S0-DTD-034

Seeing in Depth

Volume I   Basic Mechanisms

Ian P. Howard
Centre for Vision Research
York University
Toronto

Published by I Porteous, Toronto
Printed at the University of Toronto Press
2002

Published by I. Porteous
49 Dove Lane, Thornhill, Ontario, M3J 1P3. Canada

National Library of Canada Cataloguing in Publication

Howard, Ian P
Seeing in depth / Ian P. Howard.

Includes bibliographical references and index.
Contents: v. 1. Basic mechanisms / Ian P. Howard -- v. 2. Depth perception / Ian P. Howard, Brian J. Rogers.
ISBN 0-9730873-0-7 (v. 1).--ISBN 0-9730873-1-5 (v. 2)

1. Depth perception. 2. Vision. I. Rogers, Brian J. II. Title.

QP475.H68 2002 573.8'8 C2002-902810-8

Acknowledgements

We are most grateful to Antonie Howard for gathering much of the literature cited in these volumes and for her careful editing of both volumes. We thank Alistair Mapp and Hiroshi Ono for helping to write Section 17.7 and Robert Allison for helping to write Section 6.8. We also thank Alistair Mapp for proof reading the whole of the final manuscript. We thank Robert O'Shea for helpful comments on Chapter 7. We thank James Zacher for help in preparing the camera-ready document prior to printing and Martin Howard for helping in the distribution of the book. We thank Teresa Manini, and Xueping Fang for their secretarial assistance. Finally, we thank all the people who responded to our request for photographs.

Contents of Volume 1

CHAPTER

1 Introduction

1.1 SCOPE OF THE BOOK

These two volumes contain a survey of knowledge about the visual perception of the three-dimensional visual world. The primary interest is biological vision. Machine vision and computational models are mentioned only where they contribute to an understanding of the living system. The present work grew out of Howard and Rogers' *Binocular Vision and Stereopsis,* which was published in 1995. The topics have been reorganized, many new sections and chapters have been added, and the literature review has been brought up to date. The present book covers all aspects of depth perception, including a review of monocular cues to depth.

Volume I deals with the basic visual mechanisms used in depth perception. In Chapter 2 the history of the subject is reviewed to the early 20th century. Up until the 17th century, the word *optics* meant all things pertaining to vision. The study of binocular vision and space perception in general was fostered by those in the *Perspectivist* tradition, which stressed the geometrical aspects of vision. The tradition started with Euclid in the 3rd century B.C. and progressed through Ptolemy in the 2nd century A.D, Alhazen in the 10th century, Roger Bacon, John Peckham, and Vitello in the 13th century, and Aguilonius and Kepler in the 17th century. They all wrote books with titles containing either the word *Optics* or the word *perspectiva,* which formed a continuous tradition. These works were either only recently translated into English or have not yet been translated. Most visual scientists are unaware of this old tradition, which culminated in Kepler's discovery of the laws of image formation, Newton's book of *Optics,* and projective geometry. Many of the early discoveries of the Perspectivists, especially those having to do with visual perception, were forgotten after the 17th century and were rediscovered in the 19th and 20th centuries.

Throughout the book we have attempted to identify the original discoverer of a phenomenon, idea, or procedure. We would be pleased to hear about any earlier claims.

People have always been fascinated by devices that create imaginary visual worlds. The ancients had to rely on masks, puppets, and theatre. Peepshow boxes became popular in the 15th century. In the 16th century the camera obscura became popular. In the 17th century the shadow theatre was imported into Europe from the East and the magic lantern was invented. During the 18th and 19th centuries most cities in Europe and America had Panoramas, which were huge painted scenes displayed round the interior of large arenas. After Wheatstone invented the stereoscope in 1832, domestic stereoscopes became all the rage. Panoramas and stereoscopes were eclipsed by the advent of the cinema. We now have stereoscopic movies and virtual reality displays with which the viewer can interact. These display systems are reviewed in Chapter 2.

Perhaps, in the present century, synthetic worlds will become so real and the real world will become so contrived and managed that the two will be indistinguishable.

Many psychophysical and analytic procedures have been used to investigate the visual perception of depth. Chapter 3 provides only a general introduction to these procedures. Key references are provided to more detailed treatments.

Perception must start with the detection of relevant features of the environment. A stereoscopic movie camera can detect all visible features of a scene. In theory, a scene created by information picked up by movie cameras can be indistinguishable form a real scene. But cameras do not perceive anything. Perception represents the ability to respond differentially to stimuli, and to discriminate, identify, and describe them. These abilities require that stimuli be processed in complex ways, not merely detected or reconstructed. Chapter 4 is an introduction to general principles of visual coding, starting with detection and going on to discrimination, identification, and description.

Over one million axons from each eye feed into the visual cortex, more than from all the other sense organs combined. The processing of these inputs involves almost every part of the cerebral cortex. Vision is therefore the main gateway to the understanding of the central nervous system. Visual phenomena that depend on information picked up by both eyes must depend on processes occurring in the central nervous system. Chapter 5 is a review of the general physiology of the visual system.

Stereoscopic depth perception depends on the detection and processing of differences between the images in the two eyes—binocular disparities. Chapter 6 reviews the physiology of processes devoted to the detection of binocular disparity.

The fact that inputs from two eyes feed into a common mechanism gives rise to several interesting problems. Signals from the two eyes that arise from the same object must be distinguished from signals that arise from spurious superimposition of non-matching stimuli. Matching signals falling on neighbouring points on the two retinas project to the same region in the visual cortex and fuse to create the impression of one image. Non-matching images falling on the same region in the two eyes rival for access to the visual system. Chapter 7 deals with these issues.

Chapter 8 deals with the ways in which images in to the two eyes facilitate or suppress each other. Under certain circumstances, binocular images are perceived more readily and appear brighter than monocular images. Under other circumstances, superimposed, neighbouring, or successively presented binocular images engage in mutual suppression. Chapter 8 also deals with interocular transfer. A visual phenomenon shows interocular transfer when an aftereffect generated by presenting a stimulus to one eye shows when only the other eye is open. The study of interocular transfer reveals how and, to some extent, where inputs from the two eyes are combined.

When we attend to an object, the lenses of the eyes automatically accommodate to the correct distance. At the same time, the eyes converge horizontally, vertically, and by rotation about the visual axes so as to bring the two images of an object of interest into corresponding positions and orientations on the two retinas. Chapter 9 deals with these mechanisms and with defects in the co-ordinated movements of the two eyes.

As the central nervous system grows, billions of cells form appropriate synaptic connections, sometimes as many as 20,000 on one cell. This is the most complex ordered process that we know of. The study of the development of the visual system promises to be the most fruitful approach to understanding the development of the central nervous system. This is because, in the visual system, one can most easily see relationships between genetic and experiential factors. This is especially true of the development of binocular vision, which is peculiarly susceptible to sensory experience in early life.

Chapter 10 reviews the general growth of the visual system and, in particular, the pre- and postnatal growth of the binocular system. Even before the eyes open, activity arising in the eyes affects the growth of cell connections in the growing visual cortex. The study of the effects of stimuli arising in the two eyes has been particularly rewarding in young animals just after their eyes have opened. In the first place, the routing of growing axons at the optic chiasma provides a model system for investigating mechanisms of axonal guidance. Secondly, more than any other branch of developmental neuroscience, study of the development of binocular cells in the visual cortex has revealed how genetic and experiential factors interact. With the advent of high-resolution microscopes, micro-manipulation procedures, and the mapping of the human genome, we can expect major developments in our understanding of the growth of the most complex of all known mechanisms, the human visual system.

Visual functions develop in parallel with the development of the functional capacities of the visual system. Some functions mature under the guidance of genetic factors with little influence from visual activity. Other functions develop only when certain types of visual activity occur in certain critical time periods. But all visual functions become finely tuned by experience and complex relationships between them build over many years and, to some extent, over the whole lifetime. The development of visual functions is reviewed in Chapter 11, with an emphasis on the development of depth perception.

Much can be learned about the visual system by studying clinical defects and abnormalities. Defects of depth perception that result from brain damage or genetic defects such as albinism are reviewed in Chapter 12. Particular attention is paid to the signs and symptoms of loss of binocularity.

Much can also be learned about the development of the visual system by studying the consequences of early deprivation of sight in one or both eyes. Monocular deprivation within a critical period after birth severely disrupts vision in the deprived eye and usually disrupts binocular vision and stereopsis—a condition known as amblyopia. This topic has attracted a great deal of attention because of the clinical importance of amblyopia. Also, the behavioural and physiological consequences of experimentally induced monocular deprivation in animals has revealed much about the way the visual system develops and functions. These issues are reviewed in Chapter 13.

Almost all our knowledge about depth perception has come from the study of cats and primates, including humans. But there is a bewildering variety of mechanisms for the detection of depth in the rest of the animal kingdom. There are some very strange particular mechanisms and some remarkable examples of parallel evolution. Chapter 14 contains a review of seeing in depth throughout the animal kingdom, from insects to mammals.

Volume II deals with the perception of three-dimensional space. The first problem is to define the geometry of binocular space. We start by defining co-ordinate systems that can be used to specify the positions of images in each eye. We then define coordinate systems that can be used to specify the positions of points in space with respect to both eyes. In theory, one can determine the locus of points in space that project images to corresponding locations in the two retinas. This locus is known as the horopter. The horopter can be derived theoretically by ray-tracing or empirically by measuring which points appear fused or aligned. The issues turn out to be quite complex and are reviewed in Chapter 15.

Neighbouring images in the two eyes that are sufficiently similar are combined in the nervous system and passed on for processing to higher levels. The problem is to determine what stimulus features are used by the visual system to relate images in one retina with those in a corresponding region of the other retina. For example, binocular stimuli may be treated as one stimulus only if they are similar in contrast, shape, colour, and motion. One can also ask whether image matching is done only locally or both locally and globally over wide areas. These questions are discussed in Chapter 16.

Images in the two eyes may be superimposed or juxtaposed to produce a perceptual effect not evident when either image is presented alone. Any such effect is known as a cyclopean effect. Stereoscopic vision is a cyclopean effect but there are many others, such as cyclopean figural effects, cyclopean motion, and cyclopean acuity. These are discussed in Chapter 17. Another issue of cyclopean vision discussed in Chapter 17 is how visual direction sensed by each of the two eyes is unified into a coherent impression of visual direction.

Because the eyes are spatially separated, the images in the two eyes formed by a solid object differ. These differences, or binocular disparities, form the basis for stereoscopic vision. Binocular disparities can involve differences in position, orientation, texture, colour, temporal phase, or motion. In Chapter 18 we discuss the extent to which each of these differences is used as a basis for stereopsis. A type of stereopsis can arise because part of an object seen by one eye is not seen by the other eye, an effect known as monocular occlusion. This was regarded as a possible depth cue by Euclid and Leonardo da Vinci but the stereoscopic effects of monocular occlusion have only recently been demonstrated, as we show in the second half of Chapter 18.

Our ability to discriminate differences in depth on the basis of binocular disparity alone is truly remarkable. Under the best conditions, an angular disparity of only about 2 arcsec can be detected, which is equivalent to detecting a depth interval of 4 mm at a distance of 5 m. Methods for measuring stereoacuity and the factors that influence it are reviewed in Chapter 19.

Binocular disparities can be considered on a point-for-point basis. However, when registering the layout of surfaces, the visual system processes patterns of disparities, such as differences in image orientation, size, and shear. The visual system also registers spatial gradients of disparity, including linear gradients that specify flat surfaces inclined in depth, and higher-order spatial derivatives of disparity that specify curvature in depth. The geometry of disparity patterns is discussed in Chapter 20.

The eyes are separated horizontally and it is easy to see that this introduces disparities along the horizontal dimension. It has been generally assumed that only horizontal disparities are used to code depth. However, the images from an extended surface also possess vertical disparities and we now know that the visual system uses vertical disparities in a variety of ways, including the perception of absolute distance, depth scaling of horizontal disparities, and the perception of 3-D shape. These issues are discussed in Chapter 21.

An object in one location can influence the perceived spatial disposition of an object in a neighbouring location or of an object seen successively in the same location. These effects come under the heading of depth contrast and are discussed in Chapter 22.

The appearance of an object or of the way we respond to it can be influenced by its perceived distance with respect to other objects. For instance, the way one object appears to move with respect to another object is influenced by how the objects are arranged in depth. Also, stimuli that interact when seen in the same depth plane may cease to interact when separated in depth. This is a useful feature of perception because it allows us to concentrate our attention on objects in the plane of interest without being distracted by events occurring in other depth planes. For example, we can visually pursue a moving object at one distance while ignoring potentially distracting motion signals arising from objects at other distances. These issues are discussed in Chapter 23.

The impression of depth can be very compelling when only one eye is open. In Chapter 24 we review static monocular cues to depth. In theory, the angle of convergence of the two eyes specifies the distance of a fixated object. However, this information is useful only for near distances because beyond about 2 m the angle of vergence changes only slightly. The extent to which we use eye convergence to judge depth is discussed in the second half of Chapter 24.

In Chapter 25 we discuss the dynamic monocular cue of motion parallax produced by motion of an observer with respect to a 3-D display. The impression of depth created by motion parallax has a striking resemblance to that created by binocular disparity. Fundamentally, the two sources of depth information are the same.

Information about the distance and depth of an object interacts with information about other features of an object. Depth constancies are one manifestation of this interaction. Perceptual constancy refers to the ability to judge accurately a feature of the visual world that is detected by sensory components that are not constant. For example, we can estimate the size of an object in spite of the fact that its retinal image varies with the distance of the object. Also, we can judge the depth dimension of an object from binocular disparity in spite of the fact that the disparity between the front and back of the object varies with the distance of the object. These issues are discussed in Chapter 26.

Information about depth arising from a specified visual feature is known as a depth cue. Depth cues interact in many ways. Information provided by two different cues may be added or averaged or one cue may resolve the ambiguity of another cue. When information from one cue conflicts with that from another, the conflict may be resolved by weighting the cues or one cue may be ignored. In Chapter 27 we review these and other ways in which depth information is combined.

The processing of binocular disparity has temporal as well as spatial characteristics. The question of how the visual system processes signals that arrive both at different times and in different locations is discussed in Chapter 28.

Approaching objects produce images that change in the following ways. First, each image grows in size, an effect known as looming. Secondly, the images in the two eyes change in disparity over time. Thirdly, the two images differ in the way they move. The signals used in the perception of approaching objects and the way these signals are processed in the nervous system are discussed in Chapter 29.

Volume 2 ends with Chapter 30, which is an account of stereoscopic instruments and applications of stereoscopy.

1.2 BASIC TERMS AND CONCEPTS

1.2.1 Binocular vision and stereopsis

Strictly speaking, all animals with two eyes have binocular vision. Even animals with laterally placed eyes integrate the information from the two eyes to form a coherent representation of the field of view. Also, the field of view of almost all mammals has some region in which the monocular fields overlap. However, the term "**binocular vision**" is usually reserved for animals possessing a large area of binocular overlap within which differences between the images are used to code depth.

An area of binocular overlap provides several advantages to an animal. Visual detection, resolution, and discrimination are slightly better when both eyes see the stimulus (Section 8.1.1). Many complex visual tasks, such as reading and eye-hand co-ordination, are also performed more effectively with two eyes than with one, even when the visual display contains no depth (Jones and Lee 1981; Sheedy *et al.* 1986). Prehensile movements of the hand towards binocularly viewed targets show shorter movement times, higher peak velocities, and a more accurate final grasp position of the hand than do hand movements towards monocular targets (Servos *et al.* 1992). Binocular vision confers some advantage when available during only the planning stage of a motion (Servos and Goodale 1994). Convergence of the visual axes on an object of interest helps us to attend to the object and disregard objects lying nearer or further away (Section 23.5). Finally, binocular vision provides the basis for stereoscopic vision.

The term "**stereoscopic vision**" means literally, "solid sight" and refers to the visual perception of the 3-D structure of the world, when seen by either one eye or two. The various sources of visual information that could be used for the perception of dis-

Table 1.1. Sources of information for the perception of distance and relative depth.
Dashed lines connect strongly related cues.

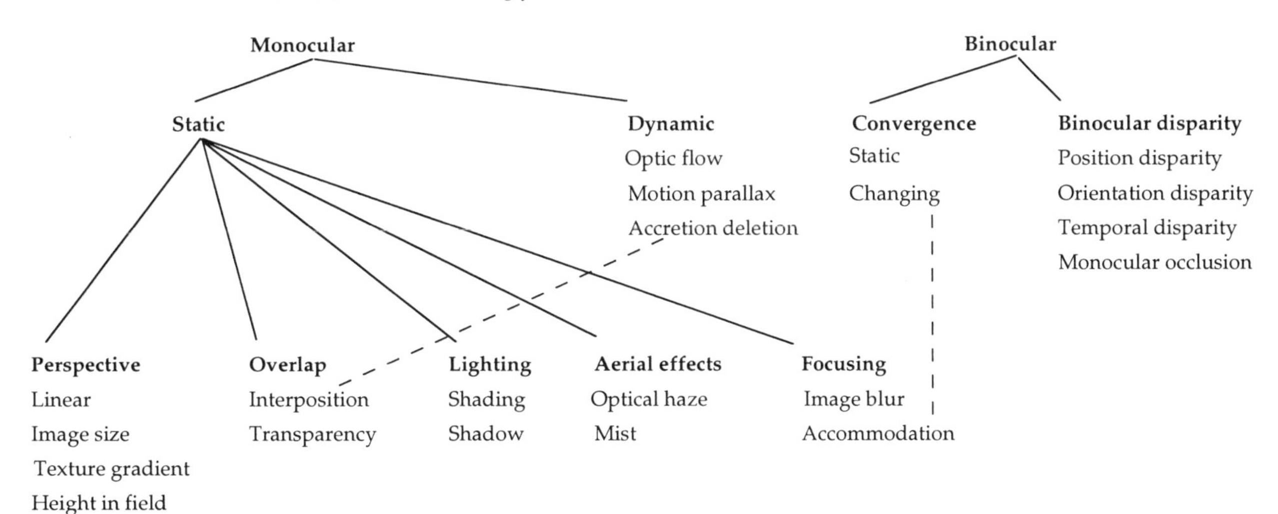

tance and relative depth are shown in Table 1.1. There are two exclusively binocular cues to depth—the vergence position of the eyes and binocular disparity. Binocular disparity, or binocular parallax, is the difference in the positions, orientations, or shapes of the images in the two eyes due to the different vantage-points from which the eyes view the world. A special type of disparity occurs when a part of an object seen by one eye cannot be seen by the other eye. We refer to this as monocular occlusion. The term "binocular stereopsis" is often used for the impression of depth arising from any binocular cue. We use the term "stereopsis" for the impression of depth arising from only binocular disparity.

Stereopsis based on disparity allows us to discriminate small differences in depth. Under the best conditions, we can detect the depth between a fixated object and a second object when the images of the objects are only 2 to 6 arcsec apart. We will see in Section 25.2 that motion parallax created by moving the head from side to side is the only other cue to depth providing this degree of precision

Monocular cues to depth are reviewed in Chapters 24, 25, and 26. Monocular cues interact in many ways with binocular cues to depth, as described in Chapter 27. In particular, binocular disparity and monocular parallax are closely related, since the successive monocular views obtained by moving the head from side to side through the interocular distance provide the same information as the simultaneous views obtained by the two eyes with the head stationary.

Binocular stereopsis helps in the performance of fine 3-D motor tasks, such as guiding a ring over a contorted loop of wire (Murdoch *et al.* 1991). Stereopsis and parallax generated by head motion help to break camouflage in stationary objects (Section 25.1.4). However, stereopsis does not add to the quality of performance of all depth-related tasks. For instance, experienced or newly trained pilots with one eye covered land an aircraft just as accurately as when they use both eyes (Lewis *et al.* 1973; Grosslight *et al.* 1978). To take another example, the task of recognizing a familiar shape when it is rotated to very different orientations in 3-D space may be performed just as well by people who lack binocular stereopsis as by people with normal vision (Klein 1977).

We will see in Chapter 14 that the eyes of some animals have a fixed angle of vergence. For these animals, the disparity between the images of an object varies in a consistent way with changes in the absolute distance of the object from the eyes. The ability to judge depth on this basis is referred to as **range-finding stereopsis**, since it works like a rangefinder. The eyes of humans and many other animals change their angle of vergence, so that the two visual axes intersect on the object of interest. Thus, the images of a fixated object have zero disparity at all viewing distances. The images of an object outside the plane of fixation have a disparity with a sign that depends on whether the object is nearer or further away than the fixated object, and a magnitude that is proportional to its depth relative to the fixated object. Under these circumstances, disparity from individual objects provides information only about the relative depth between two objects and not about the distance of either object from the viewer.

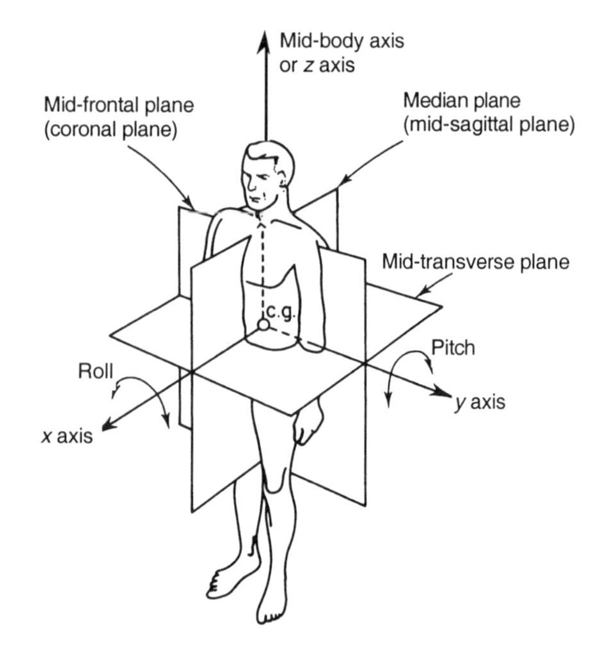

(a) The planes and axes of the human body.

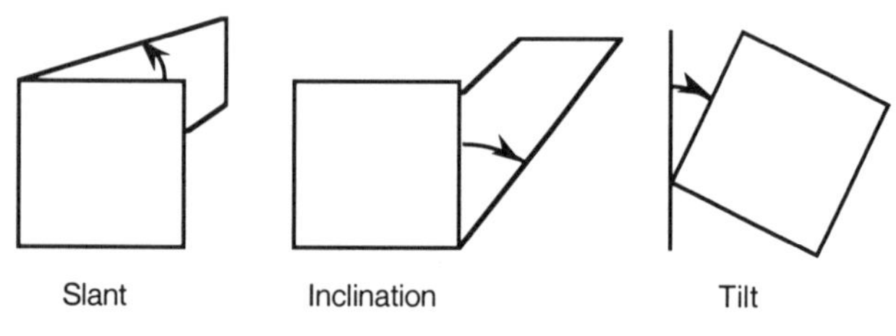

(b) Three orientations of a plane.

Figure 1.1. Definitions of planes and axes.

We will see in Section 21.3.2 that the pattern of disparity over an extended surface provides information about the absolute distance of the surface, even when no other information is available. Absolute disparity between the images of a particular object, or that between the retinal images as a whole, controls vergence movements of the eyes. The absolute distance of an isolated object could be provided by the convergence angle of the eyes. However, we emphasize throughout the book that the perception of depth by stereopsis does not depend on absolute disparities, but rather on spatiotemporal patterns of relative disparity over surfaces or between sets of objects.

The planes and axes of the human body are shown in Figure 1.1. Since this book is about depth perception, we will assume that the head is erect to gravity. We will use the term **frontal plane** to refer to any plane parallel to the mid-frontal plane of the head. We define a **slanted** line or plane as rotated about a vertical axis. An **inclined** line or plane is rotated about a horizontal axis in a frontal plane. A **tilted** line or plane is rotated about the roll axis. These definitions are illustrated in Figure 1.1.

The **distance** of an object is sometimes specified in terms of the distance between the midfrontal plane of the head and the frontal plane containing the object. The distance of an object may also be specified in terms of its radial distance from an eye or its radial distance from a point midway between the eyes (cyclopean eye).

1.2.2 Binocular stimuli and processes

Several terms referring to binocular vision are in common use, but their meanings vary from author to author. The terms "binocular vision" and "stereoscopic vision" have already been defined. The **binocular visual field** is that part of the visual field common to both eyes. A **monocular stimulus** is a distal display seen by only one eye because (a) the other eye is closed or absent, (b) the stimulus falls outside the field of view of the other eye, or (c) the stimulus is occluded to the other eye by a nearer stimulus. A **binocular stimulus** is a single distal display seen by both eyes at the same time or by the two eyes in rapid succession. The two images of an object may differ in position or shape because of the different positions of the eyes. **Dichoptic stimuli** are distinct distal stimuli, one presented to one eye and one to the other, which an experimenter can control independently. Dichoptic displays are typically created in a stereoscope, but the images of distinct objects fused by diverging or converging the eyes may also be regarded as dichoptic. Dichoptic stimuli usually differ in some defined way specified by an experimenter. The difference may be (a) a disparity of position, size, or orientation between parts or the whole of the stimuli (b) a difference in luminance, contrast, colour, shape, or motion.

The term **monoptic stimulus** is sometimes used to mean a distal stimulus seen in its entirety by one or both eyes, in contrast to dichoptic stimuli of the type in which one part of the stimulus is seen by one eye and another part by the other eye. The term **dioptic stimulus** has been used to mean a pair of identical stimuli in a stereoscope, in contrast to dichoptic stimuli that differ in some respect (Gulick and Lawson 1976). According to the Oxford English Dictionary, "monoptic" means seeing with one eye and "dioptic" means pertaining to vision. Other types of binocular images are defined in Section 15.3.5.

A binocular visual process is any process engendered by stimuli arising in the two eyes. We distinguish two basic levels of binocular processing; non-interactive and interactive. The simplest type of non-interactive binocular process is formation of a **binocular composite**. For example, a set of lines presented to one eye can be combined with a different set of lines in the other eye to form the image of a cube, which resembles the image formed when the same lines are combined in one eye, as in Figure 4.4. A second type of non-interactive binocular process is **binocular comparison** of dichoptic stimuli. For instance, one may compare the length of a line seen by one eye with the length of a line seen in another position by the other eye. Alignment of nonius lines also involves interocular comparison (Section 15.6.4). In a non-interactive binocular process, the stimuli do not interact, they are simply seen together or compared. The task is essentially the same as when both stimuli are presented to the same eye. The main difference is that changes in vergence affect the composite stimulus but not the same stimulus seen by one eye.

In an **interactive binocular process,** the images of dichoptic stimuli interact so that the appearance of at least one of them is changed in the presence of the other. In the Poggendorff illusion, collinear oblique lines appear non-collinear when a pair of vertical lines is placed between them. This illusion is evident when the oblique lines are presented to one eye and the parallel lines to the other. Thus, the dichoptic form of the illusion depends on processes occurring in the brain after the inputs from the two eyes are combined. However, the illusion is also evident when the parts of the display are presented to the same eye, so that the illusion does not depend exclusively on binocular processes and, with monocular viewing, may not be exclusively central.

An **exclusively binocular process** is a stronger form of interactive process that occurs when inputs from the two eyes are combined, and is not evident when the same stimuli are combined in one eye. An exclusively binocular process depends on a cortical mechanism activated only by binocular inputs. For example, the sensation of depth arising from static binocular disparity is an exclusively binocular process, as is binocular rivalry produced by combining distinct images in the two eyes.

The detection of binocular disparity, as a process for detecting differences in spatial location, is not exclusively binocular because the difference in location of two dichoptic stimuli is evident when the same stimuli are combined in one eye. However, the physiological properties of the two processes are not identical. Some cells in the visual cortex are specialized for detection of binocular disparity and this mechanism differs from that involved in the spatial discrimination of monocular stimuli. Thus, in these respects, the detection of disparity is an exclusively binocular process. Binocular lustre created by dichoptic combination of different luminances is an exclusively binocular effect (Section 7.3.1). Dichoptic colour mixing has some features in common with monocular colour mixing, but in other respects it is exclusively binocular (Section 7.2).

A **cyclopean stimulus** is one that is evident when two displays are combined in a stereoscope, but not evident in either monocular display (Chapter 17). A cyclopean stimulus may be a binocular composite or it may arise from binocular interactive processes.

1.3 USING THE BOOK

1.3.1 Viewing the stereograms

The act of fusing a side-by-side pair of images by diverging the eyes is known as **divergent fusion** or uncrossed fusion. The act of fusing images by converging the eyes is **convergent fusion** or crossed fusion. The stereograms presented in this book can be divergently fused with the aid of the prisms provided. The prisms should be held close to the eyes and about 12 cm above the page. The viewer should be parallel to the plane of the page and correctly oriented within that plane. Incorrect orientation shows as an elevation of one image with respect to the other. The stereograms may also be free fused by diverging or converging the eyes. In learning to free fuse, it helps if the eyes converge on a pencil point held at the correct distance between the stereogram and the eyes. For divergence, one may place a piece of clear plastic over the stereogram and fixate the reflection of a point of light seen beyond the plane of the stereogram. The correct distance can be found by observing how the images move as the pencil is moved in depth.

After some practice, readers will find that they can converge or diverge without an aid. When stereograms are free fused, one sees each eye's image on either side of the fused image. The presence of three pictures confirms that correct vergence has been achieved. Free fusion is a skill well worth acquiring, since it is often the only way to achieve fusion with displays presented at vision conferences or when you have lost your stereoscope.

A pair of images has one sign of disparity when fused by convergence and the opposite sign of disparity when fused by divergence. A change in the sign of disparity reverses the apparent depth relationships in the fused image. For stereograms in which the sign of disparity does not matter for the illustration of a phenomenon, only one pair of images is provided. When the effect depends on the sign of disparity, two stereogram pairs are provided—one pair for readers who prefer to converge the eyes, and the other for readers who prefer to diverge. Note that the provided stereoscope fuses by divergence only. Some stereograms in the book have triple images in a row. These create two side-by-side fused images with opposite signs of disparity, plus flanking monocular images. Therefore, five images are seen when the images are correctly fused. In some cases, it is instructive to compare the image formed by convergent fusion with that formed by divergent fusion. In other cases, only one of the fused images is of interest. In this case, the location of the fused image of interest is indicated by a cross for those who fuse by convergence and by two parallel lines for those who fuse by divergence.

1.3.2 Indexes and references

When a key term is introduced and defined, it is printed in bold type. In the reference list at the end of each volume, the page numbers where each reference is mentioned are entered in square brackets. This list serves as the author index. Names of people not associated with a reference are listed in the subject index. A list of portraits is provided at the end of each volume. A list of journals cited and their abbreviations is provided at the end of Volume 2.

The references are available on the Web at ftp://hpl.crestech.ca/pub/stereo vision/howard2.zip

Ian P. Howard may be contacted by e-mail at ihoward@hpl.cnrestech.ca.

Throughout the book, we suggest experiments we think could decide theoretical issues. These suggestions are printed in italics.

1.4 BOOKS ON SEEING IN DEPTH

We are not aware of any book in print devoted to the broad topic seeing in depth. Books on spatial vision such as those by DeValois and DeValois (1988) and by Regan (1991) are concerned more with contrast and acuity than with depth perception. Gibson's *The Perception of the visual world* published in 1950 was devoted largely to depth perception. There are chapters on depth perception in Volume VIII of the *Handbook of Sensory Physiology* edited by Held *et al.* (1978), the *Handbook of Perception and Performance*, edited by Boff *et al.* (1986), and *the Blackwell Handbook of Perception* edited by Goldstein (2001).

There have been several books on stereopsis and binocular vision, including Ogle (1964), Ogle *et al.* (1967), Julesz (1971), Gulick and Lawson (1976), Solomons (1978), Reading (1983a) Regan (1991a), and Howard and Rogers (1995). The national Stereoscopic Association maintains the Oliver Wendell Holmes Stereoscopic Research Library in Cincinnati and publish a bimonthly magazine called Stereo World (www.stereoview.org).

Books on vergence and binocular vision include Schor and Ciuffreda (1983), Pickwell (1989), von Noorden (1990), Scheiman and Wick 1994), and Goss (1995). Books on more specialized topics are mentioned throughout the text.

Reviews of stereopsis and binocular vision have been provided in journals by Arditi (1986), Tyler (1983, 1991a), and Patterson and Martin (1992). Books and reviews on specific topics are cited at the ends of relevant sections in the book.

2 Historical background

2.1 HISTORY OF VISUAL SCIENCE

2.1.1 The Greeks

2.2.1a *Extromission and intromission theories*
There is evidence that the Babylonians and Egyptians practised medical ophthalmology in the third and second millennia BC (Duke-elder 1961). However, there are no records about their knowledge of optics or vision.

Greek science and philosophy originated in the 6th century BC in cities spread over the Mediterranean coastline, especially that of Asia Minor and the Greek islands. This so-called Ionian period ended when Asia Minor was conquered by the Persians in 530 BC. Later, the Greeks overcame the Persians and Athens became the main centre of learning during the Classical Period from 480 to 330 BC. The centre of learning then shifted to Alexandria, the Egyptian city founded by Alexander the Great. The Alexandrian Period of Greek learning extended from 330 to 30 BC, when the last Egyptian ruler, Cleopatra, died and Egypt became a Roman province. Greek learning continued with declining vitality in Rome and in the Eastern Roman Empire of Byzantium.

The Greeks were apparently the first to enquire into the nature of vision. Many Greek philosophers, including Alcmaeon of the Pythagorean school (early 5th century BC), Plato (c. 427-347 BC), Euclid (c. 300 BC), Hipparchus (160-125 BC), and Ptolemy (c. AD 150), followed the suggestion made by Empedoclés (5th century BC) that light leaves the eye as a stream of corpuscles within a cone with its apex in the pupil. This became known as the emanation, emission, or **extromission theory** of sight. The corpuscles were believed to sense the surfaces upon which they fall, in the same way that the fingers feel an object. It was believed that the light emanating from the eyes coalesces with external light to create an effective optical medium through which images of contacted objects are transported to the mind and generate visual sensations. Several early theorists regarded the image that can be seen reflected in a person's cornea, as crucial to the visual process, presumably believing that it represented the image seen by the eye. Aristotle (c. 330 BC) refuted this idea and explained that the image in the cornea is formed by reflection. He also challenged the emanation theory by stating that, according to that theory, we should be able to see in the dark.

The emanation theory was designed to solve the problem of how the visual world is externalized and seen in its proper size. In touch, the problems of external reference and proper size were regarded as solved because the fingers touch an external object and the impression formed on the skin has the same location and size as the object. In a similar way, it was believed that light rays from the eyes sense external objects in their proper sizes. The distance of an object was thought to be sensed by the length of the light ray in the same way that the distance of a touched object is sensed by the degree to which the arm is extended. The idea that the eye is a source of light may have been inspired by the flash seen when a finger is pressed against the eye in dark surroundings—the pressure phosphene.

The followers of the atomist school founded by Leucippus (c. 430 BC) and his pupil Democritus (c. 460 BC) rejected the extromission theory. Instead, they proposed that objects continuously emit 3-D images of themselves, known as *éidola* or *simulacra*. The images move in straight lines into the eye through the intermediate translucent medium. This became known as the **intromission theory**. Epicurus (342-270 BC), of the atomist school, realized that images must shrink as they enter the eye. The atomists were then left with the problem of how the image of a given object shrinks by the correct amount for observers at different distances. The substance of the images was variously described as atoms, corpuscles, or an ephemeral substance that peeled off the object, like the skin of an onion.

Aristotle (384-322 BC), in his *De anima* and *De sensu*, rejected the idea of a substance emitted by the object, and stated that images travel to the eye as a disturbance of the transparent medium of the air, which he called the diaphanous medium. For Aristotle, light could not travel in a vacuum. However, in the *Meteorologica*, Aristotle accounted for the rainbow in terms of visual lines leaving the eye. Alexander of Aphrodisias in about AD 200 suggested that Aristotle used extromission theory in this context for mathematical convenience (see Frangenberg 1991).

Aristotle did not mention light rays but he had the idea of rectilinear propagation, since he realized that one sees double when the movement of the transparent medium evoked by a given point in space does not fall on corresponding places in the two eyes. He described how an object upon which the gaze is fixed appears double when the eyes are caused to misconverge by pressing against one eye with the finger (see Beare 1906, 1931). This is perhaps the earliest known reference to binocular disparity. Early theorists believed the crystalline lens to be the primary source of visual sensations, which were then conveyed to the brain. See Siegel (1970) and Hahm (1978) for an account of early Greek theories of vision.

Aristotle's theory was developed by his student Theophrastus (c 370-286 BC) who became Director of the Lyceum after Aristotle left Athens. We are indebted to Theophrastus for most of our knowledge of early Greek visual science. George Stratton (1917) translated his work *On the senses* into English. See Lindberg (1978) for more details about the intromission-extromission controversy.

Aristotle and other Greek philosophers placed the centre of thinking in the heart and relegated the job of cooling the blood to the brain. But in the 5th century BC, Alcmaeon of Crotona, Anaxagoras (500-428 BC), founder of the Athenian school, and Hippocrates (460-375 BC) proposed that the brain was the centre of mental activity and visual perception.

2.1.1b *Euclid*

Euclid (323-285 BC) was a Greek living in Alexandria. He was born one year before Aristotle died. His extant works include the thirteen books of the *Elements of geometry*, the *Data*, the *Phaenomena* on astronomy, the *Elements of music* and the *Optics*. The *Optics*, was written in Alexandria in about 300 BC, and is the earliest known book on the subject. Burton (1945) produced an English translation. The term "optics" is derived from the Greek word for vision. Until the 17th century, the science of optics was mainly the science of vision and of the way things appear to the human eye. It included the study of reflection and refraction because of their effects on vision. The term "optics" now refers to the physics of all forms of electromagnetic radiation, whether they are visible or not. New terms such as "physiological optics", "ophthalmology", "optometry", and "visual science" are used for the study of vision and visual perception.

Aristotle and other Greeks before Euclid had applied geometry to vision but, as far as I know, Euclid's *Optics* was the first systematic treatment. It laid the foundation for geometrical optics, leading through Ptolemy and Alhazen to Kepler. In the Greek period the mathematical approach to vision became distinct from the philosophical approach and developed a distinct terminology. People following the mathematical tradition built on the geometry of light rays became known as **Perspectivists**. They laid the foundation for the use of perspective in map making and painting, and for projective geometry and visual science. The philosophical tradition continued as metaphysics and epistemology. The two traditions are still with us, each with its own literature. There is little contact between them.

Euclid's *Optics* begins with seven definitions, or postulates. These postulates establish that light proceeds from the eye in straight lines in the form of a cone, or pyramid, with its apex centred on the eye. Only objects on which the cone of light falls are visible. Objects subtending a larger angle appear larger. Objects intersecting higher rays are seen above those intersecting lower rays, and objects intersecting rays to the left are seen to the left of those intersecting rays to the right. Today we would encompass these

Table 2.1. A selection of Euclid's theorems with equivalent statements in modern terms.

Euclid's theorem	Restatement in terms of retinal image
For a horizontal surface located above eye level, the parts further away appear lower.	More distant objects on a ceiling plane project on the retina above nearer objects.
An arc of a circle placed on the same plane as the eye appears as a straight line.	An arc in a plane containing the nodal point projects on the retina as a straight line.
When the eye approaches a sphere, the part seen will be less, but will seem to be more.	When the eye approaches a sphere, less of its surface projects on the retina but the image increases in area.
For a sphere with a diameter smaller than the distance between the eyes, more than the hemisphere will be seen.	For a sphere with diameter less than the interocular distance, the cyclopean image extends beyond half the sphere.
When the eye moves nearer to an object, the object will appear to grow larger.	The size of the image of an object is inversely proportional to the distance of the object from the eye.
When objects move at equal speed, those more remote seem to move more slowly.	The angular velocity of an object moving at constant linear velocity is inversely proportional to its distance.

postulates by setting up polar co-ordinates on the retina. The seventh postulate states that objects upon which more rays fall are seen more clearly. This postulate arose from Euclid's assumption that a fixed number of rays are emitted in the visual pyramid so that, with increasing distance, an object becomes less visible because fewer rays strike it. Eventually the object becomes invisible because it falls between rays. The same idea can be expressed in modern radiometry or in wave optics (Koenderink 1982). Theoretically, we can divide light flux into rays, each containing one photon per unit time. The amount of spatial information available to any optical instrument is the number of rays per unit area. No optical instrument can exceed the limit imposed by the discrete nature of light quanta, and the wavelength of light. Additional limitations on the spatial sampling of the image in an optical system are imposed by the optical elements of the system and by the density of receptors.

Sixty-five theorems were built on these postulates. All were concerned with geometrical relationships between the lengths and directions of light rays and the angles subtended at the eye by lines, arcs, and surfaces. Although Euclid wrote about the appearance of objects, he referred only to the geometry of distal stimuli. He described no illusions or perceptual effects arising from properties of the visual system. Today we would express almost all his theorems as statements linking the geometry of light rays to the shapes and positions of retinal images. They form part of what we call physiological optics. Euclid's theorems and proofs are still valid, except for his statements about emission of rays of light from the eye. Several of Euclid's theorems are listed in Table 2.1 with corresponding statements in terms of the geometry of the retinal image.

Today we distinguish between the geometry of retinal images (physiological optics) and accounts of visual sensations (psychophysical functions), because we know that a given retinal image produces different sensations depending on the context. Euclid knew nothing about the retinal image. He did not describe experiments or apparatus since he was concerned only with the geometry of light rays and relied on mathematical proof. He presumably made visual observations but he did not mention such things as shape constancy or aftereffects, which do not follow from his theorems.

Euclid described how a far object occludes a near object by an extent that varies with the position of the objects with respect to the horizon and their distances from the eye. He extended this analysis to explain how an eye cannot see the whole of one half of a sphere. He then described how two eyes see more of a sphere or cylinder than either eye alone when the object is smaller than the interocular distance. He was thus aware that the two eyes obtain different views of a solid object but did not conclude that this is a cue to depth. We refer to this type of difference between the two eyes' views as occlusion disparity to distinguish it from disparity in the positions of the images of the same object (Section 18.2).

Several of Euclid's theorems describe how line elements subtend different visual angles to an eye according to their relative inclinations to the line of sight and their relative distances from the eye. Theorem 6 states that parallel receding lines on a hori-

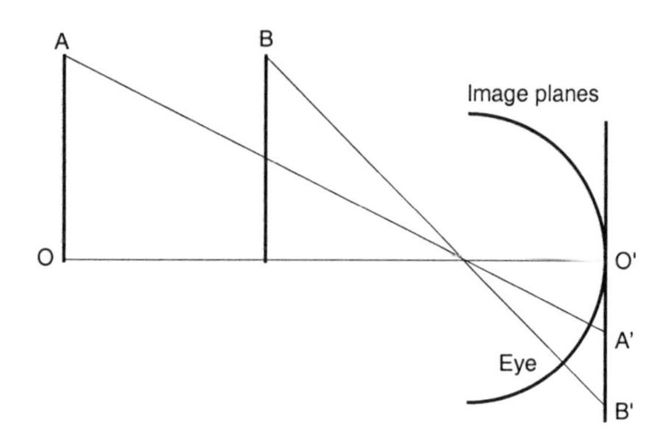

Figure 2.1. Illustration of Euclid's theorem 8.
The image produced on a flat image plane by vertical line *B* is twice at long as that produced by line *A*, twice as far from the eye. But the image produced on a spherical image plane by line *B* is less than twice that produced by a line twice as far from the eye. In deriving theorem 8 Euclid used a diagram that illustrated projection on a curved surface.

zontal surface appear to converge. Theorem 8 states that "Equal and parallel magnitudes unequally distant from the eye do not appear (inversely) proportional to their distance from the eye." Euclid's theorem is correct for a spherical image plane like the retina, as Figure 2.1 shows. Image size is inversely proportional to distance only for a flat retina (Section 24.1.1c). Several theorems deal with conditions under which circles appear as ellipses. These theorems provide a partial analysis of linear perspective (Section 2.1.6).

One of Euclid's theorems describes how a line element of a given length subtends the same angle when its ends touch a circle passing through the eye. Another describes how a fixed object subtends the same angle when the eye moves along an arc of a circle passing through the ends of the object. This theorem is a short step from proving that the horopter is a circle through the two eyes and the fixated object, but this extension of the theorem had to wait until 1804 (Section 2.2.4).

Seven of the final theorems deal with visual motion. One describes how an object moving at the same speed as the eye seems to stand still, while objects moving at other speeds appear to move. Another states that a stationary object appears to move in the opposite direction to the motion of nearby objects. We call this induced motion (Section 23.4.7). Other theorems state that, as the eye moves towards an object, the object seems to grow larger and that an object increasing in size appears to approach the eye. Finally, Euclid states that of objects moving at equal speed, those more remote seem slower.

Heron of Alexandria wrote on optics in about the year 62 AD. He explained the laws of reflection by the principle that light rays travel by the shortest path. But if any important advances in vision occurred in the 450 years between Euclid and Ptolemy all records of them have been lost (Hahm 1978).

2.1.1c *Ptolemy*

Claudius Ptolemaeus, or Ptolemy (c. AD 100-175), was a Greek-speaking astronomer, optician, and geographer living in Alexandria during the reigns of the Roman emperors Hadrian and Marcus Aurelius. Although past its heyday, Alexandria was still a great centre of learning. It had the world's greatest library, which was founded by Pharaoh Ptolemy Soter (died 283 BC). Part of it was destroyed during Caesar's siege of Alexandria during the reign of Cleopatra, the last Pharaoh of Egypt. The other part survived until 361 AD when it was destroyed by a mob after the Christian Emperor Theodosius ordered the destruction of pagan temples.

Ptolemy is best known for his work on celestial orbits, which he set out in the *Mathematike Syntaxis* or *Almagest* as it came to be known. His *Geographia*, contains procedures for map making and 27 maps (see Section 2.1.6a). He also wrote *Tetrabiblos*, a book on astrology in which he speculated about the influence of planetary configurations on human affairs.

Some time later, Ptolemy wrote a five-volume work entitled *Optics*. This work received little attention in his own time. However, after its translation into Arabic, probably in the 9th century, it became known to Arabic scholars and formed one of the foundations for Alhazen's *Book of Optics* in the 11th century. The first book of Ptolemy's *Optics* and parts of the fifth were lost before Alhazen's time. The remaining parts survive in a Latin translation of the Arabic version made during the latter half of the 12th century by Emir (Admiral) Eugene of Sicily. At that time Sicily was flourishing under Norman rule and was a meeting place for Arabic, Byzantine Greek, and Western scholars. Latin versions from the 14th century survive in Berlin, Paris, and the Bodleian Library, Oxford. Alhazen's *Book of Optics*, which was translated into Latin with the title *De Aspectibus* at about the same time as Ptolemy's *Optics*, eclipsed Ptolemy's book, which became almost forgotten until very recently. Govi published a Latin version in 1885. Lejeune (1956) produced a greatly improved annotated version of the Latin text and a translation into French (Lejeune 1989). Smith (1996) produced an English translation. The bracketed numbers in the following refer to volume number and the paragraph numbers in Smith's translation.

Historians of science have tended to select the sections of Ptolemy's *Optics* that deal with what we now call physical optics, but Ptolemy's real interest was in vision and visual perception. He adopted Euclid's geometrical analysis of visual rays and extended his investigations into visual perception and, particularly, visual illusions. Ptolemy described light as a form of energy, and realized that objects are not visible unless illuminated. He nevertheless retained the notion that light is emitted from the eye in the visual pyramid—a cone-shaped bundle of rays, which produce sensations when they strike an object. He placed the apex of the pyramid at the centre of curvature of the cornea. He insisted that light rays form a continuous bundle rather than a set of discrete rays separated by spaces, as Euclid had postulated. He pointed out that if light rays were discrete, objects would appear discontinuous and a small object that fell between two rays would become visible again if the eye moved. The rays are what we now call visual lines, and the central ray, or proper axis, is the visual axis. He argued that only the central ray falls perpendicularly on the eye to form a clear sensation, and that other rays produce blurred impressions in the visual periphery. He had no clear idea of image formation and made no mention of the optic nerves or brain.

In Book II, Ptolemy discussed light and colour and classes of visual stimuli, such as objects, sky, and shadows. He divided vision into three phases. The first involved the initial contact between the eye and rays extending to external objects. The second stage involved the immediate registration of simple visual properties such as colour, size, and distance. The third stage involved perceptual judgments derived by inference from the simple properties.

According to Ptolemy, visual flux issues forth from the eyes at great speed and strikes external objects, and feels them. The further the flux extends from the eye the weaker its capacity to sense what it touches. The direction of an object is detected by the angle the ray makes with the visual axis (II, 26). The size of an object is provided by the angle formed at the eye by the rays from the extremities of the object. The distance and slant of a near object are apprehended by the lengths of the rays that strike its surface. The distance of a far object is apprehended by the dimming of the image with increasing distance. He described how the perceived size (size constancy) of an object depends on the angle subtended by the object, its distance, and its inclination to the frontal plane (II, 52-62). He discussed lateral motion in terms of the changing visual rays intersecting an object, and motion in depth in terms of the shortening or lengthening of the visual rays (II, 76-81). He described how shading can create an impression of three dimensions in an otherwise flat surface (II, 128).

Ptolemy described several visual illusions. For example, he described how a person on a stationary boat in a swiftly flowing river perceives the boat as moving. When the person looks at the shore the river appears to move and the boat is seen to be stationary (II, 131). We refer to this phenomenon as visually induced self-motion, or vection.

He also described how a portrait of a face appears to follow an observer moving past it. He explained how the gaze of the painted face remains aligned with the visual axis of the viewer as it would if the face were real and moved with the viewer (II, 133).

Books II and III also contain an account of the geometry of binocular vision (see Section 2.2.1).

Books III and IV deal with reflection (**catoptrics**), including reflection from convex, concave, and polygonal mirrors Book V deals with refraction. Phenomena due to refraction, such as the apparent bending of a half-submerged object and the magnification of objects seen through a bottle of water, were well known to the Greeks and Romans, but Ptolemy seems to have been the first to investigate refraction quantitatively. He measured the angles of incidence and refraction by half submerging a protractor in water and aligning a point seen in the water with a second point above the water (see Delambre 1912). He did this for various combinations of transparent media, and set out the results in a table. He concluded that the ratio of these two angles is constant for a given pair of media. This is approximately true for small angles. The correct rule is that the ratio of the sines of the angles is constant for a given pair of media. Willebrord Snell, professor of mathematics at Leiden, published the correct rule in 1621. However, Thomas Harriot, scientific advisor to Sir Walter Raleigh, had stated the correct rule in unpublished notes in about 1601 (McLean 1972). The use of lenses in the ancient world is discussed in Section 2.1.3a. Ptolemy's treatment of binocular vision is reviewed in Section 2.2.1.

2.1.1d *Galen*

Galen (c. AD 129-201) was the son of an architect in Pergamon, Asia Minor, which was part of the Roman Empire. He was educated in Pergamon, Smyrna, and Alexandria. He practiced medicine in Pergamon, including healing gladiators in the amphitheatre. At the age of 32 he went to Rome where he became a friend of Emperor Marcus Aurelius and *medicus* to three succeeding emperors. Much of his extensive writing perished in a fire in the year 191. His book *De usa partium corporis humani* (On the uses of parts of the human body), which he completed in AD 175, is available in English translation. It consists of 17 books, with Book 10 devoted to the eyes.

Galen did not dissect humans, but rather pigs, oxen, goats, and tailless apes. He also carried out experiments on living animals, such as cutting nerves and the spinal cord to reveal what functions they served. Sometimes he erroneously generalized his findings to humans. He based his anatomy of the eye on Herophilus of Alexandria who described the anatomy of the eyeball and optic nerve in about 300 BC. The optic chiasma was first described by Rufus of Ephesus (c. AD 50), working in Alexandria. A spherical lens at the centre of the eye was considered to be the essential organ of vision and an extension of the brain. Galen regarded the retina as an organ that nourishes the lens. It has been suggested that early writers placed the lens at the centre of the eye because it tends to migrate there in the dead eye. Galen also proposed that each optic nerve is a hollow tube, which projects from the rear surface of the lens to the lateral cerebral ventricle on its own side of the brain. He referred to these ventricles as the thalami, meaning "inner chambers", but overlooked the nucleus that we now call the thalamus.

According to Galen, "visual spirit", or *pneuma*, is conveyed from the brain to the eye along the hollow optic nerve. The *pneuma* leaves the eye and interacts with the surrounding air to form a sentient medium, which extends to distant objects. He argued that the lens is the principal organ of sight and that visual spirit conveys visual sensations from the lens along the optic nerve to the cerebral ventricles, where it mixes with the "animal spirit". He also stated that animal spirit is generated in the base of the brain from "vital spirit" arriving from the heart. The animal spirit was believed to be stored in the ventricles, from and circulate through nerves to different parts of the body. He concluded from observing the effects of head injuries that the brain is the seat of "reason", or "mind", rather than an organ for cooling the blood, as Aristotle had proposed

Galen believed that the meeting of the optic nerves in the chiasma unites impressions from the two eyes into a single experience and concentrate the flow of visual spirit into one eye when the other eye is closed. This idea gave rise to the notion of a cyclopean eye located at the chiasma. We shall see later in this section that this idea was not overthrown until the 17th century. We now use the term 'cyclopean eye' to refer to the fact that we judge the directions of objects as if we see from a single eye midway between the eyes (Section 17.7). Galen described the six extraocular muscles, although not accurately.

Galen adopted Euclid's optics. He described binocular parallax and how each eye sees distinct parts of an object, such as a cylinder, which are then combined into a unified visual impression. However, he did not relate this to the perception of distance or solidity. He suggested that binocular vision is advantageous because it extends the field of view. Several pages of Book 10 contain a condescending explanation of why he does not describe the geometry of light rays and binocular vision. He claimed that his readers would not understand geometry and would hate him for explaining it. The little geometry he did provide is inaccurate and vague. One wonders whether it was Galen himself who did not understand geometry. Siegel (1970) reviewed Galen's theories of perception. Up until at least the 13th century, Galen's writings were treated as dogma, which inhibited observation and experimentation.

2.1.2 The Arabs

2.1.2a *Al Kindi*

After the destruction of the museum in Alexandria in AD 369 scholarly activity continued in Byzantium and Greece. The neo-Platonist school in Athens was disbanded in AD 529 by Justinian, the Emperor of Byzantium, and the scholars took refuge in Persia and Syria. The library in Baghdad founded by the Abbasid Caliph al-Maímòun became a centre for translation of Greek works into Arabic between the eighth and tenth centuries. After the westward expansion of Arabic civilization from its centres in Egypt, Damascus, and Baghdad in the 8th century, Arabic and Jewish centres of learning were established as far west as Sicily and Cordoba in Spain by al-Hakam II. Moslem religious zealots subsequently burned most of the books in the library at Cordoba. Many books that had been translated from Greek into Arabic, or had been written by Arabic scholars, were subsequently translated into Latin by such scholars as Constantinius Africanus in Salerno in the 11th century and Gerard of Cremona in Toledo in the 12th century. This produced a revival of learning in Europe in the 13th century, which laid the foundation for the Renaissance and the growth of modern science (Sharif 1966).

The first great Arab scholar was Abu Yusuf Ya'quib ibn Ishaq al-Kindi, a person of royal descent who was born in the late 8th century in the city of Al-Kufa, now in Iraq. He worked in Baghdad under the patronage of the Abbasid caliphs and died in about 866. He was an optician, musical theorist, pharmacist, mathematician, and philosopher. He wrote about 260 books, but most have not survived. His most important extant work on vision was based on Euclid and Ptolemy, and survives only in a Latin translation made in the 12th century by Gerard of Cremona, entitled *De Aspectibus*. This became a popular textbook, and its influence lasted for hundreds of years.

In line with his holistic and magical view of the universe, Al-Kindi believed that everything in the world produces rays in all directions, like a star, and this radiation binds the world into a network in which everything acts on everything else to create natural and magical effects. Al-Kindi adopted Euclid's geometrical approach to vision and conducted experiments with shadows to establish the rectilinear propagation of light. He described how a clear view of an object is built up by scanning the object with the ray of clearest vision—the visual axis. In spite of this theoretical approach, and in spite of his experiments, he clung to the emanation theory of vision. He argued that the 3-D éidola of the atomists could not account for effects of perspective, such as the elliptical appearance of a circular object viewed at an angle. He falsely reasoned that only an emanation theory could account for perspective. He did not realize that his theory of light rays solved the problem of perspective (see Lindberg 1978).

2.1.2b *Johannitius*

Hunayn ibn Ishäq (AD 808-873), known in the West as Johannitius, wrote one of the earliest known Arabic works on ophthalmology. He was a Nestorian Christian and, like Al-Kindi, worked in Baghdad under the patronage of the Caliph. He translated Greek scientific and philosophical works into Arabic, and his own works include *Ten Treatises on the Eye* and the *Book of the Questions on the Eye*. These works were very influential in Islam, and the *Ten Treatises* were translated into Latin in the 11th century. This translation contains the earliest known diagram of an eye and was the principal source of information for Western scholars about Galen's theory (Eastwood 1982).

2.1.2c *Avicenna*

Attacks on the extromission theory of vision appeared in the writings of al-Razi (Rhazes) (d. 923) and his younger contemporary al-Farabi (Alpharabius) (d. 950). These attacks continued in the works of Abu Ali al-Husain ibn Abdullah ibn Sina (AD 980-1037), known in the West as Avicenna. He was born in Bokhara and became physician and vizier to the Emir in Hamadan and later worked under the patronage of the Sultan of Isfahan in Persia. His great book *Qänün* (Canon) reviewed ancient and contemporary medical knowledge and was used in Arabic lands and in Europe for six hundred years. His writing was clearer and more systematic than was Galen's. He discussed vision in the *Canon* and in several other books, including *De anima seu sextus de naturalibus*, which still exist. He was mainly concerned with refuting the emanation theory of vision in all its forms. He adopted Aristotle's theory of visual optics and Galen's ideas on visual anatomy. See Lindberg (1976) for a discussion of Al-Kindi, Avicenna, and other Arabic scholars.

2.1.2d *Alhazen*

Of the Arabic scholars, Alhazen (Abu Ali Al-Hasan ibn Al-Hazan ibn Al-Haytham) made the most significant contributions to optics and vision (see Bauer 1912; Winter 1954; ten Doesschate 1962; Crombie 1967). He was born in about AD 965 and lived in Basra in what is now southern Iraq before he moved to Cairo, where he spent the rest of his life teaching and writing under the Fätimid caliph Al-Häkim. Egypt had come under the control of the Fatimid dynasty in 969, after the fall of the Abbasid Caliphate. Alhazen died in about AD 1040. In his autobiography he tells very little about himself, but provides a list of ninety-two of his works, more than sixty of which have survived. Sixteen works were on optics. In the seven books of his great synthetic work, the *Book of Optics* (*Kitäb al-manäzir*), he set out to examine the science of vision systematically, using mathematics and experimental observation.

Alhazen was familiar with the writings of Aristotle, Galen, and Ptolemy and with the *Conics* of Apollonios, although he did not make explicit reference to them. He used the same division of the subject into vision from rectilinear rays, vision by reflection, and vision by refraction, as in Ptolemy's *Optics*. He also used a similar tripartite division of vision into image formation, immediate perception, and inferential perception. Alhazen's work was not well known in the Arabic world for 250 years after his death, but became generally known after Kamäl al-Din Abdu'l-Hasan al-Farisi (died c. 1320) produced his *Tanqih al-manäzir* (*Revision of the Optics*) in about the year 1300 in Iran. This book competently reviewed all the subjects discussed by Alhazen and was supplemented by criticisms and new ideas (Sabra 1987a). In the 12th century, the *Book of Optics* was translated into Latin with the name *Perspectiva*, by an unknown person. Fourteen Medieval Latin manuscripts produced between the 13th and 15th centuries survive. In Basle in 1572 Risner produced the *Opticae Thesaurus, Alhazeni Arabis libri septum,* which contains the first printed version of Alhazen's book with added titles and annotations. It also contains Vitello's book *Perspectiva*, a 13th century work based largely on Alhazen. The *Opticae Thesaurus* became the principal source for optics in Europe until the 17th century. The original Arabic version of Alhazen's *Book of Optics* produced by Alhazen's son-in-

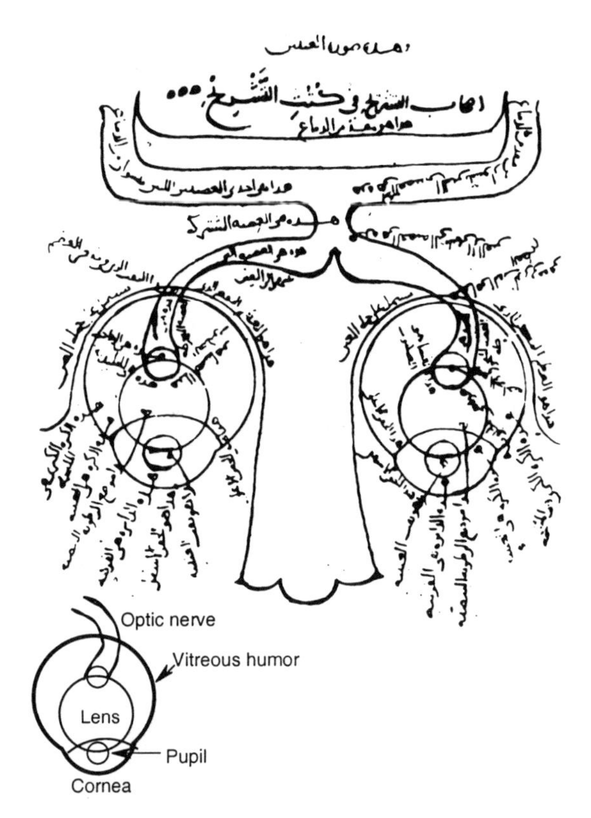

Figure 2.2. Alhazen's diagram of the visual system.
Diagram from the *Book of Optics* (*Kitāb al-manāẓir*) by Alhazen, copied by his son-in-law in 1083 from an earlier version (MS 3212 in the Faith Library in Istanbul). In the lower corner is a key to the parts.

law in 1083 was unknown to Western scholars until 1913 when Rescher made a brief announcement of its existence, but it remained unnoticed until the investigations of Krause in 1936 (see Polyak 1941). In 1989, A. I. Sabra of the Department of the History of Science at Harvard University produced an English translation of the first three books of the *Book of Optics* from the original Arabic version in Istanbul . in the following summary the numbers in brackets refer to pages in Sabra's translation.

Book I of the *Book of Optics* is devoted mainly to visual optics and the structure of the eye. Alhazen firmly rejected the emanation theory of vision and described how rays of light enter the eye from sources of light such as the sun and from objects that reflect and refract light. In the Greek intromission theory, a visible object issues a copy of itself to the eye. This raised the problem of how multiple copies can be sent to a multitude of eyes over an extended period, and the problem of how a copy of a large object can enter the pupil. In solving these problems, Alhazen adopted al-Kindi's idea that light radiates in all directions from each point of an object. Instead of an object issuing copies of itself, each point on an object emits or reflects light rays in all directions and those rays that enter the eye produce an image. This geometrical analysis of visual rays solved the problems raised by the older intromission theory and laid the foundations of modern geometrical optics, although Alhazen regarded light rays as geometrical conveniences rather than as real.

Alhazen conducted experiments with a "dark chamber" into which light entered through a small hole to form an inverted image without the aid of refraction. This was what we now call the pinhole camera, or **camera obscura** (see Section 2.1.6c). Al Kindi had also described such an instrument (Werner 1910). Also, a camera obscura was described in the *Mo Ching*, a Chinese work from the 4th century BC, and was used back then to prove that light travels in straight lines (Ronan 1978). The invention of the camera obscura is often credited to Roger Bacon or Leonardo da Vinci. The pinhole camera works because the aperture is so small that only a narrow beam of light passes from any point on the object to a given point in the image. Alhazen described how an eclipse of the sun could be safely observed by looking at the image produced by a small hole in the wall of a dark chamber, and he understood that the image is sharp only when the hole is small. He used the pinhole camera to prove that light rays travel in straight lines and pass through a small aperture without interacting. The use of the camera obscura in art is discussed in Section 2.3.1b.

Since the Moslem religion forbade dissection, Alhazen based his ideas of the anatomy of the eye on Galen. Figure 2.2 shows a diagram of the eyes and visual pathways from the earliest known copy of Alhazen's book. This, and a diagram in Book III, are the only figures in the original text. Other diagrams in Latin translations were added later (Bauer 1912).

Alhazen realized that the pupil is too large to allow the eye to work as a pinhole camera. Although he did not understand image formation by a lens he realized that, for clear vision, each point on the recipient surface in the eye must receive light from only one visual ray. He was thus confronted with the problem of how light from each object forms a distinct image on the surface of the lens without being diluted by rays from other objects.

The solution Alhazen adopted was that the surface of the cornea and the front surface of the lens are concentric, so that those rays striking the two surfaces at right angles pass unrefracted through the common centre. He proposed that only rays striking each point on the surface of the eye at right angles are allowed to pass into the sentient interior of the lens. He realized that other rays are refracted, and assumed that because of this they are weakened (p. 124). We now know that most rays falling on a given point on the cornea reach the retina, but that each ray is refracted to a point on the retina appropriate to its point of origin in the part of the visual scene upon which the lens is accommodated. There is nevertheless some truth to Alhazen's idea of the predominance of orthogonal rays because light rays normal to the retinal surface are more likely to enter the elongated visual receptors than are rays at any other angle. This is the Stiles-Crawford effect (Section 5.1.2a). Also, the compound eyes of insects work in the way suggested by Alhazen, since each ommatidium accepts light from only a narrow visual angle. In Book V, Alhazen questioned his own theory and admitted that refracted as well as normal rays are detected. He thus grappled with the problem of how the cornea and lens form a clear image by refraction. But he failed to find a solution (p. 116; see Lindberg 1976, p. 76). A solution was not forthcoming until Kepler developed a theory of image formation in his *Ad Vitellionem paralipomena* in 1604, which was published in his *Dioptrice* in 1611.

Alhazen stated that the initial process of sentience, which we now call visual processing, occurs along radial lines in the interior of the lens. The lens senses the images (forms) of objects defined by the cone of light rays that come from the surface of the object. The rays enter the lens after striking the cornea and lens at right angles. Each object produces a distinct cone of rays, so that objects are seen distinct from each other. He argued that the image must be sensed before the rays converge to a point. In the first place, sensation cannot occur at the focal point because images of distinct objects would all be fused together. In the second place, sensation cannot occur after the focal point because the image would be inverted. He believed that the rear surface of the lens is mounted on the optic nerve. Like Galen, he believed that the form (image) is conveyed along a set of hollow tubes in each optic nerve to the optic chiasma where the images from the two eyes fuse into the *sensus comunus*. He realized that the rectilinear propagation of light rays determines the spatial integrity of the initial image, but that once the "form" is conveyed into the curved optic nerve its spatial integrity depends on retention of the relative order of parts within the optic nerve, rather than on rectilinear propagation. We make a similar distinction between the retinal image formed by rectilinear propagation of light rays and topographic projection along the axons of the optic nerve.

Alhazen's description of image formation and his concept of an image were very hazy (Eastwood 1986). Sabra (1989) describes the uses that Alhazen made of the Arabic word for *form* or *image*.

Alhazen's ideas on basic visual processes, erroneous though they were in many ways, set the stage for subsequent developments in physiological optics and, ultimately, for Kepler's account of image formation.

In addition to the discussion of optics and the structure of the eye, Book I contains descriptions of several perceptual phenomena including visual masking, dark adaptation, afterimages, the dependence of acuity and colour perception on luminance (pp. 51-54), and colour mixing (p. 97). Book I also contains a description of corresponding retinal points (p. 87), diplopia of images falling on noncorresponding points, and fusion of those falling on corresponding points (see Section 2.2.2). Alhazen argued that the eye is round so that it may be moved quickly to bring the images of different objects onto the region of clearest vision—the region we call the fovea (p. 104). He explained that clear vision is built up from the separate impressions obtained as the gaze moves over the scene.

Book II contains some discussion of image formation in the eye but is concerned mainly with visual perception. Although Alhazen referred to the lens as the sentient structure he also stated that no sensation is accomplished until the image arrives in the brain (page 89). This seeming contradiction is resolved when we read that only primary sensory features of light and colour are sensed at the recipient surface, and that scanning eye movements, inferential processes, and memory are required for the detection and recognition of higher features such as shape, movement, and particular objects.

Today we say that the first stages of visual processing occur in the retina to be followed by further stages in the brain, involving complex interactions and memory. We also distinguish, as did Alhazen, between the initial, or preattentive, impression of a visual stimulus and impressions gained only after higher levels of processing have occurred (p. 209). Alhazen 's views on the roles of inference and experience at an unconscious level (p. 136) are very similar to Helmholtz's theory of unconscious inference (see Sabra 1978). Alhazen also described a level of perception involving conscious knowledge (see Q. Wang *et al.* 1994).

Book II also contains descriptions of basic visual features such as direction, distance, size, shape, continuity, motion, and transparency, and of phenomena based on these features. The phenomena include colour constancy (p. 141), additive colour mixing on a spinning top (p. 145), the role of texture gradients and the ground plane in the perception of distance (pp. 152, 179), size constancy (p. 177), and the

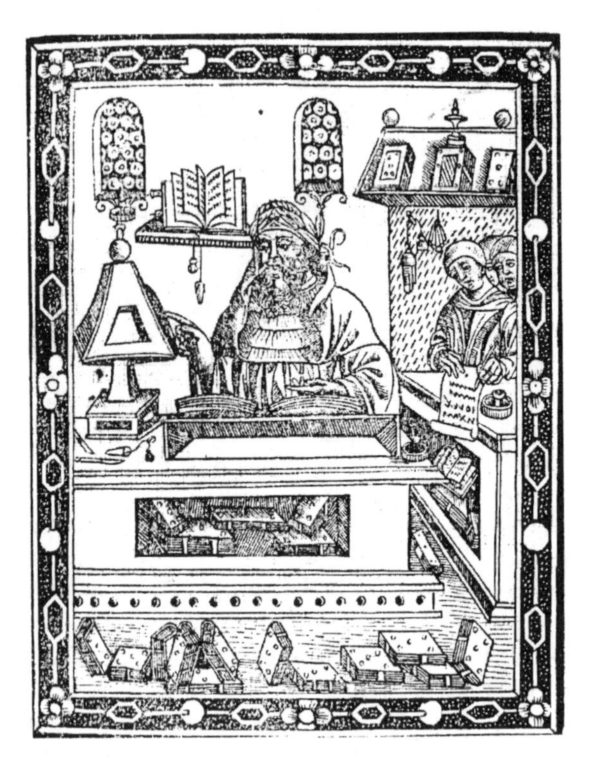

Figure 2.3. Frontispiece from *Perspectiva communis* by John Peckham (1504).
The book summarises the writings of Euclid, al-Kindi, Alhazen, Grosseteste, and Bacon.

contribution of eye movements to the perception of motion (p. 193). Book III is concerned with errors and illusions in visual perception. There are descriptions of the equal motion of the eyes (p. 229), induced visual motion (p. 261), and many other visual phenomena. The book also contains Alhazen's ideas on binocular vision, which are reviewed in Section 2.2.2.

Books IV to VII are devoted to reflection and refraction. He made paraboloid mirrors, probably on a lathe, and explained how parallel rays are brought to a single focal point by such a mirror, although Diocles had proved this in the 2nd century BC. He repeated Ptolemy's measurements of refraction, but failed to find the sine law, even though the Arabs knew Hindu trigonometry. He discussed magnification by a plano-convex lens.

He described the apparent enlargement of the moon near the horizon, an effect known as the moon illusion (see Section 26.2.6) In the *Almagest,* Ptolemy explained the moon illusion in terms of refraction of light through the atmosphere. However, in the *Optics* he explains the illusion in terms of an illusory size change induced by differences in apparent distance. Cleomedes, who probably lived in the first century AD, had offered a similar perceptual explanation (Ross 2000). Shu Hsi proposed a perceptual explanation of the moon illusion in the 3rd or 4th century AD (see Needham 1962, vol. 3, p. 226). Alhazen, also, provided an explanation of the moon illusion in terms of the relationship between apparent size and apparent distance (see Ross and Ross 1976; Plug and Ross 1994; Sabra 1987b). However, in a collection of writing on the moon illusion, Alhazen did not rule out a possible contribution from refraction (Sabra 1996). Books IV to VII of Alhazen's *Book of Optics* have not been translated into English.

Although Alhazen's *Book of Optics* inspired many commentaries and derivatives, it was not superseded until the early 17th century. Alhazen's discoveries in optics and physiological optics contained in Book I have been extensively reviewed. His discoveries in visual perception set out in Books II and III were mentioned by Priestley (1772) and reviewed by Bauer (1912). However, they have been almost totally ignored by visual scientists and, in the visual science literature, many of the phenomena discussed by Alhazen are described as having been discovered only in the last two hundred years (Howard 1996).

2.1.3 Europe to the 18th century

By the 13th century, Arabic science had declined but interest in science began to revive in Europe. At this time, the word 'perspective' was synonymous with the word 'optics'. Optics was the most important science in 13th century Europe. Light was primary because of its association with spiritual light in Christian theology. Also, in the gnostic tradition, going back to Persia, the Neoplatonists of Greece, and the 2nd century Hermetic literature of Egypt, illumination represented the divine essence that penetrated the planetary spheres to reach the Earthly sphere of mortal existence (Yates 1964).

Roger Bacon (ca 1214-1294), the "doctor admirabilis", was a Franciscan monk who studied in Oxford under Robert Grosseteste and in Paris. He was familiar with the works of Euclid, Aristotle, Galen, and Alhazen. In his *Opus Majus* of 1268 (Edited by J. H. Bridges, Oxford 1897) he mentions that people with weak eyes can use a lens for reading. His work, and especially his *Scientia perspectiva,* was based on the geometrical optics of Alhazen. The same can be said of other 13th-century scholars. One of these was Vitello (1230-70), who lived in Poland and wrote *Perspectiva.* This was the first European treatise on optics (1270). Another was John Peckham (1240-91), Archbishop of Canterbury, who wrote the book *Perspectiva communis.* The frontispiece is shown in Figure 2.3 (ten Doesschate 1962; Lindberg 1983).

The tradition of geometrical optics starting with Euclid, and continuing through Ptolemy, Alhazen, Bacon, Vitello, and Peckham was known as the *perspectivist* tradition and the practitioners were known as the **Perspectivists.** The tradition culminated in Kepler's discovery of the basic laws of optics. In the 14th century, the disastrous plague, the Black Death, brought most scientific enquiry to a halt. Scholastic implications of classical learning for Christian doctrine and the exercise of dialectical skills dominated learning in Europe.

2.1.3a *Lenses and spectacles*

The oldest known lenses were made in Egypt between 2600 and 2400 BC. They were convex-plano lenses of high quality made from rock crystal, a form of quartz. They formed part of artificial eyes placed in funerary statues during the IVth and Vth dynasties of the Old Kingdom. These statues, with their eyes, can be seen in the Louvre and the Egyptian Museum in Cairo. The lenses create the impression that the eyes follow an observer walking past the statue. Enoch and Lakshminarayanan (2000) have built a replica of one of these eyes.

Lenses dating from Greek and Roman times have been excavated. Aristophanes (c. 257-180 BC) mentioned their use in focussing the sun's rays to make fire, and Pliny (AD 23-79) mentions their use in cauterising. Seneca (c. 4 BC-AD 65) described how a ball of glass filled with water magnifies letters (Polyak 1957).

The Chinese appear to have made lenses from rock crystal or glass to focus the sun's rays as early as the 3rd century (see Needham 1962, p.118). Magnifying lenses were used in China in the 12th century for reading illegible documents and possibly for fine engraving, but do not seem to have been used as spectacles until the early Ming dynasty in the 15th century (Needham 1962, p. 119).

The foundations of the science of refraction, or **dioptrics**, were laid in Greek and Roman times. Alhazen mentioned the magnifying properties of plano-convex lenses in the 11th century. Kepler, in 1604, provided the first full account of image formation by a lens.

Spectacles were first made in about the year 1285 by an unknown person, probably a worker in glass in Pisa, Italy. The earliest written reference to spectacles is in notes for a sermon delivered by the Dominican Friar, Giordano da Rivalto, in the church of Santa Maria Novella in Florence on Wednesday, February 23rd, 1306. The earliest known work of art depicting spectacles is a portrait at Treviso of Hugh St. Cher painted by Tommaso da Modena in 1352. Since Hugh St. Cher died well before the painter was born and more than 20 years before spectacles were invented, the spectacles in the portrait are an anachronism. Rosen (1956) provides an amusing account of spurious claims that spectacles were invented in Venice, in England by Roger Bacon, in Belgium by Bacon transformed into a Walloon, and in Germany.

2.1.3b *Leonardo da Vinci*

The following passage from Leonardo da Vinci (1452-1519) has been viewed as the first suggestion that light travels as waves rather than corpuscles, as earlier writers believed. He wrote,

"Just as the stone thrown into water becomes the centre and cause of various circles, and sound made in the air spreads out in circles, so every body placed within the luminous air spreads itself out in circles and fills the surrounding parts with an infinite number of images of itself, and appears all in all and all in each part." (cited in Keele 1955)

In Lindberg's (1976) opinion this was merely an analogy describing the propagation of images and said nothing about how light itself is propagated. Be that as it may, the preceding quotation essentially reaffirms al-Kindi's principle of universal radiation. Leonardo's knowledge of optics was based on Alhazen, either directly or through Bacon, Peckham, or Vitello. Like Alhazen, Leonardo proved that light from many objects passes through each point of space by showing that many objects produce distinct images through the same hole in the pinhole camera. He then proved that light from any one object is in each location of space by showing that several images of the same object are produced simultaneously by several pinholes. He was amazed by the fact that all light rays entering the eye pass through one narrow opening without interference. He wrote,

"... here the figures, here the colours, here all the images of every part of the universe are contracted to a point. O what point is so marvellous!" (Keele 1955).

He observed the colours of the spectrum formed by light passing through a glass of water. In his *Six Books on Light and Shade* (Richter 1970, pp.67-128), Leonardo distinguished between attached shadows and cast shadows (see Section 24.4.1).

Leonardo boiled an eye in egg white until it became hard and then sectioned it. This is one of the first uses of tissue embedding. His drawings show a spherical lens, probably because the lens of a boiled eye becomes spherical, like that of a dead eye. He compared the eye with a camera obscura, believing that an inverted image is produced in the centre of the eye by light passing through the pupil, but that the rays cross again by reflection or refraction to form an erect image on the head of the optic nerve. While holding the view that vision results from light entering the eye, it seems that he believed that vision had special powers over things seen. For instance, he wrote:

"*. . . the wolf has power by its look to cause men to have hoarse voices.*" "*The ostrich and the spider are said to hatch their eggs by looking at them.*" "*Maidens are said to have the power in their eyes to attract to themselves the love of men.*" (Todd 1991).

This last statement has some basis in fact.

Leonardo observed changes in pupil size accompanying changes in illumination and noted that the pupils of nocturnal animals, such as the owl, increase enormously in the dark. In his early writings, Leonardo had the erroneous idea that the larger the pupil the larger the appearance of the object. He may have gained this impression from his observation that a dark object on a bright background appears smaller than a bright object on a dark background (Strong 1979, p. XXXII). After 1513 he refers to vision being intensified when the pupil dilates (Lindberg 1983). He failed to observe pupil changes accompanying changes in accommodation. These changes were first described by Scheiner in 1631.

Leonardo injected wax into the cerebral ventricles to obtain an accurate idea of their shape, and concluded that the optic pathways end in the posterior ventricles, rather than in the anterior ventricles as generally believed at that time (see Keele 1955).

Leonardo's contribution to drawing in perspective is described in Section 2.1.6d and his ideas on binocular vision are discussed in Section 2.2.3a.

Leonardo's writings on vision had no effect because they were in private hands until 1636. They were not studied seriously until the end of the 18th century.

2.1.3c *Giovanni della Porta*

Giovanni Battista della Porta (c. 1535-1615) (Portrait Figure 2.4) was a flamboyant collector and investigator of natural wonders, a playwright, and translator of Greek texts. His father's house in Naples was a centre for philosophers, musicians, and poets. Giovanni founded a group calling themselves *Otiosi* (Men of Leisure). Each member was required to have made a new discovery in natural science. This was the first scientific society of modern times. At the age of 23 he wrote *Magiae naturalis* (1558), a collection of wonders, recipes, and remedies. It was one

Figure 2.4. Giovanni Battista della Porta (c. 1535-1615). (From Mach 1925)

of the most popular books of its time. It was translated into English in 1658 and Dover published a facsimile edition in 1957.

In his major work on optics, *De refractione optices parte libri novem* (1593) he dealt with refraction and expounded Alhazen's view that an image is formed on the lens by perpendicular rays in the manner of a pinhole camera. However, he added his own view that this happened only after a second inversion of the image by reflection from the back of the eye, which acted as a concave mirror. He was apparently the first to give an account of binocular rivalry between differently shaped images in the two eyes. He also described two tests of eye dominance, one based on binocular rivalry and the other based on binocular parallax occurring when one or the other eye is closed (see Wade 1998b). He was aware that a lens placed in the aperture of a camera obscura improves the image and he obtained an erect image by use of a concave mirror. He understood that increasing the size of the aperture increased the illumination of the image. He failed to apply this knowledge to the eye, preferring to believe that the

Figure 2.5 Andreas Vesalius (1514-1564).
(Frontispiece in A. Burggraeve, Etudes sur André Vésale, Gand, 1841)

image is formed on the lens of the eye. He was the first to use the camera obscura for drawing, simply by tracing round the image.

2.1.3d *Benedetto Castelli (1557-1643)*

Benedetto Castelli was born in Northern Italy in 1557. He was a priest and lecturer in mathematics in Pisa and then in Rome. He was the first and closest disciple of Galileo and the only member of the Galilean school to write about vision. His *Discorso sopra la vista* was written in 1639 and printed in Bologna in 1669. An English translation was published by Ariotti (1973). Ariotti points out that Castelli is unknown and has been totally ignored outside Italy. He is not mentioned in the histories of visual science by Priestley (1772) or Boring (1942).

The treatise is in the form of a long letter to Giovanni Ciampoli, a prelate of the Roman Curia. Castelli described several well known and some novel visual phenomena. Like della Porta, he describes a camera obscura with a lens but added some of his own novel observations. He described the effects of using lenses of different focal length and noted the range of distance of the projected image over which it remained in focus. He explicitly compared the camera obscura with the eye. He wrote:

"We found by experiment that when the hole was made notably larger, there also followed confusion and fogging up of the images and when it was made very narrow the image appeared very dimmed."

He described how the afterimage of an illuminated widow frame appears large when projected onto a far surface and small when projected onto a near surface. He explained this as follows:

"...once the image is impressed on the retina it occupies a determinate area of this tunica. When we turn the eye to an object like a white wall placed ten or thirty times farther away than the first {object}, the already impressed area of the retina will be covered with an image of as large a portion of the wall as greater is the distance between the eye and the wall to the distance between the eye and the original object."

This clear statement of Emmert's law was written 242 yeas before Emmert (see Section 26.2.5). Castelli applied the same explanation to the moon illusion (Section 26.2.6). He described a size-distance illusion that occurs in drawings.

"...were a painter to draw ... two equal figures of men ... against a background in such a way that one appeared in a place far away from our eye and the other one nearer, we would then judge that the one that is represented as much farther away as, so to speak, a giant even though the two figures are of equal height."

Most textbooks on perception contain such a picture (see Figure 26.4), although nobody has cited Castelli. Also, this is a clear statement of the size-distance invariance hypothesis (see Section 26.2.2).

Castelli had a dig at classical scholarship. He declared that nature itself is "... the original book of every true knowledge of ours."

"I care not the least about ... those who do nothing other than collate diverse opinions from different volumes and ... give birth to most extraordinary monsters and to most futile chimerae of new views ... that have no other reality than in the fantasies and in those sheets of paper that they keep filling up."

Figure 2.6. Felix Platter (1536-1614).
Professor of Medicine in Basel.

Figure 2.7. Johannes Kepler (1571-1630).
(From Mach 1929)

2.1.3e *Vesalius and the development of anatomy*

Prohibition of dissection of the human body was lifted in the 14th century. Anatomists in Italy began to direct the dissection of executed criminals, victims of the Black Death, or those suspected of having been poisoned. The dissection itself was performed by an assistant. Andreas Vesalius initiated the modern study of human anatomy. He was born in Brussels in 1514 and studied medicine in Paris, Louvain, and Padua where, the day after his graduation at the age of 25, he was appointed to the chair of anatomy and surgery (see O'Malley 1964). In 1544 he became physician to Charles V, Holy Roman Emperor of Spain, parts of Italy and the Netherlands. This prevented him from engaging in further research. After the abdication of Charles in 1559, Vesalius hoped to return to his chair in Padua but was forbidden to leave Madrid by the new Emperor, Philip II. In 1564 Vesalius made a pilgrimage to Jerusalem. On the return journey, stormy weather forced him to land on the island of Zante, where he became sick and died in miserable circumstances (Portrait Figure 2.5).

The great work *De Corporis Humani Fabrica* (*On the Structure of the Human Body*) was published in 1543, when Versalius was 29 years old. It contains many fine anatomical drawings, including drawings of the eye, based on his own dissections and presented in the form of woodcuts made by a master engraver. A reprinting from the same blocks was produced by Saunders and O'Malley (1950). Vesalius could not confirm that the optic nerve was hollow, as required by Galen's theory that it transported visual spirits. In spite of this, and his critical attitude to classical anatomy, Vesalius did not question the doctrine of animal spirits, which persisted until after Kepler. He also retained a spherical lens in the centre of the eye and placed the optic nerve on the optic axis. He suggested that the retina is the sensitive organ of sight but produced no supporting evidence. Like Galen, Vesalius believed that the optic pathways project to the lining of the most anterior of the three cerebral ventricles.

It has been claimed that, in the 12th century, the Spanish-Arab, Averroes, described the retina rather than the lens as the site of image formation, but this claim has been denied by others (see Lindberg 1983). The idea was expressed by Abn Rushd, another Spanish-Arab scholar in the 13th century (Polyak 1957), and hinted at by Vesalius in 1543.

Felix Platter, (1536-1614) (Portrait Figure 2.6) was the first to state clearly that the image is formed on the retina and he produced supporting anatomical evidence in his *De corporis humani structura et usu* of 1583. However, his account lacked a clear statement of the dioptric principles of image formation. The

Figure 2.8. George Berkeley (1685-1753).
He was born in Dysert, Ireland. He became a Fellow of Trinity College Dublin in 1707. Between 1713 and 1721 he travelled in England and Italy. He then held offices in the University of Dublin and was Dean of Derry. He visited America between 1728 and 1732. In 1734 he became Bishop of Cloyne in Ireland. (Oil painting by J. Simbert, 1725. National Portrait Gallery, London)

honour of discovering those principles belongs to Kepler. Platter was Professor of medicine at Basel and was the first person to dissect the human body in a German speaking country.

Eustachio (1520-1574), in his *Tabulae Anatomicae*, was the first to recognize that the optic pathways do not project directly to the brain but first pass to the posterior part of the thalamus (the lateral geniculate nucleus), although his discovery was ignored for more than 150 years. In 1854 Louis-Pierre Gratiolet (1815-1865) discovered the optic radiations projecting from the thalamus to the occipital cortex.

2.1.3f *Kepler and visual optics*

Johannes Kepler (1571-1630) (Portrait Figure 2.7) was born of Lutheran parents in Weil. He studied at the University of Tübingen, intending to become a Lutheran clergyman. Instead, he taught mathematics at the Protestant seminary in Graz. In 1600 he became the assistant to the astronomer Tycho Brahe at the court of Rudolf II in Prague. When Brahe died a year later, Kepler became court mathematician and astrologer. He later became a mathematician in Linz, and finally in Rostock. He lived at the height of the witch-burning craze and his own mother would have been burned if he had not intervened. The horrific thirty-years war raged in Germany during the last 15 years of his life (Wedgwood 1995). See Koestler (1960) for a biography of Kepler.

In the 17th century, astronomy and mathematics were emerging from the mystical, hermetic tradition of thought with which they had been inextricably connected. Kepler's attitude to astrology was ambivalent. He cast horoscopes for the Emperor but once remarked that the "wayward daughter, astrology", had to support the "honest dame astronomy". Nevertheless, he believed that the stars and planets affected human affairs. In his *Mysterium Cosmographicum* of 1597 he stated that the planetary orbits conform to the diameters of nested Platonic solids. This theory was inspired by mystical concepts derived from the Hermetic-Cabalist (magical) tradition. But, unlike a true mystic, when Brahe produced evidence that did not fit with this theory, Kepler relinquished it in favour of his theory of elliptical orbits, as set out in his *Harmonice mundi*. This laid the foundation for Newton's theory of gravity.

Like Marin Mersenne in France, Kepler was among the first to state clearly the distinction between testable scientific hypotheses and mystical speculation, and between the use of mathematics to describe natural phenomena and its use in mystical numerology. This is clear in his famous dispute with Robert Fludd, a writer within the magical tradition (Yates 1964).

While in Prague, Kepler discovered the geometrical principles of image formation on the retina. They were set out in *Ad vitellionem paralipomena, quibus astronomiae pars optica traditur* (commentary on Vitello) which was published in 1604 and later in his book *Dioptrice* (1611). Kepler based his analysis on an approximation to the sine law of refraction, which Willibrord Snell discovered some years later in 1621. Since Vittello's book of optics was little more than a commentary on Alhazen's book of optics, one can say that Alhazen provided the foundation for Kepler's work. Kepler established the modern principles of dioptrics and put physiological optics on a firm foundation. As his starting point he considered the two problems raised by Alhazen—the problem of the inverted image in the eye and the problem that clear vision requires that each point in the image plane receives light from only one point in an object plane.

At first Kepler was disposed to accept Alhazen's solution to the inverted image problem by positing a second inversion, but he soon abandoned this idea. He concluded that a mental process was responsible for seeing the image in its correct orientation, which is tantamount to saying that he left the problem unsolved. It was left to William Molyneux, a Dublin lawyer and scientist, in his *Treatise of Dioptricks* (1692, p. 105), to propose the correct answer to the problem. The perceptual system does not have ac-

Figure 2.9. René Descartes (1596-1650).
(Engraving by W. Holl from a painting by Frans Hals in the Louvre, Paris)

cess to the absolute orientation of the retinal image but only to its orientation with respect to that indicated by other sensory modalities. Molyneux had been prompted to approach the problem in this way by reading John Locke's *Essay concerning human understanding* (1619). Molyneux posed the question about what a blind person would see if sight were suddenly restored (Section 13.1.3). Bishop George Berkeley (1709) also stated the correct solution to the problem of the inverted image (Portrait Figure 2.8). For an account of investigations of the effects of inverting the retinal image, see Howard (1982).

Christopher Scheiner (1579-1650), a Jesuit friar who lived in Bavaria, produced the first accurate drawing of an eye, showing the correct curvature of the lens and the correct position of the optic nerve. He proved experimentally that a miniature inverted image is formed on the retina (Scheiner 1619, 1630). He observed the image directly by cutting away the back of a bull's eyeball. Knowledge about the structure of the retina had to wait until the microscope was developed in the 17th century.

With regard to the second problem, Kepler rejected Alhazen's solution of allowing only normal rays to enter the eye, on the grounds that there would be no way to distinguish between normal rays and slightly refracted rays. By analyzing the paths of light rays through a lens, he arrived at the correct solution, which he then applied to the eye. Light radiating out from any point in an object plane is refracted by the lens to form a cone of light, which converges on a point in the image plane. Light radiating from any other point in the object plane converges in another point in the image plane. There is thus a one-to-one mapping of object points onto image points, and therefore no stray refracted light rays dilute the image. Kepler concluded that an inverted and left-right reversed picture of the visual scene is projected on the retina. He wrote:

"Thus vision is brought about by a picture of the thing seen being formed on the concave surface of the retina. That which is to the right outside is depicted on the left in the retina, that to the left on the right, that above below, and that below above, Green is depicted green, and in general things are depicted by whatever colour they have. So, if it were possible for this picture on the retina to persist when taken out into the light by removing the anterior parts of the eye which form it, and if it were possible to find someone with sufficiently sharp sight, he would recognize the exact shape of the hemisphere compressed into the confined space of the retina."

Kepler derived his ideas of the anatomy of the eye and the idea that the retina is the site of image formation from Platter's *De corporis humani structura et usu*. Like Alhazen, Kepler realized that optical principles could not apply beyond the retina, because the optic nerve is not straight and contains opacities. He clung to Galen's notion of visual spirits, but admitted that he knew nothing about visual processes beyond the retina, only insisting that they were not optical. Any useful optical system must change its focal length when the distance of the object changes. Porterfield (1738) introduced the term "accommodation" to describe a change in focal length. He also invented the optometer (Section 9.1.1). Kepler speculated that changes of focus in the eye are achieved by back and forth movements of the lens. In 1671, Rohault (1620-1675) constructed an artificial eye and changed focus by moving the image surface. Scheiner (1630) speculated that the eye could accommodate by changing the shape of the lens. Descartes (1664) adopted the same theory but proof had to wait for experiments conducted by Priestley (1772) and Thomas Young (1801).

2.1.3g *Descartes*

René Descartes (1596-1650) (Portrait Figure 2.9) was born of Jewish parents near Tours in France. He was educated at a Jesuit school where he formed a lifelong friendship with his fellow student Mersenne. After studying law in Poitiers he lived aimlessly in Paris on his private income. He made a sudden decision to travel and joined the Dutch Protestant army of Prince Maurice of Nassau as a gentleman volunteer. The Dutch were fighting for independence from Spain. Then, after wandering through northern

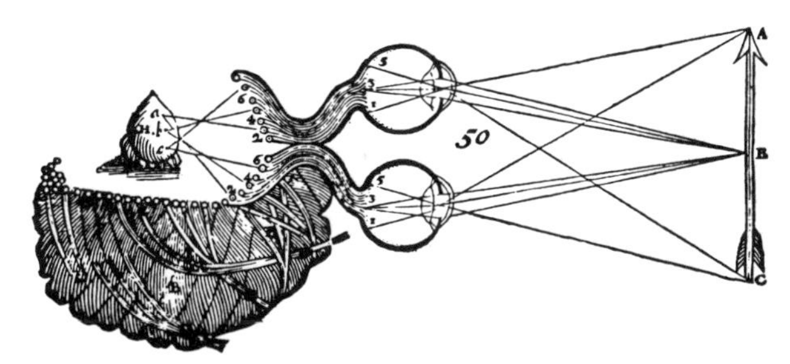

(a) Descartes (1664) used Kepler's theory of image formation. Each optic nerve projects to the lining of the ipsilateral cerebral ventricle, an idea taken from Galen. Pairs of corresponding points map onto the pineal gland to form a unified image.

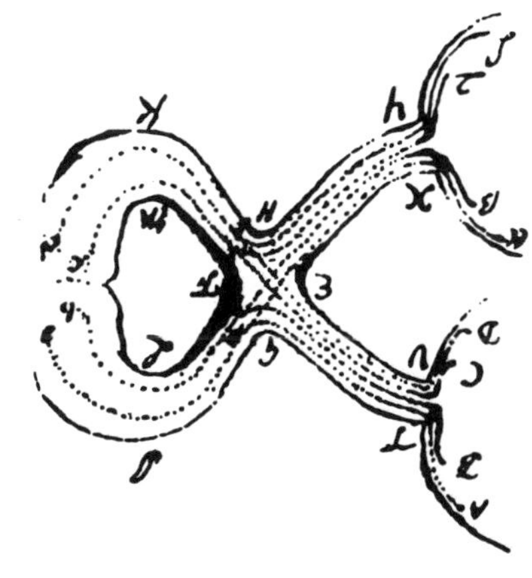

(d) Newton, in about 1706, was the first to suggest hemidecussation of the pathways. He erroneously fused corresponding fibres in the chiasma. (Unpublished manuscript, see Crone 1992)

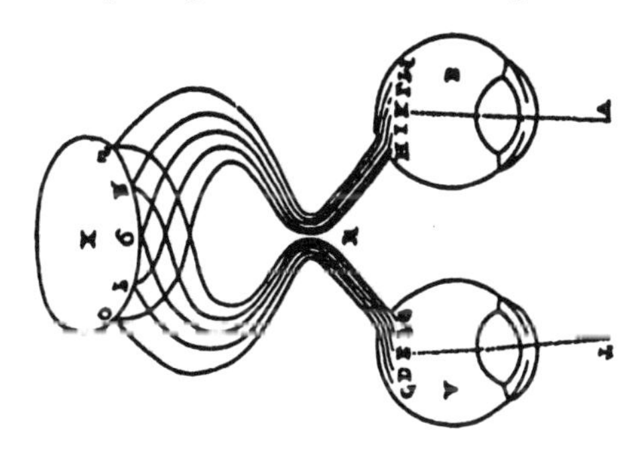

(b) Rohault (1671) adopted ipsilateral projection and combined corresponding fibres in the brain.

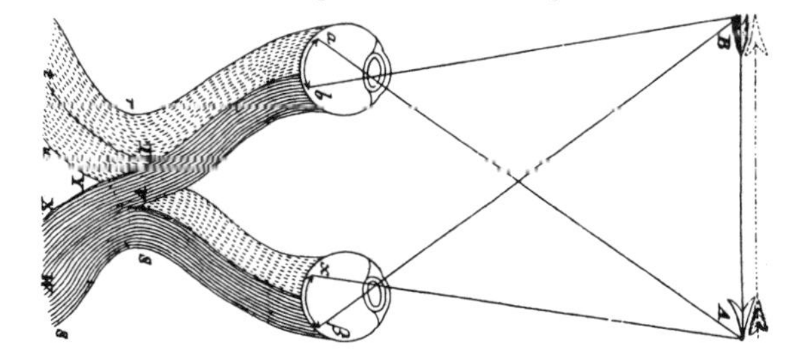

(e) Taylor (1738) corrected Newton's erroneous idea of fusion of fibres in the chiasma

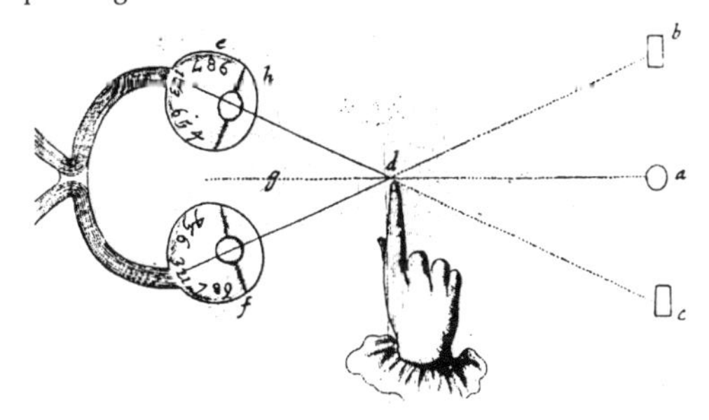

(c) Briggs (1676) adopted ipsilateral projection and described the optical projection of corresponding images.

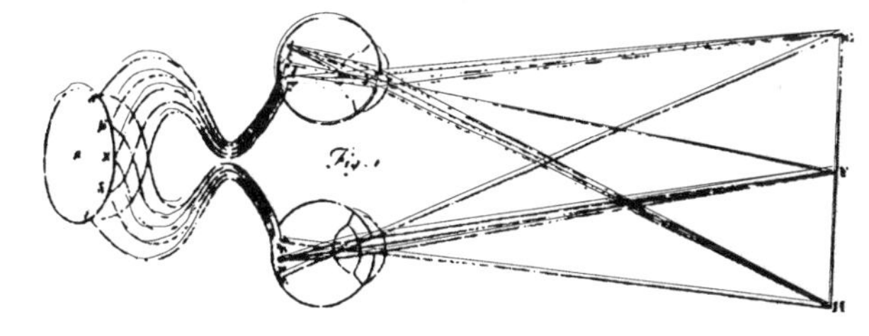

(f) Porterfield (1759) rejected Newton's hemidecussation, but showed how an object projects to corresponding retinal points.

Figure 2.10. Early drawings of the visual pathways.

Europe, he joined the Catholic army of the Duke of Bavaria. He seems to have had an easy life as a gentleman soldier during one of the most horrific of wars—the Thirty Years' War of 1615 to 1645, which left most of Germany in ruins. During this time he invented analytic geometry. After further wandering he moved to Holland to devote himself to mathematics and philosophy. In Holland, he was safe from the clutches of the Inquisition. This was the time when Galileo was brought before the Holy Office, heretics were burned at the stake, and those accused of witchcraft were burned by Catholics and Protestants alike. Nevertheless, he was very cautious about what he published. In 1649 he was reluctantly persuaded to move to Stockholm to teach philosophy to 23-year-old Queen Christina but died of pneumonia during the first winter at the age of 54. His *Traité de l'hommme* (*Treatise on Man*), published in 1664 is available in English translation (Hall 1972). Descartes adopted Kepler's ideas of image formation but retained Galen's notions of animal spirits. He believed these spirits were formed in the pineal gland. He proposed that fine filaments in the optic nerves opened valves in the walls of the ventricles and pineal gland and allowed animal spirits to flow to muscles. These mechanical ideas were inspired by his interest in clocks, animated statues, and toys. Descartes saw the human body, as an automaton controlled by the mind and soul operating through the pineal gland. The consequences of this misguided duality concept—this unfortunate division between the material and mental—are still with us.

Figure 2.11. Santiago Ramon y Cajal (1852-1934).
Professor at Valencia, Barcelona, and Madrid. He was 55 in this photograph (From Polyak 1957) .

Descartes believed that each optic nerve projects to its own side of the brain. He thus moved the site where inputs from the two eyes form a united image back from the chiasma into the brain. He proposed that each fibre in the optic nerve projects from a location in each retina to a specific location on the lining of the ipsilateral cerebral ventricle. From there, inputs from the two eyes combined in the pineal gland, as shown in Figure 2.10a. This figure seems to have been added after Descartes death (see Wade 1998a). Wade concludes that it is unlikely that Descartes had a clear idea that each point from one eye projects to the same location in the pineal gland as the corresponding point from the other eye, to form a unified impression of a single object. The French Physicist, Jacques Rohault (1671), clearly suggested that the ipsilaterally projecting corresponding fibres from the two eyes converge somewhere in the brain (Figure 2.10b). Descartes' ideas of brain structure and function were largely speculative.

Descartes, in his *La Dioptrique* (1637), described the eyes as "feeling out" a distance by the convergence of the visual axes, just as a blind man might feel out a distance with two staves, one in each hand (Section 24.6).

2.1.4 Microscopic structure of the visual system

It is not clear who first made a microscope. The Jesuit priest Athanasius Kircher (1602-1680) was perhaps the first microscopist. His *Ars Magna et Umbrae* contained observations on light and lenses and a description of his microscope. Marcello Malpighi (1628-1694) was the greatest of the 17th century microscopists and the founder of histology. He was professor of anatomy at Bologna, Pisa, and Messina and, during the last three years of his life, became physician to Pope Innocent XII. His *Pulli in Ovo* of 1673 contains descriptions of the neural groove, and the cerebral and optic vesicles. The Dutchman Anton Van Leeuwenhoek (1632-1724) made his first microscope in Delft in 1671 and continued to make microscopes of increasing quality. He studied bacteria and animal tissue, including nerve tissue. In 1674 he sent a communication to the Royal Society of London in which he reported that he could not find any canals in the optic nerve, as claimed by Galen.

Robert Hooke (1635-1703), professor of mechanics to the Royal Society, constructed the first compound microscope and published drawings of the microscopic structure of common materials in his *Micrographia* of 1667. Images in early microscopes were distorted. It was not until the early 19th century that the details of retinal structure began to be known.

In 1830, Purkinje demonstrated that shadows of retinal capillaries are visible when one looks through a pinhole. Heinrich Müller (1854), professor of anatomy at Würzburg in Bavaria, measured the parallactic displacement of the shadows for a given motion of a light source and, from this, computed the distance that the receptive layer must lie behind the vessels. This distance was the same as the anatomically determined distance between the layer containing the rods and cones and that containing the blood vessels. He thus deduced that photosensitivity occurs in the outer segments of the rods and cones. Schultze (1866) demonstrated that the fovea contains only cones. Knowing that the fovea and colour vision do not function in dim light, he deduced that cones are responsible for colour vision.

In 1873 Camillo Golgi (1843-1926) discovered that silver nitrate stains many nerve cells to reveal their fine structure. However, Golgi failed to observe synapses and believed that nerve cells form continuous networks.

Santiago Ramón y Cajal (1852-1934) (Portrait Figure 2.11) was the founder of modern neuroscience.

He was born in the village of Petilla de Aragón in the Northeast of Spain, the son of a doctor. As a young man his passion was art but, because of his father's opposition, he agreed to study medicine in Zaragoza. After some years as an army physician in Cuba he obtained academic appointments in Zaragoza, Valencia, Barcelona, and finally Madrid (see Ramón y Cajal 1937). He used the Golgi staining method to reveal the organization of retinal layers in great detail (Section 5.1.2). He formulated the idea that information is received by dendrites and transmitted to other nerve cells by the axon. He remained unsure about how information was passed from one cell to another, but rejected the idea that nerves form a continuous network. He discovered the growth cone and formulated the theory of neurotropism (Section 10.2.3).

Cajal wrote papers only in Spanish in a journal that he founded. His three-volume work *Textura del Sistema Nervioso del Hombre y de los Vertebrdos* was published between 1899 and 1904. This work became generally known only after it was translated into German by Albrecht von Kölliker and into French by Cajal's friend Leon Azoulay as *Histologie du system nerveux de l'homme et des vertébrés* (1911). Stephen Polyak visited Cajal in 1924 and was inspired to apply his methods to the primate retina, including the human retina. Polyak's monumental book *The Retina* appeared in 1941. In 1906, Cajal and Golgi shared the Nobel Prize in Physiology. In 1901, Cajal published a paper on stereopsis and binocular vision in a journal of photography in Madrid, in which he described a form of random-dot stereogram (Bergua and Skrandies 2000) (Section 30.1.4).

In 1870, Bernhard von Gudden confirmed that the human visual pathways hemidecussate. Von Gudden was an eminent neuroanatomist and professor of psychiatry in Zürich and Munich (Duke-Elder 1961). King Ludwig II of Bavaria was one of his patients. The politicians were annoyed with King Ludwig's extravagance and asked von Gudden to certify him insane and have him incarcerated in Schloss Berg. On the second day of incarceration, in June 1886, the king asked von Gudden to walk with him by lake Starnberg. Both men were later found drowned in shallow water. It is generally believed that the king killed von Gudden and then drowned himself, but this was never proved (Blunt 1970).

Figure 2.12. Painting on a Greek vase (kantharos).
From the second half of the fifth century B.C (British Museum)

Figure 2.13. Greek parallel perspective.
From an Apulian calyx-krater, 4th century BC Würzburg.

2.1.5 Discovery of cortical visual areas

Our knowledge of the brain derives from scientists such as Thomas Willis (1621-1675), Luigi Galvani (1737-1798), Paul Broca (1824-1880), David Ferrier (1843-1928), Eduard Hitzig (1838-1907), and Charles Sherrington (1857-1952). A lively account of this history is provided by Finger (2000).

Hermann Munk (1839-1912), professor of physiology in Berlin, established the occipital cortex as the site of the primary visual cortex, although he erred in placing the primary visual cortex on the lateral surface of the occipital lobe rather than at its posterior pole (Munk 1879). By studying the effects of brain lesions, Salomon Henschen (1847-1930) in Stockholm established that the visual cortex lies in and around the calcarine fissure, but was uncertain about the representation of the fovea. By observing the effects of central scotomas, H. Wilbrand (1890), in Hamburg deduced that each location in the visual cortex receives inputs from a pair of corresponding locations in the two eyes. Anatomists known as the decentralists, including D. Ferrier, C. von Monakow, and E. Hitzig, objected to the idea of a precise cortical mapping of visual space, mainly because they had observed visual defects from damage to many parts of the brain. We now know that defects arise from visual areas other than the primary visual area.

Ramón y Cajal suggested that crossed and uncrossed axons from the two eyes remain distinct in the visual pathway. In 1913, Mieczyslaw Minkowski, Director of the Brain Research Institute in Zürich, was the first to establish that they terminate in distinct layers of the LGN. Minkowski also helped to establish that the visual cortex is the main recipient of visual inputs and that the visual field is mapped in a precise and stable manner. In 1905 Alfred W. Campbell (1868-1937) of Sydney published his *Histological Studies on the Localization of the Cerebral Function*. See Polyak (1957) for a review of early studies on the visual pathways and cortex.

Clare and Bishop (1954) were the first to discover a visual area outside the primary visual cortex. This was the suprasylvian cortex of the cat, also known as the Clare-Bishop area. In the 1970's, Zeki revealed a series of visual areas in the monkey, as described in Section 5.7.

2.1.6 The discovery of perspective

2.1.6.a *Perspective in the ancient world*

In paintings from ancient Egypt, Babylonia, and Assyria, depth was represented by overlapping images, but there was no hint of any type of perspective. Human figures were drawn in front view or profile or in a combination of front view and profile. Chariots were drawn in side view with one circular wheel. Greek paintings before about 500 BC were limited in a similar way. By about 500 BC, human figures began to be drawn at an oblique angle and wheels and shields seen at an angle were drawn as ellipses, as in Figure 2.12 (Knorr 1992). In a handbook on architecture written in about the year 25 BC Vitruvius wrote that Agatharcus of Samos (525-456 BC) had invented a method for painting scenes for the theatre of Dionysus in Athens so that they appeared in depth. Such procedures were known as scenography (*skenographia*). Plutarch describes how Alcibiades kidnapped a scene painter and forced him to decorate the walls of the aristocrat's mansion (Little 1971). The practice of painting the interiors of mansions with frescos in perspective continued into Roman times. By about 425 BC, foreshortening began to show in Greek vase paintings. Half open doors were represented foreshortened with inclined top and bottom edges, but the receding edges were parallel rather than converged. Foreshortening and oblique parallel lines were also used in drawings of buildings, as shown in Figure 2.13. About 300 BC, Euclid provided some geometrical analysis of perspective in his *Optics* (Section 2.1.1b).

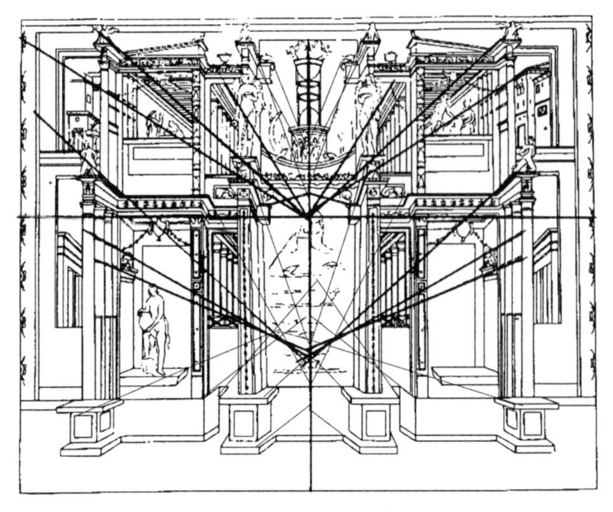

Figure 2.14. Drawing made from a mural in Pompeii. Added bold lines approximately converge on two vanishing points. Added fine lines converge on other points usually, but not always, in the midline. (Adapted from Little 1971)

Frescos discovered in Pompeii have convergent perspective, as shown in Figure 2.14. These date from before AD 79, the year that Pompeii was engulfed by the eruption of Mount Vesuvius. The added construction lines show that the artist used two vanishing points in the midline. Perhaps the upper one was designed for a standing observer and the lower one for a seated observer. Not all receding lines converge to one of these vanishing points, suggesting that artists used an intuitive procedure rather than vanishing points. Several murals in perspective have also been unearthed near Rome. Textual evidence reveals that the Romans, like the Greeks, painted stage sets in perspective (Richter 1970; Little 1971). In his handbook on architecture Vitruvius wrote,

" *after a fixed point in space had been determined for the line of sight and the diffusion of the rays, the lines of the painting were bound in keeping with natural principle to correspond to that centre, with the result that in the paintings of the stage the definite images given by an unsubstantial representation has the appearance of actual buildings in which some of the elements represented on the flat vertical panels of the stage setting seemed to recede in space and others to project outwards.*"

(a) Illusory divergence of parallel lines on the apparently receding sides of the drawing of an object.

(c) A fresco in the monastery of Decani in Yugoslavia painted in the 13th or early 14th century. Note the divergence of the sides of the cart and the haphazard perspective on the towers. (From Mijovic 1966)

(b) A fresco in the Partriarchate of Pec in Yugoslavia painted between 1335 and 1350. Note the divergence of the sides of the bench and tower. (From Ljubinkovic 1964)

(d) A fresco from the grotto of Touen Houang, China from period of the Tang dynasty (618-906). Note the divergence of the sides of the table. (From Fourcade 1962)

Figure 2.15. Medieval divergent perspective.

Opinions differ about whether the word "centre" refers to a vanishing point in the picture or the centre of the eye. Although this passage describes the general principle of linear perspective it does not provide an exact geometrical procedure for drawing in perspective.

The practice of map making influenced the development of painting in perspective. In 13th and 14th century Italy, maps of the Mediterranean were made for navigation. These so-called portolan maps consisted of stretches of coastline with place names and a superimposed lattice of radiating direction lines. In 1395 some prominent citizens of Florence formed a study group to learn classical Greek. In 1400, two members of the group, Manuel Chrysoloras and Jacopo d'Angiolo, journeyed to Constantinople in search of early Greek texts and returned to Florence with Ptolemy's *Geographia*, which was unknown in Western Europe. This 2nd century book contained eight books and 27 maps, one of which was of the whole of the known world from Sweden and Russia in the North to the Nile in the South, and from Gibraltar in the West to India in the East. The work was translated into Latin in 15th century Florence. Most importantly, the work provided three procedures for mapping lines of longitude and latitude on the curved surface of the earth onto a flat surface (Edgerton 1975). In one procedure the eye

(a) *Last Supper* by Duccio. Siena, Mus. dell' Opera del Duomo.

(b) *Jesus Before the Caïf,* by Giotto (1305). The ceiling rafters show convergent perspective but the inconsistent vanishing point is above the horizon. Edges of the dais have parallel perspective.

Figure 2.16. Perspective in 14th century Italy.

Figure 2.17. The baptistery of San Giovanni, Florence.
View from the door of the cathedral. (Photograph by Brunner in Damasch 1994)

was placed in the plane of a defined line of latitude which was drawn as a straight horizontal line on the map. The central line of longitude was then drawn as a vertical line. With the eye as the centre of projection, other lines of latitude and longitude were projected into the picture plane as curved lines. Ptolemy increased the spacing of lines of latitude as they approached the North Pole, to avoid the crowding that strict projection creates. Unlike the direction lines of the earlier portolan maps, the lines of latitude and longitude precisely indexed locations on the map. The eye is mathematically equivalent to the projection centre, and the central line of latitude is equivalent to the horizon line of a perspective transformation. Thus, Ptolemy had developed the basic principles of drawing in perspective in the 2nd century, although he applied them only to map making. The arrival of the *Geographia* in Florence in 1400 was one of the factors that contributed to the development of perspective in art in that city. It was also a major factor inspiring Columbus to sail across the Atlantic.

M. Friendly has produced an illustrated history of map making and other graphic procedures (see http://www.math.yorku.ca/SCS/Gallery/milestone/

The Latin word *perspectiva* refers to the tradition of applying geometry to vision, developed by Euclid, Ptolemy, Alhazen, and the 'perspectivists' of medieval Europe, as described in Section 2.1.1b. This tradition culminated in Kepler's discovery of the optical principles of image formation. As we shall see, practising artists developed methods of drawing in perspective without reference to the perspectivists, but the first formal description of perspective and the subsequent development of projective geometry owed something to the perspectivists.

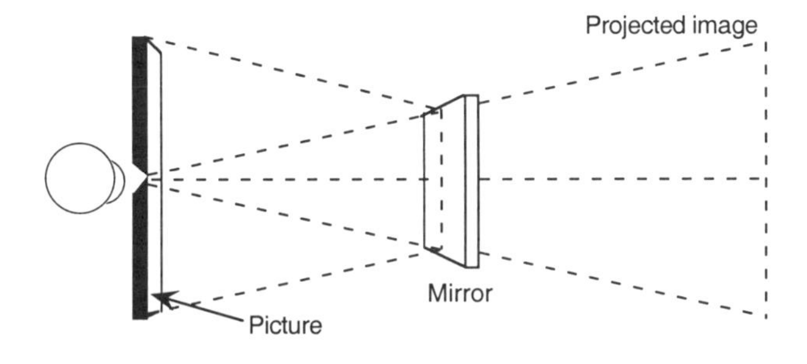

Figure 2.18. Illustration of Brunelleschi's viewing device. The observer stood behind the picture and looked through a hole at a reflection of the picture in a mirror.

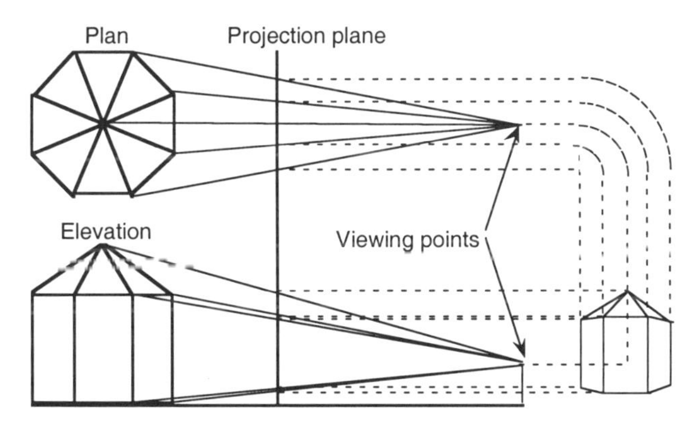

Figure 2.19. Perspective method possibly used by Brunelleschi. (Adapted from Carter 1970)

2.1.6b *Perspective during the 14th century*

During the medieval period in China and Europe, artists used overlap, height in the field, foreshortening, and oblique lines to represent depth. There was no attempt to represent parallel lines converging into the distance. In fact, parallel lines were often drawn diverging rather than converging with distance, as in the examples shown in Figure 2.15. This may be because lines in parallel perspective appear to diverge, as shown in Figure 2.15a. An artisan copying a sketch drawn in parallel perspective onto a fresco would draw the illusion. This could then serve as a model for other artisans and the effect would get larger as successive generations of pupils copied from their masters.

The use of convergent perspective did not emerge again until the early 14th century (Bunim 1940). During the 14th century, Duccio, Giotto and his pupils the Lorenzetti brothers, began to use converging lines to depict ceilings and floors, as shown in Figure 2.16. Avanzano and Giusto used the same procedures in Padua during the last quarter of the 14th century. However, while lines depicting parallel receding edges sometimes converged to a point, the same vanishing point was not used consistently for different parts of the painting. These artists followed certain basic rules that had been described by Cennino Cennini before Giotto. He wrote, "the mouldings which you make at the top of buildings should slant downwards from the edge next to the roof; the mouldings half way up the face must be quite level and even, the mouldings at the base of the building must slant upwards in the opposite direction to the upper mouldings." (see Kemp 1978).

In drawing tiled floors or ceilings in perspective, artists of the 14th century seem to have used one or two distance points (see Section 24.1.1). Distance points are points on the horizon to which lines at 45° converge. For example, the fresco in the Lower Church at Assisi has a tiled ceiling drawn in accurate perspective, even though the painting as a whole is not in accurate perspective (Klein 1961). There are still brackets exactly where the two distance points are located. It was apparently a common practice in early 14th century Tuscan workshops to draw the diagonals by rotating a piece of string attached to nails (Panofsky 1927). Thus, the artisans of the 14th century were using distance points before the use of the principal vanishing point and well before Viator gave the first written account of distance points in his *De artificiali perspectiva* of 1505. These artisans at least got as far as the Romans. Progress must have been delayed by the advent of the Black Death in 1346.

2.1.6c *Perspective in the renaissance*

A precise procedure for drawing in linear perspective was discovered by Filippo Brunelleschi (1377-1446), the architect who designed the cathedral in Florence. In about 1420 he painted the baptistery as seen from a distance of about 35 m, which is a position inside the door of the cathedral, approximately as seen in Figure 2.17. (Damasch 1994) The painting has not survived. According to Brunelleschi's biographer, Antonio Manetti (1423-1491), people stood in the cathedral doorway and, while facing the back of the painting, looked through a small hole in its centre, as shown in Figure 2.18. The hole was as small as a "lentil" on the painted side and widened to the size of a "ducet" on the viewing side. A suspended mirror produced a reflection of the painting, which filled the same visual angle as the baptistery. Instead of painted sky, burnished silver on the picture surface reflected the real sky and drifting clouds. People were fooled into believing that they were looking at the actual building.

Brunelleschi's second painting in perspective was of the Palazzo della Signoria in the Palazzo Vecchio. It was apparently larger and viewed directly with the parts corresponding to sky cut away so that

Figure 2.20. Masaccio's *Trinity* fresco.
This fresco was painted in the church of Santa Maria Novella, Florence in 1425. It is the oldest known painting with a unified vanishing point.

when viewed from the correct position, the skyline in the painting coincided with the skyline of the real buildings in the Palazzo. This painting has also not survived.

There are three theories about how Brunelleschi produced his paintings. The first theory is that, as an architect, he used the measuring instruments and geometrical constructions available to architects at that time. Instruments included rods, squares, quadrants, and mirrors. Geometrical constructions consisted of plans and elevations of buildings. In his *Life of Brunelleschi* (1550) Vasari stated that Brunelleschi drew lines from a plan and an elevation of the building he was painting to intersect planes according to what is now known as orthographic projection (see Hyman 1974). Carter (1970) suggested that Brunelleschi used the construction depicted in Figure 2.19. This resulted in the painting being a mirror image of the scene, which would explain why Brunelleschi used a mirror to exhibit the painting.

The second theory is that Brunelleschi traced the reflected image of the scene on the surface of a mirror (Krautheimer and Krautheimer-Hess 1982). For example, Lynes (1980) concluded that Brunelleschi used a polished metal sundial. The mirror inversion of the picture would not be apparent in the painting of a symmetrical building seen through an arch. But he must have realized later that the mirror image would be corrected when the picture was viewed by reflection through a hole in the picture. If he had drawn a painting that was not a mirror image it could have been viewed directly through a hole in a panel and there would have been no need for a hole in the picture. He need not have drilled a hole because a sundial would have had a hole to support the gnomon. Sundials contained a series of etched lines converging on the gnomon, which may have prompted the idea of perspective. Where there was sky, Brunelleschi could have simply left the polished surface of the mirror that reflected the real sky. Brunelleschi was a friend of the mathematician Toscanelli who used a large sundial on the cathedral in Florence to measure the altitude of the sun at noon. One problem with using a mirror is that the head of the viewer obstructs the view of the scene. Lynes suggests that Brunelleschi overcame this problem by standing back from the mirror and, with the head in different positions, marking the mirror with each point of the scene as it aligned with the reflection of the end of a rod inserted into a hole in the mirror. This could have been the same hole that supported the gnomon. Lynes also suggests that Brunelleschi painted the second picture by tracing it on the window of a church. In this case, no mirror was required.

Neither of the above two methods involves the explicit use of a vanishing point, although the paintings would contain a unified vanishing point. The third theory about Brunelleschi's method is that he used geometrical principles of perspective. We have seen that the basic ideas underlying perspective were expressed in Euclid's Optics, and fourteenth century painters came very close to the correct solution.

A few years after Brunelleschi painted the baptistery some painters in Florence began to use a unified vanishing point. For example, Masaccio, a friend of Brunelleschi, painted the *Trinity* fresco in the church of Santa Maria Novella in 1425 (Figure 2.20). It has a unified vanishing point at eye level and is the oldest known painting with a perfect vanishing point. It has been suggested that Brunelleschi drew the sketch for this painting but the evidence is not conclusive (ten Doesschate 1964).

In 1420, Ghiberti used inconsistent vanishing points in his first baptistery door. Two of the panels of the Gates of Paradise, made in the 1420's and 30's,

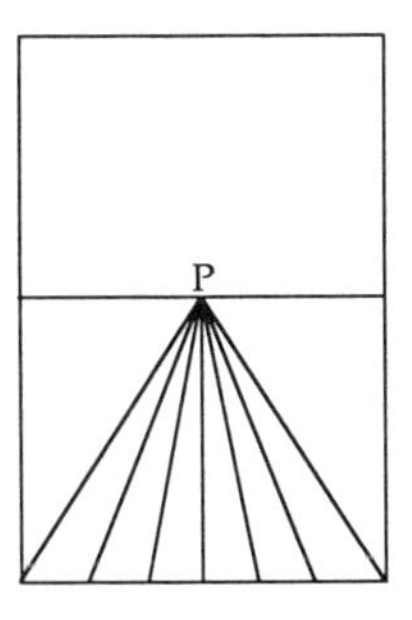

(a) A vanishing point, P, is placed on the picture plane at eye height and diverging lines, representing the images of receding parallel lines, are drawn to the base of the picture.

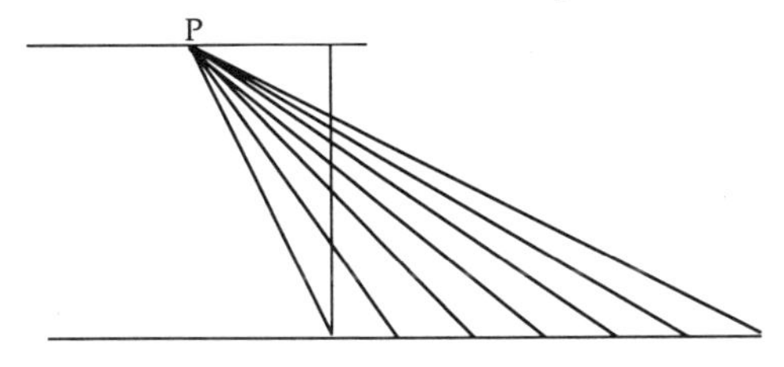

(b) Intersections of visual lines to equally spaced transversals on the ground plane are plotted on the side of the picture plane.

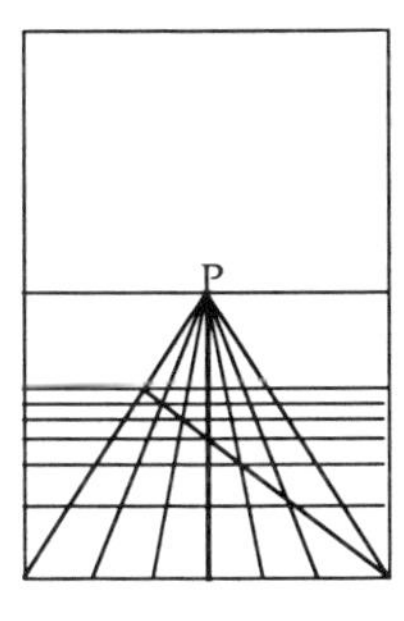

(c) The spacing of foreshortened images of transversals is checked by drawing a diagonal across the picture plane.

Figure 2.21. Alberti's construzione legittima.

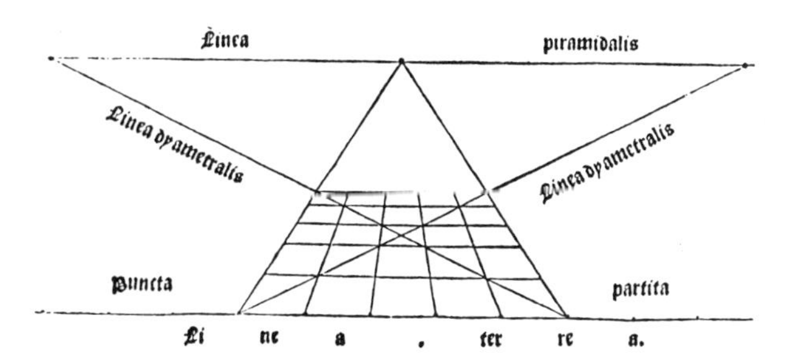

Figure 2.22. Diagram illustrating distance points.
Linea terrea is the ground line, *linea piramidalis* is the horizon line. The diagonals (*linea dyametralis*) end in the distance points on the horizon line. (From Viator's *De Artificiali Perspectiva* of 1509)

embody a central vanishing point (Parronchi 1964). Strangely, the distance point of one picture was the principal vanishing point of the other picture. The distance points and vanishing points are not on the same horizon, as they should be. Other artists continued to use inconsistent vanishing points well after 1420 (Krautheimer and Krautheimer-Hess 1982).

In 1435 Leon Battista Alberti (1404-1472) wrote *Della Pittura*. It was available only in manuscript form until it was published in Basle in 1540. The book is the first account of the geometry of drawing in perspective using a single vanishing point. Alberti's father, Lorenzo, was a banker who had been banished from Florence by a rival banking family. Leon Battista grew up in Genoa and Padua and in 1431 became architect to Pope Eugene IV. He was in Rome between 1431 and 1434 where he surveyed buildings and composed Latin letters for the pope. There is no indication that he saw ancient Roman murals drawn in perspective. He travelled with the Pope to Northern Europe and in 1434 arrived in Florence just as Brunelleschi was completing the dome of the cathedral, Donatello was completing the sculptured façade, and Ghiberti's doors were newly installed on the Baptistery. This so impressed Alberti that he devoted himself to making the art of Florence understandable to a wide audience.

Before the Renaissance, artists were tradesmen organized into guilds. Painting and sculpting were not part of the Liberal Arts, which included philosophy, grammar, dialectic, mathematics, and astronomy. Nor were they part of Mechanical Arts, which included architecture, navigation, and medicine. This attitude probably originated in Rome where artists were usually slaves. Also, Plato had condemned painting because he argued that visual perception is subject to errors, which are compounded in painting. Plato, respected only knowledge based on mathematical certainty. Alberti and renaissance artists saw perspective as a way to ground painting in mathematics and thus elevate it to the level of geometry and astronomy.

Over a period of time, leading architects and artists such as Alberti, Brunelleschi, Donatello, Uccello, Verrocchio and his apprentice Leonardo da Vinci, gathered at the home of Paolo Toscanelli on the banks of the Arno. Toscanelli was a physician and leading mathematician. Leonardo learned mathematics from him. Brunelleschi and Leonardo da Vinci were also close friends of the Franciscan monk Luca Pacioli. He too was a leading mathematician and wrote the first treatise on double entry bookkeeping. Leonardo drew the diagrams for Pacioli's book on geometry, *De Divina Proportione* in which he praised linear perspective.

Like other humanists of the renaissance, Alberti was fascinated with antiquity and read Greek and Roman authors, including Galen, Euclid, Ptolemy, and Vitruvius. He was aware of the contributions of Alhazen through the writings of Roger Bacon, John

Figure 2.23. An error in the use of three-point perspective.
The box marked S is inclined to the horizontal. Vanishing point P1 should therefore be above the horizon line. Vanishing point P2 is correct since the corresponding sides are both horizontal. (From *Perspective* by Jan Vredeman de Vries, 1604)

Peckham, and Vitello. He adopted the concept of the visual pyramid and centric ray from Galen and relied on Euclid's theorem 21 to establish that the size of an image in the picture plane is inversely proportional to the distance of the object. He became interested in cartography and applied ideas from Ptolemy's *Greographia* to construct a map of Rome. He used a grid based on polar co-ordinates with the origin on the Capitoline hill (Edgerton 1975).

In constructing drawings in perspective Alberti probably used a small box with a peephole, and a floor marked out as a square grid. The picture plane was first drawn in front view sitting on a horizontal ground plane marked off into a grid of equal squares. A vanishing point was placed at eye level on the midline of the picture plane (Figure 2.21a). Diverging lines were drawn from the vanishing point to points where the grid lines intersected the ground line of the picture plane. These lines indicated the images of receding parallel lines on the ground plane. The picture plane was then drawn as a line in side elevation. The station point was drawn at eye level at a designated distance from the picture plane (Figure 2.21b). Visual lines were drawn from the station point to each of the equally spaced transverse lines on the ground plane. Horizontal lines were drawn on the picture plane through the points where the visual lines intersected the picture plane, as shown in Figure 2.21c. The accuracy of these transversals was checked by drawing diagonal lines across the picture plane. Each diagonal should intersect the corners of all the squares that it traverses. Opposite diagonals intersect the horizon line in two distance points. The distance between a distance point and the central vanishing point equals the distance between the painter's eye and the picture (see Section 24.1.1).

Alberti made no reference to distance points and was thus ignorant of or discarded the methods used by 14th century painters. Alberti's method was known as the **construzione legittima**. But Alberti also described drawing in perspective using a glass plate or of a lattice of orthogonal threads suspended on a vertical frame, a method that became known as Leonardo's window (see Section 2.1.6d).

Jean Pélerin (pseudonym Viator) (c 1445-1524) gave the first account of distance points in his *De artificiali perspectiva* of 1509 (see Figure 2.22). However, we saw in the last section that distance points were used by painters in the 14th century. The first and second editions of Viator's book are reproduced in Ivins (1973). Viator was not an artist but secretary to Louis XI and, later, a canon of the Benedictine abbey at Toul.

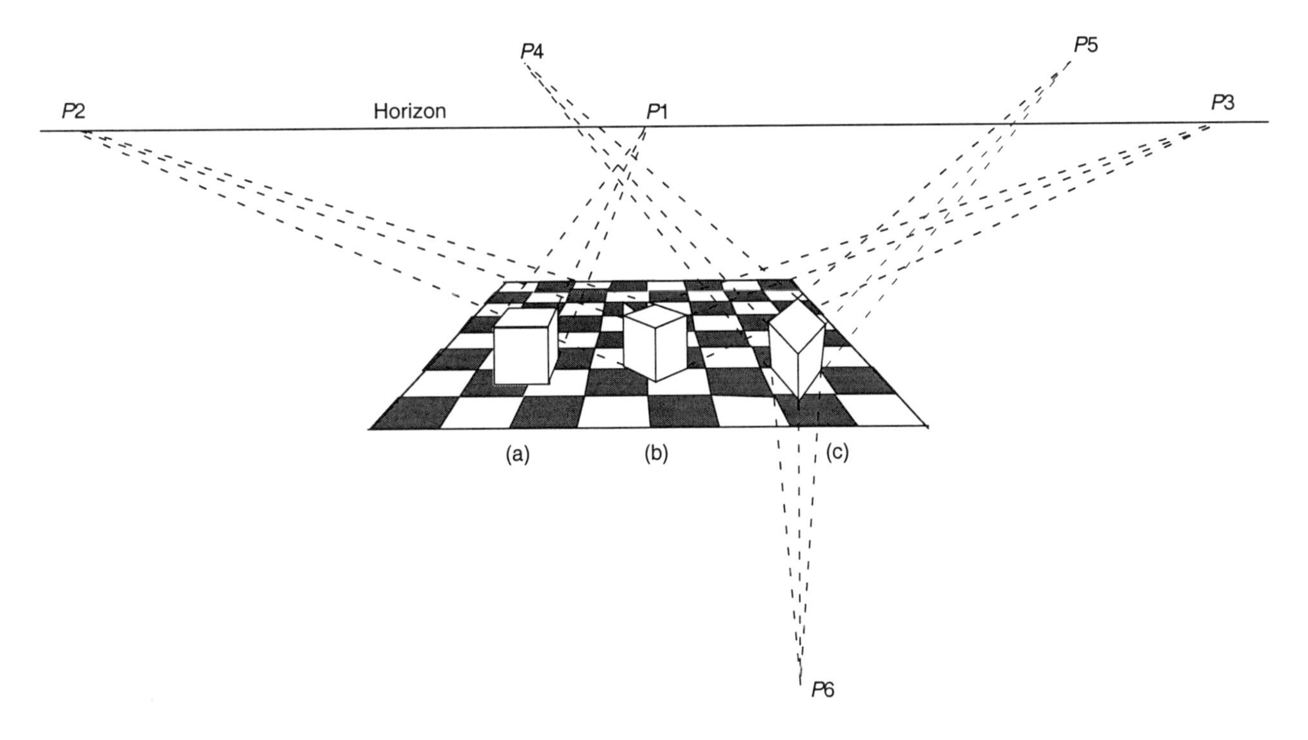

Figure 2.24. One- two- and three-point perspective.
(a) Cube on horizontal surface and orthogonal to picture plane has a single vanishing point, P1.
(b) Cube on surface at about 45° to picture plane has two vanishing points, *P2* and *P3*, on the horizon.
(c) Cube inclined forward has three vanishing points, *P4*, *P5*, *P6*, none of which are on the horizon.

The first drawings in partial or full perspective did not rely on the perspectivist tradition of geometrical optics originating in Euclid, Ptolemy, and Alhazen. Many of the painters were probably illiterate and, in any case, would not have had access to ancient texts. Alberti did consult ancient texts and his method derives mainly from perspectivist geometry rather than from practical methods used by artisans. The geometrical theory of visual optics became known as **perspectiva naturalis** because it dealt with the formation of the natural image in the eye. The geometry of perspective drawing became known as **perspectiva artificialis**, or *perspectiva practica*.

The retina is spherical but pictures are painted on flat surfaces. This has led to the mistaken notion that a drawing in accurate perspective does not produce the same image on the eye as the 3-D scene (Panofsky 1940). But in polar projection, the scene, the drawing, and the retinal image of the scene are projectively equivalent when the eye viewing the picture is in the same location as the eye of the painter. A picture may appear distorted relative to the original scene (Section 24.1.2d). But this is not due to perspective but to the visual cues indicating that the drawing is flat. When these cues are removed, the distortions are no longer present (Section 30.1.6).

Paolo Uccello (1396-1475) was an exponent of perspective. Piero della Francesca (1414-1492) wrote *De Prospectiva Pingendi*, which circulated only in manuscript form during the Renaissance (see Field 1986). Daniele Barbaro (1513-1570), Chronicler of the Venetian Republic, included large parts of Piero's treatise in his *La Practica della Perspettiva* of 1569.

Leonardo da Vinci used the *construzione legittima* until about 1500. He then became dissatisfied with it because it took no account of the curvature of the eye or of movements of objects and the eye. He was frustrated by the fact that a flat picture never looks as solid as the real object. He described how an object obscures from each eye a different part of the distant scene but did not describe any other kind of binocular disparity. In his later drawings he portrayed objects in isolation rather than in architectural settings and strove to represent motion in the form of waves, whirlpools, birds in flight, the human figure in action, and the effects of eye movements on the appearance of things (Strong 1979, p. 392). He would have been delighted to see a stereogram or a stereoscopic movie.

Albrecht Dürer learned about perspective during a visit to Bologna in 1506, probably from Luca Pacioli, an associate of Alberti. He had access to the manuscript of *De prospectiva pingendi* by Piero della Francesca, and to the manuscript version of Alberti's

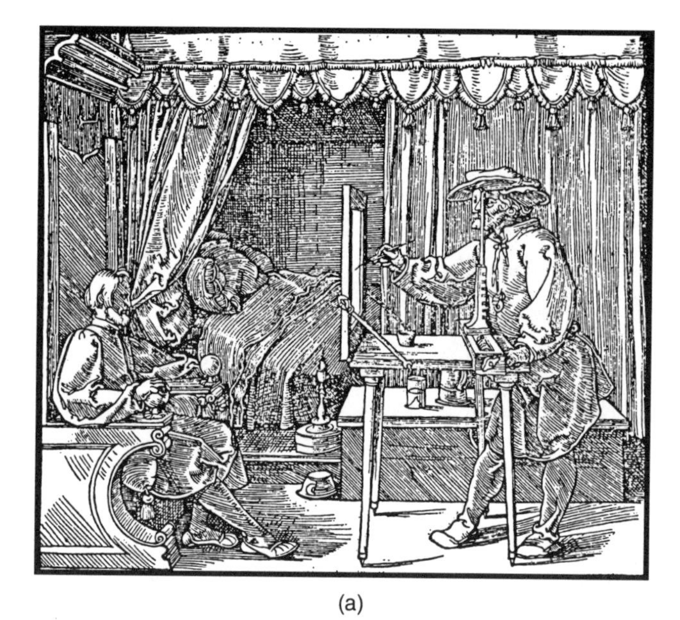

(a)

(b)

Figure 2.25. Two types of Leonardo's window.
From Abrecht Dürer 's *Unterweisung der Messung* (Nuremberg, 1538)

Della pitura. But Viator's *De artificiali perspectiva* seems to have had the greatest effect because, soon after it appeared, the perspective in Dürer's drawings and prints became more precise and some of Dürer's drawings resemble those in the first edition of Viator. In the second edition of his book, Viator reciprocated by using some of Dürer's drawings (Strauss 1977). In 1525 Dürer published the book *Unterweisung der Messung,* which remained the standard German work on perspective for some time. Dürer developed his own method, which he called *näherer Weg* or "shorter way". This was Alberti's *constuzione legittima* but with the station point moved to the projection plane. This change revealed that Dürer had an imperfect understanding of the geometry and it introduced distortions into Dürer's paintings (see Carter 1970; Ivins 1973).

Jan Vredeman de Vries (1527-1604) was a Dutch architect and painter who worked in Antwerp, Danzig, and at the court of Rudolph II Prague. He wrote illustrated pattern books for apprentice architects and engravers. In 1604 he published *Perspective,* which became the leading book on perspective in Holland. The book was reprinted by Dover in 1968. The book contains only illustrations, one of which is shown in Figure 2.23. In 1616 another Dutchman, the mathematician Marolois, produced a book that contained a geometrical analysis of perspective. This became the standard text for Dutch artists who sought a theoretical understanding of perspective.

Guidobaldo del Monte, in his *Perspectivae libri sex* of 1600 pointed out that perspective is not affected by rotation of the eye (see Frangenberg 1986). Pozzo (1642-1709), clergyman, painter, and architect,

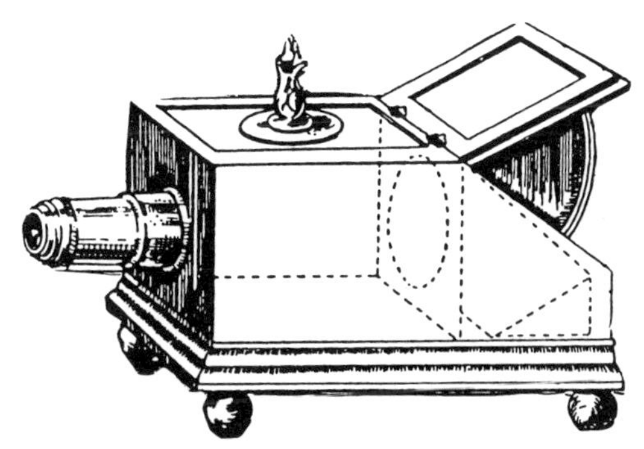

(a) Camera obscura made by Johann Zahn in 1685. The image reflects onto thin paper placed on a plate of glass.

(b) An 18th century camera obscura.

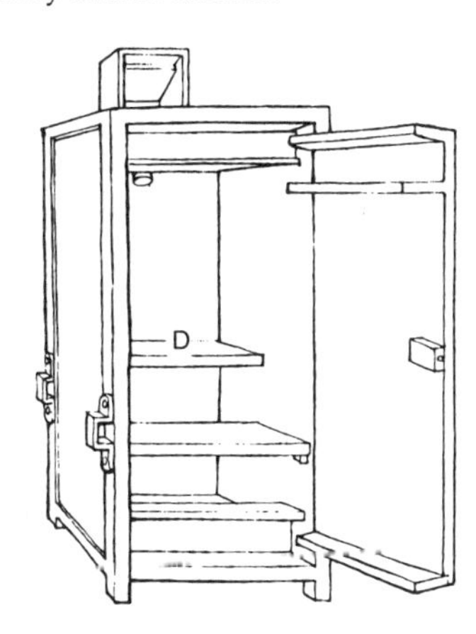

(c) An 18th century French camera obscura. A rotatable mirror on top reflected the image through a lens onto the drawing surface at D. A light trap under the seat provided ventilation. The apparatus could be carried on poles like a sedan chair.

Figure 2.26. Examples of the camera obscura.
(From Wright 1983)

wrote *Rules and examples of perspective* in 1593. An English translation appeared in 1707, which was reproduced in 1971. The English mathematician, Brooke Taylor (1685-1731), wrote *New principles of linear perspective* in 1715. This was the first geometrical analysis of perspective written in English.

In **one-point perspective**, all lines in the scene have the same vanishing point because they are either parallel to the picture plane or to each other, as shown in Figure 2.24a. In paintings, this usually means that all objects are rectilinear, on a horizontal plane, and with one surface frontal. In **two-point perspective**, all lines have vanishing points on the same horizon line because they are either parallel to the picture plane or to another common surface, but not all parallel to each other. In a typical two-point perspective painting, all objects are on a horizontal surface but some are slanted 45° to the frontal plane, as in Figure 2.24b. In **three-point perspective**, not all vanishing points are on the same horizon line because lines not parallel to the picture plane are not parallel to the same surface. In painting this usually means that some objects are both tilted and inclined to the frontal plane, as in Figure 2.24c. The vanishing points of inclined objects lie above or below the horizon (Section 24.1.1). Brunelleschi used two-point perspective in his first painting because the baptistery sits on a horizontal surface but is not rectangular. In his text on perspective, Viator drew objects at various angles but always on horizontal surfaces. He therefore used only one and two-point perspective. In his textbook, De Vries, drew inclined objects but incorrectly placed the vanishing points on the principal horizon line, as in Figure 2.23. The box marked S is clearly not horizontal and one of its vanishing points is incorrectly placed on the horizon. There is a passage in Piero della Francesca's *De prospectiva pigendi* describing 3-point perspective in the drawing of a cube placed on one of its corners (Elkins 1988).

The mathematics of perspective had its origins in Euclid and Apollonius in the 3rd and 4th centuries BC and in Pappus of Alexandria in the 3rd century AD. A full understanding of the mathematics of perspective had to wait for the development of projective geometry (see Section 3.8.2) by the French mathematicians Girard Desargues (1593-1662), Blaise Pascal (1623-1662), and J. V. Poncelet (1788-1867) and the 19th century German mathematicians Von Staudt and Felix Klein.

In the 15th century, manuals of perspective drawings of architecture and machinery began to circulate in Italy (Ferguson 1977). One of the most influential was an unpublished but widely circulated illustrated manual *Trattato di architettura* written about 1475 by Francesco di Giorgo Martini. One of the surviving copies contains margin notes by da Vinci. In 1588, Agostino Ramelli produced a book containing hundreds of technical drawings, many of novel machines. In 1556, Georgius Agricola produced *De re metallica*, the classic record of mining

Figure 2.27. Trompe l'Oeil.
An example of Trompe l'Oeil art by Domenico Rosselli in the study of the Ducal Palace of Urbino, 1476. The cabinets and their contents are painted. (Photograph by Alinari-Giraudon, in Damisch, 1994)

machinery (English translation, Dover 1950). Thus, the tradition of recording and transmitting technical information by perspective drawings was well established by the 15th century when Leonardo da Vinci began filling his notebooks with technical drawings. Denis Diderot, one of the Encyclopedists of 18th century France, produced hundreds of drawings in perspective in his *Pictorial encyclopedia of trades and industry* (English translation, Dover 1987). The tradition has continued to the present day.

For accounts of the development of perspective in art see White (1967), Pirenne (1970), Edgerton (1975), Descargues (1976), Bärtschi (1981), Wright (1983), and Kubovy (1986). Experimental studies of perspective are discussed in Section 24.1.2.

2.1.6d *Devices for drawing in perspective*

Many artists used projection devices rather than geometrical constructions to draw in perspective. Alberti's viewing box was essentially a window and he described a picture as a "window through which we look out into a section of the visible world." In his *Treatise on painting* Leonardo da Vinci described in detail a device for painting in perspective, which became known as Leonardo's window. The drawing was first made from a fixed vantage-point on a vertical plate of glass and later transferred to canvas. The device is illustrated in one of Dürer's woodcuts shown in Figure 2.25a. Leonardo wrote,
"Perspective is nothing but seeing an object behind a sheet of smooth transparent glass, on the surface of which everything behind the glass may be drawn; these things approach the point of the eye in pyramids; and these pyramids cut the said glass." (see Keele 1955).
In a variant of Leonardo's window the artist viewed the scene through a grid made of threads and transferred what was seen in each square to a corresponding square drawn on the canvas, as shown in Figure 2.25b. Pozzo drew the figures on a curved ceiling of a church by sighting from a fixed point through a horizontal grid placed below the ceiling (See Pirenne 1970).

Before the invention of photography many artists used a camera obscura, in which the scene to be painted is projected onto the canvas through a lens and small aperture. Giovanni della Porta (Section 2.1.3c) seems to have been the first to describe the camera obscura for drawing. He wrote in his *Naturiae Naturalis* (1558),

"If you cannot paint, you can by this arrangement draw with a pencil. This is done by reflecting the image downward onto a drawing board with paper. For a person who is skilful, this is a very easy matter."

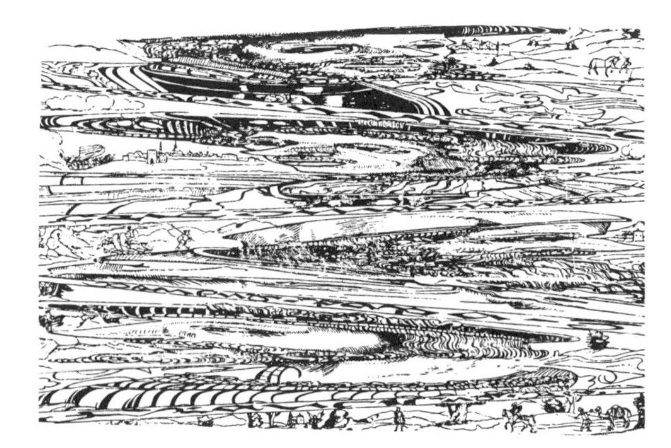

Figure 2.28. Anamorphic art.
Portraits of Charles V, Ferdinand I, Pope Paul III, and Francis I emerge when the picture is viewed obliquely from the side. (*Vexierbild* by Erhard Schön, c. 1532)

Figure 2.29. Anamorphic mirror art.
Portrait of Charles 1st, c. 1649. In Gripsholm Castle, Sweden.

Artists tended to be secretive about their use of artificial devices, presumably because they wished it to be thought that they painted by pure skill. For example, Bernardo Bellotto (1721-1790), known as Canaletto, concealed his use of the camera obscura. Philip Steadman (2001) has demonstrated that Jan Vermeer of Delft (1632-1672) probably used a camera obscura to paint his photograph-like interiors (see also Hockney 2001). Some early instruments are shown in Figure 2.26. In about 1567 Robert Boyle made a camera obscura with a lens on an extending hood and an opening in the top for drawing landscapes, like that shown in Figure 2.26b. In the instrument depicted in Figure 2.26c, the artist sits in a cabinet and traces the image of the scene reflected by a mirror on top of the cabinet. Pictures drawn with these instruments were limited by the size of the drawing surface and by the angle of view of the lens (Wright 1983). The camera obscura went out of use after it developed into the camera. Artists could then paint from photographs, which is an image in a camera obscura captured on film. See Hammond (1981) for a history of the camera obscura.

In 1806, the English scientist William Wollaston (1766-1828) patented the **camera lucida**. In a 'see through' camera lucida the artist sees the scene through two reflecting mirrors that produce an erect image. One of the mirrors is semi-transparent so that the artist can see the scene superimposed on the drawing paper. In the 'split pupil' camera lucida the artist uses part of the field of view to see the scene through mirrors or prisms and the other part to see the drawing paper. A camera lucida is much more compact than a camera obscura. Versions of the camera lucida for use on microscopes still appear in the catalogues of optical instrument makers. See Hammond (1987) for a history of the camera lucida.

2.1.6d *Trompe l'oeil and anamorphic art*

Renaissance artists were preoccupied with representing 3-D scenes on a flat surface. This directed their attention to monocular cues to depth, which can be represented in a picture, and away from binocular cues. In the tradition of painting known as **trompe l'oeil**, or fool the eye, the artist paints in perspective with the aim of fooling the viewers into believing that they are looking at a real 3-D object or scene. The tradition goes back to Roman times when wall frescos were designed to create the impression of a large room or garden. It is also represented in the device used by Brunelleschi to display his first painting in perspective. Renaissance artists painted objects in perspective on walls, ceilings, and doors in great detail and the same size as the real objects. They placed them where one would expect to see real objects, such as cabinets full of curiosities or rare books, and windows with outside views. An example is shown in Figure 2.27.

Artists usually draw in perspective on a surface orthogonal to the median plane of the head. Any drawing in perspective creates the correct image in the eye only when viewed with the picture normal to the line of sight and with the eye at the correct vantage-point. A picture drawn on a canvas slanted about a vertical axis produces the correct image only when viewed with the picture at the same angle. Viewed frontally, the picture may appear a mere jumble of lines. This simple principle gave rise to the tradition of **anamorphic art**, which flourished in the 16th and 17th centuries. Figure 2.28 shows one of the earliest anamorphic pictures, entitled *Vexierbild* (puzzle picture). It is a wood engraving produced by Erhard Schön, a pupil of Dürer, in about 1532. An

Figure 2.30. 17th century distorted room.
The four panels fold into a triangular box. When viewed from the correct point on the open side, the room appears rectangular and in correct perspective. The box was painted by Samuel van Hoogstraaten (1627-1678). (In the Municipal Museum of The Hague)

anamorphic portrait of Edward VI, painted in 1546, hangs in the National Portrait Gallery in London. Hans Holbein's *The Ambassadors*, painted in 1533, contains an anamorphic skull. The scene was first drawn on a frontal canvas and then sighting through the drawing point-by point onto a canvas or wall at a steep angle to the picture.

In cylindrical anamorphic art, the picture is drawn so that it appears in its true form only when viewed through a vertical cylindrical mirror placed on its centre, as shown in Figure 2.29. Many of these pictures depicted erotic scenes.

Another type of anamorphic art involved constructing 3-D displays that were actually one shape but appeared a different shape when viewed from the correct point. The first box of this kind seems to have been built by one of Rembrandt's pupils, Carel Fabritius of Delft (1622-1654). One of his boxes is in the National Museum in Copenhagen (Hultén 1952). The box is triangular but appears as a rectangular interior when viewed through a peephole. Figure 2.30 shows the plan of a viewing box constructed in Holland by another pupil of Rembrandt, Samuel van Hoogstraten (1627-1678) (van de Geer and de Natris 1962). The four surfaces fold to form a triangular room, which appears rectangular when viewed from the correct point. Hoogstraten travelled to Rome and Vienna, where he was patronized by the Emperor. He visited London in 1666, the year of the Great Fire, and finally settled in his native town of Dordrecht. An elegant viewing box, which he constructed at about that time, is in the National Gallery in London. This box portrays a series of connected rooms and has two viewing holes through which two scenes can be seen. Another box, probably by Hoogstraten, is in the Detroit Institute of Arts (Hultén 1952). In his manual on the art painting entitled *Inleyding tot de Hooge Schoole der Schilderkonst* Hoogstraten wrote,

"But I say that a painter whose work it is to fool the sense of sight, also must have so much understanding of the nature of things that he thoroughly understands by what means the eyes are fooled." (p. 24).

The Ames distorted room is an anamorphic structure built for scientific purposes (Section 24.1.2).

In 17th century Italy, the principle of anamorphic art was developed on an architectural scale. For example, the famous ceiling in the church of St Ignazio in Rome, painted by Andrea Pozzo, creates an impelling impression of 3-D columns with a multitude of human figures when the viewer looks up while standing on a spot marked on the floor below. The painting is actually on the curved surface of the ceiling. The Palazzo Spada in Rome contains a real arcade with misleading perspective, which looks elongated when viewed from one end and shortened when viewed from the other end. The Piazza del Campidoglio in Rome was constructed to enhance the sense of distance and size. The ancient Greeks evidently used the same principles to design stage sets (Section 2.1.6a). Stage designers and artists continue to use these principles (see Wade and Hughes 1999).

Anamorphic art is discussed in Pirenne (1970), Mastai (1976), Baltrusaitis (1977), Leeman (1976), and Kemp (1990).

2.1.7 The advent of instruments

Before the advent of psychophysics in the middle of the 19th century, investigations of the visual system

Figure 2.31. Johannes Müller (1801-1858).
Professor of Anatomy and Physiology in Bonn and Berlin. (From Polyak 1957).

Figure 2.32. Hermann Ludwig von Helmholtz.
(From Koenigsberger 1902)

were almost entirely qualitative and phenomenological. In his *Critique of pure reason* of 1781 Immanuel Kant declared that the perception of space and time is beyond the scope of experimental science. But he was soon to be proved wrong. Through the 19th century specialized instruments were invented which allowed precise measurement of temporal aspects of vision such as visual persistance, and reaction times. These included the Wheatstone's chronoscope of 1845, Plateau's stroboscope of 1833, and Volkmann's tachistoscope of 1859 (see Wade and Heller 1997).

The study of spatial aspects of vision was triggered by Wheatstone's invention of the stereoscope in 1836. The second half of the 19th century was marked by the discovery and measurement of visual illusions of all kinds. The demonstration that something as subjective as a visual illusion could be measured had a profound impact. However, a typical visual illusion, such as the Müller-Lyer illusion, is not due to a single process in the visual system and measurements are notoriously difficult to interpret. Great ingenuity is required to design stimuli and procedures that tap specified visual process.

The precise measurement of visual acuity, including stereoacuity, began in the second half of the 19th century. A task involving detection of a simple stimulus such as a single light source would seem to be immune to ambiguities. However, with the development of signal detection procedures, we now know that the subject's criterion can influence the results (see Section 3.1.1).

The precise determination of visual detection and visual discrimination thresholds had to wait for the invention of precise light sources and photometers in the early 20th century. The second half of the 20th century saw the development of the oscilloscope and computer monitor for vision research. An early example is the measurement of the contrast sensitivity function by Schade (1956).

The development of instruments in the 19th century led to the need for more precise quantitative procedures. The development of psychophysical methods by Weber and Fechner between 1846 and 1860 filled this need. Fechner believed that the entire universe was alive and conscious and believed that psychophysics would measure the soul. Since then, psychophysical methods have become more and more sophisticated (see Chapter 3).

2.1.8 Empiricist-nativist controversy

2.1.8a *The protagonists*

The empiricist-nativist controversy that flourished in Germany in the 19th century had its origin in the English empiricists such as Hume, in Kant's concept of the *a priori* status of the percept and concept of space, and the local sign theory of Hermann Lotze. The controversy acquired its most vigorous opponents in Hermann von Helmholtz and Ewald Her-

Figure 2.33. Ewald Hering.
Born in Altgersdorf, Germany in 1834. He obtained an MD from the University of Leipzig in 1858. Between 1862 and 1870 he worked in the Physiological Research Unit at the University of Leipzig and the Josephs-Akademie. In 1870 he was appointed Professor of Physiology at the University of Prague. From 1895 until his death in 1918 he was Professor of Physiology at the University of Leipzig. He was awarded the Graefe Medal of the German Ophthalmological Society in 1906. (From Polyak 1957)

Figure 2. 34. Ernst Mach (1838-1916).
Born in Turas, Moravia. He studied in Vienna and became Professor of Mathematics in Graz for 3 years, in Prague for 28 years and finally, Professor of Physics in Vienna. (Bidarchiv der Osterreichischen Nationsbibliothek)

ing. The extended battle between these two giants has been described by Turner (1993, 1994).

Helmholtz was born in the Prussian city of Potsdam in 1821 and died in 1894. His father was master at the Potsdam Gymmnasium. His mother was a descendant of William Penn. He did his doctoral work with Johannes Müller in Berlin (Portrait Figure 2.31). While serving as a military physician, he was able to conduct experiments in the Potsdam barracks. His first experiments in physiology showed that energy is conserved in metabolic processes, leaving no room for the vital force proposed by German idealist philosophers. He shared these views with his friend, the German physiologist, founder of electrophysiology, and essayist, Emil du Bois-Reymond (1818-1896). In 1850 Helmholtz was appointed professor of anatomy and physiology at the University of Königsberg, where he measured the speed of nerve conduction and invented the ophthalmoscope. He journeyed to England in 1853 and tried without success to meet Wheatstone. In 1855 he became professor of anatomy in Bonn and, in 1858, professor of physiology in Heidelberg. In the 1850's his interests shifted from nerve and muscle physiology to sensory physiology. His *Handbuch der Physiologischen Optik,* appeared in full pin 1867. This is the most influential book in visual science to have been written. In 1870 he became professor of physics at the University of Berlin. By 1875 he had abandoned sensory physiology for physics. He died of a stroke in 1894. Koenigsberger (1902) produced a biography of Helmholtz (Portrait Figure 2.32).

For Helmholtz, sensations were, "signs of external objects" learned by "practice and experience". He argued that sensations do not resemble the objects they symbolize, any more than letters of the alphabet resemble the sounds they represent. He revolted against German idealist metaphysical philosophy, as expounded by such figures as Hegel. For Helmholtz, metaphysical hypotheses were worthless if not accompanied by critical empirical investigation. Helmholtz (1910, Vol. 3, p. 533) stated:

"The fundamental thesis of the empirical theory is: The sensations of the senses are tokens for consciousness, it being left to our intelligence to learn how to comprehend their meaning. . . . The only psychic activity required for this purpose is the regularly recurrent association between two ideas which have often been connected before."

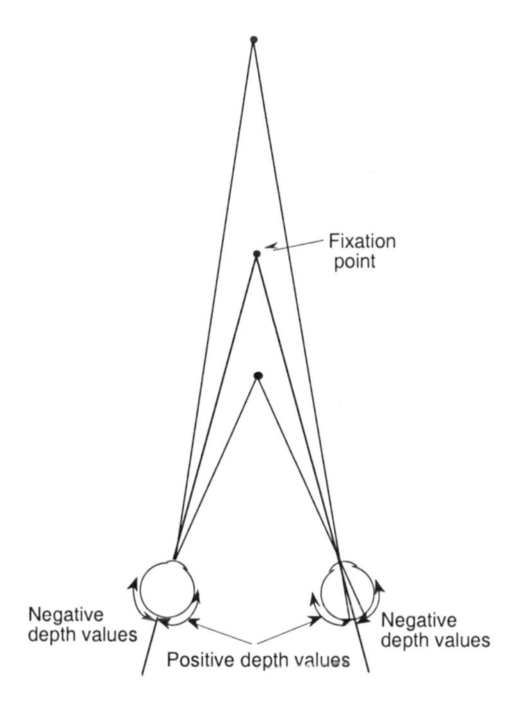

(a) According to Hering, images on each nasal retina have positive depth values, and those on each temporal retina have negative depth values.

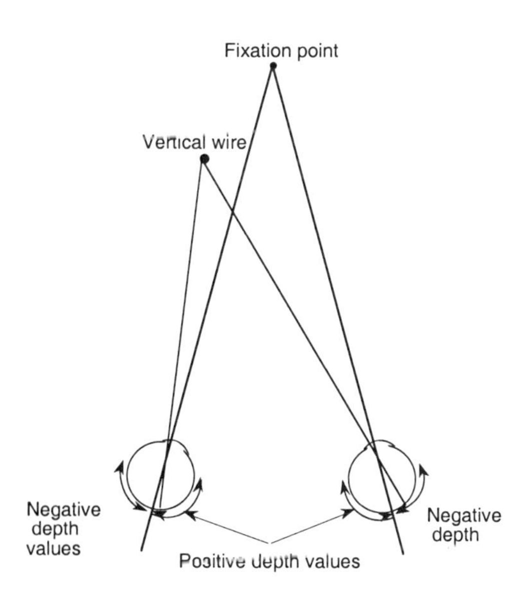

(b) Hering claimed that a near wire on the left of fixation produced an image with positive depth value in the left eye and therefore appears beyond the fixation plane. Its right-eye image has negative depth value and appears nearer than the fixation plane. Helmholtz could not confirm the effect.

Figure 2.35. Hering's theory of depth values.

Helmholtz (1910, Vol. 3, p. 2) also wrote,

". . . we always believe that we see such objects as would, under conditions of normal vision, produce the retinal image of which we are actually conscious"

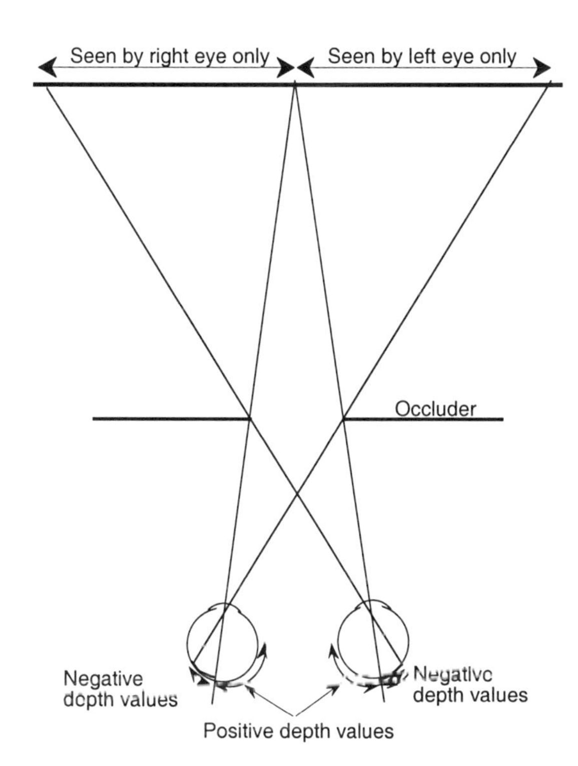

Figure 2.36. Helmholtz's refutation of Hering's theory. Helmholtz argued that, according to Hering's theory of retinal depth values, a wall seen on the right by only the left eye and on the left by only the right eye should appear to slant in opposite directions about the midline.

In attacking Hering's nativist theory of visual direction, he stated (p. 535),

"... it is even probable that the growth of the muscles, perhaps too even the efficiency of nervous transmission, is so adapted to the demands made upon it during the life of the individual, and perhaps by inheritance during the life of the species, that the requisite movements that are the most suitable become also the easiest to execute."

He weakened his case for empiricism by admitting that natural selection may also play a role. The nativist also takes this view. But he qualified this by stating, "In any event, even if this anatomical mechanism exists it is merely conducive, and not obligatory."

In part 3 of his *Handbook of Physiological Optics* Helmholtz defined and defended empiricism, but Hering, the nativist, was his real target. Others who supported the empirical approach included Wilhelm von Bezold (1837-1907), Franz Exner (1846-1926), Alfred von Graefe (1830-1899), Johannes von Kries (1853-1928), and Willibald Nagel (1848-1917).

Ewald Hering (1834-1918) (Portrait Figure 2.33) was born in Saxony. His father was a Lutheran pastor. He studied medicine in Leipzig. There is no biography of Hering. Although not a student of Johannes Müller, it is ironic that, like Helmholtz, he was very much influenced by Müller. Hering practised as a private physician between 1860 and 1865 during which time he published his first papers on binocular vision and wrote the five-volume work *Beiträge zur Physiology*. In 1865 he was appointed Professor of Physiology in the Military Medical Academy in Vienna, where he worked on respiration and binocular vision. His *Theory of binocular vison* appeared in 1867, a year after Helmholtz's *Physiological Optics*. In 1870 he succeeded Jan Purkinje as Professor of Physiology at the University of Prague where he wrote the book *Spatial sense and movements of the eye*. In 1895 he succeeded Karl Ludwig at the University of Leipzig. His final book, *Outlines of a theory of light sense*, appeared in 1920, two years after his death.

Hering acknowledged the role of learning in perception and resented being identified as a nativist. However, he insisted that basic sensory and perceptual mechanisms are a product of a long evolutionary process and occur mainly in or near the sense organs rather than at the level of cognition. He vigorously opposed what he considered to be Helmholtz's undue emphasis on learning and experience. His students and those devoted to his approach included Alfred Bielchowsky (1871-1940), Carl von Hess (1863-1923), Franz Hillebrand (1863-1926), Armin Tschermak-Seysenegg (1870-1952), and Ernst Mach (1838-1916) (Portrait Figure 2.34).

Hering was 13 years younger than Helmholtz. He was sometime deferential but usually attacked with a sarcastic and vitriolic style. A sample of that style can be found in the foreword to his book *On the Theory of the Light Sense* of 1874. Hering wrote,

"... *that modern tendency in sensory physiology, which has found its most acute expression in the Physiological Optics of Helmholtz, is not leading us to the truth, and whoever wishes to open up new avenues of research in this area, must first free himself from the theories which now prevail.*"

He argued that, just as earlier physiologists had explained troublesome phenomena in terms of vital forces, so today in treatises on physiological optics one sees invocations of the "psyche" or "inference". For Hering, this was tantamount to "spiritualism" and the idealism that Helmholtz abhorred. Hering looked forward to the day when physiological psychology, including the physiology of consciousness, would replace the descriptive tradition of "philosophical psychology" which investigated the phenomena of sensation without regard for its organic basis.

In his Founder's Day lecture at Berlin University in 1878 Helmholtz declared,

"*First of all, nativistic hypotheses about knowledge of the visual world explain nothing at all, but only assume that the fact to be explained exists, while at the same time rejecting the possibility of tracing this knowledge back to reliably established mental processes.*"

Helmholtz was irritated and frustrated by Hering's attacks but tried to maintain composure. For instance, he wrote (1910, Vol. 3, p. 557),

"*I have been obliged to make this criticism of Mr. Hering's views of the facts of the case, but I trust it will not be regarded as an expression of personal irritation on account of the attacks which he has made on my latest articles.*"

There was a strange futility about the battle between Helmholtz and Hering, and indeed about the whole nativist-empiricist controversy. It was fuelled as much by personal animosity as by scientific issues. The following examples show that, in spite of the animosity between the two men, Helmholtz came to agree with many of Hering's views and Hering came to agree with many of Helmholtz's views. The disagreement stemmed largely from Hering's vitriolic attacks and from Helmholtz's reluctance to consider low-level physiological mechanisms for such phenomena as stereopsis, colour contrast, and co-ordinated eye movements. The nativist-empiricist debate about colour vision has been reviewed by Kingdom (1997). The following is an account of their debate about three aspects of spatial vision: eye movements, visual direction, and binocular vision.

2.1.8b *The debate about eye movements*

In the debate about eye movements, Helmholtz argued that the eyes are separate organs, which, in principle, may be moved wholly independently of each other. Adherence to Donder's and Listing's laws (Section 9.2.2d) is a habit acquired by the use of the eyes of an individual during his lifetime to facilitate clear and easy visual orientation. Once acquired and ingrained, however, the facility cannot be overridden by acts of will. But such movements are anatomically possible. He argued that by using prism glasses, which produce abnormal separation of the visual axes, we can induce the eyes to perform vertical or absolute divergences. He observed that, when sleepy, he saw double images of objects, indicating that the eyes had diverged vertically or cyclorotated. When fully awake, he could not perform these eye movements voluntarily (Helmholtz 1910, Vol. 3, p. 59).

Hering held that the co-ordinated movements of the eyes are innate. He responded sarcastically to Helmholtz's observation,

"It is likely that the great Helmholtz in his dozing state, had simply failed to notice that he had allowed his head to nod to one side. This would produce the same result." (Hering 1864)

Helmholtz retorted,

"I did not make the mistake which he (Hering) attributes to me, and of which even a person with little training in observing double images could scarcely be guilty; namely, the mistake of supposing that the images were on different levels when they were really side by side, simply because my head happened to be tilted!" (Helmholtz, footnote, p. 59).

In his *Theory of binocular vison* Hering insisted that the eyes do not perform cyclovergence movements but, in his later book on *The spatial sense and movements of the eyes*, he agreed with Helmholtz that these movements exist.

Helmholtz supported the theory that the motion aftereffect is due to eye movements. It seems that he was reluctant to allow that there are dedicated motion detectors at an early stage of visual processing. Dvorák (1870), working under Ernst Mach, pointed out that this theory cannot account for the spiral aftereffect in which motion in several directions occurs at the same time (see Broerse *et al.* 1994).

2.1.8c *The debate about visual direction*

Steinbuch (1811) was the first to propose a theory of how visual directions are calibrated. He suggested that the spatial value of each optic nerve was provided by the motor response required to move the eyes to that location. In contrast, Tourtual (1827) proposed that the spatial senses were innately calibrated (see Rose 1999). Lotze (1852) coined the term 'local sign' and also proposed a motor theory of spatial calibration.

According to Hering (1864) each point in each retina has a local sign composed of three space values: its elevation, azimuth, and depth value. He named the point of fixation the core point (*Kernpunkt*) and the vertical plane containing the horizontal horopter he named the *Kernfläche*. He stated that objects lying in this plane appeared to lie on a frontal surface. In Hering's terminology, elevation and azimuth signify the direction of the point, and the point's depth value is its signed position relative to the fovea. Points on the nasal retina have positive depth values and those on the temporal retina have negative depth values (Figure 2.35a). Images with positive depth values appear beyond the convergence plane, even when only one eye is open. Images with negative depth values appear nearer than the convergence plane. The depth values of images falling on corresponding points in the two retinas are equal and opposite and the object that creates them lies in the horizontal horopter. He speculated that all points on the horizontal horopter appear on a frontal surface. He conceded that the perception of the absolute distance of an object depends on learned cues rather than on innate mechanisms.

Hering claimed to demonstrate his theory of depth values by fixating a point in the midline and observing the double images of a vertical wire nearer and to the left (Figure 2.35b). He claimed that the image of the wire on the nasal half of the left eye appears more distant than the fixation plane and that on the temporal half of the right eye appears nearer than the fixation plane (Hering 1865, p. 341).

Helmholtz attempted to observe this effect. He wrote,

"I have gazed at the pin so long and so fixedly that everything was extinguished by the negative afterimages. ··· I have never been able to persuade myself that this phenomenon occurred in the main as it ought to occur according to the Hering theory; and I never should have ventured to lay the foundation of a new theory of vision on an observation made with images that are half-extinguished in this fashion. However, I admit that I may have been unskillful. Only, Mr. Hering will have to forgive me for not being able to say that I have been convinced by this 'overwhelming proof' of the correctness of his theory, as he put it." (Helmholtz 1910, p.554)

Helmholtz (1910) argued that, according to Hering's theory, a wall should appear inclined in opposite directions when one eye sees only the nasal half and the other eye only the temporal half (Figure 2.36). But this does not happen. Hering abandoned his theory of retinal depth values after this attack and agreed with Helmholtz's that stereopsis is based on binocular disparity.

Helmholtz supported his empiricism by referring to the fact that persons with restored could not recognize simple objects (Section 13.3.3). However, he admitted that they rapidly learn to distinguish between objects in different positions, and concluded that, prior to all experience, adjacent points are recognized as adjacent (Helmholtz 1910, Vol. 3, p. 227). This was close to Hering's view.

Helmholtz asked, if perceptions are shaped by learning to conform to the objects of the world, why do we suffer from illusions? He described two classes of illusions. The first type arises because identical impressions on a sense organ can be produced by distinct distal stimuli. For example, the illusion that a stereogram viewed in a stereoscope is three-dimensional arises because it produces the same proximal stimulus as a real scene. The second type of illusion arises when a sense organ is used in an unusual way or is exposed to an unusual stimulus configuration.

Hering's experiments on binocular visual direction (Section 17.7.2) persuaded Helmholtz that directions are referred to a point midway between the eyes—the egocentre. But for Helmholtz, this relation between images and direction is learned. It is odd that neither Hering nor Helmholtz cited Ptolemy, Alhazen, Briggs, or Wells on this question.

2.1.8d *The debate about binocular vision*

The first encounter between Helmholtz and Hering occurred in 1864 when Hering was an unknown lecturer in Leipzig. The two men had independently produced a general solution of the horopter. There was some dispute over priority (see Helmholtz 1910, Vol. 3, p. 484, footnote 4) but the main dispute concerned the inclination of the vertical horopter, which Helmholtz believed was shaped by experience to lie along the ground, and which Hering described as varying at random (Section 15.7).

In general, Hering believed that binocular correspondence is innate. Helmholtz held an empirical view of binocular correspondence as described in Section 2.2.5. He stated that,

"... *any theory that assumes that fibres proceeding from corresponding places on the two retinas are united in a single fibre that transmits the impressions in the two eyes unseparated to the brain, is inadmissible and incompatible with the facts.*" (p. 539).

He reluctantly considered the possibility that corresponding fibres split into two branches, two of which unite and two of which remain distinct. See Section 7.9 for a discussion of this issue.

The debate about whether binocular correspondence is innate or learned centred on reports that people with squint develop an anomalous pattern of binocular correspondence. Followers of the nativist school allowed that anomalous correspondence can develop but argued that this does not prove that the normal pattern is learned (Section 15.4.2). Another long debate about the shape of the horizontal horopter is described in Section 15.6.2.

Helmholtz came to accept that the pattern of retinal correspondence responsible for the form of the horopter has an innate component. In spite of these concessions to Hering, Helmholtz persisted in believing that experience shapes the development of spatial perception.

The nativist views of Hering and the empiricist views of Helmholtz arose because each antagonist selected evidence to suit his theory and each looked at different aspects of the visual process. For example, Helmholtz's trichromatic theory of colour vision and the apparently conflicting opponency theory of Hering are now understood as different stages in a complex process. Both men became emotionally involved. Hering used ridicule and bombast. Helmholtz was more restrained and conceded several points to Hering but he occasionally resorted to sarcasm. Helmholtz became very frustrated by his encounters with Hering, as revealed in one of his letters to his friend, du Bois-Reymond, dated February 1865 (Kirsten 1986).

"*Mr. Hering has annoyed me considerably with his impertinent ways of judging other people's work which, in part, he has not taken the trouble to understand properly. However, I do not want to treat him in a nasty way since he is an intelligent man in his own way. Even though, at the moment, his views conflict with mine, he is working out his own viewpoint in a consistent manner. He has been, as I have heard, mentally ill and this has until now held me back from bringing him down, which he has at times deserved.*"

It may have been his frustration with Hering that caused Helmholtz to return to physics in 1875.

We will see in Chapters 10 and 11 that the interplay between genetic factors and experiential factors that governs the development of the nervous system and visual mechanisms is much more complex than either Helmholtz or Hering imagined.

2.2 HISTORY OF BINOCULAR VISION

Since ancient times, artists have struggled with the problem of how to represent 3-D space in a picture. This preoccupation with 2-D pictorial space led to an emphasis on monocular cues to distance. Euclid knew that the two eyes have different views of 3-D objects, and Aristotle noted that one sees double images when a finger presses against one eye (Section 2.1.1b). But the problem raised in peoples' minds was how we form an impression of a single visual world, despite these differences between the two images. The binocular disparities and double images were regarded as something to be overcome rather than made use of.

It is amazing that the simple facts about binocular stereoscopic vision were not appreciated until about 160 years ago. One reason for this ignorance is that, even with one eye closed, a rich variety of monocular information is available for coding depth. Thus, the importance of binocular stereopsis is not apparent with casual observation of everyday scenes.

2.2.1 Ptolemy on binocular vision

Parts of Books II and III of Ptolemy's *Optics* deal with binocular vision. The following quotations are from an English translation of Ptolemy's work on binocular vision prepared for us by Dr. Fiona Somerset of Oxford University. The full translation is in Howard and Wade (1996). I quote these sections at some length, since this is the earliest known account of the basic geometry of binocular vision and, apart from a brief review by Crone (1992), it does not seem to have been cited in the vision literature. The section and figure numbers correspond to those in Lejeune (1956). The first sections come from Book II.

27] An object appears in one location when seen with only one eye, but when seen with both eyes an object is seen in one location only if it falls on consimilar radii, namely those that have corresponding positions with respect to the visual axes. And that comes about when the visual axes converge on the object to be seen, which happens when we see things with a simple gaze and in the way which is natural when we inspect an object.

We now refer to consimilar radii as corresponding visual lines and we will use the modern term in the remaining quotations.

28] It seems too that nature sets up double vision so that we will look more and so that our viewing will be ordered and brought to a definite position. It is natural for us to turn our gaze toward diverse locations, our gaze shifts without our conscious effort with a marvellous and diligent turning motion, until both visual axes intersect on the centre of the object we wish to see, and other pairs of corresponding visual lines within the two visual pyramids are also brought into coincidence.

Here Ptolemy suggests that we have two eyes and double vision so that we will actively bring the visual axes onto the object of interest. There is no suggestion that binocular vision has anything to do with depth perception.

33] Let A be the left eye and B the right eye (Figure 2.37a). *Place two rods G and D on the perpendicular to AB. Extend to them from each eye the visual lines GA, GB, DA, DB. Let the eyes be converged on the nearest rod G.*

34] AG and BG fall on the visual axes. Of the remaining two visual lines, AD is to the left of the visual axis of the left eye and BD is to the right of the visual axis of the right eye. Thus G is seen in one location, because the two visual axes are corresponding visual lines; but D appears double since visual line AD is to the left of the visual axis of the left eye, but visual line BD is to the right of the visual axis of the right eye. Therefore when we cover the left eye, the left image will not be seen; and when we cover the right eye, the right image will not be seen.

35] If the eyes converge on D, it will come about in the opposite way. Because AD and BD are on the visual axes D will be seen as one. G will appear double because AG is to the right of the visual axis of the left eye, and BG is to the left of the visual axis of the right eye. If we cover the left eye, the image which appears on the right on visual line AG will not be seen, and if we cover the right eye, the image which appears on the left on visual line BG will not be seen.

These sections describe the essentials of binocular disparity and the difference between what we now refer to as crossed images produced by objects nearer than the convergence point and uncrossed images produced by objects beyond the convergence point.

36] If the visual axes are parallel so that they do not converge on either rod (Figure 2.37b) *both rods will be seen double according to the principles we have presented.*

37] To demonstrate this clearly the near rod at L should be painted white and the far rod at M black. Therefore objects at L and M will both be seen in two positions to the sides of their actual positions. Therefore if we cover the left eye, the images which are on the right side will not be seen; but if we cover the right eye, the images on the left side will not be seen. Visual line AL will be more toward the right than line AM, and line BL more toward the left than line BM. In this way the images on the right will be seen through the left eye, and those on the left by the right eye.

38] Again place the visual axes parallel and parallel to the line between the eyes, and place the white rod on the visual axis of the left eye and the black rod on the visual axis of the right eye (Figure 2.37c). *The distance between the rods is the same as that between the eyes. The two rods will be seen as three.*

39] Through the corresponding visual lines, each rod will be seen as one, although neither of them falls on both visual axes because the rods are placed side by side. But through the non-corresponding visual lines AM and BL, the third, middle image will be seen, composed of an image of the white rod from the right eye and an image of the black rod from the left eye. If we cover the right eye, the image of the black rod to the right side of the middle and the white image of the middle composite image will not be seen. If we cover the left eye, the image of the white rod on

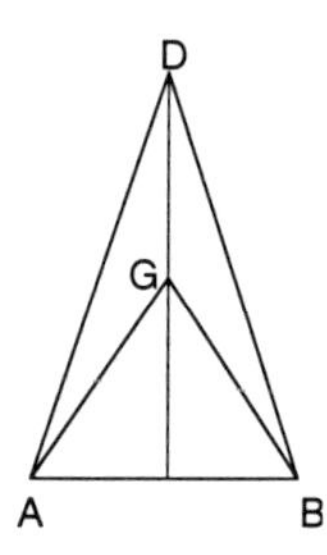

(a) Eyes at A and B are converged on D.

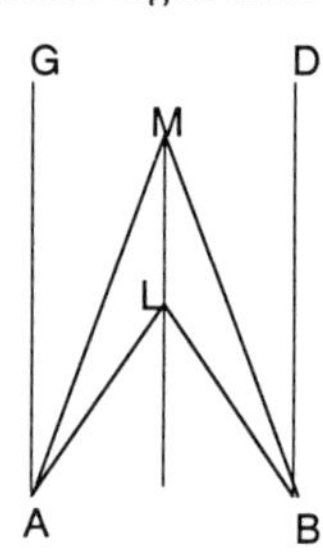

(b) The visual axes are parallel. Rods L and M are on the midline.

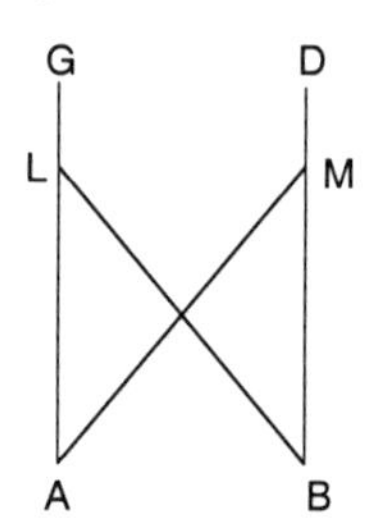

(c) Rods L amd M are on the parallel visual axes.

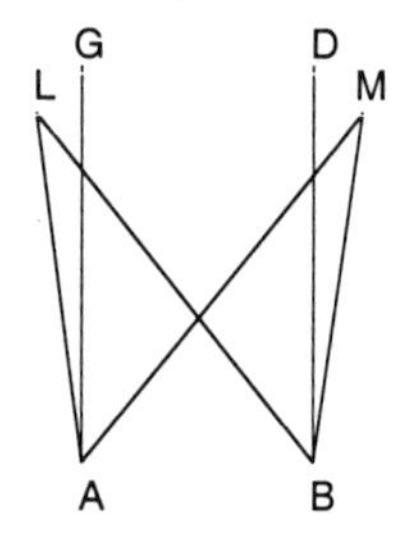

(d) Rods L and M are outside the parallel visual axes.

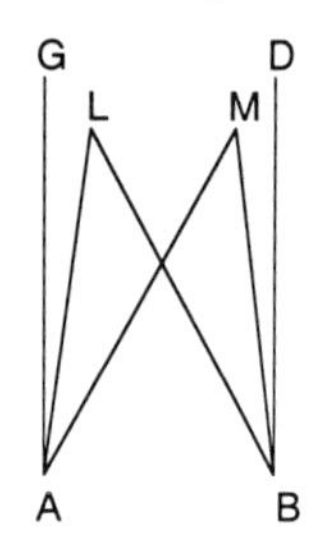

(e) Rods L and M are inside the parallel visual axes.

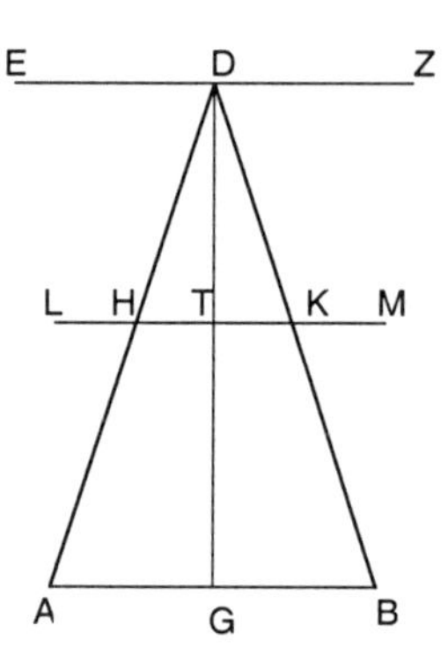

(f) Eyes at *A* and *B* converge on *D*. Objects are placed at H, T, K.

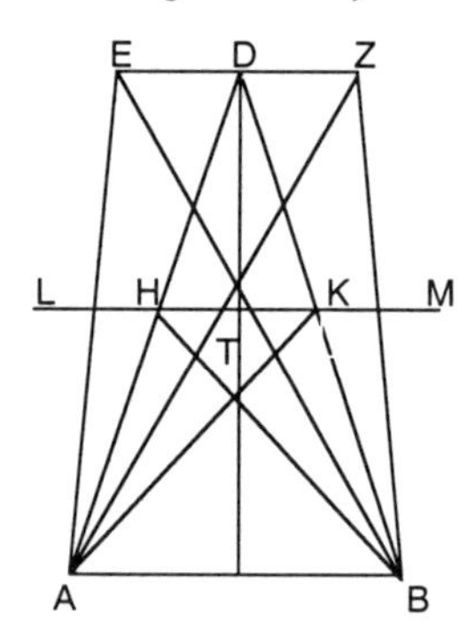

(g) Eyes converge on *D*. Objects are placed at *H*,T, K.

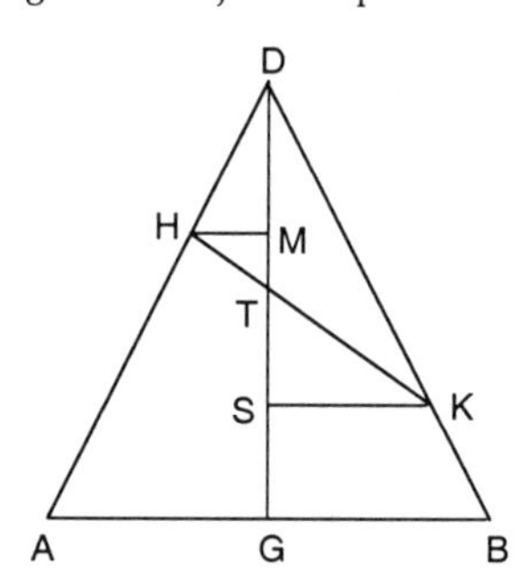

(h) Eyes converge on D. Objects K and H appear at M and S.

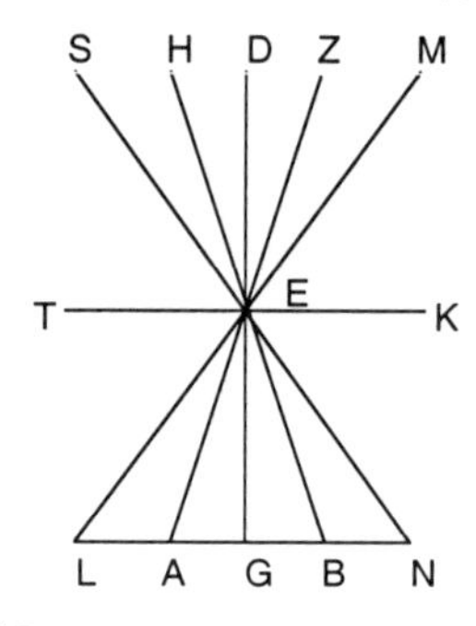

(i) Eyes at *A* and B converge on E.

Figure 2.37. Diagrams from Ptolemy's Optics.

the left and the black image of the composite middle image will not be seen.

40] For when we have joined the non-corresponding visual lines AM and BL, both rods will be seen in one position, namely that on which the two colours come together; but as concerns the remaining two visual lines falling on L and M, through the right of them will be seen the black rod, and through the left the white. Therefore when we cover the right eye, the black rod on the right and the white part of the central composite image will not be seen;

and when we cover the left eye, the white rod on the left and the black part of the central composite image are not seen. And this is demonstrated by the figure (Figure 2.37c).

Ptolemy was confused here, although the confusion may have been introduced in the translation from the Greek or from the Arabic. It is the images projected by the corresponding visual axes (*AL* and *BM*) that fuse into a composite image on the cyclopean axis. Ptolemy incorrectly formed the midline composite image from the images projected by visual lines *AM* and *BL*. It is clear from Sections 28 and 35 of Book III that he was well aware that objects anywhere on the visual axes are seen as one in the midline. The image of the black rod projected by *AM* is seen well to the right and the image of the white rod projected by *BL* is seen well to the left. He may have been misled by his diagram in which visual lines *AM* and *BL* intersect. His description of which images disappear when one eye is closed is also in error. The images would have disappeared in the way he described if he had inadvertently converged rather diverged.

41] When the distance between the rods is not equal to that between the eyes, two rods will be seen in four locations.

42] If the distance between L and M, is greater than that between the eyes (Figure 2.37d) *and the rods are outside the visual axes, the black rod will be seen in two positions on the right, since AM and BM are both to the right of the visual axes, and the white rod will be seen in two positions on the left, since AL and BL are to the left of the visual axes. Therefore when we cover the left eye, the image of the black rod on visual line AM and the image of the white rod on visual line AL will not be seen. When we cover the right eye, the image of the white rod on visual line BL and the image of the black rod on visual line BM will not be seen.*

43] If the distance between L and M, is less than that between the eyes (Figure 2.37e), *the visual axes are to the sides of L and M.*

45] These phenomena occur only by virtue of the horizontal separation of the eyes since the height and depth of the eyes are the same. Both visual axes turn until they converge on the thing to be seen. The eyes can converge horizontally to different positions; but they do not change their vertical angle of vergence, since one of the eyes is not placed higher than the other.

Ptolemy goes on to discuss size constancy, and returns to binocular vision in Book III.

26] Let us speak first about that construction in which the heads of the two visual pyramids (the eyes) *are points A and B joined by line AB and divided at the middle at point G* (Figure 2.37f). *Produce from this middle point a perpendicular GD and let the visual axes AD and BD converge on an object at point D. Under these conditions object D is seen as one and in its correct location.*

27] If from point D we draw a line EDZ at right angles to GD, anything positioned on that line, since it is at the head of (in the same frontal plane as)*point D, will appear as one and in its correct location.*

Ptolemy claims incorrectly that the locus of single vision is the frontal plane through the fixation point. Theoretically, it is a circle through the eyes and the fixation point. However, as Tyler (1997) pointed out, the difference between these loci is small for small angles of eccentricity and the empirical horopter is flatter than the theoretical horopter (Section 15.6).

28] When the line HTK is produced parallel to EDZ, and the eyes are converged on point D, an object at point T will be seen in two locations H and K. Moreover, two objects positioned in H and K will be seen in three locations, T, L, and M. They will both appear superimposed at point T as if they were one thing. In addition, they will appear separately, H at point L and K at point M. Any object on LT and TM will be seen in the same manner on HK.

Here Ptolemy asserts that an object on the midline nearer than the fixation point appears diplopic, with a separation equal to the distance between the visual axes. Objects on the two visual axes, with symmetrical convergence, are seen as a fused pair in the midline and as monocular images separated by twice the distance between the visual axes. These ideas express the fundamental principles of cyclopean visual direction, which are usually credited to Hering (1865) or Wells (1792) (Section 17.7).

29] If we converge on point T we will see D at points E and Z.

30] This may be confirmed by someone using a board on which two rods are placed. Whoever wants truly to recognize their locations may discern them by placing a finger on the thing to be seen. For his finger will land upon the object when it appears in its correct location. When the object does not appear in its correct location, his finger will not land upon it ....

31] Those objects seen by corresponding visual lines, even if there are two of them, are seen as if in one position; but if not by corresponding visual lines, even if there is only one it will be seen as if it were in two locations.

32] If we join lines AE, AZ, ZB, EB, TA, TB, BH, AK (Figure 2.37g) *any of E, D, and Z will appear in one location, since AD and BD are the visual axes, and the visual lines which converge on E and Z are corresponding visual lines because AE corresponds to BE and AZ corresponds to BZ. But H and K, will appear in one location T, since AH and BK are visual axes. Because BH and AK are non-corresponding visual lines H and K will appear at points L and M: Because visual lines AT and BT are non-corresponding, point T will appear in locations H and K.*

Here, Ptolemy restates that objects falling on corresponding visual lines are seen as one. However, he is not correct in stating that points *E* and *Z* fall on corresponding visual lines. To do so, they would have to fall on the locus of equal binocular subtense, which is a circle (the Vieth-Müller circle) passing through the eyes.

34] If the line HTK (Figure 2.37h) *is not parallel to line AB but instead AH is greater than BK; and the visual axes converge on point D, things placed on points K and H will appear on line GD, which passes through point T. But K will appear nearer than H because the line HK is inclined to the plane AB upon which the eyes are positioned. Then H will be seen at M, and K at S. Points H and K are such that perpendiculars from line GD from points M and S fall on points H and K.*

This rule about perpendiculars is somewhat arbitrary. More accurately, an object on a visual axis and its apparent location on the cyclopean axis lie on an arc centred on the point of convergence.

35] It is fitting that nature should equalize the distance between the two visual axes, and gather them in accordance with the position of the thing to be seen. Therefore the visual axes are seen as falling upon the line through the midpoint between them and the point where the axes converge. This line is equidistant from the two visual axes and the two visual axes appear to coincide with it. Objects on the two visual axes are in different directions since the visual axes are inclined to each other. The only way they can be seen as one is if they are both seen as lying on an axis midway between them. And that middle axis should rightfully be called the common axis.

Here Ptolemy explains the need to combine the distinct monocular headcentric visual directions into one. Ptolemy called the midline axis upon which objects on the two visual axes appear to lie the common axis. It is now known as the cyclopean axis.

37] Let the lines AD and BD be the visual axes (Figure 2.37f). *Objects on line EDZ appear in their correct positions; but objects on line HTK appear displaced from their true positions.*

38] It is clear that points E, D, and Z will appear in their true positions, because each of them falls on the perpendicular to the common axis at the point where the visual axes intersect and where the distance from the visual axes to the common axis is zero. Just as the distance from the visual axes to the common axis is zero so the distance between where points E, D, and Z appear and their true position will also be zero. Therefore, each of these objects will be seen in its true position.

This is a spurious proof of the incorrect conclusion that the horopter lies in the frontal plane through the fixation point. Ptolemy realized that an object appears single and in its true position when it falls at the intersection of a pair of corresponding visual lines, but failed to realize that this does not generate a planar horopter. We will see that, in the 11th century, Alhazen deduced that an object appears single when it subtends the same angle to the two eyes as that subtended by the visual axes. From this, Alhazen proved that the horopter is not a plane. However, a description of the true geometrical horopter had to wait for Pierre Prévost in 1804. It is ironic that the simple theorem of Euclid that Prévost used for this proof was well known to both Ptolemy and Alhazen.

43] We can see this more clearly if we take a black rectangular board (Figure 2.37i) *and mark off on its shorter side two points A and B separated by the distance between the eyes, and extend from the midpoint G a perpendicular GD, and draw lines AEZ, BEH, and TEK, with TEK parallel to AB. Colour GD white, TEK green, AEZ red, and BEH yellow. Place the eyes at points A and B and converge on a small object placed at point E.*

44] Lines AZ and BH fall on the visual axes, and the green line TEK will appear as one line, since it intersects the point of convergence. Red line AEZ and yellow line BEH will appear superimposed on GD; but each of them will also appear in another position, AEZ on LEM, and BEH on NES. The white line GED, will appear on lines AZ and BH.

45] When we cover eye B, green line TEK is still seen. But the white line on AZ, the yellow line on GD, and the red on LM will be hidden. The other lines will keep the positions they had when both eyes were open. All this is consistent with what has already been determined.

Ptolemy completed his treatment of binocular vision with a discussion of the binocular appearance of oblique lines and curves in the plane of regard.

Ernst Mach (1926) gave an account of Ptolemy's work on binocular single vision and reproduced two diagrams similar to Figures 2.37a and 2.37c, but I have found no other reference to Ptolemy's work on binocular vision in the visual science literature.

2.2.2 Alhazen on binocular vision

In his *Book of Optics*, written in the 11th century, Alhazen followed Galen in explaining that we have two eyes so that, when one is harmed, the other remains intact. He added that two eyes beautify the face (Sabra 1989, p. 102). Like Aristotle and Galen, he mentioned that an object appears double when a finger pushes one eye. Also, like Galen, he pointed out that when the visual axes converge on a point they lie in one plane, which we now call the plane of regard. He stated that the two eyes always move together and by an equal amount, so that the visual axes converge on the object of interest. As Heller

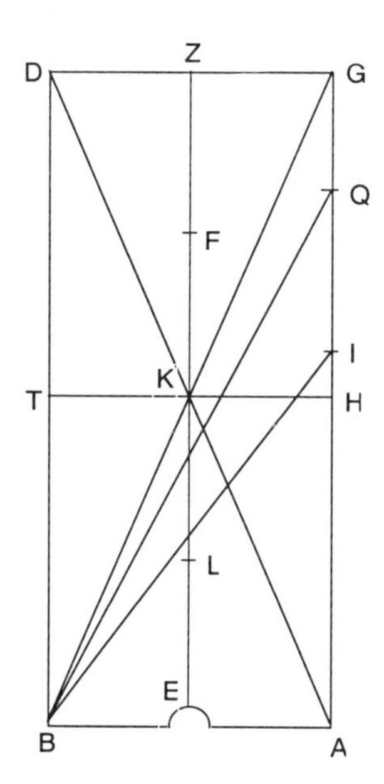

Figure 2.38. Alhazen's geometry of binocular vision.
The board is held horizontally with the bridge of the nose at *E*, the eyes at *B* and *A*, and fixation on *K*. Small objects are placed at *K*, *F*, *L*, *I*, and *Q*. Line *EZ* appears as a cross through *K*. Lines *BG* and *AD* appear as four lines with the two centre lines superimposed on the midline. Objects *F* and *L* produce double images, straddling the midline. Object *Q* produces double images, on one side of the midline. Object *I* produces double images too near together to be seen as two. Objects at *T* and *H* appear single but objects more eccentric on the same frontal plane appear double. (From Alhazen's *Book of optics*, trans. Sabra, 1989, p. 238)

(1988) pointed out, this idea lay dormant until the 19th century when Hering, without any reference to Alhazen, described the principle of equal innervation (Section 9.9.1).

In Chapter 6 of Book I and in Chapter 2 of Book III, Alhazen described the concept of corresponding points in the image planes of the eyes. He explained that visual lines from an object near the intersection of the visual axes project to corresponding points in the eyes and appear single. Images falling on non-corresponding points are seen as double. Alhazen therefore extended Ptolemy's concept of corresponding visual lines to that of corresponding points in the eye, although he had no clear idea of the image in the eye.

Like Ptolemy, Alhazen believed that the sensations evoked by lines on the two visual axes converge in what we now call the chiasma, in a point he referred to as "the centre". We now use the term "egocentre" or "cyclopean eye" to refer to the point in the head from which visual directions are judged. We do not place it in the chiasma or think of it as a place where images fuse. Also, like Ptolemy, Alhazen called the cyclopean axis extending in the midline to the fixation point the "common axis".

He stated that an object above or below the fixation point in the midline is not seen double, because its distance from the two eyes is the same and it therefore projects equal angles to the two eyes. This idea anticipates the modern concept of the vertical horopter (Section 15.7). He then discussed double images produced by an object nearer or further away than the fixation point, with both the object and the fixation point in the median plane. He invited the reader to view lines on a board extending horizontally from the bridge of the nose, as shown in Figure 2.38. This resembles the figure used by Ptolemy. He described the following visual impressions.

An object at point *I* does not appear double if it is not too far from the frontal plane *TKH* in which the eyes are converged, but an object at *Q*, well away from the plane of convergence, appears double. He thus realized that small differences in visual angle are tolerated without diplopia. We now refer to this tolerated disparity as Panum's fusional area (Section 7.1). We should call it Alhazen's fusional area! Objects closer to or farther than the fixation point (points *L* and *F*) appear double and on opposite sides of the fixated object when they are between the visual axes. They appear on the same side of the fixated object when they are outside the region lying between the visual axes (point *Q*). Here he follows Ptolemy in describing the basic facts about diplopic images and their relative order.

Small objects, *T* and *H*, in the same frontal plane as the fixation point appear single when not too far from the fixation point, but double when well to the side. He proved this by showing that the angle between an eccentric object and the median plane was not the same for the two eyes. He thus proved that the locus of fused images for a given viewing distance does not lie in a frontal line or plane, as Ptolemy believed. It is a pity that he did not go one step further and apply Euclid's geometry to show that it is a circle passing through the fixation point and the two eyes.

He followed Ptolemy in describing how the central line, *EZ*, appears as two lines intersecting in the fixation point, *K*. Finally, two lines, *AD* and *BG*, extending diagonally from each eye and intersecting in the fixation point appear as four lines, with the middle two appearing close together along the median plane of the head. He wrote,

"The reason why two of the four appear closer together is this: when the two visual axes meet at the middle object, then each of the two diameters will be perceived by the eye next to it through rays that are very close to the visual axis; thus their forms (images) *will be very close to the centre within the common nerve* (the chiasma) *and their*

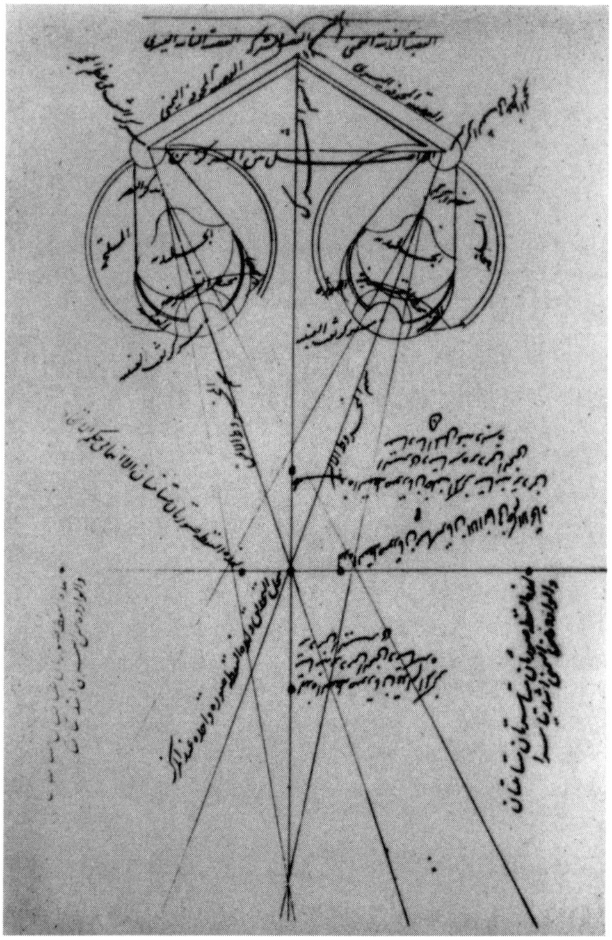

Figure 2.39. Diagram of the binocular system by al-Farisi.
From a 14th century Arabic manuscript with an English translation of *Book of reflections on the science of optics*, in the Ayasofya Library, Istanbul.

point of intersection will be at the centre itself, and thus the diameters will appear very close to the middle (the median plane)." (p. 242).

Note that Alhazen stated that lines falling "close to the visual axes" appear close to the midline. His theory predicts that lines on the visual axes appear as one on the cyclopean axis. In a work entitled *Doubts about Ptolemy*, written some time after his *Book of Optics*, Alhazen claimed that Ptolemy was wrong, both experimentally and in theory, in claiming that visual lines meeting in the point of convergence appear to lie on the common axis. Alhazen argued that they approach each other, but do not meet (see Sabra 1966). But Ptolemy was correct—when lines are drawn from the centres of the pupils they do indeed appear to coincide in the cyclopean axis. Perhaps Alhazen did not have his lines lined up with the centres of the pupils. It is not clear why Alhazen claimed that Ptolemy was theoretically incorrect when his own theory predicts the effect that Ptolemy reported. This work by Alhazen proves that he had read Ptolemy's *Optics*.

These ideas of Ptolemy and Alhazen on binocular vision have been almost totally ignored. Figure 2.10c is from *Ophthalmographia* written by William Briggs (1650-1704) in 1676. It illustrates the fact that objects on the visual axes appear in the midline. Wells gave an account of cyclopean visual direction in his *Essay upon single vision with two eyes*, written in 1792, 87 years before Hering wrote his account in 1879. None of these authors acknowledged the contributions of Ptolemy or Alhazen (Section 17.7).

Kamäl al-Din Abdu'l-Hasan al-Farisi (died c. 1320) reviewed Alhazen's writings in his *Tanqih al-manäzir* (*Revision of the Optics*) in about the year 1300 in Iran. A diagram of the binocular system from one of his manuscripts is shown in Figure 2.39. It shows how rays from a fixated object fall on corresponding points and how rays from nearer or more distant points fall on non-corresponding points. After Alhazen, there were no significant advances in knowledge about binocular vision until the 17th century.

2.2.3 Europe to the 18th century

2.2.3a *Leonardo da Vinci on binocular vision*

Leonardo da Vinci (1452-1519), in his book *Trattato della Pittura*, wrote,

"A painting, though conducted with the greatest art and finished to the last perfection, both with regard to its contours, its lights, its shadows and its colours, can never show a relievo equal to that of the natural objects, unless these be viewed at a distance and with a single eye."

He went on to describe how an object obscures from each eye a different part of the distant scene (see Figure 18.20). This is occlusion disparity that had been described by Euclid and Galen. Unlike those authors, da Vinci described it as a source of information about depth (see Keele 1955; Strong 1979). He observed that objects seen with both eyes appear rounder than when seen with only one eye.

2.2.3b *Aguilonius on binocular vision*

Franciscus Aguilonius (François D'Aguillon) was born in Brussels in 1546, the son of the secretary to King Philip II. He became a Jesuit priest in 1586 and died in Antwerp in 1617. He taught logic, syntax, and theology and was charged with organizing the teaching of science in Belgium. In 1613 he published part one of a three-part work on optics designed to synthesize the work of Euclid, Alhazen, Vitello, Roger Bacon, and others. He died before completing

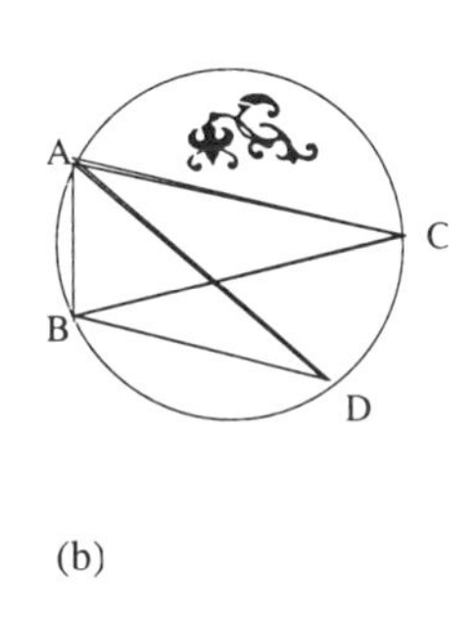

Figure 2.40. Diagrams from Aguilonius (1613).
(a) Aguilonius defined the horopter as the plane in which double images appear to lie. The eyes at *A* and *B* converge on point *C* on a frontal plane, *DE*. Images of an object at *F* project onto the plane at *G* and *H*. Those of an object at *I* project to *K* and *L*.
(b) Locus of equal subtense of visual lines. Points *C* and *D* on a circle through the eyes at *A* and *B* project equal angles to the eyes.

the other parts. The published work consists of six books with the title *Opticorum Libri Sex*, with illustrations by Rubens.

In his treatment of visual optics and perception he followed the order of topics in Alhazen's book. *Opticorum Libri Sex* appeared 2 years after Kepler's *Dioptrice* but Aguilonius did not refer to Kepler's work on the formation of the retinal image. Like Alhazen, he realized that clear vision is possible only if each object point is represented by one image point in the eye. He adopted Alhazen's theory that only light rays orthogonal to the cornea and lens surface are clearly registered.

His ideas on binocular vision were presented in Book 2. He was aware that binocular vision improves the sense of depth but did not relate this to binocular disparity. He adopted Galen's idea of the cyclopean eye located in the chiasma. One of the illustrations by Rubens shows putti dissecting the eye taken from a cyclops. In theorems 144 to 146 he clearly described how an object of appropriate size and at an appropriate distance occludes a far surface so that half the far surface is seen by one eye and half by the other eye. He also described how a larger object creates a region of the far surface invisible to either eye and a smaller object creates monocular zones separating a binocular zone (see Section 18.2.1). However, he did not relate these facts to depth perception.

Aguilonius introduced the term "horopter" to describe the locus in space within which both fused and diplopic images appear to lie. The word is derived from the Greek words "horos" meaning "boundary", and "opter" meaning "observer". Aguilonius presented the diagram shown in Figure 2.40a and wrote,

(a) A man, blind in one eye, is shown having difficulty reaching for an object.

(b) The eyes are converged on a point on the vertical plane. Disparate images of the small sphere in front of the plane are projected on the far plane, which Aguilonius defined as the horopter. (From *Opticorum Libri Sex*, Bodleian Library, Oxford)

Figure 2.41. Etchings by Rubens from Aguilonius (1613).

"Let the centres of sight be at A and B which the straight line AB connects. The optic axes AC and BC come together at C, and through C, parallel to AB runs a straight line, DE."

He called this line the horopter and the vertical plane containing it the horopter plane. He went on to state,

"The appearance of all those objects placed in the plane (of regard) *assume places for themselves. For example, F is a visible object, the optic radii AF and BF join at F, but they carry beyond the image of the object, until they site it in the horopter as in a common terminus and station, where the twin sites of H and G are placed. For an object at I, the images appear at K and L. In this way, the horopter is the terminus of all things which exist beyond and on this side of the junction of the optic radii ."* (Definition 10, p.111).

Figure 2.42. Isaac Newton.
Born in Woolsthorpe, England in 1642. He attended Trinity College of Cambridge University in 1661 and became Lucasian Professor of Mathematics 1n 1669. In 1689 he was elected to Parliament and became Master of the Mint in 1697. He was President of the Royal Society from 1703 until he died in 1727. (From Polyak 1957)

Thus, for Aguilonius, the horopter was not based on the concept of corresponding points, which he does not discuss in his book. However, in theorem 148 on page 52 he stated,

"The object on the point on the horopter when the optic axes meet is seen most clearly. In the second place, objects lying on the rest of the horopter are seen less clearly. In the third place, most imperfectly of all are seen those things which lie outside the horopter, which are seen as double."

Aguilonius went on to describe how only objects on the horopter are seen in their true location and he built an instrument to measure the spacing of double images in the horopter as he defined it. Rubens provided the fanciful illustration of this instrument shown in Figure 2.41. In the actual instrument, the vertical plane could be moved to different distances from the observer.

It is clear from Rubens' illustration, from Figure 2.41a, and from Aguilonius' account that he used his instrument to plot projected positions of disparate images rather than the locus of fused images. Aguilonius maintained that the horopter, defined this way was a frontal plane passing through the point of convergence. He probably believed this because he visually projected the double images of objects that were well outside the frontal plane. We will see in Chapter 15 that the horopter, defined as the locus of fused images, is approximately a circle passing through the point of convergence and the two eyes—the Vieth-Müller circle. Alhazen had already proved in the 11th century that the locus of fused images is not the frontal plane, although he did not establish the shape of this locus. Aguilonius had read Alhazen and cited him four times. However, he did not refer in this proof, to Alhazen's concept of corresponding points, nor to his demonstrations on cyclopean vision and the limits of fusion.

We now define the horopter as the locus in space within which an object must lie to appear fused, a definition that Aguilonius only hinted at. On page 156 of his book Aguilonius produced the drawing shown in Figure 2.40b, accompanied by the following statement;

"If objects fall upon different rays it can happen that things at different distances can be seen at equal angles. If point C be directly opposite the eyes, A and B, with a circle drawn through the three points, A, B, and C. By theorem 21 of Euclid's Third book, any other point D on its circumference which lies closer to the observer than C, will subtend an angle ADB which will equal angle ACB. Therefore, objects at C and at D are judged equally far from the eye. But this is false, because point C is farther away than D. Therefore a judgment of distance is false when based on the angles between converged axes, quod erat probandum."

At first glance, it looks as though Aguilonius discovered the geometrical horopter more than 200 years before Prévost and Vieth and Müller. However, it is clear from this quotation that he was concerned to prove that objects equidistant from an observer do not subtend equal angles to the two eyes. He thought of his circle as the locus of equal angles of binocular subtense of visual lines, rather than as the locus of zero disparity. Euclid himself had used the same theorem to prove that an object subtends the same visual angle when an eye moves round the circumference of the circle passing through the ends of the object and the centre of the eye (see Burton 1945, p 367). It would have been an easy step for Aguilonius to prove that the locus of equal binocular subtense and the locus of fused images are theoretically the same. Aguilonius did not take that step, probably because, like Euclid, he did not have a

Figure 2.43. John Taylor (1703-1772).
(Frontispiece of Taylor's *Mechanismus*, 1750)

clear conception of how light rays are projected onto the retinas. Aguilonius followed Alhazen in believing that the image is formed in the lens. The idea of a frontal-plane horopter persisted until the early 19th century, when Pierre Prévost established that the locus of fused images is a circle passing through the centres of the eyes. He used the same theorems that Euclid and Aguilonius had used to establish the locus of equal angles of binocular subtense. He was apparently unaware of the Aguilonius contribution and did not refer to Alhazen.

2.2.3c *Kepler on depth perception*

In his *Dioptrice* of 1611, Kepler explained depth perception in terms of the feeling of rotation of the eyes as they converge on an object. He probably derived this motor theory of depth perception from Alhazen. René Descartes also adopted the motor theory in his *La Dioptrique* of 1637, which contains a picture of a blind man using two sticks to triangulate distance, which Descartes described as analogous to the use of convergence by sighted persons. The motor theory of depth perception and of vision in general was further elaborated by George Berkeley in his *Essay towards a new theory of vision* (1709).

The theory stemmed from the general belief that depth could not be seen by vision alone, since the image in the eye is both two-dimensional and in the eye. The deeper philosophical position underlying these views is that vision is mediated by images that replicate the seen object—by pictures in the mind—and that these pictures are interpreted in terms of motor actions. The modern view is that vision, like other sensory processes, is mediated by coding mechanisms that process information about the perceived object, but not in an isometric form or even, necessarily, in a topographic form. The coding of depth by binocular disparity is a good example of non-topographic coding of a spatial feature. The persistence of the old view caused the long delay in the discovery of purely visual mechanisms devoted to the perception of depth. The old view still lingers on when people wonder why we do not see upside down when the retinal image is reversed or speculate that the geometrical transformation of the cortical image has something to do with shape recognition (Section 5.4.2).

The set of possible matches between the images of an array of objects is known as the Keplerian projection (Section 15.2.3). I have not been able to trace the source of this idea in Kepler's writings.

2.2.3d *Newton*

We now know that, in humans, the inputs from only the temporal half of each retina project ipsilaterally. Inputs from the nasal half of each retina cross over, or decussate, in the optic chiasma and project to the contralateral half of the brain. Isaac Newton (Portrait Figure 2.42) in his *Opticks* (1704) was the first to propose that visual inputs segregate in this way. He wrote,

"*Are not the species of objects seen with both eyes united where the optick nerves meet before they come into the brain, the fibres on the right side of both nerves uniting there, and after union going thence into the brain in the nerve which is on the right side of the head, and the fibres on the left side of both nerves uniting in the same place, and after union going into the brain in the nerve which is on the left side of the head, and these two nerves meeting in the brain in such a manner that their fibres make but one entire species or picture, half of which on the right side of the sensorium comes from the right side of both eyes through the right side of both optic nerves to the place where the nerves meet, and from thence on the right side of the head into the brain, and the other half on the left side of the sensorium comes in like manner from the left side of both eyes.*" (p. 346)

He went on to explain that this is true only of animals with frontal vision. He conceived of each optic nerve as a multitude of "solid, pellucid, and uniform capillamenta" which transmitted vibrations caused by light to "the place of sensation" in the brain. Newton believed that corresponding fibres fused just after the chiasma, so that the brain received only one nerve from each pair of corresponding retinal points, as shown in Figure 2.10d from one of his manuscripts (see Crone 1992). He thus returned to Galen's concept of the cyclopean eye. Newton stated that similar images falling on corresponding points fuse to give the impression of a single object, and are seen as double when they fall on noncorresponding points. He also realized that dissimilar stimuli falling on corresponding points rival rather than fuse. He wrote,

"... *they cannot both be carried on the same pipes into the brain; that which is strongest or most helped by phantasy will there prevail, and blot out the other.*" (quoted in Harris 1775).

2.2.3e *Others in the seventeenth and eighteenth centuries*
Other writers in the 17th and 18th centuries, including the French physicist Jacques Rohault (1671), Nicholas Malebranche (1674), and Robert Boyle (1688), clearly stated that binocular vision contributes to the impression of visual depth. Like Aguilonius, they noted that it is more difficult to reach accurately for an object with one eye than with two. Rohault noted that after losing one eye people recover the ability to judge the distances of objects, and suggested that they use parallax generated by moving the head from side to side. Perhaps he was aware of the use of binocular parallax by people with two eyes.

Sébastien Le Clerc (1679), an authority on perspective, clearly described the differences between the images of a solid object in the two eyes, but did not relate these differences to the perception of depth. Smith (1738), in his *Compleat system of opticks*, described how he sighted a distant object between the points of a vertical pair of dividers about 6 cm apart. When the dividers were placed at the correct distance, the inner pair of diplopic images fused to appear as a rod extending down the midline from the hand to the distant object. This is essentially the same effect that Ptolemy and Alhazen had observed when looking at the fused image of lines extending out from the two eyes.

William Porterfield of Edinburgh (1759) produced drawings of an object as seen by each eye (Figure 2.10f). He cited anatomical authorities in rejecting Newton's idea of hemidecussation of the visual pathways. Like Rohault (1671), he suggested that corresponding visual fibres combine in the brain.

In 1738, the ophthalmologist Chevalier John Taylor (1703-1772) (Portrait Figure 2.43) adopted Newton's idea of hemidecussation but he produced anatomical evidence that corrected the erroneous notion that the optic fibres from corresponding retinal regions fuse in the chiasma (Figure 2.10e). However, we shall see in the next section that anatomical evidence on this question was not available until 1870. John Taylor claimed to have performed the first operation for correction of squint. In fact, Johann Dieffenbach (1792-1847) performed the first successful operation.

Jean Théophile Desagulier (1683-1744) was an ardent disciple of Newton (Wade 2000). He was born in France but, when he was 2 years old, his father, fleeing religious persecution, brought him to England. He studied and taught at Oxford and in 1714 became demonstrator and curator of the Royal Society. He investigated corresponding points and colour rivalry by placing a candle on the visual axis of each eye and viewing them through an aperture. In support of Newton's notion that corresponding nerves from the eyes fuse in the chiasma, Desagulier claimed that different colours presented to corresponding regions in the two eyes rival rather than fuse. He argued that this supported Newton's idea that inputs from the two eyes rival for access to the common nerve in the chiasma. Other early studies of colour rivalry are described in Section 7.2.1.

Joseph Harris, Master of the Mint in London, who died in 1764, made drawings of crossed and uncrossed disparities arising from objects nearer and further than the point of fixation, as had Alhazen and others before Harris. He realized, as had da Vinci, that monocular occlusions contribute to the impression of depth, not only between an object and its background, but also within a single object, as revealed in the following passage from his *Treatise of optics*, published in 1775, 11 years after his death.

"*And by the parallax, on account of the distance betwixt our eyes, we can distinguish besides the front, part of the two sides of a near object not thicker* (wider) *than the said distance and this gives a visible relievo to such objects, which helps greatly to raise or detach them from the plane, on which they lie. Thus, the nose on a face, is the more remarkably raised by seeing each side of it at once. These observations, I say, are of use to us in distinguishing the figures of small and near objects; and when the breaks, prominences and projections are more considerable, we do not want them. The distances betwixt the legs of a chair, are visible many yards off, and the projection of a building is visible still farther. But as the distance is increased, different degrees of eminences, cavities, et cetera, disappear one after another.*" (p. 171)

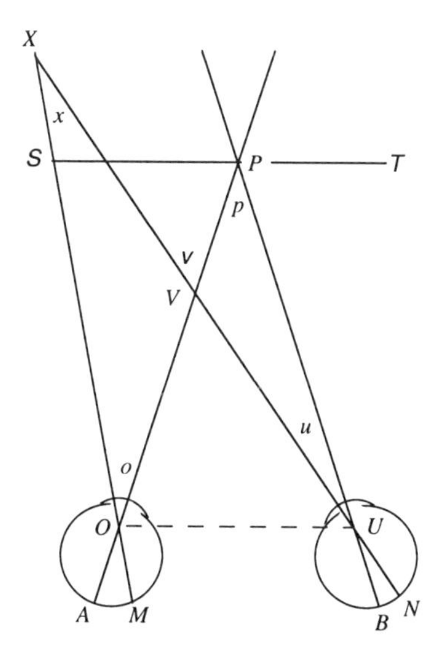

Figure 2.44. Drawing used by Vieth (1818).
Vieth used this drawing to prove that the horizontal horopter is a circle passing through the optical centre of each eye (*O* and *U*) and fixation point *P*. Points *A* and *B* represent the foveas and are therefore corresponding points. Points *M* and *N* represent the images of point X and fall on corresponding points when *M* is as far from *A* as *N* is from *B*. This is true when angles *o* and *u* are equal. But angles *o* and *u* are equal when angles *x* and *p* are equal, since they belong to similar triangles *OXV* and *UPV*. Angles *x* and *p* are equal when they fall on the circumference of a circle passing through *O* and *U*, since angles subtended on a circle by common chord *OU* are equal. Line *ST* is the frontal-plane.

It is clear from this passage that Harris realized that occlusion disparity is scaled by absolute distance. He seems to have been the first to realize this. It is not clear whether he referred only to occlusion disparity or also to position disparity. He used the term "horopter" in its modern sense as the locus of objects producing fused images. However, he believed the horopter to be a frontal surface. He also wrote; "An object that is a little out of the plane (of the horopter), may yet appear single · · · it will also shift its place by winking either eye, and looking at it with both eyes."(p. 113). This description of Panum's fusional area was written 700 years after Alhazen had made the same point and 80 years before the account provided by Peter Ludvig Panum, professor of physiology at Kiel University. The quotation also states the principle of parallactic motion, although Galen had made a similar observation.

2.2.4 The horopter

For Ptolemy in the 2nd century, the locus of objects producing fused images was the frontal plane through the fixation point (see Section 2.2.1). Alhazen, in the 11th century, proved that the locus is not a frontal plane and defined a distinct vertical locus for fused images, in the median plane of the head. Aguilonius introduced the term 'horopter' but defined it as the frontal plane in which the images of fused and diplopic images appear to lie. I do not know who first used the term 'horopter' to refer to the locus of objects producing fused images. Pierre Prévost was the first person to describe the horopter, defined in this way, as a circle passing through the centres of the eyes and the fixation point. In his *Essais de philosophie ou étude de l'esprit humain*, published in Geneva in 1804, he wrote:

"It follows from the stated law that, in the plane of the optic axes, the position of those points seen single with the two eyes is a circumference of a circle which passes through the two centres of the eyes and the intersection of their axes. I refrain here from demonstrating this proposition which is easy to deduce." (Shipley and Rawlings 1970a).

Presumably, Prévost's deduction of the circular horizontal horopter relied on Euclid's theorem that angles subtended by the cord of a circle on the circumference are equal. Prévost also described, incorrectly, a vertical horopter formed by rotating the intersection of the visual axes about the interocular axis.

Vieth (1818) was the first to specify clearly the geometry of the horizontal horopter, which he defined as the locus of objects producing fused images. He wrote,

"Firstly it is correct and established from common experience, that point P in Fig. 2 (Figure 2.44) *towards which both eyes are directed, or at which both visual axes intersect is seen singly.*

Whether the so-called corresponding points M and N, or more specifically, whether these images of a point X are equidistant or at unequal distances from A and B, the images of point P, that depends on whether the angles o and u at the pupil are equal or unequal. However, o = v - x and u = v - p. Therefore, . . . if x is equal to p, then o is equal to u.

Thus, in that condition, where the angles p and x are equal, the images M and N are equidistant from A and B, and this case occurs when X lies on the circumference of a circle, which passes through O and U and P, because all angles on this circumference are subtended by the same chord O U."

Here Vieth applied the same geometrical theorems that Aguilonius used to establish the locus of equal angles of binocular subtense, to establish that the locus of single vision—the horopter—is a circle through the fixation point and the centre of each eye. Vieth went on to write,

"Thus, if by the expression corresponding points one understands such points which lie in the same direction in both eyes, and are equidistant from A and B which seems to me the correct meaning, and one asserts one sees that thing singly whose images fall on such corresponding points, then, according to this rule, one sees that thing singly which is situated within the boundary of a sphere which passes through O, U, P, hence, not what lies in the plane S T, which one has called the horopter."

Here Vieth incorrectly generalized his principle of corresponding points in claiming that the horopter is a sphere rather than a circle. Not all angles subtended by a chord on the surface of a sphere are equal. As we will see in Chapter 15, the theoretical horopter for parallel visual axes is an infinite toroid formed by sweeping the horopter circle about the interocular axis. A. Prévost (1843) first pointed out that when the visual axes are converged, the horopter is not a surface but a horizontal circle and a vertical line in the median plane (Section 15.6).

Vieth went on to state, as had Ptolemy and Alhazen, that an object positioned between the visual axes projects images on opposite sides of the image of the fixation point, whereas an object to either side of the two visual axes produces images that fall on the same side of the image of the fixation point. Like others before him, he stated that images of an object nearer than the fixation point are crossed with respect to the fixation point and those of an object beyond the fixation point are uncrossed. These ideas are explained more fully in Section 15.3.1.

Johannes Müller (1826) produced a similar and independent analysis, but became aware of Vieth's work, which he acknowledged in his paper. Müller (1801-1858) was Professor of physiology and anatomy in Berlin. He taught Helmholtz and was one of the founders of experimental physiology. The theoretical horizontal horopter is now known as the **Vieth-Müller circle**. Müller had the horopter passing through the centres of the two lenses. He discussed binocular disparity in the context of fusion and rivalry of binocular images and concluded that the differences between the images in the two eyes were too small to be detected.

In 1843, five years after Wheatstone reported that disparate images produce a sensation of depth, Müller agreed that disparity is involved in the perception of depth. Volkmann (1836) first specified the geometrical assumptions underlying the theoretical horopter, and Helmholtz (1864) generalized the geometry of the horopter over the visual field. For more details on the history of the horopter, see Shipley and Rawlings (1970a).

2.2.5 Physiology of stereopsis

Before the 1960's, many visual scientists believed that binocular stereopsis arose from high-level cognitive processes rather than from the conjunction of visual inputs at an early stage of visual processing. This idea was motivated by the belief that only higher mammals have stereoscopic vision and by the observation that the 3-D appearance of the world does not change appreciably when one eye is closed. Helmholtz (1893, p.262) wrote,

"We therefore learn that two distinct sensations are transmitted from the eyes, and reach consciousness at the same time and without coalescing; that accordingly the combination of these two sensations into a single perceptual picture of the external world is not produced by any anatomical mechanism of sensation, but by a mental act."
He realized that stereopsis depends on the registration of disparities but argued that;

". . . the coincidence of localization of the corresponding pictures received from the two eyes depends upon the power of measuring distances of sight which we gain by experience."

This view stemmed from his empirical theory of vision and the associated theory of unconscious inference (Section 2.1.8). Wundt (1894, p. 209), who had been Helmholtz's assistant in Heidelberg, expressed the same opinion. Sherrington (1904), also, concluded from his work on binocular flicker that monocular images are processed independently, and that the final synthesis is " mental".

Ramón y Cajal (1911) proposed that inputs from corresponding regions of the two retinas converge on what he called "isodynamic cells", and that this mechanism forms the basis of unified binocular vision. This idea received experimental verification when Hubel and Wiesel (1959; 1962) reported that pairs of visual afferent fibres of the cat converge on binocular cells in the visual cortex, and that the receptive fields of each binocular cell occupy corresponding positions in the two eyes. If a binocular cell had identical receptive fields, identically positioned in each eye, it would respond optimally to stimuli with zero binocular disparity, and depth could not be recovered from its output. This was the gist of Helmholtz's argument against the idea of convergence of visual inputs.

The problem would be solved if there were cells specifically tuned to similar images in slightly different positions in the two eyes. Different cells would be optimally tuned to different disparities.

Simple as this idea is, it was not proposed until 1965. This is probably because the idea of any cortical cell being tuned to a specific stimulus feature was not in vogue until 1959, when Hubel and Wiesel discovered cortical cells tuned to stimulus orientation and movement. Hubel and Wiesel failed to find disparity-sensitive cells. But they did not have close control over the positions of the eyes, and it is not clear from their report whether they had thought of binocular cells devoted to the detection of disparity.

Jack Pettigrew produced the first evidence of disparity detectors at an early stage of visual processing. He did this work for his undergraduate thesis written in the University of Sydney in 1965. He got the idea while inspecting a Julesz random-dot stereogram and mentioned it to his supervisor, Peter Bishop, who was working on binocular cells in the cat visual cortex, but not with this particular idea in mind. Bishop, suggested to Pettigrew that he repeat the experiments of Hubel and Wiesel on the visual cortex of the cat using a Risley prism to control the disparity of the images from a single display rather than using separate stimuli for each eye. The search for binocular cells selectively tuned to different disparities was beset with the problem of ensuring that the images in the two eyes were in binocular register. Pettigrew solved this problem by paralyzing the extraocular eye muscles with gallamine and curare.

Bishop took Pettigrew's thesis to a conference in California in 1966 and showed it to Horace Barlow who had just set his graduate student, Colin Blakemore, the task of looking for disparity detectors. Barlow invited Pettigrew to work with him and Blakemore at Berkeley. The three of them confirmed the presence of disparity-sensitive cells in the visual cortex of the cat and reported their findings in 1967 (Barlow *et al.* 1967). They found that certain binocular cells in the visual cortex of the cat respond selectively to line and bar stimuli having a particular binocular disparity. Similar findings, based on work done between 1965 and 1967, were reported about the same time from the University of Sydney, Australia by Pettigrew, Nikara, and Bishop (1968). The history of these discoveries is described by Bishop and Pettigrew (1986). In 1977, Gian Poggio and his co-workers at Johns Hopkins University in Baltimore first reported disparity detectors in the primary visual cortex of the monkey. Developments after that date are described in Chapter 6.

For a detailed history of ophthalmology see Hirschberg (1982). For more details on the history of visual optics and binocular vision see Lindberg (1976), Polyak (1957), and Gulick and Lawson (1976). Wade (1987) has provided an interesting account of the discovery of stereoscopic vision.

Figure 2.45. A shadow theatre in Paris in the 1890's.
The show was designed by M. Caran d'Ache and often depicted battle scenes. Some parts of the silhouettes consisted of coloured paper. Assistants moved them across the screen while the audience viewed them from the other side. (From Hopkins 1898)

2.3 HISTORY OF VISUAL DISPLAY SYSTEMS

2.3.1 Early display systems

2.3.1a *Shadowgraphs*

More than a thousand years ago showmen travelled all over China with shadow plays. Shadows of hand-operated flat puppets were cast onto a screen of fine paper or cloth to the accompaniment of music, as shown in Figure 2.45. The audience sat in a darkened room. Shadow plays spread to all parts of the Far East and can still be seen in Java. They sometimes formed part of religious ceremonies or funerals. In the 17th century, shadow plays reached Europe through Turkey, where they were displayed in palaces and at street corners. In the 18th century there were at least four large shadow theatres in London, one with a screen 14 feet high. By the 19th century, wooden hand-operated puppets were replaced by mechanically operated metal puppets. People could buy puppet kits for use in their own homes (Thurman and David 1978).

2.3.1b *The camera obscura and magic lantern*

In the camera obscura an image of a scene is projected through a small hole or through a lens onto the wall of a darkened chamber. The camera obscura had its origin in the pinhole camera that was known from ancient times (Section 2.1.2d). During the 16th century, several people added a biconvex lens,

(a) The first Parisian panorama built on the Boulevard Montmartre in 1802. (From Bapst 1841)

(b) Panorama Marigny built on the Champs-Elysées in Paris in 1886. It exhibited the *Battle of Buzenval* and later a *Diorama of Paris*. In about 1893 it was converted into the Théâtre Marigny, which still exists. (From Architektonische Rundschau, 1886. In Oettermann 1997)

Figure 2.46. Panoramas.

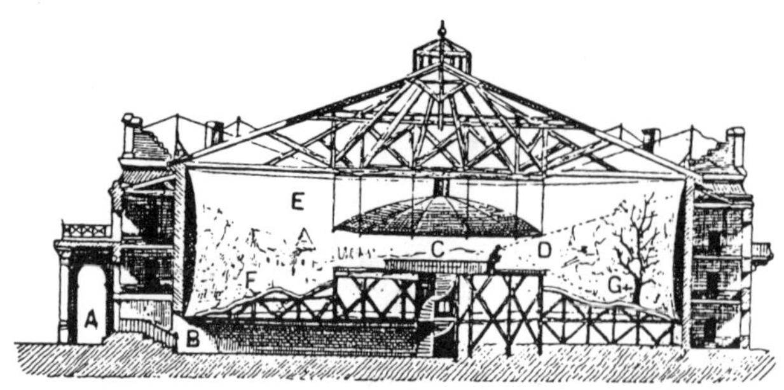

Figure 2.47. Cross section of a panorama.
People entered at A and passed through passage B up to the observation platform C. They viewed the cylindrical display, E, with a visual angle indicated by D. The floor consisted of a mock terrain that blended into the panoramic display. (From Oettermann 1997)

Figure 2.48. Painting a panorama.
The scene depicts the Battle of Gettysburg and was exhibited in New York at the end of the 19th century (From Hopkins 1898)

which allowed more light to enter the dark chamber. These included Girolamo Cardano, professor of mathematics in Milan and Daniele Barbaro, a Venetian architect, inventor of the variable diaphragm and author of *La practica della perspettiva* (1568). The instrument was used for observing solar eclipses and for popular amusement (see Hammond 1981). Della Porta (c. 1535-1615) was probably the first person to use the instrument for drawing (Section 2.1.3c). Kepler coined the name "camera obscura".

Examples of room-size instruments still exist in Edinburgh, Bristol, the Isle of Man, and in the Cliff House in San Francisco. The oldest, built in 1836, is in the museum in Dumfries, Scotland. A lens in the roof can be rotated to show a panoramic view of the surrounding countryside. In the 16th century small versions of the camera obscura with an added lens were used by artists as an aid to drawing in perspective (Section 2.1.6).

In the 17th century, these basic elements were brought together in the construction of the first projector—the magic lantern. Perhaps the first magic lantern was built by Athanasius Kircher (1601-1680), a Catholic priest. He was born in Germany and worked in Rome, where he published *The great art of light and shadows* in 1648. Painted transparent slides were projected onto a screen by a lens attached to a lantern containing a candle. He placed several slides on the rim of a disc. The Dutch scientist Christiaan Huygens also constructed an early magic lantern. After the discovery of more intense light sources at the end of the 18th century, magic lantern shows became a popular form of public entertainment. The superimposition of pictures from two projectors allowed one picture to be dissolved into another. A rear-projected picture could be made to move or loom. Mechanical slide holders allowed one part of the picture to move with respect to a fixed background (see Coe 1981). In 1798 a show called *Phantasmagoria* was opened in a reputedly haunted chapel in Paris. Pictures of human skeletons were projected onto billowing smoke to create an impression of solid writhing forms emerging from the tombs of the chapel. William Nicholson (1802) described a *Phantasmagoria* presented by a Mr. Philipsthal at the London Lyceum in about 1802. Similar methods were used to create ghosts on the theatre stage (Hopkins 1898). The magic lantern became the modern projector and, operating in reverse with the addition of light-sensitive film, the camera (Gernsheim 1969).

2.3.1c *Viewing boxes*

In about 1410, Brunelleschi had people look through a hole at a reflection of his perspective painting of the Baptistery in Florence. Since the apparatus was placed in front of the real Baptistery, people were fooled into believing they were seeing the real scene. The first viewing box containing a painted image was probably devised by Leon Battista Alberti in Florence in the 15th century. During the 17th century, peepshow boxes were constructed on the same principle of projective equivalence as the Ames distorted room. A room that appears in its correct distorted form when viewed from the wrong angle appears rectangular when viewed from the designated station point. These devices are described in Section 2.1.6c. The English artist, Thomas Gainsborough, made a peep show box in about 1781 for which he painted pictures on glass plates. Viewing boxes with slides or short movie sequences remained popular at seaside resorts until well into the 20th century. Today they are represented by flight simulators, video games in amusement arcades, and television.

2.3.1d *Panoramas and dioramas*

A panoramic display is one that encircles the viewer, at least horizontally. In 1787, Robert Barker patented the idea of a panoramic display and the word "panorama" came into the language in the 1890's, although it is not known who first used it. Robert Baker built the first panoramic display for large audiences in Edinburgh and London in 1797. Viewers stood at the centre of a cylinder, 13 metres in diameter, that was painted with a view of Edinburgh or London as seen from a tall building. Panoramas were built in many European and American cities during the 19th century. Examples are shown in Figure 2.46. The audience entered a central observation platform from below, as shown in Figure 2.47. The cylindrical canvas was illuminated by light from a window round the roof of the building or from lamps suspended in the building.

Panoramas were designed to reproduce a scene so accurately that viewers could believe they were looking at the real thing. They were an extension of the trompe l'oeil technique to fill the whole field of view. The floor surrounding the audience contained a 3-D false terrain that blended into the 2-D scene painted on the cylindrical canvas, as shown in Figure 2.47. Artists painted the scene on a series of flat canvases, which were then joined and formed into a cylinder. This introduced some distortion of perspective, which could be avoided by use of a camera lucida with a curved ruler; a device invented by a Frenchman called Gavard in 1830. At a later time the panoramic scene was made from a series of photographs. Teams of artists standing on a moveable scaffold, as shown in Figure 2.48, transferred the preliminary paintings or photographs onto the full size canvas, which was typically 300 feet long and 40 feet high. This was done by drawing square grids on the painting and on the large canvas. A long drawing instrument was used to transfer the image piecemeal. Photographs could be converted into slides and projected onto the large canvas. Different artists specialized in painting people, buildings, and landscape. The whole process took at least a year to complete. For a history of the panorama see Bapst (1841) and Oettermann (1997).

A diorama is a large display that typically fills much of the visual field but does not extend through 360°. Louis Daguerre, before he turned to photography, built the Diorama in Paris in 1822. He and Charles Bouton, painted enormous pictures, 72 feet wide and 46 feet high, with translucent and opaque paints with subjects such as *The tomb of Napoleon*, *The beginning of the deluge*, and *The Grand Canal of Venice*. These were exhibited in a large room with a mixture of reflected and transmitted light controlled by mirrors and shutters. In 1823, John Nash had a similar structure built in Park Square, London. It seated 200 people on a circular platform, which rotated periodically to reveal a different scene. The building still exists, but not the machinery. Dioramas created such realism that audiences were convinced they were observing a 3-D scene.

Painted dioramas that fill the field of view for viewers in a given position are still used as backdrops to exhibits in many museums, such as the Natural History Museum of New York. In recent times, realistic 3-D moving displays have been created in wide-angle cinemas such as those developed by the Imax Company.

Figure 2.49. Sir Charles Wheatstone.
(Engraving from a photograph in the Illustrated London News, 1868, 52, p. 145. Reproduced by permission of the Illustrated London News Picture Library)

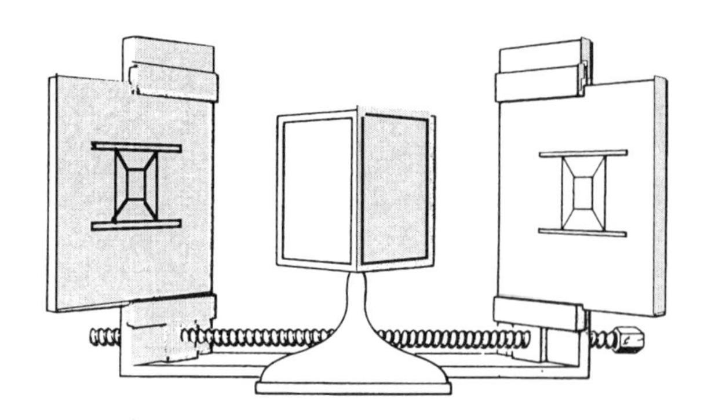

Figure 2.50. Wheatstone's first mirror stereoscope.
(From Wheatstone 1838)

2.3.2 Advent of the stereoscope

2.3.2a *Early devices for dichoptic viewing*
I have already explained how Ptolemy and Alhazen combined the images of distinct objects by converging on an object in another depth plane. In the 18th century, Desagulier (1716) added the refinement of viewing objects on the two visual axes through an aperture. This removes the unwanted monocular images. Dutour (1760) removed unwanted monocular images by placing a board between the eyes as they converged on two side-by-side displays. Reid (1764) viewed distinct objects through two tubes aligned with the two visual axes. They all used these devices to study corresponding points, binocular disparity, and binocular rivalry but none of them realized that disparities could create impressions of depth.

The first binocular optical instrument was a twin telescope built in Middleburg, Holland by the optician Hans Lippershey in 1608. A Capuchin friar, Antonius Maria Schyrleus de Rheita, was familiar with Lippershey's work and built his own binocular telescope in 1645 which he described in his book *Oculus Enoch at Eliae*, published in Antwerp in 1645.

A binocular microscope was made by the Capuchin monk, Père Chérubin d'Orléans, in 1677 and presented to the Dauphin of France. Chérubin d'Orléans also made a binocular telescope in 1671. However, in all these instruments each image was inverted so that they created a pseudostereoscopic effect in which the sign of disparity was inverted. Riddell (1853) added erecting eye pieces to a binocular microscope to produce a true stereoscopic effect (see Wade 1981). But even this instrument was not a stereoscope because the stimuli were simply 3-D objects viewed by both eyes. In principle, using the instrument was like looking at the world through two tubes.

2.3.2b *Wheatstone*
The invention of the modern stereoscope must be credited to Sir Charles Wheatstone (Figure 2.49) who was born near Gloucester, England, in 1802. As a young man he made and sold musical instruments and invented the concertina. He also wrote some scientific papers on acoustics. He contributed to many fields, including chronometry, optics, cryptography, and telegraphy, and invented many useful devices. In 1834 he became professor of experimental physics at King's College, London. He died in Paris in 1875.

Towards the end of 1832 he had two stereoscopes made by Murray and Heath, opticians in London. One was a mirror stereoscope and the other a prism stereoscope (see Gernsheim 1969). But he became involved with the electric telegraph and waited until 1838 before reporting his mirror stereoscope and his experiments with the instrument to the British Association in Newcastle-on-Tyne in the North East of England (Wheatstone 1838). However, his colleague Herbert Mayo gave an account of Wheatstone's stereoscope in his *Outlines of human physiology*, which appeared in 1833. Wheatstone called his new instruments stereoscopes. Aguilonius had used the word *stéréoscopique* in 1613 to denote binocular vi-

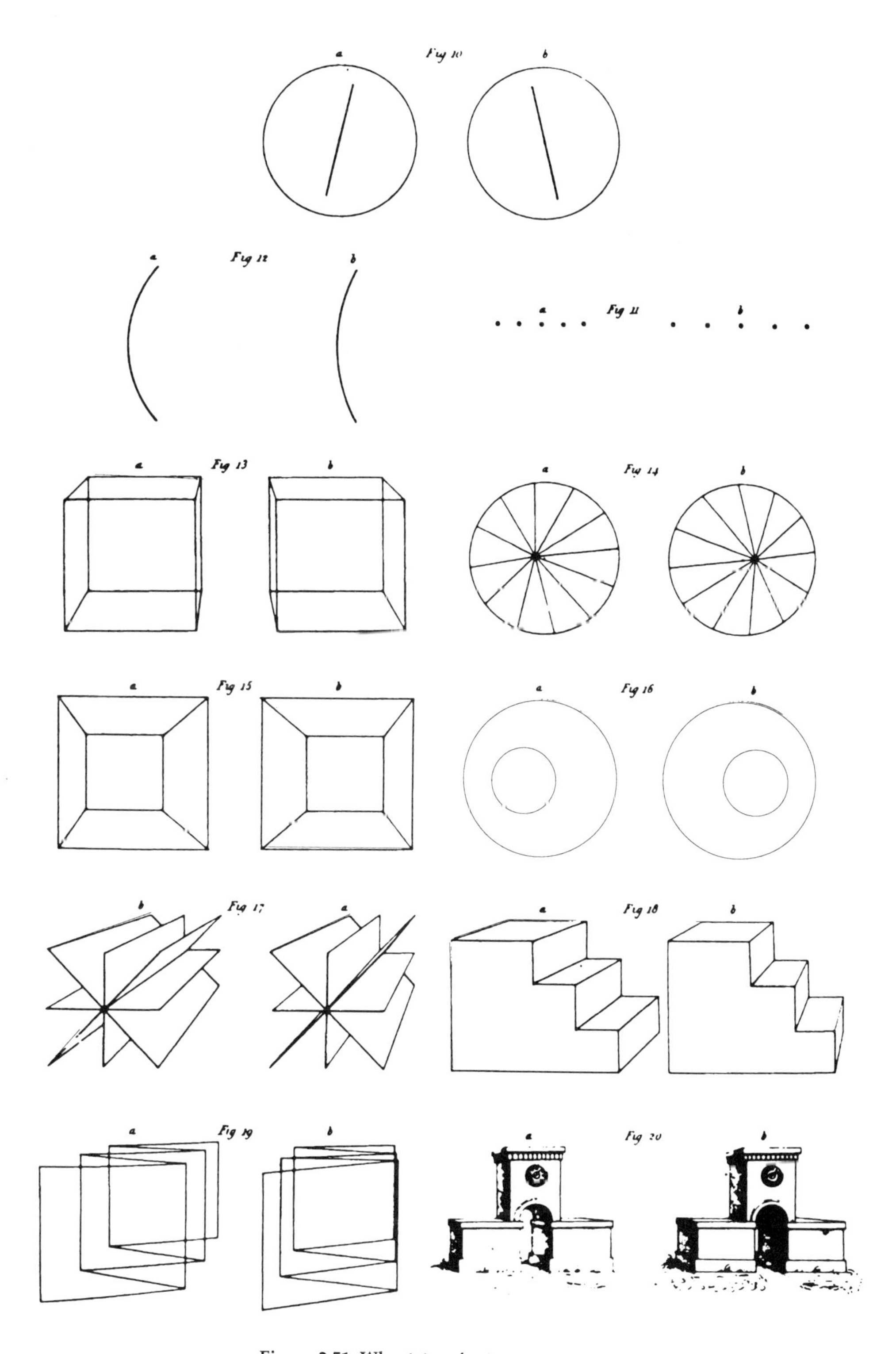

Figure 2.51. Wheatstone's stereograms.
(From Philosophical Transactions of the Royal Society, 1838)

sion and J. G. A. Chevallier had used the word *stéréoscope* to describe an instrument that created an impression of solid objects from a variety of monocular cues (see Helmholtz, Vol. 3, p. 363). Wheatstone stated the principle of the stereoscope thus,

"It being thus established that the mind perceives an object of three dimensions by means of the two dissimilar pictures projected by it on the two retinas, the following question occurs. What would be the visual effect of simultaneously presenting to each eye, instead of the object

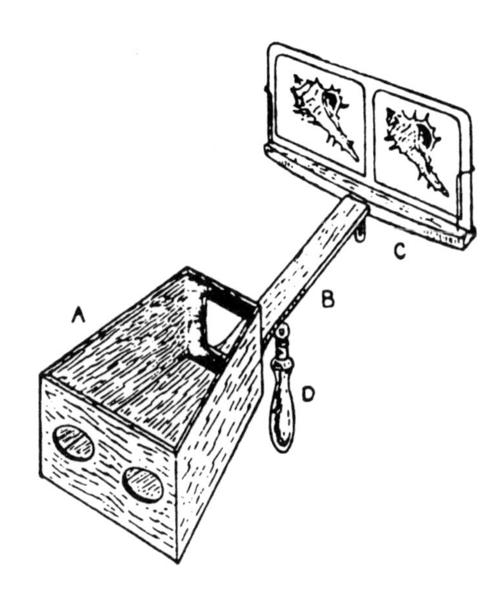

Figure 2.52. Elliot's stereoscope of 1839.
Modified by Lockett in 1912. (From Lockett 1913)

itself, its projection on a plane surface as it appears to that eye? To pursue this inquiry it is necessary that means should be contrived to make the two pictures, which must necessarily occupy different places, fall on similar parts of both eyes." (p. 373)

His mirror stereoscope consisted of two mirrors at right angles and two vertical picture holders (Figure 2.50). In a later version the two halves of the instrument could be rotated to change the angle of convergence. He described twenty pairs of pictures, or stereograms, which appeared solid when viewed in his stereoscope. These included a series of points stepped in depth, a cube, a cone, and a pyramid, as shown in Figure 2.51. He observed that all these shapes appeared flat when the pictures in the two eyes were the same and appeared in reverse depth when the pictures with disparity were reversed to the two eyes.

James Elliot, a mathematician in Edinburgh, had the idea in 1834 of constructing a stereoscope but did not construct one until 1839. He was not aware of Wheatstone's work. Elliot's stereoscope consisted of a box that allowed each eye to see one of a pair of stereographic pictures. It contained no mirrors or lenses (see Elliot 1852). Figure 2.52 shows a modified version of the stereoscope made by Lockett (1913).

The essence of any stereoscope is that it allows one to control the image in each eye separately. An experimenter can thus isolate binocular variables and study their effects quantitatively—it provides an experimenter with dichoptic control. With his new instrument, Wheatstone demonstrated the relationship between binocular disparity and depth perception. His stereoscope with adjustable arms allowed him to vary convergence while keeping disparity constant, and thus show that impressions of depth do not depend on disparity alone. The invention of the stereoscope inaugurated the modern study of stereoscopic vision.

In 1852, Wheatstone presented a paper to the Royal Society in which he described the pseudoscope, which reverses the inputs to the two eyes. This reverses the sign of disparity and makes concave surfaces appear convex. In a communication to the Microscopical Society, Wheatstone (1853) described the binocular microscopes made by Père Chérubin d'Orléans, in 1677 and by Riddell in 1853.

2.3.2c *Brewster*

Sir David Brewster (Portrait Figure 2.53) was born in Jedburgh, Scotland, in 1781, the son of the rector of the grammar school. He died in 1868. He wrote many scientific papers on optics, especially on the polarization of light, and invented the kaleidoscope. He was a scientific editor, college Principal at St. Andrews University, Principal of Edinburgh University, and secretary to the Royal Society of Edinburgh. Brewster witnessed Wheatstone's demonstration of the mirror stereoscope at the British Association meeting in 1838 and bought a model with which he began his own experiments. At a meeting of the Royal Society of Edinburgh in 1849 he described a stereoscope in which two side-by-side pictures were placed in a box and viewed through prisms made from half lenses, which fused and magnified the images. The instrument was described in his book on stereoscopes published in 1856. Brewster made his original prism stereoscope by cutting a convex lens in half and arranging each half with its vertical cut edge on the temporal side of an eye. An early version is shown in Figure 2.54.

Examples of early stereograms produced for the prism stereoscope are shown in Figure 2.55. The prism, or lenticular, stereoscope is still referred to as the Brewster stereoscope, although Wheatstone had made one in 1832. The subsequent development of stereoscopic instruments in general is described in Chapter 30.

Brewster wrote an anonymous letter to the Times in October 1856, in which he disputed Wheatstone's claim to the invention of the stereoscope and the principle upon which it was founded. Wheatstone effectively refuted Brewster and the two men engaged in an acrimonious correspondence in the newspaper. In 1859, Brewster attacked Wheatstone again. He claimed that a pair of pictures found in Lille and drawn by the 17th century Florentine artist

Figure 2.53. Sir David Brewster.
(An engraving from the Illustrated London News, 1868, 52, p. 189. Reproduced by permission of the Illustrated London News Picture Library)

Figure 2.54. Brewster's prism stereoscope.
A mirror in the hinged panel reflects light onto the stereogram.

Jacopo Chimenti were a stereogram. Wheatstone obtained photographs of the drawings and showed that they did not produce an impression of depth when placed in a stereoscope. In spite of Brewster's objections, it was generally agreed that the differences between the drawings were accidental and not related to binocular disparity. For a lively account of this debate see Wade and Ono (1985) and Ono and Wade (1985). The works of Wheatstone and Brewster have been edited by Wade (1983).

By the 1850's, the science of binocular vision boomed. During this decade, the percentage of papers in vision devoted to binocular vision increased from about 19 to 30%. About 70% of these papers were written by Germans, including Wundt, Helmholtz, Hering, Dove, Panum, von Graefe, Meissner, and Nagel (Turner 1993).

2.3.3 Stereophotography

Johann Schulze (1687-1744), a German physician, produced the first photographic images in 1725. He placed cut-out letters on a bottle containing a mixture of chalk and silver nitrate. Tom Wedgwood, son of the potter Josiah Wedgwood, produced images of botanical specimens on sensitized silver salts in 1796 (see Pollack 1977). However, these men did not know how to fix the images. Joseph Nicéphore Niépce, Cardinal of Amboise (1765-1833), produced the first photograph in about 1826, and Louis Daguerre produced his first successful daguerreotype photograph in 1837 (Portrait Figure 2.56).

The principle of first making a transparent negative from which many positives can be produced was introduced by the Englishman Fox Talbot in 1840. In 1841, Wheatstone employed two photographers, Richard Beard and Henry Collen, to help him produce the first stereophotograph, which was a portrait of Charles Babbage, the inventor of the first calculating engine. The following year, the Parisian photographer A. Claudet produced daguerreotype stereophotographs for Wheatstone, but these were not satisfactory because of the reflective surface of the prints.

The first stereophotographs were taken by moving a single camera through the interocular distance. Since long exposures were required with early film, the subject had to be stationary for a long period. In 1853, the Parisian photographer A. Quinet made the first binocular camera, which he called the Quinétoscope. In 1856 J. B. Dancer, independently, made a stereocamera in Manchester, England. It is illustrated in Figure 2.57a.

In 1896, the French company, Jules Richard, produced the "Verascope", the first mass-produced stereo camera. In 1901, Kodak introduced the Stereo

Figure 2.55. Early photographic stereograms.
The upper stereogram is one of the earliest ones produced for the prism stereoscope. It is a photograph of the Wheatstone family, taken by Antoine Claudet probably in the mid 1850's. (Reproduced by permission of the National Portrait Gallery) The lower stereogram is a view of the Great Exhibition of 1851 in the Crystal palace, London. The stereograms should be viewed with a stereoscope or with divergent fusion.

Kodak camera. These stereo cameras consisted of two cameras side-by-side. In 1853, F. Barnard produced an attachment, known as the stereo reflector (Figure 2.58) that converted a single lens camera into a stereo camera. The left-eye and right-eye views are reflected so that they project through the same lens and fall on opposite halves of the film. In the Leitz stereo attachment, prisms replaced the mirrors. The attachment for the Contax camera contains a reflecting system and two miniature lenses. The normal lens is replaced by the attachment. The distance between the reflecting system and the lens for proper separation of the half images is proportional to the lens aperture and inversely proportional to the tangent of the angular field of the lens.

For an account of further developments in stereophotography see Maude (1978).

By 1846, stereophotographs were being sold in the shop of James Newman in Soho Square, London. However, stereophotography did not arouse much interest because inexpensive stereoscopes were not available. Brewster took his prism stereoscope to Paris in 1850 and engaged the interest of the optician Jules Duboscq who built a number of them, together with a set of daguerreotype stereophotographs. These stereoscopes were shown in London at the Great Exhibition of 1851. One was made for Queen Victoria who took a great interest in the device. Within three months, nearly a quarter of a million prism stereoscopes were sold in London and Paris.

(a) First commercial stereo camera. J. B. Dancer, Manchester, 1856. It used wet collodian plates or dry collodian-albumen plates.

(b) High speed spark-drum stereo camera. Lucien Bull 1903.

(c) Edison's Kinetoscope.

(d) Ives Kromskop. Three stereopositives with superimposed red, green, and blue filters were combined to produce a coloured stereoscopic picture.

Figure 2.57. Early stereoscopic devices.
(South Kensington Science museum. Science and Society Picture Library)

Louis Daguerre.

Fox Talbot

Nicéphore Nièpce

Figure 2.56. Pioneers of photography.

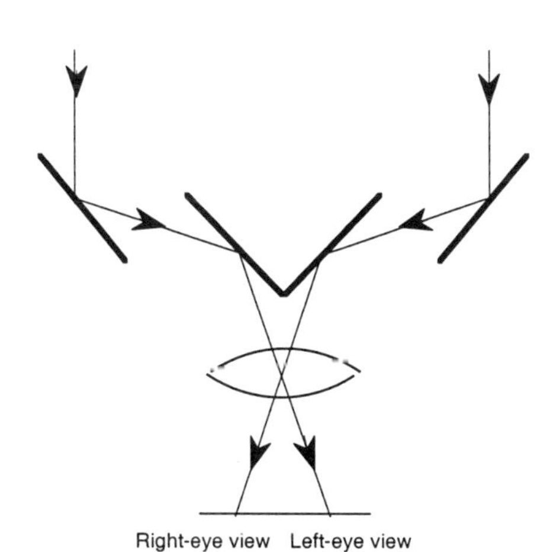

Figure 2.58. The stereo reflector.
A system of mirrors attached to a single lens camera converts it into a stereo camera in which the left-eye and right-eye views are projected to opposite halves of the film.

Duboscq patented the prism stereoscope in 1852, but the patent was successfully challenged and annulled in 1857. Stereoscopic views of the Great Exhibition of 1851, such as those in Figure 2.55, were popular because people who could not get to London were able to see the exhibits. Claudet devoted himself to the improvement of stereoscopic photography and patented a folding pocket version of the prism stereoscope in 1853. A rotary stereoscope holding 50 or 100 views was made in England in 1854. Stereoscopes made for the wealthy became very elaborate. Figure 2.59 shows a mirror stereoscope built in 1856 and a prism stereoscope of 1862.

In 1854, George Swan Nottage, a man of humble origin and limited education, founded the London Stereoscopic Company with the motto 'No home without a stereoscope'. By 1858 the company had sold over half a million stereoscopes and its travelling photographers had produced 100,000 stereoscopic photographs of famous places from many parts of the world. Nottage became Lord Mayor of London and died in 1885, a wealthy and honoured man. Stereoscopic photography was introduced into

Mirror stereoscope 1856

Prism stereoscope by Hirst and Wood 1862

Figure 2.59. Early domestic stereoscopes.
(South Kensington Science Museum. Science and Society Picture Library)

Figure 2.60. The Kaiser Panorama.
Twenty-five people viewed 50 stereoscopic slides that changed position every few minutes inside the 15-foot diameter cylinder. (From Oettermann 1997)

the United States in 1854 by William and Frederick Langenheim. They founded the American Stereoscopic Company in New York in 1861.

The idea for stereoscopic book illustrations was patented by P. B. Godet in 1857. The first book illustrated with stereoscopic photographs was Charles Smyth's account of Tenerife, published in 1858. Another early book was entitled *Stereoscopic views among the hills of New Hampshire* was published by the Bierstadt Brothers of New Bedford, Massachusetts in 1862. Viewing stereograms in a book requires an open type of stereoscope, which can be placed on the surface of the book. John Parker (1858) described a pair of prisms with a partition extending from the bridge of the nose to the stereogram. In the same month, J. B. Spencer (1858) described a similar stereoscope for use with books.

In 1859 a Mr. Bennett of London described the "Clairvoyant Stereoscope", which was an open sided hand-held prism stereoscope with a sliding picture holder. Joseph Beck (1860) patented an improved version in September 1859. Oliver Wendell Holmes, essayist and Harvard professor of medicine, designed a similar hand-held version of the prism stereoscope in 1863. In 1864, his friend Joseph Bates, added a sliding picture holder. The instrument was mass-produced for home entertainment, both in America and Europe. It is readily available in antique shops, and is still manufactured. Wendell Holmes (1859) wrote three articles in the Atlantic Monthly about stereoscopic pictures. He wrote,

"The time will come when a man who wishes to see any object, natural or artificial, will go to the Imperial, National, or City Stereographic library and call for its skin or form, as he would for a book at any common library."

The Oliver Wendell Holmes Stereoscopic Research Library is maintained by the National Stereoscopic Association (www.cincinnati.com/3Dlibrary). The Association publishes the bimonthly magazine Stereo World (www.stereoview.org).

In 1862 Henry Swan (1863) patented stereoscopic miniatures, which became known as "Swan cubes". Transparent positives were mounted on two small prisms so that they created a 3-D image when viewed from the correct position.

Ives built a device the device shown in Figure 2.57d for producing coloured stereoscopic pictures. Three pairs of stereo pictures taken through red, green, and blue filters were viewed in a stereoscope containing the same filters.

By 1862, more than a thousand professional photographers were producing stereoscopic photographs, which were sold by the million. The Keystone View Company of America absorbed its main rival, Underwood and Underwood, and eventually dominated the market. Until the advent of the cinema, the stereoscope was the optical wonder of the age. It allowed people to see the world in the comfort of their own living rooms (see Earle 1979).

In 1880 the physicist August Fuhrmann opened the Kaiser Panorama on the Unter den Linden in

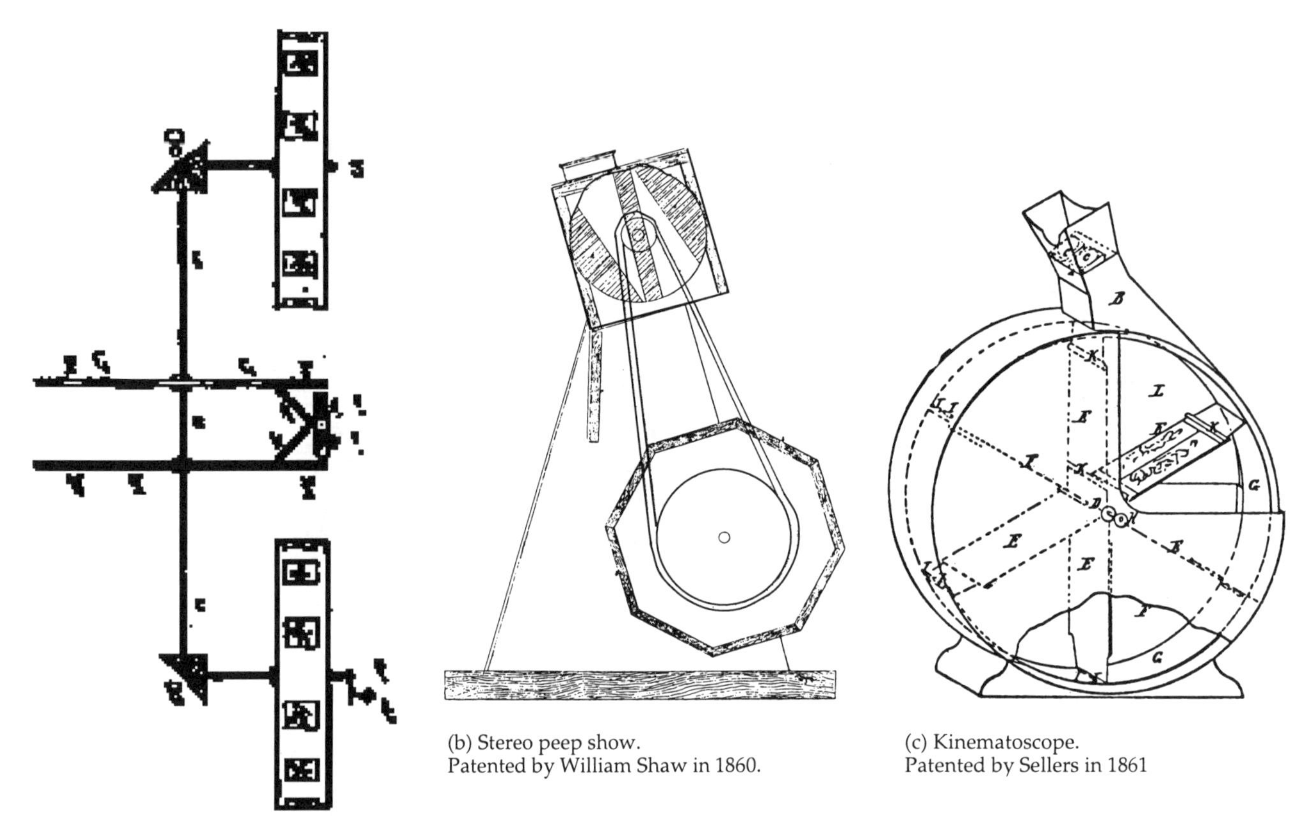

(a) Twin zoetrope.
Patented by Henri DuMont in 1859.

(b) Stereo peep show.
Patented by William Shaw in 1860.

(c) Kinematoscope.
Patented by Sellers in 1861

Figure 2.61. Early moving stereoscopes.

Berlin. It remained open until 1939. A 5-metre diameter housed 25 viewing stations, as shown in Figure 2.59. Stereoscopic slides rotated past the viewing stations at intervals of a few minutes. Many photographers were employed to collect photographs of exotic places and headline-making events. At one time there were 250 Kaiser panoramas throughout Germany. The German Kaiser, the Sultan of Turkey, and the Pope had copies of the photographs for private viewing. Several thousand of these stereoscopic pictures were published in 1915 in a book entitled *Goldenes Buch der Zentrale für Kaiserpanoramen*.

Stereoscopic peep shows lost their wide appeal with the advent of illustrated magazines. In 1992, there was a resurgence of interest in stereoscopy with the advent of the random-dot autostereogram, described in Section 30.1.4. See Judge (1950), Darrah (1964), Gernsheim (1969), and Morgan and Symmes (1982) for more details on the history and methods of stereoscopic photography.

2.3.4 Stereoscopic movies

Joseph Plateau (1801-1883) was the Belgian scientist best known in visual science for the Talbot-Plateau law and the Plateau spiral. In his Doctoral dissertation at the University of Liége in 1829 he observed that a sensation of continuous motion can be created from a series of intermittently viewed objects. Michael Faraday (1831) made similar observations using superimposed toothed wheels rotating in opposite directions. On the basis of these stroboscopic effects, Plateau developed the 'phenakistiscope' in 1833. It consisted of a disc with slits round the rim and a series of pictures in a ring concentric with the slits. The rotating disc was held in front of a mirror and the observer looked with one eye through the passing slits at the pictures reflected in the mirror. A picture from the moving sequence appeared each time a slot passed before the eye. The intermittent sequence of pictures appeared as a single moving picture. At the same time, Simon Stampfer independently developed a similar device in Vienna. A device known as the 'zoetrope' was invented by William Horner (1786-1837) in Bristol in about 1834. A series of pictures was placed on the inside of a rotating cylinder and viewed through a sequence of slits in the opposite wall of the cylinder. In 1853, Baron Uchatius mounted a rotating picture disc and sectored shutter on a magic lantern to create moving images. See Deslandes (1966) and Coe (1981) for accounts of the history of cine photography.

In 1849, Plateau proposed that a binocular version of his phenakistiscope would produce moving images in stereoscopic depth, an idea he credited to Wheatstone. However, there is no record of this device having been made. In a letter to the journal *La Lumière* in 1852, Antoine Claudet described how he had constructed a stereoscope in which one sees moving images and wrote that Wheatstone was attempting to construct a similar instrument. Although he announced that a full description of these instruments would be published, the publication never materialized (see Gernsheim 1969 and Gosser 1977). It fell to the Parisian optician, Jules Duboscq, to patent the first stereo moving picture device in 1852 (Duboscq 1857). He called it the 'stéréofantascope'. It consisted of Plateau's phenakistiscope and two mirrors, which stereoscopically combined 12 pairs of photographs, with each pair placed along the radius of a revolving disc. The radial arrangement of the pictures introduced some distortion because the pictures for the two eyes moved at different velocities. The following year Claudet (1865) took out a British patent for a similar device involving a prism stereoscope rather than a mirror stereoscope (see Gosser 1977). Czermak made a similar instrument in 1855 (Helmholtz 1910, p. 357).

In 1859, Henri DuMont, a French civil engineer, patented a series of instruments for showing moving stereoscopic images. In the version shown in Figure 2.60a pictures are arranged on the outside of two drums and viewed through mirrors and slits in discs, which rotated with the picture drums. In early 1860, William Shaw, in Middlesex, England, combined a zoetrope with a mirror stereoscope and also with a prism stereoscope to produce a moving stereoscopic peep show (Shaw 1861). In the version shown in Figure 2.60b the pairs of pictures were mounted on the sides of a rotating octagonal drum and viewed through cone shaped apertures in a rotating cylinder so that vision was blanked out between pictures. One of his displays was of a moving train. He called this instrument the 'stereotrope' and showed it at the International Exhibition of 1862. In February of 1860, Pierre H. Desvignes from Lewisham, Kent patented a similar device and also used a train as one of his pictures. He proposed the use of intermittent illumination to overcome the problem of image blur but there is no record of his having built such a device.

Coleman Sellers (1827-1907), an American engineer, combined a vertical zoetrope with a prism stereoscope in 1861, to make a stereoscopic peep show called the 'kinematoscope' (Figure 2.60c). This was an advance on earlier devices because, instead of lining a cylinder, the cards containing the pictures radiated out from a central shaft so that their motion was along the line of sight rather than orthogonal to it. This reduced image blur. The cards came into view one at a time as the shaft rotated. At the same time, an outer cylinder with a series of slots rotated between the cards and the stereo viewer so that vision was blanked out between cards. Sellers built only one model, which he kept at home as a toy (see Gosser 1977).

In about 1870, Wheatstone also constructed a stereo zoetrope with pictures arranged round the inside of a rotating cylinder. He added a pawl device, which moved the cylinder intermittently so that it was stationary when each pair of pictures was seen. He was not the first to use this important principle. He did not use a shutter system to interrupt viewing between pictures (see Gosser 1977).

Interest in stereoscopic moving images lapsed in the period after 1870, when modern cinematography was being developed. Eadweard Muybridge started his carrier as a photographer of stereoscopic views of North America. During the 1870's he developed a chronophotographic system for recording animals in motion (Muybridge 1899) It involved an array of 40 cameras, which were triggered in sequence along the path of motion. The resulting sequence of images is what is required for cinephotography. He mounted the sequence of photographs round the rim of a wheel and projected them with a magic lantern to produce a brief moving image. But the moving display lasted only about one second. He attempted to make stereo versions of these pictures. Étienne Marey developed a chronophotography system using a single camera in which a sequence of pictures could be taken in quick succession on a rotating glass plate. He used two cameras rather than 40 to produce moving stereoscopic images (Marey 1895). But his use of a glass plate severely limited the duration of the projected picture. The development of flexible film was the key to further success.

Louis Le Prince, a Frenchman living in Leeds, England, seems to have been the first to use intermittently moved flexible film to produce moving projected pictures (see Coe 1981). He patented his device in Leeds in 1888. In 1890, Frederick Varley, a civil engineer in London, working with photographer William Friese-Greene, patented the first stereo cine camera using a roll of celluloid film (Varley 1890). The unperforated Eastman film was about 17 cm wide and 7 yards long with at most five exposures per second. This low exposure frequency severely limited the quality of the moving image. Furthermore, the machine was not suitable for commercial exploitation. In 1903 Lucien Bull built the high-speed stereo camera shown in Figure 2.57b. A rotating drum triggered a series of flashes.

The first commercially viable cine projector, called the 'kinetograph', was developed by the Scottish engineer William Dickson (1860-1935), working in Thomas Edison's laboratory in New Jersey. Dickson also built a peep-show version called the 'kinetoscope', which showed tiny, brief moving pictures (see Figure 2.57c). A kinetoscope peepshow parlour opened in New York in 1894. The kinetograph involved all the essential elements of the modern system, including a sprocket wheel for intermittent advancement of a roll of 35-mm film and an associated shutter. At this stage, Edison did not believe there was any future in large projected moving pictures.

The first working cine projector was developed by Louis and Auguste Lumiére. They obtained celluloid film from Eastman and coated it in their own factory in Lyons. They opened the world's first movie house in Paris in December 1895. The first film showed them feeding their own baby. Many other inventors were involved in the maturation of cinephotography into its present form.

Edison patented a stereoscopic version of the kinetoscope, but there is no evidence that the machine was built. In 1903, Auguste and Louis Lumière exhibited, in France, an anaglyph stereo movie sequence of only a few seconds' duration. In 1915, the Famous Players Film Company (later the Paramount Picture Corporation) released three short anaglyph stereo films produced by the cinema pioneer Edwin Porter (Hayes 1989). However, there seems to be no record of where these films were shown.

Harry K. Fairall produced the first commercially successful stereo movie *The Power of Love*, which opened at the Ambassador Hotel Theatre in Los Angeles in September 1922. In the same year, a stereo film *Plastigrams* made by Jacob Leventhal and Frederick Ives was shown in New York. Also in 1922, William Van Doren Kelley exhibited his short movie *Plasticon* at the Rivoli Theatre in New York (Kelley 1924). In 1935, Metro-Goldwyn-Mayer released a series of short stereo movies called *Audioscopiks*. These were the first stereo movies with sound (Norling 1939). All these movies used colour anaglyphs. This method does not allow the use of colour in the film. In 1935 Edwin Land demonstrated a stereo film in colour using the polaroid method of separating the images. This process was developed in the 1930's by the Zeiss-Ikon Company of Germany and by Raymond and Nigel Spottiswoode who produced the first stereo film in full colour and with stereo sound for the 1951 Festival of Britain in London.

In 1922, Laurens Hammond and William F. Cassidy demonstrated a short film at the Selwyn Theatre in New York. This film used the 'Teleview' system in which alternating left- and right-eye views of a movie were projected on a large screen. Each member of the audience viewed the screen through a rotating shutter synchronized with the alternation of the images on the screen. In a modern version of this system, used by the Imax Company of Toronto, members of the audience view alternating pictures on a wide-angle screen through electrically operated liquid-crystal shutters.

A stereo cinema system using a parallax grating was developed by Semyon Pavlovich in Russia in the 1940's. This used 30,000 silver wires, weighing six tons, suspended in front of the screen. It did not require the use of viewing glasses, but had several drawbacks, including darkening of the image, image diffraction, and dependence on viewing position. Also in the 1940's, Professor Noaillon of Belgium developed a radial converging grill with wide slits, which were rendered invisible by a rapid oscillation of the grill in its own plane. Stereoscopic pictures were projected through the grill onto a screen so that a member of the audience saw alternate strips of each picture presented to the two eyes. The Frenchman Francois Savoye patented a similar system in 1942. Jennings and Vanet (1952) developed a version of this system in which an inclined cylinder of black bars and slits, tapered from top to bottom, rotated rapidly about its central axis. Alternating left- and right-eye vertical strips of two stereoscopic images were projected through the slits onto a screen inside the cylinder. The tapered cylinder projected a series of radiating zones into the audience area, and a viewer within any zone saw the alternating images correctly. The rotation speed of the cylinder was set to avoid stroboscopic effects between the grill and the projector. A small system, called the cyclostereoscope, was made for home use by A. Mattey of Paris (Blum 1983).

Projection systems using lenticular screens have been used for stereo cinematic projection, but screens for large pictures are expensive and several projectors are required. The construction and projection of lenticular-sheet stereograms are described in Section 30.1.4.

There was a boom in stereo films based on the use of polaroid glasses in the 1950's and again in the early 1980's. Stereoscopic television and video films have not been successful.

This completes this historical review. Historical backgrounds to particular topics will appear in various parts of the book.

3 Psychophysics and analysis

3.1 PSYCHOPHYSICS

3.1.1 Psychophysical methods

3.1.1a *Basic terms*

Psychophysics is the quantitative study of how people or animals detect, resolve, discriminate, identify, categorize, or describe defined stimuli. This section briefly outlines psychophysical methods with emphasis on those used in the study of depth perception. For fuller accounts see Guilford (1954), Torgerson (1958), Swets (1964), Green and Swets (1966), Carterette and Friedman (1974), Gescheider (1976), and Falmagne (1985).

A **stimulus domain** is a set of objects or events described with respect to defined features and selected values of those features, plus rules of composition and transformation of those features and values. A stimulus domain is the defined set from which stimuli used in an experiment are drawn.

A visual scientist defines a stimulus domain to investigate some property of the visual system. Stimuli may be constructed in the laboratory. For example, Stiles used two 10° patches illuminated at various luminances and wavelengths to derive the standard observer for colour discrimination. Shapes viewed at various inclinations in depth are used to investigate shape constancy. Defined visual displays may be used to investigate the effects of binocular disparity on the perception of motion in depth. On the other hand, a stimulus domain might be a set of commonly occurring objects that an experimenter believes are familiar to a group of observers, such as different makes of automobile, or different breeds of dog. An experimenter with an incomplete or incorrect knowledge of a stimulus domain from which stimuli are drawn produces incorrect data and false theories. An observer who responds to stimulus features outside the stimulus domain of interest to an experimenter produces artifactual data.

A **task domain** is the type of judgment that an observer is requested to make when presented with defined stimuli. The basic types of task in a psychophysical experiment are detection, resolution, discrimination, categorization, identification, and description. These are defined later in this section. The **response** that indicates performance on the task may consist of a key press, an adjustment of a stimulus, or sorting stimulus objects. Verbal responses can range from a simple "yes" or "no" to a description of a sensory experience. In animals and preverbal infants, responses consist of such things as eye movements, pointing, or a conditioned response.

The **basic parameters of performance** are accuracy, precision, and speed. Take the task of setting a variable stimulus to appear equal to a standard stimulus, such as setting one object to appear at the same distance as another. The **point of objective equality** (POE) is the objective distance of the standard, and the **point of subjective equality** (PSE) is the signed mean of a series of settings of the variable. **Accuracy** is the signed difference between the POE and the PSE.

Accuracy is synonymous with **constant error** and bias. **Precision** of performance in a task is indicated by the mean of unsigned deviation scores with respect to the PSE. Related measures are the standard deviation, variance, and standard error of scores. Precision is synonymous with sensitivity, variability, and variable error. The term "accuracy" is often used where "precision" is meant. Accuracy is also often used inappropriately to signify the mean of unsigned deviation scores with respect to the POE. This measure confounds accuracy and precision, and should be avoided. Accuracy and precision, are independent, or orthogonal, measures.

In assessing human performance on detection or discrimination tasks, it is useful to have some idea of the theoretical limit that can be reached by a detector. A detector that performs at this theoretical limit is known as an **ideal observer** for that task. For instance, it is possible to calculate the performance of an ideal observer for the detection of light, given the quantal nature of light and the "noise" within which the signal is presented (Barlow *et al.* 1971). The ideal observer for a particular task provides a yardstick for assessing the performance of a human observer on that same task. An ideal observer for stereopsis is described in Section 19.3.3.

In a **class A psychophysical procedure** the subject detects a discontinuity such as a light in a dark field, a boundary in a bipartite field or a break in a vernier line (Brindley 1970). A class A procedure removes virtually all ambiguity from the task because the subject performs like a null-reading instrument. A well-administered class A psychophysical experiment is the most sensitive measure we have for revealing what is detected, resolved, or discriminated. Under the most favourable circumstances, sensory thresholds are near the theoretical limit of efficiency defined by the ideal observer.

When we say an object is detected or that two stimuli are resolved or discriminated, we say nothing about what subjects experience, other than that they experience a discontinuity. We do not talk about sensory 'qualia', such as red patches and the like. We first measure defined stimuli by a standard procedure of defined accuracy and precision, which we call the physical measure. For example, we measure the photometric luminance of stimuli. We then map the physical measurements onto the probability of detection of each of a set of stimuli. If we have no instrument finer than a given sensory system, we cannot measure that sensory system. For example, people who test wool quality by touch or tea or wine by taste outperform all instruments so that it is not possible to measure what they are doing. One can measure only their repeat consistency. We can also measure precision under one condition with respect to that under another condition. Without some independent measure we cannot compare accuracies.

In a **class B procedure** the subject matches or compares two things with respect to some defined features, where the things differ in some other feature. For instance, an observer who adjusts the lengths of the two halves of the Muller-Lyer illusion to appear equal is performing in a class B experiment because the two halves of the figure still differ with respect to the arrows on their ends. In class B experiments it may be difficult to be sure which aspect of the stimuli the subject is responding to. The literature can be confused, with a welter of conflicting theories and contradictory evidence. Examples are the literature on the moon illusion (Section 26.2.6) and shape constancy (Section 26.3). Some of this confusion is due to the tendency to regard a perceptual phenomenon as due to a single process at one level in the system. For example, the apparent motion of a stationary object seen against a moving background is known as induced motion. But the effect can arise from processes occurring at at least three levels in the nervous system (Section 23.4.7). The topic of levels of perceptual processing is dealt with in Harris and Jenkin (2002).

The assumption that one can tap a particular process has worked well in class A experiments. For instance, we have the beautiful coincidence between the psychophysical spectral sensitivity curve and the physically determined absorption characteristics of extracted visual pigment. As soon as subjects are required to isolate the stimulus features being judged from among other features, they bring to bear a repertoire of sensory, perceptual, and linguistic functions and skills. We no longer have a null instrument but a knowledgeable strategist.

3.1.1b *Basic methods*

All the standard psychophysical methods have been used in the study of depth perception. In the **method of adjustment** the subject adjusts a variable stimulus until it is detected or until it matches a standard with regard to some specified feature. In the **method of limits** the experimenter increases or decreases the variable stimulus until the subject detects it or indicates that it matches a standard. The mean signed error of settings with respect to the standard is the constant error. The variability of settings about the signed mean measures the precision of the judgments, and reflects the sensitivity of the sensory system.

These methods are especially useful for measuring steady-state constant errors. For instance, to measure how an inclined surface affects the apparent inclination of a superimposed test line, subjects are asked to set the line to the apparent vertical. The mean signed error indicates the extent of the induced inclination. Errors of anticipation and habituation are avoided by averaging settings from trials in which the line starts at various angles on either side of the vertical.

The **method of constant stimuli** is used when the effect being measured is transient, or when it is important to avoid the presentation of stimuli in an ordered sequence. Suppose one is measuring the accuracy and precision of setting a variable stimulus to match a comparison stimulus. About seven values of the variable are selected that lie within the region of uncertainty on either side of the PSE. On each trial, one of these stimuli is presented briefly and the subject indicates whether, by a defined criterion, it is greater or less than a comparison stimulus. The comparison stimulus may be presented with the test stimulus or stored in the subject's memory. Subjects are asked to make a decision, even when the two stimuli look alike. The order of presentations is randomized and each stimulus is presented many times. The percentage of trials in which the variable stimulus is judged to be greater than (or less than) the standard stimulus is plotted on the *Y* axis against stimulus magnitude on the *X* axis to yield a **psychometric function**. Small differences between stimuli are difficult to detect because of noise in the visual system. As the difference increases, the probability of detection increases until it saturates at a probability of one. Therefore, the psychometric function is usually an S-shaped curve in the form of a cumulative normal distribution, or normal ogive, as shown in Figure 8.2.

The psychometric function defines the relationship between variations in a stimulus attribute and the probability of a particular psychophysical response. The value of the stimulus that yields 50% "greater-than" judgments is the PSE. The difference between the PSE and the POE is the constant error. The value of the stimulus that yields 75% of "greater than" judgments is usually taken as a measure of the upper discrimination threshold. The 25% point defines the lower discrimination threshold. The difference between the 50% and 75% points is the just noticeable difference, or **JND**.

The method of constant stimuli may also be used to measure the detection threshold for a given stimulus attribute. In this case, a single stimulus is presented with various values close to the detection threshold, and subjects report whether or not they have detected it. The psychometric function is a plot of the percentage of times the stimulus was detected against the value of the stimulus on the *X* axis. The detection threshold is usually defined as the value of the stimulus that yields 50% detection, although the actual percentage chosen is arbitrary.

Probit analysis (Finney 1971) is used to improve the fit of a cumulative normal curve to a set of data, and thus to derive more precise estimates of the parameters of the function. Both the ordinate (response probability) and abscissa (stimulus strength) are first transformed into standard scores (standard-deviation units). This has the effect of linearizing the psychometric function. Weights are then assigned to each datum point in inverse proportion to its standard error, which means that greater weight is given to points in the upper part of the psychometric function. In addition, datum points near the centre of the psychometric curve are weighted more heavily because they contribute more to the determination of the position of the curve than do points at either end of the curve. The best fitting line is then determined by weighted linear regression.

3.1.1c *Staircase methods*

Judgments about stimulus values that lie some distance from the PSE are less informative than those about stimuli near the PSE. It is therefore best to concentrate stimuli near the PSE. Stimuli should also be symmetrically arranged around the PSE so as to avoid biasing the results. A derivative of the method of limits, known as the **staircase method**, provides an efficient way of achieving these two goals. Stimuli are presented briefly in ascending or descending order until the response of the subject changes. The order of stimulus progression is then reversed until the subject's response changes again. After several repetitions of this process the stimulus values converge on the PSE. The threshold is the mean of a criterion number of trials after the judgments have reached a constant level of fluctuation. The difficulty of deciding when a constant level has been reached can be avoided by defining the threshold as the stimulus value above which 50% of the judgments are "yes". If the step size is too great, the subject merely alternates between "yes" and "no" judgments, and if it is too small, time is wasted. It is a useful strategy to start with a large step size and reduce it during the course of the experiment. A simple rule is to halve the step size when the response of the subject reverses, and double it again if a criterion number of similar responses are made in succession.

With a simple staircase, the subject can anticipate when to change the response. For instance, if several "yes" judgments have been made in succession, the subject may guess that it is time to say "no", even though no change in the stimulus has been detected. The **double-staircase method** reduces the effects of sequential dependencies. In this method, two staircase sequences are run at the same time with stimuli from the two sequences presented in random order (Cornsweet 1962). Several variations of the staircase method have been devised (Taylor and Creelman 1967; Levitt 1971; Watson and Pelli 1983).

3.1.1d *Signal detection procedures*

When a stimulus is presented on every trial, subjects may seemingly improve their performance by adopting a more lax criterion, reflected in their willingness to report a stimulus when unsure of its presence. On the other hand, subjects may seemingly perform less well by adopting a stricter criterion. Classical psychophysical methods do not distinguish between a change in the detectability of a stimulus and a change in the willingness of the subject to report its presence. The method of **signal detection** measures the separate contributions of these two factors (Green and Swets 1966).

A signal is a well-defined physical event which a sensory system, or receiver, is attempting to detect. The basic idea is that all neural discharges created by a signal are accompanied by noise arising from other stimuli or from random events in the sensory system. It is assumed that the noise level fluctuates at random around a mean value with a given variance. The sensory response generated by a given signal is also assumed to vary at random around a mean value, with the same variance as for the noise alone. It is also assume that responses on different trials are independent, and that performance is stable over a set of trials.

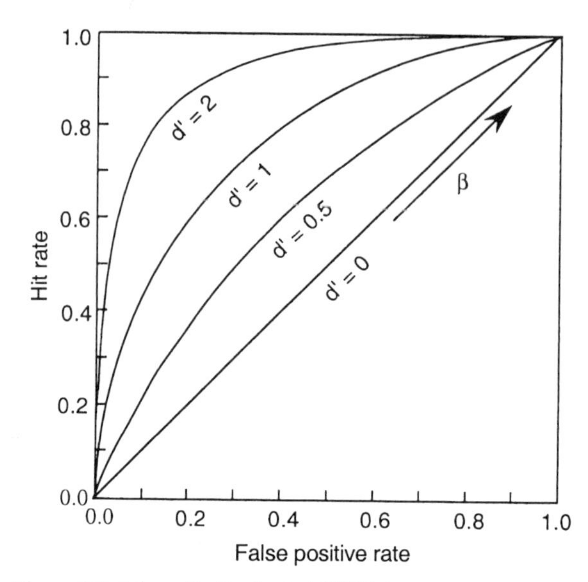

Figure 3.1. A hypothetical set of ROC curves.
Symbol d' signifies the detectability of the stimulus. Each curve is

The **detectability** of a signal with respect to the noise is defined as the difference between the mean of the distribution of the response to noise alone and the mean of the distribution of the response to the noise plus signal, divided by the variance of the distributions. Detectability is denoted by the symbol d' ("*d* prime").

Within the threshold region, a subject is necessarily uncertain about whether a weak sensation is due to a signal or to noise. The level of sensory activity above which the subject reports a signal is the **criterion level**, denoted by the symbol β. The subject's task can be described as that of estimating the likelihood that the sensory activity on a given trial arises from noise plus signal, relative to the likelihood that it arises from noise alone. The ratio of these two likelihoods is the **likelihood ratio** and forms the most efficient basis for a detection task.

The way a person responds in an uncertain situation depends on the perceived rewards and penalties (payoff) associated with each of the four types of response. These are: (1) correctly detecting a stimulus (hit), (2) saying it is present when it is not (false positive), (3) not detecting a true signal (miss), and not reporting it when it is not present (correct rejection). The payoff associated with each type of response is known as a **payoff matrix**.

The method of signal detection separates out the effects of the criterion, β, from changes in the detectability of the stimulus, d'. A series of signals of varying strength in the region of the threshold are presented at random, along with catch trials in which there is no signal but only noise in the sensory system. A record is kept of the rate at which targets are detected in a given number of trials (hit rate), and the rate at which targets are reported when none is present (false-positive rate). These data are plotted on a graph with hit rates along the *Y* axis and false-positive rates along the *X* axis. A curve, known as the **receiver operating characteristic**, or ROC curve, is fitted to the data (Figure 3.1).

When the hit rate increases in simple proportion to the false positive rate the ROC curve is a diagonal line. This signifies that the subject was merely guessing, since an improved hit rate was achieved only by an equal increase in false positives. The detectability of the stimulus is indicated by the area between the ROC curve and the diagonal, which increases with increasing separation between the curve and the diagonal. The area under the curve corresponds to the probability of detection in a two-alternative forced-choice task. It is a nonparametric measure that does not rely on assumptions about the distribution of responses. The criterion is given by the point along the ROC curve at which the subject is operating. Macmillan and Creelman (1991) have written a users guide for signal detection theory.

Signal detection methods can be used to plot the probability of response of single neurones in the visual cortex to well-defined stimuli of variable strength to yield a **neurometric function** (Parker and Newsome 1998) (See Section 6.2.1d).

3.1.1e *Forced choice method*

A simple procedure for ensuring that measurements of thresholds are not affected by changes in criterion was proposed by Blackwell (1952). Subjects are presented with two stimulus windows, either at the same time or sequentially. Only one of the windows contains a stimulus; the other is blank. Stimulus strength and the positions of the stimulus and blank are varied at random over a series of trials. On each trial, subjects are forced to say which window contains the stimulus; hence the name **two-alternative forced-choice (2AFC) procedure**. A two-alternative decision is independent of changes in criterion, since subjects are forced to choose on each trial. The percentage of correct responses is plotted on the ordinate against the value of the stimulus, to generate a psychometric function. Since the chance level of performance is 50%, ordinate values run between 50 and 100%. The stimulus value that a subject correctly identifies 75% of the time is taken as the threshold. The 75% point is the mean of the psychometric function based on the forced-choice procedure. The slope of the function indicates the rate at which performance improves as stimulus strength is increased. It is the reciprocal of the standard deviation of the distribution of responses. The standard error of judgments for each stimulus value is calculated by the equation for the standard error of a proportion. The standard error is largest when subjects are most uncertain in their judgments, which is when the test and comparison stimuli are most similar. As the percentage of correct responses increases, the standard error tends to decrease. McKee *et al.* (1985) have described this procedure.

3.1.1f *Scaling*

In another class of psychophysical methods, known as sensory scaling, subjects make categorical judgments about the magnitudes or sensory qualities of stimuli. Readers are referred to Torgerson (1958), Garner (1962), and Falmagne (1986).

3.1.1g *Phenomenological analysis*

Many perceptual phenomena can be investigated by simply asking subjects to describe what they see. Before the nineteenth century, most investigations of the functioning of the visual system relied on this method. The Gestalt psychologists used this method extensively in the early part of the 20th century. Most visual phenomena were discovered by chance observation or by an inspired guess that a given phenomenon may occur if stimuli are arranged in a particular configuration. For instance, Celeste McCollough, extrapolating from some work on chromatic aberration, anticipated that something of interest would be seen if alternating gratings were paired with alternating colours. That led to the discovery of one of the first contingent aftereffects (Section 8.3.5). Wheatstone's discovery of the stereoscope and his subsequent observations with a variety of stereograms revealed the basic characteristics of the human stereoscopic system.

Once a visual phenomenon has been discovered, the stimulus conditions that give rise to it can be established by comparing the frequency of its appearance under different experimental conditions. Powerful inferences can be made about mechanisms underlying visual processes from qualitative observations made under cleverly devised circumstances.

There are no general procedures for this type of investigation, but great care must be taken not to overgeneralize conclusions. For instance, two of the discs in Figure 3.2 look like mounds, and two look like hollows. When the figure is inverted, mounds become hollows and hollows become mounds. In most textbooks, it is concluded that the convexity or concavity of a shaded region is interpreted in a way consistent with the light source being above with respect to gravity. However, when Figure 3.2 is viewed with the head upside-down, impressions of

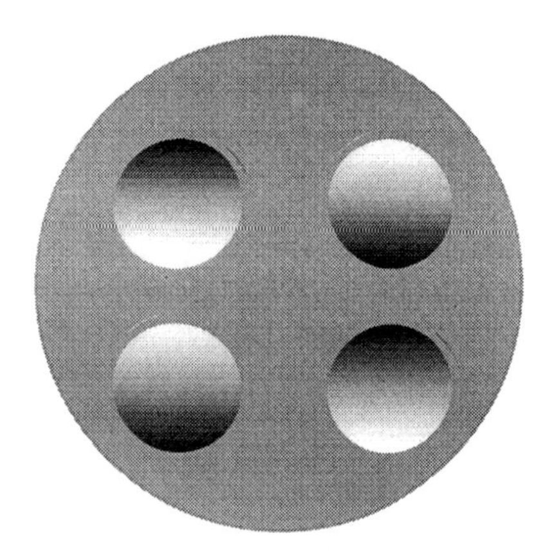

Figure 3.2. Frames of reference in shape from shading.
When the figure is inverted, the mound discs change in to hollows, and visa versa. Viewing the figure with inverted head reveals that the crucial factor is orientation of the discs to the head rather than to gravity. When the figure is viewed below the chin almost parallel with body, with head erect, the crucial factor turns out to be the retinal orientation of the discs.

convexity and concavity are determined by the orientation of the dark and light areas relative to the head, rather than to gravity. But even this is not the correct account. The figure can be viewed with head upright but with the figure at a steep angle beneath the chin so that, the part of the picture that is "top" with respect to gravity and to the head, is upside-down on the retina. Now the convexities and concavities are interpreted in terms of retinal coordinates (Howard *et al.* 1990).

It is difficult to change one feature of a stimulus without changing others. One may think that one has uncovered the function relating a perceptual effect to variations in a particular stimulus dimension. However, changes in the selected dimension may be incidental to changes in another crucial dimension that one has not considered. For example, changing the density of texture elements may affect the perceived size of a surface patch. But the crucial factor may be the change in the total number of elements in the patch rather than density. Only vigilance and imagination in designing control conditions can prevent one from falling into this type of trap; there are no foolproof rules.

Perhaps half the perception literature consists of investigators pointing out how other investigators have failed to take account of crucial stimulus dimensions and of criticized investigators arguing that these stimulus dimensions are not crucial. Even when a physical stimulus responsible for a given sensory effect has been properly identified, there is still the question of how the stimulus is processed in the nervous system. Many types of intervening processes can mediate identical responses to identical stimuli. There can be many causes of a given perceptual phenomenon. Just because an effect has a name like contrast, masking, or induced motion, does not mean it is due to one process.

3.1.2 Detection

A stimulus is detected when an observer can decide at above chance level whether it is present or absent. A persisting patch of light is detected when the mean rate at which quanta fall in the patch is discriminably different from the mean rate at which quanta fall within the surrounding region (Section 19.3.3). A brief patch of light is detected when the total number of quanta falling on each unit area of the test patch in a defined unit of time is sufficiently different from the number falling on the surround. In a typical task, a line of a given length and luminance, on a background of a different luminance, is varied in thickness until the line is detected. Because of diffraction, the image of a line is spread across several receptors, so that detection reduces to the task of detecting a luminance gradient.

For an illuminated line on a dark background, no line is so thin that it cannot be detected. This is because the luminance of the line can be increased to compensate for any reduction in width to generate a discriminable luminance gradient. One can talk about the minimum resolvable thickness of an illuminated line only if the contrast between line and surround is specified. It has been estimated from frequency-of-seeing curves that a short line of light seen against a dark background is detected at above chance level if two quanta of light are absorbed within a critical area and within a critical time period of about 10 ms. Stimulus energy is completely summed within this critical area (Ricco's law) and critical time (Bloch's law). The critical area and time vary with light wavelength and retinal location (Bouman and van den Brink 1952; Schwarz 1993).

A black line seen against a bright background must be at least 0.5 arcmin wide to be detected, however bright the background (Hecht and Mintz 1939). The visual target can also be the boundary between two unequally illuminated regions. In all cases, performance depends on the ability to detect a luminance gradient—the subject is not required to respond to any other attribute of the stimulus.

3.1.3 Resolution

3.1.3a *Width resolution*

As two initially superimposed fine lines are separated, the first impression is of a line increasing in width. This is because the distribution of activity over the set of detectors becomes broader as the lines separate, but the two peaks of excitation are not distinct enough to be discriminated. Thus, two spatially separated lines can be distinguished from two perfectly superimposed lines before they are far enough apart to be seen as two distinct lines. This type of resolution is **width resolution**. It exceeds the limits set by the Nyquist or Rayleigh criteria. In colour, width resolution shows itself as a loss of saturation as a monochromatic light is replaced by two monochromatic lights that produce the same hue as the original monochromatic light. Width resolution in stereopsis is discussed in Section 19.10.2. Geisler (1984) described an ideal observer for acuity and hyperacuity and Snippe and Koenderink (1992) developed an ideal observer for width discrimination and hyperacuity in metameric sensory systems.

3.1.3b *Resolution of stimulus separation*

Two stimuli are resolved when they are detectable as two stimuli. For temporal resolution, the excitation incurred by the first stimulus must subside sufficiently before the second stimulus is presented. The limiting factors are the speed of stimulus onset and the time constant of the sensory system. For spatial resolution, the stimulus must excite two distinct detectors at a discriminably higher level than it excites a third detector in an intermediate location. Thus, a set of detectors in a square lattice can resolve a periodic stimulus, such as a black-white grating, only if the spatial period of the grating (distance between two black bars) is at least twice the spacing of the detectors, as illustrated in Figure 5.5. This is known as the **Nyquist limit**. A related statement is that, for a diffraction-limited system, two point sources can just be resolved when the peaks of their images are separated by the radius of the inner bright regions of their diffraction patterns (Airy's disc). This is known as the **Rayleigh criterion**. For green light (540 nm) and a numerical aperture of 1.4, this criterion imposes a resolution limit of 240 nm.

The colour system has only three channels, with very wide and overlapping tuning functions. Since neither the Nyquist limit nor the Rayleigh criterion is satisfied in this system, our capacity to resolve wavelengths is zero. No matter what wavelengths are present in a patch of light, we see only one colour. The colour we see depends on the relative extent to which the different colour receptors are excited. If two lights with different wavelength components excite the three receptors in the same ratios, those lights appear identical, even though the wavelength components are discriminably different when presented one at a time. The lights are said to be metameric matches.

The visual local-sign system is metameric only locally. It has about one million channels (ganglion cells) which, at the theoretical limit, allow it to resolve a black and white grating with bars as narrow as the diameter of receptors. In other words, resolution is limited by the ability of the neighbouring receptive fields to detect differences in luminance contrast. Two or more visual stimuli falling wholly within an area of the retina where the excitatory regions of neighbouring receptive fields mutually overlap, appear as one stimulus in a position that depends on the mean or centroid of the total luminance distribution. This occurs when two short parallel lines are presented together within an area of about 2 arcmin, which is about the size of the smallest receptive fields in the retina (Badcock and Westheimer 1985; Watt *et al.* 1983). This metameric merging of stimuli occurs over larger distances in the peripheral retina, where receptive fields are larger. Metameric merging accounts for our limited grating acuity in which adjacent lines of a grating merge into a grey patch. When lines are presented to distinct locations or successively, their separate positions can be discriminated to much finer limits, just as wavelengths of light can be discriminated when coloured patches are presented in different spatial locations. Spatial discriminations beyond the Nyquist limit are referred to as hyperacuity, as we will see.

3.1.3c *Resolution of secondary features*

Resolution is more difficult to investigate in secondary spatial or spatiotemporal features—features derived from the initial coding of local sign and time. This is because all such features can be also resolved by the million-channel, local-sign system. For example, even if orientation detectors could not resolve the angle between two long intersecting lines, the lines would still be perceptibly distinct because they fall on distinct regions of the retina.

The presence of dedicated detectors for given secondary feature is revealed by electrophysiological procedures and also by the presence of metamerism (Section 4.2.7). Given that orientation is coded in the visual cortex by dedicated detectors with overlapping tuning functions, two neighbouring or intersecting short lines at slightly different orientations should metamerize their orientations—they should appear as one line at an intermediate orientation. Parkes *et al.* (2001) showed that neighbouring patches of grating differing in orientation all appear orientated at the mean orientation of the set. Perhaps metameric processes within excitatory regions of neighbouring orientation detectors can explain why closely spaced parallel lines appear wavy, and why

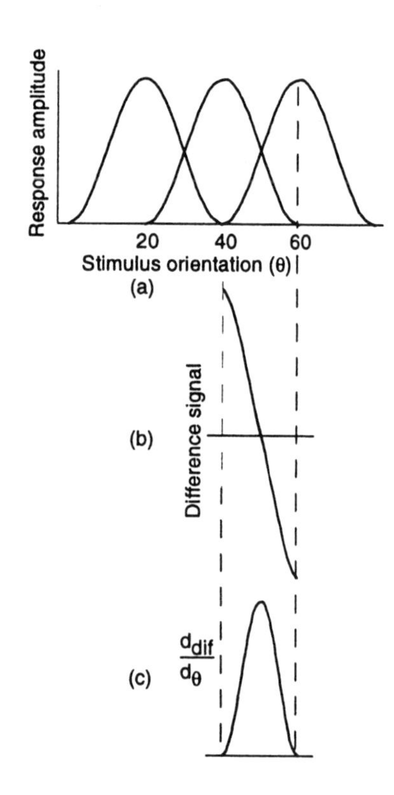

Figure 3.3. Detection and discrimination of orientation.
(a) Hypothetical tuning functions of three orientation detectors.
(b) Signed difference in firing rate of neighbouring orientation detectors as a stimulus is rotated from 40° to 60°.
(c) The orientation discrimination function derived by differentiating the difference signal. The difference signal is strongest at the peaks of the tuning functions, but the discrimination signal is strongest at the point of intersection of the tuning functions.

the orientation of a short, briefly presented line is difficult to detect (Andrews 1967).

For similar reasons, metamerism should be evident in visual motion, and there is evidence that it is. A display of short-lifetime dots moving in different directions in the same general direction appears as a set of dots moving in a common mean direction. The discriminability of a change in the mean direction of motion for a mixed display of dots was the same as for an array of dots all moving in the same direction (Watamaniuk *et al.* 1989). When the two directions of motion are widely separated and the dot trajectories do not intersect, averaging gives way to the perception of two distinct displays (23.2.2). Furthermore, an array of short-lifetime dots moving in the same direction at different speeds resembles an array of dots moving at the mean speed of the set (Watamaniuk and Duchon 1992). These results can be explained in terms of metameric processes within the motion-detection system that arise because the tuning functions of motion channels overlap. Averaging of speed does not occur when the dots have a long lifetime because the differences between the speeds of the component dots are then discriminated on the basis of relative changes in position of identifiable dots. Metamerism in the disparity system is discussed in Section 19.8.

3.1.4 Discrimination

3.1.4a *Basic features*

Two stimuli are discriminated if one is detectably different from the other, given that they have been resolved as two stimuli, either in space or in time. A metameric system with poor resolution can have exquisite discrimination. For instance, even though the wavelength resolving power of the human eye is zero, we can discriminate between many hundreds of spectral colours (or their metamers), as long as they are presented sequentially (resolved in time) or to different regions of the retina (spatially resolved).

Resolution requires the detection of a difference between the means of two overlapping and simultaneously present distributions of activity along the sensory continuum. Since the two distributions of activity are present at the same time, the detection of either mean is confounded by the presence of the other stimulus, so that performance is subject to the Nyquist limit.

Discrimination depends on the detection of a difference in the mean response of one set of detectors and the mean response of either the same detectors at a different time or of a set of detectors in a different location on the sensory surface. There is no theoretical limit to the precision with which the mean of a single distribution of excitation across a set of detectors can be registered when no confounding stimuli are present. Precision depends on the square root of the number of photons and their spatial distribution. In neural terms, the precision with which the location of a single stimulus can be registered depends on the rate of change of response across the set of detectors. The precision with which the location of a stimulus can be detected by two detectors with overlapping tuning functions depends on the steepness of the tuning functions at the point on the stimulus continuum where the tuning functions overlap. That is, it depends on the relative rate of change of the signal in each of the two detectors as the stimulus is moved over the stimulus continuum. This point is illustrated in Figure 3.3 for orientation discrimination. Resolution depends mainly on the signal-to-noise ratio, and on the tuning width and density of sensory channels along the sensory continuum. The noisiness of the individual channels seems to be less important for discrimination than for resolution (Bowne 1990).

The fine discrimination in sensory systems, compared with resolution, explains hyperacuity. Examples are detection of a change in separation between two neighbouring but distinct points, and detection of an offset between two lines (vernier acuity). Both these acuities are several times finer than the mean spacing of receptive fields. Another example is the task of setting a point midway between two other points, which has yielded thresholds of approximately 1 arcsec (Klein and Levi 1985). The key idea is that if two stimuli are separated by more than about 5 arcmin they do not metamerize or interfere with each other. In a resolution task, the stimuli are necessarily crowded together. In a hyperacuity task they are spatially separated (Geisler 1984).

The distinction between resolution and discrimination (hyperacuity) in a locally metameric spatial modality can be vividly illustrated on the skin. When the back is prodded simultaneously by two pointed objects about 1 cm apart, the stimuli metamerize into apparently one object at an intermediate position. The apparent position of the fused stimuli depends on their relative strengths. Békésy (1967) used the term "funnelling" for metameric averaging. If the objects are presented sequentially with the same separation, their distinct positions can be discriminated (Loomis and Collins 1978).

3.1.4b *Discrimination with respect to a norm*

For oppositional stimulus continua we can ask what is the least departure from the norm that can be detected. For example, we can measure the threshold for detection of the motion of a point or the colour of a patch, the curvature or tilt of a line, or the departure of a surface from the frontal plane. In these cases, only one stimulus is present. The norm serves as the comparison. A norm, such as no motion, gray in the red-green opponent system, zero disparity in a crossed-uncrossed disparity scale is an inherent value in an oppositional scale. Such norms are subject to temporary modification but are not arbitrary learned values. An experimenter may define an arbitrary norm that the subject is required to learn such as a particular velocity or colour, or the mean of a set of repeatedly exposed stimulus values. Subjects are then asked whether each of several stimuli is greater than or less than the memorized norm.

3.1.4c *Discrimination functions*

Discrimination sensitivity in a metameric system should show local maxima where the tuning functions of detectors intersect, because this is where the difference signal changes most rapidly. The well-known hue-discrimination function is an example. The local-sign system is locally metameric and it, too, should show undulating discrimination functions. There has been some debate about whether sensitivity to changes in the separation of lines shows peaks and troughs, as the distance between the lines is varied (Hirsch and Hylton 1982; Westheimer 1984a). Wilson (1986) interpreted some data from Klein and Levi (1985) as showing peaks and troughs like those in the hue-discrimination function. Regan and Price (1986) found undulations in sensitivity to changes in line orientation as the line was varied in orientation. The highest sensitivity to changing orientation occurred at the vertical and the horizontal, which suggests that the tuning functions of orientation detectors intersect at these values. This means that the peaks of orientation tuning functions are on either side of the main meridians. This raises the possibility that in any sensory system, the tuning functions of neighbouring detectors intersect where sensitivity to change is required to be highest. Another illustration of this point is that sensitivity to changes in binocular disparity (stereoscopic depth) is greatest about zero disparity, which is where the tuning functions of disparity detectors seem to overlap. This issue is discussed in Section 19.3.2.

All this suggests that a stimulus is detected most efficiently when it excites a detector at the peak of its tuning function. However, a change in a stimulus along a feature continuum is discriminated best when the stimulus falls on the steep flanks of the tuning functions of neighbouring detectors. There is physiological evidence supporting this conclusion. In the visual cortex of the cat, cells tuned to stimulus orientation responded most reliably to stimuli oriented at the peak of the tuning function, but sensitivity to changes in orientation was greatest on the flanks of the tuning function (Scobey and Gabor 1989). Motion-sensitive cells in V2 and the middle temporal area (MT) of the alert monkey had directional tuning functions with a half-width of 50° at half-height. Although each cell fired most vigorously to motion at the peak of its tuning function, it was most sensitive to changes in motion direction (as little as 1.1°) when the stimulus fell on the flank of its tuning function (Snowden *et al.* 1992).

Tanner (1956) proposed that the detectabilities of single stimuli in each of two detectors (d'_1 and d'_2) sum like vectors to predict the discriminability between stimuli presented to the two detectors ($d'_1 d'_2$). The cosine of angle θ between the vectors represents the coefficient of correlation between the noise in the two detectors. Thus

$$d'^2_{1,2} = d'^2_1 + d'^2_2 + 2d'_1 d'_2 \cos\theta$$

When cos θ is zero the detectors and the noise sources are independent and discriminability is the square root of the sum of the two detectabilities. When cos θ equals +1 there is only one sensory channel and the stimuli are discriminated only to the extent that they differ in detectability. When cos θ equals -1 the noise in two channels is negatively correlated and discriminability equals the sum of the two detectabilites. This formulation has been used to assess the extent to which the two eyes operate as distinct detectors (Section 8.1.3).

The tasks of recognition, identification, and description are discussed in Section 4.7.1.

3.1.5 Temporal thresholds

Temporal aspects of sensory processing have been studied with a great variety of procedures. Only a brief outline of these methods is provided here.

With a suprathreshold stimulus, one can measure the time required to detect it. In a typical experiment, subjects press a key as quickly as possible after a stimulus is presented. The average **reaction time** is the reaction time. The reaction time includes the time taken for the stimulus to be processed and the time taken for the response to be prepared and executed. If the same response is used for different stimuli, differences between reaction times provide a measure of differences in sensory processing time. These procedures have been used to study effects of learning on the processing time for stereopsis (see Section 19.12).

In some cases, the results of temporal processing of sensory inputs are reflected directly in a nontemporal percept. For instance, a difference in arrival of sounds at the two ears of a few milliseconds causes an apparent shift in the position of the sound source that subjects can identify by simply pointing in the appropriate direction, taking as long as they wish. Similarly, in the Pulfrich stereophenomenon, a target moving in a frontal plane appears to move in depth when image processing in one eye is delayed by introducing a dark filter in front of that eye. A very precise mapping of interocular time differences into disparities can be obtained by simply asking subjects to indicate the depth of the path of the moving target (Section 28.1).

In another temporal procedure, the duration of time for which the stimulus is presented increases gradually on succeeding trials until the subject reports either the presence of the stimulus or some defined change in the stimulus. The resulting measure is known as the **temporal threshold**. Subjects are not required to respond rapidly but are merely required to say on each trial whether or not the stimulus occurred, or in which of two windows it occurred. As the luminance intensity of a stimulus is increased, the temporal threshold becomes vanishingly small. As the stimulus is weakened, the temporal threshold increases up to a limiting value that depends on the capacity of the sensory system to integrate stimulus energy over time. This threshold is the **temporal integration time**. The capacity of a sensory system to integrate stimuli over time can also be investigated by presenting brief stimuli in succession with variable interstimulus intervals. This topic is discussed in more detail in the next section.

3.2 APPLICATIONS OF PSYCHOPHYSICS

In addition to providing information about the sensitivity of sensory systems, psychophysical procedures allow one to make inferences about how, where, and in what order, information is processed in the visual system. More specifically, they provide information about the following.

The existence of a feature detector For example, the perception of depth in a stereogram lacking all other cues establishes that there are processes devoted to the detection of binocular disparity.

The number and bandwidth of subchannels processing a feature For example, psychophysical evidence suggests that there are about seven subchannels devoted to the detection of spatial frequency (Section 5.5.3).

The location of visual processes Psychophysical procedures can prove that certain visual processes occur in the retina. For example, it has been established psychophysically that trichromacy and one form of luminance contrast occur within the retina. A visual effect that is apparent only when both eyes are open must occur at a post-retinal level (Section 17.1). However, it is difficult, if not impossible, to establish the precise site of post-retinal visual processes by psychophysical procedures alone.

The order of visual processes Psychophysical procedures that allow one to infer the order in which processes occur in the nervous system are known as **psychoanatomical procedures** (Julesz 1971). It is argued that, because one percept depends on another, it must be processed at a later stage. For example, from the fact that the perceived direction of motion of a grating in an aperture depends on the way line ends are interpreted, it has been concluded that line processing precedes the processing of motion (Section 23.2.1). But the logic may be flawed. A motion signal could be extracted first and then reinterpreted after line ends are assessed. In other words, a given visual feature can be processed in several stages; some of which precede and others follow the processing of another feature. Also, the outcome of a perceptual process could feed back to an earlier stage or to a parallel stage and affect what is occurring there. Without supporting anatomical or physiological evidence, one may not be able to draw firm conclusions about the order in which sensory qualities are processed merely from the way one perceptual phenomenon depends on another.

The logic is firm in other cases. For example, the detection of the shape of a figure contained in a random-dot stereogram must occur after binocular disparities are processed. The shape is not present in the input from either eye.

If a perceptual process occurs more rapidly than it would take for neural feedback to occur, one may conclude that it involves only simple feedforward processes.

If an effect in sensory dimension *A* occurs only for certain values of stimulus dimension *B*, then it is reasonable to conclude that the stimulus is processed for *B* before it is processed for *A*. For instance, certain visual effects, such as perspective illusions, occur with stimuli defined by luminance but not with chromatic stimuli. They must therefore occur after chromatic and achromatic channels are partitioned.

The linearity of sensory processing Psychophysical procedures can allow one to characterize nonlinearities, such as signal rectification, ceiling effects (compressive nonlinearity), and spatiotemporal interactions between stimuli. They can also reveal whether distinct visual features, such as distinct cues to depth, interact in a linear or nonlinear fashion (Chapter 27).

Convergence and equivalence of processes If two sensory inputs produce the same perceptual effect we can infer that they feed into a common neural process. For example, binocular disparity and monocular motion parallax produce similar impressions of depth, and other psychophysical evidence supports the idea that they feed into a common mechanism (Section 27.2.1).

The influence of attention and learning

Psychophysical techniques that allow one to infer the characteristics of visual processing are described in the following sections. Their uses in the study of stereopsis are described in other parts of the book.

3.2.1 Threshold summation and masking

Two sensory processes are said to be independent when the probability that one of them detects a given stimulus is unaffected by the simultaneous stimulation of the other, and when the internal noise associated with one process is uncorrelated with the noise associated with the other. When the same weak stimulus is presented to each of two independent sensory detectors, the probability that the stimulus will be detected is greater than the probability of detection when the stimulus is presented to only one detector. This is because, when more detectors are stimulated, the signal in one or other of them is more likely to exceed the fluctuating noise level than when only one is stimulated. This is known as **probability summation**. Pirenne (1943) introduced this concept into visual science. For two detectors of equal sensitivity, the probability of detecting a stimulus using both (P_b), relative to that of detecting it with either one alone (P_1 and P_2), is given by

$$P_b = (P_1 + P_2) - P_1 P_2$$

Quick (1974) provided the following simplified formula for calculating the effect of probability summation on stimulus detection when there are several detectors with different sensitivities,

$$s = \left[\sum_{i=1}^{N} \left| s_i \right|^p \right]^{\frac{1}{p}}$$

where N is the number of independent detectors responding to a set of stimuli, S_i is the sensitivity of the *ith* detector, and p is the slope of the psychometric function at the point where a stimulus is detected 50% of the time.

If the probability of detecting a stimulus by two detectors is greater than that of detecting the stimulus by only one detector by an amount that exceeds probability summation, then the two detectors are not independent. This means that signals in the two detectors converge and sum before a decision is made about the presence of a stimulus. If the noise in the two detectors is not correlated, the noise signals partially cancel out, so that noise after convergence is less than the sum of the noise in the two detectors. This improves the signal-to-noise ratio and increases the probability of detection above the level of probability summation.

Consider a defined continuously variable stimulus feature such as orientation, velocity, or disparity, detected by a multichannel sensory system. A cell within a subchannel for a given feature responds to all suprathreshold stimuli that fall within the bandwidth of that subchannel and within the receptive field of that cell. Within the bandwidth of a subchannel, signals from subthreshold stimuli converge on a common neural pathway and may summate to produce a suprathreshold stimulus—an effect known as **subthreshold summation**. By varying the relative values of two simultaneously presented stimuli over the defined sensory continuum, one can determine the stimulus range of subthreshold summation. This reveals the bandwidth of subchannels within a sensory system devoted to a given sensory feature. For example, the tuning bandwidth of orientation detectors has been inferred from the range of orientations over which subthreshold summation of two differently orientated stimuli occurs (Thomas and Gille 1979). Similarly, the spatial-frequency bandwidth of detectors has been inferred from the range of spatial frequencies over which subthreshold summation of grating acuity occurs (Wilson and Gelb 1984). The application of subthreshold summation to sensory processing of 2-D spatial contrast stimuli has been reviewed by Graham (1989). Applications of subthreshold summation to binocular vision are discussed in Section 8.1.

A briefly presented suprathreshold stimulus tends to elevate the threshold for a briefly presented test stimulus presented in the same or a neighbouring location, at the same time or in close temporal contiguity. This is known as **masking** and its applications to binocular vision are discussed in Section 8.2. In Section 7.3 we will see that inputs from the two eyes arising from stimuli with similar orientation, spatial frequency, and contrasts reinforce each other when they reach the visual cortex, while inputs from nonmatching stimuli suppress each other. Thus, binocular stimuli with similar features are combined in one way and those with differing features are combined in another way.

3.2.2 Adaptation and contrast effects

3.2.2a *Contrast and assimilation*

In many instances, the appearance of a stimulus, rather than its visibility, is affected by another stimulus presented at the same time or successively. A gray patch appears darker when next to a lighter patch than when next to a brighter patch. It is assumed that such interactions occur because of spatial or temporal overlap between the tuning functions of detectors for a particular visual feature. For instance, the distance between two parallel lines is underestimated when they are separated by a gap slightly greater than is needed to resolve them. This is an **assimilation**, or **attraction**, effect. When parallel lines are slightly further apart than about 4 arcmin they appear to repel each other (Köhler and Wallach 1944; Badcock and Westheimer 1985). This is a simultaneous **contrast**, or **repulsion**, effect. Successive contrast effects in visual location are known as **figural aftereffects**. Assimilation could be due to spatial summation of signals in the sensory mechanism. Contrast is probably due to spatial inhibition. The greater spatial range of contrast than of assimilation is presumably due to inhibitory interactions extending beyond summatory interactions.

Contrast aftereffects also occur in the spatial-frequency domain (see Howard 1982). In orientation, simultaneous contrast reveals itself as geometrical illusions such as the Hering illusion. Contrast effects in depth are discussed in Section 22.4 and those in motion-in-depth in Section 29.5. Physiological correlates of simultaneous and successive contrast effects have been revealed in the responses of single cells in the visual cortex (see Saul and Cynader 1989).

Contrast effects enhance signals associated with changes in stimulation, relative to signals associated with regions of steady stimulation. If it is assumed that contrast effects depend on the overlap of sensory channels, they provide a psychophysical procedure for measuring the physiological properties of metameric systems, such as the number and bandwidth of channels coding a given feature.

3.2.2b *Oppositional sensory dimensions*

In many sensory dimensions there are natural balance points, or **norms** (Section 4.2.8). For instance, "vertical" is a norm for orientation, "equidistance" is a norm for relative depth, and "stationarity" is a norm in a scale of movement from one direction to the opposite. Such stimulus dimensions are known as **oppositional**. It is characteristic of oppositional stimulus dimensions that prolonged inspection of a stimulus displaced from the norm causes that stimulus to appear more like the norm. The effect persists after removal of the induction stimulus and causes other stimuli on the same dimension to appear displaced accordingly.

Thus, when one looks at a tilted line, it gradually appears more vertical and a vertical line looks tilted in the opposite direction, a phenomenon known as the **tilt aftereffect**. The effect can be as large as 5° when the test stimulus is presented for 100 ms (Wolfe 1984a). Curved lines come to look straighter, and objects at different distances appear more equidistant. In **induced visual motion** a stationary spot appears to move in the opposite direction to a moving background. Successive motion contrast is known as the **motion aftereffect** (Section 23.4.7).

A persistent asymmetry of stimulation with respect to a norm signifies that there is a systematic distortion of the visual input. Rescaling responses to the sensory continuum adjusts the system to the disturbance. For example, on average, the natural world contains as many lines slanting or curving one way as lines slanting or curving another way. Even if a natural object in the world is slanted, the slant it creates in the retinal image balances out as the observer moves about and views it from different directions. A persistent slant or curvature over the whole visual field signifies that the visual system is wrongly calibrated and in need of correction. After a while, systems that detect orientation, curvature, or motion automatically adjust themselves. James J. Gibson used the term "normalization" to refer to the tendency for stimuli in an oppositional scale to regress to the norm. The prevailing mean value of stimuli on an oppositional scale is subtracted from the value of other stimuli that may be present. For example, a vertical line appears tilted in the opposite direction to a tilted surrounding frame. Thus, normalization can be regarded as an automatic calibration, or error-correcting mechanism (Andrews 1964).

The term "normalization" is also used to refer to the process of dividing the response of a feature-detector neurone by the pooled response of neighbouring cells so as to render the response of the feature detector invariant to changes in stimulus contrast (Section 4.2.8). This process implies division, rather than subtraction.

A plausible physiological explanation for Gibsonian normalization can be provided in terms of adaptation within opponent mechanisms. One must assume that the tuning functions of detectors tuned to opposite sides of the norm intersect at the position of the norm and that a stimulus appears at the norm when the distribution of activity in the set of detectors is symmetrical. For instance, after a line just off the vertical has been inspected for some time, it should appear displaced towards the vertical because the detector on that side of vertical will have become adapted, or fatigued, relative to the one on the other side of vertical. Note that inspection of a vertical line has no effect because it excites the two detectors equally. Similarly, the opponent colours, red and green, normalize towards grey, whereas an equal mixture of red and green remains grey. I suggest that the tuning functions of subchannels in sensory systems that normalize have evolved or grow so that they intersect at the position of the natural norm. This has the added advantage that sensitivity to change in the stimulus feature is greatest at this position. Normalization is due to selective adaptation, but its effect on a particular stimulus depends on the differential tuning widths and differing concentrations of detectors along a sensory continuum.

3.2.2c *Contingent aftereffects*

Sensory adaptation can result in an elevation of threshold, lasting a few minutes. There are also adaptation processes that can last days or even weeks. The best known of these are the **contingent aftereffects**. For example, a period of exposure to red vertical lines alternating with green horizontal lines produces a long-lasting aftereffect in which vertical lines look slightly green and horizontal lines look slightly red (McCollough 1965). Long-lasting adaptation effects have also been reported after exposure to single visual features such as visual motion (Favreau 1979) and visual tilt (Wolfe and O'Connell 1986). In these cases, long-lasting effects occur when test trials do not immediately follow the induction stimulus. However, these seemingly simple effects may be specific to incidental features of the induction stimulus, such as its size or the surroundings. If so, they may be best described as contingent aftereffects. There is evidence that cells in the visual cortex of monkeys adapt specifically to stimulus contingencies. However, this was shown only for linked stimuli within the single visual dimension of orientation (Carandini *et al.* 1997). The most significant contingencies are probably those between distinct stimulus dimensions.

Contingent aftereffects can be understood as mechanisms to correct for cross-talk between feature-detection systems. The orientation-colour contingent effect may be responsible for nulling the perceptual effects of chromatic aberration in the optical system of the eye, or in lenses that people habitually use. The perceptual system must recognise them as arising from some feature of the visual system rather than the environment. Other stimulus contingencies arise from recurring structures and events in the visual environment (Section 4.6.4a). For example, an approaching object produces looming of the retinal image and changes in disparity (Section 29.2.2). The visual system seems to develop specialized detectors so that perception can be focussed on these recurring stimulus complexes. The relevance of contingent aftereffects to binocular vision is reviewed in Section 8.3.5.

3.2.3 Interocular transfer

A visual effect is said to show **interocular transfer** if the stimulus inducing it is presented to one eye and the test stimulus is presented to the other eye. It is argued that, since interocular transfer of an effect arises in binocular cells, which occur first in the visual cortex, the processes responsible for the transferred portion of the effect must be cortical. This argument is not always valid. An afterimage shows interocular transfer, in the sense that an afterimage impressed on one eye is visible when that eye is closed and the other eye is open. This does not prove that afterimages are cortical in origin, it simply means that activity arising in a closed eye still reaches the visual cortex and can appear superimposed on whatever the open eye is seeing. This interpretation is confirmed by the fact that an afterimage is no longer visible when the eye containing it is paralyzed by the application of pressure to the eyeball (Oswald 1957). To prove that an aftereffect is cortical, one must show that it survives paralysis of the eye.

The test of **monocular independence** is related to that of interocular transfer. In the test of independence, opposite induction stimuli are presented to each eye and the aftereffect is tested in each eye separately. For instance, a leftward-moving textured display is presented to the right eye and a rightward-moving display is presented at the same time to the corresponding area of the left eye. A stationary test pattern is then presented to each eye in turn to reveal whether the direction of the aftereffect in each eye is appropriate to that eye's induction stimulus. If it is, then at least some aftereffect must have been generated in pathways specific to each eye. If the aftereffect in each eye is just as strong as when that eye alone is exposed to an induction stimulus, it is concluded that the processes responsible for the aftereffect are totally independent in the two eyes. If the aftereffect were generated wholly in a pathway common to both eyes, induction stimuli of equal magnitude but opposite sign should cancel when one is presented to each eye. Experiments involving this procedure are reviewed in Section 8.3.

3.2.4 Dichoptic stimuli

3.2.4a *Dichoptic composite stimuli*

I define a dichoptic composite stimulus as one in which part of the stimulus is presented to one eye and part to the other, where both parts are required for the composite percept. For instance, a display of lines presented to one eye can be combined dichoptically with a different display of lines in the other eye so that they form a composite shape not evident in either monocular image, as in Figure 3.4. The literature on dichoptic composite stimuli is reviewed in Section 8.3.6.

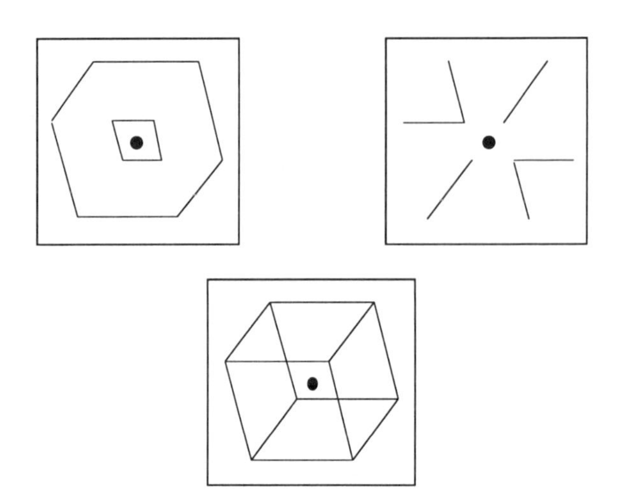

Figure 3.4. A dichoptic composite stimulus.
Fusion of the two upper displays produces the lower shape.

3.2.4b *Cyclopean stimuli*

The term "cyclopean" is difficult to define, because different authors have used it in different ways. The Cyclops was a one-eyed giant in Homer's Odyssey, and the term is used to describe a birth defect in which there is only one central eye. Hering and Helmholtz used the term "cyclopean eye" to denote the point midway between the eyes which serves as a centre of reference for headcentric directional judgments. Ptolemy and Alhazen had the same idea (Sections 2.2.1, 2.2.2, and 17.7).

Julesz (1971) generalized the term "cyclopean" to denote the processing of visual information after inputs from the two eyes are combined. This gives the term the same connotation as "central processing" as opposed to "retinal processing". In this sense, the processes responsible for coding orientation, motion, and disparity are cyclopean. Julesz also defined a **cyclopean stimulus** in a "strong sense" as a stimulus formed centrally but which is not present in either retina—it can be said to bypass the retinas. A cyclopean phenomenon is based on a cyclopean stimulus, and therefore involves processes occurring at or beyond the primary visual cortex.

Three types of cyclopean procedure exist. They all enable a stimulus to be presented to the visual cortex without there being a corresponding stimulus on the retina.

1. Paralysis of an eye For instance, if a bright light is presented to one eye and the afterimage is still visible after that eye has been pressure paralyzed, one must conclude that afterimages can be generated at a central site.

2. Direct stimulation of the visual cortex When the visual cortex is stimulated directly by an electric current, the subject sees visual phosphenes. Fortification illusions of migraine are a naturally occurring cyclopean stimulus, since they arise from a direct disturbance of the visual cortex.

3. Use of dichoptic stimuli In this procedure, which may be called the **dichoptic cyclopean proce-**

dure, distinct images are presented to the two eyes in a stereoscope to create an effect not evident, in whole or in part, in either monocular image. A dichoptic cyclopean stimulus can be an area of binocular rivalry, binocular disparity, dichoptic motion or flicker, or any combination of these features. Once synthesized, a cyclopean shape can be made to move by simply moving the dichoptic boundaries that define it. This is the main cyclopean procedure discussed in this book. Many subtypes of dichoptic cyclopean procedure exist, and are described in Chapters 8, 13, and 17.

3.2.5 Trading functions

Two or more types of stimulus may generate the same perceptual effect (Section 4.6.5). For instance, the sensation of a sound source in a particular direction relative to the body midline can be generated by a difference in time of arrival of sounds at the two ears or by an interaural difference in sound intensity. An apparent displacement of a sound source produced by changing one cue can be nulled, or titrated, by an opposite displacement produced by changing the other cue (Harris 1960). The set of null points for different values of each sensory cue defines a **trading function** in microseconds per decibel. Examples of a trading function in stereoscopic vision are provided by titration of binocular disparity against monocular parallax (Section 27.2.5) or against perspective (Section 27.3). The titration procedure is applicable only when the two cues interact in a continuous fashion. Titration of one cue against another allows one to investigate the degree of equivalence between cues and their relative efficiency. The existence of a cue trading function is evidence that inputs from two cue systems converge to produce a signal common to both.

3.2.6 Effects of attention and learning

3.2.6a. *Preattentive and focal processing*

In any complex information processing system, a choice must be made between processing all information simultaneously or devoting specialized resources to the analysis of the item of greatest importance in a particular location. Parallel processing is faster than sequential scanning but it would be biologically expensive to duplicate the complex neural circuits for in-depth analysis of distinct features at each location of the visual field. It is better to have one complex mechanism that can be applied to any location of interest. The visual system uses both strategies. Inputs over the whole visual field are rapidly processed in parallel, but only to a certain level, which has been called the preattentive level (Neisser 1967). Items of particular interest are then foveated, attended to, and processed in greater detail at the level of focal attention, which is limited to the location where attention is focussed or to particular visual features.

3.2.6b *Effects of practice*

Most investigations of the response properties of cortical cells are carried out on anaesthetized animals. With this procedure, effects due to changes of attention, motivation, or learning remain undetected. There is a growing body of evidence from work on alert animals that all these factors modify the responses of cortical neurones, even in the primary visual cortex or perhaps in the lateral geniculate nucleus. The concept of sets of cortical cells tuned to specific and fixed stimulus features must give way to a view of the cortex as a highly flexible organ in which response characteristics of cells depend on neighbouring stimuli and the activity of other centres in the brain. Some of the literature on this issue is reviewed in Section 5.5.7.

Practice leads to improvement in a variety of simple visual tasks, such as orientation discrimination, contrast detection, and Snellen acuity (see Bennett and Westheimer 1991). Prolonged practice did not improve a gap bisection task (Klein and Levi 1985) or a three-point alignment task (Bennett and Westheimer 1991), both hyperacuity tasks. Bennett and Westheimer could not find any practice effects with a grating discrimination task. On the other hand, prolonged practice improved gap bisection when the stimuli were not foveated (Christ *et al.* 1998).

It looks as though practice has more effect in the part of the visual field that is less often used for hyperacuity tasks such as discrimination of orientation and spatial intervals. However, Westheimer (2001) found no improvement in grating or gap resolution with practice in either the central or peripheral retina. Grating resolution depends on the spacing of detectors and it is difficult to see how practice could affect it. Hyperacuities, such a vernier acuity and orientation discrimination, depend on more complex neural processes that might well change with practice, especially in the peripheral retina.

Poggio *et al.* (1992) proposed that the nervous system sets up task-specific neural modules that improve performance on a repeated visual task. They illustrated this idea by a computer simulation of a neural network that manifested improved performance in a vernier acuity task when provided with appropriate feedback. Naïve observers showed a similar task-specific improvement in vernier acuity over a few tens of trials when given knowledge of results. The model replicated several features of human performance, such as the dependence of vernier acuity on the length and relative orientation of the test lines.

Fahle and Edelman (1993) reported improved performance on a vernier acuity task after observers were given prolonged practice with no knowledge of results. Improved performance was specific to the orientation of the test stimulus. Improvement on one hyperacuity task, such as vernier acuity, was found not to transfer to other tasks, such as curvature and orientation discrimination (Fahle 1997; Christ *et al.* 1998). The ability to see a shape defined by relative motion of line elements improved with practice, but, in this case, the learning was not specific to the orientation of line elements or to the direction of motion (Vidyasagar and Stuart 1993). We will see in Section 19.12 that stereoacuity also improves with practice.

In a real-life situation it is not clear what constitutes visual feedback in hyperacuity tasks. Perhaps visual feedback in the form of an error signal is not required. Improvement may occur because of an increase in response efficiency of a particular configuration of cortical cells due either to repetitive stimulation or attention-driven facilitatory processes from higher levels in the nervous system. An increase in response efficiency could involve a narrowing of the tuning functions of cortical detectors for the set of features involved in the task. It could also involve recruitment of detectors with neighbouring tuning functions, thus providing a better sampling of stimulus features. Learning, as ordinarily understood, need not be involved. It could simply be a matter of the nervous system recruiting its local resources for the performance of a repeated specific task.

According to this view, there would be a cost in terms of a performance decrement on tasks involving the use of sensory mechanisms that have been diverted to the specific task. It is difficult to see how any general learning-dependent improvement could occur in adult visual mechanisms responsible for basic tasks such as visual acuity, since we use our eyes all the time we are awake. However, it is possible that the system has evolved ways to concentrate those basic mechanisms when we repeatedly perform a specific task. Feature-detecting systems may have an inherent capacity to modify their tuning functions locally and temporarily to improve performance on specific tasks, either by error feedback or simply as a result of stimulus repetition. Attention is a process for concentrating limited resources on a selected task.

3.3 ANALYSIS OF LINEAR SYSTEMS

3.3.1 Nature of linear systems

A system is any device that transforms inputs into outputs to achieve some specified function. A function that transforms well-defined inputs into specified outputs is a **transfer function**. The aim of systems analysis is to design systems or to determine transfer functions of existing systems. One must first specify the system. Human-made systems usually have well-defined inputs and outputs and well-defined components, or modules, which can be investigated independently. Any natural system, such as the eye, has an unspecifiably large number of component systems and is itself a component in an unspecifiably large number of larger systems. One must specify the inputs that one wishes to study. In the visual system, this could be a set of stationary black-white gratings, a set of coloured patches, a set of objects at different distances, or any other stimuli that evoke responses. One must then specify the outputs. In the visual system, the output could be precision or accuracy of detection, discrimination, or recognition, or it could be eye movements, or neural activity at some specified site in the nervous system.

In a given experiment, a defined system is a 'black box' the internal structure of which can be inferred only from specified responses to a given set of stimuli. Paradoxically, it is easier to infer the order in which subroutines are executed in a nonlinear system than in a linear system because linear systems are commutative so that the same outcome can be achieved by doing things in different orders, whereas nonlinear systems are often noncommutative. In a linear system, one can independently determine the transfer functions of subsystems and combine them mathematically to predict the transfer function of the larger system. The transfer functions of in-series modules are combined by multiplication and those of parallel modules by addition. Modular systems that combine in a linear fashion are easy to construct (or to genetically programme) and malfunctions are easy to trace and treat locally. However, a linear system cannot perform operations such as multiplication or division, or any of the other nonlinear operations that are known to serve important functions in the nervous system (Section 3.4).

Many physiological systems, such as the heart, kidney, and lungs can be considered as modular units that can be made to operate in relative isolation. The visual system has certain obvious structural-functional modules, such as the two eyes, the lens, the pupil, and the extraocular muscles. Physiological and psychophysical investigations have revealed what look like modular structures in the neural structures of the visual system, such as the various cell types in the retina, the LGN, and the various in-series and parallel processing streams in the central nervous system. However, no physiological system is strictly linear and the performance of the whole system is not predictable from the responses of modules studied in isolation

One may say that the aim of visual science is to identify and characterize functional modules, and to derive their transfer functions and the rules governing their interactions. This is an ambitious enterprise. Consider the bewildering array of potential modular components that one can choose to investigate. One can select a pigment molecule, any receptor cell, amacrine cell, bipolar cell, or ganglion cell, or any collection of these retinal cells, or any synapse or collection of synapses, or the optic nerve or any of the large number of visual centres in the brain. For each component one must choose the stimuli and responses deemed to be of interest. The visual system or any of its components is sensitive to an unspecifiably large number of stimuli and responds in an unspecifiably large number of ways. For example, a retinal receptor is responsive to light, pressure, chemical changes, and electricity, and responds by changing its membrane potential, temperature, optical properties, oxygen consumption, and chemical composition. In addition, no two cells and no two eyes are exactly alike. The visual system changes over time, because of adaptation, learning, and ageing. It is also an evolving system with a history and, we hope, a future.

An investigator must decide which aspects of the system to study and at what level of generality and abstraction. There is no such thing as a complete analysis of any natural system. The visual system is what it is. The descriptions and theories that we erect are human constructs based on an arbitrary selection of some aspect of the system derived for some specific purpose and based on certain assumptions. Even when a functional description has been found that mimics some aspect of the visual system, it may not specify the physiological structures involved. The reason for this is that a given function can be implemented in many different physical systems. Constructing a functional description is like defining the algorithm of a process that can be executed by distinct machines, or hardware (see Marr 1982).

Systems fall into two main classes, linear and nonlinear, each requiring very different experimental and mathematical procedures. In very general terms, a linear system is one in which the response to input *A* plus input *B* is equal to the sum of the responses to *A* and *B* separately. This is the **principle of superposition**. Also, in a linear system, the response to a given input is the same whenever it is applied. This is the **principle of time invariance**. In practice, any system is linear only over a certain range of stimulus values. A complex system such as the eye behaves as a linear system in some respects and as a nonlinear system in other respects. Components of a system may be highly nonlinear and yet produce a linear response when working together. For example, an amplifier may be nonlinear and produce a distorted output but be linear when an error feedback signal is added. The following provides only a very general guide to systems analysis and indicates sources from which more detailed information can be obtained. The analysis of control systems is discussed in Section 3.5.

3.3.2 Fourier analysis

The fundamental assumption underlying linear systems analysis is that the transfer function of a system is fully characterized when one knows how the system responds to a set of sine-wave inputs. In a linear system, a sine-wave input produces a sine-wave output with the same frequency. The signal can be shifted in phase and its amplitude either attenuated or amplified. Over the part of the frequency spectrum for which the change in amplitude is constant, the system is said to have a flat response. A **low-pass system** attenuates responses above a specified frequency, and a **high-pass system** attenuates responses below a specified frequency. A **bandpass system** transmits inputs over only a limited band of frequencies. In practice, the response of any natural system begins to weaken and eventually stops as frequency is increased beyond a certain limit; thus, all natural systems are either low-pass or bandpass systems. Within the transmission range of a linear system, different frequencies may be attenuated by differing amounts.

In 1807, Fourier established a fundamental theorem, which is used extensively in linear systems analysis (his paper was rejected and not published until 165 years later). The core idea is that any waveform can be synthesized by combining a specified set of pure sine waves of appropriate amplitudes and in appropriate phase relationships. Note that a pure sine wave in the temporal domain extends forever; it has no beginning or end. A pure sine wave in the spatial domain extends indefinitely in space. If a complex waveform is periodic and repeats at a frequency of *F* Hz, the component sine waves include one with a frequency of *F* Hz (the fundamental) plus sine waves with frequencies that are multiples of *F*.

For example, the sinewave components of a repetitive square wave are a sine wave with a frequency, F, equal to that of the square wave, plus all odd harmonics ($3F$, $5F$, $7F$, . . .) with amplitudes decreasing in inverse proportion to frequency. Thus, the frequency components of a repetitive waveform are a series of discrete components described mathematically by a Fourier series.

If the waveform is aperiodic, the frequencies of the component sine waves vary continuously and are described mathematically by a Fourier integral. In either case, the **Fourier transform** of a signal gives the amplitude and phase of each sine wave component of the original waveform. The graph of the amplitude of each component sine wave as a function of its frequency is the **amplitude spectrum**, of the signal. The phase of component sine waves as a function of frequency is the **phase spectrum**, of the signal. For example, the Fourier amplitude spectrum of a single impulse, such as a very brief burst of sound, consists of pure tones of all possible frequencies, all of equal amplitude. The set of pure tones coincide (are in phase) at only one moment in time; at all other moments they mutually cancel because they are out of phase. A transient signal is known as an **impulse**, or delta function. A narrow vertical line is a spatial impulse, which, in the Fourier domain, is an infinite number of vertical spatial sine waves of equal amplitude, which are in phase only at that particular location. At all other locations, the sine waves cancel to a constant (dc) luminance level.

The **Fourier power spectrum** of a signal is the square of the amplitude of each frequency term in the Fourier transform. The power spectrum has peaks at the frequencies to which the system responds most vigorously.

For a mathematical treatment of Fourier analysis see Bracewell (1978) and Brigham (1974).

Applications of Fourier analysis to the visual system are discussed in Section 4.5.1.

3.3.3 Transfer functions

The **transfer function** of a system is some measure of the transmitted signal plotted against the spatial or temporal frequency of the input. One important measure of a linear system is its **gain**. The gain of a system is the magnitude of some feature of the output divided by the magnitude of the equivalent feature of the input. **Amplitude gain** is the peak-to-peak amplitude of the output divided by that of the input. For example the amplitude gain of a visual receptor is the amplitude of the generator potential divided by the intensity of light. **Velocity gain** is the velocity of some response of the system divided by

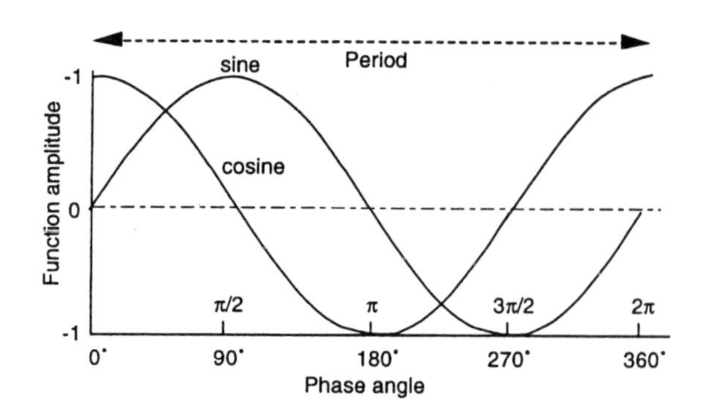

Figure 3.5. Sine and cosine functions.

the velocity of some input signal. For example, the velocity gain of the visual pursuit system is the velocity of an eye movement divided by the velocity of the moving stimulus. When gain is expressed as a ratio of output to input a value of zero indicates that there is no output and a value of 1 indicates that the output equals the input. A gain greater than 1 indicates amplification and a minus gain indicates that the sign of the input has been reversed. In particular, a gain of –1 indicates that the output is the exact opposite of the input. Gain is sometimes expressed in decibels. For amplitude gain, g, one decibel (dB) = $20\log_{10}g$. Thus, a ratio gain of 1 is equivalent to a decibel gain of zero.

A second important measure of a linear system is its **phase** shift. When a system is stimulated by a sine wave, its phase shift is defined as the phase of the output minus that of the input, indicated by degrees or radians. When the output is delayed with respect to the input we have a phase lag and when the output anticipates the input we have a phase lead. A 180° phase lead or lag brings the input and output into antiphase. A 360° lead or lag brings the input and output into phase again.

When spatially modulated signals are used we have the **spatial amplitude transfer function** and the **spatial phase transfer function**. When temporally modulated signals are used, we have the **temporal amplitude transfer function** and the **temporal phase transfer function**. The spatial amplitude transfer function is often referred to as the **modulation transfer function**, or **MTF**.

The simplest spatially modulated signal is a black and white grating in which luminance is spatially modulated according to a cosine or sine function. Over a phase interval from 0° to 360° the cosine function has a symmetrical appearance and the sine function has an asymmetrical appearance, as shown in Figure 3.5. That is why they are sometimes referred to as even-symmetric and odd-symmetric functions. The **spatial frequency** of a grating is the number of complete white-black cycles in one degree of visual angle, expressed as cycles per degree (cpd). The **spatial period** of a grating is the reciprocal of its spatial frequency, or the angular subtense of one cycle of the grating. The amplitude of a grating is the difference in luminance between the peaks and troughs. Other features of a sine wave are described in Figure 5.7.

When a sinusoidal grating of a given spatial frequency is transmitted through an optical system, its amplitude of luminance modulation is to some degree attenuated by the summed effects of light loss and dispersion. The **amplitude attenuation** produced by an optical system is the reciprocal of the gain of the system. This definition can be generalized to cover both optical systems and image-processing systems.

In any practical optical system, amplitude attenuation is complete for all spatial frequencies above a certain value. This simply means that the system cannot resolve gratings above a certain spatial frequency. The human visual system—the whole system, not only the optics—is also insensitive to gratings below a certain spatial frequency. The range of frequencies resolved by any system is its spatial bandwidth (or bandpass) at that contrast. Gratings with frequencies outside the spatial bandwidth appear as homogeneous grey patches.

A spatial transfer function of an optical or visual system is derived from its responses to sinusoidal gratings. From Fourier's theorem it follows that a visual display, however complex, in which luminance is modulated along only one dimension, can be synthesized by superimposing sets of parallel sinusoidal gratings, with suitable amplitudes, phases, and wavelengths. These sets of gratings constitute the spatial Fourier components of the display. In practice, luminance cannot be modulated about a value of zero since there is no negative light. All spatial patterns therefore contain a certain mean level of luminance, which can be regarded as a dc, or zero spatial-frequency component, which must be added to the Fourier transform.

Any 2-D visual scene can be synthesized by superimposing sets of sine-wave gratings, with each set oriented at a different angle in the plane of the display. If a set of spatial sine waves is transmitted through a spatially homogeneous linear system, the image consists of a set of sine waves with the same spatial frequencies. The amplitudes of component spatial frequencies can change by different amounts in a linear system. A linear system may also displace, rotate, or invert the image, because such transformations do not affect spatial frequency. Strictly speaking, a linear system cannot magnify or minify the input, with all frequencies scaled up or down proportionally. However, most optical systems either minify or magnify the image. This need not violate the assumption of linearity, since it is only the linear dimensions of the image that are minified or magnified, not the angles subtended at the nodal points of the optical system.

The spatial amplitude transfer function of an optical system is derived by using a photoelectric probe to measure the luminance modulation of the stimulus grating and of the image of the grating at each of several spatial frequencies within the spatial bandwidth of the system. In the eye, one uses a photometer to measure the variation in luminance modulation of light reflected from the retinal image of a grating. The ratio of the amplitude of luminance modulation of the image to that of the stimulus defines contrast transmission, or gain, which is plotted as a function of the spatial frequency of the grating. This spatial transfer function indicates how efficiently the system transmits spatial sine waves. It can be used to predict the quality of the image of any pattern.

A Fourier analysis of the stimulus pattern is first performed. Each sine-wave component is then amplified or attenuated by an amount determined by the modulation transfer function of the system. When the pattern is restored by Fourier synthesis, it defines the spatial properties of the image produced by the system. The image is the result of passing the visual display through a set of sine-wave luminance filters, each with an infinitely narrow bandwidth. For a full specification of the image, one must also know the spatial phase transfer function. The optical transfer function is derived from the amplitude and phase transfer functions and, when defined for all orientations of the image, it fully specifies the performance of a linear optical system in transmitting spatial information for a given aperture and optical axis.

The temporal properties of a system are specified by the temporal amplitude transfer function and the temporal phase transfer function. One can combine two spatial dimensions and time to produce a spatiotemporal Fourier transform. This transform specifies the unique set of drifting sinewave gratings at each orientation that are required to synthesize a given moving display. The spatiotemporal transfer function of a linear system is a complex-valued function of spatial and temporal variables.

Although a linear system as a whole can be formally described by this analysis, one need not assume that the system contains distinct components that actually carry out these operations.

3.3.4 Point-spread functions

The **point-spread function** provides an alternative measure of the optical quality of an image. Even if the image is perfectly focussed, diffraction of light by the pupil, dispersion in the optical media of the eye, and scatter of light by the surface of the retina cause each point in the distal array to be represented by a blurred disc in the image. If the effects of dispersion and scatter are ignored for a point object, one can calculate the distribution of light in the image due only to diffraction. Such a system is said to be **diffraction limited**. For a circular pupil (aperture), this turns out to be a central bright disc, known as **Airy's disc**, surrounded by bright and dark rings arising from the way light waves alternately summate and cancel round the boundary of the pupil. This is the diffraction point-spread function. The diameter of Airy's disc in radians of visual angle is given by $2.44\lambda/a$, where λ is the wavelength of the light and a is the diameter of the pupil, both expressed in the same units. For a 2.3-mm pupil and a wavelength of 550 mm, Airy's disc has a diameter of 1 arcmin of visual angle. This corresponds to about three times the diameter of a foveal cone, so that the central disc produced by a bright point of light, however small, necessarily falls across about seven cones (see Section 5.1.3).

A similar calculation can be done for a thin-line stimulus, in which case we have a **line-spread function**. A thin line in the spatial domain is a spatial impulse, or delta function in the Fourier domain. Impulses are used widely in testing communication systems, both natural and man-made, because the Fourier transform of an impulse is a set of equal-amplitude sine waves extending evenly across the whole frequency spectrum. The crests of the waves of the sine waves coincide at the location of the impulse. At all other places, crests and troughs cancel. When the eye is exposed to a thin line, it is as if a complete set of spatial sine waves were simultaneously injected into the visual system. The amplitude transfer function of the eye determines how each component is attenuated, and the Fourier transform of the amplitude transfer function of a linear system is its response to a spatial impulse (the line-spread function). Put another way, the spatial amplitude transfer function is the Fourier transform of the line-spread function. The line-spread function and the transfer function are thus equivalent representations of a linear visual system. This mathematical relationship forms the basis of many inferences about the performance of the eye, cortical cells, and the visual system as a whole. A spatiotemporal impulse is a stimulus confined both spatially and in time.

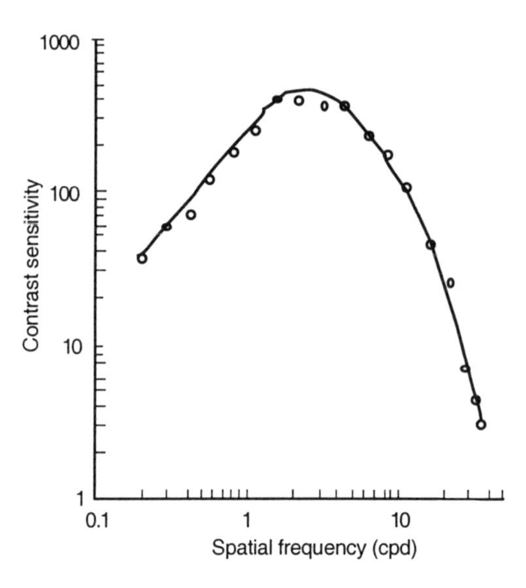

Figure 3.6. Contrast sensitivity function.
Contrast-sensitivity of human observer for a sine-wave grating, luminance 500 cd/m^2. (Adapted from Campbell and Robson 1968)

The line-spread function resulting from diffraction imposes an upper limit on the spatial frequency of a grating that can be imaged on the retina. This limit is called the **cut-off frequency** and is given by:

$$\text{Cutoff frequency} = \frac{d}{\lambda} X \frac{\pi}{180}$$

in cycles per degree. Where d is the diameter of the pupil and λ is the wavelength of the light. Aberrations other than diffraction reduce the cut-off frequency below this theoretical limit and we will see in the next section that the finest grating that a person can detect is further limited by neural factors beyond the optics of the eye.

If the spatial amplitude transfer function of a system is the same for all positions of a grating, the system is homogeneous. The eye is far from being homogeneous. If the transfer function is the same for all orientations of a grating the system is isotropic; if not, it is anisotropic. Astigmatism is an example of anisotropy in the visual system. For an introduction to linear systems analysis, see Toates (1975). For more detailed treatments see Bracewell (1978), Cooper and McGillem (1967), and Brigham (1974).

3.3.5 The contrast sensitivity function

The method for determining the amplitude of the output of a system must be adapted to the type of output that the system produces. In an optical system, the output is an image that can be measured with a physical instrument. Special methods are required to determine the modulation transfer function of the optics of the eye (Section 5.1.3).

Several methods may be used to measure the modulation transfer function of the visual system as a whole, including neural processes. The amplitude of the output can be derived from objective responses in the form of eye movements or the responses of a cell or of a set of cells in the visual system. On the other hand, it may be indicated by psychophysical judgments made by subjects under specified conditions. De Lange (1958) was the first to apply linear systems analysis to psychophysical data, in his investigations of visual flicker.

Campbell and Robson (1968) first applied these methods to psychophysical data derived from the use of spatial patterns modulated in luminance. In this application, the output of the visual system is defined as the luminance contrast required for detection of a sine-wave grating at some specified criterion for detection. The threshold contrast as a function of spatial frequency is known as the **contrast sensitivity function**, or CSF. It may be regarded as the spatial transfer function of the contrast detection mechanism of the visual system as a whole at threshold. A typical contrast sensitivity function is shown in Figure 3.6.

We will see in Chapter 19 that an analogous sensitivity function relates the threshold modulation of a depth corrugation to the spatial frequency of the depth modulations of the surface. In general, the spatial amplitude transfer function of any system is a mapping of the output of the system onto sinusoidal inputs of varying spatial frequency. The temporal amplitude transfer function is a mapping of the output onto sinusoidal inputs of varying temporal frequency. At suprathreshold contrasts, the output of the visual system is assessed by asking observers to match the contrast of gratings at different spatial frequencies (Georgeson and Sullivan 1975).

3.3.6 Signal analysis

3.3.6a *Basis functions*

Any mathematical analysis applied to input or output signals, is known as **signal analysis**. In applying such methods one does not necessarily assume that the system is linear. The first task is to decompose the stimuli being considered into a set of **basis functions**. For example, sine waves of different frequencies provide a set of basis functions by which any well-behaved function can be approximated arbitrarily closely by a set of sines and cosines, summed over a range of frequencies and phases. We can then ask whether the visual system analyses complex visual stimuli in terms of the defined basis functions. For this purpose, basis functions can be regarded as a set of filters applied to the visual input, or we can talk about a set of visual channels or coding primitives. Physiologically, a visual primitive in the space domain is the sensitivity profile of the receptive fields of a set of similar cells at the level of the visual system being considered. For example, at the level of ganglion cells, the visual primitives are the types of receptive fields of ganglion cells (Section 5.3.2). The idea can be generalized to the spatiotemporal response profiles of cortical cells (Section 5.5.4).

It is important that the set of coding primitives is complete. A set is complete with respect to a defined stimulus domain when each discriminable stimulus within that domain can be represented by a distinct weighted sum of the primitives. For example, zero crossings (regions of maximum change in luminance) do not form a complete set of visual primitives because there are textures that appear different but produce the same representations in terms of zero crossings (see Daugman 1990). Gaussian derivative functions provide an example of a complete set of primitive wavelets (Young 1987; Koenderink 1990). In a complete coding system, the number of independent degrees of freedom in the code is at least as large as the dimensionality of the stimuli.

A second important attribute of coding primitives is their linear independence. A coding process is optimally efficient when the primitives are independent, so that each captures a property of the input not captured by any other—there is no cross talk. Independent primitives are often described as orthogonal but, mathematically, primitives that are not orthogonal can be independent. In biological sensory systems, detectors for distinct sensory attributes are generally independent. For instance, the colour of a line is not affected by the line's orientation. However, sensory detectors for a given feature are not independent—they overlap. For instance, visual orientation detectors have broad and overlapping tuning functions, which undersample the stimulus dimension to produce metamerism. Such detectors are inefficient for resolution but are efficient for discrimination, as explained in Section 4.2.7.

A third attribute of good coding primitives is their ability to exploit redundancies in stimuli, and thus economize on signal transmission and processing (Barlow 1961). In a non-redundant visual world, each point varies in luminance over the full range of values in a totally random fashion and independently of the luminance of neighbouring points. Visual white noise has these characteristics. There is no way to compress the signal from such a stimulus. Since long lines are rare in a white-noise world, there would be little point in having detectors tuned to line orientation. In fact, natural visual scenes contain redundancies, since points with similar luminance tend to cluster along lines or within areas.

For an infinite homogeneous (shift invariant) detector, Fourier components are the most efficient way to transmit information about spatially redundant stimuli (Bossomaier and Snyder 1986). However, for an inhomogeneous detector of finite size, like the retina, wavelets based on oriented Gabor patches with overlapping spatial scales are well suited to exploit simple redundancies in natural images. They achieve an optimal compromise between information preservation and economy of sampling over each stimulus dimension (Sakitt and Barlow 1982; Field 1987; Olshausen and Field 1996). Little attention has been paid to the most efficient way to exploit temporal redundancy in natural images.

Basis functions used to describe visual primitives are described in Section 4.5.

3.3.6b *Convolution*

Once the sensitivity profile of a linear detector is known, the magnitude of its response to a given stimulus can be derived. The distribution of light intensity across the receptive field of the detector is plotted. At each spatial location across the surface of the detector the stimulus magnitude is multiplied (weighted) by the local value (height) of the sensitivity profile of the detector. All the values are added (integrated) to yield a single number, which represents the response of that detector to that stimulus at that instant.

Keeping the stimulus in the same location, the procedure can be repeated for each of the set of detectors that overlap the location of the stimulus. The resulting numbers plotted against the positions of the centres of the detectors yield a one- or two-dimensional spatial response profile to that stimulus over that region of the receptor surface. The spatial response profile is the **convolution function**. In general, the convolution function derived from two continuous functions $f(t)$ and $g(t)$ for $t > 0$, is:

$$f(t) * g(t) = \int_0^t f(t-u)g(u)du$$

The order of the operations makes no difference, or $f(t) * g(t) = g(t) * f(t)$. Calculation of a convolution and its inverse is eased by the fact that the Laplace transform of a convolution equals the product of the Laplace transforms of the two functions. One could convolve the spatiotemporal sensitivity functions of a set of cells with the spatiotemporal distribution of stimulus intensity to derive a function that represents the total response of the system to a stimulus over a defined time interval.

3.4 ANALYSIS OF NONLINEAR SYSTEMS

The response of a linear system to any input can be calculated when its response to each of a set of sine waves (or equivalently, its response to an impulse) is known. This follows from the fact that any signal can be represented in the frequency domain as a set of sine waves of appropriate amplitude and phase. This is not true of a nonlinear system. Types of nonlinearity in the visual system are listed in Section 4.2.9. There is no general method for characterizing nonlinear systems. However several procedures have been applied to the visual system.

In the Volterra method it is assumed that nonlinearities arise from interactions between impulses. For a single detector generating impulses at n different times, the potential interaction between impulses is proportional to the product of their intensities. The contribution of these interactions to the response of the system at time t is the product of impulse intensities multiplied by a function of the n time lags, known as a Volterra kernel. The system is characterized by the integral of these products over the set of n impulses. The zeroth-order term, kernel, indicates the response when no stimulus is present. For low amplitude stimuli the first-order term dominates the response and the kernel is the linear component of the system. A linear system has no terms higher than the first. Higher terms indicate deviations from linearity, which become more apparent as signal size increases. The method can be generalized to a system of multiple inputs.

During World War II Wiener developed the method of Wiener kernels for investigating nonlinearities in reflected sonar signals. The method is defined in terms of a system's response to white noise and can be used for systems that (1) do not change rapidly over time and (2) become linear as signal strength is reduced. First, the response of the system to Gaussian (white noise) stimuli is measured. Then the Wiener kernels are calculated. These are essentially orthogonal terms in a power series, with the zeroth-order term representing the response of the system to a static input, the first-order term the linear characteristics of the response, and higher terms representing nonlinearities. The values of the kernels depend on the stimulus. Ideally, responses to an infinitely long Gaussian white noise stimulus will allow one to predict the response of the system to any stimulus, because all stimuli are contained in white noise. In practice, one must use finite stimuli of lim-

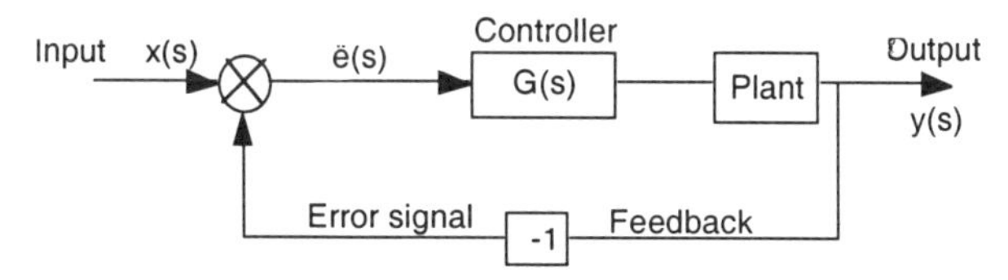

Figure 3.7. Basic elements of a linear control system.

ited spatial and temporal bandwidth (for details see Marmarelis and Marmarelis 1978).

A variant of the Wiener kernel method, with improved signal-to-noise ratio, involves testing a system with a small set of superimposed sine waves rather than white noise (Bennett 1933). Victor (1979) used this procedure to analyze responses of single neurones in the visual pathway (see Section 5.5.4).

In another approach, the stimulus consists of a spatiotemporal sequence of binary signals in the form of a grid of black and white squares modulated in time (Sutter 1992). The display is constructed from a repeating binary sequence of length 2n-1, known as a maximum length shift-register sequence, or **M-sequence**. The order of n is made sufficiently large so that the repeating sequence is not apparent. Such a sequence approximates a white noise stimulus, since all spatial and temporal frequencies are represented equally. The stimulus is easy to generate and allows for rapid calculation of the first and higher order Wiener kernels. For example, a two-dimensional M-sequence display has been used to map the spatiotemporal structure of the receptive field of a cortical cell (Reid *et al.* 1997).

The structure of the stimulus determines which aspects of the receptive field are revealed. For a linear system, a sparse stimulus is sufficient, because only the first Wiener kernel is relevant. For nonlinear systems a reasonably dense stimulus is required to reveal the higher-order kernels. For example, more details of a cell's receptive field are revealed with finer stimuli modulated at higher temporal frequencies. However, a finer stimulus has less stimulus energy and, if the stimulus exceeds the resolving power of the visual system, it becomes indistinguishable from uniform grey.

The transfer function of a linear system cannot reveal the nature or even the presence of linear subcomponents because the same overall transfer function can arise from a multitude of equivalent subcomponents. To explore the inner workings of a linear system one must probe the input-output relations for each sub-component.

For a system consisting of a cascade of linear and nonlinear components, the characteristics of the nonlinear component can uniquely determine the response characteristics of the whole system (Spekreijse and Reits 1982; Korenberg and Hunter 1986). For example, when stimulated with superimposed stimuli of frequency F_1 and F_2, a nonlinearity, such as rectification, produces **cross modulation products**, or complex harmonics, of the general form $nF_1 + mF_2$, for integral values of n and m. The relative amplitudes of these terms depend on the nature of the nonlinearities. Regan and Regan (1988, 1989) provided a mathematical analysis of these processes and applied it to the processing of binocular signals (Section 6.7.2). This method is more immune to the effects of noise than Wiener kernel methods.

Nonlinear systems may also be studied by building an analogue computer with similar nonlinear features, and have it compute outputs to defined inputs. Alternatively, a digital computer can be programmed to simulate a nonlinear system. For example, Lehky *et al.* (1992) characterized the nonlinear receptive fields of complex cells of the monkey visual cortex by recording the responses of cells to a great variety of patterned stimuli. An artificial **neural network** was then created for each neurone using an iterative optimization algorithm. The responses of the cell to some stimulus patterns defined the training set for the neural network. The cell's responses to patterns not used in the training set were predicted with a median correlation of 0.78. For discussions of neural-network methods, see Rumelhart and McClelland (1986), Hinton (1989), and Miller *et al.* (1991).

Neural network models can be used to characterize the receptive-field structure of cells, but do not indicate the function of the cells. Network models of binocular disparity processing are discussed in Section 6.8. For a discussion of nonlinear visual processes see Pinter and Nabet (1992).

3.5 CONTROL THEORY

Control theory is concerned with designing models to simulate the response of a system to specified stimuli. A model has an input stage, one or more in-series or parallel controllers with defined transfer functions, and an output that changes the state of a defined physical system, or plant. This is the forward loop of the system. Feedback loops convey error signals from the output of one or more controllers to earlier stages in the processing stream. An error signal modifies the response of a controller so as to restore the system to some goal state (Figure 3.7). In proportional error control, the response of a system is proportional to the size of a step input. With constant input, proportional control results in a steady-state error. The error may be reduced by integrating the input (integral control), but this renders the system insensitive to rapid fluctuations in input—it lowers its frequency response. Differentiation of the input improves the frequency response but renders the system insensitive to constant inputs. Models usually contain both integral and differential elements. The stability and accuracy of a control system may be improved by adding appropriate filters in the forward loop or feedback loop. Computer programmes, such as CAD software can be used to design appropriate filters. The error is sampled at defined intervals in a sampled data system, and continuously in a continuous system.

Many biological systems can be described by a differential equation. For example, the force (F) required to rotate an eye or move an arm is the sum of three forces.

1. Force to overcome elasticity—resistance that depends on position, θ.
2. Force to overcome viscosity—resistance that depends on velocity, $d\theta/dt$.
3. Force to overcome inertia—resistance that depends on acceleration, $d^2\theta/dt^2$

A coefficient is the value of a function when the variable is set at unity. The coefficient of elasticity is E, that of viscous resistance is V and that of inertia is M (mass), then:

$$F = E\theta + V\frac{d\theta}{dt} + M\frac{d^2\theta}{dt^2}$$

A first-order system contains no terms higher than the velocity term, and a second-order system contains an inertial term.

Differential equations are not easy to solve because the terms are not algebraic quantities that can be added or subtracted. The differential of an exponential function equals the value of the function so that if we convert the terms of a differential function into exponential functions then we can treat them as algebraic quantities. This is what the **Laplace transform** does. Mathematically, the Laplace transform is the convolution (sum of the products) of the function, f(θ), and an exponential function, $e^{s\theta}$ with exponent $s\theta$. The function $e^{-s\theta}$ is known as the kernel and the Laplace transform is denoted by F(s). It is the area under the curve formed by multiplying the function and the kernel at each value of θ. This process is represented by the formula:

$$\mathrm{F}(s) = \int_0^\infty e^{-s\theta} f(\theta) d\theta$$

When $f(\theta)$ is zero for negative values of θ and the real part of the complex variable s is zero, Laplace and Fourier transforms are the same.

The Laplace transforms of standard differential equations are found in a table. The overall transfer function is then obtained by algebraic procedures. The resulting Laplace transform is converted back into non-Laplacian form by again looking in the table. In general, all bounded continuous functions have a Laplace transform. In most practical cases, each function has a unique Laplace transform and each transform has a unique inverse function.

For example, the Laplace transform of an exponential function, e^{at} is

$$S(e^{at}) = \int_{t=0}^{t=\infty} e^{-st}(e^{-at})dt$$

Since s is not a function of time it can be treated as a constant.

$$S(e^{at}) = \int_{t=0}^{t=\infty} e^{-(s+a)t} dt$$

$$= \left[\frac{-e^{-(s+a)t}}{s+1}\right]_{t=0}^{t=\infty}$$

$$= 0 - \frac{-e^{-0}}{s+a}$$

$$= \frac{1}{s+a}$$

Table 3.2 shows examples of Laplace transforms of input functions and of transfer functions that describe the way the system transforms the input.

For example, let the input be a step and the transfer function be integration. In Laplace terms the input is $1/s$ and the transfer function is $1/s$. The output is the product, namely $1/s^2$, which is the Laplace transform of a ramp. Thus, a step passed through a system that integrates produces a ramp.

If a system contains several linear processing stages in series, the overall gain is the product of the component gains and the overall phase lag is the sum of the component phase lags. Also, the Laplace transform of the whole set is the product of the Laplace transforms of the separate transfer functions. The Laplace transforms of linear in-parallel transfer functions may be summed. Thus, a single transfer function can be derived for the whole linear system. The Laplace transform of the input times the

Table 3.7 Examples of functions and their Laplace transforms.

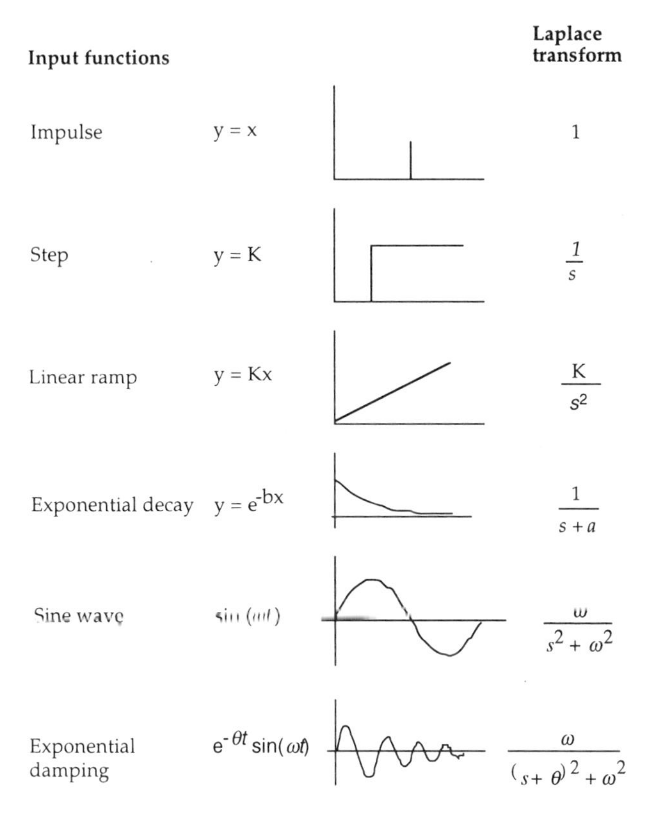

Input functions		Laplace transform
Impulse	y = x	1
Step	y = K	$\frac{1}{s}$
Linear ramp	y = Kx	$\frac{K}{s^2}$
Exponential decay	$y = e^{-bx}$	$\frac{1}{s+a}$
Sine wave	$\sin(\omega t)$	$\frac{\omega}{s^2+\omega^2}$
Exponential damping	$e^{-\theta t}\sin(\omega t)$	$\frac{\omega}{(s+\theta)^2+\omega^2}$
Transfer functions		

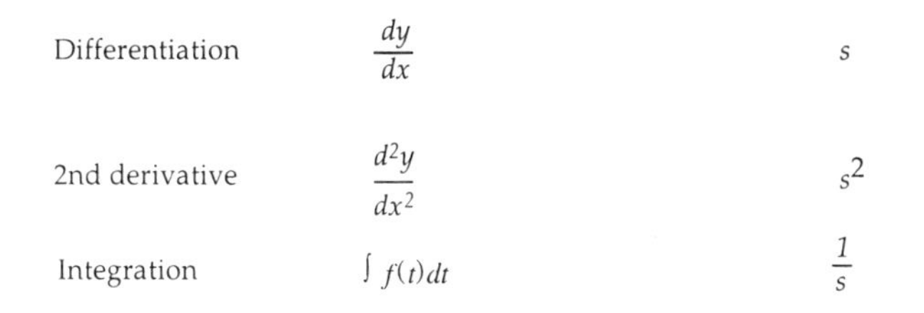

Differentiation	$\frac{dy}{dx}$	s
2nd derivative	$\frac{d^2y}{dx^2}$	s^2
Integration	$\int f(t)dt$	$\frac{1}{s}$

Laplace transform of the transfer function of a system gives the Laplace transform of the output.

When there is negative feedback between the output, y, and the input, x, an error signal e is produced with a Laplace transform of $e(s)$ equal to the Laplace transform of the input, $x(s)$, minus the Laplace transform of the output, $y(s)$. For an introduction to the Laplace transform see Grove (1991).

The **closed-loop gain,** $g(s)$ of a system is the amplitude of some attribute of the output divided by the amplitude of the same attribute of the input when the system is operating under feedback control. For example, the closed-loop gain of vergence can be defined as the vergence movement of the eyes divided by the opposed angular displacement of the images in the two eyes. Gain may also be de-

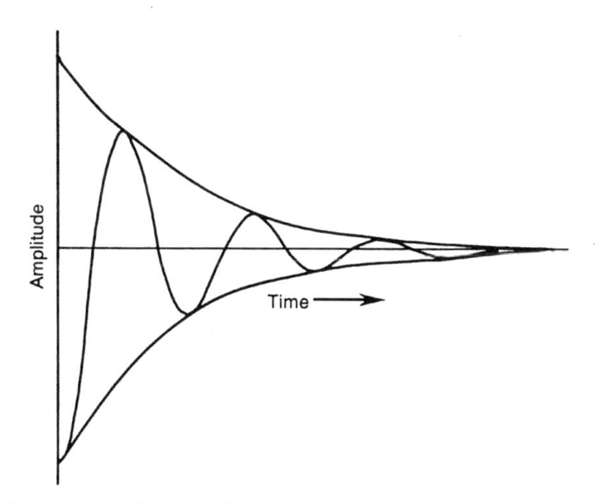

Figure 3.8. A damping function.
Representing the exponential decay of a sinusoidal oscillation.

fined in terms of other attributes of the stimulus and response, such as velocity. Gain in decibels is 20 times the logarithm of proportional gain. Thus, a proportional gain of 1 is equivalent to 0 db.

The **open-loop gain,** $G(s)$ of a system is the amplitude of the output divided by the amplitude of the input when the feedback loop is cut. There is a simple relationship between $G(s)$and $g(s)$:

$$e(s) = x(s) - y(s)$$
$$y(s) = G(s)\ddot{e}(s)$$
$$y(s) = G(s)[x(s) - y(s)]$$
$$y(s) = G(s)x(s) - G(s)y(s)$$
$$y(s)[1 + Gs)] = G(s)x(s)$$
$$y(s) = \frac{G(s)}{1 + G(s)}x(s)$$
$$But\,\frac{y(s)}{x(s)} = g(s)$$
$$Therefore, g(s) = \frac{G(s)}{G(s)+1}$$

If the feedback mechanism has the transfer function $H(s)$ then the closed-loop transfer function is

$$g(s) = \frac{G(s)}{1 + G(s)H(s)}$$

For example, if $H(s)$ is 1 and the closed-loop velocity gain of visual pursuit is 0.9 then the open-loop gain is 9. Thus the eyes will move at 9 times the velocity of the visual target if the feedback loop is cut by stabilizing the retinal image so that the motion of the

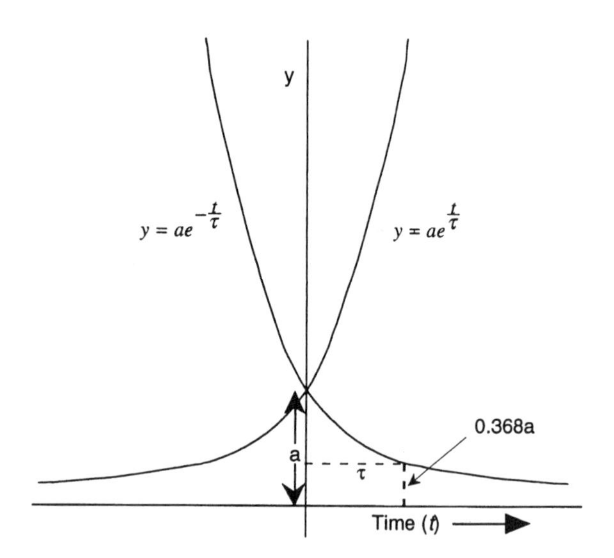

Figure 3.9. General exponential function of time.

retinal image is not reduced when the eyes move. If the closed-loop gain is 0.5 the open-loop gain is 1.

If a sinusoidal signal is applied to a linear control system the output is also sinusoidal, with the same frequency. The amplitude of the output may be larger (positive gain) or smaller (negative gain) than that of the input. Also, the gain may vary with the frequency of the input. A graph of gain against the frequency of the input to a system is known as the **gain Bode plot**.

The **phase lag** of a linear control system is the extent to which the output lags behind the input, measured by the phase angle between a sinusoidal input and the corresponding sinusoidal output. A plot of phase lag against the frequency of the input is the **phase Bode plot**. If the phase lag is a constant fraction of the period of the sinusoidal input then it is a constant time lag.

Many processes have a natural frequency of sinusoidal oscillation, the amplitude of which decays exponentially when the driving function is switched off. The amplitude, θ, of a damped oscillation of frequency ω can be described as the product of a sinusoidal function and an exponential damping function, as shown in Figure 3.8, and described by:

$$\theta = ae^{bt} \sin \omega t$$

The time taken for a system to reach a steady state after a stimulus has been switched off also depends on the damping of the system. If damping is set to a critical value, the system changes from one state to another without sinusoidal oscillations. If a system is overdamped, it reaches the steady state more rapidly than a critically damped system. An undamped system will either oscillate indefinitely, or become unstable. A system will be unstable if the gain of the feedback loop is greater than 1 or if the phase lag of the feedback loop exceeds a critical value.

The **stability** of a linear system can be investigated by plotting gain against phase on polar coordinates, a graph known as the **Nyquist diagram**. It can also be investigated by the root locus procedure (D'Azzo and Houpis 1995).

Many neural systems show spontaneous or stimulus-initiated oscillation. Two coupled oscillators are phase locked when the phase difference between them is constant. Two coupled phase-locked oscillators in antiphase are in push-pull mode and their combined output is the difference between their amplitudes. Two oscillators in phase are synchronized. A chain of coupled phase-locked oscillators can generate a travelling wave. For example, the motion of a snake is governed by a travelling wave of neural activity. Synchronized neural systems are discussed in Section 4.3.2.

Exponential functions describe the time course of many natural processes. Any natural process has a **time constant**, which indicates how rapidly it reaches a new steady state when disturbed.

The general exponential function, as graphed in Figure 3.9 is $y = ae^{bx}$. Constant a defines the intersection of the function on the y axis. Constant b defines the slope of the function. When b is positive the function describes a growth process and when negative it describes a decay process. The number e is the base of natural logarithms and equals 2.718 $\cdots$. It equals the sum of the following series:

$$e = \frac{1}{0!} + \frac{1}{1!} + \frac{1}{2!} + \frac{1}{3!} + \frac{1}{4!} \ldots$$

Raised to a power x, it becomes:

$$e^x = \frac{x^0}{0!} + \frac{x^1}{1!} + \frac{x^2}{2!} + \frac{x^3}{3!} + \frac{x^4}{4!} \ldots$$

When e^x is differentiated, the series is unchanged since each term becomes the term before it:

$$\frac{de^x}{dx} = \frac{x^0}{0!} + \frac{x^1}{1!} + \frac{x^2}{2!} + \frac{x^3}{3!} + \frac{x^4}{4!} \ldots$$

This means that the slope, or differential, of an exponential function, is everywhere equal to the value of the function, at that point.

$$y = ae^x \text{ then } \frac{dy}{dx} = ae^x$$

In other words, the rate of growth or decay of an exponential function is proportional to the present magnitude of the function. For example, if rabbits were left to reproduce unhindered, the increase in numbers would follow an exponential function because the rate at which they reproduce would be proportional to the number of rabbits.

If $y = e^x$ then $x = \log_e y$. In other words, an exponential function is the inverse of a **logarithmic function**. One can be obtained from the other by interchanging the axes.

In the time domain x equals time, t. Let $b = 1/t$ where τ is the time constant. Then an exponential function of time is:

$$y = ae^{-\frac{t}{\tau}}$$

Where a is the value of y at time zero ($ae^0 = a$, since $e^0 = 1$) and τ is the time at which $y = ae^{-1}$. When $t = \tau$,

$$y = ae^{-1} = \frac{a}{e} = \frac{a}{2.718} = 0.368a$$

Thus, after time τ the value of the exponential function has fallen to about one third of the value it had at time zero. If τ = one second, the function falls to $1/e$ of its initial value in one second. The time constant is a measure of how long it takes a natural decay process to decay. Many natural physiological and perceptual processes obey an exponential function approximately. For examples of models applied to vergence eye movements see Section 9.6.10.

3.6 TIME SERIES

The above discussion of the analysis of events over time is part of the broader subject known as **time series analysis**. A time series is a set of measurements of a defined process made over time. Time series analysis is designed to reveal trends and recurring events in a single process (univariate) or relationships between two or more processes (bivariate or multivariate). The methods also apply to series of measurements made over space.

The first task in analyzing a time-varying or space-varying signal is to distinguish between signals due to changes in the input to the system from noise fluctuations. Noise often consists of high-frequency random fluctuations, which can be removed by taking a moving average over successive sets of data points. This is tantamount to passing the data through a low-pass filter.

The effects of noise can also be reduced by applying a well-defined stimulus many times and averaging the responses over many cycles of stimulus repetition. **Signal averaging** emphasizes components in the response that are time-locked to the stimulus, and attenuates components due to noise, which average out over several cycles of stimulus repetition because they are not correlated between one repetition and the next.

Standard signal-averaging procedures overlook episodic noise such as bursts and oscillations. These events can be characterized by **phase-locked spectral analysis**, which measures the difference between responses to each cycle of a periodic stimulus and the response average. Standard signal-averaging procedures also fail to register stimulus engendered variations in response to particular cycles of a periodic stimulus. Episodic activity generated by the stimulus, but not time-locked to it, can be revealed by power spectrum analysis or as peaks in the autocorrelation function, as described below (Schiff *et al.* 1999). Long-term variations can be removed by passing the data through a high-pass filter. Signal averaging has been used extensively in the analysis of cortical activity (Section 6.7.1).

Trends can be revealed by fitting the data to a defined polynomial function by the least mean squared error procedure. When data are passed through two filters in series, the final output is obtained by the process of convolution, described in Section 3.3.6. These procedures can be found in any textbook of statistics.

The **autocorrelation function** is one of the main tools for revealing recurrent patterns in a process. The correlation coefficient between N observations made at repeating intervals of time or space and the identical set of observations is necessarily 1. This coefficient is therefore ignored. The correlation coefficient, r_1, between measurements made at interval t (x_t) and those made at interval t+1 (x_{t+1}) is first obtained. This is essentially the mean of the products of the paired deviations from the mean divided by the sample variance. It is a measure of how similar the value of the function at interval t+1 is to its value at interval t. In this case, the lag interval, k, is 1 and the coefficient measures any tendency for neighbouring events to be related. Then measurements two intervals apart (k=2) are correlated to give r_2, and so on for values of k usually no greater than $N/4$. Each of the k correlation coefficients, r_k, can be plotted as a function of lag (k) to give the autocorrelation function, or correlogram. The process should be stationary; that is, it should not contain long-term trends. For a random series of events, all coefficients for k>0 vary randomly about zero. A series containing short-term dependencies, in which neighbouring values tend to be the same, show large coefficients for small values of k. A series with a regular alternation shows coefficients that alternate in sign. The coefficients of a sinusoidal series also vary sinusoidally. Peaks in the autocorrelation function can represent the contributions of different temporal or spatial frequency components. We will encounter autocorrelation functions in Sections 6.7.1, and 8.1.3b.

The **power spectral density function** of a process represents the contribution of each sinusoidal spatial or temporal frequency to the variance of the series of measurements. The area under the whole spectral density function is the variance of the whole time series. The spectral density function is in the frequency domain and the autocorrelation function is in the time or space domain. However, they are closely related. A spectral density function based on a total variance normalized to 1 is the Fourier transform of the autocorrelation function.

In bivariate analysis, a series of measurements of one process can be correlated with measurements of a second process over lag intervals from 0 to some specified value to produce a **cross correlation function**. After allowance has been made for contaminating effects of autocorrelations in each process, the cross correlation function reveals common fluctuations in the two processes over time or space. For example, the visual system can be said to cross correlate the images in the two eyes over a spatial transformation to reveal how one image should be matched to the other (Section 16.1).

A **cross-spectral density function** can be derived from two processes. It is the Fourier transform of the cross correlation function. There are several forms of the cross-spectral density function. The cross-amplitude spectrum measures how the amplitudes of two processes are related and the cross-phase spectrum indicates how their phases are related. The **coherence function**, is the frequency-domain analogue of the squared cross-correlation coefficient. It expresses the extent to which two processes covary as a function of frequency and has been used in the analysis of synchronous neural activity (Sections 4.2.6 and 4.3.2g).

A **chaotic system** is one in which a small change in its state at time zero leads to an exponential growth in uncertainty about its future state. The Earth's atmosphere is chaotic. Cortical neural networks with recurrent excitatory and inhibitory circuits may be chaotic (Van Vreeswijk and Sompolinsky 1996). Chaos imposes a limit on our ability to predict behaviour. Short-term predictions are to some extent possible in a chaotic system, but not long-term predictions.

A **stochastic system** is one determined by a multitude of independent interacting factors. The output of a stochastic system resembles that of a deterministic chaotic system and it is difficult to distinguish between them. However, a stochastic process does not allow short-term predictions while a chaotic system does. Also, the autocorrelation functions of the two types of system differ. This type of analysis has been applied to binocular rivalry (Section 7.9).

For a general account of time series analysis see Chatfield (1997). Time series analysis can be applied to the **dynamics of nonlinear systems**. This branch of mathematics has not been applied to any of the topics discussed in this book. For an introduction of non-linear dynamics see Kaplan and Glass (1995).

3.7 BAYESIAN INFERENCE

Bayesian analysis, like any other account of perception, starts by defining a **stimulus domain (SD)**. A visual stimulus domain is a set of objects or events defined with respect to specified visible features and selected values of those features, plus the rules of composition and transformation of those features and values. The stimulus domain is not merely a stimulus that happens to be present at a given moment but rather the defined set from which particular stimuli are drawn. The notion of **frame** used in artificial intelligence is defined as a collection of features (attributes) and associated values that describe some entity in the world (Minsky 1975). The value of an attribute may be another frame. Thus frames can form a hierarchical semantic net. A stimulus domain is the complete set of frames defined on a specified set of objects or events (Section 3.1.1a).

The complete static visual stimulus domain is the set of discriminably different scenes. If the scene were broken down into a 1,000 by 1,000 pixel array with each pixel occurring at one of 10 discriminably distinct levels of luminance, the stimulus domain would contain $10^{1,000,000}$ displays. Each display would, in theory, produce a distinct neural response. A world consisting of random sequences of such arrays would contain no redundancy and no structure. There is no point in using or even thinking about such a domain except to measure a person's ability to detect whether two such displays are correlated (Section 16.1).

For a synthetic stimulus domain, such as displays on a computer monitor, the rules of feature composition and transformation may or may not conform to those occurring in natural scenes. Stimulus domains defined with respect to natural objects or events necessarily conform to natural rules of composition and transformation. However, it is only features of objects, not the objects themselves that are in a visual stimulus domain. Any natural object has an unspecifiably large number of features, many of which, such as temperature, atomic structure, and weight, are not visible. Thus, an unspecifiably large number of stimulus domains can be defined over a given set of natural objects. Scientific instruments reveal ever more properties of natural objects but we have not revealed all the properties of any object.

Each of the set of static or dynamic visual displays in a given stimulus domain has a certain probability of occurrence. The distribution of probabilities over the set of displays is the prior probability distribution of the stimulus domain, or **domain prior**, denoted by $p(SD)$.

Consider an observer who makes certain assumptions about a given stimulus domain. The set of assumptions, correct or incorrect, about the set of stimulus objects is **the observer's stimulus domain**, which I denote by OD. If some of the assumptions are correct, the observer knows something about the domain. For an observer with correct and complete knowledge of the domain, $SD = OD$. The observer may have assumptions about the relative probabilities of various scenes in the stimulus domain. The distribution of assumed probabilities over the set of displays is the observer's prior probability distribution, or **observer prior**, denoted by $p(OD)$.

A particular retinal image can arise from more than one object. For example, an inclined circle and a frontal ellipse can produce the same retinal images. Identical images produced by distinct distal stimuli are **essential ambiguities** of the stimulus domain. Image ambiguity can also arise because of noise in the visual system, such as optical opacities, light dispersion, distortions, scotomata, or eye tremor. Some of these sources of noise can be allowed for. For example, the effects of chromatic aberration can be discounted (Section 3.2.2c) and so can the effects of eye tremor. Other forms of noise, such as defects of accommodation, involve loss of information, which cannot be restored. These are essential sources of noise. However, an observer may know what type of information is missing and thereby estimate the degree of uncertainty introduced by an essential source of noise.

The distribution of the relative probabilities of obtaining a given image, I, over all displays in the stimulus domain is the **domain likelihood function**, denoted by $p(I|SD)$. Observers can derive the domain likelihood function only if they possess complete knowledge of the stimulus domain and of the transmission characteristics of the visual system. The likelihood function derived from whatever assumptions the observer makes about these things is the **observer likelihood function**, denoted by $p(I|OD)$.

A visual system has access to only the retinal image and must decide which display in the stimulus domain most probably produced that image. Consider an observer with complete knowledge of the stimulus domain and of the image forming system. Each display of the stimulus domain will have a certain probability of producing a given image. The distribution of these relative probabilities over all displays of the domain is the posterior probability distribution for a given image, or simply the **domain posterior**, denoted by $p(SD|I)$. The probability that a given image has been produced by display X rather than by display Y will depend on the relative probabilities of occurrence of the two displays, $p(SD)$, and on how likely it is that each display could produce that image, $p(I|SD)$. More precisely,

$$p(SD|I) \propto p(I|SD)p(SD)$$

Dividing by a normalizing factor $p(I)$ we obtain the basic Bayesian formula.

$$p(SD|I) = \frac{p(I|SD)p(SD)}{p(I)}$$

The normalizing factor is simply the integral of all products of likelihood functions and priors.

The domain posterior indicates the probabilities that each of the stimulus displays could produce a given image in a given visual system. The final decision about which display is present can be based on either the peak or mean of this distribution. The decision may also depend on the risks or costs involved in making particular errors of stimulus identification. For example, if an object is life threatening, it is better to falsely conclude that an ambiguous image was produced by the dangerous object than to falsely conclude that a harmless object produced it.

An observer with complete knowledge of the stimulus domain and of uncertainties introduced by visual noise can identify displays with the greatest possible accuracy, within limits set by essential ambiguities and noise. Such an observer is an **ideal perceiver** of that domain. Human observers fall short of the ideal because they have limited knowledge of the stimulus domain or of visual noise.

Bayesian inference is based on conditional probabilities. A conditional probability of event A, given a particular value of state B is expressed by $p(A|B)$. For example, we can state the probability a man being bald, given that he is 40 years old. A conditional probability of event A can be expressed over a range of values of state *B* to yield a probability distribution of event *A* over variable *B*. For example, we could state the probability of a man being bald at each age between 1 and 100 years.

Bayesian methods are useful in artificial visual systems in which the designer can fully specify the knowledge competence and the optical system of the machine. But difficulties can arise even in machine vision. If the machine is operating with natural scenes, the stimulus domain and the prior probabilities are difficult to specify. Even a simple and well-defined stimulus domain can be problematical, as illustrated by Bertrand's paradox in which two methods of measuring the probability that a randomly selected chord of a circle is longer than the side of an inscribed equilateral triangle give different results (Bertrand 1889). Albert (2000) showed that similar problems arise in defining the prior probabilities of directions of motion of a point moving at random in 3-D space (see Jepson *et al.* 1996).

A human observer's best Bayesian estimate of a given stimulus is $p(OD|I)$ derived from $p(OD)$ and $p(I|OD)$. For a human observer, the quantities $p(OD)$ and $p(I|OD)$ are, typically, difficult to estimate because the experimenter must know what the observer assumes or knows about the stimuli and about uncertainties in the visual system. But one may question whether human observers have even an approximate idea of the relative probabilities of scenes in any significant stimulus domain. Even if $p(OD)$ and $p(I|OD)$ cannot be quantified, the performance of an observer on a visual recognition task can be compared with that of an ideal perceiver.

Assuming a noise-free visual system, an ideal-perceiver analysis can be used to define the least amount of information in the retinal image required for inferring the specific displays in defined stimulus domains. For example, it has been shown that the complete 3-D metric structure of a scene can be recovered from just three views of four noncoplanar points (Ullman 1979), or from just two views of eight points (Longuet-Higgins 1981) (Section 25.5.1).

Applications of Bayesian statistics to perception are discussed in Knill and Richards (1996). Witkin (1981) used it to develop a model of the perception of shape from texture. Read (2002) developed a Bayesian model of stereopsis.

3.8 CONCEPTS OF GEOMETRY

3.8.1 Symmetry and groups

In the most general sense, a symmetrical pattern is formed by the repetition of an element over space or time according to a simple transformation rule. A linear transformation of an element creates a frieze pattern; a rotation creates circular structures; dilatation creates radial patterns; reflections create bilaterally symmetrical patterns. Most objects contain one or more of these symmetries, and we are particularly sensitive to symmetries in otherwise random patterns (Fisher and Bornstein 1982; Kahn and Foster 1981). A world without symmetries would be without pattern. Group theory is the branch of mathematics that deals with symmetries.

A group is a set of things (elements) and a binary operation that can be applied to any pair of elements in the set. The binary operation can be multiplication, addition, a rotation, a translation, or any other operation that, when applied to any pair of group elements, maps into another group element. Whatever it is, the binary operation between two elements *a* and *b* can be represented by the sign for multiplication, *ab*, or by the sign for addition, *a+b*. A group must satisfy the following axioms:

1. The operation must be **closed,** which means that it must not produce elements outside the defined set. For example, the odd integers are not closed under addition, and therefore do not form a group with addition. The even integers form a group under addition.
2. The set must contain an **identity element,** *e*. The group operation applied to the identity element and any other element leaves that other element unchanged. Thus $ae = a$. For example, *e* is 1 for the group formed by multiplication of the set of natural numbers. Zero is *e* for addition. For rotation, *e* is zero degrees.
3. Each element of the set, has an **inverse element,** a^{-1}. The group operation applied to an element and its inverse produces the identity element. Thus $aa^{-1} = e$. For example, in addition, -2 is the inverse of 2, so that 2 + -2 = 0, the identity element.
4. The operation must be **associative**. A group may or may not be **commutative**. Thus, *ab* may or may not equal *ba*.

The **permutation group** of a set of numbers represents all ways in which the set may be mapped onto itself. The permutation group of all numbers contains all other groups as subgroups (Caley's theorem). A group in which *n* repetitions of the same operation return the group to its original state is a **cyclic group** of **order** *n*. Thus, successive rotations of a line by 60° form a cyclic group of order 6 (C_6). It is a **subgroup** of the group of rotations through 30° (C_{12}). The order of any subgroup of a group of order *n* is a factor of *n* (Lagrange's theorem). If *n* is prime, there are no subgroups.

Groups with the same abstract structure are **isomorphic**. Two problems that may appear different

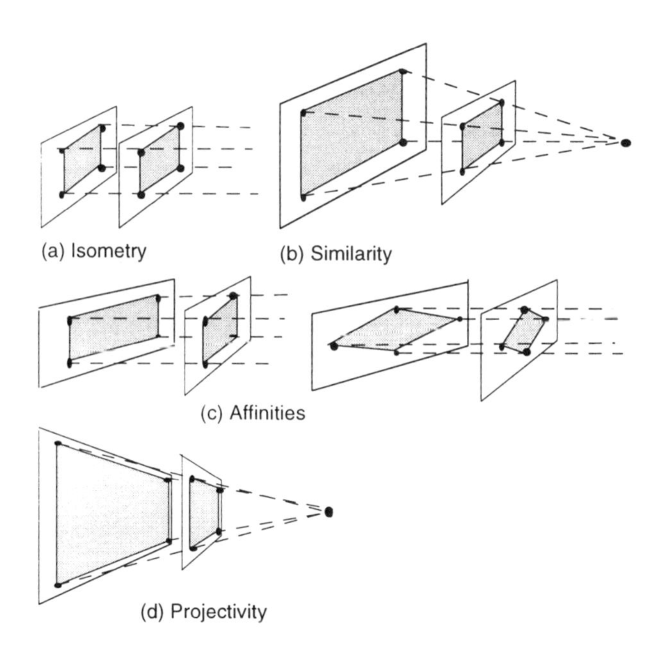

Figure 3.10. Types of transformations underlying geometries.

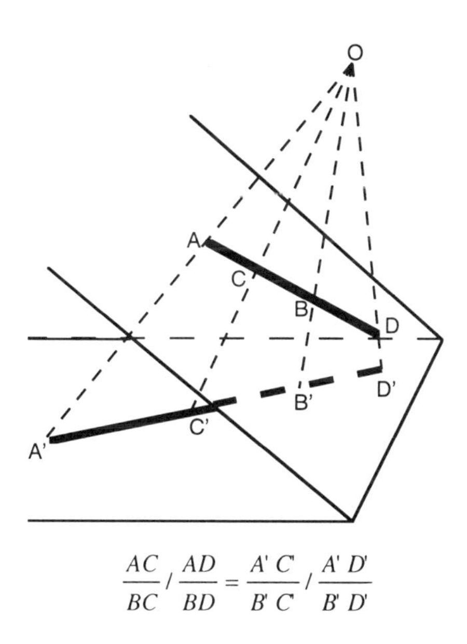

$$\frac{AC}{BC} / \frac{AD}{BD} = \frac{A'C'}{B'C'} / \frac{A'D'}{B'D'}$$

Figure 3.11. The cross ratio.

on the surface may be isomorphic. For example, the problem of why one's reflection in a mirror is reversed left to right but not top to bottom is isomorphic with the following problem. Place two pennies heads up and one penny tails up. By turning the pennies two at a time make them all tails up. Can you continue turning pairs of pennies until they are all heads up? The mirror problem is discussed in Section 4.7.3d.

Group theory is beautifully explained in Grossman and Magus (1964), Budden (1972), and Shub-

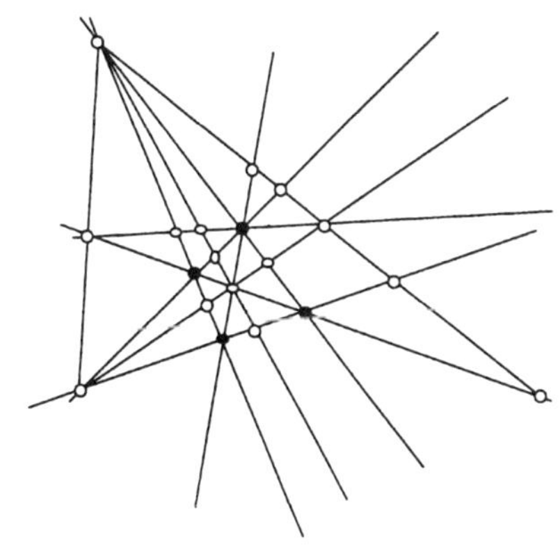

Figure 3.12. The Möbius net.

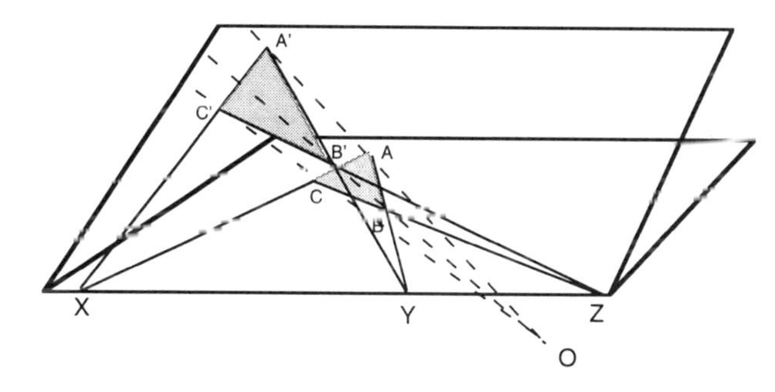

Figure 3.13. Desargues theorem.
Corresponding sides of projectively equivalent figures converge on the same straight line.

nikov and Koptsik (1974).

Groups formed by continuous, or infinitesimal, transformations are known as **Lie groups** (Hoffman 1966). A Lie group is a differentiable manifold. The local differential operators of Lie Groups are known as orbits and occur in orthogonal pairs: horizontal and vertical grids for defining translations, concentric and radial patterns for defining rotations and dilations, and orthogonal hyperbolic patterns for defining hyperbolic rotations. The operators can be commutatively combined by summation and multiplication to form a Lie algebra.

Patterns of optic flow and some processes in spatial vision, such as size and shape constancy, can be described in terms of Lie orbits (see Section 4.4.1).

3.8.2 Types of geometry

Every group of transformations has an associated geometry. Each **geometry** studies the properties of patterns under a defined group of transformations. The transformations must satisfy the group axioms of closure, identity element, inverse element, and associativity. Each of the following transformation groups forms the basis for a geometry.

Congruencies Two shapes are congruent when they occupy exactly the same space. Congruence is a null transformation and a geometry based on congruence allows one to plot shapes on co-ordinates and specify all their internal dimensions but not to move them.

Isometries Isometric geometry admits rigid motions, namely translation, rotation, and mirror reflection. It preserves shape and size, but not position or orientation, as in Figure 3.10a. In these geometries, any two isometric shapes can be brought into exact congruence by translation and/or rotation.

Similarities A geometry based on similarities admits rigid motions plus dilations/contraction. It preserves shape, but not size, position, or orientation, as in Figure 3.10b.

In the visual constancies, we readily recognize an object whatever the position, orientation, or size of its retinal image. These are the transformations that define isometric and similarity geometries, and, in some sense, the visual system must carry out these transformations. Any two similar shapes can be brought into congruence by translation, rotation, and change of size.

Affinities Affine geometry admits rigid motions, dilations/contractions, plus shear. It preserves collinearity, parallels, the ratio of lengths along a line or of segments on parallel lines, and the ratio of areas of any two figures in a plane. It does not preserve angles, size, or the ratios of non-collinear distances, as in Figure 3.10c. A fundamental theorem of affine geometry is that, any triangle in an object plane can be projected onto a triangle that is similar to any triangle in an image plane, given that one is allowed to arrange the two planes at an appropriate angle.

Shadows cast by the sun provide an example of an affine transformation, since the light rays are effectively parallel. Certain types of binocular disparity involve affine transformations of shear, as we will see in Section 20.2.8.

Projectivities Projective geometry admits rigid motions, dilations/contractions, shear, plus non-parallel projection. It preserves collinearity, concurrence, and order of points, but not parallels, lengths, or angles, as in Figure 3.10d. For example, the notion of angles between two lines has no meaning and all triangles and all quadrilaterals are projectively equivalent. Projective geometry grew out of the medieval perspectivist tradition and the development of drawing in perspective in 15th century Italy (Section 2.1.6).

Projective geometry is particularly important in vision because the retinal image is a projection of the visual scene. Consider the general case in which each point in an object plane is mapped by a straight projection line to one point in an image plane, where the two planes may or may not be parallel. When the projection lines are parallel (parallel projection) we have an affine geometry which preserves parallels and ratios of lengths, but not angles. When all the projection lines pass through a common point (central projection), we have a projective geometry which preserves collinearity, concurrence of intersecting lines, and relative order of points, but not parallels, lengths, areas, or angles. Ratios of distances along a line are not preserved but **cross ratios** of distances among four collinear points are preserved, as shown in Figure 3.11. A **perspectivity** is a mapping of a set of points through a centre of projection onto a second set. A **projectivity** is a mapping of a set of points onto another set by one or a sequence of perspectivities. A projectivity can thus be the product of two or more perspectivities.

The fundamental theorem of projective geometry is that a projectivity is determined when any three points on a line are matched to corresponding points in the image. Four points, with no three collinear, completely determine a projective transformation. To prove this, join any four non-collinear points by lines to create new points where the lines intersect. The points have a defined position, since each is collinear with two intersecting lines. By continuing to link up the newly formed points the whole surface is eventually filled with points in determined positions. This is known as the Möbius net (Figure 3.12).

Desargues' theorem states that, for triangle ABC, and its projection $A'B'C'$, the intersections of corresponding sides are collinear. The proof is easy to visualize by applying the following principles:

1. The images of any two coplanar lines intersect.
2. A line and its projected image lie in the same plane as the centre of projection.
3. Two planes intersect in a straight line.

Let triangle ABC and its image $A'B'C'$ projected from point O lie in non-parallel planes, as in Figure 3.13. From (2), corresponding sides of the triangles are coplanar with O. From (1) they must meet. But the triangles lie in distinct perspective planes that, from (3), must meet in a line. Therefore, all pairs of corresponding sides meet in the same line. The intersection of two perspective planes is the perspective axis. The theorem is true for any polygon because any polygon can be divided into triangles. For an introduction to projective geometry see Coxeter (1964). For applications to vision see Section 24.1.2.

Topological transformations Topology is rubber-sheet geometry in which a pattern may be deformed in any way as long as it is not cut or joined to another pattern. It admits rigid motions, dilations, contractions, shear, non-parallel projection, plus elastic deformation. It preserves connected neighbourhoods and edges plus relative order of points and routes between points. Metric properties such as lengths, angles, areas, or cross ratios are not preserved. In topology, one cannot distinguish between a doughnut and a coffee cup! A topological transformation preserves all points and their neighbours. Thus the mapping is **biunique** (one to one) and **continuous**.

As the distance between two points a and b reduces to zero, the distance between their images a$'$ and b' also reduces to zero. If you can go from a to b through a defined set of points in a given order then you can go from a' to b' through the images of the same points in the same order. Points in a surface remain in the surface, and those outside remain outside. Points on the boundary of a surface remain on the boundary. A surface is bounded, or closed, if it is finite but has no boundaries, like the surface of a sphere.

Jordan's theorem states that a simple closed line divides a plane into an inside and an outside. Any two points with the same **polarity** are said to be **connected**. They are said to be simply connected if there are no holes in the surface. Jordan's theorem is implicitly embedded in the figure-ground mechanism of the perceptual system, which allows us to recognize the difference between the inside and outside of a figure. This becomes difficult with complex figure like that shown in Figure 3.14a. People recognize when Jordan's theorem has been violated, as in Figure 3.14b, even though they may not be able to say why. Jordan's theorem is not true in three-dimensions, as is illustrated by the Möbius strip. The surprise generated when this strip is cut and yet remains connected indicates that our perceptual system assumes that Jordan's theorem does hold in three-dimensions.

The geometries listed above form a hierarchy, because the theorems true in any one of them are also true in those earlier in the list. Each group of transformations is a subgroup of those following it, as shown in Figure 3.15. Topology is the most general geometry because its theorems are true in all other geometries.

In any geometry, the features of patterns that remain unchanged are **invariants**. The patterns are said to be equivalent over the set of transformations that are allowed in the geometry, and are called **automorphs**. For example, in isometric geometry, the different orientations of a shape are automorphs. In projective geometry, all shapes that project the same image are automorphs. In each case, the objects form an equivalence class. An equivalence relation is reflexive (any element equivalent to itself), symmetrical (if a is equivalent to b then b is equivalent to a), and transitive (if a is equivalent to b and b is equivalent to c then a is equivalent to c).

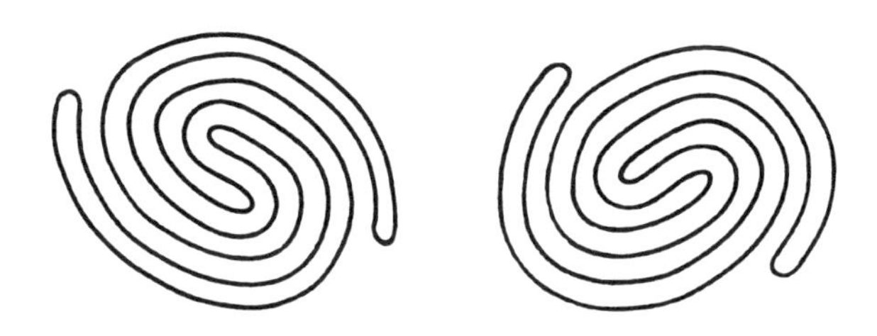

(a) The Minsky spiral. It is difficult to see that the figure on the left consists of two spirals while that on the right consists of only one. (From Minsky and Papert 1969)

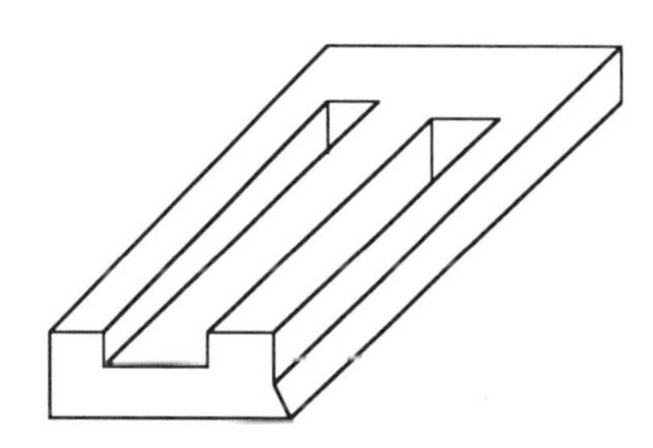

(b) An impossible object.

Figure 3.14. Paradoxical objects.

Conformal transformations A conformal transformation preserves continuity and local angles but does not necessarily preserve collinearity or parallels. Riemannian geometry on curved surfaces can be defined in terms of conformal transformations of Cartesian co-ordinates.

The projection of each half of the retina onto the surface of the visual cortex is basically a conformal transformation. So is the projection of the sense organs in the skin onto the sensory homunculus in the somatosensory cortex. Hemidecussation involves a discontinuous mapping of the nasal and temporal hemiretinas so that the retina as a whole is not mapped topologically. The way organs change in shape during embryology or evolution can often be described by conformal transformations that arise through differential rates of growth (allometric growth). For example, shapes of shells in related genera of molluscs and shapes of primate skulls (See Figure 4.10) can be described by conformal transformations. These transformations were described by D'Arcy Wentworth Thompson in his book *On Growth and Form* (Thompson 1952) and by Julian Huxley in *Problems of Relative Growth* (1932).

Our ability to recognize cartoon drawings and family resemblances between faces suggests that our

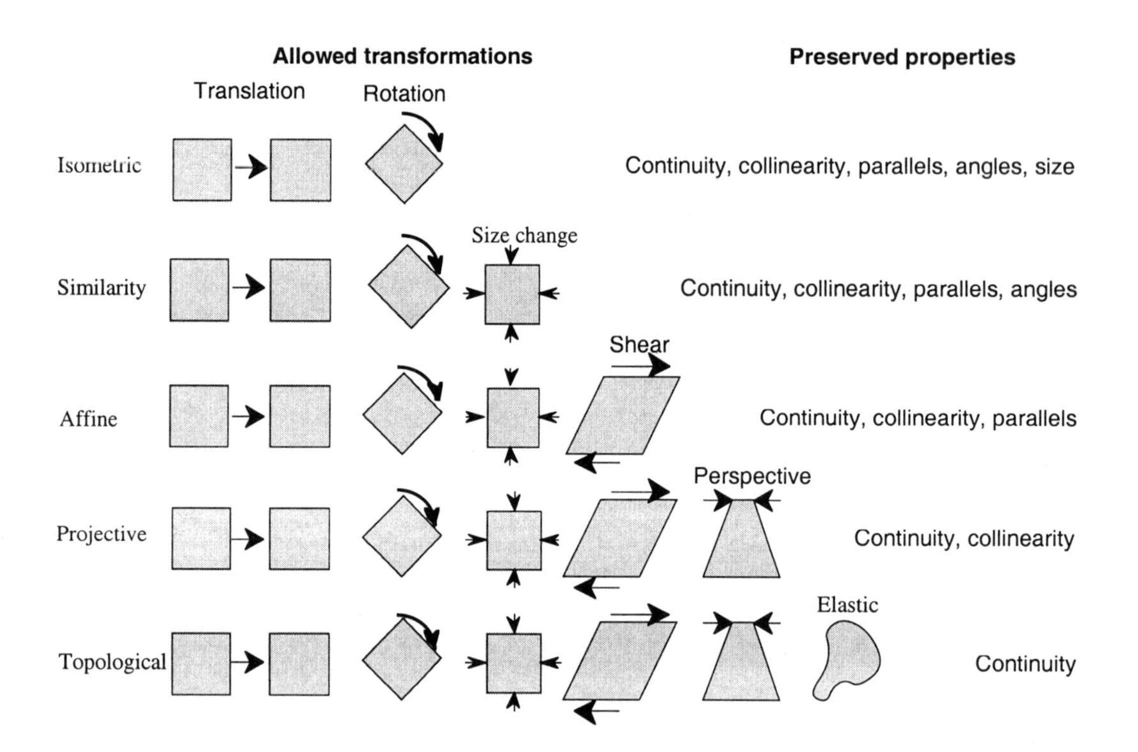

Figure 3.15. Geometries and transformation groups.

visual system is capable of carrying out conformal and topological transformations of visual stimuli.

3.8.3 Analytic geometry

3.9.3a *Homogeneous co-ordinates*

The **Cartesian co-ordinates** of a point specify its distance along each of two or three orthogonal axes. In **polar co-ordinates,** the position of a point is indicated by its distance from a central point and its radial direction with respect to that central point. Polar co-ordinates are used to specify locations of points on the retina (Section 15.2.2) or eye positions (Section 9.2.2c). Co-ordinate systems used to specify binocular disparity are described in Section 20.1. In all these co-ordinate systems there is no way to represent points and lines at infinity in the extended plane of projective geometry. For this purpose we use homogeneous co-ordinates.

Homogeneous co-ordinates were invented by Möbius in 1827. Originally they consisted of the sides of a triangle A_1, A_2, A_3, as in Figure 3.16a. The position of any point is specified by its distance from an apex of the triangle. If, at each apex, a weight is placed proportional to the distance of the point from the apex, then the position of the point is the centre of gravity, or centroid, of the triangle. That is why triangular homogeneous co-ordinates are called **barycentric co-ordinates**. The position of a point may also be specified by its perpendicular distance from each side of the triangle.

Triangular homogeneous co-ordinates are used for specifying trichromatic colour values in the CIE colour system. In this system, the X, Y, and Z axes of 3-D colour space are used to represent the absolute luminances of standard red, green, and blue wavelength components in a given colour specimen. The X, Y, Z axes may be set at 60° rather than at 90° and a plane is placed across them to form an equilateral triangle, as in Figure 3.16b. The height of this chromaticity plane above the origin represents luminosity. Any 3-D point in X,Y,Z space can be projected onto the chromaticity plane by drawing a line through the origin (O) and the point. The co-ordinates of the point in the chromaticity plane are the three perpendicular distances from each of the three sides of the triangle, namely x, y, z, as in figure 3.16b. The values of x, y, and z are normalized to add to 1. The co-ordinates thus represent the relative proportions of red, green, and blue in the colour specimen. In the centre of the triangle, there are equal amounts of red, green, and blue so the colour there is grey.

We can see from the above example, that 2-D homogeneous co-ordinates are a projection of 3-D co-ordinates into a plane. Thus, a point in 2-D homogeneous co-ordinates is specified by three numbers rather than the two numbers of 2-D Cartesian

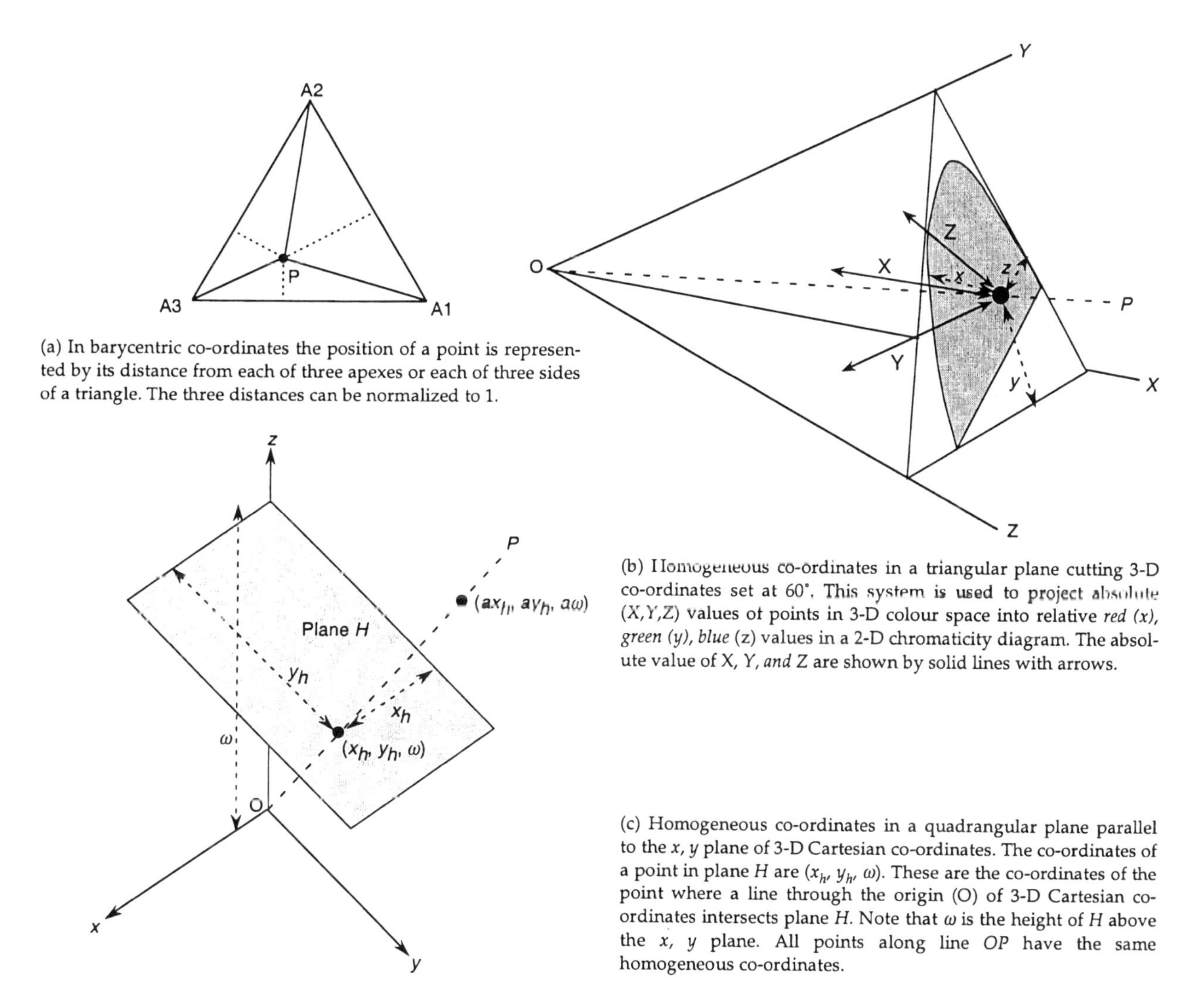

(a) In barycentric co-ordinates the position of a point is represented by its distance from each of three apexes or each of three sides of a triangle. The three distances can be normalized to 1.

(b) Homogeneous co-ordinates in a triangular plane cutting 3-D co-ordinates set at 60°. This system is used to project absolute (X,Y,Z) values of points in 3-D colour space into relative *red (x), green (y), blue* (z) values in a 2-D chromaticity diagram. The absolute value of X, *Y, and* Z are shown by solid lines with arrows.

(c) Homogeneous co-ordinates in a quadrangular plane parallel to the *x, y* plane of 3-D Cartesian co-ordinates. The co-ordinates of a point in plane *H* are (x_h, y_h, ω). These are the co-ordinates of the point where a line through the origin (O) of 3-D Cartesian co-ordinates intersects plane *H*. Note that ω is the height of *H* above the *x, y* plane. All points along line *OP* have the same homogeneous co-ordinates.

Figure 3.16 Homogeneous co-ordinates.

co-ordinates. Homogeneous co-ordinates express only relative values, since all points on a line through the origin have the same homogeneous co-ordinates. They all project to the same point in a plane. That is why they are called homogeneous co-ordinates. In general. **homogeneous co-ordinates are a projection of n-dimensional co-ordinates into an n-1 dimensional space**. Absolute values are lost, leaving only relative values.

I will now describe the homogeneous co-ordinates used to indicate movements or projections of objects in space. Let the co-ordinates of a point with respect to orthogonal 3-D Cartesian co-ordinates be (x, y, z). Let *H* be a plane containing homogeneous co-ordinates. Plane *H* can have any position and any orientation relative to the 3-D co-ordinates. It has already been noted that the chromaticity plane cuts all three axes of 3-D colour space to create a triangular plane. For representing spatial transformations it is usual to set plane *H* parallel to the *x, y* axes and orthogonal to the *z* axis of 3-D space, as in Figure 3.16c. For example, an artist sets the easel orthogonal to the depth dimension of the scene being painted. When the easel is set at an angle, the painting is anamorphic (see Section 2.1.6d). Let the height (position on the *z* axis) of plane *H* above the *x, y* plane be w. For simplicity, plane *H* is confined to the positive quadrant of space. A line through the origin (point O) of the Cartesian co-ordinates intersects plane *H* in point (x_h, y_h, ω). These are the homogeneous co-ordinates of the point. Therefore, as in any polar projection, each point in plane *H* is associated with a line passing through the origin of 3-D co-ordinates. Each line is associated with a point in a given plane.

In 2-D homogeneous co-ordinates a point is represented by the triplet (x_h, y_h, ω) where $x_h = x'\omega$, and $y_h = y'\omega$. Two points are the same in homogeneous co-ordinates if they are proportional. Thus $(x_h, y_h, \omega) = (ax_h, ay_h, a\omega)$. Scaling simply moves plane H along the z axis. We are free to give ω any value. For many purposes it is convenient to make $\omega = 1$.

Any line in plane H is defined by the intersection of a plane through O and plane H. Since any line is the locus of points that satisfy a linear equation

$$ax + by + c = 0$$

we can call the ordered set of three coefficients $[a, b,$ and $c]$ the homogeneous co-ordinates of a line. The line at infinity (the horizon) is the line $\omega = 0$. In general, points and lines in homogeneous co-ordinates are both represented by three numbers. This reflects the duality of points and lines in projective geometry. For every theorem about points there is an equivalent theorem about lines.

In homogeneous co-ordinates, geometrical theorems can be reduced to algebraic theorems. For example, the projection of a point in plane H onto a second plane H' can be represented analytically by a set of linear equations,

$$\begin{aligned} x_h' &= a_1x_h + b_1y_h + c_1\omega \\ y_h' &= a_2x_h + b_2y_h + c_2\omega \\ \omega' &= a_3x_h + b_3y_h + c_3\omega \end{aligned}$$

Theorems in projective geometry become theorems about number triplets defined by these transformations. This branch of mathematics is known as **analytic geometry**. Theorems can be easier to prove when expressed in algebraic form, but one may lose sight of spatial structures that equations represent.

3.9.3b *Transformations in matrix notation*

Suppose one wishes to move, translate, rotate, or shear a shape in a 2-D computer graphics display. In a plane defined by two-dimensional Cartesian co-ordinates, point (x, y) can be carried to point (x', y') by the following transformations:

Translation through Tx and Ty
$x' = x + Tx, \quad y' = y + Ty$

Scaling by Sx and Sy
$x' = x\,Sx, \quad y' = y\,Sy$

Rotation about the origin through θ
$x' = x\cos\theta - y\sin\theta, \quad y' = y\cos\theta + x\sin\theta$

Transformations of a plane figure can be achieved by carrying out these calculations in sequence for each point of the figure. In any transformation, directed line elements (vectors) that map into line elements with the same direction are **eigenvectors** and the scalar multiple that transforms an eigenvector into its image is an **eigenvalue**. Eigenvalues are the n roots of the equations representing the transformation.

These calculations can be carried out more efficiently by expressing the points in homogeneous co-ordinates and the transformation as a multiplication of two matrices. The homogeneous co-ordinate values of the point in its original position form a one-dimensional matrix (x_h, y_h, ω). The coefficients of the transformation equations are arranged in a 3 by 3 matrix that characterizes that transformation. Multiplying the point matrix by the equation matrix gives the matrix representing the transformed values of the point (x_h', y_h', ω'). The rules of matrix multiplication are described in any textbook on matrix algebra. These rules represent, in compact form, operations that solve a set of linear equations.

From now on the subscript h will be dropped and it will be assumed that all points are in homogeneous co-ordinates. Basic movements and rotations of a point in a plane can be represented by 3 by 3 matrices, as follows:

$$\text{Translation} \quad (x', y', 1) = [x, y, 1]\begin{bmatrix} 1 & 0 & 0 \\ 0 & 1 & 0 \\ Tx & Ty & 1 \end{bmatrix}$$

$$\text{Scaling} \quad (x', y', 1) = [x, y, 1]\begin{bmatrix} Sx & 0 & 0 \\ 0 & Sy & 0 \\ 0 & 0 & 1 \end{bmatrix}$$

$$\text{Rotation} \quad (x', y', 1) = [x, y, 1]\begin{bmatrix} \cos\theta & \sin\theta & 0 \\ -\sin\theta & \cos\theta & 0 \\ 0 & 0 & 1 \end{bmatrix}$$

These equations can be abbreviated to:

$$\begin{aligned} P' &= P \cdot T(d_x, d_y) \\ P' &= P \cdot S(s_x, s_y) \\ P' &= P \cdot R(\theta) \end{aligned}$$

Successive translations and rotations of a point can be derived by addition of the transformation matrices. Successive scalings of a point can be derived by matrix multiplication. Addition or multiplication of matrices is known as **composition**. Composition yields a single matrix that represents a set of transformations in a compact form. This provides considerable economy of computation. For small rotations, further economy can be achieved by substituting 1 for $\cos\theta$.

In general, matrix composition is not commutative—the order in which the operations are performed cannot does not affect the result. However, the similarity compositions of translation, scaling, and rotation are commutative.

The method can be generalized to three dimensions. In this case, we project 4-D Cartesian co-ordinates onto 3-D homogeneous co-ordinates. Therefore, each point is represented by four numbers (x, y, z, w), or $(x/\omega, y/\omega, z/\omega, 1)$. Transformations are now represented by 4 by 4 matrices. For example, the matrix for rotation of a point in 3-D space about the z axis becomes

$$R_z(\theta) = \begin{bmatrix} \cos\theta & -\sin\theta & 0 & 0 \\ \sin\theta & \cos\theta & 0 & 0 \\ 0 & 0 & 1 & 0 \\ 0 & 0 & 0 & 1 \end{bmatrix}$$

The projection of point $(x, y, z, 1)$ in 3-D space onto point $(x' y', z', \omega)$ in a plane normal to the z axis and distance d from the origin is represented by:

$$(x', y', z', \omega) = [x, y, z, 1]\begin{bmatrix} 1 & 0 & 0 & 0 \\ 0 & 1 & 0 & 0 \\ 0 & 0 & 0 & 1/d \\ 0 & 0 & 0 & 1 \end{bmatrix}$$

We let $\omega = z/d + 1$, which denotes the magnification of the projection (Section 30.1.1). The co-ordinates of a point in the plane are related to those of a point in 3-D space by:

$$[x', y', z', \omega] = [x/\omega, y/\omega, z/\omega, 1]$$

These procedures are used in computer graphics to move objects within a 2-D display or to create 2-D displays from 3-D scenes. Objects in the scene are typically represented by polygons. This speeds up the calculations because, to achieve a desired movement or projection of an object, one need transform only the apexes of the polygons. For more details on these methods see Hearn and Baker (1986), Foley *et al*. (1990), and Faugeras (1993).

3.8.4 Differential geometry

In differential geometry one defines a basic set of local differential operators, namely translation, rotation (*curl*), dilatation (*div*), and two types of shear (*def*) which can be combined algebraically to describe any continuous transformation of points on a surface or in a volume. It is used extensively in analyzing air flow in aerodynamics.

When we move through a visual scene the retinal image undergoes continuous transformations known as optic flow, which can be described in terms of differential geometry (Werkhoven and Koenderink 1990). These transformations inform us about our self motion and about the 3-D layout of objects in the visual scene (Chapter 25). We are very sensitive to changing patterns of optic flow, which suggests that our visual system embodies continuous space-time transformation operators. There is physiological evidence that visual centres in the brain contain cells that respond selectively to the differential operators of geometry (Section 5 4.6). We do not know which set of primitive transformation operators the visual system uses. Hoffman (1966) suggested that the system uses the differential operators of Lie groups. Koenderink (1990) has proposed that we use the similar set of basic operators of differential geometry (Section 20.2.8). Perhaps different geometrical operators are used for different purposes in different parts of the brain. Differential operators have been applied to the analysis of patterns of binocular disparity (Section 20.2.8). The brain is, among other things, a geometry engine.

Also, in differential geometry, one defines various types of curvature that can be used to describe smooth 3-D objects. Applications to visual perception are described in Sections 21.4.2c and 24.1.6.

3.8.5 Non-Euclidean geometry

In Euclidean geometry, length and angles have meaning. Accordingly, geometries based on isometries and similarities are Euclidean. Affine and projective geometries are non-Euclidean. A second characteristic of Euclidean geometry is that the axes are straight. In the non-Euclidean geometry developed by Riemann, the axes are curved. A space of positive curvature is elliptical, and the co-ordinates of a Riemannian elliptical geometry lie on the surface of a sphere or ellipsoid. A space of negative curvature is hyperbolic, and the co-ordinates of a Riemannian hyperbolic geometry lie on a hyperbolic cone (Coxeter 1961). Einstein's theory of the curvature of space-time induced by gravity is the most famous application of non-Euclidean geometry.

In any geometry, the shortest distance between two points is called a geodesic. In Euclidean space, geodesics are straight lines and all straight lines are geodesics. Parallel straight lines never meet. In Riemannian geometry, geodesics are curved, although not all curved lines are geodesics. For example, the shortest distance between two points on the surface of a sphere is an equatorial circle, or great circle, through the points. That is why aeroplanes navigate along great circles on long journeys. In Euclidean geometry, the angles of a triangle sum to 180° but in Riemannian geometry they sum to more or less than 180°. For example, on the surface of a sphere, a triangle formed by the intersections of three geodesics (great circles) sum to more than 180°. All great circles intersect, like lines of longitude on the Earth, so that geodesics in elliptical geometry cannot be parallel. For the same reason, the images of two straight lines on a spherical retina cannot be parallel. One can draw parallel (concentric) circles on a sphere, but such circles are not geodesics. Attempts have been made to construct a non-Euclidean geometry of visual space (see Section 4.4.2).

4 *Sensory coding*

4.1 STRUCTURE OF SENSE ORGANS

Most human sense organs consist of receptors arranged on a sensory epithelium supported by a mechanical structure. The epithelium and supporting structure have intricately designed mechanical properties, which serve the following four purposes.

Filtering appropriate stimuli A given sense organ is designed to respond preferentially to one type of stimulus energy, called the **adequate stimulus**. However, each sensory end organ is sensitive to several types of energy. For instance, in addition to light, retinal receptors are sensitive to pressure applied to the eyeball, vibration, and electricity. The mechanical properties of the eyeball, the orbital tissues, and the receptors normally shield the eye from forms of energy other than light. Similarly, the mechanical properties of the skull and the cochlea protect the receptors on the basilar membrane of the ear from forms of energy other than sound. The semicircular canals respond preferentially to head rotation because their sensory hair cells project into a fluid-filled annulus, and the utricles respond preferentially to head tilt because their sensory hair cells have heavy crystals attached to them.

Efficient collection of stimulus energy Structures associated with sense organs are designed for efficient transmission of energy to the sensory epithelium. For example, the three ossicles in the middle ear act as a lever that matches the mechanical impedance of air to that of the fluid in the inner ear. In some sense organs, associated muscles direct the receptors to different positions or directions. Mobile hands, eyes, and pinnae are examples.

The mechanical structure of sensory end organs ensures that stimulus energy is efficiently collected. For example, light absorption in retinal receptors is facilitated by the arrangement of pigment molecules on a folded membrane. Furthermore, the outer segment of a retinal receptor forms a waveguide in which light reflects from the inner wall, rather than escapes (Section 5.1.2a). The dense hexagonal packing of cones in the central retina and their alignment with the centre of the pupil also facilitate efficient collection of light.

Spatial distribution of stimuli The simplest sensory systems consist of a single detector that increases its rate of response monotonically as the strength of stimulation increases. The simple eyespots of some invertebrates are essentially of this type. More complex sensory systems, such as the skin, the eye, and the ear, have receptors distributed over a sensory membrane. In the skin, stimuli are applied directly to different locations so that no special devices are required to ensure that the spatial distribution of receptors corresponds to the spatial distribution of stimuli. The ear has a frequency analyzer, consisting of the response of the basilar membrane to travelling waves, which creates an ordered distribution of frequencies over the hair cells. The semicircular canals of the vestibular system are arranged in orthogonal planes so that different directions of head rotation are coded distinctly. The eye uses a lens to ensure a faithful distribution of light to an image. The fact that we have two eyes separated laterally, enables us to form images that contain binocular disparities for the coding of depth. Vergence eye movements bring the images of inspected objects in the two eyes into spatial correspondence.

Trophic functions and protection Many structures associated with sense organs, such as blood vessels and glands, serve trophic functions. Sense organs are designed so that these structures do not interfere with stimulus detection. Thus the cornea and lens do not contain capillaries, but, instead, have highly specialized trophic mechanisms. The capillaries of the retina are kept away from the fovea and are placed so that they do not form sharp shadows under normal viewing conditions. The auditory system filters out the auditory effects of blood circulation. Associated structures or muscular systems may protect the sense organ against injury or sensory overload. The eyelids, epidermis, and the muscles of the middle ear are examples.

4.2 TYPES OF SENSORY CODING

A sensory end organ transduces a spatiotemporal pattern of stimulus energy into a corresponding pattern of electrical activity, which eventually produces neural discharges in a sensory nerve. In the simplest case, the neural discharge is linked more or less directly to a response mechanism. For example, in a simple phototaxic response of a maggot, fluctuation of light intensity as the animal moves its head from side to side produces a corresponding fluctuation in the rate of discharge from a light detector, which causes the head to turn in the direction of least stimulus intensity (Fraenkel and Gunn 1940). The coding process consists essentially of a simple transduction and some amplification. No features of the stimulus are extracted, other than its time varying intensity.

A higher level of complexity is illustrated by the response of the semicircular canals of the vestibular system to head rotation, which causes the eyes to move in a corresponding direction. In this case, the time-varying response from each canal and the spatial distribution of stimulation over the three canals are processed, but nothing else.

Certain features of visual stimulation feed into more or less direct response mechanisms with a minimum of processing. For instance, changes in light intensity evoke the pupillary light reflex, and image blur evokes changes in accommodation. These responses are involuntary and do not involve perception. Like the eye, an automatic camera operates an aperture and focuses.

Visual perception involves much more than a transduction of spatiotemporal patterns of light. All visual features that we perceive are contained in the optic array, and are transduced into spatiotemporal patterns of neural discharge. In that sense we can say that all perceived visual features are coded or represented in the neural discharge. One could say equally well that visual features are coded in the pattern on the television monitor. Used in this way the terms 'coded' and 'represented' tell us nothing about processes underlying perception because they merely indicate that stimulus information has been transduced into a corresponding neural form. A perceptual mechanism, unlike a simple imaging system, involves complex filtering processes that analyze the neural signals into a multitude of simple features such as position, colour, movement, and size. These features are combined into complex and abstract forms, which are stored in memory and compared with other stimulus patterns or with patterns reconstructed from memory.

At a particular time, we process only those parts or features of the visual input to which we are paying attention, even though nonattended stimuli are transduced in the retina and reach the visual cortex. Thus, mechanisms responsible for perceptual awareness can be decoupled from the early stages of visual processing. They may also be decoupled from response mechanisms because we can perceive without responding and we can learn to respond in a multitude of ways to the same stimulus.

In the following discussion of sensory coding, I start by describing sensory transduction processes and then processes responsible for extraction of information from the neural discharge.

4.2.1 Impulse codes

Receptors and bipolar cells are analogue devices that respond to stimulus energy by a graded change in membrane potential, known as a **generator potential** (Section 5.1.2a). Short inhibitory interneurones throughout the nervous system also respond with graded potentials. The **action potential**, or nerve impulse, is the basic unit for long-range transmission of information in the nervous system. An action potential is a transitory change in the standing potential across the cell membrane of a neurone, mediated by transmembrane migration of sodium ions. The potential travels at a velocity of up to 2 m/s along unmyelinated axons, and up to hundreds of meters per second in myelinated axons. Velocity increases with increasing axon diameter. In a given neurone, all action potentials have the same amplitude and, on that account, are said to obey the **all-or-none law**.

Action potentials in the visual system are first formed in ganglion cells in the retina. The axons of ganglion cells carry action potentials to the thalamus, where they are relayed to the visual cortex of the brain. A typical cell in the visual cortex has a single axon and a rich tree of branching dendrites, which carry graded potentials from synapses to the cell body (soma). Action potentials are generated in the soma or in the initial segment of the outgoing axon and can propagate back from the soma into the dendritic tree. We will see later in this section and in Section 10.4.5 that back-propagating action potentials may play an important role in synaptic plasticity (Stuart and Sakmann 1994; Magee and Johnston 1997; Markram *et al.* 1997). There is evidence that slowly rising action potentials mediated by calcium ions are also generated in dendrites.

Because it has fixed amplitude, an action potential is a **quantized signal**. A system that transmits signals as a temporal string of quantized impulses is an **impulse-code**. An impulse-code is used for long-range transmission in the nervous system, just as it is in the digital computer. An efficient impulse-code has the following requirements:

1. It must use quantized signals of fixed amplitude. Such signals are relatively immune to the effects of noise within the system because they are stronger than all but the most severe noise.
2. Signals must get neither weaker nor stronger during transmission. An axon with a diameter of 1 mm has a resistance of about 10^{10} ohm/cm, which is about 10^7 higher than that of metal. If signals in an axon relied on passive electrical conduction, they would fade within a few millimetres. Nerve impulses are boosted to a fixed amplitude at the nodes of Ranvier, which occur along myelinated axons at intervals proportional to axon diameter.
3. Signals must not coalesce during transmission and they must arrive in the same temporal order in which they were sent. Two properties of nerve conduction ensure that these two requirements are met. First, although large-diameter axons conduct more rapidly than small-diameter axons, the velocity of conduction of impulses along a given axon is constant. Second, after an impulse has been transmitted, the cell is unresponsive, or refractory, for a few milliseconds. As a consequence, nerve impulses are discrete and cannot get out of order. Another consequence of the refractory period is that a nerve cell can transmit only up to a limiting frequency—about 500 impulses per second.

Since signal amplitude is fixed, an impulse-code system cannot use amplitude modulation for transmitting information along a single axon. In the single axon, information can be transmitted only in terms of the time of occurrence of impulses or as a temporal frequency or a modulation of temporal frequency.

In computers, an impulse code system is used to construct a digital (number) code. In a defined interval of time, an electrical impulse signifies the binary number 1, and the absence of an impulse signifies the binary number 0. In each case, one bit of information is coded. A given message is conveyed as a string of binary numbers, each of defined length (a word). A master clock defines the temporal position of each binary number in a word. Hence both time and signal strength are quantized in a digital code. In a stored message, each number in a word is placed in a defined location.

Neural impulses generated by particular stimuli do not occur in discrete time intervals within well-defined "words", so there is no digital coding in the nervous system. Thus, a sensory nerve does not transmit information about whether an impulse occurred or did not occur in a predefined time interval. However, synchrony between signals in different neurones is an important attribute of neural signalling (see Section 4.3.2). A sensory neurone attached to a single receptor transmits information about the state of depolarization of the receptor, which is a function of stimulus magnitude. The efficient encoding of this information is ensured because a generator potential is lawfully related to the logarithm of stimulus intensity, and the frequency of nerve impulses in a sensory neurone is a monotonic function of the generator potential.

The probability of response in a sensory neurone plotted against stimulus strength forms a **neurometric function** in the form of a cumulative probability curve, or ogive. The same curve defines a psychometric function obtained when probability of seeing is plotted against signal strength in a psychophysical experiment. This linkage between neural responses and psychophysical responses has been extended to responses of cortical cells and forms the basis of the growing field of **cognitive neuroscience** (Parker and Newsome 1998). The sensitivity of cortical cells tuned to binocular disparity has been compared with psychophysically determined depth threshold (see Section 6.5.1d).

The preceding statements refer to steady-state conditions within the normal operational range of a sensory system. The firing rate of many sensory neurones declines exponentially when a constant stimulus is applied to the receptor. These are **phasic cells** that transmit information about changes in stimulation, rather than about its absolute level. Cells that maintain a response to a steady-state input are **tonic cells**. Furthermore, neighbouring receptors may engage in mutual inhibition, which renders them more sensitive to spatial gradients of intensity than to regions of constant intensity. The matter is further complicated because sensory axons typically receive inputs from several receptors through a complex cellular network—its **receptive field**. In vision, this implies that the response of a retinal ganglion cell may be related to the spatial distribution or motion of stimuli within its receptive field, and not merely to the total energy falling on the receptors. Another possibility, which is discussed later, is that the temporal pattern of firing of a ganglion cell, not merely its frequency, is related to the spatial distribution of the stimulus.

4.2.2 Analogue processing

Axons are designed for long-range transmission of information in the nervous system. However, the most important coding processes do not occur in axons but at synapses, where signal transmission obeys very different rules and information from many sources can be combined using operations such as addition, subtraction, multiplication, and mutual inhibition. These operations are typically analogue, local, and nonlinear. In a computer, logical operations are performed by and-gates and or-gates, which are perfectly structured connections between binary elements. The nervous system does not possess perfect structures of this kind. Whatever logical operations it performs arise from a statistical and changeable weighting of multiple inputs at synapses (Mel 1994).

Signal transmission at synapses is graded rather than quantal. Each cortical neurone is studded with up to 200,000 synapses arranged along a complex of branching dendrites. Neurones exhibit a bewildering variety of dendritic structures. Analogue processes are involved in conveying signals along the dendrites to the cell body of the typical cortical neurone. However, spikes mediated by calcium ions also occur in dendrites. These are slower than the sodium spikes of the axon. Sodium action potentials back-propagated along axons of cortical pyramidal cells facilitate subthreshold calcium action potentials initiated by dendritic excitatory inputs (Larkum *et al.* 1999). This allows for coupling of inputs arriving at different layers of the visual cortex.

Short interneurones transmit signals in an analogue fashion rather than by neural impulses. They serve to inhibit rather, than excite, cells they impinge upon. Excitatory and inhibitory interactions at a given synapse depend on the relative locations of inputs on the dendrites, on ionic currents released by stimulation, and on the magnitude of membrane conductances.

We will see in Section 5.4.4e that glial cells, which outnumber neurones, also take part in the transmission process (Nedergaard 1994). Other influences, known as **neuromodulators**, affect the properties of neurones or synapses, rather than evoke neural impulses. For example, neurohormones circulating in the blood or cerebrospinal fluid regulate neuronal activity at target sites throughout the nervous system. Some neural centres such as the locus coeruleus project axons carrying neuromodulatory signals to most parts of the brain. Other neuromodulatory signals arise locally within particular neural circuits (Katz and Frost 1996).

Information can be transmitted at a much higher rate by an analogue neural signal than by neural spikes (de Ruyter van Steveninck and Laughlin 1996). Also, local analogue processes occurring in parallel within a large network can be performed much more rapidly than equivalent in-series digital processes executed in computers (Koch *et al.* 1986). Analogue processes are dedicated to a particular task, whereas digitally programmed processes in computers may be modified. Lack of flexibility is not a disadvantage in a system which subjects all inputs to the same limited range of transformations, and for which speed of processing is important. However, we will see in the next chapter that parallel processes within neural networks possess some flexibility, even in the primary visual cortex. Flexibility can be achieved by growth processes, by slower serial processing performed at a later stage, and through the mediation of attention mechanisms. Stimulus-contingent changes at synapses allow for some short-term flexibility in processing and for long-term learning (see Section 5.4.5). Both the extent and distribution of back-propagated action potentials from soma to dendrites are more restricted after several action potentials have occurred in sequence (Spruston *et al.* 1995; Yuste and Denk 1995).

There is a bewildering variety of dendritic patterns in cortical neurones (see Figure 5.16). Presumably, the different patterns serve distinct functions. For example, there are cells in the mammalian brainstem that act as coincidence detectors for sound localization. They respond maximally when inputs from one ear arrive at one cluster of dendrites at the same time as inputs from the other ear arrive at a distinct cluster of dendrites. The spatial separation of the two dendritic clusters increases signal strength because signals arriving on neighbouring dendrites show non-linear saturation, and hence have a high threshold, while those arriving on distinct dendrites sum linearly and have a low threshold (Agmon-Snir *et al.* 1998). Stellate cells and pyramid cells in the visual cortex have distinct dendritic branches. A similar mechanism may render these cells sensitive to coincidence of stimuli within or between the eyes.

Thus, the real work of the nervous system is achieved by nonlinear analogue processes within clusters of synapses and within the dendritic branches of each neurone. The axonal impulse-code system is simply a way to get information from one synaptic cluster to the next without loss. Computers are quantal and digital throughout, both for transmission of signals and for signal processing.

4.2.3 Monopolar and bipolar detectors

Some stimulus dimensions have direction as well as magnitude. They are said to be oppositional, or **bipolar stimulus dimensions**. For example, visual motion along a given axis is a bipolar stimulus feature, since the motion can be in either of two directions. Some receptors are **bipolar detectors** because they produce a bipolar response to bipolar stimuli. They are thus able to code both the magnitude and direction of bipolar stimuli. For instance, a hair cell in a semicircular canal of the vestibular system hyperpolarizes when the head rotates one way and depolarizes when it rotates the other way. Since there are no negative nerve impulses, neurones can produce bipolar responses only with respect to a resting state of maintained discharge. For example, the afferent fibres from the semicircular canals maintain a steady discharge that declines when the head turns one way and increases when it turns the other way.

Light intensity is a monopolar stimulus since there is no negative light. Retinal receptors do not generate bipolar receptor potentials—they all hyperpolarize when stimulated. The stimulus feature of light-on as opposed to light-off can be considered bipolar. Light-on and light-off responses are achieved in two distinct sets of bipolar cells into which the receptors feed. They are called bipolar because of their structure, not because of their function. Functionally, each bipolar cell is **monopolar** and acts as a half-wave rectifier, since it responds to alternate half cycles of a stimulus oscillating in luminance. A bipolar cell does not respond to a steady stimulus. A cell producing distinct signals for both light increase and light decrease would have to maintain a constant discharge when there is no change in illumination. Similarly, most simple cells in the visual cortex have little or no maintained discharge and act as half-wave rectifiers with respect to the spatial distribution of dark-light stimuli falling within their receptive fields. Each cell responds to only one half of the cycle of stimulation produced by a dark-light grating moving over the retinal receptive field of the cell. The full range of stimulation is covered because the cells occur in pairs with opposite spatial phase (Section 5.4.6).

There are several reasons why it is more efficient to have two sets of oppositely tuned monopolar detectors than a single bipolar detector:

1. Monopolar detectors do not need a maintained discharge.
2. Any disturbance of maintained discharge upsets the calibration of a bipolar system. For instance, alcohol upsets the maintained discharge of vestibular receptors, with well-known consequences.
3. Two monopolar detectors have double the dynamic range of a bipolar detector.
4. Outputs of monopolar detectors can be combined to produce a difference signal or ratio signal that is independent of changes in a stimulus feature to which both detectors are equally sensitive. For instance, if the output of a detector tuned to one orientation is subtracted from that of a detector tuned to another orientation, the resulting signal is independent of the overall luminance of the stimulus.
5. Nonlinearities in each detector may be cancelled when two opposed signals are combined. An example of this is provided in Section 29.3.7.

There is a price to pay for monopolar detectors; there must be twice as many monopolar as bipolar detectors, and there must be an additional opponency mechanism that combines the two half-wave rectified signals into a unitary signal. The detectors in the semicircular canals of the vestibular system are functionally bipolar and do not rectify the head-acceleration signal. Presumably, for them, the need to economize on the number of detectors and subsequent analysis outweighs the disadvantage of maintaining a discharge when the head is not moving. Also, these detectors respond to only one stimulus feature and have no need for an opponent mechanism to discount the effect of stimulus intensity. Simplicity and the consequent speed of processing are important for vestibular detectors because they control eye-stabilizing reflexes and postural responses, which must be rapid.

4.2.4 Primary coding

I refer to a coding process evident at the level of the individual receptor as a **primary coding process**, and to a stimulus feature coded by a primary sensory process as a **primary feature**. A primary coding process can involve changes in the frequency of firing in individual receptors or a specialization of different detectors for different stimulus values. Thus, in the eye, modulation of stimulus intensity is a primary feature that involves frequency coding. Position and colour are primary features that involve sets of differentially tuned receptors. Beyond the basic coding process, primary features become elaborated by neuronal processing. The basic types of primary coding process are as follows:

4.2.4a *Frequency coding*

The physical intensity of a stimulus is the energy falling on unit area of a sensory surface in unit time. When an adequate stimulus varies in intensity but not in any other respect, the response of a receptor is some monotonic, saturating function of physical intensity, modified by effects of adaptation. For a given class of receptors, the steepness of the tuning function and the position of the tuning function along the stimulus-intensity axis vary from receptor to receptor. Typically, each local group of receptors feeds into a higher level neurone and constitutes the receptive field of the higher-level neurone. The response of a higher-level neurone to a stimulus of a given area and duration depends on the total physical energy falling within its receptive field, and within the time interval over which the energy is integrated. In the psychophysical domain, the area relationship in vision manifests itself as **Ricco's law**, which states that, for a stimulus of fixed short duration and below a certain area, the detection threshold is inversely proportional to the product of the area and intensity of the stimulus. The time relationship manifests itself as **Bloch's law**, which states that, for a stimulus of a fixed area and below a certain duration, the threshold is inversely proportional to the product of time and intensity.

The effectiveness of a stimulus of a given intensity varies as a function of other attributes of the stimulus. For example, for a given physical intensity of light, the frequency of response of a retinal cone is a bell-shaped function of the frequency (wavelength) of the light. Also, the frequency of response of an auditory receptor is a bell-shaped function of the frequency of a sound of a given intensity. Finally, the effectiveness of a stimulus also varies as a function of its position with respect to the receptor or with respect to the receptive field of a detector at a higher level in the nervous system. The variation in the response of a sensory detector as a function of changes in the position of a small stimulus within its receptive field is known as the **point-spread function** of the detector. The direction from which a stimulus impinges on a receptor may also affect its power to evoke a response. For instance, a retinal cone is more sensitive to light that arrives from the direction of the centre of the pupil than to light arriving from other directions; a phenomenon known as the **Stiles-Crawford effect**.

The temporal features of the response of a single detector can code stimulus onset and offset, stimulus duration, and changes in stimulus intensity over time. It has been suggested that temporal modulation of the response train of a single receptor also plays a part in pattern recognition, as we shall see in Section 4.3.1.

4.2.4b *Response variability*

The frequency of response of a neurone anywhere in the visual system is not precisely the same when a given stimulus is repeated. Any such variance in response is called **intrinsic noise**. The ability of any detector to detect a signal depends on the ratio of signal strength to variance due to noise—the **signal-to-noise ratio**. There are two basic types of intrinsic noise. **Additive noise** is a constant level of noise added to signals of any strength. For additive noise, the signal-to-noise ratio declines as signal strength is reduced and may exceed signal strength in the neighbourhood of the stimulus threshold. **Multiplicative noise** is proportional to signal strength, so that the signal-to-noise ratio remains constant as signal strength is varied. There has been some debate about whether multiplicative noise is an intrinsic property of the way single neurones respond or whether it is due to properties of neural networks. Kasamatsu *et al.* (2001) found that the signal-to-noise ratio of cells in the visual cortex is improved when an elongated stimulus is accompanied by neighbouring aligned stimuli. Thus, multiplicative noise in the visual cortex is due, at least in part, to the balance of excitation and inhibition within the network of connected neurones (Section 5.4.7).

4.2.4c *Dynamic range*

Luminance changes by 10 log units between starlight and sunlight but visual receptors have a dynamic range of only about 3 log units. Furthermore, a ganglion cell can transmit up to a frequency of only about 800 impulses per second. At least four mechanisms serve to allow the visual system to compensate for the limited dynamic ranges of the receptors and transmission system.

1. Pupillary control of the amount of light entering the eye.
2. Nonlinear compression of response sensitivity at high luminance. This is due to such causes as saturation of bleached visual pigment and of neural responses.
3. General or local adaptation of the sensitivity of the whole response range as a function of local or mean intensity level. This process is equivalent to placing a neutral density filter over the whole or part of the visual scene.
4. Subtractive adaptation due to lateral inhibition. This reduces responses to even areas of luminance compared with responses to spatial or temporal contrast boundaries. This type of visual process is responsible for simultaneous and successive contrast.
5. Segregation of "on" and "off" ganglion cells doubles the dynamic range of the system.

4.2.4d *Labelled-line codes*

A labelled-line code depends on the type of sensory cell stimulated. For instance, three types of cone code long, medium, and short wavelengths of light, and different types of receptors in the skin code pain, pressure, and temperature. In a general sense, the different sensory modalities such as vision and audition are labelled lines, in that each responds to distinct stimuli and evokes distinct sensations. The concept of labelled line neurones is related to the concept of **specific nerve energies** associated with Johannes Müller (1843) although the idea originated before Müller (see Boring 1950, p. 80 and Rose 1999). Most stimulus continua, such as motion, orientation, and disparity, are coded by sets of detectors, with the members of each set being sensitive to a particular range of the continuum. These detection systems may be called labelled-line systems, and the distinct sets of detectors for a given stimulus feature are known as sensory **channels**. All labelled-line detectors have a bandpass tuning function, in which the firing rate at first increases and then decreases as the stimulus is varied over the sensitivity range of the detector. The tuning function is thus two-valued, which introduces an essential ambiguity into the response of any single labelled-line detector, in contrast to the unambiguous single-valued tuning functions of responses to stimulus intensity. Another reason for ambiguity in a labelled-line detector is that a change in response can be due to either a change in stimulus intensity or a change in the stimulus dimension to which the detector is tuned. There are two types of labelled-line coding.

4.2.4e *Topographic coding*

Each sensory modality has one and only one **topographic labelled-line system** by virtue of the fact that each receptor has a distinct position on the sensory epithelium and sends signals to a particular location in the central nervous system. Stimulation of a given receptor evokes an impression of a given value on a sensory continuum, which is the **local sign** of the receptor. The local sign of a sensory cell is a function of how it is connected in the central nervous system, rather than of its position in the sensory membrane. However, since sensory receptors do not change their positions, one may talk about them coding location on the sensory surface. A sensory system can devote its topographic system to only one sensory feature, which is the local sign, or topographic, feature for that sensory system. I refer to this as the **local-sign exclusion rule**. For both vision and touch, position is the local-sign feature, for audition it is frequency, and for the utricles it is direction of head acceleration.

The retinal image has a precise geometry, which can be encoded only if receptors maintain a fixed spatial order. Once the spatial attributes of an image have been coded into nerve impulses in the optic nerve, a fixed spatial ordering of those nerves is no longer required. It is only required that the specific connections that each neurone makes be preserved until spatial information is encoded into some other form, or evokes a response. The fact that the visual hemifields are represented in different hemispheres or that the mapping of the retina onto the cortex can be described as a conformal logarithmic function (Section 5.4.2) has no significance for spatial coding. Within the nervous system, it is only the connections that cells make with other cells that count; the geometry of the disposition of cells on the cortical surface is irrelevant for coding of spatial information, although it is important for economizing on the lengths of neural connections.

There has been a long debate about whether local signs are innate or calibrated through experience (see Section 2.1.8c). The debate centred on the restricted idea of a one-to-one correspondence between retinal location and perceived visual direction. But a richer set of concepts emerges when one considers the topology of visual space. Consider the following propositions regarding retinal local sign.

1. A point stimulus evokes a sensation of only one point. N point stimuli evoke a sensation of N points when they are sufficiently well separated.
2. Closely adjacent stimulus points are perceived as connected.
3. The relative order of stimulus points is preserved in the percept.
4. The greater the separation between stimulus points, the greater the perceived distance between them.
5. Collinear stimulus points are perceived as collinear.
6. A stronger version of proposition 5 is that images arising from straight lines (and only such images) are perceived as straight. This carries the corollary that a line perceived as straight when imaged on one part of the retina is perceived as straight when imaged on any other part.
7. The relative directions of stimulus points are preserved in the percept.

The first five propositions probably relate to innate mechanisms that are not modified by experience. Even Helmholtz, the arch empiricist, believed proposition 3 depends on innate processes. The last two propositions are not necessarily true and are modified by experience. Thus exposure to curved lines induces a percept of curvature in straight lines. Geometrical illusions attest to the fact that propositions 6 and 7 are subject to short-term experience.

4.2.4f *Nontopographic coding*

All labelled-line systems in a given sensory modality, other than the local-sign system, are **nontopographic labelled-line systems**. In a nontopographic system the detectors differ in terms of their tuning to a stimulus attribute (their filter characteristics), rather than in terms of their position on the sensory epithelium. One such system in the human retina is the three-cone colour system in which the receptors act as differential filters for wavelength, which is not a position-dependent stimulus feature. Another such system in the retina is the on-bipolar cell system and the off-bipolar cell system (Section 5.1.2c).

A complete set of colour channels must be present at each location of the local-sign system, which itself has only one complete set of channels. This imposes a severe constraint on the number of colour channels. If a feature other than local sign were to be coded by many channels, only a subset of these channels would be activated at a given time by a given stimulus. This would degrade the spatial resolution of the local-sign system. Also, the sensory epithelium would have to be very large. For instance, if there were many colour channels, the retina would become impossibly large. It was this logic that led Thomas Young (1802) to propose that there are only three types of receptor for colour, and that they have widely overlapping tuning functions. The only way to escape this constraint is to have an eye with one part devoted to colour coding and a distinct part devoted to spatial resolution. The mantis shrimp has this type of eye. One set of low-density ommatidia contains at least ten types of colour pigment, which filter the incoming light, and which are thus capable of resolving the chromatic spectrum. Two sets of high-density achromatic ommatidia resolve the image spatially. The three sets of ommatidia converge on the same location in space. See Section 14.2.4 for more details of these remarkable eyes.

All primary sensory features must be coded in terms of either temporal frequency or local sign or in terms of receptors with distinct tuning functions for a nontopographic feature. In the visual system, this means that receptors can signal only intensity, temporal changes in intensity, oculocentric direction (local sign), and wavelength. Coding for other features must be deferred to beyond the receptors. I now discuss deferred, or secondary, coding.

4.2.5 Secondary coding

Coding processes that require the cooperative activity of two or more receptors or of neurones at a later stage of processing are **secondary coding processes**. Visual features such as motion, orientation, and disparity are **secondary features** because they are spatial or spatiotemporal derivatives of the sensory input, and are not represented in the activity of single receptors. Colour cannot be derived from the spatial or temporal properties of the local-sign system and therefore requires its own set of primary receptors. In addition, colour has its own secondary features, such as colour opponency and constancy. Neither motion nor disparity can be coded by a single receptor. Orientation could be coded by oriented receptors but this does not happen in practice, since all visual receptors have a circular cross section.

We are able to see the motion, orientation, and depth of stimuli at each location in the visual field. There must therefore be a complete set of secondary detectors for each of these features, serving each location of the visual field. However, secondary features do not require dedicated retinal receptors because the outputs from a given set of receptors can be configured in different ways to code different secondary features. The formation of distinct channels for secondary features can occur in the retina or can be deferred until inputs reach the visual cortex. In a complex visual system, it helps to defer coding of secondary features because there is more room in the brain than in the eye for the neural processes required for channelling these features. All detectors for secondary features are labelled-line, multichannel systems.

4.2.5a *Multiple-channel systems*

Consider first the possibility that all detectors sensitive to a given feature have the same broadband tuning function that spans the range of detectable stimuli. If the detectors are to distinguish between many values of the feature, the function relating frequency of firing to values of the stimulus must be single valued. The tuning function could be S-shaped, showing saturation at high stimulus values. We have already seen that only detectors of changes in stimulus intensity are of this type. The tuning functions of all primary or secondary labelled-line detectors, for stimulus features other than intensity, show a bandpass characteristic. This means that they have bell-shaped tuning functions with a peak and two flanks, in which the response falls to zero. For instance, a retinal cone shows a bandpass response when the wavelength of a light of fixed intensity is varied, or when the position of the stimulus is varied. Also, a hair cell in the auditory cochlea shows a bandpass response as the frequency of a tone of fixed intensity varies. However, for a bandpass tuning function, pairs of stimuli lying on opposite flanks of the tuning function generate the same response, so the response is ambiguous.

The universal solution to this problem is to have sets of detectors tuned to a particular bandwidth within the feature continuum, and with the tuning functions of the different sets overlapping. Detectors tuned roughly to the same bandwidth in a sensory continuum are often referred to as a sensory subchannel. A stimulus with a particular value of the relevant feature evokes a particular combination of responses in a particular set of overlapping subchannels. The combined response thus provides a unique input for each single value of the stimulus continuum. For example, each wavelength of the visible spectrum evokes a unique response in the set of red-green-blue receptors. However, the use of the combined response of a multichannel system does not overcome all ambiguity because different mixtures of stimulus values can evoke the same combined response—the system is metameric. For example, the same colour impression can be created from different combinations of wavelengths and luminance. There is no cure for this type of ambiguity in a given detector system. However, metamerism in a detector for one spatial feature may be compensated by another spatial feature. For example, different orientations of a set of lines may not be resolvable by the orientation-detection system but may be resolved by the position-detection system.

The response of a detector in a multichannel system is ambiguous also because changes in stimulus intensity are confounded with changes in the stimulus feature to which the detector is tuned. The ambiguity can be resolved by using the difference in response of two neighbouring detectors (Section 4.2.8), or by normalizing the response of a given cell by dividing it by the mean response of neighbouring cells (Section 5.4.6).

There is another reason why it is better to cover a wide stimulus continuum with several detectors rather than with only one. The design characteristics of an efficient detector for one end of a feature continuum differ from those of a detector at the other end. For example, a visual detector for high spatial frequency (fine patterns) is fundamentally different from a detector for low spatial frequency (coarse patterns). To take another example, a detector sensitive to vertical lines has a vertically oriented receptive field, which renders it relatively insensitive to horizontal lines. Even for frequency-coded detectors of stimulus intensity, the stimulus range is typically partitioned out among several detectors with S-shaped tuning functions at different positions along the stimulus-intensity continuum. Thus, rods operate at low-intensities and cones operate at high-intensities. The following discussion is concerned with only multichannel, labelled-line systems in which each channel has a bandpass tuning function.

A single channel of a multichannel, labelled-line system can determine the value of a stimulus with a precision no finer than half the width of its tuning function. This presents no problem if there are many channels, as in the million-channel local-sign system of the eye. However, according to the local-sign exclusion rule, there can be no more than one multichannel local-sign system in a given sense organ. Having many narrowly tuned channels is not the only way to achieve good discrimination within a given stimulus continuum. The other way is to have

Table 4.1. Types of sensory code.

Primary coding (evident in the single receptor)
Frequency code (monotonic tuning functions)
Examples: Intensity and intermittency
Labelled-line code
Topographic
Example: Local sign/visual direction
Nontopographic
Example: Colour
Secondary coding (distributed at the receptors but later channelled in local labelled-lines).
Examples: Size, orientation, movement, disparity.
Higher-order coding (non-local spatial or temporal derivatives of a primary or secondary feature)
Examples: Shape, optic flow, disparity gradients.
Composite coding (dedicated detectors for combinations of features in the same modality).
Examples: Motion-in-depth, parallax, constancies.
Intersensory coding (mixed features from different modalities).
Examples: Vertical, headcentric visual direction.

a few broadly tuned channels spanning the stimulus range, with overlapping tuning functions. There is no theoretical limit to the precision with which the value of a stimulus can be detected by a two-channel system if the outputs of the channels to the same stimulus are compared.

The Germans made use of this fact during the Second World War. Their bombers navigated down the locus of intersection of two overlapping radio beams. Any departure from this locus was immediately detected by a change in the strength of the signal from one beam relative to that from the other.

For this type of system to work, the tuning functions of the channels must overlap, but the output of each channel must retain its identity. Any detector system with only a few overlapping tuning functions at each location retains the power to detect a change in the value of a single stimulus but loses the capacity to resolve the values of the feature for two or more stimuli presented at the same time to the same set of detectors. Such systems are **metameric sensory systems**. Most sensory systems are metameric, including the one that codes binocular disparity. The properties of metameric systems are discussed in more detail in the following sections. Types of sensory coding are listed in Table 4.1.

4.2.5b *Channel number and channel homogeneity*

One can ask how many subchannels span the total bandwidth of a detection system for a given sensory feature. For colour there are three, and for local sign there are one million. Estimates of the number of channels for motion, spatial scale, and binocular disparity vary between three and about 20. A related question is what proportion of the bandwidth of a detection system does the mean bandwidth of detectors occupy. This proportion is not the reciprocal of the number of channels, since channels overlap.

Another question that is often confused with the question of the number of channels serving a visual feature is the question of **channel homogeneity**. Consider a detection system in which, at each location in the visual field, there is a set of detectors that covers the detectable range of the stimulus feature. Tuning functions are homogeneous when they are similar over the area of the retina where the detectors occur. For example, the three types of cone have similar chromatic absorption functions, wherever they occur in the retina. The discrete nature of chromatic channels is revealed by humps and dips in the hue-discrimination function, since hue discrimination is best where neighbouring colour channels overlap (Section 4.2.7). Hue-discrimination functions are derived from stimuli subtending 2°. However, since the visual pigments are the same over wide areas of the retina, the peaks of the discrimination function should occur at the same wavelengths wherever the stimulus is placed. A sensory system can also be homogeneous with respect to the distribution of the different detector types over the retina. The colour system is inhomogeneous in this respect since the proportions of cone types vary from one location to another.

Suppose that motion is served by three motion-sensitive channels at each location of the visual field. The motion-detection system does not have homogeneous tuning functions because motion detectors based on small receptive fields in the central retina are sensitive to lower velocities than motion detectors based on large receptive fields in the visual periphery. Although motion detection may be a three-channel system in the sense that three detectors span the bandwidth of the system at any location, the bandwidths vary over the retina. A sensory system can be discrete locally, with a few distinct channels, but show a continuum of detector types over the retina as a whole. In a system of this type one would expect to find undulations in the discrimination function for small stimuli that stimulate a local discrete set of detectors, but a smooth discrimination function for large stimuli that stimulate a wide range of detector types over a large area. Colour may be the only visual feature that is coded by channels with homogeneous tuning functions. Visual features derived from spatial coding, such as motion, spatial scale, and orientation are grossly inhomogeneous because of the steep increase in the size of receptive-fields as one moves away from the fovea. For the same reason, channels with inhomogeneous tuning functions code binocular disparity. This question is discussed further in Section 6.5.1.

4.2.6 Feature detectors

4.2.6a *Definition of feature detectors*

A **feature detector** is a neurone whose firing rate varies as a function of a change in a defined stimulus feature. The cell's tuning function to a particular stimulus feature is measured by recording its firing rate as the feature is varied for a stimulus presented in the cell's receptive field (Section 5.4.1). A stimulus feature is **channelled** at a specific level in the nervous system when detectors for that feature exist at that level. In other words, the feature is represented explicitly at that level (Marr 1982). Light intensity is channelled at the receptor level, since a change in light intensity changes the response of receptors. Other stimulus features channelled at the receptor level are position, flicker, stimulus duration, and wavelength. Contrast and colour opponency are channelled in the retina at the ganglion cell level. Motion is channelled at the ganglion cell level in the frog and rabbit but, in primates, it is not channelled until the visual cortex. Orientation and disparity are also first channelled in the visual cortex—in the retina they are **unchannelled, or distributed**. It is inefficient to have more than the minimum number of features channelled at the receptor level because, if many specific features were encoded there, most receptors would be inactive most of the time and this would degrade acuity. It is better to have a set of receptors with more or less the same broad response characteristics so that gradients of light intensity may be detected at all locations. It is also undesirable to channel many features at the ganglion cell level because this would increase the size and complexity of the eye. It is better to channel secondary features in the central nervous system, where cells can be specialized for detection of specific secondary or higher-order stimulus features.

4.2.6b *Sensitivity of single cells*

The probability of response of single neurones in the visual cortex to well-defined stimuli of variable strength yields a **neurometric function** (Parker and Newsome 1998) analogous to the psychometric function derived from psychophysical experiments.

The threshold of the most sensitive single cells in the visual cortex corresponds to the behaviourally determined contrast threshold (DeValois and DeValois 1988; Barlow *et al.* 1987). The most sensitive cells in the visual cortex of anaesthetized cats signal differences in spatial frequency or orientation that are only slightly larger than the difference thresholds determined psychophysically (Bradley *et al.* 1987). Similarly, the orientation discrimination of monkeys determined psychophysically is similar to the sensitivity of cortical cells to changing orientation obtained in the alert monkey (Vogels and Orban 1990). In monkeys, the threshold of cells in MT and MST to the degree of coherent motion in a display of random dots is similar to the psychophysically determined threshold (Section 5.7.3b).

These facts do not mean that single cells unambiguously code variations in one stimulus feature, as we will now see.

4.2.6c *Adaptation and discriminability*

The sensitivity of retinal receptors is an inverse function of the level of retinal illumination. In addition, cells in the visual cortex exhibit contrast adaptation, in which responsivity is adjusted to the prevailing level of contrast. This prevents response saturation to high-contrast stimuli and increases dynamic range.

In addition, complex cells in the visual cortex of the monkey show pattern-specific adaptation. After a brief presentation of a grating the cells become less sensitive to gratings of similar orientation (Müller *et al.* 1999). This loss of responsivity to a persisting stimulus conserves metabolic energy and increases discriminability for orientations in the neighbourhood of the adapted orientation. Stimulus-specific adaptation improves discriminability because it reduces the correlation between signals arising from stimuli with similar orientations.

4.2.6d *Coding density and ensemble coding*

A visual system, at a particular level of analysis, in which each neurone responds to most visual features, is said to exhibit **dense coding**. Density of coding is also a function of the width of the tuning functions of cells to each component feature to which they respond. The coding is called dense because, with most natural visual scenes, each neurone would have a high probability of firing. A system in which each neurone responds to only one feature exhibits **local coding**. Most of the time, most neurones would not respond. A system in which each neurone responds to a few features exhibits **sparse coding**. Dense coding is redundant and transmits little local information, although information is present in the spatiotemporal patterns of activity. Local coding in the retina would require an impossibly large number of detectors. Sparse coding combines efficiency with reasonable capacity to carry information. Anything that increases the feature specificity or narrows the tuning functions of cells increases the sparseness of coding. For example the specificity of a cell in the visual cortex varies as a function of the responses of other cells in neighbouring locations and of recurrent influences arising from attention and learning (Vinje and Gallant 2000).

Each receptor in the retina responds to a multitude of visual features and coding is therefore dense. The only specificity is with respect to simple features such as position, luminance, and wavelength. Cells in the striate and peristriate areas show selective tuning to contrast, length, colour, motion, orientation and binocular disparity and their coding is therefore sparser. These features are more complex than those coded locally in the retina. The single cell cannot be said to code any one of these features unambiguously, since variations in firing rate may be due to a change in any one or any combination of them. At higher levels of visual processing, coding becomes sparser because cells become selective to even more complex features (Barlow 1961). For example, cells in the inferotemporal cortex respond selectively to faces. Sparse coding also economizes in energy consumption (Levy and Baxter 1996). This is important in the brain, which accounts for up to 50% of the total energy consumption of the body.

It is believed that particular stimulus features are distinguished in a sparse coding system by the cooperative activity of populations of cells tuned to that feature. For example, the orientation of a line may be uniquely coded by the output of the set of orientation-sensitive cells that the given stimulus excites. The motion of the same line could be coded by the output of the set of motion-sensitive cells that it excites. Since the two sets of cells are at least partly distinct, each stimulus feature is coded distinctly, although not necessarily in the activity of individual cells (see Abbott and Sejnowski 1999). Zohary (1992) has developed an algorithm for determining how many cells in an ensemble of cortical cells, each tuned to two stimulus dimensions, is required to match the psychophysically determined performance of the animal. Ensemble coding is discussed further in Section 4.2.8.

4.2.6e *Interdependency of feature detectors*

Feature detectors do not form a neat set of distinct parallel modules, each serving a distinct stimulus attribute. They interact in the following ways.

Cross talk For example, in the McCollough effect, the orientation of a grating can affect its perceived colour (Sections 3.2.2c and 8.3.5).

Pooled systems A given stimulus attribute can be

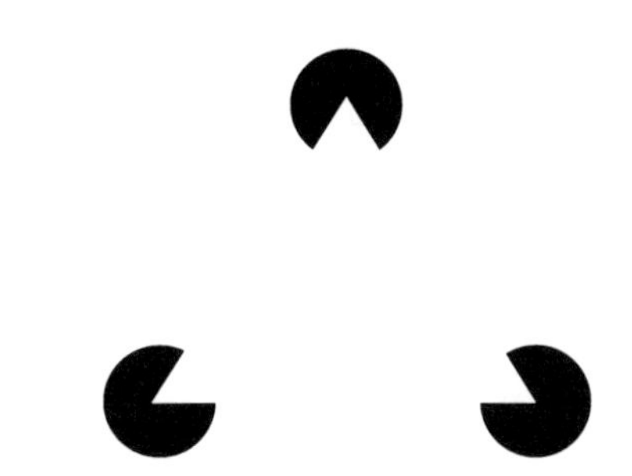

Figure 4.1. Modal and amodal completion of contours. A complete white triangle is seen. This is modal contour completion. Each disc appears complete behind the triangle. This is amodal completion. (After Kanizsa 1979)

detected by more than one sensory system. For example, motion can be detected in an edge defined by luminance, colour, texture, or disparity (Section 4.6.5). There may therefore be several motion-detection systems, which pool their outputs, or the outputs of the different types of edge detector could be pooled before motion is detected.

Mutual interpretation The interpretation of one feature can depend on the value of associated features. For example, the perception of relative motion depends on the perceived relative depth and transparency of objects (Sections 23.2 and 23.4.4). These effects must depend on complex and highly nonlinear feedback processes by which the outcome of one perceptual process influences the way another process proceeds.

Composite features Certain features are combined at an early stage of processing to form dedicated composite features. For example, motion signals are combined with stereo signals to create motion in depth (Section 29.3).

Figure-ground segregation Stimuli defined by different features interact spatially to segregate figure from ground perceptually (Section 5.5.6).

4.2.6f *Importance of inhibition*

At all levels of the nervous system, inhibitory connections between cells play as important a role as excitatory connections. Inhibitory linkages in the retina are involved in the centre-surround organization of receptive fields of ganglion cells. It is believed that these inhibitory linkages enhance contrast. Mach bands are one manifestation of this process (Ratcliff 1965; Sagi and Hochstein 1985). Inhibitory linkages also occur in the visual cortex (see Sections 4.2.6 and 5.4.7).

Hyperpolarizing inhibition is determined by the linear sum of negative and positive currents. **Shunting inhibition** reduces the excitatory response by a nonlinear increase in membrane conductance (Borg-Graham *et al.* 1998).

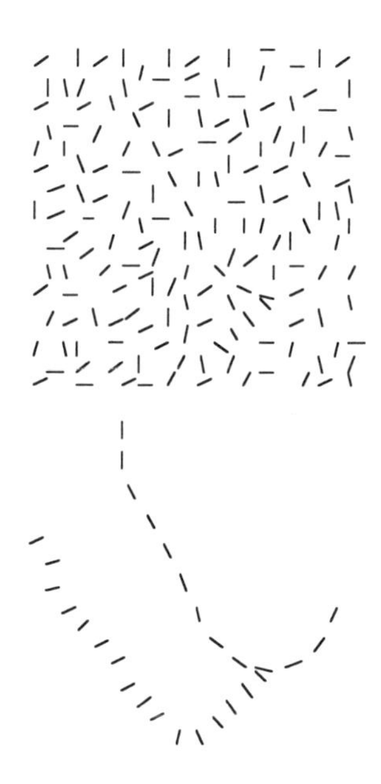

Figure 4.2. Detection of aligned sets of lines.
In the upper figure a set of aligned elements can be detected more easily than a set of elements orthogonal to the line through the set. The two sets of elements are shown in the bottom figure. (Redrawn from Field *et al.* 1993)

Inhibitory synapses near where nerve spikes are generated produce **proximal inhibition**, which is capable of blocking all excitatory responses of the cell. Inhibitory synapses on particular dendrites produce **distal inhibition**, which can selectively inhibit particular branches of the neurone or produce graded inhibition (Vu and Krasne 1992).

4.2.6g *Recurrent feedback*

Some neural connections feed forward into the central nervous system while others feed back to an earlier stage of processing. This issue is investigated by a procedure called **temporal slice analysis**. The build-up of a cell's response is recorded in 10-ms intervals after the onset of a relevant stimulus. Recurrent inhibitory loops reveal themselves as a delay in the achievement of the cell's steady-state response, whereas feedforward loops incur no such delay. This procedure has revealed that basic feature detectors in the visual system, such as those for orientation and binocular disparity, involve feedforward rather than feedback loops (Section 5.3.3 and 5.7.1). This has the advantage that feature detectors reach their steady-state response very soon after stimulus onset, allowing the animal to recognize briefly exposed or moving stimuli. It has also been argued that neural units responsive to complex features, such as cells in the inferotemporal cortex that respond selectively to faces, involve only feedforward loops (Oram and Perrett 1992). However, higher-order feedback systems must be involved when the response of cortical cells coding particular features is modified by attention or by the behavioural significance of the stimulus (Section 5.5.7).

The feature detectors discussed so far are for processing relatively simple visual features. Higher-order feature detectors are discussed in Section 4.6.

4.2.6h *Perception of figural coherence*

At a very general level, neighbouring stimuli and stimuli in temporal sequence tend to be correlated. In other words, stimuli tend to have spatial and temporal extension. One can regard cortical cells that respond to specific features, such as oriented lines or lines of a specific length, as reducing the redundancy of natural images (Rao and Ballard 1999).

The ability to perceive coherent objects is basic to visual perception. The Gestalt psychologists proposed several principles of figural organization, or visual grouping to explain how a set of isolated stimuli is perceived as a coherent pattern (see Koffka 1935). These principles are contiguity, continuity, similarity, good figure, and common motion.

Sometimes we see an object as complete when parts of its boundary cannot be seen because the object has the same luminance and texture as the background, in other words, when the object is camouflaged. In Figure 4.1, devised by Kanizsa (1979), we 'see' the camouflaged edges of a white triangle even though there is no physical contrast in those regions. This is known as **modal completion**. Edges of this type are known as **subjective contours**, cognitive contours, or illusory contours. The rule is that modal completion occurs when the object to which the edge belongs is perceived in the foreground.

Some cells in V1 and V2 of the monkey respond to subjective contours created by real contours outside their receptive fields (Grosof *et al.* 1993; Bakin *et al.* 2000). Subjective contours have been used in studies of stereoscopic vision (see Sections 7.3.2 and 23.1.3). See Kanizsa (1979), Parks (1984), and Spillmann and Dresp (1995) for reviews of this topic.

We readily perceive an object as complete when parts are occluded by other objects or by parts of the same object, an effect known as **amodal completion**. For instance, the cutout corner pieces in Figure 4.1 appear as full discs partially occluded by the corners of a triangle. The impression that one surface occludes another is indicated by the presence of a T-junction (see Section 24.3.2).

4.2.6i *Sensitivity to collinearity*

Collinear sets of dots stand out within an irregular array of dots, as shown in Figure 17.16. The stereoscopic implications of these so-called Glass patterns

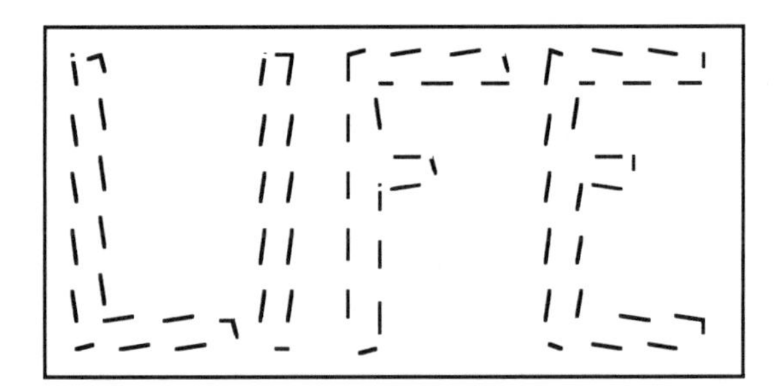

Figure 4.3. Conflict between local and global orientation. The letters are parallel to the frame but appear tilted in the direction of the line elements. (Adapted from Fraser 1908)

are discussed in Section 17.5. Morgan and Hotopf (1989) suggested that diagonal lines seen running between the intersections of regular grid patterns are due to the activation of collinearity detectors.

People are also sensitive to collinear points that traverse a 3-D random array of points, when disparity provides the only information about their collinearity (Uttal 1983; Hess *et al.* 1997a). However, people are more sensitive to collinear points lying in one depth plane than to points that traverse several depth planes.

Aligned line elements show mutual facilitation of visibility to a greater degree than elements with the same orientation, but which are not aligned (Polat and Sagi 1994). Thus, aligned line elements are easy to detect among randomly oriented line elements, as can be seen in Figure 4.2 (Field *et al.* 1993). The line elements need not fall on a straight line but can form tangents to a curve. Furthermore, the line elements may be a few degrees out of alignment and still be interpreted as belonging to the same line. However, in this case, the perceived orientation of the line may be distorted in the direction of the line elements, as can be seen in the well-known "crazy letters" illusion shown in Figure 4.3 (Fraser 1908). People can also detect aligned Gabor patches in a stereoscopic random array in which alternate patches lie in different depth planes (Hess and Field 1995).

The contrast detection threshold for a bar was improved 40% in the presence of a neighbouring aligned bar. The response of many complex cells in V1 of alert monkeys showed a similar improvement in the presence of a neighbouring aligned bar (Kapadia *et al.* 1995). People are particularly sensitive to stimulus elements that are part of a figure rather than part of a background (Section 23.1.2).

4.2.6j *Coherence detectors*

Cells in the primary visual cortex tuned to the same orientation are linked by excitatory lateral connections (Section 5.4.7). Linkages are particularly strong for orientation detectors that not only have the same orientation tuning but also have aligned receptive fields (Nelson and Frost 1985). These linked orientation detectors could form alignment detectors responsible for, modal and amodal completion. Higher levels of neural processing are also involved (Sheth *et al.* 1996; Mattingley *et al.* 1997). Thus, some cells in area V2 of the visual cortex of the monkey respond to subjective contours (von der Heydt and Peterhans 1989). Cells in the inferior temporal cortex of the monkey, which respond when the animal recognizes a shape, also respond when the shape is partially occluded. These cells respond to amodal completion—the tendency to see a partially occluded object as complete (Kovács *et al.* 1995).

Morgan and Hotopf (1989) referred to sets of connected orientation detectors as collector units. Field *et al.* (1993) called them **association fields**. A more specific name for an association field concerned with orientation is **alignment detector**. One can generalize the idea to cover associations between other feature-detectors. An association field of collinear points would be a **collinearity detector**. An association field based on the size or colour of texture elements could allow one to perceive figures defined by these stimulus features. An association field of motion detectors with similar direction sensitivities could underlie the perception of shape defined by motion (the gestalt phenomenon of common fate). A set of detectors of this type would be a **motion coherence detector**. Disparity detectors could unite points on surfaces to form a **coplanar detector** (see Hess *et a*l. 1997a). The role of temporal synchrony in stimulus binding is discussed in Section 4.3.2.

The existence of detectors for stimulus coherence seems to contradict the fact that the visual system is particularly sensitive to spatial and temporal discontinuities, such as discontinuities of contrast, motion, texture, and binocular disparity. The interaction between inhibition and enhancement in the visual cortex is complex and depends on the particular stimulus features and on contrast (Section 5.4.7).

4.2.6k *Cue-invariant detectors*

A shape can be perceptually defined by almost any feature. We can thus have shape defined by orientation, collinearity, size, colour, motion, or disparity. Each feature system could have a distinct shape-recognition mechanism, but it would be more economical if the outputs of the various detectors were to converge on one cue-invariant contour-detection mechanism (Cavanagh *et al.* 1990). Leventhal *et al.* (1998) found a few cells in V1 of the cat and monkey that respond to a boundary defined by texture, luminance, or contrast. A greater number of cue-invariant cells were found in V2. Cue-invariant cells have also been found in the inferotemporal cortex, MT and MST (Sections 5.7.2c and 5.7.3).

4.2.7 Metamerism and channel bandwidth

4.2.7a *Metameric systems*

In physiological terms, a metameric system is a set of detectors with mutually overlapping tuning functions along a particular stimulus continuum. All visual features, other than luminance and contrast, are processed by sets of detectors that are, at least to some extent, metameric. The visual local-sign system contains one million channels in the form of one million ganglion cells. Thus, the stimulus continuum of retinal position is covered by one million tuned receptors (receptive fields). This system is metameric only locally within each small region of the retina where a group of receptive fields forms a mutually overlapping set. Similarly, the frequency coding system in audition is metameric only locally within each critical band—a region along the basilar membrane over which hair-cell tuning functions mutually overlap. Detectors for motion and for orientation have overlapping tuning functions and are therefore partially metameric (Section 3.1.3c). Binocular disparity detectors are also partially metameric (Section 19.8.2). The number of channels and their degree of overlap in these cases are not known with certainty. By contrast, the receptor stage of the colour system is wholly metameric, since it has only three overlapping channels over the whole stimulus continuum.

A metameric sensory system produces an identical output to certain physically distinct mixtures of stimuli in a given stimulus dimension, but distinct outputs to the components presented one at a time. For instance, the colour system cannot distinguish between mixtures of wavelengths that stimulate the three colour channels in the same proportions, but can discriminate between the same wavelength components when they are presented one at a time or in distinct regions of the retina.

A metameric stimulus is a combination of physical stimuli within a stimulus continuum that produces a sensation of a single value within that continuum, even though the component stimuli produce distinct sensations when presented separately. A set of metamers forms an equivalence class under the operation of resolution but not under the operation of discrimination.

Inferences from the use of metameric stimuli are based on two assumptions. First, once stimuli have been combined metamerically, information about the component stimuli is lost. Second, two stimuli that produce identical sensations generate identical physiological activity at some location in the nervous system. It follows from these assumptions that the identical appearance of two metamerically matched stimuli cannot be disturbed by any change applied equally to the metamerically combined neural signals. If two metamerically matched stimuli remain matched for all possible changes applied equally to both, then the metameric process must be at the initial site of processing of that sensory system. Conversely, if some change applied equally to two stimuli disturbs a metameric match, the metameric process must be preceded by a process that detects the applied sensory change, or there is a nonlinear feedback between later and earlier stages of processing. Thus, we infer that trichromacy is achieved at the front end of the visual system from the fact that metameric colour matches continue to match for all states of adaptation of the eye (Grassmann's third law).

For a more complete analysis of this logic, see Brindley (1970). Furthermore, the shape of the metameric matching function for colour (the CIE colour chart) provides a basis for powerful inferences about the nature of the cone mechanisms responsible for trichromacy.

Similar binocular disparities within a local spatial region can average to produce an impression of one depth plane (Section 19.8.2). This means that two different mixtures of disparities could average to give the same depth impression. These would be metameric matches within the disparity system. Nobody has investigated the stability of such matches over different states of visual adaptation. However, they should change with adaptive changes occurring in the retina but not with changes occurring beyond the primary visual cortex.

Metamerism should not be confused with stimulus ambiguity. An ambiguous stimulus is one that can be interpreted in several ways. For example a frontal square is projectively equivalent to a whole family of tapered shapes. To some extent, an observer can perceive projectively equivalent stimuli as different. The equivalence is produced by the geometry of visual rays and has nothing to do with the way information is processed in the nervous system. Metameric stimuli arise because of the way stimuli are processed by the receptors and nervous system. Although many physical stimuli produce the same sensation, observes have no choice but to perceive metameric stimuli as identical. There is loss of information in metamerism but no ambiguity.

Metameric systems exhibit several related properties, which are discussed in subsequent sections of this chapter.

4.2.7b *Channel bandwidth*

In any multichannel sensory system with overlapping tuning functions one can ask how many channels cover the sensory continuum and what is the bandwidth of each channel.

Four psychophysical procedures have been used for this purpose.

Adaptation Determination of the range of stimulus values over which an adapting stimulus of a given value elevates the threshold of a subsequently exposed test stimulus.

Subthreshold summation Determination of the stimulus range over which subthreshold stimuli reduce the threshold of a superimposed test grating.

Masking Determination of the stimulus range for which suprathreshold-masking stimuli elevate the threshold of a superimposed test stimulus.

Comparison of detection thresholds with identification thresholds If two stimuli on a given sensory continuum stimulate distinct sensory channels, the identity of each stimulus should be apparent at the same contrast at which it is detected. This is because each labelled-line channel produces a distinct sensation. Two such stimuli are said to be perfectly discriminated. For example, Nachmias and Weber (1975) found that gratings of 3 and 9 cpd were perfectly discriminated at the contrast at which they were just detected.

4.2.8 Sensory opponency

Strictly speaking, an opponent system extracts the difference between inputs from two oppositely tuned detectors for a bipolar sensory continuum. However, the term "opponency" is often used to denote any sensory mechanism in which a difference signal is generated, even if the stimuli do not form a bipolar sensory continuum.

In the colour system, inputs from red and green detectors are combined in an antagonistic, or seesaw fashion, as are those from blue and yellow (red plus green) detectors. This process occurs in the retina to produce ganglion-cell receptive fields with opponent centres and surrounds. It has been claimed, but not confirmed, that inputs from ganglion cells converge in push-pull fashion on double-opponent cells in the visual cortex (Gouras 1991). Since a change in luminance affects members of each opponent pair equally, colour opponency produces signals that vary with hue, independently of changes in luminance. The luminance signal is derived by adding the inputs from red and green cone types.

The vestibular system also contains a two-stage opponent mechanism. The end organs (cupulas) in the semicircular canals operate as bipolar receptors. Inputs from matching pairs of semicircular canals on opposite sides of the head combine in a push-pull fashion in the vestibular nuclei to signal the direction of head rotation (see Howard 1982).

In the orientation domain, opponency makes orientation discrimination independent of contrast over a wide range of contrasts (Regan and Beverley 1985). Independence of contrast has also been reported for spatial frequency discrimination (Regan 1982), speed discrimination (McKee *et al.* 1986), and temporal-frequency discrimination (Bowne 1990). Furthermore, judgments of either the orientation, spatial frequency, or contrast of a grating presented with different combinations of the three attributes were just as precise as when the grating varied in only the attribute being judged (Vincent and Regan 1995). Thus, subjects precisely judged one sensory feature when the stimulus varied with respect to features not being judged. Since individual cortical cells respond to changes in many features, the ability to isolate one feature must depend on opponent processes between sets of cells. The ratio of responses of two detectors can be used for the same purpose.

A related mechanism for rendering the response of feature detectors invariant to changes in stimulus contrast is to scale (divide) the response of each cell by the pooled response of neighbouring cells. This type of automatic gain control is called **response normalization** (Heeger 1992a; Carandini and Heeger 1994). Normalization in opponent systems is discussed in Section 3.2.2b.

Detection of binocular disparity is largely independent of luminance and contrast (Section 19.5.1). However, we will see in Chapter 6 that disparity detectors in the visual cortex are not opponent detectors, since each detector responds either to crossed or to uncrossed disparity but not to both. Opponency between crossed and uncrossed disparities must be registered at a higher level in the nervous system. In Chapter 21 we will see that the extraction of other types of difference signals within the disparity system renders stereoscopic vision immune to the effects of misconvergence, image misalignment, and unequal magnification of the images in the two eyes.

Binocular rivalry is a type of sensory opponency operating between the two eyes when they are presented with distinct stimuli. This topic is discussed in Chapter 7.

4.2.9 Nonlinear visual processes

Since a linear system can only add or subtract, other forms of computation must depend on nonlinearities. The basic nonlinearity in a digital computer is the two-state transister. Computations other than addition carried out by the nervous system depend on nonlinearities at synapses. There are several types of nonlinearity in the nervous system. For some types, the system is approximately linear over a limited range of amplitude modulation. An **essential nonlinearity** is one that does not approach linearity when input amplitude is reduced. Multiplication, division, and rectification are essential nonlinear processes. The following is a list of common nonlinearities encountered in sensory systems.

Thresholds In the region of the threshold, the receptor potential produced in a receptor is a nonlinear function of light intensity and the principle of superposition does not hold. One reason for this is that additive noise in the system (as opposed to multiplicative noise that is proportional to signal strength) becomes larger in proportion to the signal as signal strength is reduced. Another reason is that stimulus events become quantal near the threshold. For instance, whether or not light is detected depends on the statistical probability of light quanta falling on a receptor and being absorbed by pigment molecules. A third reason is that sensory responses may be quantal. For instance, a synaptic response depends on the release of neurotransmitter molecules from at least one synaptic vesicle.

Beyond the receptor level, neurones produce spike potentials in an all-or-none fashion. Thus, stimulated receptors within the receptive field of a ganglion cell produce a neural spike in the ganglion cell when the combined generator potential reaches a threshold value. In the presence of uncorrelated noise, some receptors will be nearer threshold than others at any time. The resulting response variability smoothes the response of the ganglion cell to variations in stimulus strength in the threshold region (Anderson *et al.* 2000).

Compressive and accelerating nonlinearities In a compressive nonlinearity a response saturates as stimulus strength is increased. A ceiling effect is a saturation of response at high levels of stimulus intensity. The logarithmic relationship between stimulus intensity and the generator potential of retinal receptors is a nonlinear property of this type. A system may be linear over its operating range but, as the intensity of a stimulus is increased beyond a certain level, the response levels off to an asymptotic value.

In an accelerating nonlinearity, the output accelerates relative to the input. This typically occurs near the sensory threshold. Raising an input by a power greater than one introduces an accelerating nonlinearity. Several characteristics of binocular cells in the visual cortex can be accounted for if it is assumed that inputs are squared (Section 6.5).

Rectification Many neurones act as half-wave rectifiers of the input since they respond only to displacements in one direction along the stimulus continuum. Rectification arises when the stimulus dimension is bipolar and extends in both directions with respect to a norm, such as motion and binocular disparity, but for which the sensory detectors are monopolar. For example, retinal bipolar cells are half-wave rectifiers because they respond either only to stimulus onset or only to stimulus offset. Also, many motion detectors in the visual cortex respond only to motion in one direction. We will see that some detectors of binocular disparity are also half-wave rectifiers in that they respond only to crossed or only to uncrossed disparities (Section 6.5). The combined output of two half-wave rectifiers, with signs ignored, produces a fully rectified signal. Full-wave rectification is formally equivalent to squaring the inputs, since squaring removes signs. For example, if the signals from on-centre and off-centre receptive fields were combined without regard to sign it would produce a fully rectified signal that would indicate an edge whatever the sign of its contrast. A fully rectified signal of binocular disparity would signal the amplitude of a disparity without indicating its sign. The combined output of two half-wave rectifiers, with signs intact, reconstitutes the original signal. Thus the output signal from two rectifiers may be linear even though each rectifier is nonlinear (see Warland *et al.* 1997). Advantages of rectification in sensory systems were discussed in Section 4.2.3.

Multiplicative nonlinearities Multiplication of two signals is a nonlinearity since the output is more than the sum of inputs. Logical, or Boolean, operators such as AND-gates and OR-gates involve a type of multiplication based on threshold processes. A set of Boolean operators forms the basis of the **perceptron** model of visual processing (Rosenblatt 1962). A perceptron mechanism can carry out many types of computation, but cannot compute distributed functions such as figure-ground segregation. An example of an AND-gate is provided by cells in the visual cortex that respond only when both eyes are stimulated (Section 5.6).

Detection of correlation between inputs involves multiplication. For example, the Pearson correlation coefficient is the mean of the products of normalized deviations. The detection of the direction of a moving stimulus necessarily involves a multiplicative nonlinearity. For example, in the Reichardt motion detector, a linearly filtered output from one detector is multiplied by the delayed output from a neighbouring detector and then averaged over time (Reichardt 1987). People are able to detect the degree of correlation between images presented to the two eyes (Section 16.1.3).

In general, any real continuous function can be approximated by a polynomial. Polynomials can be derived by multiplying inputs from a set of detectors, and can be used to compute a broader set of functions than can be computed by a 'perceptron' (Koch and Poggio 1992).

The response of a Hebbian synapse, thought to be responsible for synaptic plasticity and learning, depends on the product of the presynaptic potential and the simultaneous postsynaptic potential (Section 10.4.5).

Hysteresis This is a nonlinear process in which the response of a system to a particular stimulus value depends on the direction from which that value is approached. For example, the response of a neurone subject to fatigue is greater for a given value of stimulation if the stimulus increased from a low value than if it decreased from a high value. As stimulus strength is slowly increased and decreased over a given range, the response-stimulus function does not traverse the same path but traces out a loop, known as a **hysteresis loop**. The binocular fusion mechanism shows hysteresis in that the disparity at which initially fused stimuli become diplopic is not the same as the disparity at which initially diplopic stimuli fuse. Binocular fusion is a bistable system in which the change of state is saltatory rather than continuous (Section 7.1.6).

Nonlinear facilitation and inhibition These types of nonlinearity arise when the response of a cell to two inputs is higher or lower than the sum of the responses to each input acting alone. Ganglion cells, relay cells in the lateral geniculate nucleus, and simple cells in the visual cortex are reasonably linear, in that their responses to complex stimuli can be predicted from the superimposition of their responses to simpler stimuli. Complex cells in the visual cortex and visually responsive cells in higher visual centres, such as the inferior temporal cortex and parietal cortex, respond in a highly nonlinear way to visual inputs. Detectors at all levels of the nervous system show inhibitory interactions, which are believed to reduce activity within homogeneous regions relative to activity at boundaries between distinct regions. The result is that contrast, rather than luminance determines activity at higher levels of the visual system.

Many detectors show subthreshold summation in which a subthreshold stimulus presented to one detector is brought above threshold by the simultaneous stimulation of a second detector. This type of nonlinear facilitation occurs in the response of binocular cells in the visual cortex to inputs from the two eyes (Section 6.3). The nonlinearity is extreme in so-called "AND" cells, which respond only to simultaneous inputs from the two eyes.

Interactions between neurones can change the gain of a neural system, or can change the time and space constants of receptive fields of sensory neurones—both nonlinear processes.

Cross-modulation products The response of a linear system to two superimposed sine-wave inputs is two sine waves with their frequencies unchanged. The output of a nonlinear system to two sine waves consists of the two sine waves (the fundamentals) plus harmonics of each sine wave, plus sums and differences of the harmonics of the two sine waves, such as $3F_1 + 2F_2$. These **Cross-modulation products** are characteristic of nonlinear systems. Each type of rectifying nonlinear system produces characteristic cross-modulation products. Thus, one can infer the type of rectifier in a given system by measuring its cross-modulation products, and consulting a catalogue. Regan and Regan (1988) developed a more general mathematical treatment involving the zoom fast-Fourier transform. This gives a very high resolution of the cross-modulation products produced by two sine-wave inputs. When applied to the analysis of evoked potentials generated in the human visual cortex by two lights flashing at different frequencies, it produces a sharp separation between stimulus-related signals and signals arising from noise, as shown in Figure 6.18. Zhou and Baker (1996) recorded from cells in visual cortical areas 17 and 18 of the cat, which responded to moving cross-modulation products (spatial beats) of superimposed sine-wave gratings, even though the spatial frequencies of the gratings were outside the tuning range of the cells. Cross-modulation products are not detected by a purely linear system, since they are not represented in the Fourier domain. Any system generating cross-modulation products must therefore involve nonlinear processing.

Second-order stimuli Stimuli defined by luminance are often referred to as first-order, linear, or Fourier stimuli to distinguish them from second-order, or non-Fourier stimuli defined by texture, disparity, or motion. They are called non-Fourier stimuli because they are not represented in the luminance Fourier spectrum of the stimulus. But this terminology is rather arbitrary since luminance-defined edges are not necessarily detected by linear processes and it is not clear that the visual system contains Fourier mechanisms in the strict meaning of the term. A second-order stimulus, such as a figure defined by texture, can be converted into one with Fourier components if it is first rectified. In the visual system, full-wave rectification can be achieved by pooling the inputs from on-centre and off-centre receptive fields. This introduces a spurious doubling of the spatial frequency of the stimulus, which can be removed by subsequent filtering at an octave lower than the spurious signals. The resulting signal can be detected by a detector of stimuli defined by luminance. For example, a Reichardt motion detector can detect first-order motion defined by motion of luminance-defined edges. However, it cannot detect second-order motion of edges defined by contrast or texture. However, after a second-order signal has been rectified and filtered it can be detected by a Reichardt detector (Chubb and Sperling 1988; Wilson and Kim 1994).

Adaptation and differentiation Sensory adaptation, or short-term modification of the response of a sensory system exposed to constant stimulation is a nonlinearity because it violates the time-invariance assumption of linear systems. Sensory adaptation amplifies transient inputs relative to sustained inputs and is equivalent to differentiating the input.

Synchronization Enhancement of responses to synchronized inputs relative to non-synchronized inputs is a nonlinear process.

Learning The long-term modification of a response to a given stimulus through learning is also a nonlinear process. Spontaneous changes in the interpretation of a visual stimulus, such as those that occur in reversible perspective and ambiguous figure-ground displays are also nonlinearities. These nonlinearities in high-level control systems are difficult if not impossible to specify mathematically.

Procedures for analyzing nonlinear systems are reviewed in Section 3.4.

4.3 TEMPORAL CODING

4.3.1 Temporal coding in single neurones

It has been assumed generally that the single neurone conveys information in terms of spike frequency and that the probability of an action potential in an axon per unit time is proportional to firing rate. This is known as the Poisson model. However, spikes are sometimes generated with precise timing. For example, the relative timing of spikes arriving at the olivary nucleus from the two ears is used for sound localization and high-precision timing of spike trains may also be used in binocular vision (Section 28.3). The Poisson model cannot account for why ganglion cells tend to respond in discrete sets of spikes rather than at a continuously varying rate or why these spike trains recur with high precision when the same stimulus is repeated (Berry *et al.* 1997). Theoretically, a discrete burst of spikes can convey up to 3.6 bits of information per spike. This is much more than is conveyed by spike frequency alone. A process that conveys information by the pattern of spikes in one or more neurones is known as a **correlation code**. For example, synchronous firing of ganglion cells is believed to carry information (Berry *et al.* 1997). Correlated inputs from the two eyes are important for binocular rivalry and stereopsis.

A group at the National Institute of Health in Bethesda proposed a radical idea of temporal sensory coding (Richmond and Optican 1987, 1990; Richmond *et al.* 1990; Gawne *et al.* 1991). They recorded from the lateral geniculate nucleus and from complex cells in the visual cortex, as alert monkeys fixated small patterns of black and white squares and rectangles (Walsh patterns) which varied along three stimulus dimensions—pattern, duration, and contrast. The spike train of a cell's response to each stimulus was smoothed to produce a spike-density profile over the first 260 ms. Components that accounted for successively smaller correlations between stimuli and the response profiles over the set of stimuli were determined by principal components analysis (a form of factor analysis). A weighted sum of these components represented the response of a cell to a particular stimulus. The first component was correlated with mean firing rate. Higher order components of the spike train presumably arose because of differential latencies of subregions of the receptive field or differential delays in recurrent inhibition from different regions of the receptive field. The effect of varying one stimulus feature depended on the value of the other features, so that the features interacted non-additively. The response was the same for many combinations of the three stimulus features—pattern, duration, and contrast. However, principal component analysis of the spike train allowed the investigators to extract some information about each stimulus feature, as well as about combinations of features. In another study with Walsh patterns, spike trains of cells in V1, V2, and V4 were found to contain a temporal code for colour and a distinct temporal code for pattern (McClurkin and Optican 1996). These investigators suggested that pattern and colour are processed in distinct multiplexed temporal-code channels, rather than in distinct labelled-line channels. This would facilitate the binding of spatially related features. Gawne (2000) found that stimulus orientation was related to spike frequency of single complex cells in monkey V1 but, that stimulus contrast was related to response latency and spike modulation. Both features could be reliably recovered from stimuli that varied in both contrast and orientation when the responses of several neighbouring neurones were pooled.

The principal components method reveals independent components only if the underlying structure of the spike train is linear. The approach seems to be justified because a nonlinear method based on neural networks did not reveal evidence of significant nonlinearities in spike trains produced by cortical neurones, at least to static stimuli (Fotheringhame and Baddeley 1997).

The approach makes two novel claims. First, that responses of single neurones contain distinct temporal components that relate to distinct stimulus features. Second, that the nervous system is capable of decoding this temporal information. The traditional view is that transmission of sensory information in the single axon is in terms of response frequency, with temporal modulation serving to indicate only temporal modulations of overall stimulus strength.

Even if the spike train of single neurones contained information about a variety of stimulus features, additional neural machinery would be required to extract it. Since there is no indication of what this would be, the approach leaves the problem of stimulus analysis in the nervous system unsolved. Furthermore, the analysis was based on only a subset of values of two or three stimulus features exposed for a fixed duration. Golomb *et al.* (1994) found a well-defined set of principal components in the spike train of cells in the lateral geniculate nucleus, but only for a stimulus of fixed duration. A typical cell in the LGN or visual cortex is also influenced by stimulus motion, colour, flicker, disparity, and size. Thus, in practice, the job of disentangling the contribution of each value of each feature to the principal components of the spike train of a single neurone becomes difficult to perform.

Tovée *et al.* (1993) applied a similar analysis to responses of cells in the primate temporal cortex to faces. They also found that the first principal component of the spike train is closely related to the mean firing rate of the cell, and accounts for about 70% of the variance. Furthermore, they found that about 85% of the information about firing rate available during a 400-ms period could be extracted during the first 50 ms of the cell's response. Almost half of it was available in the first 20 ms.

Signals from the visual cortex to the inferior temporal cortex pass through four stages. Each stage adds about 20 ms to the total latency of response in the temporal cortex (Rolls 1992). This suggests that effective information about firing frequency is extracted in about 20 ms at each stage of processing. If we assume a firing rate of 100 Hz, this means that firing frequency estimates in a single neurone are based on up to five spikes. Tovée *et al.* found that only about 19 and 8.5% of the information in a spike train was contained in the second and third principal components, respectively. Furthermore, a good part of this information was found to reflect the latency of the cell's response. They concluded that features of the response train other than latency and mean firing rate are probably not significant for cortical processing. The rapidity of visual processing is also indicated by the finding that the visual-evoked potential corresponding to a decision that a complex scene exposed for 20 ms does not contain a specified object occurred within 150 ms (Thorpe *et al.* 1996).

Heller *et al.* (1995) applied a neural net analysis to spike train inputs to cells in monkey V1. They concluded that variations in firing rate could not be resolved over intervals shorter than about 25 ms. Most information is contained in the mean firing rate in the initial short part of the input. Some information is available in modulations of response over a longer interval, but whether this information is actually used is still an open question.

To overcome the temporal resolution limitations of principal components analysis Victor and Purpura (1996) measured the spike latency, spike count, and interspike interval in 352 neurones in V1, V2, and V3 of alert monkeys during the first 256 ms of exposure to gratings varying in orientation and spatial frequency, and to checkerboards varying in contrast, element size, and texture type. Most of the cells showed evidence of stimulus-specific tuning by spike time and spike interval, and about half showed evidence of tuning by spike count.

Oram *et al.* (1999) found large numbers of stimulus-elicited precisely timed spike patterns in the LGN and visual cortex but concluded that many of these patterns arose by chance and carried no information that was not available in the spike count. Reinagel and Reid (2000) analyzed the spike-train responses of LGN cells to defined random sequences of flashing light and concluded that some cells use temporal patterns to encode temporal information. Each cortical cell receives thousands of excitatory and inhibitory inputs, each of which fluctuates over time. The resulting chaotic activity of a neural network puts constraints on the use of precise temporal patterns (Van Vreeswijk and Sompolinsky 1996).

Higher-order components in the temporal waveform do not necessarily have to be analyzed to be useful. Rather than conveying information about stimulus features to higher levels of processing, they may be involved only in low level detection of texture boundaries for the following three purposes.

Identification of surface discontinuities An evenly textured surface has a consistent texture, colour, contrast, and motion, so that detectors in that region respond with similar and synchronous response trains. The responses of the cells onto which the detectors converge would be enhanced by this stimulus synchrony because synchronous inputs summate at synapses. Hence, the postsynaptic cells would resonate to an input from an evenly textured region. At the boundary of two textured regions, there would be a discontinuity in the pattern of resonance, which could be used to locate the boundary, even before any analysis of particular stimulus features had taken place. There is evidence from responses of single cells in the visual cortex that cells with similar responses to stimulus orientation fire in synchrony (Section 5.4.6). Evidence of neuronal mass activity related to texture segregation, which is independent of the visual feature defining the textures, has been revealed in visual evoked potentials from the human scalp (Lamme *et al.* 1993; Bach and Meigen 1997). Fahle (1993a) found that a phase difference of only 5 ms between the temporal modulation of groups of spatially homogeneous points was sufficient for perceptual segregation of one group of points from the background.

Matching binocular images The tuning characteristics of the component monocular receptive fields of a binocular cell in the visual cortex tend to be similar. When the same image falls on corresponding points in the two retinas, the two response trains arriving at a binocular cell are the same and in phase. They could serve to generate a cross-correlation function even though what is correlated is not analyzed. Thus, the similarity of temporally modulated signals arriving at a binocular cell could indicate whether the eyes are properly converged on a given stimulus.

Detection of disparity discontinuities A model of this type of process has been provided by Murata and Shimizu (1993). Tootell *et al.* (1988c) reported enhanced neural activity (as reflected in the uptake of deoxyglucose) along cortical loci corresponding to borders between differently textured regions of a stimulus display. This enhanced activity was evident only when the display was viewed binocularly.

Discrimination of complex temporal stimuli Complex temporal stimuli are produced by moving stimuli. Reinagel and Reid (2000) found that a defined random series of flashes produced its own characteristic temporal pattern of discharges in an individual neurone in the LGN of the cat.

Whether or not stimulus-evoked temporal features of the spike train carry information, cortical neurones show stereotyped differences in the waveforms and repetitive firing properties of their action potentials. Neurones in the mammalian neocortex can be classified into four types according to their temporal dynamics. There are: "regular spiking", "fast spiking", "bursting", and "intrinsic-bursting" neurones (Section 5.4.4d).

Bursting cells in the visual cortex show characteristic bursts of activity containing up to about six spikes within a 25 ms period. Some pyramidal cells in the superficial layers of the visual cortex of the cat respond with synchronized bursts in response to depolarizing current. They have been called **chattering cells** (Gray and McCormick 1996). Bursts are more reliably related to the orientation and spatial frequency of the stimulus than are single spikes (see Lisman 1997). This is because single spikes can arise spontaneously, and therefore represent noise, whereas multiple spikes rarely occur spontaneously. Also, multiple bursts are more likely to trigger a postsynaptic response than single spikes, especially when synchronous bursts converge on a given synapse. Chattering pyramidal cells send collaterals into other cortical layers and other cortical regions and may play a key role in generating synchronous cortical activity, discussed in the next section.

Other aspects of temporal coding have been reviewed by Dinse *et al.* (1990a).

4.3.2 Temporal synchrony

Synchrony of firing of subcortical or cortical neurones to a given stimulus has been studied by recording from a pair of neurones, and deriving a time-averaged measure of the temporal correlation between spiking events. It is possible to measure synchrony of firing in a larger number of cells in great detail, and to distinguish between synchronous firing arising from lateral interactions and that due to direct stimulation (Gerstein and Aertsen 1985; Aertsen *et al.* 1989; Vaadia *et al.* 1995). Cells in area 17 of the cat synchronize their responses with a precision of a few milliseconds.

Synchrony of firing of subcortical or cortical cells could improve sensitivity, aid stimulus binding, serve as an attention mechanism, provide sensory information, and provide a basis for learning, as I will now explain.

4.3.2a *Improved stimulus tuning*

The responses of cells with similar tuning to orientation, motion, or spatial frequency are more closely synchronized than the responses of cells that differ in tuning (Braitenberg 1985; König *et al.* 1995). This correlated activity is believed to be mediated by lateral connections in the visual cortex (Section 5.4.7). Correlated discharge of nearby pairs of cells in the cat visual cortex is stimulus dependent and not a simple consequence of the unicellular responses. Although the optimal tuning of cell pairs to spatial and temporal frequency and velocity was found to be similar to that of the component cells, the receptive fields of cell pairs were narrower and their responses were briefer (Ghose *et al.* 1994a). This suggests that correlated discharges of neighbouring cells achieve a higher degree of spatial and temporal resolution than is achieved by single cells.

Synchrony of firing of spatially aligned ganglion cells could facilitate the firing of cortical cells tuned to stimulus orientation (Alonso *et al.* 1996; Meister 1996). Correlated activity between cells with different tuning characteristics could help to resolve the ambiguity in the response of single cells. For example, cells that respond both to dark and light bars may fire in synchrony only in the presence of a dark bar (Ghose *et al.* 1994a).

Dan *et al.* (1998) showed theoretically and experimentally that more information can be extracted from a pair of neurones that tend to fire in synchrony than from the firing rates of a similar uncorrelated pair of neurones. Synchronous visual inputs are more likely to drive cortical cells through the mediation of Hebbian synapses that act as coincidence detectors (see below).

On the other hand, Cardoso de Oliveira *et al.* (1997) found that responses of cells in MT and MST became synchronized just before an expected stimulus was presented. Firing became desynchronized as a function of stimulus contrast at the onset of a stimulus moving at up to 29°/s with respect to a stationary background. They concluded that motion direction is coded by differential rates of firing rather than by synchronization alone. Desynchronization of firing in the region of an attended moving stimulus, whatever its direction, relative to synchrony in the background could be used to segregate the stimulus from its background.

4.3.2b *Stimulus binding*

The presence of many visual areas, each devoted to specific features of a visual stimulus, raises the problem of how this distributed activity is synthesized into a single percept. This has been dubbed the "**binding problem**" (Hinton *et al.* 1986). Different visual objects are juxtaposed and may overlap and move. Thus, a given pattern of neural activity is only fleetingly related to a given stimulus object—the same cells may code features of several complex objects in rapid succession. The binding problem has two aspects. In the first place, a similar visual feature occurring over many receptive fields must be recognized as belonging to a single object. This could be called **spatial binding for a given feature**. For instance, the boundary of a large object stimulates many orientation detectors, and their outputs must be related to form a coherent percept of the object's shape. The role of spatial interactions between cortical neurones in this process was discussed in Section 4.2.6. In the second place, objects possess particular concatenations of distinct visual features that must be recognized as belonging to the same object. This could be called **feature binding for a given object**.

One possible solution to the binding problem is to converge diverse activities evoked by a given object onto a single cell devoted to recognition of that particular object. Such cells have been called **grandmother cells**. Such a mapping process requires an impossibly large number of dedicated cells and connections.

The other general solution to the binding problem is to code the responses of feature-extracting systems in different processing streams and at several levels of processing, which then become bound into a particular spatiotemporal pattern of activity (Zeki and Shipp 1988). Each neurone participates in many different patterns, and only the spatiotemporal response of a particular cell assembly is unique to a particular object. A distributed system can store more objects than there are neurones in the network.

Temporal synchrony of the pattern of neural activity evoked by a given stimulus, coupled with desynchronization between one pattern of activity and others, could allow the same network to code distinct stimuli in rapid succession and distinguish between several temporally or spatially overlapping stimuli (Milner 1974; Malsburg and Schneider 1986; Singer and Gray 1995). Wallis and Rolls (1997) have shown that a hierarchically organized system of feature extracting processes of this type can build representations of objects that are independent of position, size, and aspect.

Visual stimulation causes groups of adjacent cells in the visual cortex of both anaesthetized and alert cats to discharge in synchrony at frequencies of between 30 and 80 Hz (Gray and Di Prisco 1997). This so-called **gamma frequency** is outside the range of spontaneous background activity responsible for the **alpha rhythm** or the 4 to 7 Hz **theta waves** that occur during sleep (Gray and Singer 1989; Munk *et al.* 1996). Cortical cells fire in high-frequency bursts containing two to four spikes at intervals of 15 to 30 ms, which is approximately one period of the gamma cycle. A given visual stimulus evokes synchronous responses in different parts of the primary visual cortex, separated by up to 7 mm, between cells with non-overlapping receptive fields but similar tuning to orientation.

Synchronized firing could enhance the response of a pool of cells with different tuning characteristics when they are stimulated by the same stimulus object. Thus, two parallel bars stimulating cortical cells with distinct receptive fields evoked widespread synchronous activity when the bars moved in a common direction (see Gray *et al.* 1991). Pairs of cells with different preferences for motion direction in MT of the alert monkey responded in synchrony when stimulated by a single object moving at up to 6.7°/s, but not when stimulated by distinct moving objects (Kreiter and Singer 1996).

A set of cortical cells with overlapping receptive fields in the cat visual cortex responded in synchrony when stimulated by a single bar. However, when stimulated by two differently oriented superimposed bars, the cells segregated into assemblies according to orientation preference. The cells in each assembly were synchronized but there was no correlated firing between the assemblies (Engel *et al.* 1991). One long bar induces stronger synchrony than two smaller collinear bars in the same location (Neuenschwander and Singer 1996).

A 40 Hz component of the human visual evoked potential was stronger in response to a coherent figure than to a spatially noncoherent figure (Tallon-Baudry *et al.* 1996). Such activity could serve to bind responses to a continuous edge spanning many receptive fields. It could thus be part of a mechanism for figure-ground segregation and figural grouping. However, synchrony of firing between pairs of neurones was found to be no greater when both receptive fields lay in a figure region than when one receptive field was in the figure and one was outside it (Lamme and Spekreijse 1998).

Rodriguez *et al.* (1999) presented subjects with a pattern that could be perceived as meaningless shapes or as a face. Synchronized visually evoked potentials at the gamma frequency occurred at widely separated regions of the human scalp when the person reported seeing a face. The physical stimulus remained the same, only the percept changed. The responses became desynchronized as the person made a motor response.

It has also been claimed that cells in widely different cortical areas and in different hemispheres respond in synchrony with near zero phase lags at between 35 and 85 Hz to a stimulus to which the cells are similarly tuned (Eckhorn *et al.* 1988; Jagadeesh *et al.* 1992). This type of synchronous activity is most pronounced in response to stimuli with continuous contours and common motion (Gray *et al.* 1991; Freiwald *et al.* 1995), as one would expect of a system that helps to bind distinct features of a given object. The oscillations need not themselves code stimulus-specific information. They could serve to co-energize sets of feature detectors for the formation and consolidation of cell assemblies in the process of learning. Once a cell assembly has been consolidated, synchronous activity could activate it for the purpose of object recognition. Synchronous activity in a distributed set of cells can be evoked quickly enough to allow a familiar object to be recognized in a fraction of a second (Gray *et al.* 1991). Vaadia *et al.* (1995) found correlated spatiotemporal patterns of firing in the frontal lobe that varied according to the sensory discrimination task that the monkey was performing. The possible role of synchronized activity in binocular rivalry is discussed in Section 7.10.2.

Tempting as these speculations are, they must be treated with caution. Young *et al.* (1992) could find no evidence of stimulus-evoked synchronous activity in the 30- to 60-Hz range in the primary visual cortex or in the middle-temporal area (MT) of either the anaesthetized or alert monkey. Some signs of oscillation were found in the inferotemporal cortex, but only in the alert monkey. Ghose and Freeman (1992) could find no consistent relationship between oscillatory discharges in cortical cells and specific features of the stimulus other than stimulus strength, as reflected in the mean firing rate of cells. Engel *et al.* (1992) suggested that oscillatory responses might have been missed in these studies because the responses are not strictly periodic and contain a broad band of frequencies.

The results of some psychophysical experiments support the idea that neural synchrony is involved in figure-ground segregation. It has been claimed that a textured region can be detected within an identical textured surround when the two regions are presented sequentially at rates between 12 and 42 Hz, if the phase difference is at least 10 ms (Leonards *et al.* 1996). A set of collinear line elements was more easily detected within a set of randomly orientated elements (see Figure 4.2) when the two sets were presented asynchronously (Usher and Donnelly 1998). The interplay between depth cues is influenced by the degree of temporal synchrony between conflicting cues (see Chapter 27).

Other psychophysical evidence argues against the role of synchronized neural activity. Kiper *et al.* (1996) found that performance on a texture segregation task was not affected by whether the texture elements in the regions to be segregated were flickered in phase or in antiphase. Lee and Blake (1999a) reported that subjects could detect a textured figure that differed from the background only in temporal synchrony. A textured region in which elements reversed motion direction in synchrony stood out from the background in which reversals of direction were not synchronized. However, Farid and Adelson (2001) showed that a simple temporal bandpass filter could convert the difference between the two regions of the display into a difference in contrast. When they eliminated this factor, subjects could not detect a textured figure defined by a difference in temporal synchrony.

In spite of doubts about the use of widespread synchrony of neural activity in visual coding, there can be little doubt that synchrony of similar inputs at particular synapses plays a role in tuning cortical cells during development and in perceptual learning, as we will see at various points in this chapter and Chapter 10 (see Trotter *et al.* 1992).

4.3.2c *Synchrony of inputs from paired senses*

Auditory afferents converging in the superior olive form a delay line in which the timing of inputs from the two ears is compared. Calibration of this system requires precise matching of conduction velocities along the afferent axons. This is achieved by regulating the distances between axon nodes (see Sabatini and Regehr 1999). There is some evidence that synchrony between inputs from the two eyes is involved in stereopsis (Section 28.3).

4.3.2d *Synchronized activity and attention*

In states of reduced attention and sleep, large cell populations in the cortex engage in high-amplitude synchronous activity at less than 10 Hz. During arousal, the reticular activating system of the brain stem disrupts this synchronized activity but facilitates stimulus dependent oscillatory activity at frequencies in excess of 30 Hz (Munk *et al.* 1996). Attentional processes are probably very important in organizing neural activity into unitary patterns. Sillito *et al.* (1994) reported that synchronized activity in the visual cortex of the cat induced similar high-frequency synchronized spike potentials in relay cells of the LGN. They suggested that this feedback mechanism concentrates neural circuitry onto the stimulus. However, they found that this activity could arise from slow cortically induced covarying resting potentials rather than from covarying spike potentials (Brody 1998).

The strength and pattern of synchronized firing between widely different parts of the cat cerebral cortex have been shown to depend on the stimuli to which the cat is attending (Roelfsema *et al.* 1997; Cardoso de Oliveira *et al.* 1997). For example, cells in cat cortical areas 17, 7, and 5 fired in synchrony in the 4 to 12 Hz range when the visual stimulus was one associated with a learned action but not when the stimulus was novel (von Stein *et al.* 2000). This suggests that synchronization is associated with top-down neural processing. Perhaps the familiar stimulus attracts more attention than the novel stimulus. Fries *et al.* (2001) found that neurones in cortical area V4 of the monkey showed increased synchronization in the 35 to 900 Hz range and decreased synchronization in the < 17 Hz range when the stimulus was one to which the animal was attending. Synchronization could increase the efficiency of synaptic transmission of specific information passing from V4 to the inferior temporal cortex, where pattern information is processed.

We are still left with the problem of how the activity of cell assemblies is accessed by other neural processes (Engel *et al.* 1992). Treisman (1988) proposed that visual attention must be directed serially to each object in a display whenever more than one feature is used to distinguish an object from other objects. Crick and Koch (1990) proposed that attention based on stimulus position binds neural activity arising from diverse features of an object, and that this process is facilitated by temporal synchronization of the responses of neural centres activated by the various features of the object. Visual attention can be based on stimulus features rather than on position alone (Section 5.5.7). This calls for a process serving spatial binding for a given feature distributed over several locations.

Proponents of **feature-integration theory** suggest that visual features and location are initially processed independently and are bound by an act of attention. The theory is supported by perceptual confusions, called **illusory conjunctions**, which occur when attention is overloaded (Treisman and Schmidt 1982). For example, when subjects are asked to describe the colour of briefly exposed letters they sometimes confidently report seeing the letters in the wrong colour. Evidence from patients with a unilateral attention deficit has been equivocal, but a recent study seems to confirm that illusory conjunctions occur in the hemifield affected by a unilateral attention deficit (Arguin *et al.* 1994). Friedman-Hill *et al.* (1995) described a 58-year-old man with bilateral parieto-occipital lesions who confused the colours of simultaneously presented letters and objects, even when they were presented for several seconds. He was also unable to report the relative spatial positions of two objects displayed on a screen.

There has been some dispute about the interpretation of experiments on illusory conjunctions (Tsal 1989; Green 1991). Green (1992a) argued that a distributed process that does not require an act of attention binds visual features and topographic location directly and locally. Illusory conjugations could perhaps be due to unusual persistence of vision so that afterimages are carried from one object to another as the eyes scan the scene. **Palinopsia** is a clinical condition in which a visual perception recurs after the object has been removed, and illusory visual spread is an illusory spatial extension of a seen object (see Critchley 1955). Illusory conjunctions could perhaps result from confusions at the level of recall rather than at the level of perceptual registration.

Synchronized activity could involve feedback loops that help to sustain a pattern of neural activity in short-term memory. Tallon-Baudry *et al.* (2001) observed sustained synchronized activity between regions of the extrastriate cortex of two human subjects when they memorized a pattern and compared it with a second pattern presented after an interval of up to 2 s.

4.3.2e *Synchrony at Hebbian synapses*

Hebb (1949) speculated that learning depends on competitive reinforcement of synaptic contacts (Portrait Figure 10.6). He proposed that synaptic contacts are strengthened when activity in a presynaptic cell is temporally correlated with activity in a postsynaptic cell. Synaptic contacts are weakened when activity in the two cells is not correlated. Synapses behaving in this way are known as **Hebbian synapses.** When two presynaptic cells converge on the same postsynaptic membrane, the activity in either presynaptic cell is more highly correlated with that in the postsynaptic cell when the converging inputs are synchronous rather than asynchronous. This is because the postsynaptic membrane summates potentials from converging synchronous signals more effectively than it summates those from asynchronous signals. Thus, neural impulses in two axons converging on a simple cell in the cat's visual cortex produce a stronger postsynaptic response when they arrive within about 7 ms of each other (Usrey *et al.* 2000). The outcome is that correlated activity in two or more afferent pathways gains preferential access to the nervous system, and, over time, leads to more efficient transmission along that neural pathway.

Persisting increases in synaptic transmission are known as **long-term potentiation** (LTP). When converging inputs are persistently uncorrelated the synaptic strength of the one most highly correlated with the postsynaptic potential increases at the expense of the synaptic strength of the other. Furthermore, LTP in one group of synapses can be associated with **long-term depression** of responsiveness in neighbouring inactive synapses, mediated either by chemical diffusion or by lateral dendritic connections (Scanziani *et al.* 1996).

These processes have been studied most extensively in the hippocampus, a region of the cortex involved in memory. Synaptic connections can be reversibly strengthened or weakened in brain slices of the cat visual cortex. This is done by correlating postsynaptic responses generated by stimulation of white matter with depolarization or hyperpolarization of the postsynaptic membrane by current delivered through the recording electrode (Frégnac *et al.* 1994). Computer simulations of cortical pyramidal cells have revealed that cortical cells could, in theory, detect coincidence between single spikes in the submillisecond range (Softky 1994).

A neural net with a broad distribution of conduction and synaptic delays could, in theory, use the Hebbian rule to learn both static patterns (cell assemblies) and temporal sequences such as tunes (Herz *et al.* 1989).

A Hebbian synapse strengthens synaptic connections when converging inputs covary, and can therefore be thought of as a covariance detector responding to what is common between two inputs. Some sensory systems act as difference detectors since they respond to what is different between two inputs—they decorrelate the input (Dan *et al.* 1996). For instance, a ganglion cell does not fire when its receptive field is evenly illuminated but does fire when there is a luminance gradient across the receptive field. Ganglion cells work this way because of mutual inhibitory connections within the inner plexiform layer of the retina (Section 5.1.2c).

The advantage of this system is that messages are transmitted to the brain only about spatial or temporal changes in the proximal stimulus—the regions that are most informative. Mutual inhibitory mechanisms are responsible for contrast processes and opponent processes at various levels in a variety of sensory systems. Opponent mechanisms detect changes in one stimulus feature in the presence of changes in some other feature (Section 4.2.8). In Section 29.3, we discuss examples of opponent mechanisms in stereopsis. Barlow (1991) referred to mutual inhibitory mechanisms as anti-Hebbian, because they detect differences rather than coincidences. He suggested that Hebbian and anti-Hebbian mechanisms work together at successive levels within the processing hierarchy of the visual system—Hebbian mechanisms detect coincidences between inputs, anti-Hebbian mechanisms sharpen the distinctions between sets of detected coincidences.

Hebbian synapses are important in the development of the visual system, particularly the development of binocular vision (Section 10.4.5). For discussions of Hebbian synapses, see Clothiaux *et al.* (1991), Dan and Poo (1992), Ahissar *et al.* (1992), and

Malgaroli *et al.* (1995). Structural changes underlying memory in adult animals are reviewed by Bailey and Kandel (1993).

4.3.2f *Sources of synchronized activity*
Cortical oscillations could originate in the retina or lateral geniculate nucleus, (Ariel *et al.* 1983; Ghose and Freeman 1992). Multi-electrode recordings in the LGN revealed synchronized oscillations of 60 to 114 Hz evoked by large spatially continuous stationary or moving stimuli (Neuenschwander and Singer 1996). The oscillations originated in the retina, but it was argued that they alone could not account for oscillations at the cortical level.

A second source of synchronous cortical activity could be the subthreshold oscillation of the membrane potentials of many cortical pyramidal cells in layer 5 of the visual cortex and other cortical areas (Gutfreund *et al.* 1995; Gray and McCormick 1996).

Thirdly, synchronous activity could be generated by time delays in recurrent inhibitory loops in the cortical network (Freeman 1975). One type of subthreshold oscillation is mediated by a regenerative voltage-gated modulation of calcium conductance across the cell membrane. A second type is mediated by modulation of sodium conductance. Both types of oscillation can trigger spikes if the membrane is sufficiently depolarized. There is some evidence that oscillatory activity originates in inhibitory interneurones in cortical layer 4 with collaterals ramifying in layers 3 and 4 (Llinás *et al.* 1991). In the absence of stimulation, these oscillations have an amplitude of 5 mv and are in the 5- to 20-Hz range, but computer modelling indicates that during stimulation, and with appropriate synaptic coupling, much higher frequencies could be generated (Silva *et al.* 1991).

4.3.2g *Models of synchronized activity*
Several investigators have developed neural networks that model synchronized activity in the visual cortex (Eckhorn *et al.* 1990; Schuster and Wagner 1990; Grossberg and Somers 1991; Sporns *et al.* 1991; Wilson and Bower 1991; Niebur *et al.* 1993). Some models are based on known properties of excitatory pyramidal cells and inhibitory interneurones (Traub *et al.* 1996; Wright *et al.* 2000). Ghose and Freeman (1997) developed a model of cortical oscillations that would arise from the integration of oscillatory signals in the LGN and from intrinsic oscillations of cells in cortical layer 5.

Chawanya *et al.* (1993) developed a neural network model that simulates synchronized oscillations within and between orientation columns of the visual cortex, and in which the strength of the phase correlations between different columns reflects the length and continuity of bar-shaped stimuli (see also König and Schillen 1991). Schillen and König (1994) described a neural network model, which developed synchronized firing after a period of temporal correlation between the activation of distributed assemblies responding to different features of an object. Christakos (1994) provided a mathematical basis for analysis of synchrony in neural nets using the **coherence function**, which expresses the extent to which two processes covary as a function of frequency. It is the frequency-domain analogue of the squared cross-correlation coefficient. Ritz *et al.* (1994) based a network model of collective oscillations in the visual cortex on local inhibition between single spiking neurones. Parodi *et al.* (1996) developed a model based on the detection of differences in arrival time of integrated visual inputs as opposed to differences in spike-train frequencies (see also Gawne *et al.* 1996). Mechanisms of neural synchrony have been reviewed by Sturm and König (2001).

4.4 GEOMETRY OF VISUAL SPACE

4.4.1 Spatial vision and Lie groups

Basic transformations of visual stimuli such as the changing size or shape of the retinal image as an object moves in 3-D space, or patterns of optic flow produced by self motion can be described in terms of Lie operators (Section 3.8.1). Hoffman (1966) proposed that these operators are embedded in the visual system (see also Dodwell 1983). It may be significant that inspection of one member of any pair of Lie orbits for some minutes induces an aftereffect that resembles the other member of the pair. For example, inspection of a radial pattern creates an aftereffect of concentric rings (MacKay 1961). Gallant *et al.* (1993) found some cells in area V4 of the monkey that responded selectively to stimuli resembling one or other of the Lie orbits. The columnar organization of cells in the visual cortex resembles modified Lie orbits (Section 5.6.1). Lie operators are related to the operators of differential geometry that have been used to describe patterns of binocular disparity (Section 20.2.8).

4.4.2 Non-Euclidean geometry

For ordinary purposes we use Euclidean geometry to record the positions of objects in the world. On a cosmic scale, Einstein used non-Euclidean geometry to describe the intrinsic curvature of space-time in the neighbourhood of massive objects. One can ask whether the visual system has its own intrinsic geometry in terms of which the positions of objects are perceived. If so, this geometry should manifest itself in the way people perform psychophysical tasks and we should be able to derive the basic parameters of the intrinsic geometry of visual space.

Euclidean geometry is based on isometries and similarities. Affine and projective geometries are non-Euclidean. Therefore, the projective geometry that describes the shapes of images of objects in 3-D space is non-Euclidean (Section 24.1.1). Some people have argued that the curvature of the locus of zero binocular disparity (the horopter) indicates that visual space is curved. But the curvature of the horopter varies with viewing distance and, in any case, the shape of the horopter is not necessarily related to the way surfaces are perceived (Section 15.6.2).

The co-ordinates of Euclidean geometry are straight and Euclidean space has zero curvature. In Riemannian geometry, space is curved. A space of positive curvature is elliptical, and the co-ordinates of a Riemannian elliptical geometry lie on the surface of a sphere or ellipsoid. A space of negative curvature is hyperbolic, and the co-ordinates of a Riemannian geometry lie on a hyperbolic cone. In Euclidean geometry, the angles of a triangle sum to 180° but in Riemannian geometry they sum to more or less than 180°. For example, on the surface of a sphere, the angle of a triangle sums to more than 180°.

In Euclidean space, the shortest distance between two points (a geodesic) is a straight line but, in Riemannian geometry, geodesics are curved. For example, the shortest distance between two points on the surface of a sphere is the equatorial circle, or great circle, through the points. Since all great circles intersect, like lines of longitude on the Earth, geodesics in elliptical geometry cannot be parallel in the sense of never meeting. If we assume that the space of visual perception is Riemannian with constant curvature, then the main task is to determine the constant that determines that curvature.

While visiting the Dartmouth Institute in New Hampshire in 1945 Rudolf Luneburg was shown some data on space perception that did not conform to Euclidean geometry. He also considered the puzzling results in the Blumenfeld alley experiments (Blumenfeld 1913). In the **distance alley**, subjects are shown a pair of fixed lights on the horizon, one on each side of the median plane. A pair of test lights is then presented at different distances on the horizon plane and the subject adjusts their separation until it appears the same as that of the fixed lights. In the **parallel alley**, subjects adjust two receding lines of lights terminating in the fixed lights until they appear parallel. In Euclidean space the settings in the two tasks should match. In fact, the lights in the parallel alley are usually placed nearer the median plane than are those in the distance alley.

Luneburg (1947, 1950) concluded that the settings correspond to a non-Euclidean hyperbolic geometry—a geometry of constant negative curvature. This conclusion rests on the assumption that the geometry remains the same across these tasks. A parallel array of lights probably provides more information about absolute and relative distances than the two pairs of lights in the distance alley. For example, suppose that the distances of the lights in the distance-alley are underestimated relative to the distances of the lights in the parallel alley. Because of size-distance scaling, observers would perceive the distance-alley lights as closer together than the parallel-alley lights. An array of vertical lines in the frontal plane can appear to lie on a curved plane (see Section 15.6.2). This is not due to an inherent curvature of binocular visual space but rather to a lack of relevant distance information. When more information is added, the lines are perceived correctly.

Luneburg died in 1949 but Blank and others continued to work on the theory in the Knapp Laboratory at Columbia University until 1952 (Blank 1953, 1958). Relevant experiments were also performed by Zajaczkowska (1956), Shipley (1957a, 1957b), Foley (1964), Indow and Watanabe (1984), Rosar (1985), and Wagner (1985).

This work was motivated by a desire to find a consistent implicit geometry of binocular visual space that underlies all perceptual experience, analogous to geometries used in physics. But the formal theory is based on a restricted set of assumptions and viewing conditions and tells us nothing about visual mechanisms. The experiments are highly artificial and the geometry they reveal varies widely between individuals and between tasks. The results of a given experiment may have little to do with how we perceive spatial relations in complex visual scenes (Hecht *et al.* 1999). In any case, human space perception is subject to adaptation, contrast, figural interactions, cue interactions, and long-term recalibration. Furthermore, different visual subsystems may conform to different geometries.

There is also the problem of instructions and criteria. A person viewing railway lines vanishing into the distance may adopt a realistic criterion and report that the lines appear parallel, or they may adopt an analytic criterion and report that the images converge (Carlson 1962). Even if subjects are carefully instructed, one cannot be sure about the criterion they use because they themselves may not be fully aware of what they are doing. Also, judgments of the size and distance of objects vary with the psychophysical method (Ehrenstein 1977). The quest for a unified geometry of binocular visual space based on judgments made under conditions of reduced depth information is probably doomed to failure.

Koenderink and van Doorn (2000) used a different procedure. Observers stood in one spot on a large meadow with knee-high grass and distant trees and buildings. By remote control, observers set a horizontal pointer supported in a cube to align with a sphere, with both pointer and sphere at eye level. The pointer and sphere were at the same distance and 120° apart. Distance varied up to 25 metres. Thus, the pointer, sphere, and subject formed a triangle and the correct setting of the arrow was 30°. The curvature of Riemannian geometry can be derived from the extent to which the sum of the angles of a triangle departs from 180°. The triangle derived from the pointer settings indicated a space of positive (elliptic) curvature for near distances and a space of negative (hyperbolic) curvature for far distances. The transition occurred at about 2 metres. One crucial factor is the accuracy of the angular setting of the pointer and supporting cube to the required angle of 30° as a function of distance. Binocular disparity would probably become ineffective at about 2 metres, leaving only perspective. If the results arise from a changing bias in the local setting of the arrow to 30° they tell us nothing about the global geometrical structure of visual space.

The geometry of binocular visual space is discussed further in Section 24.1.

4.5 CODING PRIMITIVES

In signal analysis one seeks a set of **basis functions** that can describe any complex signal. For example, sine waves of different frequencies and phase provide a set of basis functions by which any well-behaved function can be approximated arbitrarily closely. We can ask whether the visual system contains a set of basis functions by which it analyses complex visual scenes. The basis functions can be regarded as a set of filters applied to the visual input, or we can talk about a set of visual channels or coding primitives. Physiologically, a visual primitive in the space domain is the sensitivity profile of the receptive fields of a set of similar cells at the level of the visual system being considered. For example, at the level of ganglion cells, the visual primitives are the spatial tuning functions of receptive fields of ganglion cells (Section 5.3.2). The idea can be generalized to the spatiotemporal response profiles of cortical cells (Section 5.4.6). The general methods of signal analysis were discussed in Section 3.3.6. Visual primitives that have been proposed for the visual system will now be discussed, starting with Fourier components

4.5.1 Fourier components

4.5.1a *The visual system as Fourier analyzer*

A linear system transmits signals of different frequencies without distortion or interactions. For many purposes, it is convenient to specify the spatial Fourier components of a visual display or of the responses of the visual system. As we saw Section 3.3, it is a particularly useful procedure when one is testing the linearity of a system.

The French mathematician Duffieux (1946) was the first to apply Fourier analysis to optical systems. The lens of the eye is a reasonably linear transmission system. But a system that merely transmits different spatial frequencies does not detect the different frequency components in complex signals. For detection, the different frequencies must stimulate distinct detectors. The system carries out a Fourier analysis, or redescription, of the input to produce output signals that indicate the amplitude and phase of each frequency component in the input.

The Dutch engineer DeLange (1954) analysed the visual response to flickering light in terms of temporal frequencies. Otto Schade (1956) was the first to apply Fourier analysis to spatial aspects of vision. The way an optical system, including that of the eye, resolves images of different spatial frequencies provides an adequate measure of its performance. But that does not mean that the eye performs a Fourier analysis of the input. It simply means that we can use Fourier analysis to assess its performance. The human ear can be said to produce at least a crude temporal Fourier analysis of sound patterns. This is because the ear has many distinct frequency channels, each responding to a narrow range of sound frequencies.

Fergus Campbell and John Robson (1968) proposed that the human visual system beyond the optics has distinct channels, each tuned to a particular range of spatial frequencies, and, as a consequence, achieves a spatial Fourier analysis of visual patterns (Section 3.3.5).

The idea that the visual system performs a Fourier analysis can be misleading. In theory, any system capable of detecting the spatial Fourier components of complex patterns efficiently must fulfil three requirements.

1. It must possess a set of independent and linear detectors each of infinite size and very narrow spatial-frequency bandwidth. Not only detectors of low spatial frequency, but also detectors of high spatial frequency must be large.

2. It must be spatially homogeneous.
3. It must encode both amplitude and phase.

The first condition is not satisfied in the visual system, since receptive fields are comparatively small and not narrowly tuned to spatial frequency. Also, there are many non-linearities in the visual system. The condition of spatial homogeneity is also not fulfilled, since receptive fields become larger and less dense in the peripheral retina. Nevertheless, a typical receptive field in the retina is maximally sensitive to a spatially periodic stimulus of a given period, and receptive fields vary in size and thus vary in their preferred spatial periodicity. The ganglion cells with different sizes of receptive field are often called spatial-frequency channels. The term "spatial-scale channels" is better because it does not suggest that the visual system performs a spatial Fourier analysis, which would require receptive fields of infinite extent.

The importance of phase can be illustrated by considering the Fourier transforms of a thin line and white noise. They have the same amplitude spectrum since they can both be decomposed into sets of equal amplitude sine waves of all spatial frequencies. They differ only in their phase spectra. Phase is random for white noise but, for a line, the peaks of the sine waves coincide at one location.

4.5.1b *Number of spatial-frequency channels*
Four psychophysical procedures have been used to determine the number and bandwidth of the so-called spatial-frequency channels in each region of the retina.

1. In the method of adaptation, one measures the range of spatial frequencies over which an adapting grating elevates the threshold of a subsequently exposed test grating (Blakemore and Campbell 1969).

2. In the method of subthreshold summation, one determines the range of spatial frequencies over which subthreshold gratings reduce the threshold of a superimposed test grating (Graham and Nachmias 1971; Sachs *et al.* 1971).

3. In the method of masking, one determines the range of frequencies over which suprathreshold masking gratings elevate the threshold of a superimposed test grating (Stromeyer and Julesz 1972; Wilson *et al.* 1983). After allowing for effects of probability summation and non-linear interactions, the results are reasonably consistent with the conclusion that there are at least six spatial-scale channels with a half-amplitude bandwidth of about 2.2 octaves for the lowest spatial-frequency channel, and about 1.3 octaves for the highest spatial-frequency channel (Wilson 1991a).

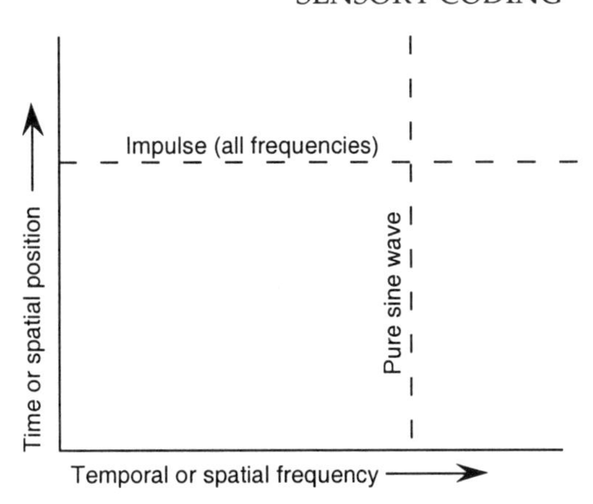

Figure 4.4. The impulse and pure sine wave.
A temporal impulse (delta function) occurs at a specific time, with energy spread evenly over the whole temporal-frequency spectrum. A pure tone contains only one frequency but extends over an infinite period of time. Similarly, a spatial impulse occurs at a specific location, with energy distributed over the spatial-frequency spectrum. A pure spatial sine wave has only one spatial frequency but extends over infinite space.

4. In the final method one compares the contrast at which two gratings are detected with the contrast at which they are discriminated. If these thresholds are the same it is assumed that the two gratings stimulate distinct spatial-frequency channels. Watson and Robson (1981) used this method and concluded that there are seven distinct spatial-frequency channels.

The spatial frequency of a visual channel depends basically on the size and internal structure of receptive fields of ganglion cells. But there is a gradual increase in the size of receptive fields with increasing retinal eccentricity. If the internal structure of receptive fields scales in the same way, there must be a continuous gradation of spatial-frequency channels over the whole retina. The important question is how many channels are present in each local region of visual space.

4.5.1c *Detection of spatial frequency and position*
Investigators have raised the question of whether spatial frequency and spatial position are coded by the same or by distinct mechanisms. At first glance this appears to be an empirical issue, requiring an experimental approach. However, at least part of the answer can be arrived at by a theoretical analysis, once the general properties of a particular system are known. Gabor (1946) applied this analysis to temporal signals in acoustic systems, rather than to spatial frequency and position in the visual system. However, the conclusions apply to both domains.

If the frequency dimension of a signal is represented on the *X* axis and the time dimension on the

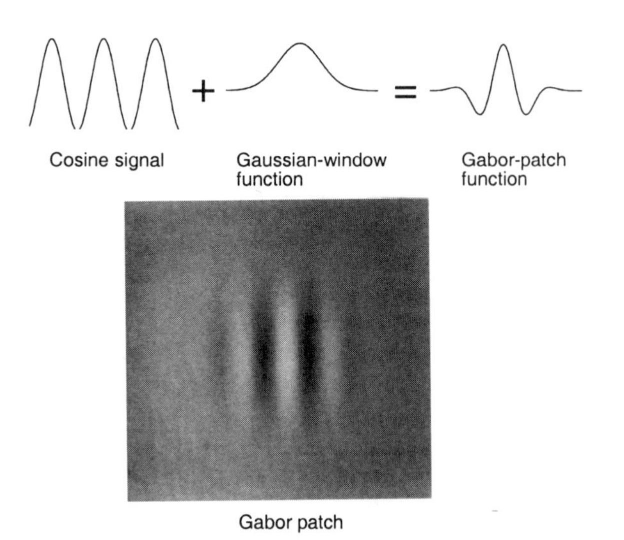

Figure 4.5. Gabor patch.
A Gabor patch results from multiplying a patch with a sinusoidal luminance profile with an aperture with a Gaussian (normal) luminance profile. (Adapted from Graham 1989)

Y axis of Cartesian co-ordinates we have what is known as **a time-frequency diagram**. Consider an impulse of sound occurring at a well-defined time but with energy distributed evenly over the whole frequency spectrum (an impulse, or delta function contains all Fourier components, all with the same amplitude). The impulse is represented in the time-frequency diagram as a horizontal line (Figure 4.4). An impulse is a useful test stimulus to apply to a linear system because, if a linear system's response to an impulse is known, one can calculate the transfer function of the system. This is because the response to the impulse of any linear system is the Fourier integral of its transfer function.

Consider, next, a pure tone of infinite duration with a well-defined frequency but an undefined temporal epoch. It is represented in the time-frequency diagram as a vertical line. Any finite signal may be regarded as intermediate between the impulse, which is determinate in time but not in frequency, and the infinite sine wave, which is determinate in frequency but indeterminate in time. A detector can be designed to extract information about the time of occurrence of events but disregard frequency, or to detect frequency and disregard time. For instance, an ideal oscillograph with uniform response over the whole frequency range and a very short time constant (quick decay of response) is an instrument of the first type. A bank of narrowly tuned oscillators, each with a long time constant (prolonged resonance), may be regarded as an example of the second type of instrument. One instrument cannot do both jobs efficiently because the design characteristics are incompatible.

The same argument applies if we substitute spatial frequency for temporal frequency and position for time. Note that the ideal spatial frequency detector is narrowly tuned to spatial frequency and has a long space constant (infinitely large receptive field), and the ideal position detector is broadly tuned to spatial frequency and has a small receptive field. Because of these opposed requirements, it is impossible to design a detector for both types of information with maximum efficiency. If both types of information are required from the same detector, there must inevitably be a compromise, which is expressed by the fact that, in any detector

$$\text{Time constant x bandwidth} > 1/2$$

A consequence of this relationship in the space domain is that, if we define the sensitivity of a detector to differences in position as *Ds*, and its sensitivity to spatial frequency as *Df*, then the product of these uncertainties cannot be less than one-half, or

$$Ds \times Df > 1/2$$

This reciprocity between two uncertainties is essentially the same as that expressed in Heisenberg's principle of uncertainty, which states that it is impossible to know both the position and frequency characteristics (mass) of a fundamental particle at the same time. The best compromise between the detection of position and spatial frequency is achieved when *Ds* x *Df* is a minimum which, in the ideal case, is 0.5. Gabor defined the characteristics of a detector for which this would be true. For a detector in the space domain, these requirements are met if the sensitivity profile of each detector is a Gaussian (normal) distribution and if, at each location, there are pairs of detectors with sensitivity profiles in quadrature. These two requirements are summarized by the expression

$$S(x) = e^{\frac{-x^2}{2s^2}} \cos 2\pi Fx \text{ (or } \sin 2\pi Fx)$$

where $S(x)$ is the normalized sensitivity profile across the *X* axis of a detector, *s* is the standard deviation of the sensitivity profile, *F* is the optimal spatial frequency to which the detector is tuned, and *x* is distance along the *X* axis (Marcelja 1980; Kulikowski 1980). A Gaussian distribution has the unique property that its Fourier transform is also a Gaussian function. Gabor pointed out that there exists a class of real-valued functions that are more general than Gabor functions and which also maximize the joint detection of position and frequency.

Two sensitivity profiles are in quadrature when one is phase-shifted 90° with respect to the other, like a sine wave and a cosine wave. A cosine function between 0° and 180° is bilaterally symmetrical, whereas a sine function is asymmetrical. A Gaussian profile multiplied by a cosine wave produces an even-symmetric **Gabor function** and a Gaussian profile multiplied by a sine wave produces an odd-symmetric Gabor function. One can think of a Gabor function as the image produced by looking at sinusoidal grating through a Gaussian window. The grating is known as the **carrier** and the window is known as the **envelope**. The scale of the Gabor function varies according to the width of the envelope. When the Gaussian envelope is very wide, the Gabor function is a sine wave (narrowband display), and when it is very narrow, the Gabor function becomes a thin line, or delta function (broadband display). An intermediate case is shown in Figure 4.5.

There is thus a family of Gabor functions extending from a sine wave to a line (Graham 1989). At intermediate values, the Gabor function is a spatially localized patch of damped sine or cosine waves, known as a **Gabor patch**. For a given spatial frequency of the sine wave, the spatial-frequency bandwidth of the patch is inversely proportional to the size of the envelope. Just as any visual scene can be decomposed into sine waves it can also be decomposed into Gabor patches derived from a Gaussian function of specified width.

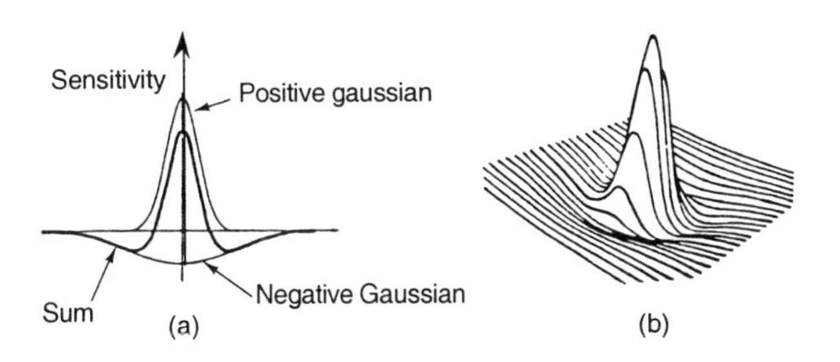

Figure 4.6. Difference of Gaussian sensitivity profile.
(a) An idealized sensitivity profile of a ganglion cell receptive field depicted as the sum of a narrow positive Gaussian distribution representing the excitatory component of the cell's response and a broader negative Gaussian distribution representing the inhibitory component of the response.
(b) A 3-D sensitivity profile of a ganglion cell. The 'volumes' of the excitatory and inhibitory regions are equal, so that when the receptive field is evenly illuminated, the firing rate of the cell is the same as in the dark. (Adapted from Rodieck 1965)

4.5.2 Gabor functions and wavelets

Elongated Gabor patches, as defined above, provide a reasonable fit to the 2-D response profiles of simple cells in the visual cortex (Jones and Palmer 1987). Furthermore, cortical cells with even-symmetric (cosine) and odd-symmetric (sine) sensitivity functions are in quadrature (their sensitivity profiles are relatively phase-shifted by 90°). There is evidence that cells in quadrature occur in pairs in the visual cortex, with different pairs tuned to different regions of the spatial-frequency spectrum (Pollen 1981). This suggests that the visual system achieves an optimal compromise between detection of spatial frequency and detection of stimulus position. This arrangement is also ideally suited for optimally deblurring images and reducing noise in the visual input.

These ideas have been generalized to two spatial dimensions and to the dimension of orientation (Daugman 1984, 1985). Thus the sensitivity profiles of the receptive fields of the basic detectors of the visual system can be described as 2-D Gabor functions with different spatial periodicities, even- or odd-asymmetric profiles, and oriented at different angles. Each detector has an orientation bandwidth of about 15°, a half-amplitude spatial-frequency bandwidth of about 1.5 octaves, and a length-to-width ratio of about 2:1.

A branch of mathematics known as **wavelet theory** has been applied in vision (Young 1987; Daugman 1990, 1991; Farge *et al.* 1993). Wavelets, are localized, self-similar, undulatory functions derived from Gabor functions. They differ with respect to size (dilation), position (translation), and phase. If the wavelets are anisotropic, like Gabor patches, or the elongated receptive fields of cortical cells, they also differ in orientation (rotation). Even with a sparse sampling of size and orientation, one can construct any complex pattern from an appropriate set of wavelets, with a resolution limited by the smallest wavelets in the set. Wavelets are efficient in that the dimensions of position, spatial scale, and orientation are detected independently with no cross talk (see Sakitt and Barlow 1982).

We shall now see that the receptive fields of cells in the visual system can also be fitted by other functions with other properties.

4.5.3 Other visual primitives

The receptive field of a ganglion cell can be considered to be an excitatory area with a Gaussian sensitivity profile superimposed on an inhibitory area with a somewhat wider Gaussian sensitivity profile. The composite sensitivity profile is known as a **difference of Gaussians** (**DOG**). The DOG profile of a ganglion cell is circular symmetric, or isotropic, as in Figure 4.6. The receptive field of a cortical cell may be represented as an elongated DOG. It is even-symmetric (cosine) when the excitatory and inhibitory components are spatially in phase, and odd-symmetric (sine), when they are 90° out of phase.

If the receptive fields of the basic units of the visual system conform to these specifications, it would be useful to use stimuli with these spatial characteristics, since they are most easily detected. Hugh Wilson and others have used DOGs extensively in the study of contrast sensitivity and stereoscopic vision, as we will see in later chapters. A DOG stimulus, unlike a sine-wave grating, has a well-defined location in space. It also has a peak, or centre spatial frequency, which can be varied by changing the width of component Gaussian distributions.

Since a Gaussian function is not a pure sine wave, it necessarily has a certain spatial bandwidth. By differentiating a Gaussian, one obtains a Gaussian with a narrower spatial bandwidth. The sixth and tenth derivatives of a Gaussian function, known as **D6** and **D10**, are often used in visual experiments, including those on stereopsis. Stork and Wilson (1990) showed that Gaussian derivatives are real-valued functions, which like Gabor functions, maximize the joint detection of position and frequency. Being real-valued they have the advantage that they require only single detectors rather than the paired detectors in quadrature required by complex-valued Gabor functions.

Other types of mathematically defined filters have been borrowed from physics and used to characterize low-level visual processing. These include zero crossings (Marr 1982), Cauchy functions (Klein and Levi 1985), dipoles (Klein and Levi 1986, Klein *et al.* 1990), and cepstral filters (Yeshurun and Schwartz 1990) (Section 16.1.4). Although each formalism has advantages, it is unlikely that the visual system conforms exactly to any one of them. Spatial filters may improve the efficiency of low level coding, but visual processes are too nonlinear to be described by any set of linear filters. Also, linear image transformations do not solve the problem of visual recognition, they simply postpone it.

4.6 HIGHER-ORDER SENSORY SYSTEMS

4.6.1 Types of processing

4.6.1a *Serial and parallel processing*

Sensory information related to space perception is combined in a great variety of ways for diverse purposes. We often distinguish between **serial processing** and **parallel processing** of sensory information in the central nervous system. In one type of serial system, referred to as hierarchical, a given stimulus attribute is processed sequentially at increasing levels of complexity. For example, visual motion is processed at successively higher levels in the nervous system. In a second type of serial system, distinct stimulus attributes are processed sequentially. For instance, colour and binocular disparity are processed serially since some colour processing occurs in the retina before binocular disparity is processed in the visual cortex. In a third type of serial system, usually referred to as serial search, stimuli in different locations or different attributes of a given object are attended to in sequence. In a parallel system, different stimuli or different attributes of a stimulus are processed simultaneously by neural mechanisms laid out in parallel. For example, inputs from rods and cones are processed in parallel in the retina.

The terms 'serial' and 'parallel' can also be used to describe the structural organization of sense organs with respect to a given task. For example, the sense organs in the joints of the arm are structurally in series with respect to judging the position of the finger in relation to the torso. I refer to serial tasks of this kind as **nested**. The two eyes are structurally in parallel. Whether sense organs are nested or in parallel has nothing to do with whether information is processed serially or simultaneously (in parallel).

4.6.1b *Types of judgment and stimuli*

A **relational judgment** requires information from two or more sensory inputs. For example, the task of judging the position of the hand relative to the torso is relational, since one cannot perform it if information from any of the joints is missing. A **multi-cue judgment**, like that of judging depth from several depth cues, is not relational since any one cue is sufficient. A judgment is **underdetermined** by an ambiguous stimulus. For example, a change in the size of a retinal image is ambiguous because it can be due to a change in object size or object distance, and perceived distance is underdetermined by changing image size. Sensory inputs are **dissociable** when judgments can be made on the basis of each one, at the same time as a judgment about the relationship between them. For example, one can judge the angle of each arm joint as well as the total posture of the arm. In binocular fusion, stimuli are **nondissociable** because the visual system has no access to the monocular components of a fused image, only to the disparity between them. The images are dissociable when the disparity is beyond the range of fusion. Sensory inputs are **independent** when a judgment based on one does not affect a judgment based on the other (Ashby and Townsend 1986 discuss types of perceptual independence). Garner (1974) distinguished between **integral** and **separable** stimulus dimensions. Integral dimensions necessarily coexist, like hue and saturation, while separable dimensions may occur independently of one another, like shape and flicker. I do not use this distinction.

In many cases, it is an advantage to have cells that respond to a particular combination of features. For instance, cells in several visual areas are jointly tuned to disparity and direction of motion. A cell of this type could be selectively responsive to motion in a particular depth plane relative to the plane in which the eyes are converged (Sections 6.2 and 23.4.5). Or it could respond to an object approaching the head along a particular trajectory (Section 29.6.2). For purposes such as comparing colours or the orientations of lines, the visual system seeks to isolate specific features, but, for other purposes, it seeks to detect the difference between or the conjoint variation of two features. An attribute of the visual world that is defined in terms of the conjunction of two or more simple features is a **higher-order feature**. A higher-order feature is **intrasensory** when the component features are detected in the same sense organ, and **intersensory** when they are detected by more than one sense organ.

Higher-order features fall into four classes; **nested sensory inputs** based on vector addition, **spatial relationships** based on differences between stimuli, **covariance functions** based on stimulus ratios, products, or correlations, and **multi-cue systems** based on summation or averaging. Each of these classes will now be considered.

4.6.2 Nested sensory systems

A nested sensory system consists of sense organs embedded in a series, or chain, of jointed body parts. For example, inputs from proprioceptors associated with the shoulder, elbow, and wrist form an **intrasensory nested system** with which we judge the 3-D position of the unseen finger in relation to the torso (see Matthews 1988). The efference copy associated with active movement at each joint can also be regarded as a sensory input. The position of the hand with respect to the torso is indicated by the sum of the arm-joint vectors, with each vector scaled by the length of the segment of the arm distal to the joint (Imamizu *et al.* 1995; Pagano and Turvey 1995). The lengths of limb segments form part of the body schema, or internal representation of the body. This schema operates largely at a preconscious level. Phantom limbs experienced by amputees illustrate that the body scheme is independent from the somaesthetic-proprioceptive system.

The sensory and motor components of the arm-joint system are depicted on the left of Figure 4.7.

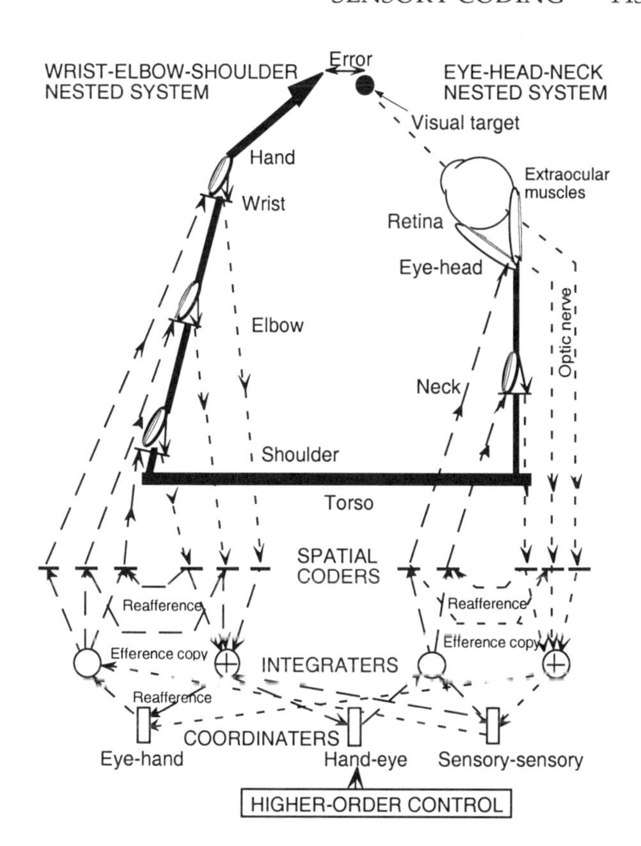

Figure 4.7. The eye-hand coordination system.
Structural components (bold) are linked by joints with muscles and sense organs. Dashed lines are sensory or motor nerves and short bars are neural centres. Spatial coders process sensory inputs and motor efference. Integrators process information from in-series sense organs or to in-series muscles. Co-ordinators relate sensory information from the arm with that from the eye-head system or relate sensory inputs with motor outputs. Examples of efference-copy and reafference signals are shown.

The hand moves in nearly straight paths, which suggests that movements are initially programmed in extrinsic co-ordinates that map the trajectory of the hand relative to the body, rather than in intrinsic joint co-ordinates (Haggard *et al.* 1995). Moreover, arm trajectories involve intersensory coding, since they are planned in terms of their visually perceived straightness rather than in terms of their actual straightness (Wolpert *et al.* 1995). Before execution, movements must be transformed into movements of muscles, which is not a simple task, since the same arm position can be achieved in many ways (see Soechting and Terzuolo 1988).

Some cells in the posterior parietal cortex of the monkey are influenced by proprioceptive inputs from two or more arm joints (Mountcastle *et al.* 1975). Some cells respond when the limb is in a particular posture (Costano and Gardner 1981). These cells could serve as sensory integrators. Some cells in the premotor cortex and basal ganglia respond to the direction of an arm movement irrespective of the muscle pattern used to implement the movement. These could serve as motor integrators. The transformation into a pattern of muscular contractions at the level of motor coders depends on both the direction of movement and the load against which the arm is working. The transformation is probably achieved by other cells in the premotor cortex, basal ganglia, and cerebellum (see Georgopoulos 1991).

An example of an **intersensory nested task** is that of judging the direction of an object with respect to the torso by vector addition of the local sign of the image, and the sensed positions of eyes in head and of head on body (see Figure 4.7). The system is subject to constant errors, especially when the eyes are in an eccentric position (Rossetti *et al.* 1994). The sense of position of the eyes is registered by efference copy (corollary discharge) but proprioception may also be involved. The sense of position of head on body is coded mainly by proprioception but efference could be involved. The spatial coders process information from each sense organ or generate the motor commands to muscles at each joint. Integrators process information from a set of nested receptors or generate co-ordinated motor outputs to the neck muscles and extraocular muscles.

Spatial information from the eyes and other sense organs is integrated at several levels in the nervous system. Some cells in the posterior parietal cortex (areas 5 and 7) of the monkey are influenced by proprioceptive signals from both the eyes and the neck, and seem to be concerned with coding the bodycentric direction of gaze (Brotchie *et al.* 1995). Visual inputs and eye-position information interact at various subcortical and cortical sites (Section 19.9). Such cells could serve as spatial integrators for the headcentric direction of visual stimuli (Andersen and Zipser 1988). Cells in the parietal lobe and premotor cortex respond to visual stimuli that occur in the region of the body surface to which the cells also respond (Section 5.7.3f).

Nested systems adapt to the growth of the body, and to unusual sensory inputs. For example, people readily adapt their pointing to visual targets seen through a displacing prism (Howard 1982). A PET scan of the human parietal cortex has revealed changes associated with such learning (Clower *et al.* 1996). When an optical system gradually distorted the visual trajectory of the hand into a curve, subjects unconsciously adapted the reaching movement to maintain an apparently straight path (Wolpert *et al.* 1995; Flanagan and Rao 1995), although the path that visually appeared straight was not actually straight (Wolpert *et al.* 1994).

A nested system is commutative if the temporal order of operations does not affect the outcome, as for rotations of the eyes, head, and torso on vertical axes. Rotations of an eye about three hypothetical gimballed axes of horizontal gaze, vertical gaze, and torsion are non-commutative, since the final position of the eye depends on the order in which the movements occur. Eye movements are commutative if they occur only about axes in a plane fixed to the head—Listing's plane (Section9.2.2).

Tasks performed by vector addition within a nested sensory system are relational and dissociable. For example, the task of judging the position of the hand relative to the torso is relational since information from all the joints is required, and is dissociable because one can judge the angle at any joint as well as the position of the hand relative to the body. The components of nested systems are typically independent in the sense that a judgment of the state of one component does not affect that of another.

Constant errors of judgments based on the components of a nested system should sum algebraically to produce the constant error of the relational judgment. For example, directional errors in the local-sign system of the retina, in the sense of position of the eye, and in the sense of position of the head on the torso should add to produce the overall error in judging the position of a visual object relative to the unseen body (Section 17.7). The variance of judgments based on the separate components should sum to produce the variance of the relational judgment. This is the **additive variance hypothesis**.

4.6.3 Stimulus relationships

Many sensory systems are devoted to the detection and discrimination of spatial or temporal relationships between two or more stimuli in the same or in different sensory modalities. The spatial attribute may be position, length, distance, orientation, curvature, alignment, aspect ratio, or any other spatial property. The task may be that of detecting, nulling, or scaling a one-dimensional difference between two stimuli. The sensory systems are structurally in parallel rather than nested. Judgments are relational, since both stimuli are necessary, but the component stimuli may or may not be dissociable. **Multidimensional scaling** deals with how we compare and classify stimulus objects that differ with respect to two or more features (see Davison 1983). Relational judgments may be intrasensory or intersensory.

4.6.3a *Intrasensory relationships*

In vision, stimulus orientation, alignment, and collinearity are examples of intrasensory relationships. Specialized detectors for these features were discussed in Section 4.2.6.

Examples of higher-level features are differences in separation, length, or orientation between two stimuli presented simultaneously. There is evidence that discrimination of such differences depends on specialized higher-order sensory systems that register simultaneous responses in cortical cells some distance apart (Morgan and Regan 1987; Kohly and Regan 2000). This involves lateral connections between cortical cells (Section 5.4.7).

Other examples of intrasensory relationships are the binocular disparities that we use for stereoscopic vision, and the differences between sounds in the two ears that we use for auditory localization.

Detection of the spatial relationships between stimuli forms the basis for pattern recognition. Recurring patterns, such as corners, T-junctions, and intersections have specialized detection systems in higher visual centres of the brain (Section 5.7.2b). Pattern perception is subject to error arising from non-linear interactions, as in geometrical illusions, tilt contrast, and figural aftereffects. It is also subject to ambiguity. For example, superimposed orthogonal gratings may be perceived as two gratings or as a checkerboard. They are more likely to be seen as two gratings under conditions that foster an impression of transparency or when they are separated in depth (Section 23.2.2).

There are also dynamic intrasensory relationships. For example, the parts of the human body move in characteristic patterns (Johansson 1973). The perception of patterns of motion is also subject to error, as in induced motion in which a stationary object appears to move when seen against a moving background (Section 23.4.7). They are also subject to ambiguity. For example, oblique, orthogonal gratings moving in opposite directions past an aperture can be seen as a coherent plaid moving in a direction and at a velocity that represent the vector sum of the component motions, or they can be seen sliding over each other (Section 23.2.2).

4.6.3b *Intersensory relationships*

A person may respond to relationships between stimuli in different sensory modalities. The task of judging a stimulus in one modality with respect to that in another is often referred to as **cross-modal matching**.

The sense organs involved in an intersensory relational task are structurally in parallel with respect to a common reference frame. The frame may be part of the body, such as the z body axis, an external reference frame, such as gravity, or a common internalized metric, such as centimetres, degrees of angle, or straightness. One can think of the stimulus attribute of each object generating a distribution of neural activity in a feature detector of a sensory system, with a mean and a variance. Intra- and intersensory discrimination depends on the least difference between the means of the two distributions that can be detected. Intersensory scaling depends on comparing two suprathreshold stimuli. Ideally, the variance of a comparison task should equal the sum of the variances of the component tasks performed separately. Constant errors should add algebraically.

In measuring intersensory discrimination, subjects should believe that the stimuli in the two sense organs arise from different objects. Otherwise the stimuli will be perceived as linked. For example, if one hears a bell ringing and sees a bell, the sound seems to come from the seen bell, even when the heard and seen bells are up to 30° apart.

Also, in measuring intersensory discrimination, the task should be performed in open-loop mode, so that error is not displayed. Otherwise, performance will improve with practice.

4.6.4 Stimulus covariance

Some stimuli covary. As one stimulus changes, the other changes in a characteristic way. The covariation may be defined by a correlation, a product, a ratio, or by a more complex relationship, such as multiplication after squaring. Judgments of stimulus covariance are relational because they cannot be performed unless both sensory inputs are present. I distinguish the following types of stimulus covariance.

4.6.4a *Stimulus covariance within a sensory system*

Stimuli can covary because of some feature of the visual system. For example, the chromatic aberration of the eye creates colour fringes of opposite colour on opposite edges of an object. These stimulus relationships provide no information about the world. In many cases, there are mechanisms for perceptually cancelling them (Section 3.2.2c). Other forms of stimulus covariance provide information about the world.

4.6.4b *Covariance between motor efference and afference*

We can regard the motor command, or **efference copy** associated with a voluntary movement as a type of sensory input. A voluntary movement also produces stimulation of sense organs. For example, when we move our eyes, the image of a stationary object sweeps across the retina. Von Holst and Mittelstaedt (1950) called these forms of sensory stimulation **reafference** to distinguished them from exafference, which is sensory stimulation that does not arise from voluntary movement. Efference copy scaled by reafference expresses the invariant relationship between voluntary motor commands and consequent sensory stimulation. Over many repetitions of voluntary movement, animals learn how efference and reafference are related (von Holst 1954). The sight of self-produced movements of the limbs plays a crucial role in infant development (Held and Bauer 1967; Bahrick and Watson 1985).

Proprioceptive reafference consists of stimuli from joint receptors, muscle spindles, or the vestibular system that are correlated with voluntary motor efference. Over time we learn the relationships between efference and proprioceptive reafference. A mismatch between the current efference-reafference relationship and the learned relationship, signifies the presence of an external stimulus or force.

The following is an example of a covariance function based on proprioceptive reafference. Motor commands for eye movements generate proprioceptive reafference, which provides an error signal for the long-term calibration of the oculomotor system. Kittens with section of proprioceptive afferents from extraocular eye muscles suffer permanent deficits in visual-motor coordination and depth discrimination (Section 13.4).

Exteroceptive reafference consists of tactile, visual, or auditory stimuli that are correlated with motor efference arising from voluntary motion of some part of the body. We also learn what stimulation to expect from a given self-produced motion. A mismatch between expected and actual stimulation signifies that part of the reafferent signal is due to external events. A persistent mismatch in either type of reafference signifies that the system is in need of recalibration. This happens as the body grows during infancy or is injured. Even a short period of mismatch between self-produced motion and reafference exteroceptor such as the eyes or ears can lead to recalibration of the system. For example, we readily learn to adapt pointing responses to visual targets viewed through displacing prisms (Howard 1982).

The following is an example of stimulus covariance based on exteroceptive reafference. The depth interval between two objects equals the ratio of the velocity of parallactic motion between the images of the objects to the velocity of sideways head movement, scaled by the distance of the nearer object (Section 25.3).

4.6.4c *Stimulus covariance and perceptual constancies*

Many stimulus features covary in a characteristic way as a function of the position, orientation, or motion of the observer. Judgments based upon such covarying features are referred to as **perceptual constancies**. They allow us to detect invariant properties of the world as we move about or as the stimulus object moves. The basic problem of perceptual constancy is to define the invariant features in stimuli viewed from different vantage-points and in different sizes and locations (see Riesenhuber and Poggio 1999). An experimenter must define these invariances before asking whether observers use them, and that is not always an easy task.

The following are examples of covariance functions that depend on the position or movement of the observer or on some other property of the observer.

1. The size of the retinal image of an object is inversely proportional to the distance of the object from an observer. Size constancy refers to the ability to use this relationship to estimate the linear size of an object at different distances (Section 26.2.2).
2. For a small spherical object moving towards an eye at a constant speed, the rate at which its image increases in size is proportional to the velocity of its approach and inversely proportional to the time to impact (Section 29.1).
3. The binocular disparity produced by two objects separated in depth by a fixed distance is inversely proportional to the square of the distance of the nearer object from the observer (Section 15.3.1).

4.6.4d *Covariance of observer-independent stimuli*

Some features of the world covary independently of the position or motion of the observer. A stimulus invariance may involve a simple association of features such as the colour, size, and noise of a bee or the colour, shape, and smell of a rose. These **taxonomic invariances** can be infinitely complex, as in face recognition, the classification of animal and plant species, and anatomical structures, social interactions, and people. They often defy analysis.

Some features of objects are related by a **transformational invariance**. For instance, the shape of an object may be transformed in some lawful way. For example, the shape of the body is transformed in a characteristic way as we grow or as species evolve.

I return to the perception of stimulus covariance in Section 4.7.3.

4.6.5 Multi-cue systems

4.5.6a *The nature of cues*

Sometimes, the same general percept is generated by different sources of sensory information. For example, a percept of self-rotation is produced by stimulation of the semicircular canals of the vestibular system or by observing a moving scene. Perceptually equivalent sources of information are often referred to as **cues**. But the term "cue" is ambiguous. It can refer to a property of the distal stimulus or of the proximal stimulus, or it can refer to one of the sensory processes that code the proximal stimulus.

For example, a cue can be an approaching object, a change in the size of the object's image, or one of the sensory processes that code changing image size. One sensory process involves registration of changing area and another involves registration of motion. These processes are independent. One can change area without stimulating motion detectors by using discrete increments or equiluminant stimuli. Also, one can create an impression of increasing area in a motion aftereffect from a rotating spiral without changing the area of the image. The motion system itself may consist of several more-or-less independent sensory channels, such as short-range and long-range motion detectors. We could call these motion channels different cues to motion.

4.6.5b *Types of multi-cue systems*

In a multi-cue system, there are two or more relatively independent types of sensory information for a judgment about a specified feature of the world. For example, judgments of the relative depth between two objects may be based on binocular disparity, perspective, accommodation, and on visual features indicating that one object overlaps the other. The sensory systems involved are structurally in parallel rather than nested, but inputs may be processed simultaneously (in parallel) or sequentially (in series) before they are combined to form the basis of a judgment or action.

Multi-cue judgments are not relational, since they may be based on any one of the cues in isolation. They are typically nondissociable. For example, we do not have separate access to depth impressions generated by each cue to depth when they are presented together, but only to a single impression of depth. Multi-cue systems can be intrasensory or intersensory. Studies of multi-cue systems have been concerned with the following issues.

1. Sensory summation and masking One issue is whether the detection threshold is lower for a stimulus seen by both eyes than for a stimulus seen by only one eye, after making allowance for the increased statistical probability of detection with two eyes (Section 8.1). A second issue concerns the circumstances under which different stimuli interfere with each other (Section 8.2).

2. Multistable percepts The issue is whether the stability of one interpretation of an ambiguous stimulus is influenced by the addition of other information. The stability of a percept is indicated by a) its latency, b) its probability of occurrence on different occasions, and c) its resistance to spontaneous change (Section 27.1).

3. Multisensory judgments of stimulus magnitude The issue is whether different sources of sensory information combine to determine the perceived magnitude of a stimulus attribute. Do they combine additively, by averaging, by a multiplicative process, by use of the most reliable information, or in some other way?

I distinguish three types of multi-cue systems.

Type 1. Common neural code systems. Different sources of information in the same or different modalities are sometimes converted into a common neural form at an early stage of processing. For example, visual motion signals and inputs from the semicircular canals converge on the same cells in the vestibular nucleus.

Type 2. Distinct continuously variable cues.

a) Where both sources of information unambiguously determine a response.

b) Where one source of information is ambiguous because it can be interpreted as arising from two distinct causes, it can be assigned to more than one covariance function. This is **covariance ambiguity**.

Type 3. Systems involving categorical judgments. For example, phonemes can vary continuously but a speaker of a given language perceives discrete categories. Similarly, strokes in written letters can vary continuously but readers see distinct letters. When a phoneme or letter stroke falls on the decision boundary between two categories there is **category ambiguity**. For example, when a sound is intermediate between a 'b' and a 'd', speakers of English hear one or other phoneme according to the context. Systems involving discrete stimulus values necessarily involve categorical decisions. For example, which of two objects overlaps the other is a discrete variable and gives rise to a categorical response of the occluding object being in front of the occluded object (Section 24.3).

Multiple cues to a given stimulus attribute may be combined in the following ways.

4.6.5c *Multi-cue averaging*

According to the central tendency theorem, the best estimate of a quantity is the mean of independent sources of information. If one source is more reliable than another, the best estimate is a weighted mean. In assigning weights to sensory cues it is important to define the judgment. A cue with high weight for ordinal judgments may contribute nothing to quantitative judgments. For example, the overlap cue to depth has high reliability for judgments of depth order but provides no information about the magnitude of relative depth. On the other hand, relative motion between the front and back of a transparent rotating object provides reliable information about relative depth but is ambiguous with respect to which is the front and back of the object (Section 24.3.3).

The defining characteristic of an averaging system is that sensory information is summed and divided by the number of inputs. If the strengths of two cues are s_1 and s_2 and their weights are w_1 and w_2, cue averaging can be represented by

$$\frac{w_1 s_1 + w_2 s_2}{w_1 + w_2}$$

Division by the sum of the weights converts them into relative values, which sum to unity. Any constant error in one cue system will distort the judgment in proportion to the number of cues involved and the weight given to each. Weights assigned to cues may vary with their magnitude or with changes in associated features of the stimuli, such as their spatial and temporal frequency. In that case, cue averaging is nonlinear (Anderson 1974).

If two sources of information that code the same quantity are combined in the most efficient way, the reciprocal of the total variance should equal the sum of the reciprocals of the component variances (Fisher 1966). Thus the reciprocal of the variance of a judgment based on averaging cues *A* and *B* (V_{AB}) should equal the sum of the reciprocals of the variances of judgments based on each cue separately, or:

$$\frac{1}{V_{AB}} = \frac{1}{V_A} + \frac{1}{V_B}$$

This means that judgments based on several cues should be more precise than those based on one cue.

In a cue-averaging system, two or more cues provide independent estimates of the same continuously variable stimulus feature. Lack of agreement between the estimates constitutes a mutual error signal. In the absence of other information, the observer cannot know which estimate is correct. For small discrepancies, the best strategy may be to use a weighted average. However, even in statistics, averaging may not be the best strategy when two estimates differ widely. It is usually better to question the validity of one of the estimates. Large discrepancies between cues are resolved by cue averaging only when the neural signals generated by the two cues are similar (multi-cue systems Types 1 and 2, involving common and distinct codes). Otherwise, large discrepancies are resolved in one of the other ways discussed below.

Disparate visual images appear in an intermediate location that depends on their relative luminances in the two eyes, even though their disparity produces depth and they appear in distinct lateral positions when the monocular images are alternated (Section 17.7.3). The apparent brightness of an uncontoured surface viewed with both eyes is a weighted mean of the brightnesses of the images seen by each eye separately (Section 8.1.3).

These examples are Type 1 multi-cue systems (common code) since stimuli are neurally combined at an early stage so that they are nondissociable. The two sources of information may be regarded as repeated measures of the same quantity. Averaging of inputs occurs most clearly in such cases and shows evidence of being reasonably linear.

In other cases, cue averaging is likely to be nonlinear. For example, the location or motion of a visual contour may be defined in terms of the visual attributes of luminance, colour, motion, texture, or binocular disparity. The literature on how precisely we discriminate shapes defined by different features is reviewed in Regan (1991b). Rivest and Cavanagh (1996) found that the precision of localization of a contour improved as more attributes were added. Rivest *et al.* (1997) found that improvement of orientation discrimination with practice transferred between bars defined by colour, by luminance, and by motion. These findings support the idea of a pooling of information from the different feature-detection systems within the same sensory modality, but the pooling is highly nonlinear.

The relative distances of objects in depth may be derived from any one or any mixture of a variety of depth cues including binocular disparity, visual parallax, and perspective (Chapter 24). It has been claimed that depth cues either add or average. For example, Bruno and Cutting (1988) reported that perceived relative depth between simple squares is based on the sum of the monocular cues of size, relative height, occlusion, and motion parallax. However, Massaro (1988) interpreted the data according to a model in which perceived depth is based on the most reliable cue. See Section 27.1.2 for more discussion of this issue.

Nonlinearity in cue averaging may arise for the following reasons.

1. The relative potencies of cues may vary in a nonlinear fashion with respect to the judged feature. For example, vergence is a cue to depth only over short distances, whereas disparity operates over larger distances and perspective over even larger distances.

2. Cues may be affected in different ways by changes in associated sensory features such as spa-

tial frequency, contrast, or colour. For example, depth cues such as overlap and disparity operate in both the luminance and chromatic domains, while others, such as shading, operate in only the luminance domain (Cavanagh 1987).

3. Experimenters sometime assume, falsely, that a given cue is absent when it is held at a value of zero while other cues are varied. For example, the size cue to depth is not absent when image size is held constant because constant image size is a cue that stimuli have the same depth.

4. The weighting function for cue averaging may vary with the level of agreement between cues. The cue-averaging system may be switched off when cues are highly discrepant. For example, Zacharias and Young (1981) suggested that the weighting of vestibular visual cues to self-rotation are averaged when they are consistent, but that one cue is ignored when they are highly inconsistent.

4.6.5d *Multi-cue trading*

The weighting and relative sensitivity of cue systems that engage in averaging can sometimes be measured by a **cue trading function** (Section 3.2.5). For example, the apparent displacement of a sound due to interaural differences in intensity may be nulled by an interaural difference in time of arrival of the sound (Harris 1960).

The existence of cue trading is evidence that inputs from two cue systems converge to produce a signal common to both. In other words, cue trading occurs in Type 1 (common-code) multi-cue systems. For example, interaural intensity differences are converted into time differences because the more intense sound is processed with shorter latency. This time difference and the time-of-arrival difference are then converted into the same position-dependent signal by a delay line in the olivary nucleus (Van Bergeijk 1962).

In other cases, integration of discrepant signals from different cue systems is achieved at a higher level of processing. For example, motion in depth produced by image expansion can be traded against motion in depth produced by changing disparity (Gray and Regan 1996). This topic is discussed in Section 29.3.7.

Cue averaging and trading occurs convincingly only between cue systems that generate similar signals at a relatively early stage of processing. Other multi-cue systems resolve large discrepancies by one or other of the following processes.

4.6.5e *Cue confirmation and percept stability*

Some multi-cue systems do not average but simply confirm or supplement one another. Cue confirmation operates most clearly in Type 3 (categorical) multi-cue systems in which the task is that of interpreting **bistable percepts** or recognizing discrete **stimulus categories**. Cue confirmation operates like voting, where the weighted values of the various cues are summed to determine the strength of a given interpretation of the stimulus. Since the alternative interpretations are discrete, the pooled voting strength of the different cues determines the **stability** of a given interpretation, not its magnitude. An interpretation is stable when it has a high probability of recurring under similar circumstances, has low latency, and does not change as the stimulus is maintained.

The reversible perspective of a 3-D skeletal cube provides an example of cue confirmation for a bistable percept. Reversals occur more frequently when the cube is viewed monocularly than when it is viewed binocularly (Howard 1961). Sperling and Dosher (1995) found that the stability of a particular depth interpretation of a bistable 3-D cube depends on the additive contributions of disparity and the relative contrasts of far and near sides. The magnitude of depth was not affected, only its sign.

An example of cue confirmation for a categorical judgment is the increased certainty about what word is spoken when we can both hear the speaker and see the lips moving. The two sources of information do not average since they are not composed of continuous variables. Under normal circumstances, they simply confirm each other, and supplement each other if one or other source is weak. Performance on this type of task involves high-level stimulus categories stored in memory. If the word heard does not prompt recall of the same word as that prompted by sight of the speaker's mouth, as in a badly dubbed movie, the system accepts the least noisy stimulus with a bias towards the word that best fits into the context of other words spoken.

It is easy to confuse a cue-confirmation mechanism with cue averaging in a Type 3 (categorical) multi-cue system. For example Trueswell and Hayhoe (1993) concluded that the magnitude of perceived depth between two test squares is an average of information from disparity and overlap (Section 27.4). However, they measured the probabilities of seeing each interpretation averaged over trials, not the magnitude of perceived depth. The averaging was done by the experimenters, not by the subjects. Thus, in depth perception, we may accept the most highly weighted cue and the one most consistent with other information and then seek for confirmation in other cues to depth. If the other cues to depth do not agree, they may be reinterpreted or ignored rather than averaged with the most reliable cue.

4.6.5f *Cue reinterpretation*

A conflict between cues to a particular stimulus feature may be resolved by a reinterpretation of the stimulus situation. This happens in a Type 2 (distinct codes) multi-cue system when a change in a stimulus feature is ambiguous, or **underdetermined**. For example, a change in the size of the image of an object could arise from motion of the object in depth, but it could also arise because the object is changing in size. Thus, a change in image size is an element in two **intersecting covariance functions**; the invariant relation between image size and distance for a given object, and the invariant relation between image size and object size for a given distance.

I propose the following rule. When there is a severe conflict between two cues to a stimulus feature, the more ambiguous cue will be reinterpreted. If neither cue is ambiguous, a conflict causes a recalibration of the cue systems rather than a reinterpretation. Cue recalibration is discussed below. Cue reinterpretation can be continuous or saltatory, as the following examples will show.

When a change in the size of the binocular images of an object matches the changing disparity between the images, we see an object of fixed size moving in depth. But if the images change in size without an equivalent change in relative disparity, then the object appears to change in size by an amount that accounts for that discrepancy. This reinterpretation resolves the conflict between cues to depth because the component of changing size that is not commensurate with changing disparity is no longer accepted as a cue to depth. There would be a residual perceptual conflict only if the object were not expected to change in size, or if other sensory information, such as tactile information, signified that it was not changing in size. The residual discrepancy may be resolved in another way. For instance, the object may be perceived as nearer than it is because a given change in disparity produces a smaller change in image size for a nearer object than for a far object. The general principle is that enunciated by Helmholtz (1910, Vol. 3, p2), "· · · such objects are always imagined as being present in the field of view as would be there in order to produce the same impression on the nervous mechanism, the eyes being used under ordinary normal conditions."

In the above example, cue reinterpretation involves reassigning a change in a particular stimulus feature from one covariance function to another, where both functions embody distinct continuous variables (Type 2 multi-cue system). In the following example, cue reinterpretation involves a change of state of a two-valued feature (Type 3 multi-cue system).

A wire cube spontaneously reverses in apparent depth when viewed for some time. When it does so, the perspective of the cube is reinterpreted to conform to its new depth. If the cube is rotating, each reversal of perspective changes the percept from that of a rigid cube rotating in its actual direction to that of a non rigid trapezoid rotating in the opposite direction (Howard 1961; Sperling and Dosher 1995).

4.6.5g *Complementary multi-cue systems*

Different sources of information may complement each other, each providing something lacking in the other. The following are some of the ways in which this can happen.

1. Disambiguation of cue attribution The ambiguity in one sensory cue can be resolved by a second cue. For example, a given otolith input may be produced by head acceleration or by head tilt. This ambiguity manifests itself in the inability of pilots flying in clouds to detect whether the aircraft is accelerating or climbing. The ambiguity may be resolved by sight of the ground or by inputs from the semicircular canals, which register head rotation but not linear acceleration.

Prior exposure to an unambiguous stimulus may bias the interpretation of an ambiguous stimulus. For example, prior exposure to a corrugated surface defined by the unambiguous cue of binocular disparity biases the perceived depth of a corrugated surface defined by the ambiguous cue of motion parallax (Section 22.6.1).

2. Disambiguation of stimulus sign Some cues are ambiguous with regard to sign but not with regard to magnitude. The ambiguity of sign may be resolved by a second cue, which is often a two-valued stimulus lacking quantitative information. For example, image blur is an ambiguous cue to lens accommodation, because an object nearer than the focal point produces the same blur as one beyond the focal point. The sign of chromatic aberration of the lens resolves the ambiguity (Section 9.1.2c). In a second example, the ambiguity of the sign of depth perspective is resolved by the two-valued cue of stimulus overlap (Section 24.3).

3. Complementary ranges Multiple cues typically extend the stimulus range of a feature-detection system. This is because one cue may be more effective at one end of the stimulus range, and a second may be more effective at the other end. For example, binocular disparity is most effective for near viewing, while perspective remains effective for distant viewing. Also, the effectiveness of a cue for a given stimulus feature may differ as a function of some other stimulus feature. For example, accommodation becomes an ineffective cue to distance for stimuli

with low spatial frequency, while perspective is relatively immune to a lowering of spatial frequency.

4. Filling in One cue may fill in for a second cue that is hidden or is not being attended to. For example, sight of someone speaking may help us to recognize a word that is difficult to hear. Multiple cues protect against loss or pathology of one cue system.

5. Provision of an error signal One cue may provide an error signal that is absent in another cue. For example, the extraocular muscles provide little or no feedback to indicate the adequacy of the vestibulo-ocular response induced by head rotation in the dark. If the eyes are open there is an error signal in the form of retinal slip velocity.

4.6.5h *Cue dominance*

More reliable cues are more heavily weighted in a conflict situation. This is known as cue dominance. In extreme cases, one source of information completely overrides conflicting information. For example, when a bell is seen in one place and heard in place less than 30° of subtense away it seems to be located where it is seen rather than where it is heard. This is the basis of ventriloquism. When a felt object is optically minified it feels smaller than when viewed normally.

4.6.5i *Cue dissociation*

When the conflict between two cues is severe, the cues dissociate and create an impression of two objects. For example, a seen bell more than 30° away from a heard bell dissociates into two, a seen bell in one location and a heard bell in a second location.

4.6.5j *Cue recalibration*

Cue conflicts may lead to long-term recalibration of multi-cue systems. This is most likely to occur when both cues are unambiguously related to a given perceptual interpretation. For example, an unusual relationship between convergence and familiar size affects the scaling of perceived depth (O'Leary and Wallach 1980). A change in perceived distances of objects, as revealed by pointing with unseen hand was produced in subjects who inspected their own feet for 3 minutes through base-out prisms (Craske and Crawshaw 1974) (See Section 24.6.6).

Summary

I have classified the ways in which sensory information is combined within and between sense organs. Roughly speaking, inputs from nested systems combine by vector addition, comparative judgments are derived by subtraction, the detection of invariants and scaling of one input by another are achieved by division or multiplication, and multiple cues to the same judgment may be combined by averaging. However, simple cue averaging probably occurs only for inputs that are in the same neural form, such as local sign from the two eyes. In other cases, distinct cues for the same judgment complement each other, strengthen a given interpretation of a stimulus, or reduce ambiguity, rather than engage in averaging. Cue conflicts are resolved in terms of reinterpretation of covariance functions, by cue dominance, or by dissociation, rather than by averaging. The ways in which sensory information and efference are combined, whether by addition, subtraction, multiplicative scaling, or averaging, depends on the physical organization of the sensorimotor systems but also on the purposes for which the information is being used. In any case, sensory information is often combined in a highly nonlinear fashion, which defies simple analysis. A fuller account of intersensory systems is provided in Howard (1997a). Interactions between different sources of information for the perception of relative depth are discussed in Chapter 27.

4.7 TYPES OF PERCEPTUAL JUDGMENT

A visual **stimulus domain** is a set of objects or events with defined visible features and selected values of those features, plus the rules of composition and transformation of those features and values. A stimulus domain may be a set of stimuli in a laboratory or a set of naturally occurring objects. The set of possible responses to defined stimuli defines the **response domain** (Section 3.1). We are concerned with responses involving some kind of judgment. They may be classified into detection, resolution, discrimination, categorization, recognition, identification, and description. They form a **task hierarchy**. Successful performance at a given level requires successful performance at all lower levels but not at any higher level. This ordered set of tasks presumably involves a hierarchical sequence of processes in the nervous system. However, performance at a higher level may modify performance at a lower level. For example, it is easier to detect or categorize familiar objects than unfamiliar objects. To describe an object one must identify (be able to name) relevant features but not necessarily the object. In other words, we can describe unfamiliar objects or objects when we have forgotten their names. Also, patients with agnosia can describe familiar objects without being able to recognize them. There is an unspecifiably large number of ways to describe any natural object because an object has an unspecifiably large number of features—it can be perceived as belonging to any of a large number of stimulus domains (see Section 4.7.3).

4.7.1 Detection, resolution, and discrimination

The basic tasks of detection, resolution, and discrimination were described in Section 3.1. All sensory systems transduce specific forms of energy falling on an array of detectors into neural responses so that spatiotemporal variations in stimulus energy may be detected and discriminated. The complexity of subsequent neural processing depends on the use to which the information is put.

For the visual system, we can think of these uses forming an evolutionary sequence, involving increasing complexity of neural processing in a hierarchy of neural centres. For example, lens accommodation uses simple attributes of the visual input, such as image blur and chromatic aberration, and is controlled by subcortical centres (Section 9.1). A more recently evolved level of control allows accommodation to be coupled to vergence eye movements. Optokinetic nystagmus, in its evolutionary simple form, is evoked by retinal motion signals extracted over a wide area and processed in the vestibular nuclei and accessory optic system. In higher mammals, the system is supplemented by more highly processed optic flow signals routed through the visual cortex (Section 23.4.5).

Neural processing for simple responses such as accommodation, eye movements, and postural control can involve fairly complex relationships between different sensory inputs, but this processing occurs with minimal conscious awareness or control. Sensory-motor learning can be involved, as when a child learns to walk, but the learning does not involve conscious knowledge of the detailed structure or contents of the visual world. Stimulus ambiguities in one sensory system are usually resolved by other sensory inputs. The tasks are those of detection and discrimination only.

More complex inputs are required for the conscious perception of the three-dimensional layout of the visual world for guiding movements and for navigation. For these purposes, elaborate neural processes occur in several hierarchical and parallel visual centres (Section 5.7.3). These processes allow us to perceive objects and features of objects as remaining the same under incidental transformations of size, distance, orientation, motion, and lighting. These are the basic **perceptual constancies** described in Chapter 26. They also allow us to perceive objects as distinct entities even when they are partially occluded, as described in Chapter 24. Finally, they allow us to perceive the movements of articulated or plastic objects, as described in Section 25.4.

4.7.2 Categorization, recognition, and identification

In a **categorization** task, subjects group, or classify, a set of stimuli according to defined criteria. In the simplest case, subjects arrange stimuli into feature-defined sets with no order within or between the sets. For example, coloured objects in one set, round objects in a second set, and moving objects in a third set. At a more complex level, subjects place the stimuli in order. The ordering can be a simply rank ordering, an equal-interval ordering, or an equal-ratio ordering. The ordering can be on a one-dimensional continuum such as size, velocity, or distance, or in a multi-dimensional feature space. Thus, in addition to discriminating between stimuli, subjects arrange them in bins or in order. Subjects need have no prior knowledge of the categories if they are shown a representative sample of stimuli in each category. In any case, they may have no explicit understanding of the criteria that determine class membership.

Even more complex stimuli and neural processes are involved in recognizing or identifying particular visual objects (Section 5.7.2). Both tasks require memory of specific stimuli. In a simple **recognition** task subjects state whether they have seen a stimulus on a previous occasion. In an **identification** task, subjects emit a distinct learned response (a name) to all stimuli in a defined category and different responses to stimuli in each of the other defined categories. Each name is arbitrarily related to the stimuli. Names may be conventional names or novel names randomly assigned to each stimulus in the set. Thus, in addition to categorizing stimuli, subjects must learn the categories and give them names. Subjects can identify a single stimulus, since they have prior knowledge of the categories.

Performance is aided by assumptions or knowledge of the world. Ways must be found to deal with ambiguous and camouflaged stimuli. We tend to classify objects into superordinate and subordinate categories, as in the Linneaus scheme for classifying species of animals and plants. These processes are simulated by **semantic nets** (Findler 1979).

A simple taxonomic or categorical schema is assumed to exist when some aspect of behaviour is contingent on the presence of a relatively simple stimulus feature or set of features. For instance, the aggressive posture of the robin red breast is contingent on it seeing a patch of red on a bird-like object. Signs of this kind are known as **releasers**. Most releasers described by the ethologists are of this simple kind. More complex taxonomic schemata underlie the ability to categorize things in terms of several features. Symbols allow things to be named, that is, expressed in arbitrary abbreviated form. A good deal of perceptual learning consists of isolating salient features or partial cues as aids to rapid identification and as the basis for an economical mnemonic system (Gibson 1967).

The inferior temporal cortex and prefrontal lobes are the regions of the brain devoted to object categorization and recognition (Section 5.7.2). In most experiments, cells in these regions in the alert monkey are found to respond to particular objects that the animal has been trained to recognize, the implication being that the cells respond to a particular concatenations of stimulus features. However, Freedman *et al.* (2001) found that cells in the frontal cortex respond to stimulus categories that the monkey has been trained to construct, even after the stimulus features defining the category boundaries have been changed.

It has yet to be shown that the response of an 'object recognition' cell depends on how an animal regards an object (what descriptive domain it is using). For example, an animal could be trained to use a box as an object to sit on or as an object to throw. Would the response of a cell vary with way that the animal intends to use the box?

Models of perception of the 'perceptron' type, such as 'Pandemonium' (Selfridge 1959), essentially embody the taxonomic level of performance. That is, percepts are defined in terms of the presence or absence of weighted elementary stimulus features. Also, most studies of concept formation have involved the study of conjunctive or disjunctive categorical concepts such as large green objects versus small red objects. It is as if zoology never got beyond Linnaeus. The mathematics underlying categorical behaviour is set theory. A set is a collection of items defined as belonging to a class, but having no internal structure. Perhaps the belief that categorical percepts are primary arose out of the belief that set theory is basic to logic and mathematics. Many aspects of perception require analysis in terms of a richer set of mathematical ideas, such as projective geometry, the calculus of variations, and group theory. These mathematical tools add the notion of internal structure to the notion of class membership.

4.7.3 Description

4.7.3a *The nature of descriptive processes*

A **representation** is something which is accepted as resembling some defined aspects of a specified thing (object, event or idea) or set of things, for some specified purpose. A **description** is a representation involving flexible, recursive composition rules and knowledge of a descriptive domain. Within a **descriptive domain**, or schema, an object is seen as a member of a set of items with which it is equivalent in some respect. This process involves memory of the set of related items and memory of the equivalence relation. When operating at the descriptive level we can:

1. Attend to one or other aspect of an object in response to centrally determined requirements or criteria, such as curiosity, knowledge, or the search for an answer to a problem.
2. Apply a process to itself to create a recursive function. For example we use language to describe the grammatical structure of a language.
3. Combine subroutines in a novel fashion to suit the demands of a complex task.

These abilities enable us to define new equivalence relations within an existing set of discriminated stimuli and thus derive ever more complex descriptive functions or clusters of such functions. This makes for great efficiency, and allows for the possibility of these processes themselves being modified as a result of other internal computational processes. Descriptive processes can have great computational depth, generality, and economy, as in the descriptive structures of science.

An implicit descriptive process is not at the conscious level and can only be inferred from behaviour. For example, a child may speak grammatically but be unaware of the rules of grammar. Also, people can recognize friends but may not be aware of the facial features that they use. An explicit descriptive process is represented at the conscious level, such as when a person derives a mathematical theorem from a set of axioms or learns the distinctive features of different makes of automobile.

The description of an object may include the nature of its parts and the way they are organized, the way the item changes over time, and what it does when handled in some way. Minsky (1961) called these descriptions "articular" descriptions to distinguish them from simple class assignment descriptions (see also Clowes 1971). Such descriptions are potentially infinite for any object, for they may be applied recursively; that is, each product of analysis can be an item of further analysis.

The descriptions that we trust the most are used to define and construct what we call reality. We then use this construct to test the validity of other descriptive systems. Generally speaking, today's scientific descriptions are based on identity matching operations such as are involved in using rulers and reading a vernier scale. These highly valid and often highly sophisticated modes of perceiving and describing the world are progressively developed in historical time. The less developed they are, the less adequate are the means of resolving contradictory experiences, and humans invent the concept of 'magic'. That is, they deliberately tolerate contradictions by saying that they exist in the world and not just in their own descriptive domains.

Even when techniques and knowledge are generally available for resolving conflicting experiences many people, who should know better, accept magical explanations of illusory effects, such as the simple conjuring tricks of mediums and stage mystics.

We can discover new descriptive domains by observing objects and events in the natural world. For example, we may walk round an object, or fly round the moon to discover its 3-D structure. Or we may construct a bridge and then test it out, or determine the gravitational constant from the swing of a pendulum. In each case, we get a 'read out' of a natural computational process in the form of a stimulus display. We need sense organs, knowledge, and a theory with which to interpret the display in relation to the question we are asking. We can use the world as a memory store in the form of objects or events that can be displayed or attended to at will. Humans are more intelligent than digital computers mainly because they have a two-way interface with the natural world. One task of any theory of perception is to understand how perceivers acquire the ability to 'consult' the natural world.

Instead of observing a real object we can observe an analogue of the object. The analogue is isomorphic with some aspect of the object. A photograph or retinal image is spatially isomorphic with an object as seen from a particular vantage-point and is said to be an 'image' of the object. If an image is blurred, faded, enlarged, projected at an angle, rotated, or stretched, it may be possible to restore or 'normalize' it to congruence with the object and apply a superimposition or 'template-matching' test to check its identity. Most computer pattern-recognition devices operate in this way. Mental images can be thought of as analogues of objects in this sense, but many good arguments have been put forward for doubting that the template-matching procedure plays a dominant role in human pattern recognition (Neisser 1967; Pylyshyn 1973). There is no need to repeat those arguments here.

The medium in which an analogue of an object is represented can be arbitrarily related to the object or event. Temporal events can be represented as spatial analogues and vice versa. For example, a phonograph record is a spatial analogue of a piece of music and a radio signal is isomorphic with a televised picture. The frequency of nerve impulses can be an analogue of the intensity of a stimulus. For any system to recognise that something is isomorphic with something else it is necessary for it to express the structure of the two things in a common medium of representation. One cannot compute over incompatible data bases.

In an analogue computer, variables in uniform analogue medium, namely voltages in an electrical network, are set up to be isomorphic with a defined physical system, such as an aerodynamic system used for designing aircraft. The sense organs achieve a general uniformity of expression by transducing all stimuli into patterns of nervous activity (see Rushton 1961). Such a uniform medium of expression provides the basic 'hardware' compatibility that any natural or artificial computer requires. However, in spite of this underlying uniformity in the nervous system, any given stimulus object may come to be represented in the nervous system in a variety of ways. For instance, an object is coded very differently by the eyes than by the sense of touch. If the perceiver is to recognise that a seen object is the same as a felt object it must converge the two coded representations into a common computable form.

When we use an analogue computer we do essentially the same thing as observing natural objects or events but speed the process up and bring it within a smaller spatial compass by simulating the feature of the world we are interested in. When we use a digital computer, we economize further by having a single machine that can simulate any well-defined feature of the world within a uniform and economical medium of expression (digital symbols).

The most powerful representational and descriptive systems are more symbolic than analogical. A system is symbolic to the extent that units of expression are arbitrarily related to the items represented. All symbolic systems retain some analogical features. For instance, expressing the formula for a circle in terms of non-complex, positive, rational numbers implies some isomorphism between the selected properties of the number space and the spatial properties of circles. Mathematics is the isomorphic mapping of aspects of real or imaginary things into a structured symbol system with rigorously defined axioms and syntax.

Natural languages do not achieve the same degree of uniformity of symbolic expression that is achieved by the formal systems of logic and mathematics. Languages are relatively loose conglomerates of specific descriptive domains, unified by a common grammar (phonemics, syntax, and inflection) but not by a common set of axioms or function rules. There is computational power within local domains but often very little between them. This is why people tolerate so many contradictions and fail to appreciate what may appear to others as obvious relationships. Humans invent and enlarge the scope of these symbolic descriptive systems in historical time.

Perceptual systems are also loose conglomerates of locally uniform descriptive systems within which there is computational power over local descriptive domains. But there is no common syntax or grammar, and no common set of axioms or rules beyond the local domain. Hence, pre-linguistic descriptive processes do not have the descriptive potential of a symbolic language although, in practice, they may outperform a symbolic system, such as language. In other words, we can often perceive things or behave with respect to them quite adequately without being able to say how we do it. Only when we can express adequately what we are doing are we able to outstrip our own pre-linguistic performance. The piecemeal specific schematic structures typical of perception are adaptive for an evolving creature, where the first priority is to develop the capability to deal with many diverse and local contingencies in a complex environment. General computational power is a luxury that can evolve later.

Symbolic systems often involve the conversion of descriptions into linear strings of symbols. Thus, what may be an underlying n-dimensional descriptive structure is reduced to an n-1 dimensional expression. As in the projection of objects into a picture plane, this reduction of dimensions leads to ambiguity. Thus, rules of syntax are needed to disambiguate the symbol string and parse it back into a higher dimensional structure, what psycholinguists call a syntax-free expression or deep structure.

Thus, perception is not language-like in the sense of using symbols, but it is language-like in the sense of having some uniformity of representational structure within local schemata (Neisser 1967;Newell and Simon 1972; Bryant 1974).

4.7.3b *The ideal perceiver*

An ideal descriptive domain is one that is fully understood, that is, one about which all answerable questions can be asked and answered. The finite groups of mathematics are prototypes of perfect descriptive domains. For example, the 17 space group of crystallography and the group that generates the set of five Platonic or regular-sided polygons are ideal descriptive domains. Linear perspective is an ideal domain, for it is possible to fully describe linear projection from three to two dimensions (Section 24.1). An ideal descriptive system may utter descriptions that a less adequate system does not understand. Also, only a perfect system can completely assess the adequacy of another system. A father is seen as an ideal perceiver to his young child, until the child discovers that the abilities and knowledge of the father are constrained.

One cannot have an ideal perceiver for an object in the world, for nobody can be said to know all there is to know about a natural object. A theory of perception maps the descriptive structures of a perceiver—not into the world—but into a description of some abstracted aspect of the world that the investigator creates. The empirical study of perception and cognition is essentially the study of the constraints of natural perceivers and thinkers and of the ways in which such constraints change with experience. Such an enterprise is always limited by the adequacy of the explicit descriptive functions that investigators possess, that is, by the adequacy of the descriptive structures of science.

The description of an ideal perceiver for a given domain is essentially the prescription of what must be done to succeed in a task that faces a perceiver with defined discrimination capacity. Such a prescription can also be regarded as a feasibility test—it establishes that the defined perceptual act is possible. It defines those aspects of the task that are attributable to the requirements of the task (given the discrimination constraints of the perceiver). Most, if not all, computer simulations of perceptual processes attempt to construct ideal perceivers that are realizations of some theoretical prescription of a task. Psychologists, studying natural perceivers, also need a prescription of the environmentally determined constraints in as perfect a form as possible, but their main concern is to determine empirically the descriptive structures that people use, with all their contradictions, confusions, and omissions. We need simulations of interesting ways in which perception fails and of systems that learn by failing.

There is an unspecifiably large number ways to describe any object. For example, consider the perceptual and conceptual descriptive domains that relate to the human body.

The **kinaesthetic body** refers to the sensory and perceptual processes that underlie our ability to judge the relative spatial dispositions and movements of unseen body parts.

The **seen body** refers to the set of rules that enable us to recognize our own hand or direct the gaze to a particular part of the body.

The **body schema** is the set of descriptive rules and stored data responsible for the sense of familiarity with our own body, and the phantom limbs that people report after amputation. This schematic structure is probably intimately connected with the

motor command centres that direct movements and body parts.
The **body image** is the set of descriptive rules, which enables us to imagine our own body.
The **known body** is the knowledge structure that allows a person to make such statements as "I have two arms, one nose and ten fingers".
The **pictorial body** enables us to recognise ourselves in drawings and pictures.
The **anatomical body** is that set of data and rules which enables the surgeon to navigate the body of a patient or pass an examination in anatomy. It is a behavioural schema because it is a collective product of human perception and intellect.

Think of the variety of descriptive domains that we can use when we look at a face. We can recognize it as a face rather than not a face. Within the domain of faces there are distinct features that allow us to recognize the species, race, sex, family, and individual. Other features allow us to recognize changes that can occur in a given face, such as aging or changes of mood or health. Finally, there are situational features that allow us to recognize the orientation, vantage-point, and motion of a face. All these features arise from the same basic anatomical structures and ways of moving that define a face. We perceive faces in all these ways without necessarily being aware of what features we are using. We recognize family resemblances and cartoon drawings of familiar people (Dodwell 1983).

The task of analyzing perceptual systems is further complicated by the fact that, for any defined task, a person may call upon more than one descriptive system for an answer. But the answers may not agree, in which case we say that the person has an illusion. For instance, the amputee with a phantom arm will decide he has two arms when he refers to his body schema, but that he has only one arm when he refers to his seen body. In the Müller-Lyer illusion, the lines appear unequal in length when 'eyeballed' but are reported as equal when end-to-end matching is applied. It is sometimes not possible to decide which description to trust.

The science of perceptual systems consists of studying the structure, growth, and adequacy of such rule systems and the relationships among them. It is a matter of empirical investigation, just how the various systems relate, and the safest initial assumption is that they function in isolation from each other.

The adequacy of peoples' perceptual schemata of these types of situation can be studied by the following procedures. In the production technique, subjects extrapolate or interpolate when presented with an incomplete stimulus. For example, they interpolate the path of an object moving behind an occluder. In what may be called the 'when-does-it-look-right' technique subjects pick out the correct stimulus from among several displays. Finally, one can measure a person's ability to adapt to anomalous experiences. The behavioural methods for teasing out the structure of perceptual schemata need enlarging and systematizing. I shall describe some examples of descriptive domains.

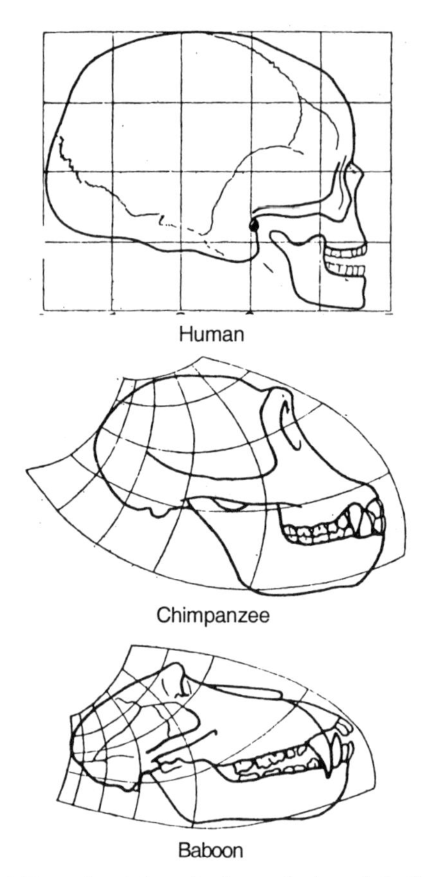

Figure 4.8. Transformations in the evolution of skulls.
(Adapted from D'Arcy Wentworth Thompson 1952)

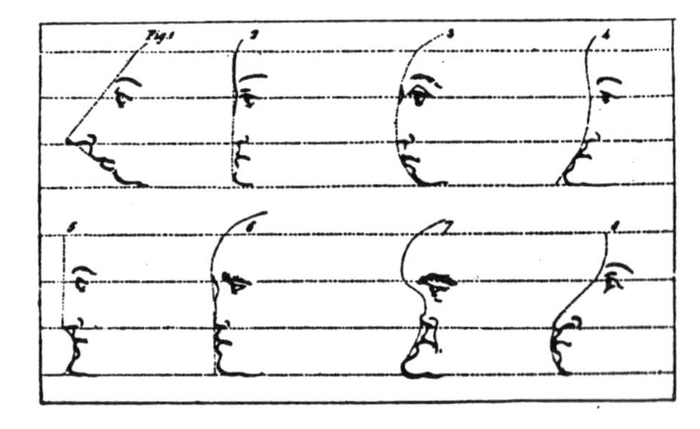

Figure 4.9. Transformations used by caricaturists.
(From *Rules for drawing caricatures* by Francis Grose 1788)

4.7.3c *Transformations*

Under this heading I discuss those structural features of classes of objects that are invariant under non-similarity transformations. These are point-for-point transformations that do not preserve length, straightness of lines, or size of angles. They are the kind of plastic transformations that are involved in growth. The classic discussion of such invariant properties of objects is contained in D'Arcy Wentworth Thompson's book *On Growth and Form*, first published in 1916. Figure 4.8 shows three primate skulls. On casual inspection, they do not seem to have a similar shape. However, if one of them is plotted on Cartesian co-ordinates, the other skulls can be derived by applying a conformal transformation to the co-ordinates.

Cartoonists have developed similar methods for producing cartoons, which stress certain essential features of the original. Artists also used such methods to explore the relationships between faces of different types so as to improve their graphic descriptive powers. Figure 4.9 is from an 18th century text on the techniques of drawing, which employed the same method used by Thompson (1952).

There are two descriptive domains in any transformation. One domain is the invariants that survive the transformation. The other is the transformation rule. This suggests an experimental approach to the question of what transformations and invariants under various transformations people can normally recognise or can be taught to recognise. But as far as I know, there are no experimental studies along these lines. In everyday experience we recognise the family resemblances between relatives, and this skill probably depends upon this kind of descriptive structure. For example, we recognize members of the Hapsburg family by the Hapsburg lip.

4.7.3d *Symmetry*

Symmetry is the regular repetition of something over a transformation. The transformation may be translation, rotation, reflection, or some other transformation in space, time, or any other dimension. Most common objects have at least one axis of symmetry. The visual system is particularly sensitive to mirror symmetry (Corballis and Beale 1970a; Barlow and Reeves 1979; Barret *et al.* 1999; Tyler 1999). See Howard (1982, Chapter 13) for a review of studies on the perception of symmetry in shape recognition. I mention here one interesting case of perception involving symmetry in three dimensions, which illustrates the importance of the concept of the group in perception and cognition.

Most people perceive that their image in a mirror as reversed from left to right but not reversed top to bottom. But people cannot understand how a symmetrical mirror can produce a seemingly asymmetrical effect. Group theory solves the problem.

Take an asymmetrical object, such as a hand. Let the X axis be between little finger and thumb, the Y axis be between finger tips (top) and wrist (bottom), and the Z axis be between palm (front) and back. Call the opposite ends of an axis its poles. Let a 180° rotation about the X axis be pitch (p), that about the Y axis be yaw (y), and that about the Z axis be roll (r). Let congruence between the poles of an axis of one hand and the poles of the same axis of another hand be denoted by 1, and non-congruence of poles be denoted by 0. Start with two identical hands in the same orientation and allow 180° of rotation about each of the three axes. If we take the axes in the order XYZ, all relative orientations of the identical hands are represented in the group Table 1. It is a group because it is closed, associative, and contains an identity element (no rotation) and an inverse element (opposite rotation) (Section 3.8.1). The symmetry of the group table indicates that the group

Starting position	Axis of rotation		
	Pitch	Yaw	Roll
010	001	111	100
001	010	100	111
111	100	010	001
100	111	001	01

Table 1. The group table for matching two identical objects. One object is rotated 180° about each axis from each of a set of starting positions. Each set of three numbers indicates matches along the X, Y, and Z axes respectively, with 1 indicating congruence and 0 indicating non congruence.

Starting position	Axis of rotation		
	Pitch	Yaw	Roll
011	000	110	101
000	011	101	110
110	101	011	000
101	110	000	011

Table 2. Group table for matching enantiomorphs. One object is rotated 180° about each axis from each of a set of starting positions.

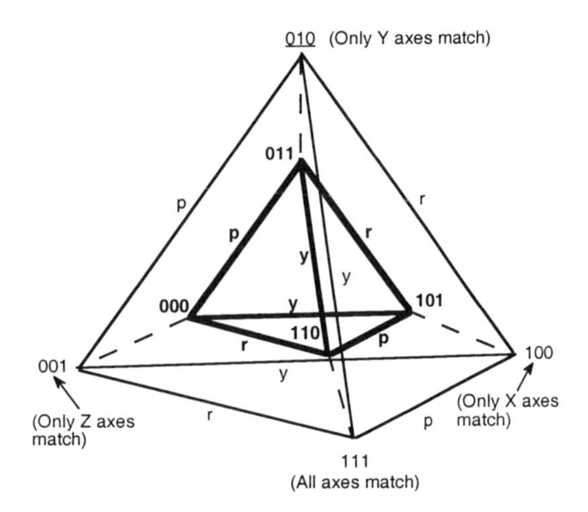

Figure 4.10. A 3-D Caley diagram for mirror-reflection problem. The outer tetrahedron indicates all ways in which an object can be matched with itself by 180° rotation about each of three orthogonal axes. The inner tetrahedron indicates all ways in which an object and its mirror reflection (enantiomorph) can be matched. Each set of three numbers indicates whether the objects match (1) or do not match (0) along their X, Y, and Z axes respectively. Each solid line represents a 180° rotation about the roll (r), pitch (p), or yaw (y) axis of one object relative to the other. Each dotted line represents reflection of one object relative to the other about an axis indicated by which of the three numbers changes.

is commutative. This means that the order of rotations does not make any difference. The group may also be represented by its Caley diagram, which in this case is a tetrahedron, as in Figure 4.10. Each apex of the diagram is a state of relative orientation of two objects and each side denotes a 180° pitch, yaw, or roll rotation of one object relative to the other.

Now take two hands of the same person or a hand and its a mirror reflection. They cannot be made congruent along all three axes. Objects of this sort are enantiomorphs. The group table is show in Table 2. No entry in Table 2 is the same as any entry in Table 1, because one can go from Table 1 to Table 2 only by reflection, which changes the match on only one axis. The two groups are said to be cosets of the larger group formed by combining them.

The two tetrahedral groups can be combined into one group based on the operations of rotations and reflection, as shown in Figure 4.10. This group is also commutative.

The answers to all meaningful questions about matching an object and its mirror reflection are contained in the group table. All arguments between people with incomplete knowledge of the group are spurious, like those between two blind men who feel different parts of an elephant. Paradoxes stem from misrepresentations of the group.

The statement that one's mirror image is reversed left-right but not up-down is misleading. It can be seen in Table 2 that an object and its enantiomorph can be matched on any pair of axes or on none, depending on whether the operation of translation, pitch, yaw, or roll is applied before the objects are compared. Thus we can walk into own reflection and match all but the front-back. Or we can imagine ourselves rotated about the X body axis (pitch) and match all but the head and feet. Or we can rotate about the Y axis (yaw) and match all but the hands. Finally, we can rotate about the Z axis (roll) and leave all axes unmatched. There is no logical priority for any one of these operations.

When we wish to identify the left and right hands of a person facing us, we naturally rotate or imagine ourselves rotated about the Y axis to bring the head-feet and back-front axes into congruence. We intuitively realize that the left and right hands are congruent only when the other two body axes are congruent. We do not need a rotation to identify the head and feet or the front and back of another person because these features have distinct shapes. If all people had blue left hands and red right hands we would not need to apply a rotation to identify left and right hands.

Young children behave differently. Even if they can identify their own left and right hands they get confused when asked to identify the hands of a person facing them and are more likely to say that the hand directly facing their left hand is a left hand. The answer is correct with the mirror image, but not with another person.

The mirror problem arises because we confuse the enantiomorphic mirror image of ourselves with the body of another person and apply the same rotation test. With the mirror image, the test reveals that left and right do not match when the other two body axes are congruent. We say the mirror has produced a left-right reversed image. We find it difficult to imagine matching ourselves with our mirror image by a translation, or a rotation about the X axis (pitch) or the Z axis (roll) because we intuitively realize that these operations bring one or both of the other body axis into non congruence. However, if we apply them, we see that it is just as valid to conclude that our mirror image is reversed front-to-back, head to toe or along all three axes as it is to say that it is reversed left to right.

People also puzzle over why the mirror image of print is left-right reversed, but not upside down. Consider a printed word on a transparent plate of glass. The print and its reflection are both normally oriented with respect to a viewer. Thus, a mirror does not reverse or invert print. We see the other side of the print in the mirror but, since the two sides are the same, this has no effect. With print on opaque paper we must either rotate the paper vertically or horizontally to see its reflection. According to which rotation we apply we see the print either reversed or inverted. It is not the mirror that causes the reversal or inversion but the rotation applied to the sheet of paper.

The mirror problem demonstrates that our descriptive structures are often an incomplete subset of more valid and complete group structures, which it is often possible to define abstractly. A well-defined group is the theory of the ideal perceiver (and thinker) for that descriptive task. The theory of the ideal perceiver can be complete within the domain of a specified group, but there are always more general group structures (theories) to be discovered. Mathematical descriptions allow us to define groups that are more complete, more comprehensive, and less subject to contradiction, than those that our perceptual system exhibits. Mathematics thus provides an abstract theory of intelligence in terms of which we assess natural intelligence. Any theory of natural intelligence is an account of how the assumptive structures that are inferred from behaviour relate to the ideal perceiver. The ideal perceiver is a system that knows the complete structure of a defined descriptive domain, and is both an account of what is perceived and of 'how' it is perceived. It does not prescribe the hardware (neurones, transistors, etc.) needed to realize a perceiver, but it can specify the operations (programme) that the perceiver carries out, as fully as one wishes.

Abstract groups have always fascinated people. This fascination is apparent in the way the Greeks, Kepler, and artists such as Escher have pondered the group of five Platonic solids. These are the only possible regular-sided solids. Such groups are fascinating because they are derived by thought and yet inform us about the world. We know more certainly that there cannot be an object in the world which fits the description "sixth regular solid" than we know any fact derived from observation. Thus we know things about the world by referring to group structures within abstract descriptive systems as well as by observation. The most powerful scientific theories are those that describe group- like symmetrical structures, such as Mendeleef's table of chemical elements, crystals, or fundamental particles, because they tell us what to look for, and what kinds of things cannot be found. There seem to be no psychological studies of group concepts, although Piaget uses the concept of groups informally in his theory. For an introduction to group theory see Budden (1972) and Shubnikov and Koptsik (1974).

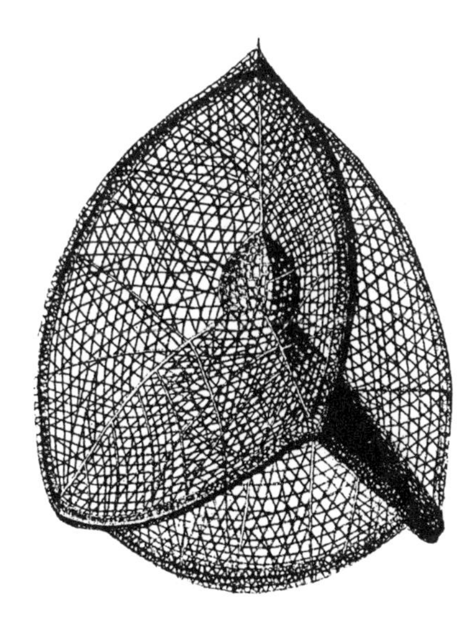

(a) Skeleton of a Radiolarian (*Callimitra agnesae*)

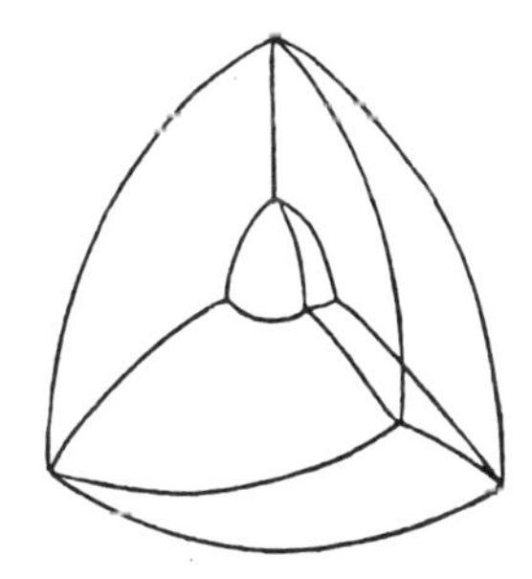

(b) Soap film formed in a tetrahedral wire frame with curved edges. The inner shape forms spontaneously in variable size.

Figure 4.11. Equilibrium structures.
(From D'Arcy Wentwoth Thompson 1952).

4.7.3e *Equilibrium structures*

The forms of natural objects have a regularity that results from the interplay of competing forces. A snow crystal has a six-fold symmetrical branching structure because of the interplay between the forces acting during its growth and certain constraints in the system. D'Arcy Wentworth Thompson (1952) showed how the forms of many animals are determined by similar principles. For instance, Figure 4.11a shows the skeleton of a radiolarian and Figure 4.11b shows the same form constructed by dipping a tetrahedral frame in a soap solution. The principle that determine these shapes is referred to as the principle of least effort, or the principle of equilibration. The shortest route on the surface of a sphere (geodesic), the brachistochrone problem (the curve of most rapid fall in a diagonal direction), and the shapes of honey comb and living cells are other examples of equilibrium systems, which can be understood (described) with the help of the calculus of variations (see Courant and Robbins 1941). These structures are essentially equilibrium states that maximize or minimize the values of certain variables. The layout of the cerebral cortex seems to be governed by the same principles (Section 10.3.1).

Another related set of structures are the spirals, as represented in the shapes of shells, horns, whirlpools and in the flight of moths round a flame. The appeal of spirals in art suggests that we appreciate the underlying orderliness of these structures, but behavioural studies seem to be lacking.

Branching structures are also well represented in nature. The perceptual recognition of different species of tree involves the building of a descriptive domain that operates over the relevant variables, such as the number and spatial disposition of branching patterns at each node and the distance between nodes. Botanists have developed formal descriptive systems for this purpose. The theory of the ideal perceiver of spirals and trees is the mathematical theory of spirals and trees (phyllotaxis). It is the job of experimental psychology to determine to what extent a human observer is or can become such an ideal perceiver.

The question of interest here is whether such descriptive structures occur in perception and behaviour. The principle is represented in the path of least effort that we tend to take when we run round obstacles. This is determined by the inertial forces acting upon the moving body and by reflexes and skills that we may not be aware of. The principle also operates in perception when we perceive the apparent path of an object as it moves behind an occluder or when we resolve ambiguous stimuli (Section 23.4.4).

4.7.3f *Causal sequences—structures of events*

When we ask "what caused 'X'", we seek a description of a structure of events in which 'X' plays the role of a necessary consequent event. In theory, there is always an infinity of such descriptions, but the context in which the question is asked usually prescribes the type of answer required. Sequences of events over time may be represented spatially, and causal relationships represented as intersections, pointers and the like, as in flow charts, family trees, switching circuits, state diagrams, and Caley diagrams (abstract state diagrams of groups).

There have been relatively few studies of the perception of causal structures. Michotte (1946) investigated the phenomenology of causal impressions. He wanted to prove that impressions of causality are direct and unlearned rather than inferred from a sequence of events, as proposed by British empiricists, such as Hume. However, Michotte did not study young children (see Olum 1956), nor the ways in which adults change their descriptions as a result of experience. Therefore, his results tell us nothing about whether impressions of causality are innate or learned. Furthermore, Michotte's experimental procedures have been called into question (Joynson 1971).

Paradoxically, the technique that Michotte used is ideal for investigating the role of experience in the perception of events, the very thing that Michotte himself did not study. The only study on this question that I know is one by Gruber *et al.* (1957) in which they showed that delaying the time between the removal of a support and the collapse of a bridge modified the impression of what caused the bridge to fall.

There are three psychophysical procedures for revealing the assumptions, or schemata, that underlay the perception of events. The first is an extrapolation procedure in which the subject observes an event sequence and executes an aiming movement towards the future position of a moving object. The second method may be called the "cognitive prism" method in which the event sequence is distorted by means of some intervention by the experimenter that is hidden from the subject. The third procedure involves asking the subject to select a correct display from an array of displays—the "when-does-it-look-right method".

Science is the enterprise of describing causal relationships and seeking the most universal invariant properties of objects and events. **Implicit physics** is a branch of psychophysics concerned with how accurately and precisely people make judgments or perform tasks that embody some natural property of physical objects. For example, most people believe that liquid in a tilted jug slants up towards the spout in proportion to the angle of tilt (Howard 1978). In this case, people assume a proportional relationship between two sensory features that does not exist—liquid remains horizontal as a jug is tilted. People's distorted understanding of the rules of perspective comes under this heading. An amusing example of naïve understanding of spatial relationships is provided in Hinton (1987). People's understanding of natural dynamics is discussed by Proffitt and Gilden (1989). Nevertheless, artisans with no formal education built elaborate devices through a process of creative imagination and trial and error (Ferguson 1977).

In other cases people believe two features are not related when they are. For example, most people say that a loop of string held in the shape of a square encloses the same area when it is pulled into a rectangle. They become confused when the string is pulled out until the area is zero. They assume that deforming a loop of string is the same as cutting a shape and reassembling the pieces. The isoperimetric principle states that the area of a rectangle enclosed by a perimeter of constant length is the product of length and width, which is greatest when the two dimensions are equal.

The history of science is replete with examples of how progress was held up because all people had an inappropriate assumption about how objects and events are related. For example, before Galileo, everyone believed that heavy objects fall faster than light objects. They also had an inappropriate mental picture of the path of a projectile, and believed that if the world rotated an object would fall at an angle when dropped from a tower. Most people still have the same inappropriate mental structures.

When invalid assumptive structures are found, we can apply the confrontation method. This is just what Galileo himself did when he became the first man in history to realize that the mass of an object does not affect the time it takes to fall a given distance. He imagined two weights linked together by a fine thread. They now become a heavy weight and should, by the assumptions made at the time, have fallen faster than if they were not joined by the thread. But, if this were so, one would have to say that one of the weights was pulling the other, which is impossible. The only way to avoid this contradiction is to assume that heavy and light weights fall in equal time (air resistance apart). Perhaps at the root of all theoretical scientific advances, there is a confrontation experience—a realization that the old formulation leads to a contradiction. Confrontation experiences are at the root of perceptual learning, that is, the acquisition of more adequate and valid perceptual structures. Confrontations are also fundamental to the scientific enterprise of constructing adequate models of perceptual processes.

4.7.3g *Feature detectors and perceptual schemata*
Dedicated neural units with selective tuning functions for a composite feature code certain higher-order features, such as the direction of approach of a visual object. A system of dedicated hardware of this type is the most efficient and rapid way of coding vital types of information that recur frequently. The extension of this idea to more complex features has given rise to the concept of the **pontifical cell**, or grandmother cell—a cell specifically dedicated to the recognition of features as complex as one's grandmother. The notion of dedicated hardware at this level of complexity has severe limitations, since it would require an explosively large number of dedicated cells, each of which would be dormant most of the time.

A much more efficient procedure at the level of complex features is one analogous to programming in a computer in which complex forms are stored as descriptions using general components and rules of composition (algorithms). Language works this way, and the higher recognition processes of perception have language-like properties. For instance, when we recognize a face we may construct a type of description by combining features from different parts of the face, an ability reflected in the way a portrait is built up by police sketch artists (see Rolls 1994; Rolls and Tovee 1995). Whereas the number of stimuli that can be encoded by local pontifical cells increases linearly with the number of cells, the number of stimuli that can be encoded by distributed descriptive processes increases exponentially with the number of cells. Rolls *et al.* (1997) have produced physiological evidence that faces are coded by a distributed process in the temporal lobe of the monkey.

Some descriptive rules underlying perception may be inborn, others are learnt at an early age and function without conscious awareness. They may be called **implicit descriptive rules**, or **schemata**. Other descriptive rules, such as those that allow to us to read, recognize different makes of automobile, or diagnose an illness, are learned deliberately and applied with conscious effort.

Implicit descriptive rules express very general properties of real world scenes, and place constraints on the interpretation of potentially ambiguous stimuli (Howard 1974). We use our accumulated knowledge of the world to consciously or unconsciously infer the object or event in the world that would most likely produce the stimulus. This process can be expressed in the formal system of Bayesian probability (Section 3.7).

The **generic viewpoint** assumption is a very general example of this principle. A given object can be viewed from different positions, in different orientations, or under different lighting conditions. These are generic variables. For certain values of a generic variable, the proximal stimulus may assume a peculiar form. For example, a cube viewed along a diagonal appears as a hexagon. The images of certain disconnected features of an object may abut when viewed from a given angle. Certain features of an object may become invisible when viewed from a given angle. The most likely object is the one that would produce the same proximal stimulus from any generic variable (Freeman 1994). Examples are provided in Section 23.1.2.

An object may appear complete when only part of it is in view. The perceptual system embodies a set of structural rules about how corners, edges, and surfaces are connected in 3-D objects, and how a changing image can signify a solid rotating object (see Section 24.2). Objects that do not conform to these rules, such as the object in Figure 3.14b, are immediately recognized as impossible. However, we are usually unable to state why the object is impossible. Without specific training, implicit rules do not provide a basis for explicit knowledge.

The perceptual system also embodies implicit rules about how motion, occlusion, depth, transparency, and shading are related in real world scenes (Stoner and Albright 1993). Implicit perceptual schemata involving stereoscopic depth are discussed in Sections 24.1 and 24.2.

Helmholtz (1910), in revolting against German idealist philosophers, stated that sensations do not resemble the objects they symbolize, any more than letters resemble the sounds they represent (Section 2.1.8). For Helmholtz, sensations were "signs that we have learned to decipher." But this ignores the distinction between stimuli and neural events that are lawful transductions of external objects, and stimuli such as words, that are arbitrary signs of the objects they denote. In the former case, the external object can be reconstructed from knowledge of the eye's optics and the filter characteristics of neural processing. If there is any ambiguity, at least a range of possible stimuli can be recovered. The development of neural processes may involve learning, but it is not learning of arbitrary signs. In the case of language, a written word can be recovered from knowledge about the visual system, but recovery of the object denoted by the word requires knowledge of the language. Helmholtz's view also ignores the possibility that the visual system is genetically programmed to interpret certain stimulus features in certain ways.

4.7.3h *The role of attention*

Selective attention dominates the processing of higher-level features. Attention is the process of concentrating limited resources on the task of immediate importance. Attention is also involved in the process of actively seeking for something in the world or in memory. A decision is first made to perform a given task. Given that the task has been well rehearsed, a decision to perform it determines the saliency of particular stimulus features related to its performance. For example, when a person drives a car, red lights acquire a particular saliency with respect to the specific response of stopping. The same stimulus has no such saliency for a passenger who, instead, may be keyed in to respond to stimuli such as the contents of a shop window. For any natural scene there is an unspecifiably large number of stimulus specificities that may be evoked with respect to different tasks. This fact is easily ignored in studies of attention involving an impoverished stimulus set. Some stimuli have an innate saliency, which breaks through whatever task is being performed. For example, a sudden movement in peripheral vision triggers an alarm response. Once salience of each stimulus has been set by the demands of a task, a lower level attention process selects the most salient stimulus from among those present. Version eye movements then direct the gaze to the most salient object and vergence movements bring the object into the plane of zero binocular disparity.

There are also processes of stimulus selection within the central nervous system. For example, we can attend to objects not imaged on the fovea. Also, in an ambiguous figure, such as Rubin's vase-profile figure or Boring's daughter-mother-in-law figure, we can change the interpretation of the figure without there being any change in the stimulus. Such changes are reflected in changes in the site of neural activity in the extrastriate cortex, as revealed by fMRI imaging (Kleinschmidt *et al.* 1998). Finally, we can use our attention processes to seek out a particular memory.

Stereoscopic depth, as an attention-getting stimulus, is discussed in Section 23.4.8. Physiological aspects of attention are discussed in Section 5.5.7. Further discussion of attention is beyond the scope of this book (see Koch and Ullman 1985).

5 *Physiology of the visual system*

5.1 THE EYE

5.1.1 General structure of the eye

The cross-section of the human eye is illustrated in Figures 5.1 and 15.32. The human eye is approximately spherical with a diameter of about 24 mm. The cornea has a diameter of about 12 mm, a radius of curvature of about 8 mm, and a refractive power of about 42 dioptres—about 70% of the eye's total refraction. The pupil and associated iris muscles situated just in front of the lens act as an aperture stop to control the amount of light entering the eye. Changes in the size of the pupil also affect the optical quality of the image. Thus, as the pupil enlarges, diffraction decreases and spherical aberration increases. Also, depth of focus decreases with increasing pupil size. The depth of focus of a 2 mm pupil is ±0.46 dioptres and of an 8 mm pupil is ±0.17 dioptres. For a given viewing distance and level of illumination, the pupil automatically adjusts in size to achieve the best compromise between these optical factors. The ocular media transmit about 75% of incoming light at a wavelength of 500 nm and about 80% at 560 nm

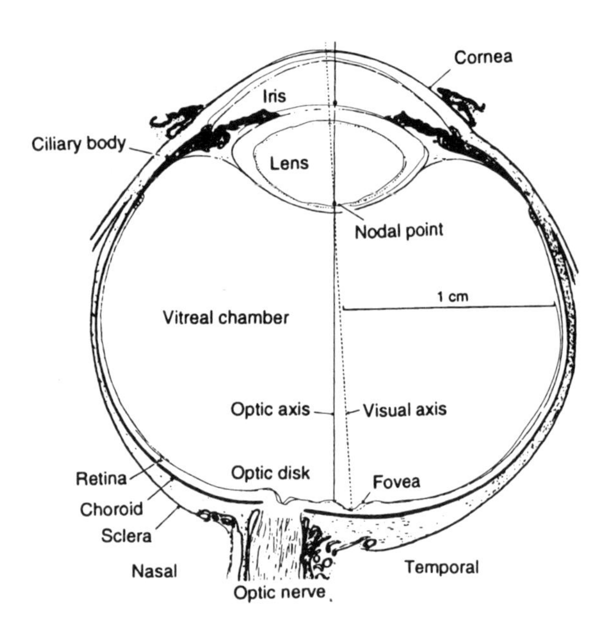

Figure 5.1. Horizontal section through the right human eye. (From *The vertebrate visual system*, by Polyak, 1957, University of Chicago Press)

(Norren and Vos 1974). Larger animals have larger eyes, except that birds tend to have unusually large eyes in proportion to their body size. Animals with larger eyes have an advantage because the size of the image increases with eye size and a large eye can house more receptors, which increases sensitivity.

The lens of the human eye has a diameter of about 9 mm and a thickness of about 4 mm. Its refractive index is about 1.4 in the core and about 1.38 in the surface regions. This gradient of refraction increases the power of the lens and reduces spherical aberration. The **visual axis** is the line joining the point of fixation and the centre of the fovea. The **optic axis** is the best fitting line through the optic centres of the four refractive surfaces of the eye. It intersects the retina about 1.5 mm from the fovea on the nasal side and about 0.5 mm above the fovea. It thus makes an angle of about 5° to the visual axis. This is known as the **angle alpha**. The optical decentration of the image of a fixated point causes the image of the point to be asymmetrical, an effect known as **coma**. Coma is compensated to some extent by an opposite decentration of the pupil. See Section 15.1 for the geometry of the visual fields and Section 9.1 for a discussion of lens accommodation.

For details of the structure of the eye, see Polyak (1957), Davson (1962), Charman (1991), and Oyster (1999). Section 10.2 deals with the development of the eye.

5.1.2 Types of retinal cells

5.1.2a *Receptors*

The fine structure of the retina is depicted in Figure 5.2. This structure was first revealed by Ramón y Cajal using the Golgi staining method, and described in a series of papers between 1888 and 1933 (see Polyak 1941). The retina is a multilayered membrane with an area of about 1,000 mm^2. It is about 250 μm thick at the fovea, diminishing to about 100 μm in the periphery. It is separated from the choroid by a pigmented epithelium, which absorbs light and prevents light that has passed through the retina from being reflected back onto the receptors. In nocturnal animals this epithelium reflects light and thus improves sensitivity. The receptors are densely packed in the outer layer—the layer furthest removed from the source of light. There are two main types of receptor—rods and cones. **Rods** have high sensitivity, an exclusively peripheral distribution, and broad spectral tuning. **Cones** have lower sensitivity, high concentration in the fovea with decreasing concentration in the peripheral retina, and three types of spectral tuning, peaking at around 450 nm (S-cones), 535 nm (M-cones), and 565 nm (L-cones). S-cones constitute 5 to 10 % of all cones. There are about equal numbers of M- and L-cones, although there is considerable variation between people (Roorda and Williams 1999). The normal range of luminance sensitivity of the human eye extends about 3 log units from roughly 10^{-7} cd/m^2 to 10^{-4} cd/m^2. The luminance of stimuli varies about 10 log units. Specialized mechanisms compensate for the limited dynamic range of the eye (Section 4.2.3).

The **retinal magnification factor** (RMF) is the linear distance on the retina corresponding to one degree of visual angle. In the human fovea it is about 0.29 mm/deg (Williams 1988).

The adult human retina has between 4 and 6 million cones with a peak density of between 100,000 and 320,000 per mm^2 at the fovea declining to about 6,000 per mm^2 at an eccentricity of 10° (Curcio *et al.* 1990). The primate fovea is a centrally placed pit about 1.5 mm in diameter, which contains a regular hexagonal mosaic of cones with a mean spacing of between 2 and 3 μm. The central fovea, which is about 0.27 mm in diameter and subtends about 1°, contains at least 6,000 cones. The human retina has 100 million or more rods, which are absent in the fovea and reach a peak density of about 160,000 per mm^2 at an eccentricity of about 20° (Osterberg 1935). Retinal receptors constitute about 70% of all receptors in the human body.

The shadows of the blood vessels of the retina become visible when one looks through an illumi-

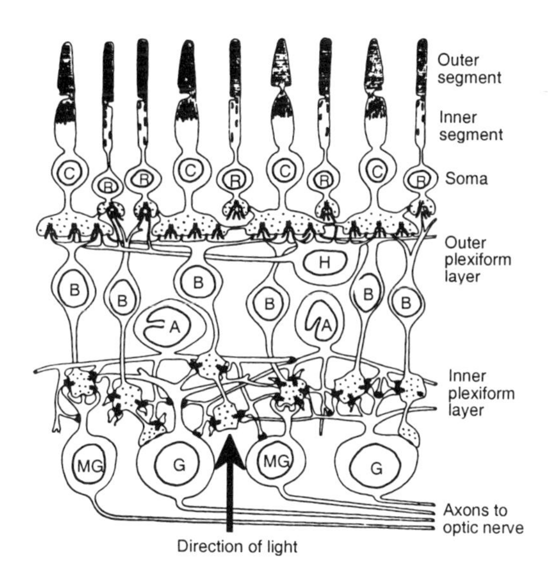

Figure 5.2. The general structure of the retina.
Rods and cones are signified by *R* and *C*. *H* is a horizontal cell in the outer plexiform layer and *A* is an amacrine cell in the inner plexiform layer. Cells marked *B* are bipolar cells. Cells marked *G* are ganglion cells and those marked MG are midget ganglion cells. (From Dowling and Boycott 1966)

nated pinhole. As the pinhole is moved from side to side, the shadows undergo parallactic motion because they are some distance in front of the receptors. Müller (1854) measured the magnitude of this parallactic motion, and, by applying his results to retinal anatomy, deduced that light is absorbed in the outer segments of rods and cones.

Each receptor has an elongated outer segment, an inner segment, a cell body, a short axon (50 to 500 μm), and a single synaptic terminal, as shown in Figure 5.2. The outer segment is about 50 μm long and consists of a membrane folded into about 750 layers to form a stack of discs. About 10^8 molecules of photopigment are packed along the membrane so that light passing through the layers stands a good chance of being absorbed. From time to time, discs are shed from the outer segment and absorbed by the pigment epithelium that lies between the retina and the choroidal layer of the eyeball (Young 1971).

The image plane of the eye's optical system is at the level of the inner segments but all the photopigments are in the outer segment. Light entering the inner segment of a receptor at the correct angle is guided into and along the outer segment by internal reflection. The two segments therefore act as a waveguide, which concentrates light quanta into the outer segment and prevents light scatter beyond the image plane of the inner segments. Their efficiency as waveguides is enhanced by the fact that their diameter is similar to the wavelength of light. If the diameter of receptors were less than about 2 μm, light would leak from one to the other (Snyder and Miller 1977). Because of these design requirements, the diameter of cones is remarkably constant over the animal kingdom and is the ultimate factor limiting visual acuity.

Most light enters the inner segment at the correct angle because the photoreceptors are aligned with the centre of the pupil—the direction from which most light rays come. This explains why we are more sensitive to light passing through the centre of the pupil than to light passing through its margin, an effect known as the **Stiles-Crawford effect** (Stiles and Crawford 1933). This mechanism reduces the effects of light scatter and of aberrations in the light that passes through the margins of the eye's optical system (see Enoch and Tobey 1981).

Some alignment of receptors is present in the retina of the neonate, and is presumably determined by the way receptors are packed together. However, an active process must control fine receptor alignment because it recovers in areas disturbed by retinal detachment (Campos *et al.* 1978). Also, receptors become less well aligned after an eye has been occluded for several days, but become aligned again after the occluder is removed (Enoch *et al.* 1979).

5.1.2b *The receptor potential*

An electrode in the form of a glass capillary 0.1 μm in diameter filled with salt solution can be inserted into a retinal receptor to measure the potential difference between the inside and outside of the cell. In the resting cell, positive sodium and calcium ions flow into the outer segment and out from the inner segment. This creates a potential difference of about -40 millivolts. When light is absorbed in the outer segment, pigment molecules isomerize (change their shape) and become catalytically active. This initiates an amplifying cascade of catalytic chemical events within the cell that results in hydrolysis of guanocine monophosphate (GMP) molecules. In the dark, unhydrolysed GMP binds to the cell membrane and opens the channels through which sodium and calcium ions enter the outer segment. Hydrolysed GMP molecules allow the channels to close, resulting in a fall in the concentration of sodium and calcium ions within the cell and hyperpolarization of the cell membrane. The membrane of the inner segment of a retinal receptor contains a sodium-potassium ionic pump that modifies the voltage changes induced in the outer segment and initiates the release of the neurotransmitter glutamate into the synaptic cleft on the cone terminal (see Yau and Baylor 1989). These events take about 0.2 s, which is similar to the integration time for rod vision measured psychophysically.

The visual transduction process in primates is studied by placing a piece of living excised retina under a microscope and drawing the outer segment of a single receptor into a micropipette 2 μm in diameter. As the receptor is illuminated with a spot of light a fine electrode picks up the membrane current.

A single response of a rod is influenced by photons absorbed over a period of about 200 ms. Vision is still possible when only one photon arrives in the rod integration time (Baylor *et al.* 1984). This represents an enormous degree of amplification (Baylor *et al.* 1987). Rods therefore have high sensitivity but low temporal resolution. The spectral sensitivity of rods determined in this way is similar to that determined psychophysically by the scotopic visibility curve.

A cone integrates photons over about 50 ms. Within the luminance range of a cone, the number of photons arriving in the integration time varies between about 100 and 10^5 (Schnapf *et al.* 1990). Cones therefore produce a graded signal with high temporal resolution.

Within the linear range of the visual system, the number of quanta absorbed is proportional to the luminance of the light. The degree of hyperpolarization of the cell membrane is proportional to the rate of absorption of light quanta by photopigment. Sensitivity is reduced after exposure to a bright light. Continued exposure to a bright light depletes the stock of unbleached pigment molecules.

Receptors are subject to noise arising from random variations in absorption of light quanta and from random fluctuations in the biochemical processes within the cell, including thermal instability of the visual pigments (Lamb 1987). Thus, a **receptor potential** is a noisy, continuously graded signal (analogue signal). Graded potentials also occur locally at synapses and within dendrites throughout the nervous system, where they are known as **postsynaptic potentials.**

Rose (1948) introduced the concept of the **quantal efficiency** of any system that detects a stimulus at low light levels. If the average number of quanta absorbed in a given time by a given detector is N, with a deviation of $\sqrt{N}$, then the smallest change in N that can be detected, ΔN is:

$$\Delta N = k_1\sqrt{N} \qquad (1)$$

where k_1 is the **signal-to-noise ratio**. It can be shown that:

$$k_2 = LC^2\alpha^2 \qquad (2)$$

Where k_2 is a constant that depends on the optical parameters, quantal efficiency, and integration time of the eye, L is the luminance of the stimulus, C is the threshold contrast, and α is the angular subtense of the stimulus. Thus, for optical or visual systems with similar optical properties and integration times, quantal efficiency is proportional to the square of the threshold contrast. Rose estimated the quantal efficiency of the human eye as 5% at threshold luminances (see also van Meeteren 1978).

5.1.2c *Bipolar cells*

The second main layer of the retina consists of bipolar cells. The synaptic region between the receptors and the bipolar cells is the **outer plexiform layer.** Each bipolar cell receives inputs from only cones or only rods. However, the two types of bipolar cell converge onto ganglion cells. There is only one type of rod bipolar cell but many types of cone bipolar cell, defined by the shape and stratification of their axon terminals in the inner plexiform layer. **Midget bipolars** in the central retina are fed by only one cone. They serve the red-green component of colour vision and high visual acuity. Midget bipolar cells are further divided into **on-bipolar cells,** which respond to light increase, and **off-bipolar cells,** which respond to light decrease. Axons of on-bipolars terminate in the inner half of the inner plexiform layer, while axons of the off bipolars terminate in the outer half. **Diffuse cone bipolars** are fed by several neighbouring red and green cones and are either on-bipolars or off-bipolars, each subdivided into six types, that differ in their temporal characteristics. Some exhibit a sustained response and some a more transient response, depending on the rate of recovery of glutamate neurotransmitter. Diffuse bipolars carry a luminosity signal and have high contrast sensitivity, but may have some chromatic specificity. Blue cones and rods feed exclusively into specialized on-bipolar cells (Kouyama and Marshak 1992).

Each cone of the central retina contacts all six types of diffuse bipolar cells and at least two types of midget bipolar cells. Furthermore, the dendritic trees of each type of bipolar cell form a complete and independent coverage of the retina. Thus, any spot of light on the retina stimulates at least one of each type of bipolar cell (Boycott and Wässle 1999). This means that the separation of visual inputs into at least 10 parallel filters, or channels, begins in the outer plexiform layer of the retina.

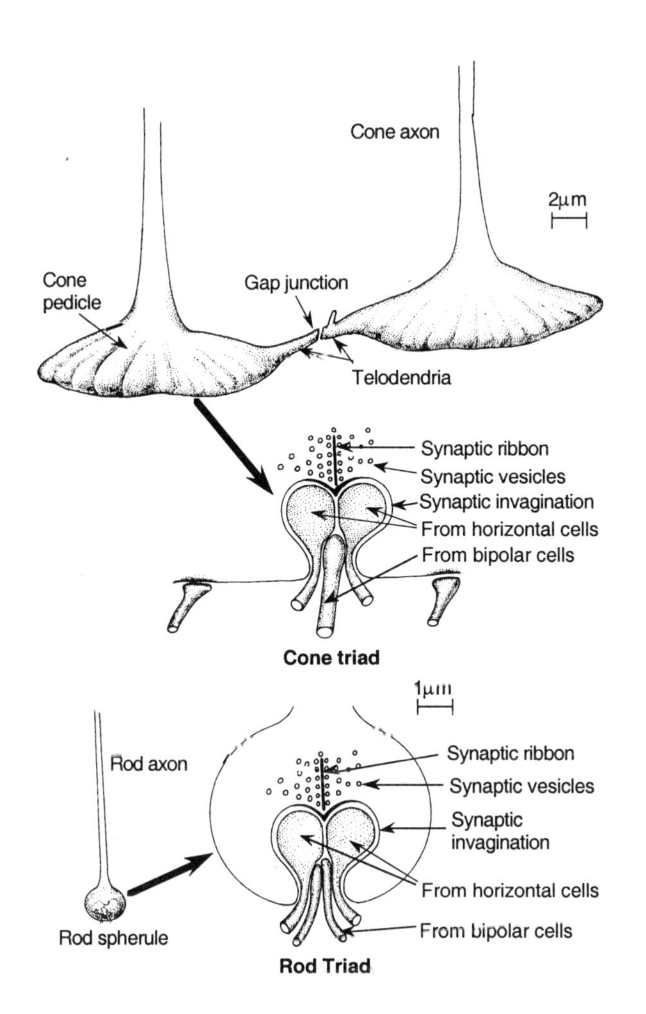

Figure 5.3. Terminals and synapses of rods and cones. (From Oyster *The Human Eye* 1999)

Each rod terminates in a spherical structure, about 2 μm in diameter, called a **spherule**. Each rod contacts horizontal cells and between 2 and 5 bipolar cells and each bipolar cell is fed by up to 45 rods (Kolb 1970). Each cone terminates in a flat synaptic **pedicle**. There are three types of synapse at the receptor level.

1. **Gap junctions** on fine radiating processes (telodendria) make electrical contact with similar processes on neighbouring rods or cones. Transmission is mediated by a family of proteins called **connexins**. For example, gap junctions connect rod amacrine cells with cone bipolar cells. Mice lacking the relevant connexin show abnormal responses in on-type bipolar cells (Güldenagel *et al.* 2001). Perhaps gap junctions open at low luminance levels to pool light energy.

2. **Synaptic invaginations**, each of which, typically, contains a single dendritic bouton from an on-bipolar cell and two boutons from horizontal cells, known collectively as a **triad synapse**. The membrane in a synaptic invagination of a rod or cone contains a synaptic ribbon, about 1 μm long, surrounded by **synaptic vesicles** containing neurotransmitter (glutamate), as shown in Figure 5.3. Rods have up to three synaptic invaginations and cones have 40 or more, each of which contacts one type of bipolar cell. Thus, a single cone can contribute to many types of bipolar cells.

3. **Fat synaptic boutons** occur only on cone pedicles. Those near an invagination, contact dendrites of on-bipolar cells. Those away from an invagination, contact dendrites of off-bipolar cells.

In the dark, the neurotransmitter L-glutamate or a similar substance is continuously released from receptors into bipolar cells (Dowling 1987). When a receptor absorbs light, the rate of neurotransmitter release is reduced. This causes on-bipolar cells to become hyperpolarized and off-bipolar cells, which form flat synaptic connections, to become depolarized. The difference is due to that fact that on-bipolar cells express a metabotropic glutamate receptor while off-bipolar cells express ionotropic glutamate receptors, namely AMPA or NMDA (Section 5.4.5b). Bipolar cells, like receptors, respond in a graded (analogue) fashion to changes in stimulus strength.

Horizontal cells run laterally in the outer plexiform layer over a distance of about 1 mm. They respond in a graded fashion to release of glutamate from rods or cones, using ionotropic glutamate receptors (AMPA and NMDA types). Horizontal cells release the inhibitory neurotransmitter GABA. One type of horizontal cell receives inputs from green cones or red cones within its dendritic field, and makes inhibitory contacts with several cone bipolar cells and with cones. A second type receives inputs primarily from blue cones or rods. For example, some horizontal cells receive inputs from red cones and send inhibitory inputs to synapses connecting green cones to bipolar cells, while other horizontal cells operate in reverse fashion.

These reciprocal connections between different types of cone were believed to form the basis for the first stage of colour opponency. However, it now seems that this is not the case (Dacey *et al.* 1996). Horizontal cells also make direct synapses with neighbouring horizontal cells of the same type so that, collectively, they form a resistive network. Signals in horizontal cells are also believed to feed back to cones in the form of a delayed depolarization, but the extent and significance of this process has been a subject of debate (Burkhardt 1993).

5.1.2d *Ganglion cells*

Ganglion cells form the third and final layer of the retina. The region between bipolar and ganglion cells is the **inner plexiform layer**. Each ganglion cell forms synaptic junctions with one or more cone bipolar cells in the inner plexiform layer. Rod bipolar cells contact on-type or off-type ganglion cells indirectly through a special type of amacrine cell. All bipolar cells use the neurotransmitter glutamate.

All-or-none **action potentials** are first formed at the inner plexiform layer. An action potential is all-or-none because it occurs only at a fixed amplitude for that nerve cell (see Section 4.2.1). Whereas a receptor potential is local, an action potential is a brief event that travels along an axon, sometimes for considerable distances. The speed of propagation of an action potential is proportional to axon diameter, which varies between 0.1 and 20 μm in vertebrates. In large axons, conduction speed reaches 120 m/s.

Amacrine cells form inhibitory connections with bipolar cells and ganglion cells within the inner plexiform layer, with dendritic fields of up to 1 mm in diameter. Amacrine cells, like ganglion cells, generate all-or-none action potentials. Thirty types of amacrine cells have been described. They differ in the type of bipolar cells they receive inputs from, which is related to their depth within the inner plexiform layer. They also differ in lateral spread of their dendritic fields and in the neurotransmitter they use (see Rowe 1991). Each type seems to fulfil a distinct function, although the functions served by most of them remains obscure. The dendritic fields of the different types overlap but the fields of each type tile the retina without overlap (MacNeil and Masland 1998).

Discharges in the optic nerve as a whole were first recorded by Adrian and Matthews (1927) in the eel. The first responses from single ganglion-cell axons were obtained by Hartline and Graham (1932) in the arthropod *Limulus*, and by Hartline (1938) in the frog. The set of receptors that directly or indirectly affects the firing of a given ganglion cell is the **receptive field** of that ganglion cell, a term introduced by Hartline (1938). Kuffler (1953) showed that ganglion-cell receptive fields in the cat retina are circular, with a concentric organization of excitatory and inhibitory regions. Receptive fields with a central excitatory region and inhibitory surround are known as **on-centre receptive fields**. They cause the ganglion cell to fire preferentially to the onset of a stimulus in the receptive-field centre. Those with an inhibitory central region and excitatory surround are known as **off-centre receptive fields**. They cause the cell to fire to stimulus offset. Ganglion cells have overlapping receptive fields because receptors supply inputs to several ganglion cells (see Chichilnisky and Baylor 1999). Those with overlapping receptive fields that are either both on-centre or both off-centre tend to fire at the same time, while those with opposite-sign centres tend not to fire at the same time (Mastronarde 1983). On-centre and off-centre receptive fields feed into pathways that remain distinct as far as the visual cortex (Schiller 1992). As luminance is reduced, the frequency of response of cells with off-centre receptive fields falls more rapidly than that of responses of on-centre cells (Ramoa *et al.* 1985).

The centre of an on-centre ganglion cell receives direct signals from several receptors through on-bipolar cells, and indirect signals from horizontal cells that originate in off-bipolar cells. The centre of an off-centre ganglion cell receives direct signals from off-bipolar cells, and indirect signals from horizontal cells that originate in on-bipolar cells. Thus, bipolar cells act on ganglion cells in a "push-pull" fashion. This mode of operation increases the dynamic range of these cells to changes in luminance (Sterling 1990). The organization of receptive fields into excitatory and inhibitory regions also ensures that the strongest signals come from local regions where luminance changes spatially or over time. These are the signals that contain useful information. The on-channel can be selectively blocked by applying aminophosphonobutyrate (APB) to an animal's retina. This impairs the animal's ability to detect light increments and reduces contrast sensitivity. However, responses to shape, colour, and movement are impaired only mildly (Schiller *et al.* 1986).

5.1.2e *Types of ganglion cells in cat*

In the cat, ganglion cells fall into three classes—X, Y, and W—each of which includes on and off types (Enroth-Cugell and Robson 1966). Type **X cells** are concentrated in the central retina and have small receptive fields with clearly organized excitatory and inhibitory regions from which signals are summed in a linear fashion. Their axons conduct at between 15 and 23 m/s and project mainly to cortical area 17. They are most sensitive to high spatial-frequency and low temporal-frequency stimuli. Type **Y cells** are more widely distributed over the retina. They have larger receptive fields that are not segregated into clearly defined regions, and show nonlinear summation of luminance distributions. Their axons conduct at between 20 and 30 m/s and project mainly to cortical area 18. They respond best to low spatial-frequency and high temporal-frequency stimuli (Pasternak *et al.* 1995). Type **W cells** have large receptive fields concentrated in the central retina, conduct at between 2 and 18 m/s, and project to areas 17, 18, and 19 and to the pulvinar.

Nerve impulses in ganglion-cell axons take longer to reach the optic nerve the greater their distance from the optic disc. However, at least for X cells, this differential delay is compensated for by the increase in axon diameter with increasing eccentricity. Axons with greater diameter conduct more rapidly (Stanford 1987).

5.1.2f *Types of ganglion cells in primates*

In primates, ganglion cells are broadly classified into **colour-opponent cells** and **achromatic cells** rather than X and Y cells.

Colour-opponent cells project to the **parvocellular laminae** of the LGN. They have small receptive fields, hence the prefix "parvo-", the Latin word for small. They constitute about 90% of the total number of ganglion cells. Each cone contributes, directly or indirectly, to at least two midget bipolar cells, one on-centre and one off-centre (Kolb 1970). Red and green cones feed into the cellular network of the retina to create four types of ganglion-cell receptive fields—red on-centre/green off-surround; green on-centre/red off-surround, red off-centre/green on-surround; green off-centre/red on-surround. The first two types correspond to inner **midget ganglion cells** and the latter two types to outer midget ganglion cells. Midget ganglion cells are also called **P cells** because they project to the parvocellular layers of the LGN. About 70% of ganglion cells in the retina of the monkey are midget cells. The receptive fields of inner and outer midget cells form distinct mosaic coverings of the retina.

Similar opponent structures are created from yellow (red plus green) and blue cones feeding into a distinct class of ganglions cells known as bistratified ganglion cells. The firing rate of a colour-opponent ganglion cell increases above baseline when the excitatory region is stimulated most and decreases below the baseline rate when the inhibitory region is stimulated most. For example, a cell with a red-excitatory centre and green inhibitory surround produces an increased response to a long-wavelength stimulus and a decreased response to a medium-wavelength stimulus. Note that single opponent cells do not respond to chromatic boundaries. We shall see later that the four types of red-green ganglion cells converge on so-called double-opponent cells in the visual cortex that produce signed difference signals related to local colour contrast.

According to the traditional view, the receptive field of a colour-opponent ganglion cell contains cones of one type in the centre and of another single type in the surround (Reid and Shapley 1992). Lennie *et al.* (1991) suggested that, at least for eccentricities up to 10°, the receptive-field centre consists of a single short-, medium-, or long-wavelength cone, and the surround contains a random mix of cone types, which feed into bipolar cells through horizontal cells. This type of receptive field organization creates at least some colour opponency. There is conflicting evidence on this issue (Lee 1996). It had been assumed that horizontal cells in the outer plexiform layer connect cones of opponent types, but recent evidence suggests that horizontal cells connect cone types of all kinds (Dacey *et al.* 1996). Amacrine cells forming inhibitory links between midget ganglion cells in the inner plexiform layer are also indiscriminate with respect to the spectral specificity of the cells they connect (Calkins and Sterling 1996).

Colour-opponent ganglion cells have small receptive fields, a low sensitivity to luminance but high sensitivity to chromatic modulation. They produce a sustained response to continued stimulation with a high degree of linearity in their temporal response, and conduct nerve impulses at medium velocity. Their small receptive fields enhance their spatial resolution, but their sustained characteristic reduces their temporal resolution. Thus, parvocellular cells in the monkey can resolve gratings up to 40 cpd but are most sensitive to temporal frequencies of only 10 Hz (Derrington and Lennie 1984). Colour opponency is reduced at scotopic levels of luminance.

Achromatic ganglion cells are called parasol ganglion cells. They have large receptive fields and project to the **magnocellular laminae** of the lateral geniculate nucleus and are called **M cells**. Achromatic cells have receptive-field centres and receptive-field surrounds made up of rods or a variety of cone types. They are thus broadly tuned to wavelength and do not show colour opponency. They are referred to as **broadband cells**. However, for some of them, the receptive field centre and surround are not in spectral balance, and colour opponency is revealed under certain conditions of stimulation (Shapley 1991). Some have on-centre receptive fields and others have off-centre receptive fields. Their large receptive fields enhance their light-collecting efficiency and render them sensitive at lower luminance. Cells with large receptive fields have higher sensitivity also because they have a higher signal-to-noise ratio. Noise is due to fluctuations in photon distribution and to spontaneous events at the photopigment and synaptic levels. When N receptors feed into one receptive field, the total noise is proportional to $\sqrt{N}$.

Magnocellular ganglion cells have large-diameter axons and consequently conduct nerve impulses rapidly. They show a transient response to continued stimulation. Their large receptive fields reduce their spatial resolution but their transient characteristic improves their temporal resolution so that they are able to respond to higher rates of flicker than the sustained colour-opponent cells. Thus, in the monkey, broadband cells can resolve gratings up to a spatial frequency of only about 10 cpd but are most sensitive to temporal frequencies of 20 Hz (Derrington and Lennie 1984). About 10% of ganglion cells are magnocellular, and recent evidence suggests that the percentage is the same in fovea and periphery (see Lee 1996).

Ganglion cells are classified in several other ways according to the organization of their receptive fields, but this topic goes beyond the scope of this book. For more details on the structure and function of the retina see Polyak (1941), Dowling (1987), Wässle and Boycott (1991), and Lee (1996).

For each cone in the central fovea of the monkey retina there are three to four ganglion cells. Each ganglion cell receives inputs from more than one cone and each cone influences more than one ganglion cell. At an eccentricity of about 15° there is one ganglion cell per cone. In the peripheral retina there are many more cones than ganglion cells (Wässle *et al.* 1990). Thus, the precision with which the retina samples the distribution of light in the image declines systematically with increasing eccentricity.

A single receptor, even in the fovea, typically contributes to the centre and surround of several ganglion-cell receptive fields. In other words, neighbouring receptive fields overlap. However, the receptive fields of a given type of ganglion cell pave the retina efficiently without too much overlap or too many gaps.

For receptive field centres arranged in a square lattice and separated by distance *a*, as in Figure 5.4a, the minimum radius of the receptive fields required for complete coverage is $\frac{a}{\sqrt{2}}$. For receptive fields arranged in a hexagonal lattice, as in Figure 5.4b, the minimum radius for complete coverage is $\frac{a}{\sqrt{3}}$. For a given type of ganglion cell, the ratio of receptive-field spacing to receptive-field diameter is close to that predicted from the most efficient coverage for a hexagonal lattice (Wässle *et al.* 1981). Thus, at every point, the retinal image is efficiently sampled for the different visual features that each of the different types of receptive field detects.

The areas of centre and surround regions of the receptive fields of both opponent and achromatic ganglion cells increase in proportion to eccentricity, with receptive fields of achromatic cells having about twice the area of those of opponent cells at any given eccentricity. Peak sensitivities of centre and surround regions are inversely proportional to receptive-field area, and hence to eccentricity, so that integrated contrast sensitivities (contrast gains) are constant over the visual field (Croner and Kaplan 1995).

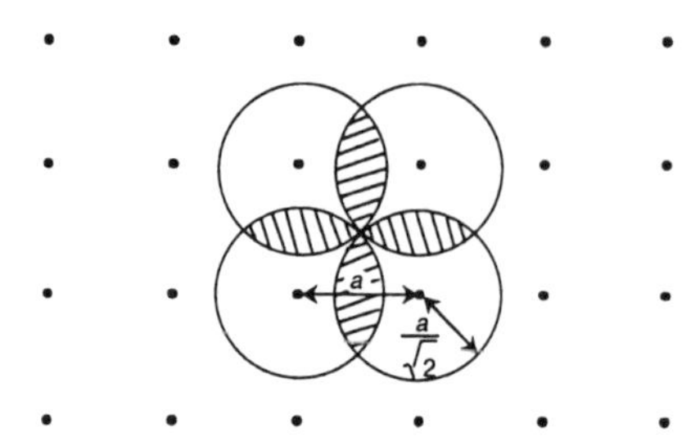

(a) For receptive fields in a square lattice, with centres *a* unit apart, the minimum radius of coverage is $a/\sqrt{2}$

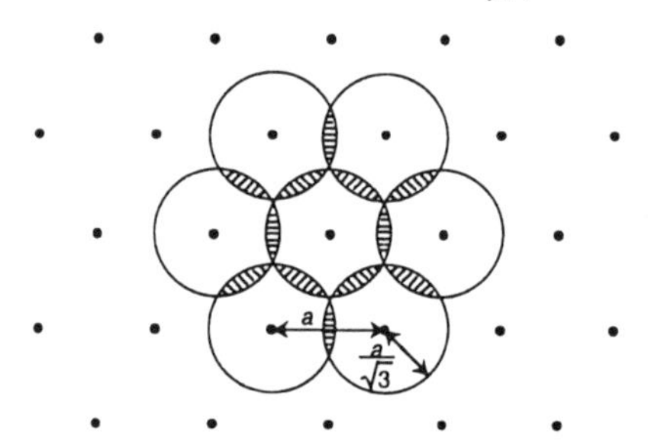

(b) For an hexagonal lattice, the minimum radius is $a/\sqrt{3}$

Figure 5.4. Coverage of ganglion-cell receptive fields.
(Adapted from Wässle *et al.* 1981)

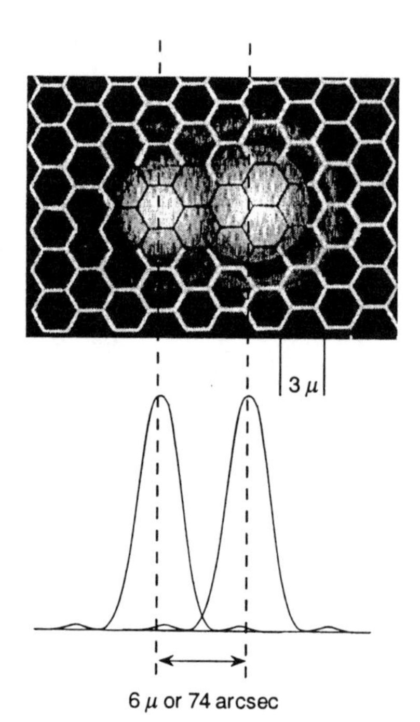

Figure 5.5. Retinal images of two point sources.
The upper diagam shows that Airrey'sdisc spreads over seven retinal cones. (Adapted from Emsley 1952)

Most ganglion cells maintain an irregular discharge in the absence of stimulation. The irregularity of the maintained discharge of ganglion cells, especially in the low-frequency range, is also evident in the power spectrum of their response to stimuli. The irregularity presumably arises from the stochastic quantal nature of light and from random variations in the initiation of neural spikes (Robson and Troy 1987).

5.1.3 Image resolution

The adult human retina has between 4 and 6 million cones, with a peak density at the fovea that is highly variable from person to person (100,000 to 320,000 cones/mm^2). The range of grating acuity predicted from these cone densities is 47 to 86 cpd. Thus, on average, the retinal mosaic is well suited to deal with the highest spatial frequency (60 cpd) transmitted by the optics of the eye (Campbell and Gubisch 1966). Cone density falls steeply with increasing eccentricity, being ten times lower 4° away from the fovea than at the fovea. The human retina has 100 million or more rods and a central rod-free area about 1.25° in diameter. The two eyes have similar numbers of cones and rods and similar photoreceptor topography (Curcio *et al.* 1990).

Like most optical systems, that of the eye suffers from diffraction, light dispersion by the optic media, spherical aberration, coma, and transverse and longitudinal chromatic aberration. If one considers only the effects of diffraction, a point of light imaged by a lens through a circular aperture creates an interference pattern consisting of a central luminous disc, known as **Airy's disc**, surrounded by concentric annuli of diminishing luminance. The diameter of Airy's disc increases as the wavelength of light increases and as pupil diameter decreases. For a pupil diameter of 3 mm and a wavelength of 550 nm, about 75% of the light from a point source is contained in Airy's disc, which spreads over about three cones in the central retina, as illustrated in Figure 5.5 (O'Brien 1951). The distribution of light in the image may also differ along different meridians because of the effects of astigmatism, and it may also be skewed because of the offset of the optic axis, an effect known as coma.n

When we consider all factors that affect the focussing of an image, the distribution of light in the image from a point source is the **point-spread function** (see Section 3.3.4). While image dispersion due to diffraction can be calculated from the pupil diameter and the wavelength of light, the point-spread function must be measured by scanning a photometer over the image. Measurements made on an excised

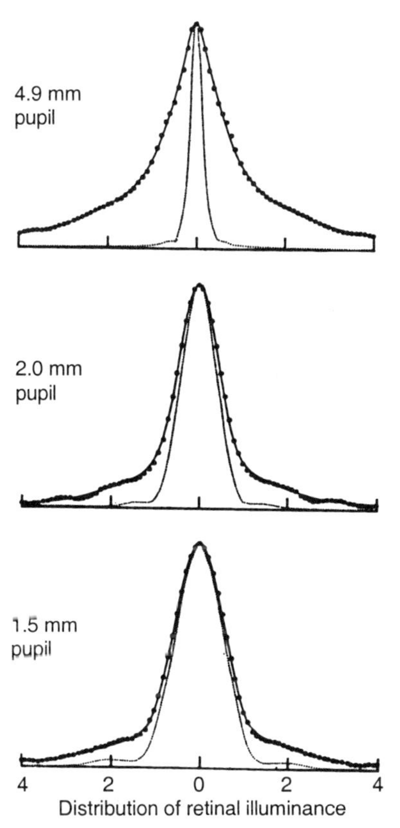

Figure 5.6. Point-spread functions of the human eye. Each curve is the normalized distribution of illuminance in the foveal image of a thin line, measured as described in the text. The narrow curve in each case indicates the calculated diffraction image of the line at the given pupil diameter. (Adapted from Campbell and Gubisch 1966)

eye of a cadaver are affected by *post-mortem* changes. In another method, an image of a bright line is formed in the living eye and photometer measurements are performed on a secondary reflected image created in space. Campbell and Gubisch (1966) used this procedure to produce the point-spread functions shown in Figure 5.6.

The wider the point-spread function the lower the ability of the eye to resolve a grating. The point-spread function depends on the diameter of the pupil in a complex way. For example, as the pupil enlarges, diffraction decreases and spherical aberration increases. For image resolution, a pupil diameter of about 2 mm achieves the best compromise between these competing factors.

However, a wider point-spread function may improve the detectability of changes in the spatial location of distinct images because more receptors are stimulated, which can provide an improved estimate of the mean position of the stimulus. With a pupil diameter of less than about 2 mm, most of the

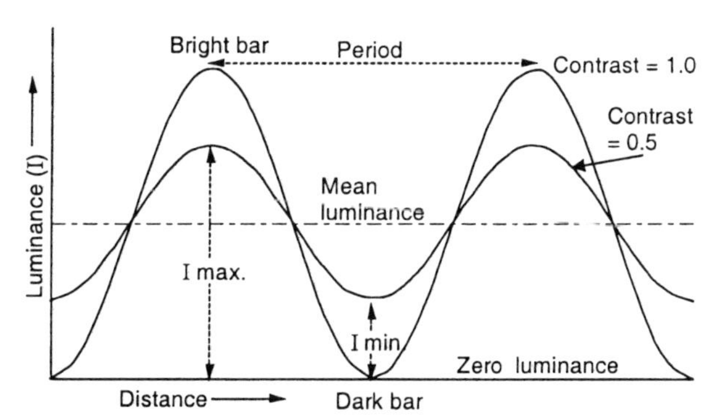

Figure 5.7. Characteristics of a sine-wave grating.
Spatial frequency is the number of luminance modulations per degree of visual angle—the reciprocal of the period. Two levels of contrast are illustrated.

$$\text{Mean luminance} = \frac{\text{Imax} + \text{Imin}}{2}$$

$$\text{Luminance modulation} = \text{Imax} - \text{Imin}$$

$$\text{Michelson contrast} = \frac{\text{Imax} - \text{Imin}}{\text{Imax} + \text{Imin}}$$

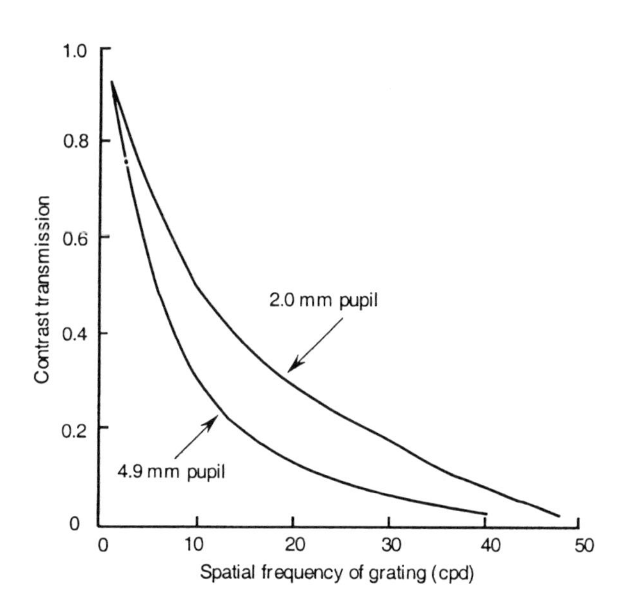

(a) Transfer functions for two pupil sizes. The curves were obtained by measuring the light reflected from the image of a grating on the fovea of a human eye. The functions result from loss of contrast due to diffraction and other optical aberrations.

light dispersion is due to diffraction, and the optical system can be regarded as linear.

The optical quality of the eye, like that of any linear optical system, can also be specified by its **modulation transfer function** (MTF). The MTF is obtained by measuring the contrast of the image of a sinusoidal grating of fixed contrast at each spatial frequency over the visible range of spatial frequencies (Section 3.3.5). The contrast and spatial frequency of a sinusoidal grating are defined in Figure 5.7. The MTF of the optical system of an eye relates the proportional loss of contrast in the image to the spatial frequency of a grating. With a small pupil, almost all the contrast in the stimulus is preserved in the image at spatial frequencies below about 5 cpd. At about 40 cpd, contrast is reduced about tenfold, and at above about 60 cpd, all contrast in the image is lost. As pupil diameter increases, the high frequency limit declines, as shown in Figure 5.8a. Modulation transfer functions can be normalized to the highest spatial frequency transmitted by an ideal diffraction-limited sysntem, as shown in Figure 5.8b. Any residual departure from the ideal system is due to factors other than diffraction, such as spherical and chromatic aberrations and light scatter in the optical media (Campbell and Gubisch 1966).

The optical quality of the image is substantially constant out to an eccentricity of about 12°, beyond which it declines with increasing distance from the optic axis (Jennings and Charman 1981).

Two stimuli can be spatially resolved only if they excite two detectors at a discriminably higher level

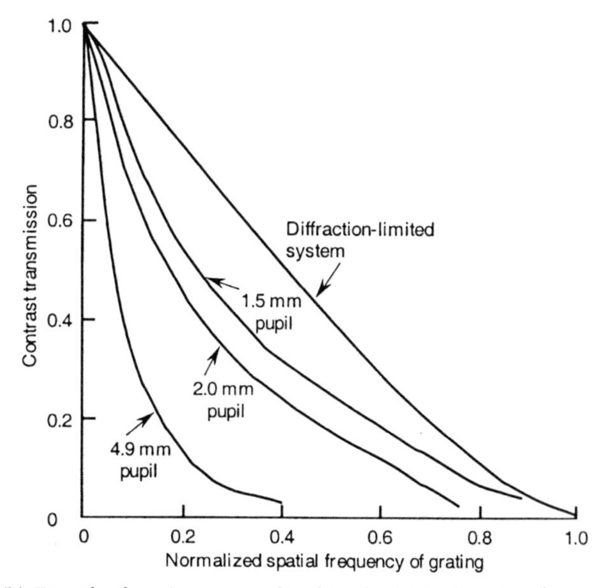

(b) Transfer functions normalized to the highest spatial frequency transmitted by an ideal diffraction-limited system with 570 nm wavelength. The ideal function falls on the same curve for all pupil diameters. The three empirical functions depart from the ideal function as pupil size is increased because of increasing dispersion of light due to spherical and chromatic aberrations.

Figure 5.8. Modulation transfer functions of the human eye.
(Adapted from Campbell and Gubisch 1966)

than they excite a detector in an intermediate location. Thus, a set of independent detectors arranged in a square lattice can resolve a periodic stimulus, such as a grating, only if the spatial period of the stimulus is at least twice the spacing of the detectors. This is the **Nyquist limit**. A stimulus with a smaller period is said to be undersampled. For a hexagonal lattice, like the cone mosaic, it is easy to prove that the Nyquist limit is $\sqrt{3}$ times the spacing of the detectors.

The smallest period in radians, v, of an extended grating that can be resolved by the optics of the eye is limited by the wavelength of the light, λ, and by diffraction, which is inversely proportional to pupil diameter, D (Westheimer 1972). Thus,

$$v = \frac{\lambda}{D} \quad (3)$$

According to the Nyquist limit, the spacing of cones in a hexagonal lattice required to match the optics of the eye is therefore,

$$v = \frac{\lambda}{D\sqrt{3}} \quad (4)$$

Image quality is best when the pupil diameter is 2.4 mm and λ = 555 nm. Putting these values in equation 4 gives a cone separation of 27.4 arcsec, which is close to the value reported by O'Brien (1951). It is an advantage to have receptors as large as possible so they capture the maximum number of photons, but if they are too large they fail to match the resolving power to the eye's optics. The advantage of having the cones touch to avoid loss of photons outweighs the disadvantage of intercone leakage (Snyder and Miller 1977).

Lord Raleigh defined a criterion for the limit of resolution of two points. For a diffraction-limited system, two point sources can just be resolved when the peaks of their images are separated by the radius of the inner bright regions of their diffraction patterns (Airy's disc). Since at least three aligned detectors are required to resolve two points, the diameter of the detectors is half the diameter of Airy's disc at the resolution limit.

The image of a grating finer than the Nyquist limit forms an interference or moiré pattern with the receptor mosaic, as illustrated in Figure 5.9. Although the grating is not visible, the interference pattern is visible because the bars of the grating come into and out of phase with the receptors at a spatial frequency lower than that of the grating.

If the spatial frequency of the receptor mosaic is f and that of the stimulus grating is $f + n$, then the interference pattern has a spatial frequency of n. This process is known as **aliasing**. However, the effects of aliasing are not normally visible because the eye is not optically capable of forming images as fine as the Nyquist limit. Thus, the optics of the eye constitutes

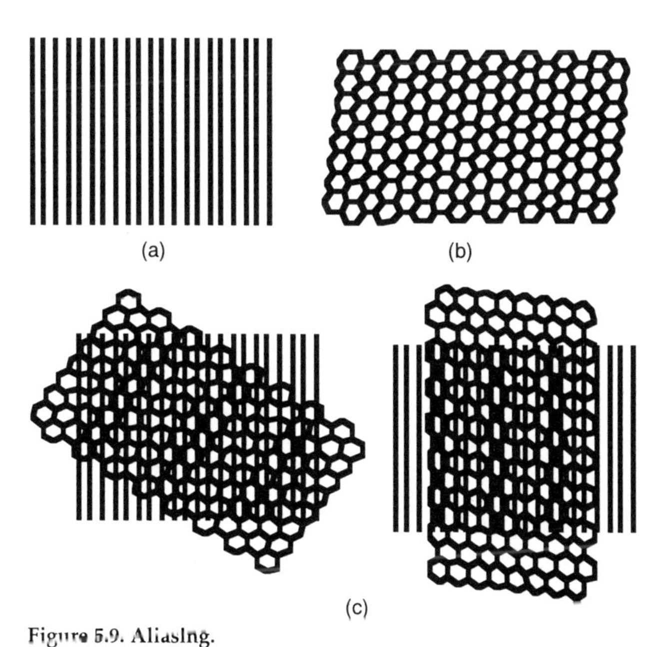

Figure 5.9. Aliasing.
A fine grating (a) projected onto a hexagonal retinal mosaic (b) produces an interference (moiré) pattern (c). The pattern is most evident when the spatial frequency of the grating is slightly higher than that of the mosaic. The two patterns in (c) demonstrate that the moiré pattern has a 60° periodicity because of hexagonal packing of foveal receptors.

an anti-aliasing filter. This is not true in the peripheral retina where gratings far beyond the cut-off frequency of the classical contrast sensitivity function may be detectable as moiré patterns produced by undersampling of the image (Snyder *et al.* 1986; Thibos *et al.* 1996).

The effects of aliasing can be made visible in the central retina by converging two laser beams on the retina to form a fine interference pattern. This pattern may be finer than the spacing of the retinal mosaic, since it bypasses the optics of the eye. By comparing grating acuity for a normal grating with that for interference fringes, Campbell and Gubisch showed that visual performance in the fovea is limited by the eye's optics rather than by the density of receptors. Because of the hexagonal packing of the receptors, the moiré pattern changes as the interference pattern is rotated through 60°. The period of the finest visible moiré pattern reveals that the mean spacing of receptors is about 0.5 arcmin, corresponding to a resolution limit of about 60 cpd, a value that tallies with anatomical determinations of the spacing of foveal cones. Above the 60-cpd limit, the interference pattern may be visible because of aliasing. The 60-cpd resolution limit for an interference pattern is somewhat finer than the finest pattern that can be resolved by the eye's optics. This means that foveal performance is limited by the optical quality of the image rather than by the spac-

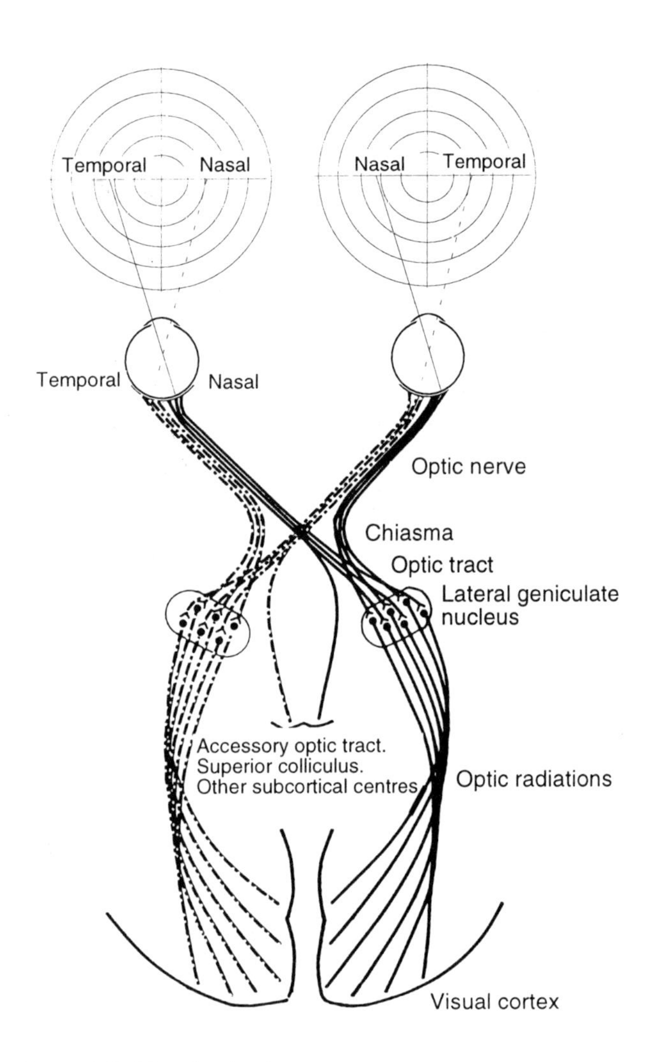

Figure 5.10. The visual pathways.
Axons from the right half of each eye (left visual hemifield) project to the right occipital lobe and those from the left half of each eye (right visual hemifield) project to the left occipital lobe. (From Howard 1982)

ing of receptors (see also Williams 1988; Hirsch and Curcio 1989; Smallman *et al.* 1996). Outside the fovea, the density of receptive fields seems to be the limiting factor in resolution (Marcos and Navarro 1997).

Haig (1993) has suggested that, in addition to providing an anti-aliasing filter, the high density of receptors relative to the resolving power of the eye's optics compensates for the degradation of spatial sampling due to the presence of three types of wavelength-sensitive cones. However, his argument seems to be based on an overestimation of the resolving power of the retinal mosaic. The close correspondence between the resolving power of the fovea and the size of foveal cones implies that each cone provides an independent sampling of spatial information, which is conveyed to the brain. This tallies with the fact that the ratio of ganglion cells to cones in the fovea is at least one-to-one.

An interference pattern formed by converging laser beams bypasses the optics of the eye but is still subject to preneural factors such as quantal fluctuations in the stimulus, opacities of the ocular media and, finally, the aperture, quantal efficiency, and density of receptors. Banks *et al.* (1987) compared the contrast sensitivity of the human observer with that of an ideal observer incorporating the preneural properties of the human visual system. They concluded that foveal performance is limited by preneural factors rather than by neural processing occurring beyond the receptor level.

5.2 VISUAL PATHWAYS AND DECUSSATION

5.2.1 Decussating pathways

The axons of ganglion cells leave the eye to form the **optic nerve**. Each optic nerve has a diameter of 3 to 4 mm and contains about one and a quarter million axons. After passing out of the retina at the optic disc, the optic nerve travels about 5 cm to end in the **optic chiasma**. In most vertebrates, most of the axons from each eye cross over to the contralateral side in the chiasma. This is known as **decussation** from the Latin *decussare*, meaning to divide crosswise. In primates and some other mammals, axons from the nasal hemiretinas decussate at the chiasma, but those from the temporal hemiretinas remain on the same side. This is known as **hemidecussation.** The nerves that emerge from the chiasma form the **optic tracts.** In primates, axons from the temporal half of the left eye join decussated axons from the nasal half of the right eye to form the left optic tract. Axons from the temporal half of the right eye join decussated axons from the nasal half of the left eye to form the right optic tract. In this way, inputs from the two eyes with similar local signs come together.

Collateral branches of optic-tract axons lead to the superior colliculus, the pretectum by way of the accessory optic tract, the hypothalamus, and other subcortical areas. In some mammals, such as the rat and tree shrew, it seems that both visual hemifields are represented in each superior colliculus (Kaas *et al.* 1974).

Each optic tract leaves the chiasma and terminates on its own side in a part of the thalamus known as the **lateral geniculate nucleus** (**LGN**). Within each LGN, inputs from the two eyes remain in distinct layers, or laminae, where they synapse with **relay cells.** Axons of the relay cells leave the LGN on each side and fan out to form the **optic ra-**

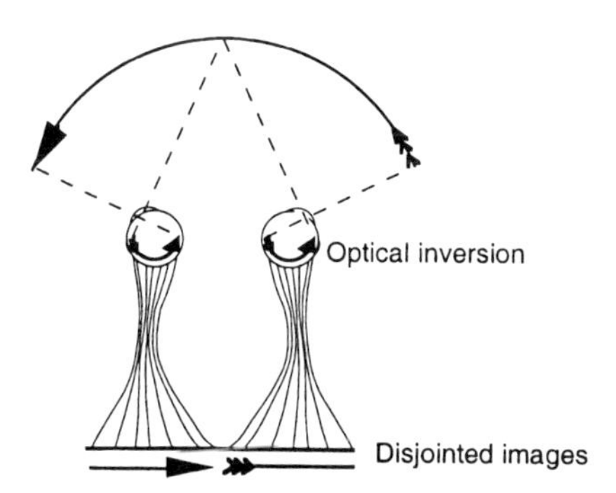

(a) Optically inverted images and undecussated pathways form a central map with the two halves topologically disjointed.

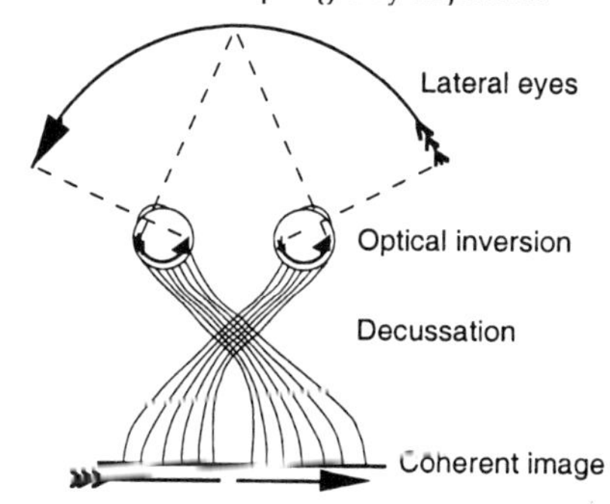

(b) Full decussation forms a coherent central map.

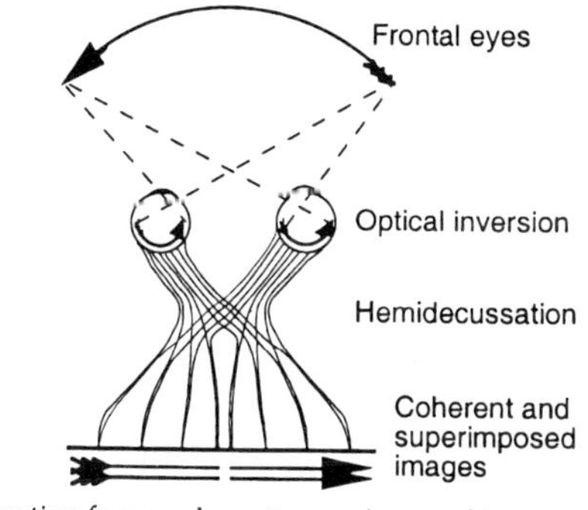

(c) Hemidecussation forms coherent superimposed images.

Figure 5.11 Types of visual pathway. (After Ramón y Cajal 1911)

diations, which course backwards and upwards to terminate in the visual cortex in the ipsilateral occipital lobe of the cerebral cortex (Figure 5.10). Thus, axons from the right half of each eye project to the right occipital lobe, and those from the left half of each eye project to the left occipital lobe. Because of the reversal of each retinal image, the left half of the visual field (left hemifield) is represented in the right cerebral hemisphere, and the right hemifield is represented in the left hemisphere.

The optic nerves in submammalian vertebrates almost fully decussate. Ramón y Cajal (1911) proposed that this primitive condition evolved to preserve the spatial integrity of the central neural map of the images from the two eyes. Because of the optical inversion of each retinal image, there is a disruption of the continuity of the central mapping across the hemispheres when the pathways are undecussated, as in Figure 5.11a.The central map is continuous when the pathways decussate, as in Figure 5.11b. The spatial integrity of the internal map is not important as such, since spatial location is coded in terms of fibre connections and patterns of firing, not in terms of spatial maps. However, transcallosal fibres connect spatially adjacent regions from opposite sides of the midline, so that visual stimuli in the midline region can be processed. These connections for decussated pathways are shorter than for undecussated pathways. It is believed that the crossing over of visual inputs to the opposite visual cortices led to the crossing over of the motor pathways so that visual inputs from a given half of space control the movements of limbs on the same side of the body. Nobody has proposed a better explanation of visual and motor decussation.

5.2.2 Hemidecussation

In most submammalian species, the visual pathways decussate and the ipsilateral projection is weak or absent. The optic nerves hemidecussate in mammals, but not in all mammals. When they do, the ratio of uncrossed to crossed fibres is proportional to the size of the binocular visual field, which depends on the extent to which the eyes are in a frontal position. This relationship is known as the **Newton-Müller-Gudden law**. Thus, the proportion of uncrossed fibres is almost zero in the rabbit, about one-eighth in the horse, one-fourth in the dog, one-third in the cat, and half in primates, including man (Walls 1963, p. 321). The weak ipsilateral projection of the visual pathways in most submammalian species is not related to the degree of overlap of the visual fields (Hergueta *et al.* 1992). This question is discussed in more detail in Section 14.1.

The Newton-Müller-Gudden law applies to the mammalian retinogeniculate pathway. Retinal projections to the hypothalamus are not topographically organized. They seem to be concerned with synchronization of metabolic activity with the day-night and seasonal cycles. The primitive condition of the retinohypothalamic pathway in non-mammalian vertebrates is one of equal ipsi- and contralateral inputs. This condition is also present in primitive mammals, such as anteaters, sloths and bats. Other nonprimate species, such as the cat and tree shrew, have a predominance of contralateral inputs and primates have a predominance of ipsilateral inputs (Magnin *et al.* 1989).

In primates, the boundary between decussating and nondecussating ganglion cells falls approximately along the midvertical meridian of the eye. In nonprimate mammals, the position of this boundary varies according to the type of ganglion cell, with some types of cell remaining fully decussated, whether they arise in the nasal or in the temporal retina (Leventhal *et al.* 1988). In some birds, such as the pigeon, hemidecussation occurs beyond the thalamus (Chapter 14).

One important function of hemidecussation is to bring inputs from a given part of the binocular visual field to the same location in the brain, as illustrated in Figure 5.11c. This allows the visual system to compare inputs from roughly corresponding regions of the two retinas with a minimum length of connections, and provides the basis for detecting binocular disparities and hence for binocular stereoscopic vision. The other function of hemidecussation is in the control of binocular eye movements. When the gaze moves over the visual scene, the eyes must move together to ensure that light from the same points in the visual scene projects to corresponding points in the two retinas.

Binocular inputs are not essential for coordinated shifts of gaze, since the eyes move through equal angles when one eye is closed (Section 9.9.1). Even opposed eye movements, which converge the visual axes on an object of interest at a particular distance, may occur when one eye is closed. However, binocular disparity is sufficient to initiate vergence eye movements, and the control of directional shifts of gaze and vergence is more precise when both eyes are open (see Chapter 9).

Stereoscopic vision is developed particularly well in mammals with frontal eyes, such as cats and primates. In these animals, visual inputs from corresponding regions in the two eyes converge on binocular cells in the visual cortex, which are tuned to binocular disparity (see Chapter 6). There are also some disparity-tuned binocular cells in some mammals with laterally placed eyes and small binocular fields, such as rabbits, sheep, and goats. Some nonmammalian species, such as certain insects, amphibians, and birds, have frontal vision and perhaps some binocular range-finding (Chapter 14).

The binocular field and the associated mechanism of corresponding points are not necessary for the perception of a unified visual field. Animals with a binocular field suffer diplopia when the mechanisms responsible for conjunctive and disjunctive eye movements are damaged, as in strabismus. Animals with laterally placed eyes, which have only a small binocular field, are less affected by diplopia. They no doubt experience a unified panoramic visual field, which may extend 360°. We humans experience a unified visual field when the nasal half of

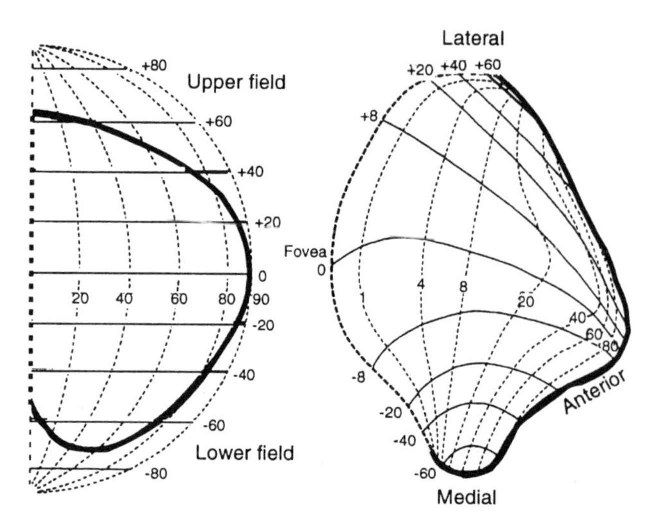

Figure 5.12. Projection of the visual field onto the LGN. Schematic view of the right hemifield of the monkey's retina and its projection onto the dorsal surface of layer 6 of the left LGN. Numbers represent degrees. The dotted lines are azimuths and the solid lines are elevations. The heavy dotted and solid lines represent the limits of the visual field. (Adapted from Malpeli and Baker 1975)

each eye's visual field is occluded. A simple way to demonstrate this is to hold up against the nose an occluder just wide enough to make the nasal limit of vision for one eye coincide with the nasal limit for the other eye. Three fingers are about the correct width. The visual field seen with such an occluder looks complete, although it is composed of only abutting monocular temporal hemifields.

Section 5.4.2 provides more details about the visual pathways.

5.3 LATERAL GENICULATE NUCLEUS

5.3.1 Structure of the LGN

In the monkey about 90% of the axons in each optic tract go to the **lateral geniculate nucleus** (LGN). In each LGN the axons segregate into distinct layers, or **laminae** where they synapse with **relay cells** (Perry *et al.* 1984). The axons of relay cells go to the primary visual cortex. The other 10% of ganglion-cell axons in each optic tract project only to other subcortical structures, such as the superior colliculus, the accessory optic system, and the suprachiasmatic nucleus of the hypothalamus (Schein and Monasterio 1987). In rodents, only about 37% of ganglion cells project to the LGN. About 95% of axons project to the superior colliculus and many of the axons reaching the LGN are branches of these axons (Martin 1986). In the cat, 77% of ganglion cells project to the LGN (Illing and Wässle 1981).

Methods for tracing connections in the nervous system are described in Section 5.4.1. The projection of the visual field onto the primate LGN, shown in Figure 5.12, was revealed by use of retinal lesions (Brouwer and Zeeman 1926; Clark and Penman 1934), and in more detail by the microelectrode recording technique (Malpeli and Baker 1975). The horizontal retinal meridian divides the LGN along its axis of symmetry, with the lower visual field represented in the medial superior half and the upper field in the lateral inferior half. The vertical meridian of the visual field divides the LGN in the orthogonal direction. The fovea is represented at the posterior, or caudal pole and peripheral regions at the anterior, or rostral pole. The number of parvocellular neurones in the macaque LGN devoted to each unit area of the visual field is 10,000 times higher in the region devoted to the fovea than in the region devoted to the far periphery (Malpeli *et al.* 1996).

The LGN of the cat contains four principal laminae—A, A1, C, and C1—and two others known as C2 and C3. Cells in laminae A and C receive their inputs from the contralateral eye and those in laminae A1 and C1 from the ipsilateral eye. The two A laminae contain similar types of cells but the cells in lamina C, originating in the nasal hemiretina are considerably larger than those in layer C1, originating in the temporal hemiretina. The axons of most X and Y cells terminate in laminae A and A1 with about 62% of all Y cells terminating in lamina A1. The C lamina receives a few X and Y cells but mainly W cells. W cells are a heterogeneous group of slowly conducting ganglion cells. Their large receptive fields have poorly defined excitatory and inhibitory regions and poor spatial and temporal resolution. In addition, the cat's LGN has two associated nuclei known as the medial interlaminar nucleus and the geniculate wing (Kaas *et al.* 1972).

Retinogeniculate pathways in the monkey have been investigated by tracing the retrograde transport of horseradish peroxidase from specific layers of the LGN to specific types of ganglion cell in the retina (Perry *et al.* 1984). They have also been investigated electrophysiologically by recording from cells in the LGN (Kaplan and Shapley 1986).

The axons of colour-opponent ganglion cells (midget cells) terminate in the four dorsal laminae in the primate LGN, where they synapse with relay cells (Figure 5.13). They are known as the parvocellular laminae, or **P laminae**, because they contain small cells. Inputs from the ipsilateral eye go to laminae 3 and 5 and those from the contralateral eye go to laminae 4 and 6. The whole visual channel,

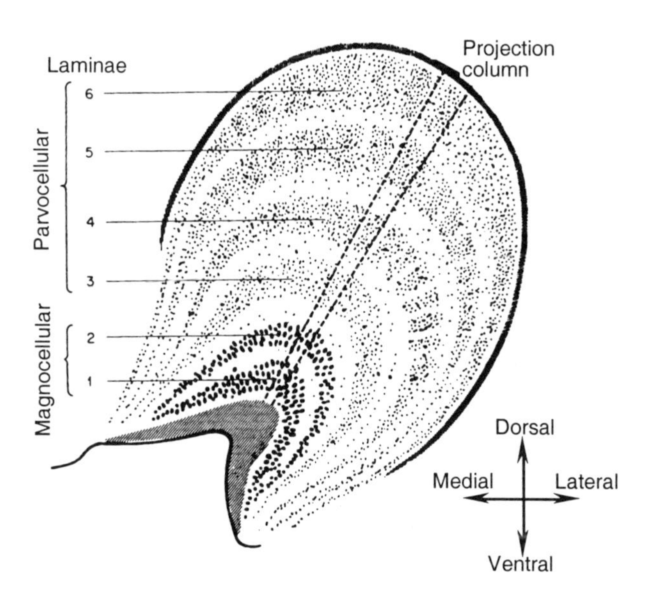

Figure 5.13. The lateral geniculate nucleus.
Lamination and projection columns in a coronal section of the LGN of a monkey. A column is defined as having 90 % of the cells with a single visual direction. (From Szentágothai 1973)

including the colour-opponent cells, the relay cells in the LGN, and the cortical pathways to which they subsequently lead, is the **parvocellular system**. Opponent ganglion cells are also called **P cells**.

The axons of achromatic ganglion cells terminate in the two ventral laminae of the LGN, known as the magnocellular laminae, or **M laminae**, because they contain large cells. Inputs from the ipsilateral eye go to lamina 2 and those from the contralateral eye go to lamina 1, as shown in Figure 5.143 The achromatic cells, the magnocellular laminae, and the pathways into which they feed are known as the **magnocellular system**. Achromatic ganglion cells are also called **M cells**.

In the part of the LGN devoted to the central retina there are four P and two M laminae. In parts devoted to the peripheral binocular visual field, there are only four laminae; two P and two M laminae. The blind spot is represented in the transition zone between the six-layered and four-layered regions (Lee and Malpeli 1994). The region devoted to the monocular temporal crescent of the visual field receives only crossed inputs and therefore contains only two laminae, one P and one M lamina (Kaas *et al.* 1972). The inputs to each lamina in the LGN are projected in systematic retinotopic order. Inputs from corresponding areas of the two eyes lie in **projection columns** running at right angles to the laminae, as shown in Figure 5.13. In each lamina, ganglion cells with on-centre receptive fields and those with off-centre fields terminate on distinct cells.

According to one estimate, magnocellular inputs reach the LGN of the monkey, on average, 17 ms before parvocellular inputs (Schmolesky *et al.* 1998). But, it has been claimed that this temporal advantage is eliminated by the time inputs reach the visual cortex, because the more numerous parvocellular inputs converge on cortical cells more than do magnocellular inputs (Maunsell *et al.* 1999).

The primate LGN also has so-called **koniocellular cells**, or **K cells**, with physiological properties similar to those of W cells in cats. They occur mainly in three ipsilateral and three contralateral layers in the LGN but some are scattered in M and P cell layers (Hendry and Yoshioka 1994). They relay inputs from blue cones to cytochrome oxidase blobs in layer 1 of V1 and to the superior colliculus. Groups of large neurones that tend to occur in K-cell layers, with properties similar to K-cells, project to V2, V4 (Section 5.7.2a), and the inferior temporal cortex (Section 5.7.2b). These pathways may be responsible for blindsight. K cells receive substantial inputs from V1 and are the only LGN cells to receive inputs from the superior colliculus (Hendry and Reid 2000).

The neuroanatomy of the LGN is reviewed in Garey *et al.* (1991).

5.3.2 LGN receptive fields

5.3.2a *Inputs to LGN receptive fields*

The receptive-field of each relay cell in the LGN is fundamentally the same as that of the ganglion cell with which it is connected. Furthermore, there is the same number of relay cells as ganglion cells, although there may be some divergence and convergence of optic nerve fibres onto LGN relay cells (Schein and Monasterio 1987).

Although any one neurone in the LGN receives a direct excitatory input from only one eye, there is evidence of extensive inhibitory and excitatory interactions in the LGN and of inputs from places other than the retina (Marrocco and McClurkin 1979; Kato *et al.* 1981; Ahlsén *et al.* 1985). All the interneurones in the LGN contain the inhibitory neurotransmitter gamma-aminobutyric acid (GABA), and all GABAergic cells in the LGN are interneurones (see Mize and Marc 1992).

According to one estimate, only about 20% of synaptic junctions found on geniculate relay neurones of the cat originate in the retina. About 50% derive from layer 6 of the visual cortex. The other 30% are composed of inhibitory (GABAergic) inputs from interneurones and cholinergic, noradrenergic, and serotonergic inputs from the perigeniculate nucleus, brainstem reticular formation, tegmentum, superior colliculus, pretectum, and locus coeruleus (Sherman and Koch 1986). Inputs to the monkey LGN are similar except that they do not include noradrenergic synapses (Bickford *et al.* 2000). According to Montero (1992), only 12% of synaptic junctions on relay cells in the cat LGN are from the retina. Of synapses on interneurones, 25% are from the retinal and 37% from the cortex. Many inputs to the LGN from sources other than the eye terminate in interlaminar spaces, where they synapse with dendritic extensions of cells in the main laminae (see Casagrande and Brunso-Bechtold 1988).

Stimuli applied outside the receptive field of relay cells in the LGN exert an inhibitory influence, which is stronger for high-contrast stimuli than for low-contrast stimuli (Solomon *et al.* 2002). It could therefoe serve as a gain-control mechanism.

5.3.2b *Corticofugal inputs to the LGN*

Axons descend from layer 6 of the primary visual cortex to the LGN. Injection of tracers has revealed that parvocellular layers of the LGN receive most of their inputs from the upper half of cortical layer 6, which is connected to cortical layers 4Cβ and 4A—the layers receiving inputs from parvocellular layers of the LGN. Magnocellular layers of the LGN receive all their inputs from cells in the lower half of layer 6, which is connected to layer 4Cα— the layer receiving inputs from magnocellular layers of the LGN (Lund and Boothe 1975; Fitzpatrick *et al.* 1994).

5.3.2c *Spatiotemporal responses of LGN cells*

Most cells in the LGN maintain a low rate of neural discharge in the dark of between 10 and 20 impulses/s (see Snodderly and Gur 1995). Only some cortical cells into which the LGN feed have a maintained discharge—the rest are silent in the dark (Section 5.4.3). Inhibitory circuits within the cortex are probably responsible for the suppression of spontaneous activity in most cortical cells.

Cai *et al.* (1997) recorded from LGN cells of cats as bright and dark bars were flashed for 13 ms on different parts of the receptive fields. Most cells had a centre-surround organization and responded with a 30 ms burst of activity above the resting level followed by a similar period of activity below resting level. The response of the receptive-field surround was typically delayed relative to that of the centre.

Cells in the magnocellular laminae have longer latencies to stimulation at the level of the chiasma, (which eliminates differences in retinal latency) than those in parvocellular layers. Also, cells in magnocellular laminae have larger receptive fields, greater sensitivity to luminance contrast, and better temporal resolution than cells in parvocellular laminae (Levitt *et al.* 2001).

In the light, relay cells in the LGN and other thalamic nuclei respond either tonically with a discharge that persists as long as the stimulus persists or phasically with a rapid burst of 2 to 10 spikes whenever the stimulus changes. Transition between these modes depends on the initial hyperpolarization of the cell membrane, which depends on inputs from the visual cortex and subcortex. Sherman (1996) suggested that relay cells fire in tonic mode when specific objects are being inspected and in arrhythmic bursts when the animal is searching for an object. Feedback from the cortex modulates pattern-specific centre-surround interactions in the receptive fields of LGN cells (Cudeiro and Sillito 1996). Cortical feedback also alters the temporal response properties of LGN cells (Marrocco *et al.* 1996). Also, feedback from the cortex through the reticular complex and perigeniculate nucleus may enhance the response of active sites in the LGN and that this creates an attentional mechanism (Crick 1984).

5.3.2d *Orientation sensitivity of LGN cells*
Most relay cells in the cat LGN are sensitive to the orientation of stimuli, especially stimuli with high spatial frequency (Daniels *et al.* 1977; Soodak *et al.* 1987; Shou and Leventhal 1989) and monkey (Smith *et al.* 1990). They tend to respond preferentially to stimuli arranged radially with respect to the fovea. About one-third of the LGN relay cells of the cat are sensitive to direction of motion. Like cortical cells, they are especially sensitive to motion of stimuli with low spatial frequency (Thompson *et al.* 1994a). Some tuning of LGN cells to orientation and motion seems to arise in the retina, it survives decortication, at least in the cat (Thompson *et al.* 1994a). However, removal of inputs from areas 17 and 18 reduces the number of LGN cells tuned to oblique orientations, which suggests that corticofugal projections have an influence (Vidyasagar and Urbas 1982). Cortical cells are more strongly tuned to orientation and direction of motion than are LGN cells. Cortical cells show differential tuning over the whole range of spatial frequencies to which they respond, rather than to only a part of that range (Thompson *et al.* 1994b).

Synaptic boutons of individual corticofugal axons are sparsely distributed over a wide region of the LGN. However, within this region, there is an elongated region of high synaptic density with an axis that is either parallel to or orthogonal to the receptive field of the parent cell in the visual cortex (Murphy *et al.* 1999). The parallel feedback could serve to enhance the orientation specificity of cortical cells by synchronizing the response of inputs from the LGN. The orthogonal feedback could enhance the motion-direction sensitivity of cortical cells.

5.3.2e *Arousal responses in the LGN*
In the cat, responses of LGN cells to light stimuli are facilitated by concurrent stimulation of the skin (Hotta and Kameda 1963). In the alert monkey, rapid eye movements, blinks, and auditory and somaesthetic stimuli produce non-specific responses in the LGN, even in the dark (Feldman and Cohen 1968). These non-retinal inputs may serve attentional processes (Sherman and Koch 1986) but their non-specificity suggests a general arousal function rather than attentional gating of specific locations or stimuli. Electrical stimulation of the mesencephalic tegmentum, an arousal mechanism in the brain stem, increases the response of relay cells in the LGN (Livingstone and Hubel 1981). This increase is particularly evident for stimuli in the centres of the receptive fields of relay cells (Hartveit *et al.* 1993).

5.3.3 Binocular responses in LGN

5.3.3a *Binocular responses in cat LGN*
Relay cells of the LGN of the cat give brisk excitatory responses to stimuli presented to one eye—the dominant eye for that cell, but respond weakly or not at all to stimuli presented to the other eye. However, stimulation of the nondominant eye for a given cell may enhance or inhibit responses to stimulation of the dominant eye (Noda *et al.* 1972). For example, a weak conditioning shock applied to the optic nerve of one eye can facilitate the response of LGN neurones to a test shock applied to the other optic nerve (Marshall and Talbot 1940; Bishop and Davis 1953). Also, the response of many relay cells is inhibited by a conditioning stimulus applied to the nondominant eye (Suzuki and Kato 1966). The direct excitatory and indirect inhibitory inputs to relay cells arise from corresponding regions in the two eyes (Marchiafava 1966; Sanderson *et al.* 1969, 1971; Singer 1970).

Guido *et al.* (1989) found that stimulation of the nondominant eye of the cat with a drifting sinusoidal grating reduced the spontaneous discharge in 29% of that eye's relay cells and produced weak excitatory responses in 25% of the cells. Some cells showed both types of response, according to the spatial frequency of the stimulus. These responses were stimulus tuned with respect to orientation and spatial frequency, and occurred in X, Y, and W cells in all LGN layers (but see C. Wang *et al.* 1994). Murphy and Sillito (1989) obtained similar results. Binocular inhibition was more common in cells receiving a dominant input from the ipsilateral eye than in those receiving a dominant input from the contralateral eye (Suzuki and Takahashi 1970). Application of bicuculline, an antagonist to the inhibitory neurotransmitter GABA, blocked the inhibitory responses but revealed excitatory responses in cells previously unresponsive to the nondominat eye (Pape and Eysel 1986; Murphy and Sillito 1989). Binocular interactions in the LGN of the cat are not disparity tuned (Xue *et al.* 1987).

Indirect influences in the LGN are postsynaptic and may be mediated by intrageniculate connections, by projections from other subcortical nuclei, such as the nucleus of the optic tract, or by corticofugal inputs from the visual cortex. About 25% of cells in the cat visual cortex project to the LGN. Some are fast conducting complex cortical cells serving binocular or monocular segments of the visual field. Others are simple cells conducting at intermediate velocities, which originate wholly in binocular cells. Thus, at least some corticofugal influences are involved in binocular vision (Tsumoto and Suda 1980).

However, there is some dispute about the role of cortical inputs to the LGN. Some investigators found that binocular interactions in the cat LGN require an intact visual cortex. Reports that interocular influences are greatest when stimuli presented to the two eyes differ in position, orientation, contrast, and movement prompted the suggestion that cortical influences facilitate transmission of signals from stimuli lying on the horopter, and are involved in binocular fusion and rivalry (Section 7.3) (Schmielau and Singer 1977; Varela and Singer 1987). Others reported that binocular interactions in the LGN do not require corticofugal inputs (Sanderson *et al.* 1971; Murphy and Sillito 1989; Tumosa *et al.* 1989a; Tong *et al.* 1992).

In the cat, responses of LGN cells to stimulation of the dominant eye are not much affected by changes in the orientation or direction of motion of stimuli presented to the nondominant eye. However, the responses of LGN cells are affected by changes in the spatial frequency of those stimuli (Moore *et al.* 1992). This suggests that corticothalamic processes balance the responses to small interocular differences in stimulus contrast, by adapting the relative contrast gains of inputs from the two eyes (Section 19.5.4). Perhaps binocular rivalry and gain control are served by different processes in the LGN. McClurkin *et al.* (1994) suggested that cortical feedback to the LGN modulates the number and temporal waveform of spikes in parvocellular neurones so as to enhance differences in response to distinct stimuli. Cortical feedback in the monkey has been found to be also responsible for modulation of the response of LGN cells to a flashing spot by a grating presented outside the receptive fields of the cells (Marrocco *et al.* 1996).

5.3.3b *Binocular responses in primate LGN*

There has been conflicting evidence on binocular responses in the primate LGN. Rodieck and Dreher (1979) reported that the response of LGN cells of the monkey to stimulation of the cells' dominant eye was partially suppressed when the nondominant eye was stimulated at the same time. But this occurred only in the magnocellular laminae. Marrocco and McClurkin (1979) found that about 13% of cells in both the parvo- and magnocellular laminae of the LGN of the monkey responded only to binocular stimulation. These studies were conducted on anaesthetized and paralyzed monkeys.

Multiple electrodes applied to the LGN of alert monkeys have revealed four types of binocular interaction to light flashes in both the parvocellular and magnocellular laminae (Schroeder *et al.* 1990). Stimulation of the nondominant eye reduced the response below the spontaneous level for one type of cell and increased it above the spontaneous level for a second type of cell. For a third type, the response was less vigorous to binocular stimulation than to stimulation of only the dominant eye. For a fourth type of cell, the response was more vigorous to binocular stimulation than to monocular stimulation. Since the researchers used multiple electrodes, they could not estimate the proportion of cells showing these different responses. The latency of interactions was too short to involve corticofugal influences. However, the later response components could have been due to cortical influences. It is unfortunate that only featureless flashes were used, since binocular interactions revealed by psychophysical procedures depend on the presence of contours (Section 7.3.2).

5.4 VISUAL CORTEX

5.4.1 Neurophysiological procedures

5.4.1a *Visualizing single neurones*

Nerve cells can be stained in a slice of neural tissue and viewed in a microscope. Neurones contain **intrinsic chemical markers**, such as a protein specific to that type of cell that takes up a specific stain or antibody. The shape and identity of the cell is evident if the protein is distributed throughout the cell. However, it may be difficult to isolate single cells from among cells of the same type. If the protein is localized in the synapse, the cells containing it will be difficult to identify unless the synapses occur in well-defined layers, such as the inner plexiform layer of the retina. Detection of an intrinsic protein can be combined with a staining procedure that reveals the whole cell. In this way one can identify the neurotransmitters or synaptic receptor proteins used by specific types of cell.

Other histochemical procedures involve staining the cell, or structures within the cell, with an extrinsic chemical agent. In the **Golgi method,** the neural tissue is stained with silver salts. Ramón y Cajal used this method to trace neural structures and connections in the retina. The method is difficult to apply and the results are unpredictable. Also, the method fails to reveal certain types of cell, such as amacrine starburst cells in the retina.

In the **Nauta method,** selected cells or axons in some structure such as the LGN are destroyed or severed in the living animal and, some time later, areas of cell or axon degeneration in corresponding areas of the visual cortex are revealed by staining (Hubel and Wiesel 1969). This staining procedure is sometimes difficult to apply.

In more recent methods, particles, proteins, or dyes are injected into chemically preserved or living cells in a cell culture through a visually guided fine pipette. In another method, known as **photofilling,** all the cells in a cell culture absorb chemical agents. Particular cells are then irradiated with a fine laser beam, which oxidizes one of the chemical agents into a fluorescent form, which diffuses throughout the selected cells. The 3-D structure of the selected cells can be recorded by a sensitive camera attached to a scanning microscope. Scanning microscopes are described in Section 30.2.3.

Chemical agents or fluorescent dyes can be injected into neural tissue where they are absorbed by the neurons and transported in either the anterograde direction to nerve terminals or in the retrograde direction to the cell body. The agent can be visualized by staining, by fluorescence, or by radioactivity.

In one procedure, the enzyme horseradish peroxidase is transported retrogradely from a site of injection in the living animal to cell bodies where it is detected histochemically. This type of procedure reveals the multiple sources of afferent fibres entering a particular neural structure, such as a region of the cortex.

Fluorescent tracer dyes, such as lucine yellow, can be injected into neurones in slices of living neural tissue. Synaptic connections between two cells can be identified by injecting the cells with different fluorescent dyes that emit at different wavelengths. Three-dimensional images of cells can be constructed by use of the confocal microscope (Section 30.2.3). Brain slices can be kept alive for long periods so that the growth of axons, dendrites, and synapses may be observed (Dailey 1964). The use of these and related methods in the elucidation of the cellular structure of the retina is reviewed in Masland and Raviola (2000).

Procedures used to study the development of the nervous system are reviewed in Section 10.1.

5.4.1b *Mapping stimulus selective areas*

There are various procedures for labelling cells that are sensitive to particular stimuli. In **autoradiography** a mixture of tritiated proline and tritiated fructose or deoxyglucose labelled with carbon 14 is injected into the eye of a living animal. The radioactive tracer gradually travels up the optic nerve and becomes concentrated in metabolically active cells in layer 4 of the contralateral visual cortex. The eye is then exposed to particular stimuli for an hour or more, which causes the radioactively labelled sugar to be selectively absorbed by neurones in the visual cortex that respond to the stimulus. The resulting patterns of radioactivity in thin sections of visual cortex are recorded on film to produce autoradiographs.

This process reveals all cells in that slice of visual cortex that respond to a particular stimulus feature, say vertical lines. The separate slices of an autoradiograph can be combined by computer reconstruction into a complete 3-D pattern of cells responding to a given stimulus feature.

Patterns of cortical activity can also be imaged in the living brain. Dyes sensitive to voltage changes associated with neural impulses are infused into cortical tissue of the living animal. An electrical stimulus is administered to a given location or the animal is exposed to specific stimuli, and the differential reflectivity of the cortical surface is recorded by photodiodes or by a video recorder (Blasdel 1992a; Orbach and Van Essen 1993). The method has a temporal resolution of 1 ms and is useful for recording spatial and temporal changes in cortical responses over a wide area. However, dyes introduce side effects. Active cortical areas reflect slightly less light than inactive areas. These changes in reflectivity can be photographed without the use of dyes. This method provides high spatial but low temporal resolution (Ts'o *et al.* 1990; Frostig 1994).

Stimulus specific regions of neural activity in the visual cortex may also be mapped by staining naturally occurring chemicals associated with neural activity, such as cytochrome oxidase (Tigges *et al.* 1992), GABA (Hendry *et al.* 1994), and protein kinase (Hendry and Kennedy 1986). Neural activity also induces genes to express mRNA transcription factors, such as Fos, Jun, and Zif268, which bind to specific DNA chromosomal sites and regulate the production of specific proteins. Immunocytochemical staining reveals the presence of these activity-dependent factors in particular neurones (Chaudhuri *et al.* 1995).

In **electrophysiological procedures** the electrical activity of neurones is recorded by an electrode. In single cell electrophysiology a fine electrode is injected through a hole in the skull and into a neurone or into the neighbourhood of a neurone. This procedure is usually carried out on the anaesthetized and paralyzed animal but it can also be applied on an alert animal since, once applied, the electrode causes no discomfort. In the procedure of visually evoked potentials, an electrode is applied to the scalp and the pooled activity of cortical cells in the underlying region of the cortex is recorded. This procedure is described in Section 6.7.1. The magnetic fields generated by neural activity in the human brain can be recorded with supercooled magnetometers (see Regan 1989a).

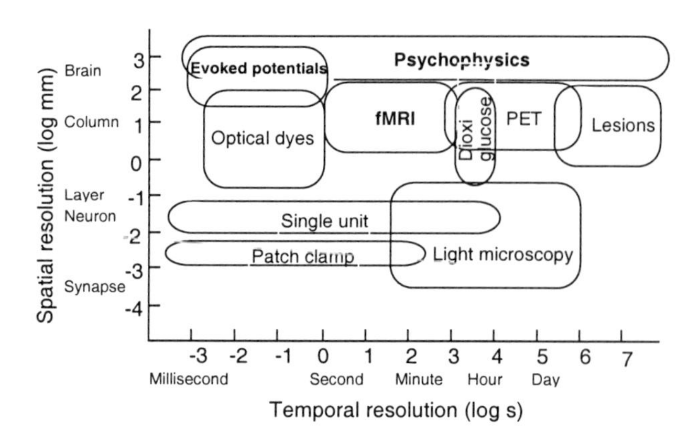

Figure 5.14. Methods for studying sensory processing. Methods printed bold are non-invasive. (Adapted from Churchland and Sejnowski 1988)

In **positron-emission tomography** (PET), a sugar containing a radioisotope, such as O15, is injected intravenously. When a positron emitted by the isotope interacts with an electron, two gamma rays are emitted in opposite directions. Each pair of rays triggers a pair of coincidence detectors on opposite sides of the head. Several pairs of detectors are placed in an annulus and the intersection of the diameters joining active detectors indicates sources of positron emission in the plane of the annulus. Several annuli are stacked to form a cylinder, which detects sources of activity in a volume of the brain. To improve sampling density, the cylinder oscillates through a small angle about its central axis. Active sites recorded in a 3-D co-ordinate system indicate regions of high glucose metabolism in the brain. Regions of high glucose metabolism correspond to regions of high synaptic activity (Eriksson *et al.* 1990). The method has a spatial resolution of up to about 1.5 mm but poor temporal resolution. It is labour intensive and expensive.

In **magnetic resonance imaging** (MRI), the brain is exposed to a strong homogeneous magnetic field. The magnetic moments of protons in elements with odd atomic weight, such as hydrogen, become aligned with the magnetic field. A pulse of radio waves momentarily perturbs their alignment. On returning to the aligned state, the atoms emit radio waves at a frequency characteristic of the chemical and physical structures in which they lie. By applying a small gradient to the applied magnetic field across the brain and by changing the orientation of this gradient, one can obtain distinct radio signals related to position within the living brain (Narasimhan and Jacobs 1964). The head must be held firmly in one position. The signals can detect transient changes in blood oxygenation levels (BOLD) due to synaptic activity (Ogawa *et al.* 1992; Menon *et al.* 1998a). After sensory stimulation the local decrease in blood oxygen peaks after about 1.5 s (Vanzetta and Grinvald 1999). A less localized change in blood flow takes several seconds to occur. This so-called **functional MRI** (fMRI) procedure is non-invasive and, although the signal level is low with rather slow dynamics, it has a spatial resolution of a few millimetres and provides a greater wealth of information than the PET scan. By analyzing images in a series of planes the folded cortical surface can be represented in a planar map.

The fMRI procedure has produced retinotopic maps of V1 and extrastriate visual areas in the human brain. This was done by recording activity produced by flickering textured patterns in the form of rotating segments and expanding rings (see Engel 1996). The maps are similar to those of the corresponding areas in the monkey brain. The boundaries between visual areas can be mapped by recording the locations where travelling waves of neural activity triggered by moving stimuli reverse in direction. The method can locate cortical areas sensitive to particular stimuli and measure changes in response associated with changes in retinal location, contrast, colour, and motion of the stimulus (Heeger 1999). The method can also detect neural activity associated with changes in attention (Kastner *et al.* 1998), and that acciated with perceptual changes in ambiguous stimuli (Kleinschmidt *et al.* 1998) and binocular rivalry (Section 7.10.2c). It can detect activity created by illusory effects such as the motion aftereffect (Culham *et al.* 1999) and subjective contours. Finally, it is used to localize neural deficits (Wandell 1999).

In a typical experiment the map of activity over the cortex in response to a baseline stimulus is subtracted from the map produced by a particular stimulus. Each stimulus is presented for a few seconds many times. At each location, the activity induced by the baseline stimulus is subtracted from that induced by the test stimulus. The difference signal is superimposed on a map of the cortex to reveal where activity has increased in response to the defined stimulus. Since the haemodynamic response lags the neural response by about one second, the method cannot be used to track rapid changes in neural response. However, it can detect onset time differences between the responses from different regions of the brain with an accuracy of about 30 ms. For example, when the subject performs a visual motor task, it can detect differences in response times between visual and motor areas that correlate with reaction times determined behaviourally (Menon *et al.* 1998b).

Signals from fMRI have multiple components. The initial dip is believed to originate from oxygen transport from capillaries associated with local neural events. Later signals originate from more global effects of deoxygenation and blood flow. Under certain conditions, the local component can be enhanced by using two stimuli, such as orthogonal gratings, which generate differential responses in independent groups of neurones. The local fMRI response can resolve orientation and ocular dominance columns in alert monkeys and humans (Kim *et al.* 2000; Grinvald *et al.* 2000).

Mapping the various visual areas of the human cortex is made difficult by the fact that they lie mainly in deep sulci, which vary in position and complexity from one person to another. Also, distinctions between areas are not easy to define. Modern computers allow cortical areas to be mapped in a variety of configurations, including a flat map of the entire cortical surface (Van Essen *et al.* 2001).

The spatial and temporal resolutions of methods of studying neural systems are depicted in Figure 5.14 (Churchland and Sejnowski 1988).

5.4.2 Visual cortical projections

The primary visual cortex in each hemisphere is mainly on the banks of the calcarine sulcus—a deep horizontal fissure on the medial surface of the occipital lobe at the caudal pole of the cerebral cortex. The primary visual cortex of subprimates is also known as Brodmann's **area 17**. In primates, it is usually referred to as **V1**. It is also known as the **striate cortex** because of the prominent stripe of Gennari it contains.

5.4.2a *Inputs to the visual cortex*

Each optic radiation projects from the main layers of the lateral geniculate nucleus to form the major input to the ipsilateral primary visual cortex. Intralaminar neurones in the LGN also project to the visual cortex.

The retinotopic order of incoming axons in the primary visual cortex of each cerebral hemisphere is preserved topographically. The central parts of the retina are represented near the caudal pole of the occipital lobe and the monocular crescents are represented more rostrally. The vertical meridian is represented along the border between V1 and V2.

The visual cortex also receives inputs from other thalamic areas including the pulvinar. The pulvinar has at least three visuotopically organized areas with connections to V1, V2, V4, MT, and the parietal and prefrontal cortex. There are also connections to auditory and somatosensory areas (Adams *et al.* 2000; Gutierrez *et al.* 2000). The pulvinar is implicated in the control of visual attention (Robinson and Petersen 1992; Levitt *et al.* 1995; Grieve *et al.* 2000). It probably processes stimuli that require rapid responses, such as visual looming, which signifies impending collision (King and Cowey 1992; Mestre *et al.* 1992).

Responses of some cells in the pulvinar are modulated by the position of the eyes, like responses of cells in the parietal cortex. These cells thus code signals in a headcentric or bodycentric frame of reference. Some pulvinar cells respond before the onset of self-initiated arm movements, even before responses occur in the primary motor cortex or posterior parietal cortex (Cudeiro *et al.* 1989).

The neocortex, including the primary visual cortex and other visual areas, receives inputs from more than 20 subcortical areas. These include the superior colliculus, hypothalamus, the locus coeruleus and pontine reticular system in the brain stem, and other centres listed in Section 5.4.5f.

The hippocampus in the old cortex receives direct visual inputs. The hippocampus and the neighbouring amygdala have reciprocal connections with visual areas in the neocortex and also connect with the pulvinar.

The claustrum is a nuclear mass in the basal forebrain with reciprocal connections with many areas of the cerebral cortex. One part contains a topographical map of the visual field and contains binocular cells sensitive to stimulus motion and orientation. In the cat, claustral afferents project to layers 1, 6, and 4 of the visual cortex. Some of those reaching layer 4 contact inhibitory interneurones and seem to be involved in end-stopping of receptive fields (LeVay 1986; Sherk and LeVay 1983).

5.4.2b *Cortical magnification factor*

The **cortical magnification factor** (M) is the distance apart in millimetres of two points on the surface of the visual cortex that corresponds to one degree of visual angle in the visual field (Daniel and Whitteridge 1961). The area of cortex devoted to the fovea is disproportionately greater than that devoted to the peripheral retina. In the macaque M is about 16 mm/° at the fovea. Also, in the monkey, more visual cortex is devoted to the lower than to the upper visual field (Van Essen *et al.* 1984; Tootell *et al.* 1988d). It has been claimed that the magnification factor is about twice as large across an ocular dominance column serving one eye than along it (Sakitt 1982). However, Tootell *et al.* (1988d) found no evidence of this anisotropy in the macaque. Since the right and left eyes are represented by alternate ocular dominance columns running mainly parallel to the vertical meridian, the binocular field has twice as much cortical representation across the horizontal meridian than along it.

In primates, M decreases with increasing eccentricity from the centre of the visual field. Cowey and Rolls (1974) estimated M for humans from impressions of light (phosphenes) evoked by electrodes implanted at various locations on the visual cortex (Brindley and Lewin 1968). At an eccentricity of 2°, M was approximately 4 mm/° and declined monotonically to 0.5 mm/° at an eccentricity of 25°. Another estimate puts M at 11.5 mm/° for the human fovea (Drasdo 1977).

The change in the magnification factor as a function of eccentricity is known as **M scaling**. For all positions in the visual field, a microelectrode must move between 2 and 3 mm over the surface of the monkey visual cortex before an entirely new region of the visual field is represented (Hubel and Wiesel 1974a). This suggests that the same number of millimetres of visual cortex is devoted to each ganglion-cell receptive field. One estimate is that 0.88 mm of human visual cortex is devoted to one ganglion-cell receptive field (Ransom-Hogg and Spillmann 1980). It follows that the amount of cortex devoted to each degree of visual field is directly related to visual acuity and inversely related to the mean size of ganglion-cell receptive fields (Rolls and Cowey 1970; Rovamo and Virsu 1979; Virsu and Rovamo 1979).

In summary, ganglion-cell receptive fields, and hence areas of the retina that can just spatially resolve two stimuli, are represented by equal areas in the visual cortex. Because receptive fields are smaller and therefore denser in the fovea than in the periphery of the retina, the fovea claims proportionately more of the cortical surface (M is greater). When allowance is made for the decrease in M with increasing eccentricity, grating resolution, vernier acuity, and stereoacuity are about the same at all eccentricities (Levi *et al.* 1985). Thus, visual hyperacuity depends on the number of processing units in the visual cortex that are devoted to the task irrespective of retinal location. A complicating factor is that receptive fields in the central retina are fine enough to provide an adequate sampling of the image produced by the eye's optics, but receptive fields in the periphery undersample the image (Section 5.1.3).

There has been some dispute about whether the variation in M arises simply from differential density of ganglion cells over the retina, or, whether each foveal ganglion cell feeds into more cortical cells than each ganglion cell from the peripheral retina (the overlap factor). In the macaque, Wässle *et al.* (1990) counted three to four ganglion cells for each foveal cone, one ganglion cell per cone at an eccentricity of about 15°, and many more cones than ganglion cells in the periphery. They concluded that the 1000-to-1 change in ganglion cell density with increasing eccentricity accounts for the change in M.

Azzopardi and Cowey (1993) came to the opposite conclusion. They used a retrograde tracer to determine directly the number of ganglion cells projecting to measured areas of the striate cortex of the macaque monkey. Foveal cones were allocated 3.3 times more cortical tissue than peripheral cones in one animal and 5.9 times more in a second animal (see also Perry and Cowey 1985).

Azzopardi *et al.* (1999) used the same method to establish that the ratio of parvocellular to magnocellular units in the macaque LGN decreased from a mean of 35:1 at the fovea to 5:1 at a cortical eccentricity of 15°. For the fovea, but not for the periphery, the LGN ratio exceeded the ratio of parvo- and magnocellular ganglion, showing that inputs from the central fovea are represented at higher levels in the visual system to a greater extent than predicted from ganglion-cell density. Popovic and Sjöstrand (2001) came to the same conclusion from an analysis of human visual resolution, ganglion-cell density, and magnetic imaging.

Bijl *et al.* (1992) produced psychophysical evidence that changes in ganglion-cell density predict the variation of contrast sensitivity over the visual field for high spatial frequencies. However, they found that, for low spatial frequencies and for the detection of localized discs, performance depends on variations in both ganglion-cell density and the degree of overlap of receptive fields, especially in the nasal hemifield.

It has also been claimed that, for the magnocellular system, the number of afferents per unit cortical area increases steeply with increasing eccentricity. Thus, each magnocellular axon from more central regions has many more cortical cells devoted to it than each axon from the periphery (Schein and Monasterio 1987). There is a constant number of magnocellular afferents per point image, defined as the area of cortex activated by a stimulus at a point in space (equivalent to the number of receptive-field centres of ganglion cells that overlap a given point on the retina).

5.4.2c *The topology of cortical mapping*

The mapping of the retina onto the cortex can be described by the conformal logarithmic function,

$$\omega = \log(z + a) \qquad (5)$$

where z is a complex number denoting the position of a point on the retina, ω is a complex number representing the position of the stimulus on the cortex, and a is a constant. A function is conformal if its first derivative (in this case the magnification factor) is isotropic (independent of orientation) and if the sizes of local angles are preserved.

Schwartz (1980) illustrated how, in a logarithmic mapping, the cortical images of squares of different sizes are transformed into images of the same size and shape, but in different locations. In a similar way, retinal images that are rotated with respect to each other are transformed into images differing in spatial phase. The recognition of size- and orientation-invariant features in the image formed by such a system reduces to the computationally simpler process of deriving translation-invariant properties. Schwartz argued that the logarithmic retinocortical mapping is part of a process for extracting size- and orientation-invariant properties of visual objects.

One problem with this theory is that retinal images differing in position are transformed into images differing in size, and this complicates the problem of deriving position-invariant features. But a more serious problem is that this theory of shape recognition is essentially a template-matching model, which is adequate for the extraction of only very simple invariant features in a highly structured input. Furthermore, it is not clear how it applies to the visual cortex. The geometrical layout of the cortical "image" is not represented in the activity of cortical cells. Cortical cells code the local sign of their origin on the retina not their local sign in the cortex. All features are coded in the central nervous system in terms of cell connections and the simultaneous and successive patterns of cell firing, not in terms of the spatial dispositions of cells. The notion of a topographic code, in the sense of the spatial arrangement of cells over a surface, ceases to have any significance beyond the retina. Spatial maps in the cortex demonstrate to an experimenter where spatial information is processed but, for the perceiver, the spatial organization of the stimulus is represented only by spatiotemporal patterns of neural connections.

Cortical mapping has four prominent features.

1. It is M-scaled. This allocates equal space in the cortex to equally discriminable regions of oculocentric space.

2. It is topographic. Lateral inhibition is required only between cells that are near neighbours because lateral inhibition serves to attenuate the response from regions of local homogeneous activity, thus accentuating the response from regions with high gradients of activity. This ensures that information regarding changes in stimulation passes to the next level of analysis, thereby economizing on information transmission (Barlow 1961). Also, pooling of spatial and other information is required more often over small regions than over larger regions. Thus, keeping spatially contiguous regions together economizes on the lengths of dendritic connections. Cortical mapping of location is discontinuous across alternating columns of visual cortex devoted to each eye (Section 5.6).

3. The mapping is optimized for uniform coverage of feature detectors. Each cortical location contains a complete set of cells tuned to the visual features of eye or origin, colour, orientation, spatial frequency, and motion (Swindale *et al.* 2000).

4. The mapping is optimally continuous over values of each feature. In each location, the different values of orientation preference and motion preference are mapped in order, with inevitable discontinuities at the end of each sequence of feature values (Das and Gilbert 1997). These mappings are discussed more fully in Section 5.6.1. Mapping continuities juxtapose cells that process similar features within a local region of oculocentric space, so that local neural processes, such as facilitation and inhibition, can be achieved economically.

The response of some cells in the visual cortex of alert cats to a stimulus in a given retinal location varies with the direction of gaze (Weyland and Malpeli 1993). Cells with this property are also found in the parietal lobe (Section 5.7.3d).

5.4.3 Cortical layers

The mammalian visual cortex is a convoluted sheet of neural tissue about 3 mm thick consisting of six main layers designated layers 1 to 6, with layer 1 at

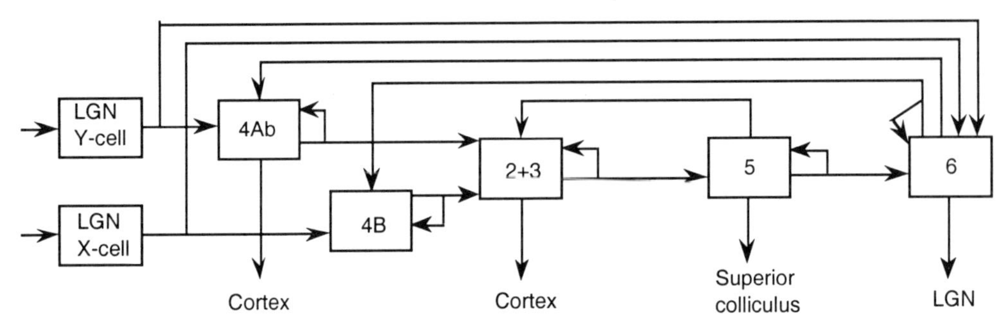

Figure 5.15. Intracortical connections of the cat's visual cortex.
The numbers refer to layers of the visual cortex. (Adapted from Gilbert and Wiesel 1985)

the outer surface and layer 6 bordering the inner white matter. More details are presented in Section 10.3.1. Layer 4 is known as the granular layer and other layers are known as extragranular layers. The white matter consists of bundles of axons that project to and from subcortical nuclei and other cortical regions. In the cat, area 17 receives inputs from X ganglion cells in the A laminae of the LGN. Area 18 receives some X-cell inputs from the A laminae and inputs from Y cells in the C laminae. Area 19 receives inputs from the C laminae and from extrageniculate cell groups (see Holländer and Vanegas 1977; Pasternak *et al.* 1995).

In primates, visual inputs from the main laminae of the LGN project to spiny stellate cells in layer 4 of area 17 (V1). Visual inputs reach other cortical areas from LGN interlaminae and other subcortical nuclei (Section 5.7). The magnocellular laminae of the LGN terminate in spiny stellate cells in the upper half of layer 4C, known as layer 4Cα. Those from the parvocellular laminae terminate in the lower half of layer 4C, known as layer 4Cβ, and in layer 4A (Hubel and Wiesel 1977; Tootell *et al.* 1988a). Cells in the middle of layer 4C receive inputs from both layers 4Cα and 4Cβ and therefore from both parvo-and magnocellular layers of the LGN (see Yoshioka *et al.* 1994). Some LGN inputs, probably W cells, project to superficial levels of layer 3 and sparsely to layer 1, at least in the monkey (Fitzpatrick *et al.* 1983).

The predominant magnocellular projection from layer 4Cα is to spiny stellate cells in layer 4B in the same 1-mm-wide vertical column of tissue (Yabuta *et al.* 2001). From there, magnocellular inputs pass to layers 2 and 3 in the same column and to neighbouring columns (Katz *et al.* 1989). Cells in layer 4Cβ project abundantly to blob and interblob regions of layers 2 and 3 and to a lesser extent to pyramidal cells in layer 4B. Layer 4B stellate cells therefore receive mainly magnocellular inputs and layer 4B pyramidal cells and layers 2 and 3 receive mainly parvocellular inputs (Sawatari and Callaway 1996; Yabuta and Callaway 1998). Cells in layer 4Cα also project sparsely to layer 5 and those in layers 4Cα and 4Cβ project sparsely to layer 6. Cells in layers 2 and 3 project to layers 5 and 6, which in turn project back into layer 4 (Gilbert and Wiesel 1979; Fitzpatrick *et al.* 1985). These relationships for the cat's visual cortex are illustrated in Figure 5.15. Layer 4B of the primary visual cortex contains a dense horizontal plexus of myelinated axons. Details of cell connections in the visual cortex have been reviewed by Henry (1991), Valverde (1991), Peters and Rockland (1994), and Callaway (1998a).

5.4.4 Cortical cell types

In the neocortex there are two main histological types of excitatory neurones and one type of inhibitory neurone. There are also many glial cells. The excitatory neurones are circular **spiny stellate cells**, most of which radiate dendrites in all directions, and **pyramidal cells** that have triangular cell bodies with a single long apical dendrite and several shorter basal dendrites (Figure 5.16a). The dendrites of both classes of excitatory cell bear a multitude of minute **dendritic spines**. A pyramidal cell has approximately 10,000 spines. Inhibitory neurones are devoid of spines and are called **smooth cells** (see Figure 5.16b). Figure 5.16c (in colour section) is a stereogram of a pyramidal cell prepared from one of Ramón y Cajal's slides.

In the macaque visual cortex there are approximately 120,000 neurones per cubic millimetre, about twice as many as in other parts of the cortex (O'Kusky and Colonnier 1982). There is about the same number of glial cells. On average, there are over 2,000 synapses per cell, and each cortical cell provides between 7,000 and 8,000 synapses to other cells. Since a given cell makes only between one and five synaptic contacts with any other cell, one must conclude that each cell connects to thousands of other cells (divergence) and receives inputs from

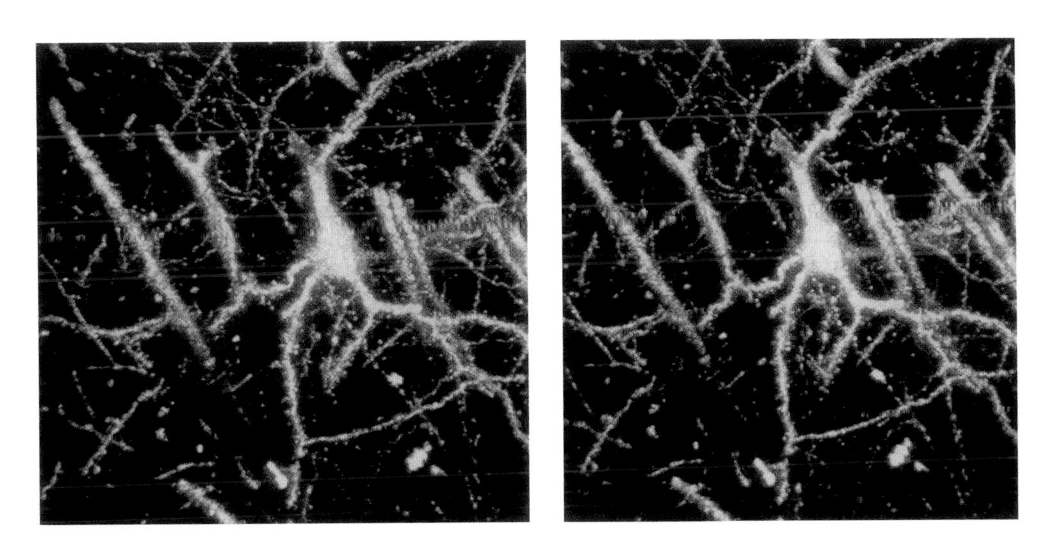

Figure 5.16b. Stereogram of a cortical pyramidal cell.
This stereogram was prepared from one of Ramon y Cajal's slides. With divergent fusion, the 3-D cell body and part of the apical dendrite can be seen.

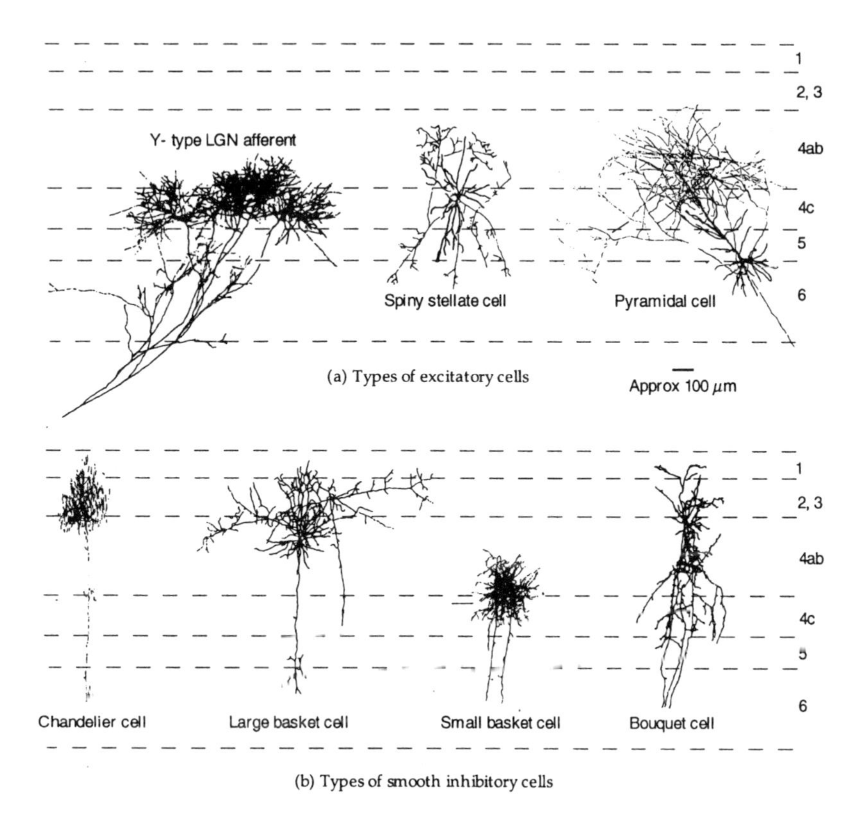

Figure 5.16a. Types of cortical cell.
(Adapted from Gilbert and Wiesel 1885)

thousands of other cells (convergence) (Salin and Bullier 1995). Details of cell types in the visual cortex are provided in Peters and Rockland (1994).

5.4.4a *Pyramidal cells*

Pyramidal cells make up about 75 % of neocortex neurones. They form excitatory (glutamatergic) synapses. They occur in all cortical layers except layer 1. The apical dendrites of pyramidal cells in layer 5 form into distinct vertical clusters, which ascend into layers 4, 3, and 2. Back-propagating sodium action potentials initiated in the soma of these cells can facilitate subthreshold calcium action potentials initiated by stimuli impinging on apical dendrites (Larkum *et al.* 1999). This produces synchronous coupling between inputs arriving at different layers in the visual cortex. Pyramidal cells of layer 4 form cone-shaped clusters of apical dendrites.

Dendritic shafts and spines of pyramidal cells receive excitatory inputs from stellate cells, neighbouring and distant pyramidal cells, the thalamus and other brain stem nuclei, and the basal forebrain. Inhibitory inputs received mainly on the axon and cell nucleus (somata) arise from inhibitory stellate cells, brain stem, and basal forebrain (claustrum).

Axons of pyramidal cells are the major output components of the neocortex. Outputs from pyramidal cells in layers 5 and 6 of the primary visual cortex project to the extrastriate cortex and to ipsilateral and perhaps also to contralateral subcortical regions (Creutzfeldt 1977). The subcortical regions include the pretectum, superior colliculus (Berman *et al.* 1975), pulvinar, LGN, and other areas of the thalamus (Gattass *et al.* 1979; Casanova *et al.* 1989; Fitzpatrick *et al.* 1994), the caudate nucleus, and the cerebellum by way of pontine nuclei (Brodal 1972). Pyramid-cell axons form the white matter of the cerebral cortex. Details of cortical efferents are provided by Swadlow (1983). Pyramidal cells in area V1 receive excitatory feedback signals from the extrastriate visual areas to which they project.

They also form excitatory networks within V1 (Johnson and Burkhalter 1997). In cortical layer 6 there are eight types of excitatory pyramidal cells. One type receives and transmits signals to magnocellular layers 4B and 4Ca and another type receives and transmits signals to parvocellular layers 2, 3, and 4CB. Other types transmit to the same layer from which they receive inputs and one type receives and transmits to all cortical layers (Briggs and Callaway 2001).

Feedback signals from extrastriate areas feed onto relatively few inhibitory GABAergic interneurones in V1. Feedback from area V5 in monkeys improves the differential responses of cells in V1, V2, and V3 to figure-ground stimuli (Hupé *et al.* 1998). Feedback connections from higher visual areas may also enhance responses to stimuli to which the animal is attending (Section 5.5.7c).

5.4.4b *Spiny stellate cells*

Spiny stellate cells occur only in cortical layer 4. Almost all inputs to the primate visual cortex from the LGN impinge on spiny stellate cells. Spiny stellate cells project horizontally within layer 4 and to a lesser extent to other layers, but their dendrites remain mainly within the same local column of cells. Only about 10% of excitatory synapses impinging on spiny stellate cells of layer 4 originate from the LGN, the rest originate from subcortical nuclei, neighbouring spiny stellate cells, and pyramidal cells, especially along recurrent axons from layer 6 (Fitzpatrick *et al.* 1985; Freund *et al.* 1989; Ahmed *et al.* 1994). Recurrent excitatory circuits are therefore very prominent in the visual cortex.

5.4.4c *Smooth inhibitory cells*

Smooth cells occur in layers other than layer 4 and are thought to be inhibitory (GABAergic) interneurones. There are three main types of GABAergic cell; chandelier cells, basket cells, and bouquet cells. It can be seen from Figure 5.16b that they differ in their patterns of dendritic arborization, but the significance of the different types is not known (Anderson *et al.* 1993). Some types of smooth cell occur only in particular cortical layers. Although most inhibitory cells operate over short distances and are known as interneurones, basket cells have axons that spread laterally up to 2 mm (Kritzer *et al.* 1992). Some GABAergic cells also slowly release neuropeptides that can either excite or inhibit other neurones. Inhibitory synapses form about 15 to 20% of synapses in the visual cortex and are most dense in layers 4A and 4B, which receive geniculocortical inputs (Douglas *et al.* 1995).

5.4.4d *Grouping by temporal dynamics*

Neurones in the mammalian neocortex have been classified into four types according to their temporal dynamics. **Regular spiking neurones** fire rapidly at stimulus onset but adapt to a steady level within about 100 ms. **Fast spiking neurones** show little or no adaptation and are usually inhibitory. **Bursting or chattering neurones** produce periodic bursts of 2 to 6 spikes. **Intrinsic-bursting neurones** fire with a burst of spikes followed by a pause and a tonic spike train (Connors and Gutnick 1990; Gray and McCormick 1996). These differences depend on the intrinsic properties of ion channels on the cell's soma and

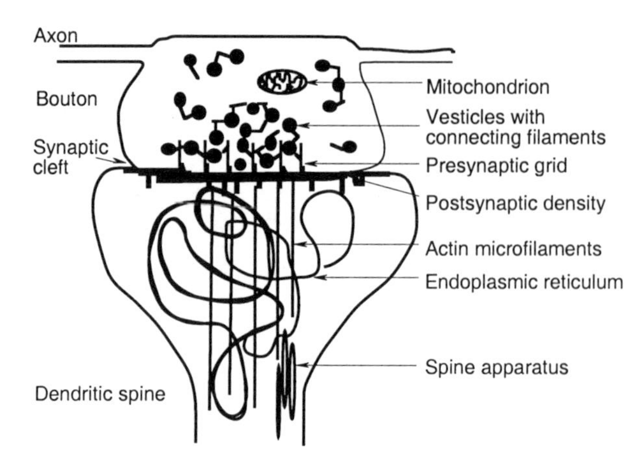

Figure 5.17. Schematic diagram of a typical cortical synapse. (Adapted from Edwards 1995)

dendrites (Solomon *et al.* 1993). They have been modelled by equations derived from the Hodgkin-Huxley analysis of time-dependent responses of ion channels on neuronal membranes (Wilson 1999). Similar differences exist in the response properties of motor neurones, such as those controlling eye movements (Section 9.11).

5.4.4e *Glial cells*

There are at least ten times more glial cells than neurones in the central nervous system of vertebrates. In the central nervous system there are two main types of glial cell: microglia, which act as phagocytes and remove dead cells, and macroglia, which consist of astrocytes and oligodendrocytes.

Astrocytes are star shaped and perform a variety of functions. Some form the protective seal between blood capillaries and the brain, known as the blood-brain barrier. Other astrocytes cover neurones in the vicinity of synapses. They remove surplus neurotransmitter from the synaptic cleft and extracellular space and help to maintain ionic balance, especially the balance of potassium ions. Reduction of the number of glial cells in the vicinity of synapses reduces synaptic efficiency (Oliet *et al.* 2001). Yet other astrocytes form a reticular network linked by gap junctions (low-resistance synapses that allow direct electrical coupling).

Astrocytes also interact with neurones by gap junctions (see Verkhrastsky *et al.* 1998). Synaptically released neurotransmitters activate receptors on glial cells and evoke waves of calcium ions. Also, glutamate released from glial cells modulates synaptic transmission between neurones and may be involved in inducing synchronous neural activity (Alvarez-Maubecin *et al.* 2000; Haydon 2001). Glial cells are involved in cortical development (Sections

10.2.2b, 10.3.2, and 10.4.2). Glial cells are also a source of neural regeneration in a damaged chicken retina (Fischer and Reh 2001).

Oligodendrocytes form myelin sheaths around axons in the central nervous system and Schwann cells form the sheaths in the peripheral nervous system (Section 10.2.3e).

5.4.5 Cortical synapses and neurotransmitters

5.4.5a *Structure of synapses*

Most neurones in the central nervous system have a cell body, or soma, from which arises a complex dendritic tree and a single axon, which branches into collaterals some distance from the soma. The soma, axon, or dendrites can form presynaptic or postsynaptic structures. However, axons form most of the presynaptic elements and dendritic spines form most of the excitatory postsynaptic elements. Inhibitory synapses are usually formed on smooth dendrites of interneurones. In the cat's visual cortex, 84% of synapses are excitatory and about 80% of these occur on dendritic spines and only 0.1% on the soma of a postsynaptic cell (see Edwards 1995).

The structure of a typical cortical synapse is shown in Figure 5.17. The axon of the presynaptic cell forms **presynaptic membranes** at a series of small swellings, or **boutons**, which are 0.5 to 1.0 μm in diameter. Each bouton contains mitochondria and a reserve pool of **vesicles**, synthesized in the Golgi apparatus and bound to contractile microfilaments of **actin**. Each vesicle is about 30 nm in diameter. A nerve impulse triggers an influx of calcium ions, which releases a cascade of chemical events involving phosphorylation of the protein synapsin by cadmodulin-dependent protein kinase II (Turner *et al.* 1999). This releases vesicles from the reserve pool, which migrate along actin filaments to the active zone at the synaptic membrane. Neurotransmitter molecules formed in the cytoplasm are pumped into the vesicle by vesicular transporter molecules embedded in the vesicle wall. Specific transporter molecules have been identified for glutamate, GABA, and other neurotransmitters (Bellocchio *et al.* 2000).

Each vesicle at the presynaptic membrane contains about 1,500 molecules of neurotransmitter. When the synapse is activated, vesicles that have migrated to the active zone near the synaptic membrane fuse with the membrane—a process known as **exocytosis** (Zenisek *et al.* 2000). Neurotransmitter molecules are thereby released into the **synaptic cleft** in less than a millisecond. Neurotransmitter molecules cross the cleft, which is about 20 nm wide, and attach to receptor molecules on the postsynaptic membrane. After synaptic transmission, glutamate molecules are rapidly absorbed from the synaptic cleft by a sodium-ion process on the presynaptic membrane.

The release of neurotransmitter from one vesicle can be regarded as one quantum of synaptic excitation. According to one view, release of neurotransmitter from one vesicle on the presynaptic membrane activates almost all the receptor sites on the adjacent postsynaptic membrane. Thus, the strength of response at a single synaptic site depends on the number of receptors at the site. According to another view, single vesicles do not saturate the postsynaptic membrane, so that response strength depends on the number of vesicles released (see Stevens 1995). Synaptic activity leads to an increase in the number of receptor molecules (see Section 10.4.4a).

The **postsynaptic membranes** of excitatory synapses occur mostly on spines distributed in great numbers along each branch of the dendritic tree of the postsynaptic cell. Each spine has a volume of about 1 μm^3. The receptor membrane on each spine contains a dense collection of receptor sites, actin-binding protein molecules, and a heterogeneous population of actin filaments, known collectively as the **postsynaptic density**. The spines can be small and blunt, elongated, or elongated with an expanded end. Actin microfilaments form the cell cytoskeleton and extend into the dendritic spines. Receptor molecules spanning the postsynaptic membrane are anchored to actin filaments by specific proteins. **Actin-binding proteins** regulate the polymerization of actin and control the arrangement of the actin filaments (Section 5.4.5c). Contraction of actin filaments transforms the blunt type into the other types during early development and during learning in the adult. Depolymerization of actin reduces the number of clusters of NMDA receptors and the number of spines containing AMPA receptors (Allison *et al.* 1998). These receptors are described in the next section. Time-lapse imaging of protein molecules in the postsynaptic density of slices of developing hippocampal tissue has revealed that these molecules are highly dynamic. Clusters of molecules form, move, remodel, and disappear on developing synaptic spines on a time scale of minutes to hours (Marrs *et al.* 2001).

Each muscle fibre receives only an excitatory input, typically from only one motoneurone, through a synapse involving the neurotransmitter acetylcholine. In contrast, each neurone in the central nervous system receives many excitatory inputs, mostly on dendritic spines, and many inhibitory inputs, the most effective of which impinge on the cell body. The resulting potentials propagate electrotonically

Table 5.2. Types of cortical synapse

VOLTAGE-GATED

Triggered by a voltage change across gap junctions.

LIGAND-GATED

Involve neurotransmitters acetylcholine, serotonin, glutamate.

Ionotropic

Triggered directly by a neurotransmitter.

AMPA type

Kainate type

NMDA type

Metabotropic

Triggered by a neurotransmitter and a cascade of secondary molecules.

INHIBITORY SYNAPSES

Involve the inhibitory neurotransmitter GABA.

to the cell body where they are integrated over a certain time period to produce action potentials in the axon. There may also be a few synaptic inputs on the axon, which modulate the action potentials.

5.4.5b *Excitatory cortical synapses*

Types of synapse are set out in Table 5.2. There are two main types. The first type is the **voltage-gated synapse** in which the pre- and postsynaptic membranes abut at **gap junctions**. Gap junctions contain well-aligned channels in the two membranes formed by proteins known as **connexins**. Small molecules and ions diffuse across these channels. The synapse is triggered directly by voltage changes produced by the presynaptic action potential passing across gap junctions. Generator potentials produced by sensory receptors activate voltage-gated synapses. Voltage-gated synapses may be unidirectional or bidirectional. They have a high threshold, but are fast acting and capable of synchronizing responses in neural tissue. They conserve the sign of depolarization of the presynaptic membrane. In invertebrates and lower vertebrates they are used in escape and warning mechanisms. They are also present in the vertebrate retina (Section 5.1.2c) and the primate neocortex (Sloper 1972). They are crucial for neurone differentiation and the growth of neuronal connections in the developing nervous system. Inhibitory interneurones in the cortex are connected by gap junctions. There seem to be two independent networks of connected interneurones—fast spiking and low-threshold spiking (Galarreta and Hestrin 2001). It is believed that these networks facilitate synchronization of neural activity.

The second main type, the **ligand-gated synapse**, is triggered by neurotransmitter molecules released from vesicles in the presynaptic membrane. Concentration of molecules in the vesicles is controlled by transporter molecules (Varoqui and Erickson 1998). Neurotransmitter molecules diffuse across the synaptic cleft and bind to receptor sites on the postsynaptic membrane. This increases the permeability of the membrane to sodium or calcium ions and, directly or indirectly, triggers a neural impulse.

Ligand-gated synapses can be experimentally activated by an **agonist**, which can be a naturally occurring neurotransmitter or an amino acid extracted from another source, and may be blocked by an **antagonist,** which can be an inhibitory neurotransmitter or a synthetic molecule.

Since ligand-gated synapses involve the diffusion of molecules across a relatively wide synaptic cleft they are much slower than voltage-gated synapses. However, they have low thresholds and act as amplifiers, since presynaptic activity releases many molecules of neurotransmitter. Some ligand-gated synapses produce an inhibitory hyperpolarization rather than an excitatory depolarization of the postsynaptic membrane. I shall now describe the two main types of ligand-gated cortical synapses.

5.4.5c *Ionotropic synapses*

In a **direct, or ionotropic ligand-gated synapse**, the neurotransmitter binds to a receptor on the postsynaptic membrane. The receptor molecule is part of the pore through which ions pass into the cell. A cylinder constructed from several glycoprotein molecules surrounds each pore. The glycoproteins round each pore differ to form distinct receptor subunits that help to determine the specificity of responses to diverse inputs to the synapse (Monyer *et al.* 1992). When the synapse is activated, the molecules lining the pore momentarily change their shape and allow ions to pass. The channel then closes and is refractory to further activation for a few milliseconds. The flow of ions can be measured by applying a patch clamp over the pore. The clamp is a fine glass pipette containing the neurotransmitter or a synthetic ligand molecule, which binds to the receptor.

There are three types of ionotropic synapses in the cortex. In all of them, the excitatory neurotransmitters are amino acids derived from glutamate. Synapses vary in the types and distributions of receptor molecules and ion channels on the postsynaptic membrane, which determine how they transform inputs into a spike-train output. Although all types of receptor respond to the same glutamate neurotransmitter, each type of receptor is identified by a specific externally applied agonist.

The first type of synapse is identified by its response to α-amino-3-hydroxy-5-methylisoxazole-4-proprionic acid, and the synapses are known as **AMPA or quisqualate synapses.**

The second type is identified by the application of **kainate** agonists, and the synapses are known as **K synapses**. Kainate agonists are extracted from plants and are more potent and more specific than glutamate—the endogenous cortical neurotransmitter. AMPA and kainate synapses open sodium- or potassium-ion channels and mediate fast excitatory transmission.

The third type of synapse is identified by the synthetic agonist N-methyl-D-aspartate, and the synapses are known as **NMDA synapses**. They are also sensitive to a number of chemical agents, such as glycine and polyamines The post-synaptic membrane contains three main types of receptor molecules known as **NR1**, which is required for synaptic activation, and **NR2 and NR3**, which modulate the response (Valtschanoff and Weinberg 2001). Activation of these receptors opens calcium-ion channels in addition to sodium and potassium channels. NMDA synapses have a prolonged response and are under strong inhibitory control. They also have the unique feature that they are blocked by extracellular magnesium ions, unless these ions are driven out by depolarization of the postsynaptic membrane by simultaneous activation of associated non-NMDA receptors on the same postsynaptic membrane. Activation of receptor molecules on the post-synaptic membrane allows an influx of calcium ions into the postsynaptic spine. This activates cadmodulin-dependent protein kinase II (CaMKII) and other enzymes, which phosphorylate membrane ion channels and thereby potentiate the response of non-NMDA receptors (Smart 1997). Activation of NMDA receptors also leads to protein synthesis and cytoskeletal changes in the dendritic spine. We will see below and in Section 10.4.5 that these stimulus-contingency features of NMDA receptors allow them to mediate growth, plasticity, and learning in the visual cortex.

These three types of synapse are subdivided into at least 16 subtypes that are identified by their responses to synthetic ligands containing different peptide subunits (Hollmann and Heinemann 1994).

The variety of synapses provides a molecular basis for diversity and specificity of postsynaptic mechanisms. The different types of synapse are differentially distributed in the mammalian brain, although receptors for NMDA and AMPA tend to occur together on the same postsynaptic membrane (Kaczmarek *et al.* 1997). In other words, they are activated by release of glutamate in the same synapse. The significance of this **cooactivation** is discussed in Section 10.4.5. Geniculocortical transmission in cortical layer 4 of the cat is mediated mainly by AMPA and K synapses, while intracortical transmission in layers 1, 2, and 3 is mediated by all three types (Larson-Prior *et al.* 1991). In the human striate cortex, NMDA receptors are most dense in layers 1 to 4C, with highest density in layer 4C. AMPA receptors are most dense in layers 1 to 3 and least dense in layers 4B and 4C. Kainate and metabotropic synapses are fairly evenly distributed. The types of synapse do not vary between cortical columns (Albin 1991).

5.4.5d *Metabotropic synapses*

In an **indirect, or metabotropic ligand-gated synapse**, the receptor is some distance from the ion channel pore and activates the pore through intermediate molecules known as **G-proteins**. The receptor structure consists of an extracellular ligand-binding region, a transmembrane region, and a cytoplasmic region within the cell that binds to G-protein molecules. (Kunishima *et al.* 2000). Activation of the receptor releases a cascade of molecules, known as a second messenger system. Metabotropic synapses fall into three classes according to the amino acid sequence of the metabotropic receptor molecules in the postsynaptic membrane. They are identified by their responses to specific synthetic agonists and antagonists (Riedel 1996). Metabotropic synapses have an onset time of hundreds of milliseconds—much longer than that of any of the ionotropic types. Cortical synapses involving the neurotransmitters norepinephrine or serotonin are also metabotropic.

5.4.5e *Inhibitory synapses*

The main inhibitory neurotransmitter in the cortex is γ-amino-butyric acid (GABA), a substance also derived from glutamate (Mize and Marc 1992; Gutiérrez-Igarza *et al.* 1996). The molecule exists in various forms, which bind to different receptor molecules. Ion channels on the postsynaptic membrane of an inhibitory synapse consist of a cluster of five receptor subunits each with an extracellular peptide, a transmembrane sequence, and a large intracellular loop (De Blas 1996). Fast acting $\mathbf{GABA_A}$ receptors are of the direct ionotropic type and $\mathbf{GABA_B}$ receptors are of the slow-acting metabotropic type (Moss and Smart 2001). Inhibitory interneurones are also connected by gap junctions, as indicated in Section 5.4.5b

Inhibition can work in two ways. In **hyperpolarizing inhibition** the change in membrane potential is determined by the linear sum of negative and positive currents. In **shunting inhibition** the excitatory response is reduced by a nonlinear increase in membrane conductance (Borg-Graham *et al.* 1998). Inhibitory synapses near the zone where nerve spikes are generated produce **proximal inhibition**, which is capable of blocking all excitatory responses of the cell. Inhibitory synapses on particular dendrites produce **distal inhibition**, which can selectively inhibit particular branches of the neurone or produce graded inhibition (Vu and Krasne 1992).

Inhibitory interneurones in the cortex work in partnership with excitatory pyramidal cells but are shorter and faster acting than pyramidal cells. There is a bewildering diversity of interneurones. Most of them are excited or inhibited by two or three neurotransmitters, which include noradrenaline, muscarine, serotonin, and glutamate (Parra *et al.* 1998). In general, inhibitory interneurones modulate the threshold for initiation of action potentials and dendritic spikes, and play a major role in controlling the activity of cortical neural networks. Short- and long-range lateral inhibitory interactions fine tune neurones to specific stimulus features, such as orientation (Section 5.4.7a). Feedforward, and feedback inhibition generates synchronous and oscillatory responses (Section 4.3.2). Inhibitory circuits are also involved in attentional gating (Section 5.5.7c).

5.4.5f *Receptors on presynaptic membranes*

Presynaptic membranes can contain ionotropic or metabotropic receptors of various types that modulate the release of neurotransmitter. Some presynaptic receptors are autoreceptors activated by the neurotransmitter released by the same membrane. Others are activated by distinct neurotransmitters released by neighbouring neurones or by cholinergic receptors triggered by afferents from centres in the basal forebrain. Some receptors involve GABA-mediated presynaptic inhibition while others facilitate neurotransmitter release. The functions of receptors on presynaptic membranes are not yet understood (MacDermott *et al.* 1999).

5.4.5g *Other cortical neurotransmitters*

The neocortex receives inputs from a variety of other subcortical areas, each involving a distinct neurotransmitter. Cortical cells contain a variety of receptors for each of these transmitters.

Afferents project from the **rostral brainstem** to most parts of the thalamus, where they constitute about 90% of brainstem afferents. They relay to all parts of the neocortex, terminating in synapses that release the neurotransmitter **acetylcholine**, which excites fast nicotinic and slower muscarinic receptors, and has indirect inhibitory effects through GABA receptors. Also, cholinergic neurones project from the **basal forebrain** to all parts of the cortex, either directly or through the thalamus. Thus, pyramidal cells and GABA interneurones in all parts of the cerebral cortex may be activated by cholinergic afferents arising from the basal forebrain or thalamus. Cholinergic afferents operating through nicotinic receptors on the presynaptic membrane promote intracolumnar inhibition while those operating through muscarine receptors reduce some forms of intralaminar inhibition (Xiang *et al.* 1998). These systems regulate the flow of information in the cortex. They seem to control selective attention and perhaps consciousness (McGehee and Role 1996; Perry *et al.* 1999). They are involved in the generation of hallucinations and cognitive disorders in clinical conditions such as schizophrenia, epilepsy, and Alzheimer's disease (see Alkondon *et al.* 2000).

Inputs from the **substantia nigra** and ventral tegmentum also project to the neocortex, especially to the prefrontal cortex. Their synapses involve the neurotransmitter **dopamine**. Five different receptors for dopamine are distributed in distinct regions of the central nervous system. Dysfunction of this system is implicated in Parkinson's disease and schizophrenia.

Fine unmyelinated axons originating in the **locus coeruleus** in the dorsal pons provide a diffuse, innervation of the cerebral cortex, largely to layer 6 (Levitt and Moore 1979). Neurones from the dorsal portion project to the visual cortex. Their synapses release the neurotransmitter **norepinephrine**. The locus coeruleus is probably associated with attention and has been implicated in cortical plasticity (Section 13.2.4e).

Axons from the raphe nucleus and **pontine reticular formation** project to all cortical areas and their synapses release the neurotransmitter **serotonin**. In the monkey visual cortex, the strongest projection is to layer 4. In the cat, there are some serotonin axons in layer 4 during the first few weeks of life but, in the adult cat, the strongest projection is to layers 1 to 3 (Gu *et al.* 1990).

5.4.5h *Synaptic changes during learning*

The synaptic events underlying learning have been studied most intensively in the hippocampus, but the results probably apply to other areas of the cortex. Electron microscopy of the hippocampus of neonate rats has revealed that numerous dendritic filopodia form synaptic contacts with axons or with other filopodia. By postnatal day 12 dendritic spine synapses develop and it seems likely that filopodia synapses give rise to spine synapses (Fiala *et al.* 1998).

The following structural changes in synapses have been associated with learning in adult animals.

1. Increase in the density of AMPA receptor sites on the postsynaptic membrane triggered by protein kinases activated by responses of NMDA synapses. These changes are believed to mediate the expression of plastic changes.

2. Increase in the area of synaptic membranes produced by curvature of the membranes, or by changes in the size or number of dendritic spines. Such changes have been observed by the electron microscope in cultured slices of rat hippocampus between 30 to 60 minutes after bursts of electrical stimulation (Buchs and Muller 1996; Toni *et al.* 1999). Laser scanning of cultured slices from the hippocampus infected with a virus expressing a fluorescent protein reveals the fine structure of dendrites. Time-lapse imaging showed that, within 60 s after 100 Hz tetanic stimulation, new dendritic spines developed and existing spines elongated. The effects were localized to within about 100 μm and were abolished with the introduction of an antagonist for NMDA receptors (Maletic-Savatic *et al.* 1999).

3. Formation of new dendritic spines and changes in the shapes of existing spines produced by formation or contraction of actin filaments in the cell. These changes have been observed within seconds of synaptic activity (Fischer *et al.* 1998). Formation and contraction of actin filaments is controlled by actin-binding proteins released from the postsynaptic density by synaptic activity (Section 5.4.5a). These morphological changes are associated with increases in synaptic transmission and reduction in latency (Yuste and Bonhoeffer 2001).

4. Formation of multiple receptor zones on the postsynaptic membrane, produced by perforations in the membrane (Edwards 1995).

5. Expression of genes leading to the production of protein molecules and the binding of proteins into protein complexes in the postsynaptic membrane (Husi *et al.* 2000).

6. It has been generally believed that neurogenesis does not occur in the neocortex of adult primates (Bourgeois *et al.* 1994). However Gould *et al.* (1999) have produced evidence that, in the adult macaque, new neurones are continually produced in the subventricular zone of the prefrontal, posterior parietal, and inferior temporal areas, but not in the striate cortex. The new neurones migrate through the white matter and extend axons.

Spontaneous release of neurotransmitter seems to be required to maintain dendritic spines on cortical neurones (McKinney *et al.* 1999). Processes of learning are discussed further in Section 10.4.5a.

Figure 5.18. David H. Hubel.
Born in Windsor, Canada in 1926. He graduated in Mathematics and Physics in 1947 and in Medicine in 1954, both from McGill University in Montreal. Between 1954 and 1958 he conducted research at the Walter Reed Army Institute of Research. In 1958 he moved to the Laboratory of S. Kuffler at the Wilmer Institute of John Hopkins Hospital. In 1959 the whole laboratory moved to Harvard University Medical School to form the new Department of Neurobiology. In 1981 he won Nobel Prize in Medicine with David Hubel and Roger Sperry.

Robinson (2001) has reviewed the history of discoveries on synaptic transmission. Reviews of recent work on synaptic transmission are contained in Cowan *et al* (2001).

5.4.6 Receptive fields of cortical cells

A neurone within the visual cortex responds, or its ongoing response is modified, when an appropriate stimulus falls within a specific retinal area. That area is defined as the **receptive field of the cortical cell**, and its position is specified by the location of its centre. Receptive field centres are represented retinotopically within each layer of area 17, although this arrangement is perturbed by local random scattering of the same order of magnitude as the size of the receptive fields at each location (Hubel and Wiesel 1977) (Portrait Figures 5.18 and 5.19).

The response of a cortical cell can be modified by stimuli that fall well outside the receptive field as defined by simple stimuli (Hammond and MacKay 1981; Nelson and Frost 1985). It was mentioned in the previous section that the receptive fields of some cortical cells expand over a period of minutes when the surrounding area is stimulated.

Each spiny stellate cell in layers 4A and 4C receives a direct input from only one eye. These cells have circular-symmetric receptive fields, which resemble those of either the parvocellular or magnocellular LGN cells that feed into them (Blasdel and Fitzpatrick 1984). Thus, their response does not depend on the orientation of the stimulus. The other excitatory cells in the primary visual cortex are pyramidal cells. Most of them have elongated receptive fields and show orientation specificity. They fall into two classes, simple cells and complex cells, according to the organization of their receptive fields, and their functional properties.

Each **simple cell** receives inputs from about 30 cells in the LGN. Some simple cells have a symmetrical (cosine) distribution of zones—an excitatory on-centre and two flanking inhibitory off-zones or an off-centre with flanking on-zones. Other simple cells have an asymmetrical (sine) distribution of zones—an on-zone flanked by a single off-zone, as depicted in Figure 5.20. The on-regions of simple-cell receptive fields show an excitatory postsynaptic potential (EPSP) to stimulus onset and an inhibitory postsynaptic potential (IPSP) at stimulus offset. Off-regions show an IPSP at stimulus onset and an EPSP at stimulus offset (Ferster 1988). The inhibitory potentials could perhaps sharpen responses to brief stimuli and improve orientation and disparity selectivity.

Antagonistic inputs from ganglion cells as well as interactions between neighbouring cortical cells could be involved in these reciprocal interactions. In the monkey, simple cells with a single excitatory zone receive excitatory inputs from only the parvocellular or only the magnocellular layers of the LGN. Some simple cells with multiple excitatory zones receive mixed parvocellular and magnocellular inputs (Malpeli *et al.* 1981).

The frequency of response of a simple cell to a stimulus that fills its receptive field is a linear sum of the frequencies of its responses to spots of light falling in each part of its receptive field. In other words, simple cells integrate luminance in a linear fashion. Most simple cells have little or no maintained discharge in the absence of stimulation. A simple cell does not respond to even illumination of its receptive field, because, with even illumination, excitatory responses are cancelled by inhibitory responses. As a dark-light grating of appropriate spatial frequency and orientation is moved over the receptive field of a simple cell the response of the cell is maximal when the bright bars coincide with the on zones and zero when they coincide with the off zones. The cell acts as a half-wave rectifier with respect to the spatial distribution of dark-light bars falling within its receptive field. The full range of stimulation is covered because the cells occur in pairs with opposite spatial phase—sine and cosine (Heeger 1992b).

Figure 5.19. Torsten N. Wiesel.
Born in Uppsala, Sweden in 1924. He obtained an M.D. at the Karolinska Institute, Stockholm in 1954 and was postdoctoral Fellow and Assistant Professor in Ophthalmology at the Johns Hopkins University Medical School from 1955 to 1959. In 1960 he moved to Harvard Medical School where he became Robert Winthrop Professor of Neurobiology. In 1983 he moved to the Rockefeller University in New York where he was Vincent and Brooke Astor Professor of Neurobiology and University President from 1992 to 1998. He is now President Emeritus. In 1978 to 1979 he was President of the Society for Neuroscience. He has received many honours including the 1981 Nobel Prize in Medicine with David Hubel and Roger Sperry, the Friendenwald Award of the Association for Research in Vision and Ophthalmology in 1975, the Karl Lashley Prize of the American Philosophical Society in 1977, the Ledlie Prize from Harvard University in 1980, and the Helen Keller Prize for Vision Research in 1996.

In spite of their basic linearity, simple cells show three types of nonlinearity—their response saturates at high stimulus contrasts, they respond more rapidly at high contrasts, and their response to superimposed orthogonal stimuli is less than their response to a stimulus in one orientation (cross-orientation inhibition). These nonlinearities could arise from a gain-control mechanism that depends on scaling (dividing) the cell's response by the pooled activity of neighbouring cells—a process called **normalization** (Carandini and Heeger 1994). Normalization makes it possible for a cell's response to critical features of the stimulus, such as motion, orientation, and disparity, to be independent of stimulus contrast.

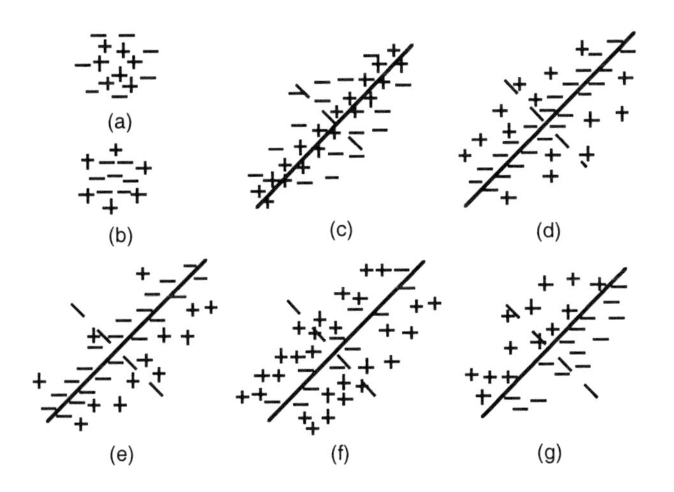

Figure 5.20. Types of receptive field.
(a) An on-centre receptive field of a ganglion cell.
(b) An off-centre receptive field of a ganglion cell.
(c)-(g) Receptive fields of simple cells in cat visual cortex. The receptive fields are shown with preferred orientations of 45°. (Adapted from Hubel and Wiesel 1962)

Complex cells are selectively tuned for orientation, spatial frequency, and direction of motion but their receptive fields are not clearly segregated into excitatory and inhibitory zones. This causes them to integrate luminance in a nonlinear fashion. A grating drifting across the receptive field of a simple cell produces a modulated response. The same stimulus produces an unmodulated response in complex cells.

Simple cells tend to be stellate cells and complex cells tend to be pyramidal cells. However, this correlation between structure and function is not perfect (Gilbert and Wiesel 1979). In Hubel and Wiesel's hierarchical model, simple cells feed into complex cells. Ghose *et al.* (1994b) revealed only polysynaptic excitatory connections from simple to complex cells and monosynaptic excitatory connections from complex to simple cells in the cat.

Some complex cells receive inputs from only the parvocellular or only the magnocellular layers of the LGN while others receive a mixed input (Malpeli *et al.* 1981). There is evidence that the phase invariance of complex cells arises from recurrent cortical inputs from other complex cells rather than from a pooling of inputs from a set of simple cells with differing phase sensitivities (Chance *et al.* 1999).

The structure of receptive fields of cortical cells is discussed in more detail in Section 5.5.4. Most cells in the visual cortex of cats and primates receive inputs from both eyes (Section 5.6.2). These binocular cells therefore have two receptive fields, one in each eye.

5.4.7 Lateral cortical connections

5.4.7a *Types of lateral connection*

Collaterals from the axons of pyramidal cells in primate cortical layers 2, 3, 4, and 5 of V1 and V2 project horizontally for up to 8 mm. This represents several receptive field diameters. Visual cortex horizontal projections are typically longer along one axis and produce spaced clusters of predominantly excitatory synapses (Rockland and Lund 1982, 1983; Rockland 1985; Gilbert and Wiesel 1985). Short-range connections up to 500 μm in area 17 of the cat are largely circular symmetric (Das and Gilbert 1999). According to one estimate, derived from horseradish peroxidase labelling, 80% of horizontal connections are with other pyramidal cells and are excitatory, while 20% are with smooth inhibitory interneurones (McGuire *et al.* 1991). However, these anatomical facts do not determine the relative strengths of inhibitory and excitatory connections because the effects of inhibition could be amplified relative to the effects of excitation.

Estimates, based on electrophysiological recording, of the proportion of cells in the visual cortex of the cat that exhibit surround suppression have varied from 10% to 80%. But some investigators use a criterion of 100% suppression while others use less stringent criteria. Also, some investigators measure only inhibition from the end of the receptive field while others measure inhibition from all round the receptive field. Walker *et al.* (2000) used a circular patch of optimally oriented grating and found that the response of 56% of simple and complex cells in the cat's visual cortex was reduced by at least 10% when stimuli were presented in the receptive-field surround. They found only a few excitatory interconnections. For binocular cells in all cortical layers, the degree of surround suppression was strongly correlated for stimuli presented to the two eyes.

Optical recording from the surface of the visual cortex in conjunction with microelectrode recording has revealed that a local stimulus induces a wide area of subthreshold activation and a much smaller area over which neural spikes are evident (Das and Gilbert 1995a). By applying a voltage-sensitive dye to the visual cortex of the monkey, one can observe the real-time spread of activity evoked by a locally applied visual stimulus. Activity was found to spread from its initial locus at a velocity of 100 to 250 μm/s to cover an area with a radial space constant of 1.5 mm along ocular dominance columns and of 3 mm orthogonal to the columns (Grinvald *et al.* 1994).

The patterns of lateral connections have been studied by retrograde tracing with 2-deoxyglucose and by correlating discharges of pairs of cells over various time delays. These methods have revealed that long-range horizontal dendrites link pyramidal cells with a similar preference for stimulus orientation (Gilbert and Wiesel 1989; Hirsch and Gilbert 1991; Ts'o *et al.* 1986).

Within layers 2 and 3 of the visual cortex of the tree shrew, lateral excitatory connections spread furthest and give off more terminal boutons in the direction of orientation tuning of cells in a given region (Bosking *et al.* 1997). Thus, cells with axially aligned receptive fields are connected more richly than cells with parallel receptive fields, which are connected more richly than cells with distinct orientation preference. These rich connections could serve to enhance the visibility of continuous edges (see Section 5.4.7b). Sincich and Blasdel (2001) found the same anisotropy in layer 3 of squirrel and owl monkeys. In primates with strong ocular dominance columns, any anisotropy of lateral connections related to axial orientation tuning is difficult to detect because it is masked by anisotropy related to eye dominance. Short-range connections of up to 500 μm decline with increasing distance and are largely independent of orientation preference (Das and Gilbert 1999).

Adorján *et al.* (1999) developed a model of orientation tuning based on intracortical excitatory connections.

Linkages are also more common between columns of the same ocular dominance than between those of opposite dominance and between blobs and blobs or interblobs and interblobs than between blobs and interblobs (Livingstone and Hubel 1984; Yoshioka *et al.* 1996). Binocular regions of the visual cortex are not connected to monocular regions (Malach *et al.* 1993).

Matsubara *et al.* (1985) found lateral interconnections in area 18 of the cat between cortical cells with orthogonal orientation preferences. These connections spanned only 2 mm and were inhibitory (GABAergic). For the majority of cells in area 17 of the cat, inhibition of a central grating by a drifting grating outside the classical receptive field is maximum when the gratings are orthogonal. For some cells, inhibition is independent of the orientation of the surround grating (Sengpiel *et al.* 1997). In contrast to the patchy arborization of long range excitatory lateral axons, arborizations of inhibitory axons are homogeneous (Kritzer *et al.* 1992).

Buhl *et al.* (1994) described three types of local inhibitory interneurones in the hippocampus of the rat. "Basket cells" synapse on the somata of hundreds of principal cells and cause rapid hyperpolarization followed by rebound. This type of cell could cause synchronous firing of large cell populations. "Axo-axonic cells" synapse only on the initial segment of the axon of the principal cells. This type seems well suited to control the discharge of principal cells. "Bistratified cells" make synaptic contact with the base and apical dendrites of principal cells and have properties like those of Hebbian synapses (Section 10.4.5).

There is thus a division of labour between different inhibitory interneurones in controlling the synchronous activity of cortical cells. Szabadics *et al.* (2001) found similar inhibitory networks in the somatosensory cortex of the rat. It is not yet known whether similar networks exist in the visual cortex.

We will see in what follows that the balance of excitation to inhibition depends on the relative positions, orientations, spatial frequencies, and contrasts of stimuli.

5.4.7b *Functions of lateral connections*

Lateral cortical connections could serve any or all of the following functions.

1. Cortical cells with collinear orientation preferences linked by long-range horizontal fibres could serve to build a mechanism sensitive to continuity of lines in a particular orientation, as discussed in Section 4.2.6 (Mitchison and Crick 1982). For example, Polat and Sagi (1993, 1994) produced psychophysical evidence that collinear stimuli show mutual facilitation to a greater degree than stimuli with similar orientation but which are not collinear. For this purpose, the linkup should be between cells tuned to collinear stimuli rather than between all cells with similar orientation.

Polat and Norcia (1996) found that evoked potentials from the human visual cortex were stronger for a set of high-contrast aligned stimuli than for a set of stimuli with different orientations. Nelson and Frost (1985) revealed a highly specific form of facilitation in the cat visual cortex between neurones with aligned orientation tuning functions. The response of complex cells in V1 of alert monkeys to a line was enhanced when a collinear line was added outside the receptive field of the first stimulus (Kapadia *et al.* 1995, 2000).

Normally, the increased response of a cell to an increase in stimulus strength is accompanied by a proportional increase in the variability of responses to repeated stimulus presentation. However, increases in responses of cells in cat area 17 to a Gabor patch in the presence of flanking collinear patches were not accompanied by an increase in response variability (Kasamatsu *et al.* 2001). Thus, collinear stimuli improve the signal-to-noise ratio, especially at threshold levels.

Cells in V2 of the monkey responded to cognitive contours indicated by disconnected but aligned boundaries of occluded objects (von der Heydt and Peterhans 1989).

Lateral connections of pyramidal cells in higher visual areas, such as the inferior temporal cortex and the temporal polysensory area (area STP), cover an area more than six times that of pyramidal cells in V1 (Elston *et al.* 1999). Presumably, this is due to the selectivity of these areas for complex stimuli.

2. Cortical cells linked by short-range excitatory connections that are independent of orientation preference could be involved in the construction of cell assemblies tuned to specific patterns, such as corners and T junctions.

3. Lateral connections can modify the response of cells according to the nature of surrounding stimuli (Nelson and Frost 1978; Gilbert and Wiesel 1990; Gilbert *et al.* 1991). There is considerable psychophysical evidence that visual acuity and shape discrimination are degraded when a test stimulus is flanked by similar stimuli (Section 8.2.4). The processes responsible must be cortical, since the effects are present when the test and flanking stimuli are presented to different eyes (Westheimer and Hauske 1975). Like stimulus interactions, inhibitory interactions between a cortical cell and its surroundings are greater when the surrounding stimuli are similar in orientation and size to the test stimulus (Kooi *et al.* 1994; Walker *et al.* 1999). Responses of a cortical cell in V1 of the alert monkey are reduced when neighbouring lines lie on the flanks of the cell's receptive field. Responses are enhanced when neighbouring lines are collinear with the cell's receptive field (Kapadia *et al.* 2000). The inhibitory interactions could enhance the visibility of a visual stimulus or textured region that differs from surrounding stimuli, as in the pop-out phenomenon described in Section 18.1.7 (Nothdurft *et al.* 2000). The facilitatory interactions could enhance the visibility of lines and help in the integration of figural information. Dragoi and Sur (2000) have modelled these orientation-specific facilitatory and inhibitory interactions.

4. Lateral interactions depend on the relative contrasts of the stimuli (Levitt and Lund 1997). Stimuli that manifest orientation or spatial-frequency specific inhibitory interactions at high contrasts show mutual facilitation at low contrasts (Cannon and Fullenkamp 1993; Kapadia *et al.* 2000). Inhibitory mechanisms must have a higher threshold than excitatory ones. The important thing at low contrasts is to see the stimuli, even though they may not be clearly discriminated from each other. Also, the visibility of a central stimulus at or below threshold is increased by the addition of suprathreshold surrounding stimuli (Knierim and Van Essen 1992; Grinvald *et al.* 1994). This is because noisy fluctuations in activity due to surrounding stimuli contribute to the response of the subthreshold central stimulus and bring it into the detectable range by a process known in physics as response resonance. This type of interaction could serve to enhance the visibility of weak parts of a patterned stimulus. Stemmler *et al.* (1995) have modelled these inhibitory and facilitatory processes.

5. Lateral connections may also explain the geometrical illusions, tilt contrast, and figural aftereffects in which a stimulus is apparently displaced away from a neighbouring stimulus (see Howard 1982, Chapter 4).

6. Lateral connections are involved in interactions between dichoptic stimuli. Thus, a cortical cell that responds only to excitatory inputs from one eye may be suppressed by inputs presented to the eye to which it does not normally respond.

7. Lateral interactions may be involved in normalization of the response of a cortical cell. Normalization is a non-linear process, which divides the response of a cortical cell by the pooled response of surrounding cells (Carandini and Heeger 1994). It renders the response of cortical cells to features such as orientation, motion, and disparity independent of stimulus contrast (Section 5.4.6). Lateral connections may also be involved in the registration of the brightness of surfaces (Rossi and Paradiso 1999).

8. Lateral pathways may also be involved in adaptations to cortical scotomata. Amputation of a finger in the adult monkey causes cells in the affected region of the somatosensory cortex to become sensitive to inputs from adjacent fingers (Merzenich *et al.* 1983). In a similar way, when corresponding areas of both retinas of the adult cat are lesioned, cells near the boundary of the deafferented cortical site immediately begin to respond to stimuli adjacent to the lesion. After a few months the whole of the silenced area begins to respond to stimuli surrounding the lesion. Silent regions in the LGN do not recover, so cortical recovery must involve changes within the cortex. Thalamocortical arbors do not extend beyond their normal range and the orientation specificity of recovered cortical cells is the same as that before the lesion. These facts suggest that inputs are conveyed to the reactivated cortical cells through horizontal axons arising from cortical cells with similar orientation preference (Darian-Smith and Gilbert 1995). Part of this effect could be due to a lowering of the threshold of horizontal cells or the removal of GABAergic inhibitory influences around the affected area (Gilbert and Wiesel 1992; Das and Gilbert 1995a; Chino 1997). But there is also evidence of ax-

onal sprouting of laterally projecting neurones (Darian-Smith and Gilbert 1994). Similarly, after a local lesion is made in one retina of the adult cat, the corresponding cortical region begins to respond to stimuli applied near the lesion, although firing rates are unusually low and transient (Schmid *et al.* 1996). This change is evident within a few hours after the induction of the retinal lesion (Calford *et al.* 1998).

A localized lesion in one eye of kittens during the critical period of development abolishes responses of corresponding cortical binocular cells to stimulation of that eye. However, after some time, binocular cells begin to respond to stimuli applied to regions round the lesioned area (Chino *et al.* 2001).

A lesion in the monocular zone of the retinal periphery creates a deafferented region in the contralateral visual cortex of adult monkeys. After several months, some cells in this region begin to respond to stimulation of the boundary of the binocular zone but most cells remain unresponsive even after a year (Rosa *et al.* 1995). Thus cortical plasticity varies with location in the visual cortex.

Occlusion of the receptive field of a cortical cell in the cat accompanied by stimulation of the surrounding area over a period of 10 minutes caused a five-fold increase in the area of the occluded receptive field (Chino *et al.* 1992; Pettet and Gilbert 1992). This type of stimulation simulates a local scotoma. A local artificial scotoma applied in one eye of a cat also led to an expansion of receptive fields in both eyes, of cortical cells serving that area (Volchan and Gilbert 1995). This suggests that the expansion is due to recruitment of previously subthreshold lateral connections. Orientation tuning and ocular dominance of the cells were not affected and receptive fields returned to their normal size when stimulated directly (Das and Gilbert 1995b). However, DeAngelis *et al.* (1995) failed to find any change in the size or structure of cortical receptive fields of adult cats during application of an artificial scotoma. They found only a short-term and reversible change in the responsiveness of cells in the surround. A small stimulus near the boundary of an artificial scotoma is perceptually displaced towards the centre of the scotoma (Kapadia *et al.* 1994). Kalarickal and Marshall (1999) produced a computational model of these processes, involving plasticity of inhibitory and excitatory lateral connections.

9. Lateral pathways could also be involved in long-term stimulus-dependent changes in cortical responses. Long-term changes in synaptic conductivity along lateral pathways in the cat's visual cortex have been induced by pairing synaptic responses with conditioning shocks of depolarizing current (Hirsch and Gilbert 1993).

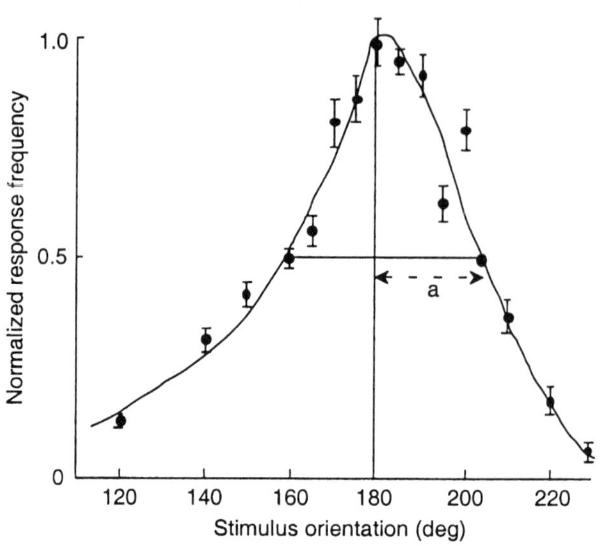

Figure 5.21. Orientation tuning function of a cortical cell. Tuning function of a simple cell in the cat's visual cortex to the orientation of a line. Distance *a* is half the width of the tuning function at half amplitude, and indicates the cell's orientation selectivity. This particular function is asymmetrical. Bars are standard errors. (Redrawn from Heggelund and Albus 1978)

5.5 STIMULUS TUNING OF CORTICAL CELLS

5.5.1 Contrast sensitivity of cortical cells

In response to a drifting grating, simple cells in the visual cortex of the cat fire at a frequency that depends on stimulus contrast. Most cells have a contrast threshold below 0.05 contrast (Skottun *et al.* 1987). After contrast has reached the threshold value, firing frequency shows an initial acceleration and then a linear dependence on contrast up to a contrast of about 0.3, after which the response saturates at a contrast of about 0.75 (Dean 1981; Albrecht and Hamilton 1982). The slope, or gain, of the contrast/response function varies from cell to cell, as does the position of the linear portion of the function along the contrast axis. For binocular cells, the slope of the contrast/response function was steeper for a stimulus presented to the dominant eye than for one presented to the nondominant eye, but was most steep when the stimulus was presented to both eyes (Chao-yi and Creutzfeldt 1984). Cortical cells also adjust their contrast sensitivity by scaling their response by the pooled response of neighbouring cells. This type of automatic gain control is response normalization (Carandini and Heeger 1994). The variance of the response of cortical cells increases with increasing response amplitude (Tolhurst *et al.* 1981). This may account for why contrast discrimination is best at low contrasts (Weber's law).

Hawken and Parker (1984) found that cells in the visual cortex of the monkey with relatively high contrast sensitivity were segregated from those with relatively low sensitivity. They identified the former with the magnocellular system and the latter with the parvocellular system. The contrast sensitivity of LGN cells is similar to that of cells in V1, but cells in MT (Section 5.7.3) have enhanced contrast sensitivity. This is probably because many magnocellular cells converge to form the receptive fields of cells in MT, which are about 100 times larger than the receptive fields of cells in V1 (Sclar *et al.* 1990).

The contrast threshold of a cortical cell as a function of the spatial frequency of a drifting grating defines the cell's contrast sensitivity function. The bandwidth and peak spatial frequency of the contrast-sensitivity functions of cortical cells are relatively insensitive to changes in contrast (Skottun *et al.* 1986a).

5.5.2 Orientation tuning

5.5.2a *Orientation tuning functions*

Both simple and complex cells respond best when a line or edge is oriented along the long axis of the cell's receptive field. The stimulus orientation that evokes the strongest response in a given cell is its **preferred orientation**. The function relating the firing rate of a cell to the orientation of a line centred within its receptive field is its **orientation tuning function**. An example from the cat is shown in Figure 5.21. The full tuning bandwidth of a cell is indicated by the width of the tuning function at half the height of its maximum response. Bandwidths of cells in the monkey striate cortex range from 6° (sharply peaked tuning) to 180° (flat tuning). The mean is about 40°, which covers about a quarter of the maximum range of 180° (DeValois *et al.* 1982a). The orientation bandwidth, determined psychophysically in humans by a masking procedure, has been found to decrease with increasing spatial frequency of a test grating from about 60° at 0.5 cpd to 30° at 11.3 cpd (Phillips and Wilson 1984). Although orientation-sensitive cells respond most reliably to stimuli oriented at the peak of the tuning function, their sensitivity to changes in orientation is greatest on the flanks of the tuning function where the change of response per unit change in orientation is greatest (Scobey and Gabor 1989).

5.5.2b *Mechanisms of orientation tuning*

Hubel and Wiesel's original idea was that the orientation selectivity of a cortical cell derives from the fact that each excitatory cortical cell receives inputs from several LGN neurones with overlapping circular receptive fields that are aligned on the retina. Confirmation of this idea has been provided by Stryker (1991), Chapman *et al.* (1991), Reid and Alonso (1995), and Ferster *et al.* (1996).

There is evidence that mutual facilitation of cells tuned to the same orientation coupled with mutual inhibition of cells tuned to orthogonal orientations sharpens the orientation selectivity of cortical cells (Sillito *et al.* 1980a; Hata *et al.* 1988; Volgushev *et al.* 1993; Crook *et al.* 1997; Eysel *et al.* 1998). One possibility is that intracortical inhibition sharpens the tuning of cortical cells for both spatial frequency and orientation by attenuating their response to low spatial frequencies (Vidyasagar and Mueller 1994). In one model, intracortical inhibition and the amplifying effects of recurrent excitation enhance the preexisting weak orientation tuning created by orientation bias in LGN neurones or by cortical convergence of inputs with aligned receptive fields (Vidyasagar *et al.* 1996). Shevelev *et al.* (1998) found that intracortical inhibition sharpens the orientation tuning of one class of cortical cells but produces stimulus dependent changes in preferred orientation and tuning width in a second class of cells, which they called "scanners".

Hirsch *et al.* (1998) found that the temporal attributes of excitatory responses of simple cells in cat area 17 to oriented stimuli matched those of inputs from the LGN. However, inhibitory responses were governed mainly or wholly by inhibitory cortical interneurones driven by stimuli of reverse contrast rather than by reduction in the output of geniculocortical synapses. This separation of excitatory and inhibitory processes could endow the visual cortex with a wider dynamic range.

Perhaps the response of cortical cells tuned to orientation is enhanced by a tendency for spatially aligned ganglion cells to fire in synchrony (see Meister 1996). Mechanisms of orientation tuning have been reviewed by Ferster and Miller (2000).

5.5.2c *Temporal aspects of orientation tuning*

Inhibitory or excitatory mechanisms, precortical or cortical that help to determine the orientation selectivity of a cortical cell could involve feedforward loops or recurrent feedback loops. Celebrini *et al.* (1993) demonstrated that the orientation selectivity of cells in the visual cortex of the monkey is present and fully formed in the first 10 ms of its response. Other investigators have found that, although optimal orientation tuning of most cells in the cat visual cortex is stable over the first 100 ms, tuning becomes sharper during this period (Volgushev *et al.* 1995).

Shevelev *et al.* (1993) found that only about 37% of cells in the cat visual cortex show stable optimal tuning during the first 600 ms. The other cells, identified as "scanners", showed systematic shifts in tuning, first in one direction and then in the other. A short build-up time is inconsistent with the idea that time-consuming feedback loops are involved in orientation tuning. The faster, feedforward, mechanism allows the animal to assess the orientation of a stimulus rapidly. Ringach *et al.* (1997) found that cells in layer 4C of the monkey visual cortex showed a unimodal response to an optimally oriented stimulus, with a single peak between 30 and 45 ms after stimulus onset and a stable orientation preference over the duration of the response. By contrast, many cells in other layers showed two peaks and their preferred orientation was not stable over time. They concluded that the broad orientation tuning of cells with a unimodal response profile arises from a direct feedforward mechanism but that the narrower tuning of cells with a multimodal response profile arises from intracortical inhibition.

These results reinforce the idea of two classes of orientation detectors. Those in the first class have rapid onset of tuning that relies on feedforward mechanisms. Cells in the second class ("scanners") rely on recurrent circuits and their orientation tuning is modified by the short-term characteristics of the visual task (Section 3.2.2).

McLaughlin *et al.* (2000) developed a neural network model of orientation tuning based on the anatomy and the feedforward and feedback responses of layer 4Cα of macaque V1.

5.5.2d *Immunity to changes in contrast*

Changes in stimulus contrast have very little effect on the orientation selectivity of cortical cells (Sclar and Freeman 1982; Ramoa *et al.* 1985). Troyer *et al.* (1998) proposed a mechanism for contrast independence. In the model, LGN inputs have an orientation-tuned component that varies with the spatial phase of the stimulus relative to that of the receptive field, and an untuned component that is not sensitive to phase. Excitatory cortical connections between cells with the same phase sharpen the tuned component, while inhibitory connections between cells in antiphase eliminate the untuned component.

5.5.2e *Detection of higher order orientation features*

The response of a cortical cell to an optimally oriented grating is reduced when a grating in another orientation is superimposed on the receptive field (DeAngelis *et al.* 1992). This is cross orientation suppression. On the other hand, the response of a cortical cell to an oriented stimulus is enhanced when a stimulus in another orientation is introduced outside the receptive field and suppressed when the other stimulus has the same orientation. Cells in V1 of the monkey that are tuned to the same orientation engage in mutual inhibition over considerable distances (Nelson and Frost 1978; Knierim and van Essen 1992; Walker *et al.* 1999). This enhances the response in regions where orientation changes, and can therefore be understood as a mechanism for detection of **orientation contrast**. Similarly, human discrimination of changes in orientation of a line is better when surrounding lines are orthogonal rather then parallel (Li *et al.* 2000). These interactions depend on the long-range lateral connection described in Section 5.4.7. Stetter *et al.* (2000) have developed a model of these processes.

Figure 5.22. Anthony Movshon.
Born in Ridgewood, New Jersey in 1950. He obtained a Ph.D. in neurophysiology and visual psychophysics from Cambridge University in 1975. In 1975 he joined the Department of Psychology at New York University, where he is now a Professor. In 1991 he became an Investigator of the Howard Hughes Medical Institute.

Cells with the same orientation preference with aligned receptive fields show mutual facilitation (Nelson and Frost 1985). This can be regarded as an **alignment detection** mechanism used to detect connected edges (Section 4.2.6).

There is physiological and psychophysical evidence for **angle detectors** in the visual cortex. These are specialized detectors tuned to the angle between lines rather than to the orientations of single lines (Sillito *et al.* 1995; Regan *et al.* 1996; Heeley and Buchanan-Smith 1996). Thus, even at an early stage of processing in the primary visual cortex, orientation detectors engage in interactions designed to respond to significant higher-order features of the visual world. In V4 most cells respond better to angles and curves than to simple lines (see Section 5.7.2b).

5.5.3 Spatial-periodicity tuning

Cortical cells are also differentially tuned to the spatial periodicity of a grating within their receptive field (Campbell and Robson 1968). The spatial-frequency tuning function of a cell is the change in firing rate as a function of the spatial frequency of a stimulus grating of fixed contrast. Such cells are said to have a preferred spatial frequency although, strictly speaking, spatial frequency cannot be mathematically specified within a restricted area. The spatial-frequency tuning functions of simple and complex cells in the cat's visual cortex have a full bandwidth at half amplitude of between 0.6 and 1.9 octaves (Movshon *et al.* 1978) (Portrait Figure 5.22). The spatial-frequency channels in the central retina of the cat seem to be spaced at half-octave intervals over a total spatial-frequency range of 2.5 octaves (Pollen and Feldon 1979). Thus, neighbouring channels overlap and about six channels cover the total range of sensitivity.

For the best response, the stimulus must fill the cell's receptive field but not extend beyond it, because stimuli outside the classical receptive field can modify the response of a cell. DeValois *et al.* (1985) found that simple cells in the visual cortex of cat and monkey, on average, responded best when about 2.5 cycles of a grating fell within the receptive field. Complex cells responded best to about 3 cycles. No cells preferred more than 7 cycles within their receptive field. Simple cells, and most complex cells tuned to a narrower range of spatial frequencies had receptive fields with more cycles.

The preferred spatial frequency of cortical cells is essentially the same for different stimulus contrasts (Albrecht and Hamilton 1982; Skottun *et al.* 1986a). The sensitivity of cells to changes in orientation and spatial frequency improves with increases in contrast but reaches a maximum at quite low contrasts (Skottun *et al.* 1987). Parker and Hawken (1985) measured (a) the spatial-frequency tuning function of cells in the visual cortex of the monkey for a high-contrast grating, (b) the change in probability of firing of a cell as a function of a change in spatial frequency and, (c) the sensitivity of cells to a change in spatial phase of a grating. The results are comparable with psychophysical measures of human hyperacuity.

Evidence reviewed in Section 4.5.2b suggests that the central region of the visual field contains about seven distinct spatial-frequency channels.

A non-Fourier stimulus is one not defined by modulation of luminance. Such stimuli are also known as second-order stimuli. For instance, a high spatial-frequency sinewave grating modulated in contrast at a low spatial frequency contains no Fourier components at the modulation frequency, and yet many cortical cells of the cat respond when the phantom spatial frequency falls within their spatial-frequency bandwidth (Zhou and Baker 1994; Marechal and Baker 1998). Responses to other non-Fourier stimuli, such as non-Fourier contours and non-Fourier motion, have been found at higher levels of the visual system. It has been proposed that non-Fourier stimuli are processed in specialized visual channels (Zhou and Baker 1996). There is ample psychophysical evidence that we perceive non-Fourier stimuli, as will become apparent at various places in this book.

5.5.4 Spatiotemporal tuning of cortical cells

5.5.4a *The spatiotemporal transfer function*

The delay between the response of an LGN relay cell and that of its associated cortical cell is about 3 ms and the monosynaptic delay between cortical cells is about 1.5 ms. Many cortical cells show a biphasic response to a brief stimulus. Cortical cells are differentially tuned to temporal features of stimuli falling within their receptive fields. Also, cells in V1 of the monkey show a lower cut off and greater diversity in their temporal-frequency bandwidth than do ganglion cells or LGN cells (Hawken *et al.* 1996). The amplitude transfer function of a cell describes the amplitude of the cell's response as a function of either the temporal or spatial frequency of the stimulus.

The phase transfer function describes the temporal or spatial phase of the cell's response as a function of the temporal or spatial frequency of the stimulus respectively. From the phase transfer function it is possible to infer the latency of a cortical cell and whether it has an odd- or even-symmetric receptive field (Hamilton *et al.* 1989). The amplitude and phase transfer functions together specify the **spatiotemporal transfer function** of a cortical cell. In a linear system, the amplitude and phase of the response is not affected by changes in contrast. While variations in contrast have little or no effect on the shapes of the amplitude and spatial phase transfer functions of cortical cells of cats and monkeys, an increase in contrast causes a temporal-frequency-dependent advance in temporal phase and a shortening of latency (Albrecht 1995).

5.5.4b *Spatiotemporal response of simple-cells*

One can think of a cortical cell as having a spatiotemporal tuning function and a spatiotemporal structure to its receptive field. Plotting the spatiotemporal structure of a receptive field using a single

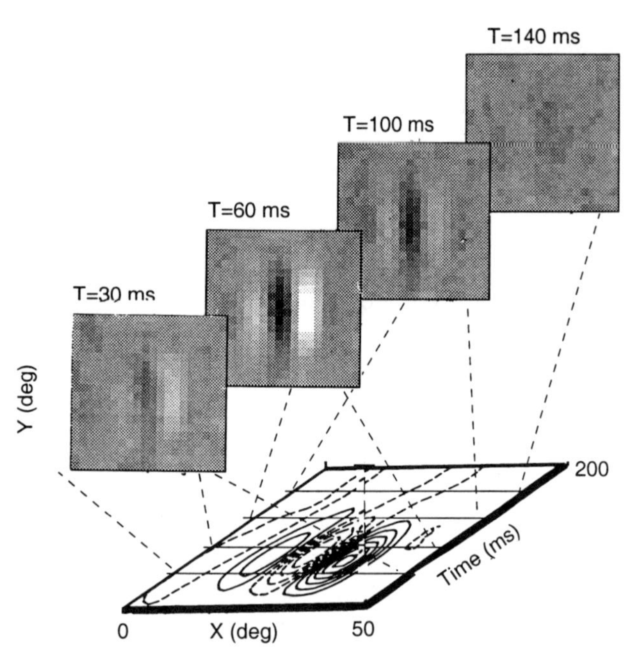

(a) Each panel is the receptive-field profile of the cell at a particular time after stimulus onset. White areas represent responses to a flashed bright bar and dark areas are responses to a flashed dark bar. The projection of the panels produces the spatiotemporal profile of the cell for one spatial dimension over time. Solid lines are boundaries of bright-excitatory, 'on' response regions. Dashed lines are boundaries of dark-excitatory, 'off' response regions.

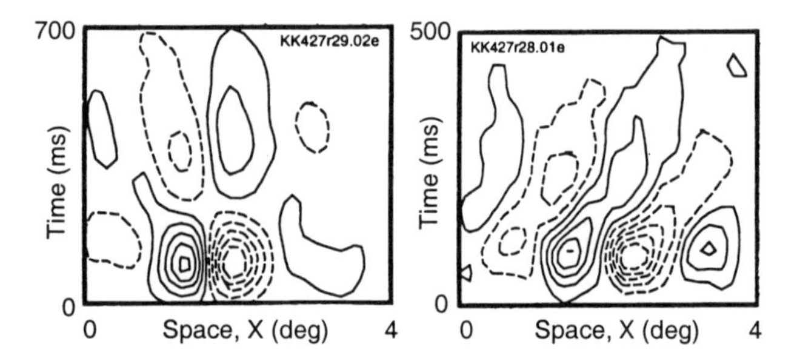

(b) Spatiotemporal response profiles of simple cells from 8-week-old cat. The example on the left shows space-time separability in which the response profile is the product of the space and time profiles. The example on the right shows space-time inseparability. This cell is selectively responsive to motion in the direction of the "tilt" of spatiotemporal regions in the receptive field profile.

Figure 5.23. Spatiotemporal response profiles of simple cells in cat's visual cortex. (Adapted from DeAngelis *et al.* 1993a)

stimulus probe is a time-consuming procedure. The so-called reverse correlation procedure is more efficient (DeAngelis *et al.* 1993a). A continuous series of briefly exposed bar-shaped stimuli is presented. The stimuli occur at different positions within the spatial confines of the receptive field and at different times within the temporal epoch over which the response of the cell persists. The responses of the cell are then cross-correlated backwards in time with the input to yield the spatiotemporal impulse response of the cell. The method works only if the cell behaves linearly, and can therefore be applied only to simple cortical cells.

An example of the spatiotemporal structure of a simple-cell receptive field is shown in Figure 5.23a. Each slice represents the momentary activity within the excitatory and inhibitory regions of the receptive field. The sequence along the time dimension represents the temporal phasing of the response. Figure 5.23b shows the plot of the receptive field along one spatial dimension on the abscissa and over time on the ordinate. In the first example, the positions of the excitatory and inhibitory regions are constant over time, and the cell is said to show space-time separability. The response of the cell is the simple product of its spatial response and its temporal response. Note that the temporal response of this cell is biphasic. In the second example, there is a spatial shift of activity within the receptive field over time. The cell is said to show space-time inseparability (space and time interact). Cells of this type show a preference for stimuli moving in the direction in which their spatiotemporal tuning function is "tilted" in the space-time domain. Cells with space-time separable response profiles are generally not selective for a particular direction of motion. After making allowance for the nonlinearity of the response of simple cells to contrast, DeAngelis *et al.* (1993b) predicted the direction selectivity of simple cells from the spatiotemporal profiles of their receptive fields, indicating that simple cells embody a linear spatiotemporal filter (see also McLean and Palmer 1989, 1994).

5.5.4c *Spatiotemporal response of complex-cells*

The spatiotemporal structure of the receptive fields of complex cells cannot be derived from responses to single spots or bars because the cells are nonlinear. However, responses of complex cells to two spots or bars have been used to examine their spatial frequency tuning and motion selectivity (Movshon *et al.* 1978; Emerson *et al.* 1992). They have also been used to explore the disparity sensitivity of complex cells, as described in Section 6.5.4.

5.5.4d *Models of spatiotemporal responses*

The so-called **energy model** has been used to account for the spatiotemporal responses of cortical cells. In this model, cortical cells summate inputs from two or more subunits, which could be simple cells. Each subunit has a linear spatiotemporal filter followed by static nonlinearities which half rectify and square the output (Adelson and Bergen 1985). In one form of the model, rectification results in one pair of subunits responding to luminance increase (on units) and the other pair to luminance decrease (off units). In each pair, one unit is 90° phase shifted in space and in time with respect to the other, to form a pair in quadrature.

Emerson (1997) produced physiological evidence for a two-subunit power model for simple cells. Gaska *et al.* (1994) tested the model by deriving the second-order Wiener kernels (Section 3.4) from responses to white noise of complex cells in the monkey visual cortex. The kernels predicted the way the cells responded to drifting gratings and the results were consistent with the energy model.

We can conclude that the power model predicts the feedforward responses of many simple and complex cells, such as the way they respond to single bars, double bars, drifting gratings, and white noise. However, the model does not account for how responses of simple and complex cells are modified by neighbouring cells and by feedback from higher centres. For example, the squaring process in the model predicts that the response should increase with the square of contrast, whereas, in fact, responses saturate at high contrast. Heeger (1992a) proposed that the output of complex cells is normalized by the mean response of neighbouring cells. Jacobson *et al.* (1993) concluded that a cascade of linear filtering followed by rectification and squaring provides a more accurate description of simple cells than of complex cells. However, they produced evidence that, even for simple cells, other forms of nonlinearity, such as surround suppression, play a significant role.

Jagadeesh *et al.* (1997) recorded intracellular synaptic potentials in simple cells in the cat visual cortex evoked by a series of stationary sinewave gratings. A linear model applied to the data predicted the direction selectivity of the cells but the tuning width indicated by the synaptic potentials was wider than that indicated by the extracellular action potentials. It thus seems that a nonlinear mechanism between the synaptic input and the output sharpens direction sensitivity. Reid *et al.* (1991) found some evidence of sharpening of directional selectivity of simple cells by nonlinear inhibitory interactions. Removal of intracortical inhibition by chemical suppression of GABA reduced directional selectivity of simple cells in areas 17 and 18 of the cat (Crook *et al.* 1996, 1997).

Other aspects of spatiotemporal coding are discussed in Section 4.3. The relationship between sensitivity of cortical cells and psychophysically determined thresholds is discussed in Sections 4.2.6 and 4.2.8.

5.5.4e *Motion sensitivity of cortical cells*

In submammalian vertebrates, such as frogs, turtles, and birds, cells tuned to direction of motion are found in the retina (Maturana *et al.* 1960; Jensen and Devoe 1983; Maturana and Frenk 1963). Motion-selective cells are also found in the retinas of rabbits and squirrels (Barlow and Hill 1963; Michael 1968). Only a few motion-selective cells have been found in the cat retina (Stone and Fabian 1966) and none has been found in the primate retina (DeMonasterio 1978). Visual evoked potentials from the retina (ERG's) and the cortex of humans reveal that most, if not all, processing of motion occurs at the cortical level (Bach and Hoffmann 2000).

Most cells in the visual cortex respond best when the stimulus moves in a particular direction, the **preferred direction**. The axis of the preferred direction of an orientationally tuned cell is at right angles to the long axis of its receptive field. Thus, the columnar organization of the axes of preferred motion corresponds to the columnar organization of preferred orientation. Some cells respond to movement in either direction along a given axis and are said to be bidirectional. Some cells in the visual cortex of the cat respond over a wide range of stimulus velocities but may be directionally selective over only high or only low velocities. Other cells respond only over low velocities (Duysens *et al.* 1987). Snowden *et al.* (1992) found that motion-sensitive cells in V2 and MT of the alert monkey had directional tuning functions with a half-width of 50° at half-height. Although they responded most vigorously to stimuli moving in the preferred direction, they were most sensitive to changes of motion when the stimulus excited the cell on the flank of its tuning function. It has been proposed that two or three temporal-frequency channels are required to account for human motion sensitivity (Smith and Edgar 1994).

5.5.5 Parvo-and magnocellular areas of V1

In 1978, Margaret Wong-Riley informed Hubel and Wiesel that she had found clusters of cells with a high concentration of cytochrome oxidase in the primary visual cortex of the monkey. Cytochrome oxidase is a metabolic enzyme found in the mitochondrial membrane of neurones. Its concentration is a sensitive indicator of neuronal activity (Wong-Riley 1979a, 1989) (Portrait Figure 5.24). Two years later, Horton and Hubel (1981) investigated the physiological properties of regions with high levels of cytochrome oxidase and found that they are centred on the ocular dominance columns and contain cells not tuned to orientation. These regions are called **blobs**, or puffs. The spaces between the blobs

Figure 5.24. Margaret Wong-Riley.
Born in Shanghai in 1941. She obtained an M.A. in Science Education at Columbia University in 1966 and a Ph.D. in Neuroanatomy with H. J. Ralston at Stanford University in 1970. She conducted postdoctoral work with R.W. Guillery at the University of Wisconsin and A. Lasansky and M. Fourtes at the National Institutes of Health. She was on the faculty of the Department of Anatomy at the University of California at San Francisco from 1973 to 1981. In 1981 she joined the Department of Cellular Biology, Neurobiology and Anatomy at the Medical College of Wisconsin, where she is now Professor. Recipient of Martin Luther King Humanitarian Award, Wisconsin Medical College, 1997.

Figure 5.25. Jonathan C. Horton.
Born in Edmonton, Canada in 1954. He obtained an A.B. from Stanford University and a Ph.D. from Harvard University with D. Hubel and T. Wiesel. In 1990 he joined the faculty at the University of California, San Francisco where he is William F. Hoyt Professor of Neuro-Ophthalmology. He received the Troutman-Véronneau Prize by the Pan-American Association of Ophthalmology in 1999 and a Lew R. Wasserman Merit Award, from Research to Prevent Blindness, N. Y. in 2000.

have a lower concentration of cytochrome oxidase and are known as **interblobs**. The centres of the blobs are about 0.4 mm apart, making about 5 blobs per mm^2 and a total of about 15,500 in the binocular visual field of the macaque (Schein and Monasterio 1987). The density of blobs is about half this value in the monocular visual cortex. The number of parvocellular cells projecting to each blob remains constant at about 110 over the visual field of the monkey. Cytochrome oxidase blobs seem to be present in all primates and form elliptical patches aligned with ocular dominance columns in primates with such columns. They also occur in area 17 of the cat (Murphy *et al.* 1995). They have also been found in autopsy specimens of human brains, where they are organized into rows about 1 mm wide, running at right angles to the 17-18 border (Horton and Hedley-White 1984) (Portrait Figure 5.25).

Cytochrome oxidase is concentrated in cortical layers 4A and 4C—layers associated with the parvocellular system—and in layer 6 which projects to subcortical nuclei, including the LGN (Snodderly and Gur 1995). The literature on cytochrome oxidase in the visual cortex has been reviewed by Wong-Riley (1994). Cells in the blobs and interblobs have the following properties:

1. Most blob cells are not tuned to orientation, while almost all interblob cells are tuned to orientation.
2. The excitatory regions of the receptive fields of blob cells are larger than those of interblob cells (Snodderly and Gur 1995).
3. Colour-opponent ganglion cells converge on so-called double-opponent cells in the cytochrome oxidase blobs (Tootell *et al.* 1988e) (Portrait Figure 5.26). Some blob cells specialize in red/green opponency, others in blue/yellow opponency. These cells produce signals related to local colour contrast, which are independent of overall luminance, since an increase or decrease in luminance affects the inputs in the same way, leaving the difference signal the same (Gouras 1991). Only a few interblob cells show some colour opponency (Livingstone and Hubel 1984). An opponency mechanism typically operates with respect to a resting discharge, which may explain why maintained neural discharges in the dark occur specifically in cytochrome-oxidase regions (Snodderly and Gur 1995). This also explains why cytochrome-oxidase cells show high metabolic activity.
4. Blob cells respond best to gratings with low contrast, low spatial frequency, and high temporal frequency. Interblob cells prefer higher contrast,

Figure 5.26. Roger B. H. Tootell.
Born in San Francisco. He obtained a B.A. in Psychology at the University of California, Santa Barbara in 1975 and a Ph.D. at Berkeley in 1985. He conducted postdoctoral work at Harvard Medical School. He gained an academic appointment in Neurobiology at Harvard Medical School, in 1988. He is now Associate Professor in Radiology at Harvard Medical School.

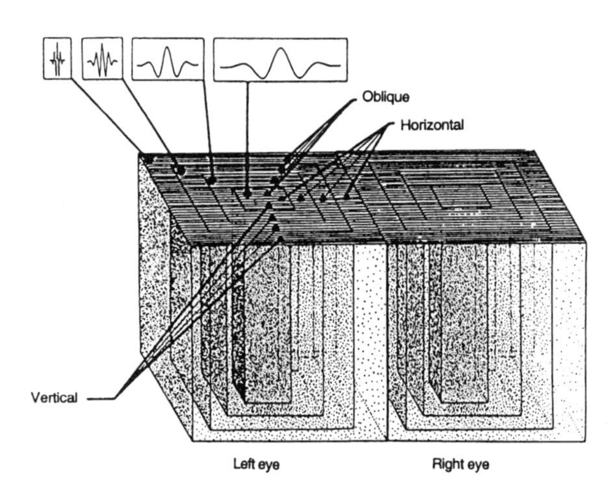

Figure 5.27. Pinwheel organization of primate visual cortex.
Each module (blob plus interblob region) is dominated by one eye. Along radii, cells are tuned to progressively higher spatial frequencies but have the same orientation preference. Orientation preference changes sequentially around the centre of the module. (From DeValois and DeValois 1988)

high spatial frequency and low temporal frequency (Tootell *et al.* 1988b; Shoham *et al.* 1997).

5. In the binocular cortex, blobs occur in rows superimposed on the centres of ocular dominance columns (Horton 1984; Ts'o *et al.* 1990). Cells in these regions respond to excitatory inputs from only one eye. Cells in intermediate regions respond to both eyes. Monocular enucleation in adult monkeys leads to a reduction of cytochrome oxide activity in ocular dominance columns corresponding to that eye (Horton and Hocking 1998a).

6. Blob cells of similar type tend to be laterally connected, as do interblob cells. Blob cells and interblob cells do not interact (Ts'o and Gilbert 1988).

The functional properties of cells in the blob and interblob regions are not sharply segregated; border cells can show both colour and orientation specificity (Ts'o and Gilbert 1988). Also, the contrast sensitivity and spatial-frequency tuning of cortical cells change gradually between blob and interblob regions (Edwards *et at.* 1995).

In the monkey, blobs receive inputs from the magnocellular layers of the LGN routed through cortical layers 4Cα and 4B and from parvocellular layers routed through cortical layers 4Cβ and 4A. The interblob regions receive parvocellular inputs through layers 4Cβ and 4A and mixed magno-and parvocellular inputs from cells in the middle of layer 4C (Lachica *et al.* 1992, Yoshioka *et al.* 1994). Both regions therefore receive mixed magnocellular and parvocellular inputs.

There are two views about the significance of the blob-interblob division of the parvocellular system. According to Livingstone and Hubel, it represents a dichotomy into a channel specialized for colour but not form and a channel specialized for colourless form. DeValois and DeValois (1988) proposed a different scheme in which the primary distinction between blobs and interblobs is in spatial-frequency tuning (see also DeValois 1991). This scheme is illustrated in Figure 5.27. Blobs contain cells tuned to low spatial-frequency stimuli (coarse detail) and the interblobs contain cells tuned to high spatial-frequency stimuli (fine detail), with a gradual transition between the two, rather than a dichotomy. They argued that the predominance of colour-coded cells in the blobs is a simple consequence of the fact that parvocellular cells operate as colour-opponent cells for low spatial frequency stimuli and as luminance-contrast cells for high spatial-frequency stimuli. This accords with the analysis provided by Ingling (Section 6.3.5). They also argued that the finer tuning for orientation in the interblobs is a consequence of the fact that cells tuned to higher spatial frequencies also have narrower tuning for orientation. The two views both end up with a similar division of cell types, but the DeValois' scheme makes this division consequent on a fundamental segregation of cells according to spatial-frequency tuning.

5.5.6 Contextual and figural responses

5.5.6a *Responses to texture-defined boundaries*

The response of a cell in the primary visual cortex is modulated by stimuli falling outside the receptive field as mapped by isolated stimuli. For example, in the primary visual cortex of the alert monkey, the response of a cell to a particular stimulus tends to be suppressed by the addition of a textured surround. The degree of suppression increases with increasing density of the surround (Knierim and Van Essen 1992). Furthermore, the response of some cells in the cat's visual cortex to an optimal stimulus centred on the receptive field is larger when surrounding lines outside the normal receptive field of the cells have a contrasting orientation or spatial frequency (Nelson and Frost 1978; DeAngelis *et al.* 1994). The orientation-specific surround effect took only 8 to 20 ms to develop (Knierim and Van Essen 1992). DeAngelis *et al.* found that, for some cells, the inhibitory effect of similarly oriented stimuli could be evoked dichoptically. This suggests that it depends on intracortical inhibitory connections. This mechanism could provide a physiological basis for the perceptual segregation of textured regions, since the response of cells of this type would be highest for texture boundaries (Section 18.1.3).

5.5.6b *Figure-ground segregation*

Zhou *et al.* (2000) found that, for about half the cells in V2 and V4 and some cells in V1 of the alert monkey, the response to an edge or line was modulated according to whether the edge or line belonged to one or other side of a figure. Some of these cells were also selective for the contrast polarity of an edge. Modulation of responses by border-ownership emerged in under 25 ms, which suggests that it does not depend on feedback from higher centres. It seems that integration of figural information starts at these early levels in the nervous system.

Zipser, *et al.* (1996) found that the response of cells in V1 of alert monkeys to a textured region filling the receptive field varied according to whether the region was differentiated from its surroundings by disparity, colour, luminance, or orientation. Response modulation had a latency of 80-100 ms, long enough to allow feedback from higher centres.

Lamme (1995) reported that neurones in V1 of the macaque monkey respond more vigorously to texture elements belonging to a figure than to the same elements belonging to a ground region. The response of cells to a textured display filling the receptive field was modulated by the colour, motion, or orientation of surrounding texture. It was also claimed that responses are modified by the disparity of surrounding elements. Lee *et al.* (1998) proposed that the late (>100 ms) part of the response of V1 neurones is influenced by feedback from higher visual centres responsible for figure-ground segregation. The early component of response of V1 neurones occurred whether or not the monkey perceptually registered the stimulus object (as indicated by saccadic refixation) but the late response was absent when the object was not registered. The late response was also absent in anaesthetized monkeys (Supér *et al.* 2001).

This latter finding may explains why Hupé *et al.* (2001) did not observe feedback in V1 with a 100 ms delay in anaethetized monkeys. They did find that the first 10 ms of response of V1 neurones to a moving bar was affected when MT was inactivated by cooling. However, one need not conclude that feedback from MT occurs with such short latency, because the bar was already moving before it entered the receptive field of the cell being recorded.

Cumming and Parker (1999) could not replicate the Lamme's modulating effect of the disparity of the surround and Rossi *et al.* (2001) found that cells in V1 responded only when a texture boundary fell in the receptive field. They concluded that V1 neurones signal texture boundaries but do not distinguish between figure regions and ground regions.

The response of cells to an even area of illumination varies with the illumination of the surrounding area in a manner analogous to changes in the perceived brightness of a gray patch as the luminance of the surround is varied (Rossi and Paradiso 1999).

The MRI response of the human lateral occipital area adapted when the same shape was presented with or without occluding contours, but did not adapt when the perceived shape of an ambiguous stimulus changed (Kourtzi and Kanwisher 2001). In other words, the area responded to perceived figure-ground features rather than to the unprocessed pattern of stimulation.

5.5.7 Effects of attention and learning

The response properties of cortical cells are usually investigated in anaesthetized animals. Such a procedure cannot detect effects due to changes of attention, or motivation. A growing body of evidence from work on unanaesthetized animals indicates that these factors modify the responses of cortical neurones. The concept of sets of cortical cells tuned to specific and fixed stimulus features must give way to a view of the cortex as a highly flexible organ in which the response characteristics of cells are conditional on simultaneous activity in other centres in the brain and on feedback from higher to lower centres. Neural activity in different parts of the brain that underlies a particular percept may be linked by response synchronization (see Section 4.3.2). Feedback from higher to lower centres must also be involved. There are four types of attention enhancement of the responses of cortical cells. The first results from general arousal, the second is location specific, the third is stimulus specific, and the fourth results from learning.

5.5.7a *Effects of general arousal*

The activity of relay cells in the LGN is enhanced in states of general arousal (Section 5.3.2). Livingstone and Hubel (1981) found that cells in the primary visual cortex of the awake cat generally show a reduced spontaneous firing rate and an enhanced response to visual stimuli compared with when the animal is asleep. Some cells in V1 and V2 respond more vigorously before saccadic eye movements but not specifically to eye movements to stimuli in the cell's receptive field (Robinson *et al.* 1980; Moran and Desimone 1985). In other words, the enhanced response at the level of the LGN and early stages of cortical processing is not location-specific or feature-specific but reflects a change in general arousal.

5.5.7b *Location-specific attention*

Increased response to stimuli to which monkeys are redirecting their attention has been noted in subcortical structures such as the superior colliculus (Robinson and Kertzman 1995) and the pulvinar in the thalamus (Robinson and Petersen 1992).

Responses of cortical cells, also, are influenced by the locus of attention, even though there is no change in the direction of gaze. Thus, a cell in V1 and V4 of the rhesus monkey responds more strongly to a grating when the animal attends to the grating rather than to a stimulus outside the cell's receptive field. The tuning curve of the cell is not affected by a change in attention (McAdams and Maunsell 1999). For a given cell, attention causes a proportional increase in its responses to stimuli in all orientations. This suggests that attention produces a multiplicative scaling of the response of a cell (Maunsell and McAdams 2000). Cells in V4 also increase their response to a stimulus when it is near a second stimulus to which the animal is attending, and the effect depends on the distance between the two stimuli (Connor *et al.* 1996).

Motion-sensitive cells in MT and MST of the monkey respond more vigorously when the animal is attending to the moving stimulus (Treue and Maunsell 1996). Many cells in the parietal lobe of the monkey respond more vigorously to a stimulus when the animal is attending to it, reaching for it, fixating it, or is about to make a saccadic eye movement towards it (Lynch *et al.* 1977; Bushnell *et al.* 1981; Robinson *et al.* 1978, 1995). Similar cells exist in the prelunate and frontal lobes (Fischer and Boch 1981; Wurtz *et al.* 1980) and, for many of these cells, the stimulus can be either a visual or auditory target in a given location (Vaadia *et al.* 1986).

In all these cases, the enhanced response is related to the location of the stimulus rather than to its identity. The PET scan has revealed that attention to a particular place or a particular object produces activation of the parietal and frontal lobes in humans (Fink *et al.* 1997). Also, in the object recognition pathway of the human brain, the fMRI response is stronger when attention is directed to an object set among neighbouring objects (Kastner *et al.* 1998). The fMRI response in V1, V2, and V3 was enhanced in humans while they were making difficult stimulus discriminations (Ress *et al.* 2000). The enhancement was related to the location of the stimuli and the difficulty of the task but persisted during intervals between stimulus presentations.

5.5.7c *Stimulus-specific attention*

Attention-selective processes in V1 and V2 may be related to shape discrimination rather than to attention to a single stimulus for purposes of fixating it or reaching for it. When monkeys were required to select a test bar in a particular orientation from among bars in other orientations, cells in V1 and V2 showed enhanced response to the bar (Motter 1993). Enhancement did not occur when monkeys attended to an isolated bar. Also, in V1 of the monkey, response to a curved line that the animal was tracking was stronger than that to an intersecting curve that was not being tracked (Roelfsema *et al.* 1998).

Attentional gating related to the identity of the stimulus has also been revealed in area V4 of the monkey (Moran and Desimone 1985). Cells responded more vigorously when the colour or luminance of a bar in the receptive field of the cell matched the colour or luminance that the animal had been trained to select in a discrimination task (Motter 1994). Cells in this area have receptive fields between 2 and 4° wide and respond to colour and spatial attributes of stimuli. When effective and ineffective stimuli were presented simultaneously within a cell's receptive field, the cell responded only when the animal attended to the effective stimulus. When the animal attended to the ineffective stimulus, the response to the effective stimulus was suppressed. An ineffective stimulus presented outside the cell's receptive field had no power to inhibit responses. In further experiments, two different effective stimuli were presented in the same receptive field of cells in V2 or V4. The cells responded most strongly to whichever stimulus the monkey attended to. Thus, a signal presumably arising from higher in the nervous system, determined which of two stimuli gained access to cells in V2 and V4 (Luck *et al.* 1997; Reynolds *et al.* 1999).

A process of attentional gating was also revealed in the inferior temporal cortex. Many cells in this area have receptive fields covering the whole retina. The response to an effective stimulus was inhibited when the animal attended to any ineffective stimulus—a stimulus to which that cell did not respond. Cells in MT of the monkey responded more vigorously to a moving stimulus when, in a discrimination task, the animal selected a stimulus moving in the preferred direction of that cell (Britten *et al.* 1996). Also, cells in MT or MST responded more vigorously when a monkey attended to a stimulus moving in the receptive field of a cell than when it attended to a stimulus outside the cell's receptive field, even though the two stimuli were identical (Treue and Maunsell 1996). Attention had a larger effect on the response of cells in the ventral intraparietal cortex (VIP) (Cook and Maunsell 2002).

When humans attended to the relevant feature in a detection task, they were better able to detect a slight change in colour, motion, or shape of a set of similar objects. At the same time, the region of extrastriate cortex showing most activity in a PET scan varied according to which feature was being attended to. Attention to shape activated the ventromedial occipital region, and attention to movement activated the inferior parietal lobe, a region corresponding to monkey MT (Corbetta *et al.* 1990).

Frith and Dolan (1997) reviewed the brain areas responsible for modification of responses according to stimulus familiarity. Mechanisms of visual attention are reviewed in Kastner and Ungerleider (2000).

5.5.7d *Learning-specific effects*

Receptive fields of cells in the somatosensory and auditory cortex undergo considerable reorganization following either intracortical microstimulation or sensory stimulation (Dinse *et al.* 1990b; Recanzone *et al.* 1993). Within the primary visual area of the cat, long-term changes in synaptic conductivity along lateral pathways have been induced by pairing synaptic responses with conditioning shocks of depolarizing current (Hirsch and Gilbert 1993) (Section 5.4.7). This suggests that these pathways are involved in stimulus-dependent changes in cortical responses. McLean and Palmer (1998) paired stimuli of specific orientation or spatial phase with iontophoretic application of glutamate or GABA to the visual cortex of cats. This produced long-term increases or decreases in the response of cortical cells to stimuli close to the orientation or spatial phase of the conditioning stimuli.

Visual stimuli having behavioural relevance as a result of learning evoke stronger responses in cortical cells. Thus, about one-third of the cells in the visual cortex of the monkey increased their response by about 20% when the stimulus was one that the animal had been trained to recognize (Haenny and Schiller 1988). A similar response increment plus a narrowing of orientation tuning was shown by about three quarters of cells in area V4. Zohary *et al.* (1994a) found a 13% increase in sensitivity of motion-sensitive cells in MT and MST of the monkey associated with a 19% improvement in the ability to discriminate directions of visual motion. Presumably, these feature-specific effects are due to feedback from higher-order visual systems. Ablation of area V4 in monkeys produced severe deficits in their ability to select less prominent stimuli from an array, and in their ability to generalize discrimination learning to new stimuli (Schiller and Lee 1991). Some cells in the inferior temporal cortex of monkeys responded more vigorously when a feature of the stimulus, such as colour, was one to which the animal had to attend to in order to solve a discrimination task (Fuster and Jervey 1981). Cells in the so-called frontal eye fields in the frontal lobes normally respond to any stimulus to which the animal is about to make a saccadic eye movement. In monkeys trained to make eye movements to targets of only one colour, cells in the frontal eye field became selectively responsive to stimuli of that colour (Bichot *et al.* 1996). Cells in the prefrontal cortex of alert monkeys that responded in the period between presentation of a stimulus and the monkey's response to the stimulus did so in a manner that varied with the type of reward received (Watanabe 1996). See Buonomano and Merzenich (1998) for a review of cortical plasticity.

A variety of simple visual tasks, such as vernier acuity and orientation discrimination improve with practice (Section 3.2.6). Ghose *et al.* (2002) found no changes in monkey V1 or V2 that could account for improved orientation discrimination with practice. Improvement of stereoacuity with practice is discussed in Section 19.12.3. Cortical plasticity in development is discussed in Section 10.4.5.

5.6 CORTICAL COLUMNAR ORGANIZATION

Lorente de Nó (1949) was the first to propose that the cerebral cortex is organized into columns. He studied patterns of synaptic linkages anatomically and showed that they run predominantly vertically from layer to layer with fewer connections running horizontally. Sperry *et al.* (1955) provided the first functional evidence of columnar organization. He found that vertical slicing of the visual cortex produced little or no effect on the ability of cats to perform fine visual discriminations. Mountcastle (1957) produced the first electrophysiological evidence of columnar organization by recording from single cells in the somatosensory area of the cat. He wrote,

"*. . . neurons which lie in narrow vertical columns, or cylinders, extending from layer II through layer IV make up an elementary unit of organization, for they are activated by stimulation of the same single class of peripheral receptors, from almost identical peripheral receptive fields. Columnar organization is a fundamental property of the whole cerebral cortex.*" (Mountcastle 1997).

The cortex develops by columnar growth from progenitor cells lining the embryonic ventricles (see Chapter 10). This lining contains a protomap of the prospective cytoarchitectonic areas (Rakic 1988).

5.6.1 Column topology

The cells in each small column of tissue running at right angles to the cortical surface of the primary visual cortex have the same tuning to stimulus orientation, although there is sometimes a reversal of orientation tuning at layer 4 (Dow 1991). These columns of similarly tuned cells are **orientation columns**. In a single column there are thousands of cells with similar orientation preference. In the cat and monkey, the orientation preference of cells rotates smoothly through its full range of 180°, as one traverses across the cortical surface through a distance of 0.5 to 1.0 mm. However, there is some random variation of orientation tuning within each column so that a random variation in tuning is superimposed on the smooth transition of tuning between columns (Hetherington and Swindale 1999).

Each cell in the visual cortex is also selectively tuned to a particular range of spatial frequencies of a grating that falls within its receptive field (Section 4.2.6). There is some dispute about the columnar organization of cells with the same tuning to spatial-frequency.

Cells in each column of the visual cortex, with a predominant input from one eye are segregated from those in neighbouring columns with a predominant input from the other eye. The two types of column form alternating stripes over the surface of the cortex known as **ocular dominance columns**.

Thus, within each column, cells have similar tuning for position, orientation, spatial frequency, and ocular dominance. They also have moderately similar temporal properties, including preferred temporal frequencies and response latencies. However, the simple cells in cortical columns of the cat visual cortex have spatial phases that vary between 90 and 180° (Pollen 1981; DeAngelis *et al.* 1999). This could allow for economy of linkages between cells in phase quadrature, as described in Section 4.5.2c. It could also allow for the formation of complex cells with responses independent of stimulus position within the receptive field, and for noise reduction.

Hubel and Wiesel (1974b) favoured the view that orientation bands are orthogonal to the ocular dominance bands. However, this so-called "ice-cube" picture of columnar organization had to be revised when it became evident that columns containing cells tuned to a particular orientation do not form continuous bands but are interspersed with blobs containing cells not tuned to orientation. In **linear zones**, iso-orientation contours run in parallel over distances of between 0.5 and 1 mm and orientation preference changes linearly with distance across the contours. The linear zones are interrupted by a variety of nonlinear zones. At ridge-shaped **fractures**, orientation preference changes abruptly by less than 90°. At **singularities**, orientation preference changes by more than 90°, either increasing with clockwise rotation or with counterclockwise rotation. At **saddle points** orientation preference rotates clockwise in either direction from one axis and counterclockwise from the orthogonal axis (Blasdel 1992b; Obermayer and Blasdel 1993).

It has been suggested that the centre of each singularity is occupied by a giant Meynert cell (von Seelen 1970; Braitenberg and Braitenberg 1979). Meynert cells are solitary giant pyramidal cells in layer 5 of the visual cortex that send richly branched apical dendrites into cortical layers 1 and 2, and other dendrites into layers 5 and 6. These are the layers in which intracortical and subcortical connections originate (Chan-Palay *et al.* 1974).

Singularities of orientation preference tend to occur along the central axis of each ocular dominance column and coincide with peaks of ocular dominance (Crair *et al.* 1997). Cytochrome oxidase blobs also tend to occur along the same axes but are not coincident with the singularities (Bartfeld and Grinvald 1992). Linear iso-orientation contours tend to be orthogonal to the boundaries between adjacent ocular dominance columns (Blasdel *et al.* 1995). Fractures tend to be either parallel to or orthogonal to ocular dominance bands (Blasdel and Salama 1986).

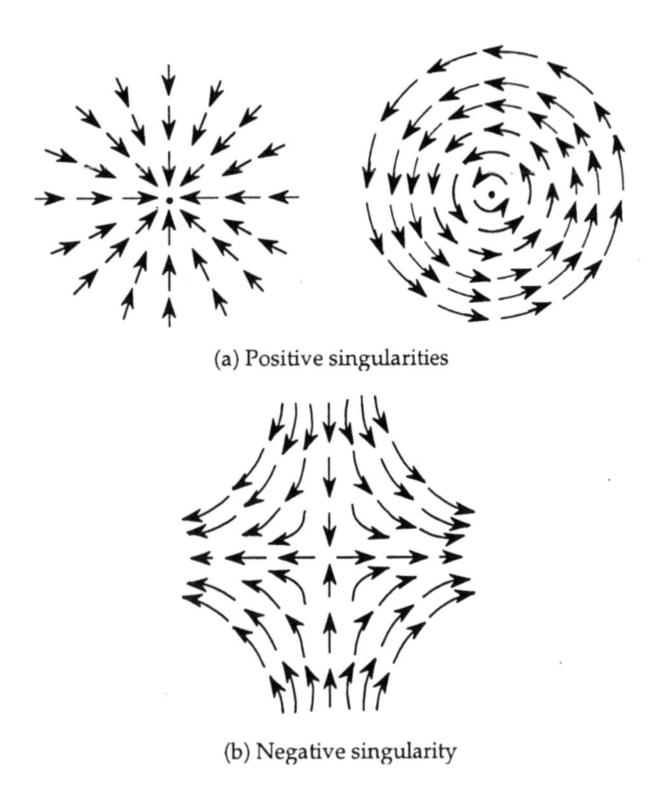

Figure 5.28. Types of singularity formed by patterns of vectors

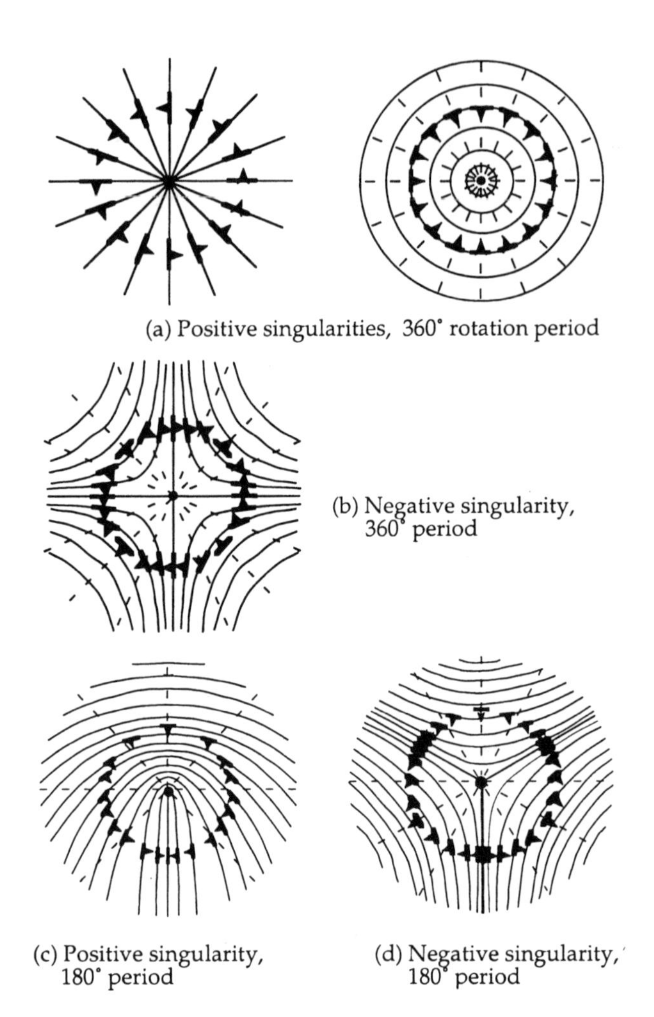

Figure 5.29. The topology of cortical columns.
Lines represent loci of receptive-field alignment. Dotted lines represent loci of equal orientation preference. Bars with an arrow represent the orientation and direction preference of a cell.

There has been some disagreement about the cortical representation of spatial frequency. According to the results of one study using optical imaging in the visual cortex of the cat, cells preferring high spatial frequency (X pathway) are segregated from those preferring low spatial frequency (Y pathway). Regions coding low spatial frequency tended to occur in the centres of ocular dominance columns (Hübener *et al.* 1997).

According to another imaging study on the cat, spatial-frequency domains, like orientation domains, are organized continuously in pinwheel patterns (Everson *et al.* 1998). Issa *et al.* (2000) produced a finer map of spatial-frequency domains in the visual cortex of the cat by combining imaging with microelectrode recording. They found a complete and continuous representation of spatial frequency within each hypercolumn (0.75 mm). The low and high extrema of the range tended to occur in distinct pinwheel singularities for orientation preference. They found no significant relationship between low spatial-frequency domains and ocular dominance columns. They argued that placement of spatial frequency extrema on the centres of orientation pinwheels ensures that all orientations are represented at these rare but important spatial frequencies. The less extreme spatial frequencies are continuously mapped round the orientation pinwheel.

The preferred direction of motion of a cortical cell is orthogonal to its preferred orientation. The combined orientation/direction preference of a cell can thus be represented by a directed line element (a vector). Maximum continuity of both orientation and direction preference would be achieved if, round each discontinuity, orientation preference cycled twice through its 180° range of values, while direction preference cycled once through its 360° range of values. In the scheme depicted in Figure 5.27 on page 205, orientation/direction preference changes in an ordered sequence round the centre of the singularity, so that cells with a particular orientation/direction preference radiate out from the centre to form a pinwheel pattern. Cells tuned to progressively higher spatial frequencies are arranged out from the centre and different blobs have different chromatic properties (DeValois and DeValois 1988).

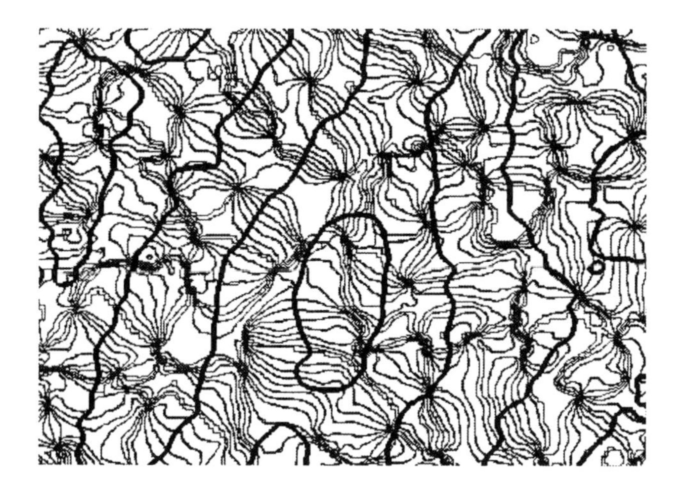

Figure 5.30. Cortical columns of monkey V1.
Iso-orientation lines (gray) are drawn at intervals of 11.25°. Black lines are borders of ocular dominance bands.
(From Obermayer and Blasdel, Copyright 1993 by the Society of Neuroscience)

Theoretically, the pinwheel pattern is only one of three patterns that can be formed from a vector field by continuous deformations involving a cyclic periodicity of 360°. Concentric and radial patterns define positive singularities, and a hyperbolic pattern defines a negative singularity, as in Figure 5.28. Consider a stimulus vector (directed line element) moving clockwise over a circular path round a singularity in a plane. If the orientation/direction preference of the cells that the vector encounters also changes clockwise then the pattern of preference forms either a radial or concentric pattern, as in Figure 5.29a. Iso-orientation loci and loci of stimulus alignment coincide in the radial pattern but are orthogonal in the concentric pattern. If stimulus preference changes counterclockwise, the hyperbolic pattern in Figure 5.29b is formed (Penrose 1979). In this pattern, iso-orientation loci coincide with loci of stimulus alignment only along cardinal directions. These are the only types of singularity with continuous transformations of vectors. However, we shall see that cortical patterns contain a discontinuity.

For cells tuned to a particular orientation, those tuned to motion in one direction are spatially segregated from those tuned to motion in the opposite direction (Bonhoeffer and Grinvald 1993). Thus, orientation preference cycles through its 180° range only once round each singularity while direction preference changes through only half its 360° range. A second orientation singularity is required to cover the other half of the range of direction preference.

Since direction preference does not change round an orientation singularity, the periodicity of a cyclic change in orientation preference becomes 180° rather than 360°. Let a stimulus vector move clockwise over a circular path round a singularity and encounter cells for which orientation preference also changes clockwise through 180°. This generates the loop pattern shown in Figure 5.29c. The tri-radius pattern shown in Figure 5.29d is generated when the cyclic path and the changing orientation preference have opposite signs. Finger prints, Zebra stripes, and stripes on several species of fish conform to this topology. If a loop is assigned a value of +1 and a tri-radius pattern a value of -1, a cyclic path that contains one of each type of singularity has a value of 0. In general, the value of any circular path over the plane is the algebraic sum of the values of the singularities contained within the cyclic path. In both the loop and tri-radius patterns all the iso-orientation loci are radial but the stimulus-alignment loci are radial only at certain points.

Electrophysiological recordings from single cells distributed in a grid over the cat's visual cortex have revealed a cyclic organization of orientation preference (Swindale *et al.* 1987; Dow 1991). Cyclic patterns of orientation-selectivity have been revealed more directly by the optical imaging of light reflected from the cortical surface (Ts'o *et al.* 1990). Blasdel (1992b) used optical imaging to reveal both loop and tri-radius patterns of orientation preference in V1 of the monkey. Regions responding to a grating presented in a particular orientation have higher absorption of red light because of the presence of deoxyheamoglobin. Bonhoeffer and Grinvald (1993) used this procedure to produce detailed maps of orientation preference in area 18 of the cat. Weliky *et al.* (1996) also used optical imaging and confirmed that orientation preference cycles once round each singularity with a 180° discontinuity in direction preference, and that two singularities are required to cover the full range of directions preference.

Thus, iso-orientation domains are subdivided into patches selective for motion in opposite directions. There is some tendency for orientation singularities to occur in pairs with opposite signs and to be connected by iso-orientation bands running at right angles to the boundary between the two ocular dominance columns that contain them (Bartfeld and Grinvald 1992; Blasdel and Salama 1986). Connections between orientation singularities of the same sign form saddle points. Figure 5.30 shows loci of iso-orientation preference superimposed on ocular dominance columns derived from optical imaging (Obermayer and Blasdel 1993).

It has been generally believed that cells at the centre of orientation singularities are not tuned to orientation. However, detailed probing within the

Figure 5.31. Nicholas Swindale.
Born in Edinburgh in 1951. He obtained a B.A. in Natural Sciences at Cambridge in 1972 and a D.Phil. with P. Benjamin in Neurobiology at Sussex University in 1976. He did postdoctoral work with C. Blakemore and H. Barlow at the Physiological Laboratory, Cambridge. He was Assistant Professor in the Department of Psychology and Physiology at Dalhousie University from 1984 to 1988. HCurrently, he is Associate Professor in the Department of Ophthalmology at the University of British Columbia, Canada.

singularities has revealed that they contain as many cells sharply tuned to orientation as do iso-orientation regions (Maldonado *et al.* 1997). The apparent lack of orientation tuning arises when responses are averaged over the singularity.

Alan Turing (1952) demonstrated theoretically that diffusion of activator and inhibitor chemical morphogens between idealized cells in a growing organism results in a variety of static or oscillatory patterns depending on the relative rates of diffusion and reactivity of the morphogens (see also Peng *et al.* 2000). These Turing patterns have recently been demonstrated in an actual medium and resemble the columnar and blob-interblob patterns of V1 and V2 (Kapral and Showalter 1995). Swindale (1980, 1982) (Portrait Figure 5.31) developed a model of the development of cortical columns based on short facilitatory and long inhibitory neural or chemical interactions between cortical cells of the same type and opposite interactions between cells of the opposite type (see also Linsker 1986). In a development of the model, orientation singularities tended to become centred in ocular dominance columns if plasticity of selective tuning is turned off first within ocular dominance columns (Swindale 1992). A neural network model of the cyclic organization of orientation tuning in the visual cortex has been described by Costa (1994).

The radial, concentric, and hyperbolic patterns, together with linear grids, form the family of differential operators (orbits) of Lie groups (Section 3.8). Gallant *et al.* (1993, 1996) found that some cells in area V4 of the monkey respond selectively to one or other of the Lie operators, although not selectively to linear grids. The response of these cells was largely invariant over changes in the position of the stimulus (Section 5.7.2b). The receptive fields of the specifically tuned cells must be built up from cells in V1 or V2, and their construction is presumably helped by the patterns of orientation and direction preference exhibited in these areas.

The topology of the visual cortex reconciles two competing requirements. First, that those cells coding a complete range of stimulus features be packed efficiently in each cortical region. Second, that cells with similar stimulus preferences occur as neighbours so as to economize on connections and facilitate the formation of receptive fields of cells higher in the visual system. The topology of ocular dominance columns is discussed in Section 5.6.2c.

Several computer models of the topology of the visual cortex have been developed. They have been reviewed by Tanaka (1991) Erwin *et al.* (1995), and Swindale (1996).

5.6.2 Ocular dominance columns

5.6.2a *Mapping ocular dominance columns*

Ocular dominance columns were first revealed by recording from single cortical cells as the electrode was moved over the cortical surface. They were then revealed anatomically in the monkey by the Nauta method, which involves tracing areas of cell degeneration in the cortex produced by selective destruction of LGN cells arising from a given eye (Hubel and Wiesel 1969) and by silver staining (LeVay et al. 1975).

Grafstein and Laureno (1973) showed that a mixture of tritiated proline and fructose injected into the eye of a living mouse gradually travels up the axons of the optic nerve through the LGN and becomes concentrated in cells in layer 4 of the contralateral visual cortex. The resulting patterns of radioactivity in thin sections of cortical tissue can be recorded on film to produce autoradiographs. Wiesel *et al.* (1974) used the same method to reveal the ocular dominance columns in the *Macaque* monkey. In a related procedure, deoxyglucose labelled with carbon 14 is injected into an animal, which is then exposed to stimuli presented to one eye for an hour

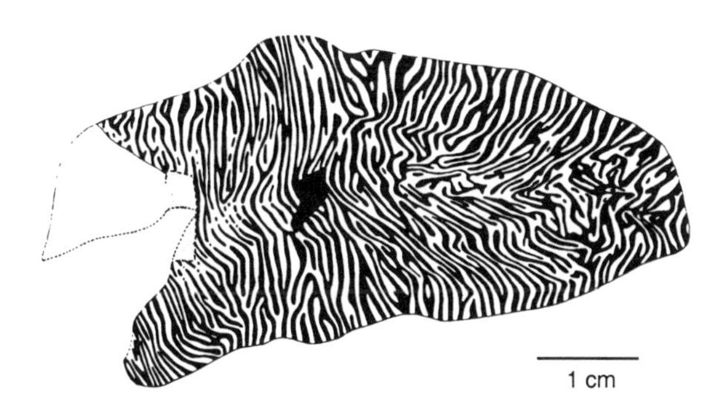

Figure 5.32. Ocular-dominance bands of monkey visual cortex. Pattern derived from a series of autoradiographs and transposed onto a 3-D map of the visual cortex. (From LeVay *et al.* Cpyright 1985 by the Society of Neuroscience)

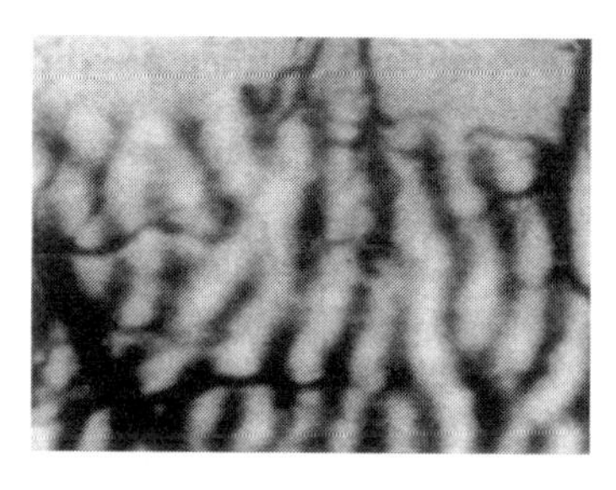

Figure 5.33. Photograph of ocular dominance bands. Taken from the visual cortex of a living monkey. Dark bands are dominated by the left eye. The boundary between V1 and V2 is across the top of the picture. Ocular dominance bands do not cross this boundary. (Reprinted with permission from Ts'o *et al.* 1990 Copyright American Association for the Advancement of Science)

or more. The labelled glucose compound is taken up by cortical neurones responding to the stimulated eye but not by inactive neurones related to the closed eye (Sokoloff *et al.* 1977).

Cells taking up radioactive tracer in the hemisphere contralateral to the injected eye are those that normally receive a strong excitatory input from that eye. Since cells with balanced ocular dominance also receive excitatory inputs from the ipsilateral eye, the dark bands produced by injection of one eye must overlap those produced by injection of the other eye.

The separate slices of an autoradiograph can be combined by computer reconstruction into a complete pattern of columns. Using data on the topography of the visual cortex, the pattern of ocular dominance columns can be transposed onto a 3-D map of the visual cortex, as shown in Figure 5.32.

Dyes sensitive to voltage changes associated with neural impulses may be infused into the cortex of an alert animal exposed to stimulation of one eye, and the cortical surface scanned by a video recorder (Blasdel 1992a). Changes in cortical reflectivity can also be photographed without the use of dyes. Figure 5.33 shows an example of an *in vivo* photograph of ocular dominance columns obtained by Ts'o *et al.* (1990). Ocular dominance columns have been revealed in post-mortem sections of human striate cortex (Hitchcock and Hickey 1980; Horton *et al.* 1990) and in fMRI recordings from the living human cortex (Menon *et al.* 1997; Grinvald *et al.* 2000).

Ocular dominance columns may also be mapped by suturing or removing one eye and, some days later, preparing slides of the visual cortex stained for natural products of neural activity. These include cytochrome oxidase (Florence and Kaas 1992; Tigges *et al.* 1992; DeYoe *et al.* 1995), GABA (Hendry *et al.* 1994), protein kinase (Hendry and Kennedy 1986), or mRNA transcription molecules that regulate gene expression (Chaudhuri *et al.* 1995).

The development of ocular dominance columns is discussed in Section 10.3.5.

5.6.2b *Properties of ocular dominance columns*

Hubel and Wiesel (1959) working with the cat, provided the first physiological evidence of convergence of inputs from the two eyes onto the same cortical cells. In the macaque monkey, they found that each local group of cells in layer 4C receives excitatory inputs from only one eye. More recent evidence from alert moneys suggests that many cells in layer 4C receive inputs from both eyes, rather than from only one (Snodderly and Gur 1995). Ipsilateral inputs may derive from pyramidal cells in layer 6 (Wiser and Callaway 1997). Signals are relayed to cells in other layers in the same vertical column of cortical tissue. Most of these cells also receive inputs from the other eye from cells in layer 4C in a neighbouring column (Hubel and Wiesel 1962).

Cells that receive inputs from both eyes are known as binocular cells. The binocular cells in each local column of cortical tissue receive an excitatory input predominantly from one eye. These are ocular dominance columns Similar columns form bands of tissue running through the thickness of the cortex known as ocular dominance bands, although they are also referred to as columns. Left-eye bands alternate with right-eye bands. The projections of the ocular dominance bands on the cortical surface are known as ocular dominance stripes.

In the visual cortex of the cat, in a region corresponding to a strip extending 12° on either side of the vertical midline, about 90% of complex cells and 70% of simple cells are binocular. Only layers 4 and

6, which receive inputs from the LGN, contain appreciable numbers of monocular cells. Beyond 12° on either side of the midline, contralateral inputs gradually become more numerous than ipsilateral inputs until the monocular crescent is reached (Berman *et al.* 1982).

Each binocular cell has two receptive fields, one in each eye, with similar oculocentric positions. The two receptive fields are similar in their tuning to orientation and spatial frequency, and many have similar tuning to direction-of-motion. They also have the same length-summation, end-stopping, and simple or complex characteristics (Hubel and Wiesel 1962; 1968; Maske *et al.* 1984; Skottun and Freeman 1984; Hammond and Fothergill 1991). The orientation preferences of monocular cells are isotropically distributed, while those of binocular cells, especially in the central retina, tend to cluster about horizontal and vertical meridians (Payne and Berman 1983).

Binocular cells that respond to excitatory inputs from either of the two eyes do so with similar short latency. This suggests that they receive direct inputs from layer-4 cells of both the ipsilateral and contralateral eyes, rather than a direct input from one eye and an indirect input from the other (Ito *et al.* 1977). Not only is the latency of excitatory inputs to a binocular cell the same, but so is the complete pattern of synaptic inputs from the two eyes, involving monosynaptic excitatory inputs and polysynaptic excitatory and inhibitory inputs (Ferster 1990). The issue of differential response latencies is discussed in more detail in Section 28.1.1.

This fundamental similarity in receptive fields allows the visual system to match the images in the two eyes, which is a prerequisite for the creation of a unified binocular field. Nevertheless, differences between the receptive fields of binocular cells do occur and, as we will see in later sections of this chapter, these differences form the basis of disparity-detecting mechanisms and stereopsis.

The width of the region of visual cortex containing a full 180° cycle of orientation preferences is roughly the same as that of an ocular dominance column. Regions containing a full set of orientation-tuned cells for each eye are called **hypercolumns**. In the monkey, hypercolumns have a diameter of 0.5 to 1.0 mm at eccentricities up to at least 15°. A similar pattern of ocular dominance columns was revealed by cytochrome-oxidase staining in the brain from a man who had been blind in one eye (Horton *et al.* 1990).

There is a surprising degree of variability between members of the same species. For example, one macaque monkey had 101 pairs of ocular dominance columns along the V1/V2 border while another monkey had 154 pairs. Also, mean column width varied between 395 and 670 μm (Horton and Hocking 1996a). It is assumed that a hypercolumn contains all the cell types required for coding a full range of visual features in the area of the visual field from which it derives its inputs (Hubel and Wiesel 1974a). Neighbouring hypercolumns receive inputs from overlapping retinal regions, and each hypercolumn is fed by the same number of ganglion cells. Excitatory and inhibitory dendritic connections extend over several columns (see Section 5.4.7).

5.6.2c *Topology of ocular dominance columns*

Columns with left-eye ocular dominance form a series of parallel bands interspersed with similar bands with right-eye dominance. The bands extend through all the layers of the visual cortex (Tootell *et al.* 1988c). In the cat, the bands in area 17 are between 0.25 and 0.5 mm wide in both the central and peripheral visual field. The bands are wider in area 18 (Shatz *et al.* 1977; Anderson *et al.* 1988). In the macaque monkey, they have been reported to be about 0.47 mm wide for the contralateral eye and about 10% less wide for the ipsilateral eye (Tychsen and Burkhalter 1997).

In each half of the visual cortex, the whole binocular visual hemifield is retinotopically mapped onto the ocular dominance bands of the left eye, and again onto the ocular dominance bands of the right eye. Along bands, this representation is approximately continuous. Across bands, it is interrupted by alternations between left- and right-eye bands. For details of how these interrupted retinotopic representations are organized see Hubel and Wiesel (1977). Even along ocular dominance bands, the change in position of receptive fields is not strictly continuous but varies as a function of changes in orientation preference (Das and Gilbert 1997).

To some extent, the ocular dominance bands follow isoeccentricity lines. However, LeVay and Voigt (1988) claimed that the major factor determining the band pattern in the cat is the tendency of bands to run across the elliptical visual cortex at right angles to the area 17/18 border. LeVay *et al.* (1985) found this tendency to be stronger in the monkey than in the cat. They argued that this is the simplest way to combine two circular monocular retinotopic maps onto an elliptical surface and minimizes anisotropy of the magnification factor across and along bands. Anderson *et al.* (1988) found only a weak tendency for the bands in area 17 to run at right angles to the area 17/18 border in the cat, although this tendency was more pronounced in area 18. They pointed out that LGN layers and the striate cortex have similar oval shapes, so that the cortical axis is not elongated along any axis by interdigitating LGN inputs.

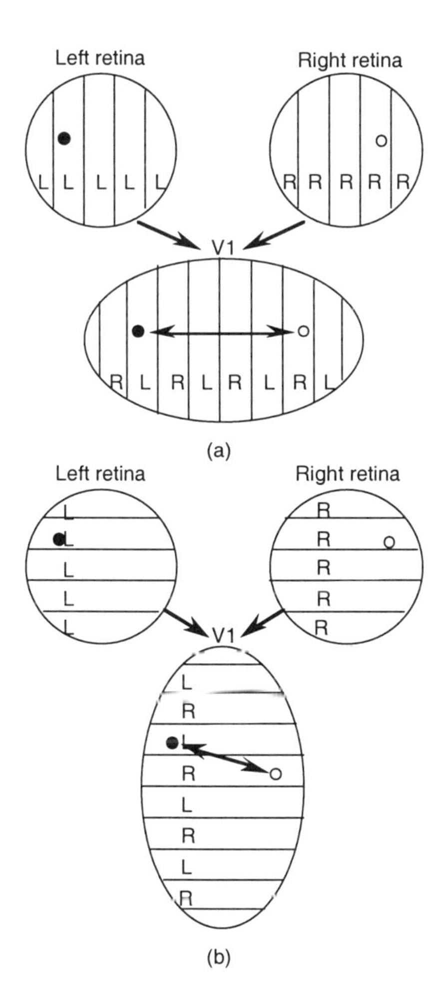

Figure 5.34 Orientation of ocular dominance bands.
The distance between points in V1 will be greater if ocular dominance stripes are orthogonal to the direction of a disparity, as in (a), than if they are aligned with the direction of the disparity, as in (b). (Adapted from Chklovskii 2000)

Jones *et al.* (1991) developed a computational model based on these differences between monkeys and cats, which accurately predicts the different ocular-dominance patterns in the two species. Similar models were developed by Goodhill and Willshaw (1990) and Bauer (1995).

Chklovskii (2000) argued that the theory that the topology of ocular dominance bands is designed to minimize cortical stretching does not explain the different orientation of bands in the parafoveal region. They proposed that, in each part of the visual field, ocular dominance bands are parallel to the direction of the most frequently occurring binocular disparity. They argued that, in the foveal region, disparities are predominantly horizontal while, in the periphery, they tend to follow isoeccentricity lines. This arrangement of the ocular dominance bands minimizes the lengths of connections between cells that register disparate images, as illustrated in Figure 5.34. This is because left-eye and right-eye bands alternate and the cortical magnification factor across the bands is about twice that along the bands (Section 5.4.2b).

5.6.2d *Ocular dominance scale*

Hubel and Wiesel (1962) introduced a seven-group **ocular dominance scale**. Cells in group 1 respond only to inputs from the contralateral eye and those in group 7 only to inputs from the ipsilateral eye. The eye not evoking a response in a given cell is known as the **silent eye**. Cells in group 4 respond equally well to inputs from either eye, and cells in the other groups have a corresponding degree of ocular dominance. In the monkey, 72% of cells in V1 fall into groups 2 to 6 (Schiller *et al.* 1976). The classification of cells in a particular animal depends on the method used to categorize the responses but, for a given method, it has been found to be stable over a period of 8 hours (Macy *et al.* 1982).

The scale of ocular dominance is based only on the excitatory effects of monocular stimulation from each eye separately. It takes no account of inhibitory connections or of the few cells that fire only in response to binocular stimulation (Grüsser and Grüsser-Cornehls 1965). The classification of cortical cells into ocular dominance columns is further complicated by the fact that the ocular dominance of some complex cells varies over time and depends on the spatial frequency and velocity of the stimulus (Hammond 1979, 1981).

Several lines of evidence show that cells in the centres of ocular dominance columns are strongly monocular while cells in intermediate positions receive excitatory inputs from both eyes. Thus, radioactive tracer injected into one eye migrates to the centres of ocular dominance columns for that eye (Horton and Hocking 1996c). Also, monocular enucleation causes loss of cytochrome oxide activity in the centres of ocular dominance columns for that eye Horton and Hocking 1998b).

5.6.2e *Dichoptic interactions*

The ocular dominance scale takes no account of the fact that almost all cells in groups 1 and 7 lying within the binocular field, and classified as exclusively monocular by Hubel and Wiesel's criterion, are affected by the simultaneous stimulation of the corresponding region in the silent eye. In fact, cells with strong ocular dominance show evidence of stronger binocular interactions of this type than do cells classified as having a balanced binocular input (Gardner and Raiten 1986). Cells with strong dominance are probably those that code large disparities.

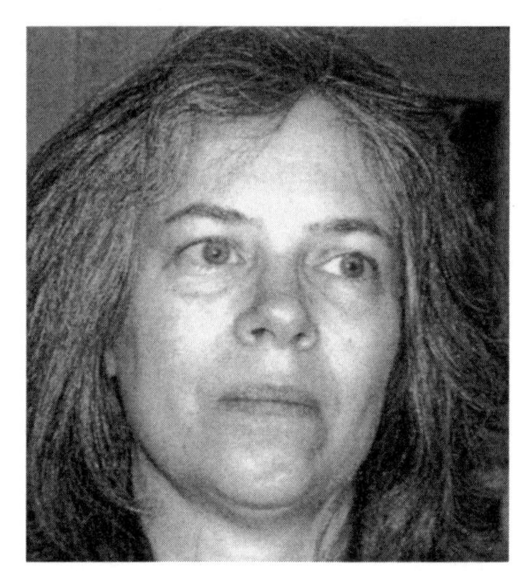

Figure 5.35. Margaret Livingstone.
She graduated from MIT in 1972 and obtained a Ph.D. from Harvard Medical School in 1979. She conducted postdoctoral work at Princeton and then with David Hubel at Harvard. She has remained at Harvard Medical School.

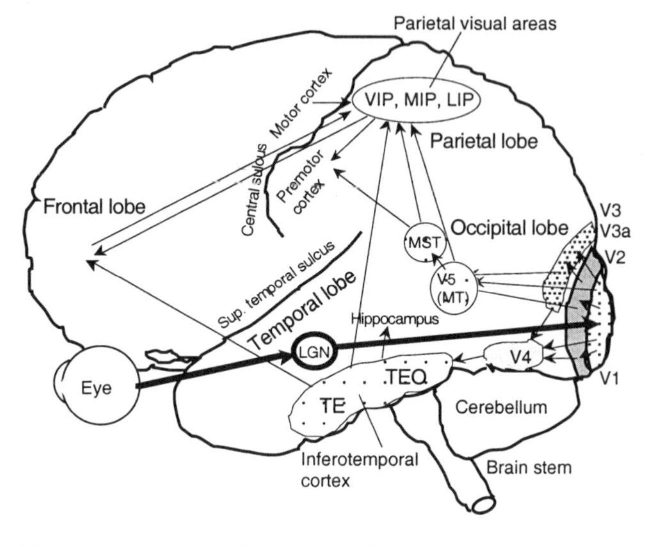

Figure 5.36. Main visual areas in the human cerebral cortex.
The dorsal stream goes from V1 and V2 through V3 to MT (V5) and MST and on to the prietal lobe. The dorsal stream goes from V1 and V2 through V4 to the inferotemporal cortex.

The effect produced by stimulation of the silent eye may be inhibitory and thus reduce the response evoked by stimulation of the dominant eye or it may involve subthreshold facilitation and lower the threshold of response to stimulation of the dominant eye (Grüsser and Grüsser-Cornehls 1965; Henry *et al.* 1969; Bishop *et al.* 1971; Poggio and Fischer 1977; Kato *et al.* 1981).

Since all direct visual inputs to cortical cells in layers 4C and 4A are believed to be excitatory, one must conclude that the suppressive effects produced by stimulation of the silent eye involve lateral connections (Section 5.4.7). Application of bicuculline (a GABA antagonist) to the surface of the cat's visual cortex caused cells that were strongly or exclusively driven by one eye to become responsive to both eyes (Sillito *et al.* 1980b).

Hammond and Kim (1996) found that, for binocular complex cells in visual cortex of the cat , the suppressive effect of stimulating the nondominant eye on the response to stimulation of the dominant eye varied with the relative orientations and directions of motion of dichoptic stimulus gratings. Some cells showed interactive effects when parallel gratings moved in opposite directions, others showed effects when orthogonal gratings moved in orthogonal directions, others when the gratings moved in either of these two ways, and still others when the moving gratings had different but similar orientations. A final group showed no or only variable suppression.

5.6.2f *Ocular dominance in New World monkeys*

The order of primates evolved at least 60 million years ago. There are two suborders, the prosimians, including tarsiers, lemurs, lorises, and galagos, and the anthropoids including New World monkeys, Old World Monkeys, and hominoids (apes and humans). The New World suborder (Platyrrhines) separated from the Old World suborder (Catarrhines) about 30 million years ago (Fleagle 1988).

Ocular dominance columns have been found in all Old World monkeys and great apes that have been studied (Hendrickson *et al.* 1978; Tigges and Tigges 1979; Sengpiel *et al.* 1996). They have also been found in at least one prosimian species, the bushbaby (*Galago*) (Glendenning *et al.* 1976). New World monkeys have eye-specific laminae in the LGN, although the parvocellular laminae are not as well defined as in Old World monkeys (Hendrickson *et al.* 1978). However, as we shall now see, not all New World monkeys have well-defined cortical ocular dominance columns.

The New World spider monkey (*Ateles ater*), has anatomically distinct ocular dominance columns in the V1, as revealed by autoradiography. They are especially evident in layer 4B and there is a good deal of overlap between them (Florence *et al.* 1986).

The tufted capuchin monkey (*Cebus apella*) has eye specific layers in the LGN. Single-unit recording has revealed that most cells in the visual cortex are binocular with a preference for one or other eye. Cytochrome-oxidase staining has revealed ocular dominance columns in the visual cortex (Rosa *et al.* 1992). Staining applied 8 months after removal of one eye also revealed ocular dominance columns. (Hess and Edwards 1987).

The squirrel monkey (*Saimiri sciureus*) has cytochrome oxidase patches and autoradiography reveals orientation columns. Early studies revealed only faint indications of ocular dominance columns (Humphrey and Hendrickson 1983; Tigges *et al.* 1984). Most cortical cells are binocular and respond selectively to disparity but fewer of them show the strong ocular dominance evident in Old World monkeys (Livingstone *et al.* 1995) (Portrait Figure 5.35). In a more recent study, involving finer anatomical resolution, ocular dominance columns were observed but they were only about 225 μm wide, the narrowest found in any animal. They were organized in a fractured, irregular mosaic and did not correspond with the distribution of cytochrome oxidase blobs, as they do in other animals (Horton and Hocking 1996b). Adult squirrel monkeys made strabismic at an early age show some evidence of ocular dominance columns in layer 4Cβ of V1 (Livingstone 1996a). Squirrel monkeys reared with one eye removed, show no evidence of ocular dominance columns (Hendrickson and Tigges 1985).

The owl monkey (*Aotes*), shows only faint ocular dominance columns (Kaas *et al.* 1976; Hendrickson *et al.* 1978). However, ocular dominance columns may be obscured by noise in autoradiography.

Adult marmosets (*Callithrix jacchus*) show no evidence of ocular dominance columns with autoradiography. Ocular dominance patches evident in 3-month-old marmosets disappear during the first year. These patches are retained in adult animals reared with one eye occluded (Spatz 1989; Sengpiel *et al.* 1996; DeBruyn and Casagrande 1981). It is not known whether any other New World monkeys show this early loss of ocular dominance patches.

It seems unlikely that genes for ocular dominance columns evolved separately in New and Old World monkeys. The emergence of columns after monocular occlusion suggests that the mechanism for their formation is present, even though they are not evident in the normal adult. Epigenetic factors such as the area of V1 and the degree convergence of visual afferents may account for variability among New World monkeys. The size of the cortex and the separation of the eyes are larger in larger animals, and ocular dominance columns seem to be confined to the larger species of New World monkeys. The squirrel monkey and the owl monkey have stereoscopic vision. Thus, well-defined ocular dominance columns are not required for stereopsis.

Figure 5.37 Alan Cowey.
Born in Sunderland, England. He obtained a degree in Natural Sciences and a Ph.D at Emmanuel College, Cambridge. After a year at the Medical School of Rochester University, he returned to Cambridge as a Lecturer in the Department of Experimental Psychology. After a further year as a Visiting Fulbright Fellow at Harvard University, he went to Lincoln College, Oxford as a Royal Society Research Fellow and a Nuffield Senior Research Fellow. He became Reader in Experimental Psychology at Oxford in 1973 and ad hominem Professor of Physiological Psychology in 1981. From 1990 to 1996 he was Director of the MRC Interdisciplinary Research Centre for Brain and Behaviour. In 1997 he became a Medical Research Council Research Professor. He has been President of the European Brain and Behaviour Society and the UK Experimental Psychology Society.

5.7 OTHER VISUAL AREAS

Axons leaving the primary visual cortex emerge from pyramidal cells above and below layer 4. Axons from layers 2 and 3 project retinotopically to extrastriate areas. The primary visual cortex also sends intracortical fibres to visual areas in the inferior temporal cortex (Gross 1973), the parietal lobe, the frontal lobe, and to several subcortical nuclei.

Visual areas surrounding the striate cortex are known as the **extrastriate cortex**, or **prestriate cortex**, although I have not found precise definitions of these terms. In primates, the extrastriate cortex includes V2, V3, V3A, and V4 in the occipital lobe. It also includes the middle temporal area (MT), the medial superior temporal area (MST), the ventral posterior area (VP), the ventral occipitotemporal area (VOP), and the ventral interparietal area (VIP) (Zeki 1974b; Wong-Riley 1979b; Maunsell and Van Essen 1983a).

At least ten visual areas have been identified in the parietal lobe. These occupy most or all of Brodmann's area 7. The inferior temporal cortex also contains many visual areas (Gross 1973). The frontal lobe contains the frontal eye fields and other areas related to visual functions. Many cells in all extrastriate visual areas are binocular (Zeki 1979). Figure 5.36 depicts the layout of visual areas.

Thirty-two distinct visual areas have been revealed in the brain of the monkey, with over 300 pathways connecting them (Felleman and Van Essen 1991; Van Essen *et al.* 1992). Together they occupy about 60% of the monkey neocortex. No doubt other visual areas remain to be discovered.

In cats, both areas 17 and 18 receive inputs from the LGN. In primates, V1 receives almost all the direct visual inputs from the main laminae of the LGN. Area V2 receives inputs from interlaminar layers only, which are fed directly from the retina or through the superior colliculus (Bullier *et al.* 1994). Areas V4, and MT receive some inputs from both the laminae and interlaminar layers of the LGN (Büllier and Kennedy 1983).

Visual inputs that bypass V1 may be responsible for **blindsight**—residual visual functions, not involving conscious awareness, that occur in destriate monkeys or in humans with striate lesions (Weiskrantz 1987; Stoerig and Cowey 1997). Most visual centres, along with other parts of the neocortex, receive inputs from several subcortical areas (see Sections 5.4.2a and 5.4.5e).

The striate cortex (V1) contains a fine and well-ordered representation of the whole contralateral visual hemifield. The visual hemifield is represented in a much less orderly fashion in each extrastriate visual area. The representation is coarser because the receptive fields are larger and there are topographic irregularities in the mapping from one visual area to another. In some visual areas, parts of the visual field are exaggerated relative to their representation in V1 and other parts are diminished or absent.

The cells within the boundaries between one visual area and another—for example, between V1 and V2—have receptive fields along either the vertical or horizontal retinal meridians. Fibres running through the corpus callosum connect regions representing the vertical meridians. These boundary regions can therefore be recognized by a sudden change in retinotopic representation.

One consequence of this juxtaposition of cells from the main retinal meridians is that retinotopic representations of visual hemifields in succeeding visual areas are mirror images of each other (Cowey 1979) (Portrait Figure 5.37). This fact has made it possible to map the borders of visual areas in the human cerebral cortex by inspection of magnetic resonance images produced by phase-encoded retinal stimuli (Sereno *et al.* 1995).

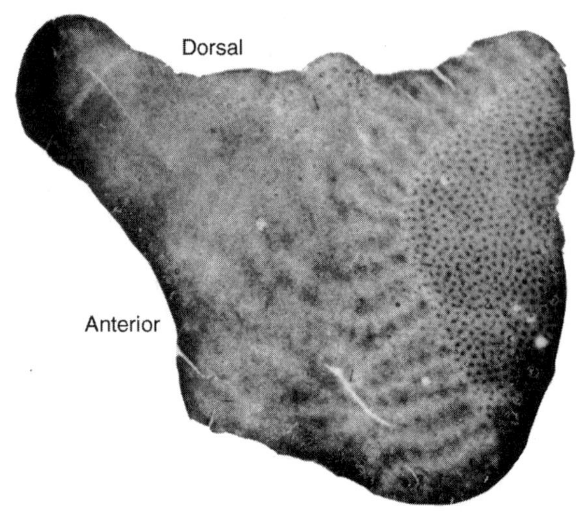

Figure 5.38. Cytochrome oxidase areas of the visual cortex. Section from layer 3 of the lateral surface of the visual cortex of the squirrel monkey, stained for cytochrome oxidase. The spotted area is V1 and the adjoining striped area is V2. (Reprinted with permission from Tootell *et al.* 1983, Copyright American Association for the Advancement of Science)

One might ask why, in some cases, cells with different tuning functions are assembled in alternating columns within the same visual area while, in other cases, they are assembled in distinct cortical areas. The answer is probably that cells are juxtaposed when lateral connections are required between them so that they can code higher-order stimulus features, such as motion in depth. Different types of cells are assembled in distinct areas when extensive local interactions between them are not required because each area is specialized for the processing of a particular feature.

As one proceeds to higher visual centres it becomes more difficult to determine what processes are occurring. Important differences between one visual area and another may not be revealed by the tuning characteristics of single cells but only by properties of larger functional units. Even if patterns of connections between cortical cells are known their functions are difficult to interpret without knowing what the system is designed to accomplish.

Most investigations of single cells at or below the level of the primary visual cortex have been conducted on anaesthetized animals. Anaesthetics probably mask effects of feedback from higher centres, even at the level of the LGN. Anaesthetics must surely affect even the basic processing carried out by cells in higher visual areas. Reliable results can probably be obtained only by using alert animals.

Another problem is that simple stimuli may not reveal differences between cortical areas. As one ascends the processing hierarchy, cells are selective to ever more specific and complex stimuli. It is difficult to decide which of an infinitely large set of stimuli to use. Investigators often employ a shotgun approach using an arbitrary set of complex stimuli.

Once the tuning specificity of cells in a given area has been determined there is still the question of how that tuning was achieved. The tuning selectivity of cells in a given visual area could be due to three processes.

1. Specifity could arise because each cell in a visual area receives converging inputs from two or more distinctly tuned cells in another area. For example, a cell tuned to two lines forming a corner may receive inputs from two cells tuned to single lines in distinct orientations.

2. Specificity of tuning in a given visual area could arise from lateral excitatory or inhibitory connections within that area. I have already mentioned the debate about whether orientation specifity of cells in V1 arises through convergence of LGN inputs or through intracortical connections (Section 5.5.2). But the same issue arises at every transition from one visual area to another.

3. Specificity of tuning may arise because of feedback from an area higher in the processing hierarchy. All three processes may operate together.

Some visual areas operate in **parallel**, each processing distinct types of information. This is presumably because the processing of each type of information requires cells with distinct properties.

Other visual areas operate in sequence to form a processing **hierarchy**. This occurs when processing carried out in a given area depends on information supplied by another area. Visual areas beyond V1 become smaller and, often, contain only a partial representation of the visual field. Receptive fields of cells in higher visual centres become larger and more specialized. Also, beyond V1, the relationship between response and stimulus contrast becomes steeper so that cells have narrower dynamic ranges and act more like on-off switches (Sclar *et al.* 1990). This suggests that the detailed information processed at each stage of a hierarchy is available from that stage but not from higher stages (Lennie 1998). Each higher stage takes only that information from the preceding stage that it requires for the processing it performs.

Thus, each hierarchical stage provides a different level of information, and information from each stage is available when needed. If higher centres are lost, we may not be able to recognise things but we can still perceive the visual world and perform basic tasks such as stimulus detection and discrimination of simple features. Thus, perception is not the end product of a hierarchical process, but different types of perceiving are possible from the outputs of each stage of processing from V1 to the highest levels.

5.7.1 Areas V2 and V3

5.7.1a *Area V2*

In primates, V2 is about the same size as V1, from which it receives about 90% of its inputs. I have already mentioned that area 18 in cats receives visual inputs from the main laminae of the LGN, while V2 in primates receives visual inputs from only interlaminar regions of the LGN.

In the macaque monkey, V2 is not partitioned into ocular dominance columns. However, it shows alternating stripes when stained for cytochrome oxidase. The **cytochrome-oxidase stripes** run over the surface of V2 approximately perpendicular to the border with V1, as shown in Figure 5.38 and extend through all the cortical layers (Tootell *et al.* 1983). Each stripe cycle is about 1 mm wide and consists of dark staining **thin and thick stripes** and a light staining **interstripe**. There are about 12 stripe cycles in dorsal V2 of the macaque monkey (Roe and Ts'o 1995). The blob regions of V1, containing mostly nonorientation-specific, colour-coded cells of the parvocellular system, project heavily but not exclusively to the thin stripes. There is also a magnocellular input to the blobs (Nealey and Maunsell 1994). The interblob regions of V1, containing the orientation-tuned cells of the parvocellular system, project to the nonstaining interstripes. Cells of the magnocellular system project from layer 4Cα to layer 4B and then to the thick stripes (Blasdel *et al.* 1985; DeYoe and Van Essen 1985; Hubel and Livingstone 1987; Livingstone and Hubel 1987). This subdivision of cell types is not complete—there are cells in each type of stripe that are tuned to colour and depth, colour and orientation, or to all three features (Ts'o *et al.* 1989; Levitt *et al.* 1994; Gegenfurtner *et al.* 1996). The pulvinar (Section 5.4.2), a thalamic nucleus, sends inputs to the thick and thin stripes of V2 but not to the interstripes (Levitt *et al.* 1995).

Within each stripe cycle, a region of visual space is remapped three times, once in the thin stripe, once in the thick stripe, and once in the interstripe. At each stripe border there is a topological 'jump back' discontinuity, like that between ocular dominance columns in V1. Adjacent stripe cycles represent adjacent regions of space and the mapping is continuous between stripes of a given type across the cycles (Roe and Ts'o 1995). In humans, the cytochrome oxidase stripes of V2 are replaced by a rather disorderly jumble of patches (see Tootell *et al.* 1996).

By combining single-unit recording with *in vivo* optical imaging Ts'o *et al.* (2001) revealed a finer functional organization within each thin, pale, and thick stripe of V2. Each stripe contains patches that are selectively tuned for colour, orientation, or disparity. Within disparity-tuned patches most cells prefer vertical stimuli. Between colour patches and disparity patches there are cells jointly tuned to colour and disparity. Within colour patches, darker regions respond to equiluminant colour and lighter regions to stimuli defined by luminance.

The tuning characteristics of cells in V2 resemble those of cells in V1, except that cells in V2 have larger receptive fields, are more likely to be binocular, and are almost all complex cells (Zeki 1978). Cells in V2, unlike those in V1, respond to subjective contours and contours defined by texture boundaries (von der Heydt *et al.* 1984; Mareschal and Baker 1998). Some cells in V2 of the alert monkey have been found to be selectively sensitive to complex line stimuli such as angles, circles, and hyperbolic and polar gratings (Hegdé and Van Essen 2000).

5.7.1b *Area V3*

In the monkey, V3 is a narrow strip of cortex along the dorsal border of V2. An area known as VP lies along the ventral border of V2 and is sometimes treated as part of V3. Area V3 is organized retinotopically and receives inputs from V1, mainly from layer 4B, and from V2. It projects to V4 and the temporal visual area in the ventral pathway, and to MT, MST, and to the posterior and ventral intraparietal areas in the dorsal pathway, as described in the next sections (Felleman *et al.* 1997a; Lyon and Kaas 2001). Human areas V1, V2, V3, and VP seem to be homologous to those in macaque.

Cells in V3 have larger receptive fields than those in V2 and only about 30° of the contralateral visual field is represented in V3 (Felleman *et al.* (1997b Most cells in V3 are orientation selective and cells with similar orientation tuning are organized into columns. About half the cells are selective for direction of motion, especially in the lower visual field. Cells in V3 prefer lower spatial frequencies and higher temporal frequencies than those in V2 (Gegenfurtner *et al.* 1997).

Estimates of the proportion of cells showing colour selectivity vary between about 15% (Baizer 1982; Felleman and Van Essen 1987; Adams and Zeki 2001) and 54% (Gegenfurtner et al. 1997).

About half the cells in both V3 and VP are tuned to binocular disparity (Burkhalter and Van Essen 1986; Felleman and Van Essen 1987). In the macaque, cells in V3 with similar joint tuning to orientation and disparity are organized into columns (Adams and Zeki 2001). This organization is well suited to the extraction of higher order orientation disparities and disparity gradients required for the perception of 3-D form. We will see that V3 projects to the posterior parietal cortex, which is concerned with processing 3-D form.

Beyond V2 and V3, visual processing is partitioned into two main pathways; a ventral pathway serving mainly the parvocellular system and a dorsal pathway serving mainly the magnocellular system but this division is not complete. Fluorescent tracers have revealed that the two major processing streams also have distinct projections to subcortical nuclei (Baizer *et al.* 1993).

5.7.3c *Area V3A*

Area V3A lies along the border of V3 but contains a distinct representation of both lower and upper visual fields, and distinct functional properties and connections (Felleman and Van Essen 1991). The border between V3 and V3A represents the vertical retinal meridian and is connected by the corpus callosum to the corresponding region in the opposite hemisphere. V3A receives inputs from V1, V2, and V3. In the macaque, cells in V3A have large receptive fields and are less selective to the speed and direction of motion than are cells in V3 (Gaska *et al.* 1988). By contrast, MRI procedures have revealed that V3A in humans is more motion selective than V3 (Tootell *et al.* 1997). Some cells in V3A of the alert monkey respond to the position of relatively small stimuli in a headcentric frame of reference, while others respond to stimuli in an oculocentric frame of reference (Galletti and Battaglini 1989).

5.7.2 The ventral pathway

The parvocellular system projects mainly ventrally to V4 and then to the inferior temporal cortex, the superior temporal polysensory area, and frontal cortex (Baizer *et al.* 1991).

5.7.2a *Area V4*

An area in the lingual and fusiform gyri in humans is the homologue of area V4 in monkeys. In primates, V4 receives some direct inputs from the LGN (Yukie and Iwai 1981; Büllier and Kennedy 1983), which may explain why a patient lacking a visual cortex retained some colour discrimination (Stoerig and Cowey 1989). But V4 receives most of its inputs from V1, about 50% of the outputs of V2, and some outputs from V3 (Nakamura *et al.* 1993).

In area V4 of the macaque only the central 35° of the contralateral visual hemifield is represented in a crude and rather disorderly fashion. This may be related to the fact that the cells have large receptive fields which, being colour coded, are confined to the central region of the retina (Gattass *et al.* 1988). There are extensive lateral connections in V4 and callosal connections are widely distributed (Van Essen and Zeki 1978). Retrograde labelling has revealed that V4 has modular compartments that receive inputs from either thin stripes (parvocellular channel) or inter-stripe regions (magnocellular channel) of V2, and V3/V3A, although there is also some cross-channel convergence. There is also modular-specific input to and feedback from the inferior temporal cortex to V4 (Felleman *et al.* 1997b).

The properties of the parvocellular system feeding into V4 and the inferior temporal cortex suggest that these structures are specialized for pattern discrimination, colour vision, and fine stereopsis. I shall deal with each of these features in turn.

Many cells in V4 of the monkey are selective for orientation although their tuning widths are broader than those of cells in V1. Regions tuned to different orientations are spatially organized, at least in the part devoted to the central retina (Ghose and Ts'o 1997). Most cells are selective for the spatial frequency of gratings and for the length and width of bars (Desimone and Schein 1987). However, cells in the awake macaque respond better to features such as angles and curves pointing in a particular direction than to simple edges or bars (Pasupathy and Connor 1999).

Some cells in V4 respond selectively to periodic patterns such as concentric gratings, radial patterns, and hyperbolic patterns that resemble Lie orbits (Section 3.8) (Gallant *et al.* 1993). Cells in V4 are selective for speed of motion (Cheng *et al.* 1997) and some are selective for motion direction. It seems that V4 processes simple shape features required for the recognition of complex shapes at higher levels in the ventral pathway.

Lesions in V4 in monkeys do not have much effect on basic visual functions, such as contrast or motion sensitivity. However, they severely disrupt performance on form discrimination tasks, such as discrimination of the relative orientation of lines or selection of an object that differs from other objects (Heywood *et al.* 1992; Merigan 1996; Schiller 1993). Similar symptoms have been noted in a human patient with lesions in V4 (Rizzo *et al.* 1992). Responses of cells in V4 are modulated in complex ways by the attentional state of the animal (Section 5.5.7c).

Positron emission tomography (PET) has revealed that blood flow to V4 increases when human subjects view a coloured display (Zeki *et al.* 1991). Patients with lesions in the ventral portions of the occipital lobe, a region considered to be the analogue of V4 in monkeys, suffer a form of colour blindness known as cerebral **achromatopsia** (Zeki 1990). Patients describe the world in terms of gray, although it seems that all three cone mechanisms are intact. One patient could see depth in a random-dot stereogram presented with red-green anaglyph filters even though he could not detect the red stereogram elements when he looked through the red filter (Hendricks *et al.* 1981). Disparity detection must depend on processes at an earlier stage than that responsible for the colour defect.

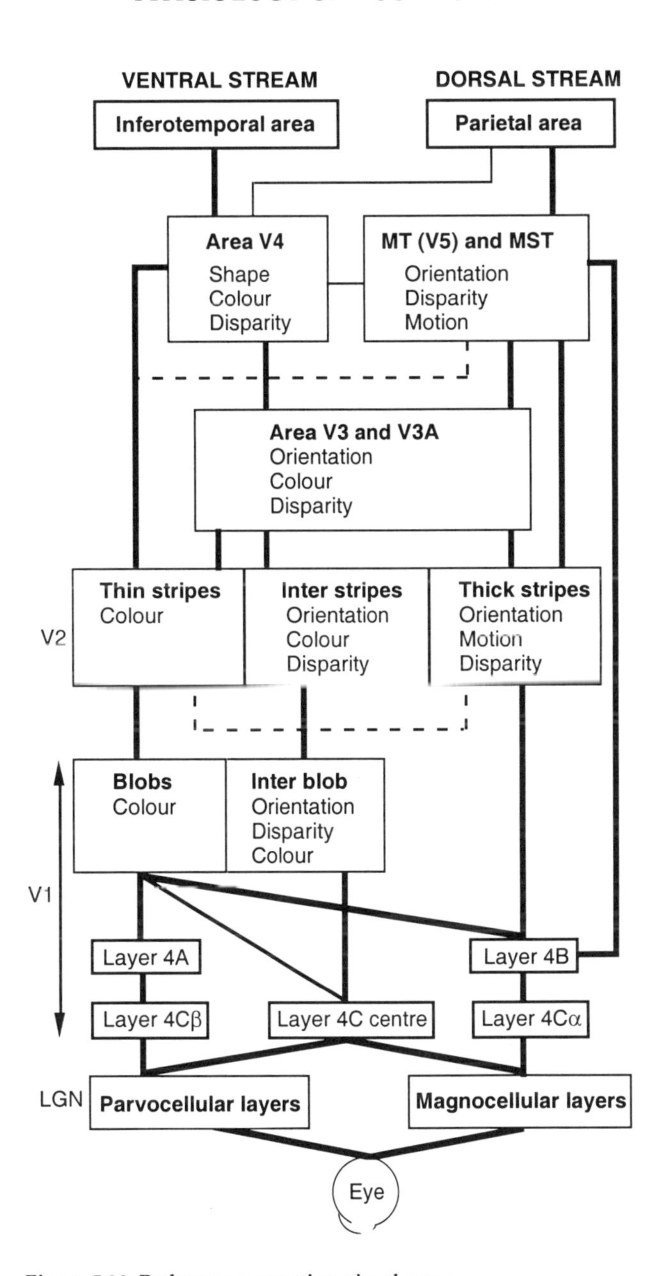

Figure 5.39. Pathways connecting visual areas.
Visual areas of the macaque, with an indication of functional specializations. (Adapted from De Yoe and Van Essen 1988)

In monkeys, V4 lesions have little or no effect on colour discrimination or on the ability to select an odd colour in an array (Heywood *et al.* 1992). Perhaps achromatopsia involves a deficit in colour appearance rather than of colour discrimination, or perhaps lesions in humans involve white matter, which is spared by experimental lesions in monkeys.

The role of V4 in depth perception is discussed in Section 6.3.4.

5.7.2b *Inferior temporal cortex*

The primate **inferior temporal cortex** (IT) receives most of its visual inputs from V4 and also provides feedback to V4. There are also some direct visual inputs from interlaminar zones of the LGN (Hernández-Gonzalez *et al.* 1994). Neurones in IT that are tuned to similar complex stimulus features are aligned in columns normal to the cortical surface, although IT has no topographic organization (Fujita *et al.* 1992; Gross 1992). All cells have large receptive fields that almost always include the fovea and extend into both left and right halves of the visual field. The lateral connections of pyramidal cells are very much more extensive than in V1 (Elston *et al.* 1999).

In the monkey, IT consists of two distinct cytoarchitectonic areas. The first is a posterior area known as TEO. The second is an anterior area known as TE. Area TEO has extensive connections with TE and the parietal lobe. Its connections to the frontal lobe are limited to areas 8, 12, and 45 (Section 5.7.3f). Area TE has fewer connections with the parietal lobe than does area TEO but more extensive connections with the frontal lobe, namely to areas 8, 11, 12, 13, and 45. (Webster *et al.* 1994). The inferior temporal cortex also connects with the medial temporal lobe, with regions in the old cortex, such as the hippocampus, perirhinal and entorhinal areas, and the amygdala. The medial temporal lobe also projects to subcortical areas.

Some areas of the temporal cortex are purely visual while others receive visual, auditory, and somatosensory inputs. It is generally believed that cells in IT respond only to visual stimuli, although fMRI has revealed that this region in the human cortex is activated by the both visual objects and objects detected by touch (Amedi *et al.* 2001). Some cells in IT respond best to particular combinations of stimulus features such as texture, shape, and colour, but they respond at different rates to a variety of stimuli (Tanaka *et al.* 1991).

Many cells in monkey IT respond selectively to a particular shape over a 2- or 4-octave change in the size of the stimulus (Ito *et al.* 1995; Sato *et al.* 1980). The cells show the same selectivity when the stimulus is moved several degrees away from the fovea or as the animal moves it gaze (DiCarlo and Maunsell 2000). The cells respond to a given shape whether it is defined by luminance contrast, by motion, or by texture (Sáry *et al.* 1993). The response to a given outline shape is not affected when the contrast polarity is changed or when the shape is mirror reversed. However, the response changes when the figure-ground appearance of the display is changed (Baylis and Driver 2001). Cells in IT therefore respond to shape rather than to lower-level features that define the shape or to the position, contrast polarity, or orientation of the shape.

Cells in IT also respond to shape defined by binocular disparity (see Section 6.3.4c). Some cells responded more strongly to solid objects than to projected images of the objects, even when only one eye was open (Gross *et al.* 1972). Janssen *et al.* (2000a) found that about 50% of cells in the lower part of TE in the macaque were selective for disparity-defined 3-D shape while very few cells in the lateral part of TE showed this selectivity.

Some cells in IT respond selectively to faces (Perrett *et al.* 1982; Rolls *et al.* 1994). The responses of some of these cells are invariant with respect to the position, size, and view of the face (Wallis and Rolls 1997; Gochin 1996). Damage to this area in humans can lead to deficits in face recognition, or **prosopagnosia** (Damasio *et al.* 1990). Lesions in the inferior temporal and medial temporal areas of monkeys impair visual discrimination learning for complex patterns. They do not affect discrimination of simple features such as differences in orientation (Holmes and Gross 1984). Lesions in TE disrupt the ability to discriminate between previously learned complex patterns. Therefore, this area is implicated in the retrieval of visual information. Lesions in TEO disrupt the ability to learn complex patterns (Mishkin 1982).

Some cells in TE respond specifically to complex stimuli that a monkey has memorized in a delayed matching task (Miyashita 1988; Miller *et al.* 1993). Similar cells are found in the inferior convexity of the monkey frontal lobe, to which IT projects. However, these cells retain responsiveness to memorized stimuli for longer periods than do those in IT (Miller *et al.* 1996). TE has reciprocal connections to the perirhinal cortex of the limbic system, an area associated with memory encoding (Naya *et al.* 2001). MRI imaging has revealed that distinct regions of the human inferior temporal cortex are active during the process of memorizing complex visual stimuli and during retrieval of memorized items (Gabrieli *et al.* 1997). The response of cells in the inferior temporal area is modified by attention (Section 5.5.7).

Figure 5.40. Peter H. Schiller.
Born in 1931. He obtained a B.A. in Psychology from Duke University in 1955 and a Ph.D. from Clark University in 1962. He conducted postdoctoral work in the Department of Psychology at MIT from 1962 to 1964. He then gained an academic position at MIT, where he now occupies the Dorothy Poitras Chair for Medical Physiology. He received an NIH Merit Award.

Figure 5.41. John Maunsell.
Born Great Baddow, Essex, England in 1955. He obtained B.Sc. in Zoology at Duke University in 1977 and a Ph.D. in Biology at the California Institute of Technology in 1982. He conducted postdoctoral work with Peter Schiller at MIT. He held academic appointments in the Center for Visual Science at the University of Rochester from 1985 to 1992. He is now Professor in the Division of Neuroscience and Department of Ophthalmology at Baylor College of Medicine, Houston. He is an Investigator of the Howard Hughes Medical Institute.

5.7.2c *Medial temporal lobe*

The inferior temporal cortex projects strongly to the medial temporal lobe, which includes the parahippocampal gyrus, hippocampus, perirhinal cortex, entorhinal cortex, and amygdala.

The parahippocampal region provides major inputs to the hippocampus. Some cells in the hippocampus respond selectively to the location of the animal in the visual environment with which it is familiar. These are known as **place cells** (O'Keefe and Nadel 1978; Eskandar *et al.* 1992; Wilson MA and McNaughton 1993; Rolls *et al.* 1998; Best *et al.* 2001). The hippocampus is also involved in other forms of memory (Wood *et al.* 1999).

Increased blood flow has been detected in the hippocampus when human subjects recognize the spatial coherence of 3-D objects (Schacter *et al.* 1995). Patients with bilateral damage to the hippocampus were unable to remember where objects had been seen (Gaffan 1994) or to verbally associate sounds with visual stimuli (Bechara *et al.* 1995).

Cells in the perirhinal cortex, entorhinal cortex, and amygdala of the monkey are selectively responsive to complex objects, faces, or familiar places (Leonard *et al.* 1985; Suzuki *et al.* 1997). Kreiman *et al.* (2000) recorded from these three areas and from the hippocampus of alert human patients with epilepsy while they discriminated between pairs of objects, such as faces, buildings, and animals. Of 427 neurones tested, 14% responded selectively to the category to which the shapes belonged.

5.7.3 The dorsal pathway

In primates, the magnocellular dorsal pathway projects to areas V3, the middle temporal area (MT), and the medial superior temporal area (MST) (Shipp and Zeki 1985; Krubitzer and Kaas 1990; Motter 1991). It then feeds into the superior temporal polysensory area (STP) and the posterior parietal cortex. Figure 5.39 shows these relationships. This system is specialized for coding low spatial frequency, fast flicker and motion, spatial location, and coarse stereopsis. These functions are associated with the analysis of the spatial positions and motions of objects and of visual motion arising from self motion, and the visual guidance of motor responses (Ungerleider and Mishkin 1982; Schiller *et al.* 1990; Lagae *et al.* 1993; Hietanen and Perrett 1996) (Portrait Figure 5.40).

5.7.3a *MT*

In monkeys, the middle temporal area is a small heavily myelinated area on the posterior bank of the superior temporal sulcus with direct inputs mainly from magnocellular cells of V1 and V2 (Maunsell *et al.* 1990). In humans, the homologous region is V5.

MT contains an irregular but complete topographic representation of the contralateral visual field, with emphasis on the lower temporal quadrant (Van Essen *et al.* 1981; Maunsell and Van Essen 1987) (Portrait Figure 5.41). The diameters of receptive fields in MT are about 10 times larger than in V1 and increase in size with increasing eccentricity (Albright and Desimone 1987). Receptive fields defined by single stimuli are surrounded by regions within which stimuli do not have a direct effect on the cell but modify the response to stimuli in the receptive-field centre (Allman *et al.* 1985). As one would expect from their magnocellular inputs, cells in MT have high contrast sensitivity and do not respond to equiluminant coloured stimuli (Tootell *et al.* 1995).

In the owl monkey, MT has distinct bands (Born and Tootell 1992). Cells in some bands respond best to motion in the same direction over a large area (zero-order motion) and are therefore sensitive to the type of visual motion produced by head rotation. The cells in other bands have centre-surround antagonistic receptive fields. The response to a central moving stimulus is inhibited by surround motion in the same direction and enhanced by surround motion in the opposite direction (Allman *et al.* 1985). These cells therefore respond best to relative motion (first spatial derivative of motion) which arises from motion parallax produced, for example, by lateral motion of slanting or inclined surfaces or of surfaces at different distances. Other cells respond best to the second spatial derivative of motion produced by motion of 3-D curved surfaces. These cells seem to be designed to detect velocity gradients in patterns of optic flow generated by self motion or to register the 3-D layout of objects from patterns of relative motion (Xiao *et al.* 1995; Buracas and Albright 1996; Treue and Anderson 1996). Morrone *et al.* (2000) used magnetic resonance imaging (fMRI) to reveal regions in the V5/MT complex of the human brain that respond to circular and radial patterns of optic flow. Other regions in the same complex respond to translatory motion.

Unlike cells in V1, cells in MT do not respond selectively to component motions of superimposed patterns moving in opposite directions (Snowden *et al.* 1991; Movshon and Newsome 1996). Lesions imposed in MT of monkeys specifically elevate motion discrimination thresholds (Newsome and Paré 1988). Most motion-selective cells in MT of the monkey are cue invariant, which means that they respond to second-order motion of contours defined by contrast, texture, flicker, or disparity (second-order motion) as well as to first-order motion defined by luminance-defined contours (O'Keefe and Movshon 1998; Albright 1992).

Bradley *et al.* (1998) trained monkeys to indicate the direction of motion of an ambiguous 2-D projection of a revolving textured cylinder (Section 25.5.1b). Many neurones in MT changed their activity whenever the direction-of-motion percept changed, even though the stimulus remained the same. Dodd *et al.* (2001) obtained similar results and showed that the correlation between responses of cells and the monkey's reports are not due to eye movements, response bias, sensory adaptation, or attention to particular locations of the stimulus.

The sensitivity of MT cells is modulated in the interval between presentation of a learned stimulus and performance of a direction-of-motion discrimination task (Seidemann *et al.* 1998).

Motion-sensitive cells of MT are also tuned to binocular disparity (Maunsell and Van Essen 1983b). The implications of this fact are discussed in Sections 6.3.2 and 29.6.2.

Area MT is also involved in the control of smooth pursuit eye movements (Newsome *et al.* 1988).

Motion-sensitive cells have been reported in the optic tectum of the pigeon (Frost *et al.* 1981) and in the suprasylvian cortex (Clare-Bishop area) of the cat (Rauschecker *et al.* 1987; Krüger *et al.* 1993).

The PET scan has revealed that V5 (MT) in the human brain is specialized for motion (Zeki *et al.* 1991). A patient with bilateral lesions that included MT was unable to experience objects moving in depth or objects moving faster than 10°/s, even though she could perceive stationary objects (Zihl *et al.* 1983). Magnetic resonance imaging (fMRI) revealed that area V5 in the human cortex is particularly active in response to 3-D stimuli undergoing rigid or nonrigid motion (Orban *et al.* 1999). This method has also revealed that increasing the directional coherence of a display of moving random dots has a similar effect on the strength of the fMRI signal from V5 of humans and on the firing rate of single cells in monkey MT (Rees *et al.* 2000).

5.7.3b *MST*

The medial superior temporal cortex (MST or V5A) receives a major input from MT. Inputs converge so that receptive fields of cells in MST, especially in the dorsal portion, are larger than those of cells in MT (Ungerleider and Desimone 1986). This suggests that the receptive fields of MST cells are constructed from the receptive fields of MT cells.

Figure 5.42. Robert H. Wurtz.
Born in St. Louis in 1936. He received his A.B. from Oberlin College and his Ph.D. from the University of Michigan with James Olds. He did postdoctoral research in the Department of Physiology, Washington University and at the National Institute of Health. In 1966 he joined the Laboratory of Neurobiology at NIMH and in 1978 he founded the Laboratory of Sensorimotor Research in the National Eye Institute. Dr. Wurtz was elected to the National Academy of Sciences in 1988, the Institute of Medicine of the National Academy of Sciences in 1997, the American Academy of Arts and Sciences in 1990. He was President of the Society for Neuroscience in 1990.

Cells in MST are sensitive to patterns of optic flow, especially to global patterns of visual motion such as translation, rotation, expansion/contraction, and rotation in depth (fanning). Responses to motion are more position invariant than in MT (Duffy and Wurtz 1991, 1997; Lagae *et al.* 1994; Graziano *et al.* 1994; Bradley *et al.* 1996). These cells could be involved in coding the direction of heading as one moves through a 3-D scene (Page and Duffy 1999). Electrical stimulation of cells in MST of monkeys biased their judgements of heading direction (Britten and van Wezel 1998). The responses of cells in MST were found to be independent of the preceding stimulus (Paolini *et al.* 2000). In other words, they were not tuned to specific changes in the flow field that might arise as the animal moves in a complex path through the environment.

Some cells in MST of the monkey respond specifically to stimuli rotating in depth, some to rotation about a horizontal axis and others to rotation about a vertical axis (Saito *et al.* 1986; Sakata *et al.* 1986). Some cells in the dorsal MST of alert monkeys respond to the tilt alone, the slant alone, or both the tilt and the slant of surfaces defined by motion (Sugihara *et al.* 2002). Some cells in the lateral-ventral region of MST are jointly tuned to motion and binocular disparity. For some of these cells the disparity preference in the centre of the receptive field differs from that in the surround (Eifuku and Wurtz 1999). They thus respond to spatial gradients of disparity and could be involved in perceptual segmentation of moving camouflaged objects.

In monkeys, the threshold of response of individual cells in MT and MST to the degree of coherent motion in a display of random dots is similar to the psychophysically determined threshold. The two thresholds vary in the same way to changes in the properties of the stimulus (Celebrini and Newsome 1994).

The motion specificity of MST cells could be due to cells receiving inputs from either subregions sensitive to distinct linear directions (direction mosaic hypothesis) or subregions sensitive to similar patterns of optic flow (vector field hypothesis). The position invariance of responses of MST cells supports the latter hypothesis, although MST cells are not entirely position invariant (Tanaka *et al.* 1989; Duffy and Wurtz 1995) (Portrait Figure 5.42). Some MST cells are influenced by the direction of gaze and by the direction of pursuit eye movements (Squatrito and Maioli 1996).

The response of motion sensitive cells in MST is substantially the same whatever feature defines the motion boundary. Thus, the preferred direction of cells was the same for motion of a dot pattern, a solid or an outline square, and of a region defined by flicker (Geesaman and Andersen 1996). This suggests that MST is involved in the detection of shape from motion. For this purpose, one would expect connections to the inferior temporal cortex.

5.7.3c *The rostral superior temporal cortex*

MT and MST are in the caudal superior temporal cortex. There has been some disagreement about the functions of the rostral superior temporal cortex. Some investigators have claimed that it is associated with spatial orientation while others have claimed that it is associated with object recognition. The area receives polysensory inputs from the inferior parietal lobe of the dorsal stream associated with spatial orientation and from the inferior temporal cortex of the ventral stream associated with object recognition. Therefore, this area is probably associated with both functions (see Karnath 2001).

The traditional view is that lesions in the right inferior parietal lobe of humans cause visual neglect, in which the patient is unable to fixate, attend to, or recall objects in the contralateral visual field. However, Karnath *et al.* (2001) found that neglect in patients with no other visual-field defects is due to lesions in the rostral superior temporal cortex or in the

Figure 5.43. Hideo Sakata.
Born in Sapporo, Japan in 1934. He graduated in Liberal Arts in 1955 and in Medicine in 1959 from Tokyo University. He obtained a D.M.S. in Physiology from Tokyo University in 1964. Between 1964 and 1973 he held academic appointments at Osaka City University, the Scrips Institute of Oceanography in San Diego, and the Johns Hopkins University School of Medicine. In 1973 he moved to the Tokyo Metropolitan Institute for Neurosciences where he became head of the Laboratory of Neurosciences. In 2000 he became Professor at Nihon University.

Figure 5.44. Vernon B. Mountcastle.
Born in Shelbyville, Kentucky in 1918. He obtained a M.D. from John's Hopkins Hospital in 1943. Since 1948 he has been at the John's Hopkins University School of Medicine. He was Professor of Physiology from 1959 to 1980 and Professor of Neuroscience from 1980 to 1991. Awards include the Lashley Prize of the American Philosophical Society, the Schmitt Prize and Medal at M.I.T., the Gold Medal of the Royal Society of Medicine, the Gerard Prize of the Society for Neuroscience, the Lasker Award, the McGovern Prize and Medal of the A.A.A.S., and the Neuroscience Prize of the N.A.S.

basal ganglia or pulvinar in the thalamus. They concluded that these three centres form a cortico-subcortical network that underlies spatial awareness. The left superior temporal cortex in humans is associated with language.

The fact that the caudal superior temporal cortex receives inputs from the ventral pathway that codes form and the dorsal pathway that codes motion suggests that it is involved in coding forms defined by motion. Vaina *et al.* (2001) showed that this region is activated in the human MRI when subjects perform discrimination tasks involving biological motion, such as recognizing a human walker from a pattern of moving light points.

5.7.3d *Ventral intraparietal area*

The ventral intraparietal area (VIP) receives inputs from MT and MST (Baizer *et al.* 1991). Cells in the VIP of the monkey are also specifically sensitive to optic flow. Some cells respond best to a stimulus moving from any azimuth direction towards a particular point on the face of the animal. In other words, they are selective to impact direction (Colby *et al.* 1993). Other cells respond to textured surfaces rotating in depth or to patterns of optic flow that simulate rotation in depth, although these cells are not as stimulus specific as those reported by Saito *et al.* in MST (Schaafsma *et al.* 1997). Like cells in MST, VIP cells show position invariance, which suggests that the vector field mechanism operates for them (Schaafsma and Duysens 1996). Also, like some cells in MST, some VIP neurones encode the direction of visual stimuli in terms of headcentric co-ordinates (Duhamel *et al.* 1997).

Sakata *et al.* (1999) (Portrait Figure 5.43) found cells in the anterior intraparietal area (AIP) of the alert monkey that responded selectively to the visual shape, size, and 3-D orientation of objects that were being manipulated or were about to be manipulated. Many of these cells were binocular. In a related area in the lateral bank of the caudal intraparietal sulcus, Sakata *et al.* found cells that were selectively responsive to the 3-D orientation of bar-like objects portrayed in a stereogram. Other cells in the same general area responded selectively to the 3-D orientation of surfaces defined only by disparity (Taira *et al.* 2000). Using fMRI, Shikata *et al.* (2001) revealed that areas within the intraparietal sulcus in humans are active when subjects discriminated the 3-D orientation of a surface defined by the monocular cue of texture gradient.

Other cells in AIP that were moderately responsive or unresponsive to visual features of objects showed selectivity for the type of hand grip involved in reaching for and grasping objects of different sizes, shapes, and orientations (Murata *et al.* 2000). Cortical lesions in the homologous area in humans cause deficits in grasping. The fMRI from normal subjects revealed that this area is specifically activated during grasping (Binkofski *et al.* 1998).

5.7.3e *Posterior parietal cortex*

The human posterior parietal cortex includes the superior parietal lobe (Brodmann's areas 5 and 7), the inferior parietal lobe (Brodmann's areas 39 and 40), and the medial and lateral intraparietal areas (MIP and LIP), and the anterior interparietal area (AIP). In the monkey, the inferior lobe contains areas 7a and 7b. Areas MIP and LIP receive visual inputs via V3A, MT, and MST, as well as auditory, somaesthetic, and vestibular inputs. There are also inputs from the superior colliculus, cerebellum, and basal ganglia, and frontal lobes. Area AIP receives inputs from area LIP and projects to the premotor cortex (Nakamura *et al.* 2001).

Visual inputs and eye-position information interact at various subcortical sites, including the superior colliculus (Sparks and Porter 1983) and pulvinar (Robinson *et al.* 1990). However, the posterior parietal cortex is the main cortical centre for coordination of visually guided movements of arms, head, and eyes in relation to an object to which an animal is attending (Bruce and Goldberg 1985).

Magnetic resonance imaging from the human parietal lobe has indicated that a change in the stimulus being attended to is associated with activity in lateral intraparietal cortex (LIP) and that a change in the response made to a given stimulus is associated with activity in the posterior intraparietal cortex (MIP) (Rushworth *et al.* 2001).

Some cells in the monkey posterior parietal cortex respond to a stimulus in a given retinal location but the magnitude of response depends on the position of the eyes in the orbits (Andersen *et al.* 1990; Duhamel *et al.* 1992). Some are influenced by proprioceptive inputs from one, two, or more arm joints (Mountcastle *et al.* 1975; Leinonen *et al.* 1979). Other cells are influenced by proprioceptive signals from the neck (Brotchie *et al.* 1995). Others are active when the animal reaches for or manipulates an object in a given location and their activity is independent of the spatial trajectory of the arm movement (Mountcastle *et al.* 1975; Hyvärinen and Poranen 1974; Bushnell *et al.* 1981) (Portrait Figure 5.44).

Areas area LIP and 7a receive visual inputs from extrastriate areas along with eye-position and head-

Figure 5.45. Richard A. Andersen.
He obtained a B.Sc. in Biochemistry from the University of California at Davis in 1973 and a Ph.D. in Physiology from Johns Hopkins Medical School in 1981. He was on the faculty of the Salk Institute in La Jolla from 1981 to 1987 and the faculty of MIT from 1987 to 1994. In 1994 he joined the Biology Division of Caltech in Pasadena where he is the James G. Boswell Professor of Neuroscience and Director of the Sloan Center for Theoretical Neurobiology. Recipient of the Spencer Award at Columbia University in 1994.

position signals, and auditory direction signals. Area LIP of the monkey seems to be specialized for coding the headcentric positions of stimuli to which the animal is about to make an eye movement (Snyder *et al.* 1998). Some cells respond preferentially to a stimulus in a given headcentric direction to which the animal is attending, others to stimuli at a given distance. Some are tuned jointly to direction and distance (Sakata *et al.* 1980; Andersen and Mountcastle 1983). Some cells respond in relation to eye movements in 3-D space (Section 9.11.2). The responses of these cells in the alert monkey do not depend directly on sensory inputs or motor outputs, but seem to encode a movement that is about to occur, even after the visual target has been removed (Eskandar and Assad 1999).

Cells in LIP project to cortical areas concerned with generation of hand and eye movements, including the MIP and frontal lobes (Andersen *et al.* 1997). They also project to the superior colliculus (Section 9.11.2). The neighbouring parietal-reach region (PRR), is concerned with planning arm movements to particular locations (Batista and Andersen 2001) (Portrait Figure 5.45).

Area 7a seems to be specialized for coding the bodycentric positions of objects in extrapersonal space, for directed reaching and navigation (Snyder *et al.* 1998). It projects heavily to the hippocampus, an area involved in spatial memory and navigation. Sakata *et al.* (1994) found many cells in area 7a that responded selectively to rotation of a slit in a given direction in 3-D space. The response of some of these cells showed periodic changes when the monkey viewed an ambiguous rotating trapezoid (Ames window) monocularly.

Neurones in the medial intraparietal cortex (MIP) are specialized for stimuli within reaching distance. They respond to either somatosensory or visual inputs or both, and code stimulus features such as direction and movement. Some maintain their response during a memory-guided response (see Colby and Goldberg 1999).

Cells with receptive fields defined in headcentric co-ordinates also occur in the anterior bank of the parietal lobe (V6) in the monkey. They have larger receptive fields than cells with similar coding properties in area V3 (Galletti *et al.* 1993).

Some cells in the inferior parietal lobe of the monkey respond best to motion in depth defined by looming, others to motion in depth defined by changing disparity, and others to motion in depth defined by either stimulus (see Sakata *et al.* 1997).

In humans, lesions in the superior parietal cortex cause defects in visual localization that affect the accuracy of reaching to visual targets in the contralateral visual field (Holmes and Horrax 1919; Ratcliff and Davies-Jones 1972; Perenin and Vighetto 1988). They may also produce astereognosis, or inability to recognize objects by touch, and asomatognosia, or denial that a part of the body is one's own.

One can think of the parietal lobe as maintaining a representation of the position of body parts in relation to external objects that are guiding behaviour. Sensory inputs and motor commands keep the representation updated (Wolpert *et al.* 1998).

The mechanisms involved in willing a part of the body to move and in the detailed planning of the movement remain mysterious.

For a review of the parietal lobes see Stein (1991), Caminiti (1995), Sakata *et al.* (1997), Galletti *et al.* (1997), and Andersen *et al.* (1997).

5.7.3f *Frontal lobes*

The distinction between ventral and dorsal processing streams extends into the prefrontal cortex. In the monkey, the posterior parietal cortex projects to the principal sulcus and arcuate regions of the frontal cortex. Cells in these regions respond to spatially localized stimuli and maintain their response when the animal is required to remember the location of the stimulus after it has been removed (Funahashi *et al.* 1989). The inferior temporal cortex projects to area 8 in the arcuate sulcus, area 12 in the inferior prefrontal convexity, and areas 11 and 13 on the orbital surface of the frontal cortex (Ungerleider *et al.* 1989; Webster *et al.* 1994).

Lesions in these areas in monkeys and humans produce deficits in the recognition of complex objects, such as faces and words. The response of cells in these regions to complex patterns is relatively independent of the size, orientation, and colour of the pattern. The cells continue to respond to a pattern during intervals when the animal is required to remember the pattern (Wilson *et al.* 1993).

Some cells in the frontal cortex responded only to "what" features when monkeys were required to remember a given object. Other cells responded only to "where" features when monkeys remembered a location. However, some cells responded to either "what" or "where" features according to the demands of the task and other cells responded to both "what" and "where" features at the same time (Rao *et al.* 1997). Thus, the parvocellular and magnocellular systems projecting to the frontal lobes have different, though overlapping, functions.

Event-related fMRI in humans has revealed that maintenance of items in spatial memory is associated with activity in frontal area 8, while selection of an item from memory is associated with activity in frontal area 46 (Rowe *et al.* 2000).

The response of some cells in the lateral prefrontal cortex on monkeys responded selectively to computer-generated images of dogs and cats that the monkeys had been trained to categorize. The cells responded to the same categories even though the stimulus boundary between the categories had been changed (Freedman *et al.* 2001).

Both the ventral and dorsal streams converge on the lateral frontal eye fields (Bullier *et al.* 1996). Cells in this region in the monkey respond in relation to eye movements that the animal is planning to make (Kim and Shadlen 1998).

5.7.3g *Premotor cortex*

In primates, the ventral intraparietal area (VIP), the lateral intraparietal area (LIP), and (MST) project to area 7b in the posterior parietal lobe, which projects to the premotor cortex, especially the ventral region. The premotor cortex contains a somatotopic representation of the body, especially of the arms, face, and mouth. Many cells are bimodal and also respond to visual stimuli placed near the region of skin to which they respond. Thus, a cell that responds to a tactile stimulus on the arm changes the location of its visual receptive field as the arm moves. This cell codes visual stimuli in armcentric co-ordinates. Similarly, a cell that responds to touch on the face changes it visual receptive field as the head moves. This cell, like some cells in the VIP, codes visual stimuli in headcentric co-ordinates (Graziano *et al.* 1997).

Bimodal cells that code visual stimuli in armcentric and headcentric co-ordinates have also been found in the putamen, a subcortical nucleus in the basal ganglia, which receives inputs from the parietal lobe and premotor cortex (Graziano and Gross 1993).

5.7.3h *Interactions between ventral and dorsal pathways*

The anatomical distinction between the ventral and dorsal cortical streams is by no means complete. There is evidence of partial convergence of parvocellular and magnocellular inputs even in the primary visual cortex (see Schiller *et al.* 1990). In addition, there are extensive interconnections between all visual areas, especially between the temporal and parietal lobes (DeYoe and Van Essen 1988).

Although V4 receives a predominant input from the parvocellular system, mainly from the magnocellular system, a few sites show evidence of inputs from the parvocellular system (Maunsell *et al.* 1990; Ferrera *et al.* 1992). The anterior superior temporal polysensory area (STP) receives inputs from both the dorsal and ventral pathways and contains cells that are jointly sensitive to the form and motion of visual stimuli (Oram and Perrett 1996).

Some cells in the monkey posterior parietal cortex (LIP) respond differentially to different shapes, even when the animal is not performing a motor task (Sereno and Maunsell 1998). This shape selectivity may depend on inputs to area LIP from V4 and the inferior temporal cortex (Webster *et al.* 1994).

One reason for interconnections between regions processing different features is that ambiguity in the location of contours defined by one feature can be resolved by reference to another feature. Thus, information flow between different systems allows one to exploit redundancies in the visual world.

5.7.4 Evidence for distinct pathways

The effects of lesions within the parvocellular and magnocellular systems of the monkey support the idea of two major processing streams. Lesions in the parvocellular laminae of the LGN produced deficits in colour vision, fine pattern discrimination, and fine stereopsis (Schiller *et al.* 1990). Damage to the parvocellular retinogeniculate pathway reduced chromatic sensitivity at all spatial frequencies and achromatic sensitivity at high spatial and low temporal frequencies (Merigan 1989). Lesions in V4 also produced colour-vision defects, although they were less severe and less permanent than those produced by LGN lesions. Lesions in the inferior temporal lobe produced deficits in object recognition (Pohl 1973).

Humans with lesions in the inferior temporal cortex cannot discriminate between shapes in a similar orientation but can discriminate between shapes that differ widely in orientation. Dijkerman *et al.* (1996) described a patient with visual form agnosia arising from damage to the inferior temporal cortex. She could not match the orientation of one disc to that of another. However, she could accurately grasp a disc displayed in different orientations, presumably through the mediation of an intact parietal system (see Section 5.7.3c).

Milner (1997) described a patient with carbon monoxide poisoning who could not discriminate between stimuli that differed in size, orientation, or shape even though she could perform skilled actions that required the registration of those stimulus attributes. The literature on the dissociation between perception and action is reviewed in Milner and Goodale (1995). Lesions in the magnocellular laminae of the LGN produce deficits in motion perception, high-frequency flicker perception, and pursuit eye movements (Schiller *et al.* 1990) as well as reduced contrast sensitivity for low spatial frequency gratings modulated at high temporal frequencies (Merigan and Maunsell 1990). Lesions in MT produce similar deficits (Dürsteler *et al.* 1987). Lesions in the posterior parietal cortex in man and other primates are associated with loss of spatial memory, disturbances of spatial attention, and defects in representing spatial relations. See Critchley (1955) and Andersen (1987) for reviews. Humans with lesions in the parietal cortex can discriminate between shapes but cannot discriminate between rotated shapes (Walsh and Butler 1996).

It has been proposed that the magnocellular system is solely responsible for coding depth and that the parvocellular system is blind to binocular disparity (Livingstone and Hubel 1988). Cells sensitive to binocular disparity are certainly present in area V3 and in MT, which are considered part of the magnocellular system (Maunsell and Van Essen 1983b; Felleman and Van Essen 1987). However, the idea that disparity detection is confined to the magnocellular system must be rejected (Section 6.3.5).

The picture that emerges is of many retinotopically coded visual areas, which process different visual features. The system is hierarchical in that

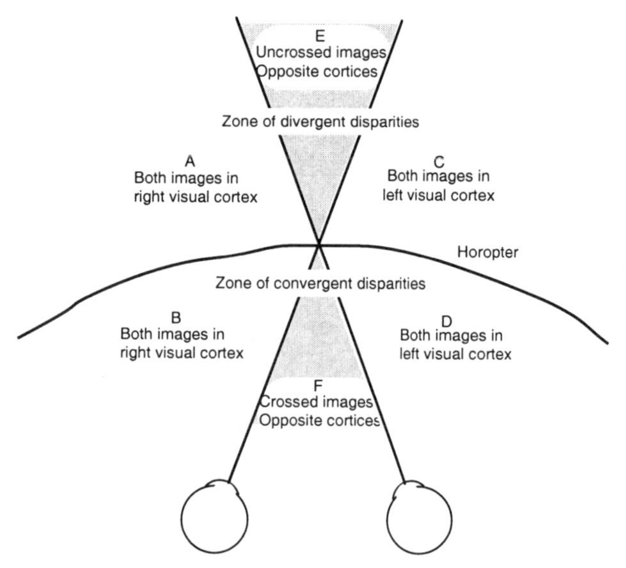

Figure 5.46. Divisions of the visual field.
The diagram indicates regions of visual space where binocular images project to the left, right, or opposite visual cortices. Regions between the visual axes generate crossed or uncrossed images according to whether they are nearer than or beyond the fixation point. Regions beyond the horopter generate images with divergent disparity and regions nearer than the horopter generate images with convergent disparity.

Figure 5.47. Bertram Payne.
Born in Egypt in 1953. He obtained a B.Sc. and Ph.D. in Zoology from the University of Durham, England, He conducted post-doctoral work at the Medical College of Pennsylvania. In 1984 he moved to Boston University School of Medicine where he is a Professor of Anatomy and Neurobiology.

each area receives its inputs from the striate cortex, sometimes by way of other areas. The system is parallel in that there is divergence into distinct areas that handle different visual features simultaneously. These areas interact through lateral connections. At the same time, centres higher in the sequence send recurrent signals back to earlier stages. Some higher centres receive visual inputs from the pulvinar or other centres before they receive inputs from V1. They could therefore modulate inputs from V1 by feedback (see Zeki 2001). The system is probably also multilayered in that the output of each processing unit could be available to consciousness.

Ascending connections terminate in layer 4, while descending connections terminate in layer 1 and possibly layer 6 (Felleman and van Essen 1991). Finally, the parallel pathways converge into processes that determine perception and action (see Grossberg 1990; Rockland and Van Hoesen 1994).

Ultimately, information from interrelated parallel streams must be combined to form unitary percepts of objects and of spatiotemporal relationships. Boundaries in the visual field defined by colour, texture, motion, and disparity typically coincide. If each feature were processed independently, some means would have to be found to keep contours defined by different features in register. Otherwise we would experience the world like a poorly printed photograph in which colour spills across edges.

This may be a pseudo problem when applied to the visual scene as a whole. Perhaps the only unitary percept of the scene as a whole is that of the crudely analyzed general features of the visual scene formed before the streams segregate. Distinct and specialized processing would be concerned only with the analysis of particular regions or features of the primitive scene. There may be no overall unitary percept derived from higher-order processing of inputs. Parts of the primitive scene not attended to may be perceived only in the crudest terms for purposes such as control of posture, detection of self-motion, control of eye movements, and redirection of attention. The ultimate representation of the perceptual world may be a high-level associative network, which 'consults' visual information provided by whichever visual centres are relevant to a given task. For example, if the task is that of discriminating between two orientated lines, the network may consult only V1. If the task is that of motion discrimination, it may consult MT, and if it is that of recognizing a visual object it would consult the inferior temporal cortex.

5.8 MIDLINE INTERACTIONS

If the nasal and temporal inputs are perfectly partitioned in the chiasma all axons from the nasal hemiretinas go to the contralateral cerebral hemisphere and all those from the temporal hemiretinas go to the ipsilateral hemisphere. Objects to the left of both visual axes (regions *A* and *B* in Figure 5.46) produce images in the right visual cortex and objects to the right of both visual axes (regions C and D) produce images in the left visual cortex. Images are said to have uncrossed disparity when the object is beyond the horopter and crossed disparity when it is nearer the horopter (Section 15.2).

Consider an object lying in the zone between the visual axes. The images of an object beyond the fixation point (region E) fall on the nasal halves of each retina and those of an object nearer than the fixation point (region F) fall on the temporal halves. In both cases, the images project to opposite cerebral hemispheres. With perfect partitioning at the chiasma, no binocular cells would receive direct inputs from the disparate images and stereopsis based on direct inputs would be impossible for such objects. In fact, stereoscopic acuity is particularly good for objects lying on the midsaggital plane near the fixation point. Therefore, there must be cortical cells serving the midline region that have receptive fields in opposite hemiretinas in the two eyes.

Evidence for this convergence of inputs has been provided from the cat. A strip of cortex at the boundary between areas 17 and 18 of each hemisphere of the cat contains binocular cells with receptive fields that overlap in the vertical midline of the visual field (Stone 1966; Leicester 1968; Blakemore 1969). Cortical cells in this region have receptive field centres up to 3° into the ipsilateral visual field, along the horizontal meridian. Along more eccentric horizontal meridians above and below the retinal equator, the receptive field centres extend up to 10° into the ipsilateral field (Payne 1990) (Portrait Figure 5.47). Cells with ipsilateral inputs near the border between areas 17 and 18 of the cat are driven by Y cells, are broadly tuned for orientation, and are strongly dominated by the contralateral eye (Diao *et al.* 1990). Two mechanisms could underlie bilateral projection to midline cells; imperfect partitioning of inputs at the optic chiasma, and interhemispheric connections projected through the corpus callosum.

5.8.1 Partitioning of hemiretinas

Linksz (1952) suggested that the nasal and temporal hemiretinas are not perfectly partitioned. One reason for this could be that ganglion cells with large receptive fields in the midline region necessarily receive inputs from receptors in both halves of the retina. Kirk *et al.* (1976a, 1976b) recorded from ganglion cell axons that crossed in the chiasm in cats. They found that, although the receptive fields of X cells did not encroach more than 0.5° across the retinal midline, those of Y cells and slowly conducting W cells encroached 15° or more over the midline.

Another reason for imperfect partitioning of visual inputs could be that decussating and non-decussating axons are incompletely segregated in the midline region. In a region extending about 1 to 2° on either side of the vertical retinal meridian in the cat, some ganglion cells project to the ipsilateral LGN and some to the contralateral LGN (Stone 1966; Nikara *et al.* 1968; Sanderson and Sherman 1971; Cooper and Pettigrew 1979a; Levick *et al.* 1981). While few if any X cells in the temporal retina project ipsilaterally, 5% of Y cells and 60% of W cells do so (Stone and Fukuda 1974). W cells have large receptive fields and project mainly to subcortical nuclei such as the medial interlaminar nucleus and superior colliculus. This suggests that they are involved more with the control of vergence than with stereopsis. However, Pettigrew and Dreher (1987) found cells in cortical area 19 of the cat, which receive inputs from W-type ganglion cells, and are tuned to zero or uncrossed disparities.

In the macaque monkey, horseradish-peroxidase labelling revealed a few ipsilaterally projecting cells in the nasal retinas and a few contralaterally projecting cells in the temporal retinas. Both were within a 1°-wide vertical strip around the midvertical meridian, which expanded to 3° in the perifoveal region (Stone *et al.* 1973; Bunt and Minckler 1977). Bilateral projection from the foveal region could explain **foveal sparing**, in which vision is preserved around the fovea following damage to one visual pathway. However, eccentric fixation may create a false impression of foveal sparing (Williams and Gassel 1962). Also, foveal sparing does not always occur; sometimes there is foveal splitting, in which the scotoma partitions the fovea.

The results of a study by Leventhal *et al.* (1988) may solve this problem. They, also, used horseradish-peroxidase labelling in the macaque and found cells in the nasal retina that projected ipsilaterally but no cells on the temporal side that projected contralaterally. From this, it follows that damage to one visual pathway should produce foveal sparing in the contralateral eye and foveal splitting in the ipsilateral eye. They cited clinical cases that conformed to this prediction.

Direct bilateral projection of the visual inputs has been tested psychophysically in humans. Subjects used one hand to press one of two keys to indicate whether a monocular target was to the left or right of fixation. The procedure was repeated with the other hand. Reaction time was 25 ms shorter when the visual target was projected to the same half of the brain as that controlling the manual response, even when the target was only 15 arcmin away from the midline (Harvey 1978). This result argues against the idea of overlap of visual projection in the midline region. It is what one would expect if the longer reaction time involves a longer route through the callosum. Lines and Milner (1983) measured simple manual reaction times and obtained an advantage of about 2.4 msec when the stimulus was in the same half of the brain as that controlling the hand. The simple reaction time is a purer measure of interhemispheric conduction time than the choice reaction time used by Harvey. I will now consider other evidence for transcallosal visual inputs.

5.8.2 Corpus callosum

5.8.2a *Callosal pathways*

Bilateral projection to midline cortical cells could be carried by interhemispheric fibres in the corpus callosum even if inputs from the retinas were perfectly partitioned in the chiasma. In each hemisphere, recipient zones for transcallosal fibres are coextensive with projection zones from which transcallosal fibres originate (Van Essen *et al.* 1982). Retinal areas near the vertical meridian project to the region between cortical areas 17 and 18 in the cat and to the region between V1 and V2 in the monkey. In both animals, the boundary region is well served by interhemispheric connections (Choudhury *et al.* 1965; Hubel and Wiesel 1967; Harvey 1980). In the monkey, 80% of transcallosal cells occur near the V1/V2 border and are from the foveal area. The other cells are from the periphery and extend 7 mm into V2 from the V1/V2 border in finger-like bands.

Callosal linkages occur between cells an equal distance from the border, which means that they interconnect visual inputs from symmetrically opposite locations in the visual fields of the two eyes (Abel *et al.* 2000). They are thus in a position to process images with large disparities spanning the midline. These disparities could serve coarse stereopsis or control vergence.

In V1, the terminals of these neurones are confined to cortical layers 2, 3, 4B, and 5. In V2, they occur in all layers (Kennedy *et al.* 1986). Transcortical inputs are also concentrated in the boundary regions between visual areas V2, V3, V3A, V4, and VP, and their presence helps to reveal the boundaries of those areas (Van Essen *et al.* 1982).

Visual areas V4 and MT in the monkey receive extensive transcallosal inputs which, although most dense near the representation of the vertical meridian, are not confined to the midline region (Van Essen and Zeki 1978; Maunsell and Van Essen 1987). Recordings in V4 of the alert monkey reveal that transcallosal influences extending more than a degree or two beyond the vertical midline are purely inhibitory. It has been suggested that these widespread inhibitory influences are concerned with colour constancy and figure-ground segregation rather than with stereopsis (Desimone *et al.* 1993).

Anatomical studies on human brains have revealed bands of transcallosal terminals along the boundary between areas V1 and V2 and in surrounding areas (Clarke and Miklossy 1990).

The concentration of transcallosal connections near the border between areas 17 and 18 seems to depend on visual experience, since transcallosal connections are severely reduced in number in cats reared with both eyelids sutured (Innocenti and Frost 1980). It seems that information carried by the callosal pathway is confined to low spatial and low temporal frequencies and high contrasts (Berardi and Fiorentini 1987; Berardi *et al.* 1987). The functions of the corpus callosum have been reviewed by Berlucchi (1972) and Kennedy *et al.* (1991).

5.8.2b *Section of the chiasma*

One approach to the issue of midline stereopsis is to study the effects of midsagittal section of the optic chiasma. This procedure severs all direct inputs from the contralateral eye to each hemisphere, so that any remaining binocular cells in the visual cortex receive their contralateral input indirectly through the callosum (Berlucchi and Rizzolatti 1968). Estimates of how many binocular cells survive in split-chiasm cats have varied. Milleret and Buser (1984) claimed that none survive, Lepore and Guillemot (1982) claimed that 30% survive, and Cynader *et al.* (1986) claimed that up to 76% survive. In split-chiasm cats there is a complete loss of binocular cells responsive to uncrossed disparity arising from stimuli in the median plane (Lepore *et al.* 1992). This is what one would expect because the images of such objects fall on the nasal hemiretinas, the inputs from which are severed in the split-chiasm cat. Split-chiasm cats performed poorly on a random-dot stereo discrimination test presented on a jumping stand apparatus, which constitutes a test of fine stereopsis (Lepore *et al.* 1986). On the other hand, Blakemore (1970a) reported the case of a boy in whom the decussating pathways in the optic chiasm were completely sectioned and who could discriminate depth in the region of the midline, as revealed by a coarse test of stereopsis involving disparities of at least 3 arcmin.

5.8.2c *Callosectomy*

A second approach to the issue of midline stereopsis is to study the effects of cutting transcallosal pathways. The physiological effects are discussed first. Cooling the border between areas 17 and 18 in one hemisphere of the cat produced a selective loss of binocularity in a significant number of midline cells in the other hemisphere (Blakemore *et al.* 1983). Transection of one optic tract also led to a loss of binocularity in the contralateral hemisphere (Lepore *et al.* 1983). Almost all disparity-tuned cells in area 19 of the cat lost their disparity tuning following section of the corpus callosum, suggesting that these cells receive their input from the contralateral eye by this route (Guillemot *et al.* 1993). Callosectomy in adult cats produced a permanent reduction in binocular cells in the region of area 17 receiving callosal inputs, extending 4° to either side of the retinal midline region (Payne *et al.* 1984a, 1984b). However, there is conflicting evidence on this point. For instance, Elberger and Smith (1985) reported that callosectomy affected binocularity and visual acuity at all retinal eccentricities when performed on cats before the age of about 3 weeks, and had no effect after that age (see also Elberger 1989, 1990). Minciacchi and Antonini (1984) also failed to find any loss of binocularity in areas 17 and 18 of unanaesthetized callosectomized adult cats. However, they did not test the cells for disparity sensitivity. Part of the effect of early callosectomy on binocularity may be due to eye misalignment that this procedure introduces (Elberger 1979). Binocular cells in visual areas beyond areas 17 and 18, such as the lateral suprasylvian area and the superior temporal sulcus, are not affected by neonatal callosectomy in the normal cat. However, they are affected by this procedure in Siamese cats, in which the visual pathways fully decussate (Zeki and Fries 1980; Elberger and Smith 1983; Marzi *et al.* 1982).

I now consider the behavioural effects of callosectomy. In cats with neonatal section of the callosum, coarse stereopsis revealed in reactions to a visual cliff was adversely affected (Elberger 1980). Mitchell and Blakemore (1970) reported a clinical case in which callosectomy led to a disruption of midline stereopsis, also measured by a test involving only coarse (large) disparities. In other studies, fine stereopsis was not affected by callosal section in the neonatal or adult cat (Timney *et al.* 1985; Lepore *et al.* 1986). Jeeves (1991) found that four subjects with partial or complete loss of the corpus callosum were able to detect only crossed disparities in the midline. They suggested that detection of crossed disparities is mediated by the anterior commissure, which was intact in these subjects. Rivest *et al.* (1994) found that three subjects with congenitally absent corpus callosum were as precise as normal subjects in adjusting two textured plates to equidistance when the plates were on opposite sides of the midline. However, one and perhaps all three subjects had intact anterior commissures and, in any case, may have performed the task on the basis of monocular cues to depth, since the plates were actually moved in depth. Transection of the part of the corpus callosum known as the splenium did not to affect midline stereopsis in the monkey (Cowey 1985) and section of all parts other than the anterior commissure was without effect in monkeys (LeDoux *et al.* 1977) and humans (Bridgman and Smith 1945). It looks as though at least some of the crucial fibres serving midline stereopsis cross in the anterior commissure. Some fibres could cross in the corpus callosum from the lateral geniculate nucleus to the opposite visual cortex. The evidence for such connections is controversial (Glickstein *et al.* 1964; Wilson and Cragg 1967).

In cats with unilateral removal of the visual cortex, midline cells tuned to fine disparity were still present in the remaining hemisphere but cells tuned to coarse disparity were lost (Gardner and Cynader 1987).

All these findings support a suggestion made by Bishop and Henry (1971) that the callosal pathway is responsible for midline bilateral integration only for coarse stereopsis, whereas fine stereopsis in the midline depends on overlap of direct visual inputs in the midline region. This issue is discussed in Section 16.2.5.

6 *Physiology of disparity detection*

6.1 DISPARITY DETECTORS

Before the 1960's many leading visual scientists, including Helmholtz, believed that binocular stereopsis did not arise from the conjunction of visual inputs at an early stage of visual processing but arises from high-level cognitive processes (Section 2.2.5). Ramón y Cajal (1911) proposed that inputs from corresponding regions of the two retinas converge on what he called "isodynamic cells" and that this mechanism forms the basis of unified binocular vision. This idea received experimental verification when Hubel and Wiesel (1959, 1962) reported that the receptive fields of binocular cells in the visual cortex occupy corresponding positions in the two eyes. If this were strictly true, and if each binocular cell had identical receptive fields in each eye, all binocular cells would respond optimally to stimuli lying along the horopter, and disparity could not be recovered from the output of such cells. Jack Pettigrew, a student of Peter Bishop, discovered disparity detectors in the cat in 1967 in Sydney (Section 2.2.5). Gian Poggio and his co-workers at Johns Hopkins University in Baltimore discovered disparity detectors in the monkey in 1977.

The search for binocular cells that are selectively tuned to different disparities was beset with the problem of ensuring that the images in the two eyes were in register. If the images are slightly out of register, a cell that is really tuned to zero disparity will appear to be tuned to a disparity equal to the image misregistration. Also, any movement of the eyes during the recording introduces artifacts. Several procedures have been used to solve this problem. In the anaesthetized animal, eye movements are controlled by paralyzing the eye muscles and attaching the eyeball to a clamped ring. A rotating mirror or a prism of variable power controls the ef-

Figure 6.1. Horace B. Barlow.
Born in England in 1921. He graduated from Trinity College Cambridge. He was a Research Fellow at Trinity College between 1950 and 1954 and a Lecturer at King's College, Cambridge between 1954 and 1964. Between 1964 and 1973 he was Professor of Physiological Optics and Physiology at the University of California at Berkeley. He then returned to the Physiological Laboratory at Cambridge University as a Royal Society Research Professor. He became a Fellow of the Royal Society of London in 1969.

Figure 6.2. Colin Blakemore.
Born in Stratford-upon-Avon, England in 1944. He obtained a B.A. in Medical Sciences from Cambridge University in 1965 and a Ph.D. in Physiological Optics from Berkeley in 1968. He also holds a Sc.D. (Cantab) and D.Sc. (Oxon). He was Lecturer in Physiology at Cambridge from 1972 to 1979 and has been Waynflete Professor of Physiology, Oxford since 1979. He is Director of the Oxford Centre for Cognitive Neuroscience. Recipient of the Robert Bing Prize from the Swiss Academy of Medical Sciences, the Netter Prize from the French Académie Nationale de Médecine, the Royal Society Michael Faraday Award, the G.L. Brown Prize from the Physiological Society, the Charles F. Prentice Award from the American Academy of Optometry, and the Baly Medal of the Royal College of Physicians.

fective direction of gaze. In the reference-cell procedure, different electrodes record responses of a test cell and a reference binocular cell, each with receptive fields in the central retinas. Changes in the response of the reference cell indicate when eye movements have occurred (Hubel and Wiesel 1970a). In a related procedure, eye drift is monitored by the response of a reference cell to monocular stimulation (Maske *et al.* 1986a). Image stability can also be indicated by responses of LGN cells of foveal origin, one from each eye (LeVay and Voigt 1988).

These procedures indicate when eye drift has occurred, but they do not specify when test stimuli have zero disparity, since the reference cell may not be tuned to zero disparity. One solution to this problem is to use the mean response of several reference cells to define zero disparity (Nikara *et al.* 1968). Another procedure is to use an ophthalmoscope to project images of retinal blood vessels onto the screen on which the stimuli are presented (Bishop *et al.* 1962; Pettigrew *et al.* 1968) (Portrait slides 3 and 4). The problem is simplified when testing is done on alert monkeys trained to converge their eyes on defined targets.

A second problem in identifying a disparity detector is to ensure that changes in the response of a binocular cell are not due to incidental changes in stimulation. For example, the stimulus in one eye may move outside the receptive field of the binocular cell when the experimenter changes the disparity of the stimuli. By measuring the response of the binocular cells for many combinations of image positions, effects of monocular position can be separated from effects of disparity (Ohzawa *et al.* 1990). Another procedure is to use random-dot stereograms, for which disparity is not related to monocular features of the stimulus. The physiology of stereopsis has been reviewed by Gonzalez and Perez (1998b) and Cumming and DeAngelis (2001).

6.2 DISPARITY DETECTORS IN CATS

Barlow, Blakemore, and Pettigrew (1967) reported from Berkeley, California that certain binocular cells in the visual cortex of the cat responded selectively to line and bar stimuli with a particular binocular disparity (Portrait Figures 6.1, 6.2, and 6.3).). Similar findings, were reported about the same time from Sydney, Australia, by Pettigrew, Nikara, and Bishop (1968) (Section 2.2.5) (Portrait Figure 6.4).

Figure 6.3. John D. Pettigrew FRS.
Professor of Physiology and Pharmacology at the University of Queensland, Australia.

Figure 6.4. Peter O. Bishop.
Born in Tamworth, New South Wales, Australia in 1917. He obtained the M.B. and B.S. in 1940 and the D.Sc. in 1967 from the University of Sydney. After serving as a surgeon during the war he studied at University College London from 1946 to 1950. He held academic appointments at the University of Sydney from 1950 to 1967 when he became Professor of Physiology at the Australian National University in Canberra. He retired in 1983. He is a Fellow of the Australian Academy of Sciences, Fellow of the Royal Society of London, and Officer of the Order of Australia. Joint winner of the Australia Prize in 1993.

Disparity-selective neurones are referred to as **disparity detectors**. The **disparity tuning function** of a cell is its frequency of firing as a function of the disparity of the retinal images. The **preferred disparity** of a cell is the disparity to which it responds most vigorously. A preferred disparity also has a sign (crossed or uncrossed) and an axis (horizontal, vertical, or oblique). The **disparity selectivity** of a cell is indicated by the width of its disparity tuning function at half its height. The narrower the tuning function, the higher the cell's selectivity. The **response variability** of a cell is the mean fluctuation in the firing rate for a constant stimulus (Crawford and Cool 1970). A binocular cell shows facilitation, summation, or occlusion depending on whether its response to a binocular stimulus is greater than, equal to, or less than the sum of responses to monocular stimuli. Binocular summation is discussed in Section 8.1.1.

The preferred disparity of a cell is measured by observing its response to stimuli presented simultaneously to the two eyes as a function of disparity. In a related procedure, the mean retinotopic position of the receptive field of a binocular cell is determined first in one eye and then in the other. The separation in degrees of visual angle between the two separately determined monocular fields is the cell's receptive field offset. It is often assumed that the preferred disparity of a binocular cell equals its receptive field offset. However, it is not easy to determine the relationship between these two measures. The receptive field offset can be measured only in cells that respond to each eye separately, but many cells tuned to disparity give an excitatory response to monocular stimulation of one of the eyes but not of the other. In any case, the disparity sensitivity of some, or all, binocular cells may depend on offsets of subunits within their receptive fields rather than on the offset of the fields as a whole (Section 6.5.4).

Each retina projects retinotopically into the visual cortex, so that, as one progresses across the cortical surface within an ocular dominance band, a systematic change occurs in the retinal location of the receptive fields in the single eye. However, in each small region of the cortical surface there is a random scatter of receptive field locations. In the cat, the variance of this scatter has been estimated to be 0.12° (Albus 1975). In a more recent study, the scatter of receptive-field positions of binocular cells within each cortical column was about half the mean size of receptive fields (Hetherington and Swindale 1999). Some of this variation in position was correlated between the two eyes. Only the uncorrelated component of variation could contribute to disparity detection. In these studies, the mean variation of receptive field position was measured within a given cortical column. There could also be variation in mean offset between neighbouring columns, which could contribute to disparity detection. If the monocular fields in a given small region are paired at random to form receptive fields of binocular cells,

Figure 6.5. Gian F. Poggio.
Born in Genoa, Italy in 1927. He received the Doctor of Medicine from the University of Genoa in 1951. He then held Fellowships in Neurological Surgery and in Physiology at John's Hopkins University. In 1960 he joined the faculty of Johns Hopkins University, in the Department of Physiology from 1960 to 1980 and as Professor of Neuroscience from 1980 until he retired in 1993. In 1989, with Bela Julesz, he received the Lashley Award of the American Philosophical Society for work on stereopsis.

then the variance of the field offsets should equal the sum of the variances of the monocular scatters. For the cat, this is approximately true (Bishop 1979). Blakemore and Pettigrew (1970) found that the positions of receptive fields of binocular neurones show greater variance in the ipsilateral retina than in the contralateral retina. Thus, the more recently evolved ipsilateral projection is less precise than the phylogenetically older contralateral projection.

The preferred disparities in the study by Barlow *et al.* were distributed horizontally over 6.6° and vertically over 2.2°. Other investigators have found that the preferred disparities of cells in the cat's visual cortex had a standard deviation of only about 0.5° for both horizontal and vertical disparities for eccentricities of up to 4°, increasing to 0.9° at an eccentricity of 12° (Nikara *et al.* 1968; Joshua and Bishop 1970; von der Heydt *et al.* 1978). These values suggest a range of disparities of only about 3°. Ferster (1981) found no cells sensitive to disparities over 1°. There are at least two reasons for these discrepancies. The first is that the apparent peak disparity to which a given cell is tuned is affected by the extent to which eye movements are controlled. The second is the accuracy with which the position of zero disparity is registered in testing.

Blakemore (1970c) found that disparity-selective cells in the cat's visual cortex were arranged in distinct columns of tissue, which he called **constant depth columns**. He also described a type of columnar arrangement in which the binocular cells were driven by receptive fields in the contralateral eye that were all in the same region of the retina and by receptive fields in the ipsilateral eye that were scattered over several degrees. The cells in such a column have a variety of preferred disparities, but they all respond to a stimulus lying on an oculocentric visual line of one eye. The column "sees" along a tube of visual space lined up with one eye. This finding should be replicated with more adequate control of receptive field mapping.

Disparity-tuned cells in areas 17 and 18 of the cat have been classified into three main types. Those in the first type have a narrow disparity tuning function centred at zero disparity and are known as **tuned excitatory cells**. Cells of the second type fire maximally to crossed disparities and are known as **near cells**. Cells of the third type respond to uncrossed disparities and are known as **far cells**. The distinction between the physiological responses of these cell types was not always clear and the cells probably lie along a continuum. Cells that could not be placed in any of these categories are described in later sections of this chapter.

Tuned excitatory cells are ocularly balanced—they respond equally well to either eye. The near and far cells show strong ocular dominance—they respond more strongly to stimulation of one eye than of the other or exclusively to stimulation of one eye (Fischer and Krüger 1979; Ferster 1981). Thus, most cells with high ocular dominance are sensitive to nonzero disparity and cells driven well through either eye are tuned to disparities close to zero or are nonselective for disparity (Maske *et al.* 1986a, 1986b; Gardner and Raiten 1986; LeVay and Voigt 1988). After ablation of areas 17 and 18, cats lost their ability to discriminate depth based on disparity although abilities such as vernier acuity and brightness discrimination survived (Ptito *et al.* 1992).

Disparity-tuned cells have also been found in area 19 of the cat. Pettigrew and Dreher (1987) found that such cells receive inputs from W-type ganglion cells and are tuned to zero or uncrossed disparities compared with the predominant crossed-disparity tuning of cells in area 17. Guillemot *et al.* (1993) found that only about 34% of cells in area 19 of the cat were tuned to disparity, compared with over 70% in area 17. Almost all cells in area 19 lost their disparity tuning following section of the corpus callosum, suggesting that these cells receive their input from the contralateral eye by this route. W cells have

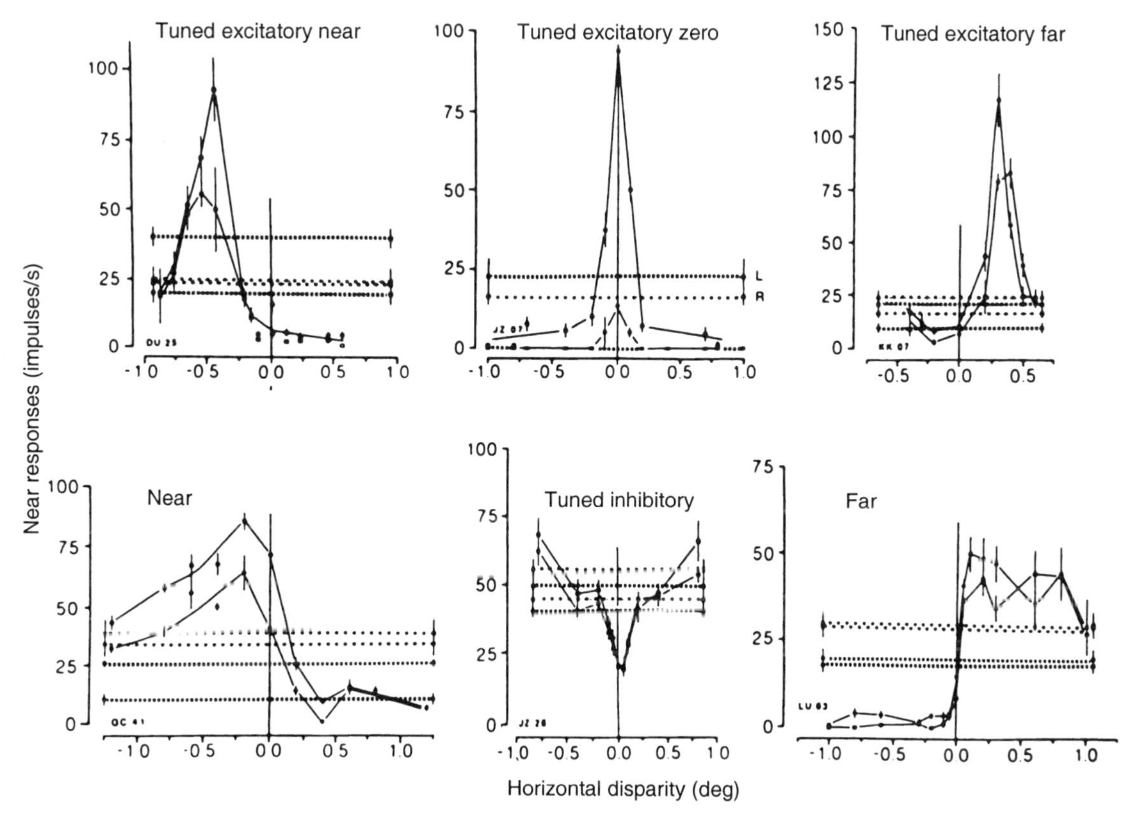

Figure 6.6. Six classes of disparity-tuned cells in monkey visual cortex.
For each function, frequency of nerve impulses is plotted for different horizontal disparities of a bright bar moving in each of two opposed directions across the cell's receptive field. For each cell, the functions for the two directions of motion are plotted separately. Vertical bars are standard errors. Finely dotted horizontal lines are responses of the left eye alone to two directions of motion. Coarsely dotted lines are responses of the left eye alone. (From Poggio 1991)

large receptive fields and may be involved more with the control of vergence than with stereopsis.

Cells in area 21a of the cat, which receive inputs from areas 17 and 18, are orientation selective and about 75% are binocular. Most of the binocular cells showed disparity tuning functions similar to that of cells in area 17, and there was no correlation between orientation tuning and disparity tuning (Wang and Dreher 1996). The tuned response of many of these cells to modulation of spatial phase between dichoptic gratings was retained for differences in orientation of the gratings of up to at least 45° (Vickery and Morley 1999).

6.3 DISPARITY DETECTORS IN PRIMATES

6.3.1 Disparity detectors in V1

Hubel and Wiesel (1970a) were the first to look for disparity-tuned cells in the anaesthetized monkey. They found no such cells in V1 but found them in V2. Their inability to find them in V1 was probably due to inadequate control of eye alignment. More recent recordings from V1 and several other visual areas of the anaesthetized monkey have revealed the same three types of disparity-tuned cell found in the cat plus an infrequent type that is inhibited by zero disparity, known as **tuned inhibitory cells**. These four types of cell have been found in V1 but are especially prevalent in V2, where at least 70% of neurones are tuned to horizontal disparity (Poggio and Poggio 1984; Hubel and Livingstone 1987). Evidence of a larger number of binocular cells in V2 than in V1 has also been obtained from visually evoked potentials in humans (Adachi-Usami and Lehmann 1983). Ablation of the foveal region of V2 in the monkey caused a severe elevation of stereo threshold (Cowey and Wilkinson 1991).

Poggio and Fischer (1977) were the first to record from disparity-tuned cells in the visual cortex of an alert animal—the rhesus monkey (Portrait Figure 6.5). These findings were extended by Gian Poggio and coworkers in Johns Hopkins University in Bal-

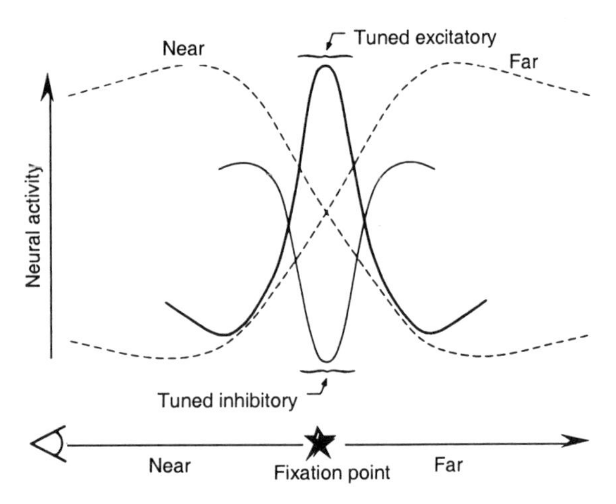

Figure 6.7. Idealized tuning functions of disparity detectors.
Tuned excitatory and tuned inhibitory cells are optimally tuned to zero disparity, and their tuning functions form a reciprocal pair. Tuning functions of "near" and "far " cells (dotted lines) are asymmetrical and do not show a well-defined preferred disparity. They, also, form a reciprocal pair. (Reproduced with permission from Poggio *et al.*, 1985, Vision Research, Pergamon Press)

Figure 6.8. Bruce Cumming.
He received a B.A. and M.D. at Oxford University and a Ph.D. with S. Judge, also from Oxford. He conducted postdoctoral work with A. Parker in the Department of Physiology at Oxford University. In 2000 he became an Investigator in the National Eye Institute at the National Institutes of Health in Bethesda, U.S.A.

timore (Poggio and Talbot 1981; Poggio *et al.* 1985, 1988; Poggio 1991). The monkey was trained to fixate a small visual target while bar stimuli were presented in an area 2° wide and in different depth planes relative to the fixation target. The problem of aligning the two visual fields was thus greatly simplified. Both bar stimuli and random-dot stereograms were used to determine the disparity tuning functions of the cells. In addition, the sensitivity of cells to changes in dichoptic correlation was determined with dynamic random-dot displays that changed from being correlated in the two eyes to being uncorrelated.

More than half the simple and complex cells in V1 were found to be disparity tuned and an increasing proportion of disparity-tuned cells was found as testing progressed into areas V2, V3, V3A, MT, and MST. About equal numbers of simple and complex cells were disparity-tuned, but complex cells were particularly sensitive to the depth in random-dot stereograms. The subfields within the receptive fields of complex cells presumably allow these cells to respond to the disparity between the microelements of the stereogram. The complex cells were also more sensitive than simple cells to changes in image correlation in a random-dot stereogram, probably for the same reason. Some cells sensitive to differences in image correlation were also sensitive to the sign and degree of disparity, while others were sensitive to differences in image correlation only when disparity was zero (Gonzalez *et al.* 1993a).

Binocular cells in the alert monkey were classified into six types, the four already described (excitatory cells tuned to zero disparity, tuned inhibitory cells, and near and far cells) plus **tuned excitatory cells,** tuned to either crossed disparities or uncrossed disparities (Figure 6.6). Another type of cell responding to stimuli moving in opposite directions in the two eyes is described in Section 29.6. Tuned excitatory cells with a narrow tuning function peaking within ± 12 arcmin of zero disparity are the most common type. All six types are inhibited by uncorrelated images and tend to have balanced ocular dominance. Tuned inhibitory neurones are suppressed by stimuli around zero disparity and most of them are excited by uncorrelated images. The tuning functions of these two types of cell are symmetrical and form reciprocal pairs. The near cells respond to crossed disparities and are inhibited by uncrossed disparities and the far cells have the opposite characteristics. The near and far cells do not have a well-defined preferred disparity and only about a third of them respond to changes in image correlation. Their tuning functions are asymmetrical and also form reciprocal pairs, as depicted in Figure 6.7. The tuned inhibitory and the near and far cells tend to have strong monocular dominance, suggesting that inputs from the weaker eye inhibit those from the dominant eye, except when the stimulus is at the appropriate depth relative to the horopter. The tuned near and tuned far cells have tuning functions with a well-defined preferred disparity, peaking at crossed or uncrossed disparities of up to 0.5°. They also have an inhibitory flank on the zero

Figure 6.9. Andrew J. Parker.
Born 1954 in Burnley, England. He obtained a BA in Natural Sciences in 1976 and a Ph.D. in 1980, both from Cambridge University. He is now University Lecturer in Physiology at Oxford University and Fellow St. John's College, Oxford. Recipient of the James S. McDonnell Foundation: 21st Century Scientist Award.

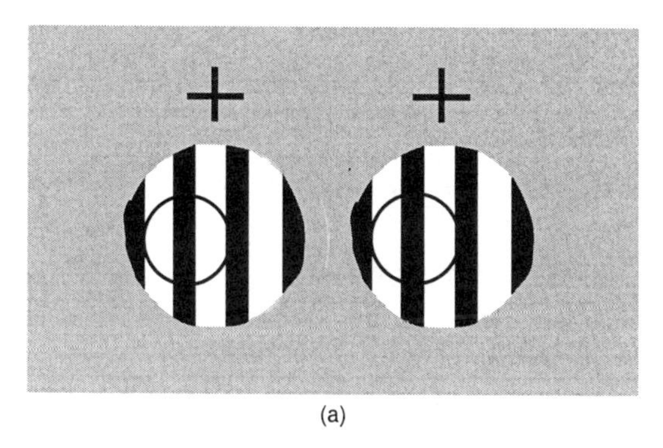

(a)

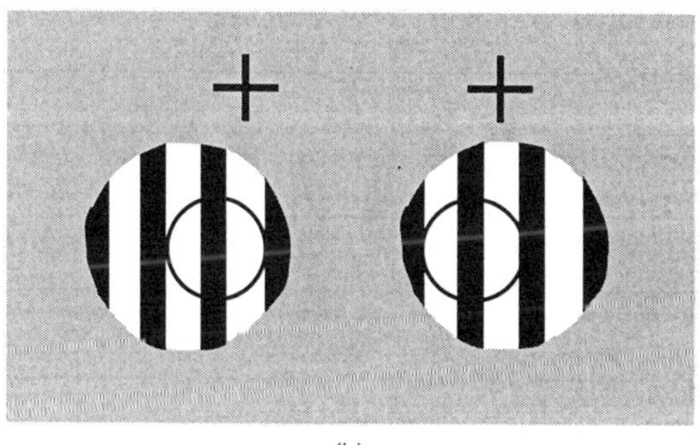

(b)

Figure 6.10. Detection of interpolated depth.
The small circles represent monocular receptive fields of a binocular cell in V1. In (a) the binocular cell receives a zero-disparity input from the gratings and the disc containing the gratings appears coplanar with the fixation cross. In (b) the receptive fields still receive a zero-disparity input but, with crossed fusion, the disc containing the grating appears in front of the fixation cross. The figure is adapted from Cumming and Parker (2000), who used a sine-wave grating.

disparity side of the tuning function. They, too, form a reciprocal pair.

The classification of disparity detectors into six classes and the scheme depicted in Figure 6.7 are abstractions in terms of standard prototypes. One should not regard them as a fixed number of exclusive types. Disparity detectors do not fall into exclusive categories and are not uniform over the visual field. Some binocular cells, particularly in V1, give the same response to all disparities. Some cells in V2 respond to both vertical and horizontal disparities (Section 6.6.4) and some are jointly tuned to disparity and motion (Section 6.6.3).

Poggio suggested that excitatory/inhibitory pairs of disparity detectors provide the physiological basis for fine stereopsis and that the near/far pairs provide the physiological basis for coarse stereopsis. This argument is based on the different widths of the tuning functions. However, we know that very fine discriminations can be achieved with a set of broadly tuned channels (Section 4.2.7). For instance, the chromatic channels are broadly tuned but achieve fine discriminations of colour. On this basis the near/far disparity channels should provide good discrimination for differences in disparity around zero because their tuning functions are steepest and overlap just at zero. The relative change in signal strength in the two channels is therefore greatest at this point, as can be seen in Figure 6.7. Fine discrimination around zero-disparity based on narrowly tuned detectors would require several detectors tuned to different disparities. See Lehky and Sejnowski (1990) for discussion of this issue.

Cumming and Parker (1999) found that the response of most binocular cells in V1 of alert monkeys changed when the absolute disparity of a random-dot stereogram was changed, leaving relative disparities constant (Portrait Figures 6.8 and 6.9). No cells were found that responded consistently to relative disparity over changes in absolute disparity. Mechanisms tuned specifically to patterns of relative disparity must reside in other visual centres.

Cumming and Parker (2000) placed identical gratings in a pair of dichoptic apertures. The apertures had a horizontal disparity equal to the period of the grating, as in Figure 6.10. This caused the fused image of the apertures and gratings to appear in front of the fixation cross. A binocular cell in V1 with receptive fields confined to the grating was exposed to a zero-disparity stimulus even though the grating appeared in front of the fixation cross. Binocular cells of this type in the alert monkey responded in the same way as they did to a grating in a zero-disparity aperture. In other words, these cells registered the local zero disparity of the grating

Figure 6.11. Rudiger von der Heydt.
Born in Rauschensamland, Germany in 1944. He studied Physics at the Universities of Göttingen, Marburg, and Munich from 1963 to 1969 and had research training in Neurophysiology with G. Baumgartner in the Department of Neurology at the University Hospital Zurich. He obtained a DR. sc.nat. (Ph.D.) from the Swiss Federal Institute of Technology in 1993. He has been Professor of Neuroscience at the Johns Hopkins University School of Medicine since 1993 and Professor in the Krieger Mind/Brain Institute since 1994. He received the Alfred Vogt Preis Award of the Swiss Ophthalmological Society in 1986 and the Golden Brain Award, Minerva Foundation, Berkeley in 1993.

Figure 6.12. Gregory C. DeAngelis.
Born in Fairfield, Connecticut in 1965. He obtained a B.Sc. in Biomedical Engineering at Boston University in 1987 and a Ph.D. in Bioengineering at Berkeley with Ralph D. Freeman in 1992. He conducted postdoctoral work at Stanford University with W. T. Newsome between 1995 and 1999. He is now an Assistant Professor of Neurobiology at Washington University School of Medicine.

whether or not the grating was perceived as displaced in depth. Cumming and Parker suggested that depth percepts arise at a higher level than V1.

Prince *et al.* (2002a) measured disparity tuning functions of cells in V1 of the alert monkey using dynamic random-dot stereograms. These stimuli contain no monocular information about depth. They also contain a broad range of orientations, which allows disparity selectivity to be measured independently of preferred orientation. Disparity selectivity varied from cell to cell in a continuous fashion rather than falling into discrete groups. Disparity sensitivity was measured by the discriminability of the maximum and minimum points on the disparity tuning function (see Section 6.5.1). Sensitivity was correlated with the degree of tuning for motion direction but not with the orientation preference or the ocular dominance of the cells. This issue is discussed further in Section 6.6.2. The responses of the cells could, approximately, be accounted for in terms of the energy model described in Section 6.5.1.

6.3.2 Disparity detectors in V2 and V3

Von der Heydt *et al.* (2000) (Portrait Figure 6.11) found cells in V1 of the anaesthetized monkey that responded to local disparities in a random-dot stereogram. However, V1 cells did not respond to the edges of the cyclopean shape defined by disparity, even though they responded to contrast-defined edges. In V2, many cells were sensitive to disparity in random-dot stereograms but many of them responded to both contrast defined edges and disparity-defined edges of a stereogram. They signalled the polarity of the depth step and the orientation of an edge. It seems that detection of a disparity-defined edge requires an extra step that is not achieved in V1 but is achieved in V2. Cells in V2 of the monkey respond to disparities lying outside their receptive fields when these disparities indicate the presence of an edge running across the receptive field (Section 23.1.3a).

In V3 of the monkey, which borders V2, about half the cells are disparity-tuned. About half of them are tuned excitatory cells centred on zero disparity. The others are tuned inhibitory cells or are tuned to either crossed or uncrossed disparity (Felleman and Van Essen 1987). Similar types of disparity-tuned cells have been found in area VP, which also borders V2 (Burkhalter and Van Essen 1986). In the macaque, cells in V3 with similar joint tuning to orientation and disparity are organized into columns (Adams and Zeki 2001). This organization is well suited to the extraction of higher-order orientation disparities and disparity gradients required for the perception of 3-D form. We will see that V3 projects to the posterior parietal cortex, which is concerned with processing 3-D form.

6.3.3 Disparity detectors in the dorsal stream

6.3.3a *Disparity detectors in MT and MST*
Disparity-tuned cells have been found in the dorsal processing stream going through MT and MST, to the parietal lobe. About two-thirds of cells tested in MT belong to the same four disparity-tuned types found in V1. Most of these cells are as sensitive to vertical disparity as to horizontal disparity (Maunsell and Van Essen 1983b). Disparity-tuned cells in MT occur in patches between 0.5 and 1 mm in diameter interspersed with regions not tuned to disparity. Cells in the same vertical column have similar sensitivity to disparity and similar preferred disparity. Across each patch there is a smooth transition of preferred disparity from crossed to uncrossed (DeAngelis and Newsome 1999). Cells in MT show disparity tuning for both drifting and stationary random-dot stereograms, although their response is generally stronger for moving stimuli (DeAngelis *et al.* 2000) (Portrait Figure 6.12).

Some cells in the lateral-ventral region of MST are jointly tuned to motion and binocular disparity. For some of these cells the disparity preference in the centre of the receptive field differs from that in the surround (Bradley and Andersen 1998; Eifuku and Wurtz 1999). They thus respond to relative or absolute spatial gradients of disparity and could be involved in perceptual segmentation of moving camouflaged objects.

DeAngelis *et al.* (1998) showed that binocular detectors in MT of the monkey are involved in stereopsis. They electrically stimulated clusters of MT cells possessing similar disparity preference. This biased the responses of monkeys in a near-far depth discrimination task in the direction of the disparity preference of the cells that were stimulated.

There is no evidence that MT is directly involved in the control of vergence eye movements. However, the intraparietal cortex is involved in vergence control (Section 9.11.2c) and motion and depth signals generated in MT feed into the parietal cortex.

Evidence reviewed in Section 7.10.2b suggests that MT is involved in binocular rivalry. Other evidence of disparity tuning in MT and MST is presented in Section 23.2.2.

6.3.3b *Disparity detectors in the parietal cortex*
Cells selectively responsive to the 3-D orientation of elongated objects or planes have been found in the caudal part of the lateral bank of the intraparietal sulcus (CIP and LIP) of monkeys (Sakata *et al.* 1999). Most cells responded to the orientation of surfaces in line and random-dot stereograms containing only disparity information about depth (Taira *et al.* 2000). In a later study it was found that most cells in area CIP sensitive to surface orientation in depth responded when the cue was either disparity or perspective or when both cues were present (Tsutsui *et al.* 2001). However, the response of these cells to disparity was stronger than their response to perspective. Some cells responded exclusively to disparity-defined depth and a few cells responded exclusively to perspective. A few responded only when both cues were present. Discrimination of surface orientation was impaired when CIP was inactivated by injection of muscimol.

This area is connected with the anterior intraparietal cortex, an area concerned with the manipulation of objects (Section 5.7.3). Disparity-sensitive neurones in LIP are also related to the control of eye movements in 3-D space (Gnadt and Mays 1995).

6.3.4 Disparity detectors in the ventral stream

6.3.4a *V4 and depth perception*
Cells sensitive to 3-D shapes occur in the ventral processing stream going through V4 to the inferior temporal cortex. Felleman and Van Essen (1987) reported preliminary data indicating that up to 50% of cells in V4 of the macaque are sensitive to disparity. Hinkle and Connor (2001) found that 80% of cells in macaque area V4 were tuned to disparities in the disparity range –1° to +1°. The tuning functions were similar to those found in other visual areas.

Lesions of V4 in monkeys produced no defects in stereopsis as tested with static or dynamic random-dot stereograms or Gaussian patches (Schiller 1993).

The tuning of some cells in V1, V2, and V4 to changes in stimulus size was found to depend on the distance of the stimulus. Some cells responded best to near stimuli while others preferred far stimuli. Some of the cells retained their distance sensitivity under monocular conditions but, for most of these cells, it was not possible to decide which depth cues were being used (Dobbins *et al.* 1998).

6.3.4b *Disparity detection in the inferior temporal cortex*
Janssen *et al.* (2000a) found that about 50% of cells in the inferior temporal cortex of the alert rhesus monkey were selective for the global 3-D structure of convex and concave disparity-defined random-dot displays depicting surfaces curved about a horizontal axis. They were not sensitive to local changes in disparity. These cells are therefore sensitive to higher spatial derivatives of disparity. Janssen *et al.* (2001) found that that most of these cells responded selectively to either disparity along the edges of a curved surface or to disparity gradients within the surface. They also found cells selective for magni-

tude and direction of curvature about a vertical axis.

Most of these cells were in the lower part of the subregion of the inferior temporal cortex known as TE. Only a few such cells were found in the lateral part of TE. Cells in both parts of TE were selective for 2-D shapes. Typically, the response of cells to a preferred 3-D shape was greater than the sum of responses to monocular stimuli. In other words, they showed binocular summation. Cells sensitive only to 2-D shapes did not show binocular summation and often showed binocular inhibition. Most of these cells were selective for either disparity gradients or disparity curvature. The response of disparity-curvature-detectors was disrupted by disparity discontinuities, such as edges and steps. The response of most cells was maintained when the stimulus was moved 3.2° in various directions. The response of all cells was affected to some degree by changes in stimulus size or curvature (Janssen *et al.* 2000b).

Uka *et al.* (2000) also found most neurons in the inferior temporal cortex of the alert monkey to be selective for both shape and disparity. Most cells were 'near' or 'far' cells; only a few were tuned to zero disparity. Neighbouring neurones had similar disparity selectivity. The receptive fields of most cells included the fovea, and disparity selectivity was reasonably constant when the stimulus was moved 2° in any direction from the fovea.

Ferraina *et al.* (2000) found cells in the monkey frontal eye fields that were sensitive to coarse disparities. These cells may be related to the planning of large vergence eye movements, since the frontal eye fields receive inputs from LIP in the parietal lobe, an area concerned with eye movements (see Section 5.7.3e), and project to the superior colliculus.

6.3.4c *Detection of shapes defined by disparity*

The inferior temporal cortex is concerned with shape recognition (Section 5.7.2b). Some cells in this area respond selectively to shapes defined by luminance, texture, or motion (Sáry *et al.* 1993). Tanaka *et al.* (2001) found cells in the inferior temporal cortex of the monkey that responded selectively to shapes in random-dot stereograms, which are therefore defined by disparity alone. The responses were the same to the same shapes defined by different patterns of random dots. For some cells, responses to disparity-defined shapes were correlated with responses to shapes defined by luminance or texture.

6.3.5 Parvo- and magnocellular disparity detectors

Livingstone and Hubel (1988) proposed that the parvocellular system is blind to disparity-defined depth. They based their conclusion on the report by Lu and Fender (1972) that depth cannot be seen in an equiluminant random-dot stereogram. This argument relies on the false assumption that the parvocellular system is wholly chromatic. The parvocellular system is not merely a colour-opponent system; it also codes high spatial-frequency stimuli defined by luminance contrast (Section 5.5.5). An equiluminant stimulus shuts off only the luminance-contrast component of the parvocellular system. In any case, stereopsis could not be confined to the magnocellular system because that system does not have the spatial resolution exhibited by the disparity system (Section19.3). The more reasonable conclusion is that the chromatic component of the parvocellular system does not code depth. Even this conclusion has to be modified, as we will see in the following and in Section 18.1.4.

The distinction between the chromatic and luminance channels is not the same as that between the parvocellular and magnocellular systems. While the magnocellular system is wholly or almost wholly achromatic, the parvocellular system is both chromatic and achromatic. The parvocellular system is structurally simple but functionally complex and can be considered to consist of four subchannels. Which subchannel is activated depends on the spatial and temporal characteristics of the stimulus (Ingling and Matinez-Ugieras 1985: Ingling 1991). Consider a ganglion cell in the red-green (r-g) opponent system. For a plain steady stimulus (low spatial and temporal frequency), responses from the red and green zones of the cell's receptive field are subtracted to yield an opponent chromatic signal. This subtractive process manifests itself as a photometric subadditivity, in which the threshold for detection of a mixture of green and red light is higher than one would predict from the thresholds of green and red lights tested separately. When the red-green stimulus is flickered, the red and green components begin to add to yield a luminance signal. The cell then loses it spectral opponency and shows photometric additivity. When a high spatial-frequency pattern is added to a steady red-green stimulus, the r-g components again begin to add. The cell again loses its spectral opponency and shows photometric additivity. Thus, the pure chromatic channel is a low spatial and low temporal frequency system. The r-g system is a colour-opponent system for low spatial and temporal frequencies, a pure luminance system for either high temporal frequencies or high spatial frequencies, and a mixed system in the middle range of spatial and temporal frequencies.

Thus, one is not likely to obtain stereopsis with equiluminant stereograms containing high spatial-frequency patterns. There is some evidence that equiluminant stereopsis occurs with low spatial-frequency patterns (Section 18.1.4). The magnocellular system has no colour opponency and cannot process high spatial frequencies. Therefore, this system will not see depth in any equiluminant stimuli or in fine patterns defined by luminance contrast.

The magnocellular or parvocellular layers in the LGN of the monkey can be destroyed selectively by injecting ibotenic acid. Lesions in the parvocellular layers severely reduced the ability to detect depth defined by small disparities in random-dot stereograms with high spatial frequency elements. The lesions had less effect on the ability to detect large disparities, especially in stereograms with low spatial frequency elements. Lesions in the magnocellular layers produced deficits in high temporal-frequency flicker and motion perception but no deficits in stereopsis, even for low spatial frequency stereograms (Schiller *et al.* 1990). This suggests that depth in low spatial frequency stereograms is detected just as well in the parvocellular system as in the magnocellular system.

Disparity in afterimages can create a sensation of depth (Section 19.9.2a). Ingling and Grigsby (1990) reported depth in afterimages of a perspective illusion and in afterimages of a reversible perspective figure. They assumed that afterimages do not arise in the transient magnocellular system and concluded that these sensations must arise in the parvocellular system. Depth sensations in these illusions are absent at equiluminance, so that if we accept Ingling and Grigsby's assumption, the sensations they observed must have arisen in the luminance component of the parvocellular system. However, this argument is weakened by evidence that afterimages do arise in the magnocellular system (Schiller and Dolan 1994).

Kontsevich and Tyler (2000) found stereoacuity to be high in a dynamic random-dot stereogram in which the luminance of each element was modulated smoothly over 300 ms. This stimulus was designed to stimulate the sustained parvocellular system. Acuity was low when element luminance was modulated back and forth in transient steps over the same interval. This stimulus was designed to stimulate the transient magnocellular system.

This evidence, and other evidence cited in Section 18.1.4, leads to the following conclusions. Only the parvocellular system processes disparity in fine patterns defined by luminance contrast or in coarse equiluminant patterns. Both the parvo- and magnocellular systems process disparity in coarse patterns defined by luminance contrast. The magnocellular system is probably most sensitive to rapid changes in disparity (see Tyler 1990).

6.4 SUBCORTICAL DISPARITY-TUNED CELLS

6.4.1 Disparity tuning in the pulvinar

Although there are binocular interactions in the LGN of the cat (Section 5.3.3), disparity-tuned cells have not been found there (Xue *et al.* 1987). Cells tuned to disparity occur in the cat's pulvinar, a subcortical nucleus closely associated with the LGN, which receives most of its inputs from the superior colliculus and visual cortex (Casanova *et al.* 1989). In the monkey, the pulvinar provides the major subcortical input to cortical area 18 (V2) (Levitt *et al.* 1995). Since cells in the pulvinar were found to be tuned to opposite directions of motion in the two eyes, they may be concerned with coding motion in depth. A person with a lesion in the left pulvinar had impaired stereoacuity (Takayama *et al.* 1994). The pulvinar has been implicated in controlling attention to salient visual stimuli (Robinson and Petersen 1992; Morris *et al.* 1997). Cells sensitive to phase disparity of dichoptic gratings have been found in the perigeniculate nucleus (Xue *et al.* 1988).

6.4.2 Disparity tuning in nucleus of the optic tract

In the cat, the response of about half the cells of the nucleus of the optic tract (NOT) that respond to moving displays are also tuned to binocular disparity; some show an excitatory response to a limited range of disparities and others show an inhibitory response (Grasse 1994). The NOT is part of the pretectum and is concerned with the control of optokinetic eye movements. Evidence reviewed in Section 23.4.5 shows that disparity signals are conveyed to the NOT in primates and serve to link the optokinetic response to the plane in depth upon which the eyes are converged.

6.4.3 Disparity tuning in the superior colliculus

Disparity-tuned cells, mostly of the tuned excitatory type, have been found in the superior colliculus of the opossum (Dias *et al.* 1991). However, stereopsis has not been demonstrated in this animal. Berman *et al.* (1975) found binocular cells in the superior colliculus of the cat that showed summation or facilitation when dichoptic images fell on corresponding receptive fields. Bacon *et al.* (1998) found that 65% of binocular cells in the cat's superior colliculus were sensitive to disparity. They found excitatory and inhibitory types tuned to zero disparity, as well as far and near cells. The superior colliculus is a subcortical nucleus with multisensory inputs, that is concerned with the direction of attention and control

Figure 6.13. Izumi Ohzawa.
Born in Hida Takayama, Japan in 1955. He graduated in Electrical and Electronics Engineering at Nagoya University in 1978 and obtained a Ph.D. Physiological Optics from the University of California, Berkeley in 1986. He continued postdoctoral research at Berkeley until he moved to Osaka in 2000. He is now Professor of Biophysical Engineering in the Graduate School of Engineering Science at Osaka University.

of saccadic eye movements to designated locations (Mays and Sparks 1980). A directional map of space based on visual inputs overlays a directional map based on auditory inputs. *Further work is needed to reveal whether this co-ordinated mapping extends to the third dimension.* Such a mapping could help to initiate vergence eye movements (see Section 9.11.2).

6.5 DISPARITY-DETECTOR PROPERTIES

6.5.1 Disparity tuning functions

6.5.1a *Tuning function characteristics*

The disparity tuning function of a cortical cell is the number of impulses per second plotted as a function of the horizontal or vertical angular separation of two optimally oriented dichoptic bars or gratings. Zero separation is defined in terms of the averaged response of several binocular cells with receptive fields in the foveal region when the eyes are converged on the stimulus. Tuning functions for orientation disparities are discussed in Section 6.6.2. A disparity tuning function, like any other sensory tuning function, has six basic features:

1. The peak amplitude of response.
2. Sensitivity, as indicated by the peak amplitude of response modulation to unit change in disparity. A related measure is the overall extent to which the firing rate is modulated by disparity. Ohzawa and Freeman (1986a) defined a binocular interaction index (BII), analogous to Michelson contrast:

$$\mathrm{BII} = \frac{R_{\max} - R_{\min}}{R_{\max} + R_{\min}} \qquad (1)$$

where R_{max} and R_{min} are the maximum and minimum points on the disparity tuning function. Prince *et al.* (2002a) pointed out that this index takes no account of variability in firing rate. They defined a disparity discrimination index (DDI):

$$\mathrm{DDI} = \frac{\left(R_{\max} - R_{\min}\right)}{\left(R_{\max} - R_{\min}\right) + 2RMS_{error}} \qquad (2)$$

where RMS_{error} is the root mean square of the variance over the whole tuning curve.

3. The tuning width, specified as the width (sometimes half width) of the tuning function at half its peak amplitude.
4. The relationship between excitatory responses and inhibitory responses.
5. The degree of symmetry of the tuning function about its peak amplitude.
6. The preferred disparity, indicated by the size and sign of disparity that evokes the strongest response. One must also consider possible interactions between the effects of horizontal and vertical disparities.
7. The resting level of response for disparities outside the tuning width. Most cortical cells have little or no resting level of discharge so that excitatory responses are usually stronger than inhibitory responses.
8. Response variability as indicated by the variability of the spike count with repetition of the same stimulus. The co-efficient of variation is the standard deviation divided by the mean spike count.

The first four features define the shape of the tuning function, the fifth specifies its position along the disparity axis, and the sixth defines its position along the response axis. This section deals with factors that determine the shape of a disparity tuning function. Section 6.5.2 deals with the number of tuning functions required to span the range of detected disparities and the homogeneity of tuning functions over the visual field.

Physiological data on the widths of disparity tuning functions were presented in Section 6.3. Behavioural data are reviewed in Section 19.6.3.

6.5.1b *Disparity tuning and monocular receptive fields*

It has been claimed that one can account for disparity tuning functions of simple cortical cells in terms of the size and strength of the excitatory and in-

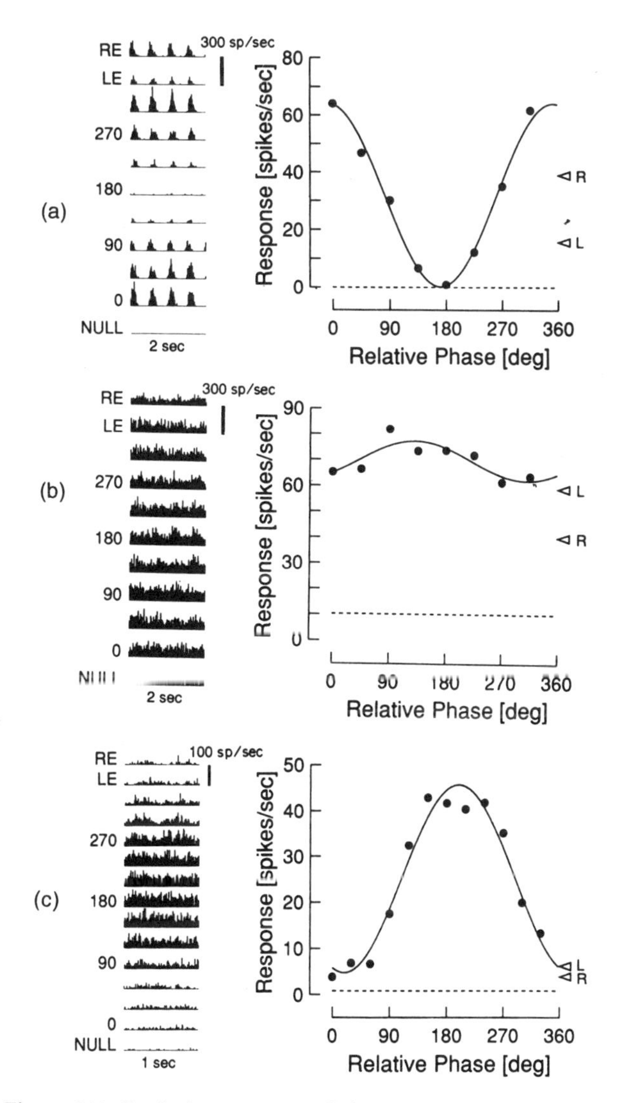

Figure 6.14. Cortical responses to dichoptic gratings.
Responses of a simple cell (a) and complex cells (b and c) in the visual cortex of a cat to a drifting sinusoidal grating presented dichoptically at various relative spatial phases. Time histograms of the responses are shown on the left, for each eye stimulated separately, and for various relative phases of dichoptic stimulation. Dashed lines represent the level of spontaneous activity. (Reproduced from Freeman and Ohzawa, 1990, with permission from Elsevier Science)

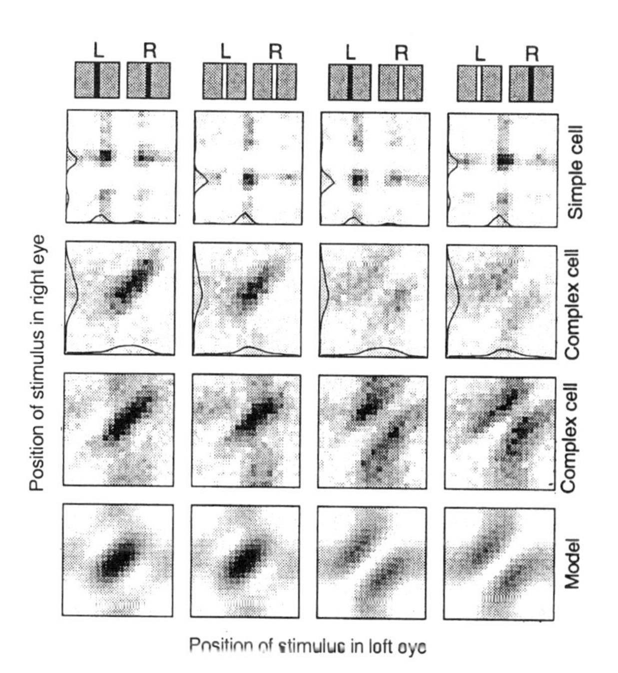

Figure 6.15. Reponses of cortical cells to disparity offset.
Each square in the top three rows shows the firing rate (higher rate—darker) of a binocular cell in cat visual cortex as a function of lateral positions of optimally oriented bars in left and right eyes. Columns show responses to bar stimuli shown at the top. In the two left columns, stimuli have the same luminance polarity in the two eyes. This causes tuning functions to have a single peak because stimuli come into register at only one position. In the two right columns, stimuli have opposite polarity. This causes tuning functions to have two peaks because there is one excitatory region in the eye with a bright bar and there are two in the eye with a dark bar (off flanks of simple cells are excitatory for dark bars). Separation between the peaks reveals the spatial period of receptive-field subunits. Profiles on the edges of some squares represent tuning functions to monocular stimuli. The bottom row of squares indicates responses predicted from a theoretical model. (Reprinted with permission from Ohzawa *et al.* 1990. Copyright American Association for the Advancement of Science)

hibitory regions of the cell's receptive fields in each eye (Bishop *et al.* 1971; Ferster 1981). Disparity-tuning functions of this type are said to conform to an **energy model**. The shapes of the monocular receptive fields of binocular cells are well described by elongated Gabor functions, which are sinusoidal modulations within an elongated Gaussian envelope (Section 4.5.2). The energy model of disparity detection predicts that the tuning function of a disparity detector will also be well described by an elongated Gabor function.

Ohzawa and Freeman (1986a) stimulated simple cells in the cat's visual cortex with dichoptic drifting sinusoidal gratings of optimal spatial frequency and orientation (Portrait Figure 6.13). Most cells responded most vigorously when the gratings in the two eyes were in a particular spatial phase and least when they were 180° away from the optimal phase (see Figure 6.14). Phase-specific interactions were absent when the gratings were orthogonal. The phase specificity of a cell's response did not depend on the degree of ocular dominance of the cell, except for a few strongly monocular cells that showed a purely inhibitory, phase-independent response to stimulation of the silent eye.

The modulation of response of a binocular cell in area 17 of the cat was the same for dichoptic gratings with very different luminance contrasts as for gratings with equal contrasts (Freeman and Ohzawa 1990). Similar results were obtained from V1 of the monkey (Smith *et al.* 1997a, 1997b). There must be a gain-control mechanism that keeps the monocular inputs to disparity-tuned cells in balance. In contrast gain-control, the contrast range of a detection system becomes adjusted to the mean level of contrast over a given area. Contrast gain-control occurs at both retinal and cortical levels (Shapley and Victor 1978; Ohzawa *et al.* 1985). Truchard *et al.* (2000) recorded responses of binocular simple cells in cat area 17 to drifting sinusoidal gratings presented dichoptically at various phases. A 10-fold increase in contrast of the grating in one eye sharply reduced gain for that eye but had only a small effect on binocular gain. Thus, most control of contrast-gain occurs in monocular pathways. However, effects due to interocular suppression do occur (Section 7.10.2b).

The important point is that binocular interactions of most simple cells could be predicted from the linear summation of the excitatory and inhibitory zones revealed by the cell's response to gratings presented to each eye separately. Cells of this type conform to the energy model. Linear summation of excitatory and inhibitory zones, revealed when each eye was tested separately, accounted for phase-specific binocular interactions in about 40% of complex cells (Ohzawa and Freeman 1986b). About 40% of complex cells exhibited nonphase-specific responses and about 8% showed a purely inhibitory influence from one eye.

Hammond (1991) agreed that most simple cells show a phase-specific response to a moving sine-wave grating of optimal spatial frequency and orientation but variable interocular phase. But he found that most complex cells do not. For a simple cell, the excitatory and inhibitory zones of the receptive field are spatially segregated. For a complex cell, the excitatory zones are coextensive for single bright and dark stimuli, and inhibition is revealed only as an interaction between two stimuli (Movshon *et al.* 1978). In either case, a phase-specific modulation of response implies that the period of zones is similar in the monocular receptive-fields of a given binocular cell. Relative phases of receptive-field zones may differ in the monocular fields of a binocular cell (Section 6.5.4).

Ohzawa *et al.* (1990) recorded responses of binocular complex cells in the visual cortex of anaesthetized cats as an optimally orientated bar was moved across corresponding receptive fields in each eye. The bars were both black, both white, or black in one eye and white in the other. The responses are shown in Figure 6.15. They developed a model of the monocular receptive fields that feed into a complex binocular cell. The receptive field for each eye consists of four subunits, each with a distinct luminance profile. A luminance profile of a subunit is the function describing how the firing of the cell is modulated above or below its resting level as a stimulus is moved over the subunit. In each eye, the fields of one pair of subunits have symmetrical (cosine) luminance profiles of opposite polarity and those of the other pair have asymmetrical (sine) luminance profiles of opposite polarity, as shown in Figure 6.16a. In each eye, the signals from the opposed symmetrical subunits mutually inhibit each other, as do those from the opposed asymmetrical subunits. Outputs from these two push-pull systems from the same eye combine through a nonlinear half-wave rectifier. The output of the symmetrical system is 90° out of spatial phase with respect to that of the asymmetrical system. The two are said to be in quadrature. A model complex cell with matching properties in the two eyes fires maximally when the stimulus occupies the same position in each eye. Such a cell has a symmetrical tuning function about zero disparity.

Ohzawa *et al.* modelled binocular cells tuned to nonzero disparity by providing receptive field subunits with a double quadrature organization, as shown in Figure 6.16b. In this case, the receptive field for the right eye is 90° phase-shifted relative to that for the left eye for all four subunits, which means that the cell fires maximally only when the stimulus in one eye is offset with respect to that in the other. Nomura *et al.* (1990) and Nomura (1993) developed a similar model.

If a cell has a periodic pattern of excitatory and inhibitory zones it could show several peaks of response as the relative phase of the images of a dichoptic bar stimulus is varied. This could account for the fact that many cortical cells have more than one peak in their disparity tuning function (Ferster 1981). Ohzawa *et al.* found that the response of some complex cells with disparity selectivity narrower than the receptive field remained constant as the stimulus was moved over the receptive field. The response of these cells showed **position invariance** in their disparity tuning.

Livingstone and Tsao (1999) used a similar mapping procedure in the visual cortex of alert monkeys. All cells showed position invariance in their disparity tuning. This was true of complex cells and also of cells with offset ON and OFF subregions. Segregation of ON and OFF regions is characteristic of simple cells but position invariance is not. They found no cells with disparity tuning that could be described as a simple consequence of the structure of

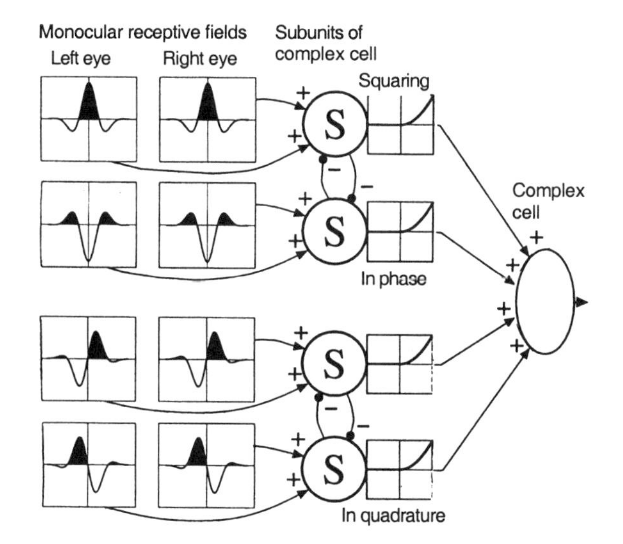

(a) Model of receptive field of a binocular complex cell tuned to zero disparity. The receptive field of the cell contains four subunits, each of which receives inputs from both eyes arranged as two mutually inhibitory pairs, one pair operating in phase, the other in quadrature (90°) phase. In the profiles of the monocular receptive fields the dark areas represent excitatory regions and the blank areas inhibitory regions. One pair of subregions receives inputs from symmetrical receptive fields and the other from asymmetrical receptive fields.

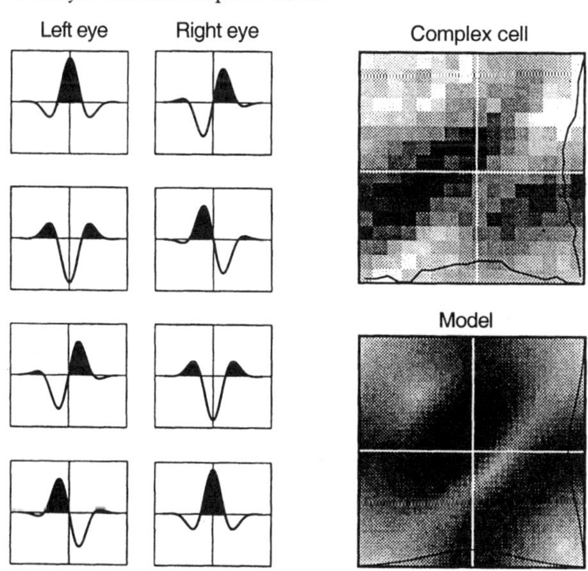

(b) Hypothetical receptive-field subunits for a binocular conplex cell tuned to nonzero disparity. The cell was modelled by receptive-field subnits with double quadrature organization.

Figure 6.16. Model of binocular complex cell.
(Reprinted with permission from Ohzawa *et a*l. 1990. Copyright American Association for the Advancement of Science)

their monocular receptive fields. This result does not support the energy model of Ohzawa *et al.*

Prince *et al.* (2002a) summarized the following lines of evidence related to the energy model.

1. The tuning functions of cells in V1 of the monkey to horizontal disparity can be adequately described by a Gabor function, as predicted by the energy model.

2. The energy model is supported by the fact that disparity tuning functions in response to stereograms with reversed contrast in the two eye are inverted (Ohzawa *et al.* 1990; Cumming and Parker 1997). The weaker response to reversed-contrast stereograms than to normal stereograms is not predicted by the energy model. However, Read *et al.* (2000) showed that this result can be explained by a simple modification of the model.

3. The energy model predicts that the response of a cell to binocular inputs from dichoptic uncorrelated random dots is the sum of its responses to each monocular display. However, the binocular response is closer to the mean of the monocular responses. However, this discrepancy can be accounted for by a simple process of response normalization, in which the response of a cell is modulated by the mean response of neighbouring cells (Fleet *et al.* 1996b).

4. Some binocular cells are more weakly tuned to disparity than would be predicted by the energy model. Weak disparity-tuning could be due to the monocular receptive fields being poorly matched in their response to one or more visual features. It could also be due to the presence of receptive-field subunits that differ in their disparity tuning.

6.5.1c *Disparity tuning and contrast polarity*

A simple disparity-detector that receives direct inputs from the two eyes does not produce a pure disparity signal—one not influenced by incidental changes in the stimulus. For example, the signal for black bars is not the same as that for white bars and simple detectors are sensitive to slight changes in object location (Qian and Zhu 1997). In the cat, complex cells produce more robust disparity signals than do simple cells. But, even for some complex cells, the sign of the disparity signal is inverted when contrast sign is reversed, such that white regions in one eye are superimposed on black regions in the other (Ohzawa *et al.* 1997).

Cumming and Parker (1997) reported that most binocular cells in V1 of the monkey produce a normal disparity tuning function to a dynamic random-dot stereogram with the same sign of contrast in the images in the two eyes (correlated dots). However, the cells produce an inverted tuning function to a stereogram with contrast sign reversed in the image to one eye (anticorrelated dots). They predicted this reversal of the tuning function by modelling the response of a binocular cell with monocular receptive fields that match in spatial and temporal properties. For each receptive-field subunit, inputs from the two eyes are summed, squared and linearly combined to form the response of a complex cell. They assumed that the cells code disparity in terms of receptive-field offset but the same predictions follow if disparity is coded in terms of receptive-field phase.

Visually, images with reversed contrast evoke rivalry, not depth (Section 16.2.8). Also, Cumming *et al.* (1998) found that human subjects saw no depth in reversed contrast stereograms like those that produced inverted tuning functions in the visual cortex of monkeys. They concluded that disparity detectors in V1 of the primate respond to a type of disparity that is not used directly for depth perception. However, other interpretations of these findings are discussed in Section 16.2.8c.

Inverted-disparity signals could have several uses. Two or more simple detectors could feed into a complex detector that accepts only same-contrast signals. Other complex cells could use reversed-contrast signals to evoke rivalry and help to indicate whether complex images in the two eyes are in proper register (have a maximum proportion of same-contrast edges). But periodic stimuli, even when in register, can contain regions of opposite contrast. Under these circumstances, complex cells that accept opposite-contrast stimuli could use them to improve their response (Ohzawa *et al.* 1997).

6.5.1d *Precision of disparity tuning*

The probability of response of single neurones in the visual cortex to well-defined stimuli of variable strength can be used to produce a **neurometric function**, analogous to a psychometric function derived from behavioural data (Parker and Newsome 1998).

Prince *et al.* (2000c) compared the sensitivity of binocular cells in V1 of alert monkeys with the ability of animals to detect depth in a dynamic random-dot stereogram. The animals fixated a zero-disparity dot and indicated whether a 0.4° central patch in the stereogram was in front of or behind a 1° wide surrounding annulus. Neural performance was assessed in terms of the probability that the response of a binocular cell to each of a range of disparities was greater or less than the response to zero disparity. On average, neuronal thresholds were about four times higher than behavioural thresholds. However, the thresholds for the best neurones were slightly lower than the behavioural threshold. It looks as though monkeys use the information from their most sensitive detectors. When the surround was absent or contained uncorrelated dots, the behavioural threshold increased more than the neuronal threshold. This suggests that the neuronal threshold depends on the local absolute disparity of the test patch, while the behavioural threshold depends on the relative disparity between test patch and surround. An increase in vergence instability when the surround was absent or uncorrelated may have been a factor. The animals were trained to fixate the central test patch but it is not clear to what extent vergence movements occurred in the two-second test periods. Energy models are discussed further in Section 6.8.

6.5.2 Number and homogeneity of detectors

Suppose that each region of the visual field contains a set of disparity detectors with the same bandwidth that together span the range of disparities detected in that region. The bandwidth of each detector as a fraction of the total bandwidth divided by the fractional overlap of the tuning functions defines the number of detectors required to span the full range of detectable disparities. As we move into the visual periphery, the mean size of receptive fields increases and detectable disparities are larger. Thus, the disparity detection system, like other spatial systems, is inhomogeneous. Although only a few disparity detectors with distinct tuning functions are required at each location, there are many types of detector over the whole binocular field.

Cormack *et al.* (1993) derived the tuning widths of disparity detectors from the threshold-elevation effect evident in the detection of interocular correlation. They first measured the **correlation-sensitivity function** for a single dichoptic random-dot display. This is the degree of correlation between dots in the dichoptic images that could just be detected, as a function of disparity. They then measured the correlation-sensitivity function of a test display with a superimposed near-threshold random-dot display of variable disparity. The degree of threshold summation was at a maximum when the two displays had the same disparity and decreased to the level of probability summation as the disparity difference increased. With further increase in disparity, the threshold increased, revealing inhibitory interactions. The derived tuning functions centred on zero disparity were symmetrical, with an excitatory central region and inhibitory surrounds. Those centred on a disparity to one side of zero were asymmetrical, with an inhibitory lobe on the side of zero disparity. The tuning widths were approximately 20 arcmin and the tuning functions showed considerable overlap. The largest disparity that could be used with this procedure was about 30 arcmin.

The discrete nature of chromatic channels is revealed by humps and dips in the hue-discrimination function, because hue discrimination is best where neighbouring colour channels overlap. Hue-discrimination functions are usually derived from stimuli subtending only 2°. Since the visual pigments are the same over wide areas of the retina, the hue-discrimination function shows the same undulations for larger stimuli. The disparity detection system is not homogeneous since the bandwidth of disparity detectors depends on the size of receptive fields, which increases with retinal eccentricity. One would therefore expect the disparity-discrimination function to show humps and dips only for small stimuli. For large stimuli, the humps and dips in different regions would not coincide and would therefore tend to cancel.

Humps as a function of disparity were not found by Badcock and Schor (1985) in disparity discrimination functions, by Stevenson *et al.* (1992) or Cormack *et al.* (1993) in dichoptic correlation-detection functions, nor by Felton *et al.* (1972) in contrast-detection thresholds. It has been concluded that disparity detection is not achieved by only three detectors in the manner suggested by Richards (1972). But this conclusion is premature. The width of the tuning functions revealed by Cormack *et al.* suggest that very few detectors are required to span the detectable range of disparities in any local region of the visual field. It is difficult to see how a local region could accommodate a large number of disparity detectors. Large displays were used in all the studies that showed an absence of humps in disparity-discrimination functions, so their absence may merely reflect the fact that the disparity-detection system is inhomogeneous over the binocular field. There could therefore be a small number of discrete detectors in each location but a continuous range of detectors over the visual field. An *experiment is needed in which disparity-discrimination functions are derived from small displays.*

Channels for the detection of spatial modulations of disparity are discussed in Section 19.6.3.

6.5.3 Dynamics of disparity detectors

6.5.3a *Response variance*

The response rate of cortical cells of anaesthetized monkeys to repetition of a given stimulus has a variance about equal to the mean response rate (Tolhurst *et al.* 1983). However, cells in V1 of alert monkeys have a much smaller response variance than those of anaesthetized monkeys, especially when effects of eye movements are minimized (Gur *et al.* 1997). Furthermore, the response variance of cells in V1 was found to be no less than that of cells in the LGN. A low response variance improves the capacity of small numbers of cells to discriminate stimuli reliably. Since most response variance was due to eye movements it was correlated between cells. Pooling of inputs at a higher cortical level would reduce the effects of uncorrelated noise but would not reduce the effects of correlated noise.

6.5.3b *Feedback loops*

The inhibitory processes involved in the creation of disparity selectivity of cortical cells could involve feedback or feedforward loops. Thorpe *et al.* (1991) found that the disparity tuning of cortical cells is fully developed in the first 10 ms of response, which suggests that only feedforward inhibitory loops are involved. Yeshurun and Schwartz (1990) have developed a model based on cepstral filters (Section 16.1.4) of how stereoscopic disparity can be derived from feedforward processing. The psychophysics of the dynamics of disparity detection is discussed in Section 19.11.

6.5.4 Position and phase-shift disparity detectors

6.5.4a *Phase disparity detectors*

Consider a binocular cell for which the receptive fields in the two eyes have the same tuning for orientation, motion, and spatial frequency, and the same distribution of excitatory and inhibitory zones. Disparity tuning of this type of cortical cell can only be due to the difference in position between the receptive fields in the two eyes. These binocular cells are **position-disparity detectors.** If the preferred disparity of a cortical cell depends on the offset of its receptive fields, one should be able to correlate the two quantities. In practice, receptive-field offset cannot be determined for all disparity-tuned cells because many of them do not respond when a stimulus is presented only to the nondominant eye. Some disparity-selective cells, known as AND cells, respond only to the joint stimulation of both eyes. Furthermore, many of the cells in which a receptive field offset can be measured are not disparity selective (von der Heydt *et al.* 1978). For a position-disparity detector there is no necessary linkage between the size of matching receptive fields and the preferred disparity of the cortical cell into which it feeds. A binocular cell with small matching receptive fields could have a large receptive field offset and therefore be tuned to a large disparity, and a cell with large receptive fields could have a zero offset and therefore be tuned to zero disparity. For instance, Pettigrew *et al.* (1968) found complex cells with large receptive fields in the cat that were narrowly tuned to disparity.

Computational studies gave rise to the idea that binocular disparity may be coded in terms of differences in spatial phase between monocular receptive fields feeding into a binocular cell (Jenkin and Jepson 1988; Sanger 1988). The receptive fields of binocular cells serving as **phase-disparity detectors**. would have the same positions, the same tuning for orientation, motion, and spatial frequency, but different distributions of excitatory and inhibitory zones. For instance, the receptive field of a cell in one eye could have a symmetric (cosine) sensitivity profile and that in the other an asymmetric (sine) profile, as illustrated in Figure 6.16b. Some binocular cells in the visual cortex of the cat have been reported to have this type of phase disparity (Freeman and Ohzawa 1990; DeAngelis *et al.* 1991) and Anzai *et al.* (1999a) found most phase-disparity detectors in the cat to have a phase disparity of 90° or less.

To make phase disparities comparable with displacement disparities, phase disparity should be expressed in terms of visual angle rather than phase angle. The angular disparity that produces a peak response equals the phase shift divided by the spatial frequency of the stimulus. Uncertainty increases with both the spatial frequency of the stimulus and with the size of the phase shift. Pooling responses over a local area or over spatial scale and orientation helps to reduce this uncertainty and produce an angular-disparity signal (Section19.8).

The maximum disparity detectable by a binocular cell is proportional to the spatial period of the zones within the receptive field. Thus, phase-sensitive binocular cells with small receptive fields are necessarily tuned to small disparities expressed in terms of visual angle. Simple cells with large receptive fields could be tuned to large angular disparities, since a simple cell with a large receptive field is essentially a scaled-up version of one with a small receptive field. The receptive field of a complex cell has several subunits within each of which there are excitatory and inhibitory detectors. The disparity preference of a complex cell with a large receptive field would therefore depend on the size and spatial disposition of the subunits within the receptive field rather than on the size of the receptive field as a whole.

A tendency for small receptive fields to have to detect small disparities and for large receptive fields to detect large disparities could account for why cells with greater retinal eccentricity are tuned to larger disparities. Cells in V1 have smaller receptive fields than those in V2 and this could account for the fact that V1 contains more cells tuned to zero disparity than cells tuned to near or far disparities, while V2 contains more near and far cells than zero-disparity cells (Ferster 1981). Psychophysical evidence on the relationship between disparity and spatial scale is reviewed in Section 19.7.

DeAngelis *et al.* (1991) mapped the receptive-field profiles of simple cortical cells of cats for each eye. By fitting the profiles with Gabor functions, they obtained the phase difference between the two eyes (see also Ohzawa *et al.* 1996). About 30% of binocular simple cells in area 17 showed substantial differences between the phases of the ON and OFF regions in the receptive fields in the two eyes. The cells had matching orientation, motion, and spatial-frequency tuning and most of them preferred orientations near vertical. Almost all cells tuned to within 20° of the horizontal had receptive fields with matching or near matching phases, which suggests that they code position disparity. Cells tuned to stimuli within 20° of the vertical had receptive fields with a wide variety of phase relationships, which suggests that they code phase disparity.

Prince *et al.* (2002b) distinguished between cells tuned to position disparity and cells tuned to phase disparity by the shapes of their disparity-tuning functions. For this purpose they used dynamic random-dot stereograms. Both types of disparity detector were found to be common in V1 of the monkey. They confirmed the distinction between the two types of detectors by measuring the disparity sensitivity of each type to sinusoidal luminance gratings as a function of spatial frequency.

6.5.4b *Hybrid detectors*

A given cell could code both position disparity and phase disparity. Anzai *et al.* (1999) measured both phase disparity and position disparity for the same simple cells in the cat's visual cortex. They measured the position disparity of a given cell with respect to a neighbouring reference cell, which was assumed to have zero offset-disparity. Statistical procedures allowed them to estimate the uncertainty of this procedure, although its validity depended on the unconfirmed assumption that position disparities of neighbouring binocular cells are uncorrelated. They found phase disparities to be mostly within a $\pm 90°$ range of phase angles. The sign of phase disparities larger than 90° becomes ambiguous (Blake and Wilson 1991). Phase disparities were mostly within $\pm 1°$ of visual angle whereas position disparities were within $\pm 0.5°$. This seems to contradict the notion that large disparities are coded by position-disparity detectors. However, it is not clear how eccentric the receptive fields were. Perhaps the larger receptive fields of the peripheral retina code large disparities. Anzai *et al.* concluded that phase disparity detectors code large disparities for low spatial-frequency stimuli and that position-disparity detectors provide a constant limit for high spatial-frequency stimuli for which phase disparities are small.

Anzai *et al.* found no correlation between the position disparities and phase disparities of binocular cells in the cat's visual cortex. The two types of disparity therefore add in some cells and subtract in others. For any binocular cell, any uncertainty in the registration of the relative positions of its monocular receptive fields would produce a corresponding uncertainty in the calibration of phase disparities of that cell. Therefore, the joint determination of the two types of disparity is subject to the same uncertainty associated with the determination of position and spatial frequency (Section 4.2.4c).

In theory, the position and phase components of a hybrid disparity detector could be measured by recording the disparity tuning of the cell to drifting sinusoidal gratings of different spatial frequencies. The phase offset can be derived by fitting a cosine function to the tuning function and the position offset is given by the slope of the function relating phase offset to the spatial frequency of the stimulus (Fleet *et al.* 1996a). For the position-disparity detector, response peaks occur at the same disparity for all spatial frequencies of the grating. For the phase-disparity detector, peaks occur when the phase difference between the gratings in the two eyes reaches a certain value, whatever the spatial frequency. This procedure has been used to investigate disparity detection in the owl (see Section 14.6.3), although its validity has been questioned (Zhu and Qian 1996).

Liu *et al.* (1992a) produced psychophysical evidence that position disparity dominates phase disparity in a stimulus when the two types of disparity are in conflict. Two cycles of a vertical cosine grating were presented in a Gaussian window. The grating and window were moved in one eye to produce a position disparity. The grating was moved relative to the window to produce a phase disparity of the grating relative to a fixed disparity of the window. Phase disparity had to be about three times larger than position disparity to reach the threshold for perceived depth. Thus, a zero position disparity restrained the effect of the phase disparity. It is not clear that the two types of disparity were processed by position-disparity detectors and phase-disparity detectors respectively. Also, one cannot conclude that phase disparity in a stimulus is processed less efficiently than position disparity, since one cannot produce a stimulus with pure phase disparity.

Erwin and Miller (1999) proposed a type of hybrid disparity detector for simple cells, in which position offset is correlated with phase displacement. In this type of disparity detector an on-region of the receptive field in one eye corresponds only with an on-region of the other eye. This means that a phase shift must be accompanied by a corresponding position shift and vice versa. They called this a "subregion correspondence" detector as opposed to a hybrid detector, in which phase and position offsets are not correlated, or as opposed to pure displacement or pure phase detectors. All their model detectors apply to only tuned excitatory disparity detectors with near balanced inputs from the two eyes, and which detect small disparities centred about zero. They argued that the correspondence model accounts for how receptive fields of tuned simple cells acquire the same orientation and spatial-frequency tuning and for the narrow distribution of preferred disparities of tuned disparity detectors in cortical areas 17 and 18 of the cat. For pure phase detectors the distribution of peak disparities would be broader for low than for high spatial frequencies. For pure position-disparity detectors the distribution would have the same width as the distribution of position shifts. For uncorrelated hybrid detectors the distribution would be broad.

Erwin and Miller argued that cells tuned to horizontal lines have smaller phase shifts than cells tuned to vertical lines because horizontal position shifts are larger than vertical position shifts, independent of stimulus orientation. Joshua and Bishop (1970) reported this type of anisotropy in the cat at an eccentricity of about 12° but not in the central field. Erwin and Miller claimed that only the subregion correspondence model accounts for the eccentricity dependent anisotropy. They stated that the model should be tested by recording phase and offset disparities of several cells simultaneously.

6.5.5 Orientation and disparity tuning

6.5.5a *Orientation and preferred direction of disparity*

The first question concerns the relation between the direction of disparity and the orientation of the stimulus. It is reasonable to suppose that a disparity-tuned cell with a given preferred stimulus orientation responds best to disparities at right angles to the receptive field axis. Maske *et al.* (1986a) found this to be true for cortical cells of the cat that lack inhibitory end zones in their receptive fields (non end-stopped cells). End-stopped cells responded to disparities along their common axis as well as, or almost as well as, to disparities at right angles to the axis, as long as the stimuli were shorter than the receptive field.

For long bar stimuli, the disparity tuning of a cell is necessarily maximum in a direction orthogonal to the bar, especially when the bar is at the preferred orientation of the cell. Prince *et al.* (2002a) showed that this linkage between the orientation preference of a cell and the preferred direction of disparity does not hold with stimuli with a broad band of orientations, such as random-dot displays.

However, with random-dot displays, Prince *et al.* found that disparity tuning is related to stimulus orientation in another way. Tuning functions for horizontal disparity resembled Gabor functions for cells with an orientation preference near vertical but resembled Gaussian functions for cells tuned to near horizontal. This can be explained as follows. A Gabor function indicates the multi-lobed response modulation across the width of the cell's oriented receptive field. A Gaussian function represents the single-lobed response modulation across the length of the receptive field.

6.5.5 *Sensitivity to different orientations*
A second question is whether, in general, horizontal disparities are detected more effectively than vertical disparities. Ohzawa and Freeman (1986a) found the degree of binocular interaction for cells preferring horizontal stimuli, as a population, to be similar to that for cells tuned to vertical stimuli. Other investigators have found that disparity-tuned cells preferring vertical stimuli respond more strongly than those preferring horizontal stimuli (Maske *et al.* 1986a).

Ohzawa *et al.* (1996) reported that about 30% of binocular simple cells in the visual cortex of the cat are sensitive to phase disparity and that almost all of these cells prefer stimuli orientated between oblique and vertical. Binocular cells that prefer stimuli near the horizontal have receptive fields that match in the phases of their on and off regions. This suggests that vertical disparities are detected by position disparity rather than by phase disparity.

Anzai *et al.* (1999a) confirmed that cells tuned to horizontal orientations have smaller phase disparities than cells tuned to vertical orientations. They did not find a corresponding anisotropy for position disparity. Barlow *et al.* (1967) found the same anisotropy using a measure of disparity that included contributions from both position and phase disparities. Others, including von der Heydt *et al.* (1978) and LeVay and Voigt (1988), found no anisotropy with a measure of disparity based on differences in location of monocular receptive fields that does not allow for any contribution from phase disparity.

In the cat visual cortex, Joshua and Bishop (1970) found that, with increasing horizontal eccentricity, the horizontal offset of receptive fields of binocular cells increased. Vertical receptive-field offset increased more gradually. Thus, the presence of an anisotropy depends on the measure of disparity and stimulus eccentricity.

Evidence reviewed in Section 21.2.2 shows that horizontal disparities are extracted locally whereas vertical disparities are extracted from large areas. Putting physiological and psychophysical evidence together suggests that phase disparities are extracted locally while position disparities are pooled.

6.5.6 Disparity tuning and eye position

6.5.6a *Disparity tuning and distance*
Trotter *et al.* (1992, 1996) recorded from disparity-tuned cells in V1 of alert monkeys as they fixated a visual target at distances of 20, 40, and 80 cm. At each distance, an array of random dots was presented with various degrees of horizontal disparity relative to the fixation target. The size of the dots and of the display, and the disparities were scaled for distance so that the retinal images were the same for each distance. The response of most cells was modulated by changes in viewing distance. For example, disparity selectivity emerged at only one distance or was sharper at one distance than at other distances. Gonzalez and Perez (1998a) obtained similar results and Roy *et al.* (1992) found cells with similar properties in MT.

The investigators concluded that signals mediating these changes in response are derived from changes in vergence or accommodation. They assumed that the pattern of retinal stimulation was the same at the different distances. However, fixation disparity may change with distance and changes in cyclovergence are known to accompany changes in vergence. Since no precautions were taken to prevent or compensate for these changes, the resulting changes in the alignment of the images may have caused the observed changes in the responses of cortical cells. Furthermore, the pattern of vertical disparities produced by a display in a frontal plane varies with distance (Section 21.4.1) and this factor may also have contributed to the observed changes in response of cortical cells.

Pouget and Sejnowski (1994) developed a neural-network model of disparity detectors modulated by vergence, which codes both relative and absolute distance.

The role of vergence in judgments of absolute distance is discussed in Sections 21.4.1 and 24.6.

6.5.6b *Disparity tuning and gaze direction*
Trotter and Celebrini (1999) recorded from binocular cells of the visual cortex of alert monkeys as they fixated a spot at the centre of a 6° by 6° random-dot display with +0.6°, -0.6°, or zero horizontal disparity. About half the cells tested showed a change in response rate when the visual display was moved 10° to left or right of the median plane. The display was

tangential to the horizontal horopter in both positions. A few cells showed a change in their preferred disparity. It is unlikely that changes in vertical disparity would affect the results because the display and the eccentricity were both small. Fixation disparity (Section 9.3.4) may have changed with changing gaze angle, although any such change should have introduced a consistent shift in disparity tuning functions.

6.6 CODING HIGHER-ORDER DISPARITIES

The disparity-tuned cells discussed so far have to do with the detection of local point disparities. Perhaps some disparity detectors in the visual system are specifically designed for the detection of patterns of relative disparity such as disparity gradients. Three types of disparity are considered in this section.

1. Disparity defined by a difference in spatial periodicity (dif-frequency disparity), which could be related to perception of slant about a vertical axis.
2. Disparity defined by relative shear of the two images, which could be related to the perception of surface inclination about a horizontal axis.
3. Second spatial derivatives of disparity, which could be related to the perception of surface curvature in depth.

Finally, we consider interactions between horizontal and vertical disparities.

6.6.1 Disparity of spatial scale

Hammond and Pomfrett (1991) reported that, for a majority of cells in the cat visual cortex, the spatial frequency evoking the best response from one eye was slightly different from that evoking the best response from the other eye. Most cells of this type were tuned to a higher spatial frequency in the dominant eye than in the other eye. Furthermore, the cells were more likely to be tuned to orientations close to the vertical than were cells with matching spatial-frequency characteristics. Hubel and Wiesel (1962) and Maske *et al.* (1984) failed to find cortical cells with different receptive-field structures, but their methods were not refined enough to reveal the crucial differences. Cells sensitive to differences in spatial frequency could provide the physiological mechanism for the detection of depth in surfaces slanting about a vertical axis (Section 21.1).

6.6.2 Orientation disparity

Bishop (1979) suggested that, as for horizontal disparities, the range of orientation disparities to which cortical cells are tuned arises from the random pairing of monocular receptive fields for which there is a random scatter of preferred orientations (see Hetherington and Swindale 1999). It is of interest that the range of orientation disparities to which binocular cells of the cat's visual cortex respond is about the same as the range of orientation disparities that the cat normally encounters.

Blakemore *et al.* (1972) measured the orientation of a bar eliciting the maximum response in single binocular cells in the visual cortex of the cat for a bar presented to each eye in turn. For many cells, the optimal orientations for the two monocular receptive fields differed. These differences had a range of over 15°, with a standard deviation of over 6°. Hubel and Wiesel (1973) were unable to confirm these findings but the eyes of their animals may not have been in torsional alignment. Nelson *et al.* (1977) replicated Blakemore *et al.*'s finding after controlling for possible contaminations due to eye torsion induced by paralysis and anaesthesia. For each cell, the widths of the orientation tuning functions for the two receptive fields were very similar. Thus, cells sharply tuned to orientation in one eye were sharply tuned in the other eye, even though the preferred orientations in the two eyes could differ. The response of a binocular cell was facilitated above its monocular level when the stimulus in each eye was centred in the receptive field and oriented along its axis of preferred orientation. As the stimuli were rotated away from this relative orientation, the response of the cell declined to below its monocular level, although this inhibitory effect was not strong. However, the tuning functions of the binocular cells to orientation disparity were no narrower than the monocular orientation-tuning functions. They argued that such broadly tuned orientation-disparity detectors could not play a role in the fine discrimination of inclination about a horizontal axis. However, fine discrimination does not require finely tuned channels when the outputs of several channels are compared (Section 4.2.7). Binocular cells with different preferred orientations in the two eyes have been reported also in area 21a of the cat (Wieniawa-Narkiewicz *et al.* 1992). The orientation disparity functions of these cells, like those of cells in area 17, could be derived from the difference between monocular orientation-tuning functions.

The above studies were performed on aneasthetized animals. Hänny *et al.* (1980) found a small number of cells in V2 of the alert monkey that were specifically sensitive to changes in the angle of inclination of small stimuli about a horizontal axis. The cells responded maximally to a line in the frontal plane. A 45° forward or backward inclination of the stimulus in the median plane, corresponding to an orientation disparity of 2°, reduced the response by half. By comparison, the smallest tuning width of cells tuned to orientation of monocular stimuli is reported to be 6°, with a mean of 40° (DeValois *et al.* 1982a). In a second experiment, Hänny *et al.* controlled for the effects of horizontal disparity by using dynamic random-dot stereograms that contained an orientation disparity but only randomly distributed horizontal disparities.

Von der Heydt *et al.* (1982) used a dynamic random-dot display containing a cyclopean vertical grating with no horizontal disparities, because the dots were dichoptically uncorrelated. They found five cortical cells that responded to orientation disparities in this stimulus. They concluded that the visual cortex of the monkey contains cells specifically tuned to orientation disparity. Such cells could provide a physiological basis for the detection of depth in surfaces inclined about a horizontal axis (Section 21.2). Inclined surfaces contain disparities between vertically oriented surface features, and such disparities would be detected by cells tuned to orientations near the vertical rather than near the horizontal.

Orientation disparity detectors tuned to orientations near the horizontal could be used to allow for orientational misalignment of the two images, as discussed in Section 21.2.2. They could also be used to detect the differential orientational disparities produced by surfaces slanted in depth about a vertical axis (Section 21.1). They could also evoke cyclovergence, as discussed in Section 9.8.5. For this latter purpose the cells would need to be sensitive to only low spatial frequency stimuli.

Bridge and Cumming (2001) pointed out that binocular cells that merely respond to the mean orientation of stimuli in the two monocular receptive fields do not qualify as detectors for orientation disparity. The response of the cell must be specific for a given orientation difference not merely the mean difference. This requires that the response of the cell to a stimulus with given orientation in one eye must vary with changes in the orientation of the stimulus in the other eye. They looked for such cells in V1 of the alert monkey. Of 64 cells, 20 responded to an orientation disparity that was not predictable from monocular orientation selectivity. However, these cells were also selective for positional disparity. Bridge and Cumming concluded that the apparent orientation selectivity of these cells arose from their sensitivity to positional disparity.

Rogers and Cagenello (1989) showed that differences in curvature of dichoptic lines evoke a sensation of surface curvature. If the receptive fields in the two eyes feeding into a binocular cell were tuned to lines of different length, the cell would be sensitive to differential curvature in the two eyes. DeAngelis *et al.* (1994) found a few cells of this type in the visual cortex of the cat.

6.6.3 Disparity and motion

Cells selectively responsive to both movement and binocular disparity occur in the pulvinar, a thalamic nucleus closely associated with the LGN (Casanova *et al.* 1989). They also occur in several visual areas of the cerebral cortex. Some cells in V1 and V2 of the monkey respond selectively to stimuli moving in a given direction in the two eyes, with some responding to only crossed-disparity stimuli and others to only uncrossed-disparity stimuli (Poggio and Fischer 1977; Poggio and Talbot 1981). These disparity-tuned motion detectors exert a disparity-dependent control over subcortical centres in the pretectum that process visual inputs for optokinetic nystagmus (Section 23.4.5).

Grasse (1994) found many cells in the pretectum in the cat that responded to moving displays were also tuned to binocular disparity; some showed an excitatory response to a limited range of disparities and others an inhibitory response. Evidence that cells jointly tuned to disparity and speed are involved in perception of relative depth is reviewed in Section 28,3,2.

Cells selectively tuned to both motion and disparity have also been found in the medial temporal visual area (MT) of the monkey, together with cells that respond to both crossed and uncrossed disparities but not to zero disparity (Maunsell and Van Essen 1983b). The cells in MT that respond best to zero horizontal disparity also respond best to zero vertical disparity. Finally, some cells in the medial superior temporal cortex (MST) of the monkey are jointly tuned to direction of motion and to the sign of disparity (Komatsu *et al.* 1988). There are more jointly tuned cells that are sensitive to crossed disparity than jointly tuned cells sensitive to uncrossed disparity. Most of the disparity sensitive cells in MST are tuned to either near or far stimuli rather than to stimuli with zero disparity. In a few of these cells the preferred direction of motion reversed as the disparity reversed. For example, a cell that responded to rightward motion for stimuli with crossed disparity responded to leftward motion for stimuli with uncrossed disparity (Roy *et al.* 1992). MST cells have large receptive fields, suggesting that they are more suitable for detecting parallactic motion of large parts of the visual field created by self-motion than for coding local depth. Saito *et al.* (1986) found cells in MST of the monkey that respond preferentially to patterns rotating in depth.

All the visual centres mentioned above are connected. Thus, visual inputs feed directly to the pretectum, from where they feed to the pulvinar. Other visual inputs feed directly to the superior colliculus. Collicular inputs feed to the visual cortex and the tectopulvinar pathway feeds to several cortical visual areas, such as areas 18 and 19, MT, MST, and the superior temporal polysensory area (STP) (Section 5.7.3).

One of the functions of this system, which bypasses the primary visual cortex, may be the execution of rapid responses to approaching objects. The collicular and tectopulvinar inputs survive in the destriate animal. Destriate rats and monkeys show responses, such as OKN and avoidance to approaching objects, which are served by this extrastriate system (Dean *et al.* 1989; King and Cowey 1992). The system is probably responsible for blindsight, or the ability of destriate patients to manifest some visual functions (Weiskrantz 1987).

Psychophysical evidence of a coupling between motion and stereopsis is reviewed in Section 23.2. In Section 29.3 we review visual mechanisms sensitive to the direction of approaching objects. It is not known whether the disparity sensitivity of cells in these regions survives removal of V1.

6.6.4 Joint horizontal and vertical disparities

Cells sensitive to vertical disparity have been reported in area 17 of the cat (Barlow *et al.* 1967) and in MT of the monkey (Maunsell and Van Essen 1983b). We show in Sections 20.3 and 21.2 that vertical disparities play a crucial role in stereopsis, in addition to their role in controlling vertical vergence (Section 9.7). In particular, differences between horizontal and vertical disparity indicate the relative deformation of the two images, which is used in the perception of the slant and inclination of surfaces in depth. Physiological evidence for this idea has been provided by Gonzalez *et al.* (1993b). They recorded from cells in V1 and V2 of alert monkeys trained to fixate a spot on a random-dot display subtending 24 by 14° as the horizontal and/or vertical disparity of the central region of the display was varied. Thirty percent of cells in V1 and 41 % of those in V2 were sensitive to both horizontal and vertical disparity. When horizontal disparity was zero, the response of most of these cells was least for zero vertical disparity and increased with increasing vertical disparity. In the presence of a fixed horizontal disparity, the response decreased with increasing vertical disparity. Similarly, in the presence of a fixed vertical disparity, the response decreased with increasing horizontal disparity. Thus, tuning for vertical disparity was centred on zero disparity. The response characteristics of these binocular cells is appropriate for the detection of a difference between horizontal and vertical disparities, which is what one would expect of a cell designed to detect deformation disparities. The behavioural significance of these results is discussed in Sections 21.1 and 21.2.

6.7 VEPs AND BINOCULAR VISION

6.7.1 Visual evoked potentials

Because of their high level of interconnectivity, cortical neurones tend to fire in synchrony. Furthermore, subgroups of cells tend to fire at different frequencies. In addition to these spontaneous firing patterns, groups of cells tend to respond together in characteristic ways in response to particular stimuli. The synchronous firing of groups of cells generates fluctuating electrical fields that can be detected either at the surface of the brain or on the scalp. Electrical fields generated by the visual cortex are known as **visual evoked potentials**, (VEPs). Pyramidal cells are the most likely source of VEPs. A pyramidal cell runs at right angles to the cortical surface and forms an electrical dipole when it generates nerve impulses. A single electrode on the scalp is affected by the activity of thousands of pyramidal cells since the meninges, skull, and scalp diffuse and average the potentials arising from the underlying area of tissue. Records can be taken only from cortical tissue that runs parallel to the surface of the brain and not from tissue within fissures.

Prominent types of synchronous activity arising from the visual cortex include alpha waves at a frequency between 8 and 13 Hz, evident in awake subjects, delta waves between 0.5 to 4 Hz that arise in the sleeping subject, and beta waves between 14 Hz and 30 Hz that arise when the subject is engaged in an attentive task. Gamma waves at even higher frequencies are detected in single cells or small groups of cells but not at the scalp.

In a typical experiment, the experimenter relates the location, magnitude, and form of VEPs to parameters of the stimulus. A well-defined repetitive visual stimulus is applied and the response is then filtered and averaged over many cycles of stimulus repetition. **Signal averaging** emphasizes components in the VEP that are time-locked to the repeating stimulus, and attenuates components due to extraneous stimuli and intrinsic noise which, being unrelated to the stimulus, average out over several

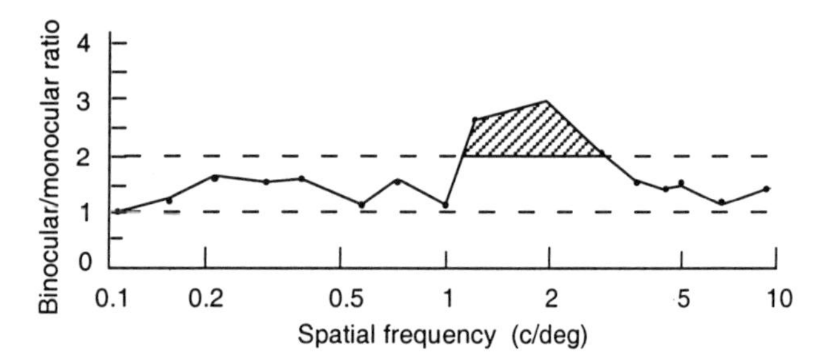

Figure 6.17 Evoked potentials and spatial frequency.
Ratio of binocular to monocular VEP as a function of the spatial frequency of a vertical grating counterphase modulated at a temporal frequency of 30 Hz. (Redrawn from Apkarian *et al.* 1981)

stimulus repetitions. Standard signal-averaging procedures overlook episodic noise such as bursts and oscillations. These events can be characterized by **phase-locked spectral analysis**, which measures the difference between responses to each cycle of a periodic stimulus and the response average. Standard signal-averaging procedures also fail to register stimulus engendered variations in response to particular cycles of a periodic stimulus. Episodic activity generated by the stimulus, but not time-locked to it, can be revealed by standard **power spectrum analysis**, as peaks in the autocorrelation function (Schiff *et al.* 1999).

In another procedure, responses of the same region to distinct stimuli are recorded or recordings are made of the responses of distinct cortical regions to the same stimulus. The degree of coherence (shared power) between these recordings as a function of stimulus frequency is used to derive a coherence function. A coherence function is the frequency domain analogue of the (squared) cross-correlation function, and its value varies between 0 and 1. In this way, one can assess the extent to which distinct stimuli evoke the same response and distinct cortical regions respond to the same stimulus.

A commonly used procedure is to identify prominent peaks and troughs in the VEP according to their amplitude, latency, and polarity. There are many sources of uncertainty in interpreting these components. For instance, Harter *et al.* (1973) reported that only the late (200-250 ms) component of the evoked response reflects the activity of binocular cells, as indicated in a greater response to identical stimulation to both eyes than to rivalrous inputs. Others have found that only the early component (100-160 ms) is correlated with stereoscopic vision (Regan and Spekreijse 1970). Results also depend on electrode position, stimulus contrast, and, as we will see below, the spatial and temporal properties of the stimulus. There is also a good deal of intersubject variability. A review of human brain electrophysiology has been provided by Regan (1989a).

The VEP has been used to reveal two properties of binocular mechanisms—summation and suppression between inputs from the two eyes on the one hand, and stereopsis on the other.

6.7.2 VEPs and binocular summation

In investigations of binocular summation and suppression using the VEP, two basic comparisons are made. First, the magnitude of the VEP evoked by monocular stimulation of each eye is compared with that evoked by binocular stimulation. Second, the response when the two eyes receive identical stimuli is compared with that when the stimuli in the two eyes are uncorrelated. The idea is that only congruent stimuli summate their inputs whereas rivalrous stimuli compete for access to binocular cells. The results are used to assess the following outcomes.

Summation A binocular VEP that is simply the sum of the monocular responses could signify that visual inputs are processed by independent mechanisms in the visual cortex. But it could also signify that the inputs from the two eyes are summed linearly by binocular cells. If summation is partial, it signifies that some inputs converge but with less than complete summation. A binocular response that equals the mean of the monocular responses (zero summation) signifies binocular rivalry, in which the input from one eye suppresses that from the other eye when both are open, or in which the two eyes gain alternate access to cortical cells. It could also arise from binocular cells that average the inputs from the two eyes.

Inhibition A binocular response that is less than the mean of the monocular responses signifies strong mutual inhibition between left- and right-eye inputs. It could occur in binocular rivalry when a strong stimulus in one eye is suppressed by a weak stimulus in the other eye.

Imbalance A monocular response from one eye that is stronger than that from the other signifies a weakened input from one eye or suppression of one eye by the other, even when the dominant eye is closed. This condition arises from anisometropia and strabismus (Section 13.5).

Facilitation A binocular response that is greater than the sum of the monocular responses indicates the presence of a facilitatory binocular interaction that one might expect in a mechanism for detecting binocular disparity. The most extreme facilitation arises in binocular AND cells that respond only to excitation from both eyes and give no response to monocular inputs.

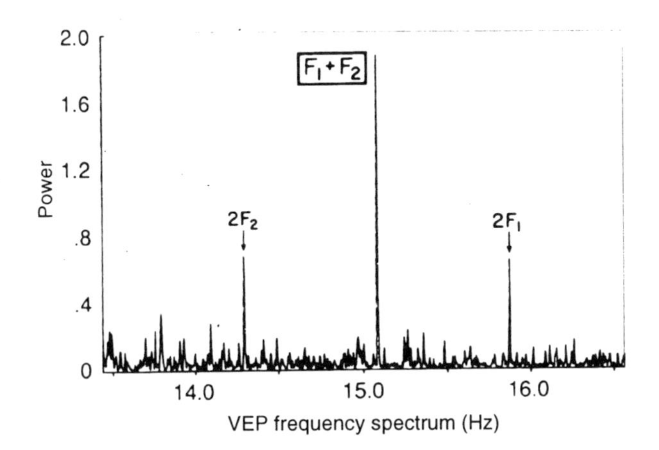

Figure 6.18. Nonlinear processing of dichoptic flicker.
One eye viewed a homogeneous field flashing at 8 Hz (F_1) with 17 per cent amplitude modulation while the other viewed a light flashing at 7 Hz (F_2) with 12 per cent modulation. The VEP spectrum was recorded at a resolution of 0.004 Hz by zoom-FFT. The $F_1 + F_2$ component in the VEP indicates a nonlinear process sited after binocular convergence. (From Regan and Regan, Can. J. Neurol. Sci, 16, 1989)

A thorough investigation of binocular facilitation of the VEP in adults was conducted by Apkarian *et al.* (1981). They presented normal adult subjects with vertical gratings of various spatial frequencies with luminance modulated in counterphase at various temporal frequencies. Binocular facilitation of the amplitude of the VEP as a function of spatial frequency for a fixed temporal frequency of 30 Hz is shown in Figure 6.17. For this subject, binocular facilitation is limited to spatial frequencies in the region of 2 cpd, which is not the region where the monocular response has its peak. Binocular facilitation was also found in a specific range of temporal frequencies of contrast modulation, generally between 40 and 50 Hz, and was higher at higher contrasts. The range of spatial and temporal frequencies within which facilitation occurred varied from subject to subject.

When a 2 log luminance difference was introduced between the images of a dichoptic reversing checkerboard pattern the amplitude of the VEP was less for binocular than for monocular presentation (Trick and Compton 1982). The effect was more evident at higher temporal frequencies of stimulus reversal. They explained the effect in terms of a relative phase shift in the inputs from the two eyes due to a difference in latency. This means that a difference in retinal illumination due to opacities in the eye could lead to a misinterpretation of VEP data, if not allowed for.

The dependence of binocular facilitation on the spatial and temporal properties of the stimulus probably explains the wide variation in the degree of binocular facilitation reported in previous studies (Cigánek 1970; Harter *et al.* 1973; Srebro 1978). Potentials evoked by diffuse flashes have been found not to exhibit binocular summation, whereas flashed patterns did exhibit summation (Ellenberger *et al.* 1978). Tests of binocular functioning that rely on the comparison of monocular and binocular VEPs are suspect if they are not based on appropriate stimulus parameters. One solution is to record evoked potentials while the stimulus is swept through a range of stimulus values—a procedure first used by Regan (1973). The amplitude of the evoked potential was recorded as the refractive power of a lens in front of an eye was varied for each of several astigmatic axes. This provided a rapid determination of the required refractive correction. In a second study, VEPs were recorded while a checkerboard pattern was optically zoomed in size, with brightness held constant (Regan 1977).

Norcia and Tyler (1985) recorded the VEP as the spatial and temporal frequencies of a grating were swept through a range of values. This gave a **spatio-temporal VEP profile**. This method is particularly useful with children too young to be tested by psychophysical procedures. In addition to providing a better basis for comparison of VEPs under a range of stimulus values, the sweep method is much faster than the presentation of different stimulus values in discrete trials.

When a vertical grating is presented to one eye and a horizontal grating to the other, the observer experiences binocular rivalry (Section 7.3). It is believed that binocular rivalry is due to competition between uncorrelated inputs from the two eyes for access to binocular cells. If so, the competition should be evident in the VEP. Spekreijse *et al.* (1972) found that the amplitude of the VEP associated with a counterphase-modulated pattern presented to one eye was strongly reduced when it was perceptually suppressed by a steady pattern presented to the other eye compared with when it was presented with a blank field in the other eye.

Since misaccommodation can cause a large reduction in the VEP, the reduced VEP from suppressed images could be due to the suppressed eye becoming misaccommodated. Spekreijse *et al.* obtained the same result when the rivalrous stimuli were presented in the context of correlated stimuli that served to keep both eyes properly accommodated.

Apkarian *et al.* (1981), also, found evidence of binocular rivalry in the VEP by showing that a vertical grating presented to one eye and a horizontal grating to the other produced a VEP that showed zero interocular summation. The binocular response fell to the monocular level when the difference of orientation of the dichoptic gratings exceeded about 20° (Tyler and Apkarian 1985). A VEP was detected in humans when a dynamic random-dot display alternated between being correlated and anticorrelated (contrast reversed) between the two eyes. The response was undiminished when the displays were composed of correlated and anticorrelated equiluminant red and green dots (Livingstone 1996b). This suggests that there are binocular cells sensitive to interocular correlation that are not involved in stereopsis, since stereopsis is weak at equiluminance (Section 18.1.4).

After a stationary pattern has been presented for some time, the VEP evoked by onset of the stimulus is attenuated. Smith and Jeffreys (1979) found almost complete transfer of this attenuation to the other eye for the CII component of the VEP, thought to originate in the prestriate cortex, but only partial transfer for the CI component, thought to originate in V1. This suggests that monocularly driven neurones are more common in V1 than in the prestriate cortex in humans, just as they are in the monkey (Zeki 1978).

Katsumi *et al.* (1986) investigated the effects of optically induced aniseikonia on the degree of binocular summation revealed in the human VEP. The stimulus was a checkerboard pattern undergoing contrast reversal at 12 Hz. When the pattern was more than 5% larger in one eye than in the other, there was no evidence of a greater VEP to binocular than to monocular stimulation. The tolerance of the stereoscopic system for aniseikonia is discussed in Section 15.8.

Another approach to using VEPs for detecting binocular interactions is to look for evidence of nonlinear interactions in the response to dichoptic flicker. Suppose that the left eye views sinusoidal flicker of frequency F_1 and the right eye views sinusoidal flicker of frequency F_2. Nonlinear processes produce harmonics of F_1 in the left eye channel and harmonics of F_2 in the right eye channel. Nonlinear processes occurring after the monocular signals are combined produce cross-modulation terms of the general form $nF_1 + mF_2$, for integral values of n and m. The relative amplitudes of these terms depend on the nature of the nonlinearities both before and after binocular convergence.

Regan and Regan (1988, 1989) provided a mathematical analysis of these processes and showed that nonlinear processing occurring after binocular convergence can be isolated from that occurring before convergence. They pointed out that, in contrast with the random-dot techniques described later, this procedure allows one to explore binocular functions even when acuity is low in one or both eyes.

Figure 6.18 shows a $(F_1 + F_2)$ component recorded by an ultrahigh resolution Fourier analysis, known as zoom-FFT. This component of the response must arise from nonlinear processes occurring after binocular convergence (Regan and Regan 1989) (see also Zemon *et al.* 1993). The $(F_1 + F_2)$ component of the VEP to dichoptic flicker is weak in stereoblind subjects (Baitch and Levi 1988). Normal infants show nonlinear responses to dichoptic flicker by 2 months of age, but the response is absent in esotropic infants, especially in those without corrective surgery (France and Ver Hoeve 1994). Thus, only subjects with normal stereopsis show nonlinear combination of monocular inputs, which arises from the way in which binocular cells combine signals from the two eyes. Presumably, the more linear addition of signals in stereoblind subjects arises from two pools of monocularly driven cells.

Electrophysiological recording from single cells in the visual cortex has revealed various forms of binocular interaction, including facilitation and inhibition (Crawford and Cool 1970, Ohzawa and Freeman 1986a, 1986b). Anzai *et al.* (1995) measured the response of single binocular cells in the visual cortex of the cat as a function of the contrast, spatial frequency, and orientation of a drifting sinusoidal grating. They applied a signal-detection analysis to derive monocular and binocular neurometric functions and contrast-sensitivity functions for each cell. These functions are analogous to behaviourally determined psychometric and contrast-sensitivity functions. They concluded that the contrast threshold is reached when the response of a small number of cells reaches a criterion level, and that the contrast-sensitivity function depends on the number and sensitivities of cells tuned to each spatial frequency. Comparison of the contrast thresholds of binocular cells for monocular stimulation with those for binocular stimulation revealed a degree of binocular advantage similar to that found in human psychophysics (Section 8.1). Anzai *et al.* concluded that binocular summation is due to binocular cells being more sensitive to binocular than to monocular stimulation, so that the criterion number of cells required for contrast detection can be achieved at lower contrasts.

Another manifestation of binocular summation is that visually evoked potentials are greater when a grating moving in one direction is presented to one eye simultaneously with one moving in the opposite direction in the other eye, compared with when either moving grating is presented alone (Ohzawa and Freeman 1988).

6.7.3 VEPs and stereopsis

There are several problems to be solved in relating changes in the VEP specifically to changes in stereoscopic depth perception based on binocular disparity. It must first be demonstrated that the response is not due to stimulus-locked eye movements. The second problem is to ensure that monocular cues to depth do not intrude. This can be done by using random-dot stereograms. The third problem is to change the depth in the stereogram without introducing unwanted motion signals. This can be done by using a dynamic random-dot stereogram, in which the dots in each monocular image are replaced at the frame rate of the display, so that there is no motion of monocular dots related to the change of depth in the stereogram. Finally, one must ensure that changes in the VEP are due to the perceived change in depth rather than to a change in the degree of correlation between the patterns of dots in the two eyes. This can be done by alternating the stereogram between equal and opposite disparities rather than between zero disparity and either crossed or uncrossed disparity. A second control is to compare the VEP evoked by a random-dot stereogram alternating in depth because of a change in horizontal disparity with that evoked by a similar change in vertical disparity. This control must be applied with caution because depth sensations can arise from certain types of vertical disparity (see Section 21.2).

Fiorentini and Maffei (1970) found that periodically reversing the contrast of dichoptic vertical gratings produced a larger VEP when the gratings differed in spatial frequency. The impression was of a surface slanted in depth about a vertical axis. The magnitude of the VEP increased as the apparent slant of the surface increased. When the dichoptic gratings fell on distinct regions of the visual field, differences in spatial frequency did not affect the VEP. They concluded that the magnitude of perceived depth determines the magnitude of the VEP. This conclusion is valid only if the VEP is not affected by spatial-frequency differences between horizontal gratings that do not produce slant.

Regan and Spekreijse (1970) presented subjects with a static random-dot stereogram, in which the horizontal disparity of the central square alternated between zero and 10, 20, or 40-arcmin. Every half-second the central square appeared to jump forward from the plane of the background and then jump back. A positive-going VEP occurred about 160 ms after each depth change and was followed by a negative-going response. The monocular stimulus produced an appearance of global motion (short-range apparent motion) for displacements less than 20 arcmin and an associated monocular VEP, but no global motion and no VEP for a 40-arcmin displacement. This same stimulus, however, produced a large VEP when viewed dichoptically. An equivalent change in vertical disparity produced a much smaller VEP. They concluded from this and from eye-movement controls that changes in the VEP were related specifically to changes in perceived depth and not to motion of parts of one of the mo-

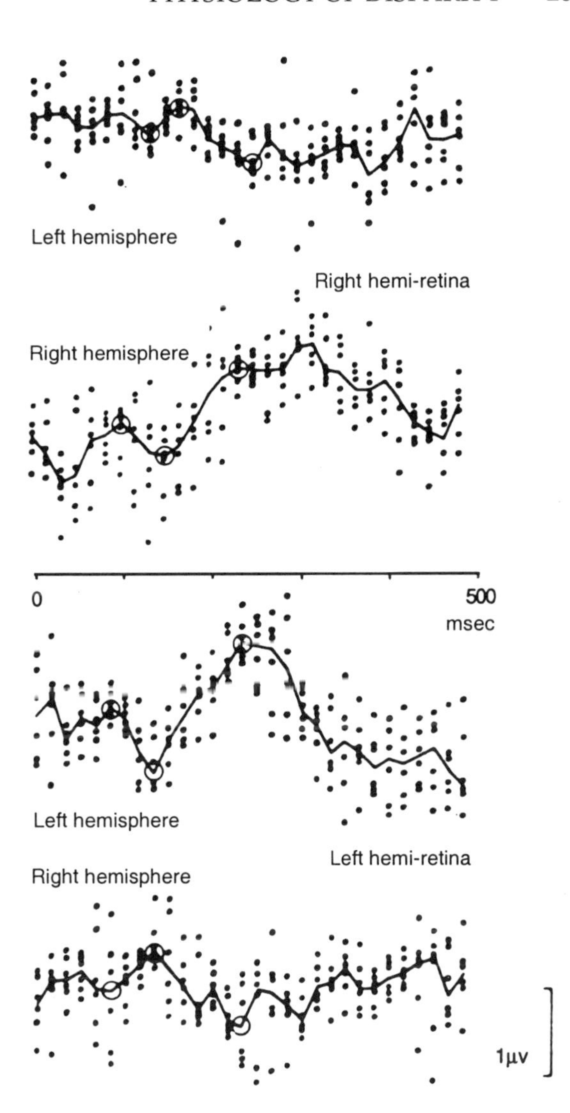

Figure 6.19. Evoked potentials and stereopsis.
Averaged evoked potentials in response to changes in depth of a random-dot stereogram presented to either the right or left hemifield and recorded from either the right or left cerebral hemisphere. With a display confined to the left visual hemifield (right hemiretinas), each change in apparent depth (indicted by circles) produced a VEP in the right hemisphere, as one would expect from the fact that the right hemiretinas project to the right hemisphere. A small mirror-image echo of the response is evident in the left hemisphere. (Reprinted from Lehmann and Julesz 1978, with permission from Elsevier Science)

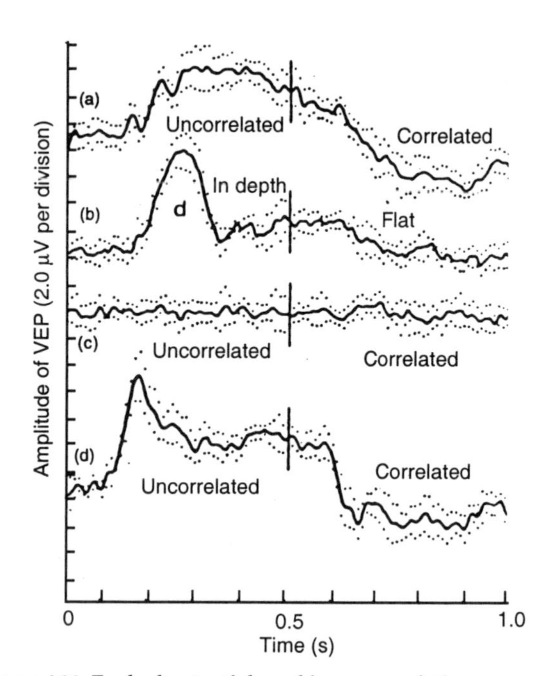

Figure 6.20. Evoked potentials and image correlation.
(a) VEP in response to a dynamic random-dot pattern alternatately correlated and uncorrelated in the two eyes.
(b) VEP in response to a cyclopean checkerboard alternating between being flat and in depth.
(c) Same stimulus as in (a) but with one eye closed.
(d) Same as in (a) but with anaglyph spectacles removed. (c) and (d) were control conditions (N=2). (From Julesz *et al.* 1980)

nocular images, to changes in disparity unrelated to depth, or to eye movements.

In the dynamic random-dot stereogram introduced by Lehmann and Julesz (1977), the dots are renewed many times per second so that any motion of the cyclopean image is not evident in either monocular image. Lehmann and Julesz (1978) used a dynamic random-dot stereogram, in which a rectangular area appeared to move out from the background and then back every half-second. With a display confined to the left visual hemifield (right hemiretinas), each change in apparent depth was followed by a VEP in the right hemisphere, as one would expect from the fact that the right hemiretinas project to the right hemisphere. There was a smaller mirror-image echo of the response in the left hemisphere (see Figure 6.19). With a display confined to the left hemiretinas, the major VEP occurred in the left hemisphere with a smaller echo in the right hemisphere. They argued that both hemispheres process cyclopean stereopsis in a similar fashion, which runs counter to some evidence from the clinical literature that damage to the right hemisphere produces a selective impairment of stereopsis (Section 12.2). There were no controls for possible eye-movement artifacts in this study.

Manning *et al.* (1992) found that the disparity threshold for detection of depth in dynamic random-dot stereograms was lower in the right than in the left visual field and that the VEP had a higher amplitude when the stereogram was presented in the right visual field. Skrandies (1997), also, obtained a larger VEP when a checkerboard dynamic random-dot stereogram was presented in the right rather than the left visual field and, like Lehmann and Julesz, found no differences between VEP's in the two cerebral hemispheres. For central stimuli, the maximum amplitude of the VEP occurred at smaller disparities than for stimuli in the peripheral field. He applied no vertical-disparity control.

The latencies of potentials evoked by a cyclopean checkerboard pattern in a dynamic random-dot stereogram fluctuating in depth were similar to those evoked by a contrast-reversing luminance-defined checkerboard. However, the cyclopean pattern evoked weaker responses than the luminance-defined pattern, presumably because there are fewer cells tuned to disparity than cells tuned to luminance contrast. Also, the spatial distribution of the potentials was different for the two types of stimuli, presumably because they are processed in distinct cortical areas (Skrandies 1991).

Stereoacuity and other forms of acuity are greater in the lower than in the upper visual field (Section 19.6.6). The magnitude of the VEP was greater for a flashed dynamic random-dot stereogram presented in the lower rather than the upper visual field (Fenelon *et al.* 1986). Evoked potentials related to stereopsis were obtained from dynamic random-dot stereograms with masking of up to 8° of the central field (Teping and Silny 1987).

Brain potentials evoked by motion in depth of the central region of a dynamic random-dot stereogram have been found in the primary visual cortex (Neill and Fenelon 1988) and in the central parietal region (Herpers *et al.* 1981).

A **dynamic random-dot correlogram** is a display of randomly distributed dots alternating between any two of the following states: being in the same positions in the two eyes (+1 correlation), being uncorrelated in position in the two eyes (zero correlation), or being in the same positions but with opposite luminance polarity (-1 correlation) (Julesz and Tyler 1976). These changes between two states are cyclopean, since they are not evident in either monocular image. A much stronger VEP was evoked by a dynamic random-dot correlogram that changed in state than by one that remained either correlated or uncorrelated (Miezin *et al.* 1981). Evoked potentials from a dynamic random-dot correlogram therefore reflect the cyclopean features of the stimulus.

Julesz *et al.* (1980) compared the amplitude of the VEP evoked by a dynamic random-dot correlogram with that evoked by a dynamic random-dot stereogram, in which alternate squares of a checkerboard pattern appeared to move in and out of the plane of the background squares. Both displays produced distinctive VEPs with a dominant latency of about 250 ms compared with when one eye was occluded (see Figure 6.20). However, the amplitude of the response to the random-dot stereogram was greater than that to the correlogram, and the waveforms generated by the two stimuli differed (see also Skrandies and Vomberg 1985).

Julesz *et al.* concluded that the greater response to the stereogram was specifically related to the appearance of depth, as opposed to the change in the degree of interocular correlation of the dots. These responses occurred only in subjects with functional stereopsis, and the authors suggested that they could be used as a simple nonverbal screening test for stereopsis. However, it is not clear whether or not the response was influenced by vergence eye movements, and there was no control for the effects of vertical disparities.

If binocular facilitation is related to stereopsis based on horizontal disparity, it should not occur for a horizontal grating, because extended horizontal gratings do not create horizontal disparities. In conformity with this expectation, Apkarian *et al.* (1981) found that the binocular response to a horizontal grating was the sum of monocular responses, while that to a vertical grating showed facilitation.

Norcia *et al.* (1985) used a dynamic random-dot display that alternated as a whole between a crossed disparity and an equal uncrossed disparity while the subject converged on a stationary point. The amplitude of the VEP increased as a linear function of the amplitude of disparity modulation, for amplitudes up to about 15 arcmin. Above this amplitude, the response first declined and then rose to a second peak at a binocular disparity of about 70 arcmin. The response to larger amplitudes of disparity alternation had a shorter latency but a greater phase lag than the response to smaller amplitudes of disparity. They argued that the two peaks in the VEP represent two disparity processing mechanisms, one for fine and one for coarse disparities (Section 11.4.3).

Kasai and Morotomi (2001) asked whether the VEP from the occipito-temporal region is affected by stimulus features that subjects were asked to detect. Subjects indicated when a dynamic random-dot stereogram contained a shape in a given orientation and depth relative to the background. Other stimuli had either wrong orientation, wrong depth, or wrong orientation and depth. A stimulus with correct orientation enhanced the VEP about 175 ms after stimulus presentation while one with correct depth had an effect at 200 ms. In both cases, the other feature could be either correct or not. This suggests that features are initially processed in distinct channels. Later VEP components were enhanced only when the stimulus had the correct orientation and crossed disparity. They concluded that these VEP components were related to the integration of form and disparity into a figure on a ground.

Gulyás and Roland (1994) used positron emission tomography (PET) to map cortical areas in ten subjects viewing random-dot displays containing texture-defined patterns or a central region with binocular disparity. Occipital, parietal, and frontal areas responded specifically to disparity, with no indication of cerebral asymmetry.

Kwee *et al.* (1999) produced functional magnetic resonance imaging (fMRI) of the brains of eight subjects viewing a 2-D display and a similar 3-D display in alternation. Four subjects exhibited bilateral activation in the intraparietal sulcus, with right hemisphere dominance. Three of these subjects also showed activation in the dorsolateral frontal lobe. Only one subject showed responses in the occipital lobe. Right parietal lobe lesions are implicated in loss of stereopsis (Section 12.2). The fMRI responses may have been related to vergence eye movements rather than to perception of 3-D structure.

Backus *et al.* (2001) used fMRI to examine responses in V1 and other visual areas of the human cortex to binocular disparity. The stereoscopic stimulus depicted two superimposed displays of random dots in two depth planes. The minimum and maximum disparities that created an impression of two depth planes were determined psychophysically. Activity in V1 increased as disparity increased above threshold levels and decreased as disparity approached the upper limit for depth perception. The relation between psychophysical and physiological measurements was particularly strong in area V3A.

Other aspects of VEPs and stereopsis are discussed in Section 11.4.3 and the whole question of evoked potentials is reviewed by Regan (1989a).

6.8 MODELS OF DISPARITY PROCESSING

(This section was written with Robert Allison)

6.8.1 Phase-disparity and energy models

Recent models have embodied the idea that binocular complex cells code the **disparity energy** at a given disparity within their receptive fields over a range of spatial frequencies (the cell's passband). These models are based on theoretical considerations and on electrophysiological recordings of simple and complex cells in V1.

fields and which matches the spatial periodicity of the receptive fields. Thus, a simple cell is most sensitive to a disparity orthogonal to the preferred orientation of the cell (the axis of the Gabor). This point was discussed in more detail in Section 6.6.2.

$$RF_r = \frac{1}{2\pi\sigma_{rx}\sigma_{ry}}\exp\left(-\frac{(x-x_r)^2}{2\sigma_{xr}^2}-\frac{(y-y_r)^2}{2\sigma_{yr}^2}\right)\cos(\omega_{rx}(x-x_r)+\phi_{rx})$$
$$RF_l = \frac{1}{2\pi\sigma_{lx}\sigma_{ly}}\exp\left(-\frac{(x-x_l)^2}{2\sigma_{lx}^2}-\frac{(y-y_l)^2}{2\sigma_{ly}^2}\right)\cos(\omega_{lx}(x-x_l)+\phi_{lx}) \quad (3)$$

6.8.1a *Simple cells*

Simple cells in cortical area 17 of the cat are the earliest cortical neurones showing binocularity. Simple cells respond to patterns of stimulation in each monocular receptive field in an approximately linear fashion with the addition of a firing threshold. Each monocular receptive field of a simple cell can be described by an elongated Gabor function, which is a sinusoidal sensitivity function (the carrier) modulated by a Gaussian window (the envelope) (Figure 6.21). Consider a vertically oriented receptive field at position (x_r, y_r) in the right eye and a similar receptive field at position (x_l, y_l) in the left eye. The sensitivity profiles across the width and height of the receptive-fields can be described by equations (3). The first term in each equation represents the overall Gaussian envelope of the receptive field, with width σ_{rx} and height σ_{ry} for the right field and width σ_{lx} and height σ_{ly} for the left field (DeAngelis *et al.* 1991). It describes the size of the receptive field. The second term reflects the transverse profile of ON and OFF regions within each receptive field. It describes the internal structure of the receptive field. The profile is modelled by a cosine function of frequency ω (in radians per degree) and phase ϕ. The binocular cell responds most strongly to a grating of spatial frequency ω_r and phase ϕ_r in the right eye and frequency ω_l and phase ϕ_l in the left eye. The preferred orientation of the cell depends on the orientation of its monocular receptive fields. The two receptive fields have the same or similar orientations. Different orientation preferences can be obtained by rotation of the profiles in equations (3).

Detectors with Gabor-function profiles achieve the minimum product of position selectivity and spatial-frequency selectivity, which makes them optimally sensitive to both spatial frequency and position (see Section 4.5.2). Other functions, such as DOGs, can achieve similar performance.

A binocular simple cell is maximally sensitive to a grating aligned with the monocular receptive

We will consider only disparities orthogonal to the Gabor axis and we will assume that the Gaussian envelopes and the preferred spatial frequencies of the monocular receptive fields are the same. Under these conditions, we can simplify the description of receptive-fields to the one-dimensional profile perpendicular to the preferred orientation of the cell.

$$RF_r = \frac{1}{\sqrt{2\pi}\sigma}\exp\left(-\frac{(x-x_r)^2}{2\sigma^2}\right)\cos(\omega(x-x_r)+\phi_r)$$
$$RF_l = \frac{1}{\sqrt{2\pi}\sigma}\exp\left(-\frac{(x-x_l)^2}{2\sigma^2}\right)\cos(\omega(x-x_l)+\phi_l) \quad (4)$$

These monocular receptive fields are linear operators, or filters, s described in Sections 3.3.6b and 4.5.2. When a stimulus, $I(x)$, is passed through a Gabor filter, $f(x)$, with width w the output, r, can be derived by multiplying the stimulus strength at each point by the sensitivity of the receptive field at that point and adding (integrating) the resulting products over the receptive field.

$$r = \int_{-w/2}^{w/2} f(x)I(x)\mathrm{d}x \quad (5)$$

This process results in a single number, which represents the strength of response of that receptive field to that stimulus. In the visual system this is represented by the frequency of firing of the cell to monocular stimuli. The response of a simple cell, r_s, can be approximated by adding the inputs from the two eyes:

$$r_s = \int_{-w/2}^{w/2} [f_l(x)I_l(x) + f_r(x)I_r(x)]dx \quad (6)$$

where l and r denote left and right eyes respectively.

If the binocular cell were to linearly sum the inputs from the two eyes, the monocular receptive fields would fully specify the response of the binocular cell to dichoptic stimulation. However, the

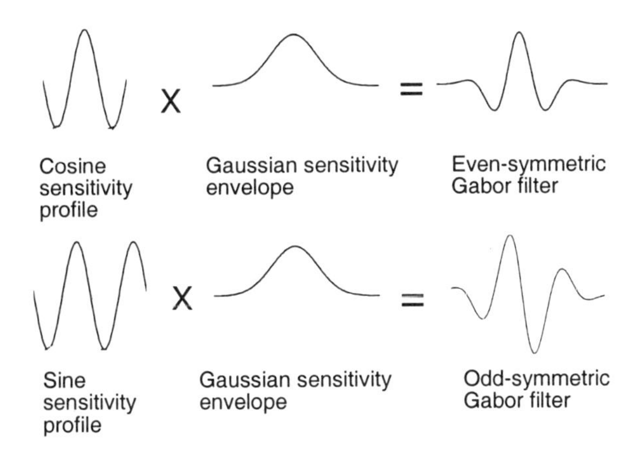

Figure 6.21. Sensitivity profiles of simple-cell receptive fields. The sensitivity profile of each monocular receptive field of a simple cell can be modelled by multiplying a cosine or sine sensitivity profile (carrier) by a Gaussian (normal) envelope. The receptive field is even-symmetric when the carrier is in cosine phase and is odd-symmetric when the carrier is in sine phase.

binocular response shows some nonlinearity. In the first place, binocular cells have a threshold and a saturating response at high levels of stimulation. Secondly, both monocular receptive fields that feed into a binocular cell respond only to light increase (ON type) or only to light decrease (OFF type) (Section 5.1.2d). No binocular cell can respond to both types of stimulus because the cells do not have a resting discharge. Thus, the output is half-wave rectified. This non-linearity could be compensated for by combining the outputs of cells sensitive to opposite contrast polarity. Another nonlinearity has been found in single-cell recordings from binocular cells. Anzai *et al.* (1999a) found that beyond the simple sum of left and right responses, the binocular receptive field contains a term proportional to the product of the left and right responses. This non-linear process can be modelled as a stage that linearly sums over the monocular receptive fields followed by another stage consisting of a static nonlinearity. The empirical results were modelled with an expansive power function with an exponent close to 2.0. Overall, the response of binocular cells shows half-wave rectified squaring. Squaring (or any multiplicative) operation introduces a term proportional to the interocular cross-correlation of the (filtered) images.

If all simple cells' receptive fields were identical in the two eyes the cells would not be sensitive to binocular disparity. The above binocular interaction would amount to binocular facilitation—firing in response to binocular stimulation would be higher than predicted from the monocular responses. Selectivity of a simple cell to non-zero disparities requires that the receptive fields differ in the two eyes.

Binocular disparity can refer to the displacement of matching image features in the two eyes with respect to corresponding retinal locations. It can also refer to the presence of dissimilar images at corresponding retinal locations. Coding of the first type of disparity requires binocular cells with similar monocular receptive fields, which look for similar image structure at non-corresponding points. However, there is some evidence that some binocular cells receive inputs from monocular receptive fields that differ in some way other than in position. This would make them sensitive to differences in the structure of images falling on corresponding retinal locations. For example, the monocular receptive fields could differ in their preferred spatial frequency, ω. This would render them sensitive to dif-frequency disparity produced by slanted surfaces (see Sections 20.2.2 and 21.1.1). Alternatively, the receptive fields could differ in preferred orientation. This would render them sensitive to disparities produced by inclined surfaces (see Sections 20.2.3 and 21.2.1). Finally, monocular receptive fields could differ in their Gabor phase (ϕ). This would render them sensitive to horizontal disparity. These are the **phase-disparity detectors** described in Section 6.5.4. The idea of phase-disparity detectors grew out of computer vision (Jenkin *et al.* 1991, Sanger 1988). The physiological evidence reviewed in Section 6.5.4 suggests that phase disparity is one of the primitive signals for stereopsis.

A position-disparity detector has monocular receptive fields with the same size, orientation, frequency sensitivity, and Gabor phase ϕ. The receptive fields vary only in the relative locations of their centres (x_r and x_l for horizontal disparity). Thus,

$$RF_r = \frac{1}{\sqrt{2\pi}\sigma}\exp(-\frac{(x-x_r)^2}{2\sigma^2})\cos(\omega(x-x_r)+\phi)$$
$$RF_l = \frac{1}{\sqrt{2\pi}\sigma}\exp(-\frac{(x-x_l)^2}{2\sigma^2})\cos(\omega(x-x_l)+\phi) \quad (7)$$

Disparity is coded by the offset of the cell's left-eye and right-eye receptive fields.

A phase-disparity detector has monocular receptive fields with the same size, orientation, frequency, and location. The receptive fields vary only in the Gabor phase of their sensitivity profiles. Thus:

$$RF_r = \frac{1}{\sqrt{2\pi}\sigma}\exp(-\frac{(x-x_0)^2}{2\sigma^2})\cos(\omega(x-x_0)+\phi_r)$$
$$RF_l = \frac{1}{\sqrt{2\pi}\sigma}\exp(-\frac{(x-x_0)^2}{2\sigma^2})\cos(\omega(x-x_0)+\phi_l) \quad (8)$$

where the monocular receptive fields are both centred on x_0 but differ in Gabor phase (Figure 6.21). Differences in phase do not correspond directly to differences in retinal disparity expressed in angular terms. Phase disparity (in radians) provides an indication of disparity only in terms of the proportion of a period. Thus a given phase disparity corresponds to a large positional disparity at low spatial frequencies and a small positional disparity at high spatial frequencies. If the spatial period of the stimulus, filtered by the Gabor receptive field, is the same as the preferred spatial period of the cell then the cell responds to a positional disparity of

$$d = \frac{\phi_r - \phi_l}{\omega} \qquad (9)$$

Typically, the Gabor receptive fields in the existing models (and in the data upon which they are based) are wide enough so that the spatial-frequency bandwidth of the resulting filters is on the order of one octave. Narrowband filters such as this are insensitive to stimuli that differ greatly from the centre frequency of the filter. Thus, one option that some models use is to simply scale the phase disparity by the preferred spatial frequency of the detector. This method will sometimes overestimate and sometimes underestimate true disparities depending on the actual spatial-frequency content in the stimulus. Pooling estimates across several disparity detector scales and phases has been proposed as a way to reduce the effects of these errors (Qian 1994; Qian and Zhu 1997).

In some computational models and computer vision implementation, estimation of the local (instantaneous) spatial frequency of the filtered stimulus has been performed and the resulting estimate used to scale the phase disparity (Fleet *et al.* 1991). Although it is unclear how this could be implemented biologically, local spatial frequency is encoded in the responses of a local population of V1 cells and this normalization may be performed at higher cortical processing stages.

Phase-based disparity detectors are also inherently limited in range. Position-based detectors are limited to detecting disparities within their receptive field boundaries but the receptive fields can be centred at arbitrary positions. In contrast, due to the quasi-periodicity of the filtered stimuli, phase-based detectors cannot distinguish phase of θ radians from one of $\theta+2\pi$ radians. Thus, for unambiguous disparity detection, phase disparities must be limited to a range of $-\pi$ to π radians. Even this range is not achievable because the phase-based detectors have a finite bandwidth and respond to spatial-frequencies greater than the preferred spatial frequency, typically up to at least twice the preferred spatial frequency. Hence a more realistic range of allowable phase disparities is $-\pi/2$ to $\pi/2$ radians.

As discussed by Fleet *et al.* (1996b), phase-based and position-based detectors could co-exist or disparity detectors could be hybrid phase-based and position-based detectors. Erwin and Miller (1999) argued, based on a hypothetical developmental model, for a strongly constrained hybrid model. Based on a developmental model that promotes correlated synaptic activity, the phase shift should depend linearly on the position shifts so that the on and off subregions of the left and right eye receptive fields lie in corresponding locations.

It has often been noted that neither binocular phased-based or position-based disparity sensitive simple cells can be considered disparity detectors. This is because their response depends strongly on monocular stimulus phase. For example, inverting the contrast of the binocular stimulus causes an inversion of the response profile (Cumming and Parker 1997). Also, a Gabor filter is not an ideal spatial frequency detector since it has a small DC response. This DC response has been noted as a possible problem in computer vision applications and a DC-cleaned Gabor filter has sometimes been used. Empirically, Cozzi *et al.* (1997) found little benefit of using DC-cleaned filters. Qian and Zhu (1997) have argued that the DC component is beneficial in that it results in a slight bias for small disparities.

6.8.1b *Complex cells and disparity energy*

The problem with simple cells as disparity detectors is that their response is a function of the local phase of the monocular signals as well as the interocular phase disparity (Qian 1994). This is because the cells respond best to signals with local phase corresponding to the spatial arrangement of their ON and OFF subregions. The energy neurone (Adelson and Bergen 1985) has been proposed as a way of achieving phase independence. The energy of a one-dimensional signal $I(x)$ over an interval $(-x_0, x_0)$ is defined as:

$$\int_{-x_0}^{x_0} |I(x)|^2 \, dx \qquad (10)$$

One way of estimating the energy of a signal at a given frequency is to combine the energy in the sinusoidal and cosine Fourier components. Since the sine and cosine components are orthogonal, their energies can be added to get the overall energy in the interval. A similar scheme can be used to combine the outputs of the Gabor filters (or related functions) formed by monocular simple cells. In this scenario, the input to complex cells is formed by squaring and combining the outputs of quadrature pairs of simple cells. Quadrature pairs consist, for example, of two detectors that differ in phase by 90° (for example, sine and cosine profiles).

Thus, the response (r) of a binocular complex cell can be modelled by summing the squared outputs of two binocular simple cells, r_{s1} and r_{s2} in phase quadrature (Ohzawa *et al.* 1990; Qian 1994; Anzai *et al.* 1999c).

$$r_q = (r_{s1})^2 + (r_{s2})^2 \quad (11)$$

Given that the disparity of the images (D) is significantly less than the width of the receptive fields:

$$r_q \approx c^2 |I(\omega)|^2 \cos^2\left(\frac{\Delta\phi}{2} - \frac{\omega D}{2}\right) \quad (12)$$

where c is a constant, $I(\omega)^2$ is the Fourier power of the stimulus within the receptive field at the preferred spatial frequency of the cell, and $\Delta\phi$ is the phase difference between the receptive fields of the component simple cells. The response is maximal when the disparity equals the phase difference divided by the spatial frequency of the cell's receptive field. This is the preferred disparity of the cell. This equation applies to receptive fields described by a general class of functions, including Gabor functions (Qian and Zhu 1997).

Anzai *et al.* (1999a, 1999b) showed that simple cells may perform the squaring operation, although other models use linear simple cells. The squaring/integration operation incorporates integration of the crossed product of the left and right images—a form of cross correlation. In contrast to the standard cross correlation procedure, the left and right images are band-pass filtered by the cell's receptive field before being multiplied. This renders the algorithm immune to image distortions smaller than the smoothing area. Also, the algorithm is computationally efficient since it is local and there is no integration across the whole stimulus, as there is in the standard cross-correlation procedure. Finally, the provision of two cross products between a quadrature pair of simple cells renders the response of the complex cell approximately the same for two black bars as for two white bars (phase independence) (Ohzawa *et al.* 1990; Qian, 1994, 1997).

The preferred disparity of a complex cell is the relative phase shift between the left-eye and right-eye receptive fields divided by the spatial frequency of the receptive-field profiles of the constituent simple cells. A narrow-band stimulus can generate phase disparities only within the range of the spatial period of the cells that it excites. In other words, small phase disparities are coded by cells with high preferred spatial frequency and large phase disparities by cells with low preferred spatial frequency.

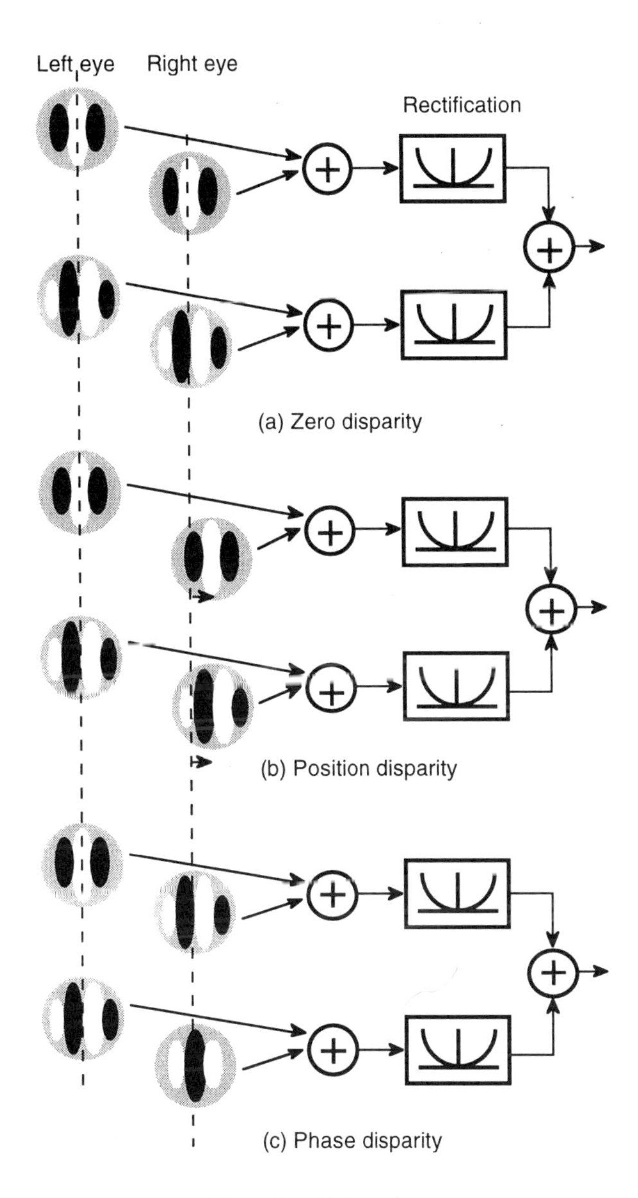

Figure 6.22. Types of disparity of binocular energy neurones. (Adapted from Fleet *et al.* 1996)

Figure 6.22 shows a binocular energy neurone that combines quadrature inputs from binocular simple cells. Disparity selectivity is modelled by giving the quadrature inputs phase (a) or position differences (b) between the monocular receptive fields. A further complication arises from the fact that simple cells are half-wave rectifiers. This can be overcome by composing each quadrature input from a combination of off-centre and on-centre cells with the same phase and disparity preference. A simple diagram of the overall model is shown in Figure 6.16a. Chen *et al.* (2001) described an extension to the disparity energy model to describe the spatio-temporal properties of disparity-selective complex cells (see also Qian 1994; Qian and Andersen 1997).

Theoretically, combining the outputs of a single quadrature pair of simple cells in a complex cell is sufficient to code the disparities at that location. However, the reliability of the estimates is still sensitive to the local phase of the signal. For complex stimuli such as random-dot stereograms, a final smoothing operation can remove high-frequency noise. The smoothing operation involves pooling disparities over an area. This has the effect of blunting the response to sharp disparity discontinuities (Qian 1994). A better approach is to use a weighted average of the responses of several quadrature pairs of simple cells (Qian and Zhu 1997). With this procedure, the final smoothing of complex-cell responses is not necessary and sharp disparity discontinuities are preserved. Pooling over a set of simple cells also reduces the effects of instantaneous spatial frequency on the disparity estimate (Fleet *et al.* 1996b) and is compatible with the larger average size of the receptive fields of complex cells compared with those of simple cells (Qian 1997).

Cozzi *et al.* (1997) analysed the performance of the phase-based disparity algorithms of Sanger (1988) and Fleet *et al.* (1991) and concluded that the algorithms are quite robust to changes in contrast, noise, interocular differences in luminance, and the spectral structure of the image.

Prince and Eagle (2000a) analyzed a model based on the energy model of Fleet *et al.* (1996a). They consider position-disparity detectors as sampling a local disparity energy function. The correct match should result in significant energy in disparity detectors tuned to the corresponding disparity. Thus, we would expect a peak in the local disparity energy function at the true disparity. As Fleet *et al.* suggested, one can use a peak-finding algorithm that interpolates between the discrete detectors to find a high-resolution estimation of the local disparity. However, there may be maxima at a number of other disparities corresponding to false image matches. To ease this correspondence ambiguity, Prince and Eagle multiplied or weighted the disparity energy function to emphasize small disparities in an operation reminiscent of the model developed by Sperling (1970). This operation enhances the contribution of small disparities and reduces the response peaks at large disparities (McKee and Mitchison 1988). This is not a disparity-gradient constraint but a simpler bias for small disparities. Use of positional-disparity rather than phase-disparity encoding allowed for sampling disparity energy over a range of 7,200°. The model could be generalized but was elaborated only for the task of disparity discrimination in a single channel tuned to a specific spatial frequency and orientation.

Prince and Eagle argued that their model could explain the phenomenon of second-order stereopsis, particularly the sensitivity of depth judgments to the disparity of the envelope of an amplitude-modulated carrier (see Section 19.7.2d). The energy detectors demodulate the carrier, and the low-frequency envelope is reflected in the pattern of activation over a set of disparities, in the shape of the disparity-energy function. This would predict strong sensitivity to the disparity of the envelope for discrimination of the sign of disparity, but subjects should still be sensitive to the carrier for discriminations on disparity pedestals or for depth estimation.

One possible problem with the disparity-energy neurone model is the idea that squared output of simple cells forms the input for complex cells. This proposal is questionable in the macaque as the required binocular simple cells appear to be rare (Livingstone and Tsao 1999). Even in the cat, complex cells may not always receive their inputs from simple cells.

Archie and Mel (2000) developed and analysed a disparity-energy model for complex cells based solely upon direct LGN inputs. They modelled a single cortical neurone with simplified morphology at the level of synaptic currents and membrane potentials. Instead of combining simple-cell subunits in quadrature pairs, the subunits are mapped onto separate branches of the dendritic tree. The excitatory LGN inputs for a given branch correspond to components of one of the four postulated subunits in the disparity-energy neurone model. The compartmentalisation provided by the separate branches provides a significant level of independence between the subunit components. The required expansive nonlinear interactions between the inputs to the cell were mediated by sodium and potassium currents and voltage dependent NMDA synaptic currents. These formed the basis for orientation and phase invariant disparity tuning typical of complex cells.

The possibility of implementing the disparity-energy neurone model either with direct LGN input or with quadrature pairs of simple cells illustrates Marr's (1982) point that the implementation of a general function is distinct from the conceptual model or the algorithm used to realize the model.

6.8.2 Neural network models

Several investigators have modelled aspects of human stereoscopic vision using neural networks (Grossberg and Marshall 1989; Becker and Hinton 1992; Grossberg and McLoughlin 1997). Artificial neural networks are only loosely based on real neural networks. The methods have been described by Rumelhart and McClelland (1986), Hinton (1989), and Miller *et al.* (1991).

Some investigators have constructed neural network models that incorporate the disparity energy model.

Lippert *et al.* (2000) used an artificial three-layer neural net, which they trained with noise patterns to learn disparity tuning similar to that exhibited by cells in the visual cortex. The disparity tuning of the network modelled that of tuned-excitatory neurones sensitive to position-disparity, phase-disparity, or to both types of disparity (hybrid type). In the hybrid-type model, the input layer consists of monocular cells representing the left-eye and right-eye receptive fields with Gabor sensitivity profiles at four scales, five phase offsets, and five position offsets relative to the centre of the receptive field. In the position-disparity model, phase-offset is constant. In the phase-disparity model, receptive-field offset is constant. The input units provide candidate subunit regions for forming the disparity-tuned response. The eight cells in the hidden layer receive the binocular inputs from these Gabor filters. The hidden layer projects to a single-output neurone. The weights of the inputs are adjusted by supervised learning, using standard backpropagation, so that the output cell has a preferred disparity and a response similar to a tuned excitatory disparity detector.

After training, disparity tuning was sharper for position-type models than phase-type models. However, for all models, disparity sensitivity peaked near the peak disparity of the teacher and generalized to novel stimuli, indicating that simple networks can learn disparity tuning from Gabor inputs. Responses were strongest and most reliable for the hybrid-type model, which was expected since it had a richer input. Sensitivity to ghost images was observed and the phase detectors exhibited the expected 180° range limitations. Thus, the authors claimed that disparity selectivity can be learned from Gabor inputs differing in phase or position, albeit, with a simple and unphysiological backpropagation learning paradigm. The model output units were much less selective for spatial frequency than are real disparity-selective complex cells. Also, the model units did not show the linear relationship between phase and frequency expected of binocular energy neurones. As a result, the networks did not show the characteristic inverse disparity tuning with anti-correlated random-dot stimuli, as described in Section 6.5.1b. The phase-networks were less precisely tuned and had a smaller range compared with the position-disparity networks.

These results must be interpreted carefully since only a single neurone was trained, with no training parameters related to spatial-frequency selectivity. Broad spatial-frequency tuning of phase-type detectors decreases the reliability of disparity estimates. A detector would have higher reliability if disparities were scaled by a measure of the actual instantaneous spatial frequency rather than by the preferred spatial-frequency of the detector. In the visual cortex there are probably multiple disparity detectors at each location tuned to the same disparity but to different spatial frequencies. Pooling responses from a set of such detectors could improve the reliability of disparity estimates. In addition, real complex cells form the substrate for much of visual processing and the demands of these processes might encourage spatial-frequency selectivity and allow more precise phase-type disparity detection.

Gray *et al.* (1998) developed a feed-forward network model that selects disparity estimates based on their reliability. Local disparity estimates from phase-based disparity-energy neurones feed into two separate pathways, a local disparity pathway and a selection pathway. The first pathway computes local disparity by spatially pooling the output of disparity-energy neurones over a local region. Competition between these disparity-sensitive cells ensures support for a unique disparity estimate for the region. The parallel selection pathway estimates the regions with the most reliable evidence for a given disparity. After training, these units become sensitive to step changes in disparity, or disparity contrast. The selection pathway gates the local disparities multiplicatively so that reliable disparity estimates are given a greater weighting when passed to the output layer. Thus, a parallel pathway determines reliability, which is then used to gate the feed-forward disparity inputs. The model demonstrates that derivation of disparity estimates can be dissociated from the determination of the reliability of those estimates. The concept of identifying and discarding unreliable disparity estimates is key to other models of stereopsis (Fleet *et al.* 1991).

7 Binocular fusion and rivalry

7.1 BINOCULAR FUSION

7.1.1 The limits of fusion

7.1.1a *Measuring fusion limits*

In the eleventh century, Alhazen noticed that images of an object continue to appear single when they do not fall exactly on corresponding visual lines (Section 2.2.2). In his *Treatise of Optics*, written in 1775, Harris wrote, "An object that is a little out of the plane of the horopter, may yet appear single." (p. 113). Wheatstone (1838) also noticed that images in a stereoscope fuse even though they do not fall exactly on corresponding points. Thus, a small point of light in one eye fuses with a similar point of light in the other eye as long the two points fall within a certain area. This area is known as **Panum's fusional area**, after Peter Ludvigh Panum, professor of physiology at Kiel, who described the first systematic experiments on the effect in 1858. The fusional range is larger for stimuli separated horizontally than for stimuli separated vertically, thus making fusional areas elliptical (Panum 1858; Ogle and Prangen 1953). However, at least part of this difference may be due to asymmetries in vergence eye movements (Mitchell 1966b).

The terms **diplopia threshold** and **fusion limit** denote the largest retinal disparity between two images for which a single fused image can be maintained. For a given direction of image separation, the diameter of Panum's fusional area is the sum of the diplopia threshold for crossed disparity and that for uncrossed disparity. The diplopia threshold is not always symmetrical; for some persons it is greater for uncrossed images while for others it is greater for crossed images. These asymmetries may be due in part to fixation disparity resulting from incorrect vergence on the fixation targets. The relationship between the fusion limit and fixation disparity is discussed in Section 9.3.4. Richards (1971a) concluded that asymmetries of the fusion limit are not due only to fixation disparity but also reflect the independent processing of crossed and uncrossed images. The size of the fusional area depends on many factors, such as retinal eccentricity, stimulus duration, the presence of surrounding stimuli, and the criterion for single vision adopted by the observer. Reported values have ranged from a few minutes of arc to several degrees.

Fusion limits are most commonly measured by the method of limits. The experimenter gradually increases the disparity between fused dichoptic stimuli until the subject reports diplopia. Then, starting with the stimuli well separated, disparity is decreased until the subject reports fusion. As we will see, there is a hysteresis effect so that the disparity at which two images appear double when initially fused is greater than the disparity at which they fuse when initially seen double. It is difficult to control vergence eye movements with the method of limits. Nonius lines indicate changes in vergence but introduce extra stimuli that may contaminate the measurements, since neighbouring stimuli can affect the fusion limit. In the method of constant stimuli, the subject aligns nonius lines just before the disparate stimuli are presented briefly. Stimuli with different disparities are presented in random order, and a psychometric function of percentage of "single stimulus" judgments against disparity is plotted. The diplopia threshold is conventionally defined as the point on the psychometric function where 50% of the judgments are "single stimulus" judgments. This method has the advantage that stimuli can be presented for too brief a period to evoke vergence eye movements. However, brief exposure introduces a temporal transient into the stimuli. Also, as we will see later, Panum's area is increased when stimuli are rapidly alternated in disparity. In a criterion-free forced-choice procedure, subjects discriminate between a binocular stimuli with a horizontal disparity and a spatially adjacent or subsequently presented pair of stimuli with zero disparity. The disparity giving 75% accuracy is generally taken as the threshold.

Finally, there is the problem of the criterion used in judging diplopia. When horizontal limits of fusion are being measured, one must ensure that subjects are judging fusion rather than apparent depth between the disparate stimuli. This is a severe problem with the forced-choice procedure because subjects tend to rely on apparent depth if that is the only difference they see. Apparent depth is not a problem for vertical limits of fusion. In monocular resolution, as two lines are moved further apart, the first sensation is of a single line becoming thicker. Similarly, a thickening of dichoptic stimuli may be noticed before fusion is lost (Heckmann and Schor 1989b). Also, as dichoptic stimuli are separated, there comes a point where edges of opposite luminance polarity are superimposed. This may evoke a sensation of binocular lustre or rivalry. If the two stimuli rival, an apparent change in position may occur. Generally, smaller fusion limits are obtained when subjects are allowed to use criteria other than diplopia than when they are required to use the criterion of diplopia.

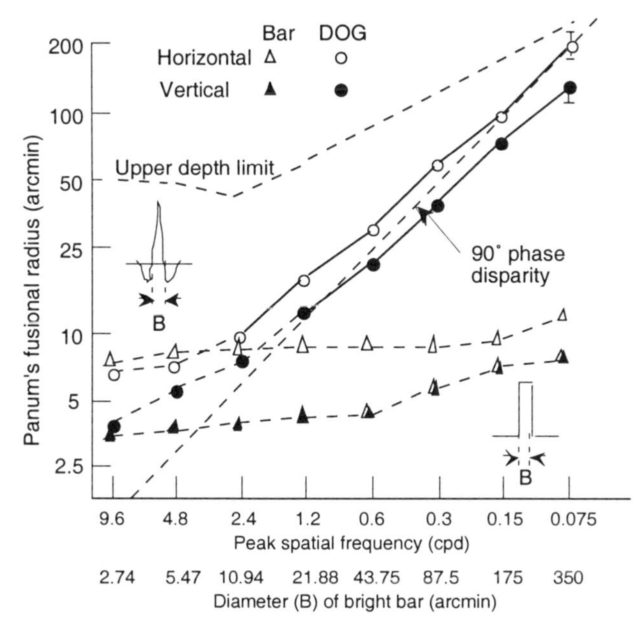

Figure 7.1. Fusion limits and spatial frequency.
Diplopia threshold (radius of fusional area) as a function of peak spatial frequency of two Gaussian patches (spatial band-width 1.75 octaves) and of the width of two bright bars. For patches with spatial frequency below about 1.5 cpd, the threshold corresponds to a 90° phase shift of the stimulus (dashed line). The fusion limit for the bars remains the same as that of the high spatial frequency patch. (Redrawn from Schor *et al.* 1984b)

7.1.1b *Fusion limits and eccentricity*

All investigators agree that the diplopia threshold increases with increasing eccentricity, although it is difficult to measure the threshold when the stimulus is more than about 10° away from the fovea. Studies reviewed by Mitchell (1966b) showed wide variations in the slope and shape of the function relating diplopia threshold to eccentricity. Palmer (1961) asked subjects to fixate between two marks 40 arcmin apart while a test spot 1.5 arcmin in diameter was presented for 10 ms with various disparities. The fusion limit was about 10 arcmin in the fovea and increased to about 30 arcmin at an eccentricity of 6°. Mitchell (1966a) reported a similar dependence on eccentricity.

Crone and Leuridan (1973) found that beyond an eccentricity of 10° the diplopia threshold increased in proportion to horizontal eccentricity. On average, the diplopia threshold was about 7% of the angle of eccentricity. Thus, a person can tolerate a 7% aniseikonia without experiencing diplopia. Ogle (1964) reported a ratio of 6%. Hampton and Kertesz (1983b) found that the diameter of the fusional area increased linearly with horizontal eccentricity with a slope of about 0.13° per degree. This is similar to the rate of increase of the magnification factor in the human visual cortex (Rovamo and Virsu 1979; Yeshurun and Schwartz 1999). There is evidence that the fusional area increases with eccentricity less rapidly along the vertical meridian than along the horizontal meridian (Ogle and Prangen 1953).

We will see later that the fusion limit for a pair of images is smaller when other images are nearby. In studying the fusion limit as a function of eccentricity, the subject fixates a binocular stimulus to hold vergence constant while the test stimuli are moved into more peripheral positions. The increase in the fusion limit with increasing eccentricity may therefore be due, at least partly, to the increasing distance between fixation stimulus and test stimuli. Levi and Klein (1990a) described a procedure for unconfounding these variables in the measurement of vernier acuity. The procedure could be adapted for measurement of stereo acuity or fusion limits. *The independent effect of increasing eccentricity could be measured by placing a zero-disparity stimulus at a fixed distance from the test stimulus as both stimuli are moved into the periphery. The zero-disparity stimulus could be an annulus around the test stimulus. The independent effect of changing image proximity could be measured by changing the separation between two stimuli on the circumference of a circle centred on the fixation point.*

7.1.2 Effects of spatial frequency and contrast

Relationships between spatial frequency and stereoscopic acuity and gain were discussed in Section 19.7.3. The relationship between spatial frequency and fusion limits will now be discussed. Kulikowski (1978) reported that the fusion limit is greater for gratings with gradual contours than for gratings with sharp contours. In more recent studies, the effects of low and high spatial frequencies on the fusion limit have been investigated directly.

Schor *et al.* (1984b) used vertical bars with a difference of Gaussian (DOG) luminance profile (spatial-frequency bandwidth of 1.75 octaves at half height) superimposed on a small fixation spot. Nonius lines were used to check for changes in vergence, which were claimed to be less than 1 arcmin. Subjects adjusted the disparity between two bars until they noticed an increase in width, a lateral displacement, or a doubling. Figure 7.1 shows that the fusion limit (radius of the fusional area) increased as spatial frequency decreased. The vertical fusion limit was consistently smaller than the horizontal limit. Below a spatial frequency of about 1.5 cpd, the limit of fusion corresponded closely to a 90° phase shift of the stimulus, indicated by the diagonal line. When the measurements were repeated with both Gaussian bars presented to one eye, the results also fell on the diagonal line. This is not surprising, because a

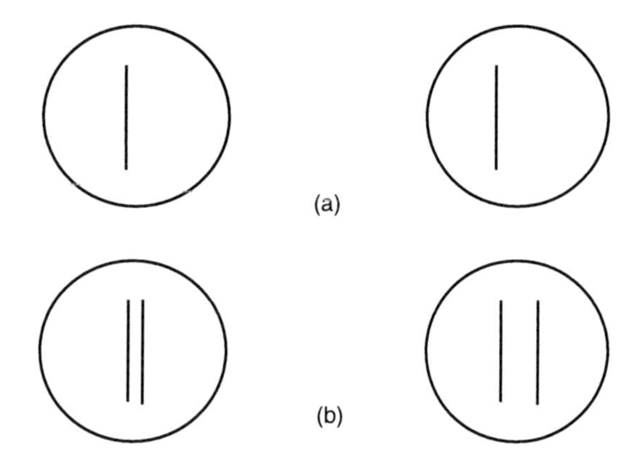

Figure 7.2. Effects of lateral spacing on fusion.
(a) The vertical lines fuse when the eyes converge on the circle.
(b) When a second pair of lines with the opposite disparity is added, the lines no longer fuse. At a distance of 30 cm, the lines have a disparity of 9 arcmin (Adapted from Braddick 1979)

90° phase shift is the Rayleigh limit for monocular resolution (Section 3.1.2). Schor *et al.* concluded that, at low spatial frequencies, the limit of binocular fusion is determined by the same factors that determine monocular grating resolution.

Both these limits are much coarser than acuity for vernier offset (Heckmann and Schor 1989b). Note that Schor *et al.* used a liberal criterion for the limit of fusion—even a slight thickening or displacement of the stimulus counted as diplopia. Perhaps the binocular and monocular limits would not match with a stricter criterion for diplopia. For spatial frequencies over about 2.4 cpd, the horizontal fusion limit levelled off to between 5 and 10 arcmin. Thus, for high spatial frequencies, the Rayleigh limit of 90° phase-shift detection ceases to be the limiting factor for diplopia resolution but not for monocular resolution. In fact, in this study and in that of Schor *et al.* (1989), the fusion limit at the highest spatial frequencies was between three and six times the width of the centre of the DOG, as depicted in Figure 7.1. Presumably the spatial resolution of dichoptic stimuli is determined by some factor other than the limit for monocular resolution.

These results emphasize the special properties of the fusion process at high spatial frequencies and pose a challenge for current theories of the underlying neural mechanism. The experiment was repeated with sharp-edged bars with widths and luminances equal to those of the bright central component of the DOG patterns. Figure 7.2 shows that the fusion limits for the bars resembled those for the narrowest Gaussian pattern, suggesting that subjects were using the edges of the bars (the highest spatial-frequency component) to make their judgments. On the other hand, Woo and Reading (1978) found diplopia thresholds for single bars were lower than monocular resolution thresholds for the same bars.

The dependence of the fusion limit on the highest visible spatial-frequency component of a stimulus might arise because, for a given contrast, high spatial-frequency stimuli have a steeper luminance gradient than low spatial-frequency stimuli. Schor *et al.* (1989) investigated this issue by asking subjects to decide which of two horizontal sine-wave gratings contained a vertical disparity. This procedure forces subjects to use any available cue, including thickening and displacement of lines, as well as diplopia. They could not use stereo depth because the gratings were horizontal. The fusion limit was measured for each of several spatial frequencies at each of several contrasts. The logic was that if spatial frequency is the crucial factor rather than the luminance gradient, then changing the contrast for a fixed spatial frequency should have no effect. But if the luminance gradient is a crucial factor, then changing contrast should have an effect, since halving the contrast halves the luminance gradient for a sinusoidal grating of fixed spatial frequency. The results showed almost no effect of changing contrast across a range of spatial frequencies from 0.4 to 3.2 cpd, a result confirmed by Heckmann and Schor (1989b). Furthermore, the fusion limit was not affected by a change in luminance gradient produced by adding a low spatial-frequency component to the sine-wave gratings, even when the added component had the higher contrast. They concluded that binocular fusion is based on information in independent spatial-frequency channels rather than on the overall luminance distribution. Linking disparate images for the detection of stereo depth can involve the overall luminance distribution of the images under certain circumstances and spatial-frequency components under other circumstances (Section 18.1.1).

Schor *et al.* argued that the fusion limit is not affected by changes in contrast because a change in contrast has the same effect on binocular cells that register fused images as on monocular cells that register diplopic images. The effect of contrast thus cancels out. Stereoacuity is adversely affected by a reduction in contrast (Section 19.5). This is presumably because the detection of disparity upon which stereoacuity is based depends only on binocular cells. Certainly, the effects of contrast and spatial frequency on diplopia detection are not the same as their effects on disparity detection.

Changes in stimulus luminance of up to 3 log units above threshold also have little effect on the fusion limit (Siegel and Duncan 1960; Mitchell 1966a).

Roumes *et al.* (1997) confirmed that the fusion limit decreases as the spatial frequency of the stimulus increases, for both crossed and uncrossed disparities. However, with a 0.3 cpd DOG superimposed on a 4.8 cpd DOG, the fusion limit was intermediate between the limits for the separate components. In other words, the low spatial-frequency component of the compound stimulus increased the fusion threshold above the limit set by the high spatial-frequency component. They used DOGs that varied in luminance in two dimensions, as depicted in Figure 4.6. This contrasts with Schor *et al.*'s finding that the fusion limit depends on the highest visible spatial-frequency component. Roumes *et al.* argued that the fusion limit is less ambiguous with their 2-D stimulus than with the 1-D Gaussian patch or bar used by Schor *et al.* They pointed out that random-dot stereograms can be fused over greater disparities than can isolated dots. The low-frequency dot clusters must play a part in determining the fusion limit of a random-dot display.

7.1.3 Fusion limits and disparity scaling

7.1.3a *Disparity scaling*

The effects of spatial scale on stereoacuity are discussed in Section 19.7.2. The related question of the effects of spatial scale on fusion is discussed here. Helmholtz (1910) noticed that disparate points are less likely to fuse when there are other objects nearby. The two lines in Figure 7.2a readily fuse when convergence is held on the surrounding circle. At a viewing distance of 30 cm the lines have a disparity of about 9 arcmin. When a second pair of lines with the same but opposite disparity was added, as in Figure 7.2b, the lines no longer fused for many subjects. Braddick (1979) independently varied the distance between the pairs of lines in one eye and the distance between disparate images in the two eyes. The crucial factor limiting fusion for a given disparity was the monocular spacing of the images rather than competing disparate images. Contaminating effects of vergence changes were avoided by having subjects align nonius lines before the displays were exposed for only 80 ms, which is too short a time for vergence movements to occur.

Braddick also showed that the reduction in the fusion limit is most evident when the closely spaced monocular images are parallel, vertically aligned, and equal in length. It is as if two closely spaced lines in one eye evoke responses in detectors of smaller spatial scale than those evoked by a single line. Thus, diplopia detection proceeds within a system of higher spatial resolution when this finer system is recruited.

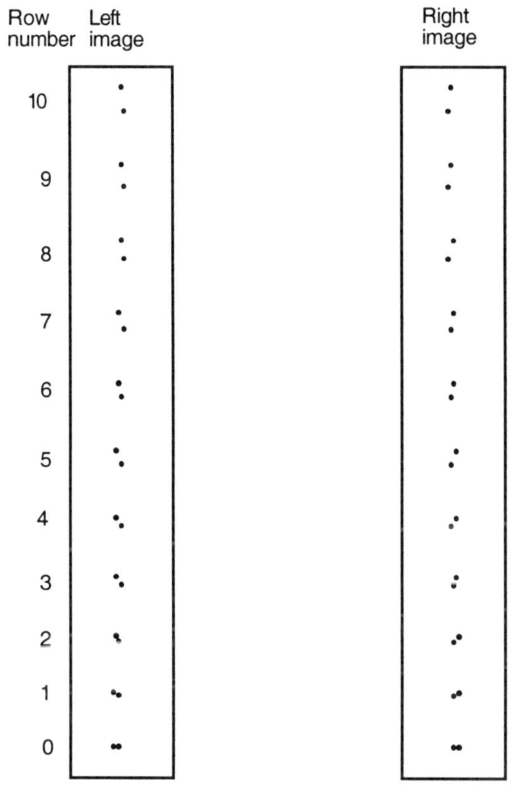

Row number	Left image	Right image
10		
9		
8		
7		
6		
5		
4		
3		
2		
1		
0		

Figure 7.3. Disparity-gradient limit for binocular fusion. Diverge or converge to fuse the columns of dots. If, at any level, the lower pair of dots in any row is fused, the upper pair fuses only if the disparity gradient is less than about 1. The vertical disparity gradient increases down the rows and may be calibrated for a given viewing distance. The disparity-gradient limit for fusion can be specified by the row number at which fusion of the upper pair of dots fails. (Derived from Burt and Julesz 1980)

Tyler (1973) quantified the limits of spatial interactions for fusion and was the first to establish a disparity-gradient limit for fusion. The disparity gradient is the difference in disparity between the images of one point and the images of a second point divided by the mean angular separation of the image pairs (Section 15.2.3). Points on a visual line of one eye have a disparity gradient of 2, and those on a line through the cyclopean axis have a disparity gradient of infinity. Burt and Julesz (1980) found that the disparity limit for maintained fusion of two points decreased as the angular separation between that pair and a fused pair of points decreased. This phenomenon is referred to as **disparity scaling** and is illustrated in Figure 7.3. Two dichoptic images do not fuse when the disparity gradient with respect to a neighbouring fused pair of images exceeds a value of about 1. Thus, in the bottom rows in Figure 7.3, the disparity gradient is steeper than 1 and, although the members of the fixated pair of dots fuse, those of the other pair remain perceptually distinct.

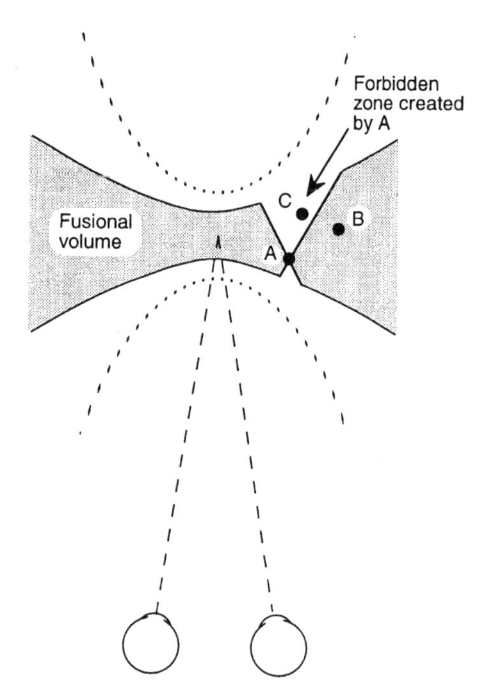

Figure 7.4. Disparity-gradient limit for fusion.
A fused object, such as *A*, creates a zone within which the disparate images of a second object, such as *C*, will not fuse. The images of *B* fuse in the presence of *A*, since *B* is outside A's "forbidden" zone. Dotted lines represent disparity limits for stereopsis. (Adapted from Burt and Julesz 1980)

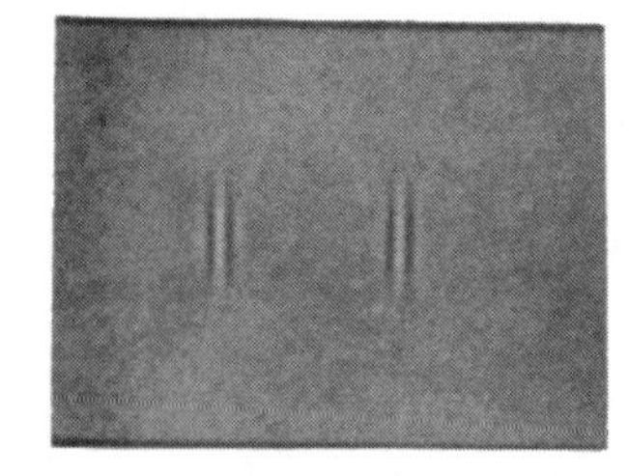
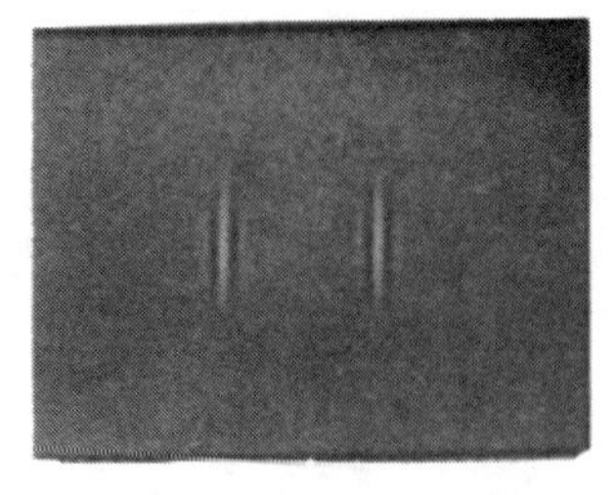

(a) A pair of D6 patches with a horizontal disparity. When fused, one patch appears in front of the other.

(b) The D6 patches are superimposed on a grating 2 octaves lower in spatial frequency. The left-hand D6 can no longer be fused.

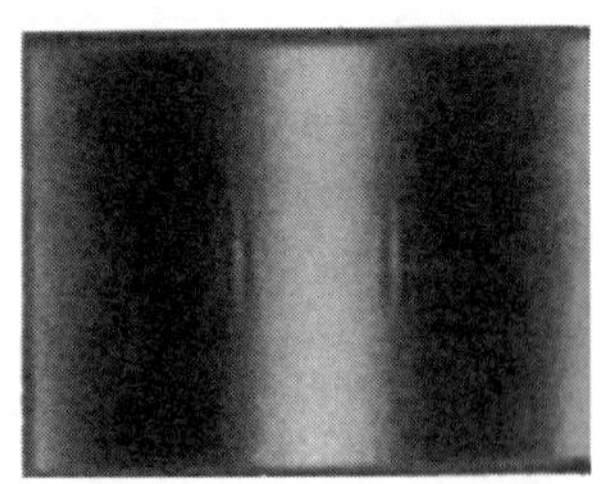

(c) D6 patches are super-imposed on a grating 4 octaves lower in spatial frequency. The patches fuse to create transparent depth. (From Wilson *et al.* 1991)

Figure 7.5. Fusion limit and superimposed spatial frequencies..

In Figure 7.3 the fused pair of dots in the bottom rows lies more or less between the unfused dots, as illustrated in the two columns of dots on the left. The spatial intrusion of the fused pair between the unfused pair could prevent the flanking pair from fusing. But the same fusion limit applies when the two disparate points are to one side of the fused points. Burt and Julesz referred to the orientation of the disparity gradient as the **dipole angle**. The dipole angle is 90° in Figure 7.3. The largest disparity gradient for which the nonfixated images could be fused was independent of the dipole angle. Given that the disparity gradient limit for fusion is 1, it follows that each fused object in the visual field creates a forbidden zone, as illustrated in Figure 7.4. Within this zone, the disparity gradient is greater than 1 and disparate images do not fuse, unless the disparity is vanishingly small. Prazdny (1985c) confirmed that the limiting disparity gradient is 1 for similar stimulus elements but found that the largest disparity gradient for fusion increased to 1.4 when the objects differed in size, and to over 2 when they also differed in luminance polarity.

Wilson *et al.* (1991) measured the effects of a background grating of one spatial frequency on the fusion limit of small vertically elongated D6 Gaussian patches, as shown in Figure 7.5. Each Gaussian patch was presented for 165 ms and had a spatial bandwidth of 1 octave with a centre spatial frequency of between 0.5 and 12 cpd. The righthand patches had zero disparity and the disparity of the lefthand patches varied. Subjects fixated between the patches and reported whether or not the left-hand patches were fused. The diplopia threshold decreased about 3.9 times when the patches were superimposed on a grating with a spatial frequency twice that of the Gaussian patches but was not affected by a grating with a spatial frequency four times that of the patches. Further tests revealed that coarse spatial scales constrained disparity processing in fine scales, but fine scales did not constrain processing in coarse scales. These effects cannot be due to changes in vergence since the stimuli were presented only briefly. They did not depend on the spatial phase of the test patches relative to that of the background grating but did depend on the test and background stimuli having the same orientation.

Wilson *et al.* concluded that binocular disparities are processed in at least three distinct spatial-

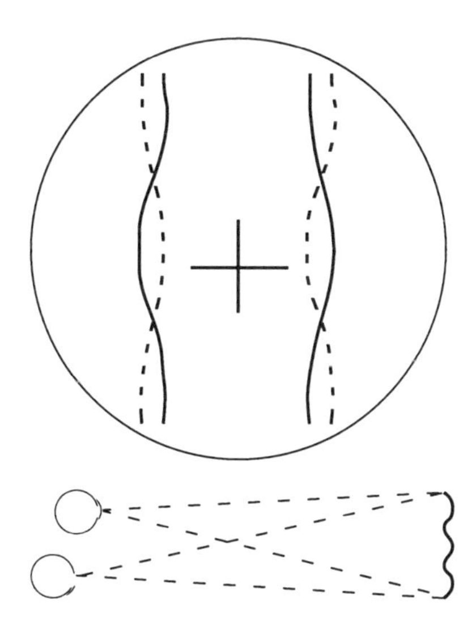

Figure 7.6. Spatiotemporal aspects of the fusion limit.
Solid wavy lines are images in one eye, dashed wavy lines are images in the other eye. Images in the two eyes were alternated at between 0.1 and 5 Hz to create a pair of lines undulating in depth, as shown below, with the sign of undulation alternating over time. The lines were 0.5° on either side of a fixation cross. (Adapted from Schor and Tyler 1981)

frequency channels, each with subchannels for near (crossed), zero, and far (uncrossed) disparities. To account for their data, they postulated that far and near cells inhibit far and near cells, respectively, in the next higher spatial-frequency channel and that zero-disparity cells inhibit both near and far cells in the next higher spatial-frequency channel. They argued that these effects could be accomplished by inhibitory feedback suppressing an appropriate subset of monocular inputs.

Scheidt and Kertesz (1993) conducted a similar experiment with an induction stimulus consisting of D10 Gaussian patterns in a 5° circular area around a central fixation point and a similar test pattern in a larger annulus around the induction stimulus. A 0.5° ring separated the inner area and the surrounding annulus. Both sets of patterns had a peak spatial frequency of 0.75 cpd, but the disparity of the induction stimulus varied from trial to trial between ±15 arcmin. When the stimuli were exposed simultaneously for 167 ms, the fusional range of the test stimuli with the Gaussian induction stimulus was reduced relative to when the induction stimulus was evenly illuminated. These results essentially confirm those obtained by Wilson *et al.* for this spatial frequency. However, when the stimuli were exposed continuously, the fusional range of the test stimulus was reduced only in the presence of an induction stimulus with uncrossed disparity. Scheidt and Kertesz proposed that interactions between fusional stimuli have a fast, wholly inhibitory, component and a slow component that is inhibitory or facilitatory depending on whether the binocular disparities in the interacting stimuli have the same or opposite signs.

The effects of the disparity gradient and of the spatial frequency of superimposed gratings are presumably aspects of the same underlying mechanism. The mechanism ensures that when stimuli of different spatial scale are crowded together, detectors of small spatial scale are devoted to the analysis of disparity between finer elements of the stimulus. This ensures that the fusion mechanism does not combine distinct parts of a dense stimulus pattern.

7.1.3b *Disparity scaling in the periphery*

Scharff (1997) found that, as eccentricity increased, an object with a given disparity had to be more widely separated laterally from a neighbouring object before its images would fuse. In other words, the critical disparity gradient for fusion decreased with increasing eccentricity. This effect may be explained by the fact that, as one moves into the periphery, resolving power decreases (receptive fields become larger) and fine disparities cease to be processed. The linkage between spatial scale and the processing of fine and coarse disparities is discussed in Sections 7.6.1 and 19.7.

7.1.3c *Disparity scaling in the blue-cone system*

Wilson *et al.* (1988) measured the horizontal and vertical diplopia thresholds for dots and lines that stimulated the blue cones only. This was done by adapting out the middle- and long-wavelength cones by a yellow adapting field. The diplopia threshold of the blue-cone system was similar to that obtained when all cone types were stimulated, and showed the same dependency on the disparity gradient when low spatial-frequency stimuli were used. In other words, the blue-cone system was subject to disparity scaling.

7.1.4 Temporal factors in fusion limits

There has been some dispute about the effects of stimulus duration on the fusion limit. Mitchell (1966a) found that the fusion limit for horizontal disparity was unaffected by increasing exposure time from 10 to 120 ms. Palmer (1961) reported similar findings. Duwaer and van den Brink (1982b) found that the diplopia threshold for vertical disparity in one subject decreased from about 10 to 6 arcmin as exposure time increased from 20 to 200 ms. Woo (1974a) found that the mean horizontal diameter of the fusional area of three subjects for a short vertical line increased from about 2 to 4 arcmin

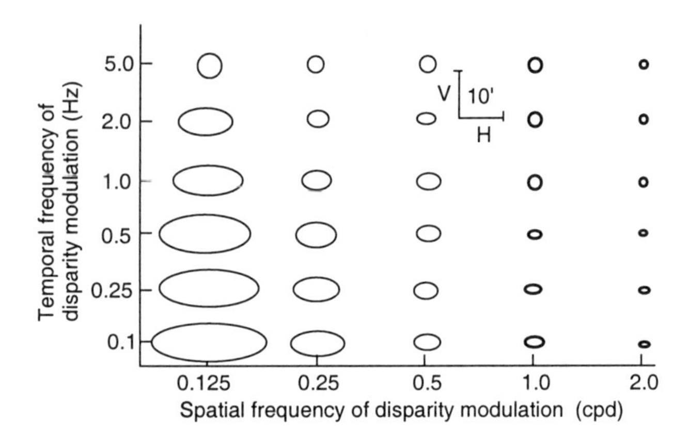

Figure 7.7. Fusional areas and spatial frequency.
Ellipses represent Panum's fusional area. Both diameters increase as the spatial frequency of disparity modulation of wavy lines shown in Figure 7.6 decreases. The horizontal but not the vertical diameter decreases as temporal frequency of depth modulation increases, especially at low spatial frequencies. (Adapted from Schor and Tyler 1981)

when the duration of exposure increased from 5 to 100 ms. The reasons for these contradictory findings remain obscure.

Woo (1974b) presented a dichoptic narrow slit for 10 ms to each eye with various intervals of time between them. The diplopia threshold was not affected until the delay was 40 ms, when the stimuli began to appear as discrete temporal events.

Schor and Tyler (1981) explored the dependence of fusional limits on the spatiotemporal properties of the stimuli. In investigating horizontal fusion limits, they presented two vertical wavy lines dichoptically with opposite phases of the waves in the two eyes. From trial to trial, they changed the horizontal disparity between the aligned peaks of the waves by changing the amplitude of the waves. The two sets of lines were placed 0.5° on either side of a fixation cross (Figure 7.6). To measure vertical fusion limits the lines were horizontal. The spatial frequency of the waviness of the lines varied between 0.125 and 2.0 cpd. The sign of disparity of the two sets of lines reversed in counterphase at between 0.1 and 5 Hz. This reduced any tendency to change convergence. They determined the amplitude of disparity modulation at which diplopia became apparent for each spatial and temporal frequency of disparity modulation. The results are shown in Figure 7.7. The fusional area increased horizontally and vertically as the spatial frequency of the waviness of the stimulus decreased. With low spatial-frequency lines, the horizontal fusion limit, but not the vertical fusion limit, decreased with increasing temporal frequency. This effect was very small with high spatial-frequency stimuli.

7.1.5 Orientation fusion limits

When superimposed dichoptic lines are rotated about their centres in the frontal plane in opposite directions, the angle at which they appear double is the fusion limit for orientation disparity. If orientation disparities were processed by a distinct mechanism, one would expect the fusion limit for orientation disparity to be independent of the length of the lines. However, if the limit depends on point disparities, the fusion limit for orientation disparity could also be independent of line length, because the point-disparity fusion limit increases with eccentricity. The measurement of the orientation fusion limit is confounded by cyclovergence, which tends to cancel orientation disparities.

Kertesz (1973) found that a larger orientation disparity was required to induce diplopia in lines subtending 2° than in lines subtending 9° when the subject fixated the centre of each line. It appears that point disparities rather than orientation disparities determine the cyclofusional limit and that the fusion limit for point disparity does not increase linearly with eccentricity for eccentricities of less than 10°. The fusion limit of orientation disparity was smaller for a set of parallel lines than for single lines. This could be because the fusion limit for crowded stimuli is smaller than for single stimuli or because gratings induce more cyclovergence than single lines (Section 9.8.5). Kertesz did not control for the effects of cyclovergence because, at the time, he did not believe it occurred.

It has been reported that the fusion limit for orientation disparity is about 2° for horizontal lines and about 8° for vertical lines (Volkmann AW in Helmholtz 1910, p. 449; Ames 1926; Beasley and Peckham 1936; Crone and Leuridan 1973; Sen *et al.* 1980). This is consistent with the fact that the fusion limit for horizontal point disparity is larger than that for vertical point disparity (Kertesz 1981). However, cyclovergence is evoked with greater magnitude by cyclorotated horizontal lines than by cyclorotated vertical lines (Section 9.8.5), and unless this is taken into account, comparison between the fusion limits for horizontal and vertical orientation disparities is not valid.

We have noticed, as did Volkmann (see Helmholtz 1910, p. 449) and O'Shea and Crassini (1982), that when a grid of horizontal and vertical lines is rotated in opposite directions in the two eyes, the horizontal lines appear diplopic before the vertical lines. This confirms that the fusion limit is greater for vertical lines (horizontal-shear disparity) than for horizontal lines (vertical-shear disparity) because cyclovergence affects both lines equally. Since the

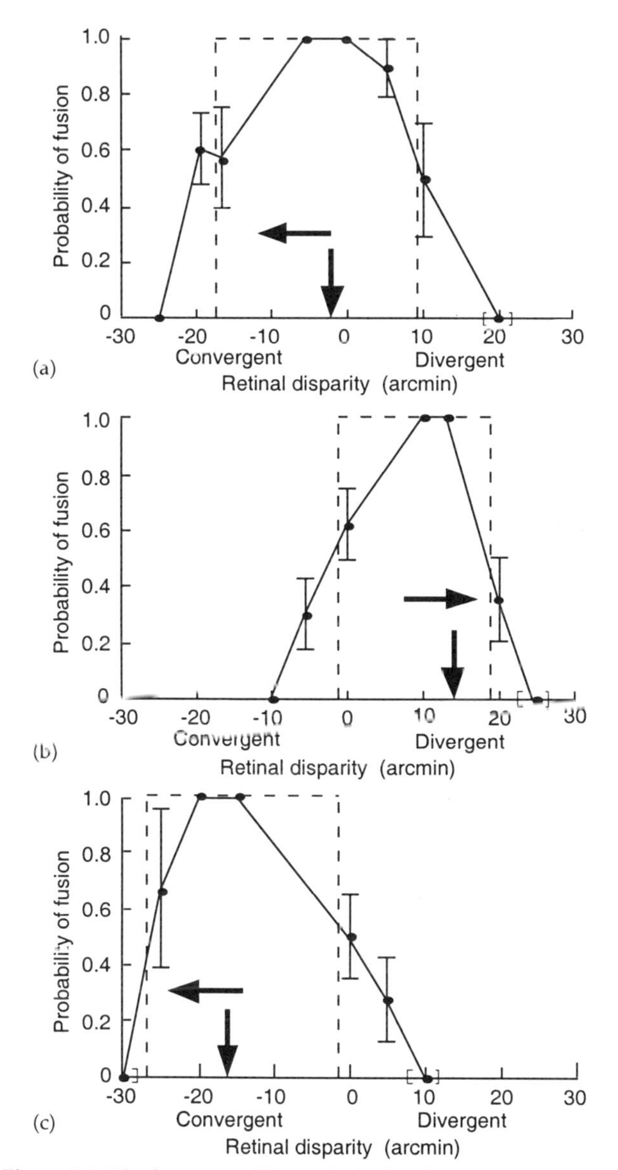

Figure 7.8. Displacement of Panum's fusional area.
A line in one eye was moved slowly into an eccentric position with respect to a fixed line in the other eye, in the direction of the horizontal arrows. At a disparity indicated by vertical arrows, a briefly exposed probe revealed the fusional limits on either side of the fixed line. Dashed lines represent the rectangular approximations to the curves. The fusional area moves in the direction of changing disparity. (Redrawn from Diner and Fender 1988)

corresponding vertical meridians of the retinas are excyclotorted about 2°, one would expect an anisotropy of fusional range if testing were done in only one direction with respect to the true vertical (Krekling and Blika 1983b). *To control for this effect and for effects of cyclovergence, experiments should be performed with stimuli rotated in both directions and with opposite directions of disparity on either side of the fixation point.* The anisotropy of fusion limits may be influenced by the fact that acuity is higher for vertical than for horizontal lines.

7.1.6 Plasticity of fusion limits

The fusion limit is smaller when diplopic stimuli are moved together until they fuse (refusion threshold) than when fused stimuli are moved apart until they appear diplopic (diplopia threshold). There has been a good deal of theorizing about the neural mechanisms responsible for this fusional hysteresis. However, hysteresis is not peculiar to binocular fusion; all psychophysical thresholds exhibit hysteresis according to the direction from which they are approached.

Fender and Julesz (1967) measured fusion hysteresis using optically stabilized images that did not move as the eyes moved. The stimulus to each eye was a single black line on a 6°-wide white surround. As the images moved apart horizontally, diplopia became apparent at an uncrossed disparity of 65 arcmin, and when they moved towards each other they fused at a disparity of 42 arcmin. When the images of horizontal lines were moved apart or towards each other vertically, diplopia thresholds were about 19 and 12 arcmin, respectively. Fusion limits with stabilized images were at least 20 arcmin smaller than with normal viewing. Fender and Julesz measured changes of vergence and claimed that these could account for the larger fusion limits in normal viewing. These fusion limits are larger than those reported by other investigators, but the line was 13 arcmin wide, which may have inflated the values.

These measurements were repeated for both crossed and uncrossed disparities of a retinally stabilized black line, but with the border of the surrounding 3° white disc unstabilized (Diner and Fender 1987). The diplopia threshold for increasing disparity, either crossed or uncrossed, was about 20 arcmin, and the refusion limit was about 10 arcmin. When an unstabilized fixation cross was added just above the line, these limits were reduced by about 5 arcmin. The range of fusion in Fender and Julesz's study, in which the disparities occurred over the entire contents of the visual field, was larger than in Diner and Fender's study, where only some of the elements were disparate. This is what one would expect from the disparity gradients in the two types of display. An overall disparity has a disparity gradient of zero, but a locally applied disparity produces a non-zero disparity gradient. Put another way, the diplopia limit is greater when there is no zero-disparity comparison stimulus in view.

Fender and his associates argued on the basis of the hysteresis effect that the fusional area becomes elongated in the direction of a gradually increasing disparity. Diner and Fender (1988) asked whether this elongation represents an overall expansion of the fusional area in both directions or an extension of the leading edge of the fusional area in the direction of movement, accompanied by a contraction of the lagging edge. In other words, does the area expand or merely shift. To answer this question, they presented a fixed vertical line stimulus to the fovea of one eye and gradually moved a test line in the other eye in the direction of increasing crossed or uncrossed disparity. Both lines were stabilized on the retina. The moving test line was replaced periodically for 2 ms by a probe line at each of several locations on either side of the fixed stimulus line. Subjects reported whether the probe line and stimulus line were fused or diplopic. When the disparity of the test stimulus was near zero, the disparity limits for the probe were approximately symmetrical about zero (Figure 7.8a). When the moving test line had an uncrossed disparity of 12 arcmin, both the left and right boundaries of the fusional area shifted in the uncrossed direction (Figure 7.8b). When the test line had a crossed disparity of 16 arcmin, the boundaries of the fusional area shifted in the crossed direction (Figure 7.8c). Diner and Fender concluded that the boundaries of the fusional area move in the direction of the overall disparity rather than expand. In fact, the data suggest that the fusional areas may contract rather than expand.

Fender and Julesz (1967) used a 3.4°-wide retinally stabilized random-dot stereogram. A central square of dots stood out in depth by a fixed disparity, and the disparity of the whole display was increased or decreased at 2 arcmin/s. They concluded that "··· for random-dot stereoscopic images there is no difference between fusion thresholds and the thresholds for stereopsis." This is a strange result since it is well known that stereoscopic depth can be perceived with diplopic images. A nonfused array of dots has a hazy rivalling appearance compared with the planar appearance of a fused array. The fusion limit for random-dot displays is ill defined. Oddly, the stereogram in the published paper consists of two uncorrelated random-dot displays. One subject saw depth in retinally stabilized stereograms when the images were separated horizontally up to 2° and vertically up to about 20 arcmin. We are not told whether the square in depth still appeared smooth or whether it took on a hazy appearance. Depth was not seen in initially unfused images until they were within 6 arcmin of each other horizontally and 1 arcmin vertically. Thus, a larger horizontal disparity limit and a larger hysteresis effect were obtained with the criterion of perceived depth using a random-dot stereogram than with the criterion of diplopia using the line target. It is not clear from this comparison whether the crucial factor is the criterion or the type of display.

Using a similar procedure, Piantanida (1986) measured crossed and uncrossed disparity limits for a retinally stabilized random-dot stereogram. The criterion was the perception of a cyclopean figure of fixed relative disparity. The range (sum of crossed and uncrossed limits) was between 68 and 150 arcmin for increasing disparity and between 46 and 96 arcmin for decreasing disparity. A small hysteresis effect was thus replicated although a cyclopean form in depth was regained at a much larger disparity than in the Fender and Julesz study. Piantanida reported that the stereograms still appeared fused after the cyclopean shape could no longer be seen. However, the criterion for fusion was the elongated appearance of the square outline of the stereogram. It was not reported whether the surface of the stereogram appeared as a flat plane or as hazy depth. Loss of fusion of the corresponding set of dots may have occurred before, not after, the loss of the cyclopean shape.

Hyson *et al.* (1983) approached the issue of fusion hysteresis with a different procedure. They presented a 9.8°-wide random-dot stereogram containing a spiral in depth with fixed relative disparity. The subject was free to change convergence between different depth planes within the stereogram. The two images were slowly separated laterally while vergence eye movements were measured. The extent to which vergence failed to keep up with the imposed disparity gave a measure of the residual overall disparity between the images. The spiral in depth could be seen for up to 10 s with up to 3° of overall image disparity. As soon as depth was lost, the eyes returned to their original converged position. The displays were then brought slowly together until the impression of depth returned, which was, on average, 2.6° in from the point where the depth had been lost. Thus, the disparity limit for maintained depth and the hysteresis effect were even larger than with the smaller stereogram used by Fender and Julesz.

These large tolerated disparities need not be regarded as extensions of Panum's fusional area, since the criterion was perceived depth, not diplopia. The subjects saw depth produced by a fixed relative disparity in a random-dot stereogram with up to 3° of overall horizontal disparity. This is equivalent to the task of registering a disparity superimposed on a disparity pedestal, as discussed in Section 19.3.2. For instance, reliable relative depth judgments were made between two lines when they were up to 2° of disparity away from the fixation point (Blakemore 1970d). Hyson *et al.* argued that, although random-dot images must fall on nearly corresponding retinal regions before depth is registered, a record of matching dot clusters may be retained over large disparities well outside the normal fusion limits, once depth has been perceived. This process would be aided if the visual system registered large dot clusters or used the edges of the stereogram.

Hyson *et al.* called this process "neural remapping". This term is misleading because it suggests that the pattern of neural correspondence has been remapped. But this is not established by these results. The relative disparity that defined the depth in the stereograms remained constant; only the overall disparity changed. Evidence reviewed in several places in this book suggests that depth is coded in terms of relative disparity not absolute disparity. It is not necessary to assume that corresponding points are remapped but only that, up to a point, overall disparities are disregarded in favour of relative disparities, especially after the relative disparities have been registered.

It was mentioned in the last section that Diner and Fender found that the fusional area for lines shifted in the direction of a slowly moving disparity. Erkelens (1988) investigated the same issue using a 30°-wide random-dot stereogram. The images were retinally stabilized for vergence movements but not for version. The subjects could thus look at different parts of the stereogram but could not change convergence appropriate to the disparity in these areas. The disparity limits for slowly increasing crossed and uncrossed pedestal disparities were measured with the criterion of perceived depth. The same limits were also measured for randomly presented static pedestal disparities. The limits for increasing disparity were similar to those for static disparities, but the limits for regaining the impression of depth were lower than those for either increasing or static disparities. Erkelens concluded that a history of perceiving fused images does not shift the disparity limit for the perception of depth, but a history of perceiving disparate images contracts the limit for that same disparity. These results confirm the hysteresis effect and Piantanida's claim that limits for regaining the impression of depth are higher than those reported by Fender and Julesz. But the results contradict Diner and Fender's claim that the disparity range for stereopsis with an increasing disparity is shifted relative to that for a static disparity. However, Diner and Fender did not investigate the refusion limit relative to the static disparity limit.

Duwaer (1983) pointed out that the disparity limit measured by the criterion of detected depth is a limit of stereo depth rather than of fusion. He found that the diplopia limit for a fixation square superimposed on a random-dot stereogram was within normal limits of about 0.3°, while depth was seen in the stereogram up to a limiting disparity of about 1.3°. He argued that the major hysteresis effect observed with random-dot stereograms does not represent a change in the fusional limits, as Fender and Julesz believed, but is due to the difficulty of regaining the correct binocular match once images have become disparate. However, Piantanida (1986) and Erkelens (1988) claimed that random-dot stereograms remain fused even after the impression of depth is lost. This is difficult to reconcile with the small fusional limits for displays with high spatial-frequency content, as reported in Section 7.1.2.

Summary

However this debate is resolved, to perceive depth in a random-dot stereogram the two images must first be linked. Fender and Julesz claim that this initial linking does not occur unless the images are within a few arcmin of being in binocular register, but Erkelens claims that it can occur with more than 1° of disparity between the images. In any case, once the disparity that defines the pattern in depth has been detected, up to 2° of overall disparity between the two images is tolerated before the sensation of depth is lost. The visual system detects relative disparities within a stereogram despite the presence of a disparity over the stereogram as a whole. It is generally agreed that, when disparity is reduced from a state of diplopia and no depth, depth in a random-dot stereogram is not perceived until the disparity has reached a lower level than that at which depth disappears when disparity is increased. Some investigators interpret this hysteresis effect as a shift in the limits of stereoscopic fusion as disparity is slowly increased, but Erkelens interprets it as a contraction of the limits of fusion due to previous exposure to unfused images.

In all the experiments on the diplopia threshold reviewed here the stimuli were lines, bars, or dots. As the images of such stimuli are separated, the contours with the same luminance polarity move further apart and contours of opposite polarity move closer together and eventually coincide. With further separation the opposite polarity contours separate. The diplopia threshold is therefore a threshold for diplopia between contours of opposite luminance polarity. To investigate the diplopia threshold for contours of the same polarity, one would have to use the stimuli shown in Figure 7.9.

7.1.7 Combining periodic patterns

Dichoptic flickering lights of slightly different frequency produce cross-modulation terms of the general form $nF_1 + mF_2$, for integral values of n and m

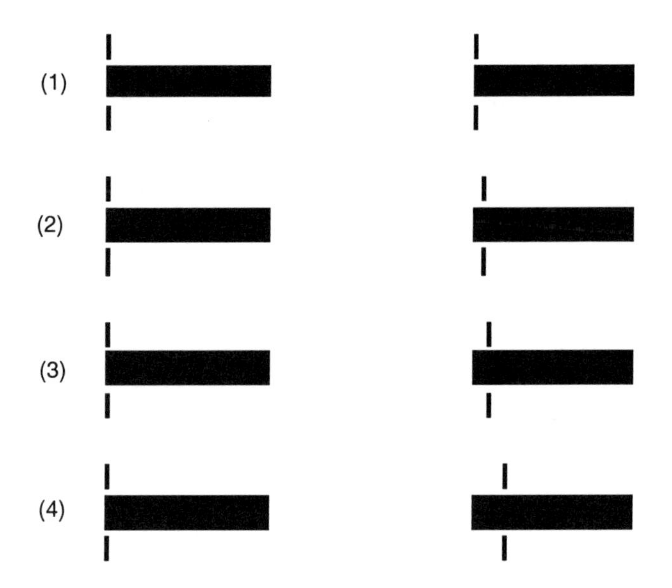

Figure 7.9. Diplopia threshold for single contours.
Display to investigate the diplopia threshold for contours of the same polarity. Each pair of patterns is fused with vergence on the vertical bars. In successive rows, disparity between the horizontal rectangles increases. At a certain disparity the left-hand edges of the rectangles become diplopic and rival.

(Section 6.7.2). The relative amplitudes of these terms depend on the nature of the nonlinearities both before and after binocular convergence. Regan and Regan (1988, 1989) showed that nonlinear processing occurring after binocular convergence can be isolated from that occurring before convergence. Only subjects with normal stereoscopic vision showed evidence of a nonlinear combination of monocular inputs, presumably arising from the way in which binocular cells combine signals from the two eyes (Baitch and Levi 1988).

One sees interference patterns (moiré patterns) when two periodic patterns with slightly different spatial frequencies are superimposed in the same eye (see Spillmann 1993). For example, a grating of 10 cpd superimposed on one of 11 cpd produces a moiré pattern of 1 cpd. There are conflicting claims about whether gratings combined dichoptically generate moiré patterns. Bryngdahl (1976) claimed that dichoptic combination of patterns of concentric rings produces a moiré pattern of vertical bars but he did not provide details or explain how he avoided binocular rivalry. Oster (1965) found no dichoptic moiré patterns. Kaufman (1974, p. 298) found that combining two offset radial patterns dichoptically did not produce the moiré pattern evident when they were combined in one eye. But these were high-contrast patterns, which therefore engaged in rivalry at each location. One would expect to see moiré patterns only if luminance were summed at each location. Low contrast gratings sum rather than rival (Section 7.3.2b) so that moiré patterns may be revealed in dichoptically combined low contrast patterns. However, we could not see moiré patterns in such stimuli.

Badcock and Derrington (1987) found that, with normal viewing, subjects were more sensitive to displacement of a 9 cpd grating relative to a 10 cpd grating than to displacement of a 10 cpd grating on its own. The rapid motion of the 1-cpd moiré pattern facilitated perception of relative displacement of the two gratings. However, subjects were less sensitive to the relative displacement of two dichoptic gratings than to the displacement of a single grating. The dichoptic stimulus did not produce a moiré pattern. Subjects saw neither depth nor rivalry in the dichoptic stimulus. Even if the gratings did rival, they were probably too similar for the change to be noticed. They concluded that the binocular mechanism cannot sum luminance variations in the two eyes.

7.2 DICHOPTIC COLOUR MIXTURE

7.2.1 Basic phenomena

Under some circumstances, a coloured area presented to one eye appears to rival an area of another colour presented to the other eye. This is **colour rivalry**. Under other circumstances, dichoptic colours combine to create a third colour. This is **binocular colour mixing**. There has been some dispute about whether binocular colour mixing ever occurs, and even those who believe that it occurs disagree about the necessary conditions.

Desargulier (1716) was perhaps the first to investigate colour rivalry. He dichoptically superimposed differently coloured pieces of silk by viewing them through an aperture. He reported colour rivalry rather than colour mixing. Taylor (1738) dichoptically viewed differently coloured glasses placed in front of candles and claimed to see colour rivalry rather than fusion. Dutour (1760), also, observed colour rivalry of patches of blue and yellow fabric combined by converging the eyes. Haldat (1806) reported that dichoptically combined glass prisms containing coloured liquid appeared in an intermediate hue.

Later in the nineteenth century, there was a controversy between those who adopted the Young-Helmholtz theory, which stipulated that all colours can be formed from mixtures of red, green, and blue light, and those who adopted the Hering or Ladd-Franklin theories, which stipulated that the sensation of yellow arises from a distinct process in the retina. The Young-Helmholtz theory predicts that yellow should arise from the dichoptic combination of red and green whereas the latter two theories do not. It is ironic that Helmholtz (1910) believed that binocular yellow was an artifact due to colour adaptation, binocular suppression, and unconscious inference, whereas Hering (1879) regarded it as due to interaction of visual inputs at a central location (for a bibliography of early studies see Johannsen 1930).

Helmholtz's reluctance to regard binocular yellow as due to a central combination of inputs arose from his belief that inputs from the two eyes are not combined physiologically. We know now that the trichromatic stage of colour processing is followed by two retinal opponent processes: one between red and green receptors and one between blue and yellow. Yellow does not arise from a distinct cone type but is formed by inputs from red and green receptors (see Boynton 1979). The theoretical significance of binocular colour mixture is still not clear.

Many early investigators have claimed to see binocular yellow. For instance, Hecht (1928) saw yellow when he combined red and green patches. Murray (1939) pointed out that the Wratten filters used by Hecht extended into the yellow region of the spectrum. Dunlap (1944) argued that binocular yellow is an artifact of adaptation of the eye to the red light and claimed to see yellow when both eyes looked at red patches for some time. He concluded that, "··· binocular color mixture can be laid away in the museum of curious superstitions." But the problem lives on. Prentice (1948) claimed to have overcome Murray's objection by using narrow-band Farrand interference filters centred on 530 mμ (green) and 680 mμ (red), neither of which extends into the yellow region of the spectrum. He obtained good binocular yellow even with brief exposure, and the fused image became more yellow with longer exposure. Others have also reported binocular yellow with narrow-band filters and appropriate controls for colour adaptation.

Hurvich and Jameson (1951) pointed out that the spectral purity of the red and green filters is irrelevant. The crucial factor is the chromatic bandwidth of the receptors, since a receptor cannot distinguish between one wavelength and another within its tuning range—the principle of univariance. The wavelengths selected by Prentice and others evoked sensations of yellowish red and yellowish green and it was therefore not surprising that they produced binocular yellow. When Hurvich and Jameson used unique red and green, which evoke the purest sensations of red and green, the dichoptic mixture was not yellow but gray. This still represents a form of binocular colour mixing that needs to be explained. The colours used by Hurvich and Jameson were close to being opponent colours that produce gray when mixed monocularly. As ordinarily understood, the opponent mechanism resides in the retina, so that the occurrence of binocular gray must depend on a distinct cortical process.

DeValois and Walraven (1967) obtained a desaturated yellow when the afterimage of a bright red patch in one eye was superimposed on a green patch in the other eye. The effect faded with the fading of the afterimage and was not present when the eye containing the afterimage was pressure blinded (Gestrin and Teller 1969).

7.2.2 Factors affecting binocular colour mixing

Dichoptic colour mixtures are affected by the following factors.

Luminance Dichoptic colour mixtures are more stable at lower than at higher luminance levels and when the luminance in the two eyes is the same (Dawson 1917; Johannsen 1930). They are also more stable when the components of the dichoptic mixture are presented on a dark rather than a light background (Thomas *et al.* 1961).

Saturation Dichoptic colour mixtures become more stable as the saturation of the colours decreases (Dawson 1917).

Stimulus duration Hering (1861) observed that prolonged inspection of a dichoptic mixture increases the stability of colour mixtures. Johannsen (1930) suggested that prolonged inspection causes the colour in each eye to become desaturated through adaptation and that this, rather than duration, is responsible for the increased stability of colour mixtures. However, dichoptic colour mixtures also seem to be more stable with very short exposure times. Rivalry is most evident with intermediate exposure durations. Thus, synchronous flicker of red and green dichoptic stimuli increased the apparent saturation and stability of binocular yellow and the best results were obtained with flash durations of less than 100 ms and interflash durations of more than 100 ms (Gunter 1951). Binocular colour rivalry does not occur with brief stimuli (Section 7.3.8). Stimulus asynchrony of more than 25 ms disrupts binocular yellow (Ono *et al.* 1971a).

Colour difference Periods of colour rivalry that occur with larger visual fields are more pronounced the greater the colour difference. Ikeda and Nakashima (1980) increased the dichoptic difference in the wavelength of a 10° test patch until the subject reported colour rivalry. The threshold difference for the occurrence of rivalry varied as a function of wavelength in a manner closely resembling the hue-discrimination curve. In other words, threshold colour differences for the production of rivalry were equally discriminable.

Sagawa (1982) asked whether the threshold for discriminating between two patches of wavelength λ and $\lambda + \Delta\lambda$ presented to one eye was affected when patches of wavelength λ were superimposed in the other eye. The idea was that if colour processing were independent in the two eyes, the addition of the dichoptic masking patches would not affect the discrimination threshold. Wavelength discrimination deteriorated in the presence of the masking stimulus, but the extent of the deterioration was largely independent of the luminance of the masking stimulus. This suggests that dichoptic masking between chromatic signals is independent of the luminance component of the visual stimulus.

7.2.3 Dichoptic and dioptic colour mixtures

Colour matches obtained under dichoptic viewing differ from those obtained with monocular viewing. Lights combined monocularly obey Abney's law, which states that the luminances of differently coloured lights add linearly. Lights combined dichoptically, whether of the same or different colours, do not obey Abney's law but produce an intermediate brightness, especially when they are similar in luminance (Section 8.1.2). Dichoptic colour matches are less saturated and more variable than similar monocular matches. The proportion of green to red required to match a spectral yellow and the proportion of yellow required to cancel blue were found to be less with dichoptic than with monocular viewing (Hoffman 1962; Hovis and Guth 1989a). Hovis and Guth (1989b) argued that less green is required for dichoptic than for monocular yellow because, with increasing luminance, the postreceptor response for green increases faster than that for red. The red/green ratio required for monocular yellow is invariant over changes in luminance because receptor inputs to the monocular opponent mechanism increase at the same rate with increasing luminance.

De Weert and Levelt (1976a) presented a dichoptic mixture of equiluminous lights of different wavelengths in a small area. Subjects adjusted the relative luminances of two lights of the same two wavelengths presented to both eyes in an adjacent small area until the two areas appeared most similar in hue. They did this for many pairs of wavelengths and derived a set of hue-efficiency functions for dichoptic mixtures. Reasonably good matches of hue were obtained between the dichoptic and dioptic stimuli with the same wavelength components. In general, a smaller amount of the wavelength component nearer the middle of the spectrum was required in the dichoptic mixture than in the dioptic mixture. A coloured patch presented to an amblyopic eye contributed less to the dichoptic colour than a patch presented to the non-amblyopic eye (Lange-Malecki *et al.* 1985).

Hering (1861) observed that dichoptic colour mixtures are more stable with small than with large stimuli and this has been confirmed more recently (Thomas *et al.* 1961; Ikeda and Sagawa 1979). With large stimuli, people experience colour rivalry rather than colour mixture. With stimuli subtending less than 2°, most subjects reported stable colour mixture (Grimsley 1943; Gunter 1951). With a display subtending 3.5°, binocular colour mixture was unstable but became stable when a fusible micropattern was superimposed on the display, as in Figure 7.10 (De Weert and Wade 1988). One could think of the textured pattern as breaking up the display into small regions and thus preventing rivalry. Binocular colour mixture is difficult to see in the presence of rivalling patterns (Dawson 1917).

Summary
Dichoptic colour mixing is a genuine phenomenon but differs in several respects from monocular colour mixing. Dichoptic colour mixing is more stable with small or textured patches than with large homogeneous patches, with flickering stimuli than with steady stimuli, and with patches of low luminance and saturation and equal luminance and chromaticity than with bright and saturated patches or patches of unequal luminance. Its occurrence implies that there must be colour mechanisms in the cortex in addition to those in the retina. The literature on binocular colour mixing was reviewed by Hovis (1989).

7.3 BINOCULAR RIVALRY

7.3.1 Introduction

7.3.1a *Basic theories of rivalry*
Before 1838, the year Wheatstone demonstrated that binocular disparity plays a crucial role in depth perception, people interested in binocular vision were preoccupied with explaining how a unified percept is formed from two images. It had been realized at least since the time of Aristotle and Euclid that the eyes have slightly different views of the world. Although a few people before Wheatstone had suggested that this contributes to the perception of

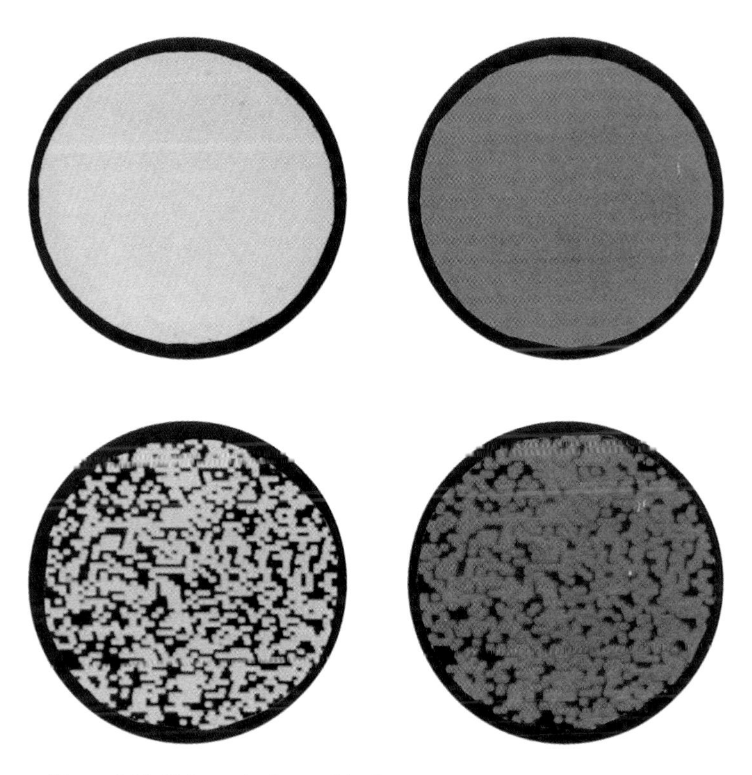

Figure 7.10. Colour rivalry and texture.
Fusion of the solid red and green discs produces unstable colour rivalry, whereas fusion of the textured discs produces stable dichoptic colour mixing. (Reprinted from de Weert and Wade 1988, with permission from Elsevier Science)

depth (Section 2.2.3), this aspect of binocular vision was generally ignored. There were two theories of how binocular images combine into a single percept.

According to the **fusion theory**, similar images falling on corresponding retinal points gain simultaneous access to the visual system to form a unitary percept while dissimilar images engage in alternating suppression. According to the **suppression theory**, both similar and dissimilar images engage in alternating suppression at a low level of visual processing.

In the fusion theory, the two images are said to fuse, although it is not always clear what this implies. Except under special circumstances, the two inputs are not simply summed, since that would make objects viewed with two eyes appear twice as bright as when viewed with one. Nor are the identities of the two signals lost in the fusion process; if they were, we would not be able to distinguish between images with crossed disparity and those with uncrossed disparity. I will take the term **fusion** to mean that similar images presented to corresponding points appear as one and are processed simultaneously rather than successively.

Even if we accept that similar images are fused and processed simultaneously, two questions remain. The first is, where in the nervous system does fusion occur? Many investigators, including Helmholtz, believed that fusion is a mental, or psychic, act. In modern terms we talk about higher, or cognitive, levels of processing. Since 1959, when Hubel and Wiesel discovered binocular cells in the striate cortex, most investigators believe that fusion of similar images occurs at this relatively low level. We will see that rivalry between nonsimilar images may not be confined to the striate cortex. The second remaining question is, what rules of stimulus combination are involved in the fusion process? Most recent work on fusion has been concerned with this question.

According to the suppression theory of binocular vision, superimposed images from the two eyes always rival by mutual inhibition, even when they are identical. Thus, in any location in the visual field, only one eye's input is seen at any one time. The dominant input varies from place to place in the visual field and alternates over time, resulting in a mosaic of alternating dominance and suppression. The fusion and suppression theories agree that binocular images rival when they are very different (Figure 7.11). The two images compete for access to higher levels of visual processing by **binocular rivalry**. The predictions of the fusion and suppression theories differ only when the images are similar. However, one cannot apply a direct test of the suppression theory when the images are similar, since any rivalry that might occur would not be visible.

In spite of this difficulty it has now been established by an indirect test that similar images do not inhibit each other in the manner required by the suppression theory (Section 7.7.1). When the images from corresponding regions in the two eyes are identical, information from both, albeit in altered forms, is passed on to higher visual processes to produce a fused image. When the images from corresponding regions differ in an appropriate way, they fuse, but the disparities are registered and produce an impression of relative depth. When the images from corresponding regions are very different, they rival so that only one of them gains access to higher stages of visual processing at any one time in any one location. Thus, similar and dissimilar binocular images are processed in fundamentally different ways.

Several investigators have noticed that rivalling stimuli form a combined percept when presented for less than 200 ms. This suggests that both similar and dissimilar stimuli are initially processed simultaneously and then segregate into those that remain fused and are processed simultaneously, and those that rival and are processed sequentially. We will also see that low-contrast dissimilar stimuli do not rival, even when exposed for some time. Furthermore, the segregation between stimuli that fuse and those that rival is not complete, even for high contrast stimuli of long-duration. Thus, some low-level features of a suppressed image can affect the processing of the dominant image and, on the other hand, the fusion process involves both inhibitory and excitatory processes. Thus, the actual processes underlying binocular vision are more complex than either the fusion theory or suppression theory suggests. The evidence for these statements is now reviewed.

7.3.1b *Basic phenomena of binocular rivalry*

When widely distinct stimuli fall on corresponding regions of the two retinas, they rival, rather than fuse—a sensation known as **binocular rivalry**. The stimulus seen at a given time is the **dominant stimulus**, and the stimulus that cannot be seen is the **suppressed stimulus**. I distinguish between contour rivalry and area rivalry.

In **contour rivalry**, a contrast-defined edge in one eye falls on the same region of the binocular field as a blank area in the other eye. For example, a small black disc on a white ground remains visible when superimposed on a larger black disc presented to the other eye, as shown in Figure 7.11a. The edge of the small disc suppresses the surrounding homogene-

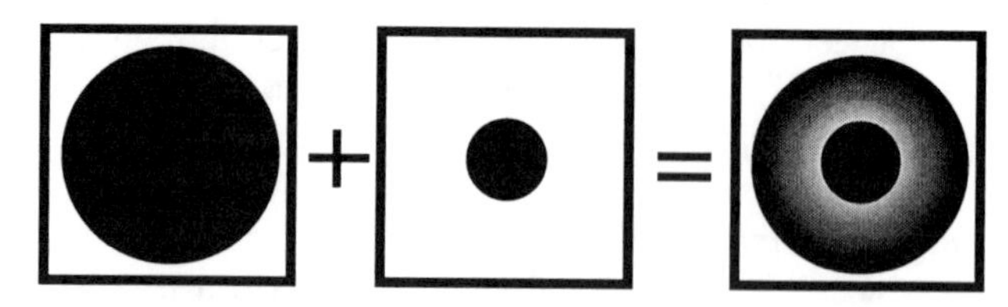

(a) A small disc in one eye suppresses the centre of a large disc in the other eye.

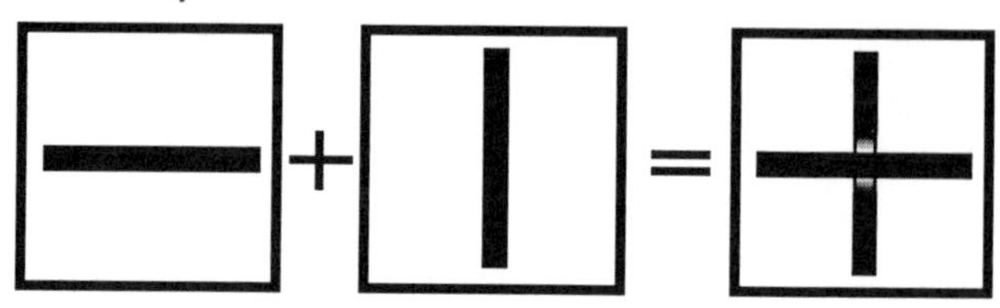

(b) Contours of one or other bar are preserved at the intersection.

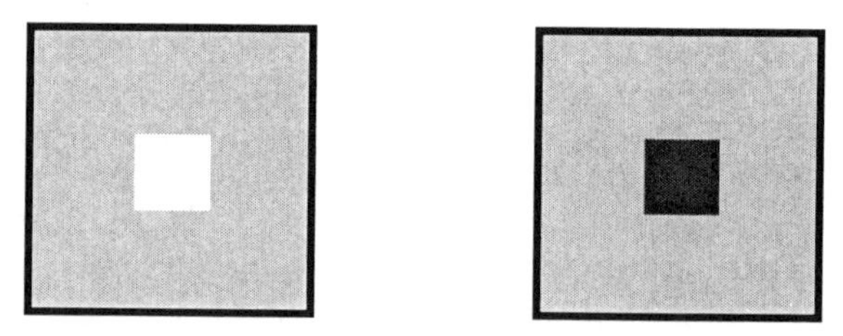

(c) Regions of opposite contrast create an impression of lustre.

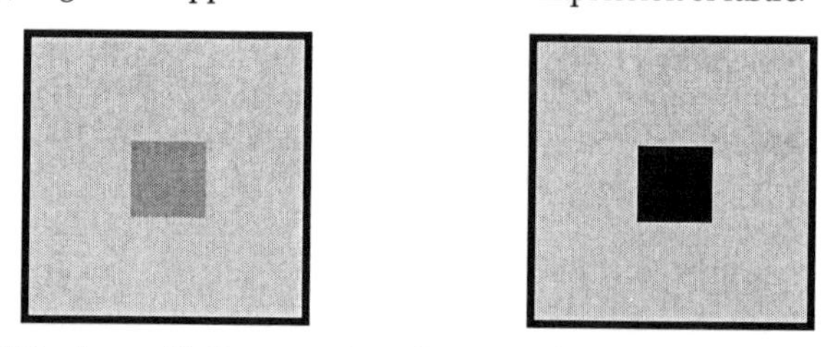

(d) Regions with the same sign of contrast fuse to an intermediate brightness.

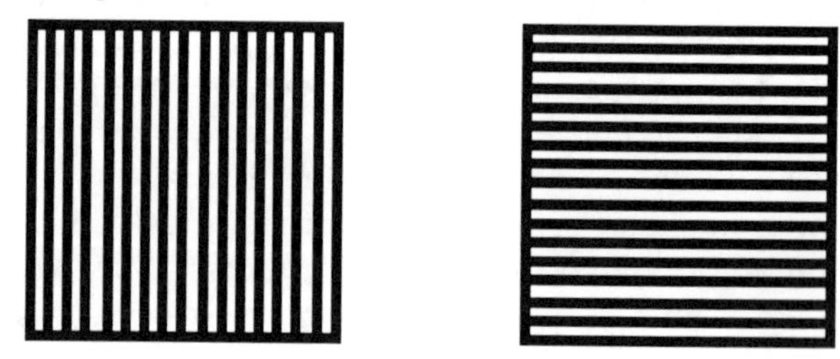

(e) Horizontal lines dominate in one location, vertical lines in another. Areas of dominance constantly fluctuate.

Figure 7.11. Basic phenomena of contours in binocular rivalry.

ous region in the larger disc. The **zone of binocular suppression** surrounding a contour in one eye presumably results from the inhibitory surrounds of receptive fields responding to the contour. A model of this process has been proposed by Welpe *et al.* (1980).

A more complex example of contour rivalry is shown in Figure 7.11b. In this case, there are two neighbouring orthogonal contours, each of which falls in the inhibitory zone of the other. They therefore compete so that sometimes the vertical bar appears complete and sometimes the horizontal bar.

In **area rivalry**, similar dichoptic regions bounded by contours of opposite contrast, as in Figure 7.11c, rival to produce a shimmering region of variable brightness. This effect is known as **binocular lustre**. Dichoptic regions that differ in luminance but which are bounded by contours with the same luminance polarity fuse to create a region of intermediate brightness, as in Figure 7.11d.

When both stimuli are small, one tends to see all of one image or all of the other in alternation. This is **exclusive dominance**. With large patterns of equal area, a part of one pattern is dominant in one area and a part of the other in another area, with these areas of dominance fluctuating in position over time. This is **mosaic dominance**. Thus, when a patch of vertical lines is presented to one eye and a patch of horizontal lines to the other, as in Figure 7.11e, one sees vertical lines in one area and horizontal lines in another, with the areas constantly shifting about. For short periods, only vertical lines or only horizontal lines may be seen.

Rivalry may not occur between brief or low-contrast images. Thus, orthogonal dichoptic patterns may appear superimposed as a plaid. A pattern with higher spatial frequency in one eye may appear to stand out in depth relative to a pattern with lower spatial frequency in the other eye (Yang *et al.* 1992). This is a figure-ground effect.

The question arises whether dissimilar dichoptic images that appear superimposed should be regarded as fused. The perceived direction of fused similar images is midway between the directions of the two monocular images (Section 17.7.3), and disparity between similar images may code depth. Dissimilar images that appear superimposed rather than in rivalry have neither of these properties, which suggests that the processes responsible for their apparent superimposition may differ from those responsible for fusion of similar images. Therefore I refer to the simultaneous appearance of dissimilar images as **image superimposition** rather than image fusion.

Inspection of any pattern for a period of time with fixed gaze leads to a loss of apparent contrast, and parts and sometimes the whole of the pattern fade completely. This is the **Troxler effect**, which is believed to be due to local adaptation. The effect is particularly evident with blurred edges, and optically stabilized images fade completely. Like binocular rivalry, Troxler fading in a complex pattern is piecemeal and fluctuates. There has been dispute about what role Troxler fading plays in binocular rivalry (Crovitz and Lockhead 1967). Liu *et al.*

(1992b) suggested that it is particularly important in rivalry at near-threshold contrasts but not at high contrasts. Prior inspection of a patterned stimulus by one eye decreased the duration of dominance of that stimulus when it was dichoptically paired with a rival stimulus (Wade and De Weert 1986).

Inspection of a moving display evokes pursuit eye movements (OKN). When dichoptic displays move in opposite directions, the eyes follow whichever stimulus is dominant. Each change in image dominance is accompanied by a change in the direction of pursuit eye movements (Enoksson 1963; Fox *et al.* 1975). This response characteristic has been used to investigate binocular rivalry in the monkey (Logothetis and Schall 1990).

Early phenomenological studies of binocular rivalry were carried out by Volkmann (1836), Wheatstone (1838), Fechner (1860), Hering (1861), Helmholtz (1910, p. 492), and Meenes (1930). Fox (1991) reviewed the more recent literature on binocular rivalry.

7.3.2 Luminance contrast and contour density

7.3.2a *Interocular differences in luminance and contrast*
Levelt (1965b, 1966) reported that the strength of a stimulus is proportional to the amount of contour per unit area and that the zone of suppression produced by a given contour widens with increasing border contrast. He proposed that the strength of a stimulus determines how long that stimulus is suppressed not how long it suppresses another stimulus (Whittle 1965; Levelt 1965b; Fox and Rasche 1969). An image with no contours is regarded as having zero strength and is believed to remain suppressed indefinitely by a patterned stimulus in the other eye. As we will see, this is not always true.

Bossink *et al.* (1993) varied stimulus strength by varying the luminance contrast, colour contrast, or velocity of a moving dot pattern. Levelt's proposition was only partially confirmed. They agreed that the strength of a suppressed image has more effect on suppression duration than does the strength of a dominant image. However, the strength of the dominant image had a significant affect on the duration for which it was dominant. Mueller and Blake (1989) found that the overall rate of alternation of rival patterns depended mainly on the contrast of the patterns in their suppressed phase but that the contrast of the patterns in their dominant phase had some effect.

A defocussed image tends to be suppressed by a clearly focussed image (Humphriss 1982). Simpson (1991) presented the horizontal arms of a Rubin cross to one eye and the vertical arms to the other eye. The surrounding frame formed a binocular fusion lock. When one eye was defocussed, its image tended to be suppressed by the well-focussed image. The area of suppression was centred on the fovea and increased as the difference in refraction between the eyes was increased.

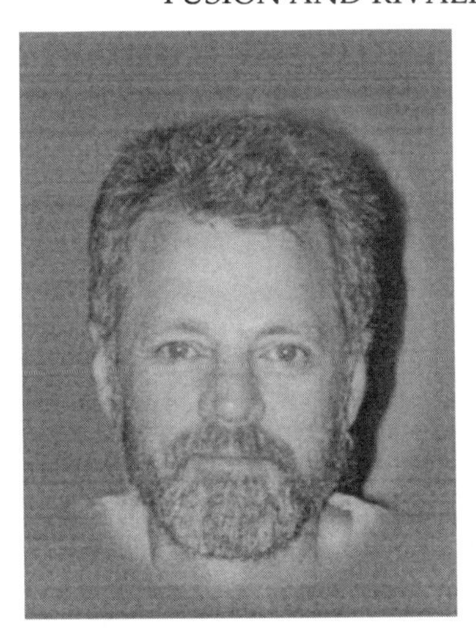

Figure 7.12. Randolph Blake.
Born in Dallas Texas in 1945. He obtined a B.Sc. in Mathematics and Psychology at the University of Texas, Arlington and a Ph.D. at Vanderbilt University, with Robert Fox. He conducted postdoctoral work at Baylor College of Medicine. From 1974 to 1988, he was Professor of Psychology at Northwestern University, Since 1988 he has been Professor of Psychology at Vanderbilt University, He Received Vanderbilt's prestigious Sutherland Prize.

Figure 7.13. Manfred Fahle.
Born in Düsseldorf in 1950. He obtained a Degree in Biology and Mathematics at the University of Göttingen in 1972. He did Graduate training in Medicine at the Universities of Mainz and Tübingen from 1973 to 1977 and postdoctoral work at the Max-Planck Institute for Biological Cybernetics, Tübingen. He was Visiting Professor at London University from 1999 to 2001. He is now the Director of the Institute for Brain Research at the University of Bremen. Recipient of the Max Planck Prize for Basic Research (with T. Poggio) in 1992.

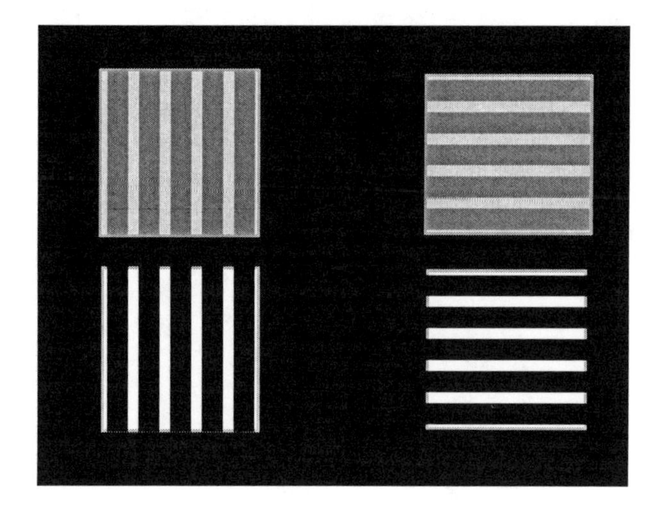

Figure 7.14. Reduced rivalry of low-contrast edges.
Fusion of the low-contrast images results in a plaid pattern. The high-contrast images produce mosaic rivalry. (Redrawn from Liu *et al.* 1992b)

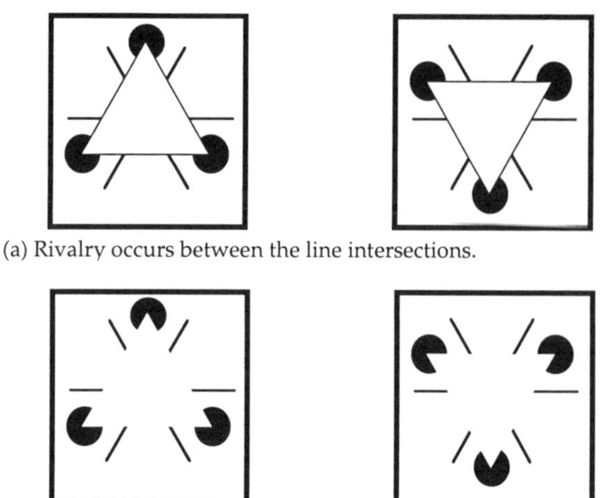

(a) Rivalry occurs between the line intersections.

(b) The triangles form a six-pointed star or show figure-ground rivalry when combined dichoptically.

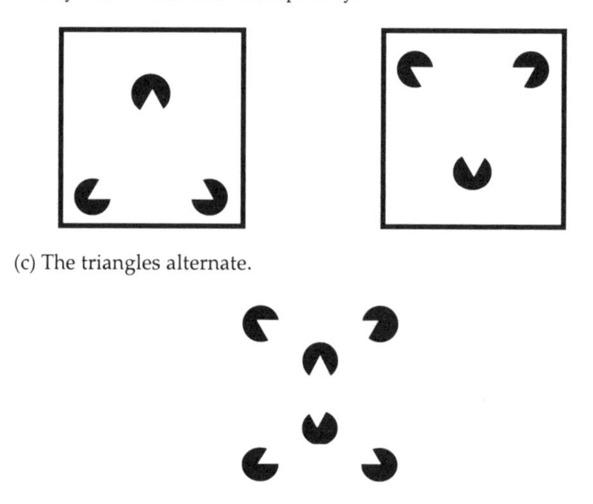

(c) The triangles alternate.

(d) The triangles alternate when combined in the same eye.
(Adapted from Bradley 1982 and Fahle and Palm 1991)

Figure 7.15. Binocular rivalry of cognitive contours

A grating near the contrast threshold can rival an orthogonal high-contrast grating in the other eye, but the low-contrast grating is visible for only short periods. The least contrast in an image that will instigate rivalry is the **rivalry contrast threshold**. The function relating rivalry contrast threshold to the spatial frequency of rivalling gratings is similar to the contrast-sensitivity function of monocularly viewed gratings (Blake 1977) (Portrait Figure 7.12). Thus, the spatial frequency for which contrast sensitivity is highest (4 cpd) requires the least contrast to initiate rivalry. However, a sine-wave grating of 4 cpd is not as dominant as a stimulus consisting of a broad mixture of spatial frequencies (Fahle 1982a). This supports the idea that rivalry occurs between distinct spatial-scale channels in the visual system. The fact that a blurred pattern is suppressed for longer periods than a sharply focussed pattern could be due to the reduction of contrast in the blurred stimulus. It could also be due to the narrowed range of spatial frequencies, which means that fewer channels are stimulated by the rival stimuli (Fahle 1982b) (Portrait Figure 7.13). The threshold for detection of a monocular flash is elevated when the flash occurs at about the same time as a sudden change in brightness in the other eye (Bouman 1955). Presumably, the change in the other eye causes that eye to become dominant and suppress the response to the test flash. Blake and Camisa (1979) found the elevation of threshold of a test flash presented to an eye when it was suppressed by the other eye to be independent of the relative contrasts of the two stimuli. They concluded that, once a stimulus is suppressed, the degree of suppression, as opposed to its duration, is independent of its contrast. The degree of suppression is also independent of the relative luminance and spatial frequency of the stimuli (Holopigian 1989). Dimming one image has no effect on the degree of suppression (Makous and Sanders 1978; Hollins and Bailey 1981).

The lower the contrast of the suppressed stimulus the more time it needs before becoming dominant (Blake and Camisa 1979). Lowering the luminance of the suppressed image has the same effect (Hollins and Bailey 1981). We must distinguish between the extent of suppression, which is not affected by the contrast or luminance of the suppressed image, and the duration of suppression, which is affected by these attributes of the suppressed image.

The fact that stimuli with high contrast or high spatial frequency tend to suppress those with low contrast or low spatial frequency helps people who wear a contact lens on one eye for hyperopia and a lens on the other eye for myopia. For near viewing,

(a) When the fine grating is oscillated slowly up and down while viewed with the nondominant eye, the dark field of the closed eye occasionally blots out the centre of the grating. The dark field appears as a gray diagonal meshwork, as depicted on the right.

(b) When one eye fixates the stationary coarse grating, the dark field of the closed eye occasionally blots out the centre of the grating and contains a low-contrast counterphase grating of the same spatial frequency, as depicted on the right. (From Howard 1959)

Figure 7.16. Binocular dominance of the closed-eye.

the image in one eye is in sharper focus than that in the other and for far viewing the image in the other eye is in sharper focus. Patients nevertheless see in sharp focus at all distances because the sharply focussed image suppresses the less well-focussed image. Under scotopic conditions the less well-focussed image is not suppressed (Schor *et al.* 1987).

The suppression of a weak image by a stronger image may explain why stereoacuity is degraded by unequal illumination or contrast in the two eyes (Section 19.5.4).

7.3.2b *Luminance contrast of both images*

High-contrast orthogonal gratings alternate more rapidly than low-contrast gratings, and continuous lines alternate more rapidly than broken lines (Alexander 1951). Alternations of rivalry occur less frequently and suppression spreads over wider areas when both images are at scotopic rather than photopic light levels (Breese 1910; Kaplan and Metlay 1964; O'Shea *et al.* 1994a).

Orthogonal dichoptic gratings just above the contrast threshold do not begin to rival for many seconds after exposure, but appear superimposed as a plaid pattern (Liu *et al.* 1992b). As the contrast of the gratings increases, the time before rivalry is experienced becomes shorter (see Figure 7.14). For a given contrast, gratings with higher spatial frequency appear as plaids longer than those with lower spatial frequency—probably because higher spatial-frequency gratings have higher contrast thresholds. This may explain why Burke *et al.* (1999) found that the plaid percept is more probable with square-wave gratings than with sine-wave gratings.

One way to think about these effects is that when both members of a dichoptic display have low contrast, neither of them has sufficient strength to suppress the other. Liu *et al.* showed that a fused aperture surrounding the stimulus significantly enhances apparent superimposition of dichoptic images. *It would be worthwhile to study rivalry in large low-contrast gratings in which any possible contribution of an aperture is minimized.*

The rate of binocular rivalry between orthogonal dichoptic gratings has been found to be higher in subjects with high stereoacuity than in those with low acuity (Halpern *et al.* 1987a). Also, the rate of binocular rivalry has been reported to be lower after alcohol ingestion (Donnelly and Miller 1995). These effects may be due to loss of contrast sensitivity associated with stereodeficiency and alcohol consumption.

7.3.2c *Rivalry between cognitive contours*

When opposed real triangles are dichoptically combined, rivalry occurs between intersecting edges, as in Figure 7.15a. Bradley (1982) observed that subjective contours (Sections 4.2.6h and 24.1.3) do not engage in binocular rivalry. When dichoptically fused, the two cognitive triangles of Figure 7.15b form a six-pointed star, as they do when they are combined in one eye. With both dichoptic and monocular viewing, the star sometimes appears to break down into superimposed triangles, which alternate with respect to their foreground-background relationship. Bradley called this figure-ground ambiguity rather than binocular rivalry. Fahle and Palm (1991) claimed that binocular rivalry is evident in Figure 7.15c. But Figure 7.15d shows that the same rivalry is evident in the same images superimposed in one eye so that this too could be put down to figure-ground ambiguity.

7.3.2d *Dominance of homogeneous fields*

It is generally believed that a featureless visual field never rivals a patterned stimulus. However, a closed eye can suppress a highly textured stimulus. Close the dominant eye and view the black and white grating of Figure 7.16a as it is oscillated up and down at about 2 Hz. A gray patch containing a diagonal meshwork pattern appears to spread out from the centre of the grating and blot it out (Howard 1959). The meshwork pattern periodically spreads and then recedes. The effect is not Troxler fading of the grating, because Troxler fading does not occur with moving stimuli. People with only one eye did not experience this effect, supporting the idea that the occluding patch is the dark field of the closed eye. When a coarse grating, like that of Figure 7.16b, is steadily fixated, a central patch of the dark field of the closed eye may still occlude the lines but, instead of a meshwork pattern, the occluded region contains a faint phase-reversed image of the grating, as depicted on the right of Figure 7.16b. Not everyone sees this image. A blank field in one eye may suppress dynamic visual noise in the other eye (Tyler personal communication).

These effects are strong violations of Levelt's proposition that the more highly patterned stimulus is dominant in binocular rivalry. The origin of the patterns visible in the field of the closed eye remains a mystery.

An homogeneous luminous field (Ganzfeld) tends to fade after it has been inspected for some time. Bolanowski and Doty (1987) found that fading did not occur when the Ganzfeld was viewed with both eyes, and concluded that fading with a monocular stimulus is due to suppression of the luminous field by the dark field of the closed eye. Gur (1991) agreed that sudden blankout in a Ganzfeld occurs only when one eye is closed and that it is due to binocular rivalry, but found that the gradual fading associated with adaptation of a stationary patterned image occurs with both monocular and binocular viewing. Rozhkova *et al.* (1982) found the same to be true of large textured displays optically stabilized on the retina. From a review of the literature on fading of afterimages, Wade (1978) (Portrait Figure 7.17) concluded that one of the major factors in the fading of monocular afterimages is rivalry between the afterimage and the dark field of the closed eye.

7.3.3 Colour

The question addressed in this section is whether binocular rivalry is affected by the colours of the rival stimuli and, in particular, whether blue cones contribute to the rivalry process. Borders defined by equiluminant blue and yellow are much less visible than those defined by red and green (Tansley and Boynton 1978; Kaiser and Boynton 1985). Since borders contribute to the strength of a stimulus, one might expect that a blue-yellow grating would be a much weaker stimulus than a red-green grating.

Periods of exclusive dominance were found to be shorter for targets of the same colour than for targets in complementary colours (Wade 1975b). Durations of exclusive dominance increased as the chromatic difference between targets increased (Hollins and Leung 1978). However, targets that varied between yellow and blue behaved like targets of the same colour. This latter finding does not prove that the blue-yellow opponent system does not affect rivalry,

Figure 7.17. Nicholas J. Wade.
Born in Nottinghamshire, England in 1942. He obtained a B.Sc. in Psychology at the University of Edinburgh and a Ph.D. at Monash University Australia with Ross Day. He conducted postdoctoral work at the Max-Planck Institute für Verhaltensphysiologie, Seewiesen, Germany. In 1970 he went to the University of Dundee, where he is now Professor of Visual Psychology. He is a Fellow of the Royal Society of Edinburgh.

because the 6-cpd gratings were probably not visible to the blue-yellow system. Stalmeier and de Weert (1988) pitted concentric black-white stripes in one eye against radial stripes of alternating colours rendered equiluminant by the flicker method in the other eye. While red-green colour pairs contributed significantly to rivalry, tritanopic colour pairs (blue-yellow) did not make an appreciable contribution. The mean spatial frequency of the radial chromatic display was about 5 cpd, which is rather high for the blue-cone system. Rogers and Hollins (1982) used 3 cpd gratings, to which blue-cones are sensitive, but still found no evidence of a contribution of the blue-cone system to rivalry.

O'Shea and Williams (1996) stimulated only the blue cones by using violet gratings on a bright yellow background, which bleached the middle- and long-wavelength cones. Since the gratings had negligible luminance contrast for the middle- and long-wavelength cones, they were essentially equiluminant. Under these conditions, orthogonal dichoptic violet gratings of 2 cpd engaged in rivalry but at a slower rate of alternation than luminance gratings. A high-contrast violet grating could be dominant over a luminance-defined grating with sufficiently low contrast. Thus, the blue-cone system can contribute to binocular rivalry and it can also contribute to fusion and stereopsis (see Section 18.1.4). However, its contribution is not evident in stimuli with high spatial frequency.

Sagawa (1981) measured the minimum luminance of a patch seen by one eye required to induce rivalry with a patch seen by the other eye as a function of their relative wavelengths. Although the contribution of the blue-yellow system was small when a red or green stimulus was presented to the other eye, it was never zero. In the suppression phase of rivalry, the sensitivity of the blue-cone system is reduced more than that of other chromatic channels (Section 7.5.2).

7.3.4 Relative velocity

A moving stimulus is dominant for longer than a similar stationary stimulus presented to the other eye (Breese 1899). The duration of dominance of a moving grating increases with its speed (Wade *et al.* 1984). However, a clear advantage of one speed over another was not observed when dichoptic displays moved at different speeds in the same direction (Blake *et al.* 1985).

Rivalry also occurs between dichoptic displays of random dots moving at the same speed but in directions that differ by more than 30° (Wade *et al.* 1984). Dichoptic displays of dots moving upward at 1.5°/s in directions that differed by up to 30° fused to produce a display that appeared to move on an inclined depth plane with respect to the circular aperture (Blake *et al.* 1985). Rivalry between oppositely moving arrays of dots is less evident when all pairs of dots in the two displays move along intersecting pathways (Matthews *et al.* 2000). Under these circumstances, the motion signal is lost as the dots intersect, which interferes with the perception of two distinct arrays. Also, intersecting dots fall on or near corresponding points for part of the time. Rivalry between orthogonal gratings moving at 1.2 or 4°/s had a similar time course as that between stationary gratings. However, depth of suppression, as measured by the detection threshold of a flashed probe, was greater for moving than for stationary gratings (Norman *et al.* 2000).

For both monocular and binocular stimuli, visual thresholds are generally lower for vertical and horizontal lines or gratings than for oblique stimuli—the so-called **oblique effect**. By analogy, *one might expect that dot patterns moving vertically or horizontally would be dominant over dot patterns moving in an oblique direction, but this question has not been investigated.*

A given stimulus is more dominant when surrounded by an annulus of moving elements, especially when elements within the target stimulus and those in the surround move in opposite directions. When the two eyes are differentially adapted to opposite directions of motion, movement aftereffects in opposite directions are seen alternating when both eyes view the same random-motion display (Blake *et al.* 1998). In this case, it is not the immediate stimuli that rival but the stimuli in the two eyes modified in different ways by adaptation.

7.3.5 Position on the retina

Each cerebral hemisphere of the monkey has more binocular cells with a dominant input from the contralateral eye (the nasal hemiretina) than binocular cells with a dominant input from the ipsilateral eye (temporal hemiretina) (Section 10.3.5). One might therefore expect a stimulus presented to the nasal retina of one eye to dominate that presented to the temporal retina of the other eye.

In conformity with this expectation, Köllner (1914) found that when a homogeneous green field was presented to the left eye and a similar red field to the right eye for about 100 msec each, the green field dominated in the left visual field (nasal half of the left eye) and the red field dominated in the right visual field (nasal half of the right eye). Thus, the left eye dominated in the left visual field and the right eye dominated in the right visual field. With longer viewing, the coloured fields began to rival. When the displays presented to the nasal and temporal visual fields were separated by a vertical black band, the colour projected to the nasal half of each eye remained dominant for prolonged periods (Crovitz and Lipscomb 1963a, 1963b).

Fahle (1987), also, showed that the pattern of alternation of rivalling stimuli varies as a function of their position in the visual field. A vertical grating in the right eye was dominant for slightly longer than a horizontal grating in the left eye when they were presented within a 20° radius around the fixation point. This was attributed to the fact that most people have a dominant right eye, but it could also be due to horizontal gratings tending to dominate vertical gratings. When the stimuli were presented more than 20° to the left of the fixation point, the left eye was dominant for about twice as long as the right eye. When they were presented more than 20° to the right, the right eye was about twice as dominant. In both cases, the temporal visual field (nasal hemiretina) of one eye tended to dominate the nasal visual field (temporal hemiretina) of the other eye.

Fahle pointed out that, in convergent strabismus, the fovea of the deviating eye competes with the dominant temporal hemifield of the other eye whereas, in divergent strabismus, the fovea of the deviating eye competes with the nondominant nasal hemifield of the other eye. This could account for why convergent strabismics develop amblyopia in the deviating eye whereas divergent strabismics do not (Section 13.4.2).

Figure 7.18. Fusion of orthogonals with brief exposure. When briefly exposed to corresponding regions of the two eyes the whole display appears composed of Xs. (Adapted from Kolb and Braun 1995)

7.3.6 Relative orientation

7.3.6a *Rivalry as a function of relative orientation*

Abadi (1976) reported that the contrast of a grating in one eye required to suppress a grating in the other eye was independent of the relative orientation of the gratings, when the angle between the gratings was larger than 20°. When the angle was less than 20°, progressively less contrast was required for suppression. Abadi concluded that suppression is due to inhibitory connections between orientation columns in the visual cortex and that these connections are strongest between columns tuned to similar orientations. There is disagreement about the role of lateral connections in the visual cortex in binocular rivalry (Section 7.10). Abadi's stimuli were at near-threshold contrasts where rivalry tends not to occur. His results may have reflected a threshold-elevation effect due to contrast adaptation rather than binocular rivalry (9.2.6).

Blake and Lema (1978) pointed out that Abadi's data could also support the conclusion that inhibitory influences are weaker between lines with similar orientation, since only a low-contrast stimulus in one eye was required to overcome the inhibitory influence of the image in the other eye. Using a detection-threshold procedure with suprathreshold stimuli, Blake and Lema found no evidence that strength of suppression varied as a function of the relative orientation of the stimuli. However, the rate of rivalry alternation has been reported to increase with increasing relative orientation of dichoptic gratings (Thomas 1978a).

The dominance of a grating in the left eye relative to an orthogonal grating in the right eye was reduced when the grating in the left eye was surrounded by an annular grating of similar orientation and spatial frequency (Mapperson and Lovegrove 1991). This effect is presumably due to masking of the left eye's grating by the surrounding grating. Bonneh and Sagi (1999) found that, at high but not at low contrasts, an array of randomly oriented Gabor patches was more dominant than uniformly oriented patches. Ooi and He (1999) reported a similar effect. These effects relate to the fact that stimuli are inhibited by neighbouring stimuli with similar orientation (Section 5.4.7). On the other hand, Bonneh and Sagi found that a high-contrast contour formed from collinear Gabor patches was more dominant than a jagged contour. This relates to the fact that the visual system is particularly sensitive to collinear stimuli (Section 4.2.6h). Other aspects of dichoptic masking are discussed in Section 8.2.

7.3.6b *Visibility of complementary rivalry zones*
Kolb and Braun (1995) constructed the dichoptic displays shown in Figure 7.18. All texture elements are orthogonal in the two images but in one region the orientation of the elements is complementary to the orientation of elements in the rest of the display. When the images were stereoscopically combined, subjects reported seeing a uniformly textured display. However, in a forced-choice procedure, they could report the position of the target region at above chance levels. Kolb and Braun concluded that this dissociation between awareness and performance is analogous to blindsight in which patients with cortical damage can discriminate visual stimuli even though they are not conscious of seeing.

Morgan *et al.* (1997) could not replicate this result and concluded that Kolb and Braun's subjects were applying a strict criterion for awareness of the vague cues that allowed them to perform at above chance on the forced-choice detection task. For example, a slight misalignment of the eyes would produce distinct asymmetrical crosses in the target region and the surround. Solomon and Morgan (1999) constructed a display containing a drifting grating defined by this type of dichoptic texture boundary. This motion is not visible to the binocular system. Subjects could not detect the motion when the display was presented for 200 ms. They concluded that motion from dichoptic texture does not have access to monocular signals. This is complementary to the finding that motion mechanisms do not have access to binocular signals (see Section 17.4.1d). The spurious signal generated by vergence instability would not operate with a moving display and this may be why static dichoptic textured regions are visible, but moving ones are not.

7.3.7 Rivalry and eye movements

A sudden motion or increase in luminance contrast of a suppressed image tends to terminate the suppression. The eyes constantly execute small saccadic movements and Levelt (1967) proposed that sudden movements of the suppressed retinal image trigger reversals of binocular dominance. However, the rate of dominance reversal of images has been found to be the same for retinally stabilized as for normal images (Blake *et al.* 1971). This shows that eye movements are not necessary for rivalry but it does not exclude a possible role for eye movements. Sabrin and Kertesz (1983) applied a more critical test of the eye-movement theory by stabilizing the image in only one eye. Rivalry still occurred but the stabilized image was suppressed for longer periods than the unstabilized image. When a motion simulating the effects of microsaccades was imposed on the stabilized image, the periods for which it was suppressed returned to normal. Thus, eye movements do affect the rate of binocular rivalry when their effects are unequal in the two eyes.

The spread of suppression around a contour into an uncontoured region in the other eye was found to be more extensive when the eyes executed vergence movements (Kaufman 1963). This effect was explained in terms of a time lag in recovery from suppression, which causes the moving eyes to leave a wake of suppression in their path.

7.3.8 Stimulus duration and temporal frequency

Dawson (1913) noticed that rivalry ceased while he rapidly blinked his eyes. Hering (1874) reported that when rivalrous stimuli are presented for a brief period they appear as two complete superimposed stimuli. Several other investigators have since noticed the same phenomenon. For instance, high-contrast dichoptic orthogonal gratings formed a grid pattern when shown for less than about 200 ms but appeared to rival in the normal way when presented for more than 400 ms (Anderson *et al.* 1978). High-contrast dichoptic orthogonal gratings did not rival when flashed on for 50 msec at a rate of 2 flashes per second (Kaufman 1963). Similarly, afterimages of orthogonal gratings did not rival when presented on a background illuminated at 2 flashes per second (Wade 1973). Thus, suppression takes time to develop and, before it develops, dichoptic stimuli slip past the suppression mechanism and reach consciousness. Another way to think about these effects is that the rivalry mechanism is less evident in the transient channel of the visual system than in the sustained channel. Subjects who continuously

Figure 7.19. Jeremy Wolfe.
Born in London, England in 1955. He received his B.A. from Princeton University in 1977 and his Ph.D. in Psychology from MIT in 1981. He held academic appointments in Psychology and in Brain and Cognitive Sciences at MIT between 1981 and 1991. In 1991 he became Director of Psychophysical Studies in the Center for Ophthalmic Research at Harvard Medical School.

Figure 7.20. Robert O'Shea.
Born in Australia in 1953. He obtained a B.Sc. and Ph.D. at the University of Queensland with Boris Crassini. He conducted postdoctoral work with Peter Dodwell, Randolph Blake, and Don Mitchell. In 1988 he went to the University of Otago in New Zealand, where he is now Senior Lecturer in Psychology.

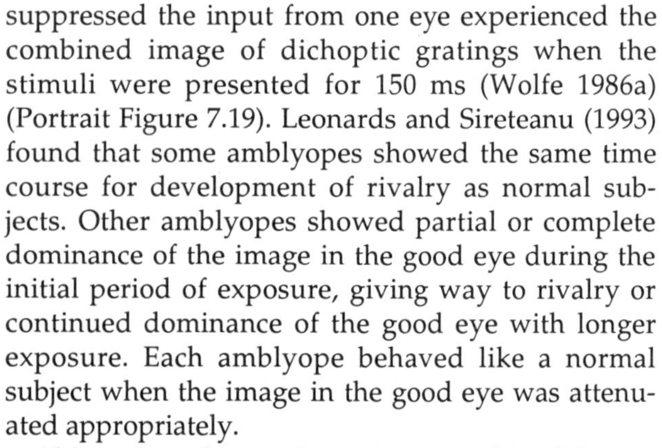

suppressed the input from one eye experienced the combined image of dichoptic gratings when the stimuli were presented for 150 ms (Wolfe 1986a) (Portrait Figure 7.19). Leonards and Sireteanu (1993) found that some amblyopes showed the same time course for development of rivalry as normal subjects. Other amblyopes showed partial or complete dominance of the image in the good eye during the initial period of exposure, giving way to rivalry or continued dominance of the good eye with longer exposure. Each amblyope behaved like a normal subject when the image in the good eye was attenuated appropriately.

Although orthogonal gratings combined into a plaid when exposed dichoptically for 10 ms, they appeared to rival when presented intermittently for 10 ms with intervals of less than 150 ms, just as they would if presented continuously (Wolfe 1983a, 1983b). Thus, the rivalry mechanism integrates over short time intervals and, once switched on, stays on for at least 150 ms, affecting the appearance of subsequently exposed stimuli.

Even though dissimilar dichoptic patterns appear as a combined image when presented briefly, Blake *et al.* (1991a) found that subjects could distinguish between a brief dichoptic image and the same images presented briefly to one eye.

O'Shea and Blake (1986) presented an uncontoured field flickering at 4 Hz to one eye and a similar field flickering at between 0.5 and 16 Hz to the other eye. These stimuli produced very few reports of rivalry. Instead, subjects reported a single field flickering irregularly at an intermediate frequency (Portrait Figure 7.20). The effect resembled that produced by superimposing the two flickering fields in one eye. Small monocular probes placed at different positions on the flickering fields in both eyes remained visible all the time. They concluded that rivalry does not occur within the transient channel of the visual system, which is most effectively engaged by flickering, uncontoured stimuli. Orthogonal dichoptic gratings, when counterphase modulated at different frequencies, were found to rival in the usual way. A grating flickering at a lower frequency created longer dominance phases than one flickering at a higher frequency.

Stereopsis occurs when stimuli with an appropriate disparity are presented alternately to the two eyes (Section 19.11.2). Rivalry, also, occurs between stimuli presented briefly and alternately to each eye, as long as the interval between them is not too great. Thus, binocular rivalry between orthogonal gratings presented in alternation to the two eyes at rates above about 20 Hz was indistinguishable from that between simultaneously presented stimuli (O'Shea and Crassini 1984). Some rivalry was apparent in stimuli alternating at 3 Hz but, below this frequency, both stimuli were seen in alternation, often with apparent motion from one to the other. Rivalry also occurred between stimuli presented alternately for 5 ms to each eye, with up to 100-ms intervals between stimuli. This is the same interval of time over which stereoscopic depth occurs with alternating stimuli to the two eyes (Section 19.11.2).

7.3.9 Rivalry and eye dominance

Eye dominance is defined in a general way as a preference for using one eye over the other. Most people are right-handed and right-eyed, but there are conflicting claims about whether eye dominance is correlated with handedness (Miles 1930; Eyre and Schmeeckle 1933; Gronwall and Sampson 1971). In animals with hemidecussating visual inputs, eye dominance has nothing to do with cerebral dominance, because each eye projects to both cerebral hemispheres.

The following three criteria have been used to define eye dominance:

1. The eye with better visual acuity, contrast sensitivity, or other measure of visual functioning. In severe cases, the weaker eye is amblyopic and is permanently suppressed when both eyes are open (Section 13.4.2).
2. The eye used for sighting when, for instance, one looks at a distant object through a ring held in both hands at arm's length with both eyes open.
3. The eye in which a rivalling stimulus is most often dominant.

There are many tests of eye dominance based on these definitions, and there is controversy about whether the different tests correlate. For instance, rivalry tests of ocular dominance have been found to be poorly correlated with sighting tests (Washburn *et al.* 1934). This literature has been reviewed by Coren and Kaplan (1973), who conducted a factor analysis on the results of 13 tests of eye dominance given to 57 normal subjects. The results revealed three principal factors, acuity dominance, sighting dominance, and rivalry dominance. Most of the variance was accounted for by sighting dominance. Another possibly important factor is the position of lateral gaze (Khan and Crawford 2001). Subjects in these studies had normal vision. In amblyopes, the principal factor in eye dominance is the eye with amblyopia.

7.3.10 Monocular rivalry and lustre

7.3.10a *Monocular rivalry*

Breese (1899) noticed rivalry between a diagonal grid of black lines on a red ground and an orthogonal grid of black lines on a green ground, when they were optically superimposed in the same eye. Sometimes only one or the other set of lines was seen, and sometimes parts of each. The colours associated with each set of lines fluctuated accordingly. Breese introduced the term "monocular rivalry" to refer to rivalry between images in one eye. Monocular rivalry between differently coloured gratings was most pronounced when they were superimposed at right angles. When the angle between them was less than about 15°, the colours and lines combined into a stable percept (Campbell and Howell 1972; Campbell *et al.* 1973). Monocular rivalry was more frequent and more complete when the colours were complementary rather than noncomplementary or black and white (Rauschecker *et al.* 1973; Wade 1975b). The rate of alternation of orthogonal gratings superimposed in one eye did not change significantly with changes in contrast (Atkinson *et al.* 1973). The rate of alternation was highest when both gratings had a spatial frequency of 5 cpd and fell off steeply on both sides of the peak (Kitterle *et al.* 1974). A grating near the peak frequency was dominant more of the time than an orthogonal grating at another frequency (Thomas 1977). Changes in stimulus size, spatial frequency, or relative orientation had similar effects on alternation rates in monocular and binocular rivalry (Andrews and Purves 1997).

The rate of monocular rivalry is considerably lower than that of binocular rivalry (Wade 1975b; Kitterle and Thomas 1980). Furthermore, monocular rivalry takes time to develop and requires constant fixation whereas binocular rivalry occurs straight away and does not require fixation.

Successive presentation to one eye of a white horizontal bar and a white vertical bar, both on black backgrounds, produced an afterimage in which the vertical and horizontal bars showed rivalry, complete with white halos at the intersections of black edges. Monocular rivalry was particularly impressive between an afterimage and a real bar. These forms of monocular rivalry appeared similar to rivalry between orthogonal afterimages impressed separately in the two eyes (Sindermann and Lüddeke 1972). When afterimages of a vertical and a horizontal bar are formed successively, there is some retention of the neural activity associated with each of the contours within the region where the afterimages intersect. The competition between these persisting neural processes is presumably responsible for the rivalry seen with monocular or dichoptic afterimages.

Monocular rivalry also occurs between two superimposed parallel gratings. For example, a 1 cpd sinusoidal grating superimposed on a 3 cpd (3rd harmonic grating) fluctuated in appearance, with the two gratings alternating in dominance. (Atkinson and Campbell 1974). The rate of fluctuation was highest when the relative phase of the gratings was 90° and least when it was 0° (peaks coincide) or 180° (peaks subtract).

7.3.10b *Monocular rivalry and monocular diplopia*
In monocular diplopia (Section 15.4.3) a stimulus presented to one eye appears double, with the image in its normal oculocentric position appearing normally bright and the anomalous image appearing dim. Bielschowsky (1898) superimposed in one eye the normal bright image of a diplopic red patch on the dim image of a diplopic green patch. Instead of producing the hue formed when two normal images of different hue are combined, the coloured patches rivalled in the manner of dichoptic patches.

Ramachandran *et al.* (1994a, 1994b) made a similar observation on a patient with intermittent exotropia. A patch of vertical lines was superimposed in the same eye on the dim diplopic image of a patch of horizontal lines. The subject experienced mosaic rivalry rather than the checkerboard pattern obtained when two orthogonal gratings of the same contrast are superimposed in the same eye.

7.3.10c *Monocular lustre*
A dark patch in one eye superimposed on a light patch in the other eye produces binocular lustre. Anstis (2000) found that a dark patch alternating with a bright patch at 16 Hz in the same eye also produces lustre, which he called **monocular lustre**. Both effects occurred only when the two patches had opposite luminance polarity—one patch had to be darker than the background and the other lighter than the background. Anstis suggested that binocular and monocular lustre arise from competition between signals in the ON and OFF visual pathways.

Burr *et al.* (1986) found that alternating orthogonal gratings to the same eye at 10 Hz produces lustrous diamond elements. Two gratings flickering in counterphase are physically equivalent to two gratings sliding over each other. Burr *et al.* suggested that this form of monocular lustre arises from competition between signals in distinct motion channels.

7.3.10d *Theories of monocular rivalry*
It has been suggested that monocular rivalry is due to eye movements that cause afterimages formed with one fixation to reinforce or attenuate components of the pattern viewed with another fixation (Furchner and Ginsburg 1978; Georgeson and Phillips 1980). But this cannot explain why Breese sometimes observed parts of each grating or why rivalry occurs between gratings at levels of contrast too low to produce afterimages (Mapperson *et al.* 1982). Others have argued that eye movements cannot be the only factor, since monocular rivalry may be observed in afterimages (Atkinson 1972; Wade 1976a; Crassini 1982).

Georgeson (1984) produced data in support of the eye-movement theory. Subjects fixated a spot as it moved to different positions on two superimposed orthogonal gratings. The grating that was perceptually dominant depended on the retinal position of the image in the preceding fixation position, as predicted by the eye-movement theory. Rates of fluctuations of monocular rivalry were independent of the relative orientation of the gratings when appropriate eye movements were made. Georgeson pointed out that afterimages are a poor stimulus for studying monocular rivalry, since they pass through various phases and are subject to fluctuations in the background upon which they are projected.

The response of a cell in the visual cortex to an optimally oriented bar or grating is suppressed by the superimposition of an orthogonal bar or grating (Section 7.10). This is known as **cross-orientation inhibition**. It is largely independent of the relative spatial phases of superimposed gratings and operates over a wide difference in spatial frequency between the superimposed stimuli. This process does not operate between gratings presented dichoptically and is therefore not the cause of binocular rivalry, but it could be a factor in monocular rivalry, as suggested by Campbell *et al.* (1973). Monocular rivalry may also be related to the Troxler effect (see Section 7.5.1.

Most investigators have concluded that monocular rivalry, like binocular rivalry, arises when a local region of the visual cortex receives, simultaneously or in rapid succession, signals that are processed in distinct orientation channels or spatial-frequency channels. Inputs in the two channels compete for access to later stages of processing. Supporters of this theory point out that the relative orientation (15°) and the relative spatial frequency (1 octave) of gratings where rivalry becomes apparent correspond to the bandwidths of orientation and spatial-frequency channels. If this theory is correct, monocular rivalry provides a method for testing the bandwidth of visual channels. This competition could be a winner-take-all process in which the dominant stimulus is seen in full strength, rather than mutual inhibition, which would entail some weakening of the dominant stimulus by the non-dominant stimulus. It seems that this issue has not been investigated.

7.3.11 Summary

When the image in one eye differs in contrast polarity, orientation, or motion direction from that in the other the two images rival. At any instant the whole or part of one image is seen and the corresponding

Figure 7.21. Effects of area on rivalry.
With large displays, the whole of one image is infrequently suppressed totally by the other image. Small displays tend to rival as a whole. (Adapted from Blake *et al.* 1992)

region of the other image is suppressed. The strength of a suppressed image has more effect on the duration of suppression than does the strength of the dominant image. The strength of an image is determined by the amount of contour it contains, and by its contrast, spatial frequency, and motion. However, the dark field of a closed eye must be regarded as having some strength, since it can suppress an oscillating grating. A stimulus on the nasal retina of one eye tends to dominate that on the temporal retina of the other eye. Low-contrast stimuli and stimuli presented for less than 200 ms do not rival, which suggests that rivalry does not occur in the transient channel of the visual system. For certain time intervals, rivalry still occurs between stimuli presented successively to the two eyes.

7.4 SPATIAL ZONES OF RIVALRY

Rivalry occurs within discrete areal units, which increase in size with increasing eccentricity and spatial frequency. Blake *et al.* (1992) referred to these units as **spatial zones of binocular rivalry**. The two procedures described in the following sections have been used to measure zones of rivalry.

7.4.1 Zones of exclusive dominance

When the image in one eye totally suppresses that in the other it is said to show exclusive dominance. Large displays, like those at the top of Figure 7.21, usually rival in a piecemeal fashion to produce a shifting mosaic of monocular images. Small displays, like those at the bottom of Figure 7.21, tend to rival as a whole. Blake *et al.* (1992) measured the percentage of viewing time that a pair of orthogonal gratings (6 cpd) in a circular patch showed exclusive rather than mosaic dominance as a function of the size and eccentricity of the patch. They noted that the area within which rivalry is exclusive is similar to the area over which suppression from a contour in one eye encroaches on a noncontoured area or on neighbouring contours in the other eye.

At the fovea, the mean diameter of the largest patch that exhibited exclusive dominance 95% of the time was 8.1 arcmin. According to Schein and Monasterio (1987), this is close to the estimated size of a cortical hypercolumn in the monkey visual cortex (the region containing one complete set of ocular dominance columns). With appropriately scaled spatial frequency, the size of the patch that generated only exclusive rivalry increased with eccentricity in a manner similar to the increasing size of receptive fields and of the cortical magnification factor (the diameter of cortical tissue devoted to each degree of visual angle).

The limiting size for exclusive rivalry depends on retinal size, not perceived size. Thus, the limiting size for exclusive rivalry between a pair of orthogonal afterimages did not vary as the apparent size of the afterimages was varied by projecting them onto surfaces at different distances (Blake *et al.* 1974).

Two low-contrast gratings showed longer periods of exclusive dominance than did two high-contrast gratings. For a given contrast, gratings with a spatial frequency of 3 cpd showed longer periods of dominance than did gratings with higher or lower spatial frequencies (Hollins 1980). Rivalry between random-dot patterns showed a similar dependency on spatial frequency (De Weert and Wade 1988). In Hollin's experiment, stimulus size was constant so that spatial frequency was confounded with the number of bars in the grating. O'Shea *et al.* (1997) confirmed Hollin's results for a stimulus subtending 2°. However, with a 0.5° stimulus, exclusive rivalry peaked at about 4 cpd. With a 4° stimulus, it peak at about 1 cpd. Peak spatial frequency for exclusive dominance was inversely proportional to stimulus area.

7.4.2 Spatial extent of the zone of rivalry

Kaufman (1963) measured the spatial extent of the zone of rivalry around a dominant contour. He presented a pair of parallel thin lines to one eye and a single orthogonal line to the other. In dichoptic viewing the single line cut across the two parallel lines and subjects reported when the part of the single line between the two parallel lines was totally suppressed. As the distance between the vertical parallel lines increased from 7 arcmin to 3.8° the percentage of time for which the orthogonal single line

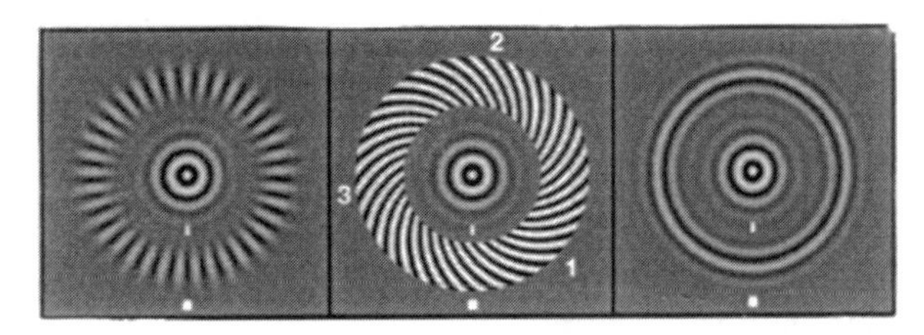

Figure 7.22. Stimuli used to measure the spread of rivalry. (From Wilson *et al.* 2001)

was visible increased exponentially from 3 to 40%, with the increase being most rapid for separations of less than 15 arcmin. The zone of suppression in a vertical direction was smaller than that in a horizontal direction. Kaufman ascribed this to a greater instability of horizontal gaze than vertical gaze.

Liu and Schor (1994) used Kaufman's procedure with elongated DOG patches with a spatial-frequency bandwidth of 1.75 octaves and a variable mean spatial frequency. Two parallel patches in one eye were dichoptically superimposed for one second on a continuously visible single orthogonal patch with the same spatial frequency. Subjects reported whether they saw a single patch between the two parallel patches. The size of the suppression zone decreased linearly with increasing spatial frequency, from about 150 arcmin at a spatial frequency of 1 cpd to about 15 arcmin at 10 cpd. At low spatial frequencies the rivalry zone was about twice as wide vertically as horizontally, at 4 cpd it was circular, and at high spatial frequencies it became a horizontal ellipse.

Schor *et al.* (1984b) had found previously that Panum's fusional area also decreased with increasing spatial frequency (Section 7.1.2) but Panum's fusion area was much smaller than the zone of rivalry at all spatial frequencies.

Liu and Schor also found that the zone of rivalry increased rapidly with increasing contrast of Gaussian patches up to a contrast of about 30%, after which it increased less rapidly. The small zone of rivalry with stimuli of low contrast and high spatial frequency could explain why orthogonal gratings fuse rather than rival when they have low contrast and high spatial frequency (Section 7.3.2).

At scotopic levels of illumination, suppression was found to spread over larger areas of gratings than at photopic levels (O'Shea *et al.* 1988). Furthermore, the size of Panum's fusional area was larger at scotopic levels. Both these effects could be due to the fact that, at scotopic levels, the excitatory regions of the receptive fields of ganglion cells become larger because inhibition from the surround becomes weaker (Barlow *et al.* 1957).

7.4.3 Temporal spread of rivalry

Large rivalling patterns produce a patchwork of zones of rivalry, with one eye's image dominant in some zones and the other eye's image dominant in other zones. The zones expand or contract, as one continues to view the pattern. Wilson *et al.* (2001) proposed that the motion of the state of suppression of one image by the other is due to spread of neural activity over lateral connections in the visual cortex. They used the stimulus shown in Figure 7.22 to measure the rate of spread of binocular rivalry. First, a low-contrast radial pattern was dichoptically superimposed on a high-contrast spiral pattern. The subject pressed a key when only the spiral pattern was visible. This triggered a brief increment of contrast at one of the eight cardinal points in the radial pattern. This caused the radial pattern to become dominant at that point and initiated a spreading wave of dominance of the radial pattern around the annulus. The subject released a key when the wave of dominance reached a designated point on the annulus. The time taken for the wave to spread from the trigger point to its final point indicated a mean speed of propagation of the wave of suppression of about 3.6°/s. The speed of propagation was 9.6°/s for concentric lines superimposed on the radial pattern. The angle between these two patterns was 45° as it had been for the radial and spiral patterns. The difference in speed probably arises from the fact that lateral connections in the visual cortex are strongest between cells with collinear orientation preference (Section 5.4.7b). The speed of propagation was, on average, about 173 ms longer for a wave that traversed the retinal midline, and which therefore traversed the corpus callosum. When the radius of the radial-plus-spiral annulus was increased from 1.8 to 3.6°, mean speed of rivalry propagation increased to about 8.3°. With fixation held on the centre of the display, the wave of rivalry for the larger display occurred in a more eccentric retinal location than that for the smaller display. When allowance was made for the cortical magnification factor (increase in size of receptive fields with increasing eccentricity), the mean propagation speed was 2.24 cm/s over the cortical surface for both sizes of display.

These results link properties of rivalry propagation to known properties of the visual cortex, such as the cortical magnification factor and preferential linkage of collinear images. This supports the notion that rivalry arises primarily between images from the two eyes in the retinotopically organized visual cortex. However, the results do not rule out some contribution from processes occurring at a higher level in the cortex.

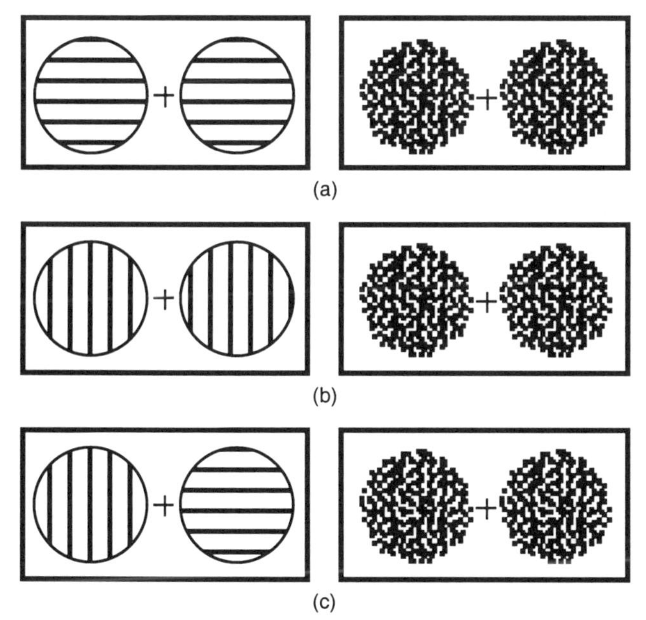

Figure 7.23. Relative orientation and synchrony of rivalry. (a) collinear images tend to rival in synchrony more than parallel but not aligned images (b) or orthogonal images (c). (Adapted from Alais and Blake 1999)

Wilson *et al.* developed a neural model of rivalry, which goes beyond previous models in providing for speed of rivalry propagation and for the effects of collinearity of stimulus elements.

7.4.4 Independence of zones of rivalry

As a general rule, when the two eyes are exposed to several distinct pairs of rivalling stimuli, the members of each pair rival independently rather than in synchrony. But there are exceptions to this rule. Connected patterns tend to appear and disappear as units, and dichoptic contour segments forming a continuous cyclopean line or pattern tend to rival in synchrony, even though the segments are in different eyes. Effects of this kind were first described by Diaz-Caneja (1928; translation by Alais *et al.* 2000) and have also been described by Whittle *et al.* (1968), Kovács *et al.* (1996), and Ngo *et al.* (2000). Also, a random mixture of red-green patches presented to one eye and the same pattern with reversed colours presented to the other eye appear either all red or all green for much of the time (Kulikowski 1992).

Alais and Blake (1999) used stimuli like those in Figure 7.23. The two images in one eye that contained collinear gratings tended to come into dominance simultaneously over random-dot images in the other eye. Synchrony of dominance was less evident with parallel gratings and less evident still with orthogonal gratings. Also, two gratings in which contrast was modulated in synchrony showed higher synchrony of dominance than gratings modulated in counterphase. Synchrony of dominance declined as the angular distance between the two rivalling patches was increased from 1 to 3°. The tendency for neighbouring aligned and synchronous images to rival in synchrony is probably due to lateral connections between cortical areas responding to similar and synchronized stimuli (Section 5.4.6).

A textured annulus placed round one of two small rival gratings increased the duration of exclusive dominance of that grating. Also, the monocular annulus was more clearly visible when the grating it surrounded was fused with a similar grating in the other eye than when it was in rivalry with an orthogonal grating (Fukuda and Blake 1992).

The boundaries of rival subregions were not determined by physically disconnected meaningful units such as words (Blake and Overton 1979). The influence of meaning on binocular rivalry is discussed in Section 7.8.2.

7.5 GENERALITY OF SUPPRESSION

According to one view, binocular suppression affects all stimuli and all stimulus features within the suppressed region (Fox and Check 1966a). Although all features of the suppressed image are removed from consciousness, there is considerable evidence that some features are suppressed more than others and that some suppressed features may affect the rivalry process and produce aftereffects.

7.5.1 Direct effects of suppressed images

7.5.1a *Rivalry, accommodation, and pupil responses*

It has been claimed that a flash of light presented to an eye in its suppressed phase of binocular rivalry evokes a weaker pupillary response than one presented to the eye in its dominant phase (Bárány and Halldén 1948; Richards 1966). Lowe and Ogle (1966) could not replicate this effect but they did find that when rival fields differ in luminance, a small constriction of the pupil occurs as the brighter field comes into dominance. This process must involve centrifugal pathways from the visual cortex to the subcortical centres controlling the pupil.

When accommodation demand differs between the two eyes, the accommodative response represents the vector average of the inputs (Flitcroft *et al.* 1992). However, when the two eyes are presented with orthogonal gratings that differ in contrast, the response at any time is determined by whichever grating is perceptually dominant (Flitcroft and Morley 1997). See Section 9.13d for more details.

7.5.1b *Threshold summation of rival stimuli*

A small near-threshold test flash superimposed on a grating presented to only one eye was detected more frequently when it was accompanied by a similar test flash presented to a corresponding region in the dark field of the other eye (Westendorf *et al.* 1982). It was concluded that a stimulus in a suppressed eye can summate binocularly with one presented in a dominant eye. But the dark field of one eye may not have been suppressed on every trial and the flash on the dark field may therefore have increased the probability of detection even if the two flashes did not summate.

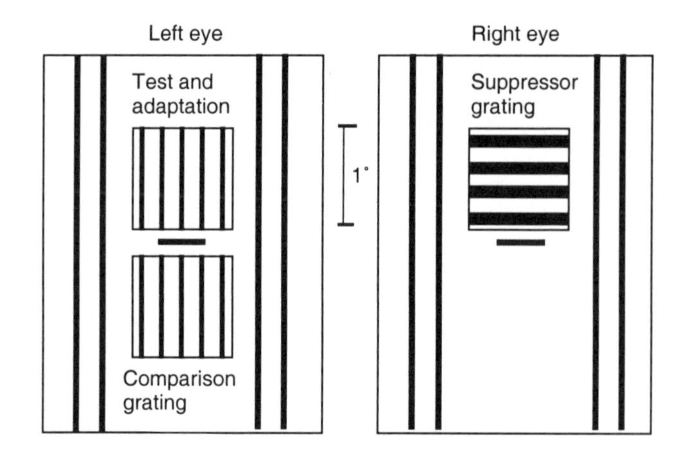

Figure 7.24. Threshold elevation from suppressed image. Stimuli used by Blake and Fox (1974b) to reveal that a suppressed grating may generate a threshold-elevation effect. While adapting, subjects fixated within the horizontal bar and the left-eye grating was suppressed by the right-eye grating. In testing, the right-eye grating was removed and subjects matched the contrast of a new comparison grating in the left eye to that of the test grating.

7.5.2 Chromatic specificity of suppression

Colour rivalry and pattern rivalry may be dissociated. For example, when two postage stamps with similar design but differing in colour and value are combined in a stereoscope the letters denoting the value of one stamp are sometimes seen in the colour of the other stamp (Dawson 1917). Also, dichoptic combination of a piece of coloured paper with printing on it with a plain piece of paper of another colour sometimes produces an impression of print with a background in the colour presented to the other eye (Creed 1935). Thus, the print dominates in one region while colours rival in the surrounding region. However, closely spaced black and coloured lines in one eye tend to be strongly dominant over an untextured patch of another colour in the other eye (De Weert and Levelt 1976a). The contours and colour are not strictly in the same location in these stimuli, so that they may not constitute an exception to the rule of nonselective rivalry.

There is evidence that the chromatic system is more affected by suppression than is the achromatic system. Smith *et al.* (1982) presented, for 20 ms, at the centre of rival orthogonal black-white gratings, small test probes varying in wavelength. When the probe fell on the dominant grating, the increment-threshold spectral sensitivity functions showed the three maxima corresponding to the absorption spectra of the cones. This is symptomatic of the chromatic system. But when the probe fell on the nondominant grating, the sensitivity function had a single broad peak near 555 nm, which is symptomatic of the achromatic system. Thus, rivalry suppression causes a greater reduction in the sensitivity of the chromatic mechanism than of the achromatic mechanism. Ooi and Loop (1994) confirmed these results for rivalry suppression and, with a more refined testing procedure, showed that loss of sensitivity is greater for blue than for red probes. The suppression of the blue-cone system during rivalry is related to the fact that blue-cones are only weakly involved in rivalry (Section 7.3.3).

Ridder *et al.* (1992) found that the change in increment-threshold spectral sensitivity was the same for all wavelengths when the test probe was superimposed on a blank field that was permanently suppressed by a grating in the other eye. In this respect, permanent suppression resembles amblyopic suppression (Section 13.4.2). Ooi and Loop (1994) also found that permanent suppression produces a more equal loss of chromatic sensitivity than does alternating rivalry, except that they found some specific loss of blue sensitivity during permanent suppression.

7.5.3 Changing the suppressed image

When a suppressed image is changed in certain ways, the suppressed phase may be rapidly terminated. Blake and Fox (1974a) presented a horizontal grating to the right eye and a vertical grating to the left eye. The horizontal grating had high contrast and was counterphase modulated at 4 Hz whereas the vertical grating had low contrast and was not modulated. The horizontal grating suppressed the vertical grating. The suppressed phase was terminated by an increase in the contrast of the suppressed image but not by a decrease in its contrast. Changes in the spatial frequency or orientation of the suppressed image did not terminate the suppressed phase. It was concluded that only energy increments terminate the suppressive process.

This evidence is not conclusive because the imposed changes may have been insufficient to overcome the flicker and larger contrast in the dominant image. Walker and Powell (1979) repeated the experiment with the left and right images equally dominant and found that changes in the contrast, spatial phase, or spatial frequency of the suppressed image terminated the suppressed phase.

Changes in spatial frequency and orientation involve local changes in stimulus energy and transient signals. O'Shea and Crassini (1981a) found that termination of suppression requires a change in orientation of a suppressed image of at least 20°. Larger changes in orientation could involve stronger transient signals than small changes in orientation. One cannot conclude that spatial frequency and orientation are analyzed as such in the suppressed image or that suppression acts selectively with respect to orientation and spatial frequency.

Freeman and Nguyen (2001) devised a procedure for controlling binocular rivalry. They presented a stationary horizontal grating to one eye. The other eye saw a horizontal grating superimposed on a vertical grating with the two gratings modulated in contrast in antiphase. As the vertical grating came into view the stationary horizontal grating was suppressed. As the horizontal gating came into view suppression ceased. They tracked the cyclic suppression of the stationary grating by measuring the visibility of a test spot superimposed upon it.

7.5.4 Eye rivalry versus stimulus rivalry

It has been generally assumed that rivalry occurs between the eyes or independently between distinct regions of the two eyes. An alternative view is that rivalry preserves a coherent pattern even though part of the pattern is in one eye and part in the other eye. One approach to this issue is to ask whether binocular rivalry can be entrained by interchanging coherent stimuli between the eyes. Blake *et al.* (1980a) reported that when horizontal and vertical rival patterns were interchanged there was no change in ocular dominance. If the dominant eye saw the horizontal grating before the interchange it saw the vertical pattern after the interchange.

Logothetis *et al.* (1996) obtained the opposite result. Oblique orthogonal gratings flickering at 18 Hz were interchanged between the two eyes three times per second. Subjects tracked changes in rivalry by pressing keys. They mostly perceived rivalry between the gratings, with dominance periods lasting several seconds, in much the same way as when each grating was presented continuously to the same eye. At higher contrasts, two of the six subjects saw the effect reported by Blake *et al.* Logothetis *et al.* concluded that the gratings competed for dominance independently of the eye to which they were presented so that patterns rivalled rather than eyes. In other words, they concluded that **stimulus rivalry** is stronger than **eye rivalry**.

I draw a somewhat different conclusion. In the period in which one grating was dominant, eye dominance switched three times per second so that there was continuity in the perceived orientation of the grating. Thus the eyes rivalled but the rivalry was entrained by the alternation of stimuli between the two eyes so as to preserve continuity of perceived form. Every few seconds subjects switched to seeing the other pattern either because of a break down in the entrainment process or because of rivalry between orientation detection systems. Other instances in which rivalry is entrained by the nature of the stimuli are discussed at the end of Section 7.8.

Lee and Blake (1999b) asked subjects to describe the frequency of rivalry rather than press keys, because it is not easy to track rapid changes in rivalry with a key. They obtained stimulus rivalry only when the orthogonal gratings were flickered at 18 Hz and abruptly interchanged between the eyes, as in the Logothetis *et al.* experiment. These are the conditions under which eye rivalry is entrained by transient signals. Mostly eye rivalry occurred when the gratings were not flickering or were interchanged gradually.

Wolfe (1984b) presented a vertical grating to one eye for some time. About 100 ms after it was removed a flashed vertical grating was presented to the same eye and a flashed horizontal grating to the other eye. Subjects saw only the horizontal grating; indicating that dominance had switched to the other eye even though this did not preserve figural continuity. Unfortunately, the test gratings were not presented the other way round.

One would expect that pattern rivalry would be more likely for coherent patterns than for noncoherent patterns. Bonneh *et al.* (2001) alternated arrays of orthogonal Gabor patches between the two eyes. Displays with patches that were closer, larger, and more uniform in orientation showed pattern rivalry. Those with small widely spaced patches with less uniform orientation showed eye rivalry.

This means that, with patterns with low spatial coherence, distinct regions of the binocular field rival independently while, with patterns with high coherence, there is interaction between rivalling regions so as to preserve pattern coherence.

Nguyen *et al.* (2001) asked subjects to indicate the location of a brief test stimulus presented to an eye in the suppressed phase of binocular rivalry produced by orthogonal gratings. Performance was not affected by whether the test stimulus and the grating in the suppressed eye were similar or not similar in colour, spatial frequency, or orientation. This supports the idea of eye suppression rather than that of stimulus-specific suppression.

7.6 EFFECTS FROM SUPPRESSED IMAGES

Binocular suppression eliminates all conscious awareness of suppressed stimuli. However, since binocular suppression depends on the relative positions, orientations, and contrasts of images in the two eyes, it is to be expected that some processing of these simple features precedes suppression. The evidence reviewed here shows that certain visual induction effects that depend on the spatial frequency, orientation, and motion of stimuli are evident while the induction stimulus is suppressed. This suggests that these induction processes occur either before the stage of binocular rivalry or bypass the rivalry process.

7.6.1 Suppression and spatial-frequency

Blake and Fox (1974b) found that a binocularly suppressed stimulus can create a threshold-elevation. They used the stimulus shown in Figure 7.24. An adaptation grating was presented continuously or intermittently to one eye for 30 s, after which the subject adjusted the contrast of a newly exposed comparison grating in a neighbouring position in the same eye to match the apparent contrast of the adaptation grating. The adaptation grating showed the familiar loss of apparent contrast, which was greater for continuous exposure than for intermittent exposure. The experiment was then repeated, but with the adaptation grating suppressed for at least half the 30-s adaptation period by a rival grating present in the other eye. The magnitude of the aftereffect was the same in both conditions, showing that adaptation occurred even when the adapting grating was suppressed. For two strabismic subjects who could suppress the adaptation grating for the full 30 s, the adaptation effect was as strong as when the grating was visible for the whole period (Blake and Lehmkuhle 1976).

Blake and Fox also showed that the spatial-frequency aftereffect (a perceived shift in spatial frequency away from that of an induction stimulus) was at full strength when the induction grating was suppressed for a good part of the induction period. They concluded that binocular suppression occurs at a site after that responsible for contrast adaptation and the spatial-frequency aftereffect. However, they measured these aftereffects only in the previously suppressed eye, and the results could be due to adaptation of monocular neurones that are unaffected by suppression (but unavailable to consciousness). Perhaps the contrast of the adaptation grating was reduced by the afterimage of the suppressor grating in the other eye. On the other hand, Crabus and Stadler (1973) had found no evidence of the figural aftereffect (apparent repulsion of a test line away from the location of a previously inspected line) when the monocular induction stimulus occurred wholly in periods of binocular suppression.

Blake and Overton (1979) overcame the problem of unsuppressed monocular neurones by showing that the threshold-elevation aftereffect in the eye opposite to that exposed to the induction stimulus was also not weakened when the induction stimulus was suppressed for a substantial part of the induction period. They also found that adaptation to a grating presented to one eye lessened the subsequent dominance of that grating when it was pitted against an orthogonal grating in the other eye.

These results strengthen the conclusion that the processes responsible for contrast adaptation and the spatial-frequency aftereffect occur before binocular suppression. Blake and Overton were convinced that contrast adaptation and the spatial-frequency aftereffect are cortical and therefore occur at least in area V1. Since the aftereffects preceded binocular suppression, they concluded that binocular suppression occurs beyond V1. They agreed that it must be processed fairly early, because it is not a cognitive process and does not occur in meaningful units (Section 7.8.2). The claim that binocular suppression occurs beyond V1 will now be assessed.

Both contrast and spatial frequency are, at least to some extent, coded at the retinal level—contrast by lateral inhibitory processes and spatial frequency in terms of receptive fields of different sizes. The two aftereffects may therefore arise in the retina or LGN. If they do, the argument for placing binocular suppression beyond V1 collapses. The belief that spatial-frequency aftereffects are cortical is based on the fact that they show interocular transfer. This, in itself, is not a convincing argument. An afterimage shows interocular transfer, in the sense that an afterimage impressed on one eye is visible when that eye is closed and the other eye opened. This does not prove that afterimages are cortical in origin, but simply that activity arising in a closed eye still reaches the visual cortex.

An afterimage is no longer visible when the eye in which it was formed is inactivated by the application of pressure to the eyeball (Oswald 1957).

Thus, to prove that any aftereffect is cortical, it must be shown that it survives pressure blinding of the eye to which the induction stimulus was presented. Blake and Fox (1972) found that the threshold-elevation effect survived retinal paralysis induced by applying pressure on the eyeball, but this still leaves open the possibility of LGN involvement. The conclusion that spatial-frequency aftereffects are cortical is supported by the finding that exposure to a grating of a given spatial frequency and orientation reduces the firing rates of neurones in the visual cortex but not of those in the LGN (Movshon and Lennie 1979).

Even if contrast and spatial-frequency aftereffects are cortical and precede binocular suppression we still do not have to conclude that the site of suppression is beyond V1. The processes responsible for contrast and spatial-frequency coding and those responsible for the combination of binocular inputs could occur sequentially within V1.

The **square-wave illusion** is an example of spatial-phase adaptation (Leguire *et al.* 1982). When a grating with a triangular luminance profile is viewed for some time it begins to look like a square-wave grating, namely, a grating with the same sine wave components but in antiphase rather than in phase. Binocular suppression of the induction stimulus severely reduces the illusion (Blake and Bravo 1985). It looks as though complex effects that depend on interactions between several spatial frequencies occur after suppression.

A disc of vertical lines appears tilted clockwise when surrounded by an annulus of lines tilted a few degrees counterclockwise. This is simultaneous tilt contrast. There is conflicting evidence as to whether tilt contrast persists when the induction annulus is suppressed by an annulus of horizontal lines in the other eye. Wade (1980) found that it did survive suppression. However, Rao (1977) found that it did not.

Vertical lines appear tilted clockwise after inspection of counterclockwise tilted lines presented in the same location. This is the tilt aftereffect. Wade and Wenderoth (1978) reported that the tilt aftereffect was not weakened when the induction stimulus was suppressed for a good part of the induction period by a rival horizontal grating presented to the other eye. These effects must be cortical because precortical sites have little or no orientation tuning (Section 5.3.2).

The McCollough effect (Section 8.3.5) occurred at full strength when the induction stimulus presented to one eye was perceptually suppressed for much of the induction period by a rival stimulus presented to the other eye (White *et al.* 1978).

7.6.2 Suppression, visual motion, and flicker

It takes a person longer to react to a moving stimulus that starts to move while it is suppressed than to one that starts in a nonsuppressed period. Presumably, the suppressed image must become dominant before its motion can be detected (Fox and Check 1968). However, we will now see that, even though motion remains unavailable to consciousness during suppression, some processing of motion signals occurs before the stage of suppression.

7.6.2a *Motion aftereffect from a suppressed image*

When a moving textured display is viewed for some time, a stationary display seen in the same location appears to move in the opposite direction. This is the motion aftereffect. The duration of the aftereffect increases with the duration of the induction stimulus. Lehmkuhle and Fox (1975) found that the duration of the aftereffect for test and induction gratings viewed by one eye was not reduced when the induction stimulus was suppressed for most of the inspection period by a stimulus presented to the other eye. Also, interocular transfer of the motion aftereffect was the same when the induction stimulus was suppressed by binocular rivalry as when it was visible for the whole inspection period (O'Shea and Crassini 1981b). Thus, some processing of visual motion must precede and survive suppression.

On the other hand, the motion aftereffect produced by a rotating display or by a rotating spiral was attenuated when the induction stimulus was suppressed (Lack 1978; Wiesenfelder and Blake 1990). Also, the motion aftereffect induced by coherent motion of a plaid was reduced when the induction stimulus was exposed to binocular rivalry (van der Zwan *et al.* 1993). The visual processes for detecting rotary, spiral, and plaid motions are more complex than those for detecting simple linear motion, and may occur in MT (Section 5.7.3). Binocular suppression seems to occur after the site of simple motion detection but before that of complex motion detection.

Alais and Blake (1998) provided other evidence for this conclusion. Gratings moving in independent directions in four neighbouring disc-shaped windows appear as one partially occluded grating moving coherently. Covering one window tends to destroy this effect. The coherence effect was destroyed in a similar way when the grating in one window was suppressed by a disc of random-dot in the other eye. Nevertheless, the random-dot stimulus suppressed the grating for shorter periods when it was accompanied by the other three discs than when it was seen in isolation.

The motion aftereffect lasts longer when a dark interval is interspersed between the induction and test stimuli. Thus a stationary test stimulus drains the motion aftereffect, presumably because it re-adapts the visual system to its normal state. In the dark, the visual system returns to its normal state more slowly. The motion aftereffect in an adapted eye also lasted longer when the stationary test stimulus in the adapted eye was suppressed by a rival stimulus in the other eye (Wiesenfelder and Blake 1992). Since a motion aftereffect can be induced by a suppressed stimulus, the inability of a suppressed stationary test stimulus to readapt the visual system suggests that induction of the motion aftereffect occurs at an earlier site than its decay. But perhaps the ineffectiveness of a stationary test stimulus is due to its being stationary. *Perhaps a suppressed stimulus moving in another direction would cause a motion aftereffect to decay.*

7.6.2b *Apparent movement from a suppressed image*

There has been some dispute about whether rivalry suppresses the generation of apparent movement. Ramachandran (1975) found that a spot superimposed on a suppressed pattern did not generate apparent motion with respect to a nearby spot presented sequentially on the dominant pattern in the other eye. On the other hand, Wiesenfelder and Blake (1991) found apparent motion under these conditions although it was not as clear as when both spots were in view. They concluded that the motion signal is weakened by suppression.

Shadlen and Carney (1986) obtained dichoptic apparent movement by presenting the same flickering gratings to the two eyes but with a 90° spatial and temporal phase shift (Section 17.4.1). Carney *et al.* (1987) presented a yellow and black grating to one eye and an equiluminant red and green grating to the other. When the two gratings flickered with a 90° spatial and temporal phase shift, apparent motion was seen in a direction consistent with the luminance component of the alternating gratings. But colour rivalry was seen at the same time. They concluded that rivalry can occur in the colour channel while motion is seen in the luminance channel. This is similar to the argument used by Ramachandran and Sriram (1972) to account for the simultaneous occurrence of stereopsis and colour rivalry in a random-dot stereogram.

To account for these effects, it may not be necessary to assume that two channels are simultaneously activated at the same spatial location. Colour rivalry may have occurred locally within areas between the contours that defined the apparent movement. In other words, the two effects may have been segregated spatially rather than by virtue of parallel visual channels serving the same spatial location.

In summary it can be stated that processes responsible for the threshold-elevation effect, the spatial-frequency aftereffect, the tilt aftereffect, and perhaps tilt contrast and contingent aftereffects occur before those responsible for binocular suppression and, to some extent, are not suppressed. It is difficult to draw firm conclusions about the site of binocular suppression.

7.6.2c *Visual beats from a suppressed image*

Carlson and He (2000) presented a red triangle facing left to one eye and a red triangle facing right to the other eye. At the same time, the luminances of the two stimuli were modulated sinusoidally at a slightly different temporal frequency. Out of phase dichoptic flicker produces a sensation of flicker at the difference frequency (visual beats). Subjects observed rivalry between the form and colour of the stimuli but continued to see the visual beats. Thus, a flicker signal penetrated the mechanism that suppressed shape and colour. Carlson and He suggested that rivalry occurs in the parvocellular system that carries colour and shape signals but that flicker and high-velocity motion signals carried by the magnocellular system survive suppression and combine.

7.7 RIVALRY AND STEREOPSIS

Galen's idea that the two images fuse in the optic chiasma held sway for about 1,500 years and was replaced in the 17th century by the view that fusion occurs in the brain (Section 2.2.3). After 1838, when Wheatstone proved that binocular disparity forms the basis of stereoscopic vision, the problem arose of reconciling fusion of images with processing of disparity. Four basic accounts of the relationship between fusion and disparity detection have been proposed. I call them the mental theory, the suppression theory, the two-channel theory, and the dual-response theory.

1. The mental theory Helmholtz (1910) objected to the idea of fusion of images at an early stage of visual processing. He wrote, "*The content of each separate field comes to consciousness without being fused with that of the other eye by means of organic mechanisms; and therefore, the fusion of the two fields in one common image, when it does occur, is a psychic act.*" (p. 499). His argument was based on the fact that black in one eye and white in the other do not produce gray, as the fusion theory would predict, but binocular lustre. This objection does not apply to the fusion of similar

images. However, he also pointed out that with very short exposures we can distinguish between depth based on crossed disparity and that based on uncrossed disparity. This means that we register which eye receives which image, which would not be possible if images fused.

This objection to the fusion theory does not hold for a mechanism containing some binocular cells tuned to crossed disparities and others tuned to uncrossed disparities. Such cells create a unified set of signals from the two inputs but preserve all the information required to code depth. In talking about a "psychic act" of fusion, Helmholtz observed that, although we do not normally notice diplopic images of objects well out of the plane of the horopter, they become visible when we make a special effort to see them. On the other hand, Helmholtz did not support the suppression theory, in which the images are processed in alternation. However, his only argument against the theory was that, " · · · *the perception of solidity given by the two eyes depends upon our being at the same time conscious of the two different images.* " (Helmholtz 1893, p. 262).

Sherrington (1904) also believed that binocular fusion results from "· · · *a psychical synthesis that works with already elaborated sensations contemporaneously proceeding.*" From his experiments on binocular flicker (Section 8.1.4) he concluded that fusion is not based on a physiological mechanism like the convergence of nerve impulses in the final common path of the motor system.

2. The suppression theory According to the suppression theory of binocular vision, rivalry is the only form of binocular interaction and operates for both similar and dissimilar images. According to one account, the position of each image is sampled intermittently and a subsequent comparison process detects disparity. According to another account, information from the suppressed image is processed for purposes of disparity detection (Kaufman 1964). In this form, the suppression theory is equivalent to the two-channel account.

3. The two-channel theory According to this theory, rivalry and fusion with stereopsis are distinct processes in separate neural channels. For instance, similar but not identical dichoptic images may simultaneously engage a channel devoted to alternating suppression and a channel devoted to disparity-detection in the same location of the visual field (Wolfe 1986b).

4. The dual-response theory According to this theory, rivalry is a state of alternating suppression occurring between dissimilar dichoptic stimuli. Similar images do not alternate, but fuse and engage the disparity-detection system. Rivalry and fusion are distinct processing modes performed by the same neural circuits. They may therefore occur at the same time in different parts of the visual field but not simultaneously in the same location. Another view is that they may occur in the same location but with one occurring in the chromatic channel and the other in the achromatic channel. The disparity-detection mechanism may involve inhibitory processes in the ocular dominance columns but not a gross alternation between the images in the two eyes.

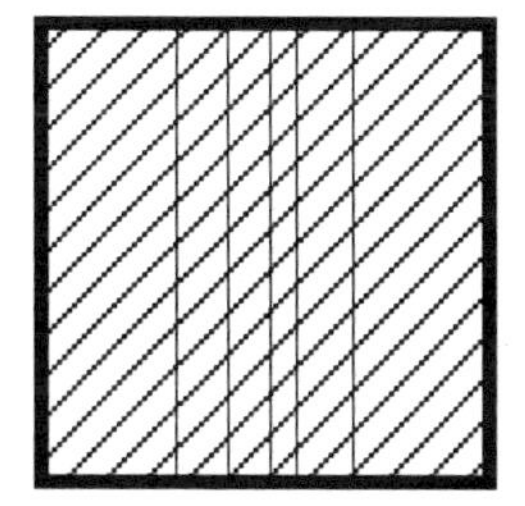
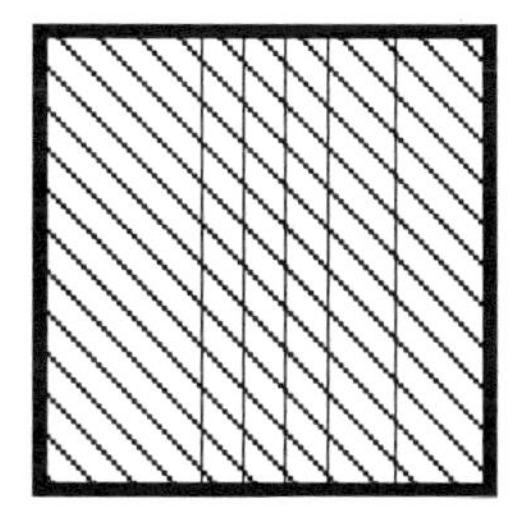

Figure 7.25. Stereopsis with rivalry.
Depth is apparent in non-rivalling vertical lines superimposed on rival lines. (Adapted from Ogle and Wakefield 1967)

7.7.1 Suppression theory of binocular fusion

According to the suppression theory of binocular fusion, rivalry occurs even between identical stimuli. According to Duke-Elder (1968, p. 684) the earliest references to the suppression theory were Porta (1593), Gassendi (1658), and Dutour (1760, 1763). Washburn (1933) and Verhoeff (1935) gave fuller accounts of the theory, although they produced no evidence other than general observations about binocular rivalry. Asher (1953) provided a lively supportive account of the suppression theory.

One cannot test the suppression theory by observing binocular rivalry because it is not possible to tell which eye is seeing the stimulus when the images are alike (see Section 17.8). The theory can be tested by measuring suppression of a test stimulus presented to an eye both in its suppressed phase and in its dominant phase. Three indicators of suppression are available: (1) the luminance or duration threshold of a test flash (Wales and Fox 1970), (2) the reaction time of a response to a test flash, and (3) success in recognizing a flashed patterned stimulus, such as a letter.

When a vertical grating was presented to one eye and a horizontal grating to the other, observers took longer to respond to a flash superimposed on the suppressed image than to one superimposed on the dominant image. When the orientation of the gratings was the same for both eyes, the reaction time to the flash was the same for either eye and the same as when the flash was superimposed on a monocularly viewed grating (Fox and Check 1966a). This suggests that neither image is suppressed when the images in the two eyes are identical.

A flash was also detected equally well by either eye when the two gratings had a disparity that produced a slanted surface (Blake and Camisa 1978). Thus, superimposed dichoptic images do not suppress each other when they differ enough to yield an impression of depth.

In another study, subjects recognized a letter superimposed on a dominant image but not one superimposed on a suppressed image (Fox and Check 1966b). A letter superimposed on either one of a pair of fused images was consistently recognized just as well as when it was superimposed on a stimulus presented to only one eye.

On the other hand, Makous and Sanders (1978) reported that a test flash presented to one eye when both eyes viewed identical patterns was not detected as frequently as when the eyes viewed different patterns, with the flash presented to the eye in its dominant phase of rivalry. This does not prove that the eyes engage in alternating rivalry when viewing identical patterns; it may simply reflect some constant mutual inhibition between identical stimuli.

When a vertical grating was presented to one eye and superimposed vertical and horizontal gratings were presented to the other eye, reaction time to a decrement in contrast in any one of the gratings was the same for all three gratings. In other words, neither of the vertical gratings was suppressed, nor was the horizontal grating that was superimposed on one of the vertical gratings. The vertical grating, being fused with the vertical grating in the other eye, kept the superimposed horizontal lines dominant (Blake and Boothroyd 1985). This demonstrates that fusion takes precedence over rivalry. Presumably a fused image supports a nonfused image only over a certain distance.

According to the suppression theory, reaction times to a monocular flash superimposed on a binocularly viewed stimulus should show a skewed distribution because the image on which the flash is superimposed is sometimes in its suppressed phase and sometimes in its dominant phase. O'Shea (1987) found the expected skewed distribution when the stimuli were rivalling but not when they were similar and fused.

7.7.2 Two-channel and dual-response accounts

According to the two-channel account, fusion and rivalry are distinct processes that can coexist in the same location in the visual field and be evoked by the same stimulus. According to the dual response account, similar images fuse, and rivalry occurs only when images have failed to fuse. Thus, fusion and rivalry are not evoked by the same stimulus and do not occur simultaneously in the same location. Evidence on the essential difference between these two accounts will now be reviewed.

Kaufman (1964) prepared a random-dot stereogram in which disparity between black dots generated a form in depth while the red and green backgrounds produced colour rivalry. He concluded that colour rivalry and fusion stereopsis occur simultaneously in the same location. The same point is illustrated by the fact that depth is evident in anaglyph stereograms in which the eyes are presented with differently coloured images. However, we do not have to conclude that the dots engaged in both rivalry and fusion-stereopsis but only that, while rivalry occurs in the chromatic channel, fusion stereopsis can occur in the achromatic pattern channel. This would be a weak version of the two-channel theory. A stronger version, espoused by Tyler and Sutter (1979) and Wolfe (1986b), is that rivalry and fusion stereopsis occur simultaneously for the same patterned stimuli. Some of the evidence for this strong version of the two-channel theory will now be considered. Other literature on the coincidence of rivalry and stereopsis is reviewed in Section 18.1.2.

Depth is apparent in disparate similar lines superimposed on a background of rivalling lines, as in Figure 7.25 (Ogle and Wakefield 1967). This suggests that rivalry and fusion can coexist in the same location. However, the effect could also be interpreted as another example of fusion taking precedence over rivalry in regions where corresponding images are located.

Julesz and Miller (1975) found that depth was still apparent in a random-dot stereogram when random-dot noise with a spatial-frequency 2 octaves higher than that of the stereogram was added to one eye. They argued that, in a given region, stimuli of one spatial frequency can generate stereopsis while stimuli of another spatial frequency generate rivalry. Blake *et al.* (1991b) used a similar stimulus but concluded that stereopsis and rivalry are generated in distinct regions. Subjects viewed a random-dot stereogram that yielded depth. When low-contrast random-dot noise was added to one eye's image, subjects saw depth but no rivalry. At intermediate contrasts, regions of rivalry and of depth were seen but not in the same place. At high contrasts, the noisy display was dominant and there was not much evidence of depth.

It has been claimed that suppressed images may contribute to stereoscopic depth. Blake *et al.* (1980a) presented a vertical grating to one eye and a horizontal grating to the other. A vertical grating was presented for 1 s to the same eye as the horizontal grating during periods when only the horizontal grating was visible. During these periods, subjects reported seeing a vertical grating slanted in depth according to the disparity between the two vertical gratings, one of which was in the suppressed eye. Blake et al. concluded that information for stereopsis is extracted before the suppression stage or that it survives suppression because it is processed in a parallel channel. However, they admitted that the similarly oriented lines in the two eyes during the 1 s test period may have temporarily suspended the suppression process, at least in those parts of the display containing near-congruent images. In view of this possibility this evidence does not provide strong support for the idea of parallel independent channels for rivalry and stereopsis in the same location.

Tyler and Sutter (1979) obtained slant from stereograms consisting of randomly changing vertical bars filtered to give slightly different spatial-frequency bands in the two eyes (Section 21.1.1). The impression of slant persisted even when the bars were uncorrelated in the two eyes and drifted at 4°/s in opposite directions. Thus rivalry existed both between the uncorrelated bars and the two directions of motion. They argued that slant arose from a primitive dif-frequency mechanism in which width disparity was detected at the same time that the uncorrelated bars engaged in rivalry. However, in the uncorrelated displays only a subset of bars rivalled at any instant, and perhaps bars that were not in rivalry carried the disparity signal.

If rivalry and disparity are processed by the same neural machinery operating in different ways, then it should take time to switch from the rivalry mode to the disparity mode. Harrad *et al* (1994) investigated this question. They presented half of a stereo target to the left eye and a grid of oblique lines to the right eye. Subjects pressed a button when the grid suppressed the stereo target. At this point, the grid replaced the other half of the stereo target for a variable period of time. Subjects judged the relative depth of the two vertical lines of the stereo target. Stereoacuity was elevated for 150 to 200 ms after the grid was removed. A control condition showed that the effect was not due to monocular masking. These are shorter intervals than the mean period for which the stereo half target would have remained suppressed if the grid had not been replaced by the other half target. In agreement with Wolfe (1986b), they concluded that the onset of a fusable target terminates suppression but that the fusion and disparity-detection mechanism takes time to become fully operational.

Summary

Most of the evidence supports the view that alternating suppression does not occur when congruent or near-congruent patterns are fused. We must therefore reject this form of the suppression theory of binocular fusion. That is not to deny that mutual inhibition of images may be involved in the fusion of similar images. The physiological evidence reviewed in Section 6.2 shows that some binocular cells are strongly dominated by excitatory inputs from only one eye and are inhibited by inputs from the other eye. The psychophysical evidence suggests that over an extended area any mutual inhibition between the two eyes when they view similar images is the same for both eyes and does not result in the alternation of suppression seen when the images are dissimilar.

The view that fusion and rivalry occur simultaneously in the same location and in distinct channels has been championed by Wolfe (1986b) and contested by Blake and O'Shea (1988) and Timney *et al.* (1989). Readers are referred to these sources for a detailed assessment of the evidence. On balance, the evidence favours the dual-response account, namely, that rivalry and stereopsis are mutually exclusive outcomes of visual processing in any given location. At least this seems to be the case for visual processing of patterned stimuli. However, it is possible that rivalry within the purely chromatic channel can proceed simultaneously with stereopsis based on disparity between patterned images in the same location. Evidence reviewed in Section 5.5.5 shows that chromatic and patterned stimuli are processed in partially distinct neural channels, even at the level of the retina and geniculate nucleus. Evidence reviewed in Section 18.1.4 suggests that stereoscopic processing is weak in the purely chromatic channel.

7.8 COGNITION AND BINOCULAR RIVALRY

So far, we have discussed binocular rivalry as if it were determined by low-level features of the stimulus plus a low-level process of alternating suppression. I now ask whether binocular rivalry is determined wholly by these "bottom-up" processes or is affected by influences descending from higher levels in the nervous system—so-called "top-down" processes.

There are two types of cognitive variables. The first type refers to the characteristics and temporary state of the observer, such as the direction of attention, expectations, and emotional state. The second type refers to high-level attributes of the stimulus, such as its familiarity, meaning, and emotional significance. Binocular rivalry was selected as a convenient tool for studying the effects of both types of cognitive factors. Each type of variable will be dealt with in turn.

7.8.1 Voluntary control of rivalry

Several investigators enquired whether the rate of alternation of rival stimuli is under voluntary control. Breese (1899) reported that subjects could influence the duration for which one or the other of two rival stimuli was seen, but he noticed that the eyes moved whenever subjects exercised this control. The more the eyes moved over a dominant stimulus, the longer that stimulus remained in view. Furthermore, a moving stimulus was dominant over a stationary stimulus. Eye movements may therefore have mediated voluntary control over rivalry.

Others have agreed that one can exercise some degree of voluntary control over the rate of binocular rivalry (Meredith and Meredith 1962; Lack 1969). However, eye movements and blink rate were not controlled and there was no objective verification of subjects' reports of the dominance of rival stimuli. Voluntary control of rivalry was still evident for stimuli viewed through small artificial pupils or after paralysis of accommodation (Lack 1971). Thus, voluntary control of rivalry is not necessarily mediated by changes in accommodation.

It has been established that a flashed test stimulus is not visible when superimposed on a suppressed image (Fox and Check 1966b). The ability of the subject to report the test stimulus can thus be used to confirm that the stimulus reported as dominant is indeed dominant. With this procedure, Collyer and Bevan (1970) obtained a 10% improvement in detection of a test flash superimposed on a given image after subjects were given 3 s to bring that image into dominance over a rival image in the other eye. However, voluntary control of dominance was not always achieved, was not continuous, and may have involved only part of the stimulus in one eye.

Ooi and He (1999) presented an array of gratings to one eye and a blank field to the other eye. An apparent motion stimulus in the blank field suppressed the grating upon which it was dichoptically superimposed. This suppression was less evident when subjects attended to that grating rather than to one of the other gratings. They concluded that attending to a dominant image can help to resist its suppression by a perturbing stimulus presented to the other eye.

Figure 7.26. Sidney R. Lehky.
Born in Chicago, 1954. B.A. and Ph.D. in Biological Sciences at the University of Chicago. with Hugh R. Wilson. Postdoctoral work at Johns Hopkins University, University of Rochester, the Salk Institute, Baylor College of Medicine, and the RIKEN Institute in Japan. Currently a research fellow in the Laboratory of Brain and Cognition at the National Institute of Mental Health, Bethesda, Maryland.

7.8.2 Binocular rivalry and meaning

Sections 7.4 and 7.5 dealt with how a suppressed image might influence certain visual effects such as binocular summation and interocular aftereffects. These processes involve only low-level features of the stimulus such as luminance, contrast, and motion. What is the evidence that high-level, or semantic, features of a suppressed stimulus are processed? Zimba and Blake (1983) addressed this question by making use of semantic priming. In this effect, prior presentation of a word shortens the time needed to decide whether a subsequently presented stimulus is a random letter string or a word semantically related to the priming word. Priming was found to operate only when the priming word was presented to an eye in its dominant phase of interocular rivalry. According to this evidence, binocular rivalry occurs before the level of semantic analysis.

The opposite conclusion was drawn from a study of rivalry in word strings. Rommetveit *et al.* (1968) presented a word such as "wine" to one eye and a typographically similar word such as "nine" to the same location in the other eye. These words were presented next to a binocularly viewed word such as "red" that made a meaningful phrase, such as "red wine", with one of the rival pair but not with the other. In a brief exposure, the semantically relevant word was more frequently reported than the irrelevant word. However, a person may be less likely to read "red nine" than "red wine" when both words are presented side by side to the same eye. Since there was no control for this possibility, the results cannot be accepted as evidence for semantic penetration of rivalry suppression. Furthermore, stimulus duration may have been too brief to allow rivalry to develop.

A person can follow a verbal message presented to one ear when a different message is presented to the other ear (Cherry 1953; Lewis 1970). However, Blake (1988) found that subjects could not read a message presented to one eye when a different message was presented to the other eye, even when the messages were in distinct fonts or when the message to be read started 5 seconds before the other message.

An erect face is seen more frequently than the same face inverted, when the two are combined dichoptically (Engel 1956; Hastorf and Myro 1959). But this may not mean that the basic rivalry process is affected by meaning. For much of the time, dominance was incomplete and parts of each face were visible. Parts of an erect face are more familiar than parts of an inverted face and are therefore more likely to form the basis of the decision about whether the face is erect or inverted. Furthermore, certain features of a face, such as an eye, are less affected by inversion than are other features, such as the nose. These factors could bias judgments in favour of an erect face quite apart from any influence of binocular rivalry. *This idea could be tested by seeing whether a binocularly viewed face consisting of an equal mixture of erect and inverted regions is more often reported to be erect or inverted.*

Ono *et al.* (1966) presented subjects with pairs of photographs of faces combined dichoptically. More rivalry was reported between the faces that had been rated as less similar on a variety of criteria, such as pleasant and unpleasant, than between those that had been rated as more similar. But it is not clear from this whether the crucial variable was similarity in terms of the semantic criteria or similarity in terms of low-level features such as the relative positions of contours.

There are also reports that the relative dominance of dichoptically combined words matched for number of letters and frequency of usage depends on the emotional impact of the words as determined by their sexual or aggressive significance (Kohn 1960; Van de Castle 1960). Such results may reflect the willingness of subjects to report a certain word rather than a greater visual dominance of one type of word over another.

Several studies reported that the personal significance of the contents of a picture affects which picture predominates when two different pictures are combined dichoptically. For example, a picture with North American content, such as a boy playing baseball, was presented to one eye and a picture with Mexican content, such as a matador, was presented to the other eye (Bagby 1957). North Americans saw more North American pictures while Mexicans saw more Mexican pictures.

There are several problems with these studies. The frequency with which one stimulus is reported may reflect the greater salience of local features of that stimulus during those periods when parts of each stimulus are visible. Because of this greater salience of partial features, one stimulus is more likely to be reported than is the other during the periods of mixed dominance. This will extend the time during which one stimulus is reported but will not necessarily reflect any basic effect of meaning on the rivalry process itself. Another problem is that recognition will be more rapid when the more familiar picture is totally dominant than when the unfamiliar picture is totally dominant.

Yu and Blake (1992) attempted to overcome these problems. They found that a face showed longer periods of exclusive dominance over a geometrical comparison pattern than did a control stimulus with the same spatial-frequency content and mean contrast as the face. They obtained the same result when the relative dominance of the patterns was assessed by the reaction time for detection of a probe flashed on the geometrical pattern. The control face may have been more dominant than the control stimulus because it was a face or because it had a coherent shape. They investigated this issue by comparing the duration of exclusive dominance of a hidden 'Dalmatian-dog' figure before it was seen as a dog and again after it was seen as a dog. The dog pattern became more dominant after subjects had been shown that it was a dog by placing a tracing of a dog over it. However, a scrambled version of the dog pattern also became more dominant after subjects had seen the same tracing placed over it. This effect must have been due to suggestion since it was not based on any objective feature of the stimuli. An upright version of the dog pattern was more dominant than an inverted version even though subjects were not aware of the dog. Whatever this set of experiments reveals about the effects of stimulus configuration, the effects are small compared with effects of stimulus variables such as contrast, spatial frequency, colour, motion, and orientation.

7.9 MODELS OF BINOCULAR RIVALRY

The simplest way to think about binocular rivalry is to suppose that it is due to mutual inhibition between competing stimuli arising in the two eyes. This can be called the **mutual suppression mechanism**. The mutual inhibition must be time dependent in the manner of a bistable oscillator (Matsuoka 1984; Sugie 1982). To account for the alternation between rival images, one must suppose that the dominant image's **potency to inhibit** the suppressed image gradually weakens but without affecting the **visibility** of the dominant image. At the same time, the suppressed image recovers its potency to inhibit the dominant image without affecting the visibility of the suppressed image. At a certain threshold point, the suppressed image becomes visible and the dominant image becomes suppressed. The two images then reverse their roles. According to this account a visible image gradually loses its inhibitory potency while the nonvisible image gains in inhibitory potency and the rate of alternation depends on the time constants of these processes and the relative visual strengths of the two images. The strength of an image is determined by its luminance, contrast, and figural complexity. The inhibitory potency of an image depends on its strength and its changing value in the duty cycle. Whether an image is visible or suppressed depends on its strength and inhibitory potency relative to the image in the other eye. Any change in the relative strengths of the two images results in a corresponding change in the relative durations of their dominance and suppression phases.

An alternative view is that an image does not compete when it is suppressed. The duration of dominance of an image depends only on the strength of the dominant image and not on its strength relative to the suppressed image. This can be called the **dominant suppression mechanism**. A third alternative is that the durations of the dominance and suppression phases depend only on the strength of the suppressed image. This view conforms to Levelt's proposition, which was presented in Section 7.3.2. I refer to this as the **suppression recovery mechanism**.

These three views generate different predictions when the contrast of the image in one eye (say the left eye) is increased while that in the other eye is kept the same. According to the mutual inhibition account, the duration of left-eye dominance should increase and that of right-eye dominance should decrease by a proportional amount, leaving the overall rate of alternation the same. According to the dominant inhibition account, the periods when the strengthened stimulus is seen should increase but the periods when the constant stimulus is seen should remain unchanged. According to the suppression recovery account, periods when the strengthened stimulus is seen should remain constant and periods when the constant stimulus is seen should be shortened. There is evidence in favour of the latter prediction and hence of the suppression recovery account of binocular rivalry (Levelt 1966; Fox and Rasche 1969). However, Mueller and Blake (1989) found that the contrast of a pattern in its dominant phases exerts some influence on the rate of alternation. Bossink *et al.* (1993), also, cited evidence that the strength of the dominant image has some effect on the duration of the dominant phase, although they agreed that the influence of a dominant image is less than that of a suppressed image. Mueller and Blake suggested that a mutual suppression mechanism is responsible for rivalry but that it contains an element that makes frequency of rivalry increase with contrast and a nonlinear element sensitive to unbalanced contrast in the two eyes.

Fox and Check (1972) measured the magnitude of suppression of concentric rings by contrast-reversed rings by measuring the threshold for detecting a test flash superimposed on the suppressed image. The magnitude of suppression, tested at several times during the suppression period, was constant. Norman *et al.* (2000) obtained the same result for rivalling orthogonal gratings. They argued that this contradicts the mutual suppression account of rivalry. But the mutual-inhibition theory, as I have formulated it, does not predict that the visibility of the suppressed image increases but only that its potency to inhibit the dominant image increases.

The converse of this finding is that a dominant image does not become visibly weaker during its dominance phase even though its inhibitory potency weakens. It is only at the point when the images change dominance-suppression roles that there is a change in visibility. In other words, the gradual change in relative inhibitory potential finally results in a saltatory, or winner-take-all, change in visibility. One must distinguish between the visibility of an image (whether it is dominant or suppressed), the strength of an image, and the inhibitory potency of an image.

The dominant suppression and suppression recovery mechanisms make the duration of the dominance phases of the two images independent, because each depends only on its own strength and not on its relation to the other stimulus. Autocorrelation procedures and fitting data to a gamma distribution have revealed that the durations of succeeding dominance phases vary randomly and independently of each other (Fox and Herrmann 1967; Walker 1975). Lehky (1995) derived a time series of dominance durations evoked by dichoptic orthogonal gratings (Portrait Figure 7.26). He then applied two tests to determine whether this time series conformed to a stochastic noisy process or to a process of deterministic chaos (Section 3.6). An autocorrelation function derived from a stochastic noise process increases linearly as a function of the dimension over which the signals are correlated, whereas the autocorrelation function derived from a deterministic chaos reaches an asymptote. Also, short-term predictions are not possible with a truly stochastic process but are possible with deterministic chaos. These two tests showed that the rivalry data conformed more to a stochastic process than to a deterministic chaotic system.

The independence and random variation of succeeding phases was the same for rivalling afterimages as for normal images (Blake *et al.* 1971; Wade 1975a). This excludes eye movements as the only cause of the random variation of dominance phases in binocular rivalry. Blake *et al.* were not justified in concluding from their data that the cause must be central. But, in any case, Wade (1977) found that the rate of rivalry between afterimages was lower than that between real images, unless the afterimages were viewed against a background flickering at below 3 Hz. This suggested that image movements due to eye tremor or image flicker increase the rate of rivalry above a baseline rate that is independent of image movement or flicker.

Since the duration of suppression depends on features of the suppressed image, Walker (1978b) argued that these features must somehow evade suppression. But this would follow only if high-level features were involved, and the evidence for this is not very convincing (Section 7.8.2). Several of the features affecting suppression such as contrast, spatial frequency, and flicker are already coded in the retina. One could explain their effect on the recovery from suppression by supposing that inputs from the suppressed eye charge a buffer mechanism until the charge breaks through the inhibitory barrier of the other eye's image. The rate of charging of the buffer depends on the firing frequency of the inputs from the suppressed eye, which in turn depends on features such as contrast, spatial frequency, and flicker. These features would not have to be processed separately, since each contributes to the undifferentiated mean rate of afferent discharge.

Dayan (1998) proposed that rivalry is based on competing high-level interpretations of the rivalling stimuli. In this model, competition occurs between competing top-down interpretations of the stimuli rather than between the inputs themselves. Perhaps top-down influences do influence rivalry under certain circumstances, but most of the evidence supports the idea that rivalry is mainly due to processes occurring at an early level.

Blake *et al.* (1990) used a sharp transient stimulus to force an eye to return to dominance whenever it became suppressed. For that eye, the dominance periods became unusually brief during the procedure and unusually long for a short time after the procedure. This result is consistent with the idea that transitions from dominance to suppression are due to a short-term adaptation or fatigue process.

A simple reciprocal suppression process cannot account for all the facts of binocular rivalry. However, a type of mutual inhibition that gives greater weight to the suppressed image and is sensitive to contrast imbalance and correlation between binocular images may provide an adequate account. Neural models of binocular rivalry based on oscillation in recurrent inhibitory connections have been proposed by Lehky (1988), Blake (1989), and Mueller (1990).

7.10 NEUROLOGY OF BINOCULAR RIVALRY

7.10.1 Rivalry at the level of the LGN

Little is known about the neural processes mediating binocular rivalry. The lateral geniculate nucleus (LGN) is the earliest stage where these processes could occur. Each cell in the LGN receives a direct input from only one eye and does not respond when only the other eye is stimulated. Nevertheless, for many LGN cells, the response is modified by stimulation of the eye from which the cell does not receive a direct input (Section5.3.3) . This modification could be mediated either by intrageniculate inhibitory connections or by descending influences from the visual cortex. There is some dispute about the role of cortical influences. Some investigators found that binocular interactions in the LGN of the cat require an intact visual cortex (Varela and Singer 1987), while others found that they do not (Tumosa *et al.* 1989a; Tong *et al.* 1992). A report that interocular influences in the LGN are greatest when the stimuli presented to the two eyes differ in orientation, contrast, and movement prompted the suggestion that these visual features are involved in binocular rivalry, since binocular rivalry is affected by interocular differences between the same features (Varela and Singer 1987).

However, subsequent experiments revealed that interocular influences in the LGN of the cat are not much affected by changes in stimulus orientation or direction of motion, but are affected by changes in spatial frequency (Moore *et al.* 1992; Sengpiel *et al.* 1995b). This suggests that interocular influences in the LGN serve to balance the responses to small interocular differences in stimulus contrast by adapting the relative contrast gains of the inputs from the two eyes. Rivalry and contrast gain could be served by the same or different classes of interactive processes in the LGN.

If inhibitory interactions responsible for binocular rivalry occur in the LGN, the degree of inhibitory coupling would have to depend on the degree of correlation between the images, as detected by feature detectors in the visual cortex. Therefore, they would have to depend, at least in part, on signals descending from higher centres. A high correlation would weaken the inhibitory coupling and lead to fusion while a low or negative correlation would produce rivalry (Lehky and Blake 1991).

Lehky and Maunsell (1996) found no difference in the responses of neurones in the LGN of alert monkeys as they fixated rival or similar gratings. This seems to eliminate the LGN as the site of binocular rivalry, at least in the monkey.

7.10.2 Rivalry at the cortical level

7.10.2a *Rivalry and the cortical VEP*

At the level of the visual cortex, Lansing (1964) reported that binocular rivalry is accompanied by changes in evoked potentials in human subjects. A 50°-wide illuminated area flashing at 8 Hz was presented continuously to the left eye, and a steady striped pattern was presented for periods of 5 s to the right eye. When the pattern was present, subjects reported that it dominated the flickering field. When this happened, there was an 82% reduction in visual evoked-potentials synchronized with the flickering field. Similar findings were reported by Lehmann and Fender (1967, 1968) and by MacKay (1968). A patient with sagittal split of the chiasma showed no change in the response to monocular flicker when the other eye was stimulated by a structured stimulus (Lehmann and Fender 1969).

Brown and Norcia (1997) presented a 2 cpd, 12°-diameter grating to each eye, either with the same orientation or along opposite diagonals. One grating oscillated at 5.5 Hz and the other at 6.6 Hz. Because of these distinct temporal labels, the VEP evoked by each grating could be recovered from the VEP spectrum. When the gratings were orthogonal, the responses from the two eyes alternately waxed and waned in synchrony with reports of rivalry. The alternation of the VEPs disappeared when the gratings were aligned. The effect was not evident with a 2°-diameter display, although Lawwill and Biersdorf (1968) had obtained similar results with orthogonal 3° gratings superimposed on a 12° steady surround. Others used only a small display and failed to find a connection between rivalry and the VEP (Cobb and Morton 1967; Riggs and Whittle 1967; Martin 1970).

Srinivasan, R. *et al.* (1999) presented a red 13° vertical grating to one eye and a blue horizontal grating to the other eye, each flickering at a distinct frequency. Neuromagnetic responses associated with a particular grating increased coherently in widely separated areas of both cerebral cortices when the subject reported that that grating was dominant. The evoked potential procedure allows one to track rivalry in animals or preverbal humans.

Kaernbach *et al.* (1999) recorded VEPs when dichoptic gratings changed from being orthogonal to being congruent. When the congruent gratings were parallel to the member of the rivalling gratings that was dominant when the gratings changed from orthogonal to parallel, subjects did not perceive a change in stimulus orientation. When the congruent gratings were orthogonal to the dominant grating, subjects perceived a change in orientation. The N1 component of the VEP, which is evoked in extrastriate cortex, changed only when subjects perceived a change in orientation. Binocular rivalry in infants is discussed in Section 11.4.1a.

7.10.2b *Rivalry and the MRI*

Polonsky *et al.* (2000) measured fMRI signals from V1, V2, V3, and V4 of the human brain while subjects experienced rivalry between a low-contrast grating and an orthogonal high-contrast grating. In each area, activity increased when subjects reported that the high-contrast grating was dominant. The increase was about half as large as that observed when the monocular images were presented alternately without rivalry. These results support the idea that the neural events underlying rivalry occur at several levels, starting at V1.

Lumer *et al.* (1998) measured brain activity with MRI while human subjects reported changes in rivalry between a grating in one eye and a face in the other. Modulations of MRI signals corresponding to perceptual changes occurred in the fusiform gyrus, the right inferior and superior parietal areas, and bilaterally in the inferior and middle frontal areas and insular cortex. Similarly, Tong *et al.* (1998) found that MRI response from the human fusiform 'face' area increased when a face presented to one eye was dominant, and the response from the hippocampus increased when the picture of a house presented to the other eye was dominant.

Lumer *et al.* found no rivalry-induced modulations of response in V1 and concluded that rivalry occurs after analysis of monocular stimuli in V1. But rivalry may not show in the MRI from V1 because overall activity in V1 is probably just as high, whichever eye is dominant, it is only a question of how the activity is distributed over binocular cells. In other words, Lumer *et al.* were probably not recording the site where rivalry occurs but rather the relative levels of activity triggered by the contents of the stimuli after rivalry had been achieved. At higher levels of the nervous system, where the perceptual contents of the rivalling images are analysed in distinct areas, the level of activity will depend on the nature and complexity of the stimulus. For example, cells that respond to faces become active when the face is dominant, just as they do when attention is voluntarily directed to a face (Wojciulik *et al.* 1998).

Tong and Engel (2001) recorded the MRI response from the region of V1 corresponding to the blind spot—a region that receives inputs from only the ipsilateral eye. A response occurred when a grating in the ipsilateral eye was dominant but not when a grating in the contralateral eye was dominant. They concluded that rivalry occurs primarily in V1.

7.10.2c *Rivalry and single cell responses*

The response of a cell in the visual cortex to an optimally oriented bar or grating is suppressed by the superimposition of an orthogonal bar or grating in the same eye (Bishop *et al.* 1973; Morrone *et al.* 1982; Bonds 1989). This is cross-orientation inhibition mentioned in Section 7.3.10. It operates over a wide difference in spatial frequency between the gratings and increases with the contrast of the superimposed grating (Snowden and Hammett 1992). Ferster (1987) suggested that the effect originates in the lateral geniculate nucleus, but other evidence suggests that it results from intracortical inhibition (DeAngelis *et al.* 1992; Morrone *et al.* 1987).

Cross-orientation inhibition is a likely candidate for the mechanism of binocular rivalry and dichoptic masking (Legge 1984a). Two findings are against this view. First, cross-orientation inhibition is as strong with monocular as with binocular viewing (Walker *et al.* 1998). Second, it is not elicited when orthogonal bars or gratings are presented to different eyes of anaesthetized cats, even though it is evident when both stimuli are presented to the same eye (Burns and Pritchard 1968; Ferster 1981; DeAngelis *et al.* 1992). This suggests that cross-orientation inhibition is generated before the signals from the two eyes are combined and is not the basis for binocular rivalry.

On the other hand, Sengpiel and Blakemore (1994) found that the response of binocular cells in area 17 of anaesthetized cats to a grating presented in its preferred orientation to the dominant eye diminished when an orthogonal grating was suddenly presented to the other eye. This suppression was replaced by facilitation when the gratings had the same orientation. In strabismic cats, interocular suppression occurred at all relative orientations of the gratings (see also Sengpiel *et al.* 1994, 1995a). Suppression was not evident in monocular cells in layer 4 of the visual cortex, before the level of binocular cells.

Sengpiel *et al.* (1995b) presented an optimally orientated drifting grating to one eye of normal cats while an induction grating of variable orientation and spatial frequency was presented intermittently to the other eye. When the spatial frequency of the induction grating was too high or too low to produce an excitatory response, it continued to produce interocular suppression, which occurred equally at all relative orientations of the two gratings. They concluded that binocular interaction is produced by the sum of facilitation for stimuli of similar orientation and by suppression that is independent of orientation. Logothetis (1998) pointed out that Sengpiel and Blakemore found a large number of cells in area 17 of the cat exhibiting orientation specific interocular suppression only when the cells had been preadapted to their preferred orientation, and that the short exposures that they used were not typical of conditions under which rivalry occurs.

Sengpiel *et al.* (1998) measured the reduction in response of cortical cells to a test grating of variable contrast presented to the dominant eye when an orthogonal grating was superimposed in either the same eye or the other eye. Same-eye inhibition caused the function relating response to the contrast of the test grating to be shifted to the right, indicating a loss of contrast sensitivity. There was also some downward shift of the function. Opposite-eye inhibition caused mainly a downward shift of the contrast-response function, indicating a reduction of response. Loss of contrast sensitivity was more evident in cortical layer 4, which suggests that inhibition underlying binocular suppression originates in part in the LGN. Reduction in response was more evident in layers outside layer 4.

Dendrites from the axons of pyramidal cells in area 17 project horizontally for up to 8 mm within layers 2, 3, and 5. This represents several receptive-field diameters. These dendrites produce spaced clusters of predominantly excitatory synapses that link cells with similar orientation preference (Section

5.4.7a). These lateral connections are therefore unlikely to serve as a basis for binocular rivalry, which occurs between stimuli differing in orientation. They could build large receptive-field units that respond to lines in a particular orientation, or they could modify the response of cells according to the nature of surrounding stimuli (Gilbert *et al.* 1991). Binocular rivalry is presumably served by shorter inhibitory connections between cortical cells (Section 5.4.7).

Logothetis and Schall (1989) trained monkeys to press one key when a display moved to the left and another key when it moved to the right. When shown dichoptic displays moving in opposite directions, the monkeys changed their response, as first one and then the other display became dominant. At the same time, the experimenters recorded the activity of motion-sensitive binocular cells in the superior temporal sulcus, probably in MT. Some cells that normally responded to a given direction of motion responded only when the monkey indicated that the stimulus moving in that direction was dominant. However, most cells remained unaffected by rivalry. In a later study, alert monkeys were presented with orthogonal gratings and responded according to whether they saw the vertical or horizontal grating (Leopold and Logothetis 1996). About 20% of cells in V1 and V2, and 38% of cells in V4 increased their activity when the orientation that the animal saw corresponded to the preferred orientation of the cell. A few cells in V4 responded when the orientation of the suppressed image corresponded to the preferred orientation of the cell. Almost all these cells were binocular cells. These results suggest that rivalry occurs between only a minority of binocular cells at these levels in the nervous system.

On the other hand, Sheinberg and Logothetis (1997) found that the response of 90% of cells in the inferior temporal cortex and superior temporal sulcus of the monkey was contingent on the perceptual dominance of the effective stimulus. The rivalling stimuli in this study were a radial pattern in one eye and the picture of an animal in the other. Thus, as one proceeds to higher levels of the ventral visual pathway—the pathway involved in pattern recognition—rivalry involves a greater proportion of neurones tuned to a particular complex stimulus. This would explain why images in their suppressed phase can influence certain visual processes occurring at lower levels of the visual system (Section 7.6). Crick (1996) has speculated that these results suggest that the seat of visual consciousness is at a higher level than V1.

Other evidence suggests that binocular rivalry involves a change in synchrony of cortical responses rather than a change in response amplitude. Fries *et al.* (1997) recorded simultaneously from many cells in area 17 of awake strabismic cats. The direction of optokinetic eye movements indicated which of two dichoptic gratings moving in counterphase was dominant at any time. Neurones that fired in synchrony to a grating presented to one eye continued to do so when that eye was dominant in the rivalry condition. The activity of neurones responsive to the stimulus that was not dominant became desynchronized. Changes in eye dominance were not associated with changes in the rates of discharge of cortical cells. They suggested that the change in firing rate found by Sengpiel and Blakemore was due to their use of anaesthetized animals and reflected the presence of rival stimuli rather than the outcome of rivalry.

A high-contrast grating in one eye suppresses the response to a low-contrast grating in the other eye, when the two gratings are set at a small angle to each other. Binocular cells in areas 17 and 18 of the cat modulated their firing rate in response to a 4.8-Hz phase reversal in the luminance of a low-contrast grating presented to one eye when a homogeneous field was present in the other eye. When a high-contrast grating was added to the other eye, however, the response modulation due to the low-contrast grating was no longer present (Berardi *et al.* 1986). These effects may have more to do with dichoptic masking, which occurs between similar patterns, than with binocular rivalry, which occurs between distinct patterns.

Miller *et al.* (2000) proposed that rivalry occurs at a high level of visual processing and that the two rivalling percepts are processed in opposite hemispheres. According to this view, rivalry is between hemispheres rather than between eyes. In support of the theory they used rivalling vertical and horizontal lines. They reported that activation of one hemisphere by caloric stimulation of one vestibular system or transcranial magnetic stimulation altered the duration of one percept relative to the other. However, O'Shea and Corballis (2000) found that binocular rivalry occurs in callosotomized (split brain) human observers and that it is similar in the two hemispheres. This suggests that rivalry occurs at a low level in the nervous system and is not due to switching between hemispheres.

Summary

No firm conclusions can yet be drawn about the physiological mechanisms underlying binocular rivalry. It probably occurs at more than one level of the nervous system and probably involves more than one process. At lower levels, rivalry is determined by simple properties of the images, such as contrast, motion, and orientation. Rivalry may also be determined by higher visual features such as figural grouping, continuity and meaning. Rivalry is a constant feature of images produced by natural scenes. Images from distinct objects outside the plane of convergence fall on corresponding retinal locations and therefore rival. Perhaps high-level rivalry processes help to preserve an impression of coherence in these rival images.

8 *Binocular summation, masking, and transfer*

8.1 BINOCULAR SUMMATION

8.1.1 Summation of contrast detection

There is said to be **binocular summation** when the task of detecting a stimulus or discriminating between two stimuli is performed better with two eyes than with one. For example, visual resolution measured with a high-contrast grating is slightly higher with binocular than with monocular viewing (Blake and Fox 1973). The suggested causes of binocular summation fall into three categories: (1) monocular-binocular difference in low-level factors such as fixation, accommodation, pupil size, or rivalry, (2) probability summation, and (3) neural summation of signals from the two eyes.

Blake and Fox (1973) thoroughly reviewed the work on binocular summation to 1972 and Blake *et al.* (1981a) extended the review to 1980.

8.1.1a *Low level factors*

The following low-level factors may contribute to binocular summation.

1. The pupil of one eye, illuminated at a given luminance, constricts more when the other eye is also illuminated compared with when the other eye is in darkness (Thomson 1947). Thus, there is binocular summation in the subcortical centres controlling pupil size.

Changes in pupil size must be taken into account when comparing monocular and binocular detection or acuity. A reduction in pupil size increases diffraction, which reduces acuity. However, a smaller pupil produces less spherical aberration, which increases acuity. The monocular-binocular difference in acuity was reduced when an artificial pupil removed the effects of changing pupil size (Horowitz 1949). Another way to eliminate effects of changes of pupil size is to illuminate the nontested eye at the same mean luminance as the eye being tested.

2. Fixation may be more steady with binocular vision than with monocular vision. The binocular superiority of acuity shows only when the images in the two eyes are in good register. Binocular Snellen acuity fell to the level of monocular acuity when fixation disparities were induced by placing a prism before one eye (Jenkins *et al.* 1992). Acuity improved when naturally occurring fixation disparities were reduced by a prism (Jenkins *et al.* 1994).

3. If the axis of astigmatism is different in the two eyes, grating acuity may be better with binocular viewing than when the grating is viewed with either eye alone. There does not seem to be any systematic evidence on this possibility.

8.1.1b *Probability summation*

Pirenne (1943) pointed out that at least part of the monocular-binocular difference in stimulus detection may be explained by the statistical advantage of having two detectors. The probability of detecting a stimulus using both eyes (P_b) relative to detecting it with either eye alone (P_l and P_r) is given by

$$P_b = (P_l + P_r) - P_l P_r \qquad (1)$$

For example, if the probabilities P_l and P_r are 0.5, P_b = 0.75—an improvement of 50%. For the same reason, one is more likely to get at least one head by throwing two coins rather than one. This is **classical probability summation**. Bárány (1946) proposed the same idea independently. Dutour had expressed a similar idea in 1763. He argued that a piece of paper seen by both eyes appears brighter than when viewed with one eye only because points not well registered by one eye may be registered by the other.

There has been some debate about the form of the probability function most appropriate for understanding binocular summation. Eriksen (1966) pointed out that formula (1) does not make proper allowance for guessing behaviour and Eriksen *et al.* (1966) proposed a version to take this factor into account. Another problem is that the calculation is invalid if judgments made with one eye are not independent of those made with the other. For instance, if the thresholds in the two eyes fluctuate together because of some central process, such as fatigue or inattention, then the assumption of independence is violated. If the spatial and temporal correlation between noise-related activity were 1 then there would be no statistical advantage in having two eyes. Even a weak correlation between noise-related activity of different neurones can restrict the statistical advantage of probability summation (Zohary *et al.* 1994b).

8.1.1c *Empirical probability summation*

Theoretical complications in defining probability summation can be side-stepped by measuring the contribution of probability summation empirically and using this to assess the contribution of other factors. This is done by measuring the effect of one stimulus on another stimulus when they are separated spatially or in time. It is assumed that probability summation occurs for well-separated stimuli but that true neural summation does not. Empirically, binocular summation is reduced to the level of probability summation when the time interval between two brief stimuli is increased to more than about 100 ms (Matin 1962; Thorn and Boynton 1974). Temporal aspects of binocular summation are discussed in more detail in Section 8.1.5.

In the preceding analysis, it was assumed that signals from the two eyes do not combine before a signal denoting the presence or absence of each monocular stimulus has been generated, and that a central decision process has access only to these independently processed signals. Now consider what might happen if the neural signals are combined before a decision about the presence of the stimulus is made.

8.1.1d *Linear summation of dichoptic inputs*

Assume a simple linear summation of neural signals and a source of internal noise that is independent of stimulus strength. Combining two weak stimuli in the same area of one eye doubles the probability of detection because signal strength doubles at the level of the generator potential within the linear range, while internal noise stays the same. Thus, the signal-to-noise ratio is doubled. When signals from two eyes are combined in the brain, trains of discrete nerve impulses combine, not generator potentials. Suppose that two stimuli are presented dichoptically and the neural signals sum linearly at a central site. If the noise in the two eyes is perfectly correlated, neural summation confers no advantage because, although signals due to the stimuli add, so do those due to noise, leaving the signal-to-noise ratio the same. If the noise in the two eyes is uncorrelated, neural summation provides an advantage because two uncorrelated noise sources partially cancel when combined. When two equal stimuli with equal but uncorrelated noise are combined, the noise level increases $\sqrt{2}$ times while signal strength doubles. If neural signals from the two eyes combine this way, then binocular signal-to-noise ratio is $2/\sqrt{2}$, so that binocular sensitivity should be $\sqrt{2}$ times monocular sensitivity (Campbell and Green 1965). With neural summation of monocular signals no classical probability summation occurs, because there are no independent decision processes.

Campbell and Green realized that their analysis rested on the assumption that noise does not arise from a closed or evenly illuminated eye. If it did, binocular performance would be twice as good as monocular performance, since the same binocular noise would be present in both cases, with the signal arising from the binocular stimulus being twice as strong as that from the monocular stimulus. Evidence reviewed in the last chapter suggests that a contour in one eye suppresses activity arising from a corresponding area of even illumination in the other eye. If this is true, then noise from an eye lacking contoured stimuli should be either attenuated or completely switched off. However, the only way to be sure that an unstimulated eye has no effect is to pressure blind it. Evidence reviewed in Section 8.1.5 suggests that noise does arise in the closed eye. It was also assumed in Campbell and Green's analysis that there is no internal noise peculiar to the channel in which the signals from the two eyes are summed.

A mechanism that sums the inputs from the two eyes can gain an advantage greater than √2 over the single-eye performance in the following three ways.

1. The binocular advantage would be 2 if there were no internal noise before the summation of signals and no severe saturation effects.
2. The binocular advantage would be greater than √2 if nerve impulses below the sensation threshold summed to a suprathreshold value at a central site. Light quanta are summed at the level of the receptor generator potential before nerve impulses are generated. Under ideal experimental conditions, noise-free stimuli are summed completely within the limits set by Bloch's law of temporal summation and Ricco's law of spatial summation (Schwarz 1993). But for this process to operate at a central site one would have to assume that a stimulus strong enough to generate nerve impulses in one eye would be subthreshold for detection at a higher level when only one eye is open.
3. There could be a facilitatory nonlinear summation of inputs from the two eyes. A binocular AND-gate mechanism works this way since it responds only to signals arriving simultaneously from both eyes. If all binocular cells were AND-gates we would see nothing unless both eyes were open. This would be superadditivity. Some binocular cells respond only when both eyes are stimulated (Grüsser and Grüsser-Cornehls 1965), and it may be these cells that determine the level of binocular summation. On the other hand, there are binocular cells that are excited by inputs from one eye and inhibited by inputs from the other. These cells would counteract the influence of the AND cells. There are reasons for believing that AND cells respond best to similar inputs from the two eyes and that inhibitory influences are strongest when the inputs differ. We will see that the advantage of binocular vision over monocular vision is greatest when the stimuli presented to the eyes are similar in shape, size, and contrast.

8.1.1e *Signal-detection theory and binocular summation*

An account of signal summation has also been provided in terms of signal-detection theory. In signal-detection theory d' is the criterion-free measure of the detectability of a stimulus, defined as the mean fluctuation of noise plus neural signal, minus the mean fluctuation of noise, divided by the common standard deviation of the two fluctuations. When two independent detectors with the same variance of signal and of noise are exposed to the same stimulus, the joint detectability of the summed stimulus d'_b is related to the detectabilities of the stimulus in each detector acting alone, d'_l and d'_r, by

$$d'_b = \{(d'_l)^2 + (d'_r)^2\}^{1/2} \qquad (2)$$

This is equivalent to saying that the precision with which the mean of a population is estimated increases in proportion to the square root of the number of observations. Under optimal conditions, this formulation gives the same √2 advantage of two eyes over one eye predicted by Campbell and Green's formulation. Green and Swets (1966) referred to this formulation as the **integration model** because it is based on the idea that signals are perfectly summed before a detection decision is made. It is a model of neural summation rather than of probability summation because, in probability summation, all signal-to-noise processing is done in each eye and only "yes" or "no" signals are finally combined. Guth (1971) argued that when the probability of correct detection for each eye (P_m), is the same, the probability of correct detection with both eyes, (P_b) is

$$P_b = P_m + d(P_m \quad 0.5) \qquad (3)$$

where d is the difference between the miss rate and the false-alarm rate. With $P_m > 0.5$, it follows from this equation that when the false-alarm rate is smaller than the miss rate, $P_b > P_m$, but when the miss rate is smaller than the false alarm rate, $P_b < P_m$. Guth also argued that whether two eyes perform better than one eye depends on the relative performance of the two eyes and on the relative frequencies of no-signal catch trials to signal trials.

In this discussion it has been assumed that neural signals from each eye reach the brain along a single channel and that signal and noise sum algebraically, at least from spatially congruent contoured stimuli. There is some support for the idea that uncorrelated external noise in the two eyes sums algebraically (Braccini *et al.* 1980). However, visual inputs are grouped into different channels defined by colour, size, and luminance-polarity, each with its own source of noise, and inputs combine by both summation and inhibition into partially distinct mechanisms in the visual cortex. We will see that several models of these processes have been proposed.

8.1.1f *Binocular summation of luminance and contrast*

Lythgoe and Phillips (1938) found that the monocular luminance threshold for detection of a white disc 12.5° in diameter was 1.4 times the binocular

Figure 8.1. Gordon E. Legge.
Born in Toronto in 1948. He obtained a B.Sc. in Physics at MIT in 1971 and Ph.D. in Experimental Psychology at Harvard in 1975. He conducted postdoctoral work with Fergus Campbell at Cambridge University. In 1977 he joined the faculty of the University of Minnesota, where he is now Professor of Psychology and Neuroscience and Director of the Minnesota Laboratory for Low-Vision Research.

threshold at all times during 20 minutes of dark adaptation. Crawford (1940a) obtained similar results for detection but found substantial binocular summation for brightness discrimination only outside the fovea. Campbell and Green (1965) measured the contrast sensitivity (reciprocal of threshold contrast) for a sinusoidal grating of various spatial frequencies. The pupils were atropinized and stimuli were viewed through artificial pupils. In monocular testing, the nontested eye viewed a diffuse field with the same mean luminance as the display in the tested eye. Binocular sensitivity was $\sqrt{2}$ higher than monocular sensitivity, in conformity with simple summation of signals from the two eyes, as mentioned previously. With spatial frequency the same in the two eyes, the ratio of monocular to binocular contrast sensitivity was constant over the visible range of spatial frequencies, a result confirmed by Blake and Levinson (1977).

In these experiments, and in others mentioned later, contrast is Michelson contrast, as defined in Figure 5.7. As the Michelson contrast of a grating is varied, its mean luminance remains constant. When the mean luminance of a monocular grating was doubled, Campbell and Green found that contrast sensitivity increased by a ratio of only 1.17. We can explain this low ratio by saying that, although the rate of neural firing increases when luminance increases, sensitivity to a difference in luminance (contrast) does not increase in proportion, because the differential threshold increases with increasing luminance. Thus, it is not signals representing luminance that are summed in binocular summation of contrast, because this would not produce the observed improvement in contrast sensitivity. Instead, signals representing contrast are derived in each eye, and it is these signals that sum. The processes of lateral inhibition in the retina generate signals related to contrast that are relatively unaffected by changes in the level of illumination The binocular advantage can be explained if it is assumed that contrast signals sum and that noise sums only by $\sqrt{2}$. There is other evidence that signals representing luminance do not sum binocularly (see Section 8.1.3).

Binocular summation is reduced when the stimuli in the two eyes are spatially separated. Thus, the detectability of a small flashed target fell to the level of summation defined by Green and Swets' integration model when the disparity of a flashed target relative to a fused stimulus was larger than about 20 arcmin. This is a disparity at about the limit of binocular fusion (Westendorf and Fox 1977). Binocular summation of low-contrast gratings, as reflected in the reaction time for detection, fell to the level of probability summation as disparity increased beyond the limits of Panum's fusional area (Harwerth *et al.* 1980). The disparity limit of binocular fusion increases as the spatial frequency of the stimulus is reduced (Section 7.1.2). The range of disparities over which binocular summation occurred showed a similar dependence on spatial frequency (Rose *et al.* 1988). It seems that binocular summation above the level of probability summation occurs only between dichoptic stimuli that are close enough to fuse.

The tuning characteristics of a binocular cell for orientation, spatial-frequency, and other stimulus features are fundamentally the same for stimuli presented to each eye. It is therefore not surprising that interocular summation of contrast sensitivity occurs only for stimuli with similar orientations and spatial frequencies (Julesz and Miller 1975; Westendorf and Fox 1975; Blake and Levinson 1977), similar directions of motion and temporal properties (Blake and Rush 1980), similar clarity of focus (Harwerth and Smith 1985), and similar wavelength (Trick and Guth 1980).

Wolf and Zigler (1963, 1965) measured the detectability of a 1° test patch at various positions on a circle with a radius of 10° around the fovea. Detectability was greater for binocular than for monocular viewing except when the test patch fell on the midvertical meridian. They argued that the two halves of a test patch on the midvertical meridian project to opposite cerebral hemispheres so that, although each hemisphere receives inputs from both eyes, it receives only half the total area. With increasing

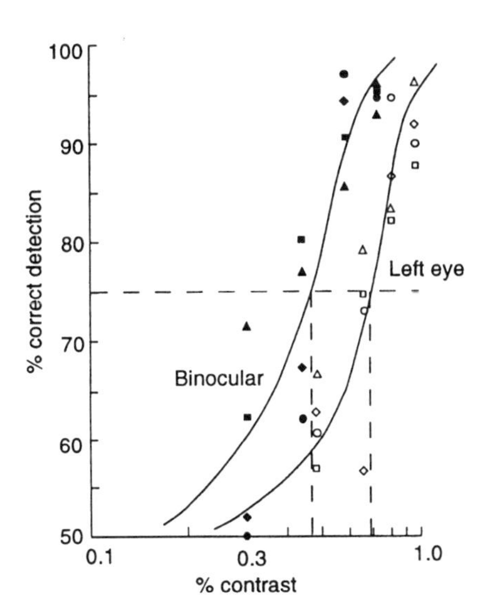

(a) Psychometric functions showing percent correct detection of a 0.5 cpd grating as a function of its contrast, for monocular and binocular viewing. For 75% detection, the monocular contrast threshold is about 1.5 times the binocular threshold, as indicated by the separation between the vertical lines. At the inflection points, both functions have a slope of about 2.

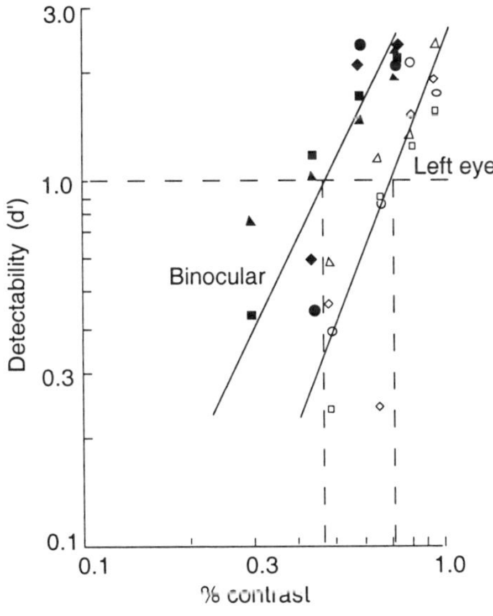

(b) The same functions plotted on log-log co-ordinates with percent detection expressed as detectability, d'. For d' of 1.0 (76% detection) horizontal separation of the functions indicates that the monocular contrast threshold is about 1.5 times the binocular threshold. Vertical distance between functions indicates that, for a given contrast, detectability of a binocular grating is about twice that of a monocular grating. (Redrawn from Legge 1984a)

Figure 8.2. Grating detection as a function of contrast.

horizontal distance from the fovea, binocular summation decreased with targets subtending 0.1° but increased with targets subtending 1.7° (Wood *et al.* 1992). Thus, binocular summation is greater when the size of the stimulus is matched to the size of receptive fields.

In summary, binocular summation of contrast is greater for stimuli that are close together in space and in time, have similar stimulus characteristics, and stimulate the same cerebral hemisphere.

Legge (1984a) (Portrait Figure 8.1) found that the monocular contrast-detection threshold for a 0.5 cpd sine-wave grating was about 1.5 times the binocular threshold (similar to the value reported by Campbell and Green). Thus, a monocular grating must have about 50% more contrast than a binocular gating to be as visible. Monocular and binocular psychometric functions had the same maximum slope of about 2 (Figure 8.2a). The horizontal separation of the two functions indicates the difference in threshold for a given % of detection. In this example, for a detection rate of 75%, the monocular threshold is 1.4 times the binocular threshold. The vertical separation of the functions indicates the % difference in detection for a given contrast. The functions in Figure 8.2b are plotted on log-log co-ordinates, and % detection is converted into the detectability measure, d'. This reveals that when both curves have a slope of 2, the ratio of binocular to monocular detectability for a given contrast is 2 to 1, as indicated by the vertical separation of the functions. For functions with slope 1, the ratio of thresholds is the same as the ratio of detectabilities. In general, the detectability (d') for a grating of contrast C presented to one eye is:

$$d' = \left(\frac{C}{C'}\right)^n \qquad (4)$$

where C' is the threshold contrast (at 76% correct detection), and n is the slope of the psychometric function, which was 2 in Legge's experiment.

Thus, when the psychometric function for detection has a slope of 2, the monocular contrast threshold is 1.4 times the binocular threshold, as predicted by neural summation. However, the detectability of a threshold binocular grating is twice that of a monocular grating at the monocular threshold. According to equation (2), neural summation predicts that binocular and monocular detectabilities should have a ratio of 1.4 to 1. The difference between the ratio of contrast-detection thresholds and the ratio of detectabilities arises because the detectability of a stimulus is not a linear function of its contrast.

Legge (1984b) proposed that the effective binocular contrast of a grating (C_b) is the quadratic sum of the monocular contrasts (C_l and C_r) or

$$C_b = \sqrt{(C_l)^2 + (C_r)^2} \qquad (5)$$

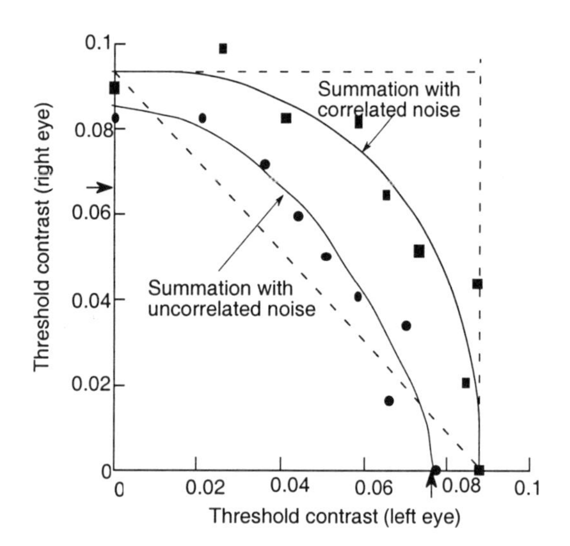

Figure 8.3. Binocular summation contours.
Detection of a dichoptic grating in the presence of uncorrelated and correlated noise. For each datum point, interocular contrast ratio was fixed and subjects adjusted the contrast of both dichoptic gratings until the fused grating was visible. Arrows on the axes indicate thresholds for monocular stimuli masked by monocular noise. The diagonal dashed line indicates perfect linear summation of binocular signals with correlated noise. Dashed lines parallel to the axes indicate loci of probability summation (N-1). (Adapted from Anderson and Movshon 1989)

When dichoptic stimuli have the same contrast and the eyes have the same threshold, the binocular contrast threshold is √2 times the monocular threshold. The quadratic summation rule assumes that the stimuli are the same except in contrast. Presumably, the model could be generalized to accommodate other differences between the stimuli by adding weighting functions to the two monocular contrasts.

Legge made the luminance of a grating different in the two eyes by placing a neutral-density filter over one eye. The degree of binocular summation decreased until, when the difference in contrast was considerable, the more luminous grating became less detectable than when the other eye was closed. This is the contrast-detection analogue to Fechner's paradox described in Section 8.1.3. The reduction in binocular summation as a function of the reduction of contrast in one eye was constant across the visible range of contrasts and spatial frequencies (Gilchrist and Pardhan 1987; Pardhan *et al.* 1989).

Anderson and Movshon (1989) used superimposed vertical sinusoidal gratings with the same luminance, phase, and spatial frequency in the two eyes. In each trial, the interocular contrast ratio was set at some value and the subject adjusted the contrast of both stimuli, keeping the ratio the same, until the grating was visible. Data for one subject are shown in Figure 8.3. The threshold contrasts of the two monocular gratings for each interocular contrast ratio define a **binocular summation contour**. For perfect linear summation of binocular signals, the contour should fall on a diagonal line, and for complete independence, with no probability summation, it should fall along lines parallel to each axis, as shown in Figure 8.3. In fact, the data fell between these two limits. The data were fitted with the following power equation:

$$\left(\frac{m_l}{\alpha_l}\right)^{\sigma} + \left(\frac{m_r}{\alpha_r}\right)^{\sigma} = 1 \qquad (6)$$

where m_l and m_r are the threshold contrasts of the left- and right-eye gratings when presented as a dichoptic pair, α_l and α_r are the contrast thresholds of each grating measured separately, and σ is a parameter inversely related to the magnitude of binocular summation. When the monocular contrasts are equal and $\sigma = 2$, the formula is equivalent to Legge's quadratic summation formula and the ratio of binocular to monocular thresholds is √2. The mean value of σ was close to 2.

Anderson and Movshon argued that, if binocular summation represents the combined action of several visual mechanisms, it should be possible to probe the contribution of each mechanism by selective masking or by adaptation. They measured the binocular summation contour when noise was added to the two dichoptic gratings. In one condition the noise was the same in both eyes (correlated) and in another condition it was uncorrelated. When contrast was similar in the two eyes, the threshold with uncorrelated noise was about √2 lower than that with correlated noise, as one would predict from Campbell and Green's formula for neural summation. These results agree with those of Braccini *et al.* (1980). Pardhan and Rose (1999) also obtained similar results and, in addition, found that binocular summation decreases with increasing levels of both correlated and uncorrelated noise.

The greater effect of uncorrelated noise can be explained as follows. Correlated noise stimulates the same zero-disparity detectors used to detect the zero-disparity grating whereas uncorrelated noise stimulates a variety of disparity detectors because the stimulus elements combine in various ways to produce lacy depth. Thus, it is easier to detect a grating in uncorrelated than in correlated noise.

Anderson and Movshon found that, as the contrasts in the dichoptic stimuli became more different, the difference between correlated and uncorrelated noise became smaller. This result can be explained in the following way. When the contrasts in the two eyes are similar, binocular cells that summate inputs from the two eyes are maximally stimulated, and signals and noise summate to give an advantage to inputs with uncorrelated noise. But when interocular contrasts differ, the summation mechanism is turned off and mutual inhibition responsible for binocular rivalry is turned on. The eye with the stronger signal now suppress the other eye so that, in the extreme case, signal and noise from only one eye are available. Under these circumstances, it makes no difference whether the noise in the two eyes is correlated or uncorrelated. Actually, as I have already argued, some such mechanism must be assumed in the Campbell and Green model to account for why noise from a closed eye does not affect the monocular contrast threshold.

Anderson and Movshon produced evidence that there are several ocular-dominance channels, each of which can be selectively adapted. They dubbed this the distribution model of binocular summation. The general form of the binocular summation contour represents the summed response of several ocular-dominance channels. The physiology of binocular summation is discussed in Section 6.7.2.

8.1.1g *Binocular summation and breaking camouflage*
Schneider and Moraglia (1994) proposed that binocular summation could enhance the visibility of a fixated object relative to that of a binocularly disparate background in the following way. Spatial-frequency components in the background, having twice the spatial period as the disparity, would be in antiphase in the two eyes and therefore binocular summation would make them less visible. The visibility of components with the same spatial period as the disparity would be relatively enhanced by binocular summation. Overall, the spectral density function for the background should show undulations with cancellation at odd multiples of the disparity and summation at even multiples. The fused image of the visual target shows binocular summation across the whole range of spatial frequencies.

Schneider and Moraglia showed experimentally that the visibility of a fused target is enhanced when its spatial frequency falls within a furrow of the spectral density function of the disparate background. In other words, spatial-frequency components of a fused target that are odd multiples of the disparity have enhanced visibility relative to a background with mixed disparities.

8.1.1h *Binocular summation of equiluminant stimuli*
Simmons and Kingdom (1998) measured monocular and binocular detection thresholds for 0.5 cpd Gabor patches that were isochromatic or red-green equiluminant. Binocular summation for both stimuli was above the level of probability summation and was particularly high for the equiluminant stimuli. They suggested that binocular summation is high in the chromatic system because binocular inhibition is weaker in the chromatic system than in the achromatic system.

8.1.2 Summation at suprathreshold contrasts

8.1.2a *Binocular summation for contrast discrimination*
Legge (1984a) used a forced-choice procedure to measure the threshold for detection of an increment of contrast in a suprathreshold grating set at various levels of contrast. Both binocular and monocular psychometric functions at suprathreshold levels had a slope of 1. As the contrast of the grating increased to 0.25 the advantage of binocular over monocular discrimination fell to zero. Thus, the amount of binocular summation for contrast discrimination decreased as the absolute level of contrast increased. This can be explained if one assumes that, at low levels of contrast, the neural effect of a stimulus is a positively accelerating function of contrast and that, at higher levels of contrast, a saturation of response occurs. There is evidence for this assumption. For a monocularly viewed grating with a contrast of up to about 0.25, the contrast-difference threshold is much smaller than the absolute level of contrast required for detection (Nachmias and Sansbury 1974).

Bearse and Freeman (1994) obtained further evidence for this conclusion. They measured orientation discrimination for one-dimensional Gaussian patches as a function of stimulus contrast and duration. Binocular performance was 66% better than monocular performance for stimuli that were both brief (50 ms) and of low contrast (8%). When either duration or contrast was increased beyond a certain level, binocular and monocular performances became equal. These results are consistent with the results of earlier experiments in which discrimination of the orientation of high-contrast gratings was found to be similar for monocular and binocular viewing (Andrews 1967). It seems that binocular energy summation occurs in the contrast or temporal threshold region, where the response of the visual system is an accelerating function of stimulus energy, and that response saturation limits binocular summation for discrimination between stimuli well above detection threshold.

8.1.2b *Binocular summation for acuity*
Given that binocular summation of contrast sensitivity is much reduced at high contrasts and that

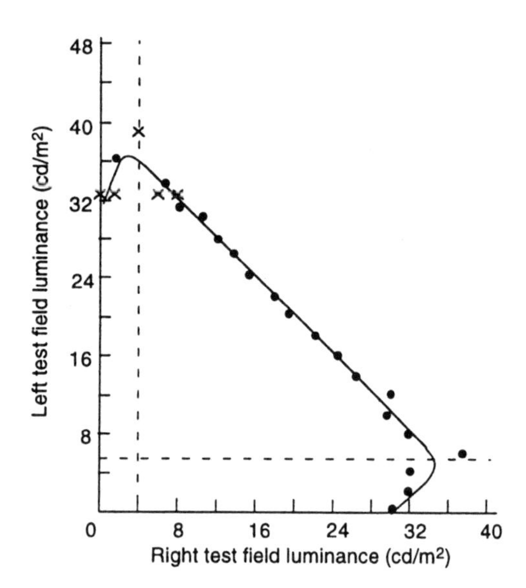

Figure 8.4. Dichoptic equal-brightness curve.
Equal-brightness curve for a 3° luminous disc presented to both eyes at a luminance of 20 cd/m² with respect to a pair of fused dichoptic discs set at various luminance ratios. Dashed lines indicate boundaries of the region in which dichoptic summation fails and Fechner's paradox is evident (N=1). (Redrawn from Levelt 1965b)

vernier acuity depends on contrast sensitivity, it is not surprising that binocular summation for hyperacuities is absent or variable at high contrasts. Home (1984) used a Landolt C and found a binocular advantage of about 40% at a contrast of 0.01 but of less than 10% at a contrast of 0.8. Banton and Levi (1991) obtained similar results for vernier acuity. Performance was between 40 and 60% better with binocular than with monocular viewing for contrasts up to about 20 times above threshold. Binocular summation declined at higher contrasts. Lindblom and Westheimer (1989) reported that binocular summation for a three-dot alignment task varied between 0 and 35% but was not much affected by an increase in dot separation up to 5 arcmin. Summation for the task of placing a high-contrast horizontal line midway between two other lines occurred only for separations of the outer lines of under 2 arcmin, probably because this task involves an element of luminance discrimination.

8.1.2c *Binocular summation for pattern recognition*
Binocular summation is near the level of probability summation for more complex visual tasks performed at a suprathreshold level of contrast, such as recognition of letters on a noisy background (Berry 1948; Carlson and Eriksen 1966; Townsend 1968; Frisén and Lindblom 1988). Cagenello *et al.* (1993) used a letter recognition task with contrasts between 0.3 and 1 and obtained a mean binocular advantage of 11% but there were wide differences between the four subjects. The mean advantage diminished as the images were made to differ in contrast.

Briefly exposed letters were recognized more accurately with binocular than with monocular presentation (Williams 1974). The advantage fell to the level of probability summation when letters fell on noncorresponding areas or were separated by intervals of more than 50 ms (Eriksen *et al.* 1966; Eriksen and Greenspon 1968). Also, this advantage was not evident in two subjects with strabismus.

As one would expect, the ability to recognize two different letters was reduced when they were superimposed dichoptically but not when they were presented to noncorresponding areas or successively (Greenspon and Eriksen 1968).

Uttal *et al.* (1995) dichoptically combined identical aircraft silhouettes, subtending 1°. One image was degraded by low-pass spatial-frequency filtering and the other by intensity averaging within local areas. Each pair of images was presented for 100 ms to facilitate binocular fusion. When two dichoptic pairs were presented sequentially, subjects could detect whether the silhouettes were the same or different aircraft with fewer errors than when two silhouettes with the same single degradation were presented monocularly. The same advantage was evident when the images with two types of degradation were physically superimposed in one eye. Thus a degraded image of an object can be seen more clearly when combined, either monocularly or dichoptically, with the same image degraded in a different way.

8.1.3 Summation of brightness

If the inputs from the two eyes sum in a simple fashion, an illuminated area should appear about twice as bright when viewed with two eyes than when viewed with only one eye. In fact, Jurin in 1755 and Fechner in 1860 observed that an illuminated area appears only slightly brighter when viewed with two eyes (see Robinson 1895; Sherrington 1904; De Silva and Bartley 1930). A bright light presented to one eye may actually appear less bright when a dim light is shone into the other eye, the effect known as Fechner's paradox.

8.1.3a *Levelt's experiments on brightness summation*
To investigate binocular brightness summation, Levelt (1965a) presented a 3° luminous disc on a dark ground to corresponding regions in each eye. The disc in one eye had a fixed luminance, and the subject adjusted the luminance of the disc in the other eye until the combined image appeared the

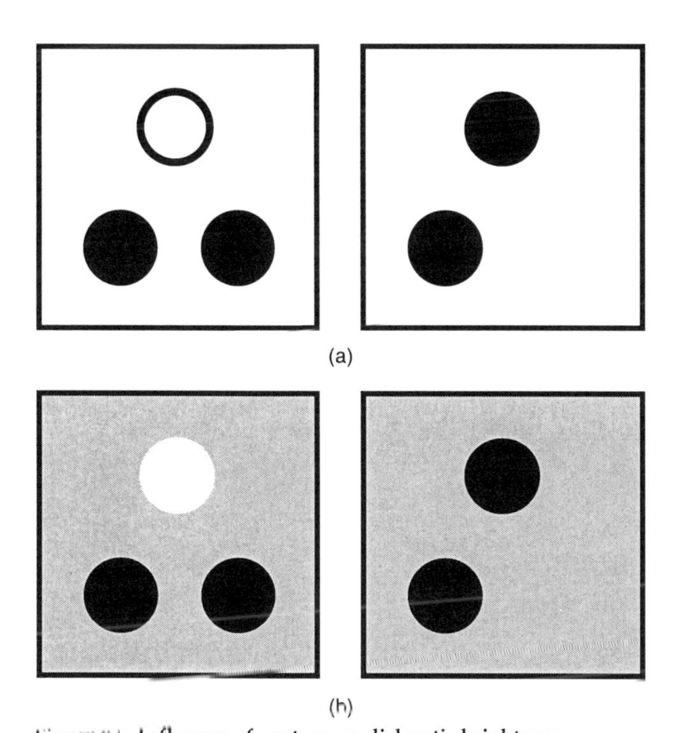

Figure 8.5. Influence of contour on dichoptic brightness.
(a) The fused black and white discs appear brighter than either the fused black discs or the monocular black disc because the strong rim round the white disc adds to the dominance of the white disc (From Levelt 1965a)
(b) The black and white discs rival because they have equally strong rims. Sometimes the fused black and white discs are as dark as the fused black discs.

same brightness as a comparison stimulus with the same fixed luminance in the two eyes. The test and comparison stimuli were presented sequentially in the centre of the visual field. The results for one subject for a comparison stimulus with a luminance of 20 cd/m^2 are shown in Figure 8.4. Over the straight part of the curve, the brightness of the comparison stimulus was equal to the mean brightness of the dichoptic test stimuli. When the test discs had the same luminance, the comparison and test stimuli were, necessarily, identical. As the luminance of the test disc was increased in one eye, it had to be decreased in the other eye by a proportionate amount to maintain the match with the comparison stimulus. Some subjects gave a greater weighting to one of the dichoptic images than to the other.

The slope of the linear function reversed when the luminance of the disc in one eye was set at or near zero, as indicated by the dashed lines in Figure 8.4. Under these circumstances, more luminance was required in the brighter disc when the dimmer disc was visible than when there was no stimulus in the other eye. This is Fechner's paradox. For the data in Figure 8.4, a luminance of about 32 cd/m^2 was required to match a monocular disc with a comparison stimulus of 20 cd/m^2, whereas a higher luminance was required when a dimly illuminated disc was visible in the other eye. Fechner's paradox can be explained by assuming that the processes underlying dichoptic brightness averaging involve both summation and inhibition (Section 8.1.5). When border contrast is similar in the two eyes, summation predominates, with greater weight given to the stimulus with greater contrast, but when the contours in the two eyes differ greatly in contrast, inhibition outweighs summation. When the contrasts are opposite in sign, inhibition becomes evident as binocular rivalry (Fry and Bartley 1933).

Inputs from a totally uncontoured region in one eye are usually suppressed by inputs from a contoured region in the other eye. When Levelt (1965b) superimposed a 2°-diameter black ring on one of the dichoptic discs, the contribution of that disc to the brightness match increased. Furthermore, in the immediate neighbourhood of a contour presented to only one eye, binocular brightness was determined wholly by the luminance in that eye. This effect of an added contour can also be understood in terms of Fechner's paradox. The influence of contour is illustrated in Figure 8.5 A white disc with a black perimeter combines dichoptically with a black disc to form a gray disc. But a black disc combines with an uncontoured white region to form a black disc, which resembles that formed by two black discs (Levelt 1965a).

One cannot be sure that a closed eye makes no contribution to a binocular match. Zero contribution from an eye can be guaranteed only if the eye is pressure blinded. This point is discussed in Section 8.1.5.

Levelt explained his results by stating that binocular brightness (B) depends on a weighted sum of the luminances of the monocular stimuli (E_l and E_r). The weights (w_l and w_r) sum to 1 and depend on the relative dominance of the two eyes and the relative strengths of the two stimuli, determined mainly by the contours they contain. Thus,

$$B = w_l E_l + w_r E_r \qquad (7)$$

This is a purely formal theory since it can describe many results if appropriate weights are selected, and there is no independent procedure for deciding the weights. Furthermore, it assumes a linear transduction of stimulus luminance and contrast into neural signals signifying brightness. Since the weights sum to 1, the formula cannot account for binocular brightness in excess of the average of the monocular luminances. De Weert and Levelt (1974)

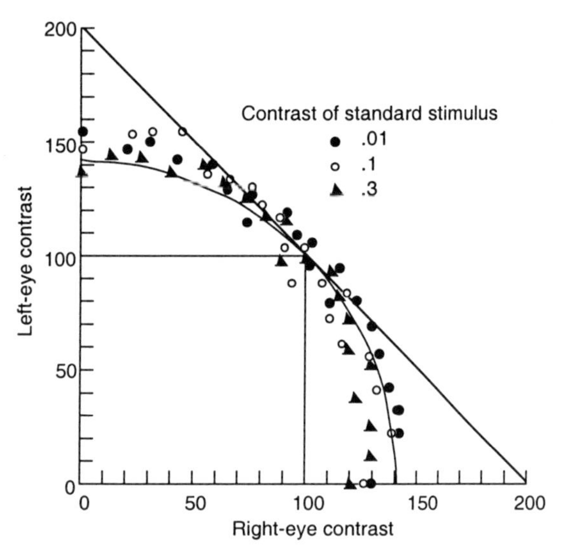

Figure 8.6. Dichoptic equal-contrast function.
Example of an equal-contrast curve for 8 cpd gratings. For each contrast of the right-eye grating the contrast of a left-eye grating was varied until the fused image appeared equal in contrast to a binocular standard grating. Both test contrasts are expressed as a percentage of that of the standard stimulus for three contrasts of the standard grating. The diagonal line is the result expected from averaging contrasts in the two eyes. The circle is the result expected from a quadratic summation rule. Horizontal and vertical lines indicate results expected when the match is determined solely by the image with the higher contrast (N=1). (Adapted from Legge and Rubin 1981)

added a parameter to the simple luminance-averaging formula to account for the fact that brightness summation is slightly better than predicted by averaging, and to account for Fechner's paradox. De Weert and Levelt (1976b) provided evidence that stimuli from the middle of the chromatic spectrum contribute more to dichoptic brightness than stimuli from either end of the spectrum.

Levelt assumed that dichoptic luminances rather than dichoptic brightnesses were averaged. Teller and Galanter (1967) held the luminance of monocular patches constant while varying their brightness, either by changing the adaptive state of the eye or the contrast between the patches and their background. In both cases the brightness of dichoptically viewed patches varied with the imposed change in monocular brightness. In particular, the level of luminance of the stimulus in one eye at which Fechner's paradox was evident did not depend on the absolute luminance of the stimulus but on its luminance relative to the luminance threshold.

8.1.3b *Other models of brightness summation*

Erwin Schrödinger (1926), as a change from his work in fundamental physics, proposed that each monocular input (f_l and f_r) is weighted by the ratio of the signal strength from that eye to the sum of the strengths of the signals from the two eyes. The binocular result, B is then given by:

$$B = f_l \frac{f_l}{f_l + f_r} + f_r \frac{f_r}{f_l + f_r} \qquad (8)$$

MacLeod (1972) added to this account by proposing that the strength of a neural signal, f, is a logarithmic transform of the stimulus contrast, as specified by

$$f = f_0 + \log\left(\frac{l}{l_0}\right) \qquad (9)$$

where f_0 is the internal noise, l is the difference in luminance across the contour, and l_0 is the threshold luminance difference. A good fit to Levelt's data in Figure 8.4 was obtained by setting $B = 1.36$, $f_0 = 0.34$, and $l_o = 2$ cd/m^2 for each eye.

Several other models of binocular brightness summation have been proposed. Engel (1967, 1969, 1970b) used a weighted quadratic sum model to account for binocular summation of brightness. In this formulation, the brightness of a binocular stimulus derived from magnitude estimations (ψ_b) is related to the brightness of monocular stimuli (ω_l and ψ_r) by the expression

$$\psi_b = \sqrt{\left(w_r \psi_r\right)^2 + \left(w_l \psi_l\right)^2} \qquad (10)$$

The weighting functions were derived from normalized autocorrelation functions of the image in each eye and reflected the amounts of contour and contrast in each image. They thus served the same function as the weights in Levelt's formula except that Engel provided a process for determining their values.

Engel's function resembles the quadratic summation model used by Legge to describe binocular summation of contrast sensitivity (equation 5), and by Legge and Rubin (1981) to describe summation of contrast in suprathreshold gratings. Legge and Rubin used the same procedure as Levelt. They presented a standard binocular grating, identical in the two eyes, and a fixed test grating in the right eye. Subjects adjusted the contrast of a test grating in the left eye until the contrast of the fused test gratings appeared the same as that of the standard grating, for different contrasts of the fixed test grating. All the gratings had the same phase and spatial frequency. An equal-contrast curve for 8 cpd gratings and three contrasts of the standard grating is shown in Figure 8.6. The results for gratings of 1 cpd were similar except that the departure from averaging

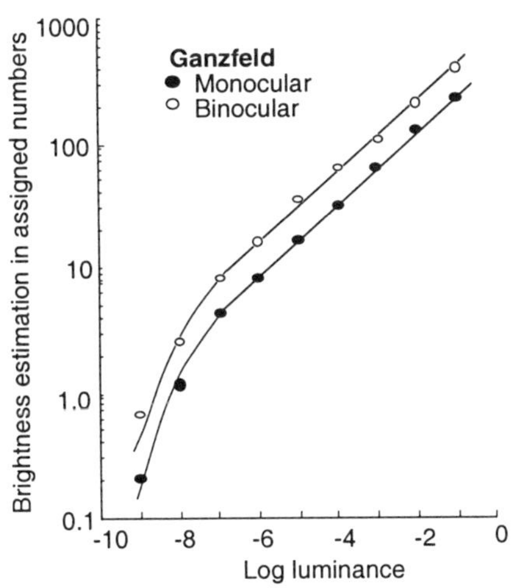

(a) Magnitude estimations of apparent brightness of a Ganzfeld presented to both eyes (upper curve) and to one eye (lower curve) as a function of log luminance.

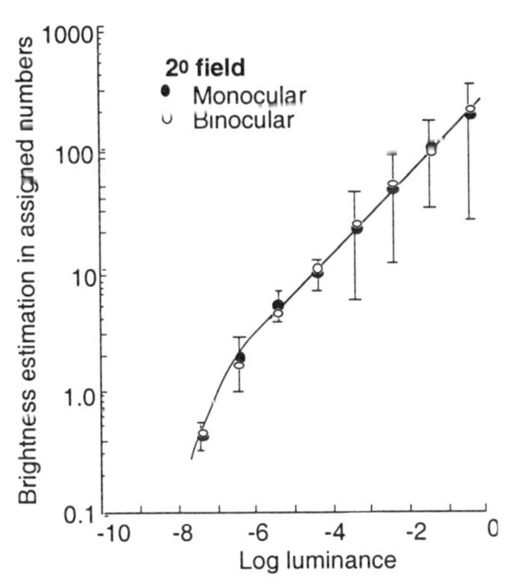

(b) Magnitude estimations of apparent brightness of a 2° spot presented to both eyes (empty symbols) and to one eye (solid symbols). Bars are standard errors (N=8). (Redrawn from Bolanowski 1987)

Figure 8.7. Dichoptic apparent brightness.

was more severe at low contrast. The results lie close to the curve representing a summation index of 2 but are well inside the diagonal, which represents an index of 1 (contrast averaging). This means that disproportionate weight is given to the grating with higher contrast.

In Levelt's averaging formula, the gratings are weighted in proportion to their contrasts. Perhaps the extra weight given to the dominant contrast arises because Legge's stimulus was a grating with many contrast borders, whereas Levelt used simple luminance discs with contours only around the edges. There is a hint of Fechner's paradox in Figure 8.6, which shows in the way some of the data points turn in towards the origin as they approach the axes. Quadratic summation implies that the contrast signal is squared in each eye before the two signals are combined with a compressive nonlinearity.

Tanner (1956) proposed that the detectabilities of single stimuli in each of two detectors sum like vectors to predict the discriminability between stimuli presented to the two detectors. According to this formulation, monocular contrasts of magnitude C_l and C_r sum like vectors to produce binocular contrast, C_b (Curtis and Rule 1978).

$$(C_b)^2 = (C_l)^2 + (C_r)^2 + 2C_l C_r \cos\text{ø} \qquad (11)$$

Cosø represents the correlation between the noise in the two eyes. When angle ø is between 90° and 120° the function reduces to averaging and accounts for Fechner's paradox. An angle of 90° signifies that contrast is detected independently in the two eyes with uncorrelated noise and the formula reduces to Legge's quadratic sum formulation. An angle of 0° signifies that binocular contrast is the simple sum of the monocular contrasts. Curtis and Rule did not propose a physiological representation of the vector-addition process. This formula contains no weightings to allow for differences between the images in the two eyes but could easily be modified to do so.

Other models of binocular brightness summation specify binocular processes that extract differences between binocular stimuli, and other processes that extract sums of binocular stimuli. For instance, in a model proposed by Lehky (1983), dichoptic stimuli with matching contours are processed in the summing channel while those with opposite luminance polarity are processed in the differencing, or rivalry, channel. Lehky also used a vector-sum formula and interpreted the angle between the vectors as the relative contributions of the summing and differencing channels. Cohn *et al.* (1981) found that stimuli in the summing channel, such as binocular increments of luminance, were selectively masked by noisy fluctuations of luminance that were correlated in the two eyes. On the other hand, signals in the differencing channel, such as a luminance increment in one eye and a decrement in the other, were selectively masked by uncorrelated noise. They argued that this evidence supports a two-process model of binocular combination.

Cogan (1987) proposed that the differencing channel receives an excitatory input from one eye and an inhibitory input from the other, and that the summing channel receives only an excitatory input from both eyes. He assumed that there are no purely monocular cells in the binocular field and that the net binocular response is the pooled output of the two channels. Sugie (1982) developed a neural network model of these processes.

Although the models proposed by Fry and Bartley, Levelt, and Engel stress the importance of contour in determining the amount of binocular brightness summation, this factor was not systematically explored. Leibowitz and Walker (1956) found that the amount of binocular summation of brightness decreased as the size of the stimulus was reduced from 1° to 15 arcmin. They attributed this effect to the increase in the proportion of contour to area, as area was decreased.

Bolanowski (1987) obtained estimates of binocular summation when all contours were removed from the visual field. He used a Ganzfeld produced by illuminating table tennis balls trimmed to fit over the eyes. Subjects rated the apparent brightness of the Ganzfeld presented for 1 s either to one eye or to both. The results for different levels of illumination are shown in Figure 8.7. When each eye received the same luminance, the apparent brightness of the binocular Ganzfeld was about twice that of the monocular Ganzfeld. Binocular summation of brightness was thus complete. When the diameter of the stimulus was reduced to 2°, binocular brightness was about the same as monocular brightness, as found by Levelt. Bourassa and Rule (1994) confirmed these results and also noted that Fechner's paradox was absent when binocular summation was measured with Ganzfeld stimuli. They also noted that summation with small stimuli with gradual borders was less than with Ganzfeld stimuli but more than with small stimuli with sharp borders.

These results can be described by Curtis and Rule's equation (11) if the angle separating the vectors increases in relation to the presence of inhibitory influences arising from sharp contours. Thus the various models of brightness summation can accommodate this result if appropriate weights are assigned to visual contours. Grossberg and Kelly (1999) developed a model of binocular brightness perception in terms of neural dynamics.

The reaction time for a button-pressing response to a flashed stimulus becomes shorter with increasing luminance. One manifestation of binocular summation is that the reaction time to a flashed binocular stimulus is shorter than the mean of the reaction times to monocular flashes (Minucci and Connors 1964). Haines (1977) reported that the reaction time to a flashed binocular stimulus was about 35 ms shorter than that to a monocular stimulus.

Another manifestation of suprathreshold binocular summation is that evoked potentials from the visual cortex are of greater magnitude when an area is flashed to both eyes rather than to one (Bartlett *et al.* 1968). A sharply focussed flashed pattern produced more binocular summation than a defocusssed pattern, except at high luminances (White and Bonelli 1970).

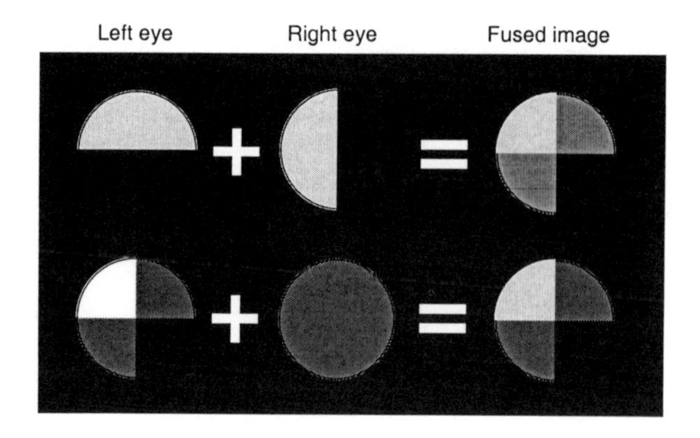

Figure 8.8. Stimuli used by Liu and Schor (1995).
Upper stimuli are dichoptic orthogonal edges which fuse to create a plaid pattern. Lower stimuli are a plaid pattern in one eye and a blank field in the other. At low contrast, the contrast of the monocular plaid pattern had to be twice that of the dichoptic patterns to produce fused images with the same apparent contrast. At high contrasts, more contrast was required in the dichoptic patterns to produce a match.

Binocular facilitation did not occur when the density of a grid pattern was different in the two eyes, nor when one eye was exposed to a grid and the other to a flashed diffuse light (Harter *et al.* 1974).

8.1.3c *Interactions between orthogonal dichoptic edges*

In the experiments reviewed so far the dichoptic stimuli were similar in orientation. In Section 7.3.2 it was mentioned that low-contrast orthogonal dichoptic gratings do not rival but rather combine to create a stable plaid pattern (see Figure 7.14). Liu and Schor (1995) assumed that, at low contrast, the contrast of each orthogonal edge in a stimulus like that shown in Figure 8.8 is preserved in the dichoptic image because the edge in one eye dominates the evenly illuminated region in the other eye. From this they predicted that the apparent contrast of a low-contrast dichoptic pattern should match that of a similar monocularly viewed pattern of twice the contrast combined with a black disc in the other eye.

Their results agreed with this prediction rather than with one based on averaging of luminance within each region of the display. At higher contrasts, the apparent contrast of the dichoptic stimulus was reduced relative to that of the monocular stimulus. This was presumably because of spreading rivalry between the orthogonal dichoptic edges. Subjects could distinguish between the dichoptic and monocular stimuli even when their contrasts matched. This could be due to the relative motion of the dichoptic images arising from instability of vergence.

8.1.4 Critical flicker fusion

The frequency at which a flickering light appears to fuse into a continuous light is the **critical fusion frequency** (CFF). With increasing luminance, the CFF increases up to a limit of about 50 Hz (Crozier and Wolf 1941). Sherrington (1904) hypothesized that, if inputs from the two eyes converge on the same cells in the same way that motor efferents converge in a final common path, then the CFF should be higher for a flickering light viewed binocularly than for one viewed monocularly. He found monocular and binocular CFF's to be about the same. He also hypothesized that the CFF should be about twice as high for flickering lights presented in phase to the two eyes than for lights presented in antiphase. Sherrington found the CFF to be only about 3% higher for in-phase than for antiphase dichoptic flicker and concluded that there is very little convergence of binocular inputs. He wrote, "The binocular sensation attained seems combined from right and left uniocular sensations elaborated independently."

Sherrington underestimated the difference between in-phase and antiphase binocular CFF. More recently, the CFF for in-phase flicker was found to be between 4.5 and 10% higher than for antiphase flicker (Ireland 1950; Baker 1970). This is due to neural summation in binocular cells, since it is higher than predicted by probability summation (Peckham and Hart 1960). Also, no significant difference between inphase and antiphase flicker sensitivity was found in subjects lacking stereoscopic vision (Levi *et al.* 1982). We now know that many inputs from the two eyes do converge on common cells. Nevertheless, Sherrington's main conclusion still stands, namely, that lights flickering in phase well below CFF do not simply sum to produce a flicker sensation of twice the frequency. On the other hand, the partial elevation of in-phase over antiphase flicker could be due to sensations of flicker arising from monocular cells. Even a few monocular cells would retain a signal of flicker after all binocular cells have ceased to register it. Thus, the CFF is not a sensitive measure of binocular summation. Furthermore, Sherrington worked at suprathreshold levels where inhibitory as well as excitatory interactions occur between inputs to binocular cells. Interocular summation of flicker sensitivity is more likely to be revealed at threshold levels of luminance.

Another factor could be the presence of contours in the stimuli. It has already been noted that dichoptic brightness summation is increased when the images contain no contours. Thomas (1956) found that the CFF with in-phase dichoptic flicker is increased by the addition of parallel lines to each image, even when the lines in one eye are orthogonal to those in the other.

The absence of dichoptic summation of flicker represents the temporal limit of the ability of binocular cells to receive alternating flashes from the two eyes, possibly because of mutual inhibitory processes evoked by stimuli in antiphase. One would expect that asynchronous dichoptic flashes below a certain frequency would not engage in mutual inhibition and would produce a sensation of double the flicker frequency. Andrews *et al.* (1996) obtained this result for alternating dichoptic flashes at frequencies of 2 Hz or less. Above this frequency, asynchronous and synchronous flashes were increasingly judged to have the same frequency.

Dichoptic interactions of flicker are also indicated by the luminance modulation of a flickering light required for the detection of flicker. This measure, expressed as a percentage of mean luminance of a sinusoidally flickering light, when plotted over a range of temporal frequencies, is the **temporal contrast-sensitivity function**, also known as a De Lange function (De Lange 1954). It is the temporal analogue of the spatial contrast-sensitivity function and has a similar bandpass shape.

Cavonius (1997) found that, with an homogeneous foveal field, flicker sensitivity increased with increasing flicker rate up to about 2% luminance modulation at about 10 Hz, and then fell rapidly to zero at a frequency of about 50 Hz. For flicker rates above about 10 Hz, sensitivity for in-phase dichoptic flicker was about 40% higher than for antiphase dichoptic flicker or for monocular flicker (Figure 8.9). At low flicker rates, sensitivity for in-phase dichoptic flicker was up to four times higher than that for antiphase flicker. This could be because of summation of neural signals arising from lights flickering in phase in the two eyes, which is **dichoptic summation of in-phase signals**, or because of summation of opposite-sign signals from lights flickering in antiphase, which is **dichoptic summation of antiphase flicker**. The summation of in-phase signals seems to be the crucial factor, because sensitivity to antiphase dichoptic flicker was about the same as sensitivity to monocular flicker with the other eye exposed to a steady field (van der Tweel and Estévez 1974; Cavonius 1979). Other evidence reviewed in the next section supports the idea of dichoptic summation of in-phase flicker but not of antiphase flicker.

8.1.5 Sensitivity to pulsed stimuli

8.1.5a *Dichoptic flashes*

Temporal sensitivity can be explored with single flashed stimuli, which may be a light spot that is

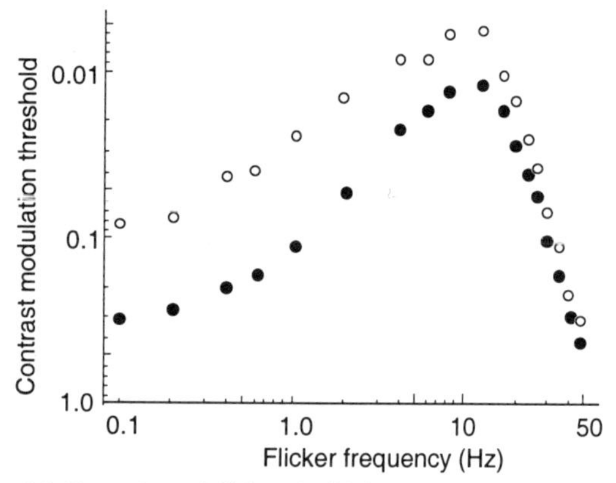

Figure 8.9. Detection of dichoptic flicker.
Threshold contrast modulation for detection of flicker of a 1° illuminated spot as a function of flicker frequency. Solid symbols are for in-phase dichoptic flicker. Open symbols are for counterphase dichoptic flicker (N=1). (Adapted from Cavonius 1979)

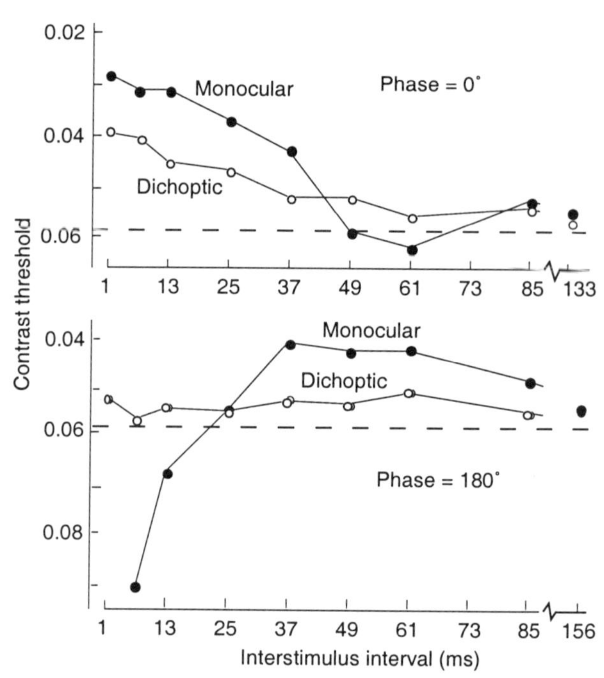

Figure 8.10. Dichoptic masking and spatial phase.
Top graph: the contrast threshold for detection of two spatially in-phase, 0.75-cpd gratings presented for 5 ms each, as a function of interstimulus interval. The gratings were presented to different eyes (empty symbols) or the same eye (solid symbols). Bottom graph: the same functions with the gratings in spatial counterphase. Dashed lines indicate thresholds for the first grating alone (N=1). (Adapted from Green and Blake 1981)

momentarily extinguished (negative polarity) or a dark spot that is momentarily increased in luminance (positive polarity). The threshold for detection of dichoptically flashed test spots that either both increased or both decreased in luminance was lower than the threshold of flashed spots that increased in luminance in one eye and decreased in the other (Westendorf and Fox 1974). Same-sign flashes were detected at a level above that of probability summation, whereas opposite-sign flashes were detected at about the level of probability summation. This provides further support for summation of dichoptic in-phase signals and independence of dichoptic anti-phase signals. When one of the flashed targets was a vertical bar and the other was a horizontal bar, flash detection was at the level of probability summation for both same- and opposite-sign flashes (Westendorf and Fox 1975).

The receptive field of a ganglion cell has a spatial sensitivity profile. For example, an on-centre receptive field has an excitatory centre with a Gaussian profile and an inhibitory surround with a wider Gaussian profile. A receptive field also has a temporal-sensitivity profile. A flash of light in the centre of an on-centre receptive field produces an excitatory discharge followed by an inhibitory phase. An off-centre field responds in the same way to a briefly darkened spot. An estimate of the durations of these phasic responses to light pulses can be obtained by measuring either the probability of seeing or the threshold for detection of a pair of flashes as the interstimulus interval is increased. The procedure is analogous to Westheimer's procedure for measuring the spatial properties of receptive fields (Section 8.2.2).

Anstis and Ho (1998) repeated Levelt's experiment using a 0.7° dichoptic test spot flickering in phase to the two eyes at 15 Hz, rather than a constantly illuminated spot. The spot presented to one eye was set at one of several luminance modulations. Subjects adjusted the luminance modulation of the spot presented to the other eye until the fused image of the two flickering spots matched the luminance of a steady gray comparison spot seen subsequently by both eyes. For light spots on a dark ground, the results were similar to those obtained by Levelt. The apparent luminance of the dichoptic flickering spot was the mean of the luminance of the component spots. At extreme values there was evidence of Fechner's paradox (see Figure 8.4). However, for dark spots on a light ground, the apparent luminance of the dichoptic spot was equal to the luminance of the spot with higher luminance. Anstis and Ho suggested that the mean weighting function for light-on-dark spots compared with the winner-take-all weighting for dark-on-light spots is due to an underlying asymmetry in the ON and OFF visual pathways. However, the state of adaptation of the eyes differed between the two conditions.

8.1.5b *Bloch's law under dichoptic conditions*
Under the most favourable conditions, the visibility of a single flash is proportional to its duration up to

about 100 ms. During this period, visibility depends on the product of intensity and duration, a relationship known as **Bloch's law**. The limiting period of temporal integration is decreased by increasing the area of the stimulus, keeping luminance constant, or by increasing the luminance of the background (Barlow 1958). Two flashed stimuli with the same size and luminance polarity presented to one eye physically sum their stimulus energy (Bouman and van den Brink 1952).

Similarly, within the limits of Bloch's law, the visibility of dichoptic flashes depends on the total energy in each flash, which is the product of the duration and intensity of the flash (Westendorf *et al.* 1972). Cogan *et al.* (1982) found that the detectability of low-contrast dichoptic flashes set within fused binocular contours was at least twice that of a monocular flash. The binocular advantage was not as large for high-contrast flashes. Binocular detectability, even for low-contrast flashes, was only 41% better than monocular detectability when the background contours were omitted from one eye. It was suggested that binocularly fused contours engage the cells responsible for binocular fusion, and which sum low-contrast stimulus energy. Contours in only one eye engage monocular mechanisms or the binocular rivalry system, which reduces the degree of binocular summation.

8.1.5c *Detection of flashes separated in time*

As the interval between two flashes is increased from zero, the mutual facilitation of the stimuli decreases to zero at about 35 ms, when the inhibitory phase of one flash coincides with the excitatory phase of the other. With a longer interval, the stimuli show mutual inhibition as the two inhibitory phases come into coincidence. When the interval reaches 100 ms there is zero interaction because the responses no longer overlap. Beyond this interval, the probability of detecting at least one of the flashes is influenced only by probability summation. This is the zero level of neural interaction. Flashes with opposite luminance polarity presented to the same eye physically cancel when they are simultaneous, show inhibitory interactions with a short interstimulus interval, and facilitatory interactions with a longer interstimulus interval (Ikeda 1965; Rashbass 1970; Watson and Nachmias 1977). These results suggest that a flash generates an initial response of one sign and a secondary response of the opposite sign. The sign of the responses depends on the polarity of the flash, and interactions between successive flashes depend on how the excitatory and inhibitory phases interact.

Matin (1962) measured the probability of detecting dichoptic flashes 35 arcmin in diameter and of 2 ms duration as a function of the time interval between them. Binocular summation was greater than predicted by classical probability summation only for interstimulus intervals less than about 100 ms. Similar results were obtained by Thorn and Boynton (1974). Note that the inhibitory effects found with monocular flashes were not reported with these dichoptic stimuli. These experiments provide evidence that similar signals falling simultaneously on corresponding points exhibit real neural summation. Blake and Fox (1973) reviewed other early experiments on this topic.

The study of interactions between flashed stimuli has been extended to flashed gratings for which stimulus alternation may involve a spatiotemporal displacement, not merely a temporal displacement. This is because two gratings of opposite luminance polarity presented in succession can be regarded as having been displaced spatially by one-half period of the grating. When a 0.75 cpd sine-wave grating was flashed to the same eye for two periods of 5 ms, there was summation up to an interstimulus interval of about 50 ms, followed by a small inhibitory effect. When the gratings were presented in the same way to opposite eyes there was similar but weaker facilitatory effect, but the inhibitory phase was absent (Figure 8.10). Gratings with opposite luminance polarity (180° spatial phase shift) presented to the same eye showed an initial inhibitory phase followed by a facilitatory phase, as shown in the figure. Opposite-polarity gratings presented dichoptically showed no facilitation or inhibition (Green M and Blake 1981). A similar result was reported for two light flashes with the same and opposite polarity (Cogan *et al.* 1990). Blake and Levinson (1977) did find dichoptic interactions between gratings of opposite polarity (180° phase shift), but these were high spatial-frequency gratings, and a slight misconvergence may have brought them into phase.

Rose (1978) found that the contrast threshold for detection of flicker or apparent movement in a sinusoidal grating between 0.5 and 7 cpd, which reversed in spatial phase at 3.5 Hz, was 1.9 times lower with binocular than with monocular viewing. This is a much greater binocular advantage than that for detection of a stationary grating or than that reported by previous workers for detection of a counterphase grating. This advantage of binocular flicker detection over pattern detection was independent of spatial frequency but was lost at temporal frequencies above 10 Hz (Rose 1980).

Responses to light onset and light offset are processed in visual channels that remain distinct at least up to the visual cortex (Section 5.1.2c). Nevertheless, these channels must interact to account for inhibitory interactions that occur when opposite-sign flashes are presented to the same eye with a small interflash interval. From the preceding evidence it seems that dichoptic interactions between transient signals of opposite sign in the two eyes do not occur at any level.

Investigators have concluded from these results that opposite-polarity stimuli arising from the two eyes are processed independently. But this is the wrong way to look at it. Think of two steady square-wave gratings presented 180° out of spatial phase to one eye. Clearly, the gratings are invisible because they physically cancel to a homogeneous gray. When the same gratings are presented dichoptically they do not physically cancel, but rival. At any instant, the dominant grating is seen just as well as when there is no grating in the other eye. Thus, the suppressed grating does not weaken the visibility of the dominant grating (Bacon 1976). From the point of view of visibility, the two gratings are processed independently, but only one of them is processed at any one time in a given location. Although opposite polarity dichoptic stimuli do not engage in simultaneous mutual inhibition, they do engage in alternating suppression, or rivalry.

The same argument can be applied to superimposed flashes of opposite polarity. They physically cancel when presented simultaneously to the same eye, and the excitatory and inhibitory phases of their neural responses interact when the flashes are presented successively to one eye. When presented dichoptically, opposite-polarity flashes rival but during the dominant phase of either one, the stimulus remains just as visible as a flash presented to only one eye. As the interstimulus interval is increased, rivalry ceases and both stimuli become visible as independent events; they do not, as with monocular viewing, engage in mutual inhibition.

It is interesting to note in this context that the threshold for detecting a low-contrast grating was lowered when it was presented just after a similar grating with a spatial phase offset of 90°, but that this facilitation was not evident when the two gratings were presented dichoptically (Georgeson 1988). The facilitation occurred between sequentially presented dichoptic gratings when they were in spatial register. According to this evidence, although the binocular detection mechanism combines dichoptically superimposed stimuli, it does not combine dichoptic stimuli in spatiotemporal quadrature.

Wehrhahn *et al.* (1990) asked subjects to decide which of two suprathreshold vertical lines, either 5 or 40 arcmin apart, was presented first. The temporal threshold for this task was lower by a factor of 1.4 when the stimulus was presented binocularly than when it was presented monocularly.

Odom and Chao (1995) measured human visually evoked potentials generated by full-field modulations of luminance at 2 Hz. A nonlinear second harmonic was evident in the VEP when the peaks were either in phase or 180° out of phase in the two eyes but was much reduced at a phase of 90°. They argued that these data support the idea of a magnocellular pathway, which sums nonlinear monocular inputs and a parvocellular pathway, which sums linear monocular inputs followed by a nonlinear stage. They produced further evidence for this model by measuring the threshold luminance modulation required for detection of flicker with dichoptic flashes at various phase differences. At 2 Hz, the threshold fell to a minimum when the phase difference was 90°. They argued that this biphasic response is evidence for the activation of both visual pathways. At 16 Hz, the threshold fell monotonically with increasing phase difference. They argued that this monotonic function occurred because only the magnocellular pathway was activated at this high frequency.

Summary

For people with normal binocular vision, binocular thresholds for luminance increments in discrete stimuli and for contrast detection in gratings are lower than monocular thresholds to a greater extent than predicted by probability summation. Binocular summation is greatest when the stimuli have similar shapes, sizes, contrasts, and locations. In other words, summation is greatest when the visual mechanisms responsible for fusion are engaged rather than those responsible for rivalry. Binocular summation of brightness is most evident when the dichoptic stimuli lack contours. Binocular summation is not evident with discrimination tasks between stimuli well above the detection threshold, presumably because of response saturation. In animals and humans that are stereoblind, binocular summation is no more than one would predict from probability summation (Section 12.3.1). This further supports the idea that in people with normal vision, near-threshold excitatory signals from the two eyes are at least partially summed when they impinge on cortical binocular cells.

There is considerable interaction between inputs from the two eyes in response to flicker, especially for low frequencies and within the modulation threshold region. In-phase flicker with similarly shaped stimuli is detected above the level of probability summation, whereas antiphase flicker is detected at or below this level. Binocular summation of

Table 8.1. Types of visual masking

Simultaneous masking
- Induction and test stimuli superimposed
- Test and induction stimuli adjacent—crowding

Successive masking
- Induction and test stimuli superimposed
  - Forward masking—1st stimulus masks 2nd
  - Backward masking—2nd stimulus masks 1st
- Induction and test stimuli spatially adjacent
  - Paracontrast—1st stimulus masks 2nd
  - Metacontrast—2nd stimulus masks 1st

stimulus energy occurs under conditions that foster binocular fusion but not under conditions that foster binocular rivalry. Light flashes or gratings with similar luminance polarity show summation when presented in quick succession either to the same eye or to opposite eyes. Flashes or gratings with opposite polarity show summatory and inhibitory phases when presented to the same eye but not when presented dichoptically. However, opposite-polarity dichoptic flashes do engage in alternating suppression. As far as visibility is concerned, they are processed in distinct channels, but these channels engage in suppressive rivalry. The binocular advantage for flicker detection seems to be greater than that for pattern detection.

8.2 DICHOPTIC VISUAL MASKING

8.2.1 Types of visual masking

An induction stimulus with near-threshold contrast lowers the detection threshold of a test stimulus (Section 8.1.1). This is threshold summation. In visual masking, a suprathreshold induction stimulus, or mask, reduces the visibility of a briefly exposed test stimulus. The mask is usually presented briefly and the test stimulus is presented either at the same time as the mask, slightly before it, or slightly after it. The mask can be a disc of uniform luminance, an edge, or a grating with sinusoidal luminance profile. The presence of masking between two stimuli is interpreted as evidence that the stimuli are detected by the same channel or by partially overlapping channels. In dioptic masking, the mask and test stimulus are presented to both eyes, while in dichoptic masking, the mask is presented to one eye and the test stimulus to the other. The main types of masking paradigm are listed in Table 8.1.

Dichoptic masking differs from binocular rivalry in two ways. First, in dichoptic masking the test stimulus is usually presented for less than 200 ms, which is too short a time for binocular rivalry to manifest itself (Section 7.3.8). Second, dichoptic masking is maximal when the test and masking stimuli have similar visual features, whereas binocular rivalry is most evident between stimuli that differ widely in shape, orientation, spatial frequency, or colour. Dichoptic masking probably occurs at an early stage in the combination of binocular signals whereas rivalry occurs later, at a stage when patterned inputs are compared (Section 8.2). Masking and rivalry could occur at different stages of processing within V1.

8.2.2 Masking without figure rivalry

Acuity is reduced when a monocular test pattern in the dominant phase of rivalry is accompanied by a distinct stimulus in the other eye (Freeman and Jolly 1994). This section reviews the circumstances under which a continuous and homogeneous stimulus in one eye reduces the visibility of stimuli in the other eye. The following evidence demonstrates that interocular suppression also occurs when binocular contour rivalry is not involved.

8.2.2a *Dichoptic interactions between featureless stimuli*

The visibility of a flash varies with the state of light adaptation of the same eye. However, most investigators have found that the overall state of light adaptation of one eye does not affect the threshold sensitivity of the other eye for a featureless stimulus (Crawford 1940a; Mitchell and Liaudansky 1955; Cogan 1989). Wolf and Zigler (1955) and Whittle and Challands (1969) found small interocular effects but their test and adapting stimuli were not devoid of visual contours. The visual potential evoked from the human scalp by a monocular checkerboard pattern reversing in contrast at 5 Hz was weaker when the other eye was closed and dark adapted than when it was adapted to a dim homogeneous light (Eysteinsson *et al.* 1993). An eye was more sensitive to a test flash within a light-adapted region in the other eye than when the other eye was dark adapted (Lansford and Baker 1969; Paris and Prestrude 1975). Adaptation of one eye to red light lowers the dark-adapted threshold for a test flash presented to the other eye by about 0.15 log units (Auerbach and Peachey 1984; Reeves *et al.* 1986).

Crawford (1940b) and Westheimer (1965) introduced the paradigm of exposing a test spot briefly on the centre of a featureless disc-shaped conditioning stimulus. The luminance threshold of the test spot was measured as a function of the duration, luminance, size, and eccentricity of the conditioning

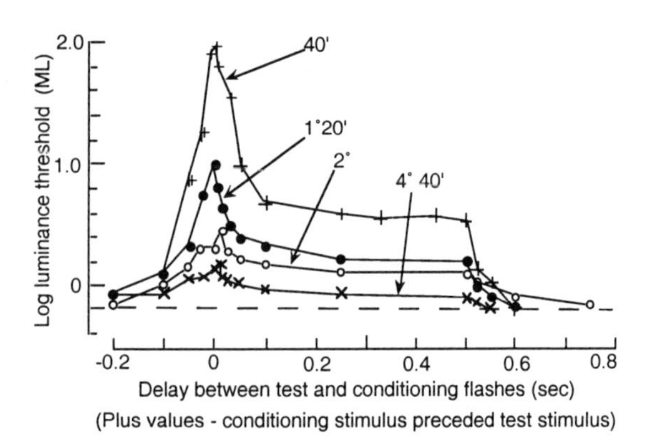

Figure 8.11 Dichoptic masking and interstimulus delay. Luminance threshold for detection of a 40-arcmin, 5-ms test flash as a function of time before or after a 500-ms dichoptic conditioning stimulus. The four curves are for different diameters of the conditioning stimulus, indicated by numbers on the curves (N=1). (Redrawn from Battersby and Wagman 1962)

stimulus. This type of procedure has revealed inhibitory interactions within the same eye (Makous and Boothe 1974). For example, Buck and Pulos (1987) found that the scotopic increment threshold for a 5 arcmin flash was increased by up to 0.6 log units when it was superimposed on a 1° photopic background in the same eye.

Inhibition of rod receptor potentials by stimulation of cones involves inhibitory interneurones in the retina, and also short-latency responses from cones occluding long-latency responses from rods as the two responses converge on ganglion cells (Gouras and Link 1966; Whitten and Brown 1973). Inhibition of rods by cones clears the post-receptor pathway so that it carries only cone signals at photopic levels. Cone-cone inhibitory interactions within the receptive fields of ganglion cells render the system particularly sensitive to luminance gradients. These types of interaction did not occur when induction and test stimuli were in different eyes (Westheimer 1967; Buck and Pulos 1987). However, in these experiments, the background stimulus was continuous. We will now see that dichoptic masking occurs with figured conditioning stimuli.

8.2.2b *Masking between figured and non-figured stimuli*
Battersby and Wagman (1962) measured the detection threshold for a 5 ms, 40 arcmin test flash presented at various time intervals before, during, or after a larger concentric illuminated disc presented for 500 ms. The results for one subject are shown in Figure 8.11. The threshold was elevated when the test flash was presented in the period between 100 ms before and 100 ms after the onset of the conditioning disc, and was maximal when the two events overlapped in time. A conditioning stimulus 4.7° in diameter had very little effect on the threshold of the test stimulus. As the conditioning stimulus was reduced in size, bringing its border closer to the test flash, the threshold for seeing the test flash became increasingly elevated during the whole period of the conditioning stimulus.

Markoff and Sturr (1971) investigated the effects of changing the size of the conditioning stimulus, with the test and conditioning stimuli presented simultaneously. A 5-ms, 3.5 arcmin test flash was presented at the same time as a conditioning stimulus exposed for 50 ms, 200 ms, or continuously. The two stimuli were presented either monocularly or dichoptically. The two top graphs of Figure 8.12 show that, with both monocular and dichoptic viewing, the luminance threshold for detection of the foveal test flash rose as the diameter of the conditioning patch increased from 10 to about 21 arcmin, after which it declined to a value that depended on the luminance of the conditioning spot. This can be explained in terms of the structure of the on-centre receptive fields of ganglion cells. A small conditioning stimulus adds to the stimulation of the on-centre and elevates the differential threshold, but, as its area increases, its edge encroaches on the inhibitory surround. When it gets larger still, its edge extends beyond the inhibitory surround, and masking declines to a level that depends on the luminance of the mask. The stimulus that produces peak masking is larger at scotopic than at photopic levels of luminance, presumably because the inhibitory surround is weaker at scotopic levels. Peak masking size is also larger with peripheral than with foveal viewing, presumably because receptive fields get larger in the periphery (see bottom two graphs in Figure 8.12).

It can be seen in Figure 8.12 that detection of a test spot in one eye is not affected by a conditioning disc wider than about 3° presented to the other eye. Therefore, as we saw in Section 8.2.2a, pure luminance masking within homogeneous areas does not occur between the eyes. One can infer that both monocular and dichoptic masking are due to interactions between the contiguous edges of the conditioning and test stimuli. We will return to this topic shortly. It can also be seen in Figure 8.12 that a conditioning disc produces very little dichoptic masking when it is visible continuously. Similar evidence was reported by Fiorentini *et al.* (1972) and Sturr and Teller (1973). One can infer that dichoptic masking is due to rivalrous interactions between stimulus onsets or offsets of contiguous edges. Chromatically selective dichoptic masking occurs with large masking flashes but only within the blue-cone system (Boynton and Wisowaty 1984).

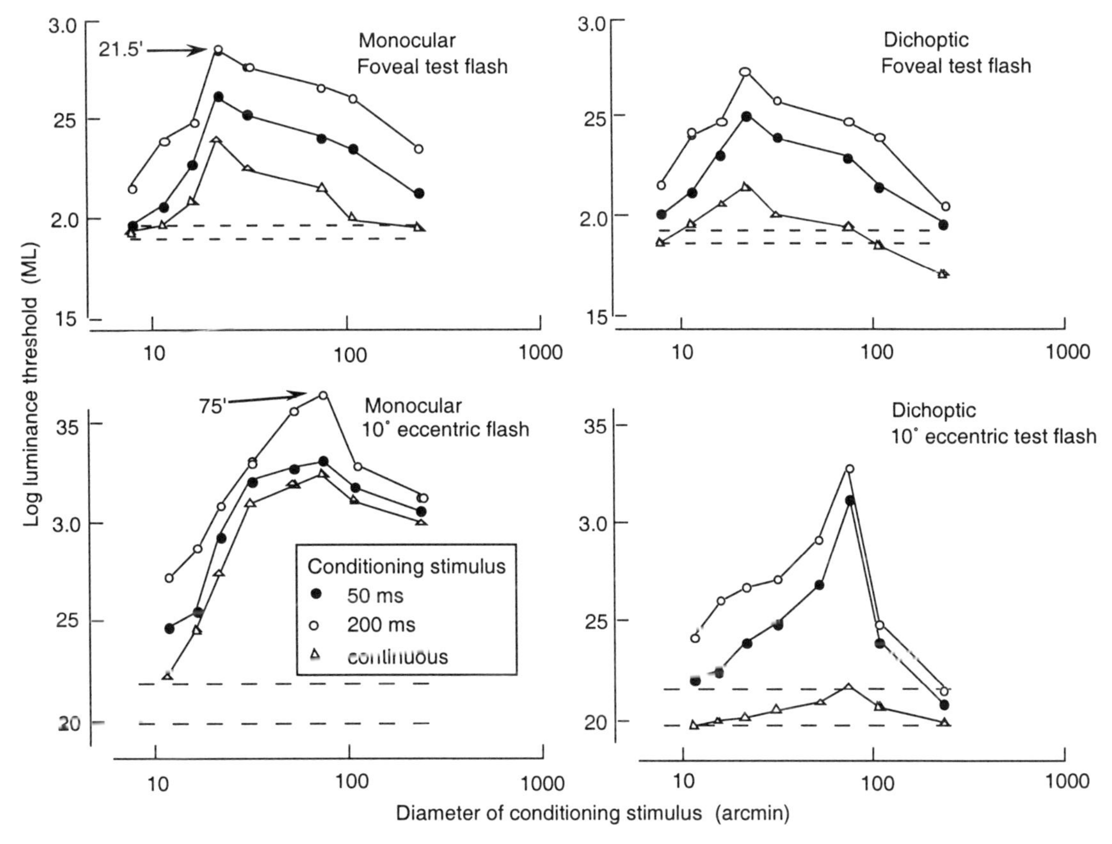

Figure 8.12. Dichoptic masking as function of stimulus size and duration.
Luminance threshold for detection of a 5-ms test flash presented with a conditioning stimulus to the same eye (left two graphs) or to the opposite eye (right two graphs). Curves in each graph are for three durations of the conditioning stimulus. The top two graphs are for a foveal test flash, the bottom two for one at 10° eccentricity. Dashed lines indicate the semi-interquartile range of the resting threshold for that position for monocular and binocular conditions (N=1). (Redrawn from Markoff and Sturr 1971)

Denny *et al.* (1991) found that contrast sensitivity for gratings with spatial frequencies over 2 cpd presented to one eye was improved when the other eye was light adapted. For spatial frequencies over 10 cpd, improved sensitivity required brighter adapting fields in the other eye. Light adapting the eye containing the stimulus had little effect on sensitivity. Also, monocular or binocular light adaptation had little effect on binocular contrast sensitivity. The same improvement in contrast sensitivity was achieved by pressure blinding the eye not containing the stimulus (Makous *et al.* 1976), which suggests that the effect is due to removal of inhibitory influences from the opposite eye rather than to facilitatory influences arising from light adapting the opposite eye. It was noted in Section 7.3.2d that the dark field of a closed eye may rival stimuli presented to the other eye.

I conclude that signals arising from rods in a fully dark-adapted eye exert a small inhibitory influence on the contrast sensitivity of the other eye. Removal of this inhibitory influence when both eyes are stimulated may contribute to binocular summation.

In Fechner's paradox, the brightness of a bright patch with a border presented to one eye declines as the brightness of a similar patch in the other eye increases from zero to the upper limit of the mesopic range—the range over which rods respond (Curtis and Rule 1980). Fechner's paradox does not occur when the eyes are evenly illuminated (Bourassa and Rule 1994). It thus seems that, with contoured stimuli, inputs from rods in one eye inhibit responses of cones in the other eye. For figured stimuli, this inhibition is least from an unstimulated dark-adapted eye and increases when a contoured stimulus within the mesopic range is introduced to the contralateral eye. Beyond the mesopic range, inputs from cones predominate and these exhibit binocular summation. There are inhibitory interactions at the retinal level (Section 8.2.2a) but these involve inhibition of rods by cones or cones by cones, rather than of cones by rods.

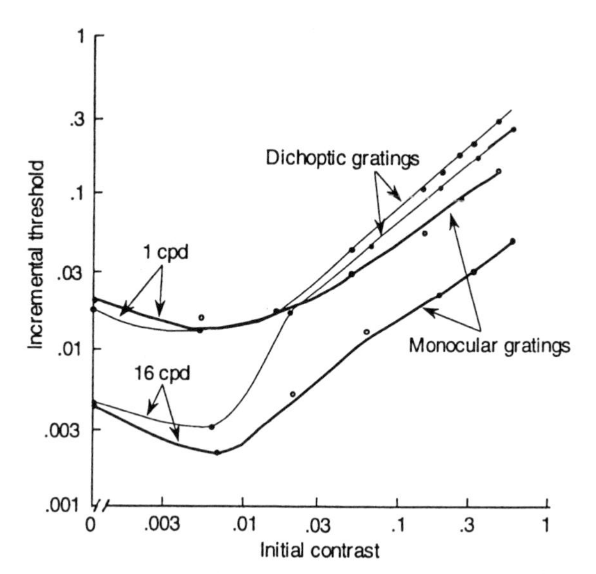

Figure 8.13. Monocular/dichoptic contrast-increment thresholds. Incremental contrast thresholds for gratings presented successively to one eye and for a grating presented to one eye and then the other eye. Each grating was presented for 200 ms with a 750 ms interstimulus interval. Both gratings had a spatial frequency of 1 cpd or 16 cpd. (After Legge 1979)

Summary

The general conclusion to be drawn from these experiments is that a large area of steady illumination in one eye does not affect the visibility of a stimulus presented to the other eye. However, the visibility of a test flash in one eye is reduced if it occurs close in time to a change in luminance in the other eye or is spatially adjacent to a contour in the other eye. In all the following experiments, both the induction and test stimuli are patterned and are usually presented for brief periods so that there are contiguous edges and contiguous temporal transients.

8.2.3 Masking with superimposed patterns

8.2.3a *Effects of contrast and spatial frequency*

In a typical masking experiment with superimposed patterns, a subject is shown a sinusoidal masking grating twice in succession to the same eye. The subject reports which masking grating has a test grating superimposed on it. The contrast of the test grating required for 75% success in this forced-choice task is its threshold contrast. When the masking and test gratings have the same spatial frequency and phase, the measurement is that of the increment threshold. When the contrast of the mask is low, the increment threshold contrast is lower than when the mask is not present. In other words, the mask facilitates detection of the test grating. As the contrast of the mask increases above about 0.3, the increment threshold contrast increases linearly, as shown in Figure 8.13. In effect, this function expresses Weber's law for incremental contrast. The results for four spatial frequencies follow the same function when the data are rescaled in units of the absolute threshold.

In dichoptic masking, the test and masking gratings are presented simultaneously to opposite eyes. Figure 8.13 shows that, for dichoptic viewing with 200-ms exposures of the stimuli, the facilitatory effect at low contrast is weaker than in monocular viewing. However, with higher contrasts, dichoptic masking is stronger than monocular masking (Legge 1979). The weak facilitatory effect represents binocular summation, discussed in Section 8.1.1. Dichoptic masking at higher contrasts represents interocular inhibition.

A high-contrast pattern in one eye and a low-contrast test pattern in the other produce Fechner's paradox—a lowering of binocular brightness relative to monocular brightness (Section 8.2.2). This inhibitory effect is not the same as binocular suppression occurring in binocular rivalry, because suppression is greatest when the dichoptic patterns are dissimilar whereas, as we will now see, masking is greatest when dichoptic patterns are similar.

Like binocular summation of simultaneously presented threshold stimuli, the masking effect of a grating is greatest when the spatial frequency and orientation of the test and mask are the same (Gilinsky and Doherty 1969). Other aspects of relative orientation in dichoptic masking are discussed in Section 7.3.6. The threshold elevation as a function of the spatial frequency of a masking grating for a given spatial frequency of a test grating is the **spatial-frequency masking function**. In general, spatial-frequency tuning functions give an indication of the bandwidth of channels tuned to different spatial frequencies (Legge 1979). However, it is difficult to compare the spatial-frequency bandwidth of dichoptic and monocular masking functions, since the two functions have very different slopes (Figure 8.14).

Harris and Willis (2001) asked whether a 1 cpd contrast-modulated grating presented to one eye masks a 1 cpd luminance-modulated grating presented to the other eye. The contrast-modulated grating was the beat pattern (moiré pattern) formed by superimposing gratings of 8 and 9 cpd. Interocular masking was as strong as when mask and test grating were presented to the same eye. The contrast-modulated grating produced results similar to those produced by a luminance-modulated mask of similar effective contrast.

Vernier acuity is degraded when the stimulus is masked by a superimposed grating or by flanking

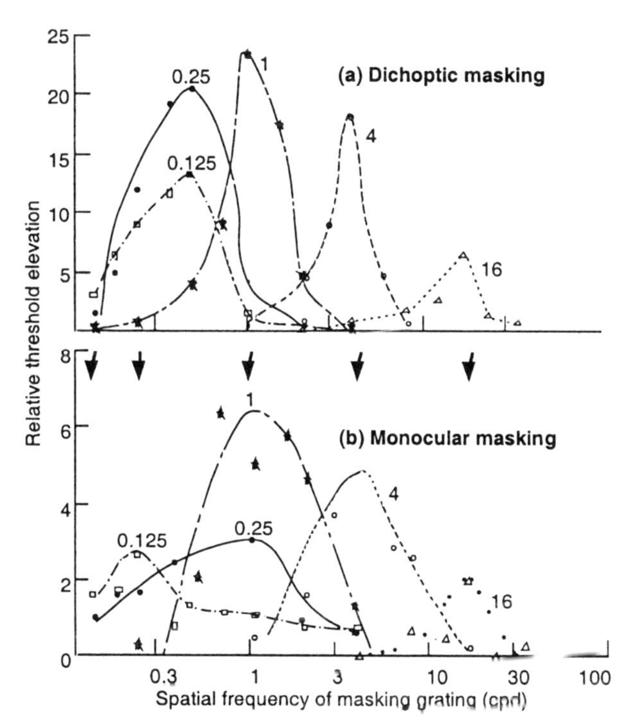

Figure 8.14. Dichoptic spatial-frequency masking functions.
(a) Each curve shows elevation of contrast threshold of a test grat-ing of a particular spatial frequency presented to the right eye, as a function of the spatial frequency of a masking grating presented to the left eye. Arrows indicate spatial frequencies of test gratings. The masking grating had a contrast of 0.19. The two gratings were presented at the same time for 200 ms. Threshold elevation is the ratio of threshold with the mask to threshold without the mask. Numbers on the curves are the spatial frequencies of the test grating (N=2).
(b) Monocular spatial-frequency masking functions obtained when mask and test gratings were presented to the same eye. (Redrawn from Legge 1979)

lines in the same eye. Mussap and Levi (1995) found that performance on a vernier target presented to one eye was similarly degraded by masks presented to the other. They argued that this demonstrates that the neural processes responsible for vernier acuity receive binocular inputs. However, it is logically possible that a neural signal arising from a vernier offset is formed before binocular convergence but that the signal is then subject to degradation by inputs from the other eye.

Visual evoked potentials recorded from the scalps of human subjects in response to a grid pattern presented briefly to one eye were reduced when a similar grid pattern was presented continuously to the other eye. This interocular suppression of the VEP was reduced when the grids in the two eyes differed in density and was positively correlated with stereoacuity (Harter *et al.* 1977).

There is a general problem with all arguments about the site of visual processes based on masking experiments. Neural processes responsible for a given visual skill may reside at many levels in the nervous system. Some processes occur in series and some in parallel, and a mask may affect any or all of these levels or may have an effect only after visual processing of a given visual feature is complete. Towle *et al.* (1980) used psychophysical tests and the VEP to measure masking of a flashed grating by a stationary grating applied to the same or the opposite eye. In the initial 100 ms, masking depended only on the relative orientations of the gratings. After about 200 ms, both relative orientation and relative spatial frequency had an effect.

8.2.3b *Spatial and temporal components of masking*
It has been suggested that masking consists of two spatial components. One component is masking by uniform illumination. The other component is masking by the pattern features of the mask.

Some evidence suggests that the first component of masking occurs in the retina. Grating detection by one eye is adversely affected by uniform illumination in the same eye but not by uniform illumination in the other eye (Blake *et al.* 1980b). The attenuation of the contrast sensitivity function (CSF) at low spatial frequencies is also believed to occur in the retina.

Yang and Stevenson (1999) questioned the view that masking by uniform illumination and the roll off of the CSF are only retinal in origin. Masking of one grating by another in the same eye declines with increasing difference in spatial frequency between mask and test gratings. Uniform illumination contains low spatial frequency components because of retinal inhomogeneities. Therefore, masking by uniform illumination should be most evident with test gratings of low spatial frequency. Also, insofar as this type of masking involves the detection of spatial frequencies, it should show interocular transfer.

Yang and Stevenson measured the monocular CSF for a grating of low mean luminance, modulated in contrast at 2 Hz. The other eye was illuminated at each of two levels of steady uniform illumination. The results are shown in Figure 8.15. Interocular masking is evident at spatial frequencies below 1 cpd. Blake *et al.* did not test below 1 cpd. There is thus a cortical component to masking by uniform illumination. Yang and Stevenson explained the smaller level of interocular masking compared with same-eye masking in terms of a post-retinal interocular gating mechanism.

The component of masking due to pattern features of the mask occurs in the cortex, where pattern features are processed (Bowen and Wilson 1994).

It has also been suggested that there are two temporal components of masking: a steady state com-

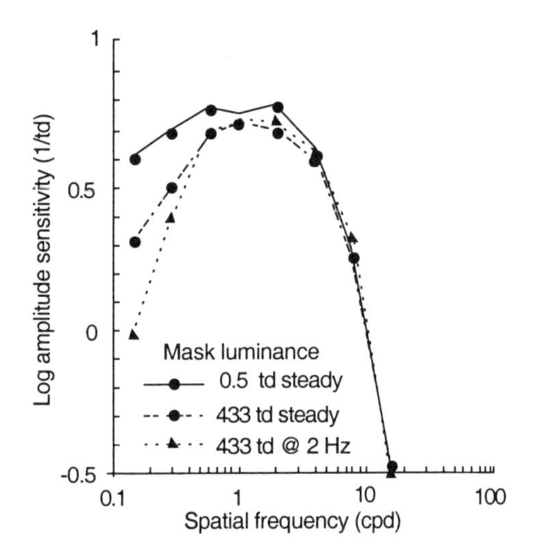

Figure 8.15 Dichoptic masking of contrast sensitivity.
A sine wave grating of mean luminance 8 trolands and contrast modulated at 2 Hz was presented to one eye and a mask consisting of even illumination was presented to the other eye. With steady even illumination, there was a loss in contrast sensitivity at low spatial frequency. The loss increased with increased luminance of the mask and was greater when the even illumination was, like the grating, modulated at 2 Hz. Results for one subject. (Adapted from Yang and Stevenson 1999)

ponent and a transient component at the onset and offset of the masking stimulus (Sperling 1965). Green and Odum (1984) found that masking of a drifting grating by a steady field of illumination showed little interocular transfer but that masking by a flickering field showed substantial transfer, like that produced by a steady patterned mask.

Yang and Stevenson (1999) proposed that the crucial factor is the similarity between the temporal features of test and mask, rather than the presence or absence of a flickering mask. It can be seen in Figure 8.15 that interocular masking is greatest when mask and test grating both flickered at 2 Hz.

8.2.3c *Masking and relative disparity*

A tone embedded in noise is easier to detect with two ears than with one, especially when the tone is presented to the two ears in antiphase. This is **binaural unmasking**. The difference between masking with inphase and antiphase tones is the **binaural masking-level difference**. Henning and Hertz (1973) revealed a visual analogue of these effects. A visual noise stimulus consisting of a mixture of vertical gratings within a narrow band of spatial frequencies, and slowly varying in contrast and phase, was presented to both eyes. A sinusoidal test grating of fixed contrast at the mean spatial frequency of the noise was superimposed on the noise. The test grating was either spatially in phase in the two eyes or in antiphase. As with auditory stimuli, the test grating was detected at much lower contrasts when in antiphase than when in phase in the two eyes, but only for spatial frequencies of less than about 6 cpd. The **binocular masking-level difference** did not depend on the temporal modulation of signal and noise (Henning and Hertz 1977).

Since vergence was not controlled, one cannot be sure whether the signal or the noise was in antiphase. In other words, we do not know whether antiphase noise is a less effective mask or whether an antiphase test stimulus is more resistant to masking. Also, Henning and Hertz did not mention whether the antiphase signal and the noise appeared in different depth planes; any misconvergence on the antiphase grating would induce either a crossed or uncrossed disparity into the noise display.

Moraglia and Schneider (1990) controlled for vergence and investigated the role of perceived depth in masking. They used a luminance-modulated Gabor patch as the test stimulus and a background area of broadband Gaussian noise as the mask. With convergence held on the background, the visibility of the superimposed test patch was augmented when it had a horizontal disparity of 13.5 arcmin. It was augmented to a lesser extent when its disparity was 40.5 arcmin. In both cases, the test patch appeared out of the plane of the noise. However, the appearance of depth was not required for augmentation of visibility because 13.5 arcmin of vertical disparity also augmented the visibility of the test patch (Moraglia and Schneider 1991). Visibility of a patch with a horizontal or vertical disparity of 67.5 arcmin was no better than with zero disparity. The augmentation effect was absent when test patch and noise were orthogonal (see also Schneider *et al.* 1989).

These results and others (Schneider and Moraglia 1992) suggest that dichoptic masking occurs at the cyclopean level within the disparity-detecting mechanism and that a stimulus is released from masking when its disparity differs from that of the mask. For disparity to help in the recognition of superimposed patterns, as opposed to their detection, all spatial frequencies required for pattern recognition must be unmasked (Schneider *et al.* 1999).

Backward masking is also reduced when the mask and the test stimulus are in different depth planes (Section 8.2.6). A depth-dependent effect has also been found for simultaneous masking of a pattern by a surrounding annulus (Fox and Patterson 1981). In a related experiment, McKee *et al.* (1994) found that a test bar in one eye is masked to a lesser degree by a superimposed mask in the other eye when a second bar, similar to the mask, is placed adjacent to the test bar. They argued that the mask becomes stereoscopically matched with the adjacent bar rather than with the test bar, which releases the test bar from masking. This suggests that binocular images are linked before the site of dichoptic contrast discrimination.

Masking of disparity detection by monocular noise is discussed in Section 19.7.4.

8.2.3d *Masking of suppressed and dominant images*
Instead of asking how a stimulus in one eye masks a stimulus in the other, one can ask how a stimulus in the suppressed phase of binocular rivalry masks a stimulus that is in either the suppressed phase or the dominant phase. Westendorf (1989) measured the reaction time to the onset of a monocular probe superimposed on one of a pair of rivalrous dichoptic patterns, both when that member of the pair was in its dominant phase and when it was in its suppressed phase. When the probe was presented in the dominant phase the suppressed image in the other eye had no effect on the visibility of the probe. When the probe was presented during the suppressed phase, however, its detection was delayed, and the delay was greater when the probe and suppressed image were identical rather than different. The greater effect with identical stimuli was ascribed to masking, since masking is greatest between identical stimuli and rivalry suppression is maximum between dissimilar stimuli. Thus, a suppressed image does not mask a probe on a dominant image but does mask a suppressed probe. It was concluded that dichoptic pattern masking occurs at a more central site than binocular suppression.

Summary
When a test pattern in one eye is superimposed on a masking pattern in the other eye for about 200 ms, the detectability of the test pattern is enhanced at low contrasts and reduced at high contrasts. The masking effect is greatest when the test and masking stimuli have the same spatial frequency and orientation. In this respect, dichoptic pattern masking differs from binocular rivalry, which is most evident when the stimuli differ. Dichoptic pattern masking is evident in the visual evoked potential. A test stimulus presented to both eyes is released from masking when the mask is in a distinct disparity-defined depth plane.

8.2.4 Dichoptic visual crowding

A stimulus is more easily detected and its features more easily discriminated when it is presented in isolation than when it is flanked by other similar stimuli. This effect is known as **crowding** or contour interaction. For example, the orientation of the letter *T* was more easily recognized for an isolated letter than for one flanked by two other letters (Toet and Levi 1992). Also, vernier acuity was better for an isolated visual target than for one flanked by other lines (Levi *et al.* 1985).

Several studies have revealed that crowding effects are evident when the test stimulus is presented to one eye and the distracters to the other eye. Detection of the gap in a Landolt C was impaired when the test stimulus was flanked by four short bars. The effect fell to zero when the distance between the bars and stimulus reached about 5 arcmin (Flom *et al.* 1963). The effect was as strong when the test stimulus and distracters were presented to different eyes.

Changes in the tilt of an isolated line were more easily discriminated than changes in the tilt of the same line flanked on both sides by other lines, even when the flanking lines had the same orientation as the test line (Westheimer *et al.* 1976). This effect also showed interocular transfer (Westheimer and Hauske 1975).

The contrast threshold for detecting a 2° diameter counterphase-modulated vertical grating was elevated by addition of an annular radial grating extending out to 20° in the same eye, especially when the radial grating was moving (Marrocco *et al.* 1985). Significant threshold elevations for both stationary and moving annular gratings were also found with dichoptic viewing.

Interocular crowding suggests that the effect is cortical in origin. This idea is further supported by the fact that crowding effects are more severe when the test and flanking stimuli are similar in size and shape (Kooi *et al.* 1994).

Crowding could be mediated by the lateral cortical connections discussed in Section 5.4.7. The spatial range of crowding is similar to the spatial range of cortical lateral connections, and both processes show a similar dependence on stimulus eccentricity (Tripathy and Levi 1994).

Tripathy and Levi presented a test letter *T* to the monocular region in the left eye corresponding to the blind spot in the right eye. Subjects reported the orientation of the test letter less accurately when three *T*s were placed in the region surrounding the blind spot in the right eye. This suggests that lateral cortical connections run from the region surrounding the blind spot in one eye into the monocular region corresponding to the blind spot in the other eye. The cortical area corresponding to the blind spot contains only monocular cells in the sense that each cell receives direct inputs from only one eye (LeVay *et al.* 1985). Nevertheless, if lateral connections run into the monocular area, this area is not strictly monocular.

8.2.5 Threshold-elevation

In the threshold-elevation effect a period of inspection of a suprathreshold masking grating elevates the contrast threshold of a subsequently exposed test grating in the same region of the same eye. The effect is observed only when the adaptation and test gratings have a similar spatial frequency and orientation. The effect shows interocular transfer of about 65% of its monocular value (Blakemore and Campbell 1969; Hess 1978). The degree of interocular transfer remained the same when the eye seeing the induction grating was pressure blinded in the test periods, showing that the effect is cortical in origin (Blake and Fox 1972). The functions relating the threshold-elevation effect to the contrast and duration of the induction stimulus were the same for the monocular aftereffect as for the transferred aftereffect (Bjorklund and Magnussen 1981). Threshold elevation produced by inspection of a line or a small number of dots did not show interocular transfer (Fiorentini *et al.* 1976). This suggests that detection of a grating involves additional processing at a higher level than detection of a line.

The elevation of contrast threshold of a test stimulus produced by an induction grating presented to the same eye was reduced when the induction grating was accompanied by a grating of a different spatial frequency presented to the other eye. This interocular effect operated only when the gratings were vertical, which suggests that it is related to stereopsis (Ruddock and Wigley 1976; Ruddock *et al.* 1979).

Adaptation to a region of high contrast reduces the apparent contrast of a subsequently seen region of lower contrast. Durgin found that this effect did not show interocular transfer and concluded that it depends on an early stage of visual processing.

All this evidence suggests that the threshold-elevation effect occurs at a site more central than that at which visual inputs are combined. The threshold-elevation effect measured binocularly was the same whether the induction stimulus was presented to one eye or the other, or alternately to the two eyes, as long as the total duration was the same (Sloane and Blake 1984). This suggests that the binocular aftereffect represents the pooled effect from binocular cells differing in ocular dominance.

Neurons in visual areas 17 and 18 of the cat respond less vigorously to a low-contrast drifting grating after they have been exposed to a similar drifting high-contrast grating (Ohzawa *et al.* 1985). Maffei *et al.* (1986) found the same level of interocular transfer of contrast adaptation in areas 17 and 18 of cats after section of either the corpus callosum or the optic chiasm. There was no transfer after both interhemispheric pathways were severed.

The threshold-elevation effect will be discussed further in Section 8.2.5a.

8.2.6 Meta- and paracontrast

In another form of successive masking, the monocular induction and test stimuli are presented only briefly and in adjacent locations rather than in the same location. For instance, a visual pattern presented for a small fraction of a second is not seen when followed by a stimulus in a nearby location when the interstimulus interval (ISI) is between 40 and 80 ms. This is known as **metacontrast** or **backward masking.** Under certain circumstances a test stimulus is masked by a stimulus that precedes it. This is known as **paracontrast** or **forward masking**. Metacontrast seems to have been first observed by Exner (1868). The early literature is reviewed in Alpern (1952). Werner (1935, 1940) used the disc-ring configuration in which a black test disc is not seen when followed by a masking black annulus. It is as if the new inner edge of the annulus desensitizes the system for subsequent processing of the edge of the disc. The two contours have opposite luminance polarity. The superimposition or close proximity of similar contours of opposite polarity seems to be a general feature of stimuli that generate metacontrast. The effects of metacontrast are evident in the response of cells in the visual cortex of anaesthetized monkeys, so that the suppression must occur early in the visual system (Macknik and Haglund 1999). Various theories of metacontrast and paracontrast have been proposed and are reviewed by Weisstein (1972).

8.2.6a *Metacontrast with cyclopean images*

Metacontrast can occur between a cyclopean shape defined by disparity and a binocular shape defined by luminance contrast, although this interdomain masking is less than when induction and test stimuli are either both cyclopean or both binocular (Patterson and Fox 1990). Lehmkuhle and Fox (1980) constructed a dynamic random-dot stereogram depicting an annular mask exposed for 160 ms and a central Landolt C test stimulus presented for 80 ms. Detectability of the gap in the Landolt C was reduced when mask and test stimulus occurred together. Detectability recovered to its value with no mask when the mask followed the test stimulus (backward masking) by more than 100 ms or preceded the test stimulus (forward masking) by more than about 300 ms. Backward masking was similar to that observed with regular contours but forward masking extended over a larger span of time than with regular contours. Masking declined as the test stimulus moved nearer to the viewer than the mask, but stayed constant when the test stimulus was moved beyond the mask (Section 23.4.2).

Metacontrast can affect the perception of depth in a random-dot stereogram. A mask consisting of a noisy 3-D array of dots reduced the accuracy with which depth in a random-dot stereogram was detected. Masking was greatest when the interstimulus interval was less than 50 ms (Uttal *et al.* 1975a). It is not clear from this result whether the masking effect depends on disparity in the mask— a 2-D mask may also produce the same effect. The results were interpreted in terms of the time it takes to perceive depth in a random-dot stereogram.

8.2.6b *Dichoptic metacontrast*

Dichoptic metacontrast can be studied by presenting the test stimulus and mask to different eyes. Werner found that dichoptic metacontrast was the same as when the stimuli were seen by the same eye (see also Kahneman 1968). Kolers and Rosner (1960) reported dichoptic metacontrast with a variety of stimuli. Like monocular masking, dichoptic masking occurs between a test flash stimulating only rods and an adjacent masking flash stimulating only cones (Foster and Mason 1977). However, there are differences between dichoptic and monocular metacontrast. Although letters are masked by a flash of light only when both stimuli are presented to the same eye they are masked by a pattern presented either to the same eye or to the other eye (Schiller 1965). For small interstimulus intervals, dichoptic masking is more pronounced than monocular masking (Schiller and Smith 1968), perhaps because of binocular rivalry. Dichoptic, but not monocular, metacontrast decreases with repetition of the stimuli (Schiller and Wiener 1963).

One way to think about dichoptic metacontrast is that the newly delivered stimulus switches dominance to the eye seeing the new stimulus before the processing of the first stimulus is complete in the other eye. Oppositely polarized luminance edges rival, even when presented simultaneously. Werner (1940) argued against explaining dichoptic metacontrast in terms of binocular rivalry on the grounds that masking due to rivalry should occur when the mask precedes the test stimulus as well as when it follows it. This is not a strong argument. When the test stimulus follows the mask, it may cause a switch of dominance to the eye containing the test stimulus. This should remove the contribution that rivalry makes to forward masking, leaving only the contribution of other factors.

The three types of cones act independently in increment thresholds (Stiles 1939). Alpern *et al.* (1970) showed that annular masking flashes are equally effective in raising the threshold of a preceding test flash detected by a given cone type, say red cones, only if the chromatic content of the flashes is adjusted to produce an equal effect on red cones (see also McKee and Westheimer 1970). It was concluded that metacontrast involves interactions at the retinal level between cones of the same type. Yellott and Wandell (1976) found that a flashed masking stimulus raised the threshold of a preceding test flash by the same amount when the flashes were presented to the same eye or to different eyes. But the inhibition was not cone-type specific in either case, which suggests some involvement of a central chromatic mechanism. However, Alpern *et al.* used a 1° test stimulus superimposed on a 9° masking disc whereas Yellott and Wandell used two 1°-wide by 3°-high bars flanking a 1°-wide test bar. Thus, there was much more opportunity for contour interactions in the latter stimulus than in the former, and dichoptic interactions are stronger when there are contiguous contours (Section 8.2.2).

The degree of masking of a 0.25° test disc, flashed on just before a slightly larger disc presented briefly to the same eye, was greatly reduced when the larger disc was immediately followed by a surrounding annulus, also in the same eye (Schiller and Greenfield 1969). Presumably the outer annulus masked the larger disc, which was then less able to mask the inner disc. When the conditioning and test stimuli were presented to one eye and the outer annulus to the other, the effect of the conditioning stimulus on the test stimulus was not weakened (Robinson 1968). Any postchiasmal effect of the annulus on the conditioning stimulus did not relieve the prechiasmal inhibitory effect of the conditioning stimulus on the test stimulus. This aspect of metacontrast does not show interocular transfer

Summary

A briefly presented test stimulus may be more difficult to see when a brief masking stimulus occurs in a neighbouring location, either just before the test stimulus (paracontrast) or just after it (metacontrast). The effect seems to be due to the close proximity of contours of opposite contrast polarity. Both forms of successive masking also occur between two shapes defined by disparity or between a disparity-defined shape and a shape defined by luminance. Both forms of masking also occur when the test and mask are presented to different eyes. Dichoptic masking may be due to rivalry between edges with opposite contrast polarity in the two eyes.

8.2.7 Transfer of chromatic adaptation

A gray patch appears green and a green patch appears a supersaturated green when superimposed on a larger area previously exposed to red light, an effect usually explained in terms of bleaching of the red receptors. DeValois and Walraven (1967) suggested that this effect is due largely to contrast produced in the small test patch by an afterdischarge from "red" receptors in the surrounding area. When a red adapting patch and a green test patch were the same size, the colour of the test patch was desaturated because of the combination of the red afterdischarge and the green of the test patch within the area of the test patch. When they placed the adapting red patch in one eye, a subsequently seen green patch in the other eye appeared gray whether it was smaller than or equal in area to the adapting patch. They argued that the contrast effect did not show dichoptically because it is retinal and that the grayness of the dichoptic test patch was produced by interocular colour combination of green from the test patch and a red afterdischarge from the same region in the other eye.

8.3 TRANSFER OF FIGURAL EFFECTS

In **figural induction** an induction stimulus affects the figural properties of a test stimulus, rather than its visibility. For instance, an induction stimulus may cause an apparent change in the orientation, size, or movement of the test stimulus. As with masking, the stimuli can be presented simultaneously or successively. Geometrical illusions are simultaneous figural effects. The tilt aftereffect and motion aftereffect are successive effects.

Interocular transfer occurs when an induction stimulus applied to one eye affects a test stimulus applied to the other. Interocular transfer has been studied for three purposes: (1) to reveal the site of processes responsible for a particular effect, (2) to investigate how inputs from the two eyes combine, and (3) to reveal effects of visual pathology on the combination of inputs from the two eyes. The first two purposes are discussed in this section and the third is discussed in Section 12.3. The physiology of binocular interactions was reviewed in Sections 7.10 and 6.7.2.

8.3.1 Experimental paradigms

Physiological studies reviewed in Chapter 6 have revealed the following types of cell in the visual cortex:

1. Binocular OR cells, which respond to excitatory inputs from either eye but no more strongly to both eyes than to either alone, some responding equally to either eye and others more vigorously to one eye than to the other.
2. Binocular AND cells, which respond only when excited by inputs from both eyes.
3. Excitatory/inhibitory cells, which receive an excitatory input from one eye and an inhibitory input from the other.
4. Monocular cells, which receive input from only one eye.

These types of cell probably form a continuum of types and their response properties may not be fixed; for instance, their response to fusible stimuli may differ from their response to rivalrous stimuli, and the excitatory/inhibitory ratio may vary as a function of stimulus strength.

Psychophysical experiments on interocular transfer of figural induction effects have been conducted to explore the role of the various types of cell in binocular processing. A major issue is the types of cortical cell required to account for the results of interocular transfer experiments. Wolfe and Held (1981, 1983) championed the view that AND cells are required in addition to OR cells and monocular cells. Moulden (1980) developed an account based on monocular cells and three types of OR cells that differ in their degree of ocular dominance, but no AND cells. Cogan's (1987) model of binocular processing involves only binocular cells. There is something arbitrary about such accounts because each explains a different set of data and could probably accommodate new data by adjusting parameters or by adding extra assumptions.

Table 8.2 summarizes the five basic experimental paradigms used in studies of interocular transfer of figural induction effects.

1. Interocular transfer paradigm In this procedure one first measures the figural induction effect with the induction and test stimuli in the same eye then with the two stimuli in different eyes. The transferred effect is expressed as a percentage of the same-eye effect. The following logic is then applied. The percentage transfer should (a) increase according to the extent to which the induction and test stimuli excite the same binocular OR cells, (b) diminish according to the extent to which the test stimulus excites unadapted monocular cells, (c) be influenced by the extent to which the different classes of cortical cells inhibit each other, (d) be influenced by the presence of post-induction effects in the closed eye during the test period, and (e) not be affected by the presence of binocular AND cells, since

Table 8.2 Paradigms used to study interocular transfer of figural effects

Paradigm	Induction stimulus	Test stimulus
1. Interocular transfer	monocular	monocular, same, or other eye
2. Monocular vs. binocular test	monocular	monocular, same, or both eyes
3. Binocular recruitment	monocular or binocular	binocular
4. Alternating monocular	alternating eyes	monocular or binocular
5. Cyclopean stimuli	cyclopean or noncyclopean	cyclopean or noncyclopean

such cells do not contribute to the strength of either the monocular effect or the transferred effect.

2. Monocular versus binocular test In this procedure the induction stimulus is presented monocularly and the test stimulus either to the same eye or to both. For example, Wolfe and Held (1981) argued that unadapted AND cells, which are activated only when both eyes are open, cause the tilt aftereffect to be less with binocular testing than that with monocular testing. The aftereffect with binocular testing after monocular induction would also be reduced through the activation of unadapted monocular cells associated with the unadapted eye. Wolfe and Held determined the reduction of the aftereffect due to this factor by measuring the amount of interocular transfer. The degree of transfer was insufficient to account for the reduction from one eye to two eyes. The extra reduction in the aftereffect was put down to the effect of recruitment of unadapted AND cells in binocular testing.

3. Binocular recruitment In this procedure the aftereffect is first measured with both the induction and test stimuli viewed binocularly and then with a monocular induction stimulus and a binocular test stimulus. The presence of AND cells should make the aftereffect with binocular induction larger than with monocular induction. Wilcox *et al.* (1990) pointed out that this argument cannot be used to infer the presence of binocular AND cells. While it is true that binocular testing after monocular induction brings unadapted AND cells into play, which reduce the aftereffect, it also brings the monocular cells of the adapted eye into play, which increase the aftereffect. Since there is no way of knowing the relative contributions of these opposed influences, it is not possible to draw conclusions about what types of cortical cell are involved in the aftereffect.

4. Alternating monocular induction In this procedure the induction stimulus is presented alternately to each eye for a period of time, and the test stimulus is then presented to either one eye or both simultaneously (Blake *et al.* 1981b). For comparison, the same procedure is followed with the induction stimulus presented intermittently to both eyes. The following logic is then applied. The alternating induction sequence adapts binocular OR cells and monocular cells for both eyes. Therefore, for these types of cells, the aftereffect should be the same for monocular and binocular testing. Since the alternating induction sequence does not adapt AND cells, the aftereffect is diluted by the activation of these unadapted cells during binocular testing. With monocular testing, the AND cells are not excited and therefore do not dilute the aftereffect. Thus, any reduction in the aftereffect with binocular testing relative to monocular testing indicates the presence of AND cells. This logic is not subject to the ambiguity of the binocular-recruitment paradigm.

5. Cyclopean stimuli A cyclopean induction or test stimulus is defined by disparities in a random-dot stereogram. Such cyclopean stimuli do not excite purely monocular cells. Both induction and test stimuli can be cyclopean or either one can be cyclopean and the other a conventional contrast-defined stimulus. The logic underlying the use of cyclopean stimuli in interocular transfer experiments becomes apparent in what follows.

There are many pitfalls in applying these arguments and the literature has become complex and rather contentious.

Any asymmetry in ocular dominance should result in more interocular transfer from one eye than from the other. A person lacking binocular cells should show neither interocular transfer nor binocular recruitment of cortically mediated aftereffects (Section 12.3).

A basic problem in all studies of interocular effects is that merely closing an eye does not stop inputs from that eye reaching the cortex. For instance, an afterimage impressed on one eye is visible when that eye is closed and the other eye opened, a fact noted by Isaac Newton in 1691 (see Walls 1953). This does not prove, as was once thought, that afterimages are cortical in origin; it simply means that activity arising in a closed eye still reaches the visual cortex (Day 1958). We now know that an afterimage is no longer visible when the eye in which it was formed is pressure blinded by pressing the finger against the side of the eye for about 30 s (Oswald 1957). Pressure cuts off the blood supply to the retina and the eye remains blind until the pressure is relieved. It is dangerous to keep the pressure on for more than about one minute.

Another problem in studies of interocular transfer of successive induction effects is that, in non-transfer trials, the same eye is used in both induction and test periods whereas, in transfer trials, the adapted eye is open during the adaptation period and closed during the test period, and the tested eye is at first closed and then opened. The sudden transition in the state of adaptation of the eyes in transfer trials could cause a spurious weakening of the aftereffect being tested. This problem can be solved by keeping both eyes in the same state of adaptation at all times. As we will see, this precaution has been applied in only one study.

There are three ways to prove that an effect is cortical in origin.

1. Show that the effect survives pressure blinding the eye to which the induction stimulus was presented.
2. Show that the effect depends on visual features that are processed only in the visual cortex. For instance, it is believed that the orientation of visual stimuli is first processed in the visual cortex so that an induction effect that is specific to the orientation of the stimulus should be cortical in origin.
3. Show electrophysiologically that the first adaptive changes to an induction stimulus arise at the level of the cortex. Other issues concerning drawing conclusions from studies of interocular transfer were discussed by Long (1979).

8.3.2 Transfer of tilt contrast

8.3.2a *Psychophysical studies*

In orientation contrast the apparent orientation of a test line in a frontal plane changes when it is intersected by or placed adjacent to a second line in a slightly different orientation. In **simultaneous orientation contrast** the stimuli are presented at the same time. In successive orientation contrast, also known as the **tilt aftereffect**, the test stimulus is presented after the induction stimulus. It is generally believed that simultaneous orientation contrast is due to inhibitory interactions between orientation detectors in the visual cortex. In the tilt aftereffect, an added factor is believed to be the selective adaptation of the orientation detectors responding to the induction stimulus, leading to a bias in the activity of the population of orientation detectors responding to the test stimulus. Inspection of an off-vertical line induces an apparent tilt of a vertical line in the opposite direction. This form of the tilt aftereffect may be due partly to orientation contrast but it is also due to normalization of the tilted line to the vertical and a concomitant shift in the scale of apparent tilt with respect to the vertical. In a contrast effect, the angle between the induction and test lines apparently increases whereas, in tilt normalization, both lines appear displaced in the same direction. See Howard (1982) for a review of this topic.

Figure 8.16. Stimuli used to measure a dichoptic tilt aftereffect. Wolfe and Held (1981) used a chevron pattern (a) as induction stimulus. In the test phase of the experiment, subjects set a chevron pattern to appear straight (b). Mouldon (1980) used tilting lines (c) as induction stimulus and subjects set lines to vertical in the test phase.

Coltheart (1973) proposed that binocular cells tuned to orientation mediate interocular transfer of orientation contrast. There seems to be general agreement that simultaneous tilt contrast does not occur when the induction stimulus and the test stimulus are presented to different eyes, with the angle between the lines less than about 10° (Virsu and Taskinen 1975; Walker 1978a). One reason for this may be that lines close together tend to rival. For larger angles between induction and test stimuli, Walker obtained a small amount of interocular effect and Virsu and Taskinen obtained transfer equal to about 60% of that obtained when both lines were presented to the same eye. Virsu and Taskinen used two intersecting lines that were 3° long, which were subject to binocular rivalry and fusion. Walker used a display free from rivalry, consisting of an annular induction grating with an outer diameter of 4.8° surrounding a 1.6° diameter test grating.

A textured surface rotating in the frontal plane about the visual axis causes a superimposed vertical line to appear tilted in a direction opposite to the background motion. When the rotating surface was presented to one eye and the line to the other, this effect remained at full strength in subjects with normal binocular vision but was significantly reduced in stereoblind subjects (Marzi *et al.* 1986). Motion-induced tilt seems to be a more strongly binocular effect than static tilt contrast.

The tilt aftereffect shows interocular transfer when the induction line or grating is presented to one eye and the test line to the other. Estimates of the extent of transfer vary between 40 and 100% (Gibson 1937; Campbell and Maffei 1971). In this case, binocular rivalry is not a factor because the induction and test displays are presented successively. The strength of the transferred aftereffect, like that of the monocular aftereffect, did not depend on whether the induction and test stimuli had the same or opposite contrasts (O'Shea *et al.* 1993). This suggests that the tilt aftereffect occurs at a level where contrast polarity is unimportant. Interocular transfer of the tilt aftereffect is positively correlated with stereoacuity and is absent in people lacking stereoscopic vision (Section 12.3.3).

When the world is viewed through prisms, the images of straight lines are made convex towards the base of the prism. After some time the lines appear straight again and, for a while after the prisms are removed, straight lines appear curved in the opposite direction. This curvature aftereffect has been reported to show between 60 and 100% interocular transfer (Gibson 1933; Hajos and Ritter 1965).

Wolfe and Held (1981) used an induction stimulus like that shown in Figure 8.16a. After exposure to the induction stimulus a test stimulus with straight lines appeared as a chevron bent in the opposite direction. Subjects adjusted a chevron test stimulus until the lines appeared straight and parallel. Wolfe and Held applied paradigm 2 (Section 8.3.1) and found that the aftereffect showed about 70% interocular transfer, but only 40% transfer when both eyes saw the test stimulus. They concluded that unadapted monocular cells were responsible for the dilution of the aftereffect in going from one eye to the other, and the recruitment of unadapted binocular AND cells was responsible for the extra dilution in going from one eye to two. Moulden (1980) did a similar experiment using a single set of parallel tilted lines, like those in Figure 8.16b, rather than a chevron pattern and obtained the opposite result, namely, more aftereffect when both eyes viewed the test stimulus than when only the unadapted eye viewed it. However, Wolfe and Held pointed out that tilt normalization might be involved in the stimulus used by Moulden. Wilcox *et al.* (1990) used both the chevron and tilted-lines stimuli and obtained the same result as Moulden. They concluded that because of the uncertain role of unadapted monocular cells, the existence of binocular AND cells cannot be established by this procedure.

In addition, Wolfe and Held applied paradigm 4 and found that, after alternating monocular adaptation, the binocular aftereffect was less than either monocular aftereffect. They argued that unadapted AND cells were also responsible for this result. The argument is more convincing in this case since there were no unadapted monocular cells. Wilcox *et al.* (1990) obtained a similar result and agreed with Wolfe and Held that this supports the idea of there being binocular AND cells. Blake *et al.* (1981b) found that gratings presented in alternation to the two eyes produced equal threshold elevations in monocular and binocular test stimuli whereas gratings presented intermittently to both eyes gave different elevations. This argues against the existence of AND cells. Wolfe and Held suggested that the stimuli used for the threshold-elevation effect, unlike those used for the suprathreshold tilt aftereffect, fall below the luminance threshold of AND cells. Wilcox *et al.* (1994) found that the difference between the monocular and binocular tilt aftereffects and that between the monocular and binocular threshold-elevation effects to be similar both at and above the contrast threshold. They suggested that Blake *et al.*'s failure to find evidence of AND cells with the threshold-elevation effect was due to the blank intervals in their intermittent exposure condition, which were not present in the alternating condition. When Wilcox *et al.* controlled for this factor they found evidence for AND cells using the threshold-elevation effect.

Wolfe and Held (1982) provided further support for the existence of AND cells. The induction stimulus was a chevron pattern defined by binocular disparities in a random-dot stereogram so that the pattern was not visible to either eye. The test stimulus was a noncyclopean chevron pattern defined by luminance, which subjects set into alignment. The aftereffect showed only when the test pattern was viewed binocularly.

Wolfe and Held argued that the effect would have been visible with a monocularly viewed test stimulus if the cyclopean induction pattern had stimulated binocular OR cells. A cyclopean induction stimulus does not adapt monocular cells. They concluded that a cyclopean image involves the stimulation of mainly AND cells that require a simultaneous input from both eyes. They admitted that the procedure might have been insensitive to the response of a small percentage of OR cells to the cyclopean image. Binocular viewing of a noncyclopean induction stimulus produced equal monocular and binocular aftereffects. Although monocular cells and binocular OR cells were adapted in this case, there should have been some advantage of binocular viewing, since it alone brought in AND cells. They suggested that the absence of a binocular advantage was due to a ceiling effect.

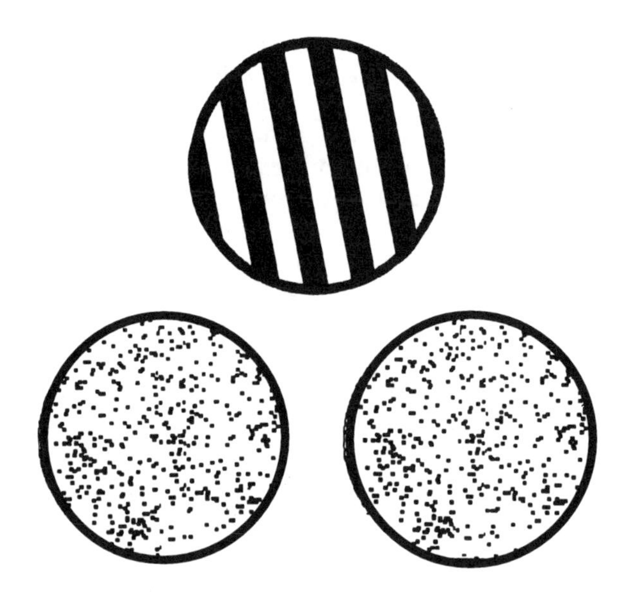

Figure 8.17. A cyclopean tilt aftereffect.
Inspection of the upper grating for about 3 minutes causes the cyclopean bar in the stereogram to appear tilted to the right. (Adapted from Burke and Wenderoth 1989)

Wolfe and Held did not examine the case in which the induction stimulus is monocular and therefore noncyclopean, and the test stimulus is cyclopean. According to their theory there should be no aftereffect because monocular stimulation cannot adapt binocular AND cells, and only AND cells are excited by a cyclopean image. Burke and Wenderoth (1989) conducted this experiment with the stimuli shown in Figure 8.17. All subjects saw an aftereffect in the vertical line defined by disparity in the random-dot stereogram after monocular inspection of tilted lines defined by luminance. Binocular inspection of the tilted lines produced a similar result. They used an induction stimulus with a single set of tilted lines whereas Wolfe and Held used a chevron pattern. Adaptation to a single set of tilted lines brings in tilt normalization—the high-level process that causes a tilted display to appear vertical. The chevron-pattern effect depends on orientation contrast—interaction between the two halves of the pattern. *Whether or not this makes a difference remains to be investigated.*

Wolfe and Held (1983) concluded from other experiments on the diminution of the tilt aftereffect when going from monocular adaptation to binocular testing that binocular AND cells are more sensitive to low than to high spatial frequencies and are not responsive to near-threshold stimuli or to stimuli that are blurred in one eye. They pointed out that stereopsis is sensitive to the same stimulus features and concluded that AND cells are involved in stereopsis.

Paradiso *et al.* (1989) obtained 92% interocular transfer of the tilt aftereffect when the test stimulus was composed of subjective contours but only 46% when it was a real bar. The same difference was found whether the induction stimulus consisted of a subjective bar or a real bar. They related this finding to the fact that von der Heydt and Peterhans (1989) had found cells responding to subjective contours only in V2 of the alert monkey, an area containing more binocular cells than are contained in V1.

8.3.2b *Physiological data on orientation contrast*

The response of a cell in the visual cortex to an optimally oriented bar or grating is suppressed by a superimposed orthogonal bar or grating presented to the same eye (Bishop *et al.* 1973; Bonds 1989). This is known as **cross-orientation inhibition**. It is largely independent of the relative spatial phases of the gratings and operates over a wide difference in spatial frequency. Cross-orientation inhibition was not elicited when the test and orthogonal gratings were presented to different eyes (DeAngelis *et al.* 1992). This suggests that the effect is generated in the visual cortex before signals from the two eyes are combined, and is not responsible for interocular transfer of orientation aftereffects.

The response of a cell in the cat's visual cortex to an optimal stimulus centred on the receptive field is larger when surrounding lines outside the cell's normal receptive field have a contrasting orientation than when they have the same orientation (Section 5.5.2). DeAngelis *et al.* (1994) found that, for some cells, the inhibitory effect of similarly oriented lines could be evoked dichoptically. This suggests that it depends on intracortical inhibitory connections. This could provide a physiological basis for interocular transfer of orientation aftereffects.

Summary

Simultaneous tilt contrast, motion-induced tilt, the tilt aftereffect, and the curvature and chevron aftereffects show about 60% interocular transfer. Considerable effort has been expended in using interocular transfer of these effects to reveal binocular AND cells. Binocular AND cells are known to exist from physiological evidence (Section 6.2), but they seem to be few in number. Perhaps the role of binocular AND cells in binocular vision will never be resolved by psychophysical means. There are so many factors to be taken into account. The same data can be explained by making different assumptions about the types of cell present, their inhibitory or facilitatory interactions, their differential luminance thresholds, their differential dependence on stimulus features such as spatial frequency, and the degree of binocular congruence of the images.

8.3.3 Transfer of the motion aftereffect

8.3.3a *Basic studies*

In the motion aftereffect, inspection of a textured display moving in one direction causes a stationary display seen subsequently to appear to move in the opposite direction. Aristotle (trans. 1931) saw the effect after looking at a flowing river, and mentioned it in his book *Parva Naturalia*. Purkinje (1825) noticed that after looking at a cavalry parade the houses appeared to move in the opposite direction. Addams (1834) experienced the landscape moving after looking at the falls of Foyers in Scotland. Purkinje and Addams believed that the effect was due to continuation of nystagmic eye movements. Wohlgemuth (1911) reviewed the early work on the motion aftereffect. Holland (1965) and Mather *et al.* (1998) have provided more recent reviews.

The motion aftereffect can be conveniently observed by inspecting the centre of a rotating spiral for about a minute and then transferring the gaze to a stationary pattern. An inwardly rotating spiral causes an apparent expansion of a stationary pattern and an outwardly rotating spiral causes an apparent contraction (Plateau 1850). The spiral aftereffect cannot be due to eye movements since it occurs in all radial directions simultaneously. The effect is generally believed to be due to the selective fatigue of one set of motion detectors and the subsequent disturbance in the balance of activity across the population of motion detectors.

The magnitude of the motion aftereffect has been measured by recording its duration (Pinckney 1964), by estimating its apparent velocity (Wright 1986), by nulling it with a real motion in the opposite direction (Taylor 1963), and by measuring its effect on the threshold for detection of motion in the adapted and unadapted directions of motion (Levinson and Sekuler 1975). In a variant of the latter procedure, Raymond (1993) measured the motion aftereffect by the elevation in the motion-coherence threshold—the percentage of coherently moving dots in a dynamic random-dot display required for detecting unidirectional motion.

A large part of the motion aftereffect shows when the inspection display is presented to one eye and the stationary test display to the other. This fact was first noted by Dvorák (1870) and has been confirmed by several investigators (Ehrenstein 1925; Freud 1964; Lehmkuhle and Fox 1976). The effect shows interocular transfer only when the induction and stationary test stimuli fall on corresponding areas of the two retinas (Walls 1953). Estimates of the magnitude of the transferred aftereffect relative to that elicited in the same eye vary between zero and 78%, with a mean of about 50% (Wade *et al.* 1993). All zero-transfer results were obtained using moving square-wave gratings and the rather insensitive criterion of duration of the aftereffect. For all other stimuli and criteria, interocular transfer was at least 40%.

The estimate of 78% was obtained by Lehmkuhle and Fox using the following procedure. During the induction period the non-adapted eye was presented with a display with the same space-average luminance as the moving display presented to the other eye. This prevented a sudden change in the state of light adaptation of the non-adapted eye in the transition from induction period to test period. Without this control, there was only 52% interocular transfer. The crucial factor may not be the presence or absence of a sudden change of luminance but the presence or absence of contours in the non-adapted eye. During the induction period the occluded non-adapted eye saw nothing whereas the illuminated non-adapted eye saw the aperture within which the moving display was presented to the other eye.

Timney *et al.* (1996) argued that conjugate eye movements induced by the moving display caused the images of the aperture to move and generate a motion aftereffect in the supposedly non-adapted eye. They supported their view with two observations. First, illuminating the non-adapted eye had no effect on the interocular transfer of the tilt aftereffect or the threshold elevation effect, neither of which involves motion. Second, when they removed all contours from the illuminated display presented to the non-adapted eye, the interocular transfer of the motion aftereffect was the same as when the non-adapted eye was occluded. It seems therefore that interocular transfer of the motion aftereffect is best reflected by the 52% value, which Lehmkuhle and Fox obtained with the non-adapted eye occluded.

8.3.3b *Motion aftereffects at different levels of processing*

If motion detectors were in the retina, one might expect some interocular transfer of the effect, since closing the eye in which the induction stimulus was presented would not prevent inputs from the adapted detectors from reaching the visual cortex. However, the motion aftereffect transferred from one eye to the other when the retina of the eye exposed to the induction stimulus was pressure blinded just after the exposure period (Barlow and Brindley 1963; Scott and Wood 1966). This proves that at least some of the motion detectors responsible for motion aftereffects in humans are at a higher level than the retina. See Section 23.2.2 for a discussion of the interocular transfer of adaptation to plaid motion in superimposed orthogonal gratings.

Raymond (1993) obtained 96% transfer of the motion aftereffect from a random-dot display when assessed by elevation of the motion-coherence threshold. A motion-coherence threshold is the least proportion of coherently moving dots in an array of randomly moving dots that can be seen as moving in a given direction. There are no sudden changes of luminance with stimuli of this kind, which may account for the high interocular transfer. Raymond favoured the view that the site of this interocular transfer is the extrastriate area MT, which is known to contain only binocular cells and to be sensitive to the degree of coherent motion (Murasugi *et al.* 1993).

Steiner *et al.* (1994) found less interocular transfer of the motion aftereffect from translatory motion than from expansion or rotation and argued that this is because the latter types of motion are processed at higher levels of the visual system.

Nishida *et al.* (1994) cited evidence that the motion aftereffect tested with a static grating involves a lower level of neural processing than that tested with a directionally ambiguous grating flickering in counterphase. In conformity with this evidence, they found that the duration of the aftereffect induced by a drifting grating showed 30 to 50% interocular transfer when tested with a static grating but almost 100% transfer when tested with a flickering pattern. However, Nishida and Ashida (2000), in agreement with Hess *et al.* (1997c), found that interocular transfer with a flickering test pattern was not complete in the retinal periphery. They also found, in agreement with Steiner *et al.* (1994), that transfer was not complete when the aftereffect was measured with a nulling method nor when the observer's attention was distracted by a secondary task.

If a dynamic test display, such as the counterphase flicker test, reflects activity at a higher level in the nervous system than does a stationary test display one should be able to use these tests to reveal motion aftereffects at different neural levels. Motion of a luminance-defined pattern is first-order motion, and motion of a pattern defined by modulation of contrast or by edges defined by texture, is second-order motion. It is believed that first-order motion is processed at a lower level in the nervous system than is second-order motion. A motion aftereffect induced by first-order motion shows with a static test stimulus but that induced by a second-order stimulus shows only with a dynamic test stimulus (McCarthy 1993). Furthermore, simultaneous adaptation to first-order motion in one direction and second-order motion in the opposite direction produced an aftereffect with a static pattern in the opposite direction to the first-order stimulus and an aftereffect with a dynamic pattern in the opposite direction to the second-order stimulus (Nishida and Sato 1995). Also, a motion aftereffect produced by a drifting grating defined by texture showed no interocular transfer when tested with a static test pattern but almost complete transfer when tested with a pattern flickered in counterphase (Nishida *et al.* 1994). See Nishida and Ashida (2001) for further discussion of interocular transfer of different types of motion aftereffects.

The motion aftereffect is greatly enhanced when the induction stimulus is surrounded by a stationary texture rather than by a blank field. This effect may be due to the fact that a display consisting only of moving elements induces pursuit eye movements, which cancel the motion of the stimulus on the retina. Enhancement of the motion aftereffect by a stationary surround does not occur when the induction stimulus is presented to one eye and the textured stationary surround to the other (Symons *et al.* 1996). This suggests that relative motion signals are required to be present in the same eye for the generation of the aftereffect.

8.3.3c *Motion aftereffects from distinct velocities*

Two lines of evidence suggest that slow motion and fast motion are processed in distinct channels.

The standard motion aftereffect does not occur for induction stimuli moving at velocities in excess of 30°/s. However, an induction stimulus moving faster than 30°/s induces an aftereffect in a test stimulus consisting of dynamic random noise (Verstraten *et al.* 1998).

Inspection of superimposed dot patterns moving at the same velocity in different directions produce a motion aftereffect that represents the vector sum of the two motion signals. When the induction stimuli move in opposite directions, the aftereffects cancel (Verstraten *et al.* 1994a). However, adaptation to two superimposed random-dot patterns moving at different velocities (4 and 12°/s) in orthogonal directions produces an aftereffect opposite the slow motion component in a static test display and an aftereffect opposite the fast motion in a dynamic test display. When the test stimulus contains both static and dynamic elements, the aftereffect is that of two superimposed patterns moving orthogonally (van der Smagt *et al.* 1999).

Dichoptic patterns moving in opposite directions engage in binocular rivalry. However, a rapid pattern in one eye and a slow pattern in the other eye do not rival but produce the impression of transparent motion of two arrays (van de Grind *et al.* 2001). They called this effect **dichoptic motion transparency**. It suggests that there are two motion channels, each with a distinct rivalry stage.

8.3.3d *Distinct motion aftereffects in each eye*
Adaptation stimuli moving in different directions and superimposed in the same eye produces an aftereffect in a direction opposite the vector sum of the induction stimuli. The aftereffects cancel when the induction stimuli move at the same velocity in opposite directions. A vector sum aftereffect also occurs in a binocular test stimulus when the two induction motions are presented to different eyes (Grunewald and Mingolla 1998). However, when only one eye views the test stimulus, the direction of the aftereffect is opposite that of the induction stimulus that was presented to that eye (Wohlgemuth 1911; Anstis and Moulden 1970). When the test stimulus consisted of stationary rivalrous dichoptic orthogonal gratings, whichever image was dominant appeared to move in a direction opposite to the motion to which that eye had been exposed (Ramachandran 1991).

These results support the idea that motion adaptation occurs in distinct directions in the two monocular sites, which then sum at a binocular site. The monocular aftereffects could be due to monocular motion-sensitive cells in the visual cortex, which selectively adapt to input from only one eye. But they could also represent adaptation in that subset of binocular cells for which that eye forms the dominant input.

Anstis and Duncan (1983) extended this paradigm as follows. A rotating textured disc was seen rotating clockwise for 5 s by the left eye, then for 5 s by the right eye, and finally rotating counterclockwise for 5 s by both eyes. The sequence was repeated 40 times. In the test period a stationary disc appeared to rotate counterclockwise when viewed by either eye alone and clockwise when viewed by both eyes. Similar eye-specific aftereffects were reported by Jiao *et al.* (1984). The binocular aftereffect must have arisen in binocular cells.

The monocular effects in Anstis and Duncan's study could not have been induced in monocular cells in a straightforward way, since each eye was exposed to equal clockwise and counterclockwise motion. Three processes have been proposed to account for these monocular aftereffects:

1. Anstis and Duncan suggested that the response of monocular cells responding to clockwise motion was inhibited by binocular cells responding to clockwise motion.

2. Tyler suggested to us that these effects can be explained by the results of his 1971 experiment in which binocular motion signals suppressed monocular motion signals (Section 19.11.4). Thus, the counterclockwise binocular motion signals do not excite monocular cells and therefore do not cancel effects of monocular exposure to clockwise motion.

3. Van Kruysbergen and de Weert (1993) distinguished between a pure monocular system, a simple binocular system, and a pure binocular system. In Anstis and Duncan's experiment, the pure monocular system for each eye was exposed to equal amounts of clockwise and counterclockwise motion and therefore did not exhibit an aftereffect. The simple binocular system received 5 s of clockwise motion from each eye separately and 5 s of counterclockwise motion from the two eyes simultaneously. Overall, the simple binocular system was exposed to clockwise motion for a longer period than it was exposed to counterclockwise motion. It therefore generated a counterclockwise aftereffect with monocular testing. The pure binocular system was activated by only counterclockwise motion and therefore exhibited a clockwise aftereffect with binocular testing. It must be assumed that the pure binocular system gives a stronger aftereffect than the simple binocular system; otherwise the two would cancel with binocular testing. Van Kruysbergen and de Weert repeated Anstis and Duncan's experiment with subjects being unaware of the conditions they were exposed to. They produced further evidence for the pure monocular system and for the two types of binocular system. In addition, their results suggest the existence of a central monocular system for each eye. These various systems are depicted in Figure 8.18 (see also van Kruysbergen and de Weert 1994).

Motion of cyclopean contours not present in either monocular image can create a motion aftereffect (Papert 1964; Anstis and Moulden 1970). This aftereffect must depend on binocular cells since only binocular cells register a cyclopean image.

8.3.3e *Interocular transfer of induced visual motion*
A stationary object superimposed on a moving background appears to move in the opposite direction, an effect known as **induced visual motion**. Swanston and Wade (1985) presented a stationary test line to one eye and a symmetrically expanding background to the other. Induced visual motion evident with ordinary binocular viewing was not present in the dichoptic condition, although binocular rivalry complicated the dichoptic judgments. Confining the moving display to an annulus surrounding a stationary disc eliminated the effects of rivalry. With this display, dichoptic induced motion occurred at about 30% of its normal value (Day and Wade 1988). It is possible, however, that part of the dichoptic effect resulted from cyclorotation of the eyes induced by the moving annulus.

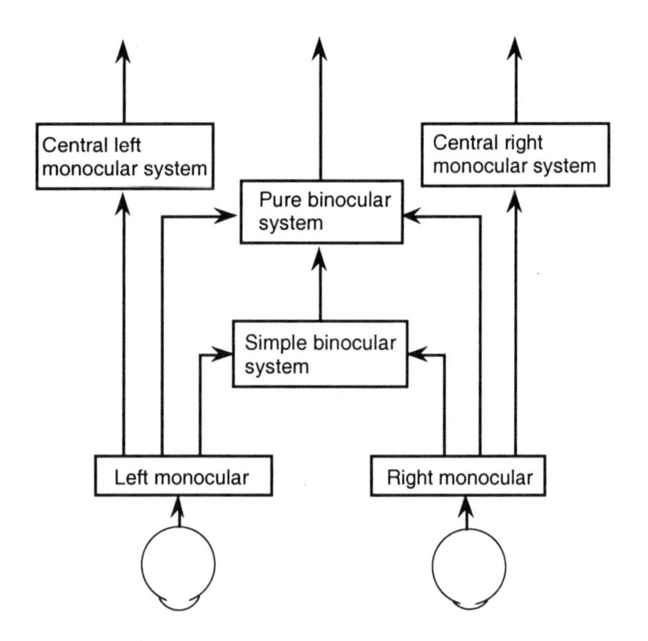

Figure 8.18. Monocular and dichoptic motion detection. Neural systems proposed by van Kruysbergen and de Weert (1993) to account for monocular and binocular motion aftereffects.

8.3.3f *Physiology of transfer of the motion aftereffect*

Physiological evidence has accumulated that motion detectors in primates are not located in the retina or LGN, but occur in the primary visual cortex and other visual areas in the central nervous system. Substantial evidence exists that motion detection at each location of the visual field is mediated by a set of cortical detectors, each optimally sensitive to a particular direction and speed of motion, but with overlapping tuning functions (see Sekuler *et al.* 1978). The motion aftereffect is believed to be due to selective adaptation of the motion detectors responding to the inspection stimulus, which causes an imbalance in the response of the set of motion detectors to a stationary stimulus. It has been reported that the motion aftereffect is not induced by inspection of a moving display that fills the visual field (Wohlgemuth 1911). This is perhaps not surprising, since the optokinetic response of the eyes would tend to null the motion of the image over the retina. Retinal motion is preserved when there is relative motion in the visual field. In any case, relative motion is a more potent stimulus than absolute motion (Snowden 1992).

It is reasonable to suppose that interocular transfer of the motion aftereffect is mediated by motion-sensitive binocular cells in V1 or at a higher level. Experiments on motion-sensitive cells in the visual cortex of unanaesthetized and lightly anaesthetized cats have provided direct evidence for this idea. For about 30 s after being stimulated by a moving grating for 30 s, the cells remained less sensitive to motion in the adapted direction and more sensitive to motion in the opposite direction (Vautin and Berkley 1977; Hammond *et al.* 1988). Cells adapted to a moving grating presented to one eye showed similar but weaker aftereffects when tested with stimuli presented to the other eye. Interocular transfer in a cell was stronger when the induction stimulus was presented to the eye that provided the dominant input to that cell than when it was presented to the nondominant eye. No transfer of the aftereffect occurred for cells classified as monocular (Hammond and Mouat 1988). Interocular transfer of direction-specific motion adaptation was most evident in simple cortical cells and showed only when the cells were adapted to motion in the preferred direction (Cynader *et al.* 1993). The interocular motion aftereffect, like the monocular aftereffect, is specific to the direction of motion of the induction stimulus. Thus, the directional tuning of binocular cortical cells must be the same for both eyes. This explains why binocular contrast sensitivity for moving gratings presented dichoptically is better than monocular contrast sensitivity only when the gratings move in the same direction (Arditi *et al.* 1981b).

Summary

The motion aftereffect shows considerable interocular transfer when the induction stimulus is confined to one eye. The extent of transfer is greater when the aftereffect is tested with a dynamic pattern with ambiguous motion rather than with a stationary pattern. A dynamic test pattern seems to tap aftereffects occurring at a higher level in the nervous system or at higher velocities. The motion aftereffect produced by second-order motion shows only with a dynamic test pattern, presumably because second-order motion is processed at a higher level than first-order motion. Distinct motion aftereffects may be generated in each eye, and seem to be at least partially distinct from an aftereffect generated by binocular viewing. Induced visual motion also shows considerable interocular transfer.

8.3.4 Transfer of the spatial-frequency shift

After inspection of a grating of a given spatial frequency, a grating of lower spatial frequency appears coarser than it normally appears and a grating of higher spatial frequency appears finer (Blakemore and Sutton 1969). This is the **spatial-frequency shift**. The aftereffect shows interocular transfer, although the size of the transferred effect relative to the ordinary aftereffect does not seem to have been measured (Murch 1972). Cyclopean shapes generated in a random-dot stereogram can also induce the spatial-frequency shift (Section 17.2). The aftereffect shows the same interocular transfer when the adapted eye is pressure blinded during the test period (Meyer 1974). Therefore transfer must be central rather than arise from activity in the closed adapted eye.

Favreau and Cavanagh (1983, 1984) obtained interocular transfer of the spatial-frequency shift with equiluminant coloured gratings, but only when the test grating was exposed for less than 400 ms or was flickering. They argued that the interocular transfer reflected the activity of colour-coded binocular cells with transient characteristics that have been found in V4 of the monkey. Transfer of other effects involving colour is discussed in the next section and the question of stereopsis with equiluminant stimuli is discussed in Section 18.1.4.

Adaptation to a high-density random-dot display reduces the apparent density of a subsequently seen display of lower dot density. This effect showed interocular transfer of about 70% (Durgin 2001).

8.3.5 Transfer of contingent aftereffects

A contingent aftereffect is one that depends on a particular combination of two stimulus features. The first contingent aftereffect to be reported involved colour and orientation. The human eye is not corrected for chromatic aberration, which means that blue light is brought into focus nearer the lens than red light. This produces colour fringes along black-white borders away from the optic axis. However, we do not see these colour fringes, presumably because the visual system applies a correction at a neural level. Colour fringes specific to the luminance polarity of edges appear when we view the world through prisms, because prisms increase the degree of chromatic aberration above its normal level. For instance, base-left prisms produce blue fringes on the right of light regions and red fringes on the left. These fringes disappear after the prisms have been worn for a few days, which reinforces the idea that the neural system compensates for them. In a footnote to a paper on adaptation to prismatically induced curvature, Gibson (1933) reported that colour fringes of opposite sign were seen for several hours after removal of prisms that had been worn for 3 days. These so-called **phantom fringes** must be neural rather than optical in origin since they show in monochromatic light (Hay *et al.* 1963). Phantom fringes represent the first-known example of a contingent aftereffect—the colours are contingent on the polarity of the edge. Adaptation to prism-induced chromatic fringes does not transfer from one eye to the other (Hajos and Ritter 1965).

Celeste McCollough (1965) discovered a contingent aftereffect that can be induced in a few minutes. Subjects viewed an orange and black vertical grating for 10 s, then a blue and black horizontal grating for 10 s. After several minutes of exposure to these alternating stimuli, a vertical achromatic grating appeared black and blue-green (the complementary colour of the vertical induction grating) and a horizontal grating appeared black and orange (the complementary colour of the horizontal induction grating). This type of contingent aftereffect is known as the **McCollough effect**. It is not an ordinary colour aftereffect obtained by gazing steadily at a coloured grating, since both colours stimulate all regions of the retina when the eyes scan the inspection stimuli. Also, the effect lasts for several hours, days, or even months, unlike aftereffects produced by single stimulus features, which last only minutes. See Stromeyer (1978) for a review of this topic.

McCollough noted that there was no colour-contingent aftereffect when the inspection stimuli were presented to one eye and the test stimuli to the other, except for one subject who reported an aftereffect in which the same rather than the complementary colours appeared in the test stimuli. This is the **positive contingent aftereffect**. Several other investigators have confirmed that the negative McCollough effect does not transfer to a nonadapted eye (Murch 1974; Stromeyer 1978). Nine out of 27 subjects tested by Mikaelian (1975) observed a weak positive McCollough aftereffect in an eye that was not exposed to the induction stimulus. Fifteen of the subjects observed the positive aftereffect in the unstimulated eye after scanning the achromatic test stimulus with both eyes for some time.

White *et al.* (1978) argued that keeping one eye closed during the induction period does not provide a fair test of interocular transfer, since the eyes receive different levels of luminance and colour. They overcame this problem by presenting the pattern component of the inspection stimulus to one eye while both eyes saw the colour component. Thus, one eye was exposed to red vertical and green horizontal gratings, as in the regular McCollough effect, while the other eye saw only the alternating colours, without the patterns. White *et al.* obtained a negative aftereffect when the eye seeing only the colour was tested. The positive aftereffect was seen when the eye that had been exposed to the patterned stimulus was pressure blinded during testing. This provides strong evidence that the figural processes responsible for the McCollough effect occurs, at least partially, in binocular cells. *The conclusion would be strengthened if the experiment were also done the other way around—exposing one eye to the regular*

McCollough induction stimuli and the other eye only to alternating achromatic gratings.

MacKay and MacKay (1975) fully partitioned the two stimulus components between the two eyes during a 20-minute inspection period. One eye saw only black and white gratings alternating in orientation while the other eye saw only a uniform green field alternating with a uniform red field. A black-white test grating presented to the eye exposed to only uniform colour appeared in the colour complementary to that of the inspection stimulus with the same orientation. However, a test grating presented to the eye exposed only to achromatic gratings appeared in the same colour as its matching inspection stimulus (positive aftereffect). Both transferred effects were less than one-third the strength of a normal McCollough aftereffect. Over *et al.* (1973) failed to observe interocular transfer, using a similar method, but with a higher rate of alternation between the pairs of right and left eye stimuli. Potts and Harris (1979) used a similar procedure and obtained the negative aftereffect when the test grating was presented to the eye previously exposed to colour, but no aftereffect when the eye exposed to colour was closed. Shattuck and Held (1975) failed to find interocular transfer of a colour-contingent tilt aftereffect. Thus, interocular transfer of the McCollough effect is not always observed and, when observed, is much weaker than the direct effect, suggesting that the aftereffect depends mainly on processes before binocular fusion.

Positive contingent aftereffects accord with a well-known property of afterimages. An ordinary coloured afterimage is in the complementary hue when projected on a light background and in the same hue when projected on a dark background or seen in the dark field of a closed eye (Sumner and Watts 1936; Robertson and Fry 1937). This dependence of the hue of an afterimage on the luminance of the background has nothing to do with binocular interactions (Howard 1960). Thus, it is not surprising that the contingent aftereffect is positive when the eye exposed to colour in the induction period is closed during the test period. Note that MacKay obtained the ordinary negative aftereffect when the eye exposed to colour in the induction period was open during the test period. *We predict that transferred contingent aftereffects will be in their negative rather than their positive form if, during the inspection and test periods, the unstimulated eye is evenly illuminated rather than being closed.*

Vidyasagar (1976) presented a red vertical grating and a blue horizontal grating alternately to both eyes. Complementary gratings (blue vertical and red horizontal) were then presented alternately to each eye separately. An aftereffect complementary to the binocular stimuli was obtained. It was argued that the complementary gratings cancelled the monocular components of the aftereffect leaving only the component arising from activation of binocular AND cells.

White *et al.* (1978) used a similar logic. Each eye was exposed to a conventional McCollough inspection stimulus with the gratings and colours changing either in phase or in antiphase in the two eyes. After in-phase inspection the binocular aftereffect was greater than either of the monocular aftereffects. After antiphase inspection the binocular aftereffect was weaker than either of the monocular aftereffects. It is surprising that there was any binocular aftereffect following antiphase inspection. These results provide evidence of binocular facilitation and cancellation of the McCollough effect, since only the binocular stimulus varied between the in-phase and antiphase induction conditions. However, the binocular facilitation was not large, suggesting that the major part of the McCollough effect is induced before the site of binocular convergence. On the basis of binocular facilitation, one would expect monocular aftereffects to be greater after in-phase than after antiphase induction, since only in-phase induction provides an effective stimulus for either binocular OR cells or AND cells. White *et al.* did not mention this point but a scrutiny of their data reveals no difference of this type. Kavadellas and Held (1977) also failed to find that monocular aftereffects after binocular induction with identical stimuli are greater than those after binocular induction with opposed stimuli. They did not compare monocular and binocular aftereffects and used a colour-contingent tilt aftereffect rather than a McCollough effect.

White *et al.* also demonstrated that aftereffects specific to each eye can be induced at the same time. Furthermore, when a black and white stimulus was presented to only one eye after a binocular induction period, the extinction of the aftereffect was largely confined to that eye (Savoy 1984). This evidence strengthens the conclusion that the McCollough effect arises before the site of binocular convergence. This conclusion is also supported by the fact that White *et al.* found the aftereffect to be at full strength when, for much of the induction period, the induction stimulus presented to one eye was perceptually suppressed by a rivalrous stimulus presented to the other.

Inspection of a red disc rotating clockwise alternating with a green disc rotating counterclockwise produces a **colour-contingent motion aftereffect** in a stationary patterned disc (Hepler 1968; Mayhew and Anstis 1972). Stromeyer and Mansfield (1970)

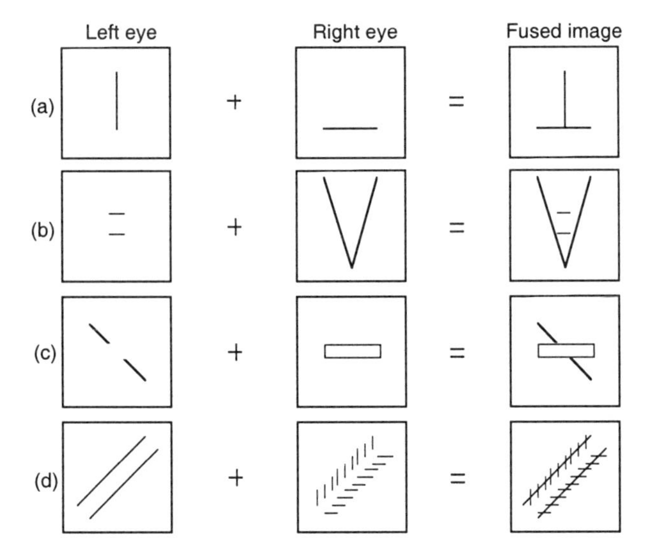

Figure 8.19. Dichoptic composite visual illusions.
Fusion of the left- and right-eye patterns produces the complete patterns on the right. In (d) binocular rivalry makes it difficult to decide whether the illusion occurs.

found that this aftereffect did not transfer to an eye that was closed during the induction period. Favreau (1978) agreed that the negative aftereffect does not transfer to an unstimulated eye, but did find a transferred positive contingent aftereffect.

In these studies a stationary test stimulus appeared to move in one direction when shown in one colour and in the opposite direction when shown in another colour. Smith (1983) found that the threshold for detecting motion of a rotating coloured spiral was elevated by prior inspection of a spiral moving in the same direction and with the same colour. This colour- and direction-specific threshold-elevation effect showed significant interocular transfer. This suggests that there are double-duty binocular cells jointly tuned to motion and colour revealed only when both the induction and test stimulus are moving, presumably because the cells do not respond to stationary stimuli.

Domini *et al.* (2000) reported **a colour-contingent depth aftereffect**. Subjects adapted to a random array of red dots distributed about one depth plane and a superimposed distribution of green dots distributed about a second depth plane. Each display had a variable disparity ($\sigma = 1.7$arcmin) and the two planes were 9.5 or 8.5 arcmin apart. The depth order of the red and green adaptation displays had a colour-specific effect on the perceived depth order of similar red and green test displays centred on one depth plane. The same aftereffect occurred when only one eye saw the colour of the dots during adaptation and only the other eye saw colour during test. Thus, the dots were matched on the basis of binocular achromatic information but the aftereffect was colour specific. This suggests that colour and disparity are processed together at a site higher than that involved in binocular matching of images.

Summary
The complex series of experiments and theoretical arguments reviewed in this section yield the following conclusions. (1) Some aftereffects show more interocular transfer than others. (2) The magnitude of an aftereffect is determined by some pooling of activity from different types of cortical cells, including monocular cells, binocular OR cells, and binocular AND cells if they exist. (3) There is some evidence that both binocular OR cells and binocular AND cells do exist. (4) Contingent aftereffects are due largely to processes occurring before binocular matching. However, a colour-depth contingent aftereffect seems to depend on processes beyond binocular matching. Evidence reviewed in Section 12.3 shows that people with defective binocular vision show less interocular transfer than those with normal vision.

8.3.6 Figural effects with binocular composites

A binocular composite stimulus is one in which part of a patterned stimulus is presented to one eye and part to the other, where both parts are required for a given visual effect. The effect produced by a binocular composite display is cyclopean only in a weak sense because a similar effect is produced when the two parts of the stimulus are combined in one eye. Witasek (1899) was apparently the first person to use binocular composite stimuli. He created a binocular composite of the Zöllner illusion shown in Figure 8.19d. Inspection of this display produces severe rivalry, but Witasek claimed that the illusion was evident after a period of practice. Ohwaki (1960) found that binocular composite illusions are much reduced in magnitude compared with normally viewed versions, and concluded that the illusions are largely retinal in origin. Day (1961) obtained similar results but concluded that the decrement of the illusions with dichoptic viewing is due to binocular rivalry. Springbett (1961) did not see the Zöllner and Hering illusions in binocular composites even in the brief moments of the rivalry sequence when both components were clearly visible. The problem is that in these illusions the lines in the two eyes overlap and thus rival. Springbett did see the Müller-Lyer illusion when the fins were presented to one eye and the connecting lines to the other. Rivalry is less of a problem in this case because the

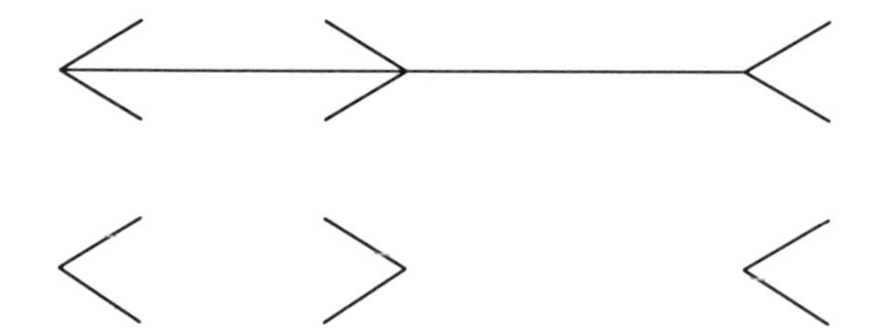

Figure 8.20. The Müller-Lyer illusion.
The illusion is evident in a figure consisting only of fins.

component lines do not overlap. He concluded that this illusion depends on processes occurring after binocular fusion. This is not a convincing argument because a Müller-Lyer illusion is evident in a figure consisting only of fins, as in Figure 8.20.

The vertical-horizontal illusion, in which a vertical line appears longer than an equal horizontal line, also survives with dichoptic lines, and in this case there is no rivalry and no intruding monocular effect (Harris *et al.* 1974).

The most thorough investigation of binocular composite illusions was conducted by Schiller and Wiener (1962). For the first three displays in Figure 8.19 the dichoptic components do not overlap and, when these were presented briefly to further minimize binocular rivalry, the illusory effects were almost as strong as with normal viewing. Very little illusion was evident in the last display, in which the component lines overlap. Schiller and Wiener concluded that these illusions depend primarily on central processes, but that the illusion is not seen in Figure 8.19d because of binocular rivalry.

Several of the geometrical illusions occur in fully cyclopean shapes generated in random-dot stereograms (Section 17.2).

8.3.7 Transfer of perceptual learning

8.3.7a *Transfer for simple visual tasks*

Practice leads to improvement in a variety of simple visual tasks, such as orientation discrimination, contrast detection, Snellen acuity (see Bennett and Westheimer 1991) and vernier acuity (Poggio *et al.* 1992; Fahle and Edelman 1993), and the ability to see a shape defined by relative motion (Vidyasagar and Stuart 1993). With the exception of one report, such training has been found to transfer fully from one eye to the other. This is to be expected because eye-specific learning would imply that learning involves only monocular cells. Learning is more likely to involve binocular cells since there are more binocular cells than monocular cells in the visual cortex. Furthermore, visual learning requires attention to the task, which implies the activity of centres beyond the primary visual cortex. Nevertheless, learning of simple visual tasks has generally been found to be specific to the shape, size, and position of the visual stimulus, which suggests that an early stage of visual processing is involved along with higher-order attention processes.

For some individuals, vernier acuity and resolution acuity improve with practice and this effect shows complete interocular transfer (Beard *et al.* 1995). Improvement in the ability to discriminate briefly flashed gratings is specific to the spatial frequency and orientation of the grating but shows complete interocular transfer (Fiorentini and Berardi 1981). Improvement in the ability to discriminate between directions of motion of dots is specific to a narrow range of directions of motion and to position but shows almost complete interocular transfer (Ball and Sekuler 1987). Improvement in identifying the orientation of oblique lines within a circular disc is orientation and position specific but shows complete or almost complete interocular transfer (Schoups *et al.* 1995a). Similarly, improvement in detecting a line that differs in orientation from a set of surrounding lines is specific to the position, size, and orientation of the stimuli but shows interocular transfer (Ahissar and Hochstein 1995, 1996; Schoups and Orban 1996). These two latter results are not surprising because monocular cells show only weak orientation tuning (Blasdel and Fitzpatrick 1984).

The only report of lack of interocular transfer of improvement in a simple visual task is one by Fahle (1994) in which he reported that improvement of vernier acuity with practice was specific to the trained eye.

8.3.7b *Transfer of pattern discrimination*

The trained ability to select one of two shapes transfers from one eye to the other in a variety of submammalian species, including the octopus (Muntz 1961), goldfish (Sperry and Clark 1949), and pigeon (Catania 1965). Transfer may be less than 100% for incompletely learned tasks or for more difficult tasks. Since the visual pathways in these species decussate fully, interocular transfer must depend on interactions between the hemispheres conveyed through the corpus callosum. In animals with hemidecussation the contralateral visual pathways from the nasal hemiretinas may be severed by section of the corpus callosum, leaving only the ipsilateral pathways from the temporal hemiretinas. Each hemisphere receives inputs from only one eye. Cats with section of the chiasm are still capable of interocular transfer of pattern discrimination (Myers 1955). The same is true of monkeys with chiasm section (Lehman and Spencer 1973).

Interocular transfer of shape discrimination in the monkey is abolished after section of the anterior commissure or of the splenium but not after section of other parts of the interhemispheric commissures (Black and Myers 1964).

8.3.7c *Mirror image transfer*

Pigeons trained to peck one of two different shapes seen only with one eye continue to select the correct shape when tested with the other eye. However, if the two shapes are right-left mirror images, the animals select the incorrect mirror-image shape when tested with the untrained eye. When trained with up-down mirror-image shapes the animals select the correct shape when tested with the untrained eye (Mello 1966). Similar effects have been found in the goldfish (Campbell 1971). In both fish and pigeon, inputs from each eye project fully to the contralateral hemisphere. Interocular transfer must therefore be mediated by the corpus callosum. In a monkey with section of the chiasm, the temporal half of each retina projects to the ipsilateral hemisphere and the nasal inputs are almost entirely lost. As in the normal pigeon, interocular transfer must be mediated by the corpus callosum. Noble (1966, 1968) found that monkeys with section of the chiasm when trained with one eye to select one of two mirror-image shapes, such as > vs. <, selected the unrewarded shape when tested with the other eye. Like pigeons, they performed correctly with shapes with up-down symmetry. Mello and Noble concluded that the indirect callosal pathway conveys mirror image representations of shape learned with one eye.

These results may be explained in a different way. Corballis and Beale (1970b) noted that pigeons tend to peck on that side of the single response key that is on the seeing-eye side. They argued that this causes the bird to switch attention from one side of the key to the other in interocular transfer. This would cause reversals because mirror-image forms possess equivalent features on opposite sides of the visual field. In support of this theory, they found mirror-image reversal to be correlated with the extent to which the birds switched sides in pecking the key.

Hamilton *et al.* (1973) noted that monkeys with section of the chiasm tend to fixate the centre of a stimulus while learning a two-choice shape discrimination. This means that only the half of the stimulus falling on the temporal retina is visible in the sectioned monkey viewing with one eye. A discrimination between a left-right symmetrical pattern would be expected to reverse in the untrained eye since the visible left half of the reinforced stimulus seen by the right eye resembles the visible right half of the nonreinforced stimulus seen by the left eye. Lehman and Spencer (1973) found that chiasm sectioned monkeys pay more attention to the nasal side of an object when looking with one eye and make interocular selections of shapes accordingly.

Campbell (1971) found that goldfish showed reversed interocular transfer of mirror-image shapes only for stimuli larger than about 15°, but that up-down mirror-image shapes transferred veridically whatever their size. Starr (1971) found that chiasm sectioned monkeys also make incorrect interocular transfer only when trained with large shapes. These results are explained if it is assumed that small shapes fall mostly in one hemisphere.

8.3.7d *Transfer of visual-motor learning*

People learn rapidly to point accurately with unseen hand to visual targets seen through prisms that displace the image up to 30°. During the learning phase, error feedback is provided at the completion of each reaching movement. During the test phase the hand remains hidden (see Howard 1982). Prism adaptation is to some extent confined to the trained arm, which suggests that it involves a recalibration of sensory or motor processes specific to one arm (Uhlarik and Canon 1971). When learning is confined to one eye, the degree of adaptation is the same whichever eye is used during testing (Hajos and Ritter 1965; Pick *et al.* 1966). Incomplete interocular transfer would indicate a change either in retinal local sign or in the sense of position of the trained eye. A change in retinal local sign is most unlikely because subjects foveated the visual target in these experiments. A change in the felt position of one eye is also unlikely since, by Hering's law of equal innervation, both eyes move together even when only one eye is open. Some degree of eye-specific adaptation has been found in subjects exposed to opposite base-up and base-down prisms (Foley and Miyanshi 1969) or opposite base-left and base-right prisms (Hajos 1968). Ebenholtz (1970) pointed out that oppositely directed prisms induce binocular diplopia, which the subject would attempt to overcome by altering the vergence of the eyes. Exposure to unusual conditions of vergence produces aftereffects, which show as errors of distance judgments (Wallach and Frey 1972; Paap and Ebenholtz 1977). Eye-specific adaptation may have arisen from these aftereffects (see Foley 1974 for a review of this issue). Prablanc *et al.* (1975) exposed the left eye and left arm to a base-right prism and the right eye and right arm to a base-left prism. Opposite adaptive shifts of pointing were induced in each arm irrespective of the eye used in testing. This shows that arm-specific adaptation is more likely to occur than eye-specific adaptation when the system is presented with the option.

Mann *et al.* (1979a) argued that people with alternating strabismus learn to dissociate the sense of position of the two eyes. They found that alternating strabismics do not display interocular transfer of prism adaptation. It is odd that prism adaptation showed no interocular transfer in these subjects, since other evidence suggests that adaptation involves processes specific to the trained arm. Perhaps alternating strabismics have a very labile sense of eye position, which they recalibrate in preference to other sensory components involved in visual-motor coordination.

9 *Vergence eye movements*

9.1 ACCOMMODATION

About 70% of the refractive power of a human eye is in the surface of the cornea, since this surface separates two media of very different optical density. This fixed focal length system brings the images of objects at any distance onto a plane not far from the retina. The ciliary muscles adjust the curvature of the lens, and hence its refractive power, to bring images of objects at a particular distance into clear focus. This response is known as **accommodation**. The refractive index of the lens is about 1.406 at the centre and about 1.386 in the outer regions. This gradient increases the power of the lens and reduces spherical aberration. The refractive power of the lens can change by about 10 dioptres. This changes the refractive power of the eye as a whole from about 59.6 dioptres to 68.2 dioptres (Davson 1962, Vol. 4, p. 105). One dioptre refracts light by 0.57°. A correctly focussed eye is **emmetropic**. An incorrectly focussed eye is **ametropic.** An eye that needs extra refraction (positive lens) to bring a distant point into focus is **hypermetropic,** or far sighted. An eye that needs decreased power (negative lens) to bring a distant object into focus is **myopic**, or near sighted. Accommodative range declines with age to about 1 dioptre at age 70 (Evans 1997). Also, with increasing age, the phase lag of accommodation to a target oscillating in depth increases (Heron *et al.* 1999).

There are four types of accommodation.

Tonic accommodation refers to the resting state of accommodation in the dark (Section 9.1.2a).

Proximal accommodation is evoked by the apparent distance of an object (Section 9.1.2b)

Blur accommodation is evoked by blur of the retinal image of an attended object (Section 9.1.3).

Convergence accommodation is triggered by a change in horizontal vergence (Section 9.5.2).

Developmental aspects of accommodation are discussed in Sections 10.2.1b and 11.2.1. Several models of accommodation have been developed (Toates 1970; Hung and Ciuffreda 1988).

9.1.1 Measuring accommodation

In subjective methods of measuring accommodation, the optometrist finds the lens that brings a defined target at a specified distance into clearest focus for the person tested. The accommodative capacity of the eye can be determined by standard eye charts with a precision of about 0.25 dioptres. In the laser speckle optometer a low-energy laser beam is reflected off a slowly rotating drum. The drum scatters the coherent laser beam to form a speckle image in front of the drum. The perceived direction of motion of the speckle pattern relative to that of the drum varies with the refractive state of the eye. The clarity of the pattern is independent of the refractive state of the eye and therefore does not act as an accommodative stimulus. The vernier alignment optometer and the stigmatoscope provide other subjective procedures (see Bennett and Rabbetts 1989).

Objective procedures for measuring accommodation include the following (see Howland 1991).

Purkinje-image method In this procedure, changes in the shape of the front surface of the lens are indicated by changes in the size of the third Purkinje image of a point of light reflected from the lens surface. This method is not affected by small changes in the direction of gaze. It is thus suitable for measuring changes in accommodation that accompany changes in vergence (Krishnan *et al.* 1977).

Fincham coincidence optometer A beam of light from a slit is projected into the eye through a collimating lens. The parallel beam is slightly off axis so that the image of the slit falls on the fovea for an emmetropic eye but to one or other side of the fovea for an ametropic eye. Light reflected from the retina is passed through the same collimating lens and split into two halves. One half is viewed directly and the other through a Dove prism, which reflects it to the opposite side of the optic axis. The displacement required to bring the two half images into vernier alignment is a measure of refractive error.

Infrared optometer The images of two narrow slits illuminated by infrared light are made to coincide on the retina when the eye is accommodated at infinity. Reflections of the images are formed on a photocell and their separation as a function of accommodative distance is measured. This is the most precise method for measuring accommodation and can record rapid fluctuations in accommodation (Campbell 1956) (Portrait Figure 9.1).

Retinoscope A small mirror reflects a point source onto the front of the patient's eye. The ophthalmologist observes this image through a hole in the centre of the mirror. As the mirror is rotated from side to side the image of the point of light sweeps across the patient's pupil. The parallactic motion of the image is nulled when the point source is brought to the anterior focal point of the eye by the addition of appropriate lenses. In static retinoscopy, the patient fixates a distant target and the lens is assumed to be relaxed to its far point of accommodation. In dynamic retinoscopy, the patient fixates a near target. Dynamic retinoscopy is rather unreliable and is not often used (Whitefoot and Charman 1992).

Partial coherence interferometry (PCI) The eye is illuminated by a split beam of coherent light. The two beams reflect off the various surfaces in the eye. The distance between a given pair of surfaces is measured by introducing a delay between the beams and observing the interference pattern produced in their reflected images. Distances between specified surfaces can be measured with a precision of between 0.3 and 10 μm (Drexler *et al.* 1998).

Electrophysiology Neural activity can be recorded in or near the ciliary muscles.

9.1.2 Tonic and proximal accommodation

9.1.2a *Tonic accommodation*

Under open-loop conditions the balance of sympathetic and parasympathetic innervation to the ciliary muscles brings the system to a state known as **tonic accommodation**, or resting focus. There are three ways to open the visual-feedback loop: viewing through a pin-hole, dark viewing, and viewing a homogeneous stimulus or one with low contrast.

On average, the eyes become about 1.5 dioptres myopic in the dark, a condition known as **dark focus** or night myopia. Nevil Maskelyne, the Astronomer Royal, was the first to notice night myopia in 1789 (see Rosenfield *et al.* 1993). He needed an extra dioptre lens when observing stars at night. The eyes also become myopic when viewing a low contrast scene, a condition known as **empty field myopia**. Pilots become myopic when viewing the empty sky. These forms of myopia are due in part to the resting accommodative state of the lens and in part to the absence of chromatic aberration as a stimulus for accommodation (Campbell and Primrose 1953).

Tonic accommodation is about 1.6 dioptres on average but varies from person to person (Leibowitz and Owens 1975). It is reasonably consistent over time for a given person (Miller 1978; Mershon and Amerson 1980; Owens and Higgins 1983) but becomes more variable after a period in total darkness (Krumholtz *et al.* 1986). The magnitude of dark accommodation decreases with increasing age (Whitefoot and Charman 1992), and is less in myopes than in emmetropes or hypertropes (Goss and Zhai 1994).

The stimulus contrast at which the eyes first show empty field myopia—the accommodation contrast threshold—is higher for stimuli of high rather than low spatial frequency (Ward 1987).

With high luminance, accommodation is most accurate when the distance of the target corresponds to the resting state of accommodation. Over-accommodation occurs for more distant targets and under-accommodation for nearer targets (Leibowitz and Owens 1975). As luminance is reduced, accommodation is pulled more towards the position of dark focus. At scotopic levels, it remains close to the position of dark focus at all viewing distances. Corrective lenses can compensate for the adverse effects of misaccommodation at low luminance (Johnson 1976). Dark focus shows some relationship to a person's refractive error (McBrien and Millodot 1987).

Figure 9.1. Fergus Campbell.
Born in Glasgow in 1924. After training in Medicine in Glasgow he became a Lecturer in Physiology in Cambridge University in 1953. In 1983 he was appointed to a personal Professorship of Neurosensory Physiology at Cambridge. He died in 1993. He received the Tillyer Medal of the Optical Society of America in 1978.

The luminance at which the eyes adopt a state of dark focus is higher than that at which they adopt the state of dark vergence. When the luminance is too low to evoke an accommodative response with monocular viewing, a change in vergence with both eyes open may evoke a change in accommodation (Jiang *et al.* 1991). The subject of dark focus has been reviewed by Rosenfield *et al.* (1993, 1994).

After viewing a near stimulus for a several minutes, the state of dark accommodation is modified by up to about 0.5 dioptres in the same direction. The aftereffect lasted 5 minutes after 5 minutes of adaptation and several hours after 1 hour of adaptation (Tan and O'Leary 1986). Monocular and binocular viewing produce similar aftereffects in both eyes (Fisher *et al.* 1987, 1988b). The aftereffect lasts longer after maintained near than after maintained far accommodation (Ebenholtz 1983, 1991). It decays more rapidly in the dark than in an evenly illuminated field (Schor *et al.* 1986b; Wolfe and O'Connell 1987). The aftereffect is weaker and decays more rapidly than adaptive changes in the resting state of vergence (Section 9.3.1 (Fisher *et al.* 1990).

It has been claimed that, in the dark, accommodation is partly under voluntary control, since it varies with instructions to think of a far object or a near object. It also varies with knowledge of the nearness of unseen surrounding surfaces (Provine and Enoch 1975; Malmstrom and Randle 1976;

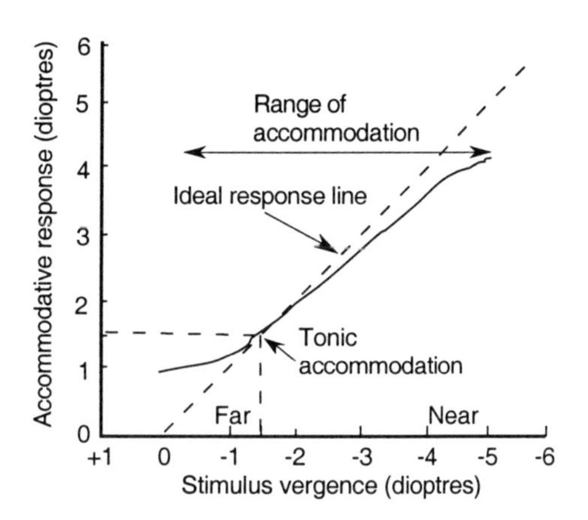

Figure 9.2. A typical accommodation response/stimulus curve. (Adapted from Ward and Charman 1985)

Rosenfield and Ciuffreda 1991b). However, these changes may have been evoked by changes in vergence through the mediation of vergence accommodation (Miller 1980; Rosenfield *et al.* 1994).

9.1.2b *Proximal accommodation*

Accommodation induced by changes in the apparent distance of an object without any actual change in distance is **proximal accommodation.** Proximal accommodation is not evident when there is a strong blur cue to accommodation (Morgan 1968). It is revealed when depth of focus is increased by viewing through a pinhole (McLin *et al.* 1988; Kotulak and Morse 1995) or when blur is under open-loop control (Kruger and Pola 1987). Proximal accommodation increases when the depth cue of motion parallax or disparity is present (Takeda *et al.* 1999).

9.1.3 Accommodation evoked by image blur

A person must make a voluntary decision to focus on a particular object in a particular depth plane. Normally, this involves fixating the object with both eyes. Once that decision has been made, accommodation is controlled by error signals derived from blur of the image of the selected object. However, a person can learn to misaccommodate voluntarily on an object that is being fixated and attended to.

9.1.3a *The blur stimulus*

An image out of the eye's plane of focus has **defocus blur** whereas a poorly focussed photograph has **target blur**. Defocus blur is under feedback control, since it is removed when the eyes accommodate. Target blur is not under feedback control and therefore remains as the eyes accommodate. A sudden change in target blur induces a transient accommodative response, but the eyes soon return to their previous state (Phillips and Stark 1977). Presumably, the way the image changes during fluctuations in accommodation allows the visual system to distinguish between the two types of blur. The accommodation system registers defocus blur of the image of the attended object and uses it as a negative feedback signal to control accommodation.

For a given state of accommodation, the range of distances in object space within which an observer is unable to detect image blur due to misaccommodation is the **depth of field**. It can be expressed in linear measure or in dioptres. The range of distances in image space within which an image appears in focus is the **depth of focus**. In linear terms, depth of field and depth of focus differ, but in terms of visual angle or dioptres, they are the same.

The depth of field of an eye can be calculated from viewing distance and pupil size if optical aberrations are ignored and a value of visual acuity is assumed. However, optical aberrations have a considerable effect so that accurate calculations are impossible (Campbell 1957). Depth of field can be determined psychophysically by asking the observer to fixate one object and report when the image of another object becomes blurred. A more refined psychophysical procedure involves measuring resolution acuity with a grating at different out-of-focus distances (Ogle and Schwartz 1956). However, the accommodation mechanism can respond to blur that cannot be detected psychophysically (Kotulak and Schor 1986c). Therefore, a more stringent method is to determine the minimum out-of-focus blur of an image which evokes a change in accommodation.

An object at a distance of x m in front of an eye has a **stimulus vergence** of -1/x dioptres. To bring the image of the object into correct focus requires an accommodative response of +1/x dioptres. However, for a given size of pupil there is a distance beyond which all objects are in focus at the same time—the depth of focus extends to infinity. As an object is approaches, with pupil size held constant, stimulus vergence increases and depth of focus decreases. Figure 9.2 shows a typical accommodation/stimulus curve for a fixed pupil size. The 45° dashed line represents the response required for perfect focus. The response for which the accommodative state equals stimulus vergence is the equilibrium state of the system, which is usually identified with dark focus. For stimuli nearer than the equilibrium distance, accommodation is usually less than that required for a perfectly focussed image, and for larger distances it is usually more than required. These are steady-state errors of accommodation that reflect a tendency to return to a state of equilibrium.

Campbell (1957) found that depth of field decreased as pupil diameter was increased from 0.75 to

7 mm, with retinal illumination and viewing distance constant. Light entering the margins of a large pupil is relatively ineffective because the receptors are aligned with the centre of the pupil. When this Stiles-Crawford effect was allowed for, depth of field became inversely proportional to pupil diameter. The decrease in depth of focus with near viewing can be cancelled by a decrease in pupil size.

For a pupil diameter of 2.5 mm, depth of field is about 0.3 dioptres. This means that errors of focus within about 0.3 dioptres do not cause detectable blurring of the image and would not be expected to evoke an accommodative response. The depth of focus represents the dead-space within which accommodation can change without generating an error signal. As pupil diameter decreases, depth of focus increases and the steady-state error of accommodation increases. With a very small aperture, accommodation falls into the state of dark focus at an accommodation distance of about 1 m and objects at all distances are in focus (Hennessy *et al.* 1976). The accommodation/vergence curve then becomes flat (Campbell 1957; Ward and Charman 1985).

By Weber's law, a given blur added to a sharp edge is more noticeable than the same blur added to a less sharp edge. It follows that depth of focus decreases with increasing spatial frequency (boundary sharpness) of a stimulus (Marcos *et al.* 1999). Accordingly, accommodation is more accurate for a grating with high spatial frequency than for one with low spatial frequency (Charman and Tucker 1977). According to the contrast sensitivity function (Section 3.3.5), contrast sensitivity peaks at between 3 and 5 cpd. Owens (1980) found that the accuracy of steady-state accommodation for a high contrast grating was also optimal at these same spatial frequencies. Amblyopes show reduced contrast sensitivity and reduced accommodative response over most of the spatial frequency range (Ciuffreda and Hokoda 1983). Thus, with a grating stimulus, accommodation and detection rely on the same contrast-detection mechanism. However, Kotulak and Schor (1987) found that changing the spatial frequency, luminance, and contrast of a DOG stimulus affected visual detection and vergence in different ways.

Depth of field increases as a linear function of log luminance and as a linear function of contrast (Campbell 1957). This is probably due to the parallel improvement in visual acuity with increasing luminance and contrast. At very low levels of luminance or contrast, accommodation reverts to the state of dark focus (Johnson 1976). Fincham (1951) concluded that accommodation is mediated primarily by the cones. Depth of focus decreased when chromatic aberration was diminished by an achromatising lens (Campbell 1957). Thus, people more easily detect an out of focus image when chromatic fringes are eliminated. Equiluminant stimuli provide poor stimuli for accommodation (Wolfe and Owens 1981).

Reduction of the luminance contrast of the stimulus reduces the accuracy of accommodation to step changes in the distance of a grating (Bour 1981). However, accuracy of steady-state accommodation is affected only slightly by reduction of contrast at suprathreshold levels of luminance or contrast (Wolfe and Owens 1981; Tucker *et al.* 1986).

When there are two overlapping stimuli at different but neighbouring distances, as when an object is seen through a mesh, accommodation usually settles at an intermediate depth (Mandelbaum 1960; Rosenfield and Ciuffreda 1991a). This suggests that accommodative stimuli are averaged over a local area, but what that area is has not been determined. Analogous effects occur in vertical vergence (Section 9.7).

Accommodation occurs with a mean latency of about 300 ms and, once started, takes about 750 ms to reach a steady-state (Campbell and Westheimer 1960; Tucker and Charman 1979). Latency and response times are similar for monocular and binocular viewing, but far-to-near responses are faster than near-to-far responses (Heron and Winn 1989).

9.1.3b *Steady-state accommodation*

When the gaze is fixed on a small target, the accommodative state of the eye fluctuates with an amplitude of up to about ±0.25 dioptres. A low-frequency component of up to 0.6 Hz can be distinguished from a high-frequency component of between 1 and 3 Hz (Campbell *et al.* 1959; Charman and Heron 1988). Fluctuations of accommodation in the two eyes are correlated in phase and amplitude, which shows that they do not arise only from instabilities in the ciliary muscles (Campbell 1960). The high-frequency component is correlated with the frequency of the arterial pulse and is therefore not under neural control or visual stimulus control. The power of the low-frequency component is related to the quality of the retinal image and its frequency is related to the frequency of stimulus oscillation in depth (Winn 2000). This shows that low-frequency fluctuations are under feedback control from the retina. Blur of the retinal image must be the crucial stimulus. The amplitude of fluctuations in accommodation is maximum at the resting state of accommodation. This is probably because tension in the accommodation system is least at this point, leaving the lens more free to oscillate (Denieul 1982; Kotulak and Schor 1986b).

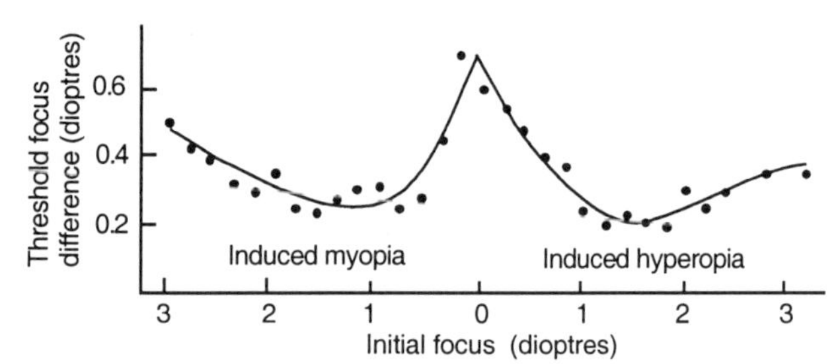

Figure 9.3. Detection of changes in image blur.
Threshold for detection of a change in blur of a visual object moving back and forth along the visual axis at 2 Hz as a function of the initial blur of the object. The eye was homatropinized and a 3 mm artificial pupil was placed before the eye. (Adapted from Campbell and Westheimer 1958)

Microfluctuations of accommodation could provide error signals required to keep the eyes reasonably well focussed on an object. If the system monitors the changing blur produced by neurally controlled fluctuations of accommodation, it could apply an appropriate corrective response.

Campbell and Westheimer (1958) paralyzed accommodation, applied a 3 mm artificial pupil, and moved a test object back and forth along the visual axis at 2 Hz. Subjects adjusted the amplitude of motion until they could just detect a change in target blur. Target excursions were centred on various pedestal values of blur between plus and minus 3 dioptres. Sensitivity to changes in blur was within the range of changes produced by microfluctuations of accommodation, which supports the idea that these fluctuations contribute to the control of steady-state accommodation. Figure 9.3 shows that the threshold for detection of a change in blur was lowest when blur was initially about 1 dioptre away from minimum. This suggests that the edge-detection mechanism detects an edge best at the point of minimum blur (the peak of its blur tuning function) but is most sensitive to a change of blur on the flanks of the tuning function (see Section 3.1.3b). Walsh and Charman (1988) paralyzed accommodation and measured the threshold for detection of sinusoidal changes in image focus of a grating as a function of temporal frequency, pupil size, and mean position of focus. They also found that maximum sensitivity to focus change occurred when the image was slightly defocussed. Watt and Morgan (1983) found a similar off-centre minimum in the threshold for detection of a difference of blur between two simultaneously presented blurred edges.

Charman and Tucker (1978) measured the accuracy of steady-state monocular accommodation as a function of the spatial frequency of a sinusoidal grating target. At a spatial frequency of 0.4 cpd, the steady-state error was similar to that observed in an empty field. As spatial frequency increased to 30 cpd, accommodation accuracy improved substantially. They explained the results in terms of a model in which image variations caused by fluctuations in accommodation become more detectable as spatial frequency is increased.

All this evidence supports the idea that a steady-state error of accommodation improves the effectiveness of the error signal arising from microfluctuations of accommodation.

Pupil constriction increases the depth of focus and thereby reduces the change in image blur as the lens changes its focal length. If microfluctuations of accommodation are to provide an effective error signal they should increase in amplitude as the pupil constricts. Gray *et al.* (1993a) found that the amplitude of microfluctuations of under 1 Hz increased as pupil diameter was reduced below 2 mm by artificial pupils. The amplitude of microfluctuations above 1 Hz was not affected. Also, increasing pupil diameter above 2 mm did not affect the amplitude of microfluctuations. They concluded that steady-state accommodation is controlled by low-frequency fluctuations of accommodation.

At low levels of luminance, high spatial frequencies become undetectable and larger fluctuations of accommodation are therefore required to produce a discriminable change in image contrast. Gray *et al.* (1993b) found that low-frequency microfluctuations of accommodation increased near the luminance threshold (below 0.004cd/m^2).

9.1.3c *The sign of accommodation*

Static blur indicates the magnitude of misaccommodation but not its direction. This is because an out-of focus image is blurred by an equal amount on either side of the plane of focus. Under natural viewing conditions, cues such as perspective, overlap, parallax, and disparity indicate whether the eyes are under or over accommodated. In the absence of such cues, dynamic or static information inherent in the misaccommodated image could indicate the sign of misaccommodation. Whatever cues are used, they are not available to conscious control, since without training, subjects cannot manually bring a stimulus into focus when the ciliary muscles are paralyzed with homatropine (Campbell and Westheimer 1959; Troelstra *et al.* 1964).The following mechanisms could provide a signed signal for accommodation.

Hunting An initial response could be made at random and then corrected if in the wrong direction. I have mentioned the spontaneous fluctuations in accommodation of a few tenths of a dioptre at frequencies up to 3 Hz. Campbell and Westheimer

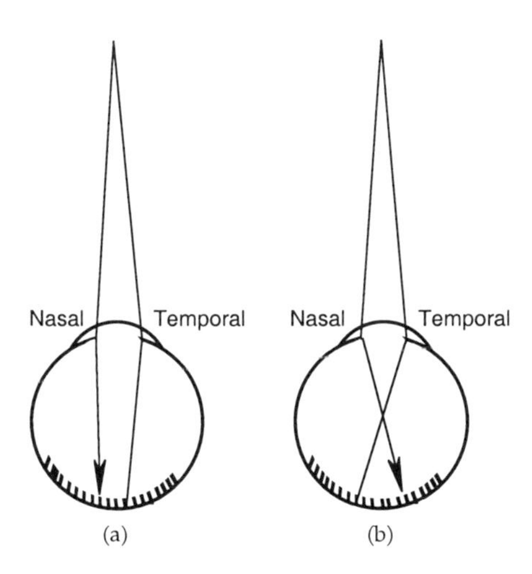

Figure 9.4. The Stiles-Crawford effect and focus sign.
Assume that the receptors are aligned with a point on the nasal side of the pupil. (a) For an under-accommodated image, light rays on the nasal side are better aligned with receptors than those on the temporal side. (b) For an over-accommodated image, temporal light rays are better aligned with receptors than nasal rays.

(1959) found that subjects made many initial errors in responses to an out-of-focus image when cues to the direction of misaccommodation were eliminated.

Troelstra *et al.* (1964) tried to eliminate all cues to the direction of misaccommodation and measured the initial responses to a horizontal line as it stepped randomly in depth every 400 ms. Two of three subjects responded in the correct direction. They were presumably using some cue that had not been successfully eliminated. Half the initial responses of the third subject were in the wrong direction, indicating that he was hunting. Hunting is therefore a possible mode of operation when other cues are not available.

Dynamic error feedback The microfluctuations of accommodation in the 2 Hz range may provide directional information within the 0.3 s reaction time of a response. This would not work for large changes in accommodation, because microfluctuations of blur would not be detected when superimposed on an image well out of focus. Responses to step changes of less than 1 dioptre have been found to be in the correct direction (Fincham 1951). Kotulak and Schor (1986d) developed a computational model of accommodation controlled by error signals arising from microfluctuations of accommodation.

The Stiles-Crawford effect For objects illuminated with monochromatic light, accommodation in response to introduction of a lens occurred only when subjects executed small lateral eye movements (Fincham 1951). For a slightly eccentric hypermetropic image, the rays nearer the fovea are more orthogonal to the retina while, for an eccentric myopic image, the rays furthest from the fovea are more orthogonal. Since orthogonal rays are a more effective stimulus (Stiles-Crawford effect), this should allow a person to detect whether the focal plane of an image is in front of or beyond the retina. In other words, the peak of the effective intensity distribution of an eccentric image moves across the blur circle of the image as focus changes from behind to in front of the retina.

The same type of mechanism could operate with a centred image if the receptors were aligned with a point to one side of the centre of the pupil (see Kruger *et al.* 1997). In fact, the peak of the Stiles-Crawford effect is usually displaced nasally, making light from the nasal side of the pupil more effective than light from the temporal side. Figure 9.4 illustrates that, under these circumstances, the side of the blur circle that is detected most effectively is on the left of an under-accommodated image and on the right of an over-accommodated image.

Chromatic aberration The visual system could rely on some static asymmetrical feature in the defocussed image such as chromatic aberration. Longitudinal chromatic aberration is due to shorter wavelengths being more strongly refracted than longer wavelengths, up to a difference in refraction of about 2 dioptres. Transverse chromatic aberration is due to the axial misalignment of the pupil, lens, and fovea. It causes blue light from an object point to fall more to the nasal side than red light. Both effects produce colour fringes, which vary according to whether the image is under- or over-accommodated. Thus the refractive state of the eye could be derived by comparing image quality in the three types of cone. Flitcroft (1990) showed theoretically that colour opponent cells in the visual cortex could perform this task. What is the evidence that we make use of chromatic aberration?

About 60% of people were unable to accommodate in monochromatic light, for which there is no chromatic aberration (Fincham 1951). Rucker and Kruger (2001) found that some subjects could accommodate to a grating that stimulated only the blue cones, but not as well as to a black-white gating.

Campbell and Westheimer (1959) had subjects view a high-contrast test object through a lens placed with its second principal focus in the pupil. This eliminated changes in illumination and image size as the distance of the test object was varied. The accommodative mechanism of the eye was paralyzed by homatropine. The target was suddenly displaced by 0.5 or 1.0 dioptre and subjects adjusted a control to restore the target to focus. When familiar with the control they were able to make an initial movement in the correct direction on every trial. They performed at chance levels when the chromatic and spherical aberrations of the eye were corrected (see also Kruger and Pola 1986).

An equiluminant red-green border or red-green grating did not elicit appropriate accommodation (Wolfe and Owens 1981) even though the grating was visibly out of focus (Switkes *et al.* 1990). Perhaps chromatic-aberration fringes are distorted or not detectable with equiluminant stimuli. Changes in the magnitude or direction of either form of chromatic aberration did not affect static accommodation (Bobier *et al.* 1992). This indicates that there is no preferential focussing on a particular wavelength.

Stone *et al.* (1993) monitored accommodation as subjects viewed a sinusoidally moving grating in which longitudinal chromatic aberration was doubled, removed, neutralized, or reversed. Doubling had little effect but removing the aberration reduced the gain and increased the phase lag of accommodation. Reversing aberration severely disrupted accommodation. These effects were most prominent for a grating with a spatial frequency between 3 and 5 cpd (see also Kruger *et al.* 1995). Kruger *et al.* (1997) obtained similar results using an open-loop accommodative stimulus. However, most subjects performed above chance with a monochromatic stimulus, which suggests that they were using an achromatic directional cue, possibly arising from a decentred Stiles-Crawford function, as described above. Kruger *et al.* (1995b) modulated the longitudinal chromatic aberration of long-, middle-, and short-wavelength components of the retinal image of a 3 cpd grating to simulate its motion from 1 dioptre behind the retina to 1 dioptre in front of the retina. This evoked appropriate accommodative changes, demonstrating that chromatic contrast at luminance borders is an accommodative stimulus. The accommodative response to a grating moving sinusoidally in depth showed increased gain and reduced phase lag as the spectral bandwidth of the illumination increased from a monochromatic wavelength of 550 nm (Aggarwala *et al.* 1995).

Changing image size Normally, when an object moves closer its image becomes larger. Cues to changing distance induce accommodative changes when the effects of blur are eliminated by viewing through a pinhole (Section 9.1.2b). Hennessey *et al.* (1976) found no effect of object distance with pinhole viewing but their stimuli were stationary. Campbell and Westheimer (1960) observed that, with normal pupils, the accommodative response was less variable when a step change in blur was accompanied by a change in image size. (Kruger and Pola 1986). found that changing image size enhanced the accommodative response to sinusoidal changes in image blur. Thus, step or smooth changes in image size seem to induce accommodation. The primary effect of a change in image size may be a change in vergence, produced by an apparent change in the distance of the stimulus. The accommodative change may be mediated by the linkage between vergence and accommodation. It has not been proved that changing image size plays a crucial role in controlling accommodation.

Whatever mechanism or mechanisms are used to sign the direction of accommodation, there is evidence that they fail for images more than 2 dioptres out of focus (Fincham 1951).

9.1.3d *Response to unequal accommodative demand*

It is generally agreed that the accommodative response, like the pupillary response, is approximately consensual when one eye is closed. However, the accommodative response when only one eye is open is 20% less than when both eyes are open (Ball 1952).

A need for unequal accommodation arises with eccentric fixation, when the image of the fixated object is nearer to one eye than to the other. Rosenberg *et al.* (1953) reported that their two subjects showed an appropriate difference in accommodation in eccentric fixation but Spencer and Wilson (1954) found that only one of their subjects showed a difference in the appropriate direction. The other subject showed a similar difference in the wrong direction.

Marran and Schor (1998) argued that the states of focus of the stimuli in these studies were not sufficiently different in the two eyes and may have been rivalrous, and that responses were measured successively rather than simultaneously. After allowing for these factors they measured the response of the two eyes to lens-induced aniso-accommodative stimuli of between 0.3 and 3 dioptres at various viewing distances. The mean aniso-accommodative response was 0.75 dioptres for the largest stimulus value. For four of seven subjects the response became more consensual with increased viewing distance.

Flitcroft *et al.* (1992) modulated accommodation demand by sinusoidally changing the power of lenses placed before the eyes. The phase of modulation differed between the two eyes by 90 or 180° or the modulation was applied to only one eye. In both humans and monkeys the accommodative response represented the vector average of the inputs. Thus, when the modulation was out of phase 180° there was no response and when the stimulus was modulated in only one eye, the response was much reduced compared with that in normal binocular viewing. When the two eyes were presented with

Table 9.1. Terminology for horizontal, vertical, and torsional vergence.

Type of vergence movement	Name of vergence movement
Opposed horizontal rotation	*Horizontal vergence*
Visual axes moving inwards	Convergence
Visual axes moving outwards	Divergence
Opposed vertical rotation	*Vertical vergence*
Right visual axis up, left axis down	Right supravergence (left infravergence)
Left visual axis up, right axis down	Left supravergence (right infravergence)
Opposed torsional rotation	*Cyclovergence*
Vertical meridians rotate top in	Incyclovergence
Vertical meridians rotate top out	Excyclovergence

modulated orthogonal gratings, stimulus averaging did not occur but, rather, the response at any time was determined by whichever grating was perceptually dominant (Flitcroft and Morley 1997).

9.1.4 Neurology of accommodation

Averaged blur signals from an area of the retina are transmitted to the visual cortex and then to the parietotemporal area (area V5) in monkeys (Jampel 1960) or to the suprasylvian area of the parieto-occipital cortex in cats (Bando *et al.* 1984; 1996). Cells in these areas respond to accommodative stimuli other than blur, such as changes in binocular disparity and motion in depth. The cerebellum, which receives inputs from the superior colliculus and from the visual cortex via the pons (Glickstein *et al.* 1972) is also involved in the control of accommodation (Hosoba *et al.* 1978; Zhang and Gamlin 1998).

Cortical areas controlling accommodation in both monkeys and cats project to the superficial and intermediate layers of the rostral superior colliculus, where microstimulation evokes accommodation responses (Sawa and Ohtsuka 1994). The accommodation-related area in the superior colliculus projects to the pretectum and mesencephalic reticular formation. These are brainstem areas involved in the joint control of accommodation, vergence, and pupil diameter (Judge and Cumming 1986). Commands are finally sent to the Edinger-Westphal nucleus in the midbrain from where motor commands are sent through the sympathetic and parasympathetic pathways of the oculomotor nerve (IIIrd cranial nerve) to the ciliary ganglion and ciliary muscles (see Ciuffreda 1991). The rate of discharge of cells in this nucleus increases linearly as accommodation is increased from infinity, but it also varies with accommodation velocity. Some of the cells also show some activity related to the state of vergence (Gamlin *et al.* 1994). The joint control of vergence and accommodation is discussed further in Section 9.11.2.

9.2 EYE MOVEMENTS IN GENERAL

9.2.1 Types of eye movement

Eye movements serve three basic functions: stabilization of the image on the retina as the head moves, fixation and pursuit of particular objects, and convergence of the visual axes on a particular object.

For reviews of the literature on eye movements see Carpenter (1988). Procedures for measuring eye movements are reviewed by Collewijn *et al.* (1975), Young and Sheena (1975), and Eizenman *et al.* (1984). For a discussion of the anatomy and mode of action of the extraocular muscles see Büttner-Ennever (1988).

9.2.1a *Eye movements for image stabilization*
Image stabilization is achieved by the **vestibuloocular response** (VOR), a conjugate eye movement evoked by stimuli arising in the vestibular organs (semicircular canals, utricles, and saccules) as the head moves (see Howard 1986). When the eyes are open, the VOR is supplemented by **optokinetic nystagmus** (OKN) evoked by the motion of the image of the visual scene (see Howard 1993a.) Both these responses are involuntary and were the first types of eye movement to evolve.

9.2.1b *Eye movements for fixation and pursuit*
Eye movements for fixation and voluntary pursuit of particular objects evolved in animals with foveas. Voluntary rapid eye movements (saccades) allow the gaze to move quickly from one part of the visual scene to another. Voluntary pursuit maintains the image of a particular object on the fovea. During

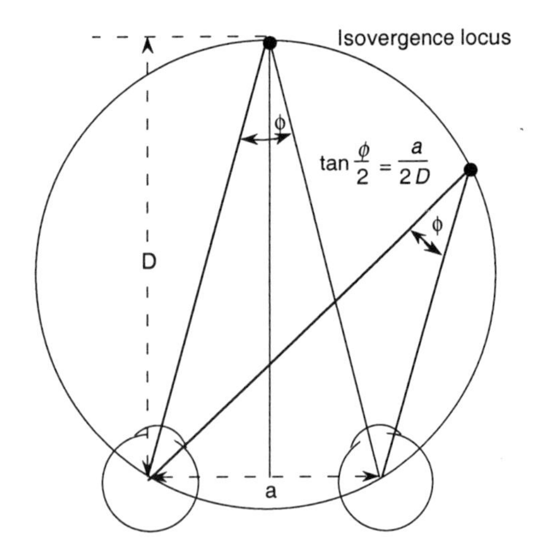

Figure 9.5. Angle of vergence.
D is the distance from the interocular axis to where the isovergence locus cuts the median plane.

Figure 9.6. Clifton Schor.
Born in 1943. He obtained a Ph.D. in Physiological Optics with Merton Flom and Lawrence Stark from the University of California at Berkeley in 1972. He is Professor of Optometry, Vision Science and Bioengineering at the University of California at Berkeley. Recipient of the Garland Clay Award and Glenn Fry Award from the American Academy of Optometry.

fixation, the eyes exhibit so-called **physiological nystagmus**, consisting of a mixture of slow drifts and microsaccades with a mean standard deviation of about 0.1° (Steinman *et al.* 1973; Ott *et al.* 1992). A person can reduce the amplitude of these movements by voluntary effort (Steinman *et al.* 1967). Marshall and Talbot (1942) suggested that the constant motion of the retinal image produced by fixation tremor improves visual acuity. If image motion is stopped, the images of all objects fade from view after a short time. Eye movements are clearly needed for continued vision, but acuity and contrast sensitivity are not adversely affected by image stabilization before the image fades (Tulunay Keesey 1960; Gilbert and Fender 1969). Furthermore, imposed image motions at velocities up to 2.5°/s have been found to have no effect on acuity (Westheimer and McKee 1975, 1977). A person makes fewer microsaccades when detecting fine detail, which would be counterproductive if microsaccades improved acuity (Winterson and Collewijn 1976; Bridgeman and Palca 1980). It seems that physiological nystagmus has not evolved to improve acuity. The stability of binocular fixation is discussed in Section 9.6.3.

A movement of an eye considered singly is a **duction**—abduction when the eye moves temporally, adduction when it moves nasally, and torsion when it rotates about the visual axis. A combined movement of the two eyes in the same direction is a conjugate movement, or **version**. A version can be in a horizontal, vertical, or an oblique direction. A conjugate rotation of the eyes around the visual axes is ocular torsion or **cycloversion**.

9.2.1c *Disjunctive eye movements*

In the third basic type of eye movement, the eyes move through equal angles in opposite directions to produce a disjunctive movement, or **vergence**. In horizontal vergence, each visual axis moves within a plane containing the **interocular axis**. In vertical vergence, each visual axis moves within a plane that is orthogonal to the interocular axis. The eyes also move in opposite directions around the two visual axes. This is **cyclovergence**. Combined rotations of the eyes about two or three axes are also possible. Table 9.1 presents the terminology of vergence. Until recently, it was believed that primates were the only animals with vergence eye movements. Hughes (1972) recorded vergence in cats, but found a good deal of individual variation. Some lateral-eyed animals, such as the chameleon, the pigeon, and the rabbit are capable of converging the eyes (Chapter 14).

Horizontal vergence occurs when a person changes fixation from an object in one depth plane to one in another depth plane. There is a weak coupling between viewing distance and cyclovergence, and changes in vertical vergence and cyclovergence occur with changes in the direction of gaze, especially at near distances. Selection of the target is under voluntary control. For horizontal vergence, the initial response brings the images of the visual target near to corresponding points without further visual feedback. This is the open-loop phase of vergence. In the final phase, all types of vergence come under control of visual error signals arising from hori-

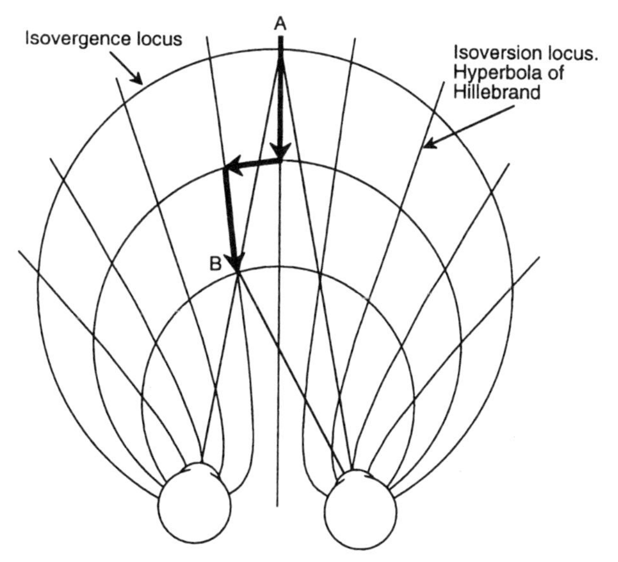

Figure 9.7. Loci of isovergence and isoversion.
Loci of constant vergence for changing version are circles through the centres of rotation of the two eyes. Loci of constant version for changing vergence are hyperbolas of Hillebrand. As the gaze moves from A to B, both vergence and version change.

Figure 9.8. Glenn Fry.
He graduated in Physics and Psychology from Davidson College in 1929 and obtained his Ph.D. in Psychology from Duke University in 1933. In 1933 he became Assistant Professor of Applied Optics at Ohio State University, where he become Director of the Optometry School and Regents Professor.

zontal, vertical, or rotation disparity. With a little practice a person can gain voluntary control of horizontal vergence in the absence of a fixated object. There is no record of people gaining voluntary control of vertical vergence or cyclovergence.

The angle between the visual axes is the **vergence angle**. When a person fixates a point at infinity, the visual axes are parallel and the vergence angle is zero. The angle increases when the eyes converge. In Figure 9.5 it can be seen that, for symmetrical convergence, the angle of horizontal vergence, ϕ, is related to the interocular distance, a, and the distance of the point of fixation from a point midway between the eyes, D, by the expression

$$\tan\frac{\phi}{2} = \frac{a}{2D} \qquad (1)$$

Thus, the change in vergence per unit change in distance is much greater at near than at far viewing distances. About 70% of a person's normal range of vergence is used within one metre from the eyes.

An **isovergence locus** is the path traced by the point of fixation when version changes with vergence held constant. For a fixed elevation of gaze, the isovergence locus is a circle passing through the fixation point and the centre of rotation of each eye. Note that the Vieth-Müller circle, or theoretical horopter, described in Section 15.5, intersects the nodal points of the eyes rather than their centres of rotation. If the eyes change their elevation, the shape of the isovergence locus depends on the co-ordinate system used to specify eye movements. In the Helmholtz axis system the isovergence locus is a toroidal surface formed by rotation of the isovergence circle round the line joining the centres of rotation of the two eyes. In the Fick axis system the isovergence locus is the isovergence circle and a vertical line lying in the median plane (Schor *et al.* 1994) (Portrait Figure 9.6).

The loci of constant version for changing vergence are known as **hyperbolas of Hillebrand**, as shown in Figure 9.7 and specified by:

$$-x^2 + y^2 + 2xy\cot 2\phi = 1 \qquad (2)$$

where x is the distance of the point of convergence from the median plane, y its distance from the interocular axis, and ϕ the angle of version with respect to the point midway between the eyes (Fry 1950) (Portrait Figure 9.8). Equation (1) holds for asymmetrical convergence if D is defined as the distance to the point where the locus of isovergence cuts the median plane, rather than the distance to the point of fixation. With increasing viewing distance, the hyperbolas become asymptotic to straight lines through the midpoint of the interocular line.

Just before, during, and just after a vergence response to a large disparity step there is a decrement in the detectability of a visual target or of the displacement or change in disparity of a target (Manning and Riggs 1984; Hung *et al.* 1989, 1990) (Portrait

Figure 9.9. George K. Hung.
Born in Shanghai, China, in 1947. He obtained a B.Sc. in Mechanical Engineering in 1970 and Ph.D. in Physiological Optics in 1977, both from Berkeley. He joined the faculty at Rutgers University in 1978, where he is now Professor of Biomedical Engineering .

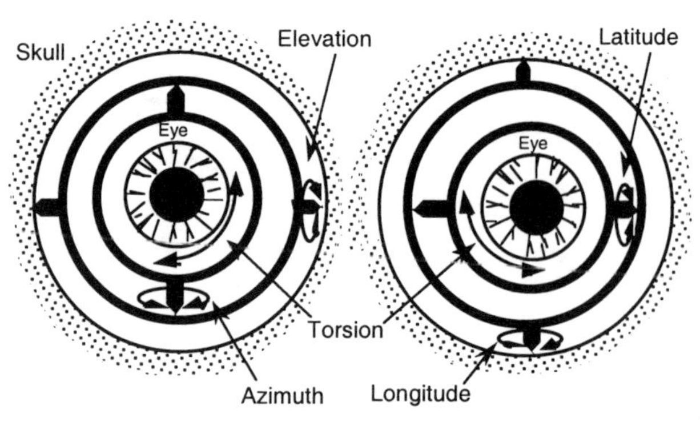

Figure 9.10. Axis systems for specifying eye movements.
In the Helmholtz system the horizontal axis is fixed to the skull, and the vertical axis rotates gimbal fashion. In the Fick system the vertical axis is fixed to the skull. (From Howard and Templeton 1966)

Figure 9.9). This loss of sensitivity is analogous to that associated with saccades (Volkmann *et al.* 1978; Matin 1974). Suppression probably helps the viewer to disregard the instability of the retinal images during vergence and may also help eliminate the effects of spurious disparity signals during the execution of large vergence movements.

Reviews of vergence have been provided by Alpern (1969), Toates (1974), Schor and Ciuffreda (1983), Carpenter (1988), Collewijn and Erkelens (1990), and Judge (1991). The development of vergence in the child is discussed in Section 11.2.6.

Maddox (1893) identified the following four types of horizontal vergence.

1. Tonic vergence
2. Proximal vergence
3. Accommodative vergence
4. Fusional, or disparity-induced vergence

These various forms of vergence are described in later sections of this chapter.

9.2.1d *Units for accommodation and vergence*
Accommodation in dioptres is the reciprocal of the viewing distance in metres. Thus, 2 dioptres of accommodation are required to bring an object at a distance of 0.5 metres into clear focus. To make horizontal vergence commensurate with accommodation, vergence can be specified in **metre angles.** One metre angle of convergence is the convergence required for binocular fixation of an object at a distance of 1 metre in the median plane. The vergence angle in metre angles is the reciprocal of the distance of the fixation point in metres. The vergence angle in degrees corresponding to a metre angle of M, for an interpupillary distance a in metres, is $2 \tan^{-1} a M/2$. Thus, the convergence in degrees corresponding to 1 metre angle varies with interpupillary distance.

In clinical practice, convergence is specified by a third measure known as **prism dioptres.** A 1-dioptre prism displaces the visual axis by 1 cm at a distance of 1 m. It follows that the angle of convergence in dioptres is the interocular distance in centimetres divided by the viewing distance in metres. Thus, a person with an interocular distance of 6.5 cm must exert 6.5 dioptres of convergence when fixating an object in the midline at a distance of 1 m. Measurements of vergence in either metre angles or prism dioptres are not applicable at near distances.

9.2.2 Co-ordinate systems for eye movements

The centre of rotation of an eye is not at the centre of the eye and is not fixed with reference to the orbit (Park and Park 1933). In other words, an eye translates a little as it rotates. For most purposes, however, it can be assumed that the human eye rotates about a fixed centre 13.5 mm behind the front surface of the cornea. The direction of gaze is specified with respect to the median and transverse planes of the head. The straight-ahead, or **primary position**, of an eye is not easy to define precisely, because the head and eye lack clear landmarks. For most purposes, the primary position of an eye may be defined as the direction of gaze when the visual axis is at right angles to the plane of the face. An eye moves from the primary position into a **secondary position** when the visual axis moves from the primary posi-

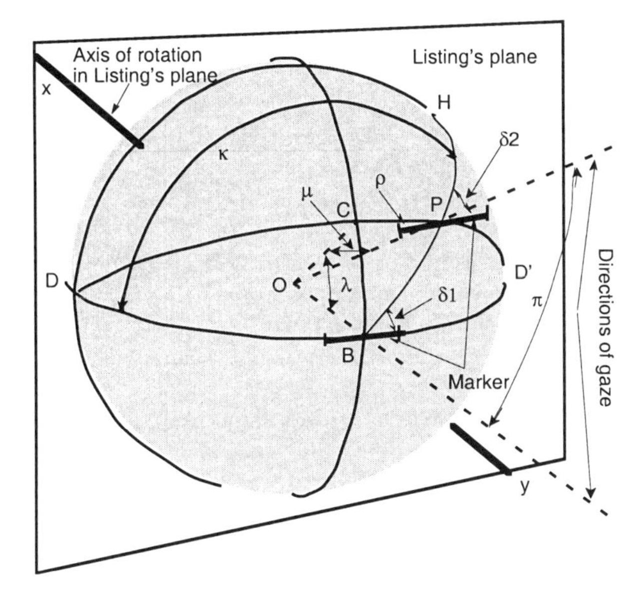

Figure 9.11. The geometry of eye movements.
The direction of gaze is assumed to have moved from the primary position OB to an oblique position OP through an angle of eccentricity π, along the meridian BH, which is at an angle κ to the horizontal meridian DBD'. This is equivalent to it having occurred about axis xy in the equatorial frontal plane (Listing's plane). The horizontal marker between the small vertical bars initially makes an angle $\delta 1$ with the meridian along which the eye moves. According to Listing's law, this angle remains constant ($\delta 1 = \delta 2$). The eye can also be regarded as having moved on Helmholtz axes, through an angle of elevation λ and an angle of azimuth μ. In the Helmholtz system, for angle δ to remain constant, the eye and marker must undergo torsion through angle ρ relative to the final plane of regard DCD'. (From Howard and Templeton 1966)

Figure 9.12. Zoï Kapoula.
Born in Greece in 1955. She obtained a Diploma in Philology and Psychology at Aristotle University of Thessalonika, Greece, 1977. Ph.D. in Experimental Psychology at the University René Descartes, 1982. She conducted postdoctoral work at Durham University, Johns Hopkins Hospital, and the National Institutes of Health. She is Research Director at the National Center for Scientific Research (CNRS, France) and Director of the "Binocular Vision and Oculomotor Adaptation group at the Laboratory of Physiology of Perception and Action at the Collège de France, Paris.

tion in either a sagittal or a transverse plane of the head. An eye moves into a **tertiary position** when the visual axis moves into an oblique position.

I assume that each eye rotates about a fixed centre. The position or rotation of an eye may be specified using any of four co-ordinate systems. The choice is arbitrary although, for a given purpose, one system may have practical advantages.

9.2.2a *Helmholtz system*

In the Helmholtz, the horizontal axis about which vertical eye movements occur is fixed to the skull. The vertical axis about which horizontal movements occur rotates gimbal fashion about the horizontal axis and does not retain a fixed angle to the skull. The direction of the visual axis is expressed in terms of elevation (λ) and azimuth (μ) (Figure 9.10). Torsion is a rotation of the eye about the visual axis with respect to the vertical axis of eye rotation.

9.2.2b *Fick system*

In the Fick system, the vertical axis is assumed to be fixed to the skull, and the direction of the visual axis is expressed in terms of latitude (θ) and longitude (ϕ). Torsion is rotation of an eye about the visual axis with respect to the horizontal axis of eye rotation. The Fick system is the Helmholtz system turned to the side through 90° (Figure 9.10).

9.2.2c *Perimeter system.*

The perimeter system uses polar co-ordinates based on the primary axis of gaze—the axis straight out from the eye socket and fixed to the head. Eye positions are expressed in terms of the angle of eccentricity of the visual axis (π) with respect to the primary axis, and of the meridional direction (κ) of the plane containing the visual and primary axes with respect to the horizontal meridian of head-fixed polar co-ordinates.

These three systems are the same co-ordinate system, simply anchored to the head in different ways (Fry *et al.* 1945). A specification of eye position can be transformed between the three systems by the following equations:

$$\tan\lambda = \frac{\tan\theta}{\cos\phi} = \sin\kappa\tan\pi \qquad (3)$$

$$\sin\mu = \sin\phi\cos\theta = \sin\pi\cos\kappa$$

9.2.2d *Listing's system.*

Listing proposed that any rotation of an eye occurs about an axis in a plane known as **Listing's plane**. Helmholtz called this **Listing's law**. Listing's plane is fixed with respect to the head and coincides with the midfrontal, or equatorial, plane of the eye when the eye is in its primary position (plane *HD'D* in Figure 9.11). Elevations and depressions of the eye occur about a horizontal axis in Listing's plane, lateral movements occur about a vertical axis, and oblique movements occur about intermediate axes. More precisely, any unidirectional movement of an eye can be described as occurring about an axis in Listing's plane that is orthogonal to the plane within which the visual axis moves (plane *OPB*). The extent of an eye movement is the angle between the initial and final directions of gaze (the change of the angle of eccentricity π). The direction of an eye movement is the angle between the meridian along which the visual axis moves and a horizontal line in Listing's plane (δ or its supplement κ).

Torsion in the Helmholtz system is the angle (ρ) between a horizontal marker on the eye and the plane within which the visual axis moves (*D'PD*). In Fick's system, torsion is the complement of angle ρ. These two torsion angles are a function of angle δ and the angle of eccentricity, π, and are not constant, even when the eyes obey Listing's law. Thus, when an eye moves into a tertiary position, it shows no torsion in Listing's co-ordinate system (assuming Listing's plane remains fixed), but it has a considerable degree of torsion in the Fick or Helmholtz systems. Any torsion occurring when the eye is in a primary or secondary position of gaze has the same angular value in all co-ordinate systems. The Helmholtz and Fick systems, being three-axis systems, can be used to specify torsion whereas Listing's system was not designed for this purpose, because it is only a two-parameter system. See Haslwanter (1995) for a discussion of the mathematics of eye movements about three axes.

Listing's law may be tested by impressing a short horizontal reference line on the eyeball in its primary position and seeing whether the angle between the line and the meridian along which the eye moves (angle δ_2) remains constant. The law appears to be correct for conjugate eye movements (Quereau 1954; Fry 1968; Ferman *et al.* 1987a) but, when the angle of horizontal or vertical vergence changes, the law does not hold and torsion, as defined by a change in the angle δ_2, does occur (Allen and Carter 1967).

Listing's plane is traditionally defined as fixed to the head. However, geometrically, a convergence of the eyes produces an outward rotation of Listing's plane in each eye. The two planes swing out about the centres of the eyes like saloon doors. Each degree of convergence of one eye rotates Listing's plane in that eye by between 0.5 and 0.9° (Mikhael *et al.* 1995; Somani *et al.* 1998). Kapoula *et al.* (1999) reported that the temporal slant of Listing's plane during convergence is more consistent among subjects for vergence evoked by disparity than for accommodative vergence, and is most pronounced when both vergence stimuli are present (Portrait Figure 9.12). For a given convergence, Listing's law holds for changes in version with respect to rotated Listing's planes. This is the **modified Listing's law**. The same rotation of Listing's plane occurs in each eye when the eyes converge through a given angle, asymmetrically or symmetrically (Steffen *et al.* (2000).

The outward rotation of Listing's plane with convergence predicts the fact that the eyes become progressively more extorted during downward gaze shifts and intorted during upward gaze shifts (Section 9.8.4).

The relationships between vergence, torsion, and gaze elevation hold for both static positions of gaze and during changes in gaze (Minken and van Gisbergen 1996). Also, cyclovergence induced by changing cyclodisparity and that induced by changes in vergence add linearly, which suggests that the two responses are controlled by distinct systems (Hooge and van den Berg 2000). Theoretical models of these relationships have been developed by Mok *et al.* (1992) and by Minken and Van Gisbergen (1996).

There is also evidence that during conjugate saccades the eyes undergo transient changes in torsion followed by a slow torsional drift in the opposite direction (Enright 1986a; Ferman *et al.* 1987a; Straumann *et al.* 1995). The effects of eye torsion on the vertical horopter are discussed in Section 15.7. Listing's law, even in this modified sense, does not hold in other circumstances, which are discussed in Section 9.8.4.

9.2.2e *The mechanism of Listing's law*

The eyes do not obey Listing's law during sleep, showing that they are not mechanically constrained to move this way. However, they obey the law when they move in the dark, so visual feedback is not required. This suggests that movements obeying Listing's law are neurally programmed (Nakayama 1975). Since Listing's plane is fixed to the head, eye-movement commands would have to be referred to the head rather than to co-ordinates fixed to the eye. The superior colliculus controls saccadic eye movements in oculocentric co-ordinates, which simplifies the coordination of eye movements with auditory targets and with movements of the arm (van Opstal

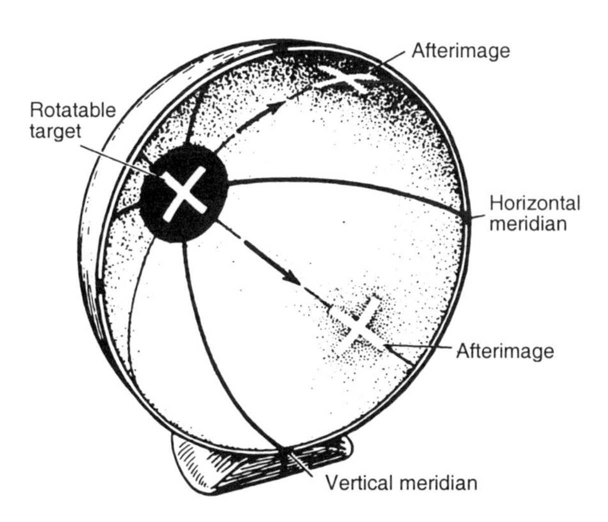

Figure 9.13. Demonstration of Listing's law.
An afterimage of a cross is impressed on the eye in its primary position. The afterimage retains the same orientation with respect to the meridian along which the gaze moves, as indicated on a hemisphere concentric with the eye. This indicates the validity of Listing's law. (From Noorden 1990)

et al. 1991). The implementation of Listing's law must be downstream from the superior colliculus.

Recent evidence suggests that the eyes move according to Listing's law because of the way the extraocular muscles are inserted on the eye rather than because of central neural control. Each lateral rectus muscle attaches to the eye at a point forward of the equator. If this were the only point of attachment, the point of tangency of the muscle on the globe would slide upward as the eye elevates and contraction of one lateral rectus would cause a torsional motion of the eye.

Miller and Robins (1987) showed that, near the point of tangency, each lateral rectus passes through a sleeve of connective tissue attached to the eye's orbit. This sleeve acts as a pulley, which causes the point of tangency to remain approximately constant as the eye elevates or depresses. The direction of action of the muscle thus remains fixed with respect to the head. The attachment of the pulley to the orbit contains smooth muscle, which could perhaps adjust the direction of action of the muscle (Demer *et al.* 1997). The other extraocular muscles are probably inserted in a similar way (Simonsz *et al.* 1985; Raphan 1998).

9.2.2f *Consequences of Listing's law*

Movement of the eyes according to Listing's law has the following consequences.

1. It reduces the degrees of freedom of eye movements and simplifies motor control. However, there are other ways in which this could be done. For example, torsion as defined in the Helmholtz coordinate system could be held constant.

2. It prevents a build up of eye torsion as the gaze moves over a circular path.

3. It economizes on the total amount of eye movement in a change of gaze. However, Listing's law in its modified form does not achieve this purpose because, when the eyes are converged, they move with respect to distinct planes. Tweed (1997) concluded, on the basis of a computer simulation, that Listing's law in its modified form achieves the best economy of eye movement compatible with maintaining the retinal meridians in torsional correspondence. The extent to which Listing's law is modified is to some extent under neural control (Section 9.8.4). This issue is discussed in more detail in Section 16.2.11.

4. Helmholtz noted that Listing's law has the important consequence that, as the gaze travels along any line in the visual field, the retinal image of the line remains self-congruent. This means that the line continues to stimulate cortical orientation detectors tuned to the same orientation (see Figure 9.13). It follows that an afterimage of a line imposed on the normally vertical meridian of an eye remains congruent with a circle parallel to the midvertical meridian of a sphere centred on the eye. These circles project as hyperbolic arcs in a frontal plane. The primary position of gaze may be specified as the centre of the set of hyperbolic arcs (Nakayama 1978). However, since Listing's law is formulated with respect to the head, the law does not ensure that the images of visual lines remain self-congruent over combined rotations of the eyes and head.

5. Van Rijn and van den Berg (1993) suggested that the modified version of Listing's law ensures that lines orthogonal to the plane of regard fall on corresponding vertical meridians. However, Tweed (1997) pointed out that this account overlooks the fact that corresponding vertical meridians are excyclorotated about 2°.

6. Listing's law in both its forms ensures that corresponding horizontal meridians fall within the binocular plane of regard. However, it does not ensure that corresponding epipolar lines are aligned between the two eyes (Section 20.1.2a).

The eyes of the chameleon obey Listing's law (Sandor *et al.* 2001). Their eye muscles differ from those of primate eyes and they do not use disparity for judging distance. One must therefore assume that the eyes of chameleons obey Listing's law so as to economize on the magnitude, and hence the speed, of eye movements. *We need to know how widespread Listing's law is in different species of animal.*

9.3 TONIC VERGENCE

9.3.1 Dark vergence

In the dark, a person's eyes adopt a characteristic state of vergence known as **dark vergence**. Dark vergence is measured by placing a person in the dark and exposing a pair of nonius lines for 100 ms, which is shorter than the latency for vergence. The lines are adjusted between exposures until they appear aligned. Repeated exposure of the lines at one distance may induce a change in vergence through the mediation of accommodation cues, and this can affect the results (Jaschinski-Kruza 1990). Other methods for measuring dark vergence are described by Rosenfield (1997).

It is believed that dark vergence depends on tonic efference arising from the vestibular system, the muscles of the neck, the eyes, and proprioceptive feedback from the extraocular muscles. The distance of the point of dark vergence varies between 0.62 and 5 m for different observers, with an average value of about 1.2 m. It is consistent over time for the same individual if the conditions of testing are constant (Owens and Leibowitz 1980; Fisher *et al.* 1988a). However, dark vergence can be altered temporarily. Maddox (1893) noted that it increases after a period of increased convergence demand induced by looking at fusible targets through base-out prisms. This effect has been confirmed many times (Rosenfield 1997).

Helmholtz (1910) observed that the eyes tend to diverge when elevated. In the dark, the eyes diverge from the position of dark vergence when the gaze is elevated and converge from that position when the gaze is depressed (Heuer and Owens 1989). See Section 9.3.4 for related effects of vertical gaze shifts. After the eyes have been held in an elevated or depressed posture for a few minutes, the position of dark vergence with horizontal gaze is temporarily biased towards the previously maintained state (Heuer *et al.* 1988).

In the dark, the eyes also take up a characteristic state of accommodation known as tonic accommodation, or **dark focus** (Section 9.1.2). Dark focus and dark vergence do not correspond to the same distance (Owens and Leibowitz 1980).

In deep sleep, anaesthesia, or death the eyes assume a posture known as the **anatomic position of rest**. This state represents the state of the muscles when they lack tonic innervation (Alpern 1969). The anatomic position of rest is more divergent than dark vergence, although there is no generally agreed value for this position (see Owens and Leibowitz 1983).

9.3.2 Strabismus

9.3.2a *Types of strabismus*

A person who cannot converge the visual axes on the intended object is said to have a squint, **strabismus** or **heterotropia**. Properly aligned eyes are said to be **orthotropic**.

The **angle of deviation** from the orthoptic state in prism dioptres is measured by having the patient fixate a point with the non-deviating eye and observing the direction of gaze of the deviating eye. The direction of gaze is measured by moving a point of light round a perimeter until its reflected image is centred on the cornea. The angle of deviation can be measured with the fixation target near or far. There are many types of strabismus. In divergent strabismus, or **exotropia**, the visual axes are directed outward from the intended point. When the angle of deviation is the same for near as for far viewing, it is a basic exotropia. When the exotropia is combined with **convergence insufficiency** the angle of exotropia is larger with near than with far viewing, and when combined with **divergence excess** it is larger with far than with near viewing. In convergent strabismus, or **esotropia**, the visual axes are directed inward, and in vertical strabismus the two axes do not lie in the same horizontal plane. Esotropia is about three times more frequent than exotropia. In **unilateral strabismus** one eye is consistently used for fixation when both eyes are open. This type of strabismus is often accompanied by some loss of visual function in the deviating eye, a condition known as amblyopia (Section 13.4). In **alternating strabismus** the person sometimes fixates objects with one eye and sometimes with the other. Alternating strabismus occurs in only about one third of esotropes but in about 80% of exotropes. Patients with **intermittent exotropia** are able to control the deviation of the eye when making an effort to fuse the images. Since amblyopia is associated with unilateral strabismus, the prevalence of unilateral strabismus in esotropes accounts for why they are much more likely to develop amblyopia (Friedman *et al.* 1980).

In **comitant strabismus** the deviation is the same in all directions of gaze, although it may vary according to the angle of convergence and from day to day. Comitant strabismus is due to a defect of the vergence mechanism rather than to a defect of particular extraocular muscles and is sometimes reduced when refractive errors are corrected optically. In **noncomitant strabismus** the deviation varies with the angle of gaze. It is due to a paresis in one or more of the extraocular muscles arising from damage to muscle or to oculomotor nerves. Noncomitant strabismus arising from paresis is also known as paralytic strabismus. The noncomitance is particularly evident when patients attempt to move the eyes in the direction of action of the paretic muscles.

In **large-angle strabismus** the angle of deviation is larger than 15 dioptres, otherwise it is **small-angle strabismus**. Deviation of one eye of up to 8 dioptres has been called **microstrabismus**, or microtropia. This condition is discussed in Section 9.3.4f.

In small-angle strabismus, the visual system may manifest anomalous retinal correspondence (ARC), defined as a shift in corresponding regions in the two retinas. The angular extent of anomalous correspondence is the angle of anomaly or angle A. In harmonious anomalous correspondence, angle A equals the angle of strabismus and there is no movement of the deviating eye when the nondeviating eye is covered (the cover test). In inharmonious anomalous correspondence, angle A is less than the angle of strabismus. Anomalous correspondence is discussed in Section 15.4.2.

About 5% of people have some form of strabismic anomaly of 5 dioptres or more. In one survey of over 3,000 patients, 74% of those tested had comitant strabismus, 10% were paretic, 8% had decompensated heterophoria, and 6% had convergence insufficiency (Stidwill 1997). About 65% of cases of strabismus develop before the age of three years, with a mean age of onset of about 30 months (Graham 1974; Stidwill 1997).

9.3.2b *Visual defects in strabismus*

With early onset strabismus the strabismic eye suffers from **amblyopia**. This involves a loss in acuity and other visual functions, as described in Section 13.4.2. In strabismus, images falling on corresponding retinal points are dissimilar and therefore engage in rivalry. Also, images from the same objects fall on noncorresponding points and create diplopia. Strabismics of long standing avoid both these disturbing effects by totally suppressing the image in the strabismic eye when both eyes are open. Clinical **suppression** can be studied in strabismic monkeys (Wensveen *et al.* 2001).

Stereoscopic vision is degraded or totally absent in large-angle strabismus of early onset. Binocular vision may be restored when strabismus of late onset is corrected. Some patients with early onset esotropia cannot fuse similar images even when the strabismus is corrected with prisms. They report that, as the images approach each other, they jump over each other to produce the opposite sign of diplopia. This condition is known as **horror fusionis**.

Small-angle strabismus (microtropia) does not usually present a cosmetic problem and there is often some residual stereopsis with tests such as the Howard-Dolman test and Titmus test—the less the deviation the greater the probability of some stereopsis (Cooper and Feldman 1979; Rutstein and Eskridge 1984). There has been some dispute about whether small-angle strabismics show evidence of stereopsis when tested with random-dot stereograms (Cooper and Feldman 1981; Henson and Williams 1980).

9.3.2c *Directional preponderance and latent nystagmus*

Strabismus developing before the age of 1 year is known as early onset strabismus and is usually associated with reduced gain of pursuit eye movements in the temporal direction and directional preponderance of optokinetic nystagmus for stimuli moving in the nasal direction with monocular viewing (Section 23.4.5). Sensitivity to visual motion is symmetrical in strabismic amblyopes so that their asymmetrical OKN cannot be due to a defect in motion sensitivity (Schor and Levi 1980). Also, eye rotations in the dark are normal so that the defect is not due to muscle imbalance. The defect is most likely due to the incomplete development of binocular vision in strabismics, which releases the temporonasal preponderance of subcortical centres from the counterbalancing influence of cortical inputs (see Section 23.4.5).

Strabismus of early onset is also accompanied by **latent nystagmus**, which occurs when one eye is closed, and sometimes when both eyes are open (Dell'Osso *et al.* 1983). In its simplest form, latent nystagmus is a spontaneous conjugate jerk nystagmus, which occurs only when one eye or the other is closed. In strabismic amblyopia latent nystagmus is larger when the affected eye is open than when the normal eye is open. The slow phase of the nystagmus is in the temporonasal direction, so that the direction of the nystagmus changes as a cover is moved from one eye to the other. In most cases the latent nystagmus is also evoked with the quick phase in the direction of gaze when both eyes are open, or in the direction of the fixating eye when the other eye is open but in a deviated position due to strabismus. Patients with alternating strabismus can control the direction of latent nystagmus by changing the eye used for fixation. Some patients who cannot control the direction of the nystagmus when both eyes are open may do so in the dark by imagining that they are looking with one eye or the other (Dell'Osso *et al.* 1979; Kommerell and Mehdorn 1982). Latent nystagmus can be regarded as a spontaneous manifestation of the temporonasal preponderance characteristic of congenital strabismus. Directional preponderance and latent nystagmus occur also in naturally strabismic monkeys (Tychsen and Boothe 1996).

Poor vergence control and fixation instability have been implicated in dyslexia (see Buzzelli 1991). However, the evidence is equivocal (Evans *et al.* 1994). Moores *et al.* (1998) could find no relationship between the ability to control vergence across saccades and dyslexia.

9.3.2d *Aetiology of strabismus*

The aetiology of strabismus is unclear. About 1% of the population has congenital infantile esotropia which may not be evident at birth but becomes evident during the first six months (Nixon *et al.* 1985). There is evidence of a genetic factor (Schlossman and Priestley 1952; Graham 1974). Strabismus of early onset may be associated with an uncorrected refractive error of over +3 dioptres that requires the patient to accommodate excessively for near vision. This evokes excessive convergence through the mediation of accommodative convergence. The esotropia usually begins as an intermittent crossing of the eyes and progresses to a constant deviation. Early correction for the hyperopia may restore binocular fixation and stereopsis in cases of accommodative esotropia (Wilson *et al.* 1993). It has been proposed that a basic cause of strabismus is a defect in the disparity feedback mechanism controlling vergence (see Kerr 1998). Early monocular occlusion in cats leads to misalignment of the occluded eye, which becomes a permanent strabismus when both eyes are allowed to see (Quick *et al.* 1989). Strabismus often accompanies congenital cataracts. More information on strabismus is provided by Lennerstrand *et al.* (1988) and von Noorden (1990).

9.3.2e *Treatment of strabismus*

It has been claimed that normal retinal correspondence and improved acuity and stereo acuity can be achieved in patients with microtropia by periodic occlusion of the good eye and refractive correction (Cleary *et al.* 1998; Houston *et al.* 1998).

Patients with intermittent exotropia, unlike those with other forms of large-angle strabismus, tend to retain stereopsis. The degree and frequency of deviation tends to increase with age and may be improved by surgery (Yildirim *et al.* 1999).

Large-angle comitant strabismus and noncomitant strabismus of muscular origin may be partially or completely corrected by surgically adjusting the extraocular muscles or by injection of a neurotoxin into selected muscles (Scott 1981). However, only a small proportion of patients with early onset strabismus show evidence of stereopsis, even when the surgery is performed before the age of 2 years (Pratt-Johnson and Tillson 1983) (Section 13.3.3a).

It has been claimed that strabismus may also be cured by wearing prisms for a period of up to 6 months This method was used widely in the nineteenth century, especially in Europe. However, Alpern and Hofstetter (1948) reported that constant wearing of prisms can lead to an increase in the angle of squint. Although some clinicians still advocate the use of prisms, debate about their effectiveness continues (see von Noorden 1990).

Estimates of the success of therapy for strabismus in restoring some binocular vision have varied from 5% or less (Flom 1963; Dobson and Sebris 1989) to over 50% (Wick and Cook 1987).

Corrective surgery can lead to the onset of amblyopia in patients who were not amblyopic before surgery (Pratt-Johnson and Tillson 1983). In most cases post surgical exercises reduce the severity of amblyopia (Murray and Calcutt 1990). Success rates, as reflected in restoration of binocular vision, are higher when surgery is performed before the age of 24 months (Keenan and Willshaw 1992). For such patients, restoration of some stereopsis depends more on the duration of the strabismus than on the age at which corrective surgery was performed (Birch *et al.* 2000a). Correction applied even in older children reduces the post-saccadic drift evident before surgery (Inchingolo *et al.* 1996).

Strabismus of late onset is not associated with amblyopia, and corrective surgery can be expected to restore binocular vision in most cases. There has been some debate about whether restoration of stereopsis becomes less probable as surgical treatment is delayed. A recent study of 25 patients with acute-onset esotropia with mean onset age of 12 years indicates that the restoration of stereopsis does not depend on the period of delay before surgery (Ohtsuki *et al.* 1994).

A patient may not be sure when a strabismus developed. Certain visual defects associated with early but not late onset esotropia, such as asymmetry of visual pursuit and dissociated vertical deviation (DVD), may indicate whether a strabismus is of early or late onset (Demer and von Noorden 1988, Schor *et al.* 1997). Corrective treatments have a cosmetic value even if amblyopia is not cured or binocular vision restored.

9.3.3 Phoria

9.3.3a *Types of phoria*

Phoria is a latent strabismus revealed only when the eyes are **disassociated**, that is, when no fusible stimuli are in view. It may be regarded as the open-loop vergence error. **Orthophoria** is the condition of zero phoria. The eyes can be disassociated by closing or covering one eye, by displacing one image with a vertical prism so that the images in the two eyes no longer coincide, or by presenting overlapping but rivalrous dichoptic stimuli. A horizontal phoria may involve an inward (**esophoria**) or outward (**exophoria**) deviation of an eye. Horizontal phoria is conventionally measured while one eye is accommodated on a distant object, because it is believed that effects of accommodation on vergence are least with far accommodation. A deviation of the left eye upward or downward is a left hyperphoria or left hypophoria, respectively, and similar deviations of the right eye are either a right hyperphoria or a right hypophoria. A torsional deviation of an eye is a **cyclophoria**: top-inward is an incyclophoria and top-outward, an excyclophoria.

It takes about 20 seconds for an eye to come to rest in its position of phoria after it has been covered (Schor 1979a). When the magnitude of phoria varies with the eccentricity of gaze, it is known as **anisophoria** (Friedenwald 1936). An essential anisophoria is due to paresis of one or other extraocular muscle in one eye and optical anisophoria is due to optical magnification of the image in one eye by a spectacle lens.

A horizontal phoria of up to 4 prism dioptres (about 2°) is considered normal. A small degree of exophoria is normally present and is referred to as **physiological exophoria**. It increases with age, especially with near vision (Freier and Pickwell 1983). A phoria is called compensated when not accompanied by other symptoms and uncompensated when accompanied by symptoms such as headache, eye strain, blurred vision, or problems with stereopsis. The presence of these secondary symptoms bears no simple relationship to the size of the phoria (Evans 1997).

The sign of phoria can vary with fixation distance. Phoria measured with distant targets is known as **distance phoria**. An esophoria that is greater for far than for near viewing is a **divergence weakness** and an esophoria that is greater for near viewing is a **convergence excess**. An exophoria that is greater for near viewing is a **convergence weakness** and an exophoria that is greater for far viewing is a **divergence excess** (see Section 9.5.3b).

The tonic state of vergence, and hence the direction and magnitude of a phoria, can be changed temporarily by holding the eyes in an extreme position of divergence or convergence (Section 9.3.5). Subjects exposed to 24 hours of monocular occlusion showed increased esophoria, especially for near vision, but normal phoria was re-established after one day of binocular vision (Bross 1984).

9.3.3b *Measures of phoria*

Phoria may be measured objectively by the **cover test**. In a simple qualitative version of the cover test the patient fixates a spot and the clinician observes the change in position of each eye as a cover is placed in front of it. To quantify a phoria, the clinician increases the power of a prism placed before the deviating eye until a change in position of that eye is no longer observed when the eye is alternately covered and uncovered. The power of the prism indicates the degree of phoria in prism dioptres, and the orientation of the prism indicates the direction of phoria. This method is limited by the smallest deviation of an eye that a clinician can detect which, on average, is about 2 prism dioptres (Romano and von Noorden 1971).

Eye position is measured objectively when greater precision is required. In the clinic, the rotation of an eye is measured by the displacement, with respect to the centre of the pupil, of a light spot reflected off the cornea. A prism can be used to bring the light spot in the deviating eye into the same position relative to the pupil as the light spot in the undeviating eye. The setting of the prism indicates the magnitude of deviation (Krimsky 1972). When measurements are made from photographs, the estimated ocular rotation per millimetre of light displacement is known as the **Hirschberg ratio**. This method has been used with children (Hasebe *et al.* 1998) and monkeys (Quick and Boothe 1989).

In so-called subjective tests for tropia and phoria, the patient is required to align visual targets. Given that the patient is able to make the judgments, subjective methods are at least as precise as objective methods.

In the first type of subjective procedure, known as the **alternate cover test**, the examiner alternately covers each eye of the patient, who reports the direction of apparent movement of a test object. In exophoria, the object appears to move with the occluder and in esophoria it appears to move against the occluder. The prism power required to null the motion of the test object indicates the magnitude of phoria.

In a second type of subjective procedure a single visual target is introduced to both eyes in such a way that the two images are **disassociated** and the fusional response is disengaged. For instance, in the **Maddox-rod test** the patient views a point of light directly with one eye while the other eye views the same point through a set of high-power cylindrical prisms arranged as a grating. Depending on their orientation, the prisms spread the point of light into either a horizontal or a vertical line. The power of a wedge prism required to bring the point and line

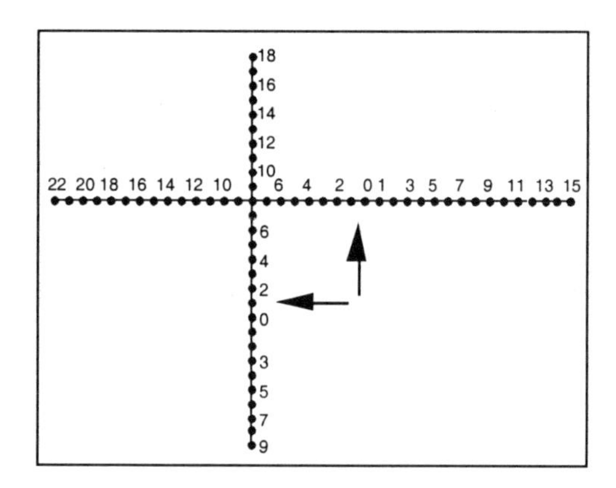

Figure 9.14. A tangent scale used in the Maddox wing test. The display is placed in a stereoscope, one eye viewing the arrows and the other eye the scales. The subject indicates which numbered dots are aligned with the arrows.

back into superimposition for the patient indicates the degree of phoria—horizontal phoria when the line is vertical and vertical phoria when it is horizontal. In the **Hess-Lancaster test** a red spot seen on a screen by one eye is moved by the patient until it appears superimposed on a green spot seen by the other eye. The colour separation is achieved with red and green filters. This is repeated for several angles of gaze to determine whether the tropia or phoria is comitant or noncomitant.

In the third type of subjective measure of tropia or phoria, disassociated dichoptic stimuli are presented in a stereoscope so that the fusional response is disengaged. The **Maddox wing test** is essentially a Brewster stereoscope in which one eye sees calibrated horizontal and vertical lines in the form of a cross and the other eye sees a vertical and a horizontal arrow. The patient reads off the position of each arrow on the appropriate scale to indicate horizontal and vertical phoria (Figure 9.14). Cyclophoria requires an annular scale. There must be nothing else in view that could lock vergence. Such a device is known as a **phorometer**.

Other stereoscopic devices used in orthoptic practice to measure and treat tropia and phoria derive from Hering's haploscope, The essential features are shown in Figure 9.15. The subject's head is fixed so that the centre of rotation of each eye is above the centre of rotation of one of the horizontal arms of the instrument. The visual targets are mounted on the ends of arms and reflected into the eye by mirrors set at 45° to the median plane. The accommodative distance of each target is adjusted by moving it along the arm. Horizontal phoria is indicated by the angular position of the arm required to bring disassociated targets into alignment. There is also a control for varying the vertical position of each target. In the clinic, haploscopes are known as amblyoscopes, synoptophores or troposcopes, depending on the manufacturer.

Schroeder *et al.* (1996) found that some tests of phoria correlate highly, others less so. Differences between tests are probably due to the degree of control of accommodative convergence and vergence adaptation.

For a review of orthoptic procedures used in the diagnosis and treatment of anomalies of binocular eye movements see Evans (1997), Goss (1995), and Griffin and Grisham (1995).

9.3.3c *Relation between dark vergence and phoria*

If phoria represents the tonic state of an eye when relieved from fusional demand it should be the same as dark vergence. We will see in the next section that, with dissociated targets, the accommodative state of the eyes can evoke a change in vergence. Thus, a difference between dark vergence and phoria could arise if the accommodative state of an eye in the dark differs from that when phoria is measured (Owens and Tyrrell 1992). Maddox (1893) assumed that accommodation is at rest for distant targets. According to this assumption, phoria measured with a distant target (distance phoria) should equal dark vergence. However, in the dark, the eyes are accommodated at about 1.5 dioptres (Section 9.1.2). Thus, in measuring distance phoria, accommodation must relax from its dark state. This produces accommodative divergence, and results in phoria being more exophoric than dark vergence (O'Shea *et al.* 1988).

For most subjects, phoria and dark vergence were highly correlated when measured with open-loop accommodation produced by viewing through 0.5 mm pupils (Rosenfield and Ciuffreda 1990). The difference between phoria and dark vergence could also be due to differing states of proximal vergence (Section 9.4) in the two conditions of measurement. Rosenfield and Ciuffreda found that this was a factor in a few of their subjects.

9.3.4 Fixation disparity

The tendency of the eyes to drift in the direction of a phoria may manifest itself as a slight deviation from the intended state of vergence when both eyes are open. This deviation is known as **fixation disparity**. The power of a prism before one eye required to reduce fixation disparity to zero is referred to as **associated phoria**, that is, phoria measured when the fusional reflexes are engaged (Ogle *et al.* 1967, p.

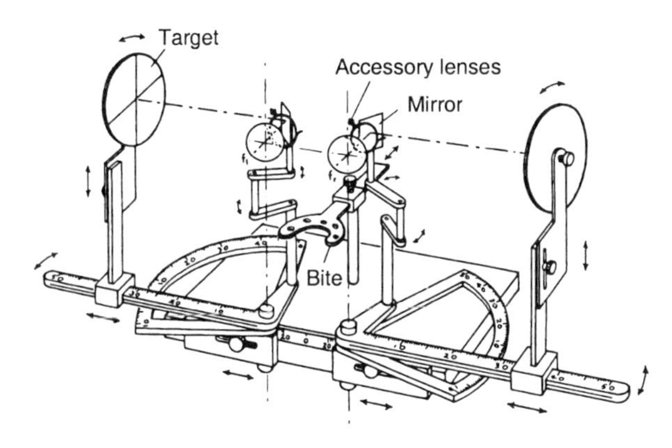

Figure 9.15. Essential components of a haploscope.
(From von Noorden 1990)

108). The point where fixation disparity is reduced to zero should be the point of oculomotor balance.

The phenomenon of fixation disparity was first mentioned by Hofmann and Bielschowsky (1900), who referred to it as fixation lag. It was observed independently by Judd in 1907 and by Lau in 1921. In 1937 Bielschowsky emigrated from Europe to the United States where he was director of the Dartmouth Eye Institute, until his death in 1940. Kenneth Ogle, who worked in the same institute, first used the term "fixation disparity". Most people have some fixation disparity. It is usually less than 6 arcmin with foveal fixation but can be as great as 20 arcmin with peripheral visual targets (Wick 1985).

The eyes tend to diverge when elevated and converge when lowered, as reflected in the state of dark vergence (Section 9.3.1). Similarly, elevating the eyes reduces esophoria and increases exophoria and lowering the eyes has the opposite effect. Also, fixation disparity becomes more eso when gaze is lowered and more exo when gaze is elevated (see Jaschinski *et al.* 1998).

The simplest view of fixation disparity is that the images of an object a person is attempting to fixate do not fall exactly on corresponding points so that the horopter does not pass through the fixation target. The viewer does not notice a fixation disparity, because the disparate images fall within Panum's fusional area. However, fixation disparity can have a slightly adverse effect on stereoacuity (Cole and Boisvert 1974) (Section 19.3.2).

9.3.4a *Nonius measurement of fixation disparity*

The **Mallett test** is a commonly used clinical test of fixation disparity. The letters, OXO, each subtending about 20 arcmin, are placed on a 1.5° disk superimposed on a page of print and viewed by both eyes. Nonius lines disassociated by polarized light are placed one above and one below the letters. The deviation of each nonius line with respect to the cross indicates the displacement of the point of fixation in each eye. Vertical fixation disparity is measured with the display turned through 90° (Mallett 1964). The magnitude of fixation disparity is indicated by the degree of prism-power required to bring the nonius lines into alignment as the subject fixates the cross.

In the **Sheedy disparometer** pairs of nonius lines with variable positive or negative offset are brought, one pair at a time, into a central aperture. The subject selects the pair of nonius lines that appear aligned (Sheedy 1980). Binocularly viewed letters surround the central aperture.

In the test used by Ogle the subject views a binocular display of letters, and fixation disparity is indicated by the distance that one nonius line has to be moved to appear aligned with another nonius line. In a more recently developed test, the subject judges the alignment of an arrow seen by one eye with one of a series of vertical coloured lines seen by the other eye (van Haeringen *et al.* 1986).

Fixation disparity may be measured by these procedures with a precision of about 2 arcmin. The Mallett test gives a more stable measure than the Sheedy test (Pickwell *et al.* 1988). Also, measures obtained with the Mallett test are typically half the magnitude of those obtained with the Sheedy test. This may be because the Mallett test provides a central fixation stimulus while the binocular lock stimulus is parafoveal in the Sheedy test.

Factors that could affect the size and precision of measurements of fixation disparity include the length and separation of the nonius lines and the size and position of the fusion lock. Ukwade (2000) found that, with a central fusion lock, the precision of nonius alignment, which reflects the stability of vergence, decreased when the vertical separation of the nonius lines increased beyond 20 arcmin. Also, for small nonius separations, precision decreased as the fusion-lock moved into the periphery (see also Wildsoet and Cameron 1985). As the fusion lock becomes more eccentric it becomes less effective, although its effectiveness also depends on its area. Overall, the size and variability of measurements were least when both central and peripheral fusion locks were provided and when the nonius lines were less than 20 arcmin apart.

The test-retest correlation of fixation disparity using a laboratory nonius procedure was 0.8 at viewing distances of 40 and 82 cm but only 0.55 at a distance of 26 cm (Jaschinski-Kruza 1993). In any subjective test of fixation disparity there is a danger

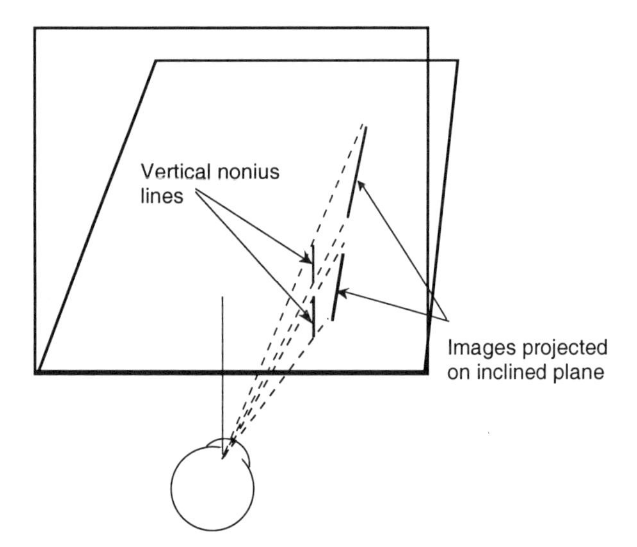

Figure 9.16. Nonius lines projected onto inclined surface.
Images of off-centre vertical nonius lines projected onto an inclined plane appear inclined and tilted.

that the act of attending to the nonius lines may increase the error in fixation or even induce it, since nonius lines do not provide a stimulus for bifoveate fixation (Verhoeff 1959). We will see below that fixation disparity varies as a function of the degree of forced vergence induced by base-in or base-out prisms.

Measurements of fixation disparity do not seem to vary significantly when the eyes move into positions of eccentric gaze (McKee *et al.* 1987). Tests of fixation disparity are described in Goss (1995).

9.3.4b *Reliability of the nonius procedure*

The nonius procedure relies on the assumption that the apparent location of an image in an eye is not affected by surrounding stimuli, and on the related assumption that objectively aligned dichoptic lines fall on corresponding retinal meridians. But the following stimulus factors can influence nonius measurements.

1. The alignment of two nonius lines presented to one eye is equivalent to a vernier acuity stimulus. Carter (1958) found constant errors in monocular vernier alignment (vernier bias) of up to about 1 arcmin. Vernier bias increases with increasing separation of the test lines. The bias differs between the eyes and, when measured with both eyes open, is the algebraic sum of the monocular biases (Jaschinski *et al.* 1999). Vernier bias, also referred to as nonius bias, should be measured and allowed for in the accurate determination of fixation disparity by the nonius method.

2. A nonius line seen against a moving background appears displaced in the opposite direction (Section 9.8.2).

3. A nonius line exposed on a real inclined surface is affected by the shear in the image of the surrounding texture elements (Section 9.8.5).

4. Shimono *et al.* (1998) showed that a pair of objectively aligned nonius lines on a horizontally disparate region of a random-dot stereogram can appear laterally displaced relative to a similar pair of nonius lines on the zero-disparity region of the stereogram. Thus, nonius lines can indicate misalignment of corresponding meridians when there is no misalignment. They also showed that, when the gaze shifts from one depth plane to another in a random-dot stereogram, vergence might change by up to about 6 arcmin without affecting the apparent alignment of a pair of nonius lines. Thus, nonius lines may fail to indicate a real change in alignment of meridians. These effects were weakened when the nonius lines were separated from the dots of the stereogram.

5. A monocular nonius line appears to lie on a stereoscopic surface. When the stereoscopic surface is inclined or slanted in depth the nonius line may appear tilted in the frontal plane (Ono and Mapp 1995). An off-centre vertical monocular line is projectively equivalent to a tilted inclined line, as illustrated in Figure 9.16. Domini and Braunstein (2001) showed that the apparent tilt of a vertical off-centre monocular line superimposed on a stereoscopically inclined textured surface closely conforms to the projective equivalence. The effect increases with increasing inclination of the surface and increasing eccentricity of the nonius lines. There is no effect for a central nonius line. However, the other effects listed here operate with central nonius lines.

6. Kertesz *et al.* (1983) reported that measures of vertical vergence obtained with the Purkinje eye tracker did not always agree with nonius measurements (Section 9.7).

7. It has been reported that the form of the forced vergence curve measured by nonius alignment can be influenced by volition (Garzia and Nicholson 1988).

8. We will see in what follows that nonius lines may be subject to shifts in binocular correspondence induced by a period of forced vergence. For this reason, the nonius method underestimates the effects of forced vergence on fixation disparity.

These cases indicate that nonius lines are affected by adjacent stimuli. It is important in using the nonius method to separate the nonius lines from other stimulus elements.

9.3.4c *Fixation disparity and fixation shift*

The monocular components of fixation disparity are the deviations from the intended point of fixation attributable to each eye. In some people, fixation disparity involves an equal and opposite deviation of each eye while, in others, one eye is deviated more than the other. Ogle *et al.* (1967) concluded that deviation of the non-dominant eye accounts for most of a fixation disparity. The contribution of each eye may vary with the state of vergence (Reading 1992, 1994). It seems that imbalance of monocular components is not related to defective binocular vision (Irving and Robertson 1991). However, degrading the image in one eye with a lens or a light-scattering filter increased the contribution of that eye to fixation disparity (Irving and Robertson 1996).

Hebbard (1962) measured the position of each eye by reflecting a beam of light off a mirror mounted on a contact lens. He assumed that, with one eye occluded, the open eye accurately fixates a target, and the change in position of an eye that occurs when the other eye is opened is the contribution of that eye to fixation disparity. The combined shift in fixation for the two eyes—the **fixation shift**—was assumed to equal the fixation disparity. He concluded from tests on one subject that the nonius and objective methods give essentially the same result. But the logic of the objective method is open to doubt, since a monocular target may not be accurately foveated. For example, a central scotoma in one eye forces the subject to use a point adjacent to the scotoma for fixation (see Section 15.4.2). Fogt and Jones (1997) found Hebbard's fixation shift method to be less reliable than objective measurement of fixation disparity using scleral search coils.

Contrast bands (Mach bands) along a black-white border increase in width with increasing eccentricity. Remole (1984, 1985) measured the width of the contrast bands in a vertical border as a function of the eccentricity of the border for each eye separately. Similar measurements made with dichoptic viewing provided a measure of fixation shift as a function of the magnitude of forced vergence. The fixation shift was found to be much larger than fixation disparity measured by nonius alignment. Remole suggested that the nonius offset arises from two effects; a fixation shift, or shift in tonic vergence, and a subjective displacement of the two nonius lines, which operates in the opposite direction. A subjective displacement of nonius lines is equivalent to a change in binocular correspondence, which tends to reduce the disparity created by forced vergence. The relation between fixation shift and fixation disparity changed with changes in forced vergence. The two measures were correlated between subjects only for forced vergence of over 5 dioptres (Remole *et al.* 1986). It seems that corresponding points defined by nonius alignment (equal visual directions) are not the same as those defined with respect to the central fovea, as used in monocular fixation. Kertesz and Lee (1987) also found considerable differences between fixation disparity measured with nonius lines and an objective measure of the fixation shift, and concluded that fixation disparity cannot be derived from uniocular changes in fixation. They also concluded that the nonius method is an unreliable measure of fixation disparity.

Fogt and Jones (1998a) found that, in several subjects, fixation disparity derived from use of nonius lines was less than that derived by measurement of the positions of the two eye by scleral coils under the same stimulus conditions. For all subjects, the objective method and the nonius method gave similar results with no forced vergence. For some subjects the nonius method progressively underestimated fixation disparity relative to the objective method as forced vergence was increased. Like Remole, they concluded that nonius lines underestimate the change in tonic vergence because nonius lines are subject to a change in binocular correspondence of up to about 1°, which operates in the opposite direction to the effects of forced vergence. The underestimation occurred only when the nonius lines were within about 3° of the fixation target (Fogt and Jones 1998b). In other words, forced vergence induced a temporary and local distortion of the horopter in the neighbourhood of a forced vergence target.

Other evidence of temporary changes in binocular correspondence is provided in Section 15.4.1.

9.3.4d *Fixation disparity and phoria*

Ogle regarded fixation disparity as an expression of muscle imbalance due to heterophoria. If this is so, the magnitude and direction of the two effects should be correlated. Some investigators reported a correlation (Ogle *et al.* 1949; Ogle and Prangen 1951, 1953; Saladin and Sheedy 1978; McCullough 1978) but others found only a weak correlation (Palmer and von Noorden 1978). Jampolsky *et al.* (1957) reported that, for far fixation, fixation disparity was correlated with esophoria but did not increase with increasing exophoria. For near fixation, fixation disparity was a function of both esophoria and exophoria.

One might expect that disassociated phoria (measured monocularly) and associated phoria (prism required to null fixation disparity) should be equal. However, Ogle *et al.* (1967, p. 108) found that associated phoria is usually less than disassociated

Figure 9.17. John Semmlow.
Born in Chicago, in 1942. He obtained a BSEE degree from the University of Illinois in Champaign in 1964 and a Ph.D in Bioengineering at the University of Illinois Medical Center in Chicago in 1970. He has held faculty positions at the University of California, Berkeley and the University of Illinois, Chicago. He currently holds a joint position as Professor of Surgery at the Robert Wood Johnson Medical School and Professor of Biomedical Engineering at Rutgers University.

phoria. The following factors may account for this difference.

1. Fixation disparity is influenced by tonic imbalance in the accommodative vergence system, in addition to imbalance in the vergence system. When accommodation was made open loop by viewing through a pinhole, fixation disparity and phoria became more closely related (Semmlow and Hung 1979). See also Semmlow and Heerema (1979a), Semmlow and Hung (1980), and Schor and Narayan (1982) (Portrait Figure 9.17).
2. Accommodation has a deadspace in the form of tolerated blur and vergence has a deadspace in the form of Panum's fusional area (Hung 1992a).
3. We will see in the next section that associated phoria may be reduced under the conditions of forced vergence used to measure it (see Schor and Narayan 1982).

Since phoria is related to the position of dark vergence (tonic balance), one would also expect fixation disparity to be related to dark vergence. In conformity with this expectation it has been reported that, as vergence deviates from the position of tonic balance, fixation disparity increases in the direction of tonic balance (Owens and Leibowitz 1983; Jaschinski-Kruza 1994). In general, when a person converges or diverges outside the position of tonic balance, the eyes are pulled back towards it within the limits of Panum's fusional area, although fixation disparity is not necessarily zero at the position of dark vergence. Also, fixation disparity is more stable for a target at the position of dark vergence (Jaschinski 1997). Linear modelling suggests that the rate of increase of fixation disparity with increasing distance from the position of dark vergence is proportional to the open-loop gain of vergence (Hung GK 1992a At near viewing distances, the position of dark focus has a small effect on fixation disparity (Jaschinski 2001). There is evidence that subjects select a viewing distance for a near-vision task that minimizes fixation disparity (Jaschinski 1998).

We will now see that the tonic state of vergence changes after vergence has been maintained outside its normal tonic state for some time and takes time to return to its normal value.

9.3.4e *Fixation disparity and Panum's fusional area*
When the magnitude of fixation disparity is increased by forced convergence or divergence, the maximum fixation disparity before diplopia is seen corresponds to the radius of Panum's fusional area. Under normal viewing conditions, fixation disparity does not usually extend as far as the radius of Panum's fusional area (Duwaer and van den Brink 1981a). Panum's fusional area is enlarged when the high spatial frequencies are removed from the stimulus by optical blurring. Normally, fixation disparity also increases (Hebbard 1964). However, people with a flat forced-vergence curve, signifying a very adaptable state of tonic vergence, do not show this dependency of fixation disparity on Panum's fusional area (Schor *et al.* 1986a). Panum's fusional area also increases with eccentricity. Ogle *et al.* (1967) found that fixation disparity did not increase while a square, serving as the fused stimulus, became larger so that its edges became more eccentric in the visual field. Other investigators have reported that fixation disparity does increase with increasing eccentricity of the fusional stimulus (Carter 1964; Francis and Owens 1983). The answer to this apparent conflict seems to be that some people show a dependence on eccentricity and some do not. Thus, for people with an adaptable state of tonic vergence, fixation disparity and the forced-vergence function are independent of the eccentricity of the fusional stimulus (Saladin and Carr 1983; Schor *et al.* 1986a).

The total range of crossed and uncrossed disparities over which depth is seen in a random-dot stereogram has been found to decrease from about 23 arcmin for subjects with no associated phoria (fixation disparity measure by prism cancellation) to about 11.5 arcmin for subjects with 4 dioptres of phoria (Jiménez *et al.* 2000a).

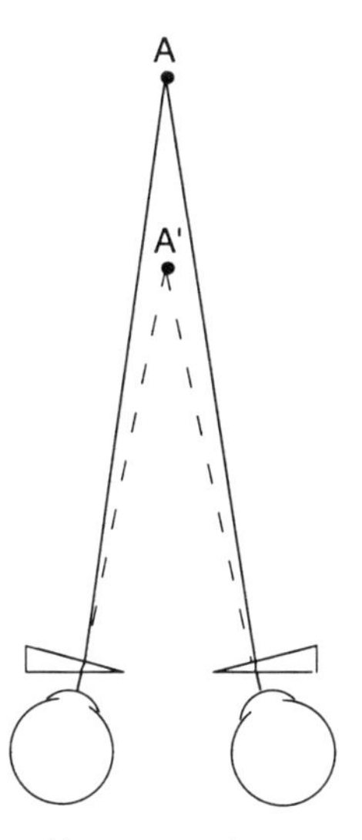

Figure 9.18. The effect of base-out prisms on vergence.
Base-out prisms cause object *A* to appear at *A'* and increase required convergence. Base-in prisms have an opposite effect.

For detailed reviews of fixation disparity, see Ogle *et al.* (1967), Schor (1983a), and Sethi (1986a). The neurology of tonic vergence adaptation is discussed later in this chapter.

9.3.4f *The monofixation syndrome*
Some patients have a permanent fixation disparity of up to 8 prism dioptres, well beyond the normal range of 8 arcmin, but not evident to casual inspection. Jampolsky (1956) used the term **small angle esotropia** and later **fusion disparity** (Jampolsky 1962). Helveston and von Noorden (1967) used the term **microtropia** and suggested that the condition arises because fixation is established at the border of a foveal scotoma in one eye (Section 15.4.2). Parks (1969), also, claimed that a foveal scotoma is the common factor in this type of fixation disparity. He used the term **monofixation syndrome** for the inability to bifixate because of a central scotoma in the deviating eye, associated with normal fusional vergence, retention of peripheral fusion, and gross stereopsis. Patients show normal peripheral fusion, as evidenced by visual testing and by the presence of binocular visually evoked potentials (Struck *et al.* 1996). Variable features include, a history of large angle strabismus, anisometropia, amblyopia, and phoria. The eyes are cosmetically straight and the condition does not respond to treatment.

9.3.4g *Summary*
Fixation disparity is a deviation of one or other of the visual axes from the intended point of convergence, too small to cause diplopia. Fixation disparity is measured by the offset of spatially separated dichoptic targets in the presence of fusible stimuli. The correlation between fixation disparity and phoria varies with viewing distance and with the position of dark vergence. Fixation disparity, like Panum's fusional area increases as the spatial-frequency content of the stimulus is reduced. Also, in some subjects, fixation disparity, like Panum's fusional area, increases with increasing stimulus eccentricity.

9.3.5 Tonic vergence adaptation to prisms

It has been known since the time of Maddox that viewing the world through base-in or base-out prisms, even for a few minutes, leads to a shift of tonic vergence lasting minutes or hours. The shift is revealed by the position of dark vergence, by phoria, or by fixation disparity after the prisms have been removed (Alpern 1946). Dynamic vergence adaptation to disturbed visual feedback is discussed in Section 9.9.3.

9.3.5a *Forced vergence curves*
When base-out prisms are placed before the eyes, as in Figure 9.18, more convergence is required to fixate a given object. Base-in prisms decrease vergence demand. The function relating fixation disparity on the y axis to vergence demand (prism power) on the x axis is the **forced-vergence curve.** (Ogle *et al.* 1967). The y-intercept indicates fixation disparity in the absence of prisms and the x-intercept indicates the prism power required to reduce fixation disparity to zero. People vary widely in the form of the forced-vergence curve. Ogle described the four basic types of curve illustrated in Figure 9.19. Type I is sigmoid, showing an accelerating degree of fixation disparity as prism power is increased from zero, one way or the other. Type II shows an accelerating change of fixation disparity to decreased vergence demand induced by base-in prisms, but no change to increased vergence demand induced by base-out prisms. Type III shows changes in response to increased vergence demand, but not to decreased demand. Type IV shows little change to either increased or decreased vergence demand. People showing this type of curve have a flat forced-vergence curve and are said to "eat the prism" (Ogle and Prangen 1953; Schor 1979b). For such people it is pointless to use prisms of fixed power for the correction of fixation disparity or phoria. This may explain why prismatic correction of fixation disparity has little or no effect on stereoacuity (Rutstein 1977). The state of tonic vergence is said to adapt to the vergence demand in those parts of a forced-vergence curve that do not change with changed demand. Beyond a certain prism power the images no longer fuse for all types.

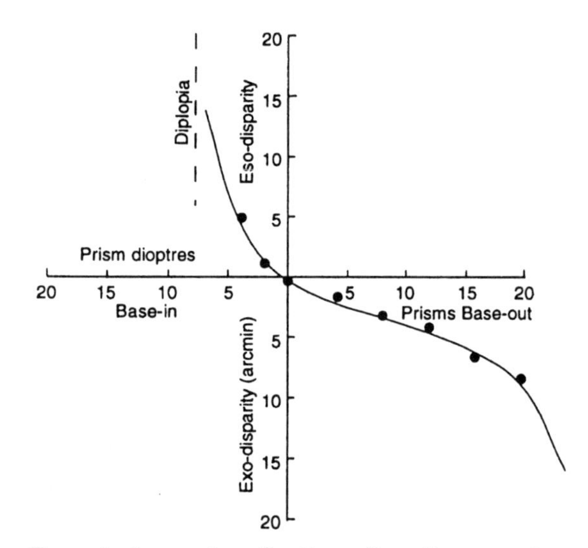

Type I. Increasing fixation disparity as prism power increases either way from zero.

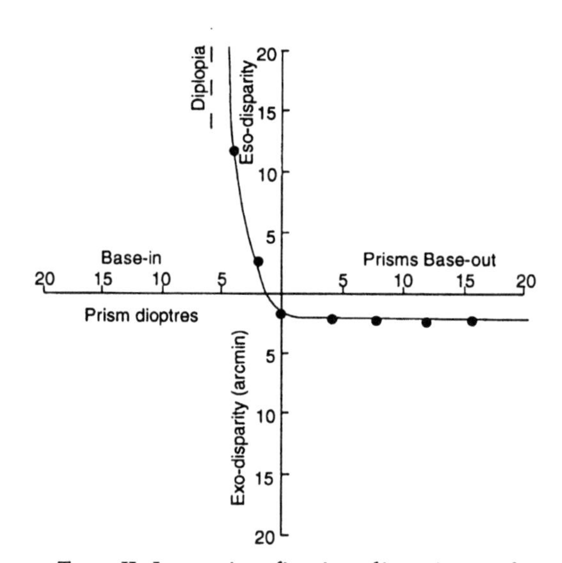

Type II. Increasing fixation disparity to decreased vergence demand. No change to increased demand.

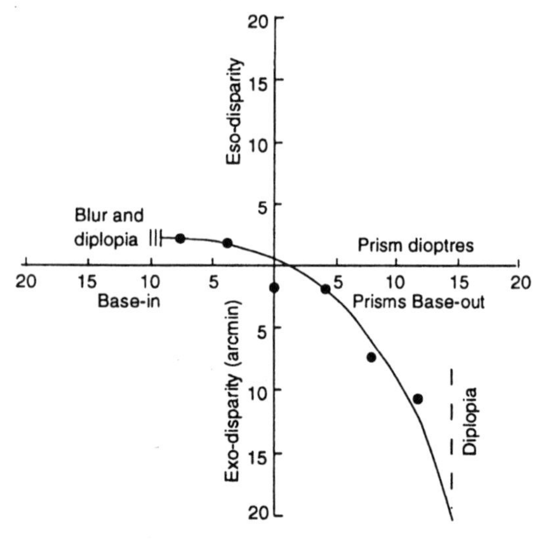

Type III. Changes to increased vergence demand but not to decreased demand.

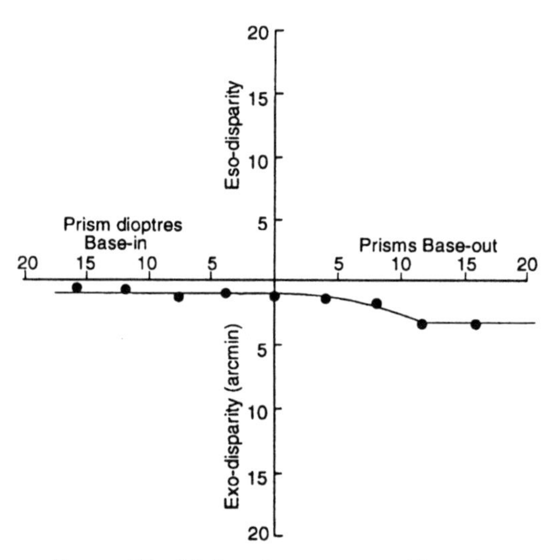

Type IV. Little change to either increased or decreased vergence demand.

Figure 9.19.Basic types of forced-vergence curves.
(Adapted from Ogle *et al.* 1967)

Haase (1980) proposed that the re-emergence of phoria during prism correction is due to the fact that the full extent of phoria is not revealed in the initial test. There is a degree of latent phoria that is revealed only after the initial phoria has been relieved. Prism power must be increased several times at intervals of several months before the full phoria is revealed. It has been claimed that heterophoria and heterotropia treated in this way show permanent correction after a five-year follow up, although the prisms must be continued to be worn to prevent relapse (Lie and Opheim 1990).

Adaptation of the vergence system to increased vergence demand has been reported up to 3 prism dioptres vertically and 10 dioptres horizontally (Carter 1963, 1965; Henson and North 1980). The form of the forced-vergence curve remains reasonably constant over months and years (Mitchell and Ellerbrock 1955; Cooper *et al.* 1981) although, for some subjects, the function shows some displacement (Daum 1983).

Fixation disparity and phoria, measured just after prisms are removed, can begin to change within the first minute of exposure to base-out or base-in prisms (Schor 1979a, 1979b). One must allow adequate intervals between tests of phoria to avoid aftereffects. The adaptability of the extraocular muscles is also shown by the fact that, when the eyes have been held in a state of eccentric version for a few minutes, it takes several seconds before they return to their normal state of relaxed version (Ludvigh *et al.* 1964). Vergence adaptation has also been demonstrated in the monkey (Morley *et al.* 1988).

Several investigators have demonstrated that, as exposure time to prisms is increased, vergence adaptation becomes more complete and takes longer to return to its preadapted state after the prisms are removed (Mitchell and Ellerbrock 1955). Even after phoria has returned to its preadapted state, the effect of a prolonged period of forced vergence is still evident in the rate of re-adaptation to prisms (North *et al.* 1986). Sethi (1986b) reported that naturally occurring phorias decay during 4 hours of monocular viewing and that the rate of this decay is correlated with the rate of adaptation to increased vergence demand. She concluded that phoria represents the natural adapted state of the vergence system. The magnitude of vergence adaptation declines with increasing age (Winn *et al.* 1994).

People adapt to large vergence demands more easily when the prismatic displacement is introduced gradually (Sethi and North 1987). Vergence adaptation is more complete when the stimulus is in the centre of the visual field rather than the periphery (Carter 1965; McCormack *et al.* 1991; McCormack and Fisher 1996). Vergence adaptation, as revealed in the change of dark vergence, occurs after a period of maintained convergence in the dark but is smaller and dissipates faster than vergence adaptation produced by maintained vergence on a visual stimulus (Ebenholtz and Citek 1995).

When prism power is increased beyond a certain limit in either direction, diplopia becomes apparent. The diplopia limit varies with the state of tonic adaptation of the extraocular muscles. For instance, diplopia occurred when a visual target was viewed through 3-dioptre prisms, which forced vertical divergence. However, after viewing a visual target for between 3 and 10 minutes through 6-dioptre prisms, the diplopia seen with 3-dioptre prisms was overcome and vertical fixation disparity sometimes returned to its normal value (Ogle and Prangen 1953). While fusional limits change with change in vergence demand, the difference between the upper and lower fusional limits (fusional amplitude) remains constant (Stephens and Jones 1990).

Patel *et al.* (1999) investigated the dynamic effects of convergence sustained for periods up to 90 s. Sustained convergence for 30 s or more reduced the peak velocity of open-loop divergence by about 25%. The velocity of convergence was not affected. This supports the idea that convergence and divergence are controlled by distinct pathways.

9.3.5b *Vergence adaptation, distance, and accommodation*

Much larger changes in fixation disparity are produced by prisms than are introduced by changing the distance of the visual target (Jaschinski 1997). This is because the change in vergence demand produced by prisms is not accompanied by a change in accommodation demand. After subjects inspected a haploscopic display, set at maximum tolerated accommodative value in one direction and maximum tolerated vergence value in the other direction, a change in tonic vergence but no change in tonic accommodation occurred (Kran and Ciuffreda 1988). A change in tonic accommodation (dark focus) occurs only when vergence and accommodation are congruent or when vergence is open loop.

Ogle *et al.* (1967) reported that about 25% of patients tested in their clinic had a forced-vergence curve that varied according to whether vergence was near or far. This same phenomenon occurred in about 40% of a sample of normal adults (Wick 1985). The most frequent change was from a type I curve with near convergence to a type II curve with far convergence, although Saladin and Sheedy (1978) found equal frequencies of type II at near and far.

With a distant visual target, there is greater vergence adaptation to base-out prisms, which increase vergence demand, than to base-in prisms, which decrease vergence demand. With a near target, adaptation produced by base-out prisms is reduced to a magnitude similar to that produced by base-in prisms (North *et al.* 1990). At least two factors could contribute to these effects. Base-in prisms cause a near target to appear more distant and this decreases the demand on the vergence system because of the action of proximal vergence. Secondly, accommodative convergence creates a greater vergence demand with a near target than with a far target.

Fixation disparity also changes when positive or negative lenses are placed before the eyes. The lenses change accommodation, which induces a corresponding change in the resting state of vergence. Furthermore, prolonged exposure to a particular state of accommodation changes the resting state of accommodation, the resting state of vergence, and the magnitude of phoria (Schor 1983a). The reciprocal coupling between accommodation and vergence is discussed in Section 9.5. Vergence demand may also be increased by viewing the world through a telestereoscope, which effectively, increases or decreases the interocular distance. The effects are not quite the same as those produced by prisms. Prisms add a constant amount to required vergence over the whole range of distance, whereas increased vergence demand produced by a telestereoscope is inversely related to distance (Figure 9.24).

Hesler *et al.* (1989) generated forced-vergence curves with accommodation rendered open loop by an artificial pupil. Relative to the closed-loop condition the curves were more exophoric on the forced convergence side but were unchanged on the forced divergence side. Semmlow and Hung (1979) obtained conflicting results so that the precise contribution of accommodative vergence to fixation disparity remains to be elucidated.

The tonic state of the eyes is clearly not fixed but adapts to the current level of vergence, more rapidly and completely in some people than in others. Hung (1992b) developed a mathematical model of vergence adaptation.

9.3.6 Noncomitant vergence adaptation

9.3.6a *Noncomitant adaptation*

A change in vergence demand is comitant when it is the same for all directions of gaze and noncomitant when it varies with the angle of gaze. When the eyes are converged on a near point directly ahead, all points in an oblique position on a frontal surface have a vertical disparity when corresponding points are specified in terms of horizontal lines of longitude (Section 20.2.8). Thus, in terms of these co-ordinates, when a person moves the gaze from a straight-ahead position to an oblique point on a frontal surface, the eyes must execute an appropriate vertical vergence to bring the images of the newly fixated object into vertical correspondence (see Figure 9.20). They must also execute movements with different azimuth angles. For instance, a vertical vergence of three prism dioptres is required to fixate a point 24° up and 24° to one side on a frontal plane, at a distance of 33 cm (Ogle and Prangen 1953). Furthermore, the required vergence for a given direction of gaze varies with the distance of the surface.

If corresponding points and eye movements are specified by a co-ordinate system involving horizontal lines of latitude, vertical vergence does not occur when the gaze moves to an oblique position. But whatever co-ordinate system we use to specify corresponding points and eye movements, the human oculomotor system must make horizontal and vertical vergence movements that are appropriate to the direction of gaze as well as to the distance of gaze. Schor *et al.* (1994) found that when the gaze is directed to a target without error feedback, the visual axes intersect with an error of no more than 0.25° for any direction or distance of the target. This suggests that vergence movements are pre-programmed for direction and distance.

Noncomitant vergence demand is induced by spectacle lenses with unequal magnification in the two eyes—a condition known as optical aniseikonia (Section 15.8). This happens when the two eyes require different amounts of optical correction. When the gaze is directed away from the optic axes of a spectacle lens, the lens acts like a prism, which increases in power as a function of the angle of gaze. If the lenses do not have the same power the eyes must move different amounts to maintain fusion. A person can compensate for the effects of spectacles by turning the head so that the eyes look through the centres of the lenses, and lenses can be made that optically correct for aniseikonia. However, most people do not need to compensate in either of these ways because they adapt to the noncomitant vergence demand by noncomitant vergence. Thus, a person used to reading with unequal lenses learns to elevate the visual axis of one eye relative to that of the other to bring images of an object onto the fovea. A person who has made such an adjustment of vergence shows a phoria that depends on the direction of gaze when tested with disassociated viewing. This is noncomitant phoria, or anisophoria (Ellerbrock and Fry 1941, 1942; Ellerbrock 1948; Allen 1974). The dynamic aspects of adaptation to aniseikonia are discussed in Section 9.9.3.

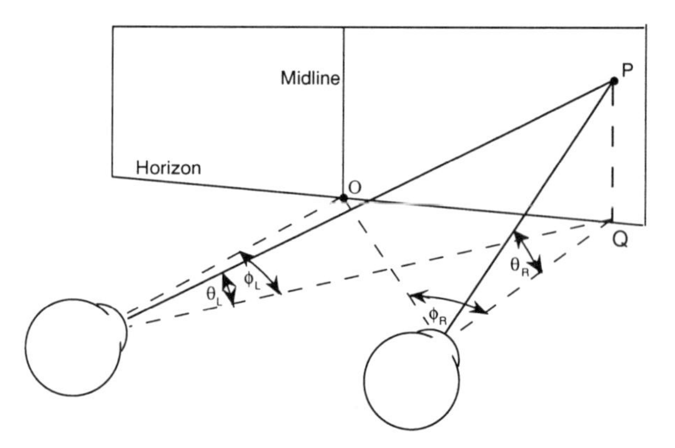

Figure 9.20. Unequal versions needed for oblique gaze. When the gaze moves from a straight-ahead at O, to an oblique position, P, the eyes must execute a vertical vergence movement to bring the images of P into vertical correspondence. This is because *PQ* subtends a larger angle, θ, to one eye than to the other eye. The eyes must also execute unequal horizontal movements, since angle ϕR is larger than angle ϕL.

9.3.6b *The adaptive field*

When a person holds the gaze in one direction while adapting the state of vergence to a fixed, prism-induced, disparity, the resulting change in phoria occurs in about 30 minutes. Although the change in tonic vergence is maximal for that direction of gaze, it also shows when the eyes look in neighbouring directions along the same meridian. The eye positions over which a locally applied change in vergence spreads is called the **adaptive field** (Henson and Dharamshi 1982). Schor *et al.* (1993a) found that changes in phoria after adaptation to a fixed direction of gaze were constant over an 18°-wide field, showing that the adaptive field is at least this wide.

Noncomitant adaptation of tonic vergence to a gradient of disparity, such as that produced by spectacles, takes much longer than comitant adaptation to constant disparity (Sethi and Henson 1984). With noncomitant adaptation, a specific degree of adaptation must be applied at each direction of gaze along a given meridian. Schor *et al.* (1993a) investigated this process by having subjects adapt to two separated targets with opposite prism-induced vertical disparities. They used vertical rather than horizontal vergence because it is not affected by accommodation. Differential adaptation of vergence was greater with greater lateral separations of the targets or with smaller imposed disparities. Thus, the steeper the disparity gradient between targets, the more difficult it became to acquire noncomitant vergence.

Schor *et al.* proposed a two-mechanism model of tonic vergence adaptation. The first is a global mechanism of rapid onset, which generalizes to all eye positions along a given meridian. The second is a slower local mechanism that adapts to distinct disparities in the visual field. The local mechanism shows some spread, so that adaptation to objects with distinct disparities does not occur when the objects are too close together.

9.3.6c *Meridional specificity of vergence adaptation*

It seems that noncomitant adaptation of tonic vergence is specific to the meridian along which the disparity gradient is presented. Thus, Maxwell and Schor (1994) adapted subjects to two different vertical disparities presented along either the horizontal or vertical meridian. Vergence adaptation, as revealed in post exposure phoria, was noncomitant only for the meridian along which the different disparities had been displayed. It was comitant along the orthogonal meridian. The noncomitant component built up more slowly than the comitant component. McCandless *et al.* (1996) described a model of the neural processes involved in noncomitant adaptation of vertical vergence.

Schor *et al.* (1993b) had subjects converge on stationary targets at different positions along an optically induced disparity gradient. This resulted in a position-dependent phoria and a differential movement of the eyes during visual pursuit, but it did not produce disjunctive saccades. The mechanism responsible for independent adaptation of saccadic amplitude in the two eyes (Section 9.9.2) must be independent of that responsible for adaptation of static vergence and visual pursuit. Gleason *et al.* (1992) developed a model of the processes responsible for adaptation of static vergence and disjunctive pursuit.

Schor and McCandless (1997) exposed subjects to vertical disparities, which varied according to whether the gaze moved in the midsagittal plane (horizontal vergence and vertical version), the frontal plane (horizontal and vertical version), or the transverse plane (horizontal vergence and version). The adaptation of vertical vergence, as measured by dark phoria, was specific to each of the three orthogonal planes. The results were modelled with a matrix, which associated pairs of eye position signals with a weighted output driving vertical vergence.

Maxwell and Schor (1996) placed either a base-up or base-down prism before one eye according to whether the head or the whole body was pitched up or down, deviated to left or right, or rotated about the visual axis left or right. In each case, 60 minutes of alternating changes in head or body position and prism deviation resulted in a head-position-dependent change in vertical vergence, which persisted when the prism was removed. The implication is that, in normal circumstances, changes in vertical vergence compensate for the oculomotor effects of vestibular and neck proprioceptive stimuli arising from movements of the head or body. Subjects can simultaneously adapt vertical vergence in an eye-position-specific manner, which varies with head position (Maxwell and Schor 1997). Thus adaptation of vertical vergence takes the positions of both the eyes and head into account.

Maxwell and Schor (1999) placed dove prisms before the eyes, which rotated the images of a scene in opposite directions about the visual axes. When subjects tilted the head 45° in one direction the dove prisms introduced a cyclodisparity that triggered incyclovergence and when the head tilted 45° in the opposite direction the prisms triggered an excyclovergence. After one hour of alternating exposure to this coupling between head position and cyclodisparity, subjects demonstrated a head-position dependent change in cyclophoria amounting to up to 13% of the cyclodisparity in the training session.

9.2.6d *Distance cues for vergence adaptation*

Schor and McCandless (1995a, 1995b) asked subjects to attempt to fuse a cross containing various degrees of vertical disparity coupled with variations in various visual cues to distance, including overlap, looming, relative size, and parallax. A 2-hour adaptation period did not induce any change in vertical vergence as indicated by post-exposure phoria. Association of vertical disparities with different values of horizontal vergence as a cue to distance did induce adaptation of vertical vergence. It was concluded that adaptation of vertical vergence occurs in response to associations between vertical and horizontal oculomotor activity but not in response to associations between vergence and monocular visual cues to distance.

9.4 PROXIMAL VERGENCE

The type of vergence that Maddox called voluntary vergence is usually referred to as **proximal vergence**. It is evoked by stimuli that give the impression of being nearer or further than the point of convergence, in the absence of disparity or accommodation cues. For instance, increasing the size of the projected image of a playing card was found to evoke a change in vergence (Ittelson and Ames 1950; Alpern 1958). Predebon (1994) observed vergence induced by the familiar size of a monocularly viewed object. Transient proximal vergence was evoked when the images of an isolated square changed sinusoidally in size at a fixed distance, or varied in disparity but not in size (Erkelens and Regan 1986). The response to a combined change in size and disparity was the linear sum of the two component responses.

Short latency (80 ms) transient convergence was elicited by an expanding dot pattern projected onto an 80° by 80° screen. Divergence was elicited by a contracting display (Busettini *et al.* 1997). The responses were similar when the dots changed in density but not in size. They were also similar with monocular viewing and when the display was confined to the temporal hemifields of both or one eye. In the latter case, the single eye saw lateral motion in the opposite direction to that in which the eye moved. This demonstrates that vergence was not simply the sum of two monocular pursuit movements. Thus, radial motion specifically triggers vergence in a machine like fashion. Radial motion detectors in MT or MST are probably involved (Section 5.7.3a).

Yang *et al.* (1999) found that the magnitude of the short latency vergence response to a looming display was proportional to the vergence angle before the response was triggered. With parallel gaze (equivalent to infinite viewing distance) the response fell to near zero. Schapero and Levy (1953) found that proximal vergence shows only at distances below 2 m and increases as the distance of the stimulus is reduced below this value. This inverse relationship with viewing distance would allow an observer moving through a natural scene to converge correctly on a distant object and ignore looming signals arising from nearby objects.

Depth cues could drive proximal vergence indirectly by evoking a change in accommodation (see Takeda *et al.* 1999). McLin *et al.* (1988) found that the ratio of vergence to a change in stimulus size resembled the ratio of vergence to changing accommodation (AC/A ratio) and concluded that size changes evoke accommodation directly and vergence indirectly. Wick and Currie (1991) compared vergence and accommodative responses to prisms and lenses with those initiated by targets at different distances. They concluded that proximal vergence can be initiated independently of proximal accommodation.

A change in disparity could contribute to the divergence produced when the size of an object increases. The edges of an object lying in the frontal plane acquire an uncrossed disparity as they get further from the fovea, because they become progressively more distant from the concave horopter. An *experiment should be conducted with an object changing size within the curved plane of the horopter or within a frontal plane at the abathic distance, that is, the distance at which the horopter lies in the frontal plane (Section 15.6). Furthermore, according to this explanation, an object at a far distance where the horopter is convex should induce divergence when it is made larger.*

Subjects made large and rapid changes in vergence when they looked back and forth between two frontal-plane horizontal fluorescent rods seen at different distances in dark surroundings (Wick and Bedell 1989). It was claimed that, because the rods were horizontal, the only cues to depth were the relative thickness and height of the rods in the field. However, accommodation cues were not eliminated. Enright (1987a) found that subjects converged when an apparently near part of the drawing of a cube was monocularly fixated, and diverged when an apparently far part was fixated. In this case the cue to depth was provided only by perspective and the display was free from artifacts present in the displays used by other investigators. Vergence movements were also elicited in the closed eye when the gaze of the open eye changed from a part of a painting that depicted a near object to a part that depicted a far object (Enright 1987b) (Portrait Figure 9.21).

Rosenfield *et al.* (1991) measured proximal vergence evoked with accommodation cues eliminated by having subjects look through pinholes, and with disparity cues to depth eliminated by inducing a vertical disparity between the images in the two eyes. A letter chart viewed in the laboratory was the target for distances up to 6 metres, and objects such as buildings seen out of the window were targets for distances of up to 1,500 metres. Vergence and accommodation changed linearly with increasing distance of the target, up to about 3 metres, after which vergence remained constant.

Ringach *et al.* (1996) found that pursuit of a patch on a monocular display of dots, which created the impression of a 3-D rotating sphere (kinetic depth effect), induced the same pattern of divergence and convergence as pursuit of an LED moving in an actual 3-D orbit. Observers could not produce this pattern of eye movements without an appropriate stimulus and could not entirely suppress vergence movements evoked by the kinetic depth display.

Vergence induced by an attempt to fixate one's unseen finger in the dark would also be proximal vergence, although people do not perform this task accurately, and changes in vergence seem to be independent of changes in accommodation in the dark (Fincham 1962). Voluntary changes in vergence to imagined objects in the dark also fall under the heading of proximal vergence (McLin and Schor 1988). Such vergence movements are unreliable and poorly correlated with the distance of the imagined object (Erkelens *et al.* 1989b).

The mechanism that signs the direction of accommodation does not work for images out of focus more than 2 dioptres (Fincham 1951). Furthermore, disparities of more than about 4° do not evoke vergence (Section 9.6.1). Other depth cues, such as perspective and motion parallax, operate at distances outside these limits. Thus, vergence movements to objects at a distance far removed from the plane of convergence are, by definition, proximal vergence. Proximal vergence and proximal accommodation bring accommodative and disparity cues to within a range in which they can help to converge the visual axes accurately onto the desired depth plane. Schor *et al.* (1992) have devised a feedback control model of these processes.

North *et al.* (1993) assessed the relative contributions of accommodation, disparity, and proximal cues to vergence by introducing three types of cue discordance. The accommodation cue was made discordant relative to the other cues by having subjects view a display at 4 m through -0.75 dioptre lenses. The disparity cue was made discordant by having subjects view the display through prisms. The proximal cue was made discordant by having subjects view the display through both lenses and prisms. They measured the immediate change in phoria and fixation disparity after a period of exposure to each stimulus. Discordance of disparity or of the proximal cue relative to the other cues had similar effects while discordance of accommodation

Figure 9. 21. James Enright.
Born in Baker, Oregon in 1932. He obtained a B.A. at University College of Los Angeles in 1957 and a Ph.D. with E. W. Fager at the Scripps Institute of Oceanography in 1974. He performed postdoctoral work with J. Aschoff at the Max Planck Institute fur Verhaltensphysiologie, Erling-Andechs, Germany. He joined the faculty in the Department of Oceanography at University College in San Diego in 1966, where he is now Professor.

had only about one third the effect of a discordance of one of other two cues. Hung *et al.* (1996) commented that, in this experiment, an absence of stimuli in the central field created a bias against accommodation. They found that proximal cues had little effect on vergence in the presence of adequate disparity and accommodation cues.

9.5 ACCOMMODATION AND VERGENCE

Horizontal vergence and accommodation normally occur together and are accompanied by a change in pupil diameter. The pupil constricts with near vergence/accommodation to compensate for the narrowed depth of field and increased spherical aberration caused by near accommodation. The pupil dilates with far vergence to reduce diffraction and improve retinal illumination. The three concomitant changes are known as the **near-triad** response. The literature on this topic is reviewed in Semmlow and Hung (1983) and Kotulak and Schor (1986a). The following account concentrates on the coupling between horizontal vergence and accommodation.

9.5.1 Accommodative convergence

A change in accommodation is normally accompanied by a change in vergence—an increase in accommodation evokes convergence and a decrease evokes divergence. This is known as **accommoda-**

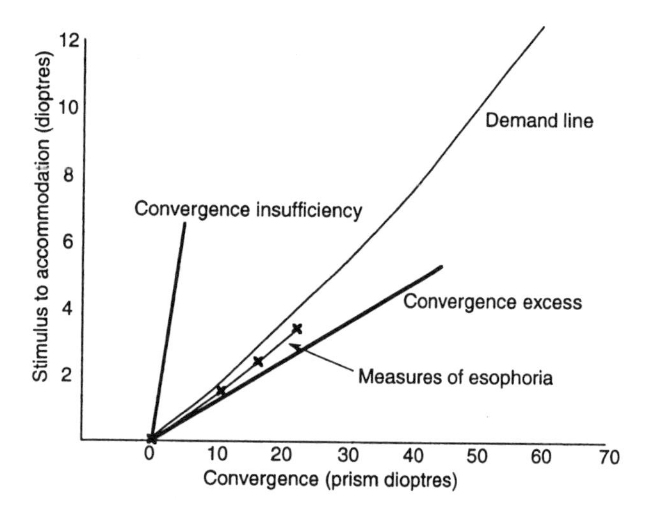

(a) The demand line is the convergence that should accompany each degree of accommodation. The bold lines represent convergence insufficiency or excess. The crosses represent phoria at each of several accommodations.

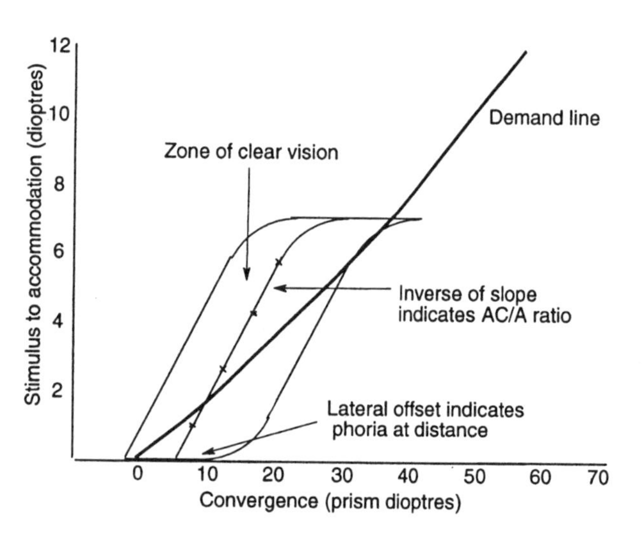

(b) The width of the zone of clear vision represents the range of vergence over which accommodation remains tolerably precise. The height indicates the range of accommodation. Phoria is the offset of the centre line from the demand line. The slope inverse indicates the AC/A ratio.

Figure 9.22. Charts of accommodation-convergence.

tive convergence (AC). The response may be evoked by a binocularly fused stimulus that is out of the plane of accommodation. Since accommodative convergence is evoked by accommodation rather than by disparity, it can be evoked by a stimulus presented to only one eye. In fact, Müller (1843) first discovered accommodative convergence by noting the convergence of the eyes as a stimulus was moved along the line of sight of one eye with the other eye closed. He concluded that the link between the two responses is not immutable, since he observed that one can learn to change accommodation without changing vergence. Donders (1864) coined the term "relative convergence" for convergence without a change in accommodation and the term "relative accommodation" for accommodation without a change in vergence. Maddox (1893) and Morgan (1944) concluded that a change in accommodation always involves a change in vergence but that vergence can change without a change in accommodation. Vertical vergence and cyclovergence are not evoked by misaccommodation since there is no natural linkage between image blur and vertical or torsional misalignment of images.

Figure 9.23. Stuart Judge.
Born in Edinburgh in 1947. He obtained a B.A. in Physics and Mathematics in 1969 and a Ph.D. in Communication and Neuroscience in 1976, both from Keele University. He is a Lecturer in Physiology at Oxford University and Fellow and Tutor in Physiology at St Anne's College Oxford.

In the absence of a fusional lock, changes in accommodation produce an approximately linear change in vergence over a 5-dioptre range of accommodation (Alpern *et al.* 1959; Martens and Ogle 1959; Flom 1960a). However, significant deviations from linearity do occur (Westheimer 1955). The amplitude of accommodative convergence (AC) evoked by a 1 dioptre change of accommodation (A) is the **AC/A ratio**. The ratio is reasonably constant over time for a given person (Martens and Ogle 1959; Flom 1960b). A low ratio means that the plot of vergence against accommodation is steeper than the demand line, as illustrated in Figure 9.22a. This signifies that convergence or divergence is insufficient for the required accommodation, a condition known as convergence weakness (see Section 9.3.3). A high ratio means that the plot of vergence against accommodation is less steep than the demand line and signifies that convergence or divergence is excessive for the required accommodation, a condition known as convergence excess (Daum 1989).

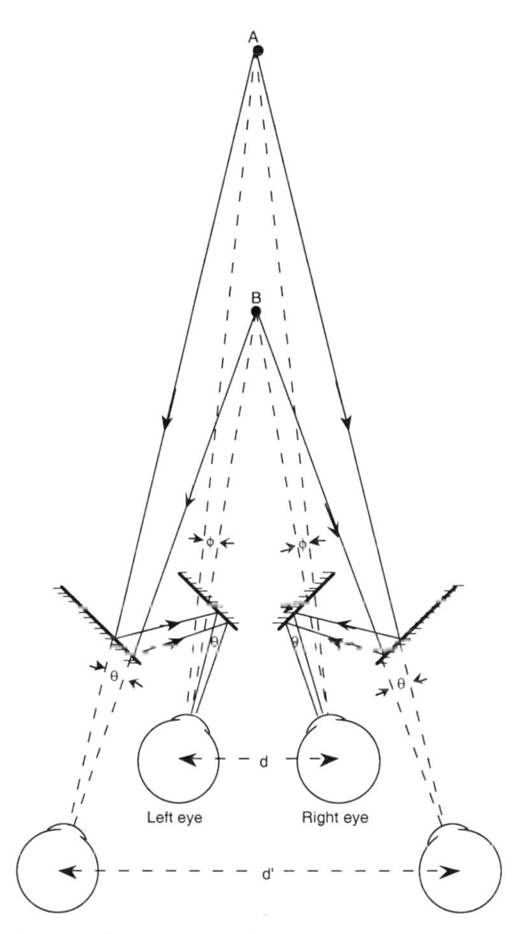

Figure 9.24. A telestereoscopic device.
The effective interocular distance is increased from d to d'. This increases the convergence required when changing fixation between two near objects, such as A and B (angle θ is larger than angle ϕ). The device also increases the path length of the light rays, which decreases the amount of accommodation. The ratio of convergence to accommodation is therefore increased.

Accommodative convergence can be measured by the change in alignment of a covered eye produced by placing negative lenses in front of the viewing eye. The AC/A ratio in degrees of vergence per dioptre of accommodation varies from about 2.4 at high luminance to 1.6 at low luminance (Alpern and Larson 1960). Ogle *et al.* (1967) proposed that a more realistic measure is the change in fixation disparity with binocular viewing as a function of lens-induced changes in accommodation for each of several states of prism-induced vergence. The two methods do not necessarily produce the same result (see Judge 1985) (Portrait Figure 9.23). Hung (1997) has proposed that these differences are due to nonlinearities arising from the fact that both accommodation and vergence have stimulus thresholds. Larson (1982) described a procedure for measuring accommodative convergence using a display containing a fusion stimulus.

Since accommodative state is difficult to measure in routine clinical practice, the AC/A ratio is measured with reference to the stimulus to accommodation, it being assumed that the accommodative response is proportional to the stimulus. This is the **stimulus AC/A ratio**, as opposed to the **response AC/A ratio** obtained when accommodation is measured (Alpern *et al.* 1959). The two measures do not always give the same result (Ripps *et al.* 1962).

The gain of accommodative convergence can be expressed as the ratio of the actual vergence change to the ideal vergence change required to fuse the target (Judge and Miles 1985; Judge 1987). This measure provides direct comparisons between subjects with different interocular distances.

Convergence is more effectively evoked by disparity than by misaccommodation. For instance, the dynamics of the initial portion of a vergence response were not improved when a pure disparity stimulus was supplemented by an accommodation stimulus, although an accommodation stimulus did improve the velocity and precision of the final stages of the vergence response (Semmlow and Wetzel 1979). Furthermore, when vergence was evoked in the monkey by disparity alone, its phase lag was smaller and its velocity higher than when it was evoked by misaccommodation alone (Cumming and Judge 1986). Accommodation seems to provide a moderate contribution to vergence only when the movement is nearly complete (Hung *et al.* 1983). The contributions of accommodation and disparity to vergence are discussed further in Section 9.5.3.

The measurement of the stimulus AC/A ratio is not severely affected by how long one eye is covered during testing (Rosenfield *et al.* 2000). Also, the response AC/A ratio is not affected by adaptation of vergence to base-in or base-put prisms (Rainey 2000).

Viewing through base-in or base-out prisms does not alter the change of vergence required per unit change in accommodation, it simply adds or subtracts a constant vergence at all accommodation distances. However, when objects are viewed through a telestereoscope, as in Figure 9.24, it is as if the eyes are further apart. The closer the stimulus is to the eyes, the greater the vergence demand. This increases the required change in vergence per unit change in accommodation. As predicted, 30 minutes viewing through a telestereoscope increased the AC/A ratio and decreased the CA/C ratio (see next section) (Judge and Miles 1985). The AC/A ratio does not change when a target is viewed through a telestereoscope because a simple change in tonic vergence can cater for the fixed change in vergence demand (Bobier and McRae 1996).

Drugs, such as homatropine, that increase the effort necessary to produce a given state of accommodation, increase the AC/A ratio (Chin and Breinin 1967). Drugs, or viewing through an artificial pupil, that reduce accommodative effort have the opposite effect (Hermann and Samson 1967).

There have been competing claims about whether the AC/A ratio is affected by orthoptic exercises (Hofstetter 1945; Manas 1958). Flom (1960c) controlled for potential artifacts due to repeated testing and for increases in near accommodation and proximal vergence following orthoptic exercises. He found that 30 minutes of orthoptic training per week over eight weeks given to 94 patients with exophoria increased the mean AC/A ratio by 0.41. A control group receiving repeated testing but no orthoptic training showed no change in the AC/A ratio.

Vergence responses of strabismics and some amblyopes without strabismus are similar to those of people with normal vision but with one eye closed (Kenyon *et al.* 1980a). Thus, strabismics show accommodative vergence but not disparity vergence.

Myopic children show enhanced accommodative convergence (Jiang 1995). A child who is esophoric must relax accommodation to maintain single vision. The resulting image blur during near work could induce axial growth of the eye and hence myopia (Gwiazda *et al.* (1999). Schor (1999) has developed a model of these processes.

9.5.2 Convergence accommodation

A change in horizontal vergence, however it is evoked, is accompanied by an appropriate change in accommodation, a response known as **convergence accommodation** (CA). The change in convergence accommodation per unit change in convergence is the **CA/C ratio**. Convergence accommodation is measured when vergence is changed in the absence of the blur cue to changing accommodation. Blur can be removed by viewing through pinholes, because an object seen through a pinhole remains in clear focus whatever its distance and whatever the accommodative state. However, some blur remains with pinholes of a practical size. Accommodation can also be made open loop, by using a difference of Gaussian visual target with a centre spatial frequency of 0.2 cpd (Tsuetaki and Schor 1987). The pinhole method has revealed that accommodation is a linear function of convergence. For young adults, accommodation in dioptres is approximately equal to vergence in metre angles, so that the gain of convergence accommodation is 1. The gain is smaller in older and more presbyopic subjects (Fincham 1955; Fincham and Walton 1957).

A speckle interference pattern formed on the retina by two laser beams provides an effective stimulus for studying convergence accommodation, since such a pattern is independent of the state of accommodation. Using this procedure, Kersten and Legge (1983) found that the average accommodation of the two eyes is linearly related to vergence angle over the eyes' accommodative range, with a mean CA/C ratio of 0.91. The CA/C ratio was almost as high at scotopic levels. Accommodation was found not to vary with changes in the angle of gaze. The gain of convergence accommodation may also be measured by the ratio of the actual change in accommodation to the ideal change required to accommodate the new visual target.

Horizontal convergence also increases the horizontal radius of curvature of the cornea, especially in young people (Löpping and Weale 1965). This effect is believed to be due to tension induced in the cornea by contraction of the medial rectus muscle.

9.5.3 Relation between AC and CA

9.5.3a *AC-CA linkage*

Convergence accommodation and accommodative convergence are two aspects of the same functional unity of vergence and accommodation. Commands for the two responses are issued concurrently and interact reciprocally. However, the eyes begin to change vergence before they begin to accommodate, since the skeletal rectus muscles respond more rapidly than the autonomic ciliary muscles (Allen 1953).

In spite of this functional unity, accommodation and vergence assume independent resting states in the dark (Owens and Leibowitz 1980). There is conflicting evidence about whether the resting states of vergence and accommodation, although different, are correlated (see Gray *et al.* 1993c). However, when allowance is made for differences in the AC/A ratio between subjects, dark accommodation may be predicted from dark vergence (Wolf *et al.* 1990; Jiang and Woessner 1996). There is probably also some uncorrelated noise in the two systems. In darkness, vergence and accommodation return to their respective resting states because there is insufficient stimulation to activate the feedback loops in each system or the crosslinks between the systems.

People with an unusually high AC/A ratio tend to have a lower than normal CA/C ratio. Abnormally high AC/A ratios and low CA/C ratios are accompanied by low adaptability of the resting state of accommodation and high adaptability of tonic vergence. Unusually low AC/A ratios are accompanied by high adaptability of accommodation and low adaptability of tonic vergence (Schor 1986, 1988;

Schor and Horner 1989; Polak and Jones 1990). With increasing age, the CA/C ratio decreases substantially, probably because of the decrease in the range of accommodation. At the same time, the AC/A ratio increases moderately (Bruce *et al.* 1995; Rosenfield *et al.* 1995a), although Ciuffreda *et al.* (1997) found that the increase occurred only after the age of 45 years and only for stimulus AC/A ratio.

A telestereoscope, increases the effective interocular separation and the required change in convergence per unit change in accommodation. This decreases the required gain of AC/A and increases the required gain of CA/C. Periscopic spectacles that bring the visual axes to the midline, reduce the interocular separation and reduce to zero the change in vergence involved in accommodating at different distances. Miles *et al.* (1987) found that 30 minutes exposure to a telestereoscope, produced a mean shift of 37% in the AC/A gain, which returned to normal over a period of about 4 hours with normal viewing (Portrait Figure 9.25). Exposure to periscopic spectacles had very little effect on AC/A gain. Exposure to base-out prisms that increase convergence demand by a constant amount at all distances caused a predicted downward shift in the AC/A curve and upward shift in the CA/C curve, rather than changes in gain. Base-in prisms that reduced vergence demand by a constant amount shifted the CA/C curve downward but had no effect on the AC/A curve. Thus, the reciprocal couplings between accommodation and convergence are subject to adaptive changes, although the effects are asymmetrical. These changes could be due to error-sensing feedback in the reciprocal control loops, but muscular fatigue could also contribute. Exposure to virtual-reality displays that place unequal demands on vergence and accommodation also changes the AC/A and CA/C ratios (Eadie *et al.* 2000).

There has been some dispute about whether the linkage between accommodation and vergence is served by tonic controllers, by phasic controllers, or by both. Evidence of a linkage involving both types of vergence and accommodation has been reported by Ebenholtz and Fisher (1982) and Rosenfield and Gilmartin (1988a, 1988b). However, more recent evidence suggests that the linkage receives inputs from only the phasic system (Schor 1992; Jiang 1996).

At frequencies of stimulus oscillation below about 0.1 Hz, accommodation does not respond to changes in vergence and vergence does not respond to changes in accommodation. As stimulus frequency is increased to about 0.5 Hz, the CA/C and AC/A ratios increase in a nonlinear fashion (Schor and Kotulak 1986). The values of both ratios are subject to fatigue (Schor and Tsuetaki 1987).

Figure 9.25. Frederick Albert Miles.
2000 Born in Grimsby, England in 1939. He obtained a B.Sc. in Animal Physiology from the University of Leeds in 1962 and a D.Phil. in Neurophysiology from the University of Sussex in 1971. He was a lecturer at the University of Sussex from 1966 to 1971. In 1971 he moved to the Laboratory of Neurophysiology at NIMH, Bethesda, where he is now Chief of the Section on Oculomotor Control in the Laboratory of Sensorimotor Research of the National Eye Institute. He received the Golden Brain Award of the Minerva Foundation in 2000.

Models of the interactions between accommodation and vergence have been developed by Schor and Kotulak (1986), Polak and Jones (1990), Schor *et al.* (1992), and Schor (1992). The symbols and units used in these models are described in Section 3.5. Schor and Kotulak's model, shown in Figure 9.38, contains separate controllers for phasic and tonic vergence and accommodation. The cross links serving accommodative convergence and vergence accommodation occur after the phasic controllers but before the tonic controllers. Hung (1997) developed a non-linear model. One nonlinearity is the dead space in the accommodation system, which arises because of tolerated blur. A second nonlinearity is the dead space in the vergence system that arises because of Panum's fusional area. The model accounts for differences in the AC/A ratio determined by the phoria and fixation disparity methods.

Reviews of accommodative vergence and vergence accommodation have been provided by Alpern (1969), Morgan (1968), Ciuffreda and Kenyon (1983), and Fry (1983). Semmlow and Venkiteswaran (1976) dealt with dynamic aspects of accommodative vergence.

9.5.3b *Vergence-accommodation charts*

For a given interpupillary distance there is a required degree of vergence and accommodation for each viewing distance. When plotted on a chart, as in Figure 9.22a, the required accommodation for each vergence angle is known as the **demand line**. The actual vergence for various degrees of accommodation for disassociated targets can be plotted on the same chart and indicates phoria. A similar plot with associated targets represents fixation disparity. The same chart can be used to plot the range of vergence before loss of fusion and the insufficiency or excess of convergence and of divergence in relation to accommodation. The range of accommodation and the range of vergence possible without excessive error in either is known as the **zone of clear single binocular vision**. Figure 9.22b shows a representation of this zone (Fry 1939). The width of the zone represents the range of vergence within which accommodation remains tolerably precise, and is approximately constant at all levels of accommodation. With training, people can learn to dissociate vergence and accommodation, and thereby broaden the zone of clear vision (Hofstetter 1945; Heath and Hofstetter 1952). The ranges of accommodation and vergence that can be achieved without discomfort is known as the **comfort zone**. The righthand boundary represents the limit of convergence, and the slope of this boundary indicates that the person can achieve higher degrees of vergence when aided by near accommodation, (the AC/A ratio). The situation is complicated by the fact that strong convergence evokes an excessive accommodative response through the mediation of convergence accommodation (Semmlow and Heerema 1979b). The distance along the vergence axis between the demand line and the limit of vergence is the **vergence reserve**. Phoria for each value of accommodation is represented by a line running parallel to the sides of the zone of clear vision.

Vergence/accommodation charts are used in clinical practice (see Hofstetter 1983; Goss 1995).

9.6 VERGENCE EVOKED BY DISPARITY

Maddox gave the name "fusional vergence" to vergence evoked by binocular disparity. Horizontal disparity induces horizontal vergence, vertical disparity induces vertical vergence, and cyclodisparity induces cyclovergence. The stimulus for a change in the state of horizontal or vertical vergence is the absolute disparity of the object of regard—its disparity with respect to the horopter. The stimulus for cyclovergence is the cyclodisparity of the contents of the binocular field. We will see in Chapter 22 that stereoscopic depth is derived from relative disparities and is largely immune to changes in absolute disparity. Thus, the disparity signal for vergence is not the same as that for stereopsis.

9.6.1 The range of vergence

The angle of vergence changes about 14° when the gaze is moved from infinity to the nearest distance for comfortable convergence at about 25 cm. Vergence changes about 36° when the gaze moves to the nearest point to which the eyes can converge. About 90% of this total change occurs when the eyes converge from 1 m. Vergence changes of 14° have been recorded in the cat (Stryker and Blakemore 1972).

The fusional range of vergence, or fusional reserve, is the range of disparity of binocular images over which a visual target remains fused when the subject is allowed to change vergence. The fusional range overestimates the range of vergence, since it includes Panum's fusional area. The range of fusional vergence is most commonly tested in the clinic by the use of Risley prisms, which change the lateral alignment of dichoptic targets without changing their optical distance. The prisms are first set so that the dichoptic images are easily fused. To maintain a fused image, the subject gradually diverges when the prisms are adjusted in a base-in direction and converges when they are adjusted in a base-out direction. The limits to which vergence can be forced before fusion is lost defines the **break fusional range**. With respect to an initial viewing distance of 50 cm, the fusional range extends about twice as far in the convergence direction as in the divergence direction. The fusion limit for divergence is typically the position of parallel gaze and is less variable than the limit for convergence (Feldman *et al.* 1989). Once the target has become diplopic, the prisms are adjusted in the opposite direction until the images fuse again. This procedure measures the recovery fusional range, which is typically about half the break fusional range. Divergence ranges are usually assessed before convergence ranges because convergence may induce vergence adaptation. However, the response may be biased whichever order of testing is adopted (Rosenfield *et al.* 1995b).

For two 3.5° vertical lines 2° to left and right of a fixation point, the mean fusional amplitude for four subjects was about 3.5° for divergence and 6.5° for convergence relative to dark vergence (Jones and Stephens 1989). The fusional range of monkeys is similar to that of humans (Boltz and Harwerth 1979; Harwerth *et al.* 1995). Kertesz (1981) measured the vergence component of the fusional amplitude. For one subject, vergence amplitude was 8.3° for a vertical line 5° long, 9.7° for a line 10° long, and 25° for a line 57° long. Divergent amplitudes were about one third as large, although it is not clear what reference vergence was used. The range of vertical vergence is discussed in Section 9.7.1.

Figure 9.26. Han Collewijn.
Born in Amsterdam in 1935. He obtained a degree in Medicine in 1960 and a Ph.D. in Neurobiology in 1963, from the University of Amsterdam. He then held a Research Fellowship at California Institute of Technology. In 1967 he obtained an appointment in the Department of Physiology at Erasmus University in Rotterdam, where he remained until he retired in 2000.

Figure 9.27. Casper Johannes Erkelens.
Born in Batenburg, the Netherlands in 1950. He obtained a B.Sc. in Physics from the Catholic University of Nijmegen in 1973 and was a high school teacher between 1973 and 1978. In 1983 he obtained a Ph.D. in Dental Physics from Utrecht University. He conducted postdoctoral work in the Department of Physiology at Erasmus University in Rotterdam and joined the faculty there in 1985. In 1992 he was appointed Professor in the Department of Physics of Man at Utrecht University. He was Director of the Helmholtz Institute at Utrecht University between 1994 ad 2000.

Feldman *et al.* (1989) compared the break and recovery fusion ranges using five clinical tests: including Risley prisms, a vectograph, and a random-dot stereogram. Each test gave good repeat reliability but correlations between the tests varied from zero to +0.94. Tests with targets of a similar size and degree of detail were correlated most highly.

The nearest distance at which convergence can be maintained is the **near point of vergence**. The accepted norm is 10 cm (von Noorden 1990). **Convergence insufficiency** is a term used to describe decreased fusional convergence. The criteria and instruments used for its assessment vary considerably (Daum 1988). With increasing age there is an increased incidence of convergence insufficiency. This trend accelerates after the age of 60 (Pickwell 1985).

As horizontal vergence is forced further from the position of dark vergence, the eyes tend to return to that position. This tendency manifests itself as an increasing fixation disparity, although this does not show in people who adapt their tonic vergence to changed vergence demand (Section 9.3.5).

Disparity in a vertical sine-wave grating cannot induce vergence movements larger than one half-cycle of the grating. In a random-dot display the visual system has difficulty finding corresponding images when disparity exceeds the mean separation of the dots, unless there are well-defined dot clusters. Nevertheless, Frisby and Mayhew (1980b) and Mowforth *et al.* (1981) claimed that vergence is initiated by disparities that exceed the dot separation in random-dot displays, when low spatial-frequency clusters are not present. Each eye was shown a 5° by 5° random-dot display filtered with circular Gaussian filters that had centre spatial frequencies of 1.75, 3.5, or 7.0 cpd. In response to 28 arcmin of disparity, appropriate vergence movements occurred with all three spatial frequencies, even when the disparity far exceeded the mean spacing of texture elements. However, their 7-cpd texture contained identifiable clusters of texture elements extending over at least 1° of visual angle. A disparity of 56 arcmin induced vergence oscillations with only the 7-cpd texture. In this case, the disparity of the texture clusters was too large and system relapsed into an oscillation.

Erkelens and Collewijn (1985b) (Portrait Figures 9.26 and 9.27) used a random-dot stereogram with a fixed disparity of 36 arcmin in a central region. Vergence was recorded as the whole of one eye's display was moved laterally with respect to the other eye's display. The eyes followed with an increasing lag until display separation reached between 1 and 2°. Above this value the images no longer fused and depth in the stereogram was no longer seen, although some vergence was evident up to a separation of about 4°, when vergence suddenly returned to a value that was independent of the stimulus. When the movement of the displays was reversed, vergence was reactivated at a smaller angle of separation than that at which it broke down. The stimulus contained a wide range of spatial frequencies and the displays were separated gradually. Erkelens and Collewijn explained the hysteresis effect by the fact that the eyes returned to a more convergent position after vergence broke down, which shifted the separation of the images in the convergent direction.

It is not surprising that random-dot displays with an overall disparity of over 4° do not evoke vergence. In the natural world, vergence movements are made between clearly distinct near and distant stimuli, where the initial disparity of the target stimulus exceeds 4°. In making such movements, we are presumably aided by cues to depth other than disparity, such as perspective.

Convergence anomalies such as convergence insufficiency, fixation disparity, and convergence inaccuracy may be alleviated by refractive correction (Dwyer and Wick 1995). Refractive correction improves the ability to detect binocular disparities that provide the error signal for vergence.

In orthoptic therapy, patients are trained to fuse targets in a synoptophore. There is general agreement that training is effective in improving the range of near vergence but is less effective in increasing far vergence (Daum *et al.* 1988). For example, the range of vergence, especially in the direction of convergence, was increased by orthoptics training for 10 minutes per day over a period of a few weeks, although the effect had mostly dissipated 6 months later (Daum 1982b).

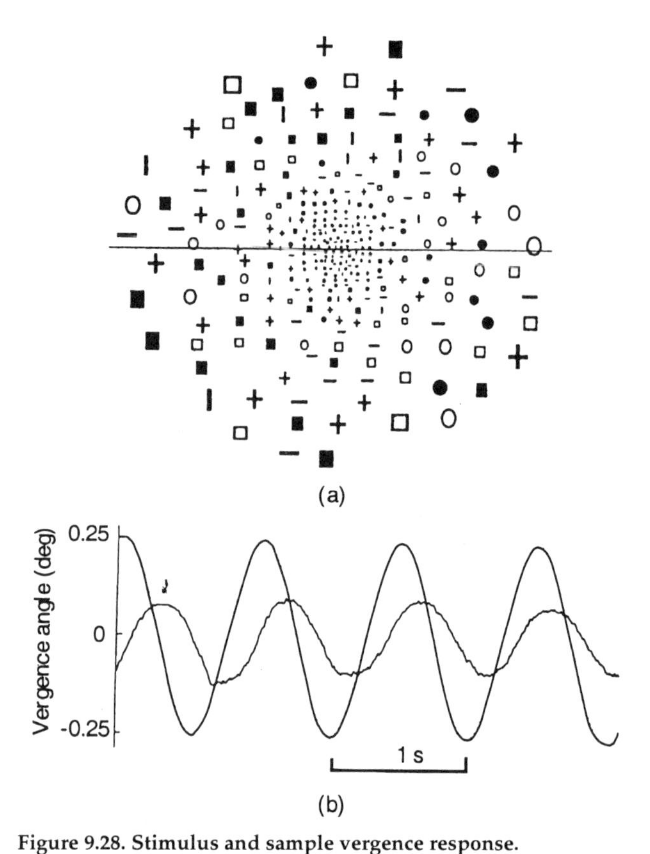

Figure 9.28. Stimulus and sample vergence response.
(a) Subjects looked at the line as the horizontal disparity of the display was modulated at 1.0 or 0.1 Hz with amplitude 0.5°.
(b) A sample response (solid line) superimposed on the sinusoidal stimulus modulation. (From Howard *et al.* 2000)

9.6.2 Disparity threshold for vergence

Duwaer and van den Brink (1981b) determined the smallest disparity required to initiate vertical vergence, as indicated by displacement of nonius lines. Since vertical vergence is not under voluntary control it provides a clean measure of the disparity/vergence threshold. For stimuli presented for 2 s at eccentricities of up to 4°, vergence was initiated by a disparity of about 1 arcmin. The threshold was higher for stimuli presented for only 160 ms at eccentricities greater than 4°. These disparity thresholds were much smaller than the disparities at which singleness of vision was lost. *The disparity/vergence threshold should be determined with an objective method of recording eye movements, and compared with that used for coding of depth.*

The ability to detect a correlation between dichoptic dynamic random-dot displays is affected by the proportion of uncorrelated dots and by luminance contrast (see Section 16.1.3). Stevenson *et al.* (1994) enquired whether the power of random-dot displays to evoke horizontal vergence is affected by the same factors. A dynamic random-dot display was set at various disparities within a circular aperture. This stimulus contained no monocular stimuli to vergence. A long horizontal line anchored vertical vergence and cyclovergence. Subjects tried to keep two vertical nonius lines aligned for 60 s after the display of dots was introduced, either with a constant disparity or with a 15-arcmin sinusoidal modulation of disparity. In spite of subjects' effort to keep the nonius lines aligned, the eyes showed vergence responses to the dot pattern. The velocity of these responses for a given disparity, although low compared with uninhibited vergence, showed a similar dependence on interocular correlation and stimulus contrast as did the detection of correlation (Section 16.1.3). Stevenson *et al.* concluded that vergence uses the same stimulus information as that used for detecting image correlation. In a random-dot display, cross-correlation provides the only information about disparity. Other sources of information, such as contour matching or colour matching, are available in real-life visual scenes and no doubt also contribute to the control of vergence.

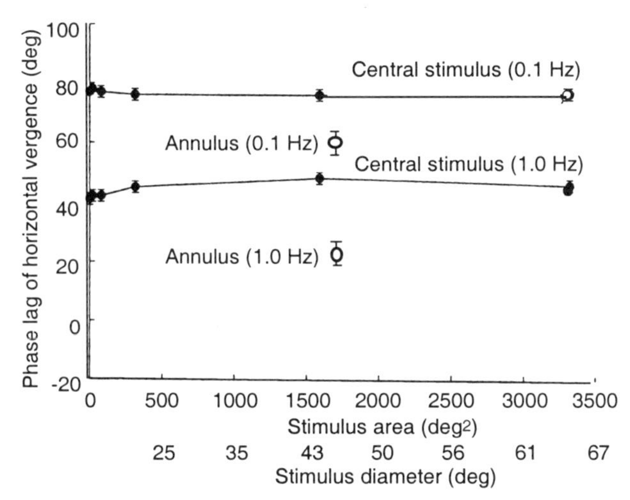

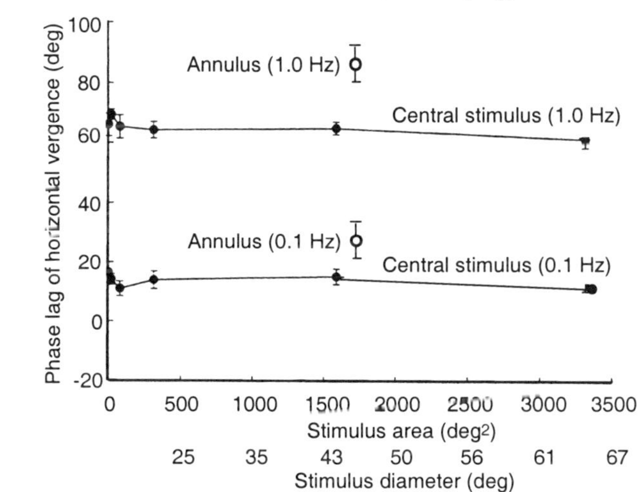

Figure 9.29. Horizontal vergence as a function of stimulus area. The horizontal disparity of the stimulus in Figure 9.27 was sinusoidally modulated through a peak-to-peak amplitude of 0.5°. The annular display with inner diameter 45° and outer diameter 65° had the same area as a central display with diameter 45°.

9.6.3 Stability of vergence

The small random movements of an eye are almost as large with binocular fixation as with monocular fixation (St. Cyr and Fender 1969). The combined motion of the two eyes produces a corresponding variation of disparity for all objects in the binocular field. Motter and Poggio (1984) found that for about 60% of the time the eyes of a monkey were misconverged by more than 7 arcmin in both the horizontal and vertical directions when the animal was fixating a small target. Vergence instability increased as the visual subtense of the target increased from 0.5 to 1°. The larger stimulus provided a less well-defined fixation target, and this, rather than stimulus size, may have caused the difference (see next section).

I predict that vergence stability will improve if extra texture elements are added to a central texture element of fixed size. This should be especially true for vertical vergence because vertical vergence is evoked by the summed disparity over a fairly large area (Section 9.7).

Steinman and Collewijn 1980 measured eye and head movements as subjects fixated a target while rotating the head from side-to-side through 20° at between 0.25 and 5 Hz. In one subject, the gain of eye movement relative to head movement was 0.87 in one eye but only 0.66 in the other. Even though vergence errors were up to 3°, the scene appeared stable and binocularly fused for all subjects. Stereoacuity and fusion of random-dot stereograms were not much disturbed by head rotation up to 2 Hz (Patterson and Fox 1984a; Steinman *et al.* 1985).

These results were not confirmed by others. Duwaer (1982a) found that head oscillations of 20° at 0.66 Hz produced vergence shifts of between 5 and 13 arcmin, as indicated by an afterimage method. Ciuffreda and Hokoda (1985) used a nonius procedure and found similar small vergence errors when subjects oscillated their heads through 20° at 4 Hz.

Motter and Poggio suggested that a dynamic feedback process insulates stereoscopic vision from the effects of fixation tremor. A neural model of this process, called a shifter circuit, was proposed by Anderson and van Essen (1987). The physiological evidence for this process is not convincing (see Gur and Snodderly 1997). We argue in Section 19.9.2 that such a mechanism is not required because even large disparity changes applied evenly over the visual field do not produce sensations of changing depth (Erkelens and Collewijn 1985a, 1985b; Regan *et al.* 1986a). All that is required is that the stereo system registers only first or second spatial derivatives of disparity (Section 20.2.7). For the same reason, stereopsis is not disturbed by naturally occurring or experimentally imposed fixation disparities (Fender and Julesz 1967).

9.6.4 Stimulus interactions

9.6.4a *Stimulus integration area*

While we can converge voluntarily on a specific object in a particular depth plane, our ability to do so precisely is limited by the area over which disparity information is integrated to produce a vergence signal. I define the **stimulus integration area** for vergence as the area of an isolated stimulus above which the gain of vergence in response to a temporal modulation of disparity does not increase. It can also be defined as the smallest separation between two stimuli with differing disparities for which convergence on one is not affected by the other.

Howard *et al.* (2000) measured the gain and phase lag of horizontal vergence to sinusoidal modulations of horizontal disparity of an isolated dot and of isolated textured displays of various sizes (Figure 9.28). It can be seen in Figure 9.29 that vergence gain was just as high and phase lag was just as low for a dot as for larger displays modulated in disparity at 0.1 or 1 Hz. Thus vergence tracking is as precise for a dot as for a large display. Therefore, stimulus integration area is very small for vergence tracking of an isolated stimulus.

The integration area for vertical vergence is several degrees of visual angle in diameter (Section 9.7). The area is larger for cyclovergence (Section 9.8.5).

9.6.4b *Stimulus integration depth*

Disparity information for the control of vergence may also be integrated over a range of disparities in a given small region. I call this **stimulus integration depth** for vergence control. Stimulus integration area times integration depth defines the stimulus integration volume. *Disparity integration depth has not been measured for any type of vergence.*

Mallot *et al.* (1996c) measured the vergence state after two planes of random dots 18 arcmin in front of and beyond a prefixation target were presented for 230 ms. When the planes contained equal numbers of dots with the same contrast, the eyes remained converged midway between them. When one plane contained more dots or dots with higher contrast, the eyes moved to a point nearer that plane. Thus vergence was elicited by a disparity signal derived from a weighted mean of the disparities in the two planes, with greater weight given to the plane containing more dots or higher contrast dots. Since neither stimulus area nor disparity separation between the planes was varied, this experiment does not indicate the area or range of disparities over which averaging occurs.

9.6.4c *Mixed disparities*

It is not difficult to fuse the images of a central object in the continued presence of horizontal disparities in stimuli in the peripheral visual field (Ludvigh *et al.* 1965). However, peripheral stimuli affect central fixation in four ways.

1. Horizontally disparate images near a centrally fixated visual object can induce fixation disparity.

2. The range of horizontal vergence within which a centrally placed stimulus can be held in a fused state is increased slightly by the addition of more peripheral stimuli, but only if the added stimuli are in a nearby depth plane. Peripheral stimuli with disparities above about 0.5° did not contribute to the maintenance of extreme positions of vergence in fixating a central stimulus (Jones and Stephens 1989).

3. Large, bold, horizontally disparate images presented suddenly to the parafoveal region can induce temporary diplopia in a small centrally fixated object (Winkelman 1951).

4. Disparity modulation of a large display can induce vergence instability. For example, a 0.25° sinusoidal modulation of horizontal disparity in a random-dot display induced about 0.15° modulation of horizontal vergence in subjects trying to hold fixation on a central stationary spot. The effect was greater for larger random-dot displays and for central displays compared with displays displaced to one side of the fixation target (Stevenson *et al.* 1999).

These results suggest that, under certain circumstances, mixtures of horizontal disparities are integrated over a certain area to provide the signal for vergence. Popple *et al.* (1998) measured the integration area for composite stimuli in the following way. They exposed, for 250 ms, a random-dot stereogram containing a central disc with 12.5 arcmin of crossed disparity. Subjects initially converged on a central dot in the plane of the surround. The magnitude of the initial vergence response to the disc, as assessed by a forced-choice nonius procedure, increased as the diameter of the disc was increased up to about 6°. This suggests that the integration area for the initial vergence response is about 6°. No conclusions can be drawn about the integration area for maintained vergence.

People cannot fuse the images of a central object in the continued presence of a large vertically disparate display in the periphery (Section 9.7).

9.6.5 Effects of stimulus position

The velocity and magnitude of horizontal vergence induced by disparity steps or ramps has been found to be greater for stimuli presented in the centre of the visual field than for those presented 3° in the periphery (Hung *et al.* 1991). Howard *et al.* (2000) found that disparity modulation of a textured display confined to the peripheral retina produced a response with lower gain and greater phase lag than modulation of a display of equal area confined to the central retina (see Figure 9.29). A stimulus with a given disparity became less effective in maintaining an accurate state of vergence as it moved into the peripheral visual field (Francis and Owens 1983; Hampton and Kertesz 1983a). Furthermore, a large disparate stimulus was more effective in evoking vergence than was the same stimulus with the central 10°-wide region occluded by an artificial scotoma (Boman and Kertesz 1985).

Differences between centrally and peripherally evoked vergence are probably due to the loss of acuity, including stereoacuity, with increasing eccentricity (Section 19.6.1). Another factor is that Panum's fusional area increases in the periphery so that, with more eccentric stimuli, the tendency of the eyes to return to the position of dark vergence induces a larger fixation disparity without loss of fusion (Ludvigh and McKinnon 1966). Thus, the peripheral retina tolerates larger disparities without triggering vergence than does the central retina. Another reason may be that peripheral disparity detectors have a smaller internal-loop gain (vergence signal per unit steady-state disparity) than do those serving the central retina. A fourth factor could be the increasingly transient character of disparity detectors as one moves into the peripheral retina.

The greater effectiveness of central stimuli in evoking vergence is advantageous because vergence is designed to bring the images of objects of greatest interest onto corresponding retinal points so that residual disparities can be coded as depth. Cyclovergence is evoked just as effectively by peripheral stimuli as by central stimuli (Section 9.8.4). But the purpose of cyclovergence is to bring the whole visual scene into register, not particular images.

9.6.6 Vergence latency

Human vergence eye movements made in response to a 2° step change in disparity of a small stimulus have a mean latency of between 130 and 250 ms, a mean velocity of about 10°/s, and take about 1 s to complete (Rashbass and Westheimer 1961a). Similar values have been reported for the monkey (Cumming and Judge 1986). However, vergence latencies of less than 60 ms have been reported in monkeys in response to step disparities in textured stimuli subtending 40 by 40° (Busettini *et al.* 1996a). These short latency responses were not due to the independent response of each eye to the motion of its retinal image, because both eyes responded even when image motion was confined to one eye. The initial acceleration of the vergence response was greater when the stimulus was presented just after the animal had made a conjugate saccadic eye movement. Postsaccadic enhancement depended at least in part on visual stimulation arising from a saccade. Thus, it occurred after a rapid shift in the retinal position of the stimulus with no eye movement. Similar short latency vergence occurs in humans, although the latency is about 25 ms longer than in the monkey (Busettini *et al.* 2001).

Conjugate saccadic eye movements are not affected by new sensory information arriving between 80 ms before the saccade starts and its completion (Becker and Jürgens 1975). In contrast, the duration of vergence movements is long enough to allow visual information to guide the response to its goal. Because of internal delay, any error correction in the initial rapid phase of a vergence movement would be effective only after the eyes had moved to a new position. This would lead to instability. Error correction is used to bring the eyes onto the target during the slower final phase of vergence.

There has been some debate about the relative latencies of convergence and divergence. Krishnan *et al.* (1973b) found that convergence had a longer latency than divergence (250 ms versus 210 ms) but Semmlow and Wetzel (1979) found the reverse (180 ms versus 200 ms). Hung *et al.* (1997) agreed with Semmlow and Wetzel. Krishnan *et al.* determined the start of a response by linear extrapolation from the constant velocity phase. This overestimates the latency of convergence because convergence has a higher velocity than divergence. Convergence has a shorter rise-time to peak velocity, a shorter time constant, and shorter total duration than does divergence (Zuber and Stark 1968; Mitchell 1970; Krishnan *et al.* 1973a; Hung *et al.* 1997).

The phase lag of vergence in response to a sinusoidal modulation of disparity is much smaller than one would predict from the 160 ms latency to a step stimulus. This difference cannot be due to the temporal regularity, and hence predictability, of the sinusoidal stimulus compared with a step, because regularly spaced steps have the same latency as irregularly spaced steps (Rashbass and Westheimer 1961a). Although sinusoidal modulations of disparity produce shorter latency responses than irregular modulations, the difference is not large enough to account for the difference between sinusoidal and step stimuli (Zuber and Stark 1968; Krishnan *et al.* 1973a). But the difference between phase lag and latency is a mystery only if one believes that vergence is controlled by a single-channel linear system. In a nonlinear system the phase lag to a smoothly varying stimulus cannot be predicted from that to a discretely varying stimulus. Large and small disparities tap distinct channels in the disparity-coding system and these channels may have different latencies (Section 6.3.1).

Latency has been found to be less to 4° disparity steps repeating at regular 2-second intervals than to steps occurring at irregular intervals (Yuan *et al.* 2000). When subjects initiate and therefore anticipate the motion of a visual target, a vergence movement starts even before the stimulus moves (Erkelens *et al.* 1989a). However, when subjects manually tracked a vergence stimulus moving sinusoidally in depth at

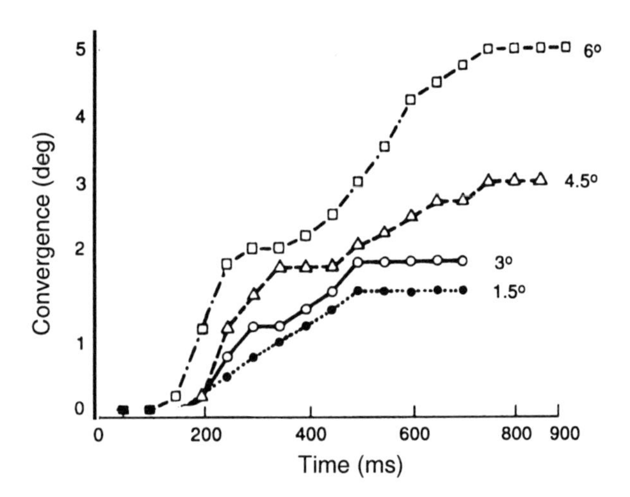

(a) Time course of symmetrical closed-loop vergence to disparity steps up to 6°. (From Westheimer and Mitchell 1956)

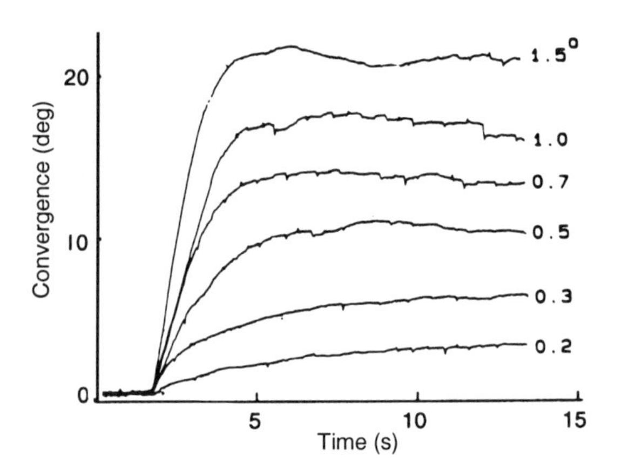

(b) Time course of symmetrical open-loop vergence to disparity steps up to 1.5°. (From Pobuda and Erkelens 1993)

Figure 9.30. Time-course of closed- and open-loop vergence.

various frequencies, vergence lag was no smaller than when the target was tracked by vergence alone (Koken and Erkelens 1993).

9.6.7 Vergence velocity

Figure 9.30a shows the time course of symmetrical convergence evoked by step changes of disparity of a small target from an initial vergence angle of 1.6° charted by Westheimer and Mitchell (1956). Figure 9.30b shows convergence to open-loop steps in disparity from an initial value of about 2.5° obtained by Pobuda and Erkelens (1993). Vergence velocity to each step was initially constant and then declined to produce a constant maintained angle of vergence. Both the initial velocity and the magnitude of maintained vergence increased with increasing size of the disparity step. The linear function relating response

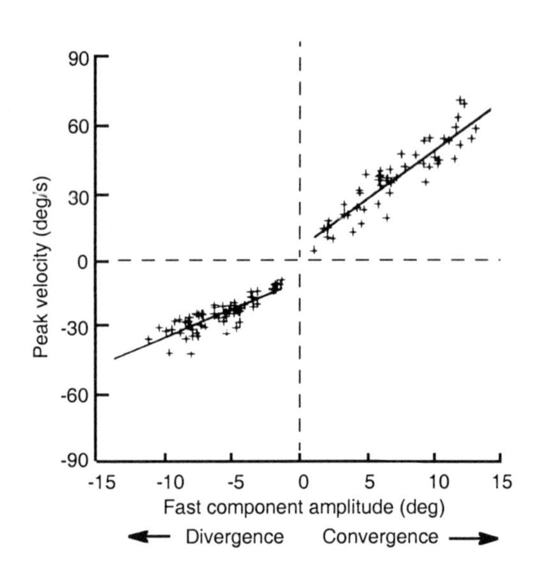

Figure 9.31. Main sequence for symmetric vergence. Peak velocity as a function of amplitude for symmetric vergence (N=1). The linear regression line is steeper for convergence than for divergence. (Adapted from Hung *et al.* 1997)

velocity to response amplitude is known as the main sequence for vergence, by analogy with the main sequence for saccades. The main sequence for vergence has been found to be constant under a variety of viewing conditions (Hung *et al.* 1994). This suggests that vergence movements are controlled by a processor that depends on only response amplitude. Most investigators agree that the rise in velocity with increasing amplitude is steeper for convergence than for divergence (see Figure 9.31) (see Hung *et al.* 1997). Neurophysiological data are consistent with this finding (Section 9.11.2).

The dependence of vergence velocity on the initial disparity of the stimulus can also be investigated by an open-loop procedure, in which vergence and disparity are linked so as to keep disparity constant. With steps of up to about 1°, velocity of open-loop vergence was approximately proportional to the size of disparity (Rashbass and Westheimer 1961a). One subject showed a 10°/s increase in velocity for each 1° increase in disparity. In the initial phase of a normal closed-loop response, before feedback had time to be effective, vergence velocity was approximately proportional to disparity up to a value of about 4°. This suggests that the open-loop transfer function of the vergence system involves the conversion of the initial disparity signal into a velocity signal. Mays *et al.* (1986) suggested that, as the response proceeds, the velocity signal is integrated into the position signal required to maintain the eyes in their new state of vergence. The greater the eccentricity of the tar-

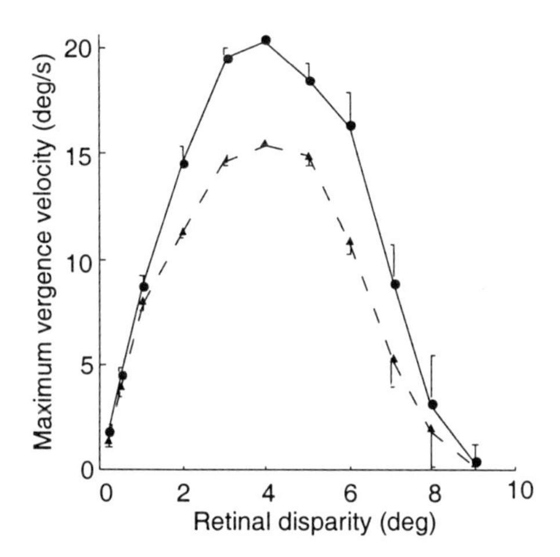

Figure 9.32. Vergence velocity and open-loop disparity. Peak vergence velocity as a function of steps of crossed disparity between retinally stabilized images of a vertical bar (dotted line) and a random-dot stimulus (continuous line). Bars are standard deviations (N=5). (Redrawn from Erkelens 1987)

gets between which vergence movements are made, the lower the velocity of vergence in the initial phase of the response (Schor *et al.* 1986c).

Erkelens (1987) used open-loop crossed-disparity steps from 0.25 to 10° for a vertical bar, a cluster of random dots, and the inner region of a random-dot stereogram. For all stimuli, maximum velocity of convergence increased steeply with increasing disparity up to about 3° and continued to increase up to a disparity of about 4°, after which it declined to zero at about 9° (Figure 9.32). Although vergence velocity differed widely between subjects, the shape of the function relating velocity to disparity was much the same for all subjects. When large angles of convergence were reached, velocity decreased until vergence saturated between 25° and 35°. Responses to disparities of more than 2° were not sustained. Convergence was more rapid than divergence.

Vergence responses to steps larger than 1.5° were found to be biphasic (Westheimer and Mitchell 1956). The initial fast response undershoots the target. Saccades, also, tend to undershoot the target. Consistent undershooting is an adaptive strategy because it ensures that only the magnitude of the secondary response need be computed. Westheimer and Mitchell suggested that the second, slower, component represents accommodative vergence. However, since both components of biphasic vergence movements obey the main sequence (Alvarez *et al.* 1998), the second component may be slower simply because it is smaller.

Busettini *et al.* (1996a) found a similar dependence of vergence velocity on the size of disparity steps in large textured displays. However responses to uncrossed disparity steps greater than about 2° tended to be in the wrong direction (convergence), probably because the sign of disparity in a dense textured display becomes ambiguous with large disparities. Convergence must be a default response to an ambiguous stimulus. Vergence responses were faster in the period just following a 10° saccade. Version is also speeded up after a saccade and Busettini *et al.* produced evidence that these effects depend mainly on the motion of the image produced by a saccade.

Pobuda and Erkelens (1993) explored the relationship between vergence velocity and stimulus velocity by comparing closed-loop responses to a stimulus that changed in disparity by 8° in 1 s, either smoothly or in two, four, or eight steps. The pattern of changing vergence was approximately a low-pass filtered version of the change in disparity with a lag of one reaction time. Vergence was not sensitive to the velocity of changing disparity as such, since the time course of the overall response was similar in the smooth and stepped conditions. A model of the vergence system based on these features of the response is discussed in Section 9.6.10.

Symmetrical vergence to a target moving along the midline in natural visual surroundings kept up with the stimulus with an accuracy of about 98% for stimulus velocities up to about 40°/s (Erkelens *et al.* 1989a). At higher stimulus velocities, vergence progressively failed to keep up with the stimulus, but performance was better when the movement of the stimulus was controlled by the subject rather than by the experimenter. When vergence was evoked by a voluntary to-and-fro motion of the subject's head and upper torso with respect to a stationary target, vergence velocity kept to within about 10% of stimulus velocity up to a velocity of 100°/s. Erkelens *et al.* did not move the subject's head passively, so we do not know whether the extra gain was due to the voluntary motion of the head or to a contribution from vestibular stimulation. These vergence velocities reported by Erkelens *et al.* are very much higher than those reported by Ludvigh and McKinnon (1968) for an isolated bar under the control of the experimenter.

The peak velocity of voluntary changes of vergence between two fixed stimuli within the median plane of the head increased from about 50°/s for vergence changes of 5° to about 200°/s for vergence changes of 34° (Erkelens *et al.* 1989b).

Execution of a large number of fast saccades or vergence movements leads to reduction in vergence velocity and accuracy, presumably because of fatigue (Yuan and Semmlow 2000). Appropriate vergence movements were evoked when the stimuli in

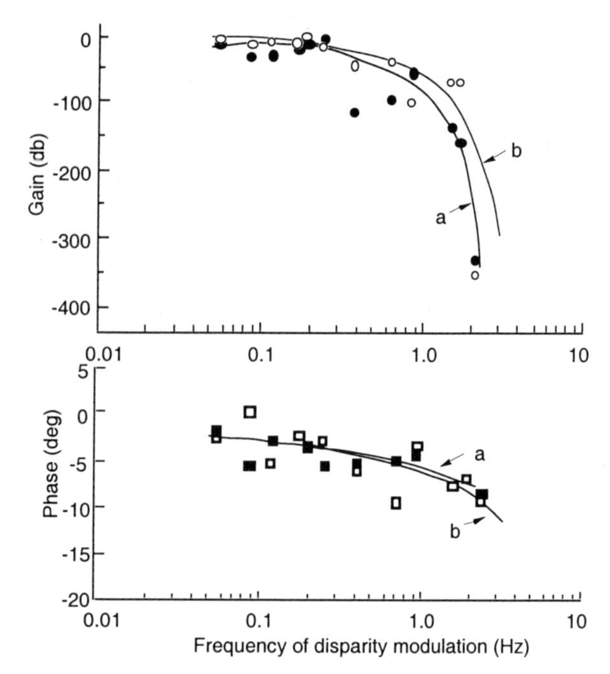

Figure 9.33. Gain and phase lag of vergence.
Gain and phase lag of vergence to predictable (lines marked a) and unpredictable (lines marked b) sinusoidal changes in disparity as a function of temporal frequency of disparity modulation. The amplitude of the stimulus was 1.75 metre angles (N=1). (Adapted from Krishnan *et al.* 1973a)

the two eyes differed in luminance by up to 1.6 log units, although vergence velocity decreased as the difference increased further (Mitchell 1970).

9.6.8 Vergence gain and phase lag

9.6.8a *Closed-loop gain*
For a sinusoidal variation in stimulus disparity, vergence gain is defined as peak amplitude of vergence divided by peak amplitude of disparity modulation of the external stimulus. This is the closed-loop gain. The gain and phase lag of vergence of a young subject made in response to both predictable and unpredictable sinusoidally changing disparity of amplitude 3.5° in a pair of vertical lines at a mean distance of 1.75 m are shown in Figure 9.33. Gain was close to 1 (zero decibel loss) for frequencies up to about 1 Hz and fell off above 1.5 Hz (Krishnan *et al.* 1973a).

Erkelens and Collewijn (1985c) measured the gain and phase lag of vergence evoked by sinusoidal oscillation of a 30° by 30° display of random dots containing a 15° by 15° diamond-shaped area with a crossed disparity of 36 arcmin with respect to the background. For stimulus amplitudes between 1° and 5° the gain was between 0.8 and 1 at a frequency of 0.25 Hz (Figure 9.34). Gain fell with increasing

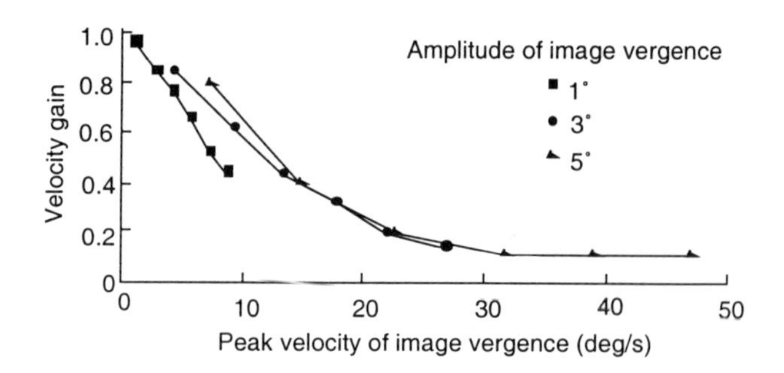

Figure 9.34. Vergence gain as a function of stimulus velocity.
Velocity gain of horizontal vergence as a function of peak velocity of image vergence for three amplitudes of image vergence (N=4). (Adapted from Erkelens and Collewijn 1985c)

stimulus amplitude at a greater rate for larger amplitudes. Gain was more closely related to stimulus peak velocity of image oscillation than to frequency or amplitude. The change of closed-loop gain with changing amplitude demonstrates that the system is nonlinear. The phase lag of vergence was about 20° at a frequency of 0.25 Hz and increased to about 100° at a frequency of 1.5 Hz in much the same way for all amplitudes. This suggests that phase lag results from a constant delay of about 250 ms.

The influence of stimulus area on vergence gain was discussed in Section 9.6.4.

9.6.8b *Open-loop gain*
The open-loop gain of vergence is the amplitude of vergence divided by the amplitude of initial stimulus disparity. The dependence of vergence amplitude on initial stimulus disparity can be investigated by presenting the stimulus for a duration less than response latency. In this way, the response is not affected by changes in disparity that occur after the response commences. The response is open loop during this initial period and the gain is the open-loop gain of the initial vergence response (Jones 1980). The peak amplitude of vergence to a 200-ms flashed stimulus increased nonlinearly with increasing stimulus disparity with respect to the position of tonic vergence, reaching a maximum at a disparity of between 2 and 3° (Figure 9.35).

The initial open-loop component of vergence occurs for stepped stimuli that are visible for only 50 ms. The magnitude of the initial response is independent of stimulus duration (Semmlow *et al.* 1993).

An open-loop vergence gain can also be obtained by coupling the stimulus disparity to the vergence response so that disparity remains constant during the response. In this case, the constant disparity acts as a persisting error signal, which drives the response to much larger amplitudes than the disparity, as we will see in the next section.

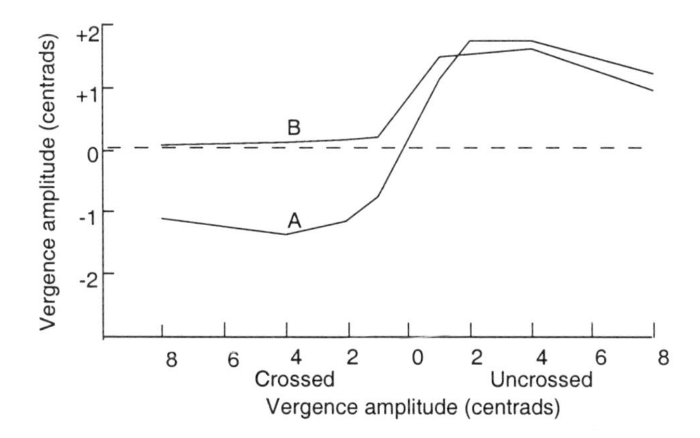

Figure 9.35. Vergence amplitude and disparity.
The peak amplitude of vergence in response to a 200-ms flashed stimulus as a function of binocular disparity. Curve A is from a subject showing a symmetrical response to crossed and uncrossed stimuli. Curve B is from a subject who responded only to uncrossed disparities. A centrad is 1/100th of a radian and is approximately equal to one dioptre for small angles. (Redrawn from Jones 1980)

9.6.9 Trigger and fusion-lock components

9.6.9a *Basic facts*

Vergence can be triggered by stimuli with large disparities, or by stimuli on opposite sides of the midline that project to opposite cerebral hemispheres (Winkelman 1953). However, these responses are transient when the images are not similar in shape. Vergence is maintained on an object to within about 2 arcmin only when the images are similar and fall within Panum's fusional area (Riggs and Niehl 1960). I will refer to transient and sustained components of vergence as the **trigger component** and the **fusion-lock component** respectively.

The two components are seen most clearly in responses to open-loop disparity. For instance, Erkelens (1987) found that open-loop disparities of up to 2° caused the eyes to converge between 15 and 25° and remain in the converged position for as long as the stimulus lasted. Thus, for open-loop disparities up to 2° the response was sustained. Open-loop disparities of 2 to 5° drove the eyes to a convergence of up to 35°, but the eyes drifted back to a vergence of less than 5°. In this case, the initial response was transient. For disparities larger than 5°, the eyes were driven to a less extreme position and the response was also transient and sometimes did not occur. The disparity at which the response became transient was the same as that at which the images were no longer fused. Thus, transient vergence is initiated by disparities well outside the fusional range and vergence is maintained by disparities small enough to provide a fusional lock.

Erkelens also found that a transient response to a large open-loop disparity reduced response velocity to other stimuli presented subsequently, but only when they had similar disparities. Thus, the short-term adaptive process underlying the transience of the trigger component to a given disparity is restricted to a given range of disparities.

9.6.9b *Functions of trigger and fusion-lock vergence*

Transient vergence to large disparities serves useful functions. In the first place, the initial high velocity response to a step change in disparity rapidly brings the angle of gaze close to the target plane where small fixation errors can be easily detected. A second rapid vergence response may occur when the initial response does not bring fixation near the target plane. Alvarez *et al.* (2000) produced evidence that second rapid responses are evoked by the sense of eye position rather than by a visual error signal.

When a person decides to change convergence from one plane to another, the large disparities in the target plane trigger an appropriate transient response. Once the images from objects in the target plane have been fused, large disparities arising from objects in other depth planes should be ignored until a decision has been made to change convergence again. If large disparities in nontarget depth planes were not ignored, the vergence system would never be able to settle down into a desired state. Subjects have no difficulty in making appropriate vergence movements to voluntarily selected parts of random-dot stereograms, even though this causes nonselected elements to become disparate (Erkelens and Collewijn 1991). Subjects can also track a chosen part of a stereogram when the selected disparity is coupled to their eye movements in open-loop control.

After locking onto a target the vergence system should continue to respond to small disparities due to movements of the target or vergence drift. If disparities are averaged over a local spatial region the eyes may not converge precisely on an object when there are neighbouring objects in nearby depth planes. This should not matter because small vergence movements would not cause images of a fixated object to fall outside Panum's fusional area, and would not disturb the relative disparities in the visual scene. Vergence changes between neighbouring objects should help to build a representation of the 3-dimensional structure of the local region.

9.6.9c *Vergence to dissimilar stimuli*

Vergence can be triggered by stimuli that differ in shape, but the response is not maintained in the absence of matching features that provide an error sig-

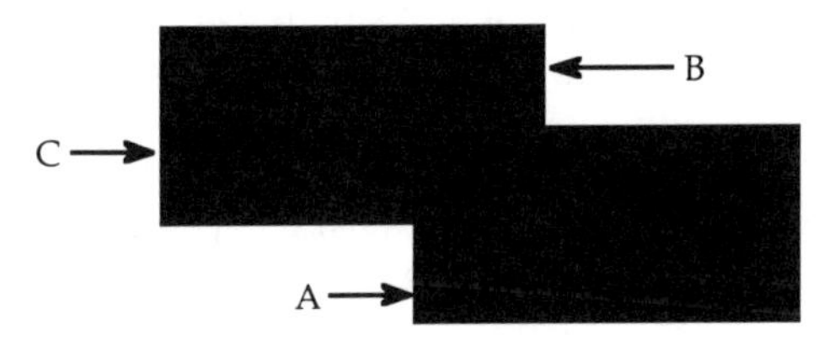

(a) For images with the same luminance polarity, vergence is evoked by matching edges C and A rather than by non-matching edges A and B.

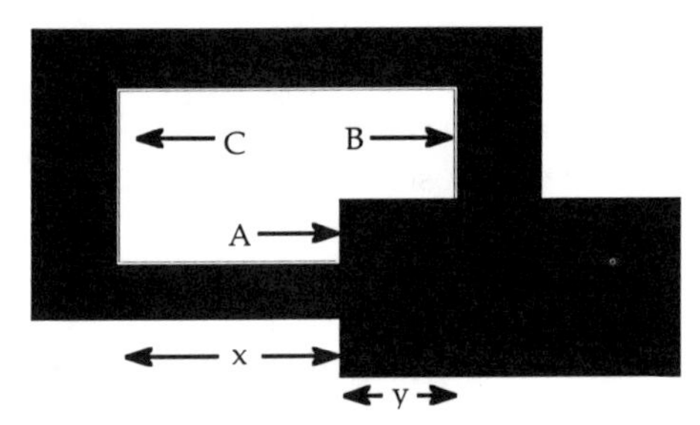

(b) For images with opposite luminance polarity, vergence may be evoked by matching edges A and B rather than by non-matching edges A and C.

Figure 9.36. Vergence to correlated and anticorrelated edges.

nal for fine vergence control (Westheimer and Mitchell 1969; Jones 1980). For instance, dichoptic vertical lines with 2° of horizontal disparity evoked sustained vergence, but the response was transient when one of the lines was horizontal. The response to dissimilar stimuli is transient even for disparities that normally evoke a sustained response. Brief exposure to a pair of similar stimuli and long exposure to a pair of dissimilar stimuli evoke the same transient response, which is not maintained because neither pair of stimuli is fusible. The similar stimuli do not fuse because of the brief exposure, and the stimuli with long exposure do not fuse because they are dissimilar.

Westheimer and Mitchell used only one pair of dichoptic images and their subjects may have converged voluntarily. Jones and Kerr (1972) overcame this problem by a vergence competition paradigm in which one eye saw a foveal stimulus and the other eye saw stimuli on both sides of the fovea. Thus, the subject had the choice of converging or diverging. They presented a pair of similar and a pair of dissimilar dichoptic images briefly at the same time. A transient response was just as likely to be triggered by the dissimilar pair as by the similar pair (Jones 1980; Jones and Kerr 1972; Semmlow *et al.* 1986).

Transient vergence to dissimilar dichoptic stimuli, such as horizontal and vertical lines, may be triggered by the common low spatial-frequency components of the stimuli. Edwards *et al.* (1998) investigated this issue using the vergence competition paradigm. Subjects were presented for 500 ms with a Gabor foveal patch in one eye and a pair of Gabor patches in the other eye, one 2.5° to the left and the other an equal distance to the right of the fovea. The dichoptic pair of images with the highest mean luminance contrast determined the direction of vergence, even when the images differed in contrast. Vergence was also determined by the pair of images that contained the lowest spatial frequency. Thus, transient vergence is determined by dichoptic images with the highest combined contrast within a low-pass spatial-scale channel.

Pope *et al.* (1999a) used the same competition paradigm. They found that, although an orthogonal pair of Gabor patches or a pair with opposite luminance polarity could induce vergence, there was a bias in favour of responding to images that matched in orientation or polarity. As the initial disparity of the images increased from 2.5 to 5°, the vergence system became increasingly indifferent to whether the images matched in orientation or luminance. The system was most selective for orientation at a spatial frequency of 2 cpd.

The above evidence suggests that transient vergence is evoked by a low spatial-frequency signal derived from the contrast envelope of the stimuli. Sato *et al.* (2001) asked how similar in size the contrast envelopes of dichoptic Gabor patches with 3.8° of crossed disparity must be to evoke transient vergence when in competition with dichoptic light patches with 3.8° of uncrossed disparity. The probability of vergence to the Gabor patches decreased as the ratio of the sizes of the envelopes increased from 1:1 to 1:8. Initially, the luminance-defined gratings (carriers) in the dichoptic Gabor patches were orthogonal and their spatial frequencies were scaled by envelope size. This ensured that there was no stimulus for sustained vergence. However, it was found that changing the relative envelope size had the same effect whatever the relative orientations or spatial frequencies of the carriers. Thus, transient vergence is determined by the relative overall sizes of the dichoptic stimuli. The system does not require the detailed structure of the stimuli to be matched.

Masson *et al.* (1997) measured the initial 100 ms of vergence responses of monkeys and humans to arrays of 2°-wide discs that were black on white in one eye and white on black in the other eye. Vergence occurred in the opposite direction to that evoked by images with the same contrast polarity. Both normal and reversed vergence occurred with a latency of 60 to 80 ms. The reversed-contrast display produced no impression of depth.

These results fit with the fact that reversed-contrast images produce inverted-disparity signals in V1. These inverted disparity signals can be explained in terms of reversal of contrast phase in the receptive fields of disparity detectors (Section 6.5.4). However, Masson *et al*, used overlapping 2°-wide discs, so that reversed vergence may have been evoked by disparity between matching edges of the type shown in Figure 9.36 (Howard 1997b). One would have to assume that closer matching edges have greater control over vergence than more distant matching edges or non-matching edges. Moreover, predictions are difficult with random multiple-disc displays. Masson *et al.* found that reversed vergence became less evident as the density of the discs was reduced. Small, low density reversed-contrast discs evoked vergence in the same direction as matching discs. These latter displays produced an impression of depth, and vergence occurred with a latency of 110 millisecond, which suggests the involvement of a higher-level neural mechanism.

Transient horizontal vergence is also evoked when dichoptic stimuli are up to about 3° out of vertical alignment (Mitchell 1970). A large vertical misalignment takes the images out of the range of the horizontal fusion-lock mechanism.

A similar distinction has been made between transient stereopsis evoked by dissimilar stimuli and sustained stereopsis evoked by similar stimuli (see Section 19.11.5).

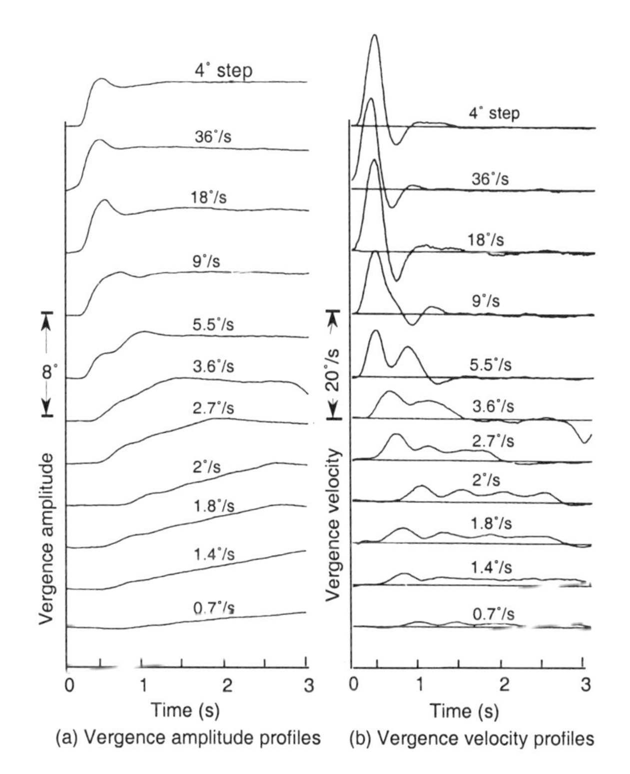

Figure 9.37. Time course of vergence.
(a) Time courses of vergence as a function of the velocity of a 4° change in disparity of two vertical lines. The velocity of the change in disparity is indicated above each curve.
(b) Velocity profiles of the responses. (N=3).
(Adapted from Semmlow *et al.* 1986.)

9.6.9d *Dynamics of trigger and fusion-lock vergence*

Semmlow *et al.* (1986) investigated transient and sustained vergence quantitatively. They varied the disparity between two vertical-lines by up to 4°, either in one step or in a ramp of constant velocity from 0.7 to 36°/s. Up to a ramp velocity of 1.4°/s, subjects tracked the changing disparity smoothly, which was taken as evidence that vergence was controlled by the sustained component. The smooth response exhibited proportional control in which vergence velocity was proportional to stimulus velocity. At higher velocities, the smooth response was interspersed with rapid responses (Figure 9.37). The ratio of peak velocity to amplitude of these rapid interludes was the same as that of the transient response to a 4° disparity step (they fell on the same main sequence). This suggests that the rapid interludes were transient responses. Their constant main-sequence characteristics persuaded Semmlow *et al.* that they are pre-programmed, or ballistic, movements based on a sampling of accumulated disparity in the ramping stimulus, in contrast to the proportional control of sustained responses. The main sequence takes account of only the first-order component of the system. Alvarez *et al.* (1999) developed a procedure for revealing the second-order components of the transient and sustained systems. A model of these process is described in Section 9.6.10. In the preceding section it was reported that Erkelens found a continuous change in vergence velocity as a function of disparity and argued against the idea of two distinct feedback dynamics in the sustained and transient systems.

The steplike responses to rapid disparity ramps reported by Semmlow *et al.* could be due to failure of vergence to keep up with the stimulus. The stimulus for vergence is image disparity, or stimulus disparity minus vergence. With rapidly changing disparity ramps, vergence lags the stimulus and disparity error accumulates to a level that triggers a transient response, which in turn restores the disparity error to within range of the sustained mechanism. With gradual ramps, disparity error does not accumulate to the level required to trigger a transient response. Analogously, voluntary pursuit of a target moving in the frontal plane is interrupted by catch-up saccades when the target moves too rapidly. Alternating fast and slow vergence does not

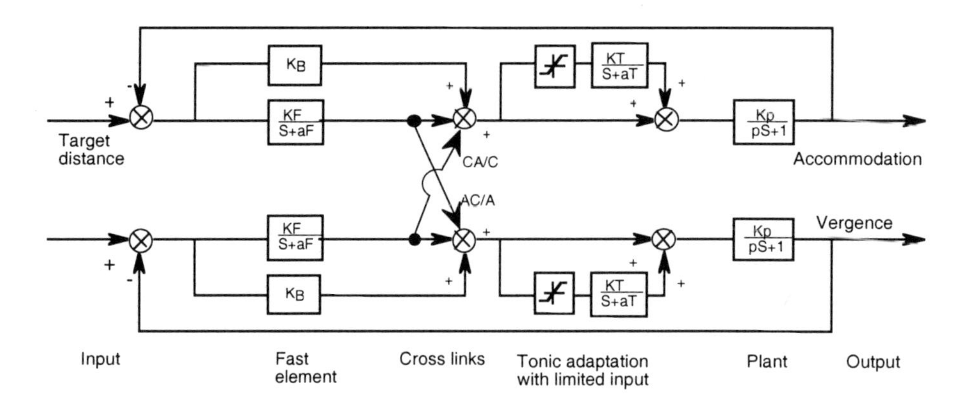

Figure 9.38. Model of vergence accommodation.
Mutual interactions between vergence and accommodation systems occur between the phasic and tonic neural integrators in the feedforward paths. Inputs to the tonic integrators have a saturation limit which could produce amplitude-dependent nonlinearities of the AC/A and CA/C ratios. Transfer functions are indicated by Laplace transforms. (Redrawn from Schor and Kotulak 1986)

occur when the stimulus ramp is presented in open-loop mode (Erkelens 1987; Pobuda and Erkelens 1993). Under these conditions, the disparity error is constant, since the movement of the stimulus is coupled to that of the eyes.

9.6.9d *Vergence asymmetries*

Figure 9.35 shows the vergence amplitude/disparity profile of a person who responded only to a line stimulus with uncrossed disparity. Other subjects responded only to a crossed disparity. Jones (1977) found that in about 20% of subjects with otherwise normal binocular vision the vergence response to briefly exposed disparities was asymmetrical; some did not respond to crossed disparities while others did not respond to uncrossed disparities with respect to the resting state of vergence. These subjects showed anomalous asymmetries in stereopsis produced by brief stimuli, like those reported by (Richards 1971) (see Section 12.1.1). However, there was only a loose relationship between the two types of asymmetry. With longer stimulus durations, asymmetrical responses showed only as slight differences in latency and velocity.

The disparities used by Jones were outside the fusion range. Fredenburg and Harwerth (2001) found asymmetries in the initial vergence response to briefly exposed Gabor patches with disparities of 30 arcmin or less. Although vergence asymmetry was correlated with performance on a stereo discrimination task, subjects with vergence asymmetries showed normal stereopsis for fine disparities. They concluded that vergence and stereopsis share an initial disparity-selective mechanism but that the two responses are subsequently processed by distinct mechanisms.

9.6.10 Models of horizontal vergence

General methods in the construction of models of sensorimotor systems were described in Section 3.5. Rashbass and Westheimer (1961a) developed the first model of the vergence system. It was a linear model with continuous feedback and a 160 ms delay element in the feedback loop. Krishnan and Stark (1983) also developed a linear model under continuous sensory control, unlike the ballistic saccadic system. Vergence can be maintained on a target with reasonable accuracy in two ways. A neural integrator could derive the final state of vergence from the integral of response velocity. Alternatively, the system could have a high internal-loop gain, defined as eye velocity per unit image disparity, which would allow small error signals to return vergence close to its desired state. Krishnan and Stark's model incorporates two parallel controllers. The first has fast, transient (derivative) dynamics with an eye-velocity output proportional to the instantaneous magnitude of disparity one reaction time earlier. This component contributes to the initial response only. The other has slow, tonic dynamics (time constant 15 s) with a leaky output related to the integral of eye velocity. The leaky output accounts for the slow drift of the eyes back to a resting state in the dark. Each controller has its own internal-loop gain, and the two are combined with a pure delay of 160 ms.

One weakness of the model is that it produces too slow a response to step inputs when the internal-loop gain of the integral controller is set low, and produces oscillations when it is set high. Performance could be improved by making the controller responsive to the predicted position of the visual target, but this would work only for predictable

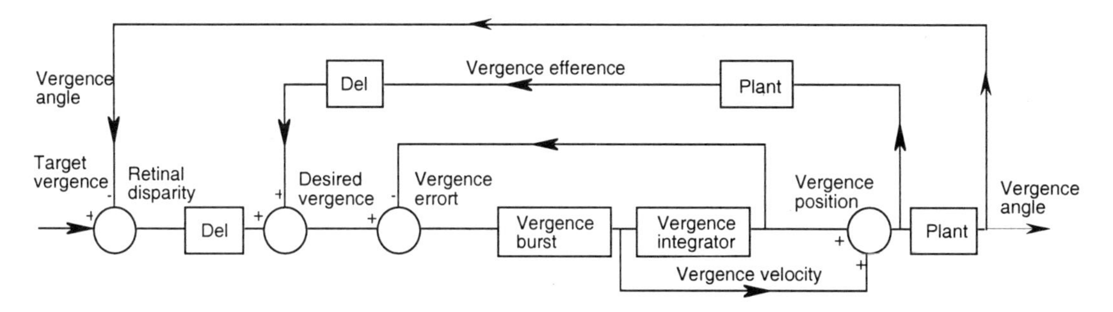

Figure 9.39. Outline of a model of the vergence mechanism.
A desired angle of vergence is derived by adding retinal disparity to the present vergence angle. Feedback from cells carrying velocity and position signals (burst-tonic neurones) passes through an internal model of the oculomotor plant and a compensating time delay (Del). The difference between desired vergence state and the feedback signal provides a vergence-error signal, which drives vergence-burst neurones. Burst neurones provide a velocity signal to oculo-motor nuclei and, by integration, a vergence position signal which maintains the desired final state. (Adapted from Zee and Levi 1989)

movements and, as has already been pointed out, vergence stability does not depend on whether the stimulus is predictable. Perhaps the phase lag of one of the control elements, when operating in a continuous tracking mode, is less than would be predicted from the latency of vergence in response to disparity steps, and this may account for the otherwise puzzling stability of the real vergence system (Rashbass 1981).

Krishnan and Stark's model deals with the nonlinear asymmetry between divergence and convergence by an appropriate asymmetry in the internal-loop gains. The compressive nonlinearities of the vergence system in the form of saturating levels of velocity and amplitude, and nonlinear interactions between vergence and version are not incorporated into the model. Nonlinearities arising from high-level control of vergence, as when a person decides to respond to one of several disparities, are also not in the model.

Schor and Kotulak (1986) developed a model that incorporates interactions between the vergence and accommodation systems (Figure 9.38). It contains an integrator with a short time constant that accounts for the initial response and an integrator with a long time constant that accounts for adaptive changes in tonic vergence. Nonlinear saturation elements turn off the integrators when the output reaches a certain value, so as to prevent overshoot.

Zee and Levi (1989) proposed the model shown in Figure 9.39, which incorporates the contribution of the saccadic system to vergence and the adaptive plasticity of the vergence system (Section 9.3.5). Cova and Galiana (1995, 1996) proposed a neural model of interactions between version and vergence that are discussed in the next section. Patel *et al.* (1997) developed a neural network model of horizontal vergence, which takes into account the nature of the input disparity signal and the motoneurone output signal. The model incorporates adaptive nonlinear control involving both position and velocity signals and explains both open-and closed-loop responses.

Pobuda and Erkelens (1993) proposed that vergence signals are processed through several parallel channels each with a gain element and a leaky integrator conferring lowpass characteristics. The gain of each channel is specific to a particular range of disparity amplitudes. As an eye movement in response to a given disparity progresses, control passes from the channel sensitive to large disparities to that sensitive to small disparities. The channels are insensitive to the rate of change of disparity. Pobuda and Erkelens also proposed that the overall lag of the system is comprised of a delay of between 80 and 120 ms in the vergence-processing loop, plus a lag in the mechanical plant. The lag in the processing loop is less than the 160 ms assumed in the other models and accounts for the small phase lag in response to sinusoidal stimuli, which other models do not explain (Section 9.6.8). The model also incorporates a slow integrator, like that proposed by Schor (1979a), to account for adaptation of tonic vergence. Although each channel is linear, their combined action introduces a nonlinearity, since the gain of the response varies with stimulus amplitude.

A "dual-mode" model of disparity control (Figure 9.40) has been developed from that of Zee *et al.* by Semmlow *et al.* (1986), Hung *et al.* (1986), and Horng *et al.* (1998a). A fast initial pre-programmed response is produced by a pulse-step mechanism with dynamics that are independent of stimulus duration or

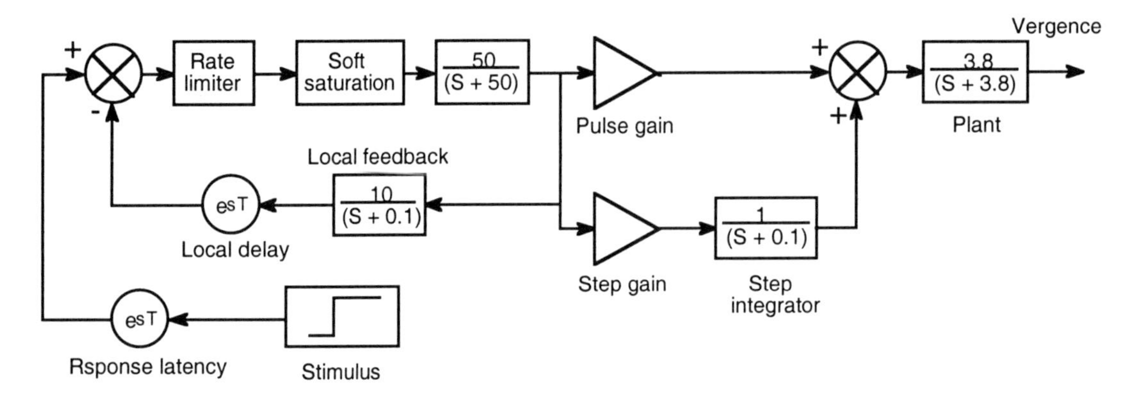

Figure 9.40. Model of initial component of vergence.
(Adapted from Horng *et al.* 1998a)

size, since the response does not depend on external error feedback. The initial response is followed by a slower response produced under feedback control. The oculomotor muscles are represented by a first-order plant with a mean time constant of 265 ms. The output of the pulse generator is derived from the difference between the stimulus step and a delayed internal feedback signal. The model also contains a nonlinear rate-limiter and an amplitude saturation element. The pulse signal feeds directly to the plant and also drives a leaky step integrator. The sum of the pulse and step signals drives the muscle plant.

For an open-loop disparity signal, the theory predicts a series of step-like responses, as a series of vergence movements is triggered by the persistent disparity signal. Semmlow *et al.* (1994) found the step-like response to an open-loop signal, which they claimed supported their model. The initial pulse response was isolated experimentally by using a 4° step stimulus that lasted only 100 ms before being replaced by darkness. Since this is shorter than the 160-s latency of vergence, the response occurred in the dark. The pulse response was found to be absent when the eyes diverged (Horng *et al.* 1998b).

Eadie and Carlin (1995) have reviewed models of the vergence system.

Summary

Vergence can be triggered by nonfusible stimuli or by stimuli outside the fusional range, but it cannot be maintained in a given state unless the fusion-lock mechanism is engaged. In a normal environment we see a multitude of disparate stimuli, each capable of triggering a vergence movement. When we decide to change fixation from one distance to another we disengage the fusion-lock mechanism, allow the new visual object to trigger an appropriate vergence response, and re-engage the fusion-lock process. If a new fusible stimulus does not materialize after the trigger response, vergence is not maintained in the new position and the response is transient. If a fusible stimulus comes into view, the fusion-lock mechanism is engaged and the response is sustained. There is some debate about the different dynamics of the two types of vergence. The neurological evidence bearing on this question is reviewed in Section 9.11. *Sustained and transient responses have not been studied in vertical vergence or cyclovergence.*

9.7 VERTICAL VERGENCE

When the principal horizontal retinal meridians are coplanar, the locus of zero vertical disparity is the median plane of the head and the plane of regard (Section 15.5.2). Consider a point object, *P*, in an oblique position, as in Figure 9.20. Because the object is nearer one eye than the other, the distance between one image and the plane of regard subtends a larger angle than the same distance for the other image (Ogle 1939c). When the gaze moves from straight-ahead to point *P*, vertical vergence is required to bring the images of *P* into vertical correspondence. For instance, a vertical vergence of three prism dioptres is required to fixate a point 24° up and 24° to one side on a frontal plane, at a distance of 33 cm (Ogle and Prangen 1953). The required vergence for a given direction of oblique gaze varies with the distance of the surface. Thus, we are constantly changing our vertical vergence as we shift our direction or distance of gaze (Section 9.7.3). However, the range of vertical disparities over the visual field is only about 2.5° compared with a range of horizontal disparities of about 14°.

Many people have a vertical phoria that manifests itself as an elevation of one eye relative to the other when there is no fusional stimulus. A vertical

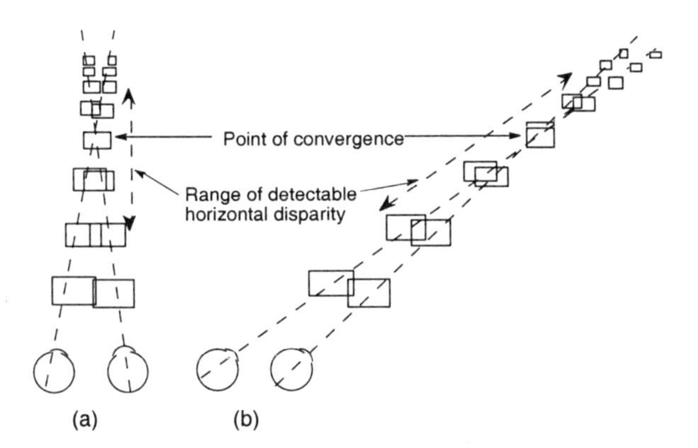

Figure 9.41. Disparity as a function of distance.
(a) Horizontal disparity increases in proportion to the distance from the plane of convergence.
(b) Along an oblique line of sight, vertical disparity decreases with increasing distance and horizontal disparity increases with increasing distance from the point of convergence. Horizontal disparity changes more rapidly than vertical disparity.

eye misalignment in the presence of a fusional stimulus is a vertical fixation disparity.

There are conflicting claims about whether objective methods of measuring vertical vergence give the same results as those obtained by alignment of horizontal nonius lines. Duwaer *et al.* (1982a) found that scleral search coils gave essentially the same results as a nonius method for a stimulus consisting of a vertical line intersected by two horizontal lines subtending 2°. Kertesz *et al.* (1983) reported that results obtained with the Purkinje eye tracker did not always agree with nonius measurements for a stimulus consisting of a square subtending 8°.

9.7.1 Range of vertical vergence

The smallest vertical disparity required to initiate vertical vergence, as indicated by displacement of nonius lines, is about 1 arcmin for stimuli presented for 2 s at eccentricities of up to 4°. The threshold disparity was higher for stimuli presented for only 160 ms at eccentricities greater than 4° (Duwaer and van den Brink 1981b). Under all conditions, the disparity threshold was much smaller than the disparity at which singleness of vision was lost.

The **vertical fusion range** is the vertical offset of dichoptic targets over which images remain fused. It includes the contribution of Panum's fusional range (nonmotor component) and the range of vertical vergence (motor component). It has been reported as 1.25° for a luminous disc subtending 0.12°, as 2.1° for a disc subtending 2° (Ellerbrock 1949a), and up to between 5.5 and 6.9° for a patterned display subtending 57° (Kertesz 1981; Duwaer 1982b). Boltz *et al.* (1980) measured vertical disparity that could be tolerated before loss of depth in a random-dot stereogram with 54 arcmin of horizontal disparity but unspecified size. The tolerated vertical disparity for both humans and monkeys was about 1.5° for a viewing duration of 500 ms. Sharma and Abdul-Rahim (1992) tested 60 subjects with a single Snellen letter and obtained a mean tolerated vertical offset of 2.3 dioptres (about 1.2°) with a range of between 1 and 5 dioptres. A low estimate could arise if objects other than the dichoptic stimuli are visible. All these figures must be doubled to obtain the total range of vertical fusion on both sides of zero disparity. The vertical fusion range was reduced when the stimulus was blurred or reduced in luminance (Ellerbrock 1949b, 1952). Duration of exposure is another important factor. Ellerbrock (1949a) found that it took 2 minutes for vertical vergence to reach its final value. Panum's fusional area contributes only between 2 and 10% to the range of vertical fusion (Duwaer 1982b). Orthoptic exercises have little effect on the vertical fusion range (Kutstein *et al.* 1988).

The **range of vertical vergence** is the range of actual vergence movements. It is best determined by measuring the movements of the eyes in response to vertical disparities impressed on dichoptic images presented in a stereoscope. Kertesz (1981) used scleral coils to measure vertical vergence evoked by the stimulus shown in Figure 9.44. For one subject, the maximum vergence in one direction was 1.9° for a disc subtending 5°, 3.5° for a 10° disc, and 5.2° for a 57.6° disc.

Hara *et al.* (1998) used a cross stimulus and obtained a mean maximum vertical vergence of 1.42° for a far stimulus and of 2.13° for a near stimulus. Thus the amplitude of vertical vergence was greater at near than at far. Under normal circumstances, vertical disparities occur only in near stimuli in eccentric positions.

The magnitude of horizontal disparity increases with the depth of the object from the point of convergence, as shown in Figure 9.41. Thus, horizontal disparity provides information about relative depth between points at each location in the visual field. The horizontal disparity of an object at infinity is 14° relative to an object at 25 cm. Images with a horizontal disparity of more than about 0.5° cannot be fused and images with a disparity of more than about 2° cannot be detected by disparity detectors. Outside the 2° range, the horizontal convergence mechanism must therefore use cues to depth other than disparity to bring an object of interest within range of the disparity detection system. Even when the eyes are converged on an object, the volume of space around that object may contain objects with a wide variety of horizontal disparities. For most pre-

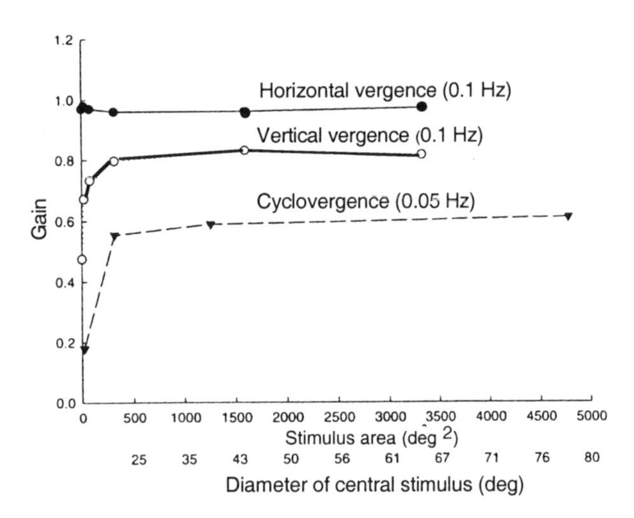

Figure 9.42. The effect of stimulus area on vergence gain.
The disparity of the stimuli was modulated sinusoidally at the frequency indicated on each curve. Top two curves from Howard *et al.* (2000). Bottom curve from Howard *et al.* (1994).

cise detection of relative disparities in the region of interest, one needs to be able to converge on one object and ignore neighbouring disparities. This means that horizontal vergence must be controlled by disparities in a selected local region.

Vertical disparity does not vary with distance for objects viewed in the normal way in the horizontal plane of regard or in the median plane. Therefore, it does not provide information about relative depth in these planes. But vertical disparity increases with eccentricity and decreases with absolute distance within each quadrant of the visual field. Even in extreme eccentricities, vertical disparities are only about 1.5°—within range of the disparity-detection system. Thus, unlike horizontal vergence, vertical vergence does not need to be evoked by cues to depth other than disparity.

For a series of objects at different distances along a line of sight in the median plane, only the horizontal disparity of the objects changes with distance. A voluntary choice must be made about which object to converge the eyes upon. For an oblique line of sight, both horizontal and vertical disparities change with distance but horizontal disparities change much more rapidly than vertical disparities. Vertical vergence will be sufficiently precise if it is evoked by the mean detectable vertical disparity in a fairly large region. It is not necessary to have attentional control over the stimulus evoking vertical vergence.

9.7.2 Effects of stimulus area and position

Howard *et al.* (2000) measured the gain and phase lag of vertical vergence for m-scaled random textured displays ranging in size from a small dot to

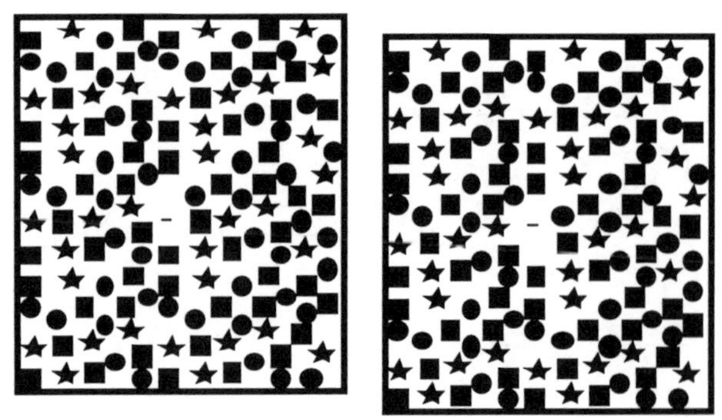

(a) The central horizontal lines cannot be fused when the vertically disparate surround is fused.

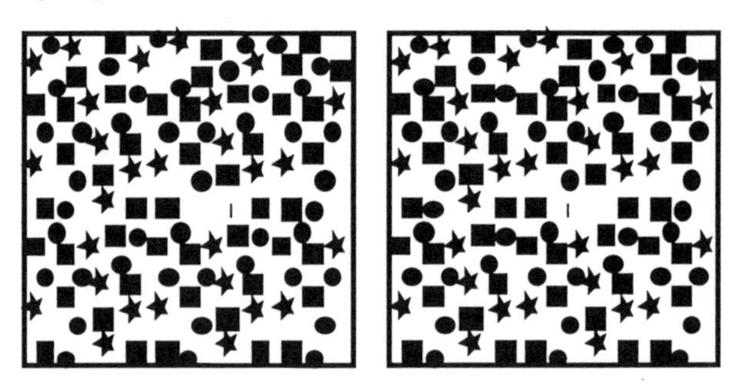

(b) The central vertical lines are easy to fuse when the surround has a horizontal disparity.

Figure 9.43. Fusing central stimulus with disparate surround.

65°. The vertical disparity of the display was modulated sinusoidally through a peak-to-peak amplitude of 0.5° at frequencies of 0.1 and 1.0 Hz. It can be seen from Figure 9.42 that the gain of the response increased as the diameter of the display increased to 20°, above which it remained reasonably constant. Phase lag decreased slightly as stimulus diameter increased to 20°. A central stimulus with a diameter of 45° produced a response with higher gain than a peripheral stimulus of the same area (an annular stimulus with inner diameter 45° and outer diameter 65°). Thus, a central stimulus is a more effective stimulus for vertical vergence than a peripheral stimulus of the same area. The figure also shows how horizontal vergence and cyclovergence are affected by stimulus area.

Vertical disparity in a surrounding display induces persistent vertical diplopia in a fixated target with zero vertical disparity (Burian 1939; Houtman and van der Pol 1982a). In Figure 9.43 it is impossible to fuse the central horizontal lines when the surround has a vertical disparity. Central vertical lines with zero horizontal disparity are easily fused in the presence of horizontal disparity in surrounding texture elements.

Stevenson *et al.* (1997) found that subjects could not hold horizontal or vertical vergence on a fixation

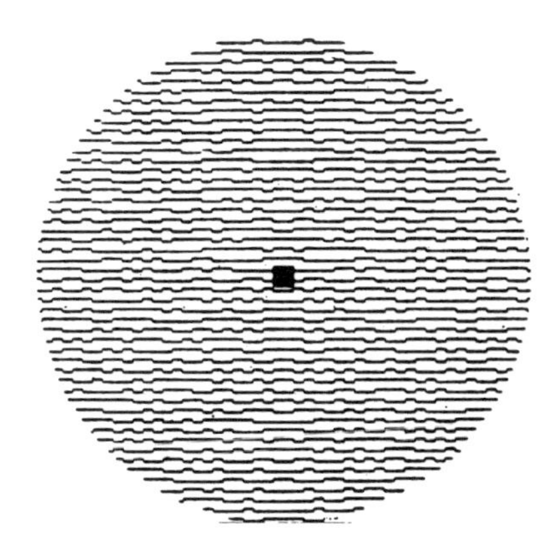

Figure 9.44. The range of vertical disparity.
Stimulus used by Perlmutter and Kertesz (1982).

target when the vertical and horizontal disparities of a 7.5° textured surround were simultaneously modulated up to 40 arcmin at 0.125 Hz. Induced vertical vergence was the same whether subjects attended to the stationary target or to the modulated surround. Induced horizontal vergence was small when subjects tried to fixate the stationary spot but had a gain of about 0.85 when they attended to the surround. This demonstrates that both horizontal and vertical vergence are driven by a weighted mean of competing signals from a certain area. However, people have some control over which of two competing stimuli is used to drive horizontal vergence but no control over which stimulus drives vertical vergence. This difference is presumably related to the fact that vertical disparities do not change as abruptly over the visual field as horizontal disparities.

Stevenson *et al.* (1999) found that induced horizontal and vertical vergence in subjects trying to fixate a stationary point decreased with decreasing area or increasing eccentricity of a textured surround modulated in horizontal and vertical disparity through 0.25° at 0.5 Hz. The decrease with increasing eccentricity was similar to the change in the cortical magnification factor.

Allison *et al.* (2000) asked whether a vertically disparate textured background continues to prevent fusion of a central target when background and target are placed in different depth planes. They measured the threshold for detecting diplopia in central horizontal lines as a function of their vertical disparity. The introduction of a zero-disparity background reduced the diplopia threshold from about 0.9° to about 0.2°. As the horizontal disparity of the background relative to the central target increased to 6° the diplopia threshold approached the value obtained in the absence of the background. Decorrelating the two background images also reduced the effect of the background on the diplopia threshold. We can thus say that vertical disparity signals are averaged over an area and over a certain depth.

Schor *et al.* (1994) found that when the gaze is directed to a target without error feedback, the visual axes intersect with an error of no more than 0.25° for any direction or distance of the target. This suggests that horizontal and vertical vergence are preprogrammed for direction and distance.

9.7.3 Dynamics of vertical vergence

Houtman *et al.* (1981) used scleral search coils to measure vertical vergence in response to a display of letters subtending 11°. An open-loop 35 arcmin vertical displacement of the image in one eye produced a saturation level of vergence of about 40 arcmin, which occurred at a velocity of about 15 arcmin/s. Responses were obtained to closed-loop sinusoidal image displacements at frequencies up to 1 Hz. For a stimulus amplitude of 33 arcmin response gain fell from near 1 at 0.03 Hz to about 0.3 at 1 Hz. Gains were lower for an amplitude of 65 arcmin. Responses away from the resting position were slower than those towards the resting position (Houtman and van der Pol 1982b).

Perlmutter and Kertesz (1982) used the stimulus shown in Figure 9.44, which subtended 8.5°. An open-loop step of vertical disparity of 14.8 arcmin produced a vertical vergence with a reaction time of 180 ms and a velocity of 39.6°/s. The final amplitude of 54 arcmin was maintained for 250 ms. The velocity of vertical vergence was proportional to the magnitude of disparity, as Rashbass and Westheimer (1961a) found for open-loop horizontal vergence. For sinusoidal modulations of vertical disparity, the open-loop gain was about 4 at frequencies up to 0.4 Hz and fell to 1 at about 0.7 Hz. For closed-loop modulations of disparity of 9.3 arcmin, the gain was about 1.1 up to a frequency of 0.4 Hz and fell to 1 at 0.9 Hz. Gain decreased as the amplitude of disparity modulation increased from 3.3 to 28 arcmin, showing that the mechanism is nonlinear. The phase lag increased from about 30° to 90° as frequency increased from 0.1 to 0.9 Hz and was the same for predictable (sinusoidal) as for unpredictable modulation of disparity, demonstrating that the mechanism does not involve a predictor. The stimulus shown in Figure 9.44 consisted of 50 notched horizontal lines with an interline spacing of 10 arcmin. At every multiple of 10 arcmin of vertical disparity many segments of the horizontal lines come into correspondence and provide a stimulus for vertical fusion. This stimulus is therefore unsatisfactory for studying vertical vergence.

Howard *et al.* (1997) used scleral eye coils to measure the gain and phase of vertical vergence. The 65° textured displays were m-scaled to homogenize visibility and were aperiodic to avoid spurious binocular matches, as shown in Figure 9.28. The vertical disparity of the display oscillated through peak-to-peak amplitudes of between 18 arcmin and 4° at frequencies between 0.05 and 2 Hz. The results are shown in Figure 9.45. Gain was near one when the stimulus oscillated through 18 arcmin at 0.1 Hz or less. As the amplitude of stimulus oscillation increased, vergence gain decreased at all frequencies, which is evidence of a nonlinearity. Gain declined with increasing stimulus frequency but was still about 0.5 at 2 Hz for an amplitude of 18 arcmin. Vertical vergence is designed to compensate for small disparities changing at moderate frequencies.

Howard *et al.* found that phase lag increased from less than 10° at a stimulus frequency of 0.05 Hz to between 100 and 145° at 2 Hz. Phase lags reported by Perlmutter and Kertesz were similar. Overall, the dynamics of vertical vergence resemble the dynamics of horizontal vergence and cyclovergence.

Boman and Kertesz (1983) claimed that the amplitude of vertical vergence decreases and its reaction time increases in the presence of a horizontal disparity but that horizontal vergence is not affected by the presence of a vertical disparity. This claim is suspect because the stimulus had a strong horizontal/vertical anisotropy—it contained several prominent vertical lines, which would provide a horizontal fusional stimulus at several horizontal disparities but no such horizontal lines. Stevenson *et al.* (1997) found that the amplitude of an oblique vergence is the sum of the horizontal and vertical components measured separately, although they used only a limited range of vergence amplitudes.

Step changes in vertical vergence involve disjunctive saccades (Bush *et al.* 1994; Van der Steen and Bruno 1995). Furthermore, subjects show rapid adaptation of disjunctive saccades to unusual patterns of vertical disparities (Section 9.9.3).

Left-over-right vertical vergence evokes a top-to-left cycloversion and vice versa (Enright 1992b; van Rijn and Collewijn 1994; Mikhael *et al.* 1995). This suggests that vertical vergence is controlled mainly by the oblique muscles. When the head tilts to one side, the eyes counterroll conjugately a few degrees in the opposite direction (Collewijn *et al.* 1985). This reflex in resonse to pitch of the head in lateral-eyed animals is greater than counterolling in frontal-eyed animals. It helps to keep the image erect. Counterrolling is accompanied by a vertical divergence of the eyes. The eye on the side of the lower ear is elevated relative to the other eye (Kori *et al.* 2001).

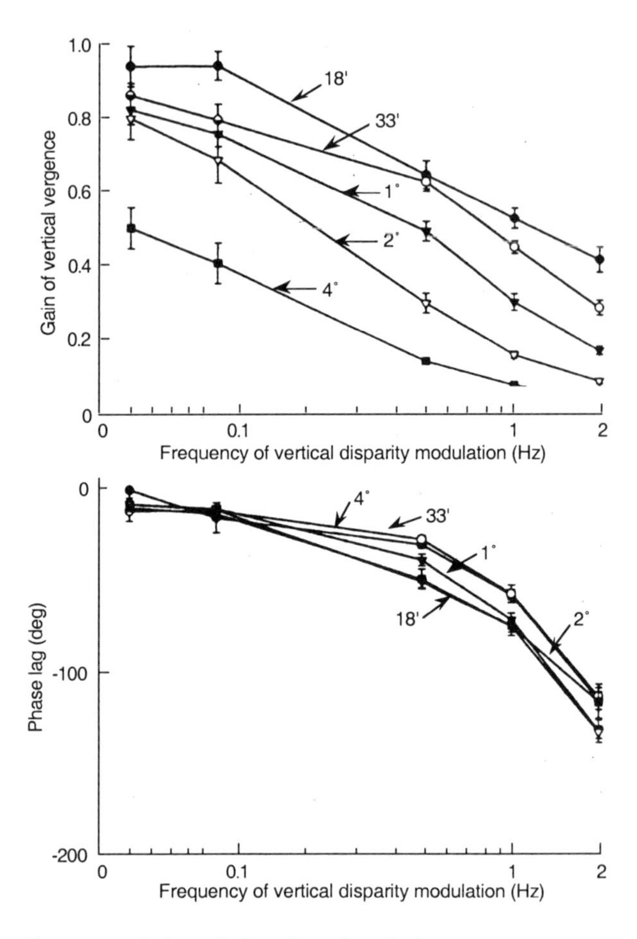

Figure 9.45. Gain and phase lag of vertical vergence.
Gain and phase lag of vertical vergence in response to sinusoidal modulation of vertical disparity of a large textured display at various frequencies and amplitudes. Mean of 4 subjects. (Adapted from Howard *et al.* 1997)

Dissociated vertical deviation (DVD) that accompanies congenital esotropia is an exaggerated form of this pattern of eye movements (Van Rijn *et al.* 1997; Cheeseman and Guyton 1999). It seems that it manifests itself when the eye movement system is not under binocular control (Brodsky 1999). People with normal binocular vision may show asymmetrical vertical phorias that resemble mild dissociated vertical deviation (Van Rijn *et al.* 1998). The vertical vergence component of DVD partially damps nystagmus associated with DVD (Guyton *et al.* 1998).

9.8 CYCLOVERGENCE

9.8.1 Types of torsional response

Cycloduction is the torsional state of a single eye indicated by the dihedral angle between the median plane of the head and the plane containing a specified meridian of the eye and the visual axis. The usual reference meridian is the normally vertical meridian as indicated by selected landmarks on the eyeball or by the apparent orientation of the after-image of a vertical line. Cycloduction is designated incycloduction or excycloduction, depending on whether the eye is rotated top towards or top away from the median plane.

Cycloversion is the equal component of the eyes' cycloductions; levocycloversion when the eyes rotate top to the subject's left and dextrocycloversion when they rotate top right. Cycloversion occurs under the following circumstances.

1. Rotation of a textured display around the fixation point induces optokinetic cycloversion (Brecher 1934; Cheung and Howard 1991).
2. Small amplitude cycloversion is evoked by a large display of lines tilted from the vertical in the frontal plane (Goodenough *et al.* 1979).
3. Cycloversion accompanies vertical vergence. Left-over-right vertical vergence evokes a top-to-left cycloversion and vice versa (Mikhael *et al.* 1995).
4. Rotation of the head about the roll axis induces an opposite **counterrolling** of the eyes of up to about 8° (Collewijn *et al.* 1985; Ferman *et al.* 1987b). In the dark, the response is evoked by signals from the vestibular system and has a velocity gain of about 0.6. In the light, optokinetic and vestibular signals combine to increase velocity gain to about 0.72 (Leigh *et al.* 1989).
5. Cycloversion can be evoked voluntarily after an extended period of practice (Balliet and Nakayama 1978).

Cyclovergence is the difference between the eyes' cycloductions. Cyclovergence is designated incyclovergence or exocyclovergence depending on whether the relative rotation is top-in or top-out respectively. Cyclovergence is zero when two horizontal nonius lines on opposite radii of a dichoptic display in the frontal plane appear collinear. It is best to use horizontal rather than vertical nonius lines as a reference because, under normal circumstances, corresponding horizontal meridians are parallel while corresponding vertical meridians have a positive declination of about 2° (Helmholtz 1910, vol. 3, p. 408).

Stimulus cyclodisparity, or **declination**, is the signed relative rotation of dichoptic stimuli in the frontal plane in external co-ordinates (Ogle and Ellerbrock 1946). **Inclination** refers to the slant with respect to the vertical of an object in a sagittal plane of the head, signed positive top away. For interpupillary distance a and observation distance d, a line with inclination i projects as a pair of images with declination θ. It is shown in Section 15.7 that:

$$\tan\frac{\theta}{2} = \frac{a \tan i}{2d} \qquad (4)$$

For small values of θ

$$\theta = \frac{a \tan i}{d} \quad \text{or} \quad i = \tan^{-1}\frac{\theta d}{a} \quad \text{in radians}$$

The image **cyclodisparity** of a pair of dichoptic images is their relative orientation with respect to corresponding retinal meridians, designated positive if the left-eye image is rotated clockwise and the right-eye image counter-clockwise with respect to the nearest pair of corresponding retinal meridians. Dichoptic images have zero cyclodisparity if they are parallel to a pair of corresponding meridians. It is assumed that they are parallel to corresponding meridians when they appear parallel.

The declination of vertical corresponding meridians causes the vertical horopter to be inclined, top away, by an amount that varies with viewing distance (Section 15.7). The inclination of the vertical horopter may also be affected by cyclovergence that accompanies a change in horizontal vergence (Amigo 1974). It follows from these definitions that, for horizontal lines, image cyclodisparity equals stimulus cyclodisparity minus cyclovergence. For vertical lines, image cyclodisparity equals stimulus cyclodisparity minus cyclovergence minus the declination of corresponding vertical meridians.

Various investigators over the last 100 years, including Hering (see Ogle and Ellerbrock 1946), Verhoeff (1934), Kertesz (1972), and Krekling (1973a) have denied that cyclovergence occurs and the response is not mentioned in most textbooks. In some cases, investigators changed their minds when they used more effective stimuli, particularly stimuli subtending a visual angle in excess of 25° and containing many horizontal and vertical elements.

Cyclovergence is evoked by cyclodisparity (Kertesz and Sullivan 1978) and as a component of horizontal vergence (Allen and Carter 1967) and of vertical vergence (Enright 1992b; Cheeseman and Guyton 1999). **Cyclophoria** is a torsional misalignment of the eyes in the absence of cyclofusional stimuli (Wick and Ryan 1982). **Cyclotropia** is a torsional misalignment of the eyes in the presence of fusional stimuli (Ruttum and von Noorden 1983).

9.8.2 Measurement of cyclovergence

The following psychophysical methods have been used to measure cyclovergence.

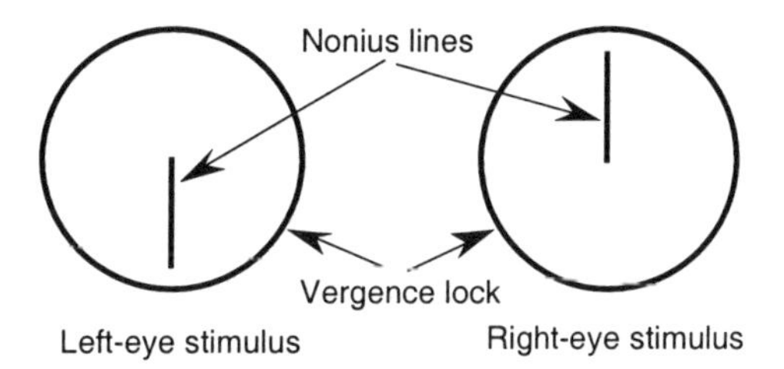

Figure 9.46.Volkmann discs.
Angular misalignment of the two nonius lines in the binocular image provides a measure of the declination of corresponding vertical meridians. In most people the corresponding meridians are relatively extorted about 2°.

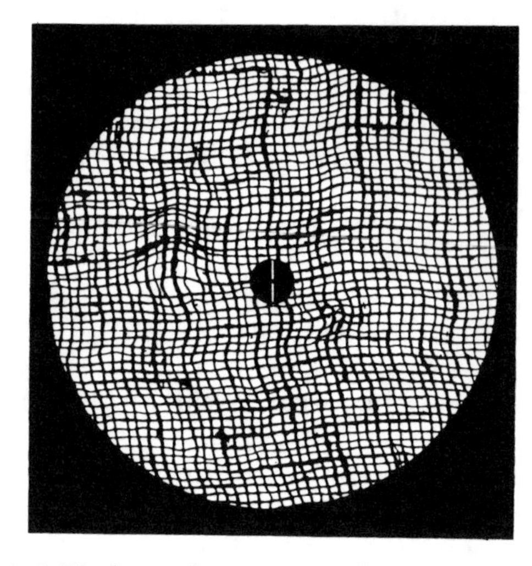

Figure 9.47. Display used to measure cyclovergence.
A dichoptic textured display subtending 75° was disjunctively cyclorotated through various amplitudes and frequencies. Cyclovergence was measured objectively and by aligning the central nonius lines. (From Howard et al. 1993)

9.8.2a *Setting a line in the median plane to vertical*

The observer rotates a test line in the median plane of the head until it appears vertical. The method is based on the assumption that a line seen binocularly appears vertical if and only if its images fall on corresponding retinal meridians, and any error in the vertical setting is due to cyclovergence, according to formula (3). However, this assumption is faulty. The phenomena of slant contrast and normalization discussed in Section 22.4 demonstrate that a line does not necessarily appear vertical when its images fall on corresponding meridians. Also, the slanting-line method gives different results depending on whether or not a reference plane is provided in the form of a circle round the test line (Harker 1960). Another problem is that a slanted line is a stimulus for cyclovergence and may therefore contaminate the results. This problem is at least partially overcome by presenting the test stimulus briefly, after the stimulus for cyclovergence has been removed. Ellerbrock (1954) further minimized the effect of the test stimulus by setting two points rather than a line into the apparent frontal plane. Amigo (1974) used a similar procedure to investigate the vertical horopter. Hampton and Kertesz (1982) compared the settings of a brief sequentially presented test line with an objective measure of cyclovergence. The perceived slant of the test line was less than that corresponding to the residual cyclodisparity in the line, and therefore did not indicate the degree of cyclotorsion. A sequential test stimulus may overcome one problem, but does not overcome the problems of slant contrast and normalization.

9.8.2b *Nulling cyclodisparity in dichoptic stimuli*

In one form of this method dichoptic lines are rotated in opposite directions in the frontal plane until they fuse into one, or appear collinear. This method was first used by Meissner in about 1854 (Le Conte 1881) and was also used by Volkmann (see Helmholtz 1910). Cogan (1979) pointed out that this measure of cyclovergence does not agree with that based on judgments of apparent vertical, since a line does not necessarily appear vertical when the images in the two eyes appear collinear. He showed that, on average, a line was set within 3° of true vertical in the median plane whereas two dichoptic images in the frontal plane, one red and one green, had to be incyclorotated by an amount corresponding to an inclination of 31° to fuse into a single image. Although setting two dichoptic images into collinearity may be the better procedure, it has its own problems. Superimposed dichoptic images tend to exhibit binocular rivalry and it is difficult to detect cyclodisparity once the images lie within Panum's fusional area. Also, a slanted line is a stimulus for cyclovergence and may therefore contaminate the results.

9.8.2c *The nonius method*

In the nonius method, cyclovergence is indicated by the angle through which a horizontal line presented to one side of the fixation point in one eye has to be rotated in the frontal plane to appear parallel to a horizontal line presented in the opposite field of the other eye. A binocular circle surrounding the lines holds horizontal and vertical vergence steady. This stimulus display is known as Volkmann discs (see Figure 9.46). This is the torsional equivalent of nonius methods used to measure horizontal and vertical vergence. Subjects read off the torsional deviation of the eyes on a scale (Sen *et al.* 1977).

Hofmann and Bielschowsky (1900) were the first to use Volkmann discs systematically. They re-

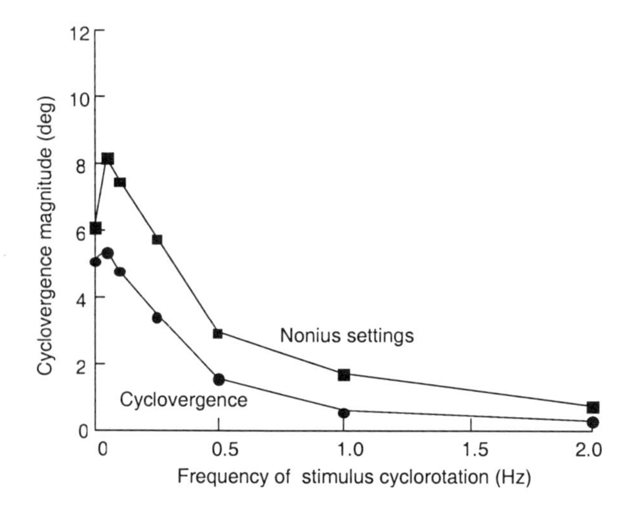

Figure 9.48. Measures of cyclovergence.
Objectively measured cyclovergence and nonius-line settings as a function of the frequency of cyclodisparity modulation of the display in Figure 9.47 (N=3). (Adapted from Howard *et al.* 1993)

corded cyclovergence of about 5°, induced by disjunctive rotation of dichoptic textured displays through 8°. Verhoeff (1934) also used a nonius method and found cyclovergence to be a slow response with magnitudes of up to 6° induced by an 8° disjunctive rotation of textured patterns. He also found that a greater amplitude of cyclovergence is induced by textured patterns than by simple line patterns, and more by cyclodisparity of horizontal lines than of vertical lines. Hermans (1943) used a nonius method to determine the position of the vertical horopter. Although subject to artifacts, the nonius method is the only satisfactory psychophysical method for measuring cyclovergence.

Howard *et al.* (1993) superimposed a pair of nonius lines on the centre of a 75° dichoptic textured display (Figure 9.47). The two images had a static cyclodisparity of 12° or were rotated sinusoidally in antiphase through an amplitude of 12°, at frequencies between 0.05 and 2 Hz. Subjects nulled the apparent offset of the nonius lines in the static display and nulled their rocking motion in the dynamic display. Figure 9.48 shows that, for static cyclodisparity, the nonius setting was slightly higher than the magnitude of cyclovergence measured objectively, using scleral search coils. When the display rotated back and forth, the nonius lines appeared to rock through a greater amplitude than predicted from the magnitude of cyclovergence. With one eye closed, the remaining nonius line appeared to rock in a direction opposite to that of the rotating surround. This is the well-known phenomenon of induced visual motion. When both eyes were open, the two monocularly induced motion effects combined with the effects of cyclovergence to create the large apparent rocking motion of the nonius lines. Thus, the nonius method is a reasonably valid measure of cyclovergence, but only for static or slowly changing disparities.

9.8.2d *Objective recording of cyclovergence*

Methods for recording eye movements, such as electrooculography and the Purkinje eye tracker, do not record eye torsion. Photographs of the iris or episcleral blood vessels must be analysed frame-by-frame (Howard and Evans 1963). Video records of the two irises can be subjected to autocorrelation to yield a continuous record of cyclovergence. Kertesz (1972) used the photographic method but failed to find cyclovergence because his stimulus was too small. Crone and Everhard-Halm (1975) and Hooten *et al.* (1979) used the photographic procedure with a more adequate stimulus and obtained clear evidence of cyclovergence.

A scleral search coil mounted on an annular contact lens was developed by Collewijn *et al.* (1985), and is available from Skalar Medical in Delft. When the coil is placed on an eye within an oscillating magnetic field, a voltage proportional to the sine of the torsional position of the eye is generated (Robinson 1963; Collewijn *et al.* 1975; Ferman *et al.* 1987b). This method provides a low-noise signal that continuously registers the torsional position of an eye to within a few minutes of arc. The only drawback is that the contact lenses can be worn for only about 30 minutes at one time. Kertesz and Sullivan (1978) used the scleral-coil procedure and obtained a cyclovergence of 3.5° to a ±5° step of cyclodisparity in dichoptic patterned displays subtending 50°.

9.8.3 Dynamics of cyclovergence

Howard and Zacher (1991) used scleral search coils to measure the gain and phase lag of cyclovergence as a function of the frequency and amplitude of cyclodisparity of the 75° dichoptic display shown in Figure 9.47. This stimulus contains a broad range of spatial frequencies, has both vertical and horizontal elements to act as cyclofusional stimuli, and is a good stimulus for keeping horizontal and vertical vergence constant. A regular grid pattern is not suitable because the eyes tend to misconverge on such stimuli. Since the display was circular and the surroundings black, there were no stationary lines to provide a cyclofusional anchor. The dichoptic displays were rotated in counterphase about the fixation point to give peak-to-peak amplitudes of disjunctive cyclodisparity of 2, 6, or 12°, at frequencies of from 0.05 to 2 Hz. The gain of cyclovergence was defined as the mean peak-to-peak amplitude of

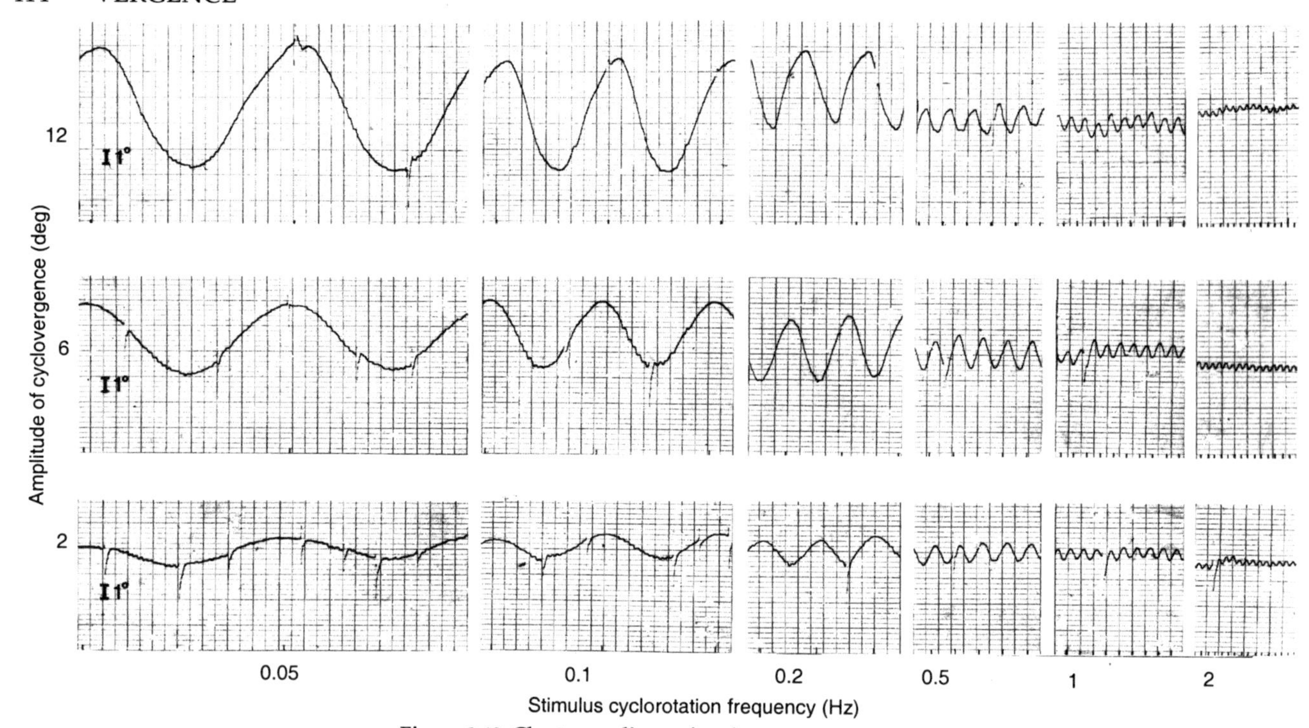

Figure 9.49. Chart recordings of cyclovergence.
Cyclovergence for different frequencies and amplitudes of cyclo-disparity of the stimulus shown in Figure 9.47. The traces represent the difference between the opposed cyclorotations of the eyes. Sharp impulses are blinks. (From Howard and Zacher 1991. Reprinted with permission of Springer-Verlag, Heidelberg)

cyclovergence divided by the peak-to-peak amplitude of the cyclodisparity of the stimulus. A sample of recordings is shown in Figure 9.49. Gain declined with increasing stimulus frequency. Phase lag was imperceptible at a frequency of 0.05 Hz, and increased with increasing frequency of cyclodisparity, reaching values of over 100° at a frequency of 2 Hz (Figure 9.50).

Cyclovergence of 2 or 3° maintained for periods up to 150s took about 5s to decay in the dark (Taylor *et al.* 2000).

Cyclovergence is designed to cope with cyclodisparities of low frequency and amplitude, as indicated by the fact that, in Figure 9.50, gain is highest and phase lag lowest at a frequency of 0.05 Hz and amplitude of 2°. For one young adult, the gain of cyclovergence reached 0.91 at this frequency and amplitude. Van Rijn *et al.* (1992) obtained a maximum gain of only 0.2, but they used only one frequency of stimulus rotation of 0.2 Hz and a stimulus diameter of only 28°. In a recent paper, they obtained a mean gain of over 0.4 when they used a 48°-wide display oscillating at 0.15 Hz (van Rijn *et al.* 1994a). We will see in the next section that the gain of cyclovergence declines rapidly as stimulus area is reduced.

The dependence of cyclovergence gain on the amplitude of stimulus cyclorotation demonstrates that the system is nonlinear, because the gain of a linear system is independent of stimulus amplitude for a given frequency. For a stimulus amplitude of 6°, the function relating gain in decibels to frequency has a slope of 20 db/decade for the five highest frequencies (gain in decibels is 20 times the log of response amplitude divided by the logof stimulus amplitude). This is the value expected of a first-order system. However, the phase lag at the corner frequency corresponding to a gain of -3 db was much smaller than expected from either a first-order or second-order system.

The high gain and low phase lag for low stimulus frequencies and small amplitudes is what one would expect of a system designed to correct for slight rotary misalignments of binocular images. Misalignments may be produced by cyclophoria or by torsional drifts of the eyes that occur as the gaze moves over a 3-D scene. Cyclovergence does not have to deal with rapid external events. Disjunctive cyclorotations of the distal visual stimulus as a whole occur only under very special circumstances as, for instance, when one observes an isolated vertical line changing its inclination. Under normal circumstances, orientation disparities produced by inclined lines or surfaces, occur in the context of other objects that are either upright or inclined in another way.

Spontaneous variation in cycloversion during fixation had a standard deviation of about 0.2°, which is higher than the standard deviation of be-

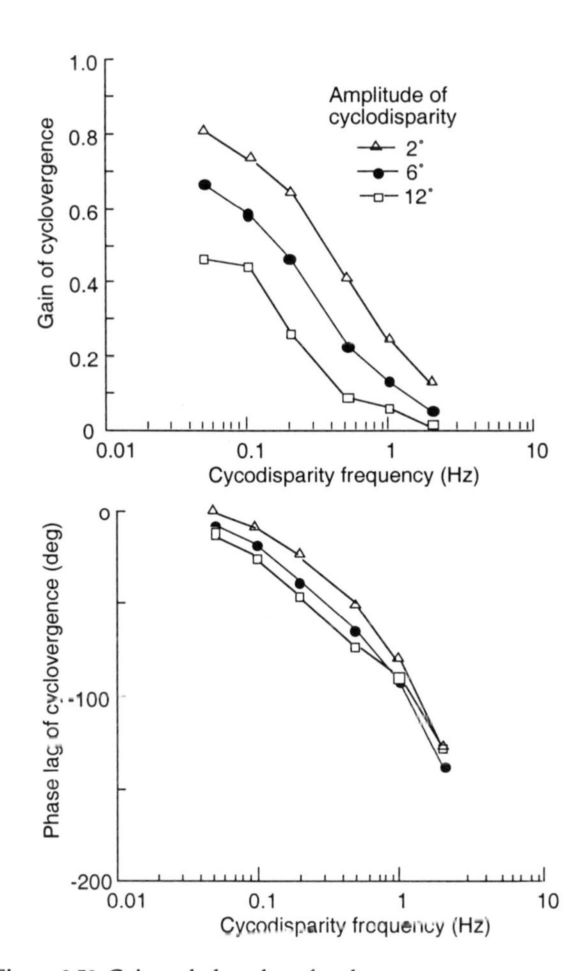

Figure 9.50. Gain and phase lag of cyclovergence.
Gain and phase lag of cyclovergence as a function of frequency of cyclodisparity for three amplitudes of disparity (N=3). (From Howard and Zacher 1991)

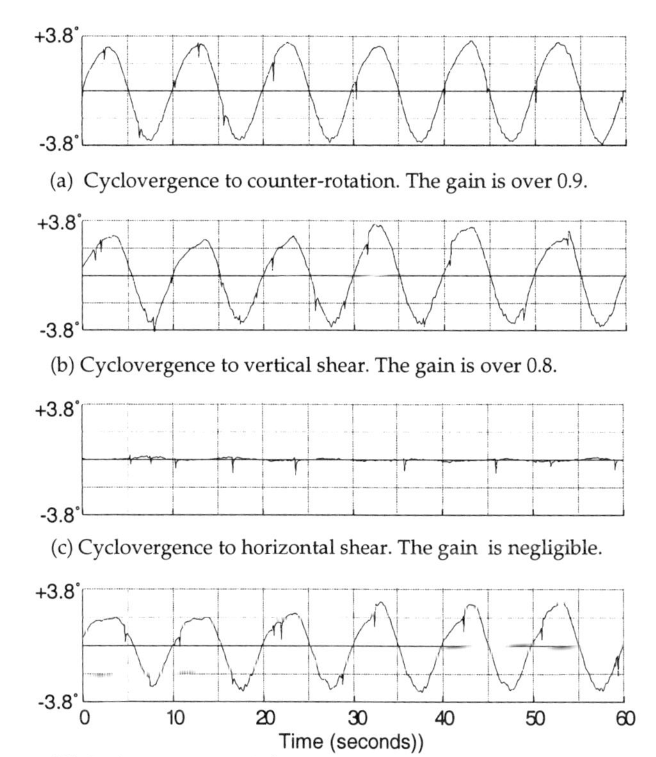

Figure 9.51. Cyclovergence to rotating and shearing patterns. (Redrawn from Rogers 1992)

tween 6 and 8 arcmin for conjugate horizontal and vertical movements of the eyes (Ferman *et al.* 1987b; Ott *et al.* 1992). Spontaneous changes in cyclovergence had a standard deviation of about 6 arcmin with fixation on an isolated point in the dark, a stimulus lacking error feedback. Spontaneous changes were only about half this value when a textured background was present (Enright 1990; van Rijn *et al.* 1994a).

9.8.4 Cyclovergence and angle of gaze

When the eyes are converged and the gaze elevates, the eyes execute an incyclovergence, and when the gaze is lowered they execute an excyclovergence (Section 9.2.2). Converging between two points of light at eye level or a horizontal change of gaze produces no significant cyclovergence (Minken and van Gisbergen 1994). Models of these responses have been developed by van Rijn and van den Berg (1993) and by Minken *et al.* (1995).

Minken *et al.*'s model accounts for effects of vergence on cyclovergence. Thus, Mok *et al.* (1992) found that with the eyes converged 30°, 30° of gaze elevation produced 5° of incyclovergence and 30° of gaze depression produced 5° of excyclovergence. The same results were obtained when subjects converged to remembered targets in the dark. Porrill *et al.* (1999b) also found that cyclovergence accompanying changes in vergence and gaze elevation conformed Minken *et al.*'s model.

Changes in cyclovergence with gaze elevation represent a departure from Listing's law and serve to keep the horizontal meridians of the two eyes in alignment, as explained in Section 9.2.2f. To some extent, these eye movements can be explained in terms of the way muscles are attached to the eyeball. However, there must also be some neural control because the magnitude of cyclovergence induced by a given elevation of the eyes for a given angle of horizontal convergence can be increased or decreased about 1.5° by exposing subjects to an unusual degree of cyclodisparity during gaze elevation (Schor *et al.* 2001). The change generalized, but not fully, to angles of gaze not used during training.

Changes in cyclovergence induced by changes in horizontal vergence with elevated asymmetric gaze show less intra- and intersubject variability than associated changes in cycloversion (Ivins *et al.* 1999).

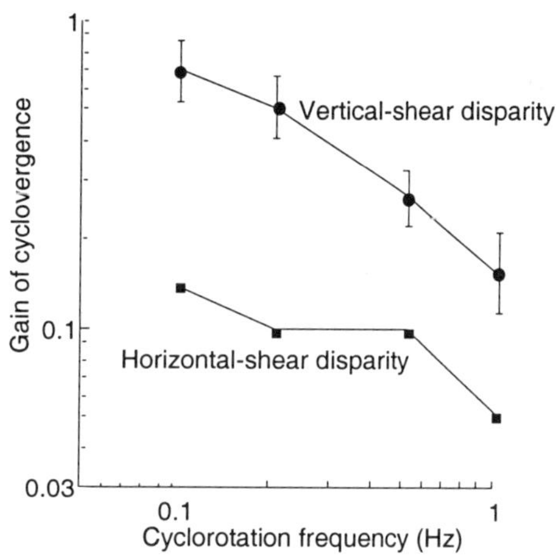

Figure 9.52. Cyclovergence anisotropy.
Gain of cyclovergence as a function of the frequency of disjunctive cyclorotation of vertical lines and of horizontal lines (N=4). (Redrawn from Rogers and Howard 1991)

Precise control of cyclovergence is more important for binocular vision than precise control of cycloversion. The stimulus used in these experiments was an isolated point of light, which provided no error feedback. Evidence reviewed in the next section suggests that cyclovergence is modified in the presence of a large visual display.

Cyclovergence is also required to correct for a tendency for the eyes to become misaligned as the head is tilted to one side. An unusual pairing of head tilt and cyclodisparity can lead to an adaptive change in cyclovergence (Section 9.3.6c).

9.8.5 Visual stimulus for cyclovergence

(This section was written with Brian J. Rogers)

9.8.5a *Horizontal and vertical shear disparities*

An orientation disparity between the images of lines in the horizontal plane of regard can be due only to eye misalignment, whereas an orientation disparity from a vertical line may be due to inclination of the line in depth. It would be a good strategy if cyclovergence were evoked only by disparities in horizontal elements, leaving residual disparities in vertical elements as cues for inclination. This strategy would prevent the occurrence of diplopia in horizontals and reduce useful disparities to values that can be detected most effectively.

Ogle and Ellerbrock (1946) claimed that more cyclovergence is evoked by cyclodisparity of verticals than of horizontals. However, they measured cyclovergence by setting a line to the vertical, which we have already seen is subject to artifacts. The more reliable nonius method revealed cyclodisparity of horizontals to be the more effective stimulus (Nagel 1868; Verhoeff 1934; Crone and Everhard-Halm 1975).

Rogers and Howard (1991) used scleral search coils to measure the gain of cyclovergence to textured patterns filling the binocular field. The patterns were subjected to either, (1) equal and opposite rotation between the two eyes, (2) equal and opposite vertical shear, (3) equal and opposite horizontal shear, or (4) equal and opposite deformation. The vertically shearing patterns created cyclodisparities along the horizontal meridians but not along vertical meridians and the horizontally shearing patterns created the opposite pattern of cyclodisparities.

Figure 9.51a is the record for one subject in response to opposite sinusoidal rotation of densely textured dichoptic patterns at a frequency of 0.1 Hz. Cyclovergence has an amplitude gain of over 0.9. The gain of cyclovergence for vertical-shear disparity (Figure 9.51b) was only slightly lower than that for rotation disparity. In contrast, the gain for horizontal-shear disparity (Figure 9.51c) was negligible. Figure 9.51d shows a gain of over 0.6 to a deforming pattern that created the same magnitude and direction of cyclodisparities along horizontal meridians as the rotating pattern, but the opposite direction of cyclodisparities along vertical meridians (Rogers and Howard 1991; Rogers 1992). Figure 9.52 shows that cyclovergence evoked by vertical-shear disparity has a higher gain than that evoked by horizontal-shear disparity over a range of frequencies of cyclorotation of the stimuli. This result was confirmed by van Rijn *et al.* (1994b) and by Taylor *et al.* (2000).

These results demonstrate that cyclovergence is driven primarily by vertical-shear disparity along horizontal meridians and minimally by horizontal-shear disparity along vertical meridians. Overall vertical-shear disparity is the more reliable indicator of eye misalignment because it does not occur under other circumstances. An overall horizontal-shear disparity is produced by an inclined plane. It is best not to have cyclovergence evoked when viewing an inclined plane because the response would eliminate the disparity cue to inclination.

Frisby *et al.* (1993) questioned whether vertical-shear disparity is a more effective stimulus for cyclovergence in real, rather than stereoscopic surfaces. They measured cyclovergence using vertical nonius lines while the subject viewed a surface inclined at a large angle to the vertical. Differences in the orientation of the nonius lines were equivalent to 50% of the horizontal-shear disparities created by the inclined surface. Since an inclined surface does not create a net vertical-shear disparity, it ought not to evoke cyclovergence, according to the results of Rogers and Howard and van Rijn *et al.*

Rogers and Bradshaw (1999) repeated the experiment using both 20 and 80° surfaces, which changed their inclination at 0.1 Hz. Cyclovergence was measured by scleral search coils. The results showed no significant cyclovergence and confirmed that cyclovergence is not driven by horizontal-shear disparity even when created by a real inclined surface. They found no misalignment of horizontal nonius lines, which confirms that the eyes do not change their cyclovergence to an inclined surface, but vertical nonius lines did appear misaligned. They suggested that this effect was due to the nonius lines appearing as part of the inclined surface. The images of a line on an inclined surface have an orientation disparity, but we do not see tilted images but one line inclined in depth. If the top half of the line is seen by one eye and the bottom half by the other eye and if each half is seen in the plane of the surface, the lines will appear to differ in orientation, which will affect the nonius alignment. The misalignment of vertical nonius lines reported by Frisby *et al.* was probably an artifact and not due to cyclovergence. Artifacts in the use of nonius lines were discussed in Section 9.3.4b.

The stimulus for cyclovergence could be the horizontal gradient of vertical disparities along horizontal meridians. Otherwise, cyclovergence could be evoked by detectors specifically sensitive to orientation disparities between horizontal elements, which are the same angular magnitude at all eccentricities.

DeBruyn *et al.* (1992) tested this latter possibility with a display consisting of dynamic, horizontal random-noise gratings, which were either correlated or uncorrelated between the two eyes. The gratings were cyclorotated at 0.1 Hz in opposite directions in the two eyes and torsional eye movements were recorded with scleral search coils. Correlated noise gratings, which created both orientation disparities and a horizontal gradient of vertical disparity were an effective stimulus for driving cyclovergence. Uncorrelated noise gratings, which contained no point disparities, produced no cyclovergence, even though the change in orientation of the bars of the grating could clearly be seen in either monocular image.

9.8.5b *Effects of stimulus area*

Howard *et al.* (1994) used scleral search coils to measure the gain of cyclovergence as a function of the area and position of the stimulus. In one set of conditions, circular textured displays with diameters of 5, 20, 40, and 75° were used and in a second set of conditions, black discs occluded the central 5, 20, or 40° of the 75° circular display. Each display was cyclorotated through a peak-to-peak amplitude of 12° at 0.05 and 0.2 Hz. It can be seen from Figure 9.53a that a 5° display evoked weak cyclovergence, and that the gain of the response improved as the diameter of the display increased from 20 to 75°. Kertesz and Sullivan (1978) also reported an increase in the gain of cyclovergence for one subject as stimulus diameter was increased from 10° to 50°. On the other hand, it can be seen in Figure 9.53b that the gain of cyclovergence was not reduced when the central 40° of the stimulus was occluded. The gain of optokinetic torsional nystagmus induced by conjugate rotation of the display was severely reduced by occlusion of the central 40°, and previous studies have shown that the same is true of horizontal optokinetic nystagmus (Howard and Ohmi 1984).

One may conclude that cyclovergence requires a large stimulus, but this does not have to be in the centre of the visual field. If cyclovergence is driven by point disparities, then this might explain the need

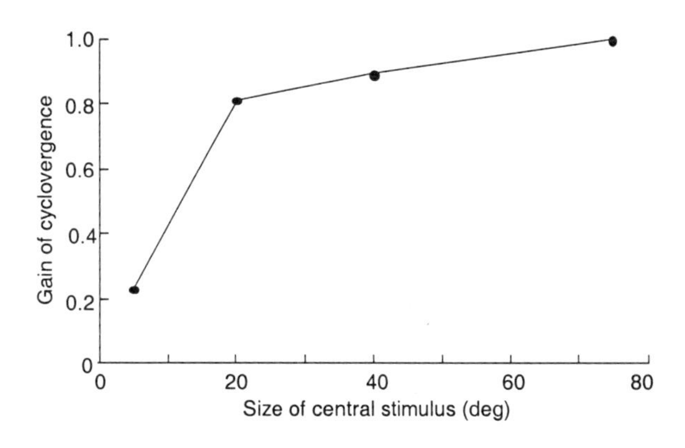

(a) Gain of cyclovergence normalized to the gain of the largest display as a function of display diameter.

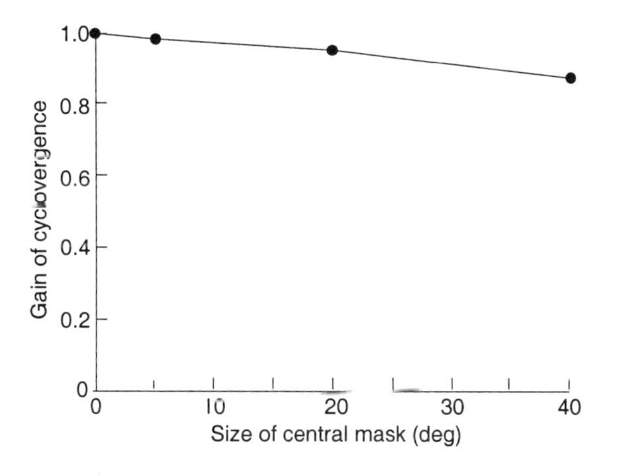

(b) Gain of cyclovergence normalized to the gain of the largest display as a function of the diameter of a central black occluder superimposed on the display.

Figure 9.53. Cyclovergence and display area.
The images cyclorotated disjunctively at 0.05 Hz. The gain for the largest display was 0.64 (N=4). (From Howard *et al.* 1994)

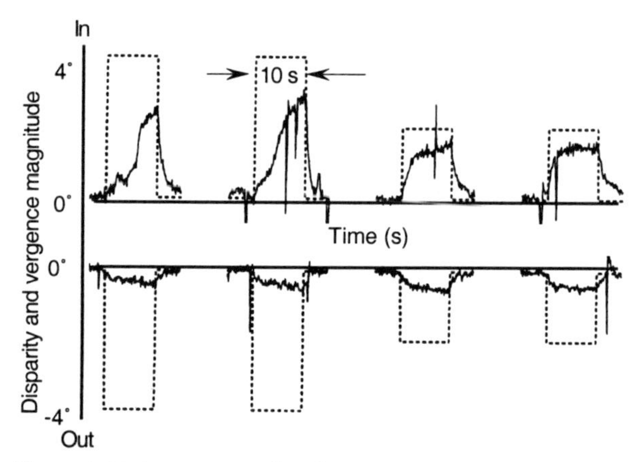

Figure 9.54. Asymmetry of cyclovergence.
Records of cyclovergence of four subjects in response to a ±4.6° or ±2.3° step in vertical-shear disparity of a 60° textured display. The stimulus (dotted lines) was maintained for 10s. The magnitude of incyclovergence was greater than that of excyclovergence in all subjects. (From Howard and Kaneko1994).

for a large stimulus and the indifference to occlusion of the central retina, since point disparities in a display rotating about the visual axis increase linearly with stimulus eccentricity. If cyclovergence is driven by orientation disparity, the preceding results demonstrate that it is driven only by detectors with large receptive fields. This makes sense, since the only purpose of cyclovergence is to keep the images of the main horizontal features of a sccene orientationally aligned. Residual disparities in vertical elements code differential inclinations of particular objects, especially those in the centre of the visual field.

9.8.5c *Asymmetry of in- and excyclovergence*
Howard and Kaneko (1994) found incyclovergence was larger than excyclovergence in their four subjects (see Figure 9.54). No subjects were found with the opposite asymmetry. Taylor *et al.* (2000) reported the same asymmetry in their two subjects. This asymmetry may be due to the natural predominance of horizontal surfaces below eye level.

9.9 VERGENCE-VERSION INTERACTIONS

9.9.1 Hering's law of equal innervation

9.9.1a *Law of equal innervation*
When fixation changes rapidly between two points differing in both depth and direction, vergence is combined with a saccade. When the eyes track a spot moving slowly between points differing in depth and direction, vergence is combined with slow pursuit. The following discussion is about how vergence combines with these two types of version.

Movements of a single eye, or ductions, are specified in terms of magnitude, direction, and velocity. Co-ordinated movements of the two eyes are specified in terms of version and vergence components. Version is an equal movement of the eyes in the same direction, and vergence is an equal movement in opposite directions. A circle passing through the centres of rotation of the two eyes and the fixation point is the path traced by the fixation point as vergence changes with version held constant. A hyperbola of Hillebrand is the path traced by a change in vergence with version held constant (see Figure 9.7). Mathematically, any movement of the two eyes may be described as the sum of a version and a vergence. Let θ be the version component with movements towards the right signed positive and let μ be the vergence component with convergence signed positive. For a pure vergence, the movement of each eye equals $\mu/2$. In general (after Ono 1980)

$$\theta + (\frac{\mu}{2}) = \text{rotation of the left eye} \quad (5)$$

$$\theta - (\frac{\mu}{2}) = \text{rotation of the right eye} \quad (6)$$

Alhazen, in the 11th century, proposed that the movement of one eye is accompanied by a movement of the other eye of equal amplitude and velocity, either in the same or in the opposite direction. This idea was developed by Hering (1868) and stated the **law of equal innervation.**

Hering wrote, "The two eyes are so related to one another that one cannot be moved independently of the other: rather, the musculature of both eyes reacts simultaneously to one and the same impulse of will."(Hering 1868, p. 17). hering did not mean that one eye cannot move whiel the other eye remains statioanary. He went on to write:

"*. . . it is possible for us to move both eyes simultaneously about different angles and with different speeds . . . and even to move one eye outward or inward while the other remains still. We are able to do this, not because we simultaneously give each eye a special innervation, but because in these movements each eye recveives two different innervations. One is a turning movement of both eyes to the right or left and the other is inward or outward turning of both eyes. Since thse two innervations of the two eyes work together in one eye and conversely in the other, the resultant movement in each eye must necessarily be different.*"

Consider the idealized case, in which the gaze shifts from point A to point B, both lying on the visual axis of the left eye. This movement may be decomposed into an equal version signal for both eyes of θ and a vergence signal which, in this case, moves the right eye through angle $-\mu/2$ and the left eye through angle $\mu/2$. By equations (4) and (5) they cancel for the left eye and add for the right eye. The right eye therefore does all the moving. People can be trained to move an occluded eye horizontally while keeping the other eye fixated on a stationary target (Manny 1980).

Hering's law would be a tautology if it merely stated that all eye movements may be described as the sum of a version and a vergence component. It is clear from Hering's use of the phrase "equal innervation" that he was thinking of component neural processes not merely mathematical components.

It is not the amplitude or velocity of the movements of the two eyes that are equal in Hering's law, but the amplitude and velocity of the vergence component in each eye and of the version component in each eye. One can erect the hypothesis that, when version and vergence are executed simultaneously, eye velocity is a linear sum of the velocities of each component movement acting alone. Empirical evidence bearing on this version of Hering's law is discussed in Section 9.9.2. Hering's law should perhaps be called the **law of equal component innervations**.

Hering's law requires that, at any instant, the visual object that evokes version is also the object that evokes vergence. Chaturvedi and van Gisbergen (1998) presented two target objects at the same time in different positions in 3-D space. Whichever object evoked the first saccadic response was almost always the object that evoked the vergence response. When the saccadic response was to a compromise position, vergence showed a similar compromise.

9.9.1b *Neurology of Hering's law*

Hering's law implies that there are two neural centres, one for vergence and one for version, and that each centre sends the same the innervation to the two eyes. The simplest assumption is that these innervations are combined linearly in the final common path so that the movement of each eye is the algebraic sum of the innervations from the two centres. But Hering's law does not require a linear combination of version and vergence signals; it only requires that the version signals remain equal and the vergence signals remain equal. For instance, it does not forbid version signals from being attenuated when combined with vergence signals.

Conjugate eye movements of all types in the horizontal plane are organized in the paramedian pontine reticular formation, or **PPRF**. Conjugate eye movements in the vertical plane are organized in the mesencephalic reticular formation, or **MRF**. These nuclei receive inputs form the superior colliculus, vestibular nucleus, and the frontal eye fields and project monosynaptically to motoneurone pools in the oculomotor nuclei. Each motoneurone pool innervates almost exclusively one extraocular muscle in one eye. Those in the trochlear nucleus innervate the contralateral superior oblique muscle, those in the abducens nucleus innervate the ipsilateral lateral rectus, and motoneurone pools in the oculomotor nucleus innervate the ipsilateral medial rectus, inferior oblique, inferior rectus, and contralateral superior rectus (Evinger 1988). Moschovakis *et al.* (1990) recorded from premotoneurones responsible for vertical saccades and found that they branch to innervate all the motoneurone pools that move both eyes conjugately. Moschovakis (1995) traced the axonal terminations of premotor lead burst neurones responsible for initiating horizontal and vertical conjugate saccades. In each case, the axons terminated in the set of motoneurones that controlled the movements of both eyes. This provides a basis for Hering's law at least for conjugate saccades.

Smooth pursuit eye movements seem to be organized by separate commands to the eyes. King and Zhou (1995) found that the pursuit movement of each eye during the initial open-loop 100 ms was controlled only by motion of the visual target in that eye, whether the eye movement was conjugate or disjunctive. In their experiment, any contribution of disparity-induced vergence to disjunctive pursuit must have had a latency longer than 100 ms but presumably obeyed Hering's law.

Hering believed that axons from distinct motoneurones for version and vergence combine in the muscles. The distinct axons could converge on the same muscle fibres, or different muscle fibres could be devoted to each component. Hering believed in the second possibility and thought he detected opposed contractions in the muscles of the stationary eye during a change of fixation along that eye's visual axis. Tamler *et al.* (1958) claomed to have detected these contractions electromyographically in humans. Others found no changes in electrical activity of muscles of a stationary eye under these conditions (Breinin 1955; Blodi and Van Allen 1957; Allen and Carter 1967). Co-contractions could be due to slight saccadic movements when vergence changed rapidly along the line of sight of one eye, or to torsional movements that accompany changes in vergence. More recent evidence suggests that co-contraction of lateral and medial recti during convergence is due to inappropriate version signals arising in the abducens nuclei, which are overcome by appropriate signals arising from centres controlling vergence (Gamlin *et al.* 1989).

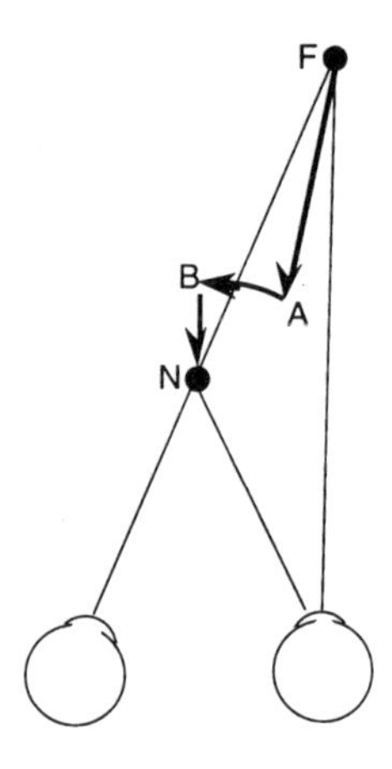

Figure 9.55. Hypothetical path of vergence-version.
Hypothetical path of the point of fixation from F to N. The eyes first converge to A, then execute a conjugate saccade to B, and finally complete the vergence movement to point *N*.

Enright (1980) revealed that there is also a slight lateral translation of an eye when vergence changes along that eye's visual axis. These torsional and translatory movements are probably incidental to the version and vergence components and have no functional significance.

O'Keefe and Berkley (1991) produced evidence of a coupling of the movements of the two eyes mediated by proprioception. Infusion of a paralytic agent into the muscle capsule of one eye in the anaesthetized cat reduced the spontaneous changes in position of both eyes. Without the paralytic agent, neither passive movement of one eye nor application of a local anaesthetic into the muscles of one eye had any effect on the other eye. It was suggested that the eyes are coupled by signals arising in proprioceptors in the extraocular muscles and that these signals are gated by efferent signals.

9.9.2 Vergence-version additivity

9.9.2a *Basic studies*

Hering's law does not stipulate that version and vergence components are combined linearly but only that any nonlinearity applies equally to the two eyes. However, one can hypothesize that, when version and vergence are executed simultaneously, eye velocity is a linear sum of the velocities of each component movement acting alone. This could be called the strong version of Hering's law. This hypothesis can be tested under two conditions: when vergence is combined with slow voluntary pursuit, and when it is combined with conjugate saccades.

People can pursue a point of light executing a closed path in a horizontal plane—a task requiring coordination of version and vergence. However, when the target moves at more than about one cycle per second, the vergence component of the tracking motion breaks down leaving only the version component (Rashbass and Westheimer 1961b). The dynamic limits of the two systems therefore differ. Semmlow *et al.* (1998) found that the velocity of slow vergence tracking is more variable than slow version tracking and that the variability of combined vergence-version tracking is predictable by linear combination. They, also, concluded that at least slow version and vergence are controlled by different mechanisms with distinct dynamics.

Additivity of vergence and slow version has also been investigated by introducing step changes in the disparity of a target that subjects were pursuing slowly along a horizontal track in the frontal plane (Miller *et al.* 1980). The velocity of an eye for which the version and vergence signals were opposite in direction was the linear sum of the component velocities. In this case, the component innervations were delivered to distinct extraocular muscles. When the version and vergence signals drove the eye in the same direction, an 11% loss in additivity occurred. In this case, the combined innervations impinged on the same muscles, and one could explain the loss in additivity in terms of a compressive nonlinearity at the neuromuscular junction. In both cases, the innervations themselves may well have been combined linearly in the oculomotor nuclei from independent command centres.

According to the traditional account, a version made in response to a stepped stimulus is normally saccadic and occurs with a latency of about 200 ms and a velocity of up to 600°/s. Vergence has a latency of about 150 ms and a peak velocity of about 20°/s. Accordingly, in a rapid eye movement involving both vergence and version, vergence should start before version, after which the required version should be achieved quickly, leaving the rest of the slow vergence to be completed. If we assume for the sake of simplicity that the stimulus is stepped along the visual axis of the left eye and vergence is inhibited during the time that version is occurring, the resulting eye movements should be as depicted in Figure 9.55 At the start, pure vergence causes binocular gaze to track along a hyperbola of Hillebrand. Then version kicks in and gaze moves along an isovergence locus. Finally, the residue of vergence again carries the gaze along a hyperbola of Hillebrand back onto the visual axis of the left eye. Note that the law of equal component innervations holds throughout this sequence. With a sequential programming of version and vergence, the amplitudes and velocities of the eyes are always equal, not merely the amplitudes and velocities of the component commands.

Several investigators have obtained records of eye movements that showed the two components occurring sequentially in the general manner just described (Alpern and Ellen 1956; Westheimer and Mitchell 1956; Yarbus 1967). Findlay and Harris (1993) found this type of movement only when the images of the dichoptic target were in the same retinal hemifields. Ono and Nakamizo (1978) found this pattern only when vergence and accommodation were disassociated or when the target moved down the visual axis of one eye. In other cases, Ono and Tam (1981) found multiple saccades, especially for large vergence changes.

Pickwell (1972) also found that the relative magnitudes of version and vergence components depended on whether the stimulus was aligned with the left or right visual axis. This asymmetry was explained in terms of a displacement of the cyclopean eye (Section 17.7.4) towards the dominant eye. Barbeito *et al.* (1986) confirmed this asymmetry but produced evidence that displacement of the egocentre is not the only causative factor. Furthermore, Enright (1998a) found that the extent of the asymmetry was not correlated with the position of the egocentre (Section 17.7.2).

Other investigators found no evidence of a clear separation in time between vergence and version components (Erkelens *et al.* 1989b). Enright (1996b) found that most subjects could make smooth vergence movements between targets in the median plane or between targets lying on the visual axis of one eye. Whether or not this is regarded as breaking Hering's law depends on how the law is interpreted.

When an accommodative change is induced in one eye by a lens, with the other eye closed, the closed eye moves but the open eye moves only slightly or not at all (Müller 1826; Kenyon *et al.* 1978; Enright 1992a). Saida *et al.* (2001) found that both eyes showed an initial response to the introduction of the lens before one eye, but that the open eye stopped moving while the closed eye continued to move. However, under open-loop conditions, in which motion of the visual target was optically stabilized in the open eye, both eyes moved equally through half the amplitude shown by the closed eye in the closed-loop condition. This suggests that, with normal viewing, the open eye maintains steady gaze on the target by error feedback from retinal slip. When this error feedback is removed by opening the feedback loop, both eyes share in the vergence response induced by the change in accommodation.

Recent evidence, which I now review, suggests that the control of vergence, especially asymmetrical vergence, is more complex than a simple sequential combination of rapid version and slow vergence.

In the first place, the saccadic component is not switched off while vergence is occurring. There is thus the question whether the saccadic and vergence components combine additively when they occur together. Ono *et al.* (1978) found that during the convergence phase of an eye movement evoked by a target that stepped in both lateral direction and depth, the differences of amplitude and velocity between the two eyes were much greater than predicted from a linear addition of component velocities. A similar supra-additivity of components occurred when a step change in the lateral direction of gaze was superimposed on a tracking vergence made in response to a target moving slowly in depth (Saida and Ono 1984). Similarly, when accidental microsaccades occurred during a vergence movement, the difference in velocity between the two eyes was greater than predicted by simple addition (Kenyon *et al.* 1980b). Collewijn *et al.* (1997) concluded from their own data that version and vergence components of composite eye movements can be executed separately during part of the movement but that in most eye movements the two components overlap in time and interact in a nonlinear fashion.

There is the related question of crosstalk between version and vergence (Ono 1983). Is the response to a stimulus requiring pure version devoid of a vergence component, and is a required pure vergence devoid of a version component?

9.9.2b *Vergence intrusions in saccades*

Smith *et al.* (1970) detected a significant lack of synchrony between the eyes in conjugate saccades. Bahill *et al.* (1976) noted that subjects with normal vision sometimes execute unequal saccades in the two eyes, but they regarded disconjugate saccades as anomalous. Williams and Fender (1977) found that, in spite of claims to the contrary, voluntary saccades are almost completely synchronized in the two eyes. Saccadic disconjugacy is more marked in young children than in adults (Fioravanti *et al.* 1995).

Collewijn *et al.* (1988a) found that, in horizontal saccades along a locus of isovergence, the motion of the abducting eye had a larger amplitude, higher peak velocity, and shorter duration than that of the adducting eye (towards the nose). As a result the eyes diverged transiently by up to 3° and a postsaccadic drift of the eyes was required to bring the gaze onto the target. The magnitude of transient divergence increased with increasing magnitude of horizontal saccades and with increasing angle of vergence (Collewijn *et al.* 1997).

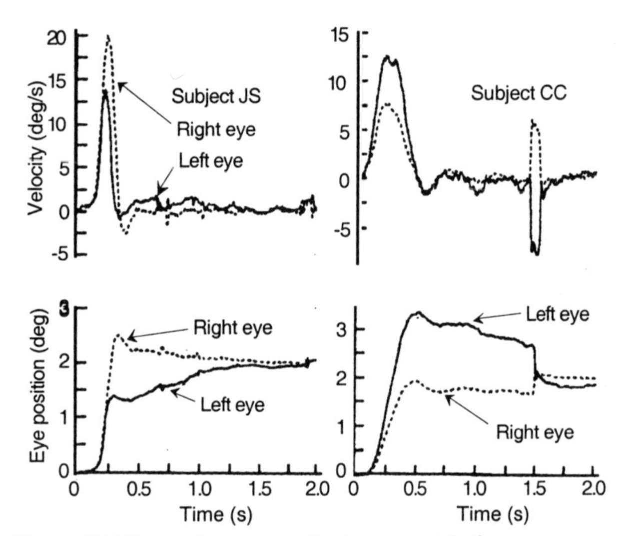

Figure 9.56. Dynamic asymmetries in symmetrical vergence. Upward displacements of both traces signify convergence, opposite displacements signify version. Upper traces show left and right eye velocities and lower traces show eye positions. The stimulus was a 2°-high dichoptic vertical line that stepped from 7° to 9° of convergence. (From Horng *et al.* 1998b)

Vergence intrusions into saccades could be due to any of the following causes.

1. They could be due to the fact that the elastic resistance of an eye to nasalward motion is greater than to temporalward motion (Collins *et al.* 1981). Zee *et al.* (1992), also, ascribed intrusions of vergence into horizontal saccades to asymmetries in the mechanical properties of the lateral and medial recti muscles.

2. Phoria could occur when fusion is lost temporarily during a saccade (Kapoula *et al.* 1987). It is more marked in strabismics, especially in strabismics lacking stereopsis (Kapoula *et al.* 1997). Furthermore, strabismics lacking stereopsis do not adapt their relative saccadic amplitudes to imposed aniseikonia, in the way that people with normal vision do (Kapoula *et al.* 1995; Bucci *et al.* 1997).

3. The horizontal horopter, defined as a circle through the nodal points of the two eyes, does not coincide with the locus of isovergence, defined as a circle through the centres of rotation of the two eyes. This means that a target on the locus of isovergence would have a crossed disparity when the eyes are in the straight-ahead position. This would induce convergence rather than divergence, but perhaps the horizontal horopter was closer than the isovergence locus at the viewing distances used in experiments involving saccades within isovergence loci.

4. Saccades into the lower visual field were found to be associated with transient convergence while saccades into the upper field were associated with much smaller transient convergence or a small transient divergence (Enright 1989a; Collewijn *et al.* 1988b; Zee *et al.* 1992). The stimuli were positioned so that a vergence change was not required. A post-saccadic corrective vergence brought the eyes onto the newly acquired target. Vergence intrusions could be due to the general tendency for objects below eye level to be near and those above eye level to be far away. However, the situation is complicated by the fact that the vertical horopter is inclined top away, which means that isovergence targets in the upper half of the visual field have a crossed disparity and those in the lower half an uncrossed disparity (Section 15.7). Vergence movements associated with vertical saccades may simply be induced by these disparities.

5. Signals to the two eyes from the oculomotor nuclei could have different dynamics. They could differ in latency, rise time, or duration. Sylvestre *et al.* (2002) measured vergence intrusions into saccades made by monkeys between eye-level targets at optical infinity. This procedure should have eliminated factors (3) and (4). Saccades with a symmetrical velocity profile showed vergence intrusions consisting of divergence followed by convergence. Saccades of larger amplitude had larger peak velocities of both the conjugate and the disconjugate components. Saccades with more complex velocity profiles showed oscillatory vergence intrusions. They concluded that the intrusions arise from differential dynamics of the conjugate and disconjugate oculomotor systems.

6. The neural system for vergence could interact with that for version.

9.9.2c *Saccadic intrusions into vergence*

Enright (1984, 1986b) reported that when subjects make rapid changes in fixation, involving both vergence and version, they make short-latency saccadic movements, which are unequal in magnitude in the two eyes. Between 40 and 70% of the required vergence was achieved by these asymmetrical saccadic movements. Enright looked for saccades only in asymmetrical vergence movements. Levi *et al.* (1987) found saccades of unequal amplitude and, in a few cases, saccades in opposite directions in symmetrical vergence, especially in the early part of divergence movements. Corrective conjugate saccades often occurred at the end of vergence movements. Horng *et al.* (1998b) reported that the two eyes show dynamic asymmetries in the initial phase of symmetrical vergence that are corrected in the final phase, either by slow or saccadic versions, to bring the eyes into their final position of symmetrical gaze. Examples of rather large asymmetries are shown in Figure

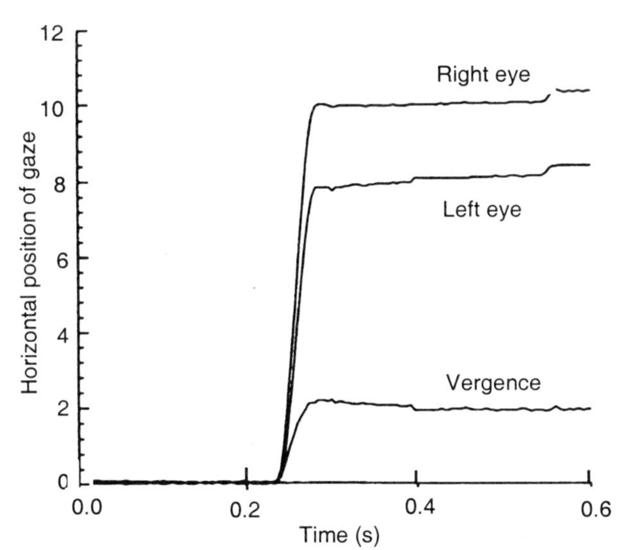

Figure 9.57. Unequal saccades in anisometropia.
Record of unequal saccades made by 12-year-old boy adapted to anisometropic spectacles, after spectacles were removed. Visual target was 10° to the right of initial fixation. Unequal saccades created spurious vergence. (Adapted from Oohira *et al.* 1991)

9.56. Similar movement patterns have been found in the monkey (Maxwell and King 1992). Out of five subjects, van Leeuwen *et al.* (1998) found saccadic intrusions in symmetrical vergence in only the two subjects with strong monocular dominance, as revealed by a sighting test.

It has been reported that approximately 95% of a divergence of 11° when combined with a version of 45° was accomplished by unequal saccades. Only 75% of a similar convergence movement was accomplished by unequal saccades (Erkelens *et al.* 1989b; Collewijn *et al.* 1995). Unequal saccades had the same duration in the two eyes, but one was slower than the other and neither had the same main-sequence dynamics (ratio of amplitude to peak velocity) as a regular conjugate saccad. Enright (1992a) reported that, although the pulse of innervation for the saccades in the two eyes is the same, the step component that determines the final position is generated independently in the two eyes. Saccadic intrusions into vergence to off centre targets brought one or other eye onto the target, so that a final asymmetrical motion of the eyes was required for binocular foveation (Enright 1998b). Collewijn *et al.* (1997) observed symmetrical rather than asymmetrical final vergence, but this may be because they averaged responses over several trials.

Unequal saccades occur particularly frequently in accommodative vergence with monocular viewing and with binocular viewing in strabismic patients who lack disparity vergence (Kenyon *et al.* 1980b). Unequal saccades during vergence could be due to nonlinear interactions between vergence and version signals in motoneurones or in the oculomotor muscles (Zee *et al.* 1992).

Saccadic latency is longer when targets lie in different depth planes rather than in the same depth plane (Honda and Findlay 1992). Hung (1998) found that, on average, vergence was initiated 36 ms before saccadic version when binocular disparity provided the only cue to depth between the targets. Under natural viewing conditions, there was no significant difference between the latencies of the two response components. This suggests that the delay in the saccadic component is due to increased time for stimulus processing rather than to increased time for organization of the response.

Hung concluded that the path followed by the point of gaze during a composite eye movement depends on the asymmetry of saccadic amplitudes, the relative latencies of version and vergence components, and differences in response times of the two eyes.

9.9.3 Adaptation to aniseikonia and paresis

9.9.3a *Adaptation to saccadic dysmetria*

A saccade that lands on the intended target is orthometric. One that misses its target is dysmetric—hypometric when it undershoots the target and hypermetric when it overshoots the target. It has been known for some time that the saccadic system adapts to dysmetria imposed by displacing the target to a different position along the path of the eye's movement while the saccade is in midflight. The displacement of the target is not visible and, initially, the subject overshoots or undershoots the target and a secondary saccade is required. After a few practice trials with a constant displacement, the amplitude of the first saccade becomes modified so as to bring the gaze close to the displaced target (McLaughlin 1967; Henson 1978). It is easier to correct for hypometria than for hypermetria (Miller *et al.* 1981; Deubel *et al.* 1986). The induced change in saccadic gain is limited to saccades within ±30° of the direction of the saccade used in training (Deubel 1987). Albano and Marrero (1995) found that dysmetric adaptation of one eye, with the other eye closed, transferred to the other eye. When both eyes were open and the visual target in only one eye was displaced intrasaccadically, the main effect was a conjugate adaptation of saccadic amplitude. However, the adaptation period was less than 30 minutes and the presaccadic visual display did not signal that unequal saccades were required.

Chaturvedi and van Gisbergen (1997) displaced a visual target intrasaccadically along an isovergence locus, in one direction for saccades in a distant depth plane and in the opposite direction for saccades in a near depth plane. The saccadic system acquired simultaneous changes in gain of opposite sign for saccades in the same direction but in different depth planes. Opposite gain adaptation was also achieved for saccades in the same frontal plane, but only when the directions of the two saccades were sufficiently different.

9.9.3b *Vergence adaptation to aniseikonia*

In the preceding studies the saccadic gain of both eyes changed in the same way. More recently, it has been shown that the vergence system can adapt to dysmetrias of opposite sign in the two eyes. Adaptation of tonic vergence to aniseikonia was discussed in Section 9.3.5.

The images of a surface in the frontal plane have a built-in pattern of horizontal and vertical disparities, which increases with eccentricity and diminishes to zero as viewing distance increases. It can be seen in Figure 9.20 that the angles of azimuth and elevation through which the right eye must move from a midline point *O* to acquire the visual target *P* in the upper righthand quadrant of the headcentric visual field are larger than those through which the left eye must move. Saccades to eccentric targets seem to be centrally programmed to perform the appropriate degree of disconjugacy. The saccadic disconjugacy was reduced after several hours during which subjects wore prisms that nulled the vertical disparity (Ygge and Zee 1995).

Spectacles worn to correct unilateral myopia of refractive origin produce aniseikonia—they enlarge the image in one eye relative to that in the other (Section 15.8). This is because the spectacle lenses are offset from the cornea and do not move with the eyes; contact lenses do not produce this effect. An off-axis object seen through spectacle lenses produces images with different angles of eccentricity in the two eyes. People who wear spectacles to correct for refractive anisometropia develop compensatory asymmetries in saccadic eye movements (Erkelens *et al.* 1989c; Lemij and Collewijn 1991a). Monkeys show the same adaptive response (Oohira and Zee 1992). Figure 9.57 shows a record of unequal saccades made by a 12-year old boy who had worn spectacles for 7 years to correct an 11-dioptre myopia in one eye. The acquired saccadic asymmetry compensated for almost all the optical aniseikonia. After 3 months of wearing corrective contact lenses, which do not produce aniseikonia, saccades became almost equal in size (Oohira *et al.* 1991).

Even short periods of exposure to aniseikonia can produce appropriate adaptation of relative saccadic amplitudes in the two eyes. For instance, Lemij and Collewijn (1991b) found that after 1 hour of exposure to a 2-dioptre lens in front of one eye, the amplitudes of saccades in the two eyes differed by the amount required for binocular acquisition of eccentric targets along the horizontal and vertical meridians. More powerful lenses caused larger adaptive effects up to a limit of 6 dioptres. Six hours of adaptation to 6-dioptre lens, which caused one image to be magnified 12%, produced appropriate saccadic asymmetries. Bucci *et al.* (2001) obtained a persistent saccadic disconjugacy after 15 minutes during which time subjects made horizontal or vertical saccades to points in a random-dot display magnified in one eye by only 2%. Disparities produced by images differing in size by 2% would be within Panum's fusional limit.

Disconjugacy of vertical saccades can be induced by opposite motion of identical dichoptic visual targets just after completion of vertical saccades (Kapoula *et al.* 1996a). However, saccadic disconjugacy is not induced by postsaccadic motion of dichoptic random-dot targets that are spatially uncorrelated in the two eyes (Kapoula *et al.* 1990). Thus postsaccadic drift is not induced by opposed motion in the absence of a recognizable disparity at the completion of the saccade.

Horizontal saccades became disjunctive after subjects made repeated saccades over a period of 15 minutes to a flashed eccentric target with horizontal disparity (Bucci 2000). The subjects had no visual error feedback.

Bush *et al.* (1994) projected random-dot patterns to the two eyes with an 8% difference in size at a viewing distance of 33 cm. Saccades to a superimposed target shown to one eye immediately produced saccades of unequal magnitude of between 4 and 7.5%. This occurred in both humans and monkeys. Thus, a stronger than usual pattern of disparity can induce unequal saccades without a period of adaptation or learning. But this is not saccadic adaptation, since the effect did not persist with monocular viewing.

Van der Steen and Bruno (1995) also obtained immediate disjunctive saccades in response to a similar display with near viewing. However, with far viewing, it took several minutes to obtain disjunctive saccades. This effect was saccadic adaptation, since it persisted for some time in open-loop conditions. With far viewing, disparities in the frontal plane diminish to zero and disjunctive saccades are therefore not normally required. Bruno *et al.* (1995) developed a model of saccadic adaptation.

Subjects rapidly adapted relative saccadic amplitudes when one image was made larger than the other or when the subjects looked back and forth between two points in a random-dot stereogram (Eggert and Kapoula 1995; Kapoula *et al.* 1996b). Kapoula *et al.* (1998) made a peripheral target disparate by placing an 8% magnifier in front of one eye. Even though the target was presented for only 100 ms on each trial and there was a 1 s interval before the saccade was initiated, subjects learned to make appropriate disjunctive saccades with 15 min of training. The effect of training persisted when the target was no longer disparate (Kapoula *et al.* 2000). Saccadic disconjugacy produced by 10% aniseikonia was reduced in the presence of strong monocular cues indicating that the visual display was frontal (Bucci *et al.* 1999).

Adaptation of conjugate saccades to imposed dysmetria is specific to the visual meridian along which training occurred (Deubel *et al.* 1986). Nonconjugate saccadic adaptation to aniseikonia has also been found to be specific to the main orthogonal and oblique directions within which training was applied (Lemij and Collewijn 1992).

Averbuch-Heller *et al.* (1999) placed a base-out or base-in prism in front of one eye so that fixation of a central visual target required 5° of convergence or divergence. For 15 minutes, subjects made 20° saccades from the central target seen through the prisms to targets seen to left or right outside the prism. Subjects developed a strong saccadic disconjugacy that persisted during subsequent monocular viewing. This was accompanied by changes in the peak velocities of both eyes and in the relative velocities of the eyes. These features of adaptation are symptomatic of a binocular saccadic-vergence interaction and cannot be explained by a purely monocular saccadic adjustment. Some subjects developed some saccadic disconjugacy in a non-trained direction along the same horizontal meridian, which could be explained only by monocular adaptation. Averbuch-Heller *et al.* concluded that both binocular saccadic-vergence interaction and a monocular component were involved.

Munoz *et al.* (1999) induced large errors in vergence by exposing subjects to a 4° disparity step superimposed on a 16°/s disparity ramp in a pair of dichoptic vertical lines. Between five and ten convergence or divergence training trials produced an increase in the velocity and magnitude of response to a test stimulus consisting of a simple 4° step. The effect faded after about 5 responses to the test stimulus. Similar changes were evident when the test step was exposed for only 100 ms and removed before the eye movement began. This shows that the adaptation was in the initial open-loop phase of the vergence response. Takagi *et al.* (2001) induced adaptive changes in convergence to a stimulus that stepped from 2 to 1 m. In the 30-minute training session the stimulus stepped from 2 to 1 m and then, within the 200 ms open-loop reaction time to this stimulus, it stepped to either 0.7 (signal increase) or to 1.4 m (signal decrease). After-signal-increase training, the velocity of the initial response to a 2 to 1 m step was increased and after signal-decrease training initial velocity was decreased.

9.9.3c *Adaptation of pursuit to aniseikonia*

Schor *et al.* (1990) enquired whether adaptations of the saccadic and pursuit systems to aniseikonia are independent. In the pursuit condition, subjects maintained vergence for two hours on a pair of horizontal dichoptic lines that moved up and down at 10°/s through 20°, giving a frequency of 0.25 Hz. In the saccade condition, subjects followed step vertical displacements of the lines, which occurred every half-second for 2 hours. Horizontal vergence was maintained on a fixed vertical line. The image of the horizontal line in one eye was magnified 10% so that it moved further than the image in the other eye. The training produced a 7.3% asymmetry of pursuit amplitude but only about 1% of saccadic asymmetry. Saccadic training produced a 6% asymmetry of saccadic amplitude but only about 2.5% of pursuit asymmetry. In other words, adaptation effects were largely specific to the type of eye movement used in training. No adaptation occurred when visual error feedback was withheld during and just after each eye movement.

9.9.3d *Oculomotor adaptation to paresis*

Saccades and voluntary pursuit adapt to partial paralysis of one eye (Optican *et al.* 1985). In one study, a man with sudden onset of paralysis of the left abducens nerve developed saccadic hypometria in his left eye. His other eye happened to be strabismic. Saccadic accuracy of the left eye, which he preferred to use for fixation, recovered after a while. When the paretic eye was patched, the normal eye showed saccadic overshooting (Kommerell *et al.* 1976).

A similar recovery of saccadic orthometria was reported in a patient with sudden onset of right third-nerve palsy. During the 6-day recovery period, the left eye was patched (Abel *et al.* 1978). Experimentally induced paresis in the horizontal recti of monkeys causes the affected eye to be hypometric. With the good eye patched for 6 days, the paretic eye regained orthometria but became hypometric again when the patch was switched (Optican and Robinson 1980). This capacity to recover from the effects of paresis seems to depend on the integrity of the cerebellum, since lesions of the cerebellum induce permanent saccadic hypermetria (Optican 1982).

In these studies, the normal eye was not used or was kept patched during the recovery period and whatever adaptation occurred for the paretic eye was also found in the patched eye—adaptation was conjugate. The visual system apparently treats both eyes alike if there is no visual information to indicate that they should be treated differentially. Snow *et al.* (1985) produced the first evidence that the saccadic system can adapt to paresis differentially in the two eyes. In six monkeys, they weakened the tendon insertions of the medial and lateral recti in one eye, causing saccades for this eye to be hypometric to a greater or lesser extent. After 30 days of binocular viewing, the ratio of saccadic amplitudes in the two eyes returned to normal in all the animals, although recovery was faster in those animals with less initial hypometria. After recovery of balance in the amplitudes of saccades, saccadic durations were longer and peak velocities smaller in both eyes. Furthermore, when the recovered operated eye was patched, its hypometria returned.

It was concluded from these two facts that recovery of saccadic balance depends on changes in neural control and not merely on recovery of strength in the operated eye. Viirre *et al.* (1988) disturbed saccadic balance in monkeys by recession (surgical reinsertion) of a rectus muscle in one eye. After a period of binocular viewing, the deviation of the operated eye disappeared and saccades in both eyes became orthometric in most of the animals. This adaptation did not occur after severe weakening of the muscle.

The experiments described in this and the previous section demonstrate that as long as there is appropriate visual feedback the saccadic system can compensate for aniseikonia or for an unbalanced muscular system by changing the balance of innervations to the two eyes. Binocular visual inputs also seem to be required just for the regular maintenance of binocular coordination of saccades. Simply patching one eye in normal monkeys for 6 days produced hypermetria and postsaccadic drift in that eye, which rapidly cleared up when both eyes were open (Vilis *et al.* 1985; Viirre *et al.* 1987).

Summary

Although version and vergence are distinct eye movements, they sum in a nonlinear fashion when they overlap in time. Vergence movements can intrude into saccadic eye movements to targets that require only version. This may be due to loss of a fusional signal during a saccade or to asymmetries in the mechanical properties of eye muscles. Saccades of different amplitudes in the two eyes may intrude into vergence eye movements. In spite of these complexities and qualifications, Hering's law remains as a fundamental statement of how frontal-eyed animals, such as cats and primates, move their eyes. They do not generally move the eyes independently like a chameleon. Also, the two visual axes intersect at the point of fixation, and a closed eye moves in version or in vergence along with the open eye.

People automatically preprogramme the relative motions of the eyes when making saccadic eye movements to targets in different directions and at different distances. The system is rapidly modified by the prevailing pattern of disparities and shows long-term adaptation when subjects are exposed for some time to unusual patterns of disparity. Adaptation is largely specific to the type of eye movement involved.

9.10 VERGENCE-VESTIBULAR INTERACTIONS

When a person rotates in the dark, the eyes execute conjugate reflex movements known as the vestibuloocular response (VOR). The eyes move in the opposite direction to the head at about the same velocity (slow phases) with periodic rapid return movements (quick phases). Stimuli for VOR arise in the semicircular canals of the vestibular system. On each side of the head there are three canals in approximately orthogonal planes. Each canal is maximally sensitive to rotation of the head in its own plane. VOR can be elicited in a horizontal or vertical direction or about the visual axis (torsional nystagmus), depending on which canal is in the plane of head rotation. The VOR is present in neonates and is basically under the control of centres in the vestibular nucleus and cerebellum (see Howard 1986 for a review).

When the eyes are open, VOR is supplemented by optokinetic nystagmus (OKN) evoked by motion of the image of the visual scene (Section 23.4.5). In addition, a vestibuloocular response is evoked by linear motion of the body in the dark (Smith 1985; Baloh *et al.* 1988). This is referred to as **linear VOR** (LVOR). The stimulus in this case arises in the utricles and saccules. These are the vestibular organs sensitive to linear acceleration. This response is also supplemented by OKN. The two types of VOR and OKN together stabilize the retinal image of the stationary surroundings as the head rotates or translates.

9.10.1 Linear VOR and viewing distance

With linear sideways self-motion, the angular velocity of eye movements required for image stability is zero for objects at infinity and increases for nearer objects. For perfect image stability at viewing distance D, the angular eye velocity, θ, for a linear displacement, L, of the head is given by

$$\theta = \tan^{-1} \frac{L}{D} \quad (7)$$

With eyes open, any inadequacy in the linear VOR (LVOR) is compensated by OKN or by visual pursuit, which is naturally scaled for viewing distance because the angular velocity of a stationary object relative to the head is inversely related to viewing distance. Paige (1989) found that eye velocity increased as the visual stimulus was brought nearer, but not rapidly enough to compensate for the reduction in distance.

Several procedures have been used to reveal whether LVOR is intrinsically scaled for viewing distance without help from visual pursuit. One procedure is to use imaginary visual targets. Gresty *et al.* (1987) found that the velocity of LVOR in the dark increased when a linear component was added to a rotation of the head about a vertical axis, and increased still further when subjects imagined that they were looking at a near visual object.

A second procedure is to record the LVOR just after the visual target has been switched off. Schwarz and Miles (1991) measured the velocity of LVOR in monkeys in response to linear sideways acceleration of the body, for the first 200 ms after switching off a fixation target at one of several distances between 16 and 150 cm. The velocity of LVOR was inversely proportional to viewing distance. However, the scaling of velocity to distance was not perfect and showed considerable individual variation. The mean gain (ratio of eye velocity to that required for image stability) was 0.74 at 16 cm and 1.25 at 159 cm.

Paige (1991) asked human subjects to fixate visual targets at various distances while their bodies were oscillated up and down. The gain of vertical LVOR continued to be related to viewing distance for some time after the target was switched off. The coupling between LVOR gain and viewing distance was affected when subjects viewed the visual targets through prisms that increased the required vergence, but not when they viewed the targets through lenses that changed the required accommodation. Paige *et al.* (1998) found that the effect of viewing distance on LVOR gain after the lights were put out declined as the frequency of head rotation increased to 4 Hz.

It seems from this evidence that LVOR is inversely scaled for viewing distance without help from visual pursuit. The scaling could arise from visual cues to distance seen before the movement starts, from the state of vergence during the movement, or from perceived or imagined distance.

Busettini *et al.* (1991) found that the velocity of the initial short-latency and open-loop portion of OKN evoked by linear motion of a textured display past the stationary animal was also inversely related to viewing distance. However, they found that the gain of closed-loop OKN remained constant. They assumed that this effect was evoked by depth cues such as vergence and accommodation but there could perhaps have been a direct neuromuscular effect of vergence state on eye velocity in the period before visual feedback became operative. They suggested that, at near viewing, the increase in the peak velocity of OKN is offset by a non-linear speed saturation in the LVOR system.

Similar results have been obtained with human subjects although the dependence of the initial velocity of OKN on viewing distance varied with stimulus velocity and other factors (Busettini *et al.* 1994). In the same laboratory, Inoue *et al.* (1998) found that some cells in MST of the alert monkey increased their response to visual motion as vergence increased, while other cells increased their response as vergence decreased.

Shelhamer *et al.* (1995) found that the gain of LVOR in the dark increased when human subjects imagined that they were looking at a near visual object, even though this did not significantly increase vergence. However, LVOR gain was not affected when subjects increased vergence by means of auditory feedback. This suggests that the gain of linear VOR is governed by a central signal related to perceived distance rather than by efference or afference signals associated with the vergence state of the eyes.

Signals responsible for VOR ascend from the vestibular nucleus to the oculomotor nuclei along the medial longitudinal fasciculus and the tract of Deiters. In the monkey, neurones in the tract of Deiters increased their response to linear head motion when the animal changed fixation from a far to a near stationary object (Chen-Huang and McCrea 1998).

9.10.2 Rotary VOR and viewing distance

The velocity of eye movements required to stabilize the image of a stationary object when the head rotates about the midbody axis (RVOR) varies inversely with viewing distance. This is because, when the head rotates, the eyes translate due to their offset from the axis of head rotation. For a head rotation of θ, the rotation of an eye, ϕ, required to stabilize the image of a stationary object at distance D is

$$\phi = \theta + \tan^{-1}\frac{d\sin\theta}{D} \quad (8)$$

where d is the distance from the axis of head rotation to the centre of rotation of the eye. For a distant object the effect of translation is negligible but, for an object at the near point, eye velocity required for image stability is about double that required at infinity. Thus, the gain of RVOR (eye velocity divided by head velocity) required for image stability increases from 1 at infinity to 2 for near vision. In illuminated surroundings, the distance scaling of RVOR could be achieved by OKN. The scaling could be achieved in the dark only if VOR were linked to assumed distance or to vergence.

Biguer and Prablanc (1981) measured RVOR gain during co-ordinated movements of the eyes and head to an eccentric visual target. For a near target, the RVOR component of the eye movement had a higher gain than when the target was far. This was still true when the target was switched off just before the head started to move, showing that visual error signals during the movement are not necessary for modulation of RVOR. Biguer and Prablanc concluded that the modulation depends on visual distance cues seen before the movement started. Hine and Thorn (1987) measured the velocity gain of RVOR while subjects rotated the head from side to side through 30° and converged on a point of light at various distances. When the target was visible the gain was accurately modulated by distance. Modulation continued less adequately after the target was extinguished (see also Paige *et al.* 1998). The gain of RVOR was not affected by lenses that changed accommodation but was affected by prisms that changed vergence. They concluded that vergence can provide the signal for modulation of RVOR.

Viirre *et al.* (1986) found that, in the monkey, eye velocity increased as the linear component of head motion was increased and as the distance of a stationary visual target was decreased. They concluded that this modulation of RVOR is not visually mediated because it occurred in the first 20 ms of the start of head movement, which is below the latency of OKN, and also occurred at frequencies of head rotation beyond the range of OKN. Presumably, the distance of the visual target was assessed before the start of the head motion.

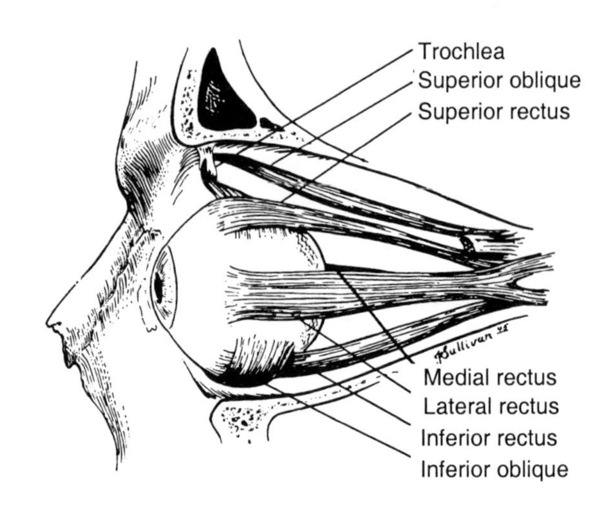

Figure 9.58. Extraocular muscles of the left eye.
(From Cogan 1956 *Neurology of the ocular muscles,* Courtesy of Charles C. Thomas, Publisher, Springfield, Illinois)

Snyder *et al.* (1992) rotated monkeys at between 30 and 500°/s for 40 ms at various times just before or during vergence eye movements between targets at different distances. The gain of RVOR increased linearly with increasing vergence angle. The latency of RVOR was shorter than the latency of vergence, and the response showed some of modulation of gain appropriate to the visual target on which the gaze was not yet directed. This suggests that the signal for modulation of RVOR is derived from the central motor command related to the shift of attention to the new vergence target rather than from proprioceptive feedback from extraocular muscles.

9.11 NEUROLOGY OF VERGENCE

9.11.1 Oculomotor nerves and nuclei

The three pairs of extraocular muscles shown in Figure 9.58 receive their innervation from three cranial nerves: the third (oculomotor), the fourth (trochlear nerve), and the sixth (abducens) nerve. Each nerve originates in a brainstem nucleus of the same name. Together, the three nuclei are called the oculomotor nuclei, which is confusing because it is also the name of one of them. The superior rectus and superior oblique muscles receive an exclusively contralateral innervation from the oculomotor and trochlear nuclei respectively. The other muscles receive an exclusively ipsilateral innervation; the inferior rectus and inferior oblique from the oculomotor nucleus and the lateral recti from the abducens nucleus (Porter *et al.* 1983).

Each extraocular muscle has an outer **orbital layer** and an inner **global layer**. Each layer has two main types of muscle fibres. The first consists of singly innervated fibres each receiving one motor axon that ends in a cluster of neuromuscular junctions at a restricted locus on the nerve fibre (*en plaque* endings). These fibres are fast acting. The second type consists of smaller, multiply innervated fibres with many neuromuscular junctions distributed over the whole length of the fibre (*en grappe* endings). These fibres are slow acting but can maintain constant states of tonic contraction. It is not known whether these multiple endings derive from one or several motor axons. The two types of muscle fibres form distinctive subtypes within both the orbital and global layers (see Spencer and Porter 1988; Porter and Baker 1992). Afferent nerve fibres innervate sensory cells in specialized muscle fibres (muscle spindles) and Golgi tendon organs. There are also distinct types of efferent nerve fibres, some have a mean diameter of 2.5 μm and are unmyelinated and others have a mean diameter of about 9 μm and are myelinated (Alpern and Wolter 1956). Many of the smaller axons are either sensory or motor and innervate blood vessels.

In the oculomotor nuclei, neurones projecting to the rectus muscles are segregated into three groups: A, B, and C. Cells in group C have smaller cell bodies and thinner axons than those in groups A and B and project to smaller and slower muscle fibres (Porter *et al.* 1983). It has been suggested that group-C cells form a distinct pathway for vergence (Jampel 1967; Büttner-Ennever and Akert 1981). However, Keller and Robinson (1972) recorded the activity of cells in the abducens nucleus of the monkey and showed that no type of eye movement was the exclusive product of a particular set of oculomotor neurones. By their own admission, their electrode may have failed to record the activity of small cells exclusively devoted to vergence. We cannot be sure that integration of version and vergence signals is achieved in the oculomotor nuclei rather than involving distinct efferent fibres.

A disparity of about 4° generates the most rapid vergence movements, and vergence is evoked even by disparities of up to 9°. These are larger disparities than those to which cortical cells have so far been found to respond (Section 6.3). Symmetrical vergence movements within the midsagittal plane are instigated by stimuli that project images to opposite cerebral hemispheres. There is evidence that the detection of large disparities between such images depends on interhemispheric connections routed through the corpus callosum (see Section 5.8). This idea is supported by the fact that a patient with section of the corpus callosum failed to produce vergence movements to targets in the visual midline but responded when the images were projected to the same hemisphere (Westheimer and Mitchell 1969). The importance of the callosal pathway for the control of vergence is also indicated by the misalignment of the eyes in callosectomized cats (Payne *et al.* 1981).

9.11.2 Neural signals for vergence

9.11.2a *Signals in brain-stem nuclei*

It was commonly believed until the early 1950s that, while convergence required active contraction of extraocular muscles, divergence resulted from passive elastic tension in the muscles. Electromyography revealed that the lateral recti contract when the eyes diverge, and that all muscles are in a state of active contraction even when the eyes are in the primary position of gaze (Breinin and Moldaver 1955). Distinct neural mechanisms for convergence and divergence have now been identified (Section 9.11).

Conjugate horizontal eye movements require the simultaneous contraction of the lateral rectus muscle of one eye and the medial rectus muscle of the other. These movements are organized in premotor neurones in the paramedian pontine reticular formation (PPRF) whence signals pass to motor neurones in the ipsilateral abducens nucleus. The abducens nucleus contains lateral rectus motoneurones that innervate the ipsilateral lateral rectus muscle. It also contains abducens internuclear neurones whose axons cross the midline, ascend in the medial longitudinal fasciculus (MLF), and terminate in the medial rectus subdivision of the contralateral oculomotor nucleus (Baker and Highstein 1975; King *et al.* 1994; Moschovakis 1995). During a lateral conjugate eye movement, equal ipsilateral and contralateral signals arise in each abducens nucleus. The internuclear circuit is specific to conjugate eye movements. Damage to the MLF creates a defect in conjugate movements while leaving vergence movements intact, a defect known clinically as **internuclear ophthalmoplegia** (Evinger *et al.* 1977). Other defects specifically impair vergence, leaving version intact (Jampel 1967).

There has been some dispute about where signals for conjugate eye movements originate. It has been generally believed that they originate in premotor neurones in the PPRF. However, Zhou and King (1998) found that almost all premotor neurones in the PPRF of monkeys controlled movements of either only the ipsilateral or only the contralateral eye. Only about 5% of PPRF cells coded conjugate saccades. Even in the abducens nucleus, only about 66% of cells exhibited binocular response characteristics. About 27% coded saccadic movements of only the ipsilateral eye and 7% coded only movements of the contralateral eye. The variable convergence of ipsilateral and contralateral premotor neurones onto motor neurones in the abducens nucleus may provide a basis for adaptation of binocular eye movements (Section 9.3.5).

Horizontal vergence eye movements involve the simultaneous activation of either both medial recti or both lateral recti, and therefore require a different neural circuit. Until recently, nothing was known about the neural processes controlling vergence. Earlier studies failed to find cells in the oculomotor nuclei of the monkey that discharge only in association with vergence eye movements (Keller and Robinson 1972; Keller 1973). Mays and Porter (1984) agreed that most cells in the oculomotor nuclei of the monkey carry signals for both conjugate and disjunctive eye movements, although not all neurones participate equally in the two types of eye movement.

Schiller (1970) found a few cells specifically related to vergence in the caudal region of the monkey's oculomotor nuclear complex. During vergence, medial rectus motoneurones in the oculomotor nucleus displayed signals related to both the position and the velocity of the eye, just as they do for conjugate eye movements (Gamlin and Mays 1992). Motoneurones of the superior oblique muscles in the trochlear oculomotor nucleus also discharge during vergence movements (Mays *et al.* 1991). Their discharge is presumably related to cyclovergence, which is known to accompany vergence.

In a more detailed study, Mays (1984) found that the firing rate of cells in the mesencephalic reticular formation of the monkey, an area just dorsal and lateral to the oculomotor nuclei, was related to vergence angle in an approximately linear manner and was not affected by conjugate eye movements. These cells are referred to as **vergence-angle cells**. Most of them increased their firing rate specifically during convergence and a few specifically during divergence. Most of them responded during a change in vergence or during a change in accommodation, but some responded only to changes in vergence and some only to changes in accommodation (Zhang *et al.* 1992). Vergence-angle cells responded with monosynaptic latencies to antidromic stimulation of medial rectus motoneurones in the ipsilateral oculomotor nucleus but not in the contralateral oculomotor nucleus (Zhang, Y. *et al.* 1991). Thus, convergence does not involve efferent pathways that cross the midline.

Vergence-angle cells have also been found in a region dorsal to the nucleus of the third nerve in alert monkeys trained to track a visual target as it moved sinusoidally to-and-fro along the visual axis of one eye at 0.1 or 0.2 Hz (Judge and Cumming 1986). On average, the firing rate of these cells increased or decreased by 16 spikes/s for each degree of change in vergence. Like the cells described by Mays, most of them increased their firing rate during convergence rather than during divergence. There was no response when the eyes moved in the same direction and most cells responded in the same way with both eyes open as with only one eye open. The cells had a mean phase lag of 34° relative to the stimulus. This is greater than the lag of 16.8° reported for cells in the oculomotor nuclei. The delay between the firing of a cell and the start of a vergence movement varied between 35 and 70 ms. When a cell was electrically stimulated, a response was initiated with a mean delay of about 30 ms.

Mays *et al.* (1986) discovered a new class of vergence-related cells in the same area, just dorsal and lateral to the oculomotor nucleus, and in a more dorsal area extending into the pretectum. These cells responded with a burst of activity just before and during vergence movements, which alert monkeys had been trained to make to stimuli that stepped in depth or moved along a depth ramp. The firing-rate profile of these cells was related to the velocity profile of vergence, and the total number of spikes in a burst was related to the size of vergence. These **vergence-burst cells** responded in the same way when the animal tracked a depth ramp, thus maintaining the stimulus in a state of near zero disparity, from which it was argued that these cells respond to eye velocity rather than to the velocity of changing disparity. On average, the response of a cell preceded the eye movement by 22 ms.

Some cells showed a burst of activity related to response velocity and a tonic response proportional to the angle of maintained vergence. The tonic response showed only for larger movements, presumably because only a large movement brought the eyes to a position of gaze for which a sizeable tonic innervation was needed to prevent them from drifting back to their resting state. These are **vergence burst-tonic cells**. Most of the burst and burst-tonic cells responded only to convergence, but a small number of divergence burst and burst-tonic cells were found. Burst cells presumably form the neural substrate of transient vergence (Section 9.6.9).

During convergence, signals from the abducens nuclei to the lateral and medial recti decrease in frequency, but not to zero. The signals to the medial recti are therefore inappropriately small and those to the lateral recti inappropriately large. The effects of these signals must be nulled at the level of the motoneurones by signals arriving from neural centres controlling vergence (Gamlin *et al.* 1989).

In many people with phoria, the tonic imbalance between the eyes returns to its pre-exposure value during a period of exposure to base-in or base-out prisms (Section 9.3.5). Morley *et al.* (1992) recorded from cells in the region dorsal to the oculomotor nucleus in the alert monkey, before and after the animal had been exposed for some time to visual targets at one of various accommodation and vergence distances. Only a few of the cells retained the same relationship between firing rate and vergence angle, so whatever mechanism is responsible for the adaptation of phoria to changed vergence demand must lie outside this region.

The cerebellum is involved in the control of both version and vergence (Keller 1989). Patients with cerebellar dysfunction have poor ocular alignment combined with esophoria and, in many cases, esotropia. They also show disconjugate saccadic dysmetria (Versino *et al.* 1996). There is conflicting evidence about how they show reduced adaptation to prism-induced phoria (Milder and Reinecke 1983; Hain and Luebke 1990). Cells have been found in the posterior interposed nucleus (IP) of the monkey cerebellum that increase their activity during either divergence or far accommodation, both when the eye movements are elicited by misaccommodation or by disparity (Zhang and Gamlin 1998).

The nucleus reticularis tegmenti pontis (NRTP) receives afferents from the frontal eye fields in the frontal lobes and has reciprocal connections with the cerebellum. It is involved in the control of conjugate saccadic and pursuit eye movements. Some cells in the NRTP increase their activity either as vergence and accommodation move to far viewing or to near viewing (Gamlin *et al.* 1996). Microstimulation of these cells produces a far response or a near response. Their activity is not related to conjugate eye movements (Gamlin and Clarke 1995). Thus, this nucleus is part of the corticocerebellar pathway controlling vergence and accommodation.

9.11.2b *Vergence control by the superior colliculus*

The deeper motor layers of the rostral superior colliculus are implicated in the joint control of vergence, accommodation, and pupil diameter of the near triad response. Cortical areas involved in control of accommodation and vergence project to the rostral superior colliculus and from there to the pretectum and other subcortical areas controlling accommodation and vergence (Judge and Cumming 1986; Ohtsuka and Nagasaka 1999). Lesions of the pretectum produce vergence defects (Lawler and Cowey 1986). Stimulation of a saccade-producing site in the rostral superior colliculus of the alert monkey not only stopped a visually evoked saccade but also reduced a pure vergence response or a vergence response associated with a saccade (Chaturvedi and Van Gisbergen 1999, 2000). This could be due to stimulation of brainstem **omnipause neurones** that are active between saccades or because stimulation of the colliculus evoked both convergence and divergence responses at the same time (Guitton 1999).

9.11.2c *Cortical control of vergence*

Cells in the suprasylvian area of the parieto-occipital cortex (Clare-Bishop area) in cats respond to accommodative stimuli, changes in binocular disparity, and motion in depth (Bando *et al.* 1984, 1996; Toyama *et al.* 1986a). These are all stimuli for vergence. Electrical stimulation of cells in the caudal part of this area evokes vergence movements (Toda *et al.* 1993). Activity of some cells is correlated with the peak velocity of vergence, while that of others is correlated with the peak velocity or amplitude of accommodation, and that of others with both responses (Takagi *et al.* 1993). The amplitude and velocity of vergence movements were reduced by bilateral lesions in the suprasylvian area. Unilateral lesions affected only the contralateral eye, resulting in asymmetrical vergence (Takada *et al.* 2000).

Cells in the medial superior temporal cortex (MST) of the monkey respond just before a vergence response. Takemura *et al.* (2001) recorded from MST cells in alert monkeys 50 to 110 ms after horizontal disparity steps of various sizes were applied to a random-dot pattern. Disparity tuning functions of the cells resembled those of cells in V1, as described in Section 6.3.1. The mean disparity-tuning function of all the cells fitted the curve describing vergence magnitude as a function of disparity step size. Takemura *et al.* concluded that the magnitude, direction, and time course of the initial vergence response to disparity steps are determined by the sum of activity of disparity-sensitive cells in MST.

Responses of cells in the area LIP on the lateral bank of the intraparietal sulcus of the monkey (Section 5.7.3d) are related to rapid changes of gaze in 3-D space (Gnadt and Mays 1995). These movements involve both saccadic and vergence components. Each cell responded differentially to the position of a visual target in a frontal plane and this differential response was independent of the distance of the frontal plane from the monkey. However, some cells responded best when the stimulus remained in the same frontal plane as the initial fixation, others when the target was stepped towards the monkey, and others when it stepped away. For many cells, depth tuning could be evoked by either an accommodative cue to distance or by disparity, but most cells responded best when both cues were available. The cells responded in a similar fashion after the eyes had moved in response to targets that had been switched off before the movement began. Thus, responses were related to the difference between current eye position and the desired eye position rather than to the position of retinal images at the time of the response. This means that information about changes in eye position, derived from proprioception or motor outflow, feeds into these cells. Humans are able to make combined vergence-version eye movements to remembered targets after intervening changes of fixation (Krommenhoek and Van Gisbergen 1994). These cells in LIP were therefore generated a premotor signal for rapid eye movements in 3-D space relative to the fixation plane.

Many disparity-tuned cells in area LIP of the rehesus monkey project to the motor map of the superior colliculus (Gnadt and Beyer 1998). They also project to the frontal eye field and Ferraina *et al.* (2000) found cells in the frontal eye field that respond to coarse disparities. Jampel (1960) elicited vergence responses in the monkey by stimulation of the frontal eye field.

Positron emission tomography (PET) has revealed that, when human subjects visually pursue an approaching object with vergence eye movements, activity occurs bilaterally in the temporo-occipital cortex, the left parietal lobe, and in the right fusiform gyrus (Hasebe *et al.* 1999).

It has already been mentioned that the nucleus reticularis tegmenti pontis (NRTP) in the brain stem receives afferents from the frontal eye fields in the frontal lobes of the cerebral cortex. Both centres are involved in the control of saccades and pursuit eye movements (Gamlin *et al.* 1996). The NRTP is also involved in the control of vergence and accommodation and Gamlin and Yoon (2000) have now found a region in the frontal eye fields, just anterior to the saccade-related region, which is involved in the control of vergence and accommodation. Thus, the frontal eye fields are involved in the control of all forms of voluntary eye movements, conjugate and disconjugate.

9.11.3 Neurology of cyclovergence

Almost nothing is known about the neurology of cyclovergence. The rostral interstitial nucleus of the medial longitudinal fasciculus (MLF) is involved in the control of torsional and vertical saccades, and the interstitial nucleus of Cajal is involved in slow torsional and vertical gaze control, including torsional and vertical vestibuloocular responses. Eye torsion and head deviation are induced by stimulation of this nucleus in monkeys (Westheimer and Blair 1975) and in humans (Lueck *et al.* 1991). The nucleus of Cajal receives inputs from the paramedian pontine reticular formation (PPRF), a centre concerned with the coordination of all rapid eye movements, and projects to the oculomotor and trochlear nuclei. Neurones in the right nucleus respond when the eyes rotate clockwise from the point of view of the animal. Those in the left nucleus respond when the eyes rotate counterclockwise. Cells in the same nucleus respond when the eyes execute vertical saccades. Both up and down movements are represented in both left and right nuclei (Vilis *et al.* 1989). Microstimulation of cells in the interstitial nucleus of Cajal in the monkey induces conjugate saccadic torsional eye movements, which obey the same clockwise-counterclockwise rule (Crawford *et al.* 1991).

Unilateral inactivation of the interstitial nucleus of Cajal in monkeys leads to a loss of rapid torsional eye movements in the ipsilateral direction, a torsional displacement of the eyes (and of Listing's plane) to the contralateral side, and an ipsilaterally beating torsional nystagmus with a vertical component. Unilateral inactivation also causes a slowing of rapid downward movements, indicating a failure of the neural integrator (Crawford and Vilis 1992). Inactivation of both interstitial nuclei of Cajal restores torsional balance and removes the torsional nystagmus, but leaves a vertical nystagmus (Helmchen *et al.* 1998).

Unilateral inactivation of the MLF produces a contralaterally beating torsional nystagmus. Bilateral inactivation leads to a loss of saccadic components of torsional and vertical eye movements (Suzuki *et al.* 1995).

10 *Development of the visual system*

10.1 METHODS

The basic structures and mechanisms of the visual system begin to develop in the early embryo and become fully functional several months after birth. Initially, development of the visual system is genetically programmed. However, even before birth, spontaneous discharges in sensory nerves affect the formation of neural connections in the visual pathways. After birth, stimulus-dependent neural activity determines the precise combination of pathways from the two eyes and fine tuning of the whole visual system. Visual mechanisms are modified by growth processes extending throughout infancy. These growth processes include: (1) changes in the size and shape of the eyes and their setting in the head, (2) modifications of accommodation and vergence, (3) changes in the size of the retina, (4) myelination of the visual pathways and visual cortex, and (5) changes in the distribution of dendrites and a reduction in the density of synaptic contacts. Contemplation of the myriad of complex factors that regulate the development of the visual system induces a sense of awe.

The study of the development of the visual system involves a variety of anatomical and physiological procedures, some of which can be applied in the living animal (*in vivo*), while others are applied in cultures of living cells taken from the brain (*in vitro*).

Cells or whole organs can be transferred from one location to another location in the same animal or from one animal to another, using procedures developed by Roger Sperry (1951). These procedures reveal whether chemical agents responsible for the growth of particular tissues are specific to the tissue or to the cellular environment in which the tissue grows. These methods are particularly applicable in the lower vertebrates, such as fish and amphibia, which possess remarkable powers of regeneration.

The fate of migrating neurones can be followed by labelling them with tritiated thymidine. Migration of differentiating neurones and cell lineages derived from a particular progenitor cell can also be traced by infecting progenitor cells with a retrovirus that expresses a green fluorescent protein. The virus transfers from a progenitor cell to its offspring but not to other cells. Growing cells containing the fluorescent protein can be viewed directly and continuously by time-lapse videomicroscopy (Okada *et al.* 1999).

Recently there has been a dramatic increase in the use of genetically manipulated mice to study the development of the nervous system. In the **transgenic procedure** an exogenous gene is introduced into the genome of a mouse zygote, which leads to overexpression of a particular protein. This procedure reveals the roles of particular proteins in development and the roles of regulatory mechanisms that govern the expression of proteins. The transgenic procedure is limited by the availability of recombination agents that effect the expression of particular genes.

In the **gene knockout procedure** a particular gene product is removed by the process of homologous recombination or a gene is introduced that expresses a toxin that inhibits production of specific proteins. This procedure may prevent specific cells from developing. For example, the retinas of a transgenic mouse containing a toxin-producing gene linked to the red opsin gene were devoid of cones. Animals with an inactivated gene may develop compensatory mechanisms that mask the normal effects of the gene. Also, knockout animals may die before the effects of transgenic procedures can be investigated.

A natural mutation may also knock out a particular gene. The study of natural growth defects, such as albinism, can reveal the mechanisms underlying normal development.

Bringing together of anatomical, genetic, physiological, and behavioural procedures promises a rapid advance in our knowledge of the development and diseases of the nervous system.

10.2 SUBCORTICAL DEVELOPMENT

10.2.1 Development of the eye

10.2.1a *General development*

The retina develops from an outgrowth of the forebrain, known as the **optic vesicle**. After making contact with the overlying ectoderm, the vesicle invaginates to form the optic cup, which is attached to the brain by the **optic stalk**. The lining of the optic cup forms the retina and the overlying ectoderm is induced to form the lens and cornea. The ventral surface of both cup and stalk invaginate to form the optic fissure. The sides of the fissure fuse and axons from the retina begin to enter the optic stalk, which becomes the optic nerve. Many genes responsible for the development of eyes have been identified. Some of these genes, such as *Pax6*, occur in many phyla, including flies, fish, birds, and mammals and cause similar visual defects when absent (Oliver and Gruss 1997).

The volume of the eyeball of the adult human is about three times that of the eye of the newborn infant. The eye grows proportionately less than the body as a whole, which increases in volume about 20-fold. The corneal surface increases about 50% while the area of the retina approximately doubles from 590 mm^2 to 1250 mm^2 (Scammon and Wilmer 1950). The axial length of the eye increases from about 15.5 mm at birth to its adult value of 24.5 mm, which is reached at about the age of 13 years, with about half the increase occurring in the first 2 years (Larsen 1971). This means that the infant eye requires about 85 dioptres of refraction to focus an image compared with about 60 dioptres in the adult eye (Lotmar 1976). Another consequence of the small size of an infant's eye is that 1° of visual angle corresponds to between 0.18 and 0.2 mm on the retina, compared with 0.29 mm on the adult retina (Hamer and Schneck 1984). Thus, for a given distal stimulus, the area of the retinal image in the infant eye is considerably less than that in the adult eye.

10.2.1b *Emmetropization*

In many neonatal vertebrates, including primates, the eyes of most individuals are too short in relation to their optics. As a result, the eyes of human neonates are hypermetropic when refraction is measured with a cycloplegic drug, which paralyzes the intraocular muscles. This means that the image of a distant object lies beyond the retina. With noncycloplegic refractions, the eyes of human neonates tend to be myopic (Howland 1993). The normal adult eye is emmetropic, so that the image of a distant object is focussed on the retina without accommodative effort. The development of emmetropia with increasing age is known as **emmetropization**. In the first 3 years the curvature of the human cornea decreases by about 5 dioptres from an initial mean value of about 48 dioptres. In about the first 12 years the axial length of the human eye increases by about 7 mm from an initial mean value of about 15.5 mm. Similar changes occur in the rhesus monkey (see Bradley *et al.* 1999). These changes are partly under genetic control and partly controlled by visual feedback (Troilo and Wallman 1991).

The axial length of the eyes of young cats and chickens increases to compensate for an experimentally induced refractive error (Wallman and Adams 1987; Schaeffel *et al.* 1988). Since compensation occurs in chickens in which the eyes are unable to accommodate, it must depend on a mechanism that detects the sign of defocus of the image (Schaeffel *et al.* 1990). However, chickens raised with a positive lens in monochromatic light showed the same corrective elongation of the eye as animals raised in white light (Schaeffel and Howland 1991; Wildsoet *et al.* 1993). Thus, chromatic aberration is not necessary to sign the defocus of the image. Presumably, other cues to the sign of defocus are sufficient. Since defocussed images induce compensatory eye growth, high spatial frequencies are not required

Figure 10.1. Carla J. Shatz.
Born in New York City in 1947. She obtained a B.A. in Chemistry from Radcliffe College in 1969 and a Ph.D. in Neurobiology from Harvard Medical School with D. Hubel and T. Wiesel in 1976. She conducted postdoctoral work with Pasko Rakic in Harvard Medical School. In 1978 she joined the faculty at Stanford University, Department of Neurobiology, where she became Professor of Neurobiology in 1989. In 1992, she moved to the Department of Molecular and Cell Biology at Berkeley, where she became Professor of Neurobiology and an Investigator of the Howard Hughes Medical Institute. In 2000, she became Nathan Marsh Pusey Professor of Neurobiology in the Department of Neurobiology at Harvard Medical School. Honours include the Golgi Award from Fidia in 1992, the Silvo Conte Award from the National Foundation for Brain Research in 1993, the Charles A. Dana Award for Pioneering Achievement in Health and Education in 1995, the Alcon Award for Outstanding Contributions to Vision Research in 1997, and the Bernard Sachs Award from the Child Neurology Society in 1999. She was elected to the American Academy of Arts and Sciences, in 1995 to the National Academy of Sciences, in 1997. Dr. Shatz was President of the Society for Neuroscience, 1994-95

(Schaeffel and Diether 1999). For some reason, a positive lens induces an eye to increase in length but a negative lens has little effect (Schaeffel and Howland 1991).

Of eight monkeys reared with a 9-dioptre contact lens on one eye, one showed no effect, one developed 3 dioptres of axial myopia, and five developed axial hyperopia of up to 3.5 dioptres in that eye. After the lens was removed, both types of refractive error diminished (Smith *et al.* 1994). This suggests that primates also have a visually controlled emmetropization mechanism that compensates for refractive errors up to 3.5 dioptres (see also Hung *et al.* 1995).

Accommodative error produced by interactions between accommodation and convergence has been implicated in the development of myopia in humans. Myopic children show enhanced accommodative convergence (Jiang 1995). A child who is esophoric must relax accommodation to maintain single vision. The resulting image blur during near work could induce axial growth of the eye and hence myopia (Gwiazda *et al.* 1999). Schor (1999) has developed a model of these processes.

Even in the adult, an excessive amount of near work, such as reading, causes the eyes to become increasingly myopic (Flitcroft 1998; Blackie and Howland 1999). There is also evidence that the axial length of the human eye decreases with advancing age to compensate for a decrease in the depth of the anterior chamber and the refractive power of the cornea and (Grosvenor 1987).

A change in the axial length of the eye of between 2 and 20 μm, corresponding to a change of between –0.036 and –0.015 dioptres, occurs as emmetropes or myopes accommodate (Drexler *et al.* 1998). This change is believed to be caused by contraction of the ciliary muscles. Long-term changes in axial length in those engaging in near work could arise from the cumulative effect of these changes.

In addition, there is evidence that loss of contrast in the image over long periods increases the axial length of the eye (Bartmann and Schaeffel 1994). When the eyelids of a young monkey were sutured so that only diffuse light entered the pupil, the eye developed an increased axial length and a consequent axial myopia (Wiesel and Raviola 1979; Tigges *et al.* 1990). Other investigators could not replicate this effect in monkeys (von Noorden and Crawford 1978) but Wallman and Turke (1978) obtained similar effects in chickens. Hoyt *et al.* (1981) found axial myopia to be associated with early eyelid closure in human infants, although the effect may arise from mechanical or thermal effects, since patients with corneal opacification did not suffer from myopia.

In contrast to the general finding that form-deprivation induces elongation of the eye (myopia), some studies have reported hyperopia after less severe deprivation or in association with strabismic amblyopia (see Kiorpes and Wallman 1995).

Schaeffel and Howland (1988) and Flitcroft (1998) have developed models of emmetropization. The topic is reviewed by Young and Leary (1991) and Schaeffel and Howland (1995).

10.2.2 Development of the retina

10.2.2a *Structural development of the retina*

The precursors of retinal cells develop from the inner layer of the optic cup—an outgrowth of the forebrain. Ganglion cells differentiate first, followed by cones, horizontal cells, bipolar cells, and lastly rods. Cells develop first in the central retina. By about the 30th embryonic day in the cat, the axons of ganglion cells begin to grow into the optic stalk and segregate into distinct bundles, or fascicles, separated by glial cells (Shatz and Sretavan 1986; Okada *et al.* 1994) (Portrait Figure 10.1).

The development of the human retina has been recorded from 13 weeks of gestation. At this age the mosaic of foveal cones is identifiable with a density of about 14,000 cones per mm^2 (Hendrickson and Yuodelis 1984; Hendrickson and Drucker 1992; Diaz-Araya and Provis 1992). By 24 weeks of gestation, cone density in the central retina is approximately 38,000 per mm^2, compared with an adult value of over 100,000 per mm^2. Cone density is inversely related to cell-soma diameter. The increase in cone density in the fovea is due to migration of cells towards the fovea from a circumferential region of undifferentiated cells rather than to cell division within the central region (Diaz-Araya and Provis 1992). The mature retina contains a regular mosaic of different types of cell. The differentiation and distribution of cell types seems to be controlled by (1) molecular markers produced by early cells, (2) tangential dispersion of differentiating cells, and (3) elimination of incorrectly positioned cells (see Cook and Chalupa 2000).

One week after birth the human peripheral retina resembles that of the adult but the macular region covering about 5° of the central retina is very immature. There is a foveal depression, but all cell layers extend across it rather than being parted as in the adult retina. The principal mechanism for formation of the foveal depression seems to be migration of ganglion-cell bodies in the inner retinal layers away from the centre (Kirby and Steineke 1992). At 26 weeks of gestation the inner segments of rods and cones are rudimentary and there are no outer segments until 36 weeks of gestation. Foveal cones of the one-week-old infant have inner and outer segments that are only about one-sixteenth the length of adult segments. In the neonate peripheral retina, the inner and outer segments of rods and cones are 30 to 50% of adult length. The outer segments of cones continue to elongate for up to 5 years. Foveal cones become narrower with age, from 5 to 7.5 microns wide at birth to 1.8 to 2.2 microns in the adult (Yuodelis and Hendrickson 1986). The decrease in diameter means that a foveal cone subtends between 1.5 and 2.2 arcmin at the nodal point in the neonate eye and about 0.5 arcmin in the adult eye.

The larger receptor aperture in the infant eye severely reduces the effective contrast of high spatial-frequency images (Banks 1988). The retina of an 11-month-old infant is similar to that of the adult in both periphery and fovea (Abramov *et al.* 1982). The retina of the monkey shows a similar development, but the fovea is more advanced at birth than in the human (Samorajski *et al.* 1965; Hendrickson and Kupfer 1976).

The visual field is smaller in human infants under 8 weeks of age than in the adult (Section 11.1.4), but to what extent this is due to optical factors rather than maturation of the retina is not known (Schwartz *et al.* 1987).

Axons of young ganglion cells are guided to the optic disc by a variety of chemical growth factors and by adhesion molecules of the extracellular matrix, such as integrin and cadmodulin. Once they reach the optic disc, the diffusing growth factor netrin-1 guides them into the optic nerve (Deiner *et al.* 1997).

10.2.2b *Functional development of retina*

The development of functional retinal synapses can be followed by immunocytochemical labelling of the synaptic glycoprotein, SV2. This protein is correlated with the presence of synapses as revealed by the electron microscope (Okada *et al.* 1994). Synaptic development follows a foveal to peripheral progression and occurs for rods before cones at the same retinal eccentricity.

Ganglion cells and amacrine cells develop before photoreceptors or bipolar cells (Robinson 1991). Even at this stage of development, optical recording using a calcium-sensitive dye has revealed slow spreading oscillations of concentration of intracellular calcium in a proportion of ganglion and amacrine cells (Wong *et al.* 1995). Calcium waves also propagate between glial cells. This activity presumably helps in the development of the tangential retinal network even before the development of structures responsible for activating the optic nerve. Stimulated spikes and waves of intracellular calcium influence the motility and morphology of the growing nervous system (Gu and Spitzer 1995).

Mammalian ganglion cells generate spontaneous neural activity well before birth. Bursts of neural activity are generated in the rat retina by at least embryonic day 17 (Galli and Maffei 1988). In the cat retina, electrical stimulation induces spikes in about a third of ganglion cells by embryonic day 30 (5 weeks before birth). By day 55 almost all ganglion cells are capable of generating repetitive discharges. Discharges are abolished by application of tetrodotoxin, which indicates that they are sodium mediated (Skaliora *et al.* 1993). In the cat, between the time of eye opening at post-natal day 7 and post-natal day 21, the intermittent bursts of spontaneous activity in ganglion cells change to a more regular spontaneous discharge. Just after eye opening, only a few ganglion cells respond to light but by postnatal day 10 they all respond (Tootle 1993).

In the adult retina, ON and OFF bipolar cells are segregated in distinct strata in the inner plexiform layer. In the developing retina, bipolar cells are not segregated and ganglion-cell dendrites ramify throughout the plexiform layer. Tootle (1993) claimed that ganglion cells in the developing retina are either ON or OFF types. However, more recent evidence suggests that immature ganglion cells in the ferret retina show both On and OFF responses to a flash of light (Wang *et al.* 2001).

10.2.3 Growth of optic nerve and tract

The axons of ganglion cells grow out from the eye to the chiasma and then to the LGN and other subcortical centres. In mammals, axons from the foveal region develop first and occupy the central region of the optic nerve and tract while axons from a series of concentric rings of increasing eccentricity occupy the outer regions of the optic nerve and tract (Walsh and Guillery 1985; Walsh 1988). Secondary axons grow from the LGN along the optic radiations to layer 4 of the visual cortex. Cells from each part of the retina are destined to connect with specific locations in the LGN and cortex. See Holt and Harris (1993) for a discussion of retinal markers.

10.2.3a *Axonal growth*

The tip of a growing axon forms a **growth cone** from which extend web-like **lamellipodia** and finger-like **filopodia**. These extensions are several micrometers long and form and retract on a minute by minute time scale. Actin filaments are assembled by polymerization at the distal end, and flow at a velocity of about 100 nm/s to the proximal region, where the polymer chain is disassembled. Behind the actin filaments microtubules extend down the length of the axon. As the axon grows the microtubules extend into or retract from the lamellipodia and filopodia. Substances required for axon growth are transported in intracellular vesicles to the tip of the growing axon (Martenson *et al.* 1993).

Pioneer axons growing along the optic nerve and tract show periods of advance, when growth cones are elongated and have few filopodia, interspersed with pauses in which growth cones spread out and project filopodia. As the axons reach "decision regions" in the chiasma or LGN, the periods of advance are short and the pause periods are prolonged for one hour or more. During the pauses, filopodia 'seek out' the appropriate pathway in the chiasma or the correct target cells in the LGN (Bovolenta and Mason 1987). When a pioneer axon encounters an obstacle, its growth cone collapses and lateral extensions develop from this axon and from trailing axons in the same fasciculated bundle, some of which develop into new growth cones (Davenport *et al.* 1999). These events can be observed by time-lapse photography of fluorescently labelled growth cones (Halloran and Kalil 1994; Mason and Wang 1997).

Five main agents guide, accelerate, or retard the growth of axons.

1. Structure of the cellular environment Cartilage and other tissues form physical barriers, and extracellular spaces in the embryonic neural tissue form channels through which neurones migrate or axons grow.
2. Cell adhesion molecules Protein molecules spanning the membrane of the growing axon fall into at least three families: **integrins, immunoglobulins**, and **cadherins**. Each molecule has an extracellular domain and a cytoplasmic domain linked by a transmembrane segment. The extracellular domain determines specific connections outside the axon and the cytoplasmic domain interacts with the cell's cytoskeleton to determine the shape and motility of the growing cell.

Integrins tend to bind with similar (homophilic) or dissimilar (heterophilic) molecules on adjacent cell membranes, causing the cells to adhere. Immunoglobulins, and cadherins bind with **glycoproteins**, such as netrin, laminin, fibronectin, and tenascin, secreted by astrocytes (a type of glial cell) on a fibrous matrix known as the **extracellular matrix** (Sanes 1989). This matrix is substantially reduced after growth is complete. Specific cell-surface receptors bind to specific glycoproteins (Reichardt 1992). Three subfamilies of cadherins have been identified but many more may remain to be discovered. The extracellular domain of the molecule varies within subfamilies and the cytoplasmic domain varies between subfamilies. Each type of cytoplasmic domain triggers a distinct type of reaction in the cell. Wu and Maniatis (1999) identified 52 human cadherin genes in three clusters, each with a constant region coding a cytoplasmic domain and a variable region coding extracellular domains. They speculate that a great variety of cadherins is created during neurogenesis by rearrangement of variable DNA regions relative to constant regions, in the same way that the immune system generates a multitude of antigens.

We will now see that extracellular glycoproteins attract growth cones, help axons adhere to the substrate, and generate signals that travel to the interior of the growth cone and to the soma of the growing neurone (Palecek *et al.* 1997).

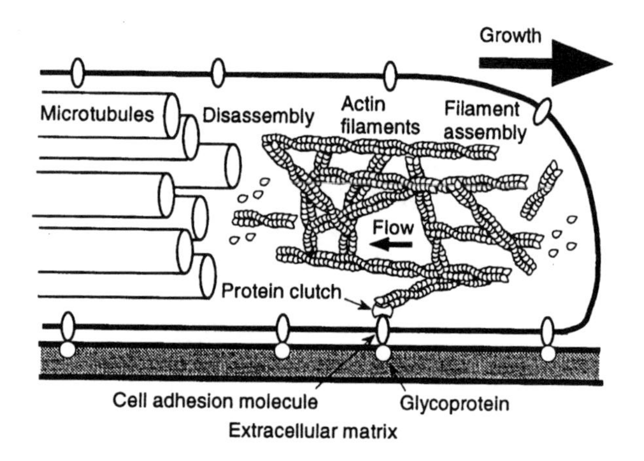

Figure 10.2. Advance of a growth cone.
Cell adhesion molecules embedded across the cell membrane of the growth cone attach to glycoproteins on the extracellular matrix. Protein molecules link the cell adhesion molecules to actin filaments in the growth cone. As the growth cone advances, the retrograde flow of the filament network is attenuated and the mictotubules move forward with the growth cone. (Adapted from Lin and Forscher 1995)

Signals are carried by cell surface receptors from the extracellular binding site to the cytoplasmic domains inside the cell (Challacombe *et al.* 1996). These signals trigger activity in a wide range of small protein molecules, such as members of the rho family (Luo *et al.* 1997). These molecules seem to control the backward circulation of actin filaments from their assembly point at the tip of the growth cone to the lace where they are disassembled. The turnover time of actin filaments in hippocampal neurones is about 44 s (Star *et al.* 2002). When the growth cone advances, this circulation is attenuated because the actin filaments become anchored to the cytoplasmic domains of cell surface molecules, which in turn are anchored to the extracellular matrix by their extracellular domains (Lin and Forscher 1995; Suter *et al.* 1998). This molecular clutch mechanism regulates the formation and movement of filopodia and lamellipodia and determines the direction of axon growth. The processes are depicted in Figure 10.2.

Activation of NMDA receptors leads to the emergence of new spines while activation of AMPA receptors inhibits the mobility of actin filaments and causes growth-cone spines to round up and become more stable and regular (Fischer *et al.* 2000). Low-frequency stimulation, which induces long-term depression of synaptic activity also stabilizes actin activity through the mediation of NMDA receptors. (Star *et al.* 2002). NMDA and AMPA receptors are defined in Section 5.4.5. We thus have mechanisms for experience-dependent modulation of actin filament dynamics and spine formation.

Signals conveyed by cell surface receptors from glycoproteins to the interior of the cell also travel to the soma of the growing cell where they control gene expression of proteins involved in morphogenesis. Antibodies of the glycoproteins inhibit axonal growth (Cohen *et al.* 1986). Glycoproteins may also have enzymatic or proteolytic properties enabling them to modify the extracellular medium through which growth cones move (Pittman 1985).

Some glycoproteins on the extracellular matrix are attractants while others are repellents and create exclusion zones from which growing axons are deflected (Pini 1993). Some attract some axons and repel others (see Tessier-Lavigne and Goodman 1996). The filopodia of a growth cone observed under the microscope *in vitro* collapse completely when chemical repellents are applied (Fawcett 1993).

The type of glycoprotein most active in a given location may change over time (Cohen *et al.* 1986). Exposure to a growth factor can affect subsequent responses to the same factor (Diefenbach *et al.* 2000). Also, as an axon grows from one cellular environment into another, the type of extracellular glycoprotein to which it responds changes because of intrinsic changes within the growth cone (see Dodd and Jessell 1988; Song *et al.* 1997). Thus, growth cones encounter a complex spatiotemporal pattern of chemical influences as they migrate through the extracellular matrix (see Letourneau et al. 1994). XE "{Letourneau et al. 1994["

Some extracellular glycoprotein molecules, such as laminin and fibronectin, appear only during neurogenesis in mammals but remain in animals, such as fish and frogs, which are able to regenerate the optic nerve. Other molecules, such as tenascin, remain in the adult mammalian brain. Interactions between extracellular molecules and cell adhesion molecules seem also to be involved in long-term potentiation (Section 10.4.1) underlying learning in the adult brain (see Jones 1996; Mueller 1999).

3. Neurotrophins As an axon approaches to within about 300 μm of its target cell the action of glycoproteins secreted by the extracellular matrix is switched off and the growth cone comes under the control of neurotrophins secreted by the target cell. It is not known whether this change is mediated by the neurotrophins or by an intrinsic timing mechanism in the growing axon. Neurotrophins and their ligands are discussed in further in Section 10.4.1b.

It seems that neurotrophins are sufficient to guide retinal axons to their destination in the tectum of amphibians, such as the axolotl. The axons reach their destination even when forced to travel an unusual route and in the absence of impulse activity (Harris 1984).

There is some evidence that growth cones are also attracted by neurotransmitters secreted by target cells (Zheng *et al.* 1994). Ultimately, the chemoaffinity between ganglion cells and their target regions in the visual cortex must be determined by the topographic expression of specific genes. Evidence of genetic transcription factors has been found in the retina and optic tectum of birds (Yuasa *et al.* 1996).

4. Intrinsic factors Proteins that foster axon growth are formed within growing neurones. Specific genes express these proteins. Production of a given protein ceases when its gene is switched off. For example, the bcl-2 protein is produced by the bcl-2 gene in growing ganglion cells. Over expression of this gene in adult mice allowed severed ganglion cell axons to regenerate (Chen *et al.* 1997). The GAP-43 protein, also known as neuromodulin, is present in developing and regenerating neurones and interacts with extracellular growth factors (Zhu and Julien 1999). This protein activates members of the G protein family within the growing axon (Vancura and Jay 1998).

Neuronal growth also depends on a supply of intracellular calcium ions. Nerve growth factors trigger a transient increase of calcium ions in the growth cone, which evokes a signalling system for cone guidance, and extension of filopodia. Neuronal growth stops if the store of calcium ions is depleted by application of a blocking agent (Takai *et al.* 1998).

5. Action potentials Spontaneous or stimulus-induced neural activity is involved in the development of synaptic contacts in the LGN and cortex (see Section 10.4.4). For example, blockage of sodium-dependent excitatory activity by tetrodotoxin (TTX) disrupts synaptic development (Shatz 1990a). However, TTX did not affect the growth of axons to their destination. Verhage *et al.* (2000) found that all neurotransmitter release is stopped throughout the brain of mice lacking the *munc 18-1* gene. Although the brainstem in these mice lost its neurones by embryonic day 18, the neocortex resembled that of normal littermates at birth. However, soon after birth, the cortical neurones of the defective mice degenerated. Thus, it seems that, at least in mice, neural activity involving release of neurotransmitter is not required for initial formation of cortical layers. However, neural activity is required for the refinement and maintenance of cortical synapses. It is possible that neural activity not involving neurotransmitter release is involved in axonal guidance. For example, waves of calcium ions spread over the embryonic cortex through electrical coupling over gap junctions rather than by synapses (Section 10. 4.4b). Spontaneous neural activity in the developing LGN is discussed in Section 10.2.4b.

In amphibia, axon growth may depend on spontaneous neural activity involving potassium ions rather than sodium ions. When McFarlane and Pollock (2000) blocked potassium channels, developing ganglion cells of the frog *Xenopus* grew aberrantly in the optic tract and optic tectum. Thus, different forms of neural activity may be involved in different groups of animals.

10.2.3b *Segregation of axons at the chiasma*

When ganglion-cell axons reach the chiasma, they segregate into those from the temporal hemiretinas that remain on the same side, and those from the nasal hemiretinas that decussate to the contralateral side. Two mechanisms have been proposed to account for this segregation. In the first, growing axons respond to structural or chemical signals as they approach the chiasma. In the second, axons from the temporal retina grow at random to one side or the other, and those taking the wrong route are subsequently eliminated.

Sretavan (1990) injected a fluorescent dye into the optic tract (postchiasma) of embryonic mice that retrogradely labelled axons in the optic nerve (prechiasma) according to whether they were destined to cross or remain on the same side. Chan and Guillery (1994) labelled retinal regions of embryonic rats. These procedures revealed that ganglion-cell axons have a retinotopic order as they leave the retina, although axons from different classes of ganglion cell intermingle. When axons reach the chiasma they loose their retinotopic order. Axons from dorsal and ventral retinal regions intermingle with those from nasal and temporal regions, although younger axons still occupy the centre of the tract (Walsh 1988). Thus, the partition mechanism that decides whether a given axon decussates or not has nothing to do with the relative positions of axons (Dräger 1985).

During the first stage of development in the mouse, uncrossed axons from the ventrotemporal retina arrive at the chiasma before crossed axons from the other eye and grow directly into the ipsilateral optic tract without approaching the midline (Marcus and Mason 1995). In later stages, crossed and uncrossed axons arrive at the chiasma at the same time. Axons from the temporal retina at first follow the same route as those from the nasal retina as they grow towards the midline of the chiasma, but at this point the temporal fibres bend sharply towards the ipsilateral side. Thus, in the mouse, there is an age-dependent change in the process of segregation at the chiasma.

Meissirel and Chalupa (1994) found that uncrossed axons from the temporal hemiretina of the monkey took an ipsilateral route through the chiasma by embryonic day 36. This is before crossed axons from the nasal hemiretina reach the chiasma. Thus, the first uncrossed axons do not require pioneer crossed axons from the opposite eye. They are presumably guided by a specific chemical. By embryonic day 42, there was a clear segregation of crossed and uncrossed axons in the optic tract, with the early uncrossed axons occupying the lower region. These pioneer uncrossed axons may be lost at a later age, since the deeper layers of the adult optic tract consist of crossed fibres.

As retinal axons approach the chiasma they become defasciculated and spread out. As they leave the chiasma they become bundled again to form the optic tract. Time-lapse video microscopy of living axons labelled with fluorescent dyes in cats and mice has revealed that axons pause several hours in the chiasma. The growth cones of uncrossed axons become highly branched with filopodia. After seemingly feeling out their environment they turn towards the ipsilateral optic tract before reaching the thin raphe of cells along the chiasmatic midline (Sretavan and Reichardt 1993; Godement *et al.* 1994; Marcus *et al.* 1995). They are presumably deflected by a chemical signal, which forbids them from crossing the midline (Godement *et al.* 1990).

Injection of a monoclonal antibody for glial cells has revealed a palisade of radial glial cells straddling the chiasmatic midline of the embryonic mouse. Growing ganglion cell axons enter the radial paliside and contact glial cells from which they pick up cell surface molecules with opposing effects on crossed and uncrossed axons (Colello and Guillery 1992). Also, in the embryonic mouse, a V-shaped pattern of CD44- neurones develops in the neuroepithelium at the site of the future chiasma a day or two before ganglion-cell axons arrive at the site. These neurones express neurotrophins and cell surface molecules, which either promote or inhibit axon growth (Sretavan *et al.* 1994). Their effects on growing axons can be seen when retinal explants are cocultured with cells from the chiasmatic midline of embryonic mice (Wang *et al.* 1995). When these epithelial neurones were removed by a specific antibody, subsequently arriving ganglion-cell axons failed to decussate (Sretavan *et al.* 1995). Specific inhibitory molecules have also been revealed in cultures of growing ganglion cells from the embryonic rat, an animal in which only a few axons from the ventrotemporal margin of the retina project ipsilaterally (Wizenmann *et al.* 1993).

The retinas, optic stalk, and various regions of the growing nervous system of many animals, including insects, fish, and mammals, express a family of **Slit** ligands known as Slit1, Slit2, and Slit3, which repel growing axons. The axon receptors for these ligands are known as **Roundabout**, or Robo receptors. Slit expressed by midline glial cells prevents ganglion-cell axons from crossing the midline in the fruit fly *Drosophila* (Kidd *et al.* 1999) and in the Zebrafish (Fricke *et al.* 2001). Slit also prevents ganglion cells from innervating inappropriate brain areas in these animals. In mice, Slit ligands are not involved in axon divergence in the chiasma. They form an inhibitory system that guides axons to their destinations in the diencephalon (midbrain), as described in Section 10.3.1b.

GAP-43 (neuromodulin), secreted by growing axons, may be involved in axon divergence at the chiasma. Disruption of the gene for GAP-43 causes many axons in the mouse chiasma to be misrouted into the ipsilateral optic tract and to form abnormal connections in the LGN and superior colliculus (Zhu and Julien 1999).

Removal of one eye in early fetal mice and ferrets causes axons that would remain uncrossed to accumulate at the chiasma (see Taylor and Guillery 1995). It seems that uncrossed axons are first inhibited from crossing and then depend on the presence of crossed axons from the other eye for their entry into the uncrossed pathway. Sretavan and Reichardt (1993) found that monocular enucleation did not affect the routing of uncrossed axons although it did abolish the pause in growth at the chiasma. The reason for this conflicting evidence remains obscure. Other evidence suggests that axons arising in the nasal and temporal retina do interact at the chiasma. Cultured nasal growth cones of the chick retina are just as likely to grow on nasal as on temporal axons, whereas temporal growth cones grow only on temporal axons (Bonhoeffer and Huf 1985). Temporal growth cones in culture collapse when they contact nasal axons. This suggests that nasal axons produce an inhibitor specific to temporal growth cones (Raper and Grunewald 1990).

It has been reported that many axons from the temporal hemiretinas of the fetal rat, ferret, and cat take the wrong route at the chiasma and are subsequently eliminated (Jeffery 1984, 1990). This provides a second mechanism for axon segregation at the chiasma. However, there is some uncertainty about this mechanism, since other evidence from mice, ferrets, and cats suggests that during early fetal stages all axons from the temporal hemiretinas take the ipsilateral route but, at a later stage, decussate and remain decussated. The late decussation could occur because the chemical signal forbidding decussation of temporal axons fades or because late-arriving axons are insensitive to that signal (see Sretavan 1990; Reese and Baker 1992; Baker and Reese

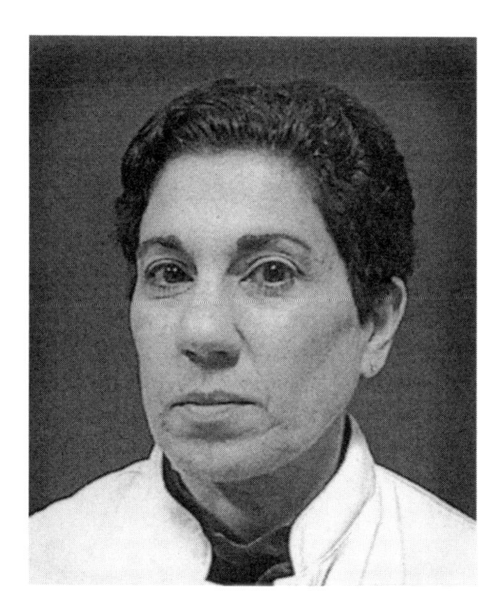

Figure 10.3. Patricia A. Apkarian (van der Veldt)
Born in the United States in 1949. She obtained a B.Sc. in Psychology from Michigan State University in 1971 and a Ph.D. in Visual Sciences from the Smith-Kettlewell Institute, University of the Pacific, San Francisco in 1979. She conducted postdoctoral work at the Smith-Kettlewell Institute and the Netherlands Ophthalmic Research Institute, Amsterdam. Since 1993 she has been a Staff Scientist in the Department of Clinical Neurophysiology, Amsterdam and member of the Faculty of Erasmus University Medical Centre, Department of Physiology, Rotterdam. In 1982 she received the Garland W. Clay Award from the American Academy of Optometry.

1993). Decussated temporal axons of cats are of a specific type, but their function remains obscure.

In fetal monkeys the adult pattern of almost complete hemidecussation is evident even before the development of ocular dominance columns and before the period of ganglion cell death (Chalupa and Lia 1991). In all mammals, axons from the nasal retinas fully decussate.

As optic fibres emerge from the mammalian chiasma to form the optic tract, crossed and uncrossed axons form crossed and uncrossed pairs. Axons segregate according to the type of ganglion cell (parvocellular, magnocellular, W-cells) from which they originate. Axons of each type acquire a retinotopic order. Those from the upper and lower halves of the retina segregate into anterior and posterior segments of the tract respectively. This segregation of dorsal and ventral regions is controlled by chemical factors carried on glial cells (Reese *et al.* 1994). These processes are controlled by regulatory genes that express patterns of proteins in overlapping longitudinal domains in the forebrain (Marcus *et al.* 1999). The cellular protein GAP-43 present in ganglion-cell axons is involved. Development of the optic tract is disrupted in rat embryos lacking this protein (Kruger *et al.* 1998).

Thus, three transformations of axons occur in the region of the chiasma. (1) Axons form into crossed and uncrossed pairs. (2) They segregate according to cell type. (3) Each cell type re-establishes a retinotopic order. Within this transition zone the glial cells change from an interfascicular organization to the radial organization typical of the diencephalon (Reese *et al.* 1994). Ganglion cell axons then enter the lateral geniculate nuclei where they segregate into distinct laminae (Section 10.2.4). The development of the chiasma was reviewed by Guillery *et al.* (1995).

10.2.3c *Achiasmatic animals*

Williams *et al.* (1991, 1994) identified an autosomal recessive mutation in some sheep dogs that causes all retinal axons to project to the ipsilateral LGN. The optic chiasma is thus eliminated—the animals are achiasmatic. The nasal retinal fibres terminate in ipsilateral layer A with the same topographic arrangement as that in the contralateral LGN in the normal dog. The temporal fibres project normally to the superimposed ipsilateral layer A1. Since the nasal fibres have not crossed the midline the nasal projection is mirror-image reversed with respect to the temporal projection and the two maps are congruent only along the vertical midline. Williams *et al.* argued that this reversed mapping could be explained if it is assumed that there is a fixed position-dependent chemoaffinity between retinal axons and LGN cells and layers. Thus, the selection of target cells in the LGN is controlled by the retinal position from which the axons originate rather than by their eye of origin. These dogs manifest spontaneous nystagmus associated with head oscillations. They exhibit the rare condition of see-saw nystagmus. This is a disjunctive vertical nystagmus in which each eye intorts as it rises and extorts as is moves down (Dell'Osso and Williams 1995).

Apkarian *et al.* (1994) described two achiasmatic children (Portrait Figure 10.3). Each optic nerve projected fully to the ipsilateral visual cortex as revealed by the absence of visual evoked potentials from the contralateral cortex. The VEP results were confirmed by magnetic resonance imaging (MRI). The children, like the achiasmatic dogs, exhibited congenital nystagmus with components of see-saw nystagmus (Dell'Osso 1996). The children lacked stereoscopic vision (Apkarian 1996). A similar case in a 35-year old man was reported by McCarty *et al.* (1992). Albinism involves the opposite type of defect in which temporal retinal axons decussate instead of projecting to the ipsilateral cortex (Section 12.5.2).

10.2.3d *Competitive survival of ganglion cells*
Ganglion-cell axons grow towards their target areas in subcortical nuclei such as the superior colliculus and lateral geniculate nucleus. The destination of a given axon seems to be determined by the retinal region from which it originates, since ganglion cells from a transplanted area of the protoretina in the toad still grow towards the destination appropriate to the original site (Fraser 1991). We will see later that other factors determine the precise way in which ganglion cells synapse with target cells in the LGN.

In the cat, ganglion cell axons start to grow in the optic nerve on the 19th embryonic day. By the 39th day the optic nerve contains about 600,000 axons. Between then and the second week after birth, the number of axons declines to the adult number of about 160,000 (Lam *et al.* 1982; Ng and Stone 1982; Williams *et al.* 1986).

In the monkey, by the 95th embryonic day the optic nerve contains about 2.85 million axons compared with 1.6 million in the adult. Axons are lost most rapidly between the 95th and 120th embryonic days, which is just when retinal terminals segregate into distinct layers in the LGN (Rakic and Riley 1983). The surplus of optic nerve axons is due to overproduction of ganglion cells rather than to axonal branching (Perry *et al.* 1983; Sefton 1986). In both normal cats and in cats with one eye removed there is close agreement between the number of ganglion cells and the number of optic nerve fibres at each stage of development (Chalupa *et al.* 1984). A similar loss of motor axons innervating muscles has been noted during early development (Cowan 1973).

Competition for the neurotrophic growth factor secreted by target cells with which ganglion cells make synaptic contacts determines which optic-nerve axons survive and which die. The growth factor binds to receptor molecules on the surface of the growing axon (see Allendoerfer *et al.* 1994). Signals are then transported rapidly down the axon to the cell soma, where they regulate the expression of genes that control the production of proteins required for cell growth and survival (Spencer and Willard 1992). In the absence of the growth factor the neurone dies. Death of ganglion cells in tissue culture is prevented when the culture contains target cells or a growth factor derived from target cells. In the growing visual system, cell death is promoted by removal of the growth factor (see Raff *et al.* 1993). Death is also promoted in cells with severed axons because they are cut off from the growth factor (Bray *et al.* 1992). Ganglion cells of an eye in which action potentials were blocked by tetrodotoxin died at the normal rate but those of the non-deprived eye died at a reduced rate (Scheetz *et al.* 1995). Ganglion cells in tissue culture respond to the neurotrophic growth factor but only when they are in an active state (Meyer-Franke *et al.* 1995). Cell growth and survival thus depend both on the growth factor and activation of the cell. Some growth factors are highly specific, and ensure that only appropriate synaptic contacts survive (Korsching 1993).

Removal of one eye reduces the number of cells competing for central connections. Thus, when hamsters and rats had one eye removed *in utero*, the optic nerve from the remaining eye had about 20% more axons than that of an eye of a normal animal (Jeffery and Perry 1982; Sengelaub and Finlay 1981). In cats with one eye removed prenatally, the remaining eye had about 180,000 ganglion cells compared with 150,000 in an eye of a normal cat, and the receptive fields of cortical cells were smaller in monocularly enucleated cats than in normal cats (Chalupa *et al.* 1984; Stone and Rapaport 1986).

In the developed retina of the monkey, the density of ganglion cells has been reported to be 300 times higher in the foveal region than in the far periphery (Perry and Cowey 1985). However, more recently, the density of ganglion cells has been estimated to be 1,000 higher in the fovea (Wässle *et al.* 1990). The processes responsible for the development of this differential density are not fully known. The loss of ganglion cells and their segregation into areas of different density are accompanied by differential growth of the retinal surface.

10.2.3e *Myelination*
Myelin is a fatty substance secreted as an insulating sheath around each axon by cells formed from a type of glial cell known as oligodendrocytes. The production of oligodendrocytes is under the joint control of a growth factor and electrical activity in the axons (Barres and Raff 1993). Ganglion-cell axons begin to myelinate after the period of axonal loss and the process mostly occurs postnatally. Axon diameter and conduction velocity are correlated with the thickness of the myelin sheath. When myelination is prevented by X-ray irradiation, axons do not increase in diameter, showing that growth of axon diameter depends on some factor derived from the myelin sheath (Colello *et al.* 1994). In humans, myelination proceeds from the brain towards the eye. Axons from the fovea myelinate before those from the peripheral retina. Myelination of subcortical visual pathways is complete by the third postnatal month (Yakovlev and Lecours 1967) and of the geniculocortical pathways by the seventh month (Magoon and Robb 1981). Myelination of the cerebral cortex system is not complete until early adulthood.

10.2.4 Development of the LGN

10.2.4a *LGN lamination*

The structure of the lateral geniculate nucleus (LGN) was described in Section 5.3.1. In the cat, retinal afferents from the contralateral nasal retina invade the LGN by the 32nd day of gestation, which is about 10 days after the first ganglion cells develop and about 5 weeks before birth. Afferents from the ipsilateral temporal hemiretinas invade the LGN about 3 days later than those from the contralateral nasal hemiretinas (Shatz 1983). This may be a consequence of the recent evolution of the non-decussating pathway. The visual effects of this asymmetrical development are discussed in Section 23.4.5. Regions of the LGN corresponding to the central retina (medial portions of each lamina) develop before those corresponding to the peripheral retina (lateral portions of each lamina) (Sretavan and Shatz 1987). Inputs from the two eyes in the cat are at first intermingled and become almost fully segregated into distinct layers by about the 54th day of the 64-day gestation period. In the ferret, inputs from on-centre and off-centre receptive fields segregate into distinct layers in the third post-natal week. Dendrites in the LGN of the neonate cat bear large numbers of spines and growth cones.

After about 4 months postnatally, the growth cones disappear, the number of spines decreases, and arborizations become restricted to the laminae appropriate to the eye of origin (De Courten and Garey 1982). Similar changes in humans extend over a longer period (Garey 1984).

Corticogeniculate and geniculocortical interconnections also develop in the pre- and postnatal periods. These interconnections developed even after retinal afferents were removed as they started to invade the geniculate nucleus (Guillery *et al.* 1985).

Injection of radioactive tracers into the eyes of monkey foetuses has revealed that all LGN cells are formed by the 64th day of gestation and their segregation into six laminae occurs between the 64th and 110th day (Rakic 1976). The LGN of the neonate monkey has the same general morphology and laminar structure as that of the adult. Magnocellular and parvocellular neurones are clearly distinguishable on the basis of their responses to visual stimulation, although immature cells have a lower spontaneous rate of firing and longer latency than do those of adult cells. Ganglion cells in the eye of the embryonic monkey diverge into magnocellular and parvocellular types soon after their last mitotic division. The two types of axon project to distinct alternating laminae of the LGN. Innervation of parvocellular laminae begins before innervation of magnocellular laminae (Meissirel *et al.* 1997). In a study of 53 human brains, geniculate cells increased rapidly in size during the first 6 to 12 months of postnatal life and then more slowly before reaching their full size at the age of 2 years. As in monkeys, cells in parvocellular layers develop faster than do those in magnocellular layers (Hickey 1977).

The prenatal segregation of visual inputs into eye-specific laminae in the LGN seems to involve an initial coarse segregation into similar cell types; for example, X, Y, and W types in the cat and magnocellular and parvocellular types in primates (see Casagrande and Condo 1988). In the cat, this is followed by growth of a terminal arborization and elimination of inappropriate axonal side branches (Sretavan and Shatz 1986a). In each LGN lamina, inputs from one eye competitively eliminate inappropriate synaptic contacts from the other eye (Shatz 1990b). In the macaque monkey, Snider *et al.* (1999) found no evidence of retraction of inappropriate axonal side branches in the neonatal LGN. They concluded that the formation of eye-specific projections depends on selective loss of whole ganglion cell axons.

10.2.4b *Lamina formation and neural activity*

Lamina formation in the LGN depends on the presence of retinogeniculate afferents. Thus, in the tree shrew, laminae do not develop following bilateral enucleation (Brunso-Bechtold and Casagrande 1985). But inputs from one eye are not sufficient. Thus, when Sretavan and Shatz (1986b) removed one eye in cats at embryonic day 23, before axons had reached the LGN, axons of the remaining eye developed a normal morphology and loss of axonal branches, but projected diffusely rather than being confined to normal eye-specific laminae.

There is a growing body of evidence that segregation of laminae in the LGN depends on bursts of synchronized nerve impulses arising in the prenatal retina (Meister *et al.* 1991). By the Hebbian rule, described in Section 10.4.5, synapses firing in synchrony in a given region reinforce each other and suppress activity at synapses firing out of phase with the dominant input. It is unclear how this competitive process propagates to form well-defined layers. Some of the evidence for the dependence of cell segregation on bursts of synchronized impulses will now be reviewed.

The segregation of inputs from on-centre and off-centre ganglion cells in the ferret LGN coincides with the onset of distinct spontaneous firing patterns from the two types of receptive fields. During the first 2 postnatal weeks both types of ganglion cell fire in a similar way. But, as GABA inhibitory circuits develop in the retina, the firing rate of on-centre ganglion cells declines relative to that of off-centre cells. This difference in firing rate allows Hebbian synapses (NMDA synapses) for the two types of ganglion cell to form distinct layers. Segregation does not occur after blockage of NMDA receptors (Fischer KF *et al.* 1998). The development of distinct temporal patterns of spontaneous activity through the growth of specific inhibitory circuits at specific times may be involved in other developmental processes.

When sodium-mediated action potentials in ganglion cells of the cat were unilaterally blocked by infusion of tetrodotoxin between embryonic day 45 and birth, the development of eye-specific layers in the LGN was severely disrupted by the proliferation of inappropriate dendritic growth across what would normally be eye-specific layers (Shatz and Stryker 1988). Dendritic morphology was affected less by bilateral blockage of action potentials (Dalva *et al.* 1994). Ganglion cells in the cat retina become capable of generating action potentials by embryonic day 30 and the subsequent increase in sodium-mediated action potentials coincides with the period of innervation of the LGN by ganglion cell axons (Skaliora *et al.* 1993). Ganglion-cell action potentials are also required for the postnatal maturation of LGN synapses but, at that stage, inputs arising from visual experience are more effective than electrically evoked potentials in nonseeing kittens (Kalil 1990).

In the ferret, an animal with well-developed stereoscopic vision, the projections to the LGN are mapped retinotopically at birth but their segregation into distinct laminae is not complete until a week or two after birth (Jeffery 1989). In the neonate ferret the ipsilateral projection to the LGN arises from only the temporal hemiretina, as in the adult animal, but the contralateral projection arises from the whole retina and only later becomes confined to the nasal hemiretina (Jeffery 1990).

The same is also true of retinotectal projections to the superior colliculus of the rat (Land and Lund 1979). Initially, axon terminals in the LGN from the two eyes overlap and then segregate into eye-specific layers. During this prenatal period spontaneous discharges in ganglion cells evoke excitatory postsynaptic currents in both NMDA and non-NMDA synapses in the LGN (Mooney *et al.* 1993). When spontaneous neural activity is pharmacologically blocked in one retina of the ferret while LGN laminae are developing, some axons from the active eye invade laminae that would normally receive inputs from the inactive eye (Penn *et al.* 1998). However, only a few axons from the active eye invade the monocular crescent of the inactive eye, showing that the coarse projection of axons to the LGN of the ferret does not depend on spontaneous retinal activity (Cook *et al.* 1999).

In monkeys, segregation of LGN laminae is severely disrupted by prenatal removal of one eye, although some segregation of magno- and parvocellular regions is still evident (Rakic 1981). Development of LGN laminae seems not to depend on postnatal visual experience, since monkeys deprived of vision in one eye from birth to 27 weeks develop a substantially normal LGN. This is so in spite of the fact that this type of deprivation leads to a reduction in the number of binocular cells in the visual cortex (Blakemore and Vital-Durand 1986a).

Competitive synaptic interactions generated by spontaneous neural discharges are probably also responsible for the fact that ganglion cells with on-centre and off-centre receptive fields segregate in the LGN and establish distinct connections with relay cells (Hahm *et al.* 1991). Early monocular deprivation permanently reduces the efficiency of synaptic transfer in the LGN (Section 13.2.1).

10.2.4c *Functional development of the LGN*

Within the first 4 weeks after birth, cells in the LGN of kittens showed low rates of maintained discharge to general illumination, weak and long-latency responses to flashed stimuli, and absence of surround inhibition. By 4 weeks the spatial characteristics of the cells' receptive fields achieve the adult form. After 4 weeks the temporal response of cells becomes more biphasic and there is a large decrease in response latency and response duration (Cai *et al.* 1997). Cells of the X-type showed mature response properties before Y-type cells (Daniels *et al.* 1978). Early geniculate responses are mainly excitatory, inhibitory circuits mature later (Ramoa and McCormick 1994).

The spatial resolution of a cell in the LGN is indicated by the highest spatial frequency of a drifting high-contrast grating that evokes a response in the cell. The spatial resolution of LGN cells with receptive fields at an eccentricity of more than 10° is much the same in the neonate monkey as in the adult. However, for the foveal region, LGN cells in the neonate monkey could resolve only up to about 5 cpd compared with 35 cpd in the adult (Blakemore and Vital-Durand 1986b). In the neonate monkey, the temporal resolution of LGN cells is low but rapidly improves in magnocellular cells over the first 2 months (Hawken *et al.* 1997).

The development of the LGN has been reviewed by Casagrande and Brunso-Bechtold (1988). The evolution of the mammalian visual pathways has been reviewed by Henry and Vidyasagar (1991).

10.3 DEVELOPMENT OF VISUAL CORTEX

10.3.1 Growth of cortical areas

10.3.1a *General divisions of the brain*

During development, cells become more specialized, in that the types of cell into which they may develop become more restricted. The zygote is a non self-renewing, **totipotent** cell because it gives rise to every type of cell but does not renew itself (unless it forms fraternal twins). Cells in different regions of the developing embryo become specialized to produce the cells appropriate to that region. They are referred to as self-renewing **multipotent stem cells**. They divide to form other multipotent cells or differentiate into more specialized **progenitor cells**. The progenitor cells for the central nervous system become further specialized into **neuroblasts**, which form neurones and **glioblasts**, which form oligodendrocytes and astrocytes.

The development of the vertebrate central nervous system starts with the establishment of the **neural tube** that runs dorsally along the anterior-posterior axis of the embryo. The **neural crest** forms above the neural tube. The neural tube is lined with a neuro-epithelium of multipotent stem cells that form the brain, spinal cord, oculomotor nerves, retina, and iris. Neural crest cells form the autonomic nervous system, most sensory nerves, and the cornea, sclera, and ciliary muscles of the eye. The stem cells of the neural tube migrate to the inner surface of the tube when they divide. Between divisions they synthesize DNA as they migrate up to then back from the outer surface of the tube. This cyclic movement is known as the elevator movement. Some stem cells differentiate into neuroblasts or glioblasts and migrate to the outer layer of the tube. The remaining stem cells divide at different rates in different parts of the tube to form the forebrain, midbrain, hindbrain, and spinal cord. The shapes of brains of different vertebrate species can be modelled by a computer simulation of differential growth rates in the neural tube (Fujita 1990). The forebrain is further subdivided into the telencephalon and diencephalon. The telencephalon forms the 6-layered neocortex, the 3-layered archicortex (hippocampus), the olfactory cortex, and the basal ganglia. The diencephalon forms the thalamus and colliculi. The adult human brain constitutes about 2% of the total body mass but consumes about 20% of the energy.

The anterior-posterior axis and the dorso-ventral axis of the growing neural tube is determined by protein signalling molecules derived from the neighbouring notochord (presumptive backbone). One such molecule is known as the **sonic hedgehog**, or Shh. The same molecule also determines the differential rates of growth of different parts of the brain (Britto *et al.* 2002). It has also been implicated in regulating cell division and survival in the embryonic retina, optic nerve and cerebellum. Thus, this molecule acts at different times and in different places to perform the same basic function of regulating cell division and survival. Other protein signalling molecules probably also perform their functions at different times and places.

The mammalian visual cortex, described in Section 5.4, is part of the convoluted surface of the neocortex. In primates, and particularly in humans, the neocortex is enlarged relative to other parts of the brain, mainly because of an expansion of the cortical surface (Hofman 1985). The neocortex of primates is about five times the volume of that of insectivores after allowance has been made for general increase in brain size (Barton and Harvey 2000). It contains areas for each sensory modality as well as motor areas and areas associated with a variety of cognitive and emotive functions. All cortical areas have six layers with an average total thickness of about 0.26 cm, which is similar in all mammals with brain volumes greater than about 3 cm^3 (Hofman 1989). Cells in the deepest layer 6 project to the thalamus, those in layer 5 project to subcortical nuclei other than the thalamus, and those in upper layers 2 and 3 project to other cortical areas. Layer 4 is the main recipient layer. All cortical areas have the same basic cellular constituents and show evidence of a radial columnar organization.

Across vertebrate species, brain weight increases as a power function of body weight (Hofman 1989). The cerebral cortices of small mammals, such as the mouse, are smooth, or lissencephalic. If brains of increasing size were to remain similar in appearance, the surface area would increase as the two-thirds power of the volume. In fact, the area of the mammalian cerebral cortex increases almost in proportion to its volume. This means that the brains of larger mammals must fold in order to fit in an economically sized skull—they become gyrencephalic. Thus, with species of increasing size the cerebral cortex becomes increasingly folded into **gyri** (convolutions) and **sulci** (grooves).

During the first half of the gestation period the human cerebral cortex has very few gyri or sulci. Among the first sulci to appear are the parieto-occipital and calcarine sulci (Polyak 1957). As the cortex grows, convolutions become increasingly complex.

At birth the human brain as a whole is only one-quarter of its mature volume, which is not reached until the age of about 2 years (Sauer *et al.* 1983). The volume of the adult human brain can vary between about 1180 and 1625 cm^3, and that of the neocortex between about 574 and 829 cm^3, without any correlated variation in intelligence (Filipek *et al.* 1994).

Before birth, the visual cortex grows more rapidly than other parts of the brain and at birth has reached about half its mature volume, which it reaches about 4 months after birth. The adult visual cortex occupies about one-thirtieth of the cortical surface. The visual cortex of the human neonate is between 1.4 and 1.7 mm thick, compared with between 2.1 and 2.5 mm in the adult (Wong-Riley *et al.* 1993).

The number of ganglion-cells, n, and therefore the number of cells in the LGN, increases with increasing eye size. To maintain constant resolution, the number of cortical processing units (hypercolumns) should increase in proportion to n. But to maintain constant angular resolution, the number of cells in each hypercolumn should increase by $n^{1/2}$. Therefore, the overall number of cortical cells should increase in proportion to n times $n^{1/2}$, or $n^{2/3}$. Stevens (2001) found that, over 23 primates including humans, the number of neurones in V1 is proportional to the 2/3 power of the number of LGN cells.

Cell density in the human visual cortex is over 1 million per mm^3 at 2 weeks of gestation. It decreases to about 90,000 per mm^3 at birth and then to about 40,000 per mm^3 at 4 months postnatally, after which it remains stable (Leuba and Garey 1987). This loss of cell density is due to overall growth, since there seems to be no loss of neurones in the visual cortex with aging. On average, there are over 13 billion neurones in the human neocortex (Braengaard *et al.* 1990). Neurogenesis continues in various regions of the adult cortex of the macaque, but not in the visual cortex (Gould *et al.* 1999). Neurogenesis does not seem to occur in the post-natal human cerebral cortex (Korr and Schmitz 1999). The cerebral cortex is reviewed in Mountcastle (1998).

10.3.1b *Intrinsic specificity of cortical areas*

Studies involving tissue culture, cell lineage, and cell transplantation have demonstrated that the embryonic nervous system exhibits laminar and regional specificities. These specificities arise from families of protein molecules expressed by specific groups of genes. These proteins in the cortical plate form density gradients and areal patterns before the arrival of thalamocortical afferents. For example, there are at least 25 homeobox genes expressed in the embryonic forebrain. These genes express proteins with a characteristic sequence of amino acids and are involved in specifying the body plans of many different species. In addition, several nonhomeobox transcription factors are expressed in specific regions and layers of the developing cortex (Bulfone *et al.* 1995).

Genes common to the whole cerebral cortex probably determine the general formation of cortical layers and columns. Particular progenitor neural cells are genetically programmed to produce proteins specific to a particular cortical area at a certain period, in the absence of influences from the surroundings of the cells. For example, progenitor cells from the presumptive limbic system express a specific genetic factor after they have been isolated *in vitro* (see Section 10.4.1a). Specification of different cortical areas and cell types in vertebrates is controlled by **neurogenic genes** similar to those controlling neurogenesis in invertebrates.

Cells destined to form a particular type of cell generate a transmembrane protein expressed by a gene called *Notch*. The protein acts on a receptor protein expressed by the gene *Delta* in neighbouring cells. This inhibits those cells from forming the same type of cell. A second mechanism depends on asymmetric cell division in which daughter cells express different quantities of regulatory proteins to form a chemical gradient (Lewis 1996).

An important family of regulatory proteins consists of EphA and EphB **protein tyrosine kinase receptors**, which tend to bind selectively to ligands ephrin-A and ephrin-B, respectively. Because both the ligand and its receptor are anchored to cell membranes, they are able to guide particular cells to particular locations. There are several numbered subtypes of each receptor-ligand class. These molecules are involved in the development of many body tissues in vertebrates, including the neuromuscular system and the visual system.

Donoghue and Rakic (1999) used DNA probes to examine the distribution of these molecules in slices of macaque visual cortex at days 65, 80, and 95 of the 165-day gestation period. They found that some members of the EphA-ephrin-A class of molecules are expressed in different patterns in the embryonic cortical plate, cortical laminae, and protocytoarchitectonic cortical zones before the arrival of thalamocortical afferents. For example, when cells are proliferating in the ventricular zone and migrating to the cortical plate (day 65), EphA6 is expressed in the region of the cortical plate corresponding to the future visual cortex and EphA3 is expressed in the region corresponding to layer 4 of the future extrastriate cortex. Other EphA-ephrin-A molecules are expressed or become differentially distributed in the cortical plate only after the arrival of thalamocortical axons (day 80-95). For example, ephrin-A5 becomes concentrated posteriorly and ephrin-A3 anteriorly during this period. In some cases, cells that become linked by Eph receptor and ephrin ligand molecules break apart in about 30 minutes and then repel each other. The break occurs because the ephrin molecules are cleaved by a transmembrane protease that is coexpressed with ephrin-A2 (Hattori *et al.* 2000).

Complementary gradients of EphA receptors on growing ganglion-cell axons and ephrin-A ligands are involved in establishing retinotopic mapping in the superior colliculus (Brown *et al.* 2000).

Ephrin-B1 appears in the neuroepithelium at the onset of neurogenesis and declines when neurogenesis is complete. It is expressed on radial glial cells and forms a high to low density gradient between the ventricle and pial surface (Stuckmann *et al.* (2001).

Cadherins are a superfamily of transmembrane proteins mediating a wide range of selective interactions between cells during neurogenesis and in the adult brain. Their role in guiding the axons of the growing visual pathways was described in Section 10.2.3a. Their role in cortical neurogenesis is not fully understood but they presumably help to label cortical domains so that incoming axons can recognize them. Some of them are expressed in the cerebral cortex of mice lacking thalamocortical projections (Miyashita *et al.* 1999). Thus, these intrinsic molecular markers specify the basic organization of the visual cortex independently of visual inputs. The differential expression of other cadherins and other genetic growth factors is not evident until after thalamocortical axons have invaded the developing cortex (Nakagawa *et al.* 1999). These factors are responsible for the influence that thalamocortical inputs have on the development of the visual cortex.

Various regions of the growing nervous system of many animals, including insects, fish, and mammals, express **Slit** ligands (Slit1, Slit2, Slit3) and their Roundabout, or Robo axon receptors. They form a general inhibitory system that prevents ganglion–cell axons from invading inappropriate areas of the thalamus, hypothalamus, and superior colliculus (Niclou *et al.* 2000; Ringstedt *et al.* 2000). The different types of Slit ligands are expressed in different places and at different times during development (Erskine *et al.* 2000).

The ephrins and Slits form two of four major families of molecules involved in axon guidance. The other two families are the **netrins** and **semaphorins**. The neurotrophins and extracellular adhesion molecules discussed in Section 10.4.1 also help to specify cortical connections. All these guidance molecules create the immense complexity of the central nervous system. The complexity depends on genetic regulation of their expression at different phases of development, and context-sensitive interactions between the different guidance systems (Yu and Bargmann 2001).

10.3.1c *Influence of thalamocortical inputs*

For some progenitor cells, there is a critical period during which they can adopt the morphological and chemical features characteristic of a cortical area into which they have been transplanted (see Levitt *et al.* 1997). Thus, the detailed structure of intracortical connections, in particular the receptive-field structure of neocortical sensory areas, seems to be determined by thalamic sensory inputs rather than by intrinsic properties of progenitor cells.

By ablating somatosensory inputs and the visual cortex of neonate hamsters Frost and Metin (1985) induced visual afferents to innervate the growing somatosensory cortex. Visual stimulation in the mature animals revealed cells in the somatosensory cortex with well-defined retinotopically organized receptive fields resembling those in the normal visual cortex. Similarly, late embryonic cells destined to form part of the visual cortex of the rat, when transplanted into the developing somatosensory area, formed the afferent connections and architectonic features characteristic of the somatosensory cortex (Schlaggar and O'Leary 1991). Visual afferents of neonate ferrets have been induced to innervate the auditory thalamus and auditory cortex. Cells innervating the auditory cortex exhibited characteristics of cells in V1. They showed orientation and direction selectivity and simple and complex receptive field organization, although their receptive fields were larger and they were less sensitive than cells in V1 (Roe *et al.* 1992).

Thus, the receptive-field structures of different areas of the neocortex are determined by the identity of cells that migrate into them from the thalamus rather than by the identity of cells ascending from the ventricular zone. However, the corticocortical and efferent connections of the visual cortex are not much affected by cross-modal manipulations (see Sur *et al.* 1990). Thus, efferent connections of the visual cortex are specified before the development of cortical layers or thalamic inputs.

Adult fish and amphibia retain the capacity to restore cortical connections after the optic nerve has been cut (Sperry 1951). Ganglion-cell axons regenerate and find their correct connections in the tectum. This process does not involve the creation of new ganglion cells (Beaver *et al.* 2001). If the cut ends of amphibian optic nerves are grafted onto the ipsilateral optic tract they form connections with the ipsilateral visual cortex. After recovery, the animals exhibit reversed optokinetic nystagmus and mirror-image reversal of movements in response to a prey

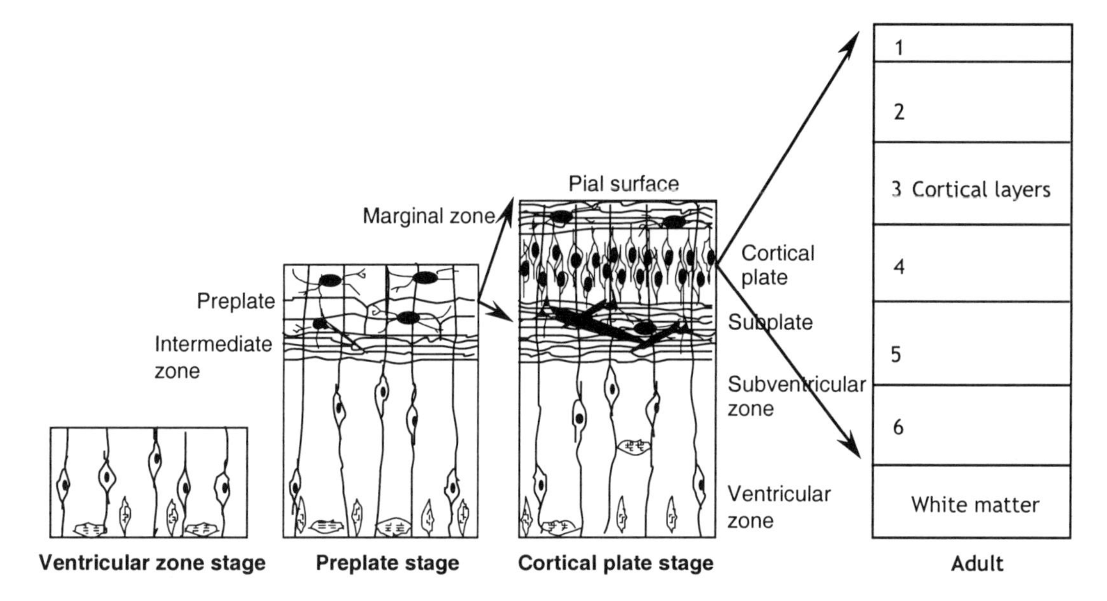

Figure 10.4 Development of cortical layers.
Progenitor cells divide in the ventricular zone and migrate along glial cells to form the preplate. The preplate then differentiates to form the marginal zone, cortical plate, and subplate. Cells from the ventricular zone continue to migrate through the cortical plate to form layer 6 of the mature cortex and then the other layers in reverse order. The intermediate zone and subplate transform into the white matter, and the other zones disappear. (Adapted from Uylings *et al.* 1990)

object (Sperry 1945). The visual cortex must treat the crossed inputs as if they originated in the eye that normally innervates that cortical hemisphere.

10.3.2 Formation of cortical layers

10.3.2a *Development of cortical layers*
In humans, the first cortical neurones develop at about 40 days after conception. Neurogenesis is complete by about the 125th day, halfway through gestation (Rakic 1988). The first cells to develop in the neuroectoderm are self-renewing **multipotent stem cells**. Each cell expresses two surface proteins, each of which binds to a different growth factor in the extracellular matrix. According to which growth factor the cell encounters, it forms one of two types of **unipotent progenitor cells**, known as glioblasts and neuroblasts (Davis and Temple 1994; Park *et al.* 1999). Glioblasts differentiate into various types of glial cells, and neuroblasts differentiate into various types of neurones. However, exposure of glioblasts to the appropriate growth factor can cause them to revert to multipotent stem cells, which may then develop into neurones (Kondo and Raff 2000).

A developing neurone first produces many neurites with growth cones. The growth cone of the neurite destined to become the axon is largest and extends numerous lamellipodia and filopodia with a high turnover of actin filaments (Bradke and Dotti 1999). The other neurites become dendrites.

Cell division occurs in a neuroepithelium, known as the **ventricular zone**, which lines the lateral ventricles. Specialized glial cells known as **radial glial cells** develop before neurogenesis starts. These cells have a short process extending to the ventricular plate and a fine radial fibre extending to the pial surface. These cells disappear or become astrocytes after neurogenesis is complete. Neurone progenitor cells migrate radially along each glial cell fibre. The migrating cells and the glial cell along which they migrate derive from the same stem cell to form a clonal unit (Noctor *et al.* 2001). The cells migrate through an intermediate zone toward the pial surface to form the **preplate**. Postmitotic cells in the preplate develop into the **marginal zone** at the pial surface, the underlying **cortical plate**, and the **cortical subplate**. Postmitotic cells migrating from the ventricular zone into the cortical plate form the six layers of the mature cortex (Figure 10.4). Details of cell migration are discussed in the next section.

In all mammalian species, cells first form the deepest layer 6. More superficial layers develop in an inside-out order from cells that migrate in successive waves from the subplate through layer 6 (Rakic 1974; Luskin and Shatz 1985). There is thus a correlation between a cell's birthday and its laminar position. In the human, cortical layers V and VI appear at 20 weeks of gestation and all layers are differentiated by 28 weeks (Yan *et al.* 1992). The extracellular protein reelin secreted by Cajal-Retzius cells in the

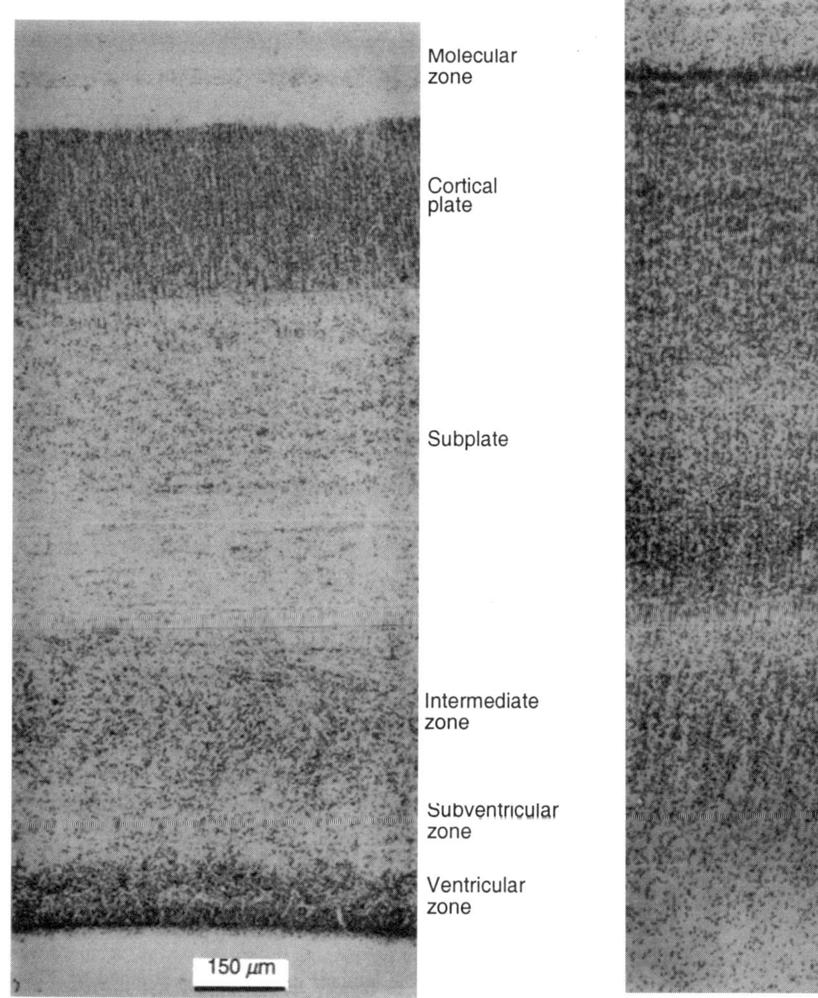

(a) Photomicrograph of a Nissl-stained section through the visual cortex of 17-week human foetus. Neurones develop in the ventricular zone, lining the cerebral ventricle. They migrate through the intermediate zone and cortical subplate to the cortical plate, just below the pial surface of the brain. The six layers of the visual cortex are not yet visible, but develop in the cortical plate.

(b) Photomicrograph of the visual cortex of the human neonate. The six layers are formed and the subplate has transformed into largely neurone-free white matter. (Reprinted from Yan *et al.* 1992, with permission from Elsvier Science)

Figure 10.5. The development of the human visual cortex.

marginal zone is essential for this process. In mice lacking the gene that encodes reelin, the preplate fails to divide into the marginal zone and cortical subplate (Sheppard and Pearlman 1997).

When cells first reach the cortical plate they have a simple structure. By the 30th week of gestation most cortical neurones have complex dendritic fields (Yan *et al.* 1992). Cortical zones of the 17-week human foetus are shown in Figure 10.5a. A section of the human visual cortex at birth is shown in Figure 10.5b. When the lamination of the visual cortex is complete, the intermediate zone and subplate form the virtually neurone-free white matter, and glial cells change into stellate astrocytes.

10.3.2b *Cell lineage and migration*

Migrating neurones can be followed by labelling them with tritiated thymidine. Cells derived from a particular progenitor cell can be traced by infecting the progenitor cell with a retrovirus that transfers only to the cell's offspring. The virus expresses a green fluorescent protein (GFP), which can be viewed continuously by time-lapse videomicroscopy.

At first, most progenitor cells divide symmetrically to produce more progenitor cells. The rate of subdivision is particularly high in the visual cortex. For example, in the mouse, each initial cell produces about 250 further progenitor cells. If we assume that each progenitor cell produces the cells in one cortical column then the final number of progenitor cells determines the number of columns. A brain factor gene may be responsible for the evolutionary expansion of the cerebral cortex by the way it controls the number of times a progenitor cell divides symmetrically to form other progenitor cells (Tao and Lai 1992). The number of times each progenitor cell divides to form neurones determines the number of cells in each cortical column. This number must be fairly constant over mammalian species, since the thickness of the cortex does not vary.

During the stage of neurogenesis, progenitor cells change to an asymmetrical, or stem-cell, mode of division in which one daughter remains a progenitor cell tethered to the preplate and the other differentiates into a neurone or glial cell, which undergoes no further cell division. Late in neurogenesis, progenitor cells divide symmetrically into two differentiating cells and the preplate becomes almost depleted of progenitor cells (Caviness *et al.* 1995).

The retinas of adult fish, amphibia, and birds contain stem cells that are capable of regenerating damaged retinal neurones (Fischer and Reh 2001). Multipotent stem cells occur in various regions of the brains of adult rodents, including the hippocampus, spinal cord, and lining of the ventricles. Application of growth factors induces these cells to differentiate into neurones or glial cells both *in vivo* and *in vitro* (Reynolds and Weiss 1992; McKay 1997). Pharmacological suppression of neurogenesis in the hippocampus of the adult rat impaired associative learning (Shors *et al.* 2001). Stem cells in the ventricular lining of the brain of the adult mouse divide very rarely in normal animals but divide rapidly after spinal cord injury (Johansson *et al.* 1999).

Gould *et al.* (1999) have produced evidence that, in the adult macaque, new neurones are continually produced in the subventricular zone of the prefrontal, posterior parietal, and inferior temporal areas, but not in the striate cortex (see Section 5.4.4f). Thus, the idea that new nerve cells cannot be formed in the adult mammalian nervous system must be revised. Stem cells from the human embryonic forebrain propagate *in vitro* and differentiate to form neurones or glial cells, but it is not known whether stem cells are present in the central nervous systems of adult primates (Carpenter *et al.* 1999; Korr and Schmitz 1999).

Pyramidal cells, spiny stellate cells, astrocytes, and oligodendrocytes derive from distinct progenitor cells, which originated from the same multipotent stem cells (Parnavelas 1999). It seems that most inhibitory interneurones (smooth stellate cells) do not develop from the ventricular zone but originate in the ventral telencephalon, a region that also gives rise to the basal ganglia. From there they migrate tangentially into the intermediate zone of the developing cortex (see Parnavelas 2000). Neurones differentiate first and glial cells differentiate during late embryonic or early postnatal stages.

Nadarajah *et al.* (2001) observed two types of radial cell migration in slices of embryonic mouse brain: locomotion and translocation. In locomotion, neurones are guided to their final destination by radial glial cells. In translocation, the neurone extends a leading process along which the nucleus of the cell migrates. Some cells migrate most of the way to their destination and then switch to translocation.

Some migrating neurones move laterally, sometimes changing their direction of movement (O'Rourke *et al.* 1992). Recent evidence suggests that these cells are progenitors of inhibitory interneurones (Parnavelas 2000). Lateral spread is caused more by movement of progenitor cells than by tangential migration of differentiating postmitotic cells (Kornack and Radic 1995; Reid *et al.* 1995). Thus, some progenitor cells migrate laterally and, as they migrate, divide symmetrically to produce a series of evenly spaced non-migratory daughter progenitor cells. Each progenitor cell that migrates radially divides asymmetrically two or more times to form radial clusters of cells of the same type; excitatory cells, or glial cells (McConnell 1995a). The radial clusters from the set of related progenitor cells form horizontal rows of 'cousin' cells within the cortical laminae.

The timing of cell differentiation is controlled by an interaction between factors intrinsic to the developing cells and extrinsic factors in the form of chemical signals from other cells. For example, stem cells, which normally produced neurones or glial cells in high density cell cultures, differentiated into smooth muscle cells when cultured at low density (Tsai and McKay 2000). Presumably, this effect depends on differential rates of diffusion of a chemical across cell membranes. The effect that a given chemical agent has depends on when it is applied. For example, sites on progenitor cells that are receptive to an epidermal growth factor increase in number during embryonic development. If a retrovirus is used to introduce extra receptor sites in the early stages, the progenitor cells behave like those in later stages and differentiate into glial cells rather than neurones (Burrows *et al.* 1997).

A progenitor cell in its early phase of cell division, when transplanted into an older host, behaves like the cells of the host (Gray and Sanes 1991). A progenitor cell in its late phase of cell division retains its initial identity when transplanted so that, if it was destined to migrate to layer 6, it continues to do so when transplanted into the visual cortex of an animal at the stage when cells were migrating into layers 2 and 3 (McConnell and Kaznowski 1991). Thus, the fate of a young neurone seems to be determined by cell-cell interactions during a critical period of cell division.

Axons growing in cultured slices of immature visual cortex from the ferret arborized within their appropriate layer. Neural activity arising from outside the cortex is thus not required. However, inhibition of neural activity in the cell culture decreased the specificity of cell arborizations (Dantzker and Callaway 1998). Therefore signals intrinsic to the visual cortex are involved in axonal guidance. It seems that, at an early stage, signals from neurones in an already complete cortical layer instruct the progenitor cell to begin forming cells for the next layer (see McConnell 1995b).

The subject of cell lineage and migration in the developing nervous system has been reviewed in Moody (2000).

10.3.2c *Arrival of afferents from the LGN*

Although geniculocortical afferents reach the cortical subplate well before their target cells in layer 4 are formed, they do not begin to migrate towards layer 4 until well after all cortical layers have formed (Lund and Mustari 1977; Rakic 1988). During this so-called **waiting period** LGN axons extend widespread branches within the subplate (Ghosh and Shatz 1992a). The resulting interactions between afferents and subplate cells are involved in selection of target cells in the cortical layers. The nature of these interactions is discussed in Section 10.3.3a. At the end of the waiting period, the subplate cells begin to die off along with the lateral connections of LGN neurones within the subplate (see Shatz *et al.* 1991).

As we will see later, it has been suggested that subplate neurones grow towards the thalamus and guide afferent fibres to their proper destination in cortical layer 4. When visual afferents reach their destination in cortical layer 4, they lose their growth cones, branch, and form synapses. It is believed that axon growth is arrested by a molecular stop-signal emitted by the cells of layer 4. These processes can be followed in cultured explants from the embryonic or postnatal cortex of the rat (Molnár and Blakemore 1999). After the cortical layers have formed, the cells in the various layers develop at a uniform rate (Lund *et al.* 1977).

10.3.2d *Loss of synaptic density*

Just before and after birth there is a rapid increase in the density and total number of synapses and in the thickness of layers in the primate visual cortex and in the cortex as a whole (Rakic *et al.* 1986).

In the rat, the synaptic density of the visual cortex increases up to the third postnatal week and then decreases to adult levels by day 25. Activity dependent release of cell adhesion molecules at NMDA synapses has been implicated in these developmental changes in the rat (Butler *et al.* 1999).

In the macaque monkey, synaptic density within the visual cortex increases exponentially to the third postnatal month, after which it decreases, at first slowly and then more rapidly to reach its adult value after about 5 years (Bourgeois and Rakic 1993). In humans, this increase is most rapid between the ages of 2 and 4 months but continues to the postnatal age of about 8 months, when mean synaptic density reaches about 25,000 per neurone. After about 8 months a massive but slow loss of synapses occurs to reach the adult level of about 10,000 synapses per neurone, which is reached by about the age of 11 years.

There is no evidence of neurone loss during this process but, because of increasing brain volume, cell density declines and cortical layers become thinner during the first few postnatal months in both humans and monkeys (O'Kusky and Colonnier 1982; Garey and de Courten 1983; Huttenlocher and de Courten 1987; Zielinski and Hendrickson 1992). Loss of synapses is related to the development of cortical dominance columns in which inputs compete for synaptic access to binocular cells. It is thus part of the process by which neural networks develop in response to maturational and environmental demands.

10.3.3 Development of cortical connections

10.3.3a *Interlayer connections*

Spiny stellate cells of cortical layer 4 are the primary recipients of visual afferents. In the cat, the projections of layer 4 cells to cortical layers 5 and 6 are complete by postnatal day 15. Connections within layer 4, which are the most complex, mature in 20 days. Connections to layers 2 and 3 take even longer to mature.

Callaway (1998b) labelled neurones in living brain slices from V1 of prenatal macaque monkeys. Connections between the main cortical layers developed with a high degree of selectivity, suggesting that they are genetically controlled. In cultured blocks of cortex from 10-day-old rats, axons grew most effectively on a membrane prepared from the cortical layer containing their target cells. Target cells must produce a layer-specific growth factor (Castellani and Boltz 1997). For example, layer 6 neurones form connections in layer 4 but bypass layer 5. The appropriate connections develop *in vitro* so that extrinsic influences are not involved. However, the connections were less specific when intrinsic spontaneous activity was suppressed by tetrodotoxin (Dantzker and Callaway 1998). Thus, although layer specific connections are determined in part by molecular markers on target cells, spontaneous neural activity must also be involved.

Connections between cells within the main layers show evidence of early exuberance followed by elimination of superfluous connections, suggesting that correct local connections are identified by neural activity.

10.3.3b *Intercolumn connections*

Cortical cells with similar orientation tuning are linked by fibres of pyramidal cells running parallel to the cortical surface in layers 2 to 6 (Section 5.4.6). Long-range fibres extend up to 6 mm and form fine branches distributed in repeating clusters corresponding to the repeating orientation-selective columns (Gilbert and Wiesel 1979; Rockland and Lund 1982; Luhmann *et al.* 1986).

In the cat, crudely clustered horizontal connections develop from layers 2 and 3 before the arrival of connections from layer 4, but the fine tuning of horizontal connections occurs after the arrival of inputs from layer 4 (Callaway and Katz 1992). Hata *et al.* (1993) traced the development of horizontal connections by a correlational analysis of spike trains recorded simultaneously at different lateral locations in the visual cortex of kittens, a procedure developed by Perkel *et al.* (1967). During the first 2 postnatal weeks, these connections were wholly excitatory; inhibitory linkages developed by the fourth week. Also, the connections were widespread between cells with very different orientation preferences and did not show the clustered distribution evident in the adult. By the seventh postnatal week, the connections became confined to a radius of about 600 μm and to cells with similar orientation tuning. At the same time, the clustering pattern emerged. The process involves the growth and elimination of synaptic connections rather than cell death (Callaway and Katz 1990). The pattern of clustering is less precise in cats binocularly deprived during the first few weeks of life (Callaway and Katz 1991). Hata *et al.* also found that correlated firings developed first in layer 4 but by the seventh postnatal week the frequency of firing in layer 4 declined to the low level typical of the adult and was overtaken by firing rates in other layers.

Dalva and Katz (1994) studied the development of connections in cultured slices of visual cortex from the ferret. The slices were perfused with a form of glutamate that remains inactive (caged) until photolyzed by ultraviolet light. By recording from a given cell as neighbouring cells were activated by a laser they were able to map out patterns of lateral connections. They confirmed that local connections are overproduced before birth and subsequently decline as long-range connections develop.

When action potentials in the visual cortex of ferrets were silenced by infusion of tetrodotoxin from post-natal day 21, the horizontal connections did not develop the adult clustering pattern. Binocular enucleation did not prevent the development of the initial crude clustering (Ruthazer and Stryker 1996). It seems that the initial clustering depends on spontaneous neural activity at the level of the LGN or cortex. The fine tuning of the clustering pattern probably depends on visual experience.

Burkhalter *et al.* (1993) used a fluorescent dye to observe the development of lateral connections in a series of post-mortem human visual cortices from 24 weeks of gestation to 5 years postnatal. Lateral connections first emerged at 37 weeks of gestation, after the development of radial connections at right angles to the cortical surface. As in the cat, lateral connections showed first in layer 4B, but also in layer 5 and then in layer 6. Fibre density increased rapidly, leading to the formation of a uniform plexus at 7 weeks postnatal. The patchiness typical of lateral connections in the adult cortex emerged after the 8th postnatal week. Longer-range lateral connections in layers 2 and 3 did not emerge until the 16th postnatal week and reached their adult form by the 15th month. These long-range connections were patchy to begin with and remained patchy. Layer 4B is associated with the magnocellular system and layers 2 and 3 with the parvocellular system. This evidence suggests that the magnocellular system develops before the parvocellular system.

Long-term changes in synaptic conductivity along these lateral pathways in the cat's visual cortex have been induced by pairing synaptic responses with conditioning shocks of depolarizing current (Hirsch and Gilbert 1993). The results suggest that these pathways are involved in stimulus-dependent changes in cortical responses. Long-range lateral connections could serve a variety of functions as listed in Section 5.4.7.

10.3.3c *Transcortical connections*

The projection of transcortical axons from area 17 to area 18 is established before birth in the cat. However, these axons are evenly distributed between the various cortical layers before the age of about 20 weeks, after which they arise mainly in layers 2 and

3. During the same period, cells projecting from area 17 to area 18 form into clusters by elimination of projections from intercluster zones (Price *et al.* 1994). These gross maturational changes do not depend on visual experience, since binocular deprivation up to 28 weeks of age did not stop them (Price and Blakemore 1985). However, binocular deprivation and blockage of afferents from the LGN reduced the density and precision of projections from area 17 to area 18 (Caric and Price 1999).

10.3.3d *Transcallosal connections*

Transcallosal connections are present at birth in the cat but the zone in each hemisphere from which they originate (**callosal efferent zone**) and the zone to which they project (**callosal terminal zone**) are spread out over areas 17 and 18.

Transcallosal axons in mice seem to be guided through the corpus callosum by two midline populations of glial cells, one above and one below the corpus callosum. Growing axons *in vivo* and *in vitro* are guided between the two populations of glial cells by repellent molecules (Shu and Richards 2001).

In the terminal zone, developing callosal afferents migrate along radial glial fibres, which guide them to their destination in the various layers of the cortex (Norris and Kalil 1991). The growth cones advance rapidly within the callosal tract with only brief pauses with extensions of filopodia. They then extend radially through the cortical layers until they reach the target region. Next, they pause and produce transitory branches and show repeated cycles of advance and withdrawal as if sensing out their environment (Halloran and Kalil 1994). The glycoprotein netrin secreted by the extracellular matrix has been implicated in the guidance of commissural nerve fibres (Serafini *et al.* 1996).

During the first postnatal week of the cat, the callosal terminal zone in each hemisphere becomes restricted to the boundary between areas 17 and 18—the region corresponding to the vertical midline of the visual field. During the first 3 postnatal months, the callosal efferent zone becomes less densely populated and also restricted to the midline region (Innocenti 1981). Dark rearing and binocular lid suture reduce the number of callosal neurones and produce an abnormally narrow distribution (Innocenti *et al.* 1985). In monocularly deprived and strabismic cats, the efferent zone remains densely populated and distributed beyond the midline region (Lund *et al.* 1978; Lund and Mitchell 1979a; Innocenti and Frost 1979). These results suggest that balanced binocular inputs are required for the thinning and regional restriction of callosal neurones. See Boire *et al.* (1995) for discussion of this issue.

For an eye with divergent squint, the median plane of the head projects to the temporal retina and hence to the medial bank of the ipsilateral lateral gyrus. For an eye with convergent squint the midline projects to the nasal retina and hence to the contralateral gyrus. In strabismic cats, the callosal connections are displaced in accordance with the sign of the strabismus, in an attempt to bring the field of the deviating eye into correspondence with that of the normal eye (Lund and Mitchell 1979b). However, Berman and Payne (1983) found that cats raised with convergent or divergent strabismus showed an expanded callosal efferent zone in both hemispheres and callosal terminal zones that, although wider than normal in some animals, were in their normal position. They thus found no support for the idea that abnormally placed callosal zones occur in the contralateral hemisphere for divergent squint and in the ipsilateral hemisphere for convergent squint. The functions of transcallosal connections were discussed in Section 5.8.

10.3.4 Development of cortical cell specificity

Braastad and Heggelund (1985) recorded from cells in area 17 of kittens between the ages of 8 days and 3 months. At 8 days the cells had receptive fields of both the X and Y type, spatially organized into excitatory and inhibitory zones. The receptive fields were large, and only about 40% of them showed tuning to stimulus orientation, compared with more than 90% at 4 weeks. As cells in areas 17 and 18 mature, their velocity preference shifts to higher velocities, they acquire greater orientation specificity, and their receptive fields become smaller (Buisseret and Imbert 1976; Milleret *et al.* 1988).

Pettigrew (1974) found direction-selective cells in the visual cortex of visually inexperienced kittens in the first 4 weeks after birth but he did not find cells with orientation or disparity specificity However, others have found cells tuned to orientation, especially to the vertical and horizontal, in 1-week-old dark-reared kittens (Blakemore and Van Sluyters 1975; Buisseret and Imbert 1976; Frégnac and Imbert 1978).

Freeman and Ohzawa (1992) reinvestigated this question using gratings. Even in 2-week-old kittens, they found many simple and complex cells in area 17 that responded to stimulation of either eye and were tuned to orientation and spatial frequency, although the responses were weak and unstable. Some of the cells were tuned to binocular disparity, although the proportion of such cells was lower than in the adult cat. The cells showed evidence of both excitatory and inhibitory binocular interactions.

Freeman and Ohzawa could not determine what proportion of cortical cells was responsive, because unresponsive cells could not be detected. Between the second and third week, the vigour of the responses increased and there was a substantial increase in the proportion of binocular cells. By the fourth week, the tuning of cortical cells was similar to that of the adult cat. They concluded that the development of the basic physiological apparatus for stereopsis in the cat predates visual experience and must therefore be genetically determined.

One must be careful not to exclude the possible influence of visual stimulation arising through the eyelids before the eyes open. Krug *et al*. (2001) found that gratings presented through the eyelids of ferrets 2 weeks before eye opening, evoked responses in the LGN and striate cortex. The responses in the cortex varied with the orientation of the grating.

Kittens reared in a visual environment of only vertical lines or only horizontal lines show a preponderance of cortical cells tuned to the orientation to which the animals were exposed (Hirsch and Spinelli 1971; Blakemore and Cooper 1970). Selective experience has little effect on the columnar organization of layer 4 of the visual cortex but, outside this layer, columns are maintained only for cells sensitive to the experienced orientation (Singer *et al.* 1981). There has been some debate about whether cells tuned to unstimulated orientations die off leaving only stimulated cells or whether the orientation preferences of unstimulated cells change to match the orientation to which the kittens have been exposed (see Stryker *et al.* 1978; Howard 1982). We will see in what follows that the maintenance and fine tuning of the stereoscopic system also depends on visual experience during early infancy.

Artificially induced synchronous stimulation of the visual pathways of 15 day-old ferrets reduced the specificity of orientation tuning of cells in the visual cortex (Weliky and Katz 1997). The induced stimulation acted like noise to weaken the naturally occurring synchronous activity generated by stimulation of aligned receptive fields.

The involvement of visual activity is indicated by the fact that suppression of NMDA synapses in the visual cortex of neonate ferret prevented the development of orientation selectivity but had no effect on the tuning of cells in the adult animal (Ramoa *et al.* 2001).

Miller (1994) elaborated a model of the development of the spatial properties of simple-cell receptive fields and of cortical columns from activity-dependent competition between ON- and OFF-centre visual inputs. Wimbauer *et al.* (1997) extended this model to the development of spatiotemporal properties of cortical cells through competition between both ON and OFF inputs and temporally lagged and non-lagged inputs.

Wiemer *et al.* (2000) applied an algorithm developed by Kohonen (1995) to simulate the development of the columnar representation of orientation, ocular dominance, and disparity in a hypercolumn of the visual cortex. The model incorporates lateral excitatory and inhibitory interactions and stimulus-induced plasticity resulting from a Hebbian learning rule. The simulator was exposed to stereo pictures of natural scenes. The resulting orientation and ocular dominance columns resembled those in the primate visual cortex. In addition, a wide range of disparities was represented in each region of constant orientation. Wiemer *et al.* suggested that each hypercolumn of the visual cortex is a module that encodes stimuli within a defined volume of space.

10.3.5 Ocular dominance column development

The organization of ocular dominance columns in the visual cortex was described in Section 5.6.2. Hubel and Wiesel (1963) found most cells in the visual cortex of visually inexperienced kittens between the ages of 1 and 3 weeks were responsive to inputs from either eye. However, they could be grouped into ocular dominance columns, according to which eye produced the larger response, in much the same way as in the adult cat.

LeVay *et al.* (1978) found no evidence of ocular dominance columns in neonate cats, but they became evident 3 weeks after birth and resembled the adult pattern by 6 weeks. Others have found some signs of ocular dominance columns in visual areas 17 and 18 in neonate cats (Albus and Wolf 1984; Blakemore and Price 1987a). Crair *et al.* (1998) found that cells in area 17 of neonate cats respond more strongly to stimulation of the contralateral eye than of the ipsilateral eye. The strength of ipsilateral responses increases to that of contralateral responses at about 3 weeks of age. Thus, it seems that segregation of ocular dominance columns starts from a condition of contralateral dominance rather than from a condition of equality. Crair *et al.* (2001) found anatomical evidence of ocular dominance columns in cats by postnatal day 14 but not by day 7. Thus, ocular dominance columns in the cat emerge before the onset of the critical period of cortical plasticity.

In the ferret, afferents from the LGN do not reach the visual cortex until at least postnatal day 9. Ruthazer *et al.* (1999) recorded the development of ocular dominance columns in the ferret (*Mustela putorius*) by injecting radioactive proline into one eye. The bands became evident about five weeks after birth and reached adult form by about nine weeks. Crowley and Katz (2000) used anterograde transport of tracers injected into eye-specific layers of the LGN and revealed ocular dominance columns by postnatal day 16. This is before the onset of visually induced responses and therefore before the period of visually induced cortical plasticity. Spontaneous discharges from the LGN could be involved. Furthermore, the ocular dominance bands developed when one or both eyes were removed during this period. Neither of these investigators determined electrophysiologically whether these bands responded normally to their respective eyes.

In the monkey, segregation of ocular dominance columns begins during the second half of gestation, although projections from the two eyes overlap extensively in layer 4C during the first 3 postnatal weeks (Wiesel and Hubel 1974; Rakic 1976). Ocular dominance columns and cytochrome oxidase blobs can be seen in the autoradiograph of the neonate monkey brain before the animals have been exposed to light (Des Rosiers *et al.* 1978; Horton and Hocking 1996c). Also, bilateral retinal ablation in midterm monkey embryos, before efferent axons had established synaptic connections in the cortex, did not prevent development of normally spaced cytochrome oxidase blobs and interblobs (Kuljis and Rakic 1990).

It thus seems that neither light stimulation nor spontaneous neural discharges are required for the development of the basic chemoarchitecture of the visual cortex. The initial development of ocular dominance columns probably depends on eye-specific molecular cues. But perhaps these molecular cues serve only as a guide for crude segregation of visual inputs. Neural activity is involved in the final shaping of ocular dominance columns and other cortical connections (see Section 10.4.6).

Stimulus-dependent changes in the degree of light absorption by the cortical surface revealed that, in the first $3^1/2$ weeks, ocular dominance columns of the cortex of the monkey increase in width by about 20%. This is about twice the rate at which the periodicity of the distribution of orientation preferences increases (Blasdel *et al.* 1995). At $3^1/2$ weeks the singularities in the orientation system show a tendency to lie in the centres of the ocular dominance columns, and the iso-orientation contours run orthogonally across the borders of ocular dominance columns, which is the adult pattern (Section 5.6.2). The full adult pattern of ocular dominance is established in monkeys by the age of six weeks (LeVay *et al.* 1980; Tychsen and Burkhalter 1997).

By the sixth postnatal day, V1 of the macaque contains an adult-like proportion of disparity-sensitive cells, although the cells are poorly tuned to spatial frequency and are not very responsive (Chino *et al.* 1997). The cells mature in the first 4 weeks, by which time the monkey begins to show evidence of stereopsis.

In adult cats and primates, the geniculocortical afferents from the two eyes project to distinct cells in layer 4C. From there, inputs from both eyes project to binocular cells in other layers (Section 5.4.3). In cats under 3 weeks of postnatal age, inputs from the two eyes are completely mixed in layer 4. Segregation into distinct cells is not complete until between 8 and 10 weeks of age (LeVay *et al.* 1978). This transition from a state of uniform binocular innervation to one of monocular dominance is believed to involve a phase of exuberant proliferation and homogeneous distribution of synaptic terminals followed by selective pruning or withdrawal of inappropriately connected dendrites. It has also been suggested that this pruning process is accompanied by expansion and maturation of appropriately connected dendrites. The process could also involve development of inhibitory connections.

Antonini and Stryker (1993) used immunohistochemical procedures to trace growth processes at the cellular level. They revealed that widely extending but immature branches of geniculocortical afferents are eliminated at the same time that other branches grow in length and complexity and segregate into patches according to their eye of origin. Mechanisms involved in the development of ocular dominance columns are discussed in the next section.

The metabolic enzyme cytochrome oxidase is present in the cortical plate of the human brain by the 26th week of gestation. When stained, it is a sensitive indicator of neural activity (Section 5.5.5). Cytochrome-oxidase blobs are evident in the human visual cortex by the 24th postnatal day and are well organized by the fourth month. Cytochrome-oxidase stripes in V2 are weakly evident in the human neonate (Wong-Riley *et al.* 1993). In contrast, both blobs and stripes are clearly evident in the visual cortex of the neonate macaque, although the distribution of the enzyme is not the same as in the adult animal (Horton 1984; Kennedy *et al.* 1985; Horton and Hocking 1996c).

Factors controlling the development of ocular dominance columns are discussed in Section 10.4.6.

10.4 GROWTH OF CORTICAL NEURONES

The role of cell adhesion molecules in guiding the growth of the visual pathways was discussed in Section 10.2.3. The role of protein tyrosine kinase receptors in specifying cortical areas was discussed in Section 10.3.1b. The neurotrophins are a third class of chemicals that determine cell connections in the growing nervous system.

10.4.1 Neurotrophins

When a growing axon comes within about 300 μm of its target cell in the LGN or visual cortex it is guided by diffusing **neurotrophins** secreted by the target cell (see Korsching 1993; Lindsay *et al.* 1994). Neurotrophins are a family of proteins, which includes four related proteins; nerve growth factor (NGF), brain-derived neurotrophic factor (BDNF), and neurotrophins NT-3 and NT-4/5. NGF is restricted to specific areas of the central nervous system while the other factors are more widely distributed. In addition, neurotrophic factors are produced by glial cells and fibroblasts.

The first neurotrophin, nerve growth factor, was discovered by Rita Levi-Montalcini and Stanley Cohen, working in the laboratory of Victor Hamburger in St Louis. Levi-Montalcini and Cohen received the Nobel Prize in 1986 (see Robinson 2001).

Each growing nerve axon produces one or more cell-surface proteins, each of which is a receptor for a specific neurotrophin. These protein molecules are known as **tyrosine kinase receptors**, or Trk receptors (Barbacid 1994; Lewin and Barde 1996). They are expressed by *trk* genes at the appropriate time during development. Low affinity Trk receptors bind to more than one neurotrophin while high-affinity receptors bind with a specific neurotrophin (Rodriguez-Tébar *et al.* 1990). After binding, the Trk receptor undergoes dimerization (bonding of two similar molecules) and autophosphorylation. These changes initiate a signal cascade in the axon, which spreads to the cell nucleus. Activation of Trk receptors in the cell membrane is required for cell survival, while activation of Trk receptors in vesicles within the cell facilitates cell differentiation (Zhang *et al.* 2000).

Some neurotrophins are released into the extracellular matrix while others remain bound to cell membranes (Katz and Callaway 1992). Some attract growing axons towards a particular target cell while others prevent axons from growing into particular regions (Chao 1992; Lindsay *et al.* 1994).

In addition to their role in attracting growth cones, neurotrophins affect cell survival, differentiation, and synaptogenesis (see Allendoerfer *et al.* 1994; Pimental *et al.* 2000). They trigger selective genes to express specific growth-promoting proteins at specific times and in specific regions of the growing nervous system. For example, *trk* genes express Trk tyrosine kinase proteins in growing axons and the *bcl-2* gene expresses a protein that promotes the growth of ganglion-cell axons. Over expression of this protein prolongs the period of cell growth (Chen *et al.* 1997). Similar processes are probably also responsible for the development of collateral connections that cortical cells receiving visual inputs (spiny stellate cells in layer 4) make with cells in layers 3 and 5, and that cortical output cells (pyramidal cells) make with layers 2, 3, and 5. Also, BDNF regulates the expression of one of the receptors for the neurotransmitter dopamine (Section 5.4.5g) in the central nervous system (Guillin *et al.* 2001).

Once growing afferents reach their target cells they compete for neurotrophins, which determine whether a given cell survives, and also modulate axonal and dendritic arborization. A cell gains competitive advantage in accessing neurotrophin by virtue of both spontaneous and stimulus dependent activity in the cell. Administration of NGF or BNDF to slices of embryonic rat brain increased the amplitude of stimulus-evoked synaptic responses and BDNF increased the frequency of spontaneous responses (Carmignoto *et al.* 1997). Administration of NGF or BDNF also reduced neuronal death (Hofer and Barde 1988). Production of neurotrophins is increased in the neighbourhood of neuronal activity and neurotrophins enhance transmitter release in presynaptic neurones containing the appropriate receptor molecules (Thoenen 1995). Neurotrophins potentiate the response of *in vitro* NMDA receptors and rapidly elevate their intracellular calcium levels (Jarvis *et al.* 1997).

Different neurotrophins are expressed at different times during development. For example, NT-3 is synthesized transiently in the retina, LGN, and visual cortex of the neonate rat and regulates the early development of neurones that express the receptor Trk C. In the cat, levels of NT-3 peak during the critical period for formation of ocular dominance columns but this growth factor is absent in the adult cat (Lein *et al.* 2000). There are strains of mice that lack NT-3 or Trk C. Also, Trk C can be inhibited by transgenic or retroviral introduction of a modified form of the gene for Trk C (Das *et al.* 2000). In all cases, there is a reduction in the number of all types of cell in the retina, suggesting that NT-3 serves to establish the number of progenitor cells from which all retinal cells develop. The rate of synthesis of NGF and BDNF in the rat is low during the first 10 postnatal days when axons invade the cortex, but is high during the third and fourth weeks when visual inputs from the two eyes compete for access to cortical cells (Schoups *et al.* 1995b). This point is discussed further in Sections 10.4.6d and 16.2.4d.

Neurotrophins and their Trk receptors vary from layer to layer of the developing visual cortex. For example, BDNF is confined to cortical layers 5 and 6 in the neonate cat but spreads to layers 2 and 3 and then to layer 4 by the end of the critical period for formation of ocular dominance columns (Lein *et al.* 2000). The effects of neurotrophins *in vivo* and on cultured slices of mammalian cortex suggest that they regulate the development of specific dendritic patterns in each cortical layer (McAllister *et al.* 1995; Liu *et al.* 1996; Gallo *et al.* 1997; Szebenyi *et al* 2001). In the living tadpole of the frog *Xenopus*, BDNF applied to the retina decreased arborizations of ganglion cells in the retina. But BDNF applied to the optic tectum increased ganglion-cell arborizations in the tectum (Lom and Cohen-Cory 1999). Thus, the same growth factor can have opposite effects in different regions. The specific receptor TrkB is required for the formation of ocular dominance columns (Cabelli *et al.* 1997).

Neurotrophins can interact with each other at particular synapses. For example, the development of dendritic arbours in cortical layer 4 is enhanced by BDNF and inhibited by NT-3, while dendritic arbours in layer 6 are enhanced by NT-3 and inhibited by BDNF (McAllister *et al.* 1997).

Ultimately, the chemoaffinity between visual afferents and their target regions in the visual cortex must be determined by the topographic expression of specific genes. Evidence of genetic transcription factors (proteins that bind to specific DNA sequences and control the expression of specific proteins) has been found in the retina and optic tectum of birds (Yuasa *et al.* 1996).

Thus, neurotrophins are involved in all aspects of **synaptogenesis**, which include the initial guidance of axons to their target cells, the arrest of the growth of the cones, the development of stable and specific synaptic connections between neurones, and the development of specific patterns of axonal and dendritic arborization. Neurotrophins peak in the neonatal period of neural development and are involved in activity-dependent synaptic plasticity during this period (see Section 10.4.6d). They are also active in the adult nervous system, especially in areas, such as the hippocampus, that are involved in learning. For example, BDNF promotes long-term potentiation (LTP) in the hippocampus by increasing the density of synaptic vesicles (Tyler *et al.* 2001).

Cholinergic afferents arising from subcortical structures innervate the developing cerebral cortex. Removal of these afferents retards neuronal development and alters cortical cytoarchitecture in the visual cortex of the cat, possibly through interactions between cholinergic afferents and neurotrophins (Hohmann and Berger-Sweeney 1998).

See Fitzsimonds and Poo (1998), McAllister *et al.* (1999), and Huang and Reichardt (2001) for reviews of neurotrophins.

Neuroligin is a protein localized on the postsynaptic membrane of developing excitatory synapses. It is a ligand for **neurexins**, which are presumably localized on the presynaptic membrane. Scheiffele *et al.* (2000) showed that addition of neuroligin to a culture containing pre- and postsynaptic neurones induced the development of synaptic vesicles. Removal of neuroligin arrested synaptic development.

10.4.2 Role of glial cells

Astrocytes, a type of glial cell, constitute about half the cells in the human brain (Section 5.4.4e). In the adult brain they surround synapses and are believed to provide metabolic support and clear surplus ions and neurotransmitter molecules from the synaptic cleft. Newer evidence suggests that astrocytes release their own glutamate neurotransmitter molecules that regulate synaptic transmission over the neurone synapse with which they are associated (Araque *et al.* 1999).

The role of radial glial cells in guiding the migration of neurones to their proper cortical layers was discussed in section 10.3.2. These glial cells disappear after neurogenesis. In young animals astrocytes are involved in guiding the growth of the visual pathways, dendritic growth within the cortex, and synaptogenesis. During the critical period, astrocytes produce a variety of cell adhesion molecules, including fibronectin and glial growth factors required for cortical plasticity (Lemke 2001).

Müller and Best (1989) transplanted living astrocytes from neonate kittens into one hemisphere of the visual cortex of adult cats and dead astrocytes into the other hemisphere. After 4 to 8 weeks of monocular deprivation, a change in the ocular dominance of cells occurred only in the hemisphere with living neonate astrocytes.

The period of synaptogenesis in the developing central nervous system coincides with the period when astrocytes develop (Ullian *et al.* 2001). Cultured retinal ganglion cells have a much higher level of synaptic activity when astrocytes are present. The astrocytes increase the influx of calcium ions and the number of vesicles released at the presynaptic membrane. Ganglion cells cultured for 14 days in the presence of astrocytes produced many more synapses than cells cultured without astrocytes. The critical factor is cholesterol secreted by the glial cells (Mauch *et al.* 2001). During the period of massive synaptogenesis the cholesterol available from neurones must be supplemented by that provided by glial cells.

Oligodendrocytes—the second type of glial cell—form myelin sheaths, which insulate axons.

10.4.3 Role of subplate neurones

Columnar and laminar organization is influenced by subplate neurones located in the cortical subplate below cortical layer 6 in the developing cortex. Even before lamination is evident in the embryonic cortex in rats, cats, and ferrets, first preplate and then subplate neurones send pioneer axons down the internal capsule to the LGN and superior colliculus (McConnell *et al.* 1989; De Carlos and O' Leary 1992; Allendoerfer and Shatz 1994). Axons growing from the LGN meet the preplate and subplate axons in the internal capsule and the two sets of axons form an intimate topographic association, which is maintained until LGN axons reach the cortical subplate (Molnár *et al.* 1998).

Subplate neurones also send axons into the cortical plate (the developing cortex), primarily into layers 1 and 4. Ghosh and Shatz (1992b, 1994) disabled the subplate neurones in a particular region of the visual cortex of 1-week-old kittens by local application of kainic acid. The cortex above the affected region failed to develop ocular dominance columns. The effect seems to be specifically related to the loss of subplate neurones, because application of kainic acid directly to layer 4 did not produce this deficit. Loss of subplate neurones produced an increase in brain-derived neurotrophic factor in the affected region (Lein *et al.* 1999). An oversupply of growth factor disrupts formation of ocular dominance columns (see Section 10.4.6d).

At an early stage of development, visual afferents form temporary synapses with subplate neurones. Subplate neurones receive spontaneous excitatory inputs from the retina and transmit these signals to cells in the cortical plate (Friauf and Shatz 1991). Spontaneous retinal activity transmitted to subplate neurones plays a crucial role in guiding the incoming axons to target cells of the growing visual cortex. Blockage of spontaneous activity by application of tetrodotoxin in embryonic cats severely disrupts the pattern of cell connections (Catalano and Shatz 1998). Removal of both eyes in monkeys at embryonic day 67, after formation of layers 5 and 6 but before formation of other layers, did not affect synaptic density in V1, or formation of cortical layers. However, it did disrupt the detailed structure of synaptic contacts on dendrites and neurone shafts (Bourgeois and Rakic 1996). Most subplate cells die in the early postnatal period.

Subplate neurones and their axons can thus be regarded as a temporary scaffold for the development of geniculocortical connections. Ablation of subplate neurones in cat foetuses disrupts the development of axons from the visual cortex to the LGN (McConnell *et al.* 1994). The subplate therefore aids in the development of both geniculocortical and corticothalamic pathways. Even before axons of subplate neurones reach the LGN, visual afferents start to grow towards the internal capsule, so they are not guided by subplate neurones in this early phase of their growth. Recent evidence suggests that the thalamic reticular nucleus and the perireticular nucleus help in the initial guidance of geniculocortical and corticothalamic axons. These two thalamic nuclei are particularly large when these connections grow (Mitrofanis and Guillery 1993).

10.4.4 Role of neural activity

10.4.4a *Neural activity and gene expression*

Long-term changes in neuronal function depend ultimately on proteins produced by specific genes. Activity in a variety of cells induces a particular group of genes, known as **immediate-early genes**, to produce mRNA transcription molecules. These molecules bind to specific immediate-early genes and induce them to produce specific proteins. Some of these transcription molecules, such as Fos, Jun, AP-1, and Zif268, are produced in neurones when the neurones are stimulated. The concentration of these molecules in cells of the visual cortex can vary over a period of a few hours depending on the degree of stimulation. For example, the concentration of Zif268 in the visual cortex is decreased in deafferented or dark-reared animals and recovers when the animals are re-exposed to light (Chaudhuri *et al.* 1995; Kaminska *et al.* 1996). The activity of singing in adult canaries stimulates the expression of BDNF, which in turn enhances the survival of new neurones (Li *et al.* 2000). Activation of NMDA synapses has been implicated in protein synthesis in developing synapses (Scheetz *et al.* 2000).

The broader topic of structural changes underlying memory in adult animals was discussed in Section 5.4.5h and is reviewed by Bailey and Kandel (1993) and Edwards (1995). Neuronal plasticity in the visual cortex has been reviewed by Frégnac and Imbert (1984) and by Rauschecker (1991).

10.4.4b *Retinal neuronal domains*

Mammalian ganglion cells generate spontaneous neural activity well before birth and even before maturation of receptors. Bursts of neural activity are generated in the rat retina by at least embryonic day

Figure 10.6. Donald O. Hebb.
Born in Nova Scotia, Canada in 1904. He obtained a BA in Psychology from Dalhousie University in 1925. He was a school teacher for several years before obtaining a Ph.D. in Psychology from Harvard University in 1936. In 1937 he became a Fellow of the Montreal Neurological Institute with W. Penfield and in 1939 he obtained an academic appointment at Queens' University in Canada. In 1942 he moved to the Yerkes Laboratory in Florida and in 1947 he was appointed Professor of Psychology at McGill University in Montreal.

17 (Galli and Maffei 1988) and in the cat retina by embryonic day 30 (Skaliora *et al.* 1993).

Optical imaging in neonate ferrets reveals that waves of spontaneous firing of ganglion cells spread over each of a varying mosaic of local areas of the retina at intervals of at least 50 s (Feller *et al.* 1996; Wong 1999). The discharges are blocked specifically by acetylcholine antagonists. Since acetylcholine occurs only in starburst amacrines, the spontaneous firing must originate in this subclass of amacrine cells, which synapse with ganglion cells and other amacrine cells. Different functional classes of ganglion cells develop their own firing patterns. The local waves of spontaneous firing produce near-neighbour correlations of firing of cells in the developing LGN and visual cortex and presumably help in the formation of central connections. The retinal waves of activity cease just before the eyes open (Wong and Oakley 1996).

10.4.4c *Cortical neuronal domains*
Spontaneous neural activity is involved in guiding afferent axons through the subplate to their target cells in the visual cortex (Section 10.4.3). In addition, domains of co-ordinated neuronal activity over distances of between 50 and 100 μm occur spontaneously in slices of cortex from the neonate rat. In each neuronal domain, waves of intercellular calcium propagate from a trigger cell at the centre. Propagation occurs over gap junctions between neurones and between glial cells rather than over synapses, which are not yet functional (Yuste *et al.* 1995). Domain activity is elicited by infusion of inositol triphosphate, a molecule involved in the second-messenger system of metabotropic glutamate synapses (Section 5.4.5b) (Kandler and Katz 1998). The cortical neuronal domains are elongated in a radial direction, which suggests that they help in the formation of cortical columns.

Garaschuk *et al.* (2000) used a two-photon scanning microscope (Section 30.2.3c) to observe slices of rat neocortex. They observed spontaneous oscillations involving intracellular calcium ions spreading over the longitudinal axis of the whole cortex at intervals of several minutes. These oscillations depend on glutamate receptors but it is not clear whether they involve normal synaptic transmission. They could be involved in the growth of long-range horizontal cortical connections.

After synapses have matured, a new type of spontaneous activity emerges, which is abolished by antagonists of chemical synaptic transmission (O'Donovan 1999).

Evidence discussed in Sections 13.2.2 and 13.2.3 reveals that the development of ocular dominance columns in the visual cortex depends on exposure of the two eyes to similarly patterned stimuli. The crucial factor may be synchrony of firing of inputs from the two eyes as they converge on binocular cells. The next section describes how synchronous activity of inputs converging on a Hebbian synapse improves synaptic conductance.

10.4.5 Hebbian synapses

10.4.5a *The Hebbian synapse*
Hebb (1949) proposed that synaptic contacts strengthen when activity in pre- and postsynaptic cells is correlated, and weaken when it is uncorrelated (Portrait Figure 10.6). A similar idea had been proposed by Ariens-Kappers *et al.* in 1936. Synapses behaving in this way are called **Hebbian synapses** (Section 4.3.2d). When activity in two presynaptic cells converging on a single cell is synchronous, it is more highly correlated with that in the postsynaptic cell than when it is asynchronous. This is because the postsynaptic membrane summates potentials from converging synchronous inputs more effectively than from asynchronous inputs. The outcome

is that correlated activity in two or more afferent pathways leads to **long-term potentiation** (LTP) of the transmission efficiency of that pathway. When converging inputs are persistently uncorrelated, the synaptic strength of the one more highly correlated with the postsynaptic potential eventually increases while the other suffers **long-term depression** (LTD). It is believed that these changes form the basis for synaptic plasticity during development and for learning in the adult.

The crucial events underlying neural plasticity at Hebbian synapses involve the coactivation of NMDA and non-NMDA receptors (AMPA and kainate receptors). When the two types of receptor occur on the postsynaptic membrane of the same cell they are said to be **colocalized**. Each receptor on a postsynaptic membrane consists of several glycoprotein molecules surrounding a pore (Section 5.4.5b). The glycoproteins round each pore differ to form distinct receptor subunits, which help to determine the specificity of responses to diverse inputs to a given synapse. The distinct subunits differentiate during early development, apparently in response to specific presynaptic inputs (Sheng *et al.* 1994; Gottmann *et al.* 1997).

Conductance at non-NMDA receptors (Section 5.4.5b) depends only on the presynaptic release of the neurotransmitter glutamate. By contrast, the response of NMDA receptors to glutamate is blocked by magnesium ions unless the postsynaptic membrane is depolarized. The postsynaptic membrane is depolarized by the activation of fast acting non-NMDA receptor sites (particularly AMPA sites) on the same postsynaptic membrane. Backpropagating action potentials within the postsynaptic dendrites may be involved (Markram *et al.* 1997, Section 4.2.2). Synapses containing only NMDA receptors cannot respond, since the blocking magnesium ions are not removed. They are known as **silent synapses**. Silent synapses become active once AMPA receptors are introduced into the postsynaptic membrane.

During LTP, presynaptic glutamate release following removal of magnesium ions by postsynaptic depolarization triggers a voltage-dependent response in the NMDA receptors. This produces a graded release of postsynaptic calcium ions into the cell. When the level of calcium ions exceeds a critical level it triggers an accumulation of **calcium-calmodulin-dependent kinase II** (CaMKII) molecules and other protein kinases (Cristo *et al.* 2001) in cytoplasmic structures abutting postsynaptic receptors (Lisman 1989; Asztély and Gustafsson 1996).

Calcium calmodulin kinase (CaMKII) is an enzyme that remains active for about 30 minutes after the trigger stimulus has ceased. This is because of its ability to autophosphorylate. The protein acts as a molecular switch. Experience-dependent cortical plasticity in the visual cortex is absent in mice lacking the gene responsible for autophosphorylation of CaMKII (Glazewski *et al.* 2000). Furthermore, introduction of the enzyme into *in vitro* hippocampal cells enhanced synaptic transmission (Shirke and Malinow 1997). There is conflicting evidence about whether inhibition of CaMKII reverses LTP after it has been induced (Chen *et al.* 2001).

The accumulation of CaMKII produces an increase in the permeability of the cell membrane to calcium and a persisting increase in the number and conductance of AMPA receptors (GluR1) on the postsynaptic membrane (Mayer *et al.* 1984). Inactive synapses on cells within about 70 μm of an active Hebbian synapse also manifest long term potentiation (LTP) (Engert and Bonhoeffer 1997). Even a short period of activation of an NMDA synapse can lead to increased transmission efficiency lasting weeks.

Stimulation that produces concentrations of calcium ions below the critical level triggers long-term depression (LTD) of synaptic conductance (Singer 1990; Kirkwood and Bear 1994a; Ghosh and Greenberg 1995; Scanziani *et al.* 1996). The critical level of stimulation for the transition between LTP and LTD is itself increased after a period of increased activity and decreased after a period of decreased activity (Kirkwood *et al.* 1996). Long-term depression is accompanied by a decrease in the number of AMPA receptors on the postsynaptic membrane.

Thus, the crucial event in LTP is an increase in AMPA receptors on the postsynaptic membrane, and the crucial event in LTD is a decrease in AMPA receptors. During LTD, AMPA receptors are removed from the synaptic membrane by becoming attached to proteins in a pit, which pinches off and forms a capsule in the cytoplasm of the cell. This process is known as **endocytosis**. During LTP, AMPA receptors return from the cytoplasm to the synaptic membrane. Other types of receptor, including GABA receptors, show the same dynamic balance between endocytosis and membrane insertion. The internalization of receptors does more than regulate synaptic transmission. For example, endocytosis of neurotrophins affects the tropic functions of the cell (Carroll *et al.* 2001).

There is debate about whether LTP involves presynaptic as well as postsynaptic changes. The usual method of measuring neurotransmitter release is to record postsynaptic potentials electrophysiologically, a method that does not clearly distinguish between pre- and postsynaptic changes. Long-term potentiation is not induced by either postsynaptic depolarization alone nor by presynaptic stimulation alone (Otsu *et al.* 1995).

Patch clamp recordings from the presynaptic membrane of cultured hippocampal synapses has revealed that repetitive correlated firing of pre- and postsynaptic neurones results in rapid and persistent enhancement of presynaptic excitability, lowered spiking threshold, and reduced variability of spike frequency (Ganguly *et al.* 2000). A fluorescent marker of presynaptic activity has revealed enhanced neurotransmitter release after LTP (Zakharenko *et al.* 2001). Inhibition of the presynaptic enzyme, protein kinase, abolished these changes without affecting postsynaptic LTP. It thus seems that there are independent presynaptic and postsynaptic processes involved in LTP. Both depend on correlated activity at NMDA synapses.

Pharmacological methods allow one to visualize the release of neurotransmitter from presynaptic vesicles at individual synaptic boutons (Malgaroli *et al.* 1995). These methods have revealed that local modifications of presynaptic neurotransmitter release contribute to LTP in NMDA synapses. This process seems to be mediated by the post-synaptic release and diffusion of the gas nitric oxide through the post-synaptic membrane (Montague *et al.* 1994; Wu *et al.* 1994). There has been some dispute about the role of nitric oxide in synaptic plasticity. In some studies, LTP was prevented by inhibition of nitric oxide synthesis (Schuman and Madison 1991) while others have reported that sensory-dependent plasticity resulting from monocular deprivation in kittens is not affected by inhibition of nitric oxide synthesis (Reid *et al.* 1996). More recent evidence suggests that there is more than one way to synthesize nitric oxide. Animals in which the genes for both types of synthesis are deleted do not show synaptic plasticity (Hölscher 1997).

Glutamate synapses of the metabotropic type (Section 5.4.5b), which involve a cascade of second messengers in the receptor, are also implicated in long-term potentiation (Bortolotto *et al.* 1994; Riedel 1996). Application of a metabotropic glutamate agonist to the visual cortex potentiates the response of NMDA synapses (Wang and Daw 1996). The second messengers of metabotropic receptors may be involved in memory consolidation. Metabotropic receptors are particularly prevalent during the critical period for formation of ocular dominance columns (Reid 1995). However, metabotropic receptors are not involved in the long-term depression of responses of cortical cells to stimulation of a visually deprived eye (see Section 13.2.4b).

Neurotrophins and the neurotransmitters γ-aminobutyric acid (GABA) and norepinephrine are other factors that have been implicated in the development of the visual cortex. This topic is discussed in Section 13.2.4.

Astrocytes may also be involved in transmission at NMDA synapses. Glutamate released at non-NMDA synapses stimulates neighbouring astrocytes to release the amino acid D-serine into the synaptic gap of NMDA synapses. The gene that makes the enzyme for D-serine is active in these astrocytes. Application of an enzyme that degrades D-serine reduced transmission at NMDA synapses in rat brain slices (Wolosker *et al.* 1999).

We can summarize the action of a Hebbian synapse by saying that an excitatory input too weak to activate NMDA receptors above a critical level decreases the efficiency of activated synapses. An excitatory input above the threshold for activation of NMDA synapses increases the efficiency of simultaneously activated synapses. We thus have a mechanism for detecting coincident and correlated activity in inputs converging on a common cell. In particular, we have a mechanism for establishing use-dependent neural networks sensitive to binocular disparity. Berns *et al.* (1993) developed a computer model of this process.

Other structural changes underlying learning were described in Section 5.4.5h.

10.4.5b *Hebbian synapses and the visual system*

Most investigations of the role of NMDA synapses in neural plasticity have used *in vitro* preparations of cells from the hippocampus, a region of the old cortex, or palaeocortex, implicated in spatial memory (Fazeli 1992; Sherry *et al.* 1992). Inhibition of NMDA synapses in neonatal rats triggers a wave of cell death in the hippocampus and in the parietal and frontal lobes (Ikonomidou *et al.* 1999). Investigations of *in vitro* slices of visual cortex of rats and mice have revealed involvement of NMDA synapses in neural plasticity in that area (Kirkwood *et al.* 1993; Kirkwood and Bear 1994b; Weiss *et al.* 1993). Furthermore, low-frequency electric shocks delivered to the white matter underlying both the hippocampus and the visual cortex produce a long-term increase in synaptic conductance, except when NMDA receptors are inhibited by an antagonist (Kimura *et al.* 1989). However, the precise role of NMDA receptors in cortical plasticity in the visual cortex is still a subject of debate and many details remain to be worked out (Fox and Daw 1993; Kaczmarek *et al.* 1997).

The post-synaptic membrane of NMDA synapses contains receptor subunits **NR1** and **NR2**, each with further subtypes. Subtype NR2B predominates in the neonatal rat brain and is later supplemented or replaced by subtype NR2A. This switch is accompanied by a shortening of response times. The switch is delayed by light deprivation but is rapidly resumed with the onset of light (Quinlan *et al.* 1999). There is thus an experience-dependent regulation of NR2A synthesis, and hence of response times of NMDA synapses (Philpot *et al.* 2001).

Cortical plasticity in the visual cortex of the rat decreases during the first 35 postnatal days, as evidenced by the fact that long-term potentiation in response to stimulation of geniculocortical axons becomes increasing difficult to induce. Application of brain-derived neurotrophic factor restores synaptic plasticity (Sermasi *et al.* 1999).

10.4.6 Development of ocular dominance columns

10.4.6a *NMDA synapses*

The following lines of evidence suggest that the segregation of distinct ocular dominance columns depends on NMDA synapses activated by the neurotransmitter glutamate (Section 5.4.5b) (Constantine-Paton *et al.* 1990). In the first place, the density of NMDA synapses increases abruptly in all layers of the visual cortex of the kitten during the period when inputs from the two eyes segregate into ocular dominance columns. After this period it declines to the adult level (Bode-Greuel and Singer 1989; Fox *et al.* 1989; Czepita *et al.* 1994). Susceptibility to long-term potentiation in cultured slices of tissue taken from the visual cortices of rats at different ages coincides with the period of ocular dominance segregation (Kirkwood *et al.* 1995).

The postsynaptic membranes of mature NMDA synapses also contain AMPA receptors—the receptors are colocalized. However, NMDA synapses in the immature cortex lack AMPA receptors. For this reason they have a high threshold and are called silent synapses. During the first postnatal month, active synapses acquire AMPA receptors and the response threshold declines (Feldman and Knudsen 1998). At the time of formation of ocular dominance columns, postsynaptic activity in NMDA synapses in the visual cortex of the rat is more pronounced than in the adult. (Carmignoto and Vicini 1992). High activity of NMDA synapses is therefore associated with cortical plasticity. Infusion of an antagonist for NMDA receptors in the cat visual cortex profoundly suppressed responses to visual stimulation (Miller *et al.* 1989a).

Roberts and Ramoa (1999) used histochemical and *in vitro* patch-clamp procedures to measure the post-synaptic receptor subunits of NMDA synapses in the developing visual cortex of the ferret. Before the period of ocular-dominance plasticity, there is a preponderance of NR2B receptor subunits. After eye opening, subunits NR1 and NR2A increase rapidly but NR2B increases beyond its initial level only after day 30, the period of greatest ocular dominance plasticity, when the animal is most sensitive to monocular deprivation. At the same time there is a decrease in the decay time of receptor mediated synaptic currents. At the period of greatest ocular dominance plasticity, NMDA receptors become both more numerous and faster acting. Glutamate synapses that are selectively sensitive to the synthetic ligand quisqualate are also particularly numerous at the time of maximum cortical plasticity in cats (Dudek and Bear 1989).

The experience-dependent co-ordinated mapping of ipsilateral and contralateral visual projections onto the tectum of developing frogs (*Xenopus*) is disrupted by application of antagonists to NMDA (Scherer and Udin 1989).

NMDA receptors are involved in changes in ocular dominance induced by early monocular deprivation (see Section 13.2.4b).

10.4.6b *Other neurotransmitters*

All intracortical excitatory synapses and cortical synapses receiving inputs from the sense organs involve the neurotransmitter glutamate. But the cortex also receives inputs from a variety of subcortical areas, each involving a distinct neurotransmitter, such as acetylcholine, noradrenaline, dopamine, and serotonin (Section 5.4.5g). Cortical cells contain a variety of receptors for each of these transmitters, which are believed to modulate the responses of cells. Application of the neuromodulators acetylcholine and noradrenaline to the visual cortex of anaesthetized neonate kittens facilitates stimulus-induced changes in ocular dominance and orientation selectivity (Greuel *et al.* 1988). Also, noradrenaline is involved in the adaptive response of cortical cells to monocular deprivation (Section 13.2.4e).

In the visual cortex of kittens between 30 and 80 days of age, cells receptive to the neurotransmitter serotonin are concentrated between cytochrome oxidase blobs (Kojic *et al.* 2000). Application of serotonin increases synaptic plasticity in these regions but decreases plasticity in the blobs. These location-specific effects of serotonin may therefore promote the development of cortical columns. Serotonin is expressed by neurones originating in the raphe nucleus (Section 5.4.5f).

10.4.6c *Competition for neurotrophins*

It is generally believed that the development of ocular dominance columns depends on competition between visual afferents for access to a neurotrophic growth factor produced by cortical cells. A cell gains competitive advantage by virtue of spontaneous and/or stimulus dependent activity in the cell. Neuronal activity increases the local production of neurotrophins, which enhance transmitter release in presynaptic neurones containing the appropriate receptor molecules (Thoenen 1995). In a given region, competition drives binocular cortical cells to be dominated by one eye. Intracortical excitatory interconnections bias neighbouring cells to have similar eye dominance. Long-range inhibitory connections bias more distant cells to have opposite eye dominance. If one eye is closed during the critical period in the first few months of life, the bands corresponding to the open eye expand at the expense of those corresponding to the closed eye. Lack of correlated inputs from the two eyes weakens the tendency to form connections between binocular inputs in favour of the tendency to form connections between the correlated inputs from neighbouring regions of the same eye (Section 13.2).

Stimulation of one eye combined with an electrophysiologically induced increase in postsynaptic activity increased the ocular dominance of cells in the visual cortex of anaesthetized cats. The change lasted several hours and occurred most readily during the critical period of development but was also present in adult cats (Frégnac *et al*. 1988).

Swindale (1981) found no evidence of ocular dominance columns in 20-week-old dark-reared cats. This suggests that visual stimulation is required in addition to spontaneous discharges at an earlier age. Autoradiography failed to reveal any ocular dominance columns in the visual cortex of kittens after both eyes had been injected with tetrodotoxin between the ages of 14 days and 8 weeks. Nearly all cortical cells were found to be well driven by stimuli in either eye (Stryker and Harris 1986). Simply rearing kittens in the dark did not have this effect. Tetrodotoxin does not arrest general growth of dendrites or produce non-specific growth. Instead, it interferes with segregation of afferents into eye specific clusters (Antonini and Stryker 1993). According to this evidence, ocular dominance segregation in cats depends on inputs from the eyes, but these inputs can be the spontaneous firing of ganglion cells and need not arise from visual stimulation.

LeVay *et al.* (1980) found ocular dominance columns in dark-reared monkeys, which suggests that their initial development does not require visual experience. I have already mentioned that ocular dominance columns begin to develop prenatally in monkeys but not in cats. This development could depend on spontaneous neural activity.

Crowley and Katz (1999) found ocular dominance columns in ferrets in which both eyes had been removed in early in life. Spontaneous firing of cells in the LGN or cortex may have been involved in their development. However, spontaneous activity in the LGN is synchronized mainly between layers—within-layer synchrony is required to generate eye-specific signals. Crowley and Katz concluded that column segregation depends on eye-specific molecular cues. But even if these molecular cues exist, they may guide only a coarse segregation of visual inputs. Neural activity is involved in the final shaping of cortical connections, as we shall now see.

Neural activity regulates the production of BDNF which in turn controls the production of cell proteins required for the growth of axons and synapses. Thus, dark-rearing or blockage of neural activity during the critical period of column formation reduces the level of BDNF in the LGN and visual cortex (Schoups *et al.* 1995b). Blockage of neural activity in one eye of the cat leads to a reduction of BDNF in layer 4 neurones of the visual cortex that receive inputs from the blocked eye (Lein and Shatz 2000).

Infusion of excess of neurotrophins BDNF or NT-4/5 into the visual cortex of the cat during the critical period removes the basis for competition and ocular dominance columns do not develop in that region (Cabelli *et al.* 1995). In addition, excess of these neurotrophins during the critical period maintains, or even restores, the responsiveness of binocular cells to stimulation of a deprived eye (Gillespie *et al.* 2000). Removal of the cortical subplate in the developing brain causes an increase in BDNF, which disrupts column formation (Lein *et al.* 1999). Infusion of BDNF has a hypertrophic effect on afferents from both the deprived and non-deprived eyes of monocularly deprived kittens, but only during the critical period for column formation (Hata *et al.* 2000).

Infusion of neurotrophins NGF, BDNF, or NT-4/5 also counteracts the effects of monocular deprivation in the visual cortex of rats (Lodovichi *et al.* 2000). Formation of ocular dominance columns is also disrupted by inhibition of the ligand, TrkB, for these neurotrophins (Cabelli *et al.* 1997). It seems that neurotrophins that activate TrkB during the critical period promote connections between cortical cells even though there is no correlated activity arising from the two eyes.

The fourth neurotrophin, NT-3, or its ligand TrkA, has no effect on the formation of ocular dominance columns because it is expressed at an earlier stage of development.

Other studies on the role of neural growth factors in the development of ocular dominance columns are discussed in Section 13.2.4d.

Bienenstock *et al.* (1982) and Swindale (1982) developed mathematical models of the development of ocular dominance based on the idea of competition between inputs from the two eyes and excitatory and inhibitory interactions between cortical cells. Bienenstock *et al.*'s model emphasizes competition in the temporal domain rather than the spatial domain and has been implemented by Blais *et al.* (1999). Models of these processes have also been proposed by Malsburg (1979), Goodhill (1993), Miller *et al.* (1989b), Bauer (1995), Shouval *et al.* (1996), Elliott and Shadbolt (1996, 1999, 2002), Elliott *et al.* (1997), Harris *et al.* (1997), and Erwin and Miller (1998).

10.4.7 Induction of ocular dominance columns

10.4.7a *Tectal chemoaffinities*

In normal fish and amphibia, inputs from each eye fully decussate to form a retinotopic map in the contralateral tectum. Inputs from the ipsilateral eye reach the tectum by way of the isthmus and postoptic commissures (Chapter 14). Although there are no ocular dominance columns in the tecta of normal fish or amphibia, inputs from two eyes can be forced to project directly to the same tectum and form ocular dominance columns. This is possible because severed optic nerves of fish and amphibians regenerate to form normal connections in the contralateral tectum (see Sperry 1951). Ipsilateral connections, which are normally routed through intertectal commissures, are also restored in the frog, as long as the direct fibres regain their normal route (Gaze and Keating 1970). If only part of the goldfish retina is removed, axons from the remaining part connect only with their proper target cells in the tectum (Attardi and Sperry 1963; Jacobson and Gaze 1965). This suggests that tectal neurones produce location-specific chemical markers. However, when the caudal half of the goldfish tectum is removed or separated from the rostral half, the rostral half acquires a complete compressed projection of the entire contralateral retina (Gaze and Sharma 1970; Yoon 1971; Cronly-Dillon and Glaizner 1974). When the two halves of the fish tectum are rejoined, the retinal projection again spreads over the whole tectum (Yoon 1972). Thus, tectal chemoaffinities specify the topographic order of the retino-tectal mapping but not necessarily specific connections. We will now see that tectal chemoaffinities are not necessarily eye specific. Axons from both eyes can be made to compete for the same local connections in the tectum

10.4.7b *Ablation of the tectum*

When one tectum in the goldfish is removed, the axons that normally innervate it grow back to innervate the remaining tectum in spatial register with the innervation from the eye that normally innervates that tectum. Autoradiography revealed regions of alternating eye dominance in the doubly innervated tectum (Levine and Jacobson 1975). Initially, the invading ipsilateral projection spread homogeneously over the tectal surface and became segregated into eye specific columns about 50 days after unilateral tectum ablation (Springer and Cohen 1981). When the eye that normally innervates the remaining tectum was removed, the fish showed reversed visuomotor behaviour after reinnervation from the ipsilateral eye was established (Easter and Schmidt 1977). The ocular dominance columns did not form when retinal neural activity was blocked by injection of tetrodotoxin (Boss and Schmidt 1984). This suggests that column formation in fish depends on activity-dependent competition between axons from the two eyes.

Similarly, when one tectum of a frog embryo is ablated, the severed optic nerve grows into the ipsilateral tectum along with inputs from the intact contralateral eye. In each location in the tectum, the ipsilateral axons displace established synapses from the contralateral eye to form ocular dominance columns (Law and Constantine -Paton 1980). The regenerated nerves establish correct retinotopic connections in the tectum but cause a mirror-image reversal of prey catching to stimuli presented to the recovered eye. Development of mirror-reversed behaviour is more rapid when one tectum and its ipsilateral eye are removed so that the remaining tectum receives inputs from only one eye (Misantone and Stelzner 1974). In other words, recovery is more rapid when the restored inputs do not have to compete with inputs from the other eye. Ocular dominance columns also form in the tectum of the frog when severed optic nerves are allowed to regrow (Straznicky *et al.* 1980).

10.4.7c *Transplanted third eye*

An eye bud can be transplanted to form a third eye near one of the eyes of a frog embryo (Constantine-Paton and Law 1978; Law and Constantine-Paton 1981).

The optic nerves of the natural eye and the transplanted eye grow to innervate the same tectum where they form 200 μm-wide ocular dominance columns resembling those of mammals. The initial projection of inputs to the appropriate region of the tectum is presumably due to chemoaffinity. However, the subsequent segregation into ocular dominance columns results from conflict between two eye-specific topographic mappings. Each eye seeks to preserve connected neighbourhoods and to minimize contacts with inputs from the other eye. The width of connected neighbourhoods could be related to regions of dendritic arborization. Column formation could depend on local synchrony of spontaneous neural firing. Action potentials in the optic nerves are required for the formation of ocular dominance columns in three-eyed frogs (Reh and Constantine-Paton 1985).

10.4.7d *Construction of a composite eye*

Fawcett and Willshaw (1982) constructed a composite eye from two nasal halves or two temporal halves from opposite sides of the head of *Xenopus* larvae. As with double innervation from a transplanted eye, the inputs from the composite eye formed rostrocaudal eye-specific columns in the tectum.

10.4.7e *Ablation of temporal retina*

Ide *et al.* (1983) removed the temporal two thirds of an eye bud of the embryonic frog *Xenopus*. The remaining nasal portion grew into a normal-appearing eye. In the mature frog, each half of the retina projected to the entire contralateral tectum in stripes alternating with stripes from the other half of the retina. In this case, the inputs were from the same eye of the same animal so that the factor determining segregation of inputs was neither eye of origin nor animal of origin. The factor was probably the greater synchrony of firing between near neighbour inputs, since abolition of neural activity by application of a neurotoxin prevents the induction of ocular dominance columns in goldfish (Meyer 1982).

11 *Development of visual functions*

11.1 DEVELOPMENT OF BASIC FUNCTIONS

11.1.1 Acuity and colour sensitivity

11.1.1a *Development of contrast sensitivity*
The contrast-sensitivity function is the reciprocal of contrast threshold plotted against the spatial frequency of a grating with a sinusoidal modulation of luminance. (Section 3.3.5).

Contrast sensitivity can be measured in monkeys by training them to discriminate between a grating and a blank field. This procedure revealed that sensitivity to spatial frequencies below 5 cpd approached adult levels by the 20th week, and sensitivity to higher spatial frequencies continued to improve until the 28th week, when the contrast-sensitivity function acquired its adult form (Boothe *et al.* 1980).

The contrast-sensitivity function of human infants has been measured by behavioural procedures (Atkinson *et al.* 1977; Banks and Salapatek 1978) and by use of visual evoked potentials (Pirchio *et al.* 1978; Norcia *et al.* 1986, 1990). Norcia *et al.* obtained higher contrast sensitivities in infants than the other investigators. At low spatial frequencies, the contrast threshold declined from 7% contrast at 2-3 weeks to 0.5% at 9 weeks. For adults, the contrast threshold determined by the frequency-swept VEP for spatial frequencies below 1 cpd was 0.32-0.22% (Norcia *et al.* 1988, 1990) (Portrait Figure 11.1). However, the shapes of the functions were similar in all these studies, and so was the developmental trend towards greater contrast sensitivity and the extension of sensitivity to higher spatial frequencies, as shown in Figure 11.2.

Movshon and Kiorpes (1988) concluded that group averaging of data in earlier studies led to the erroneous conclusion that the shape of the contrast sensitivity function changes during development. Their analysis revealed that the falloff in sensitivity at low spatial frequencies, typical of the adult, is present even in 1-month-old infants. Gwiazda *et al.*

Figure 11.1. Anthony M. Norcia.
He received his B.A. in Psychology from the University of Minnesota and his Ph.D. in Psychology from Leland Stanford Jr. University. He is now a Senior Research Scientist at the Smith-Kettlewell Eye Research Institute in San Francisco.

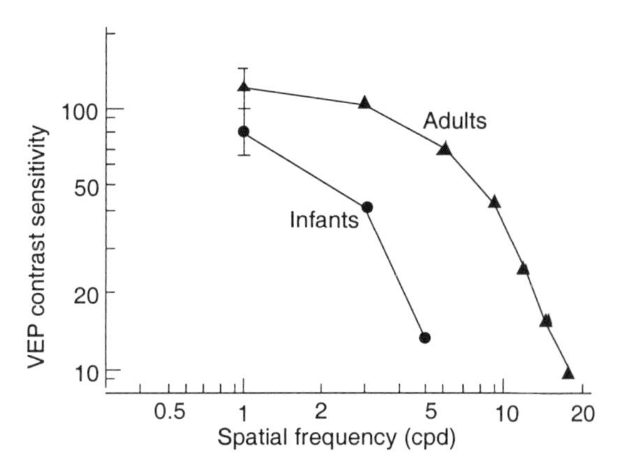

Figure 11.2. Contrast sensitivity from evoked potentials.
Contrast-sensitivity function of six adults (top curve) and four 25- to 28-week-old infants. Contrast sensitivity was measured by the amplitude of the VEP, in response to gratings of various spatial frequencies counterphase modulated at 12 Hz for the infants and at 15 Hz for the adults. (Adapted from Norcia *et al.* 1986)

(1997) used preferential looking and operant procedures to track the development of contrast sensitivity. Between the ages of 2 and 4 months they found a one octave loss of sensitivity for a 0.38 cpd grating, coupled with a continuing gain in sensitivity at higher spatial frequencies, as shown in Figure 11.3. They proposed that the low frequency loss is due to the development of lateral inhibition.

From an early age, contrast sensitivity and acuity are higher in the central visual field than in the peripheral field. Thus, at all ages between 10 and 39 weeks, visual acuity and contrast sensitivity, as revealed by visual-evoked potentials, were about 2.3 times higher for a grating confined to the central 2° of the visual field than for one confined to an annulus between 8 and 16° into the peripheral field (Allen *et al.* 1996).

In a study of 241 children, Scharre *et al.* (1990) found that contrast sensitivity had not reached adult levels by age 7 years, a result confirmed by Gwiazda *et al.* (1997).

11.1.1b *Development of grating acuity*
Grating acuity is the highest spatial frequency of a sinusoidally modulated luminance grating with maximum contrast that can just be resolved. It is indicated by the angular subtense of one period of a grating with the highest detectable spatial frequency. Grating acuity improves as the retina, LGN, and visual cortex mature. Three behavioural criteria have been used to assess grating acuity in young infants. These are, (1) the occurrence of optokinetic nystagmus in response to a moving grating, (2) the pupil response to a grating presented against a background with the same space-average luminance, and (3) the child's tendency to look at a patterned stimulus. These procedures do not always produce the same results. However, it is agreed that acuity for high-contrast gratings improves from approximately 100 arcmin at 1 month, to 20 arcmin at 3 months, to 10 arcmin at 12 months, and to the adult value of 1.5 arcmin (40 cpd) at 5 years of age (Dobson and Teller 1978; Jacobson *et al.* 1982; Mayer and Dobson 1982; Banks and Stevens 1985; Cocker *et al.* 1994).

Maurer *et al.* (1999) tested grating acuity in 28 infants immediately after their vision had been restored by removal of cataracts and insertion of contact lenses at between 1 week and 9 months of age. Their acuity was no better than that of neonates but showed signs of improvement after only 1 hour and continued to improve over the 1-month test period. Thus patterned visual inputs are necessary for the postnatal development of acuity.

Norcia and Tyler (1985) recorded evoked potentials from the infant's scalp in response to a frequency-swept, counterphase-modulated grating and obtained estimates of grating acuities of 13 arcmin (4.6 cpd) at 1 month and 3 arcmin (20 cpd) at about 12 months, as shown in Figure 11.4. The mean adult acuity determined by this procedure was about 32 cpd (Norcia *et al.* 1990). These acuities are consid-

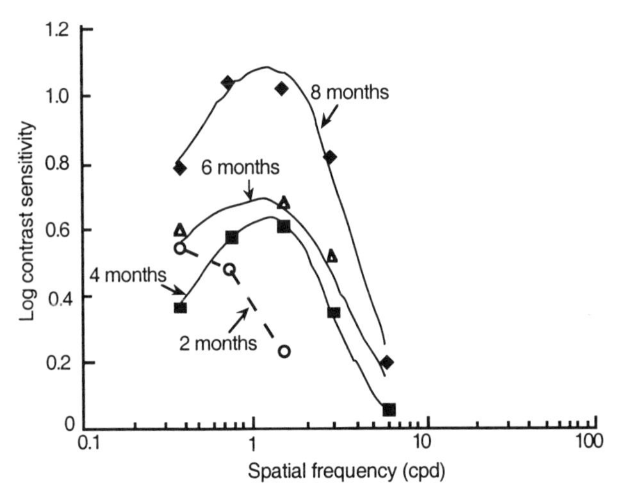

Figure 11.3. Contrast sensitivity functions in infant at four ages. (Adapted from Gwiazda *et al.* 1997)

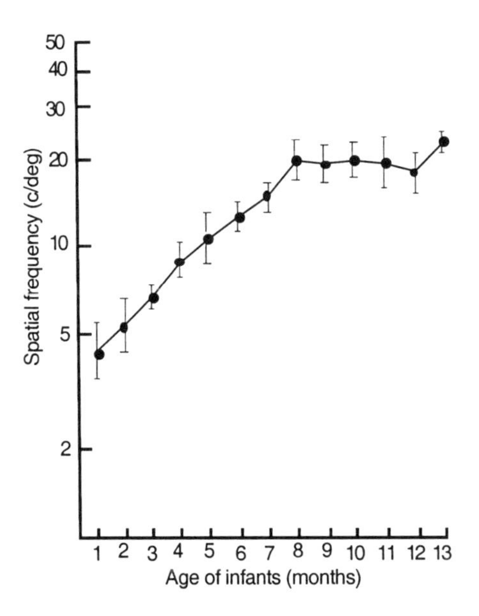

Figure 11.4. Sweep VEP grating acuity as a function of age. Mean grating acuity for 197 infants with 95% confidence bands derived from the sweep VEP procedure. (Adapted from Norcia and Tyler 1985)

erably better than those reported from behavioural studies or other evoked-potential studies referenced in their paper. The **frequency-swept VEP** provides a finer determination because it is fast and therefore immune to adaptation effects. It is also immune to the effects of probability summation (Section 8.1.1). Grating acuity obtained by this method was only slightly better with binocular viewing than with monocular viewing, and was similar in the two eyes (Hamer *et al.* 1989).

11.1.1c *Development of vernier acuity*

Vernier acuity is a hyperacuity because it is finer than the mean spacing of retinal cones—unlike resolution acuity it is not subject to the Nyquist limit (Section 3.1.2b). For this reason, vernier acuity and other hyperacuities are much finer than resolution acuity. Several investigators have enquired whether this superiority of hyperacuity is present in young children.

Manny and Klein (1985) measured vernier acuity in infants from 1 to 14 months of age. In a two-choice preferential looking procedure, the child's direction of gaze was recorded while one of two horizontal lines was replaced with a line with a vernier offset. In a tracking procedure, the infants eyes were observed as the vernier target moved from place to place. Vernier acuity improved to a mean value of about 2 arcmin at age 11 months, compared with a mean adult value of about 0.25 arcmin. Acuity was higher and less variable with the tracking procedure than with the two-choice procedure, perhaps because the tracking procedure included a motion signal and an accompanying sound.

Shimojo *et al.* (1984) used a preferential looking procedure to measure vernier and grating acuities of infants aged 2 to 9 months. For vernier acuity, infants were presented with a square wave vertical grating of 1.1 cpd and the same grating with a horizontal offset in the bars which oscillated up and down in time with a beep sound. The motion of the offset was designed to increase the probability that the infants would attend to the stimulus. It was assumed that the motion would not be detected unless the offset were detected (see Skoczenski and Aslin 1992). For grating acuity, a square wave vertical grating of variable spatial frequency was presented with a gray field. Vernier acuity was inferior to grating acuity at 2 months of age but, by 9 months, vernier acuity was 2.5 arcmin while grating acuity was only 5 arcmin. Vernier acuity developed in parallel with stereoacuity as reported in Held *et al.* (1980). Similar trends were reported by Shimojo and Held (1987).

Vernier acuity for offsets in multiple bars may be degraded relative to that for a single bar because of crowding effects. Zanker *et al.* (1992) used a two-choice preferential looking procedure in which a single vertical bright bar containing two offset segments was paired with a straight line. Vernier acuity increased from about 25 arcmin at 2 months of age to the adult value of about 0.2 arcmin at 5 years of age. Grating acuity developed more slowly from about 8 arcmin at 2 months to about 1 arcmin at 5 years. Grating acuity was better than vernier acuity during the first year, after which vernier acuity became progressively better than grating acuity.

Thus, in spite of variations in stimuli and procedures, all investigators agree that vernier acuity changes from being worse to being superior to grating acuity. The change-over age is somewhere between 3 and 12 months.

Carkeet *et al.* (1997a) measured vernier and resolution acuities in children aged from 3 to 12 years. The vernier stimulus was a bright vertical line with several offsets along it length. The resolution stimulus was a vertical dashed line with variable density of dashes. Vernier acuity was always superior to resolution acuity and improved more rapidly with age. Between age 3 and 12, mean vernier acuity improved from 15 arcsec to 6 arcsec—2.5 times. Mean dashed-line resolution improved from 70 arcsec to 48 arcsec, a much smaller difference. Carkeet *et al.* concluded that, at all ages in their sample, vernier acuity was finer than expected from the density of retinal cones.

11.1.1d *Causes of visual immaturity*

The visual performance of human neonates may be limited by the immaturity preneural factors. These include the optical quality and aperture of the eye and the capacity of the retina to sample the image. The sampling capacity of the retina depends on the photon-capture efficiency of the receptors and the spacing of the receptors. On the other hand, the limiting factor could be neural. Possible neural factors include noise in the sensory transduction process (see Skoczenski and Norcia 1998), the size and tuning characteristics of receptive fields of ganglion cells and cortical cells, spatial blurring of cortical receptive fields because of lack of inhibitory connections and the presence of inappropriate synapses, and low cortical magnification. Skoczenski and Aslin (1995) measured the effects of adding Gaussian noise on vernier acuity in 3-month-olds, 5-month-olds, and adults. They concluded from their results that reduction in intrinsic blur due to optical or neural factors accounts for improvement in vernier acuity between 3 and 5 months, but that later improvement depends also on the development of post-retinal mechanisms for processing high spatial frequencies.

Banks and Bennett (1988) estimated the contrast sensitivity and grating acuity for two ideal observers, one based on preneural characteristics of the neonate human fovea and one with characteristics of the adult fovea. This analysis predicted a difference in the grating acuities of the infant and adult of 2 octaves compared with an actual difference of 3.5 to 4.5 octaves. They concluded that immaturity of preneural structures does not fully account for the poor visual performance of the neonate (see also Jacobs and Blakemore 1988). Williams and Boothe (1981) had reached the same conclusion in their studies of monkey vision. Wilson (1988) carried out a similar analysis from which he concluded that preneural factors play a greater role in limiting visual performance in neonates than was found in the Banks and Bennett study.

An ideal observer model predicts that vernier thresholds are inversely proportional to the square root of photon capture, while grating acuity is inversely proportional to the fourth root of photon capture (Geisler 1984). Banks and Bennett (1988) argued that an improvement of photon capture with age would explain why vernier acuity improves more rapidly than resolution acuity. However, Carkeet *et al.* (1997) found that developmental changes in both vernier and resolution acuities were larger than could be accounted for in terms of improved photon capture. They concluded that the development of central neural processes is involved in improvement of acuity.

Amblyopes, especially strabismic amblyopes, are like young children in having low vernier acuity relative to resolution acuity, presumably because of deficits in central processing (Section 13.4.2).

Such analyses are valuable but can be no better than the set of assumptions and the data on which they are based.

11.1.1e *Development of chromatic sensitivity*

In the first month or two, infants are very poor at discriminating between different colours or between achromatic and chromatic stimuli, especially at the blue end of the spectrum (Varner *et al.* 1985; Adams *et al.* 1991). By 12 weeks, most infants can discriminate hues involving cones sensitive to short, medium, and long wavelengths. Morrone *et al.* (1993) found no visual evoked responses to purely chromatic stimuli before 7 to 8 weeks of age, and concluded that luminance sensitivity develops more rapidly than chromatic sensitivity in young infants. Banks and Bennett (1988) proposed that the poor colour discrimination in infants is due to poor general visual sensitivity rather than to a specific immaturity of the colour system, such as an absence of different cone types. Allen *et al.* (1993) produced evidence for this idea by showing that the ratio of luminance sensitivity to chromatic sensitivity, as indicated by visual-evoked potentials, is the same for 2- to 8-week-old infants as for adults. They found the same constant ratio when they reanalyzed the data of Morrone *et al.* Brown *et al.* (1995) came to the

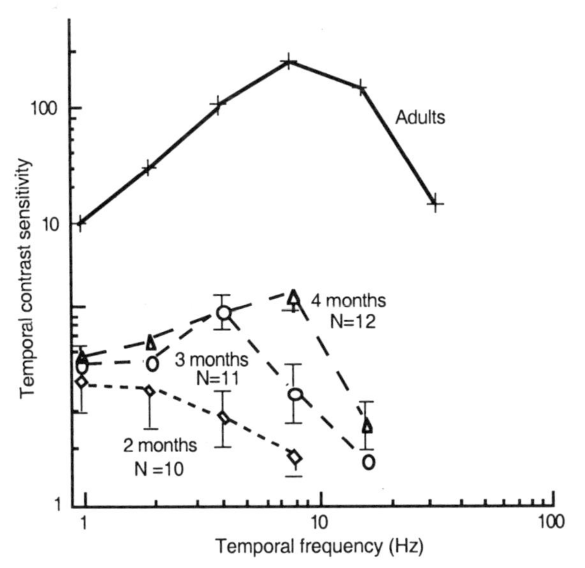

Figure 11.5. Temporal contrast sensitivity for infants and adults. The stimulus was a uniform flickering field. (Adapted from Rasengane *et al.* 1997)

same conclusion after using optokinetic nystagmus to compare colour and luminance contrast sensitivities in infants and adults. The earlier literature on colour sensitivity in infants was reviewed by Brown (1990).

11.1.2 Orientation sensitivity

Braddick *et al.* (1986) found no evidence of cortical cells tuned to orientation in the human neonate, as indicated by the VEP generated by changes in the orientation of a grating. Evidence of orientation tuning showed at the age of 6 weeks. Binocular cells have the same orientation preference for both eyes. This does not seem to depend on visual experience because the receptive fields had matching orientations in kittens raised so that both eyes did not see at the same time (Gödecke and Bonhoeffer 1996).

Behavioural tests have revealed that 6-week-old human infants can discriminate between lines in opposite oblique orientations (Maurer and Martello 1980; Held 1981). More recently, discrimination between oblique gratings has been shown in human neonates. When newborn infants were shown two opposite oblique static gratings side by side they preferred looking at the one which they had not seen previously (Atkinson *et al.* 1988; Slater *et al.* 1988). In other studies, 3-month-old infants could discriminate between gratings tilted 45° and 15°, and 4-month-old infants could discriminate between gratings tilted 45° and 22° (Bomba 1984; Bornstein *et al.* 1986). The older literature on the development of the ability to categorize orientations is reviewed in Howard (1982).

In a preferential looking task, infants under 14 weeks of age could detect a patch differing in luminance contrast from its surround but could not detect a patch containing oblique lines set in a surround of lines oriented in a different direction (Atkinson and Braddick 1992).

11.1.3 Flicker and motion sensitivity

11.1.3a *Development of flicker sensitivity*

The highest frequency of flicker of a high-contrast stimulus that produces a modulated response in the visual system is the **critical flicker frequency**, or CFF. The CFF for the electroretinogram (ERG) is adult-like in human neonates (Horsten and Winkelman 1962). The CFF for the cortical visual evoked potential (VEP) increases rapidly after the first month and is adult-like by 5 months (Apkarian 1993).

The **temporal contrast-sensitivity function** (TCSF) is the contrast threshold for detection of flicker of a sinusoidally flickering stimulus as a function of temporal frequency. The adult TCSF is typically bandpass for a uniform field or a low spatial-frequency grating flickering in counterphase and lowpass for a high spatial-frequency grating. Flicker sensitivity falls off steeply above 10 Hz to reach a CFF at about 50 Hz (Watson 1986).

Using a preferential looking procedure, Regal (1981) found that the CFF for a uniform field reached adult levels of over 50 Hz by about the 12th week. Rasengane *et al.* (1997) used a similar method to measure the TCSF for a uniform field. Peak sensitivity increased from between 1 and 2 Hz at 2 months, to 4 Hz at 3 months, to 8 Hz at 4 months, as shown in Figure 11.5. Thus, peak sensitivity shifted to higher frequencies with increasing age, causing the function to change from low pass to band pass. The CFF increased from 13 Hz at 2 months to 22 Hz at 4 months, still well below the adult value. None of the infants responded to 32 Hz, the highest frequency tested. Nevertheless, while sensitivity to high flicker rates matures at an early age, sensitivity to lower temporal frequencies was found not to reach adult levels until 7 years (Ellemberg *et al.* 1999).

The form of the TCSF depends on the spatial frequency of the flickering stimulus. Swanson and Birch (1990) found that contrast sensitivity increased between 4 and 8 months for a 1 cpd grating flickered at between 8 to 17 Hz but did not change for flicker rates of 2 to 4 Hz. Thus the TCSF at 1 cpd changed from being lowpass at 4 months to being more

Figure 11.6. Janette Atkinson.
Born in Cheshire, England in 1943. She obtained a BSc. in Psychology from Bristol University in 1965 and a Ph.D. from Cambridge University in 1971. She conduced postdoctoral work at Johns Hopkins University and was a Research Associate at Cambridge from 1972 to 1983. Between 1983 and 1993 she was Senior External Scientist of the Medical Research Council at Cambridge. She has been a Professor at University College, London since 1993.

bandpass at 8 months. At 0.25 cpd the TCSF remained bandpass. Hartmann and Banks (1992) found a similar change in the TCSF from lowpass to bandpass for a 0.5 cpd grating between the ages of 1.5 and 3 months.

11.1.3b *Development of motion sensitivity*

To establish that infants are sensitive to visual motion one must demonstrate that they respond specifically to motion rather than to flicker or change of position. Aslin and Shea (1990) found that 6-week-old infants could distinguish between stationary stripes and stripes moving at 9°/s relative to a stationary surround. At 12 weeks, they could distinguish stripes moving at 4°/s. The velocity threshold did not vary with the width of the stripes, which indicates that the infants were judging motion rather than simple flicker. One-month-old infants could not discriminate direction of motion over a wide range of velocities, although they could discriminate between stationary and moving patterns (Wattam-Bell 1996a). They could also discriminate between incoherent and coherent motion of random-dot patterns although, even at 3 months of age, detection required that 50% of the dots moved coherently compared with only 5-7% for adults (Wattam-Bell 1994). Thus, the ability to detect motion seems to develop before the ability to discriminate motion direction. Dannemiller and Freedland (1993) showed that 14-week-old infants detect motion in standing-wave line stimuli, which allow sensitivity to motion to be distinguished from sensitivity to changes in position (see also Bertenthal and Bradbury 1992).

Wattam-Bell (1992) used a preferential-looking procedure to measure the maximum displacement (d_{max}) of a random-dot pattern that allowed subjects to discriminate between directions of motion and to discriminate between coherent and incoherent motion. Between 8 and 15 weeks of age, the oldest age tested, d_{max} increased for both tasks, but the increase was greater for slow than for rapid movements. The same developmental trends were evident in the magnitude of the VEP triggered by motion of checkerboard patterns (Wattam-Bell 1991). Sensitivity to motion develops for low velocities before it develops for high velocities. Infants aged 8-10 weeks also responded to second-order motion defined by a grating of random flicker sweeping over a stationary random-dot display (Atkinson 2000, p. 81) (Portrait Figure 11.6).

The occurrence of optokinetic nystagmus (OKN) in human neonates demonstrates that they are sensitive to uniform visual motion. This measure indicates that the threshold for detection of uniform motion is constant at about 3°/s for infants between 12 and 18 weeks of age (Banton and Bertenthal 1996). Sensitivity to relative motion, especially shear, continues to develop during this period. A contaminating factor may be that OKN is under subcortical control in the young infant (Section 23.4.5) while relative motion detection, as revealed by preferential looking, presumably depends on cortical processes (Banton and Bertenthal 1997).

Hamer and Norcia (1994) used evoked potentials to measure displacement thresholds for a high-contrast, 1 cpd grating oscillating from side to side at 6 Hz. In 12-week-old infants the threshold was ten times the adult value, even though contrast sensitivity was already half the adult value. The oscillation threshold was still over four times that of the adult in 1-year-old infants.

Some directional sensitivity in neonates is suggested by the fact that the VEP is much stronger for stimuli moving nasally than in a temporal direction (Norcia *et al.* 1991). This directional asymmetry is also revealed in an asymmetry in monocular optokinetic nystagmus in neonates (Section 23.4.5) and esotropes (Section 13.4.5), although asymmetry of the VEP does not seem to be correlated with asymmetry of OKN (Kommerell *et al.* 1995). Dobkins and Teller (1996) provided psychophysical evidence for directional sensitivity in 3-month-old infants.

The development of motion sensitivity has been reviewed by Wattam-Bell (1996b). The development of sensitivity to motion in depth is discussed in Section 11.3.2b.

11.1.4 Development of the visual fields

In animals and human infants, the extent of the visual field is tested by getting the subject to fixate on a central flashing target and then observing whether the eyes move to a flashed target presented at different eccentricities. For details see Sireteanu (1996). The limits of vision may be influenced by optical factors as well as by retinal factors.

The size of the visual field increases with age. In kittens, the visual field reaches it adult extent by about 10 weeks of age (Sireteanu and Maurer 1982). In humans under 2 months of age, the binocular field is smaller than that of the adult, especially along the vertical meridian (Schwartz *et al.* 1987). The adult level is reached at between 6 and 12 months of age (Mohn and van Hof-van Duin 1986; Lewis and Maurer 1992).

Visual functions in the temporal half of the monocular visual field (nasal hemiretina) develop before those in the nasal hemifield (temporal hemiretina). For example, visual acuity revealed by preferential looking was higher in the temporal than in the nasal visual field of infants between 2 and 11 months of age (Sireteanu *et al.* 1994). For some time after first opening their eyes, kittens oriented towards stimuli in the temporal visual field of an eye, but ignored stimuli in the nasal field (Sireteanu and Maurer 1982). Similarly, human infants below 2 months of age oriented their gaze towards an isolated light presented 30° into the temporal monocular field but failed to orient towards a light only 15° into the nasal field (Lewis and Maurer 1992). A similar procedure revealed that after the age of 2 months both hemifields and the binocular field expand rapidly until the age of 8 months and then more slowly until 12 months (Mohn and Van Hof-van Duin 1986).

Although the nasal hemifield remains smaller than the temporal hemifield, the two hemifields become more similar in size with increasing age. However, even in the adult monkey, there are more cortical binocular cells with a dominant input from the contralateral eye (nasal hemiretina) than cells with a dominant ipsilateral input (temporal hemiretina) (LeVay *et al.* 1985). The cortical magnification factor (linear extent of cortical tissue devoted to each visual angle) is proportional to the density of ganglion cells (Rovamo and Virsu 1979). Since the nasal retina has a higher density of ganglion cells than the temporal retina (Curcio *et al.* 1990) the magnification factor is also higher for the nasal retina. The mature nasal hemiretina also remains more sensitive than the temporal hemiretina. Thus, the decrease in vernier acuity with increasing eccentricity of the stimulus is steeper for the temporal than for the nasal hemiretina (Fahle and Schmid 1988). Reaction times are shorter for stimuli presented to the nasal hemiretina (Payne 1967). Other aspects of hemifield asymmetry are discussed in Section 7.3.5.

The development of spatial vision has been reviewed by Mohn and Van Hof-van Duin (1991).

11.2 GROWTH OF THE OCULOMOTOR SYSTEM

11.2.1 Accommodation

The development of accommodation has been studied by measuring the refractive state of the eye with a retinoscope, with the ciliary muscles paralyzed by a cycloplegic drug (Section 9.1.1). In a second method, which does not require a cycloplegic drug, refraction is measured while the child is feeding and alert. The child views a dimmed retinoscope light that is a poor accommodative stimulus (Mohindra 1975). Several investigators have reported that the human infant is about 2 dioptres hypermetropic relative to the average adult. However, retinoscopy in young infants is unreliable, so the reported hypermetropia may be an artifact (Banks 1980). Severe untreated hypermetropia in infants can lead to amblyopia (Ingram and Walker 1979).

The development of accommodation can be studied by measuring the accommodative state of the eye while a large pattern is moved to different distances. The infrared optometer is the most precise method for measuring changes in accommodation (Section 9.1.1) but cannot be used in infants because observers must maintain fixation. With infants, refraction is measured with dynamic retinoscopy, which gives the sign of the refractive error but takes time to operate; or by photorefraction, which accurately measures the instantaneous refractive state of both eyes (Howland and Howland 1974). It is important to keep the size of the retinal image constant to ensure that any lack of accommodation is not due to the stimulus falling below the resolution threshold.

Newborn infants show signs of changing accommodation to large conspicuous objects nearer than about 75 cm (Brookman 1983; Howland *et al.* 1987; Hainline *et al.* 1992). Beyond 8 weeks, accuracy of accommodation increases until adult levels are reached between 16 and 20 weeks (Brookman 1983). Banks (1980) concluded that the accuracy of accom-

Figure 11.7. Richard N. Aslin.
Born in Milwaukee, Wisconsin in 1949. He received his B.A. in Psychology from Michigan State University in 1971 and his Ph.D. in Child Psychology at the University of Minnesota in 1975 under the guidance of Philip Salapatek. From 1975 to 1984 he was a member of the faculty in Psychology at Indiana University, Bloomington. In 1984 he moved to the University of Rochester, where he is now a Professor of Brain and Cognitive Sciences and a member of the Center for Visual Science.

modation is poor in young infants, largely because of their poor resolution. By 6 months, infants show a range of accommodation similar to that of the adult (Braddick *et al.* 1979).

Several investigators have reported that children below 4 years of age tend to be astigmatic, with a vertical axis of astigmatism. Ophthalmologists call this an "against the rule" astigmatism. However, there is some dispute on this point (see Saunders 1995). In older children and adults, any astigmatism tends to be along a horizontal axis. This known as "with the rule astigmatism" (Dobson *et al.* 1984; Gwiazda *et al.* 1984; Howland and Sayles 1984).

The depth of field is the range of distances within which an object is in focus for a given state of accommodation. Depth of field is inversely proportional to pupil diameter and to the size of the eye. Depth of field is greater in infants than in adults. This is because the pupils of infants under 2 months of age are, on average, between 1 and 2 mm smaller than those of adults, and because infant's eyes are smaller than adult eyes (see Boothe *et al.* 1985; Green *et al.* 1980). Thus, image quality is less affected by misaccommodation in the infant eye than in the adult eye. In any case, image quality is not as important for the infant because high spatial-frequencies cannot be resolved. Accommodative convergence is present in 2-month-old infants; younger infants have not been tested (Aslin and Jackson 1979). (Portrait Figure 11.7).

11.2.2 The pupillary response

The pupils of normal adults constrict about 29% more with binocular illumination than with monocular illumination (ten Doesschate and Alpern 1967). Birch and Held (1983) used this fact to investigate the development of binocularity in infants (Portrait Figures 11.8). They found that the pupil responded more to binocular than monocular illumination by the age of 4 months, and the differential response was adult-like by the age of 6 months. However, Shea *et al.* (1985) found reduced but significant levels of binocular luminance summation in the pupillary response of 2-month-old infants and in stereoblind adults (see also Sireteanu 1987). These results suggest that the development of the pupillary response is to some extent independent of the development of stereopsis.

11.2.3 Optical alignment

The pupils of neonate cats show a strong divergence with respect to the images of a point of light reflected from the corneas. In human subjects this indicates a divergent strabismus. By the end of the second month the divergence of the kitten's pupils changes into a slight convergence (Sherman 1972). Olson and Freeman (1978a) measured the angle between the visual axes of kittens by plotting the receptive fields of cells in cortical area 17 and referring the results to photographs of the pupils. They concluded that the visual axes are aligned even during the first two postnatal months when the pupils are divergent. This suggests that the change in the alignment of the pupils is accompanied by a medial migration of the area centralis. If we assume that each optic axis is normal to the pupil and passes through the entrance pupil, this process entails a reduction in the angle between the pupillary axis and the visual axis (angle lambda) from about 25 to 15°. Thus, as the eyes rotate inward during the first two months, the area centralis migrates medially. These processes occur in both normally reared and dark-reared cats, except that normally reared cats reach a steady state much earlier than dark-reared cats (Von Grünau 1979a). It seems that the basic changes are maturational. Visual inputs are required only to terminate the changes. Olson and Freeman also found that the intorsional angle between the slit pupils of the cat increases during the first 2 months to a final value of about 14°. In dark-reared animals this process continues through the third month to reach a mean value of 24°. Human neonates tend to be divergent. It is not known whether there is a medial migration of the fovea in humans but the angle

Figure 11.8. Richard Held.
Born in New York in 1922. He obtained a B.Sc. in Engineering from Columbia University in 1944 and a Ph.D. in Psychology from Swarthmore College in 1948. After postdoctoral work at Harvard University he joined the Psychology Department at Brandeis University in 1953. In 1963 he became Professor of Experimental Psychology at MIT. After retiring in 1994 he became Professor Emeritus at MIT and Director of Research in the Department of Vision Science at the New England College of Optometry.

lambda has been estimated to change from 7.9° at birth to 5.08° in the adult (London and Wick 1982).

11.2.4 Visual pursuit

Ultrasound imaging reveals a variety of slow and fast eye movements in the human foetus between 16 and 42 weeks of gestation (Birnholz 1981). Gratings moving at up to about 40°/s evoke optokinetic nystagmus in neonate infants (Kremenitzer *et al.* 1979). Optokinetic nystagmus (OKN) of human infants below the age of about 3 months shows the directional asymmetry typical of animals lacking stereoscopic vision (Section 23.4.5). The time course for the development of symmetrical OKN for large, high contrast stimuli is similar to that for the development of stereopsis (Westall 1986; Lewis *et al.* 2000). People lacking stereopsis tend to retain asymmetry of OKN.

It has been claimed that newborn infants show occasional smooth pursuit eye movements to a single object moving at up to about 20°/s (Kremenitzer *et al.* 1979). However, Aslin (1987) did not observe voluntary pursuit eye movements before 6 to 8 weeks of age. Von Hofsten and Rosander (1996) measured eye and head tracking of an oscillating visual display by infants between 1 and 3 months of age. Over this period, gain was fairly constant but phase lag decreased from 170 to 70 ms and latency from 860 to 560 ms. Human adults cannot suppress OKN when there are no stationary objects in view but can readily do so when a single stationary object is present. One-month-old infants showed no suppression of OKN in the presence of a stationary object but 2-month-old infants showed some evidence of suppression (Aslin and Johnson 1996). The literature on the development of conjugate saccadic and pursuit eye movements has been reviewed by Hainline (1993).

11.2.5 The saccadic system

New-born infants shift their gaze in the direction of a suddenly presented visual object, especially when the initially fixated object disappears (Harris and MacFarlane 1974). Aslin and Salapatek (1975) found that 1-month-old infants executed saccades from a central position to a target up to 30° in a horizontal or oblique direction but only up to 10° in a vertical direction. The probability of a response decreased with increasing distance of the target and when the initial fixation target remained in view. The directional accuracy of saccades was not specified but they were grossly hypometric and were followed by smaller corrective saccades. Saccadic latency was very variable but was less than 500 ms. Adult latencies vary between 120 and 350 ms.

Hainline *et al.* (1984) recorded the eye movements of adults and of infants between 14 and 150 days old as they scanned a set of textured patterns. At all ages, the peak velocity of saccades was proportional to saccadic amplitude, in accordance with the so called main sequence. However, infants executed slower saccades and showed greater instability of gaze. In the 2 to 14-week age period, infants increased the accuracy of saccades to particular stimulus features (Bronson 1990).

11.2.6 Vergence

Bursts of synchronous and conjugate rapid eye movements have been seen in several human foetuses after 30 weeks of gestation (Birnholz 1981). Vergence eye movements in human infants have been investigated by the corneal reflex method, in which the position of the image of a light reflected by the cornea is measured with respect to the centre of the pupil (Wickelgren 1967). The method is accurate to only about 5°, and indicates the position of the eye's optic axis rather than of the visual axis—the two can be several degrees apart (Slater and Findlay 1975a). Even allowing for this factor, one cannot obtain an accurate calibration of the instrument since an infant cannot be asked to converge on targets at known distances. However, the method is suitable for detecting changes in convergence as a visual target is moved in depth.

Slater and Findlay (1975b) found evidence of visually evoked changes in vergence in human neonates. However, the responses were unstable and occurred only within a limited range of target distance. Hainline and Riddell (1995) recorded static vergence to targets at distances between 25 and 200 cm in infants between 17 and 120 days old. Some 20-day-old infants showed evidence of stimulus-evoked changes in vergence but their response to a defined stimulus was more variable than was that of older infants.

Aslin (1977) used a photographic method to record vergence movements to a luminous cross as it approached or receded along the midline between 15 and 57 cm from the infant. One-month-old infants showed evidence of vergence in the appropriate direction but, in 3-month-old infants, vergence was more likely to occur and was more closely matched to the speed of the target. The ability to correct for a prism placed before one eye was not consistently present until 6 months of age.

The position of dark vergence—the position to which the eyes return when not subject to vergence demand—is more convergent in infants than in adults. Thus, the position of dark vergence of infants between 5 and 20 weeks of age was, on average, at a fixation distance of 35 cm compared with 120 cm for adults (Aslin *et al.* 1982). Most children are orthophoric but there is a decrease in the incidence of exophoria and an increase in the incidence of esophoria with age (Walline *et al.* 1998).

Thorn *et al.* (1994) measured the development of binocularity in human infants between 2 and 21 weeks of age. Ocular alignment was determined by observing the deviation of the first Purkinje image of a light spot reflected from the cornea (the Hirschberg test). Prism and cover tests of strabismus cannot be used with young infants. Convergence was determined by visual examination and by a test of binocular fusion involving preferential looking between a fusible pair of gratings and a rivalrous pair of gratings. The few infants that were not orthotropic during the first post-natal month were exotropic, and almost all infants were orthotropic by the fourth month. Convergence began to show at 6 weeks of age but full convergence did not occur until between the 13th and 17th week. Infants showed evidence of binocular fusion between the 12th and 16th week, and there was a high correlation between the age of onset of convergence and that of binocular fusion.

Accommodative vergence has been observed in 2-month-old infants but the magnitude of the response was not measured (Aslin and Jackson 1979). Only a weak contribution of accommodative vergence to vergence in the young infant is to be expected. In infants, as in adults, the resting state of vergence (dark vergence) is not related to the state of accommodation in the dark (dark focus) (Aslin and Dobson 1983).

Vergence movements are designed to bring images with large disparities into the disparity range where fine disparities can be detected. Infants do not need to move their eyes to detect coarse disparities. Birch *et al.* (1983) found that infants over 6 months of age, with fully developed stereoacuity, are insensitive to errors of vergence of up to 1.4°. Children had to reach an average age of 4.1 months before they could distinguish depth in stereograms in which the disparity was allowed to reach 1.4°. From this, they concluded that the development of stereoacuity in the infant is limited by the maturation of disparity-detecting neurones and not by maturation of the vergence system. The argument depends on the assumption that the younger children would also tolerate vergence errors of up to 1.4° if the neural system were mature.

11.3 DEVELOPMENT OF DEPTH PERCEPTION

11.3.1 Procedures

The development of visual perception in preverbal human infants has been studied by observing the following pieces of behaviour.

1. Reaching The accuracy of reaching to objects in different directions or at different distances is measured.

2. Avoidance behaviour The infant is presented with a stimulus such as a visual cliff or an approaching object and its reactions are recorded.

3. Preferential looking Some stimuli are naturally attractive to human infants. For instance, when presented with two stimuli with a minimum of distractions, infants spend more time looking at the more brightly coloured stimulus or the one which moves, flashes, or has higher contrast. They also prefer a 3-D display to a flat display. In the preferential-looking procedure introduced by Fantz (1965), the infant is presented with two stimuli side by side, and a record is kept of the time the infant spends gazing at each. In a refinement of the method, the person watching the infant's eyes does not know which stimulus has been presented and follows a forced-choice, bias-free procedure in deciding

Figure 11.9. Albert Yonas.
Born in Cleveland, Ohio in 1942. He obtained a B.A. in Psychology from The University of Michigan in 1964 and a Ph.D. in Psychology from Cornell University in 1968. He joined the faculty of the Institute of Child Development at the University of Minnesota in 1968, where he is now Professor and Director of the Centre for Research in Learning.

whether the infant is looking at one stimulus or the other (Teller 1979).

4. Dishabituation In a variant of the preferential-looking procedure, known as dishabituation, the infant is shown a given stimulus for some minutes or until signs of interest are no longer evident. The stimulus is then changed in a defined way, and the extent to which interest is restored is determined by observing movements of the infant's eyes or by recording indications of arousal such as increased heart rate. The stimulus is then changed in the reverse direction and, as a final control, each stimulus is flashed off and on again.

5. Operant conditioning All the preceding procedures make use of a built-in response and therefore require a minimum of preliminary training. In operant conditioning, the subject is first trained to respond to a reinforced stimulus while another stimulus differing in some crucial respect is not reinforced. A subject who learns to make differential responses to the two stimuli is deemed to have the sensory capacity to discriminate them. It is notoriously difficult to design stimuli that differ in the factor being studied and not in irrelevant factors.

6. Pursuit eye movements The ability of an infant to detect a stimulus may also be revealed by moving the stimulus and observing whether the infant's eyes track it. Optokinetic nystagmus (OKN), or the involuntary tracking movement of the eyes in response to large moving stimuli, is present in the neonate (Shea *et al.* 1980).

11.3.2 Perception of distance

11.3.2a *Accuracy of reaching movements*

The earliest identifiable responses of the arm are the neck-tonic reflex evoked by rotation of the head, the traction reflex evoked by pulling the arm, and the grasp reflex evoked by touching the palm (Twitchell 1970). None of these innate reflexes is evoked by visual stimulation. White *et al.* (1964) outlined a normative developmental sequence of visually guided reaching in human infants. In the first month, infants do not attend to objects within arm's reach, and arm movements are unrelated to vision. In the second month, infants attend to near objects and become interested in their own arms. The first visually directed swiping movements of the arm develop, but the child grasps an object only if the hand touches it. In the third month, swiping gives way to directed arm movements, and the child looks back and forth between object and hand. By the fourth and fifth months the combined action of the arms comes under visual control and gives way to the ability to reach for and grasp an object. Claims that infants only a few days old reach for visual objects and occasionally grasp them have not been confirmed (Bower *et al.* 1970b; Dodwell *et al.* 1976).

Nine-week old infants cannot reach for an object across the body midline but, by 18 weeks, infants behave like adults on this task (Provine and Westerman 1979). Co-ordinated bimanual reaching depends on the development of contralateral reaching.

Reaching movements that are without visual control once initiated are called **visually triggered movements**. Responses that are modified during execution by visually perceived error are called **visually guided movements**. A successful reaching movement to an isolated object without sight of the hand requires information about the distance of the object, which can be provided only by accommodation or vergence. A seen hand can be guided to an object by the use of binocular disparity and lateral offset between hand and object; absolute estimates of distance and direction are not required for visual guidance.

There have been several studies on the development of the reaching response in infants and conclusions have been drawn about the extent to which reaching and grasping signify that the infant has depth perception. For instance, 5-month-old infants moved the arm forward and made grasping movements with the hand when a virtual object was within reach but not when it was out of reach (Gordon and Yonas 1976; Bechtoldt and Hutz 1979). Granrud (1986) found that 4-month-old infants reached for a nearer object more consistently when looking with two eyes than with one. Furthermore, the superiority of binocular over monocular reaching was correlated with a preference for looking at a random-dot stereogram with relative depth rather than at one depicting a single surface. In another study, 5-month-old infants reached for an approaching object specified only by binocular information whereas 3.5-month-old infants failed this test (Yonas *et al.* 1978) (Portrait Figure 11.9). Infants between the ages of 18 and 32 weeks directed ballistic movements of the arms towards the virtual position of an object viewed through prisms (von Hofsten 1977). In other words, they moved their arms to the point where the eyes were converged.

When reaching to grasp an object, adults start to close the hand before touching the object and start to close earlier for smaller objects. This skill could depend on an estimate of the absolute distance of the object or of the relative visually perceived distance between the hand and the object. Compared with adult, infants between 5 and 9 months of age started to close the hand closer to the time of contact. Even, 13-month-old infants did not react differently to different sizes of object (von Hofsten and Rönnqvist 1988). Infants 4 weeks old showed signs of adjusting the orientation of the reaching hand to the orientation of a rod but this skill was more precise in 34-week-old infants (von Hofsten and Fazel-Zandy 1984).

Infants aged 6 to 20 weeks fixated on a solid object for a longer period when it was 30 cm distant than when it was 90 cm distant (McKenzie and Day 1972).

It is not clear in any of these studies that performance depended only on disparity cues to depth as opposed to vergence, and monocular cues. Nor is it clear to what extent the infants were judging the absolute distance of an object or the relative distance between the object and the hand.

11.3.2b *Cliff avoidance*

Most young mammals show a natural avoidance response when confronted with a visual cliff (Gibson and Walk 1960; Walk and Gibson 1961) (Portrait Figure 11.10). In one study, kittens used binocular cues in selecting the shallower of two steps by the age of 5 weeks (Timney 1981). In another study, human infants at the age of 2 months discriminated between the shallow and deep sides of a visual cliff, as indicated by the heart rate (Campos *et al.* 1970). However, several monocular cues to depth were available in these displays, so one cannot conclude anything about the development of binocular stereopsis in humans.

Figure 11.10. Eleanor J. Gibson.
Born Eleanor Jack in Preoria, Illinois in 1910. She obtained a BA at Smith College, and a Ph.D. at Princeton University in 1938. She held academic appointments at Smith College until 1949 when she moved to the Psychology Department at Cornell University, where she remained until she retired. She received the APA Distinguished Scientist Award in 1968, the G. Stanley Hall Award in 1970, and the Howard Warren Medal in 1977.

11.3.2c *Avoidance of approaching objects*

It has been claimed that avoidance responses to symmetrically expanding shadows occur in human infants less than 6 weeks old (Ball and Tronick 1971; Bower *et al.* 1970a; Nánez 1988). Avoidance responses were not observed when shadows expanded asymmetrically or contracted. Yonas *et al.* (1977) found no evidence of avoidance responses to symmetrically looming shadows, or even to real approaching objects, in infants under about four months of age. However, they did find that infants under 4 months of age followed a visual target that rose in the visual field. They suggested that the responses observed by Ball and Tronick were due to the infant attempting to keep the gaze fixed on the top of the approaching object as it rose higher in the visual field, and not to the movement of the object in depth. Dunkeld and Bower (1980) challenged this assertion and produced evidence that infants between 3 and 4 weeks of age show avoidance responses triggered specifically by the approach of an object, under conditions where responses to rising edges are controlled for. The avoidance response to a real approaching object in the human infant has been said to emerge between the second and fourth month (Peiper 1963). Yonas *et al.* (1978) found that infants began to respond to an approaching object in a stereoscope somewhere between the third and fifth month. Four-week old infants showed more defensive reactions, in the form of blinking responses and backward head movements, when an array of dots on a rear-projection screen expanded compared with when it contracted. An array of dots moving incoherently did not elicit defensive reactions at any age (Nánez and Yonas 1994).

11.3.2d *Development of size constancy*
Ittelson (1951) presented adults with half-size, normal-size, and double-size playing cards one at a time under conditions in which the only cue to distance was the size of the retinal image. The larger card appeared nearer and the smaller card appeared more distant than the normal card. Ittelson concluded that familiar size can be used to perceive distance. There has been considerable dispute about whether this was a true perceptual effect or simply the result of a cognitive decision.

Yonas *et al.* (1982) reported that, for adults, a monocularly viewed photograph of a face reduced in size appears more distant than an enlarged face at the same viewing distance of 32 cm. Ovals with the same sizes did not show this effect. When the same displays were presented monocularly to 7-month-old infants they reached more often for the large face than for the small face. They showed no differential preference for ovals or for binocularly viewed faces. Infants 5 months old showed no differential preference for faces. Thus, effects of familiarity with particular objects on perceived distance seems to develop between the ages of 5 and 7 months.

11.3.3 Perception of 3-D form

11.3.3a *Development of shape constancy*
Shape constancy refers to the ability to recognize an object when it is viewed from different vantage-points. Several procedures have been used to determine the age at which shape constancy develops. Bower (1966b) conditioned infants between 40 and 70 days of age to make a head movement when shown a rectangular board slanting 5° in depth. Training generalized well to the same object in different orientations but poorly to a frontal trapezoid that projected the same image as the slanted board or to a slanted trapezoid. Day and McKenzie (1973) found that infants between 6 and 16 weeks old habituated their fixation when a cube was presented repeatedly in different orientations. They did not habituate to photographs of a cube in different orientations. This suggests that they had shape constancy for a real cube but not for a 2-D representation lacking parallax and binocular disparity cues to depth. However, to establish shape constancy in infants one must demonstrate, not only that they ignore differences in the orientation of an object, but also that they can discriminate one shape from another. Cook *et al.* (1978) used the same habituation procedure and found that 12-week-old infants could distinguish between a cube and an L-shaped object and between a cube and its photograph. However, they showed no evidence of distinguishing between a cube and a truncated pyramid.

Caron *et al.* (1978) habituated fixation duration of five groups of 80-day-old infants: group 1 to a frontal square, groups 2 and 3 to 30° and 60° inclined squares respectively, group 4 to an inclined trapezoid that projected a square image, and group 5 to a frontal trapezoid. The post-habituation stimulus was always a frontal square. The pattern of results showed that the infants distinguished between a frontal square and a projectively equivalent inclined trapezoid. They were therefore not responding only to the shape of the retinal image. Also, the infants distinguished between a frontal square and a frontal trapezoid. Finally, they were sensitive to all changes in the inclination of the stimulus in going from habituation trials to test trials. However, the results did not demonstrate conclusively that the infants perceived real shape over changes in orientation. In a second study, Caron *et al.* (1979) habituated one group of 12-week-old infants to a square inclined at various angles and a second group to a trapezoid inclined at the same angles. Both groups were then tested with a frontal square and a frontal trapezoid. The results indicated that the infants perceived the real shapes of the inclined stimuli—that they had shape constancy.

Slater and Morison (1985) habituated two newborn infants to a square presented in different orientations. Subsequently, the infants preferred to look at a different shape rather than at the square in a new orientation. This is the only evidence that visual experience is not required for shape constancy.

Infants a few days old seem to be capable of recognizing a familiar face (Slater and Kirby 1998; Atkinson 2000, p. 53)

11.3.3b *Perception of motion-defined contours*
A shape defined by random dots on a background of similar dots is completely camouflaged until the dots defining the shape move with respect to the other dots. This is known as shape from motion. Kaufmann-Hayoz *et al.* (1986) found that 3-month-old infants could discriminate between two motion-defined shapes. Also, after habituating to a motion-defined shape, they looked longer at a novel stationary luminance-defined shape than at one similar to the habituated shape. To this extent, perception of shape was cue invariant. However, they showed no transfer when habituated to a luminance-defined

shape and tested with motion-defined shapes. Perhaps the novelty of motion overwhelmed any preference for one shape over the other.

When a textured surface moves over a stationary textured surface, texture elements of the background surface are occluded on the leading edge and emerge on the lagging edge. This is known as the deletion-accretion cue to the depth order of the two surfaces (Gibson *et al.* 1969). Granrud *et al.* (1984) presented 5-month-old infants with a computer-generated display representing one random-dot surface moving over another. The only cue to depth order was the accretion-deletion of dots. The infants reached to the apparently nearer surface, showing that they had used the accretion-deletion cue.

An accretion-deletion display like that used by Granrud *et al.* contains a second cue to depth order. Where the two surfaces overlap, the texture of the nearer surface has an associated moving edge, while the texture of the more distant surface has no associated moving edge (Yonas *et al.* 1987a). Craton and Yonas (1988) presented 5-month-old infants with a display representing a near surface moving over a more distant surface, in which the only cue to depth order was that one surface had a moving edge associated with its texture elements. The texture elements of both surfaces were kept away from the edge so that there was no accretion-deletion. The infants showed a reaching preference for the apparently nearer surface.

11.3.3c *Perception of rotating 3-D shapes*

Dynamic changes in perspective produced by rotating an object can be a cue to the 3-D structure of the object. The kinetic depth effect (KDE) demonstrates the power of this information. In one form of this effect, the silhouette of a twisted 3-D wire frame rear-projected onto a screen appears flat when the frame is stationary but its 3-D structure is perceived when the frame rotates.

Evidence suggests that infants are sensitive to the 3-D structure of objects when the depth cue is motion parallax before they become sensitive to static monocular depth cues, such as perspective and familiar size. Owsley (1983) used habituation tests to reveal that 4-month-old infants could not distinguish between a real cube and a wedge-shape object angled so that it projected the same image as the cube. Adults could make this discrimination. However, the infants could distinguish between two objects that were rotating about a vertical axis.

Kellman (1984) habituated one group of 16-week-old infants to a 2-D videotape of a 3-D wire frame rotating about each of two axes. A second group of infants was habituated to static views of the object taken from the motion sequence. The first group remained habituated when tested with the same object rotating about a new axis but they became dishabituated when presented with a new rotating object. The second group dishabituated to both test objects. Thus, the addition of motion parallax facilitates the perception of 3-D form in 16-week-old infants, just as it does in adults (Wallach and O'Connell 1953; Kellman and Short 1987). This could be due to perception of 2-D kinetic features of the rotating shapes, such as changing intersections, rather than of their 3-D shape.

Arterberry and Yonas (1988) obtained similar results using a dynamic random-dot display in which the shapes of the test objects (a cube and an indented cube) were defined wholly in terms of relative motion of the dots. However, it is still possible that infants were responding to differences in the 2-D flow patterns rather than to the 3-D shapes created by the motion. Two further pieces of evidence suggest that they were indeed responding to 3-D shape. First, the shape-discrimination task could be performed by 4-month-old infants with high disparity-sensitivity but not by infants with low sensitivity (Yonas *et al.* 1987b). Second, 8-week-old infants could discriminate between the two cubic forms only if they saw the whole display rather than only the region containing the greatest difference in optic flow (Arterberry and Yonas 2000).

Shaw *et al.* (1986) eliminated the possibility of using 2-D flow information. Infants were habituated to a silhouette of a rotating 3-D object in which depth was specified only by transformation of linear perspective. When tested with real stationary objects containing all cues to depth except perspective transformations, 4-month-old infants showed no preference for a novel object over the object to which they had been habituated. Six-month-old infants showed a preference for the novel test object. Thus, by 6 months of age, infants can transfer depth information from the cue of dynamic perspective to a task in which only other depth cues are available.

The perception of depth from dynamic perspective, as in the kinetic depth effect, relies on the perceiver's assumption that the object is rigid. A nonrigid flat object that deforms as it rotates can simulate dynamic perspective produced by a rotating rigid object. For a rigid object there is an invariant relationship between the deformation of the retinal image and the speed and axis of rotation of the object. Eleanor Gibson *et al.* (1979) asked whether children between the ages of 11 and 16 weeks perceive this invariant property of solid objects. The looking time of infants was habituated to a foam rubber square or circular disc undergoing rigid rotation.

Habituation was maintained when the objects were moved rigidly about a different axis but not when the objects were cyclically deformed. Habituation was not maintained when infants were habituated to motion of one rigid shape and tested with a different shape moving in the same way. Also, the children continued to recognize a shape as the same after it had stopped moving. We do not know from these results how different dynamic perspective must be before the perception of a rigid object gives way to the perception of a deforming object.

11.3.3d *Perception of depth from shading*
When a flat representation of a surface, such as that shown in Figure 3.2, is viewed in a frontal plane, the disc in which the light part is above the dark part tends to appear convex and the disc in which the dark part is above tends to appear concave. When the figure is turned round, the disc appears to reverse its convexity and concavity. Hershberger (1970) showed that chickens reared in cages in which the illumination came from below behaved as if they still assumed that light came from above when selecting a 'convex' object to peck at. This suggests that the 'light-from-above' reaction is genetically determined. This may be related to the dorsal light reaction in which fish lacking vestibular sense organs swim upside-down when the fish tank is illuminated from below. Seven-month-old, but not five-month-old human infants revealed by their reaching behaviour that they perceived convex and concave shapes on the basis of direction of shading. Thus, this ability in humans, even if not innate, develops very early (Granrud *et al* 1985).

The light-from-above-assumption could refer to 'above' with respect to gravity or 'above' with respect to head or retina. Yonas *et al.* (1979) enquired which frame of reference is dominant for 4, 5, and 7-year-old children. The children pointed to the convex shape when presented with photographs of a shaded concavity and a shaded convexity. The head was inclined 90° and the shading axis oriented either vertically (only the gravity frame relevant) or horizontally (only the head frame relevant). The responses of four-year-olds most often conformed with the assumption that light came from the top of the head. However, the five- and seven-year olds used the two frames of reference with about equal consistency. Howard *et al.* (1990) showed that adults use the head frame of reference and, more specifically, a retinal frame of reference (Section 24.4.2).

11.3.3e *Sensitivity to pictorial depth*
In the Ames window demonstration, a tapered 2-D representation of a window presented to one eye with its short side nearer can appear as a rectangular window slanting in the opposite direction. This is because the far side of a slanted rectangular window projects a smaller image. Reaching movements to the Ames window revealed that 7-month-old but not 5-month-old infants were subject to the same perspective illusion (Kaufmann *et al.* 1981). The 5-month-olds reached correctly to the nearer edge of a slanted rectangular window. Thus, infants could perceive in depth by the 5th month, but could not decode the pictorial cues present in the Ames window until the 7th month.

11.4 DEVELOPMENT OF STEREOACUITY

11.4.1 Preverbal stereo tests

11.4.1a *Preferential looking*
Held *et al.* (1980) used the preferential-looking procedure in which infants viewed line stereograms through crosspolarizing filters. Displays in 2-D and 3-D version were placed side by side. By 4 months of age, infants could distinguish a display with zero disparity from one with 1° of disparity and, by 5 months, stereoacuity had reached 1 arcmin, which was the limit of the apparatus.

The mean grating acuity of 106 infants below the age of 5 months differed between the two eyes by 1 octave. Although the results were very variable, this difference fell to about 0.5 octaves by the age of 9 months. The superiority of binocular over monocular visual acuity was evident after the sixth month (Birch 1985).

Preferential looking with random-dot stereograms revealed that stereopsis emerged between the 8th and 13th weeks in 11 infant Rhesus monkeys (O'Dell and Boothe 1997). As with the emergence of other visual functions, this is considerably earlier than in the infant human.

Birch *et al.* (1982) tested 128 human infants between the ages of 2 and 12 months using two vertical bars with either crossed or uncrossed disparity relative to a central bar. Stereoacuity was defined as the smallest disparity for which an infant showed at least 75% preferential looking at a disparate stimulus rather than at coplanar bars. Stereopsis began to show by the age of 3 months. Preferential looking revealed that three-quarters of the infants had discriminated a 1 arcmin of crossed disparity in the outer bars by the age of 5 months, but only a third of them discriminated an uncrossed disparity by this age. They concluded that crossed-disparity detectors develop before uncrossed-disparity detectors. This conclusion is valid only if the infants remained converged on the central rod in both cases. Infants showed a similar sequence in the development of a preference for looking at non-rivalrous stimuli rather than at rivalrous stimuli (Birch *et al.* 1985).

Infants below the age of about $3^1/2$ months preferred to look at dichoptically combined orthogonal gratings rather than at gratings with the same orientation in the two eyes. (Shimojo *et al.* 1986; Gwiazda *et al.* 1989). Shimojo *et al.* concluded that the infants saw the orthogonal grating as a fused grid because they found that infants of this age prefer grids to gratings. They suggested that prestereoscopic infants see a grid because they lack the binocular suppression mechanism responsible for binocular rivalry. But their results could also be explained as a preference for rivalrous stimuli in infants because they are known to prefer changing stimuli to steady stimuli. This would suggest that prestereoscopic infants do have binocular rivalry.

11.4.1b *Dishabituation*

Appel and Campos (1977) found that 2-month-old infants showed a heartbeat arousal response when a flat random-dot display was changed into one having depth defined by disparity. The infants may have responded to changes in the monocular image rather than to changes in disparity or perceived depth. Atkinson and Braddick (1976) overcame this problem by using a random-dot stereogram that contained no monocular forms. Two out of four 2-month-old infants showed evidence of discriminating between 2-D and 3-D displays.

11.4.1c *Operant conditioning*

Feldman and Cooper (1980) rewarded children between the ages of 24 and 35 months for pointing to a random-dot stereogram depicting an object in depth, by showing them a colour-sound cartoon. Ciner *et al.* (1989) first rewarded the child with a food object for pointing to a black ring, which appeared on either the left or the right side. A random-dot stereogram depicting a ring in depth was then presented on either the left or the right, and a zero-disparity random-dot display was presented on the other side. Disparity in the test stereogram was increased until the child pointed consistently to it. Ciner *et al.* (1991) used this procedure with 180 children and found an improvement in mean stereoacuity from 250 arcsec at 18 months to 60 arcsec at 5 years of age. Improvement was most rapid around the age of 30 months, and was accompanied by a large decrease in intrasubject and intersubject variability.

Similar results were obtained in a later study on 136 children, involving both preferential looking and operant conditioning, although the five-year-olds in this study reached a mean level of only 29 arcsec (Ciner *et al.* 1996). Birch and Hale (1989) used a similar procedure with a group of 76 normal infants between the ages of 19 and 60 months. They reported a mean stereoacuity of 77 arcsec at 19 to 24 months and 40 arcsec at 31 to 36 months. For more discussion of operant procedures in studies of the development of depth perception in animals see Mitchell and Timney (1982).

11.4.1d *Pursuit eye movements*

Fox *et al.* (1980) used visual pursuit in the first systematic study of the development of stereopsis. They tested 40 infants between $2^1/2$ and 6 months of age using a dynamic random-dot stereogram containing a 10° by 5° cyclopean pattern with a disparity of 45 or 134 arcmin. An infant was deemed to have stereoscopic vision if its eyes followed the moving cyclopean pattern. It is not clear whether the eye movement was voluntary pursuit or optokinetic nystagmus. The motion of the cyclopean pattern was not visible to either eye alone, and the stimulus contained no monocular cues to depth. By this criterion, stereopsis emerged between the ages of $3^1/2$ and 6 months.

With monocular viewing, optokinetic nystagmus shows a directional asymmetry in children before they have developed stereopsis, and in adults lacking stereopsis (see Section 23.4.5). Directional asymmetry of OKN could therefore be used as a test of stereopsis.

11.4.2 Standard stereo tests

Standard tests of stereoscopic vision can be applied once the child can speak. Stereoacuity norms for several standard tests have been reported for children between the ages of 3 and 6 years, by which age performance is still below adult levels (Simons 1981b). Romano *et al.* (1975) used the Titmus stereotest to trace the progress of stereoacuity in 321 children between the ages of 1.5 and 13 years. Stereoacuity increased with age until it reached 40 arcsec (the best the instrument could measure) by the age of 9 years. Cooper *et al.* (1979) traced the development of stereoacuity of 112 children between the ages of 3 and 11 years using the Titmus, TNO, and Randot stereo tests. Adult levels of performance were reached by the age of 7 on all the tests, but the Randot test produced less variable results.

Cooper and Feldman (1978a) obtained better scores for children between 2 and 5 years of age with a random-dot stereogram than with the traditional Titmus fly test or TNO test. They administered the random-dot test with an operant training procedure in which the children were rewarded for correctly reporting whether or not there was depth in the display. It is not clear whether the superior performance on the random-dot test was due to the test itself or to the increased attention and motivation provided by the training and rewards.

Heron *et al.* (1985) found that children reached adult levels of performance by about the age of 7 years but that they showed higher stereoacuity with the Frisby stereotest than with the TNO or Titmus tests (see also Broadbent and Westall 1990). Performance on none of these tests was affected by whether the children were tested normally or with motivation. Fox *et al.* (1986) tested children between the ages of 3 and 5 years on the Howard-Dolman test set up as a game with rewards. Performance of the 5-year-olds was higher than previously reported and close to the adult level. However, measurements were not taken without the rewards and attentional aids, so we cannot tell whether these were responsible for the better performance of these children. The results of these behavioural procedures do not allow one to distinguish between the child's use of binocular disparity and vergence changes. This could be done with brief stimulus presentations.

The distance between the eyes of the neonate is only about two-thirds that in the adult. Other things being equal, the minimum discriminable binocular disparity is directly proportional to interocular separation (Section 19.2.1), so this factor alone accounts for a 50% improvement of stereoacuity with age. Also, as the interocular distance increases, the system that maps disparity onto the perception of relative depth must be recalibrated.

11.4.3 VEPs and development of stereopsis

The logic for recording visual evoked potentials (VEPs) from the surface of the scalp to investigate binocular functions was outlined in Section 6.7.1. Amigo *et al.* (1978) reported that binocular facilitation of the VEP first shows in normal infants at about the age of 2 months but is still below adult levels at 5 months, from which they concluded that the VEP can be used as a test of cortical binocularity. Penne *et al.* (1987) obtained similar results in a longitudinal study of three infants. However, Amigo *et al.* tested each subject at only one spatial frequency, which was 3 cpd for adults and for infants varied between 1 and 3 cpd, according to age. Penne *et al.* used a fixed spatial frequency of 0.36 cpd. This makes it difficult to compare across ages, and they may have missed the spatial frequency that evokes the best response (Section 19.6.3). Shea *et al.* (1987) recorded the VEP in response to temporally modulated checkerboard patterns with low and high spatial frequencies and found that most infants below the age of 10 months showed binocular summation of about 145% compared with the adult value of less than 100%. They did not test infants under 2 months of age.

The enhanced binocular VEP in infants may represent the summed response of two monocular pools of neurones rather than the activity of binocular neurones (Nuzzi and Franchi 1983). It may also represent activity generated by the development of excitatory inputs to binocular cells during the critical period for development of stereopsis (Leguire *et al.* 1991). As the child gets older the neurones from the two eyes develop inhibitory interactions as part of the growth of binocularity. These inhibitory interactions reduce the level of binocular facilitation of the VEP. Many people with abnormal binocularity have an unusually large interocular suppression, and it is argued that lack of binocular facilitation of the VEP is due to abnormal interocular suppression rather than to loss of binocular cells.

The first signs of VEPs specifically related to depth in dynamic random-dot stereograms occurred in infants between the ages of 10 and 19 weeks, several weeks after the fist VEPs evoked by a random-dot correlogram or flickering checkerboard (Petrig *et al.* 1981; Skarf *et al.* 1993). Birch and Petrig (1996) used both the VEP and preferential looking to assess the early development of binocular fusion of dynamic random dots and the development of stereopsis in random-dot stereograms. Both measures revealed an abrupt onset of fusion and stereopsis at between 3 and 5 months followed by rapid development to near adult levels by between 6 and 7 months.

In a test of general binocularity but not specifically stereopsis, VEPs were recorded from the scalps of infants between the ages of 4 and 36 weeks. They were shown random-dot patterns alternating at a rate of 1.9 Hz between being correlated and uncorrelated in the two eyes, and a control pattern, which alternated between two uncorrelated states. The VEPs of most infants under 2 months of age showed the same response to the test as to the control stimulus. By the third month, all infants except one with a strabismus showed a distinct time-locked response to the test stimulus but not to the control stimulus (Braddick *et al.* 1980). A similar procedure used in a longitudinal study, revealed that the median age for the first VEP evidence of binocularity was 91 days (Braddick *et al.* 1983).

There has been some dispute about when binocular rivalry develops in the infant. Behavioural evidence suggests that it develops by the age of about 3 months (Section 11.4.1a). On the other hand, the visual evoked potentials from the two eyes of human infants between the ages of 5 and 15 months showed no evidence of binocular rivalry when the infants were presented with orthogonal dichoptic gratings (Brown *et al.* 1999) (see Section 7.10.2a). Endo *et al.* (2000) recorded from single cells in V1 of monkeys during the first 4 weeks of life. The responses of the cells showed evidence of interocular suppression when the animals were shown dichoptic orthogonal gratings. Suppression was stronger than that in adult monkeys. Endo *et al.* concluded that young monkeys experience binocular rivalry before the emergence of stereopsis at between 4 and 6 weeks.

Visual evoked potentials in amblyopia are discussed in Section 13.5.2.

11.4.4 Stereoacuity in the aged

Stereoscopic acuity has been reported to remain constant between the ages of 8 and 46 years (Hofstetter and Bertsch 1976). However, in several studies reviewed by Owsley and Sloane (1990) and Brown *et al.* (1993) stereoacuity was found to decline in subjects over 50 or 60 years of age. The decline occurs in aged people with no clinical neurological defects (Hoffman *et al.* 1959). In their own study, Brown *et al.* used a Howard-Dolman test to measure stereoacuity in four groups of subjects with mean ages of 24, 45, 56, and 64 years. The mean stereo threshold was about 16 arcsec for the first three groups and increased to 27 arcsec for the oldest group of subjects. Wright and Wormald (1992) found that of 728 people over the age of 65, only 27% had full stereopsis and 29% had no stereopsis, when tested with the Frisby stereotest.

This loss of stereoacuity after age 60 years could be due to a decrease in the optical quality of the retinal image. For human subjects between the ages of 20 and 70 years, there is a decline in the modulation transfer function of the optics of the eye. This is determined from the image produced by a laser beam reflected from the retina (Guirao *et al.* 1999).

There could also be some loss of contrast sensitivity due to changes in the central nervous system. A difference in visual acuity between the eyes can cause a loss of stereoacuity at any age (Lam *et al.* 1996). Reduced retinal illumination does not seem to be the cause of the reduction of stereoacuity with age (Yap *et al.* 1994). Greene and Madden (1987) found that when correlations between visual acuity, contrast sensitivity, and stereoacuity were taken into account, only contrast sensitivity differed significantly between younger and older subjects (see also Schneck *et al.* 2000).

Increased instability of gaze could be another factor in loss of stereoacuity with age. Fixation disparity was found to increase with age in the direction of exophoria but this was not associated with a decrease in stereoacuity (Yekta *et al.* 1989). In people over 60 years of age there is an increasing incidence of convergence insufficiency associated with exophoria for near vision (Pickwell 1985). The normally accepted near point of vergence is 10 cm but an increase up to the normal reading distance of 25 cm would have no practical consequences for stereopsis (Pickwell and Hampshire 1981).

Norman *et al.* (2000) investigated the effects of aging on the ability to discriminate between 3-D shapes. For shapes defined by disparity, older observers performed as well as young observers, except for surfaces containing large disparities and high spatial frequencies. Older observers performed poorly with shapes defined by motion parallax, especially when texture elements survived for only two frames. Overall, it seems that older observers have difficulty with steep disparity gradients and rapid temporal sequences.

11.5 BINOCULAR CORRESPONDENCE

The basic pattern of binocular correspondence is laid down before birth. However, long-term experience with small-angle misalignment of the visual axes due to strabismus leads to an adaptive shift in the pattern of correspondence. This anomalous correspondence is discussed in Section 15.4.2. But, even when there is no obvious strabismus or anisometropia, some flexibility in the pattern of binocular correspondence is needed to compensate for subclinical differences between the eyes. Even with a stable pattern of binocular correspondence the growing child must constantly recalibrate the way in which disparities are coded into relative depth. This is because the interocular distance increases about 60% from birth to adulthood, with 36% of this increase occurring in the first 6 years (Aslin 1988). A linear rescaling is required since the disparity produced by a given depth interval at a given distance is proportional to the interocular distance (Section 15.3.1).

Shlaer (1971) conducted the first experimental study of flexibility in binocular correspondence. He raised kittens with prisms that introduced 2 or 4 prism dioptres of vertical disparity into the images of the two eyes. At 4 months, binocular cells of the visual cortex were found to have developed a compensatory shift in the vertical alignment of their receptive fields in the two eyes.

The normal visual environment presents us with a persistent asymmetry of the disparity field. We see more outwardly sheared images (top outwards) than inwardly sheared images, since there are more ground surfaces than ceiling surfaces. In adults, the corresponding vertical meridians are extorted about 2° with respect to each other when the horizontal meridians are aligned. This relative shearing of the corresponding vertical meridians causes the vertical horopter to be inclined top away (Section 15.7). The shear of the vertical meridians could be a developmental adaptation to the predominance of outwardly sheared images. Two lines of evidence favour this hypothesis.

When kittens are exposed to prisms that disjunctively rotate the images in the two eyes through a small angle, the visual system adjusts by altering the orientation tuning of receptive fields so that they correspond with the imposed cyclorotation of the images. Shinkman and Bruce (1977) fitted 1-month-old dark reared kittens with goggles that produced a total of 16° of torsional misalignment of the two images. After 12 weeks, a full complement of cortical binocular cells was found and the preferred orientations of these cells for stimuli presented to each eye in turn were found to be relatively rotated by the amount of the induced optical rotation. Kittens showed incomplete adaptation of orientation selectivity to 24° of torsional misalignment of images (Bruce *et al.* 1981a).

Kittens exposed to 32° of misalignment showed a permanent disruption of binocularity and stereopsis (Isley *et al.* 1990; Shinkman *et al.* 1992). The capacity of cortical cells to accommodate their orientation tuning to imposed image cyclorotation was still present in kittens exposed to prisms after being reared in the dark until the age of 3 months, but not in those dark-reared until the age of 4 months. The cortical cells of the latter group of kittens resembled those of visually deprived animals (Shinkman *et al.* 1983).

Crewther *et al.* (1980) found very little evidence of compensation of orientation tuning in binocular cells of kittens in which one or both eyes had been surgically rotated about the visual axis soon after birth. However, in most cases, the eyes were rotated more than 16° and surgical modification of the extraocular muscles may be more disruptive than optical rotation of images (Section 12.4).

Hänny and von der Heydt (1982) reared one set of normal kittens in an environment in which visible contours were confined to a floor plane below eye level. This environment produced only outwardly sheared images. Another set of kittens was reared in an environment in which contours were confined to a ceiling plane above eye level. This environment produced only inwardly sheared images. At 4 months of age the binocular cortical cells of the kittens had preferred orientations that differed in the two eyes in accordance with the type of disparity experienced. In another study from the same laboratory, kittens were reared with lenses that magnified each image by 9° along a particular axis. The axes of magnification were set at an angle of ±45° in the left and right eyes. This created gradients of both positional and orientational disparity, which could not be corrected by vergence. At 4 months, the cortical binocular cells of these animals had adjusted the relative positions and orientations of their receptive fields in the two eyes in directions that compensated for the imposed disparities (Dürsteler and von der Heydt 1983).

A normal environment has stimuli with a variety of orientational disparities and these may be detected by binocular cortical cells whose receptive fields in the two eyes vary in their orientational selectivity (Section 6.6.2). The results described here could be explained by a preferential survival during development of cortical cells tuned to the average orientational disparity in the visual environment. The other possibility is that the interocular orientation preferences of cortical cells shift to accommodate asymmetrical inputs. In either case it seems that early visual experience helps to shape the pattern of binocular orientational correspondence.

The development of the eye is reviewed in Mann (1964) and Robinson (1991). the evolution of the visual system is reviewed in Cronly-Dillon and Gregory (1991). The development of the visual system is reviewed in Purves and Lichtman (1985), Salapatek and Cohen (1987), Lam and Shatz (1991), Simons (1993), Daw (1995), Vital-Durand *et al.* (1996), Slater (1998) and Atkinson (2000). The development of binocular vision has been reviewed by Aslin and Dumais (1980), Yonas and Owsley (1987), Timney (1988), and Held (1991).

12 *Pathology of binocular vision*

12.1 STEREOANOMALIES

12.1.1 Stereoanomalies with brief stimuli

Richards (1971b) asked observers to report whether lines on each side of a fixation cross were in front of, behind, or coplanar with the cross. The lines were presented for 80 ms with disparities of between zero and 4°. Subjects also matched the perceived depth of the flashed target with a continuously visible stereoscopic probe seen subsequently. For subjects with normal stereoscopic vision, matched depth at first increased with increasing crossed or uncrossed disparity and then decreased as disparity increased towards 4° (Figure 12.1). Stereoanomalous subjects perceived either all crossed or all uncrossed disparities at the same depth as the zero-disparity stimulus.

Richards concluded that perceived depth depends on the pooling of inputs from three classes of broadly tuned disparity detectors, one class tuned to crossed disparities, one to uncrossed disparities, and one to zero disparities. Physiological evidence supports this idea (Section 6.2). He also concluded that some people lack either the crossed or uncrossed disparity detectors and consequently fail to detect either crossed or uncrossed disparities. Richards estimated that up to about 30% of the population have a stereoanomaly of this type (Richards 1970). We will see in what follows that the idea that stereoanomalous subjects lack one or other class of disparity detector must be modified, since it has been found that the anomaly depends on the test used to diagnose it.

Stereoanomalies can be specific to the sign of luminance contrast in the visual target. Thus, observers who confuse either crossed or uncrossed disparity stimuli with zero-disparity stimuli, reverse the sign of their confusion when the stimulus is changed from dark bars on a light background to light bars on a dark background (Richards 1973). This does not fit with the idea of a simple loss of either crossed or uncrossed disparity detectors.

12.1.2 Stereoanomalies with long exposures

Effects of stimulus duration on the stereo threshold are discussed in Section 19.11.1a. People classified as stereoanomalous when tested with briefly exposed stimuli may perform normally when tested with a longer exposure. For instance, about 30% of subjects could not detect depth created by 1° of crossed or uncrossed disparity in a dynamic random-dot stereogram exposed for 167 ms. However, all but one of the subjects performed perfectly when allowed to look at the stereograms for as long as they wished (Patterson and Fox 1984b). Similar results were obtained by Newhouse and Uttal (1982) and by Tam and Stelmach (1998). The stereoanomalous observers may have succeeded with long exposure by simply converging or diverging the eyes and thus converting a disparity that they could not detect into one that they could detect. However, they also performed perfectly when a stereo target was impressed on the eyes as an afterimage. It was concluded that stereoanomalies revealed with flashed stereograms arise because subjects require time to process disparity information, and not because they lack a basic stereo mechanism. Foley and Richards (1974) trained a stereoanomalous person to discriminate between crossed disparities, uncrossed disparities, and zero disparity with stimuli presented for as long as the subject wished. After this training, the stereoanomaly that was revealed with a flashed target was considerable reduced. The role of learning in perception of depth in random-dot stereograms is discussed in Section 19.12.

12.1.3 Stimulus-specific stereoanomalies

Stereoanomalies can be specific to particular locations in the visual field. Richards and Regan (1973) developed stereo perimeter tests to investigate this question. In one test, a luminous vertical bar oscillating in depth at 2 Hz between 0 and 0.4° of crossed or uncrossed disparity was placed in different positions in the visual field while the subject fixated a stationary point. The subject was deemed to have stereoscopic vision in a given region if the target was seen to move in depth rather than from side to side. In a second test, a target with ±0.4° of disparity was flashed on for 100 ms at different positions. An observer with apparently normal vision and normal stereopsis with stationary stimuli had large areas in the visual field within which motion in depth created with uncrossed disparities could not be detected and other regions in which motion in depth with crossed disparities could not be detected. Other subjects were found to behave in a similar fashion.

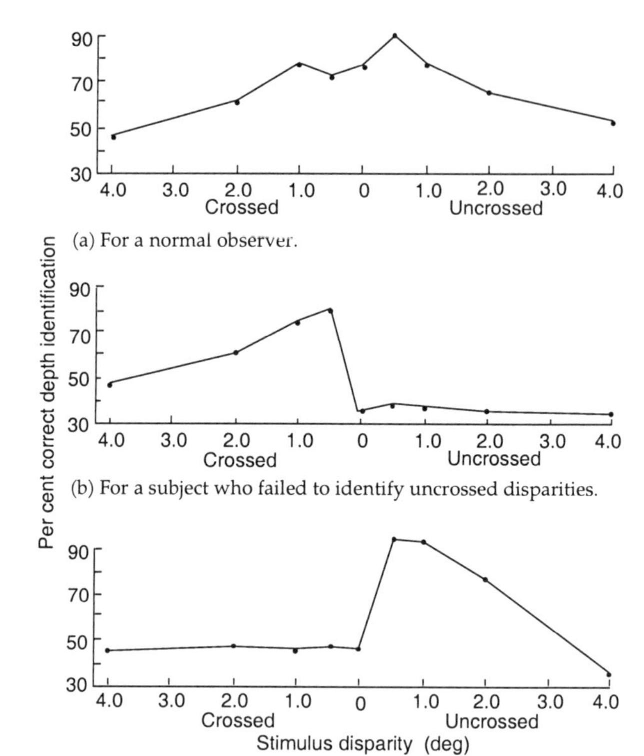

(a) For a normal observer.

(b) For a subject who failed to identify uncrossed disparities.

(c) For a subject who failed to identify crossed disparities. (Redrawn from Richards 1971b)

Figure 12.1. Depth judgments and disparity magnitude. Percent correct identification of the depth of two lines relative to a fixation cross as a function of disparity in the lines. The lines were flashed on for 80 ms.

Thus, people with normal stereoscopic acuity for stationary targets may be blind to motion in depth (see also Section 29.4.4). The opposite may also be true. About half of a group of people classified as stereoanomalous with static displays could judge depth in moving displays, such as a rotating cylinder, in which depth was defined by disparity (Rouse *et al.* 1989).

12.2 BRAIN DAMAGE AND STEREOPSIS

12.2.1 General effects of brain damage

Complete loss of stereoscopic vision and depth perception is not a common symptom of brain damage. However, there have been several reports of soldiers suffering from head injuries for whom the world appeared to lie in a single frontal plane, like a picture (Riddoch 1917; Holmes and Horrax 1919; Critchley 1955). Lesions of the right parietal lobe have been implicated in this disorder. Many patients with focal lesions of the cerebral hemispheres have some impairment of stereopsis although they do not notice the defect in their ordinary lives. Stereoscopic defects are commonly associated with other visual defects (Danta *et al.* 1978). Cerebral anoxia or coma can induce a period of cerebral blindness from which the patient gradually recovers sight. Bodis-Wollner and Mylin (1987) monitored this recovery in two patients with cerebral blindness by recording the brain potentials evoked by monocular stimuli and by binocular presentation of a dynamic random-dot stereogram. Binocular responses recovered more slowly than monocular responses. One of the patients recovered stereopsis at about the same time that she showed evidence of binocular function in the evoked potentials.

After removal of areas 17 and 18, cats lost their ability to discriminate depth based on disparity even though other abilities such as offset acuity and brightness discrimination survived (Ptito *et al.* 1992).

Cowey (1985) trained monkeys to discriminate the relative depths of two black rods. A tenfold increase in the stereoscopic threshold occurred after the part of V1 corresponding to the central 5° of the visual field was removed, and a variable but smaller increase occurred after removal of a similar region in V2. This effect of damage to the central visual field is to be expected since the test was one of fine stereopsis. A 50% increase in the stereo threshold after removal of the inferotemporal cortex was unexpected. Detection of a depth plane with a large disparity in a random-dot stereogram was unaffected by removal of the central 5° of V1. Foveal vision is clearly not essential for this coarse task. Removal of the central area of V2 or of most of the inferotemporal cortex slightly impaired performance on this test. Monkeys were unable to perform the task at all after extensive damage to the rostral superior colliculus and pretectum, which are subcortical regions associated with the control of eye movements. Subsequent tests revealed that these monkeys suffered from diplopia, which suggests that they were unable to control vergence.

12.2.2 Asymmetrical effects of brain damage

It is generally believed that the right hemisphere, which processes inputs from the left hemifield, is specialized for visuospatial tasks, such as visual localization, judgments of orientation, and depth perception, whereas the left hemisphere is specialized for language (Kimura and Durnford 1974; B. Milner 1974; Gazzaniga and LeDoux 1978). However, Birkmayer (1951) reported that of 70 brain-injured patients with impaired depth perception, 76% had left-sided damage. Furthermore, Rothstein and Sacks (1972) reported that patients with left parietal lobe lesions showed a greater stereoscopic deficit on a standard Titmus test than those with lesions in the right parietal lobe. However, only two of their ten patients had left-side damage. Lehmann and Wälchli (1975) also used the Titmus test but failed to find differential effects of left and right hemisphere damage in neurological patients. The Titmus test does not test for disparities of less than about 40 arcsec.

Danta *et al.* (1978) found that stereoscopic defects in a sample of 54 patients were more likely to be associated with damage to the temporal, parietal, and occipital lobes in the right hemisphere than in the left hemisphere (see also Ross 1983). Lesions in the left hemisphere associated with stereoscopic deficits were found to lie preferentially in the frontal and temporal lobes. The question of the lateralization of stereoscopic defects as assessed by standard tests of stereopsis is far from settled. I now consider the lateralization of defects assessed by random-dot stereograms.

Several investigators have reported that patients with disease of the right cerebral hemisphere (serving the left visual field) more often fail to perceive depth in random-dot stereograms than do normal subjects or patients with disease of the left hemisphere (Carmon and Bechtoldt 1969; Benton and Hécaen 1970; Hamsher 1978; Ptito *et al.* 1991). In the last three of these studies it was found that patients with right or left hemisphere disease performed at the same level on standard stereoscopic tests in which the forms were visible monocularly. However, Ross (1983) found both types of stereopsis to be equally affected by right hemisphere damage. Rizzo and Damasio (1985) found no laterality effects for either type of stereotest but found impairment to be greater with damage to the parietal lobe than to the temporal lobe. Lehmann and Julesz (1978) found no difference between the visual evoked potentials recorded from the left and right hemispheres when subjects were presented with random-dot stereograms (Section 6.7.3).

It has been reported that people with normal vision are better able to identify a cyclopean form in a random-dot stereogram when it is presented for 120 ms in the left visual field (right hemisphere) rather than the right visual field (Durnford and Kimura 1971). However, several cyclopean shapes were presented, and shape identification rather than stereoscopic vision may have been the factor responsible for the field asymmetry. Julesz *et al.* (1976a) used only one cyclopean shape in a dynamic random-dot stereogram and found no hemifield differences in the stimulus-duration threshold for detection of depth or in the limiting eccentricity at which a stereo target could be detected. Pitblado (1979) obtained a left-field (right hemisphere) superiority in the recognition of cyclopean shapes when the dots comprising the stereogram were small but, with large dots, performance was better in the right visual field projecting to the left hemisphere.

It has been proposed that stereopsis based on cyclopean forms in random-dot stereograms is more localized in the right hemisphere and stereopsis based on regular stereograms, in which the forms are visible in the monocular field, are more localized in the left hemisphere. Even if this were true, the crucial factor may be the relative spatial frequencies of the stimuli rather than whether or not they are cyclopean. Another possibility is that hemisphere-specific deficits reported with random-dot stereograms are due to aspects of the task other than stereopsis, such as form perception, reaction time, or defective convergence. *One way to test this would be to see whether such patients can see cyclopean shapes that are not defined by horizontal disparities (they could be defined by texture rivalry or vertical disparities) and whether they can see depth in random-dot stereograms in which the outlines of the forms are provided in the monocular images.*

The clinical category of hemisphere damage does not allow one to draw conclusions about the specific site of a deficit, and there is the problem of being sure that clinical samples are matched for factors such as age, intelligence, and motivation.

12.3 ABNORMAL INTEROCULAR TRANSFER

An induction stimulus may affect a test stimulus in one of three ways. The test stimulus can have its threshold reduced (threshold summation), or increased (threshold elevation), or some feature of the test stimulus, such as its orientation, motion, or spatial frequency, may be changed perceptually. Each of these effects can be induced in a test stimulus presented at the same time as or just after the induction stimulus. Most of these effects are manifested, although in reduced degree, when the induction stimulus is presented to one eye and the test stimulus to the other. In other words, they show interocular transfer.

Binocular summation and masking were discussed in Chapter 8 and interocular transfer of figural effects in Sections 3.2.3 and 8.3. This section is concerned with the extent to which people with defective binocular vision show binocular summation, binocular masking, and interocular transfer of aftereffects. The degree of interocular transfer has been taken as a measure of binocular interaction in the visual cortex. Given that an induction effect is cortical, any interocular transfer is assumed to reflect the extent to which the induction and test stimuli excite the same binocular cells in the visual cortex.

In normal subjects, the reduced size of the transferred effect relative to the same-eye effect is assumed to be due to dilution of the effect by unadapted monocular cells fed from the unadapted eye or by binocular AND cells that require simultaneous inputs from both eyes. If this logic is correct, a person lacking binocular cells should show no interocular transfer or binocular recruitment of cortically mediated induction effects. In practice, there are many pitfalls in applying this logic, and the literature has become complex and rather contentious.

12.3.1 Binocularity and binocular summation

A near-threshold stimulus is more likely to be detected when it is presented to two eyes rather than to one eye. Binocular summation could be due to neural summation, which is the process whereby subthreshold excitatory signals from the two eyes are summed when they impinge on cortical binocular cells. But the contribution of neural summation can be determined only after allowance has been made for the fact that detection based on the pooled output from two independent detectors shows a $\sqrt{2}$ advantage over that based on the output of a single detector. This effect is probability summation. It is best determined by measuring interocular effects under conditions in which neural summation is unlikely to occur, for instance, when the stimuli in the two eyes are separated spatially or presented at slightly different times (Section 8.1.1).

In people with normal binocular vision, binocular luminance-increment and contrast-detection thresholds for overlapping and simultaneous dichoptic stimuli are lower than monocular thresholds to a greater extent than predicted from probability summation (Section 8.1.1). It is therefore believed that true neural summation occurs. Neural summation, like the response of cortical cells, is greatest for stimuli with the same orientation and spatial frequency (Blake and Levinson 1977).

Cats reared with alternating monocular occlusion do not show behavioural evidence of binocular summation (von Grünau 1979b). In monocularly deprived cats the VEP in response to a temporally modulated grating was smaller when both eyes were open than when only the normal eye was open (Sclar *et al.* 1986). In people with severe loss of binocularity, binocular thresholds are simply what one would predict from probability summation, even with well-matched stimuli (Lema and Blake 1977; Westendorf *et al.* 1978; Levi *et al.* 1980; Blake *et al.* 1980c).

Dichoptic interactions of flicker are investigated by determining the magnitude of luminance modulation of a flickering light required for detection of flicker. This measure, plotted as a function of temporal frequency, is the temporal contrast sensitivity function. For subjects with normal binocular vision, sensitivity for in-phase binocular flicker in the region of 10 Hz is about 40% higher than that for antiphase flicker (see Section 8.1.4). At a flicker rate of 0.1 Hz, sensitivity for in-phase flicker is up to four times higher than for antiphase flicker (van der Tweel and Estévez 1974; Cavonius 1979). No significant difference between in-phase and antiphase flicker sensitivity occurred in stereoblind subjects (Levi *et al.* 1982).

A spatially uniform patch flickering in the two eyes at slightly different frequencies creates the appearance of a rhythmic modulation of luminance at a frequency equal to the difference in frequency between the two patches. This visual beat phenomenon is a simple consequence of nonlinear binocular luminance summation as the two flickering patches come into and out of phase. Three subjects with alternating strabismus and normal visual acuity and three stereoblind strabismic amblyopes failed to see dichoptic visual beats, thus providing more evidence that binocular neural summation is absent in stereoblind people (Baitch and Levi 1989).

The visually evoked potential also reveals that binocular summation is reduced in stereoanomalous observers (Section 13.5.2).

12.3.2 Binocularity and dichoptic masking

A briefly exposed suprathreshold test stimulus, such as a black and white grating, is more difficult to detect (1) when superimposed on a similar suprathreshold stimulus (simultaneous masking) or (2) when presented just before or just after a similar grating (successive masking, or the threshold-elevation effect) (Campbell and Kulikowski 1966). The test stimulus is said to be masked by the adapting stimulus, or mask.

In dichoptic masking, the mask is presented to one eye and the test stimulus to the other. This is called dichoptic masking rather than rivalry because, unlike rivalry, it occurs optimally between dichoptic stimuli of similar shape. In normal subjects, and at high contrasts, simultaneous dichoptic masking has been reported to be stronger than monocular masking (Legge 1979). Successive dichoptic masking was about 65% as strong as monocular masking (Blakemore and Campbell 1969; Hess 1978).

Dichoptic masking increases as the gratings in the two eyes are made more dissimilar in contrast (Legge 1979). Since the contrast of signals in an amblyopic eye is attenuated relative to that of signals in the good eye, one could argue that suppression of an amblyopic eye is an expression of the same mechanism that causes interocular masking in normal eyes.

Harrad and Hess (1992) tested this idea by measuring dichoptic masking with sinusoidal gratings in subjects with various kinds of amblyopia, after allowing for the difference in contrast sensitivity between the eyes. Most amblyopic subjects showed abnormal masking as a function of spatial frequency and as a function of interocular differences in contrast, when normalized to their deficit. The different types of amblyope showed specific types of masking abnormality. Only anisometropic amblyopes showed normal masking functions after allowance was made for interocular differences in contrast sensitivity. They concluded that other types of amblyopia cannot be understood in terms of the normal mechanism of dichoptic masking, and must involve more than a simple loss of contrast sensitivity in the affected eye.

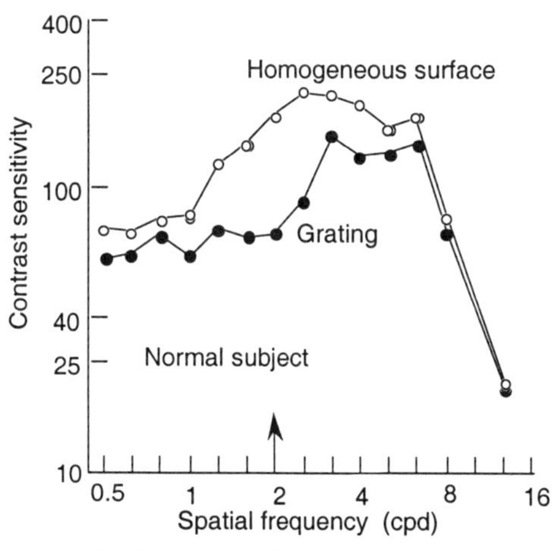

(a) Contrast sensitivity for a grating presented to the right eye of a normal subject, with the left eye viewing an homogeneous surface (upper curve), or a 2 cpd grating, 0.5 log units above threshold (lower curve).

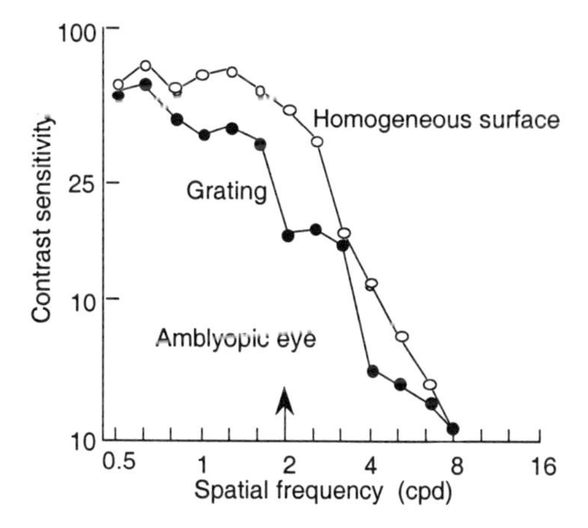

(b) Contrast sensitivity of an amblyopic eye. The non-amblyopic eye views an homogeneous surface (upper curve), or a 2 cpd grating (lower curve). Arrows show spatial frequency of mask. (Adapted from Levi *et al.* 1979b)

Figure 12.2. Dichoptic masking in an amblyopic eye.

There has been some dispute about whether dichoptic masking is present in stereoblind people with an early history of strabismus. Ware and Mitchell (1974a) found no dichoptic masking in two stereoblind subjects, whereas Lema and Blake (1977) found some in three of their four stereoblind subjects. Anderson *et al.* (1980) also found interocular transfer of the threshold-elevation effect in seven stereoblind subjects, especially from a nonamblyopic eye to a normal eye, although to a lesser extent than in normal subjects. Hess (1978) tested one strabismic amblyope, with some residual stereopsis, who showed no interocular transfer of the threshold-elevation effect and another, with no stereopsis, who showed full transfer. In a later study he found that amblyopes show no threshold-elevation in the amblyopic eye after binocular adaptation, and concluded that threshold-elevation and amblyopic suppression occur at the same cortical level (Hess 1991).

Levi *et al.* (1979b) found that subjects with abnormal binocular vision and amblyopia showed a normal level of simultaneous interocular masking at suprathreshold levels of contrast (Figure 12.2). These same subjects failed to show interocular subthreshold summation. This suggests that in people with defective binocular vision, inhibitory interactions responsible for masking still occur between the left- and right-eye inputs to binocular cells, but that excitatory interactions responsible for subthreshold summation are absent. This is in accord with the evidence reviewed in Section 13.5.2.

An important factor that may help resolve some of the conflicting evidence about interocular transfer of the threshold-elevation effect, and of other aftereffects that are mentioned later, is the spatial frequency of the stimulus. Amblyopes tend to show a selective loss of contrast sensitivity for high spatial frequencies. Selby and Woodhouse (1981) found that amblyopes showed almost normal interocular transfer of the threshold-elevation effect with low spatial-frequency stimuli to which the normal and amblyopic eye were equally sensitive. However, they showed little or no transfer with high spatial-frequency stimuli for which there was a difference in sensitivity in the two eyes. Some of these amblyopes had stereo vision, as tested on the Titmus test, and some did not, but their stereoscopic performance was not related to their degree of interocular transfer. It was concluded that stereopsis and interocular transfer of the threshold-elevation effect are not mediated by the same mechanism. However, the Titmus test involves high spatial-frequency stimuli, and it is not high spatial-frequency stereoacuity that one would expect to be related to interocular transfer of an aftereffect tested at a low spatial frequency. *Perhaps a relation between the two functions would be found if subjects were tested for stereoscopic vision with low spatial-frequency stimuli.*

A second but related factor in interocular transfer is the position of the stimulus. Even though stereoscopic vision may be lost in the central retina where both fine and coarse disparities are processed, it may be retained in the peripheral retina where only coarse disparities are processed (Section 13.5.1). Binocular subthreshold summation and interocular transfer of the threshold-elevation effect were reduced or absent in strabismic and anisometropic amblyopes for stimuli confined to the central visual field. These effects were also absent in the visual periphery for anisometropes, but strabismics showed considerable interocular summation and transfer of the threshold-elevation effect in the peripheral field (Sireteanu *et al.* 1981). One would expect binocular vision in anisometropes to be disturbed more in the periphery than in the fovea because a differential magnification of the two images produces disparities that increase with eccentricity. In strabismics, image displacement is the same over the whole visual field but affects peripheral vision less than foveal vision because the periphery has larger receptive fields.

12.3.3 Binocularity and the tilt aftereffect

Inspection of an off-vertical line or grating induces an apparent tilt of a vertical line in the opposite direction. When the induction and test lines are presented at the same time, the effect is known as tilt contrast. When the test line is presented after the induction stimulus, it is known as the tilt aftereffect. In people with normal binocular vision the tilt aftereffect shows interocular transfer when the induction line is presented to one eye and the test line to the other (Section 8.3). Estimates of the extent of interocular transfer have varied between 40 and 100% (Gibson 1937; Campbell and Maffei 1971).

Since, in primates, orientation is first coded in the visual cortex, the interocular transfer of the tilt aftereffect has been used as a test of normal binocular functioning.

Some investigators found the extent of transfer to be positively correlated with stereoacuity (Mitchell and Ware 1974) while others found no such correlation (Mohn and Van Hof-van Duin 1983). Subjects with strabismus acquired before the age of 3 years or with loss of stereopsis for other reasons showed little or no interocular transfer of the tilt aftereffect (Movshon *et al.* 1972; Ware and Mitchell 1974a; Banks *et al.* 1975; Hohmann and Creutzfeldt 1975; Mann 1978).

In these experiments the induction stimuli were gratings with a spatial frequency of about 7 cpd, which were tilted about 10° to the vertical and subtended 3° or less. Maraini and Porta (1978), used a 20°-wide grating with a spatial frequency of only 0.5 cpd, and obtained a high level of interocular transfer of the tilt aftereffect in alternating strabismics. Moreover, although consistent strabismics (esotropes) showed no transfer from the dominant to the nondominant eye, they showed good transfer in the opposite direction, albeit less than that in subjects with normal vision. Buzzelli (1981) obtained a normal level of interocular transfer of the tilt aftereffect in both directions in a mixed group of 23 strabismics. He used a grating 14° wide with a spatial frequency of 2 cpd.

Inspection of a textured surface rotating in a frontal plane about the visual axis causes a superimposed vertical line to appear tilted in a direction opposite to the background motion. When the rotating surface was presented to one eye and the vertical line to the other, this effect remained at full strength in subjects with normal binocular vision but was reduced in stereoblind subjects (Marzi *et al.* 1986).

12.3.4 Binocularity and the motion aftereffect

In the motion aftereffect a stationary display appears to move in the opposite direction to a previously inspected moving display. In people with normal binocular vision the motion aftereffect transfers at least 50% to an eye that was not exposed to the induction stimulus (Section 8.3.3).

Wade (1976b) found no interocular transfer of the motion aftereffect produced by a rotating sectored disc, in six stereoblind adults who had strabismus from early childhood, and a little transfer in 11 subjects whose strabismus had been surgically corrected (see also Hess *et al.* 1997c). Six subjects with mild strabismus and some stereoscopic vision showed some transfer of the aftereffect. Using a 10°-diameter rotating sectored disc, Mitchell *et al.* (1975) also found no interocular transfer of the motion aftereffect for subjects who were stereoblind because of childhood strabismus or anisometropic amblyopia. For subjects with some stereoscopic vision there was a positive correlation of 0.75 between the amount of transfer and stereoacuity. Mohn and Van Hof-van Duin (1983) used a similar 10°-diameter sectored disc and found no interocular transfer of the aftereffect in stereoblind subjects. In subjects with some stereo vision, they found no correlation between the amount of transfer and stereoacuity.

Keck and Price (1982) used an 8°-wide moving grating to test three groups of subjects (1) those who had central scotomata but some peripheral stereoscopic vision, (2) those with alternating strabismus but no stereoscopic vision, and (3) those with a consistent strabismus, anomalous correspondence, and no stereoscopic vision. Subjects in all groups showed less transfer of the motion aftereffect than subjects with normal vision, but transfer was absent only in the third group. O'Shea *et al.* (1994b) found that interocular transfer of the motion aftereffect was significantly reduced in 10 strabismics of early onset with a stimulus confined to the central 2.8° of the visual field but not with a stimulus confined to an annular region between 20 and 40° of eccentricity.

Raymond (1993) obtained 96% transfer of the elevation of the motion-coherence threshold. This threshold is the least proportion of coherently moving dots in an array of randomly moving dots that can be seen as moving in a given direction. Raymond favoured the view that area MT is the site of this interocular effect, which would account for its high level of interocular transfer. McColl and Mitchell (1998) confirmed the high interocular transfer of the coherent motion threshold and found that while stereodeficient subjects showed very little transfer of the conventional motion aftereffect they showed about 90% transfer of the coherence motion aftereffect. This supports the idea that the motion threshold aftereffect is processed at a higher level in the visual system than the simple motion aftereffect.

Summary

There seems to be general agreement that little or no interocular transfer occurs in stereoblind subjects. However, some transfer has been found in certain types of stereoblind subjects, and there is considerable controversy about the correlation between the degree of transfer and stereoacuity. The conflicting findings from different laboratories could be due to different clinical samples, different diagnostic tests, or different stimuli used to measure interocular transfer. One potentially important factor is the position of the stimuli. There is more interocular transfer of the threshold-elevation effect for peripheral than for central stimuli, and some stereo vision is retained in the peripheral retina when it is lost in the central retina. A related factor is the spatial frequency of the stimuli. Amblyopes show more interocular transfer of the threshold-elevation effect for low spatial-frequency stimuli to which both eyes are equally sensitive than they show for high spatial-frequency stimuli to which the amblyopic eye is relatively insensitive.

In future studies of interocular transfer of the motion and tilt aftereffects in people with visual defects, special attention should be given to the spatial frequency and size of stimuli used to test stereoscopic vision and interocular transfer.

12.4 BINOCULARITY AND PROPRIOCEPTION

Sensory receptors exist in the extraocular muscles and/or in the extraocular muscle tendons in a variety of animals, including humans (Bach-y-Rita 1975; Richmond *et al.* 1984). Inputs from these receptors enter the brain along the ophthalmic branch of the trigeminal nerve (fifth cranial nerve). Responses to stretching of extraocular muscles have been recorded in cells of the superior colliculus (Donaldson and Long 1980), the vermis of the cerebellum (Tomlinson *et al.* 1978), the visual cortex (Buisseret and Maffei 1977), and the frontal eye fields (Dubrovsky and Barbas 1977). The response of about 40% of relay cells in the LGN of the cat to drifting gratings was modified when the eye was passively rotated to different positions (Lal and Friedlander 1990). This suggests that an afferent eye-position signal gates the transmission of visual signals through the LGN.

The proprioceptive-visual gating hypothesis is supported by the fact that prolonged monocular paralysis in the adult cat causes a bilateral reduction of responses of X-cells in the LGN (Garraghty *et al.* 1982). This effect is immediately reversed by removal of proprioceptive inputs from the mobile eye (Guido *et al.* 1988). This suggests that effects of monocular paralysis are due to inhibition of X-cells by asymmetrical proprioceptive inputs from the two eyes.

Kittens with unilateral or bilateral section of proprioceptive afferents suffer permanent deficits in visual-motor coordination (Hein and Diamond 1983). For example, Fiorentini *et al.* (1985) reported that depth discrimination in a jumping-stand test is affected by unilateral section of the ophthalmic nerve in adult cats (Portrait Figure 12.3). However, Graves *et al.* (1987) found this to be true in only some cats. Significant changes in depth discrimination in cats occurred only when a unilateral section of proprioceptive afferents was performed between the ages of 3 and 13 weeks, or bilateral section between the ages of 3 and 10 weeks (Trotter *et al.* 1993). Inputs from proprioceptors in the extraocular muscles have also been implicated in the long-term maintenance of binocular alignment of the eyes (Lewis *et al.* 1994).

Trotter *et al.* (1987) severed the proprioceptive afferents of kittens either unilaterally or bilaterally at various times in the first few months after birth. This did not cause strabismus or interfere with the movements of the eyes. After the operation some of the kittens were reared with normal binocular experience and some in darkness. One month after unilateral section of proprioceptive afferents during the critical period, both the seeing cats and the dark-reared cats showed a severe reduction in the number of binocular cortical cells, which was still present $2^{1}/2$ years later. Unilateral section of the nerve had no effect when performed during the first month after birth or in the adult cat. Bilateral section of the nerve had no effect on binocular cells no matter when it was performed.

More recently, Trotter *et al.* (1993) recorded from cells in the visual cortex of adult cats in which proprioceptive afferents had been severed unilaterally when the animals were between 5 and 12 weeks of age. The stimuli were moving sine-wave gratings with dichoptic phase (horizontal disparity) set at various values. In the cells of operated cats, the range and stability of disparity tuning and the degree of binocular suppression were reduced below the level of cells in normal cats.

Figure 12.3. Adriana Fiorentini.
Born in Milan in 1926. She received a B.Sc. in Physics from the University of Florence in 1948, and the libera docenza in Physiological Optics in 1956. She worked in various capacities at the Istituto Nazionale di Ottica in Florence until 1968, when she moved to a Professorship at the Istituto di Neurofisiologia in Pisa, to commence her long collaboration with Professor Maffei. She retired in 1992 but continues her research.

Maffei and Bisti (1976) surgically deviated one eye of kittens soon after birth and occluded both eyes at the same time. The reduction in the number of cortical binocular cells was about the same as that produced by induced strabismus when both eyes were allowed to see. They concluded that asymmetrical movement of the two eyes, even in the absence of vision, is sufficient to disrupt binocularity in cortical cells. Others failed to replicate this effect (Van Sluyters and Levitt 1980). Maffei and Bisti's conclusion was also challenged on the ground that monocular paralysis in kittens leads to a reduction in X cells in the LGN and the apparent loss of binocular cells was secondary to this (Berman *et al.* 1979). Monocular paralysis produces loss of LGN X cells even in the adult cat (Brown and Salinger 1975). But this claim has also been challenged by those who found no loss of X cells in the LGN after monocular paralysis (Winterkorn *et al.* 1981). This issue remains unresolved.

A report that surgical deviation of one eye in adult cats leads to a loss of binocular cells (Fiorentini and Maffei 1974; Maffei and Fiorentini 1976b) was not replicated by Yinon (1978) but has been supported by further evidence (Fiorentini *et al.* 1979).

Others have concluded that eye motility plays a role in cortical plasticity but only in combination with abnormal visual inputs. Thus, many cortical cells recovered their response to stimulation of a deprived eye only when the normal eye was both pressure blinded and had its extraocular muscles paralyzed (Crewther *et al.* 1978).

When anaesthetized kittens with their eye muscles paralyzed were exposed to a patterned display for 12 hours they did not show any reduction in the number of binocularly activated cortical cells (Freeman and Bonds 1979). However, monocular exposure did reduce the number of responding binocular cells when the eyes were not paralyzed or were moved mechanically by the experimenter while they were paralyzed. Moving the eyes mechanically in darkness for 12 hours had no effect. They concluded that cortical plasticity depends on a combination of nonmatching visual inputs and eye-movement information, presumably arising in proprioceptors in extraocular muscles.

Buisseret and Singer (1983) came to the same conclusion after finding that neither monocular occlusion nor induced strabismus led to a change in the binocularity of cortical cells in the kitten when proprioceptive afferents were abolished by bilateral section of the ophthalmic nerve.

Cortical cells that had lost their capacity to respond to an eye that had been occluded responded to stimuli from that eye within minutes after the good eye was blinded by application of pressure and an anaesthetic block was applied to the extraocular muscle afferents of the good eye. Neither procedure was effective alone (Crewther *et al.* 1978). Thus, both proprioceptive and visual afference seem to play a role in maintaining a normal eye's suppression of a deprived eye.

Recovery of visual functions after a period of binocular deprivation is aided by ocular motility. Thus, 6-week-old dark-reared kittens showed some recovery of orientation selectivity of cortical cells when allowed to see, but not when allowed to see with their eye muscles paralyzed (Buisseret *et al.* 1978; Gary-Bobo *et al.* 1986). Buisseret *et al.* (1988) selectively severed the extraocular muscles of 6-week-old dark-reared kittens so that the eyes could move only horizontally or only vertically. After the kittens were given a period of visual experience, orientation-selectivity of cortical cells became predominantly tuned to the direction opposite that of the allowed eye movements. The influence of eye-muscle proprioception on visual functions has been reviewed by Buisseret (1995).

On balance, this evidence suggests that eye proprioception plays a key role in cortical plasticity and in the development of depth perception, but just how this is accomplished remains a mystery.

12.5 ALBINISM

12.5.1 Basic characteristics of albinism

Albinism is a group of genetically determined disorders affecting the synthesis of melanin in the retinal pigment epithelium. It occurs in all mammalian species. There are two main types of the disorder: **oculocutaneous albinism**, characterized by absence of pigment throughout the body, and **ocular albinism**, in which hypopigmentation is restricted to the eye. There are many subtypes, and the severity of the deficit depends on at least eight genes (Kinnear *et al.* 1985; Abadi and Pascal 1989). About 1 in 17,000 people has oculocutaneous albinism and about 1 in 50,000 have ocular albinism, one form of which is linked to the X chromosome and occurs only in males (Jay *et al.* 1982). For one type of albinism the affected gene encodes tyrosinase, the enzyme involved in synthesis of melanin. Introduction of this gene into albino mice allowed the visual pathways to develop normally (Jeffery *et al.* 1994). However, similar symptoms arise in human oculocutaneous albinos, who do not lack this gene. In these tyrosinase-positive cases, lack of melanin must be due to other factors—probably lack of a substrate (Witkop *et al.* 1970).

In all forms of albinism there is an absence of ocular pigmentation and a reduction of the number of rods, especially in the central retina. The absence of melanin in the pigment epithelium behind the retina causes the ocular fundus, or concave interior of the eye, to appear orange-red and renders the choroidal blood vessels visible through the ophthalmoscope.

Many albinos also have loss of pigmentation in the iris, giving the eyes a characteristic pink appearance. Lack of pigment allows light to enter the eye through the sclera and iris and to reflect from the eye's internal surfaces, causing excessive illumination and glare. Albinos typically avoid the resulting visual discomfort and exposure to excessive doses of ultraviolet light by keeping away from bright lights, a response known as **photophobia**. The yellow macular pigment, which reduces the effects of chromatic aberration, is also absent in albinos.

Albinos also tend to have astigmatism and high refractive errors, especially myopia. They show malformation of the fovea. The ganglion cell layer is present over the fovea, and central cones resemble those found normally in the parafoveal region (Fulton *et al.* 1978). The visual pathways of albinos are also deficient. These defects are accompanied by strabismus, congenital nystagmus, and a variety of visual defects including reduced acuity and impaired or absent binocular fusion.

Stereoscopic vision is either absent or deficient in albinos (Guo *et al.* 1989; Apkarian 1996). In a mixed group of 18 human albinos, 9 showed some evidence of stereopsis when tested with a variety of stereo tests, including a random-dot stereogram, although only a simple pass-fail criterion was used (Apkarian and Reits 1989). The severity of visual defects is correlated with the degree of pigment deficit (Sanderson *et al.* 1974).

12.5.2 Abnormal routing of visual pathways

12.5.2a *Abnormal routing in albinos*

In the present context, the most significant visual defect in albinism is the unusual structure of the visual pathways. Lund (1965) first revealed the unusually small number of uncrossed axons in the optic tract of albino rats. From the earliest developmental stage in albino rats, cats, and ferrets, uncrossed axons at the chiasma are reduced in number compared with those in normal animals, and tend to originate in the peripheral retina (Guillery 1986). Many fibres from about the first 20° of the temporal hemiretinas erroneously decussate. The part of the nasal retina closest to the midline also gives rise to some misrouted ganglion cells, but axons from the most nasal parts of the retina have been found to be routed normally in several species of albino animals (Sanderson *et al.* 1974). Also, cortical cells may receive inputs from the ipsilateral eye through the corpus callosum (Diao *et al.* 1983). Evoked potentials in humans reveal that they also have a reduced number of uncrossed inputs to the visual cortex (Creel *et al.* 1974).

In albino mammals of a variety of species, including monkeys, cells in the LGN receiving crossed inputs form into enlarged layers according to the eye of origin but with abnormal fusions between the layers. Cells receiving uncrossed inputs tend to form into segregated islands rather than distinct layers (Gross and Hickey 1980; Sanderson *et al.* 1974). A post-mortem study of the LGN of a human albino revealed abnormal fusions of the four parvocellular layers and of the two magnocellular layers in the region of the LGN that is normally six-layered (Section 5.3.1). In the normal LGN there is a small two-layered region devoted to crossed inputs from the monocular crescent in the far periphery of the visual field. In the albinotic LGN this two-layered region is greatly extended because of the unusual number of crossed inputs (Guillery *et al.* 1975; Guillery 1986).

The terminals of the abnormally routed visual inputs are arranged in a normal retinotopic order in the visual cortex but on the wrong side of the brain. The topological representation of the visual field normally shows a sudden reversal at the visual midline, which is represented at the border between areas 17 and 18. In the albino ferret this point of reversal occurs in the ipsilateral field up to 30° away from the visual midline (Thompson and Graham 1995). The misrouted inputs thus map a part of the ipsilateral rather than contralateral visual field and in mirror-reversed order (Kaas and Guillery 1973). This produces an unusual location of visually evoked potentials recorded from the scalp (Creel *et al.* 1978, 1981; Boylan and Harding 1983; Apkarian *et al.* 1984). The VEPs show delayed ipsilateral latency, reduced ipsilateral amplitude, or both (Guo *et al.* 1989). Retinal projections to the superior colliculus also appear to be reversed (Collewijn *et al.* 1978).

One basic cause of abnormal routing of axons is a lack of tyrosinase, the enzyme involved in melanin synthesis. It has been suggested that abnormal routing is due to absence of melanin in the developing eyestalk. However, normal visual pathways develop in animals in which melanin is confined to the retina (Colello and Jeffery 1991). The lack of melanin in the retina reduces the number of retinal cells specified to remain uncrossed (Marcus *et al.* 1996). The chemical markers that determine whether or not an axon crosses the chiasma are not known but cells that cross tend to be produced later than those that remain uncrossed. Lack of melanin seem to interfere with the spatiotemporal pattern of cell development in the retina, leading to a differential delay in the time of arrival of different axon types at the chiasma (Jeffery 1997).

12.5.2b *Abnormal routing in Siamese cats*

The visual inputs of Siamese cats are misrouted in a similar way to those of albinos (Guillery 1969; Guillery and Kaas 1971; Kalil *et al.* 1971). There are two types of visual projection; the Boston pattern and the Wisconsin pattern. Evidence of a similar distinction has been produced in the visually evoked responses of human albinos (Carroll *et al.* 1980).

In the Boston pattern, the mirror-reversed projection corrects itself to produce an essentially continuous representation of the visual field in each hemisphere (Hubel and Wiesel 1971; Shatz and LeVay

1979). Optic nerve fibres from up to 20° into the temporal retina of each eye cross aberrantly in the chiasma and terminate in the wrong LGN. In the visual cortex, the aberrant representation of the ipsilateral visual field is inserted between the normal representations of the contralateral visual fields in areas 17 and 18. This causes the cortical representation of the vertical meridian to be displaced up to 20° away from its usual location along the border between areas 17 and 18. The transition from the cortical region containing only contralateral projections to that containing only ipsilateral projects is diffuse in Siamese cats rather than fairly sharp as in normal animals (Cooper and Pettigrew 1979b). Also, the transcallosal fibres, which usually originate and terminate along the border between areas 17 and 18, acquire a wider distribution with a peak occurring in the region representing the vertical meridian rather than along the border between areas 17 and 18 (Shatz 1977a). This suggests that transcallosal fibres grow to connect corresponding regions of the visual field rather than similar architectonic regions of the cortex.

In the Wisconsin pattern the reversed projection is not corrected but there is intracortical suppression of the anomalous inputs along with all other inputs from the same LGN lamina (Kaas and Guillery 1973). The vertical retinal meridian projects to its usual location along the border between areas 17 and 18 (Kaas and Guillery 1973) and transcallosal fibres originate and terminate in substantially the same way as in normal cats (Shatz 1977b).

The superior colliculus of Siamese cats receives an abnormally large representation of the ipsilateral visual field and the part devoted to the fovea is shifted about 7° contralateral to its normal location (Berman and Cynader 1972). The superior colliculus of Boston-type Siamese cats has many binocular cells that obtain most of their ipsilateral input through the corpus callosum (Antonini *et al.* 1981). Some of these binocular cells have disparity tuning functions resembling those of cortical cells tuned to coarse disparity in normal cats (Bacon *et al.* 1999). Strabismic cats, also, retain binocular cells in the superior colliculus (Section 13.2.1d).

Siamese cats have few binocular cells in cortical areas 17, 18, and 19 (Guillery *et al.* 1974; Di Stefano *et al.* 1984). Nevertheless, they show good interocular transfer of learning to discriminate visual forms (Marzi *et al.* 1976). This could be because they retain a considerable number of binocular cells in the lateral suprasylvian area (Clare-Bishop area), an area involved in interhemispheric transfer of form discrimination (Berlucchi *et al.* 1979). Thus, Siamese cats of the Boston type have numerous binocular cells in the Clare-Bishop area, which show selectivity for motion in depth similar to that of cells in the normal animal (Toyama *et al.* 1991). The suprasylvian area is rich in transcallosal connections and callosectomy abolishes the binocularity of cells in this area in Siamese cats but not in normal cats (Marzi *et al.* 1980). Thus, the ipsilateral inputs to the suprasylvian area of Siamese cats are routed through the corpus callosum.

The abnormally routed visual projections in albinos and Siamese cats are associated with a disturbance of inputs to the oculomotor system controlling vergence. This may explain why albinos and Siamese cats usually have strabismus and spontaneous nystagmus. There is also a complete disruption of mechanisms for detecting disparity, so that albinos and Siamese cats have little or no stereoscopic vision (Packwood and Gordon 1975).

Although both albinism and the Siamese condition are genetic in origin, there has been some dispute about which defect is primary; the misrouting of visual inputs, the strabismus, or the lack of binocular cells in the visual cortex. Cool and Crawford (1972) argued that strabismus is the primary cause although they reported that while all Siamese cats lack binocular cells, some do not have strabismus. Misrouting of visual inputs is the most probable cause of the Siamese condition. In Siamese cats, lack of pigment in the embryonic eye stalk causes abnormal positioning of axons in the developing optic nerve, which in turn disrupts routing at the chiasma (Webster *et al.* 1988).

12.5.2c *Achiasmatic animals*

Williams *et al.* (1991, 1994) identified an autosomal recessive mutation in sheep dogs that causes all retinal axons to project to the ipsilateral LGN. The optic chiasma is thus eliminated—the animals are achiasmatic. The nasal retinal fibres terminate in ipsilateral layer A with the same topographic arrangement as that in the contralateral LGN in normal dogs. The temporal fibres project normally to the superimposed ipsilateral layer A1. Since the nasal fibres have not crossed the midline the nasal projection is mirror-image reversed with respect to the temporal projection and the two maps are congruent only along the vertical midline. This condition is discussed in more detail in Section 10.2.3c.

12.5.3 Congenital nystagmus

All animals with mobile eyes show involuntary pursuit movements in the same direction as a large moving visual display, interspersed with saccadic return movements. This reflex response is optokinetic nystagmus or OKN (see Howard 1993a for a review). In mammals, the subcortical nuclei controlling OKN receive direct inputs from only the nasal hemiretina of the contralateral eye, conveyed by axons that decussate in the optic chiasma. In higher mammals, outputs from the visual cortex descend to the subcortical nuclei and counterbalance the inherent directional asymmetry of the subcortical mechanism (Section 23.4.5). These cortical outputs derive from binocular cortical cells that normally receive both decussated and undecussated axons. The OKN system is held in symmetrical balance by the interplay of these two systems. In the albino, the excessive number of decussating axons upsets both the subcortical and cortical components of OKN and the balance between them, and this results in spontaneous nystagmus.

All types of spontaneous nystagmus of genetic origin are known as congenital nystagmus. Congenital nystagmus consisting of a conjugate, involuntary oscillation of the eyes, usually in the horizontal direction, is a universal feature of albinism. Nystagmus associated with albinism is due to a misrouting of the visual pathways, which reveals itself in an asymmetry in the visual evoked potentials from the two sides of the brain. Visual evoked potentials do not show this asymmetry with congenital nystagmus not due to albinism (Apkarian and Shallo-Hoffmann 1991).

An interesting feature of all forms of congenital nystagmus is that, unlike acquired nystagmus, it is not accompanied by oscillopsia, or perceived oscillation of the visual world. People with congenital nystagmus learn to suppress or ignore the retinal motion signals that arise during nystagmus. They assess the stability of the visual world on the basis of information taken in when the eyes momentarily come to rest between nystagmic sweeps. They experience oscillopsia if the retinal image is artificially stabilized (Leigh *et al.* 1988). In congenital nystagmus there is usually a position of gaze for which nystagmus is minimal or absent. This so-called null position typically shifts in a direction opposite to the motion of a moving display. Also, congenital nystagmus is usually reduced when the patient converges on a near object (Dickinson 1986). Thus, patients may reduce their nystagmus either by looking at an object with the head to one side, so as to bring the gaze into the null position, or by voluntary convergence. Prisms or surgical rotation of the eyes may help by bringing the null angle of gaze into the primary, or straight-ahead position.

Patients with congenital nystagmus may show no OKN in response to visual motion, a response with unusually low gain, or so-called reversed OKN. In reversed OKN the slow phases, which normally compensate for the motion of the stimulus, occur in the opposite direction to that of the stimulus. The slow phases often have an accelerating velocity profile instead of the constant velocity profile typical of normal slow phases (Dichgans and Jung 1975; Halmagyi *et al.* 1980; Yee *et al.* 1980). Reversed OKN occurs only in response to stimuli moving along the meridian in which the congenital nystagmus occurs (Abadi and Dickinson 1985). The reversal of OKN is presumably caused by the mirror-reversed projection of the abnormally routed cortical and pretectal inputs.

The vestibuloocular response (VOR) in albinos with congenital nystagmus has an unusually low gain when the head is oscillated at a low frequency, although gain may be normal at high frequencies. Albinos show weak or no VOR in response to caloric stimulation of the vestibular organs since this is equivalent to a low-frequency rotation of the head. Furthermore, optokinetic afternystagmus (OKAN) is absent and postrotatory nystagmus is unusually brief in people with congenital nystagmus (Demer and Zee 1984). This suggests that congenital nystagmus involves a defect in the velocity-storage mechanism common to OKN and VOR.

In summary, albinos suffer from optic glare, poorly developed retinas and poorly developed visual pathways. They also have instability of gaze, refractive error, and astigmatism. These factors contribute to loss of visual acuity and stereopsis, which is so severe in some albinos that they are classified as partially or completely blind.

13 *Effects of visual deprivation*

13.1 EFFECTS OF DARK REARING

13.1.1 Physiological effects of dark rearing

13.1.1a *Loss of general responsivity*

Cats and monkeys reared in total darkness show no obvious changes in the number, size, or staining characteristics of cells in either the retina or LGN (Chow 1973; Hendrickson and Boothe 1976). Also, the sensitivity of LGN cells in the cat to spatial frequency, orientation, or motion is not affected by dark rearing (Zhou *et al.* 1995; Mower *et al.* 1981a). However, dark rearing disrupts the normal development of cells in the visual cortex. Cats reared in the dark for the first 6 months showed a 50% reduction in the number of synapses in area 17 that received inputs from the LGN (Turlejski and Kossut 1985). Even though some cortical cells responded to stimulation in either eye, the cats were permanently stereoblind. A substantial number of cortical cells in the binocularly deprived kitten become unresponsive, weakly responsive, or respond erratically. The cells apparently do not die or lose their synaptic connections, since they recover their responsivity when an excitatory amino acid is applied locally (Ramoa *et al.* 1987).

The expression of brain-derived neurotrophic factor (BDNF) in the visual cortex increases just after young rats open their eyes. Keeping infant or adult rats in the dark or suppression of ganglion-cell activity by intraocular injection of tetrodotoxin decreased the expression of BDNF and its ligand. The growth factor returned to normal levels after visual inputs were restored. Expression of the nerve growth factors (NGF) and NT-3 did not depend on visual inputs (Castrén *et al.* 1992; Schoups *et al.* 1995b).

In normal animals, lateral connections develop between cortical cells with similar orientation preference (Sections 5.4.7 and 10.3.3). Cats binocularly deprived for at least the first 4 weeks of life develop abnormal clusters of horizontal connections (Callaway and Katz 1991). Bilateral enucleated fetal monkeys show a progressive reduction in the size of V1 accompanied by an increase in the size of extrastriate cortex (Dehay *et al.* 1996). Also, monkeys reared

Figure 13.1 Max Cynader.
Born in 1947. He obtained a B.Sc. from McGill University in 1967 and a Ph.D. from MIT in 1972. After postdoctoral training at the Max Planck Institute in Germany he held academic positions in the Departments of Psychology and Physiology at Dalhousie University in Halifax, Canada. In 1988 he became head of the Ophthalmology Research Group in the University of British Columbia. He is a Fellow of the Royal Society of Canada and recipient of the Killam Prize and the Gold Medal in Health Sciences from the Science Council of British Columbia.

with both eyes sutured have far fewer neurones in the corpus callosum that terminate in areas 17 and 18, and their distribution is severely restricted (Innocenti and Frost 1980).

Dark-reared monkeys and monkeys reared with the eyelids sutured show a reduction of visually responsive cells in the posterior parietal cortex (area 7) accompanied by an increase in the number of cells responsive to somatosensory inputs and of cells showing only spontaneous activity (Hyvärinen *et al.* 1981; Carlson *et al.* 1987). Subsequent visual experience failed to rectify these deficits fully.

13.1.1b *Loss of stimulus specificity*

During the first 3 or 4 weeks after birth, normal and dark-reared kittens have a similar number of cortical cells tuned to orientation and movement, and both possess binocular cells, of which some are tuned to disparity. After that time, cells tuned to orientation, motion, and disparity increase in number and stimulus specificity in normal animals. By contrast, in dark-reared animals, the number of tuned cells decreases and the number of cells with non-specific tuning increases (Pettigrew *et al.* 1968; Pettigrew 1974; Buisseret and Imbert 1976; Frégnac and Imbert 1978; Braastad and Heggelund 1985; Czepita *et al.* 1994).

Cells in area 18 also lose their stimulus specificity with dark rearing (Singer and Tretter 1976) and there are far fewer complex cells in area 18 of dark-reared cats than of normal animals (Blakemore and Price 1987b). Cells that retain their sensitivity to orientation tend to be monocular and tuned to the vertical or to the horizontal, thus resembling orientation tuned cells of the neonate (Buisseret *et al.* 1982).

Blockage of all neuronal activity in the visual cortex by infusion of tetrodotoxin in infant ferrets totally suppressed the development of cells with orientation selectivity (Chapman and Stryker 1993). Presumably, the retention of some selective cells in binocularly deprived animals is due to spontaneous visual inputs.

The cortical cells of kittens that have been dark reared until 5 weeks of age rapidly regained some stimulus specificity when sight was restored (Imbert and Buisseret 1975). Even after 12 months of dark rearing, cells recovered some stimulus specificity for orientation, although not for direction of motion (Cynader *et al.* 1976) (Portrait Figure 13.1). After the age of 12 weeks, lack of visual experience does not affect the orientation specificity of cortical cells in the cat's visual cortex (Buisseret *et al.* 1982).

It thus seems that cortical cells of cats develop some stimulus specificity in the first few weeks of life in the absence of visual experience. However, visual experience is required during a critical period for the maintenance and further development of stimulus specificity, especially to high spatial frequencies and fine binocular disparity.

13.1.1c *Effects on ocular dominance columns*

Swindale (1981) found a permanent loss of anatomically defined ocular dominance columns in area 17 of 30-week old dark-reared cats or cats reared with the lids of both eyes sutured. Ocular dominance columns were normal in cats kept in the dark after 6 weeks of normal vision. Visual experience of between 48 and 128 hours during a 30-week period of darkness was sufficient for normal development of ocular dominance columns in area 17. Ocular dominance columns developed in cats allowed binocular vision after 6 weeks of darkness but there was no recovery after 25 weeks of darkness (Swindale 1988). Mower *et al.* (1985) found no recovery of ocular dominance columns in cats dark-reared for between 9 and 16 weeks. However, restoration of binocular vision produced some recovery of orientation selectivity and binocularity. Dark rearing did not prevent development of ocular dominance columns in area 18 of cats (Swindale and Cynader 1986).

LeVay *et al.* (1980) found ocular dominance columns in 7-week-old dark-reared monkeys. Ocular dominance columns begin to develop before birth in monkeys but not in cats (Section 10.3.5).

13.1.2 Behavioural effects of dark rearing

Dark-reared cats have profound deficits in visually guided behaviour such as obstacle avoidance, paw placement, directed jumping, and visually elicited blinking. However, a large moving display elicits optokinetic nystagmus (Vital-Durand *et al.* 1974). After return to light conditions, visually mediated behaviour gradually recovers (Van Hof-van Duin 1976a; Timney *et al.* 1978; Mitchell and Timney 1982). However, the cats remain deficient in learning complex visual discriminations, and visual acuity does not return to normal (D.C. Smith *et al.* 1980).

Cats raised with both eyelids sutured show more severe disruption of behaviour and less evidence of recovery after restoration of sight than do cats raised in darkness (Mower *et al.* 1982). Lid suturing allows diffuse light to enter the eyes and this must be responsible for the greater severity and permanence of deficits. We will see that dark-reared cats have a prolonged period of cortical plasticity compared with normally reared cats. Visual inputs must accelerate termination of the phase of cortical plasticity in young cats, even when the stimulus is diffuse light.

Monkeys raised for 7 weeks with both eyelids sutured had reduced contrast sensitivity in both eyes. However, the deficit was much less than that produced in the occluded eye of a monkey raised with only one eye sutured. Furthermore, sensitivity to visual flicker and spectral sensitivity were not much affected by bilateral eye sutures. None of the binocularly deprived monkeys had binocular vision as assessed by binocular summation of grating detection or by their ability to detect depth in a random-dot stereogram (Harwerth *et al.* 1991). Dark-reared monkeys showed rapid recovery in a forced-choice preferential looking task and in the ability to detect a moving object. However, recovery was much slower or non-existent in the ability to recognize complex objects such as faces or in tasks requiring visually co-ordinated behaviour, such as hand placement or obstacle avoidance (Regal *et al.* 1976; Carlson 1990). It looks as though binocular suturing produces more permanent damage to higher centres of visual processing, such as the parietal cortex, than to early levels of processing.

There is some evidence that eyelid suturing or corneal opacification lead to elongation of the eye and axial myopia (see Section 10.2.1).

13.1.3 Recovery of sight in humans

In the late seventeenth century, the Irish lawyer and scientist, Molyneux, wrote to the empiricist philosopher, John Locke, to ask how a man with congenital blindness who has recovered his sight would see the world. Would he be able to distinguish a cube from a sphere without touching them? This has become known as Molyneux's question. Molyneux suggested that a person with restored sight would not be able to do so and Locke agreed with him. In 1728 William Chesselden, surgeon to Queen Caroline and ophthalmologist at St. Thomas's hospital, London, reported that a young man on whom he had performed a cataract operation could not name shapes until after a long period of learning. When he first saw, the patient reported that objects seemed to touch his eyes.

Von Senden (1960) reviewed 66 reports of cases between 1920 and 1931 in which bilateral cataracts were removed. Useful sight was restored in only a few cases. Gregory and Wallace (1963) described recovery of visual function in a man who had bilateral cataracts removed when he was 52 years old. His cataracts had developed at the age of 10 months, although he may have had some residual vision. Basic visual functions were not tested. Although it took him many months to recognize objects, such as small-case letters that he could not recognize by touch, he soon learned to recognize capital letters that he had learned to recognize by touch. However, he did not learn to read by sight although he could tell the time. Depth impressions were not evoked in perspective drawings. This patient was periodically depressed but seems to have been one of the minority who learn to make some use of their restored sight. Ackroyd *et al.* (1974) described a woman who had bilateral cataracts removed when she was 27 years old. She had been able to see until she was 3 years old. Six months after the operation her electroretinogram and visual-evoked cortical potentials were within normal limits as were her luminance threshold and dark adaptation. She could detect and locate large objects, especially if they moved, but could not recognize simple visual patterns. She regarded the operation as a failure and reverted to the life of a blind person. Apkarian (1983) described a 12-year-old girl who had her sight restored after being effectively blind since the age of 3 months. After several months of training she showed some development of visual acuity, could recognize simple objects and point to objects. However, she reverted to the behaviour of a blind person when at home. A 34-year-old man had sight restored after 30 years of blindness and although he regained some visual functions he reverted to the behaviour of a blind man (Carlson *et al.* 1986).

Mioche and Perenin (1986) tested 13 adults who had bilateral cataracts removed between 4 months and 7 years of age. Their contrast sensitivity,

Figure 13.2. Gunter Konstantin von Noorden.
Born in Frankfurt am Main in 1928, He obtained an M.D. from J.W. Goethe University, Frankfurt am Main, in 1954. He was a resident and then Assistant Professor of Ophthalmology at the State University of Iowa from 1957 to 1963, From 1963 to 1972 he was Associate to full Professor at the Wilmer Institute of Johns Hopkins Hospital and the University of Baltimore, Maryland, From 1973 to 1995 he was Professor and Director of the Ocular Motility Service at Baylor College of Medicine, Houston. He is now Clinical Professor of Ophthalmology at the University of Southern Florida, Tampa. He was President of the American Association of Pediatric Ophthalmology, the International Strabismological Association, and the American Association of Research in Vision and Ophthalmology. Co-recipient of the Hectoen Gold Medal of the American Medical Association, the Franceschetti Prize from the German Ophthalmological Society, the Proctor Award from ARVO, the Bowman Medal from the Ophthalmological Society of the UK, the Alcon Research Award, the Jackson Lecture Award from the American Academy of Ophthalmology, and the A. von Humboldt Research Prize.

particularly at high spatial frequencies, was markedly impaired for both stationary and drifting gratings in both the central and peripheral retina. Their temporal modulation sensitivity was also impaired over the whole temporal frequency range. Patients with neonatal cataracts were more severely affected than were those with cataracts that developed later.

Children with congenital cataract in one (Pratt-Johnson and Tillson 1989) or both (Tytla 1993) eyes failed standard clinical tests of stereopsis after removal of the cataract, even when it was removed after only a few months. However, some of the children with binocular cataracts showed evidence of coarse stereopsis when allowance was made for their amblyopia by testing with large stimuli, and when allowance was made for strabismus by optically aligning the images (Tytla *et al.* 1993). A 1-day old baby had a monocular cataract removed and was given occlusion therapy. Eight years later the child had good acuity in both eyes and stereoacuity of 50 arcsec (Gregg and Parks 1992).

13.2 MONOCULAR DEPRIVATION

Monocular deprivation may be induced experimentally by any of the following methods:

1. Occluding the cornea or suturing the eyelids of one eye. With lid suturing the retina is illuminated by diffuse light.
2. Creation of an artificial strabismus by surgically deviating an eye. **Tenectomy** involves simple section of the tendon of a muscle at its point of insertion on the globe. **Myectomy** is a more severe procedure involving removal of the whole of one or more extraocular muscles.
3. Optical deviation of the input to one or both eyes by prisms.
4. Optical induction of aniseikonia by a magnifying lens in front of one eye or by atropine.
5. Immobilization of an eye by induction of muscle paralysis.

These experimental procedures are designed to mimic naturally occurring visual amblyopia in humans caused by disorders such as strabismus, anisometropia, aphakia, and cataracts.

When cats or primates are subjected to a disruption of normal visual experience in one eye during a critical period in early life, binocular cells of the visual cortex develop abnormal patterns of ocular dominance, and stereopsis is deficient or lost. Profound changes occur only when monocular deprivation is applied during a critical period early in life. However, even a brief period of monocular deprivation in adult monkeys leads to a temporary reduction in excitatory and inhibitory neurotransmitters and their synthetic enzymes in cells in the visual cortex associated with the deprived eye (see Hendry and Jones 1986).

13.2.1 Subcortical effects

13.2.1a *Retinal effects*

Retinal ganglion cells are anatomically and functionally normal in cats reared with monocular occlusion (Sherman and Stone 1973; Cleland *et al.* 1980). They are also normal in cats reared with convergent or divergent strabismus induced by tenectomy in one eye (Cleland *et al.* 1982; Gillard-Crewther and Crewther 1988). Cats with 4 years of postnatal monocular deprivation had a normal electroretinogram as indicated by flash- or pattern-evoked electrical responses from the retina (Baro *et al.* 1990). However, the receptive fields of X-type ganglion cells were found to be unusually large in cats reared with bilateral convergent strabismus induced by myectomy of both lateral recti (Chino *et al.* 1980). There is also loss of contrast sensitivity in X ganglion cells serving the central retina of cats reared with convergent or divergent strabismus induced by myectomy of the lateral rectus and oblique muscles of one eye (Ikeda and Wright 1976; Ikeda and Tremain 1979; Chino *et al.* 1980). Thus, defects at the level of the retina or LGN occur only when muscle tissue is removed so that reinsertion of the muscle is not possible.

In the monkey, monocular occlusion for 24 months after birth led to some decrease in the size and density of retinal ganglion cells, whereas deprivation for 12 months had no effect (von Noorden *et al.* 1977) (Portrait Figure 13.2).

Monocular occlusion of human subjects for one week produced a flattening of the Stiles-Crawford function, which manifested itself as an increased sensitivity to light entering the eye through the periphery of the pupil and a decrease in resolution of low-intensity gratings (Birch DG *et al.* 1980). This was probably due to a change in the alignment of cones.

13.2.1b *Physiological effects in the LGN*

Effects of early monocular occlusion or enucleation are evident in the lateral geniculate nucleus, although not in such a severe form as in the visual cortex. The laminae into which inputs from the two eyes segregate are present in the LGN at birth (Section 10.2.4). Their development depends on competitive interactions between retinogeniculate projections from the two eyes before birth. Rakic (1981) removed one eye from monkey foetuses in the second and third month of gestation. One year after birth the LGN lacked the normal laminar structure and the visual cortex lacked ocular dominance columns.

In the cat, monocular suturing within the first 4 postnatal weeks leads to a reduction in the size of relay cells in the binocular laminae of the LGN served by the deprived eye (Wiesel and Hubel 1963b; Hickey *et al.* 1977). The reduction in cell size is about 20% for LGN relay cells projecting to area 17 of the cat and up to 60% for those projecting to area 18 (Garey and Blakemore 1977). Total blockage of retinal activity in one eye by tetrodotoxin for one week produced severe effects throughout the LGN, including the monocular laminae. There was some loss even when the other eye was sutured (Kuppermann and Kasamatsu 1983). Thus, spontaneous activity from a sutured eye is better than no input. While these procedures produce a reduction of cell size in the LGN without loss of cells, monocular enucleation leads to cell death in the LGN (Kalil 1980).

Figure 13.3. Yuzo M. Chino.
Born in Tokyo in 1943. He obtained his B.Sc. in Psychobiology at St. Norbert College, Wisconsin and his Ph.D. in Visual Neuroscience at Syracuse University. He held an academic appointment at the Illinois College of Optometry from 1972 to 1985. He is now Professor at the University of Houston College of Optometry.

Strabismus induced in kittens by myectomy of the lateral rectus and superior oblique also leads to a reduction in the size of cells in LGN laminae receiving inputs from the deviated eye, in proportion to the degree of amblyopia (Tremain and Ikeda 1982). However, strabismus induced by tenectomy (tendon section) produces no such effect. Also, loss of visual acuity is more severe for an eye made strabismic by myectomy than by tenectomy (Mitchell *et al.* 1984). These differences probably arise because severed tendons reattach to the eye after a few days whereas a myectomized eye remains without muscle attachment (Crewther *et al.* 1985). Another factor may be the loss of proprioceptive inputs in myectomy (Section 12.4). In kittens reared with convergent strabismus induced by monocular myectomy there is a specific loss of functional cells in LGN laminae served by the periphery of the temporal retina (Ikeda *et al.* 1977). It was suggested that this is because, for an eye deviated nasally, the peripheral temporal retina is hidden by the nose and receives an impoverished visual input.

Monocularly deprived monkeys and monkeys reared with esotropia induced by tenectomy show shrinkage of LGN cells serving the deprived eye, especially in the parvocellular layers (von Noorden and Middleditch 1975; Crawford and von Noorden 1979a; Tigges *et al.* 1984). Normal cell size recovered in five-year-old monkeys that had been reared for 30 days with induced esotropia (Crawford and von Noorden 1996). Post-mortem analysis of the LGN of a human strabismic amblyope revealed loss of cell size (von Noorden and Crawford 1992).

In the cat, ganglion-cell axons arising from a visually deprived eye develop abnormal patterns of termination in the LGN. Many X cells have unusually broad terminal fields in lamina A and the terminal fields of many Y cells are greatly reduced (Sur *et al.* 1982).

Several investigators have found that monocular deprivation affects Y cells more than X cells (Sherman *et al.* 1972, 1975; LeVay and Ferster 1977; Sireteanu and Hoffmann 1979; Mangel *et al.* 1983; Friedlander and Stanford 1984; Garraghty *et al.* 1989).

Others have found that the relative numbers and spatial properties of both X and Y cells are not affected by monocular occlusion in the cat (Shapley and So 1980; Derrington and Hawken 1981) and in the monkey (Blakemore and Vital-Durand 1986a). Brown and Salinger (1975) claimed that chronic immobilization of one eye led to substantial loss of X cells but not of Y cells in the LGN. The effects of severing proprioceptive inputs from the extraocular muscles are reviewed in Section 12.4. Levitt *et al.* (2001) found no evidence of loss of cells in the LGN of monocularly deprived monkeys.

In addition, in cats reared with one eye sutured, relay cells in the LGN show evidence of reduced metabolic activity, as reflected in the decreased level of the metabolic enzyme cytochrome oxidase (Wong-Riley 1979a). These effects are postsynaptic because there is no change in the size of presynaptic terminals. The same symptoms occur in monkeys with the lens in one eye removed but with illumination of the retina at normal levels, which shows that lack of patterned stimulation rather than lack of light is the crucial factor (von Noorden and Crawford 1977).

Action potentials still arise in the retina of an occluded eye, and it seems that these play a part in maintaining some normality of function in the LGN. The development the LGN in the cat is severely disturbed when these action potentials are abolished for several weeks by the application of tetrodotoxin (Archer *et al.* 1982). Changes in the LGN produced by monocular paralysis do not occur after intraventricular injection of the neurotoxin 6-hydroxydopamine (Guido and Salinger 1989). The following evidence demonstrates that the effects of monocular deprivation in the LGN are due to lowered level of inputs from the deprived eye relative to those from the normal eye.

Relay cells in the LGN of the cat and dog that receive inputs from the monocular crescent of the deprived eye retain their normal size, metabolic activity, and proportion of Y cells (Guillery and Stelzner 1970). Furthermore, relay cells of a monocularly occluded eye develop normally when the corresponding retinal region of the nonoccluded eye has been lesioned (Guillery 1972; Sherman and Wilson 1975; Wong-Riley 1979b).

Monocularly deprived kittens with the visual cortex removed showed the same loss of Y cells in the LGN as did kittens with the visual cortex intact. Thus, this deficit is not due to corticofugal influences (Zetlan *et al.* 1981). Cats and monkeys reared with both eyes in total darkness show no obvious changes in the number, size, or staining characteristics of LGN cells (Chow 1973; Hendrickson and Boothe 1976). These facts demonstrate that the reduced size of LGN relay cells in monocularly deprived animals is due to competitive interactions with inputs from the normal eye rather than to a simple absence of visual inputs. Delivery of the neurotrophin TN-4 into the visual cortex prevents the shrinkage of LGN cells in the monocularly deprived ferret (Riddle *et al.* 1995). Also, blockage of NMDA receptors in the kitten visual cortex renders the LGN immune to the effects of monocular deprivation (Kleinschmidt *et al.* 1987; Bear and Colman 1990). These findings suggest that activity-dependent competition for a limited supply of growth factor is responsible for selective shrinkage of LGN relay cells.

Changes in the size of cells in the LGN are closely correlated with changes in the ocular dominance columns of the visual cortex when these are measured in the same animals (Vital-Durand *et al.* 1978). As long as the eye has not been occluded for more than about 6 weeks, the cells in the LGN recover to full size soon after the occlusion is switched to the other eye (Dürsteler *et al.* 1976).

13.2.1c *Functional effects in the LGN*

There is some controversy regarding the effects of monocular deprivation on the functional properties of cells in the LGN. The spatial contrast sensitivity of X cells in layers of the LGN receiving inputs from the strabismic eyes of cats has been found to be lowered, but only for gratings with high spatial frequency (Jones *et al.* 1984a; Chino *et al.* 1994a) (Portrait Figure 13.3). A similar defect occurred in cats raised with monocular occlusion (Maffei and Fiorentini 1976a; Lehmkuhle *et al.* 1980). Also, X cells in the LGN of strabismic cats show lowered efficiency of signal transmission and unusually long latency, especially in layers innervated by the deviating eye (Cheng *et al.* 1995).

Signal transmission in LGN parvocellular units was normal in monkeys reared with esotropia induced by myectomy at between the ages of 20 and 30 days (Sasaki *et al.* 1998). Levitt *et al.* (2001) found that LGN cells of monocularly deprived monkeys

Figure 13.4. Ruxandra Sireteanu.
Born in Romania 1945. She trained as a Biophysicist at the University of Bucharest, Romania (1968) and obtained a PhD in Biophysics from the Scuola Normale Superiore in Pisa in 1976. She held postdoctoral Fellowships at the Institut d'Anatomie, Université de Lausanne and the Department for Comparative Neurobiology, University of Ulm from 1976 to 1978. She was a senior investigator at the Max-Planck-Institute for Psychiatry in Munich from 1978 to 1982 and at the Max-Planck-Institute for Brain Research in Frankfurt from 1982 to 1999. From 1991 to 1999 she was Professor for Neuroscience at the Department for Zoology, University of Mainz. She is now Professor of Biological Psychology at the University of Frankfurt and Head of the Research Group for Psychophysics and Neuropsychology at the Max-Planck-Institute for Brain Research in Frankfurt. She received the award of the Heinz-and Helene-Adam-Stiftung for "Excellence in Research in Ophthalmology" in 1991 and the Prize of the Bielschowsky Society for Research in Strabismus in 1994.

showed only a minor loss in sensitivity and latency to stimulation of the deprived eye.

13.2.1d *Effects on the superior colliculus*

In mammals, the superior colliculus is a paired structure in the midbrain involved in guiding saccadic eye movements. It is the homologue of the optic tectum of non-mammalian vertebrates. The superior colliculus receives direct visual inputs as well as inputs routed through the visual cortex. The cells have large, non-specific receptive fields and many are binocular. Unlike cells in the visual cortex, many cells in the superior colliculus retain their binocularity in cats reared with artificial strabismus or alternating occlusion of the eyes. They may receive their ipsilateral inputs through the corpus callosum. However, in strabismic cats and in cats with sectioned right medial and rectus muscles, binocular cells in the superior colliculus contralateral to the normal eye are heavily dominated by that eye (Gordon and Gummow 1975). This dominance of the normal eye was not apparent in strabismic cats that had been forced to use the deviating eye (Gordon and Presson 1977). Cortical inputs to the superior colliculus of cats are absent from an eye sutured from birth (Hoffmann and Sherman 1974).

13.2.2 Cortical effects of deprivation in cats

13.2.2a *Cortical effects of induced strabismus in cats*

In one of the first experimental studies of the effects of abnormal visual experience on the development of the visual cortex, Hubel and Wiesel (1965) reared kittens from birth to the age of 3 months or more with a surgically induced divergent deviation of one eye. This permanently reduced the number of binocular cells in area 17 to about 20% compared with 80% for a normal cat. However, there were roughly equal numbers of monocular cells that responded only to the left eye or only to the right eye. Thus, the ocular dominance columns remained roughly symmetrical. It seems that binocular cells with a left-eye dominance were converted into left-monocular cells and those with a right-eye dominance were converted into right-monocular cells.

All forms of binocular interaction, both excitatory and inhibitory, are drastically reduced in strabismic cats (Hoffmann and Schoppmann 1984; Chino *et al.* 1994b).

Similar results were observed in cats reared with prisms that disrupted the alignment of the visual axes (Bennett *et al.* 1980) and in kittens in which extraocular muscles were cut and reinserted in another position (recession), rather than severed (Sireteanu *et al.* 1993a) (Portrait Figure 13.4). Cats reared with anisometropia produced by a high-power lens before one eye also develop a preponderance of cortical cells responding to the normal eye and reduced contrast sensitivity in the originally defocused eye (Eggers and Blakemore 1978).

Cats with a natural strabismus also have abnormal ocular dominance columns (von Grünau and Rauschecker 1983). In cats, convergent strabismus has a greater effect on cortical cells than divergent strabismus (Yinon *et al.* 1975; Berman and Murphy 1982; Freeman and Tsumoto 1983; Mitchell *et al.* 1984).

Cells in area 17 driven by the deviating eye of strabismic cats show reduced sensitivity to high spatial frequencies, loss of contrast sensitivity, broadened orientation tuning, and loss of temporal resolution (see also Crewther and Crewther 1993). Strabismus has a particularly strong deleterious effect on the temporal properties of area-18 cells in the cat (Chino *et al.* 1988).

Löwel (1994) found that the ocular dominance columns in area 17 were unusually wide in cats reared with divergent strabismus induced by monocular tenectomy. Tieman and Tumosa (1997) obtained the same result in cats reared with alternating monocular occlusion. However, other investigators have found that, although the ocular dominance

Figure 13.5. Jack Crawford.
Born in North Georgia, USA in 1933. He obtained a Ph.D. in Physiological Psychology at the University of Georgia in 1962 and conducted postdoctoral work at the University of Mississippi Medical School. He is Professor and holder of the Frederic B. Asche Chair in Ophthalmology at the University of Texas Medical School in Houston.

columns of the deprived eye of cats and monkeys were reduced in width, there was no overall change in the combined width of left-eye and right-eye columns (see Crawford 1998) (Portrait Figure 13.5).

Most cells in areas 18 of strabismic cats lose their binocularity, although binocular cells sensitive to motion in depth have been found to survive in area 18 (Cynader *et al.* 1984).

In normal kittens, horizontal axons in the visual cortex connect cortical columns with similar orientation selectivity and cells in these connected columns tend to respond in synchrony (Section 5.4.6). In kittens reared with strabismus, response synchronization was normal between cell groups served by the same eye but unusually weak between cell groups served by different eyes (Löwel and Singer 1992; König *et al.* 1993). The crucial factor seems to be the lack of correlated inputs from the two eyes, which weakens the formation of connections between cells responding to the two eyes in favour of the formation of connections between cells responding to neighbouring regions of the same eye. This could account for the abnormality of ocular dominance in monocularly deprived animals (Goodhill and Löwel 1996). However, in strabismic cats, lateral connections in each hemisphere and interhemispheric callosal connections were still between inputs with similar orientation preference (Schmidt *et al.* 1997).

In normal animals, dendrites of spiny stellate cells tend to remain within their own ocular dominance column, because dendrites from an ocular dominance column of the other eye are at a competitive disadvantage. The terminal arbours of inputs from a deprived eye on spiny stellate cells are smaller and make fewer synapses per bouton than in a normal cat. The terminal arbours of the non-deprived eye are larger than normal and make more than the normal number of synapses (Friedlander *et al.* 1991).

Because of their reduced input from the deprived eye, spiny stellate cells serving a deprived eye are invaded by inputs from the non-deprived eye. This causes the dendrites of these spiny stellate cells to spread to an unusual degree into the adjacent ocular dominance columns of the non-deprived eye. The stellate cells serving the deprived eye thus receive their dominant input from the non-deprived eye (Kossel *et al.* 1995).

Changes in the upper supragranular cortical layers precede and may direct the subsequent reorganization of thalamocortical connections associated with strabismus and other forms of monocular deprivation. Trachtenberg *et al.* (2000) found that 24 hours of monocular deprivation produced loss of responses from the deprived eye in only cortical layers outside layer 4. Longer periods of deprivation were required to produce effects in layer 4. The short-term loss could arise (1) because cells receiving inputs from the deprived eye lose their capacity to influence cells in the extragranular layers, or (2) from changes in lateral connections between cells in extragranular layers.

Trachtenberg and Stryker (2001) produced evidence for the second possibility, although it does not rule out a contribution from the first possibility. They observed large losses in horizontal connections between ocular dominance columns in the upper layers of the visual cortex of kittens exposed to only 2 days of strabismus.

Chino *et al.* (1991) noted that cats reared with a 15 dioptre base-in prism before one or both eyes had a reduced proportion of cortical cells tuned to vertical contours. However, a 15-dioptre prism introduces strong curvature of vertical lines and this may have contributed to the reduced number of cells tuned to verticals. The effect did not occur when the non-deviating eye was sutured, which indicates that it is due to rivalry of inputs rather than eye deviation alone. Chino and Kaplan (1988) found a reduced sensitivity to vertically oriented gratings when recording from X-cells in the LGN of strabismic cats. The behavioural correlate of this effect is known as the "vertical effect" (Section 13.4.2d).

Figure 13.6. Takuji Kasamatsu.
Born in Nishinomiya, Japan in 1939. He obtained a B.Sc., M.D., and PhD. in Neurophysiology at Osaka University, 1959 to 1976. He was a Senior Research Fellow at California Institute of Technology, Pasadena from 1975 to 1984. Since 1984 he has been a Senior Scientist at the Smith-Kettlewell Eye Research Institute, San Francisco.

13.2.2b *Cortical effects of monocular occlusion*

Rearing kittens with one eye occluded also severely reduces the number of cortical cells responding to stimulation of the occluded eye. However, as we will see later, complete or partial recovery may occur if sight is restored early enough (Wiesel and Hubel 1963a; Blakemore 1976; Olson and Freeman 1978b). Similar results have been obtained in the mouse (Gordon and Stryker 1996). The terminals of afferents from the deprived eye in cortical layer 4 are smaller, have fewer spines, and are reduced in number (Tieman 1984). However, 7 days of monocular deprivation in kittens had no effect on the density of presynaptic vesicles of either excitatory or inhibitory synapses on the remaining terminals (Silver and Stryker 1999, 2000). The loss of cortical cells responding to the deprived eye is particularly severe in cortical layers other than layer 4 (Shatz and Stryker 1978).

Singer *et al.* (1979) sutured one eye of kittens and rotated the seeing eye in its socket through 180°. In this case, there was no shift of ocular dominance of cortical cells to the seeing eye. It seems that a spatial mismatch between a rotated eye and the other sensory-motor systems renders the eye incapable of competing with inputs from a sutured eye.

In the cat, potentials recorded from the scalp in response to flashed gratings presented to a previously deprived eye were severely reduced in magnitude (Baro *et al.* 1990). Exposure of a flickering grating to a given retinal region of a cat's deprived eye evoked potentials from the corresponding area of V1. The potentials had an abnormal waveform and abnormally long latency (Kasamatsu *et al.* 1998a) (Portrait Figure 13.6). For the normal eye, responses from a given cortical area to whole field stimulation were weaker than responses to a local stimulus because of lateral inhibition. For the deprived eye, the effects of whole-field and local stimulation were the same, showing an absence of lateral inhibition. Also, for the normal eye, but not for the deprived eye, the effects of local stimulation were evident in remote cortical sites. Thus, it seems that lateral interactions in the visual cortex serving a deprived eye are greatly reduced.

Short periods of monocular experience in dark-reared kittens produce a greater loss of cortical cells driven from the closed eye than does continued binocular deprivation (Olson and Freeman 1980b). Even three hours of monocular experience on the 30th day of dark-reared kittens produced some reduction in the number of binocular cells. A longer period of monocular experience was required before a shift in ocular dominance was observed (Schechter and Murphy 1976). One day of monocular visual experience in dark-reared kittens produced some reduction in the number of cells responding to stimulation of the closed eye (Olson and Freeman 1975). The loss was more severe after 2 to 3 days of monocular experience and was complete after 10 days (see also Movshon and Dürsteler 1977). In another study, responses from electrodes in the visual cortex of alert, 5-week-old kittens showed evidence of reduced response to stimulation of an eye that had been occluded for only 6 hours (Mioche and Singer 1989). After 12 hours of occlusion, there was a complete absence of excitatory responses from stimulation of the occluded eye. After the occluder was switched to the other eye, cortical cells slowly recovered their responsiveness to stimulation of the previously occluded eye.

The duration of monocular deprivation can influence the physiological effects. Anatomical procedures have revealed that monocular deprivation over a few days leads to shrinkage of cortical connections of afferents from the deprived eye in kittens. However, deprivation over a period of weeks is required before expansion of connections of afferents from the nondeprived eye is evident (Antonini and Stryker 1996).

The binocularity of cells in the cat's suprasylvian visual area is also disrupted by monocular occlusion (Spear and Tong 1980) though not by strabismus (von Grunau 1982).

Cells in the visual cortex in Ferrets raised with one eye occluded showed only rudimentary orientation tuning (White *et al.* 2001). Horizontal connections between cells tuned to the same orientation were less extensive than normal. Kittens raised with one eye occluded and the other exposed to only vertical lines possessed binocular cells with an orientation preference for vertical lines. Cells preferring other orientations were driven by only the eye that was occluded (Rauschecker and Singer 1981).

Sherman and Spear (1982) reviewed the early literature on the effects of visual deprivation on the cat's visual system.

13.2.2c *Types of deprivation compared*

Cats reared in complete darkness retain more binocular cortical cells than do monocularly deprived cats (Wiesel and Hubel 1965a, 1965b; Blakemore and Van Sluyters 1975; Kaye *et al.* 1982). Also, the morphology of geniculocortical synapses is normal in binocularly deprived monkeys but severely disrupted in monocularly deprived monkeys (Antonini and Stryker 1998). Thus, the effects of monocular occlusion are not due simply to disuse. They involve competitive suppression of inputs from the deprived eye by those from the normal eye. Strabismus leads to a reduction in the number of binocular cells but leaves an equal numbers of left- and right-eye monocular cells (Hubel and Wiesel 1965). In this case, the two eyes receive similar inputs but they fall on non-corresponding retinal points. Monocular occlusion, in which only one eye receives a normal input, leads to a reduction in the number of cortical cells responding to the deprived eye.

Tumosa *et al.* (1989b) compared the effects of monocular occlusion, unequal alternating occlusion, and equal alternating occlusion on the activation of ocular dominance columns in areas 17 and 18 of the cat, as indicated by the uptake of deoxyglucose. The ocular dominance columns of a given eye were reduced in width and responsivity in proportion to the degree of deprivation of that eye in relation to the other eye. Although the columns of the less deprived eye increased in width, the intensity of activation, as shown by the density of deoxyglucose uptake, was reduced relative to that of a normal eye. They concluded that cortical synapses of the less deprived eye become distributed less densely over wider areas.

13.2.2d *Removal of nondeprived eye aids recovery*

Visual acuity in a strabismic eye was lower in kittens reared with both eyes open than in kittens reared with the non-deviating eye sutured (Whittle *et al.* 1987). Recovery of function in an occluded eye is more rapid if the good eye is removed rather than simply occluded (Hoffmann and Lippert 1982). Recovery of function in a previously occluded eye involves an increase in the number of cortical cells responding to that eye and some restoration of tuning functions of their receptive fields. Thus, in kittens after 4 weeks of monocular occlusion, the proportion of cortical cells responding to the occluded eye increased to near normal levels within a few hours after enucleation of the nonoccluded eye. However, the receptive fields of these binocular cells remained abnormal. Even after 92 weeks of monocular occlusion, 22% to 40% of cortical cells began to respond to the occluded eye after the good eye was removed (Kratz *et al.* 1976; Smith *et al.* 1978; Spear *et al.* 1980). Blakemore and Hawken (1982) confirmed this effect but found it occurred only if the kitten had a period of binocular vision before one eye was occluded.

Recovery of a previously occluded eye after enucleation of the good eye takes only hours whereas recovery after switching the occluder to the other eye takes days, as we will see later. It seems that removing the eye removes its inhibitory influences on cortical cells and allows dormant excitatory inputs from the deprived eye to recover. However, recovery for most cells is not immediate, as it would be if it involved only a simple removal of inhibition from the good eye. Furthermore, temporarily pressure blinding the good eye restored input to only a few cortical cells (Blakemore *et al.* 1982), although it has been claimed that this procedure works only when afferents from the extraocular muscles are also paralyzed (Crewther *et al.* 1978). Removal of all inputs from the good eye must allow some restorative processes to occur in the weakened, but still present, excitatory inputs from the deprived eye. Some of these restorative processes occur almost immediately; others take some time. We will see later that restoration of physiological functions in a deprived eye after removal of the good eye is accompanied by some restoration of visual function. However, merely occluding the good eye after early monocular occlusion does not lead to restoration of binocular cells.

13.2.2e *Responses from a deprived eye*

Singer (1977) (Portrait Figure 13.7) found that most cortical cells of a monocularly deprived cat did not respond to electrical stimulation of the optic nerve of the deprived eye, except for some responses with unusually long latency. However, when both eyes were stimulated, most cells showed short-latency inhibitory effects that depended on intracortical circuits. Many binocular cells in the visual cortex of monocularly deprived kittens responded when the

Figure 13.7. Wolf Singer.
Born in Munich in 1943. He obtained an M.D. with O. Creutzfeld at the University of Munich in 1968. His postgraduate work was in Psychophysics and Animal Behaviour at the University of Sussex and at the Max-Planck-Institute for Psychiatry. He obtained his Habilitation for Physiology at the Medical Faculty of the Technical University, Munich in 1975. In 1980 he became Professor for Physiology at the Technical University, Munich and in 1982 the Director at the Max-Planck-Institute for Brain Research, Frankfurt am Main. He received the Ernst Jung Prize for Science and Research, in 1994, the Zülch Prize in 1994, the Hessischer Kulturpreis in 1998, and the Körber Prize for the European Sciences in 2000.

optic nerve of the deprived eye was stimulated with synchronized volleys or with simultaneous conditioning stimulation of the non-deprived eye (Tsumoto and Suda 1978). In another study a substantial proportion of cortical cells displayed evidence of both excitatory and inhibitory inputs from the deprived eye when both eyes were stimulated simultaneously by a large phase-varying grating (Freeman and Ohzawa 1988). Animals deprived of vision in one eye for more than a year had very few cortical cells that betrayed any evidence of inputs from the deprived eye.

After application of a localized lesion in one eye binocular cells serving that region become unresponsive to stimuli in the lesioned eye. However, after some time, binocular cells began to respond to stimuli applied to regions round the lesioned area (Chino *et al.* 2001). Similar effects occur in the monkey (Rosa *et al.* 1995) (Section 5.4.7b).

13.2.2f *Loss of sensitivity in the nondeprived eye*

In strabismic cats, cells in the visual cortex driven by the nondeviating eye show some loss in contrast sensitivity (Chino *et al.* 1983). Also, behavioural tests reveal some loss of contrast sensitivity in the nondeviating eye in both strabismic cats (Holopigian and Blake 1983) and strabismic humans (Levi and Klein 1985; Dobson and Sebris 1989). Loss of sensitivity in a nondeprived eye may be due to persistent blur of the retinal image arising from instability of gaze. However, Tumosa *et al.* (1989b) reported that cortical synapses formed by a nondeprived eye do not increase in number but become spread less densely over wider areas. Thus, loss of sensitivity in a nondeprived eye may be due to loss of synaptic density.

Freeman and Bradley (1980) claimed that amblyopic humans have higher than normal hyperacuity in the nonamblyopic eye but Johnson *et al.* (1982) could not confirm this.

13.2.3 Cortical effects of deprivation in primates

Monocular deprivation has been produced in infant monkeys by surgically or optically induced strabismus, by suturing or occluding one eye, or by blurring the image in one eye (induced anisometropia). In each case, the response of binocular cells to stimulation of the deprived eye is reduced or absent (Carder *et al.* 1991). In prism-reared monkeys, inhibitory connections between surviving binocular cells are less affected than excitatory connections (Smith *et al.* 1997c). The various forms of monocular deprivation also cause a contraction of ocular dominance columns for the deprived eye and an expansion of those for the seeing eye (Baker *et al.* 1974; LeVay *et al.* 1980; Movshon *et al.* 1987). These changes can be seen in the autoradiograph of the visual cortex of a monocularly deprived monkey (Des Rosiers *et al.* 1978).

Hendrickson *et al.* (1987) observed some shrinkage of ocular dominance columns of a deprived eye in monkeys reared with experimentally induced anisometropic amblyopia. Horton *et al.* (1997) did not observe column shrinkage in a monkey with natural anisometropic amblyopia, although the time of onset of the anisometropia was not known. Horton and Stryker (1993) observed no column shrinkage in the visual cortex of a man in whom anisometropic amblyopia had been detected at age $4^{1}/_{2}$ years. These results suggest that amblyopia can be induced after the critical period for induction of column shrinkage. Changes in ocular dominance do not involve any significant change in the overall density of neurones or of synapses within cortical layers (O'Kusky and Colonnier 1982).

In monkey striate cortex, monocular deprivation also produces a severe reduction in the size of cytochrome-oxidase blobs centred in ocular dominance columns dominated by the deprived eye (Trusk *et al.* 1990; Wong-Riley 1994; Horton *et al.* 1999). Autopsy

Figure 13.8. Mark Firman Bear.
Born in Alexandria, Virginia in 1957. He obtained a B.Sc. in Psychology from Duke University in 1979 and a Ph.D. in Neurobiology from Brown University with F.F. Ebner in 1984. His postdoctoral work was at the Max-Planck Institut für Hirnforschung, in Frankfurt with Dr. W. Singer and at Brown University with L.N Cooper. He joined the faculty of Brown University in 1986, where he is now Professor of Neuroscience, Fox Professor of Ophthalmology and Visual Sciences, and Investigator of the Howard Hughes Medical Institute. He received the Alfred P. Sloan Award in 1987.

specimens of human brains from cases of monocular enucleation showed the same effect (Horton and Hedley-White 1984). The *COI* and *cyt b* genes, located on the mitochondria, express RNA transcription factors that are concerned with synthesis of cytochrome oxidase. The concentration of these transcription factors is also reduced in the visual cortex of monocularly deprived monkeys (Hevner and Wong-Riley 1993; Kaminska *et al.* 1997). There is thus a direct link between visual activity and gene expression.

Kiorpes *et al.* (1998) found that surgically induced strabismus in monkeys disrupted binocularity more than lens-induced anisometropia. Both deficits reduced the optimal spatial frequency tuning of cortical cells responding to the affected eye, especially for cells serving the foveal area. They found no effect on the contrast sensitivity of cortical cells, but the behavioural loss of contrast sensitivity suffered by amblyopes might depend on the relative number of cortical cells responding.

Early monocular deprivation of various types in the monkey leads to a reduction or loss of visually evoked potentials associated with asynchronous dichoptic visual flicker (Baitch *et al.* 1991). Nine hours of monocular occlusion in adult humans did not reduce the cortical potentials evoked by a textured stimulus presented to the deprived eye but did temporarily increase the response from the seeing eye (Tyler and Kaitz 1977).

Monocular deprivation for more than a year can lead to complete loss of visual functions in the deprived eye in infant macaque and squirrel monkeys (Sparks *et al.* 1986; Wilson and Nevins 1991). The loss seems to occur in the nasal visual field (temporal retinas) before it is complete in the temporal visual field (Tieman *et al.* 1983a; Wilson *et al.* 1989). This may be related to the fact that the temporal retina has a greater density of ganglion cells than the nasal retina (Section 11.1.4).

The following evidence suggests that normal development of binocular cells depends on competition between inputs from the two eyes in innervating binocular cells. An eye receiving patterned visual inputs gains a competitive advantage over a closed or deviated eye, which disrupts the development of binocularity. The evidence also suggests that changes in ocular dominance of cortical cells involve the balance between pre- and post-synaptic activity in cortical cells in layer 4. Finally, the following evidence suggests that intracortical inhibition is also involved in this process.

13.2.4 Monocular deprivation and synaptic activity

13.2.4a *Intracortical inhibition and plasticity*

Intracortical inhibition is mediated by the neurotransmitter GABA, the effects of which can be removed by application of its antagonist, bicuculline. Seven days of monocular deprivation in kittens produced less than the expected shift in ocular dominance when the cortex was infused with bicuculline during the period of deprivation (Ramoa *et al.* 1988). Within 30 s of intravenous injection or cortical application of bicuculline in monocularly deprived cats, between 40 and 60% of cortical cells tested began to respond again to excitatory stimulation of the deprived eye, although not to the point of gaining dominance over binocular cells (Duffy *et al.* 1976; Burchfiel and Duffy 1981; Sillito *et al.* 1981; Mower and Christen 1989). Blakemore *et al.* (1982) argued that part but not all of this effect is due to the fact that bicuculline raises the general level of excitability of cortical cells to weak excitatory inputs. This could cause an increase in the probability of correlated activity between weak inputs from the deprived eye and activity from the normal eye. Thus, GABAergic intracortical inhibition contributes to adaptive responses of binocular cells to monocular deprivation. However, Silver and Stryker (2000) found that 7 days of monocular deprivation in kittens produced no detectable change in a biochemical indicator of activity at inhibitory presynaptic sites in cortical layer IV. Perhaps the site of the changes is postsynaptic.

13.2.4b *Role of NMDA receptors*

Hebbian synapses (NMDA synapses) are involved in the development of binocular cells and in neural plasticity in general (Sections 10.4.4 and 10.4.5a). It has also been claimed that they are involved in the effects of monocular deprivation.

Dark-rearing kittens to 6 weeks of age arrested the normally occurring loss of NMDA synapses after the critical period and thereby extended the period during which visual experience influenced the formation of ocular dominance columns (Fox *et al.* 1992; Czepita *et al.* 1994). Subsequent exposure to light for 10 days allowed NMDA to decrease down to adult levels.

When NMDA synapses were inhibited by cortical infusion of a specific antagonist, monocularly deprived kittens failed to show an ocular dominance shift (Bear *et al.* 1990) (Portrait Figure 13.8). Furthermore, cortical cells became less responsive to stimulation of the normal eye (Czepita and Daw 1996). Also, infusion of an antagonist prevented a switch of dominance of cells in that hemisphere of kittens after the occluder was moved to the other eye. Cells in the visual cortex of the untreated hemisphere showed the usual transfer of dominance after reversal of occlusion (Gu *et al.* 1989).

Antagonists of NMDA synapses applied in early development affect other types of synapse and lead to a general loss of responsivity and loss of orientation and direction selectivity of cortical cells in cats and ferrets (Miller *et al.* 1989a; Daw 1995; Ramoa *et al.* 2001). Therefore, loss of ocular dominance plasticity after blockage of NMDA receptors could be due to these general deficits rather than a specific effect of loss of NMDA synapses on cortical plasticity. Kasamatsu *et al.* (1998b) found considerable ocular dominance plasticity in monocularly deprived kittens receiving continuous cortical infusion of an NMDA antagonist. The major effect of the antagonist was loss of orientation tuning. They concluded that inhibition of NMDA synapses disrupts synaptic transmission rather than experience-dependent cortical plasticity.

Roberts *et al.* (1998) developed an antagonist that acts specifically on the NR1 receptor subunit of NMDA synapses (Section 10.4.5a). This antagonist decreased ocular dominance plasticity in monocularly deprived ferrets without affecting other visual functions. Thus, loss of NMDA synapses affects cortical plasticity. However, the procedure used by Roberts *et al.* did not block all NMDA receptors. Perhaps their complete removal would have more widespread effects.

Activation of NMDA receptors triggers a signalling cascade involving cAMP and the enzyme protein kinase A. Inhibition of this enzyme in the cat's visual cortex during the critical period blocks ocular dominance shifts following monocular deprivation (Beaver *et al.* 2001).

Metabotropic receptors are also numerous in the visual cortex during the postnatal period. They are also involved in long-term depression of responses in other parts of the brain. However, application of a metabotropic antagonist had no effect on the induction of long-term depression by low-frequency stimulation of slices of rat visual cortex. Furthermore, rats lacking metabotropic receptors showed normal long-term depression at NMDA receptors (Sawtell *et al.* 1999). Thus, metabotropic receptors do not seem to be involved in the long-term depression of responses to stimulation of a visually deprived eye.

13.2.4c *Effects of blockage of postsynaptic activity*

A shift in ocular dominance to the open eye during monocular deprivation in does not occur for regions of the cat visual cortex in which all postsynaptic activity has been blocked by agents such as glutamate, tetrodotoxin, or muscimol (Shaw and Cynader 1984; Reiter *et al.* 1986). Reiter and Stryker (1988) found that cortical cells of kittens become more responsive to inputs from the closed eye than to those from the open eye when postsynaptic activity is blocked. Furthermore, Rittenhouse *et al.* (1999) found greater loss of response of binocular cells to a deprived eye after eyelid suturing, which leaves spontaneous retinal activity intact, than after complete blockage of inputs with tetrodotoxin.

Thus, the crucial changes during monocular deprivation are not due merely to absence of inputs from one eye but rather to long-term depression caused by imbalance between pre- and postsynaptic activity at NMDA synapses (Hata and Stryker 1994) (Section 10.4.6).

13.2.4d *Effects of nerve growth factors*

Evidence reviewed in Section 10.4.6 shows that the formation of ocular dominance columns depends on competition between cortical afferents for a neurotrophin (NGF, BDNF, NT-3, or NT-4/5 described in Section 10.4.1) secreted by target cells in the visual cortex. Brain-derived neurotrophin (BDNF) is produced in postsynaptic cells of the visual system during the critical period for development of ocular dominance columns. Its synthesis is regulated by neuronal activity. It binds to ligand molecules on afferent axons and acts as a retrograde messenger enhancing synaptic efficiency by increasing release of a synaptic transmitter in presynaptic neurones (Thoenen 1995; Liu *et al.* 1996). Growing axons from the two eyes compete for a limited amount of BDNF, and only those receiving a sufficient amount are maintained.

The expression of mRNA for synthesis of BDNF increases in the visual cortex of the rat around the time when the eyes open (Bozzi *et al.* 1995). Also, deprivation of pattern vision during, and even after the critical period, results in a decrease in BDNF in the visual cortex of the cat (Lein and Shatz 2000). Dark-rearing or blockage of neural activity during the critical period of column formation also reduces the level of BDNF in the LGN and visual cortex of the rat (Schoups *et al.* 1995b).

Monocular deprivation in the monkey reduces the expression of cell proteins (GAP-43 and SCG10) in the LGN. These proteins are induced by BDNF and control the growth of axons and synapses (Higo *et al.* 2000). Ocular dominance columns do not develop when excess neurotrophins BDNF or NT-4/5 is applied to the developing visual cortex of cats or monkeys (Cabelli *et al.* 1995; Hata *et al.* 2000). Excess of these neurotrophins during the critical period maintains, or even restores, responsiveness of binocular cells to a deprived eye (Gillespie *et al.* 2000).

When monocular deprivation in young rats was accompanied by cortical infusion of extra NGF there occurred no shrinkage of LGN neurones and no shift in ocular dominance in the visual cortex (Berardi *et al.* 1993a, 1993b; Domenici *et al.* 1993). Also, the tuning characteristics of cortical cells did not change and the animals showed no behavioural evidence of amblyopia in the deprived eye (Domenici *et al.* 1991; Maffei *et al.* 1992). In monocularly deprived rats, visually evoked responses in the ipsilateral visual cortex are reduced, but this reduction does not occur after ventricular infusion of NGF (Yan *et al.* 1996). Furthermore, injection of NGF during the critical period facilitates recovery of binocularity and visual acuity following a period of monocular deprivation in the kitten (Carmignoto *et al.* 1993; Fiorentini *et al.* 1995). Oddly, cortical infusion of NGF in the visual cortex of the adult cat restored the capacity of the cortex to manifest an ocular dominance shift with monocular occlusion (Gu *et al.* 1994). For a review of NGF receptors, see Meakin and Shooter (1992) and Gu (1995).

Other investigators have found that brain-derived neurotrophic factor (BDNF) and neurotrophin NT-4/5 (trkB ligands) are more important than NGF (trkA ligand) in the development of ocular-dominance columns in cats, but they used more local application of neurotrophins and anatomical rather than electrophysiological assessments (Cabelli *et al.* 1995; Riddle *et al.* 1995). Also, NGF appears to be more important for the rat than for the cat.

Several forms of neural plasticity depend on the transcription factors CREB (calcium/cAMP response element binding proteins). Activity of these proteins is regulated by neuronal activity and neurotrophic factors. The proteins in turn regulate transcription of specific DNA targets. Mice with reduced CREB levels have impaired long-term memory. Also monocular deprivation in mice induces CREB mediated expression of the gene *lacZ* in the visual cortex. This induction is reduced following the critical period (Pham *et al.* 1999).

13.2.4e *Noradrenaline and cortical plasticity*

Catecholamines, particularly noradrenaline, have been implicated in the control of neural plasticity in the developing cortex. Noradrenergic axons are among the first to innervate the cerebral cortex, and reach peak levels in the second postnatal month. In the adult brain, noradrenergic axons originating in the locus coeruleus in the dorsal pons seem to provide a diffuse, non-specific innervation of the central nervous system (Levitt and Moore 1979).

After the cortex was depleted of catecholamines by intraventricular injection of 6-hydroxydop-amine, monocularly deprived kittens retained the normal proportion of binocular cells in area 17, and the cells of the seeing eye did not become dominant (Kasamatsu and Pettigrew 1979; Shirokawa *et al.* 1989). Perfusion of noradrenaline into the visual cortex of kittens over a period of one week caused a shift in ocular dominance toward the contralateral eye but only when the kittens were allowed visual experience (Kuppermann and Kasamatsu 1984). Cortical cells showed a shift in ocular dominance in anaesthetized kittens exposed to monocular stimulation for 20 hours, but only when the cortex was directly infused with noradrenaline (Imamura and Kasamatsu 1991). Also, kittens that had been monocularly deprived for one week beginning at age 4 weeks showed accelerated recovery of the normal pattern of ocular dominance when the cortex was infused with noradrenaline. Recovery was delayed when noradrenaline was depleted by application of 6-hydroxydopamine (Kasamatsu *et al.* 1981).

This evidence suggests that catecholamines are required for cortical plasticity during the critical developmental period. However, intraventricular injection of 6-hydroxydopamine produces severe side effects, including epileptic seizures. The locus coeruleus is a major source of catecholamines, and injection of 6-hydroxydopamine into this structure also caused a depletion of cortical noradrenaline and an associated loss of cortical plasticity.

Others failed to find any loss of cortical plasticity in monocularly deprived kittens after depletion of

Figure 13.9. Nancy E. J. Berman.
Born in Paducah, Kentucky in 1946. She obtained a B.A. in Psychology and Biology from Lawrence University, Wisconsin in 1968 and a Ph.D. in Psychology and Brain Science from MIT with P.H. Schiller in 1972. She did postdoctoral work in the Department of Anatomy at the University of Pennsylvania with P. Sterling and in the Department of Physiology and Biophysics at Washington University, St. Louis with E.G. Jones and N.W. Daw. In 1976 she obtained an academic appointment in the Medical College of Pennsylvania. In 1987 she was appointed Professor of Anatomy and Cell Biology in the Department of Anatomy and Cell Biology at the University of Kansas School of Medicine.

cortical noradrenaline (Bear and Daniels 1983; Frégnac and Imbert 1984; Daw *et al.* 1985). In another experiment, no differences occurred in the short-term effects of noradrenaline on responses in the visual cortices of kittens and adult cats (Videen *et al.* 1984). In these experiments, noradrenaline was depleted by lesions of the locus coeruleus or by neonatal injection of 6-hydroxydopamine, rather than by injection at the time of monocular deprivation. There is considerable restoration of the noradrenaline system with the former two procedures (see Kasamatsu 1991).

Bear and Singer (1986) suggested that this conflicting evidence could be resolved in another way. They found that depletion of both noradrenaline and acetylcholine led to a loss of cortical plasticity, in kittens even though loss of either neurotransmitter alone had no effect. They also cited other evidence, which suggests that the loss of cortical plasticity reported in the earlier studies occurred because local application of 6-hydroxydopamine depletes both norepinephrine and acetylcholine rather than only noradrenaline. Imamura and Kasamatsu (1989) agreed that the plasticity enhancing effects of the noradrenergic system are enhanced by the cholinergic system under certain circumstances.

One could argue that the effects of noradrenaline in kittens might be due to general impairment of cortical development rather than to a direct effect of the neurotransmitter. But this argument does not apply to experiments performed on the adult animal. Electrical stimulation of the locus coeruleus in combination with deprivation led to a loss of binocular cells in adult cats. In other words, induced release of noradrenaline restored plasticity in the visual cortex of cats after the critical period (Kasamatsu *et al.* 1985). Intracortical perfusion of noradrenaline had the same effect (Kasamatsu *et al.* 1979). Noradrenaline modulates synaptic activity by stimulating a second messenger system involving adenosine monophosphate (cAMP). Stimulation of the cAMP system with agents other than noradrenaline also restores cortical plasticity in adult cats (Imamura *et al.* 1999).

This evidence suggests that changes in patterns of postsynaptic intracortical excitation and inhibition play a dominant role in cortical plasticity. There is also evidence of a loss of geniculocortical afferents from a deprived eye of cats (Thorpe and Blakemore 1975). Attenuation of responses of LGN relay cells serving the deprived eye presumably has some effect on the deprivation-induced changes in ocular dominance.

13.2.5 Effects of binocular dissociation

Continued exposure to prisms that completely dissociate the images in the two eyes also leads to loss of binocular cells in kittens (Blakemore *et al.* 1975; Smith *et al.* 1979) and infant monkeys (Crawford and von Noorden 1979b). Rearing cats with both eyes occluded but with one receiving modulated light through the closed lid did not induce a shift of ocular dominance to the stimulated eye (Singer *et al.* 1977). It seems that changes in ocular dominance are due to binocular asymmetry of patterned stimuli. When both eyes receive patterned inputs that are temporally or spatially dissociated, there is no change in ocular dominance but there is a loss of binocular cells and stereopsis fails to develop.

13.2.5a *Temporal dissociation of visual inputs*

One way to dissociate visual inputs is to reverse the occluder between the eyes on alternate days so that the eyes see the same stimulus but not at the same time (Hubel and Wiesel 1965; Blakemore 1976). This procedure affects both eyes equally and, although it induces stereoblindness, it does not lead to loss of visual acuity in either eye or to an imbalance in the number of cells responding to each eye (Blake and Hirsch 1975).

Tieman *et al.* (1983a, 1983b) occluded each eye of kittens for a variable proportion of each day until they were 4 months old. The two eyes were occluded for the same length of time each day, or one eye was occluded for twice or eight times as long as the other. All groups showed a severe loss of binocular cortical cells. The greater the imbalance of eye exposure the higher the percentage of cells that responded only to the more experienced eye. Cortical cells of kittens with a balanced input had relatively normal receptive fields. Cells responding to the less experienced eye showed a poorer response to visual motion, were more broadly tuned to orientation, and had smaller receptive fields than cells responding to the more experienced eye. These deficits were most pronounced in the hemisphere ipsilateral to the less experienced eye, which receives inputs from the temporal hemiretina. Also, the nasal visual field of the less experienced eye became restricted in size and the cats ignored novel objects presented in that hemifield of the more deprived eye when both eyes were open or when only the more deprived eye was open (Tumosa *et al.* 1982, 1983). There was no recovery after restoration of normal binocular vision.

Visual inputs from the two eyes do not have to be precisely synchronous for normal development. Blasdel and Pettigrew (1979) found that cats developed normally if stimuli to the two eyes were alternated more frequently than 10 Hz. The animals were restrained during visual exposure. When kittens were allowed to move about while wearing alternating shutters that exposed each eye for longer than 0.5 s, depth discrimination was disrupted and there was a reduction in the number of binocular cells in the visual cortex (Altmann *et al.* 1987).

Figure 13.10. Nigel Daw.
Born in London, England, in 1933. He obtained a BA in Mathematics from Cambridge University in 1956 and a Ph.D. from Johns Hopkins University in 1967. Between 1969 and 1992 he held academic appointments in the Department of Physiology and Biophysics at Washington University in St. Louis. Since 1992 he has been Professor of Ophthalmology, Visual Sciences and Neurobiology at Yale University. He received the Friedenwald Award from the Association for Research in Vision and Ophthalmology in 1995.

13.2.5b *Spatial dissociation of visual inputs*

Monkeys seeing through base-in prisms that completely dissociated the images in the two eyes from age 30 to 60 days lost almost all binocular cells in the visual cortex and had no stereopsis (Crawford and von Noorden 1980; Crawford *et al.* 1983). These deficits were still present 3 years after the prisms were removed (Crawford *et al.* 1984; Crawford *et al.* 1996a, 1996b). Since this treatment affected both eyes equally, there was no amblyopia or strabismus. Three monkeys were raised in a cylinder lined with vertical stripes and with 13.5 dioptre (15°) base-in prisms on each eye (von Noorden and Crawford 1981). Disruption of cortical binocularity was as severe in these monkeys as in those raised with the same prisms but in a normal visual environment. It was expected that the repetitive pattern of stripes would have provided sufficient fusible stimuli to allow binocular cells to develop. However, a 13.5 dioptre prism introduces a nonlinear prismatic displacement and severe curvature of vertical lines and of horizontal lines above and below the horizon. These secondary distortions would have been opposite in the two eyes and would have prevented fusion. Cats reared with binocular lid closure but with flickering diffuse light presented through the closed lid of one eye did not develop greater ocular dominance for that eye (Wilson *et al.* 1977).

All these forms of deprivation have one thing in common; they reduce the frequency with which binocular cells are activated simultaneously by similarly patterned inputs. Hebb (1949) proposed that the efficiency of synaptic transmission is increased when subsets of presynaptic inputs are correlated (Section 4.3.2d). A corollary of this rule is that synaptic efficiency is lessened when the inputs are persistently uncorrelated. Binocular cells in the visual cortex seem to be good examples of the Hebbian model (Clothiaux *et al.* 1991).

Malach and Van Sluyters (1989) challenged this simple view. They exposed 4-week-old kittens to a 2-day period of monocular deprivation and then allowed them binocular vision, but with the previously deprived eye deviated. These animals showed some recovery of the number of cells driven by the deprived eye, although not to the level attained by animals allowed normal binocular vision. To recon-

Figure 13.11. Ronald S. Harwerth.
Born in Atwood, Kansas in 1939. He obtained a B.Sc. from the University of Houston in 1962 and a Ph.D. in Biomedical Sciences from the University of Texas, Houston in 1971. In 1970 he joined the faculty of the College of Optometry at the University of Houston, where he is now Professor of Physiological Optics. He received the Glenn A. Fry Award in 1980 and the Garland W. Clay Award in 1983 from the American Academy of Optometry. He became John and Rebecca Moores Scholars Professor in 1996.

Figure 13.12. Brian Timney.
Born in Sunderland, England in 1947. He obtained a B.A. in Psychology from Edinburgh University in 1969 and a Ph.D. with R. Morant at Brandeis University in 1973. Between 1973 and 1978 he conducted postdoctoral work at Queens University and Dalhousie University. In 1978 he gained an academic appointment at the University of Western Ontario, where he is now Professor of Psychology.

cile these results with the Hebbian model, one must assume that the correlation between inputs from misaligned eyes is higher than that between inputs from a closed eye and an open eye. The fact remains that normal binocular vision is required to maintain or restore the full complement of binocular cells.

13.3 THE CRITICAL PERIOD

The effects of monocular deprivation are permanent if applied early in life and maintained beyond a critical age. The period during which visual deprivation can have permanent and irreversible effects is the critical period. The critical period may not be the same for different types of visual deprivation or for different visual functions, and varies from species to species (Berman and Daw 1977; Harwerth *et al.* 1986b) (Portrait Figures 13.9, 13.10, and 13.11).

The physiological processes responsible for the onset and end of the critical period are not known. There is evidence that the development of cortical inhibition is involved. Premature enhancement of inhibition by application of diazepam in young rats triggered the onset of the critical period. Also, mice lacking the enzyme for the inhibitory neurotransmitter GABA showed reduced cortical plasticity when monocularly deprived during the critical period (Feldman 2000). Perhaps a critical level of cortical inhibition is required for the detection of synchrony between the inputs from the two eyes.

13.3.1 Critical period in cats

The eyes of cats are closed until about 10 days after birth. The period during which cortical cells are susceptible to monocular occlusion begins in the 4th week, remains high until the 8th week, and ends at about the 12th week (Hubel and Wiesel 1970b; Dews and Wiesel 1970). A similar critical period was reported in rats and ferrets (Fagiolini *et al.* 1994; Guire *et al.* 1999; Issa *et al.* 1999). At 4 weeks, binocular cells of kittens deprived of vision in one eye for only 12 hours spread over 2 days showed a massive shift in ocular dominance to the non-deprived eye. However, the normal pattern of ocular dominance was restored after 7 weeks of normal binocular vision (Malach *et al.* 1984). During the period of peak susceptibility, at about age 40 days, eye closure for 3 or 4 days induced a permanent loss of binocular cells and a deficit in depth discrimination (Timney 1990) (Portrait Figure 13.12). Also, allowing dark-reared kittens to see with one eye for a few hours during the peak of the critical period produced a distinct shift in ocular dominance to the open eye, which was evident two days later (Peck and Blakemore 1975; Olson and Freeman 1980a). The critical period for effects of strabismus is similar to that for monocular occlusion (Levitt and Van Sluyters 1982).

Figure 13.13 Donald E. Mitchell.
Born in Launceston, Tasmania in 1941. He obtained a B.Sc. in Optometry in 1962, a B.A. in Physiology from the University of Melbourne in 1965, and Ph.D. in Physiological Optics from Berkeley with G. Westheimer in 1968. He conducted postdoctoral work in the Department of Psychology at Florida State University, Tallahassee. In 1970 he joined the faculty of Dalhousie University, Canada where he is now Professor of Psychology. He is a Fellow of the National Vision Research Institute of Australia.

In other studies, some effects of monocular occlusion on ocular dominance were found when applied between the fifth and seventh months (Cynader and Mitchell 1980; Cynader *et al.* 1980). Daw *et al.* (1992) agreed that the critical period for cells in layer 4 of the visual cortex of the cat extends to the seventh month but found significant shifts in ocular dominance in other layers after monocular deprivation applied in the eleventh month. The critical period lasts longer for binocular cells in area 17 than for those in the extrastriate cortex (the lateral suprasylvian area) (Jones *et al.* 1984b). It has also been reported that the critical period for loss of binocularity ends before the critical period for changes in receptive-field structure (Berman and Murphy 1982).

In normal cats, depth discrimination thresholds assessed on a jumping stand are much better for binocular than for monocular viewing. Cats monocularly deprived by lid suture for more than 31 days from the time their eyes opened showed no binocular superiority. Cats monocularly deprived at 4 months of age showed no loss in binocular performance (Timney 1983).

13.3.1a *Postponement of critical period*

For kittens reared in the dark for up to 10 months, subsequent monocular occlusion still caused a marked shift in ocular dominance of cortical cells towards the seeing eye (Timney *et al.* 1980; Mitchell and Timney 1982; Mower and Christen 1985). Cynader (1983) found some effects of monocular occlusion in cats that had been dark reared for 2 years. After dark rearing, the period of susceptibility to monocular deprivation lasted about 6 weeks. Thus, the critical period of cortical flexibility occurs at a later age if the animal is kept in the dark. But the darkness must be complete. Monocular deprivation produced only slight cortical changes in kittens reared with binocular sutures that allowed diffuse retinal illumination (Mower *et al.* 1981b). Six hours of exposure to a normal visual environment in 6-week old kittens that were dark-reared for the first 5 months of life was sufficient to eliminate the delaying effects of dark rearing (Mower *et al.* 1983). Thus, the processes responsible for termination of the critical period must depend on visual stimulation but the visual stimulation need not be of long duration, and diffuse illumination seems to be sufficient.

Evidence from the rat suggests that prolonged plasticity with dark rearing is due to the prolonged expression of the activity-regulated gene, *cpg 15*. In normally reared rats, expression of this gene peaks during the critical period and then declines (Lee and Nedivi 2002).

Eight days of monocular occlusion starting at age 45 days produced a larger shift of ocular dominance to the nondeprived eye in kittens reared with normal vision than in kittens with induced strabismus during the first 45 days (Mustari and Cynader 1981). It was suggested that strabismus causes the ocular dominance columns to become separated and thus less vulnerable to effects of interocular competition in a subsequent period of monocular deprivation.

13.3.1b *Effects of reverse suturing*

Visual acuity of a deprived eye shows some recovery after sight has been restored but the extent of recovery declines exponentially as the period of deprivation is increased (Mitchell 1988a) (Portrait Figure 13.13). When occlusion is switched to the other eye at an early enough stage, processes leading to the dominance of the first eye are reversed. The longer reverse occlusion is maintained, the greater the shift in dominance to the previously occluded eye (Van Sluyters 1978). For instance, when monocularly deprived kittens were reverse sutured at 5 weeks of age, ocular dominance switched almost completely to the previously deprived eye within 3 weeks (Movshon and Blakemore 1974). It seems that the same cells that were dominated by the previously open eye become dominated by the newly open eye (Olson and Freeman 1978b).

In another study, reverse suturing of the eyes in a kitten at 5 weeks produced a complete switch in ocular dominance after a few weeks. Also, the orientation specificity of receptive fields in the initially deprived eye improved while the specificity of receptive fields in the newly deprived eye decreased (Movshon 1976). At the same time, the previously deprived eye improved from an initial state of functional blindness and the newly sutured eye became functionally blind. Reverse sutured animals lacked binocular cells and were stereoblind. Reverse occlusion also leads to recovery of orientation tuning as revealed in optical imaging of cortical activity (Kim and Bonhoeffer 1994).

If the reversal of occlusion is delayed for too long, the reversal of ocular dominance does not occur and behaviour mediated by the previously occluded eye, such as visually guided reaching, cliff avoidance, and jumping across gaps, remains permanently impaired (Van Hof-van Duin 1976b). Reverse suturing after 14 weeks of monocular deprivation did not reverse the pattern of ocular dominance (Blakemore and Van Sluyters 1974).

Kittens reverse sutured for 10 days after a week of monocular deprivation showed full recovery of cortical responses to stimulation of the originally deprived eye coupled with a reduction of response to stimulation of the nondeprived eye (Antonini *et al.* 1998). However, afferent arbors serving the originally deprived eye in layer 4 of area 17 showed only partial recovery. There was thus a dissociation between physiological and anatomical recovery.

A period of reverse occlusion in the cat was more effective in promoting recovery of ocular dominance and geniculate cell morphology when distributed over several sessions than when provided continuously in one session (Crewther *et al.* 1983). Simply allowing an animal to see with both eyes after a period of monocular occlusion may lead to some recovery but is not as effective as reverse suturing, which forces the animal to use its deprived eye (Mitchell *et al.* 1977; Mitchell 1988a).

Mitchell *et al.* (2001) exposed kittens between 5 and 9 days of age to 6 days of monocular deprivation. For the first 4 to 8 days, recovery of acuity in the deprived eye was more rapid when both eyes were open than when only the previously deprived eye was open. After this time, recovery was more complete with reverse suturing. This early advantage of binocular recovery seems to be in conflict with the neural competition theory of neural plasticity. Mitchell *et al.* explained the effect in terms of the fact that presynaptic activity in binocular cells from a reverse sutured eye is less than the activity when both eyes are open. Presynaptic activity below a certain threshold produces a weakening of synaptic strength (Section 10.4.5).

Although reverse occlusion leads to some recovery of a previously deprived eye the recovery is at the expense of a loss of visual capacity in the other eye. Furthermore, when sight is restored to both eyes after a period of reverse suturing, the reverse-sutured eye loses its newly gained capacity and the other eye recovers. Thus, the gain achieved by reverse suturing is not permanent in the cat (Mitchell 1988b).

13.3.2 Critical period in monkeys

By the sixth postnatal day, V1 of the macaque contains an adult-like proportion of disparity-sensitive cells, although the cells are poorly tuned to spatial frequency and are not very responsive (Chino *et al.* 1997). The cells mature in the first 4 weeks, by which time the monkey begins to show evidence of stereopsis.

Monocular occlusion in the monkey from the first week of life causes a strong shift in ocular dominance of cells in V1 to the open eye and a shrinkage of cortical columns devoted to the deprived eye (Horton and Hocking 1997). Monocular enucleation in Rhesus monkey at one week of age eliminated ocular dominance columns in layer 4Cβ leaving a uniform autoradiograph and cytochrome oxidase staining (Horton and Hocking 1998b). Columns belonging to the missing eye were much reduced in layers 4Cα and 4α. Monocular enucleation at 12 weeks produced only mild shrinkage of ocular dominance columns of the missing eye. Monocular enucleation at any age did not affect cytochrome oxidase stripes in V2. Monocular occlusion from an age of between 7 and 14 months had no influence on ocular dominance columns in cortical layer 4C, but there was still some shift of ocular dominance in cells of the upper cortical layers (Blakemore *et al.* 1978; LeVay *et al.* 1980). Monocular deprivation in the adult monkey has no effect on ocular dominance.

Two weeks of prism-induced strabismus reduced the disparity sensitivity of cells in V1 more severely when applied at the age of 2 weeks than when applied at the age of 6 weeks (Kumagami *et al.* 2000).

Reverse suturing after monocular occlusion for the first 24 days led to a reversal of dominance to the other eye (Swindale *et al.* 1981). A reversal of ocular dominance was found even after 90 days of monocular occlusion (Crawford *et al.* 1989) but no significant recovery of visual function was found after 19 months (Harwerth *et al.* 1984).

The critical period for monkeys seems to be

longer than that for cats. Anisometropia induced in the neonate monkey by placing a 10-dioptre lens in front of one eye for 30 days produced little effect in the defocused eye. When continued for 60 days or more it produced persistent amblyopia in the defocused eye as revealed by loss of contrast sensitivity for high spatial frequencies (Smith *et al.* 1985b).

The critical period in the monkey depends on the type of visual function. Thus, loss of scotopic sensitivity occurred only when monocular deprivation was initiated before 2 months of age. Spectral and contrast sensitivities were affected when deprivation was initiated before 5 months of age. Loss of binocular vision occurred even when deprivation was initiated as late as 25 months (Harwerth *et al.* 1990). Fusional vergence movements can recover fully after the eyes of monkeys are realigned following early surgically induced esotropia, even though stereoacuity may remain deficient (Harwerth *et al.* 1997). Clearly, some visual functions mature earlier than others or are less affected by deprivation than others. It seems to be a general principle that the critical period for functions processed at lower levels of the nervous system ends earlier than that for functions processed at higher levels.

13.3.3 Critical period in humans

Monocular deprivation in humans occurs as a consequence of monocular enucleation, unilateral cataracts, strabismus, or anisometropia in which the two eyes have differing refractive errors. Infants with severe strabismus over several years never recover stereoscopic vision, even when the strabismus is corrected by surgery.

Monocular vision or strabismus that develops in an older child or adult does not produce any permanent loss of binocular functioning when the condition is corrected. E.L. Smith *et al.* (1980) had six human adults with normal vision wear a monocular occluder for 12 days. This produced a small decrease in stereoacuity for a period of up to 2 hours, which was probably due to a temporary increase in fixation disparity.

The critical period for development of amblyopia in humans seems to end earlier than for development of binocularity. Jacobson *et al.* (1981) found that infants with esotropia from shortly after birth developed differences in acuity between the two eyes at a mean age of 20 weeks. One infant with esotropia from 10 months of age developed signs of amblyopia 4 weeks later. According to this evidence, the critical period for the development of amblyopia extends from the age of about 4 months to at least 12 months.

13.3.3a *Correction for strabismus*

Banks *et al.* (1975) obtained an estimate of the critical period for normal visual development in humans by testing binocular functioning in 24 adults in whom strabismus in excess of 10° had been surgically corrected at various ages. Congenital esotropes who had corrective surgery before the age of 3 years tended to develop more normal binocular functions than those having the correction at a later age (see also Enoch and Rabinowicz 1976). For subjects in whom strabismus developed after the age of 4 years, surgery produced complete restoration of binocular functioning no matter when it was performed. Banks *et al.* concluded that the sensitive period during which normal binocular inputs are required for binocular functioning in humans is between the ages of 1 and 3 years (see Section 9.3.2 for other studies).

Birch *et al.* (1990) assessed the effects of surgery and eye-occlusion therapy on acuity, eye alignment, and stereopsis in a group of 84 esotropes under 1 year of age. Three patients who responded adequately to optical and occlusion therapy achieved near normal stereopsis and acuity. Their esotropia probably arose from anisometropia. The remaining patients required surgery which, in about 88% of cases, resulted in good eye alignment, although some patients required secondary surgery. None of the patients undergoing surgery developed postoperative random-dot stereoacuity better than 200 arcsec and only 35% showed this level of performance (see also Atkinson *et al.* 1991). However, most of the patients showed normal development of acuity in both eyes after surgery followed by occlusion therapy.

Birch *et al.* (1995) found that about 40% of infantile esotropes had some stereovision when tested at 5 years of age. Those who had been surgically corrected at under 8 months of age had foveal stereoacuity of about 60 arcsec. Those corrected between 9 and 12 months of age had stereoacuity of between 60 and 200 arcsec, and those corrected between 13 and 16 months had stereoacuities in excess of 200 arcsec. However, in a later study, restoration of some stereopsis in infantile esotropes under 24 months of age was found to depend more on the duration of strabismus rather than on the age at which correction was applied (Birch *et al.* 2000a).

In a study of 82 cases of surgical correction, binocular vision, as indicated by binocular fusion and coarse stereopsis, was not restored unless surgery was performed in the first year (Deller 1988). In another study, five of seven patients in whom early onset esotropia was corrected during the first 19 weeks achieved stereoacuity of between 40 and 400 arcsec when tested 2 years later (Wright *et al.* 1994).

Figure 13.14. Lynne Kiorpes.
Born in New York in 1950. She obtained a BA at Northeastern University in 1973 and a Ph.D. in Psychology with D. Teller from the University of Washington in 1982. After postdoctoral work at Washington University she obtained an academic position at New York University, where she is now an Associate Professor.

Thus, early surgery may lead to some restoration of stereopsis but, in any case, it is justified by the amelioration of amblyopia and by the cosmetic improvement (see Smith *et al.* 1991).

Mohindra *et al.* (1985) found that all but three of 19 esotropic infants with prismatic correction showed evidence of stereopsis during the first two years, but that none of the 19 infantile esotropes had stereopsis after the age of 6 years whether they had worn a prismatic correction or not. Thus, the full adverse effects of infantile esotropia are not evident until after the second year.

13.3.3b *Correction for cataract*

There has been a good deal of debate about the best age to correct congenital cataract by surgery. A survey of 231 cases at the Wilmer Institute between 1925 and 1943 revealed that good vision (20/70 or better) was achieved in only 9% of cases when surgery was performed before the age of 2.5 years and in 69% when it was performed after that age (Owens and Hughes 1948; Bagley 1949). According to these results, it is better to delay surgery for congenital cataract. The exact ages at operation were not specified in these studies, but it is unlikely that operations were performed on children under 6 months of age. In a more recent study, eight infants with congenital unilateral cataracts developed reasonable acuity in the affected eye when they were operated on before the age of 40 days (Beller *et al.* 1981). In another study, five of 10 children developed good acuity when cataract surgery with lens replacement was performed before the age of 4 months. There was no recovery of binocular vision, presumably because, after surgery, the good eye was occluded until the age of 6 months. There was little recovery of acuity in two children operated on after the age of 4 months (Robb *et al.* 1987). Pratt-Johnson and Tillson (1981) removed bilateral or unilateral congenital cataracts from six infants in the first year of life, and found that they all developed reasonably good acuity in the affected eyes. Wright *et al.* (1992) found that five of thirteen patients who had surgery for monocular congenital cataracts before the age of 9 weeks developed vergence eye movements. One patient developed stereoacuity of 200 arcsec. Many infants with bilateral or unilateral cataract have associated visual defects, such as nystagmus, which persist after the cataracts have been removed.

13.4 AMBLYOPIA

13.4.1 Types of amblyopia

When monocular deprivation is maintained over a long period during early development, the deprived eye manifests a combination of symptoms known as **amblyopia**, literally "blunt vision". The defect has been known for at least 200 years. For instance, Thomas Reid, Professor of Moral Philosophy in Glasgow, reported in 1764 that 20 people with strabismus had defective sight of one eye.

All forms of amblyopia involve a loss of contrast sensitivity for high spatial frequency gratings presented to the central retina, and weak or absent stereopsis. Although misaccommodation may be one cause of amblyopia, spectacles cannot rectify the symptoms. Amblyopia does not involve retinal defects—receptors are oriented normally, foveal pigment density is normal (Delint *et al.* 1998), and the electroretinogram is normal (Hess and Baker 1984). Evidence reviewed in Section 13.2 shows that the physiological effects of monocular deprivation are largely confined to the visual cortex. Psychophysical evidence, also, shows that loss of contrast sensitivity in amblyopia is largely of cortical origin (see Kiorpes *et al.* 1999) (Portrait Figure 13.14).

Different types of amblyopia are defined according to their aetiology. **Deprivation amblyopia** is due to loss of form vision due to a cataract, ptosis, or retinal disorders. **Anisometropic amblyopia** is due to unequal refraction in the two eyes resulting from unequal eye growth (axial anisometropia) or corneal defects. The greater the degree of anisometropia, the greater the depth of amblyopia and the greater the loss of binocularity (Rutstein and Corliss 1999). Re-

Figure 13.15. Dennis M. Levi.
He obtained a Diploma in Optometry from Witwatersrand School of Optometry in South Africa in 1967. From the University of Houston he received an O.D. in 1971 and a Ph.D. in 1977. He joined the faculty of the College of Optometry at the University of Houston in 1972, where he is now Professor of Optometry and Physiological Optics. In 1996, he became the Cullen Distinguished Professor of Optometry. He received the Glenn Fry Award and the Garland Clay Award from the American Academy of Optometry.

fractive error due to pathological dilation of one pupil (Adie syndrome) can produce amblyopia (Firth 1999). **Strabismic amblyopia** is due to early misalignment of one eye. **Meridional amblyopia** is due to an uncorrected astigmatism and affects vision only for images oriented along the astigmatic axis.

When early disruption is applied equally to both eyes of kittens, by alternate occlusion or by prismatic dissociation of visual inputs, both eyes develop normal or near-normal acuity. However, there is still a reduced number of binocularly driven cells and loss of stereopsis (Blake and Hirsch 1975). Similarly, in a sample of 114 human infants with untreated alternating esotropia of early onset, grating acuity, as measured by preferential looking, remained normal in both eyes although stereopsis was lost (Birch and Stager 1985). However, Day *et al.* (1988) found evidence of a small equal loss of grating acuity in both eyes of 1-year-old alternating esotropes as assessed by cortical potentials (VEP) evoked by a swept spatial frequency grating alternating in phase (Section 11.1.1b). The preferential looking procedure used by Birch and Stager may not have been sensitive enough to detect the small difference recorded by Day *et al.*

Strabismic amblyopia is more prevalent among esotropes (inward deviated eye) than among exotropes (outward deviated eye). The effects on the visual cortex of cats are also greater for induced esotropia than for induced exotropia (Section 13.2.2a). The following three reasons for this difference have been suggested. (1) Exotropia develops more slowly than esotropia. (2) Amblyopia does not occur in alternating strabismics and exotropia is more likely to be alternating than unilateral. (3) In esotropes, the fovea of the deviating eye competes with the dominant temporal hemifield of the other eye. On the other hand, in exotropes the fovea competes with the nondominant nasal hemifield of the other eye (Buckley and Seaber 1982; Fahle 1987).

The subject of amblyopia has been reviewed by Levi (1991) (Portrait Figure 13.15) and Ciuffreda *et al.* (1992). As we shall now see, different types of amblyopia produce different symptoms.

13.4.2 Loss of contrast sensitivity and hyperacuity

The symptoms of amblyopia include a reduction in contrast sensitivity, sometimes for all spatial frequencies and sometimes for high spatial frequencies only (Hess and Howell 1977; Bradley and Freeman 1981). There is also a loss of grating resolution acuity (Harwerth *et al.* 1983a; Kratz and Lehmkuhle 1983) and impaired performance on vernier acuity and other hyperacuity tasks (Levi and Klein 1982a; Bradley and Freeman 1985a; Bedell *et al.* 1985).

Loss of visual acuity in a strabismic eye has three basic causal components; (1) amblyopia of the deviating eye, (2) decentering of the image of the attended object in the deviating eye, and (3) suppression or masking of the image in the deviating eye by the image in the normal eye. The first defect is evident when only the deviating eye is open. The other two effects are evident only when both eyes are open. Freeman *et al.* (1996) assessed the percentage loss of acuity of the deviating eye relative to acuity of the normal eye in nine small-angle strabismic subjects as 34% from amblyopia, 44% from image decentering, and 20% from binocular suppression. Instability of gaze, that can accompany amblyopia, does not contribute to the loss of contrast sensitivity (Higgins *et al.* 1982).

13.4.2a *Effects of luminance and duration*

As the level of illumination decreases, cones, which are concentrated in the central retina, lose their sensitivity more rapidly than rods, which are more numerous in the periphery. In strabismic, but not anisometropic amblyopia, loss of contrast sensitivity is less evident at low levels of luminance. This seems to be because strabismic amblyopes use their relatively normal periphery under scotopic conditions (Hess *et al.* 1980) (Portrait Figure 13.16).

Figure 13.16. Robert Hess.
Born in Australia in 1950. He studied optometry at the Queensland Institute of Technology in Brisbane from 1968 to 1970. He obtained an M.Sc. at Aston University in Birmingham, England in 1971 and a Ph.D. with E. Howell at the University of Melbourne in 1976. He held research positions at the University of Cambridge from 1980 to 1990. In 1990 he moved to a Chair in Ophthalmology at McGill University, Montreal. He received the Champness Medal from the Worshipful Company of Spectacle Makers, London in 1983, the Edridge-Green Medal from the Royal College of Surgeons, London in 1989, the Bobier award from the University of Waterloo in 1994, and a D.Sc. from Aston University in 1998.

In spite of loss in contrast sensitivity, gratings just above threshold appear to have the same contrast in the amblyopic eye as in the non-amblyopic eye of strabismic amblyopes. For anisometropic amblyopes, a higher contrast is required before gratings in the two eyes appear to have the same contrast (Hess and Bradley 1980). Contrast discrimination thresholds for stimuli applied to one eye are not much affected by amblyopia (see Levi 1991).

In an amblyopic eye, acuity becomes worse in proportion to the illumination level of the good eye (von Noorden and Leffler 1966) and vision may be totally suppressed when the good eye is open.

Vernier acuity in the amblyopic eyes of strabismics is more dependent on stimulus duration than that in normal eyes (Rentschler and Hilz 1985) although both eyes show a similar dependence on stimulus duration for stimuli equated for loss of contrast sensitivity (Demanins and Hess 1996a).

13.4.2b *Effects of stimulus area and eccentricity*

In a normal eye, detection of high spatial frequencies is confined to the central retina. Contrast sensitivity functions for the amblyopic eyes of esotropic amblyopes were reduced for high spatial frequency sine wave gratings presented within the central 5° of the retina. With more severe amblyopia, the deficit spread over a larger range of spatial frequencies and eccentricities (Thomas 1978b).

Hess *et al.* (1980) reported that strabismic amblyopes show loss of contrast sensitivity mainly in the central retina, while anisometropic amblyopes show loss of sensitivity in both the central and peripheral retina. After allowing for effects of spatial scale, Bradley *et al.* (1985) found that amblyopia was not related to retinal locus in seven of nine amblyopes. However, the type of amblyopia in these subjects was not clearly specified.

In the normal eye, vernier acuity develops more slowly than resolution acuity, and the decline of vernier acuity with increasing eccentricity is much steeper than the decline of resolution acuity. In strabismic amblyopia the spatial displacement of the images in the two eyes is the same over the whole visual field but affects central vision more than peripheral vision because receptive fields are smaller in the centre. Thus, the centre of an eye with strabismic amblyopia resembles that of the periphery of a normal eye. The lack of image registration in the central field of strabismic amblyopes leads to a loss of cortical binocular cells for that region of the visual field. Thus, although the central visual field of strabismic amblyopes has the same density of ganglion cells, it has fewer cortical processing units devoted to it, so it is spatially undersampled at the cortical level (Levi and Klein 1985) or subject to more intrinsic noise (Hess *et al.* 1999b). This could account for why hyperacuity, which is a function of the central retina, is affected more than resolution acuity in strabismic amblyopes of both early and late onset (Birch and Swanson 2000).

In anisometropes with aniseikonia, the displacement of one image relative to the other increases with eccentricity. As a result, the foveal region is affected less than the peripheral retina. There is thus a more balanced loss of hyperacuity and resolution acuity in anisometropes.

Optimal detection of depth requires a stimulus containing about 10 periods of a grating. As the spatial frequency of a grating is decreased, a larger stimulus is required for optimal detection (Hoekstra *et al.* 1974; Howell and Hess 1978). Thus, in comparing contrast sensitivities at different spatial frequencies, one must take the size of the stimulus into account. Hagemans and Wildt (1979) found that contrast sensitivity for low spatial frequencies increased more rapidly with increasing width of the grating for the amblyopic eye than for the normal eye of strabismic and anisometropic amblyopes.

With large stimuli, the amblyopic eye was sometimes more sensitive to gratings of low spatial frequency than the normal eye. Field size had little effect for gratings of high spatial frequency. See also Katz *et al.* (1984).

13.4.2c *Hemifield differences in amblyopia*

The nasal hemifield (uncrossed inputs from temporal hemiretina) is more susceptible to the effects of early monocular deprivation or esotropia than the temporal hemifield (crossed cortical inputs) in cats (Sherman 1973; Ikeda and Jacobson 1977; Singer 1978; Bisti and Carmignoto 1986) and humans (Sireteanu and Fronius 1990). Thus, cells in area 17 of esotropic cats show a marked loss of response to stimuli presented in the nasal visual field of the deviated eye (Kalil *et al.* 1984). This could be a consequence of the fact that nasal hemiretinas develop more rapidly than temporal hemiretinas and to the fact that the stimulus fixated by the normal eye of an esotrope falls on the dominant nasal hemiretina of the amblyopic eye. In anisometropic amblyopes, acuity loss diminished symmetrically with increasing eccentricity (Sireteanu and Fronius 1981). Hemifield differences are discussed in Section 11.1.4. Some loss of visual sensitivity may be present in mild form in the nondeprived eye (Section 13.2.2f).

13.4.2d *Orientation discrimination in amblyopia*

Orientation discrimination thresholds for amblyopic eyes are unusually high for gratings with high spatial frequency (Skottun *et al.* 1986b). In a normal eye, orientation discrimination declines with increasing retinal eccentricity of the stimulus. However, this loss can be compensated for by a proportional increase in the length of the test line. Similarly, poor orientation discrimination for a single line by an amblyopic eye can be compensated for by increasing the length of the line (Vandenbussche *et al.* 1986).

In Section 13.2.2a it was mentioned that early strabismus leads to a selective loss of cortical cells tuned to vertically orientated stimuli. While subjects with normal vision show greater acuity for vertical gratings than for gratings in other orientations (see Howard 1982), strabismic amblyopes show a selective loss of contrast sensitivity for vertical gratings, an effect known as the "vertical effect" (Sireteanu and Singer 1980). The vertical effect seems to be a consequence of strabismus rather than of amblyopia since it does not occur in monkeys in which amblyopia is induced without misalignment of the visual axes (Harwerth *et al.* 1983b). The vertical effect occurs only for high spatial-frequency stimuli in unilateral strabismics (Kelly *et al.* 1997; Sharma *et al.* 1999). It occurs in both eyes although it is not as strong in the non-deviating eye. The vertical effect is probably due to the fact that strabismus induces disparity between vertical contours but not between horizontal contours.

13.4.2e *Anisometropic and strabismic amblyopia*

In anisometropic amblyopia, the above symptoms may result from loss in contrast sensitivity due to image blur arising from defects in refraction. In these cases, differences between the two eyes disappear when the stimuli are equated for visibility (Ingram and Walker 1979). Thus, in humans with anisometropic amblyopia, deficits on hyperacuity tasks, such as vernier acuity, are proportional to losses in resolution and contrast sensitivity (Levi and Klein 1982b, 1983). However, there is conflicting evidence on this point (Kiorpes *et al.* 1993).

In strabismic amblyopia, hyperacuities, such as vernier acuity, are more severely affected than can be accounted for by loss in contrast sensitivity (Swindale and Mitchell 1994). This is true in cats, monkeys, and humans (Kiorpes 1992; Murphy and Mitchell 1991; Levi and Klein 1982b, 1990b; Levi *et al.* 1994a, 1994b). In many strabismic children, contrast sensitivity is almost the same in both eyes, but the deviating eye has a severe deficit in hyperacuity tasks and in tasks requiring the recognition of letters presented closely together (Howell *et al.* 1983). It seems that strabismic amblyopes have difficulty isolating spatial relationships for pattern identification, although they can distinguish a simple form from its background.

Wilson (1991b) modelled these differences and concluded that the fovea of a strabismic eye suffers from spatial undersampling, as indicated by loss of grating resolution, and position uncertainty, as indicated by reduced vernier acuity, but no change in receptive field size. The fovea of an anisometropic amblyopic eye suffers from a loss of sensitivity but no undersampling or position uncertainty.

13.4.3 Spatial distortions

Other symptoms of amblyopia include confusion between neighbouring stimuli (Pugh 1958), visual distortions of length and direction (Hess *et al.* 1978; Bedell and Flom 1981; Fronius and Sireteanu 1989; Lagreze and Sireteanu 1991), and defective motion, and shape discrimination (Watt and Hess 1987). Spatial phase discrimination is also defective (Pass and Levi 1982; Bennett and Banks 1987; Kiper 1994). For instance, strabismic amblyopes need higher than normal contrasts to discriminate the relative phases of a grating composed of a fundamental and its third harmonic (Lawden *et al.* 1982).

Spatial distortions in most anisometropic amblyopes can be explained in terms of loss of contrast sensitivity, and most anisometropic amblyopes perform normally on a contour integration test (Hess and Demanins 1998). But spatial defects in strabismic amblyopes cannot be explained in this way (Rentschler and Hilz 1985; Fronius and Sireteanu 1989; Hess and Holliday 1992b; Hess and Field 1994). Some strabismic amblyopes show a loss of hyperacuity but not spatial distortions while others show only spatial distortions. Furthermore, hyperacuity defects are greatest in the central retina while distortions are greatest in the periphery (Demanins and Hess 1996b).

These defects cannot be accounted for in terms of loss of retinal receptors or ganglion cells, because there is no evidence that amblyopes show such losses. Masking and adaptation studies have revealed that spatial-frequency channels of amblyopic eyes have normal bandwidths and tuning functions, except that they require higher contrasts for their activation (Hess 1980). The defects in strabismic amblyopia must have something to do with how spatial information is coded in the visual cortex. Three causes have been suggested.

13.4.3a *Spatial undersampling*
It has already been mentioned that diplopia in the central field of strabismic amblyopes leads to a reduction in the number of cortical cells responding to the deviated eye for that region of the visual field. Thus, the central visual field of the deviated eye of strabismic amblyopes has fewer cortical processing units devoted to it, so it is spatially undersampled (Levi and Klein 1985). Loss of spatial sampling at the cortical level should affect pattern acuity more than contrast sensitivity or grating acuity. For example, Levi *et al.* (1987) found that, in strabismic but not anisometropic amblyopes, 3-line bisection acuity was reduced to a greater extent than could be accounted for in terms of loss of grating acuity. This was also true of the periphery of a normal eye and, as in the normal eye, the pattern acuity of the strabismic eye improved as the number of samples in the stimulus was increased. Thus, the centre of the deviated eye of a strabismic amblyope resembles the periphery of a normal eye.

Wang *et al.* (1998) used a three-line bisection test with lines composed of a variable number of dots scrambled about the mean position with a variable standard deviation. An ideal observer analysis provided separate estimates of spatial uncertainty and sampling efficiency (number of dots used to total number of dots). Spatial uncertainty was elevated about 10-fold in both anisometropic and strabismic amblyopes relative to that in normal subjects. However, loss of spatial integration efficiency was found only in strabismic amblyopes. They suggested that this is due mainly to loss of fine-scale detectors at the cortical level.

A regular lattice of independent detectors can resolve a grating only if the spatial period of the grating is at least twice the spacing of the detectors—the Nyquist limit (Section 5.1.3). A finer grating is said to be undersampled and forms a moiré pattern with the detectors, which may be visible even though the grating is not (see Figure 5.9). This is known as aliasing. Aliasing is not normally evident in the central retina because the optics of the eye are not capable of forming images as fine as the spacing of the cones. It can be made visible by forming a laser interference image on the retina that bypasses the optics (Section 5.1.3).

A grating at twice the Nyquist limit should appear tilted by 90°. If the visual cortex of amblyopes has reduced spatial sampling, this misperception of orientation should be evident at unusually low spatial frequencies. Sharma *et al.* (1999) projected a laser interference grating on the retinas of three strabismic amblyopes. Both the amblyopic and non-amblyopic eyes showed reduced contrast sensitivity at low spatial frequencies, which, for two of the subjects, was most pronounced for vertical gratings (see Section 13.4.2d). The amblyopic eyes showed loss of contrast sensitivity at higher spatial frequencies, and there was severe misperception of the orientation of gratings with spatial frequencies between 20 and 65 cpd, well below the 120-cpd limit set by the retinal receptors. These results suggest that inputs from an amblyopic eye are severely undersampled at the cortical level. Spatial scrambling of detectors alone would not produce misperception of orientation. However, the misperception of orientation found by Sharma *et al.* was not quite the 90° value predicted by pure undersampling. Thus, the visual cortex of strabismic amblyopes may suffer from undersampling and disarray. It may not be possible to dissociate the two effects because random loss of cortical units responding to an amblyopic eye would result in an irregular detector array, which would produce scrambling.

13.4.3b *Spatial scrambling*
Perhaps the cortical local sign mechanism is spatially scrambled in strabismic amblyopia. Thus, there may be topographical disarray, or misregistration of the positions of the receptive fields of cortical cells relative to their neighbours. Strabismic amblyopes show distortions when they attempt to draw letters or gratings (Pugh 1958; Hess *et al.* 1978).

Sireteanu *et al.* (1993b) found that drawings of memorized circles of dots made by amblyopes using the defective eye were distorted in size and shape. However, drawings of complex scenes that remained visible did not show equivalent distortions and most amblyopes do not report seeing distortions in natural scenes.

Demanins *et al.* (1999) argued that a spatially scrambled array of detectors with a stimulus near the Nyquist limit should have near normal orientation discrimination because a subset of detectors would fall below the Nyquist limit. However, the discrimination of the direction of motion of a grating should be defective because motion directions are averaged over each location. For an undersampled array, orientation and motion discrimination should be equally defective. In a sample of eight strabismic amblyopes they found three who showed evidence of spatial scrambling with high spatial-frequency gratings and one who showed evidence of undersampling.

13.4.3c *Defective contour integration*

Collinear line elements are more easily detected than non-collinear elements (Section 4.2.6). This facilitates the detection of connected contours and is thought to depend on lateral connections in the visual cortex. Perhaps these connections are poorly developed in strabismic amblyopes (Polat 1999). Hess *et al.* (1997b) found that 10 out of 11 strabismic amblyopes had reduced ability to detect aligned micropatterns (Gabor patches) in a random field of micropatterns when using their deviating eye. The defect showed in only a few anisometropic ambylopes in their sample, which included patients who had undergone treatment. Chandna *et al.* (2001) found that the majority of untreated anisometropic amblyopes showed defective contour integration in their amblyopic eye. When *Hess et al.* introduced spatial jitter into their stimuli, normal subjects performed like strabismic amblyopes. They concluded that the fundamental defect is spatial disarray of orientation detectors rather than poor contour integration. However, the distinction between these defects is not well defined. Kovács *et al.* (2000) used similar stimuli and found that the poor performance of strabismic amblyopes was not due to loss of acuity or contrast sensitivity.

The visibility (determined by the contrast threshold) of a small Gabor patch is enhanced when it is flanked by two collinear Gabor patches and reduced when flanked by orthogonal patches. In a mixed group of amblyopes, collinear facilitation, measured psychophysically and by evoked potentials was reduced or replaced by inhibition (Polat *et al.* 1997). This suggests that long-range cortical connections are disrupted in amblyopia.

However, other investigators have reported that amblyopes have normal abilities in some orientation detection tasks. Levi and Sharma (1998) found that strabismic amblyopes showed normal visibility enhancement for aligned texture elements. Mussap and Levi (1999) found that three strabismic amblyopes had a normal ability to detect orientation-defined textured regions in regular arrays of Gabor-patches. However, amblyopes showed some deficit in detection of a dotted line masked by random-dot noise (Mussap and Levi 2000). They concluded that texture segmentation based on orientation is spared in amblyopia but that contour grouping based on orientation is not. It is not clear whether these different results arise from differences in testing procedures or from the existence of multiple visual mechanisms.

Monocularly deprived cats had difficulty with form discrimination tasks, such as distinguishing between an upright and inverted triangle, with the deprived eye. There was some loss of transfer of learning of simple discriminations from the deprived eye to the normal eye (Ganz *et al.* 1972).

Levi *et al.* (1999) found positional jitter did not degrade the ability of normal subjects or amblyopes to recognize letters constructed from Gabor patches. However, reduction in the number of patches defining the letters affected amblyopes more than it affected normal subjects. This suggests that the critical factor limiting pattern identification in the strabismic eye is not positional uncertainty but rather the increased number of features of a pattern required for identification. They concluded that the stimulus is undersampled at the stage where features are integrated. In a second experiment from the same laboratory Sharma *et al.* (2000) found that, compared with normal subjects, amblyopes severely underestimated the number of distinct Gabor elements in a larger array exposed for 200 ms. The visibility of the stimuli was equated for the two groups of subjects. Amblyopes also underestimated the number of gaps in an array. If the defect were due to simple undersampling of the stimulus array, they would have overestimated the number of gaps because they would perceive gaps where none existed. Sharma *et al.* concluded that amblyopes have a high-level limitation in the number of stimulus features that can be identified in a limited time.

Concentric or parallel patches appear non-parallel when they contain lines tilted in opposite directions, as in the Fraser spiral illusion. Popple and Levi (2000) found that the magnitude of the illusion in amblyopic eyes differed from that in normal eyes. However, normal eyes and amblyopic eyes showed the same magnitude of illusion when the rows of patches were shortened. They concluded that amblyopes fail to integrate spatial information over large areas.

13.4.4 Temporal resolution and motion detection

13.4.4a *Loss of flicker sensitivity*

There has been some dispute about whether sensitivity to flicker is reduced in amblyopia. There is evidence that amblyopia in the cat affects the temporal sensitivity of the low spatial frequency transient Y channel more than that of the high spatial frequency sustained X channel (Sherman and Spear 1982; Swindale and Mitchell 1994). Some human amblyopes show normal flicker sensitivity, some show reduced sensitivity, and others show enhanced sensitivity in the affected eye (Manny and Levi 1982). Bradley and Freeman (1985b) concluded that flicker sensitivity can be severely deficient in the amblyopic eye at one spatial frequency but can appear normal if the stimulus includes those spatial frequencies that are detected normally by the subject. They found contrast sensitivity in amblyopia to be highly dependent on spatial frequency, with high spatial frequencies being most affected, but to be largely independent of temporal frequency. They concluded that losses in temporal sensitivity are a consequence of losses in spatial contrast sensitivity.

Human strabismic amblyopes show longer than normal reaction times to the onset of a foveal spot of light in the affected eye (Mackensen 1958; von Noorden 1961; Hamasaki and Flynn 1981). Human brain potentials (VEP's) had longer latency and smaller amplitude when a reversing checkerboard pattern was presented to the amblyopic eye than when it was presented to the normal eye. However, the latency and amplitude of the VEP to motion onset of a checkerboard were the same for the amblyopic eye and normal eye (Kubová *et al.* 1996). This suggests that the motion pathway, probably involving mainly the magnocellular system, is relatively spared in amblyopia.

Sokol and Nadler (1979) reported a reduction in the amplitude of the electroretinogram (ERG) elicited in an amblyopic eye by a patterned stimulus, although not by an unpatterned flashing light. Nonlinear components of the VEP in response to dichoptic flicker are less evident in the stereoblind (see Section 6.7.2).

13.4.4b *Loss of motion sensitivity*

There has been some debate about whether amblyopia selectively affects motion detection (see Hess and Anderson 1993). Displacement thresholds are elevated in human amblyopes, most markedly in the fovea. Displacement sensitivity in the fovea of strabismic amblyopes, like that in the normal fovea, is improved by the addition of a stationary reference stimulus. However, motion sensitivity in the fovea of anisometropic amblyopes is degraded by the addition of a reference stimulus, as is that in the normal periphery. Thus, in this respect, the motion sensitivity of strabismic amblyopes, but not of anisometropic amblyopes, resembles that of the normal peripheral visual field (Levi *et al.* 1984). The threshold for detection of motion of a vertical grating was elevated in the deviating eye of patients with early-onset esotropia, even in the absence of defective nystagmus (Shallo-Hoffmann *et al.* 1997). Strabismic amblyopes had a weaker than normal motion aftereffect (Hess *et al.* 1997c).

Patients lacking static stereopsis due to infantile or late onset esotropia may perceive motion in depth created by opposite motion of dichoptic stimuli (Maeda *et al.* 1999). They presumably use difference of motion signals rather than change of disparity signals (Section 29.2.2).

13.4.4c *Asymmetry of motion detection*

In normal subjects, sensitivity to motion of small displays presented to one eye is higher for centripetal than for centrifugal motion (Mateeff *et al.* 1991). The gain of optokinetic nystagmus is also higher for stimuli moving centripetally. Thus, the gain of horizontal OKN for a normal eye in response to a textured display filling one half of the visual field is higher when the stimulus moves towards the fovea than when it moves away from the fovea (Ohmi *et al.* 1986). In patients with early-onset esotropia, sensitivity to centripetal motion was normal but a centrifugal bias was evident in the nasal hemifield (Fawcett *et al.* 1998).

We will see in the next section that, in early-onset esotropes, monocular optokinetic nystagmus evoked by a display filling a large part of the visual field has a higher gain when the display moves nasally rather than temporally. This asymmetry could be due to an asymmetry in motion detection at the cortical level or to a defect in the transmission of motion signals to subcortical centres controlling OKN (see Section 23.4.5).

Tychsen and Lisberger (1986) reported that, for two esotropes of early onset, monocularly viewed small line stimuli appeared to move more rapidly when moving nasally than when moving temporally. However, Brosnahan *et al.* (1998) found that, in early onset esotropes, a grating appeared to move more slowly when moving nasally, with fixation on

Figure 13.17. Murray S. Sherman.
Born in Pittsburgh in 1944. He obtained a B.Sc. in Biology, from the California Institute of Technology in 1965 and a Ph.D. in Anatomy from the University of Pennsylvania in 1969. In 1972 he obtained an academic appointment in the Department of Physiology at the University of Virginia. In 1979 he was appointed Professor of Neurobiology, SUNY at Stony Brook, where he is now Leading Professor of Neurobiology. He is the Dr. Lee Visiting Research Fellow of Christ Church College, Oxford University.

a stationary point. At least part of the asymmetry in perceived velocity could be due to the inability of subjects to inhibit pursuit eye movements when the grating moved nasally. Roberts and Westall (1990) found no directional asymmetry in perceived velocity in esotropes. Schor and Levi (1980) found that, in strabismic and anisometropic amblyopes, the contrast sensitivity function for detection of motion of a grating moving nasally to be the same as that for a grating moving temporally. Shallo-Hoffmann *et al.* (1997) reported that, in the deviating eye of early-onset esotropes, the motion detection threshold was elevated more for a grating moving nasally, and in the non-deviating eye it was elevated more for a grating moving temporally. The relation between asymmetry of perceived velocity and OKN asymmetry is clearly not yet understood.

Nasotemporal asymmetry in response to moving gratings can be detected in evoked potentials recorded from the visual cortex. No asymmetry is evident in human neonates, perhaps because early motion detectors are not directionally selective. A strong asymmetry, evident at 2 months of age, changes to symmetry by 6 to 8 months. The VEP of infantile esotropes is normal just after onset of strabismus but soon reverts to a marked asymmetry (Norcia *et al.* 1991; Norcia 1996; Birch *et al.* 2000b). Asymmetric evoked potentials associated with asymmetrical OKN in patients with infantile esotropia are correlated with loss of stereoacuity and bifoveal fusion (Fawcett and Birch 2000). The asymmetry of the VEP is reduced if surgery is done before 2 years of age (Norcia *et al.* 1995).

13.4.5 Motor symptoms of amblyopia

Amblyopic eyes produce a smaller pupillary response than normal eyes (Brenner *et al.* 1969). Also, amblyopic eyes show abnormal accommodative responses, including high steady-state error associated with abnormally large depth of focus resulting from reduced contrast sensitivity, reduced range of accommodation, and defective vergence accommodation (see Ciuffreda and Hokoda 1983).

Stimuli presented to an amblyopic eye may evoke abnormal eye movements. People with strabismic and anisometropic amblyopia show eccentric fixation and instability of gaze when attempting to fixate an object with the affected eye. Gaze is stable for fixation with the normal eye or both eyes (Schor and Hallmark 1978; Ciuffreda *et al.* 1980). Amblyopic eyes show saccadic hypometria (undershooting) and saccadic disconjugacy, sometimes for vergence eye movements in only one direction and sometimes for those in both directions (Maxwell *et al.* 1995).

Amblyopes show disturbances of optokinetic nystagmus (OKN) especially for monocular OKN evoked by stimuli moving in the temporal direction (Schor and Levi 1980; Schor 1983b; Sparks *et al.* 1986; Hartmann *et al.* 1993). With monocular viewing, voluntary pursuit eye movements of esotropes are also defective in the temporal direction (Schor 1975; Tychsen *et al.* 1985; Bedell *et al.* 1990; Kiorpes *et al.* 1996). These eye-movement defects, especially directional preponderance of OKN, are most evident in people with early, rather than late, onset esotropia (Demer and von Noorden 1988; Westall *et al.* 1998). Also, with early onset strabismus, directional preponderance of OKN is less likely to be confined to the deviating eye (Steeves *et al.* 1999).

Not all stereoblind amblyopes showed a strong directional preponderance of OKN but all amblyopes lacking interocular transfer of threshold elevation showed directional preponderance (Westall *et al.* 1989). Defects of eye movements may persist even when corrective surgery is conducted as early as 20 weeks after birth and after some stereopsis has been restored (Aiello *et al.* 1994; Wright 1996). The relationship between defective stereopsis and pursuit eye movements is discussed in more detail in Section 23.4.5 and by Kommerell (1996).

Disparity-induced vergence is absent in amblyopia, even though vergence in response to changes in accommodation is normal (Kenyon *et al.* 1981).

Strabismic amblyopes show systematic errors in pointing to visual targets when using their amblyopic eye (Fronius and Sireteanu 1994). Cats reared with strabismus show a displacement in jumping to a platform when tested with the deviating eye. Kittens younger than 4 months overcome this deficit with practice (Olson 1980). An extended period of monocular deprivation in the kitten also produces deficits in visually guided paw placement, which are more severe the longer the deprivation (Dews and Wiesel 1970).

13.4.6 Development of amblyopia

13.4.6a *Amblyopia and neural competition*

Loss of binocularity in monocularly deprived animals results from competition between inputs from the two eyes for access to cortical cells. Several lines of evidence suggest that amblyopia can be understood in the same way.

1. Visual performance is more severely degraded by monocular deprivation than by binocular deprivation. In the cat, both forms of early deprivation have more severe effects on temporal and spatial resolution than does ablation of area 17 in a normally reared adult cat (Lehmkuhle *et al.* 1982). At least part of the adverse effect of deprivation in the cat must therefore involve the extrastriate area and possibly other higher visual areas. The extrastriate area of the cat receives direct visual inputs (LeVay and Gilbert 1976). Area 17 is involved in the visual performance of dark-reared cats, since contrast sensitivity is degraded by ablation of this area in dark-reared cats (Lehmkuhle *et al.* 1984).

Cats raised with one eye sutured failed to respond to objects presented in the binocular field of the deprived eye. However, after removal of the whole visual cortex, the cats responded to stimuli in any part of the visual field of either eye (Sherman 1974) (Portrait Figure 13.17). The subcortical centres mediating the responses of decorticate cats are apparently immune to the effects of monocular deprivation.

2. In monocularly deprived animals a high proportion of cells in the monocular region of the visual cortex have normal receptive-fields (Wilson and Sherman 1977). Also, contrast sensitivity in human anisometropic amblyopes is normal outside the binocular field (Hess and Pointer 1985). Inputs to cells serving the monocular fields do not have to compete with inputs from the other eye.

3. In Siamese cats almost all cortical cells are driven only by the contralateral eye, and monocular deprivation has little if any effect on the receptive field properties of cortical cells in these animals (Berman and Payne 1982; Berman *et al.* 1989).

4. In monkeys with esotropia induced surgically on the 6th postnatal day, acuity in both eyes remained normal for four weeks, after which the acuity of the deviated eye began to deteriorate relative to that of the nondeviated eye (Kiorpes and Boothe 1980).

5. Smith E. L. *et al.* (1992) reared monkeys for between 30 and 90 days with base-in prisms. This caused a severe loss of binocular cells but no shift in ocular dominance, since cells responsive to either left or right eye were retained. This procedure did not induce amblyopia. Also, amblyopia in the same monkeys was not produced in a subsequent period, in which one eye was sutured, even though amblyopia was produced by monocular suturing applied at the same time in monkeys reared with normal vision. Thus, amblyopia is a result of a shift in ocular dominance, which excludes one eye from access to cortical cells. An eye that retains access to a substantial number of cortical cells is not amblyopic even though all binocular cells are absent.

6. Accelerated restoration of acuity in a previously occluded eye occurs after enucleation of the good eye in the cat (Hoffmann and Lippert 1982; Smith and Holdefer 1985). Monkeys raised for 4 years with induced strabismus showed some recovery of visual function in the deviated eye after the normal eye had been removed. In the deviated eye of one monkey, grating acuity improved from 0.28 to 6.3 cpd and sensitivity to flicker increased by 25 Hz over an 11-month period after removal of the non-deviating eye (Harwerth *et al.* 1986a). The same type of recovery has been noticed in amblyopic humans after loss of the nonamblyopic eye (Vereecken and Brabant 1984).

7. Visually evoked cortical potentials (VEPs) recorded in human strabismic amblyopes are reduced in amplitude and show longer latency with stimulation of the amblyopic eye than with stimulation of the normal eye. Patients with alternating strabismus showed normal VEPs for whichever eye was used for fixation at a particular time, and a reduced response for whichever eye was not used (see Franceschetti and Burian 1971). There is also reduced interocular summation of the VEP in monocularly deprived kittens (Sclar *et al.* 1986) and in human amblyopes (Srebro 1978; Sokol and Nadler 1979).

Patients with small-angle amblyopia and anomalous correspondence showed some evidence of interocular summation of the VEP but those with large-angle strabismus and binocular suppression showed no summation of the VEP (Campos 1980; Campos and Chiesi 1983). These studies did not control for effects of changing accommodation. However Barris *et al.* (1981) obtained reduced amplitude, increased latency, and reduced interocular summation of the VEP of amblyopes in response to a sinusoidally drifting laser speckle produced as an interference pattern. Such a stimulus bypasses the optics of the eye.

It can be concluded that, as a binocular cell develops, the receptive fields in the two eyes become matched in their orientation and motion selectivity, and that this matching process depends on visual experience. To allow this to happen, binocular cells must retain a higher degree of plasticity than monocular cells. With monocular deprivation, the matching process is disrupted and inputs from the deprived eye fail to guide the tuning properties of cortical cells.

13.4.6b *Recovery from amblyopia*

A traditional treatment for amblyopia is to patch the good eye. This is known as **reverse patching**. The idea is that the weak eye has more chance to recover when the good eye does not suppress it. Evidence from animal studies cited earlier shows that a previously deprived eye recovers to a greater extent when the good eye is covered. However, evidence from reverse patching in cats suggests that the deprived eye recovers only at the expense of the good eye (Murphy and Mitchell 1987).

A similar reciprocal effect has been observed in children with one eye patched and there is some risk of children developing a double amblyopia if the good eye is occluded continuously for too long (Odom *et al.* 1981). Furthermore, recovery of an amblyopic eye when the good eye is patched is unstable. For example, when both eyes of a cat were finally opened after a period of reverse occlusion, the relative performance of the eyes tended to revert to the prepatching state (Mitchell 1988b).

In a recent study, 90% of a group of 64 human strabismic and anisometropic amblyopes under 7 years of age showed some improvement of acuity after various regimens of patching the good eye over a period of about 4 months. About 70% of them showed a doubling of acuity in the amblyopic eye. Children older than 7 years showed less improvement after therapy. However, 67% of all those that improved showed some subsequent loss of acuity in the year following cessation of therapy (Rutstein and Fuhr 1992).

Birnbaum *et al.* (1977) reviewed 23 studies on improvement of visual acuity by amblyopia therapy. Although the methods were not described and there were wide variations in success rate, there was no evidence that therapy was more successful in those under 7 years of age. Adult amblyopes show some improvement of vernier acuity with practice. However, as in normal vision (Fahle 1997), the improvement is specific to the task and to the orientation of the stimulus (Levi *et al.* 1997).

It has been suggested that the best regimen for optimal performance of both eyes is one in which the good eye is patched for about half the time, with binocular vision allowed for the other half (Mitchell *et al.* 1986). However, a retrospective study of 317 patients over a period of 15 years revealed no difference between the effects of part-time and full-time occlusion therapy applied before or after one year of age (Ross-Dimmest and Morris 1990; Simmers *et al.* 1999). In another regimen, known as penalization, the image in the good eye is optically blurred so as to reduce suppression of the image in the amblyopic eye (see Fahle 1983).

Surgical correction for strabismus produces alignment of the two eyes in most cases but does not cure amblyopia. In fact there is a risk that surgery will induce amblyopia (Pratt-Johnson and Tillson 1983). In one study, four out of 20 patients were amblyopic before strabismic surgery but 16 of the 20 were amblyopic after surgery. Post-surgical exercises over a mean period of 4 years reduced the severity of amblyopia in most of these patients. By comparison, in a group of 20 adults with early onset esotropia that had not been surgically corrected, only three were amblyopic (Good *et al.* 1993). It seems that many people with early onset large angle esotropia develop normal vision in each eye (Murray and Calcutt 1990).

Strabismic amblyopia has been reviewed by Mitchell (1988c) and the effects of different regimens of occlusion on recovery from early monocular deprivation in kittens and their relevance to humans has been reviewed by Mitchell (1991). Therapies for human amblyopia have been reviewed by Garzia (1987).

Summary

The spatial distribution of defects in contrast sensitivity and acuity over the visual field of an amblyopic eye depends on the spatial frequency of the stimulus, the severity of the visual deprivation, and whether the amblyopia is due to strabismus or to anisometropia (Hess and Pointer 1985). The nasal hemifield (uncrossed cortical inputs) is more susceptible to the effects of deprivation than the temporal hemifield (crossed cortical inputs) in both cats and humans. This could be a consequence of the fact that nasal hemiretinas develop more rapidly than temporal hemiretinas. In anisometropic amblyopia, deficits on hyperacuity tasks are proportional to those in resolution and contrast sensitivity. In strabismic amblyopia, hyperacuities are more severely affected than resolution acuity or contrast sensitivity. The central visual field of strabismic amblyopes is affected more than the peripheral field because the periphery has larger receptive fields. This might explain why hyperacuity, which is a function of the central retina, is affected more than resolution acuity in strabismic amblyopes. In anisometropes with aniseikonia, the peripheral field is affected more than the central field. Spatial distortions that occur in some strabismic amblyopes are uncorrelated with loss of contrast sensitivity or hyperacuity and probably reflect both undersampling and scrambling of topographic representation in the visual cortex. Flicker detection can be defective in some amblyopes but the defect may not show at all spatial frequencies. Other amblyopic symptoms include instability of gaze and visual pursuit and errors in pointing.

Amblyopia is not due to defects in the retina or in the LGN. Several lines of evidence suggest that it results from inputs from the eye with more normal visual experience gaining greater access to cortical cells than those from the deprived eye. There are conflicting claims about procedures designed to treat amblyopia.

13.5 AMBLYOPIA AND STEREOPSIS

13.5.1 Amblyopia and stereoacuity

Humans with early strabismus, anisometropia, or uniocular cataract suffer partial or complete loss of binocularity (Levi *et al.* 1979a; Hess *et al.* 1981). People with severe strabismus fail tests of stereopsis whether or not their strabismus is accompanied by amblyopia (Cooper and Feldman 1978b). Thus, the crucial factor in loss of stereopsis is the strabismus rather than the amblyopia. However, amblyopes with stereoscopic vision show raised contrast thresholds for the detection of depth in random-dot stereograms (Wood *et al.* 1978). There is usually only a partial loss of binocular cells with anisometropic amblyopia and with strabismus of less than 3°. In these conditions, loss of stereopsis and binocular summation of threshold stimuli is confined to high spatial-frequency stimuli and is thus most evident in the foveal region. Thus, for strabismic amblyopes, stereopsis and binocular summation can be normal for low spatial-frequency stimuli and the loss is therefore not evident in the visual periphery (Holopigian *et al.* 1986). The same is true of people with alternating strabismus (Sireteanu 1982). People with severe anisometropic amblyopia suffer loss of stereoacuity. However, a patient with astigmatic anisometropia that reduced acuity only for horizontals in the affected eye had normal stereopsis (Peli 1983). Stereopsis depends on good acuity for verticals and this was not affected in this patient.

It is shown in Section 29.4.4 that stereopsis is not a unitary ability. A person can be blind for motion in depth while possessing normal stereoscopic acuity for static objects, and selective loss of one type of stereopsis can be confined to one area of the visual field. Static and motion-in-depth stereopsis are also differentially affected by early strabismus (Schor *et al.* 1983). Kitaoji and Toyama (1987) found that many subjects in whom the angle of strabismus was between 2 and 5° showed a loss of both static and motion-in-depth stereopsis in the central field but not in the peripheral field. Subjects with strabismus angles between 6 and 10° had lost static stereoscopic vision in the whole field although many of them had motion-in-depth stereopsis in the peripheral field.

McColl *et al.* (2000) found that some stereodeficient subjects could discriminate depth produced by disparity between the envelopes of Gabor patches (nonlinear disparities) even though they could not detect that produced by the linear disparity of the carrier grating within the envelope. These subjects also failed standard tests of stereopsis. The experimenters concluded that some stereodeficient subjects can process nonlinear disparities. However, the nonlinear disparities are necessarily on a coarser spatial scale than the linear disparities contained in the carrier grating. This may be why stereodeficient subjects can detect them.

Thompson and Nawrot (1999) found 10 strabismic amblyopes to be defective in both stereoacuity and in depth discrimination based on monocular motion parallax, even when the motion parallax stimulus was presented to the eye with normal acuity and contrast sensitivity.

13.5.2 Amblyopia and binocular suppression

Strabismics do not fuse corresponding images, and they therefore experience diplopia. They also suffer from a symptom called **confusion** in which all binocular images fall on non-corresponding regions in the two eyes and undergo rivalry. Many people with strabismus of early onset overcome these symptoms by suppressing vision in the deviating eye, although they can see with that eye when it alone is open. This is known as **strabismic suppression.** When both eyes are open, amblyopes have better access to information presented to their normal eye than to that presented to the amblyopic eye. Also, stimuli presented to the normal eye are not affected by competing stimuli presented to the amblyopic eye.

For subjects with normal vision, vernier acuity for a target presented to one eye is degraded by the presence of a similar target with a fixed horizontal offset presented to the other eye. For amblyopes, vernier acuity was not affected when the target was presented to the good eye and the competing target was presented to the amblyopic eye (McKee and Harrad 1993). Electrophysiological evidence of strabismic suppression has been reported in V1 and MT of the monkey (Thiele *et al.* 1997).

It is commonly assumed that strabismic suppression is an extreme form of suppression observed in normal observers during binocular rivalry (Dale 1982). However, the following evidence suggests otherwise.

1. Suppression is stronger in normal rivalry suppression than in strabismic suppression (Holopigian *et al.* 1988).

2. In normal subjects, similar images show binocular facilitation and suppression occurs only between very dissimilar images, whereas strabismic suppression occurs between both similar and dissimilar images (Schor 1977). For instance, when normal observers view dichoptic vertical gratings rotated out of alignment by a few degrees they see a fused image of a slanting surface. Gratings rival only when misaligned by many degrees. Strabismics suppress the image in the deviating eye even when the gratings have a very similar orientation. Thus, people with normal vision have a suppression mechanism for strongly dissimilar images and a fusion mechanism for similar images, whereas strabismics have only a rivalry mechanism for both similar and dissimilar images. Cells in the visual cortices of strabismic cats and monkeys did not exhibit binocular facilitation to similarly orientated drifting gratings, but stimulation of the deviated eye suppressed the responses of cells to stimuli in the non-deviated eye, whatever the relative orientation of the stimuli (see Sengpiel and Blakemore 1996).

In animals with normal vision, binocular facilitation develops because stimuli with matching features tend to fall on or near corresponding points. This produces synchronized neural activity, which strengthens short-range facilitatory neural connections through the Hebbian-synapse mechanism. The lack of synchrony between non-matching stimuli leads to the formation of long-range inhibitory lateral connections. In the strabismic animal, stimuli with matching features do not fall on corresponding points and therefore do not produce synchronous activity. Thus, the short-range binocular facilitation mechanism fails to develop in strabismics, leaving only the long-range inhibitory mechanism.

3. Suppression during normal binocular rivalry causes a greater reduction in sensitivity of the chromatic mechanism than of the achromatic mechanism (Section 7.3.3). By comparison, strabismic suppression involves an equal loss of sensitivity in the two mechanisms (Smith *et al.* 1985a).

14 *Seeing in depth in different species*

14.1 FRONTAL VISION

This chapter is concerned with depth vision in animals other than cats and primates.

Fully **frontal eyes** have one pair of visual axes that intersect. In **lateral eyes**, the principal visual axes are directed in opposite directions and never intersect. Some animals, such as the chameleon, are able to direct their visual axes either laterally where they do not intersect, or frontally where they do intersect. They thus have some frontal vision even though they do not have true frontal eyes. Some animals, such as the pigeon, are bifoveate and have two visual axes in each eye—a principal axis that lies close to the optic axis and projects to a central fovea, and a second visual axis that projects to a fovea lying in the temporal retina (see Figure 14.13) The angle between the two visual axes can be as large as 50°. Such animals are classified as lateral eyed, because their principal visual axes point laterally and never intersect. However, they have frontal vision with respect to their secondary visual axes, because these axes do intersect. Many animals do not have foveas. For instance, the rabbit has a horizontal region of high acuity in each retina, known as the visual streak. In such animals, frontality is indicated by the degree of overlap of the two visual fields.

When threatened, many lateral-eyed animals, such as the pigeon, rabbit, and horse, turn their eyes outward to gain panoramic vision at the expense of losing binocular overlap. When they wish to get a better view of something ahead of them they converge their eyes to increase binocular overlap at the expense of some loss of panoramic vision.

In most submammalian vertebrates, the visual pathways decussate, although there is usually a weak ipsilateral projection. Hemidecussation of the optic nerves occurs in mammals, but not in all mammals. When it does occur, the ratio of uncrossed to crossed fibres is proportional to the size of the binocular visual field, which in turn depends on the extent to which the eyes are frontal. This relationship is known as the **Newton-Müller-Gudden law**. In the rabbit, the proportion of uncrossed fibres is almost zero, in the horse it is about one-eighth, in the dog one-quarter, in the cat one-third, and in primates, including humans, one-half (Walls 1963, p. 321). The ferret is unusual in having large irregular projections from the centre of the ipsilateral eye in V2 with no matching inputs from the contralateral eye to V2 (White *et al.* 1999).

Some ipsilateral projections have been found in the visual pathways of all non-mammalian vertebrates that have been studied, except two minor groups of reptiles including the crocodiles (Ward *et al.* 1995). These ipsilateral projections innervate the hypothalamus, tectum, pretectum, and thalamus and are far fewer and show more interspecific variation than do contralateral projections. In general, the number of ipsilateral projections in non-mammalian species is not related to the degree of overlap of the visual fields. For example, the visual fields of the

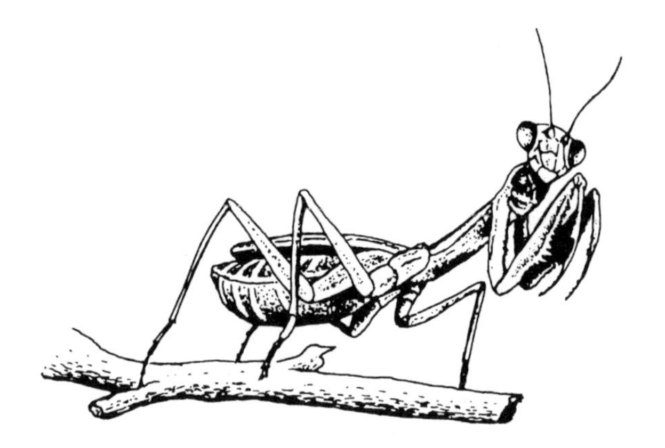

Figure 14.1. The praying mantis (*Mantis religiosa*).

lizard *Podacaris* overlap by about 18° and those of the snapping turtle *Chelydra* overlap about 38° even though the ipsilateral visual projections are more extensive in *Podacaris* than in *Chelydra* (Hergueta *et al.* 1992). Nor is the extent of ipsilateral projections in non-mammalian vertebrates obviously related to whether the animal is a predator or nocturnal, and it seems to be unrelated to taxonomic order (see Ward *et al.* 1995). In many submammalian species, such as the salamander, ipsilateral fibres develop early in life and remain fairly stable. In frogs and toads, ipsilateral fibres emerge at metamorphosis. In birds, ipsilateral fibres exhibit an initial exuberant growth followed by pruning, as in some mammals (Section 10.3). In spite of having sparse ipsilateral projections to the thalamus, we shall see that many submammalian species achieve binocular inputs onto cells of visual centres by intertectal connections or by hemidecussation at a level beyond the thalamus.

Stereoscopic vision requires some binocular overlap but animals with some overlap of their visual fields do not necessarily have stereoscopic vision. Fully frontal eyes are not required for stereopsis although, as far as I know, all animals with fully frontal eyes have stereopsis. We will see that some form of stereopsis based on binocular vision is found in a variety of submammalian phyla, including insects, amphibians, and birds, as well as in lateral-eyed mammals such as the rabbit and goat.

Several reasons have been proposed to account for the evolution of fully frontal eyes, hemidecussation, and fully developed stereopsis, with the consequent loss of panoramic vision. Collins (1922) and Le Gros Clark (1970) suggested that it was an adaptation to arboreal life in which accurate judgments of distance are required to enable animals to leap from branch to branch. However, many arboreal mammals, such as opossums, tree shrews, and squirrels, do not have frontal eyes. The squirrel, for instance, has only about 20% binocular overlap. In small mammals, binocular overlap correlates with length of skull rather than with arboreality (Cartmill 1974). The second possibility is that frontal eyes evolved in predators to enable them to judge the distance of prey and to detect camouflaged prey (Cartmill 1974; Julesz 1971). Binocular stereopsis is very effective at revealing camouflaged objects. It may also enhance the detectability of a target by differentially filtering the images of target and background (Schneider and Moraglia 1994). Camouflage may also be broken by motion parallax but at the cost of the predator having to move. Visually controlled predation is characteristic of many living prosimians and small marsupials, and these animals show primate-like specialization of the visual system. In a review of this question, Hughes (1977) proposed that the degree of binocular overlap is correlated with the evolution of visual-motor control of the forelimbs. In primates, all these factors probably created selective pressure for the evolution of frontal eyes and stereopsis.

14.2 INVERTEBRATES

14.2.1 Insects

14.2.1a *Praying mantis*

In many insects the visual fields of the two compound eyes overlap providing the possibility of binocular stereopsis (Horridge 1978). It has been reported that predatory insects such as dragonfly larvae, tiger beetles, praying mantis, and water scorpions rarely catch prey when one eye has been removed (Maldonado and Rodriguez 1972; Cloarec 1978). This suggests that they use binocular cues to relate their prey-catching activity to the distance of the target. This question has been investigated in the praying mantis (genera *Tenodera* and *Sphodromantis*). These insects have compound eyes between about 4 mm and 8 mm apart with a central forward-looking region of high-density ommatidia and a 70°-wide binocular field (Figure 14.1). When a mantis with one eye occluded is presented with a fly, it centres the prey in its visual field by a saccadic movement of the head. When both eyes are open the head moves to a compromise position so that the images of the prey are equally displaced horizontally from the centre of each eye (Rossel 1986). The mantis cannot centre a fly in both eyes simultaneously because the eyes cannot converge. When a vertical disparity between the images of an object was introduced by prisms, the head also took up an intermediate position in the vertical plane (Rossel *et al.* 1992). The mantis is also capable of pursuing moving prey with

Figure 14. 2. Samuel Rossel.
Born in Schaffhausen, Switzerland in 1948. He obtained a Diploma in Biology from the University of Zürich in 1976 and a Ph.D. with G.A. Horridge from the School of Biological Sciences at the Australian National University at Canberra. His post-doctoral work was with R. Wehner at the University of Zurich. In 1987 he joined the Department of Neurobiology and Animal Physiology at the University of Freiberg, where he is now Professor of Behavioural Physiology.

visually guided smooth movements of the head (Rossel 1980). (Portrait Figure 14.2).

When the prey is within a critical range of distance, the mantis strikes at it with its forelegs, with an accompanying lunging motion of the middle and hind legs. The large African mantis (*Sphodromantis viridis*) has an interocular separation of 8 mm and its optimal striking range extends from about 20 mm to about 60 mm from the head. The movements of the forelegs are adjusted to the distance of the prey and the attack succeeds when the prey is within 15° of the midline of the head (Corrette 1989). When base-out prisms were placed in front of the eyes the distance of strike initiation was modified accordingly. A strike occurred at that distance for which the binocular disparity of the target was the same value as without the prisms (Rossel 1983). This suggests that the mantis uses binocular disparity rather than monocular cues to distance, since prisms do not affect monocular cues. There is other evidence that the mantis does not judge the distance of a prey object on the basis of image size (Rossel 1991). Accommodation is not a cue, since compound eyes do not accommodate.

Mantids use motion parallax to manoeuvre between stationary objects, such as the stalks of plants. Before moving they create a motion signal related to the distance of the object by executing side-to-side head movements (peering). If the target is moved during the head movement the animal is deceived about the distance of the target (Poteser and Kral 1995). Motion parallax seems also to help the animals to distinguish between an object and its background.

Motion parallax cannot be used for coding the distance of a moving prey object. Mantids use binocular disparity for this purpose. The immobility of the compound eyes means that each value of image disparity is uniquely related to a particular absolute distance of the object. A typical prey object in the middle of the striking range subtends about 20°. If only one prey object is within range, the mantis need only register its mean direction from the two eyes, and its distance by extracting the difference between the horizontal positions of the two images. Rossel (1996) presented African mantids with two 4°-wide by 20°-high bar targets separated by 30°. The animals moved the head to centre one of the targets and tracked it as it moved vertically. They were thus able to match the images in the two eyes from the same object and avoid matching noncorresponding 'ghost' images from different objects. They matched the appropriate images using a simple nearest neighbour rule. When one eye could see only one object and the other eye could see only the other object, the animals centred the head between the images. Thus, they matched noncorresponding images when there were no other matches available. When the distance between the targets was reduced to 9° the animals failed to respond selectively to one target. Thus, mantids possess a very coarse mechanism for selectively matching binocular images. A coarse system is all that is required because prey objects usually subtend 20°. The visual acuity of the eyes is much finer than this limit.

Mantids still responded accurately to the distance of a target when the images were prismatically separated vertically by up to 15° (Rossel *et al.* 1992). This suggests that horizontal disparity is registered more or less independently of vertical separation.

Mantids use disparity only as a range-finding mechanism rather than a relative distance mechanism. If one eye is occluded during development, the animals perform accurately when the occluder is removed (Mathis *et al.* 1992). Thus, the growth of the mantid stereoscopic mechanism does not require binocular visual experience. Little is known about the anatomy or physiology of the mantis brain.

14.2.1b *Ants, grasshoppers, locusts, and dragonflies*

The isolated head of a bulldog ant (*Myrmecia gulosa*) snaps its mandibles to an approaching object (Via 1977). It seems that the response is triggered when an image of appropriate size falls on a small number of ommatidia. The animal cannot distinguish between a large far object and a near small object. Only a few animals responded when one eye was occluded, and their responses were more variable. The visual fields overlap by about 60°.

Grasshoppers (*Phaulacridium vittatum*), when jumping horizontally from a platform to a surface, adjust the jump to the distance of the surface. They perform almost as well with one eye occluded and most likely use motion parallax (Eriksson 1980).

Before jumping across a gap, locusts move the anterior part of the body from side to side over several cycles—an activity known as 'peering'. Given that the animal registers either the speed or the amplitude of head motion, the motion of the image of an object indicates the distance of the object (Wallace 1959; Collett 1978). When the target was oscillated in phase with the body, the locust overestimated the distance of the target. When the target was oscillated out of phase with the body, the locust underestimated distance (Sobel 1990). Thus, the locust adjusts the length of its jump as a function of the magnitude of image motion resulting from body motion, suggesting the use of absolute parallax (Section 25.2). However, the adjustment is the same for image motion against the head as for image motion of the same relative speed in the same direction as the head, showing that the sign of the motion is ignored. The sign can be ignored because, with no eye rotation, images always move against the head under natural circumstances. Taking the sign of the motion into account would complicate the control process unnecessarily.

Even without registering the speed or amplitude of head motion, relative image motion produced by objects at different distances indicates their relative distance. The relative motion also helps to distinguish between objects and to define the boundaries of objects. Collett and Paterson (1991) found that locusts are drawn to edges defined by relative motion within a textured display and tend to jump to the side of a motion-defined edge where image velocity is higher, which normally indicates that it is the nearer side.

Locusts possess neurones known as lobular giant motion detectors (LGMD), one on the left and one on the right of the visual system. Both neurones feed into a single descending contralateral motion detector (DCMD) whose fast-conducting axon projects to thoracic centres controlling jumping and flight. These neurones respond particularly well to looming arising from an object on a collision course. They also encode the velocity of an object's approach (Rind 1996; Gabbiani *et al.* 1999; Gray *et al.* 2001).

Predatory dragonflies (Libellulidae) sit and wait until an insect flies by then launch and pursue the prey. By analysing video film frame-by-frame Oldberg *et al.* (2000) found that dragonflies fly directly to the prey by steering so as to reduce the movement of the prey's image on the eye. They concluded that distance to the prey is coded by the angular velocity of the prey shortly before launch. Success did not require parallax induced by head movements.

14.2.1c *Bees*

Bees use motion parallax generated by flight to judge the distances of objects (Lehrer *et al.* 1988; Egelhaaf and Borst 1992). They can maintain a flight path down the central axis of a narrow tunnel. They do this by balancing the speeds of the motion signals arising from textures on the two sides of the tunnel, independently of the contrasts or relative spatial periodicities of the patterns (Srinivasan and Zhang 1993). Bees regulate their speed of flight by holding the average angular velocity of the images constant. Thus, they slow down when flying in a narrow tunnel and when approaching a surface upon which to land (Srinivasan *et al.* 1996). The motion detectors have small receptive fields and are insensitive to the direction of motion. As in the parallax-detecting system of the locust, detection of direction is unnecessary because images always move counter to the body during flight.

When flying in a textured cylindrical tunnel, fruit flies (*Drosophila*), also, regulate their speed of flight by holding the image angular velocity constant, independent of the temporal frequency at which texture elements pass. When the diameter of the tunnel changed suddenly, the flies adjusted their airspeed by changing the angular velocity of the image, which they kept constant (David 1982). They did not detect gradual changes in the distance of the visual display, which suggests that they use motion parallax to detect step changes in distance.

Bees have a second set of motion-detectors with large receptive fields that control optokinetic responses that stabilize yaw, pitch, and roll of the body during flight. The detectors are designed only to bring the rotary visual motion signal to zero, which occurs when the body is stable in flight. They are therefore not required to produce a well-calibrated speed signal (Srinivasan *et al.* 1999).

There is thus a distinction between a motion-detecting system designed to control linear motions of the body, which produces a directionally unsigned speed signal unconfounded by spatiotemporal frequency, and a system designed to null rotations of the body, which produces a good directional signal but confounds speed and spatiotemporal

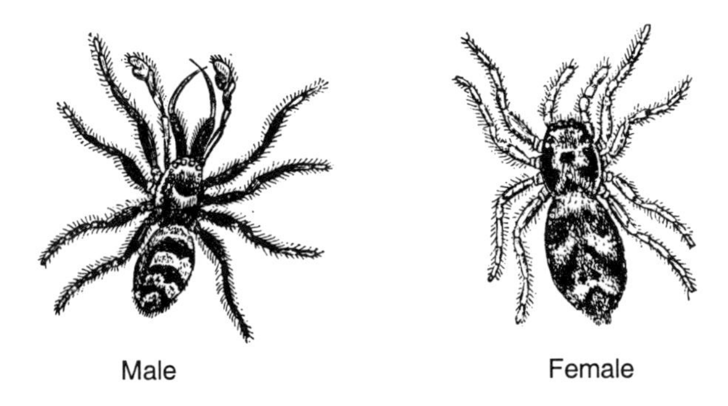

Figure 14.3. The jumping spider (*Salticus scenicus*). About 4 times natural size

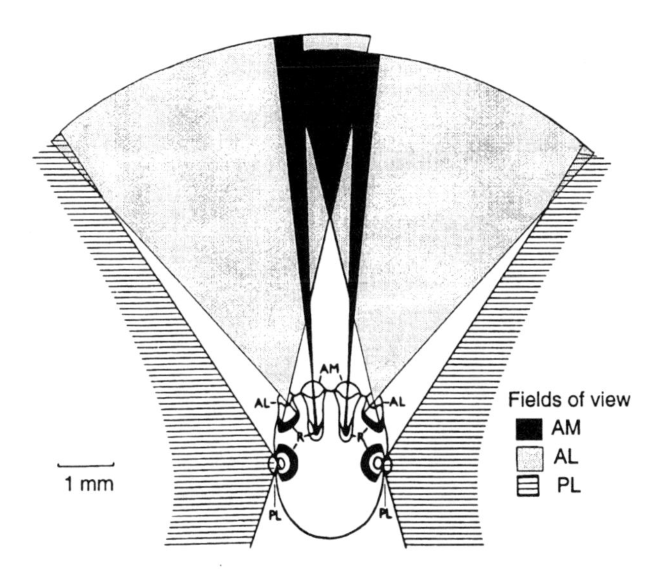

Figure 14.4. Visual fields of a jumping spider (Salticidae). The jumping spider has four pairs of simple eyes; one anterior median pair (AM), a small median pair of unknown function (not shown), an anterior-lateral pair (AL), and a posterior-lateral pair (PL). The retinas of the AM eyes move conjugately, sweeping their visual fields from side to side within the fields of the AL eyes. Fields of the AL eyes overlap in front. (From Forster 1979)

frequency. Models of these motion detectors have been proposed by Srinivasan *et al.* (1999). *It would be worthwhile to investigate the generality of this distinction throughout the animal kingdom.*

Optic flow due to body rotation is in the same direction in the two eyes, while that due to forward body motion is in opposite directions in the two eyes. Krapp *et al.* (2001) found that the response properties of movement-sensitive cells in the optic lobe of the fly are well suited to distinguish between relative binocular motions produced by different body motions (see also Dahmen *et al.* 1997).

Bees use motion signals to estimate the height of objects over which they are flying. Srinivasan *et al.* (1989) trained bees to collect food at one of several artificial flowers that differed in size, position, and height above the ground. They could discriminate a difference in height of 2 cm, whatever the size or position of the flower. With a featureless ground, the bees must have used the motion of the image of the flower. Their improved accuracy when the ground was textured suggests that they could also use motion parallax between flower and ground.

Bees also use motion signals to estimate distance flown. After returning to the hive they indicate the direction of a source of food to other bees by the direction of a waggle dance with respect to vertical. The distance of the food is indicated by the shape and duration of the dance (von Frisch 1993). Bees derive their estimate of distance flown in terms of the total angular optic flow over time produced by the visual surroundings (Srinivasan *et al.* 2000).

Miniature flying robots could perhaps perform like bees (see Srinivasan and Venkatesh 1997).

14.2.1d *Moths*

European hawk moths (*Macroglossum stellatarum*) collect nectar while in hovering flight. They maintain a constant distance from a flower swaying in the wind. Farina and Zhou (1994) concluded that they did not use stereopsis because the flower was too near. When presented with a simulated patch of flowers in front of a structured background, moths preferred flowers that were more distant from the background. This suggests that they use motion parallax between the flower and the background. However, they were able to maintain a constant distance from a 'flower' in the form of concentric rings moving back and forth without a background. Under this condition they were probably using speed of motion of the image of the rings. The gain of the response increased with increasing density of the rings up to a density beyond which there was a loss of resolution.

Wicklein and Strausfeld (2000) recorded from single cells in the optic lobes of the hawk moth (*Manduca sexta*). Some cells responded selectively to an approaching or receding blank disc but not to outward or inward motion of a revolving spiral. They were identified as being sensitive to changing size. Other cells responded to both stimuli and were identified as sensitive to optic flow. Similar types of cell have been found in the pigeon (Section 14.6.1).

14.2.2 Jumping spiders

Jumping spiders (family Salticidae) have four pairs of simple eyes; one principal anterior-median pair, a small median pair with no known function, an anterior-lateral pair, and a posterior-lateral pair (Figure

Figure 14.5. Michael Francis Land FRS.
Born in Dartmouth, England in 1942. He obtained a Science tripos at Cambridge in 1963 and a Ph.D. in Zoology and Neurophysiology at University College, London in 1967. He held academic appointments at University College, London, Berkeley, the University of Sussex, and the Australian National University between 1966 and 1984. He has been Professor of Neurobiology at the University of Sussex from 1984. Elected Fellow of the Royal Society of London in 1982. Recipient of the Frink Medal of the Zoological Society of London,

14.3). Their coverage of the visual field is depicted in Figure 14.4. The retinas of the principal median pair of eyes move sideways behind the stationary lenses. This enables the animal to move the small nonoverlapping visual fields of its median eyes across most of the larger visual fields of the anterior-lateral pair of eyes, which partially overlap in the midline. These movements can be of three types. They can be saccadic movements toward an object presented in the larger visual fields of the lateral eyes. They can be slow movements in pursuit of a moving object. They can also be regular side-to-side scanning movements of between 0.5 and 1 Hz across an object, combined with a slower 50° rotation of the retinas about the visual axes. All these movements are conjugate, so the two visual axes remain approximately parallel (Land 1969) (Portrait Figure 14.5). This system presumably provides a mobile region of high-resolution, which helps the animal to distinguish between prey, other jumping spiders, and other objects. The retinas of the anterior-median eyes are unusual in having four layers of receptors. The peak wavelength sensitivity of the receptors in each layer is matched to the wavelength of light that is brought to focus in that layer by chromatic aberration.

When a prey object enters the visual field, the spider first orients its body to centre the prey. Then the spider chases the prey and catches it on the run or, in some species, stalks the prey and jumps on it from a distance of 3 cm or more. During stalking, the spider reduces its speed of approach as it gets near the prey. In this way it does not exceed the prey's threshold for detection of an approaching object (Dill 1975). Just before jumping, it attaches a web filament to the substrate.

Removal of the median eyes of the jumping spider does not affect orienting behaviour, showing that orientation is controlled by the lateral eyes. However, a spider with only one median eye ignores stationary prey and jumps short of the target. Animals with both anterior-lateral eyes removed still orient the body but no longer chase moving prey. It seems that the principal median eyes with scanning retinas are mainly responsible for judging the distance of the prey, especially short distances. These animals do not accommodate and do not use disparity, since the visual fields do not overlap.

Distance could be signalled by the temporal delay between detection of the prey by one median eye and its detection by the other eye as the eyes scan from side to side. For a fixed velocity of scanning with parallel visual axes, this delay is approximately inversely proportional to the distance of the scanned object. It seems that in *Trite planiceps* the fixed anterior-lateral eyes contribute to judgments of distances greater than about 3 cm, since their removal affects jumping accuracy at these distances (Forster 1979). Since the visual fields of these eyes overlap it is possible that binocular disparity is used. The eyes do not move so that any disparities could be coded to signal absolute distance.

14.2.3 Scanning eyes and motion parallax

The only other known animals with scanning retinas are certain small crustacea (copepods) such as *Copilia* and *Labidocera* (Land 1988). The female *Copilia* has lateral eyes, each containing only three or four photoreceptors. The two sets of receptors oscillate laterally in counterphase, each within the focal plane of a lens (Exner 1891; Gregory *et al.* 1964). *Copilia* is planktonic and lives at a depth of about 200 m, although it may move to the surface at night. It is therefore not exposed to stationary objects and its eyes are presumably designed to detect moving objects. Scanning eye movements occur when the animal is swimming but can be elicited in a tethered animal by moving visual stimuli (Downing 1972).

Certain crustacea, such as the mantis shrimp, and some molluscs, such as the carnivorous sea snail

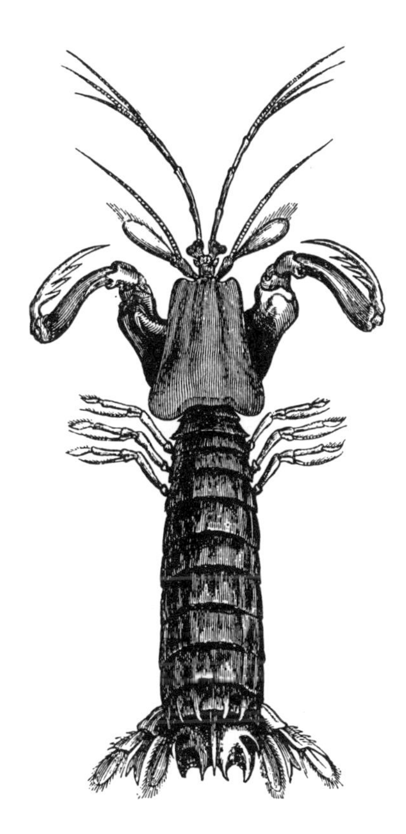

Figure 14. 6. The mantis shrimp (*Squilla mantis*).
The shrimp can be 18 cm long. (From Brehms *Thierleben* 1878)

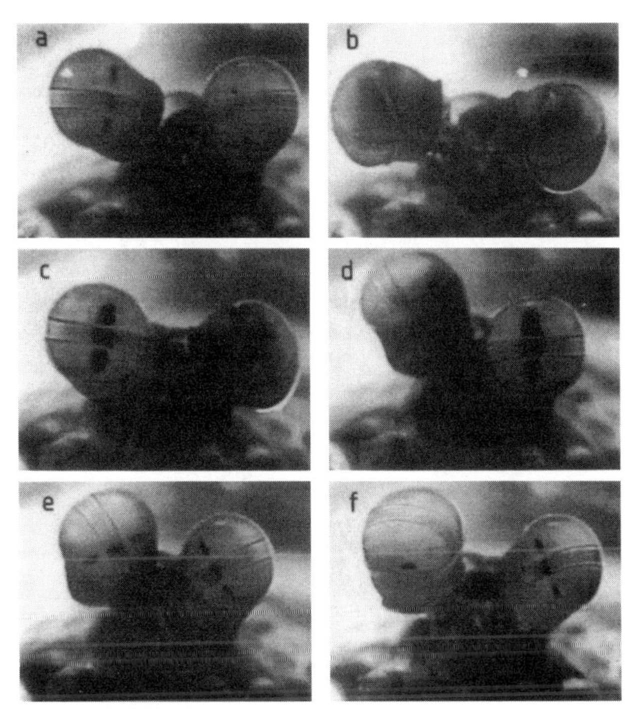

Figure 14.7. Compound eyes of the mantis shrimp.
Bands of ommatidia are flanked by upper and lower hemispheres. Six frames of a videotape show how the eyes move independently through large angles, horizontally, vertically, and torsionally. Dark regions in each eye contain ommatidia directed to the same point in space, which could therefore serve as a basis for monocular stereopsis. (From Land *et al.* 1990)

(*Oxygyrus*), execute continuous scanning movements of the whole eye. Scanning movements enable the animal to economize on the number of receptors, and it is characteristic of animals with scanning retinas or eyes that they have very few light detectors. Scanning eye movements must be slow enough to allow an image to dwell on each receptor long enough to be detected. The larger the angle of acceptance of the receptors (the coarser the resolution), the faster the eye can scan over a stationary scene before the dwell time for each detector falls below an acceptable limit of between 15 and 25 ms (Land *et al.* 1990). When the centre of rotation of an eye is some distance from the optical nodal point, scanning movements entail a translation of the vantage-point and differential motion parallax, which could code absolute and relative distances.

Flies of the family Diopsidae have long eye stalks. The increased interocular separation could serve to improve binocular stereopsis, but the eye stalks also provide a large translational component when the eyes scan and could thus sense motion parallax. Long eye stalks may also serve as a sexual attractant in males (Collett 1987). See Wehner (1981) for a review of spatial vision in arthropods.

14.2.4 Crustacea

14.2.4a *Shrimps and crayfish*

Mantis shrimps, are predatory marine crustaceans (order *Stomatopoda*) inhabiting shallow tropical waters (Figure 14.6). They live in burrows and emerge to stalk small animals, which they strike or spear with two large limbs. Their large mobile compound eyes possess many remarkable features. Each eye has a central horizontal band of six rows of ommatidia flanked by dorsal and ventral hemispherical regions containing several thousand ommatidia. Within each of the three regions, there is a group of ommatidia directed to the same location in space. These "pseudopupils" show as three dark areas in Figure 14.7. Exner (1891, p. 89) suggested that this arrangement provides range-finding stereoscopic vision in each eye. The upper and lower converging regions of each eye are separated vertically, so that stereopsis would be based on vertical disparities between images in the two regions. If this were true, these animals would be unique in two respects. They would have monocular disparity-based stereopsis and stereopsis based on vertical rather than horizontal disparities.

(a) Most crabs have widely separated, short eye stalks.

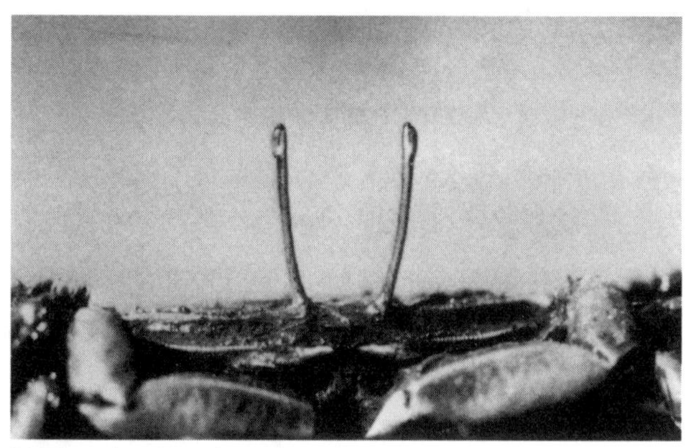

(b) Three families of the genus *Brachyura* have narrowly spaced, long eye stalks.

Figure 14.8.. Eye stalks of crabs.
(From Zeil *et al.* 1986)

The visual field of the central band of ommatidia is a few degrees high and about 180° wide. It is involved in colour vision and analysis of light polarization. The band contains at least 10 types of ommatidia, each with either a distinct filter pigment or a photosensitive pigment with a distinct spectral absorption. The rest of the eye is monochromatic (Marshall *et al.* 1991; Goldsmith and Cronin 1993). A colour system with this many channels could resolve wavelength components in a given location—an ability denied animals with trichromatic vision in which each receptor serves both colour vision and spatial resolution (Section 4.2.7).

The eyes of the mantis shrimp execute slow vertical and torsional movements through at least 60°. These movements are slow enough to allow the image to dwell on each ommatidium for at least 25 ms—long enough for detection (Land *et al.* 1990). The eyes also execute tracking movements in response to movements of a visual target. The two eyes move independently (Figure 14.7). Conjugate movements are not required if stereopsis is achieved in each eye.

The latency of the defence reflex of the crayfish (*Procambarus clarki*) decreases with increasing velocity of an approaching object. The critical variable determining latency seems to be the time required for the object to expand a specific amount. The mean discharge rate of optic nerve interneurones was a linear function of the velocity of an approaching object (Glantz 1974).

14.2.4b *Crabs*

There are two basic designs of visual system in semiterrestrial crabs (Brachyura). Most families, including the Grapsidae and Xanthidae, have widely separated, short eye stalks, as in Figure 14.8a. These crabs live among vegetation and rocks, high up in the intertidal zone. Three families of mostly nonpredatory Brachyura (Ocypodidae, Goneplacidae, and Mictyridae) have narrowly spaced long eye stalks, as shown in Figure 14.8b. These animals live in burrows on open beaches and mud flats in the lower intertidal zone or below the water line. Their long eye stalks are like mobile periscopes and allow the animals to see long distances in all directions while they are in their burrows or partially submerged. Their eyes are vertically elongated and have greater vertical than horizontal acuity. They also have a zone of high acuity round the eye's horizon. Crabs with widely spaced eyes lack these zones of high acuity.

Zeil *et al.* (1986) suggested that eyes with long eye stalks and high vertical resolution evolved in crabs to facilitate depth estimation. If the eyes maintain a constant angle of horizontal gaze, the distance of an object on flat terrain is indicated by the height of the image in the compound eye. The visual axis of a given ommatidium intersects a flat terrain at distance $D = H/\tan\theta$, where H is eye height and θ is the angle between the visual axis of the ommatidium and the horizon plane. Resolution of distances should improve with increasing eye height and with increasing vertical resolution of the eye. Any object that projects an image above the horizon plane of the eye is larger than the crab, whatever its distance. This mechanism requires that the eyes maintain a fixed angle of horizontal gaze with respect to the flat terrain.

The overlapping visual fields of eyes with long stalks could provide the basis for stereoscopic vision (Barnes 1968). For *Ocypode ceratophthalmus,* which has an eye separation of 2 cm and a horizontal resolution of 0.8 arcmin, the stereoscopic range would be about 30 cm. There is no evidence in support of this suggestion.

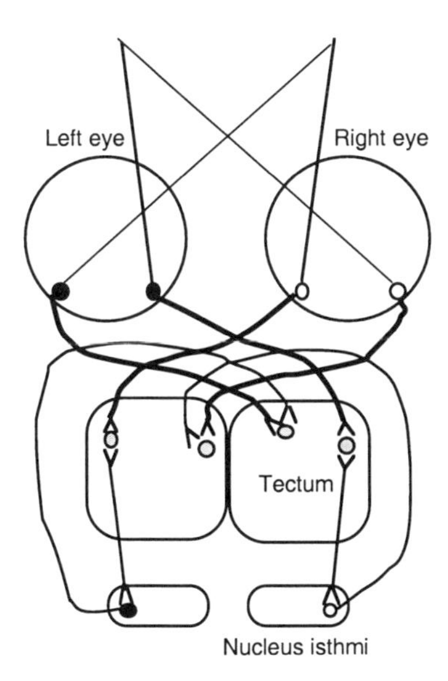

Figure 14.9. The visual pathways of a frog.
Inputs from each eye (bold lines) innervate binocular cells of the contralateral tectum directly. Binocular cells in the tectum receive their inputs from the ipsilateral eye from the contralateral tectum via the nucleus isthmi.

14.3 FISH

In bony fish (teleosts) the majority of optic nerve fibres project to the optic tectum—the homologue of the superior colliculus of mammals (Guthrie 1990). From the tectum, pathways project to the tegmentum, to three centres in the diencephalon (thalamus, preoptic area, and pretectum), and to visual centres in the telencephalon. There are also direct visual inputs to mesencephalic nuclei, the pretectum, and the thalamus (Schellart 1990).

Bony fish have a small binocular field. Most if, not all, visual inputs project contralaterally to the tectum. A few uncrossed inputs occur in the tecti of goldfish (Springer and Gaffney 1981). Ipsilateral inputs may also reach the tectum by way of the tectal commissure. Recordings from neurones in the tectal commissure revealed only large poorly defined monocular receptive fields (Mark and Davidson 1966). Some investigators found no binocular cells in the tectum of fish (Schwassmann 1968; Sutterlin and Prosser 1970; Fernald 1985). However, a few binocular cells with large receptive fields and low sensitivity have been found in the tecti of goldfish (see Guthrie 1990) and trout (Galand and Liege 1974).

The larvae of some deep-sea fish, such as *Idiacanthus fasciola*, have eyes on the ends of long rods, which are absorbed at maturity (Walls 1963). The rods hinge about their base and are probably concerned with scanning rather than with binocular vision. Hammerhead sharks also have eyes very far apart but this too seems unconnected with binocular vision, since their eyes are lateral.

The grating resolution of fish varies between 4 and 20 arcmin, according to species, compared with about 1 arcmin for humans (Douglas and Hawryshyn 1990; Nicol 1989). Fish have the optokinetic nystagmic eye movements (OKN) found in all animals with mobile eyes. Some teleost fish have foveas (Walls 1963) and some, such as the African cichlid fish (*Haplochromis burtoni*), move their eyes independently to foveate particular objects. Some voluntary saccadic movements involve a co-ordinated change in vergence when the visual object is in the binocular visual field (Schwassmann 1968; Fernald 1985). Insofar as vergence movements are evoked by image disparity, binocular cells must be involved. It is not clear what function these movements serve.

Douglas *et al.* (1988) trained goldfish (*Carassius auratus*) to distinguish a 5-cm disc from a 10-cm disc at a fixed viewing distance. The animals could still distinguish discs that were separated in depth so that they subtended the same visual angle. The fish thus exhibited size constancy, which implies that they take distance into account when judging the size of an object. They still showed size constancy when one eye was occluded, so that monocular cues to depth are sufficient. Nobody has produced evidence of binocular stereopsis in fish.

14.4 AMPHIBIA

Binocular vision has been explored in amphibia, both in the anurans (frogs and toads) and in the urodeles (salamanders). The visual system of anurans has been reviewed by Grüsser and Grüsser-Cornehls (1976) and by House (1982).

14.4.1 Frogs and toads

14.4.1a *Visual projections in frogs and toads*
Frogs of the family Ranidae, such as *Rana pipiens*, live both in water and on land and have laterally placed eyes with a visual field of almost 360° and a frontal binocular field up to 100° wide. Fite (1973) claimed that the binocular field widens to about 160° above the head but other investigators have revised this value down to 60° (Grobstein *et al.* 1980). Toads (*Bufo bufo*), also, have panoramic vision with a large binocular field extending above the head.

The optic fibres from each eye of frogs project retinotopically to the contralateral optic tectum (mesencephalon) through the lateral and medial divisions of the optic tract. The deeper layers of each tectum contain binocularly driven cells with direct inputs from the contralateral eye and inputs from the ipsilateral eye routed through the contralateral tectum by way of the nucleus isthmi and postoptic commissures, as shown in Figure 14.9 (Fite 1969; Keating and Gaze 1970a; Raybourn 1975; Gruberg and Lettvin 1980). There are also axons in the nucleus isthmi that terminate in monocular regions of the contralateral tectum (Glasser and Ingle 1978).

It had been claimed that inputs from only the central region of the binocular field of frogs project to binocular cells in both optic tecti (Gaze and Jacobson 1962). However, subsequent investigations have revealed that the whole binocular field of the frog projects to binocular cells in each tectum (Keating and Gaze 1970a). Thus, there is a complete representation of the binocular field in each tectum. In each tectum, ipsilateral inputs from the whole of one eye are in spatial correspondence with contralateral inputs from the whole of the other eye. During development, the contralateral map develops first and does not require visual experience. A coarse ipsilateral map develops at the end of metamorphosis but its refinement requires visual experience (Brickley *et al.* 1998). In contrast, the halves of the binocular field in mammals project to different cortical hemispheres. Inputs from both eyes of the frog can be forced to grow directly into the same tectum (see Section 10.4.7).

Optic fibres from each eye also project to three areas of the thalamus (diencephalon), which also receive inputs from the optic tectum. The dorsal anterior thalamus contains binocular cells that receive direct inputs from both ipsilateral and contralateral corresponding regions of the binocular field (Székely 1971; Keating and Kennard 1976). Visual inputs also go from the dorsal thalamus to the forebrain (telencephalon), an area that also receives somatosensory inputs (Kicliter and Northcutt 1975). Some cells in the telencephalon are binocular but little is known about their function (Liege and Galand 1972).

Some binocular cells in the optic tectum of *Rana pipiens* have small receptive fields and respond best to spots that are dark below and light above (Finch and Collett 1983). Other cells have very large receptive fields and respond to horizontal boundaries, dark below and light above. The cells with small receptive fields are tuned to zero vertical disparity and, on average, to 1.7° of horizontal disparity. The eyes of the frog do not change their vergence and are about 1.5 cm apart. Cells registering 1.7° of disparity were stimulated maximally by an object at a distance of 50 cm. The largest disparity tuning was 3.4°, which corresponds to a distance of 25 cm. This suggests that the binocular cells of the tectum are not responsible for the estimation of distance within the 25-cm snapping zone. Gaillard (1985) conducted a similar study in *Rana esculenta* and agreed that tectal cells with small receptive fields are not tuned to disparities within the snapping zone. However, they found a group of cells with receptive fields about 5° in diameter tuned to disparities of between 0.25° and 12°, corresponding to distances between 5 and 300 cm. These neurones would therefore be able to detect distances in the snapping zone.

Figure 14.10. Thomas Stephen Collett.
Born in London, England in 1939. He obtained a B.A. in Psychology in 1960 and a Ph.D. in Zoology 1964, both from University College London. He joined the faculty of the University of Sussex, England in 1965 where he is now Professor in the School of Biological Sciences.

Frogs of the family Pipidae, such as *Xenopus laevis*, are adapted to a completely aquatic life and feed on prey swimming above them. Because of this, their eyes migrate to the top of the head during metamorphosis. Each eye migrates 55° nasally and 50° dorsally with respect to the major body axes. As a result, the lateral extent of the binocular field increases from 30° to 162°. At the same time, the proportion of each tectum devoted to the binocular visual field increases from 11 to 77% and there is a corresponding increase in the intertectal commissures in the nucleus isthmi responsible for the ipsilateral retinotectal projection (Beazley *et al.* 1972; Grant and Keating 1989a).

During the period of eye migration in *Xenopus* there also is a great increase in dendritic branching of axon arbours in the tectal laminae receiving isthmic projections. These branches are at first widely distributed but are gradually transformed into compact arbors typical of the adult tectum (Udin 1989). Although visual inputs are not required for the development of isthmotectal projections and their increased branching, they are required for the development of compact arbors, which correspond topographically to the contralateral inputs. Thus, in Pipid frogs reared in the dark, the contralateral retinotectal projections develop normally, but the mapping of the intertectal system relative to the contralateral system shows signs of disorder (Grant and Keating 1989b). However, a normal intertectal mapping is restored when sight is restored, even in frogs deprived of vision for 2 years after metamorphosis (Keating *et al.* 1992). Thus, in the development of normal binocularity *Xenopus* has no critical period for visual experience like that found in mammals (Section 13.3). Furthermore, a severed optic nerve of the adult frog regenerates and the axons re-establish their proper topographical distribution on the tectum and their proper distribution of distinct cell types in different layers within the tectum (Keating and Gaze 1970b).

14.4.1b *Adaptations to eye rotation*

When one eye of the Pipid frog *Xenopus* at the tadpole stage was rotated through 90° or 180°, a new set of intertectal connections developed that brought the retinas back into spatial correspondence, but only when the animal was allowed to see (Keating and Feldman 1975; Grant and Keating 1992). Intertectal development was disrupted and adaptation to eye rotation did not occur in *Xenopus* larvae continuously exposed to 1 Hz stroboscopic illumination (Brickley *et al.* 1998). Thus, in *Xenopus*, interactions between signals from the two eyes guide the development of intertectal connections so that tectal neurones receive inputs from corresponding visual directions. The precise mapping of crossed ipsilateral projections requires visual experience (Section 14.4.1a). After eye rotation, a normal coarse mapping of the ipsilateral inputs develops under the control of genetic factors. About 2 weeks after metamorphosis, axons sprout new connections according to activity-dependent cues and old terminations are gradually withdrawn (Guo and Udin 2000).

When the Ranid frog *Rana pipiens* was reared from the tadpole stage with one eye rotated 180°, there was no evidence of remapping of central connections, even after both eyes were opened (Jacobson and Hirsch 1973; Skarf 1973). It appears that Pipid frogs compensate for eye rotation while Ranids do not. Presumably, visually guided plasticity in Pipids arises because they must adapt their visual functions to the migration of their eyes to the top of the head. When one eye of *Xenopus* at the tadpole stage was replaced by an opposite eye from another tadpole, about half the animals developed intertectal connections with proper spatial correspondence, but the other animals developed abnormal intertectal connections (Beazley 1975). Visual experience is required for the development of normal binocular vision in mammals (Section 13.2). However, like Ranid frogs, mammals do not compensate for rotation of one eye through a large angle (Yinon 1975).

14.4.1c *Depth discrimination in frogs and toads*

If prey is seen within a range of 1 or 2 metres the frog moves forward in a series of jumps, making detours around objects if necessary. Once the prey is within about 5 cm it is captured by a flick of the tongue. The distance at which jumping gives way to snapping depends on the size of the animal (Grobstein *et al.* 1985). Frogs usually orient themselves so as to bring the prey close to the body midline but can catch prey without first orienting the eyes or head, even when the prey is 45° or more from the midline. Within a distance of about 15 cm, frogs have been found to select from several prey items on the basis of their linear sizes rather than their angular sizes. This suggests that they have size constancy within this range (Ingle and Cook 1977).

Unlike frogs, toads first orient themselves to centre their prey in the midline and then capture it with only a flick of the tongue (Fite 1973). Like frogs, toads do not converge their eyes, so their ability to snap at prey is not derived from vergence movements (Grüsser and Grüsser-Cornehls 1976; Grobstein *et al.* 1980). As in frogs, inputs from corresponding areas of the two retinas converge in the thalamic and midbrain visual areas.

Frogs and toads have good depth discrimination, as revealed by their ability to avoid objects and jump through apertures, over gaps, down steps, and onto objects (Ingle 1976). Toads make a detour around a fence when the prey is too far away to be caught through the fence. They make for gaps formed by overlapping barriers at different distances only if the gap is sufficiently wide—a task that requires them to detect the depth separation between the barriers (Lock and Collett 1979).

A frog's ability to catch prey is not much affected by severing the optic tract of one eye, so monocular cues to distance are sufficient for this purpose (Ingle 1972). Like toads, frogs probably use accommodation (Collett 1977). Anurans, like fish, accommodate by moving the lens forwards or backwards rather than by changing its shape. Two protractor lentis muscles attached between the lens and cornea move the lens about 150 μm to achieve a 10 dioptre change in refraction. Section of the oculomotor nerves serving these muscles did not affect the accuracy of prey catching when both eyes were open. But, with one eye occluded, frogs and toads consistently underestimated the distance of prey (Douglas *et al.* 1986).

Figure 14.11. European Salamander (*Salamandra salamandra*).

Stereoscopic vision based on disparity has not been found in frogs, but there is evidence of disparity-based stereopsis in toads, as we will now see.

When a negative lens was placed in front of one eye of a toad, with the other eye closed, the animal undershot the prey by an amount predictable from the assumption that it was using accommodation for judging distance. However, when both eyes were open, negative lenses had no effect, which suggests that toads use disparity when both eyes are open. This conclusion is strengthened by the fact that base-out prisms in front of the eyes caused toads to undershoot the target (Collett 1977; Collett *et al.* 1987) (Portrait Figure 14.10). Thus, toads rely on accommodation when one eye is closed and on binocular disparity when both eyes are open. This conclusion was confirmed by Jordan *et al.* (1980) who showed that drug-induced contraction or relaxation of the accommodation muscles had little effect on distance judgments of toads with both eyes open but produced severe undershooting in monocular toads. Since the vergence position of the eyes of toads is fixed, the size and sign of binocular disparity between the images of an object signifies the absolute distance of the object from the animal. This is the information that a toad needs when catching prey.

Collett and Udin (1988) revealed another monocular cue used by the frog *Rana pipiens*. They placed frogs on a transparent sheet of plastic and a prey object on a second surface at various distances below the plastic sheet so that its image was unusually elevated on the frogs' retinas. The animals snapped increasingly short of the prey as the distance of the prey below the plastic sheet was increased. Since the eyes of a frog are normally a fixed distance above the ground, the angle of elevation of the image of a prey object is related to its distance. This is the same mechanism used by some crabs, as described in Section 14.2.4b.

14.4.2 Salamanders

About two thirds of living Salamander species belong to the family Plethodontidae (Figure 14.11). Most Plethodontids have projectile tongues, which constitute the fastest feeding mechanism among vertebrates. The animals wait in ambush until a prey object appears and then project the tongue with great accuracy both in direction and distance. The tongue extends up to two-thirds of their body length within 10 ms. They can catch flies on the wing, implying that they register the temporal as well as the spatial characteristics of the stimulus. Some Plethodontids have frontal eyes with a 90° binocular field.

Salamanders have three optic tracts. From the chiasma the basal tract extends to the peduncle, the marginal tract to the optic tectum, and the medial tract to the anterior thalamus. In most amphibians, each optic tectum receives a direct input from the contralateral eye and an indirect input from the ipsilateral eye through the nucleus isthmi. This relayed ipsilateral input involves a delay of at least 30 ms. Tongue-projecting salamanders are unique among amphibians in having a large direct ipsilateral projection to the tectum in addition to the indirect input relayed through the nucleus isthmi (Roth 1987). Since the direct tectal pathway is more rapid than the indirect pathway, tongue-projecting salamanders are able to process binocular inputs more rapidly than other amphibians. In 14 species of Plethodontids, the degree of ipsilateral innervation of the tectum was found to be related to the degree of frontal vision (Rettig and Roth 1986).

The Bolitoglossini subgroup of Plethodontids have the most frontally placed eyes and the ipsilateral projection along the marginal optic tract innervates the entire tectum, and is equal to the contralateral projection. Thus, each tectal hemisphere receives a complete projection from both eyes within the binocular visual field. The ipsilateral and contralateral projections are in topological correspondence and project onto binocular cells. The eyes have a fixed convergence and the monocular receptive fields are in closest correspondence for stimuli at about the maximum distance over which the tongue projects (Wiggers *et al.* 1995). All this suggests that tongue-projecting salamanders judge distance by a range-finding disparity mechanism, but the details of disparity coding in the tectum remain to be discovered.

Salamanders change their accommodation as they turn to fixate a prey object (Werner and Himstedt 1984). Luthardt-Laimer (1983) found that salamanders (*Salamandra salamandra*) with one eye removed caught prey as well as did binocular animals but those with one eye occluded often missed the prey. She concluded that animals with only one eye rely on accommodation but that, for some unknown reason, this option is not available to monocularly occluded animals.

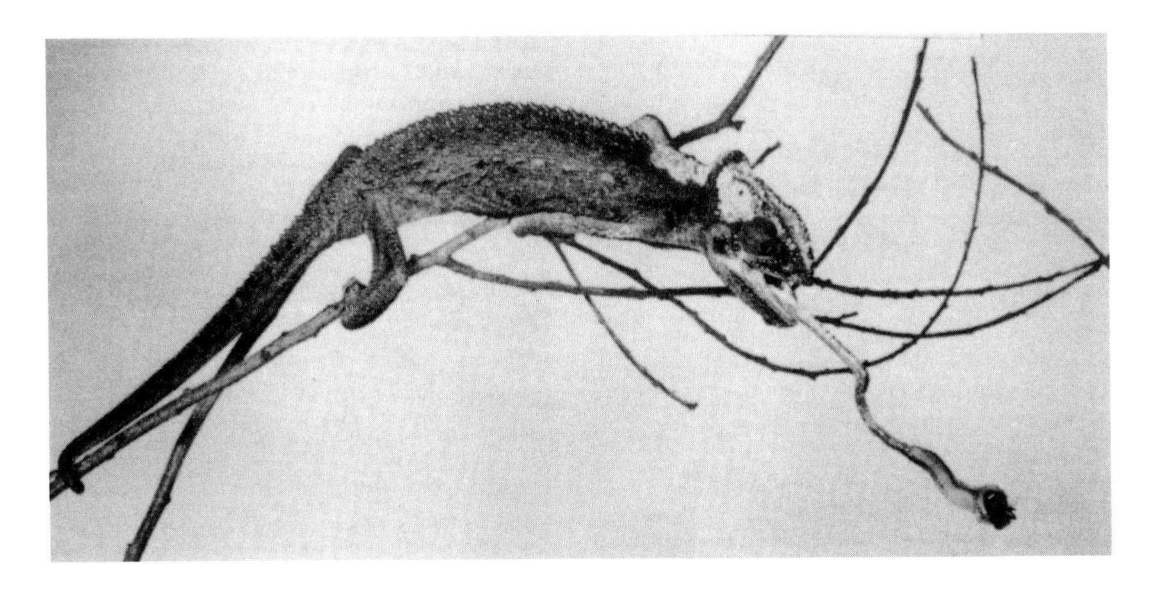

Figure 14.12. The African Chameleon (*Chamaeleo chamaeleon*)
(From the Larousse Encyclopedia of Animal Life, Hamlyn 1967)

14.5 REPTILES

Chameleons are arboreal lizards that live in shrubs (Figure 14.12). A chameleon spends most of its time sitting on a branch waiting for a small insect to come within range of its sticky tongue. Chameleons have an all-cone retina and a well-developed fovea. The laterally placed eyes are mounted in turret-like enclosures and can move 180° horizontally and 80° vertically (Walls 1963; Frens *et al.* 1998). Like the eyes of primates, the eyes of the chameleon obey Listings law (Section 9.2.2d). Optokinetic nystagmus and the vestibulo-ocular response together achieve image stability with a gain of about 0.8 as the body moves (Gioanni *et al.* 1993). When a chameleon is waiting for prey, large spontaneous saccadic movements occur at random intervals independently in the two eyes (Mates 1978). When a prey object has been detected, both eyes fixate it, although the vergence angle is not clearly related to the distance of the object (Ott *et al.* 1998). A moving prey object is pursued partly by movements of the head and partly by synchronous saccadic eye movements (Ott 2001). When the prey is in the midline of the head the chameleon shoots its tongue out at great speed up to 1.5 times the length of its body.

Harkness (1977) filmed the motion of the tongue of a chameleon and found that the distance moved was related to the distance of the prey. Convergent prisms placed before the eyes did not cause the animals to undershoot the target. Also, the precision of convergence is very poor. It seems that chameleons do not use the vergence of the eyes as a rangefinder. However, when a negative lens was placed in front of one eye with the other eye closed the animals undershot the prey. When a positive lens was used they overshot the prey. It therefore seems that chameleons use accommodation for judging distance. The accommodation response of chameleons is unusually rapid and precise. The following features of the visual system improve the precision of accommodation.

1. The unaccommodated lens has negative power, a feature not known in any other animal. A negative lens combined with a positively powered cornea forms a telephoto lens that creates an unusually large image for the size of eye. This improves the precision of accommodation by improving spatial resolution (Ott and Schaeffel 1995). A telephoto lens also increases the range of accommodation, which is unusually large (45 dioptres) in chameleons.
2. Chameleons dilate the pupils when aiming at prey. Pupil dilation reduces the depth of field and therefore makes it easier to detect when an object is not in the focal plane.
3. During scanning saccadic eye movements, only one eye is accommodated appropriately, but attention switches from one eye to the other at intervals of approximately 1 s. Just before tongue release, both eyes accommodate on the prey, which improves the precision of accommodation (Ott *et al.* 1998).
4. Like many birds, fish, and other lizards, chameleons have convexiclivate foveas in which the sides of the foveal pit are convex. Animals with well-

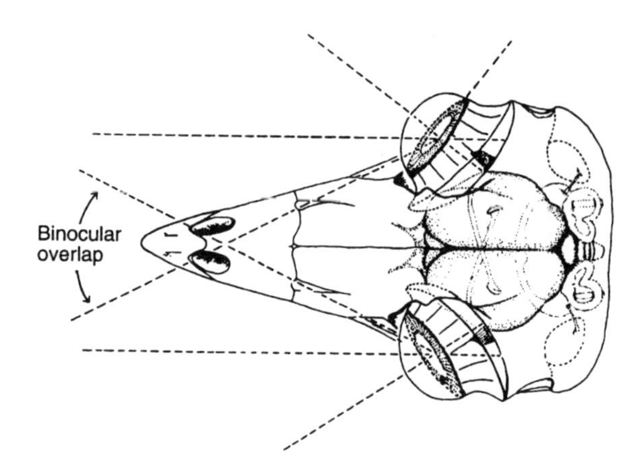

Figure 14.13. Diagram of the head of a bifoveate bird.
In each eye the lateral visual axis projects to the central fovea, and the frontal visual axis projects to a fovea in the temporal retina. (From Pettigrew 1986. Reprinted with permission of Cambridge University Press)

developed binocular vision, such as primates, have concaviclivate foveas. Harkness and Bennet-Clark (1978) showed that the pattern of pincushion distortion produced by refraction at the surface of a convexiclivate fovea is a sensitive indicator of both the direction and magnitude of image misfocussing.

A telephoto lens brings the nodal point of the chameleon eye well forward of the eye's centre of rotation. This creates motion parallax when the eye is rotated, which could provide a monocular depth cue in addition to accommodation (Land 1995).

Kirmse *et al.* (1994) proposed a novel mechanism for locating prey. The chameleon fixates a prey object with one or other eye so that its image falls on the fovea. The chameleon turns to bring the object into the median plane of the head and places the eyes in a fixed position of divergence of between 17 and 19°. Thus, just before the tongue strikes, the distance of each image from the fovea is a function of the distance of the prey from the Chameleon. Because the eyes are diverged, the image of a distant target falls nearer the fovea than does the image of a nearer target. This is not a disparity mechanism because it does not involve comparison of the positions of images in the two retinas. This explains why chameleons can aim the tongue accurately with one eye occluded. Perhaps chameleons base distance estimates on accommodation as was proposed by Harkness, as well as on the lateral position of images. There is no evidence that chameleons use binocular disparity.

The turtle *Chinemys reevesi* has extensive ipsilateral visual projections to the thalamus, and its visual fields overlap about 48°. On the other hand, the predatory turtle *Trionyx cartilagineus* has no ipsilateral visual projections, even though its visual fields overlap about 67° (Hergueta *et al.* 1992). Stereoscopic vision has not been demonstrated in these species.

14.6 BIRDS

14.6.1 The pigeon

The domestic pigeon (*Columba livia*) has laterally placed eyes with the optic axes set at an angle of about 120°. It has a total visual field of about 340° and a frontal binocular region with a maximum width of about 27°, which extends from 90° above the beak to about 40° below it (Martin and Young 1983). In the sample of birds tested by Martin and Young the visual field was widest 20° above the plane of the beak whereas, in the sample tested by Martinoya *et al.* (1981), it was widest 45° below the beak (see also McFadden and Reymond 1985; Holden and Low 1989). These differences may reflect differences in eye convergence. In any case the binocular field is well placed to help the bird peck seeds on the ground and locate a landing site.

The binocular region of the visual field is served by a specialized area in the temporal retina containing a high density of receptors, known as the **area dorsalis**. The area dorsalis projects 10 to 15° below the eye-beak plane and is distinct from the more laterally placed fovea used for lateral monocular viewing, as shown in Figure 14.13. The cones in the region of the area dorsalis contain red oil droplets whereas those in the region of the lateral fovea contain yellow pigment. Resolution of a high-contrast grating presented to the lateral fovea is between 1.16 and 4.0 arcmin, compared with resolution of about 0.8 arcmin for humans (Blough 1971). Peak contrast sensitivity of the lateral fovea occurs at a spatial frequency of about 4 cpd (Nye 1968). The acuity of the frontal fovea is only slightly higher (Hahmann and Güntürkün 1993).

When the head of the pigeon is restrained, each eye is capable of executing saccades of up to 7° amplitude to stimuli presented in the lateral field, with each eye moving independently of the other. These movements bring the image of a laterally placed object onto the lateral fovea of one eye. Both eyes execute co-ordinated vergence movements to stimuli presented at near distances in the frontal visual field. These movements bring the images onto the area dorsalis in both eyes. Vergence is most effectively evoked by stimuli 25° below the beak, the position occupied by objects on the ground at which the pigeon pecks. Before the head makes a pecking motion

Figure 14.14. Barry J. Frost.
Born in Nelson, New Zealand. He obtained a B.A. at the University of Canterbury, New Zealand and a Ph.D. from Dalhousie University, Canada in 1967. He conducted postdoctoral work at Berkeley. Since 1969 he has been at Queens University, Canada, where he is now Professor of Psychology and Biology. He is a Fellow of the Royal Society of Canada and of the Canadian Institute for Advanced Research. He received the James McKeen Cattel award and the Alexander von Humboldt Research Prize.

it moves forward in two successive movements, with the eyes converging on the ground (Goodale 1983; Bloch *et al.* 1984). The final ballistic movement occurs from a distance of 55 mm with the target centred on the area dorsalis about 10° below the eye-beak axis. The eyes close during the final peck.

Like all birds, the pigeon has two main visual pathways. One leads to the contralateral pretectum and optic tectum (the analogue of the superior colliculus in mammals), then to the ipsilateral nucleus rotundus and nucleus triangularis in the thalamus, and on to the ectostriatal complex and surrounding extrinsic recipient areas in the telencephalon. This is probably the phylogenetically older pathway since it is fully decussated and is fully developed by the time of hatching. Different regions of the nucleus rotundus of the pigeon are specialized for different functions. Cells in one region respond to changes in illumination over the whole visual field. Those in a second region are colour coded, those in a third region respond to differential motion. Those in a fourth region respond to motion in depth signalled by monocular looming. They are particularly sensitive to objects on a collision course with the head (Wang *et al.* 1993).

The second pathway of the avian visual system starts with fully decussated inputs to the optic thalamus (homologue of the mammalian LGN). From there, a hemidecussated pathway passes to a centre on the dorsum of the diencephalon known as the **wulst** (German for "bulge"). This system is not fully developed in the pigeon until about 17 days after hatching. This suggests that visual inputs are required for its maturation (Karten *et al.* 1973). This system contains cells with small receptive fields that could code the shapes and positions of stimuli (Emmerton 1983).

Frost *et al.* (1983) found very few binocular cells in either the pigeon's optic tectum or the wulst (Portrait Figure 14.14). However, binocular cells in the wulst may have been missed because of misalignment of the eyes. Binocular cells have been found in the wulst of other birds.

There is conflicting evidence about stereoscopic vision in the pigeon. Martinoya *et al.* (1988) found that pigeons could discriminate between points of light presented one at a time at different distances. However, pigeons failed to discriminate between a pair of simultaneously presented stimuli in distinct depth planes from a subsequently presented pair in one depth plane. These findings suggest that they use a range-finding mechanism based on vergence rather than binocular disparity. Pigeons can be trained to peck at one of two simultaneously presented panels at different depths. This procedure has revealed that pigeons have a depth acuity of between 0.8 and 1.8 arcmin (McFadden 1987). The animals could not perform the task when one eye was occluded, which suggests that they use binocular disparities.

14.6.2 Hawks

Hawks, like many predatory birds, have two foveas in each eye. The central foveas are used for sideways looking at distant objects and the temporal foveas serve for binocular viewing of nearer objects. The two foveas in each eye of the American kestrel, or sparrow hawk (*Falco sparverius*), are 36° apart and the visual fields overlap about 58° when the eyes are converged (Frost *et al.* 1990). The eyes move independently when the central foveas are used and conjugately when the temporal foveas are used. Pettigrew (1978) suggested that each tectum controls the contralateral eye and that conjugate movements are controlled by projections from binocular cells in the wulst back to the tectum. The wulst of the kestrel has a much greater representation of the temporal foveas than of the central foveas.

Hawks have an impressively large number of receptors per degree of visual angle. For instance, the fovea of the American kestrel, has about 8,000 receptors per degree and the red-tailed hawk (*Buteo*

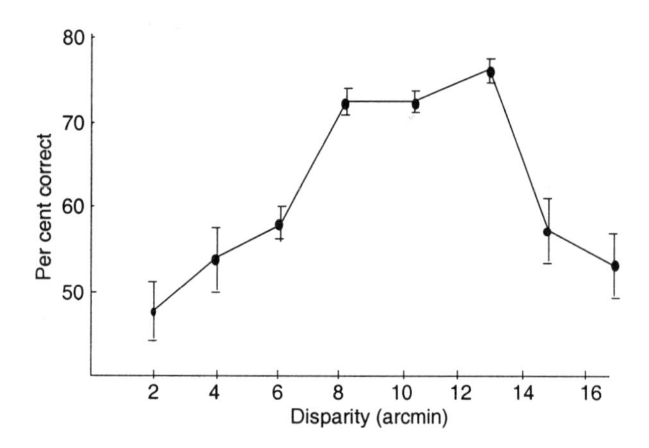

Figure 14.15. Stereoacuity in a sparrow hawk.
A hawk's percentage correct detection of depth in a stereogram, as a function of disparity. Each point is based on between 50 and 350 trials. Bars are standard errors. (From Fox *et al*. 1977)

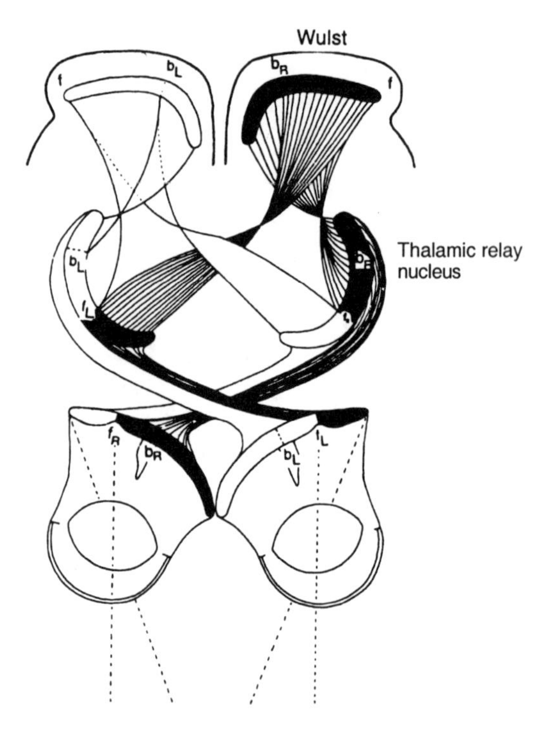

Figure 14.16. The visual pathways of the owl.
The optic nerves decussate fully as they project to the optic nucleus in the thalamus and to the optic tectum. The pathways fron the thalamus to the Wulst hemidecussate so that corresponding inputs from the binocular field are brought to the same destination in the wulst. The fovea for the frontal field is denoted by *f*, and the limits of the binocular field by *b*. (From Pettigrew 1986. Reprinted with permission of Cambridge University Press)

jamaicensis) about 15,000 per degree, compared with about 7,000 in the rhesus monkey (Fite and Rosenfield-Wessels 1975). Shlaer (1972) found that image quality in the African eagle (*Dryotriorchis spectabilis*) is up to 2.4 times greater than in the human eye. The American kestrel can detect a grating with a spatial frequency of 160 cpd. The upper limit for humans is about 60 cpd (Fox *et al.* 1976). However, the cone density and focal length of the kestrel's eye are not sufficiently greater than those of the human eye to account for its greater acuity.

Snyder and Miller (1978) suggested falconiform birds achieve high acuity because the highly concave foveal pit acts as a negative lens which, together with the positive lens of the eye, constitutes a telephoto lens system, which effectively increases the focal length of the eye. Another possibility is that the effective density of receptors is increased when they lie on the oblique walls of a concave fovea as opposed to a surface normal to the incident light.

Kestrels have a well-developed wulst containing binocular cells (Pettigrew 1978). It is larger than the wulst in pigeons but not as large as that in owls.

Fox *et al.* (1977) trained a sparrow hawk to fly from a perch to a panel containing a dynamic random-dot stereogram in preference to one containing a similar 2-D random-dot pattern. The stereogram was created by the colour-anaglyph method, with the falcon wearing red and green filters. The percentage of correct responses of the trained animal was recorded as the disparity in the stereogram was varied. It can be seen from Figure 14.15 that performance peaked for a disparity of about 12 arcmin. This suggests that sparrow hawks have full-fledged stereoscopic vision enabling them to detect the 3-D shapes of objects, and not merely a vergence or disparity range-finding mechanism serving the detection of absolute distance.

14.6.3 Owls

Owls use auditory signals to locate the direction of prey. They turn the head rapidly to face the direction of the sound source. They then swoop down and capture the prey, even in total darkness. The mechanism that detects the direction of sounds has been investigated in some detail (Konishi 1993) but it is not clear how owls detect the distance of an object from sound. Perhaps they rely on knowledge of surfaces in their surroundings acquired through vision. There is evidence that owls use their eyes for judging distance when there is some light.

The tawny owl (*Strix aluco*) has a binocular visual field with a maximum width of 48° and a total visual field of 201° (Martin 1984). The retina has a single, temporally placed fovea which, in the great horned owl (*Bubo Virginianus*), has about 13,000 receptors per degree of visual angle (Fite and Rosenfield-Wessels 1975). The visual pathways of owls decussate totally at the chiasma and project retinotopically to the optic tectum. Each cell in the optic tectum responds both when a visual stimulus falls within its receptive field and when a sound originates from the same direction. An auditory map of space in the inferior colliculus projects onto the visual map in the visual tectum (Knudsen 1982). Because the eyes are virtually immobile in the head the spatial correspondence of the visual and auditory maps is constant. However, the pattern of correspondence must change as the animal grows. The pattern is also subject to modification in animals raised wearing prismatic spectacles (Feldman *et al.* 1996).

The full decussation of the visual inputs to the tectum suggests that this system codes direction rather than distance. Figure 14.16 shows the hemidecussated bilateral projection from the optic nucleus in the thalamus to the wulst. In the owl, the wulst is particularly well developed and contains cells with properties very similar to those of cells in the mammalian visual cortex (Pettigrew 1986; Bagnoli *et al.* 1990; Porciatti *et al.* 1990). Many binocular cells are tuned to binocular disparity, and their receptive fields in the two eyes have similar tuning for orientation and motion. The cells are also arranged in ocular dominance columns (Pettigrew and Konishi 1976a; Pettigrew 1979). Efferents from the wulst project to the optic tectum in a manner analogous to the projection of cortical efferents to the superior colliculus in mammals (Casini *et al.* 1992). Like cats and monkeys, owls reared with one eyelid sutured fail to develop binocular cells and their eyes become misaligned (Pettigrew and Konishi 1976b).

Wagner and Frost (1993, 1994) measured the disparity tuning functions of cells in the wulst of the anaesthetized barn owl (*Tyto alba*) to sine wave gratings and one-dimensional visual noise. Although lesser peaks in the disparity tuning function varied as a function of the spatial frequency of the gratings, the main peak occurred at a characteristic disparity for a given cell, which was independent of spatial frequency. The characteristic disparity sensitivity of different cells ranged between 0 and 2.5°. One-dimensional visual noise also produced a peak response at a disparity of 2°. They argued that a disparity-detection mechanism based on spatial-phase differences (Section 6.5.4) produces tuning functions that vary with the spatial frequency of the stimulus, whereas tuning functions of a mechanism based on position offset are independent of spatial frequency. In the owl, the peak response of most cells tuned to disparities of less than about 3 arcmin did vary with spatial frequency. They concluded that an offset mechanism codes large disparities and a phase-difference mechanism codes small disparities.

Zhu and Qian (1996) questioned these conclusions. They pointed out that the disparity tuning of simple binocular cells is not frequency independent for either type of disparity coding—only complex cells are distinguishable in this way.

Nieder and Wagner (2000) recorded from binocular cells in the wulst of the alert barn owl as it observed a random-dot stereogram with variable disparity. About 75% of cells were tuned to disparity. The tuning functions did not form discrete categories and cells with similar tuning functions did not form columns.

Owls respond to depth in a random-dot stereogram but not when the contrast polarity of the stereogram is reversed in the two eyes (Nieder and Wagner 2001). Short-latency binocular cells in the wulst showed side peaks in their response to a particular disparity and responded to disparity in contrast-reversed stereograms. However, long-latency binocular cells showed a single peak in their response and responded only to disparity in stereograms with matching contrast polarity. Nieder and Wagner concluded that disparities are processed hierarchically by an initial stage in which local disparities are processed and a second stage in which global properties are processed. These issues are discussed in more detail in Section 16.3.

The close similarity between the disparity-detecting mechanism of the wulst and that of the mammalian visual cortex seems to be the product of a remarkable process of parallel evolution. Further work is needed to reveal how general this mechanism is among birds.

The eyes of an owl are very large in proportion to its head and they move, at most, about 1° (Steinbach and Money 1973). It is not known whether these eye movements include vergence. On the other hand, the owl can rotate its head at high velocity through at least 180°. Since the eyes are locked in a more or less fixed position of vergence, binocular disparity could code absolute depth as well as relative depth. In the young bird the eyes are at first diverged. They take up their adult positions in the second month of life. When binocular vision is disrupted during the first 2 months the eyes fail to become properly aligned. Microelectrode recordings from binocular cells in the optic tectum of strabismic owls revealed that the monocular receptive fields were also misaligned (Knudsen 1989). Thus, the neural pattern of binocular correspondence is innate and presumably guides the alignment of the eyes in the young bird.

Van der Willigen *et al.* (1998) demonstrated stereopsis in the barn owl (*Tyto alba*). Two animals could discriminate between a homogeneous random-dot stereogram and one containing an inner region in depth. They could also discriminate between two

Figure 14.17. The Meercat (*Suricata suricatta*).
(From The Larousse Encyclopedia of Animal Life, Hamtyn 1967)

stereograms containing different relative depths. They responded selectively to relative depth when absolute depth was varied. The upper disparity limit for depth discrimination was 41 arcmin.

Van der Willigen *et al.* (2002) showed that owls trained to select a given sign of relative depth in a random-dot stereogram transferred the learning to relative depth based solely on motion parallax. Thus, owls process the two depth cues in an equivalent manner.

The barn owl apparently uses accommodative effort for judging near distances. It has an accommodative range of over 10 dioptres, which is larger than that of any other species of owl (Murphy and Howland 1983). Accommodation, but not pupil constriction, is always identical in the two eyes (Schaeffel and Wagner 1992). When wearing a plus or minus lens in front of one eye and an occluder over the other, barn owls made corresponding errors in pecking at a nearby object (Wagner and Schaeffel 1991).

14.7 MAMMALS

Many mammals, including rats, dogs, cats, pigs, sheep, and humans avoid a cliff even when they are very young. This shows that they detect depth (Section 11.3.2b). This ability could depend on monocular cues such as perspective and motion parallax.

Stereoscopic vision is particularly well developed in felines (Fox and Blake 1971; Mitchell *et al.* 1979) and primates (Bough 1970; Cowey *et al.* 1975; Sarmiento 1975; Julesz *et al.* 1976b; Harwerth and Boltz 1979b; Harwerth *et al.* 1995; Crawford *et al.* 1996c). These mammals have foveas, frontal vision, hemidecussating visual pathways, and vergence eye movements. Also, visual inputs from corresponding regions in the two retinas converge on cortical cells tuned to disparity (see Chapter 6). However, the visual fields of all mammals, except the Cetaceae (whales and dolphins) overlap to some degree.

Bats and marine mammals such as Dolphins locate objects in 3-D space by reflection of emitted sound, or sonar (Au 1993; Griffin 1958; Norris 1968).

The topic of depth vision in animals has been reviewed by Walls (1963), Hughes (1977), Collett and Harkness (1982), and Pettigrew (1991). I now review evidence for depth vision and stereopsis in mammals other than cats and primates.

14.7.1 Bats

Most insectivorous bats have very poor eyesight and locate their prey by echolocation while flying at night (Simmons and Stein 1980; Thomas *et al.* 2000). In one study, big brown bats (*Eptesicus fuscus*) could discriminate a distance of 0.6 cm at 80 cm (Masters and Raver 2000). Different species of bat emit distinct high-pitched sounds which enables them to recognize their own kind (Parsons and Jones 2000). The small desert bat, *Macrotus californicus*, captures prey on the ground and when illumination is sufficient it shuts off its echolocation calls and uses vision. The visual acuity of this bat is better than 4 arcmin compared with acuities of between 15 and 60 arcmin for bats that locate prey only by echolocation (Bell and Fenton 1986). Furthermore, the eyes of the desert bat are more frontal with a larger binocular overlap than the eyes of other bats. This suggests that they use stereoscopic vision for prey location but the relevant tests have not been done.

14.7.2 Meerkats and rodents

The slender tailed meerkat (*Suricata suricatta*) is a frontal-eyed social carnivore related to the mongoose (Figure 14.17). They live in burrows in the dry South African veld. Nothing is known about its visual pathways or visual cortex. However, meerkats possess stereoscopic vision that seems to depend on binocular disparity. An animal trained to jump to the closer of two visual displays with both eyes open could discriminate relative depths corresponding to disparities of about 10 arcmin (Moran *et al.* 1983).

Thus, among frontal-eyed mammals, binocular stereoscopic vision has been demonstrated in the primates, cats, and the meerkat. As we will now see, there is some evidence for binocular stereopsis in some lateral-eyed mammals.

Rats avoid the deep side of a visual cliff as soon as the visual system becomes functional. Depth avoidance in monocular infant rats was similar to that in binocular infant rats, showing that they do not, or need not, use binocular disparity (Lore and Sawatski 1969). After some preliminary training, adult rats can jump from one platform to another across a gap and adjust the force of their jump to the width of the gap (Russell 1932). Rats discriminated a difference of 2 cm over a gap width of between 20 and 40 cm. They tended to swing the head before jumping, which suggests that they used motion parallax. Removal of one eye reduced accuracy. Rats raised in darkness for the first 100 days and then trained to jump one distance were able to adjust to another distance on the first jump (Lashley and Russell 1934). Greenhut and Young (1953) found only a weak relationship between force exerted and jump distance and that rats required practice to achieve reasonable accuracy in jumping across a gap.

Mongolian gerbils (*Meriones unguiculatus*) adjust their jump to the width of a gap. Before jumping they bob the head up and down. These movements generate motion parallax between objects at different distances and also motion of the image of the target platform or motion of the eyes as they fixate the platform. Gerbils are more likely to bob the head when other depth cues are not available or as the width of the gap is increased (Ellard *et al.* 1984). Gerbils overjumped when the target platform was smaller than the one they were used to and underjumped when the platform was larger. This suggests that they use image size as a cue to distance. However, they could compensate for changes in the size of the jumping platform by increasing the frequency of head bobbing. Varying the distance between the target platform and the background behind the platform did not affect jumping accuracy (Goodale *et al.* 1990). This suggests that relative motion produced by head bobbing is not used, but rather the ratio of head motion to the motion of the image of the platform. It is not clear whether gerbils use binocular disparity as a depth cue.

The rabbit has laterally placed eyes and, when crouched in the "freeze" position, it has 360° panoramic vision with a frontal binocular area of about 24°. At least 90% of optic-nerve fibres decussate in the chiasma (Giolli and Guthrie 1969). The retina does not have a fovea but there is a relatively high concentration of receptors along an extended horizontal region known as the visual streak, which receives the image of the visual horizon. There is also a high concentration of receptors in the posterior part of the retina, corresponding to the binocular region of the visual field. This region is about 30° wide and extends from about 35° below eye level to about 160° above eye level when the eyes are in their most converged position. The rabbit uses this part of the retina when performing visual discrimination tasks (Hughes 1971). When the animal is in the freeze position, the eyes diverge and corresponding parts of the two retinas have a horizontal disparity of about 18°. The rabbit is able to match corresponding images in the two eyes by converging the eyes but, when it does so, its visual fields no longer meet behind the head (Hughes and Vaney 1982). Convergence eye movements of up to about 18° occur when the rabbit approaches a visual display (Collewijn 1970; Zuidam and Collewijn 1979). Thus, the rabbit has two modes of viewing—one that preserves panoramic vision but loses all binocular correspondence, and one that achieves binocular correspondence within the binocular field at the expense of some loss in panoramic coverage.

The region of the rabbit's visual cortex corresponding to the binocular visual field contains binocular neurones tuned to binocular disparity. At first it was reported that the component receptive fields in the two eyes were not similarly tuned for stimulus size and orientation (Van Sluyters and Stewart 1974a; Choudhury 1980). However, when allowance was made for the fact that the eyes must converge 18° to bring corresponding images into register, binocular cells were found to have matching tuning functions in the two eyes (Hughes and Vaney 1982). The binocular cells of the rabbit show little evidence of the developmental plasticity found in the binocular cells of cats and primates (Van Sluyters and Stewart 1974b).

Rabbits can perform visual discrimination tasks for which they must integrate binocular information (Van Hof and Steele Russell 1977). They can also execute voluntary eye movements (Hughes 1971). *However, there do not seem to have been any direct tests of stereopsis in rabbits.*

14.7.3 Ungulates

Ungulates are the hoofed mammals. The even-hoofed order, Artiodactyla, includes bovines, goats, and sheep. The odd-hoofed order, Perissodactyla, includes horses, tapirs, and rhinoceroses. Ungulates possess cells in the visual cortex that are sensitive to binocular disparity. For instance, cells in V1 of sheep are tuned to small or zero disparities and those in V2 respond to disparities up to 6° (Clarke *et al.* 1976). Disparity-selective cells occur in the new born lamb, although with broader tuning than in the adult (Clarke *et al.* 1979). An autoradiographic study of projections to the visual cortex failed to reveal evidence of ocular dominance columns in sheep and goats (Pettigrew *et al.* 1984). Binocular cells extend up to 20° on both sides of the midline (Clarke and Whitteridge 1976). Sheep have widely spaced eyes, which produce four times the disparity at a given distance than that produced by cats' eyes. Sheep have good depth discrimination but the extent to which this depends on binocular stereopsis is not known.

Neonatal lambs and goats are much more advanced than neonatal cats and monkeys and, a few hours after birth, are able to follow the mother and avoid obstacles and cliffs. Nevertheless, brief periods of monocular deprivation in the lamb cause marked shifts in ocular dominance of binocular cortical cells (Martin *et al.* 1979).

Although there is no behavioural evidence of stereopsis in sheep or goats, such evidence exists for the horse. Timney and Keil (1994) trained two horses to select a display containing a square in real depth. After learning to perform that task the horses were shown random-dot stereograms viewed through red-green anaglyph filters. On most trials they selected the stereogram that depicted a square in depth. The mean disparity threshold for detection of depth was 14 arcmin. They also showed that horses are sensitive to perspective as a cue to depth (Timney and Keil 1996).

References Volume 1

Abadi RV (1976) Induction masking—a study of some inhibitory interactions during dichoptic viewing Vis Res **16** 299–75 [292]

Abadi RV, Dickinson CM (1985) The influence of preexisting oscillations on the binocular optokinetic response Ann Neurol **17** 578–86 [498]

Abadi RV, Pascal E (1989) The recognition and management of albinism Ophthal Physiol Opt **9** 3-15 [495]

Abbott L, Sejnowsli TJ (1999) Neural codes and distributed representations: foundations of neural computation MIT Press, Cambridge, MA [122]

Abel LA, Schmidt D, Dell'Osso LF, Daroff RB (1978) Saccadic system plasticity in humans Ann Neurol **4** 313–18 [425]

Abel PL, O'Brien BJ, Olavarria JF (2000) Organization of callosal linkages in visual area V2 of macaque monkey J Comp Neurol **428** 278-93 [232]

Abramov I, Gordon J, human infant Science **217** 265–7 [436]

Ackroyd C, Humphrey NK, Warrington EK (1974) Lasting effects of early blindness A case study Quart J Exp Psychol **26** 114-24 [501]

Adachi-Usami E, Lehmann D (1983) Monocular and binocular evoked average potential field topography: up Hendrickson A, Hainline L, Dobson V, LaBossiere E (1982) The retina of the newborn per and lower hemiretinal stimuli Exp Brain Res **50** 341-6 [239]

Adams DL, Zeki S (2001) Functional organization for macaque V3 for stereoscopic depth J Neurophysiol **86** 2195-203 [220, 242]

Adams MM, Hof PR, Gattass R, Webster MJ, Ungerleider LG (2000) Visual cortical projections and chemoarchitecture of macaque monkey pulvinar J Comp Neurol **419** 377-93 [183]

Adams RJ, Courage ML, Mercer ME (1991) Deficiencies in human neonates' color vision: photoreceptoral and neural explanations Behav Brain Res **43** 109–14 [470]

Addams R (1834) An account of a peculiar optical phenomenon seen after having looked at a moving body Lond Edin Philos Mag J Sci **5** 373–4 [347]

Adelson EH, Bergen JR (1985) Spatiotemporal energy models for the perception of motion J Opt Soc Am A **2** 284-99 [202, 266]

Adorján P, Levitt JB, Lund JS, Obermayer K (1999) A model for the intracortical origin of orientation preference and tuning in macaque striate cortex Vis Neurosci **16** 303-18 [196]

Adrian ED, Matthews R (1927) The action of light on the eye J Physiol **63** 378-90 [168]

Aertsen AMHJ, Gerstein GL, Habib MK, Palm G (1989) Dynamics of neuronal firing correlation: modulation of "effective connectivity" J Neurophysiol **61** 900-17 [131]

Aggarwala KR, Kruger ES, Mathews S, Kruger PB (1995) Spectral bandwidth and ocular accommodation J Opt Soc Am **12** 450-55 [361]

Agmon-Snir H, Carr CE, Rinzel J (1998) The role of dendrites in auditory coincidence detection Nature **393** 298-72 [115]

Aguilonius F (1613) Opticorum libri sex Plantin Antwerp [52]

Ahissar E, Vaadia E, Ahissar M, Bergman H, Arieli A, Abeles M (1992) Dependence of cortical plasticity on correlated activity of single neurons and on behavioral context Science **257** 1412–15 [135]

Ahissar M, Hochstein S (1995) How early is early vision? Evidence from perceptual learning In Early vision and beyond (ed TV Papathomas, C Chubb, A Gorea, E Kowler) pp 199-206 MIT Press, Cambridge Mass [354]

Ahissar M, Hochstein S (1996) Learning pop-out detection: specificities to stimulus characteristics Vis Res **36** 3487-500 [354]

Ahlsén G, Lindström S, Lo F-S (1985) Interaction between inhibitory pathways to principal cells in the lateral geniculate nucleus of the cat Exp Brain Res **58** 134–43 [178]

Ahmed B, Anderson JC, Douglas RJ, Martin KA, Nelson JC (1994) Polyneuronal innervation of spiny stellate neurons in cat visual cortex J Comp Neurol **341** 39-49 [188]

Aiello A, Wright KW, Borchert M (1994) Independence of optokinetic nystagmus asymmetry and binocularity in infantile esotropia Arch Ophthal **112** 1580-3 [526]

Alais D, Blake R (1998) Interactions between global motion and local binocular rivalry Vis Res **38** 637-44 [303]

Alais D, Blake R (1999) Grouping visual features during binocular rivalry Vis Res **39** 4341-53 [299]

Alais D, O'Shea RP, Mesana-Alais C, Wilson IG (2000) On binocular alternation Perception **29** 1437-45 [299]

Albano JE, Marrero A (1995) Binocular interactions in rapid saccadic adaptation Vis Res **35** 3439-50 [423]

Albert MK (2000) The genetic viewpoint assumption and Bayesian inference Perception **29** 601-8 [102]

Albin RL, Sakuraai SY, Makowiec RL, Higgins DS, Young AB, Penny JB (1991) Excitatory amino acids, GABAA and GABAB binding sites in human striate cortex Cerebral Cortex **1** 499-509 [191]

Albrecht DG (1995) Visual cortex neurons in monkey and cat: effect of contrast on the spatial and temporal phase transfer functions Vis Neurosci **12** 1191-210 [201]

Albrecht DG, Hamilton DB (1982) Striate cortex of monkey and cat: contrast response function J Neurophysiol **48** 217-37 [198, 201]

Albright TD (1992) Form-cue invariant motion processing in primate visual cortex Science **255** 1141-3 [224]

Albright TD, Desimone R (1987) Local precision of visuotopic organization in the middle temporal area (MT) of the macaque Exp Brain Res **65** 582-92 [223]

Albus K (1975) A quantitative study of the projection area of the central and paracentral visual field in area 17 of the cat. I The precision of the topology Exp Brain Res **27** 159-79 [237]

Albus K, Wolf W (1984) Early postnatal development of neuronal function in the kitten's visual cortex: A laminar analysis J Physiol **348** 153–85 [454]

Alexander LT (1951) The influence of figure-ground relationships on binocular rivalry J Exp Psychol **41** 376–81 [289]

Alhazen Kitäb al-manäzir (Book of optics) In The optics of Ibn al-Haytham 2 volumes (Translated by AI Sabra) Warburg Institute, University of London 1989 [15, 16]

Alkondon M, Pereira EFA, Eisenberg HM, Albuquerque EX (2000) Nicotinic receptor activation in human cerebral cortical interneurons: a mechanism for inhibition and disinhibition of neuronal networks J Neurosci **20** 66-75 [192]

Allen D, Banks MS, Norcia AM (1993) Does chromatic sensitivity develop more slowly than luminance sensitivity? Vis Res **33** 2553–62 [470]

Allen D, Tyler CW, Norcia AM (1996) Development of grating acuity and contrast sensitivity in the central and peripheral visual field of the human infant Vis Res **36** 1945-53 [468]

Allen DC (1974) Vertical prism adaptation in anisometropes Am J Optom Physiol Opt **51** 252–9 [384]

Allen MJ (1953) An investigation of the time characteristics of accommodation and convergence of the eyes Am J Optom Arch Am Acad Optom **30** 393–402 [390]

Allen MJ, Carter JH (1967) The torsional components of the near reflex Am J Optom Arch Am Acad Optom **44** 343–9 [370, 411, 419]

Allendoerfer KL, Shatz CJ (1994) The subplate a transient neocortical structure: its role in the development of connections between thalamus and cortex Ann Rev Neurosci **17** 185-218 [458]

Allendoerfer KL, Cabelli RJ, Escandón E, Kaplan DR, Nikolics K, Shatz CJ (1994) Regulation of neurotrophin receptors during the maturation of the mammalian visual system J Neurosci **14** 1795-811 [442, 456]

Allison DW, Gelfand VI, Spector I, Craig AM (1998) Role of actin in anchoring postsynaptic receptors in cultured hippocampal neurons: differential attachment of NMDA versus AMPA receptors J Neurosci **18** 2723-36 [189]

Allison RS, Howard IP, Fang X (2000) Depth selectivity of vertical fusional mechanisms Vision Res **40** 2985-98 [409]

Allman JM, Meizin F, McGuinness EL (1985) Direction- and velocity-specific responses from beyond the classical receptive field in the middle temporal visual area (MT) Perception **14** 105-29 [223, 224]

Alonso JM, Usrey WM, Reid RC (1996) Precisely correlated firing in cells of the lateral geniculate nucleus Nature **383** 815-19 [132]

Alpern M (1946) The after-effect of lateral duction testing on subsequent phoria measurements Am J Optom Arch Am Acad Optom **23** 442–7 [381]

Alpern M (1952) Metacontrast: historical introduction Am J Optom Arch Am Acad Optom **29** 631–46 [340]

Alpern M (1958) Vergence and accommodation: can change in size induce vergence movements? Arch Ophthal **60** 355–7 [386]

Alpern M (1969) Types of eye movement In The eye (ed H Davson) Vol 3 pp 65–174 Academic Press, New York [368, 391]

Alpern M, Ellen P (1956) A quantitative analysis of the horizontal movements of the eyes in the experiment of Johannes Müller Am J Ophthal **42** 289–96 [421]

Alpern M, Hofstetter HW (1948) The effect of prism on esotropia—a case report Am J Optom Arch Am Acad Optom **25** 80-91] [374]

Alpern M, Larson BF (1960) Vergence and accommodation IV Effect of luminance quantity on the AC/A Am J Ophthal **49** 1140–9 [389]

Alpern M, Wolter JR (1956) The relation of horizontal saccadic and vergence movements Arch Ophthal **56** 685–90 [429]

Alpern M, Kincaid WM, Lubeck MJ (1959) Vergence and accommodation: III. proposed definitions of the AC/A ratios Am J Ophthal **48** 141-8 [388, 389]

Alpern M, Rushton WAH, Torii S (1970) Signals from cones J Physiol **207** 463-75 [341]

Altmann L, Luhmann HJ, Greuel JM, Singer W (1987) Functional and neuronal binocularity in kittens raised with rapidly alternating monocular occlusion J Neurophysiol **58** 965–80 [514]

Alvarez TL, Semmlow JL, Yuan W (1998) Closely spaced fast dynamic movements in disparity vergence J Neurophysiol **79** 37-44 [399]

Alvarez TL, Semmlow JL, Yuan W, Munoz P (1999) Dynamic details of disparity convergence eye movements Ann Biomed Engin **27** 380-90 [403]

Alvarez TL, Semmlow JL, Yuan W, Munoz P (2000) Disparity vergence double responses processed by internal error Vis Res **40** 341-7 [401]

Alvarez-Maubecin V, Garcia-Hernández F, Williams JT, Van Bockstaele EJ (2000) Functional coupling between neurons and glia J Neurosci **20** 4091-8 [188]

Amedi A, Malach R, Hendler T, Peled S, Zohary E (2001) Visuo-haptic object-related activation in the ventral visual pathway Nature Neurosci **4** 324-30 [222]

Ames A (1929) Cyclophoria Am J Physiol Opt **7** 3–38 [278]

Amigo G (1974) A vertical horopter Optica Acta **21** 277–92 [411, 412]

Amigo G, Fiorentini A, Pirchio M, Spinelli D (1978) Binocular vision tested with visual evoked potentials in children and infants Invest Ophthal Vis Sci **17** 910–15 [483]

Andersen RA (1987) The role of the inferior parietal lobule in spatial perception and visual–vestibular integration In The handbook of physiology The nervous system (ed F Plum, VB Mountcastle, ST Geiger) Vol IV pp 483–58 Am Physiol Soc, Bethesda MD [229]

Andersen RA, Mountcastle VB (1983) The influence of the angle of gaze upon the excitability of the light-sensitive neurons of the posterior parietal cortex J Neurosci **3** 532-48 [227]

Andersen RA, Zipser D (1988) The role of the posterior parietal cortex in coordinate transformations for visual-motor integration Can J Physiol Pharmacol **66** 488-501 [144]

Andersen RA, Bracewell RM, Barash S, Gnadt JW, Fogassi L (1990) Eye–position effects on visual memory and saccade–related activity in areas LIP and 7a of macaque J Neurosci **10** 1176–96 [227]

Andersen RA, Snyder LH, Bradley DC, Xing J (1997) Multimodal representation of space in the posterior parietal cortex and its use in planning movements Ann Rev Neurosci **20** 303-330 [227, 228]

Anderson CH, Van Essen DC (1987) Shifter circuits: a computational strategy for dynamic aspects of visual processing Proc Nat Acad Sci **84** 6297–301 [395]

Anderson JC, Martin KAC, Whitteridge D (1993) Form function and intracortical projections of neurons in the striate cortex of the monkey *Macacus nemestrinus* Cerebral Cortex **3** 412-20 [188]

Anderson JD, Bechtoldt HP, Dunlap GL (1978) Binocular integration in line rivalry Bull Psychonom Soc **11** 399–402 [294]

Anderson JS, Lampel I, Gillespie DC, Ferster D (2000) The contribution of noise to contrast invariance of orientation tuning in cat visual cortex Science **290** 1968-72 [127]

Anderson NH (1974) Algebraic models in perception In Handbook of perception (ed EC Carterette, MP Friedman) pp 215-98 Academic Press, New York [148]

Anderson P, Mitchell DE, Timney B (1980) Residual binocular interaction in stereoblind humans Vis Res **20** 603–11 [491]

Anderson PA, Movshon JA (1989) Binocular combination of contrast signals Vis Res **29** 1115–32 [322]

Anderson PA, Olavarria J, Van Sluyters RC (1988) The overall pattern of ocular dominance bands in cat visual cortex J Neurosci **8** 2183-200 [214]

Andrews DP (1964) Error-correcting perceptual mechanisms Quart J Exp Psychol **16** 104-111 [85]

Andrews DP (1967) Perception of contour orientation in the central fovea Part I: short lines Vis Res **7** 975–7 [80, 323]

Andrews TJ, Purves D (1997) Similarities in normal and binocularly rivalrous viewing Proc Nat Acad Sci **94** 9905-8 [295]

Andrews TJ, White LE, Binder D, Purves D (1996) Temporal events in cyclopean vision Proc Nat Acad Sci **93** 3689-92 [329]

Anstis SM (2000) Monocular lustre from flicker Vis Res **40** 2551-6 [296]

Anstis SM, Duncan K (1983) Separate motion aftereffects from each eye and from both eyes Vis Res **23** 161–9 [349]

Anstis SA, Ho A (1998) Nonlinear combination of luminance excursions during flicker simultaneous contrast afterimages and binocular fusion Vis Res **38** 523-9 [330]

Anstis SM, Moulden BP (1970) After–effect of seen movement: evidence for peripheral and central components Quart J Exp Psychol **22** 222–9 [349]

Antonini A, Stryker MP (1993) Development of individual geniculocortical arbors in cat striate cortex and effects of binocular impulse blockade J Neurosci **13** 3549–73 [455, 463]

Antonini A, Stryker MP (1996) Plasticity of geniculocortical afferents following brief or prolonged monocular occlusion in the cat J Comp Neurol **369** 64-82 [507]

Antonini A, Stryker MP (1998) Effect of sensory disuse on geniculate afferents to cat visual cortex Vis Neurosci **15** 401-9 [508]

Antonini A, Berlucchi G, Di Stefano M, Marzi CA (1981) Differences in binocular interactions between cortical areas 17 and 18 and superior colliculus of Siamese Cats J Comp Neurol **200** 597-611 [497]

Antonini A, Gillespie DC, Crair MC, Stryker MP (1998) Morphology of single geniculocortical afferents and functional recovery of the visual cortex after reverse monocular deprivation in the kitten J Neurosci **18** 9896-909 [517]

Anzai A, Bearse MA, Freeman RD, Cai D (1995) Contrast coding by cells in the cat's striate cortex: monocular vs binocular detection Vis Neurosci **12** 77-93 [260]

Anzai A, Ohzawa I, Freeman RD (1999a) Neural mechanisms for encoding binocular disparity: field position versus phase J Neurophysiol **82** 874-90 [252, 254, 265, 267]

Anzai A, Ohzawa I, Freeman RD (1999b) Neural mechanisms for processing binocular information I. Simple cells J Neurophysiol 82 891-908 [267]

Anzai A, Ohzawa I, Freeman RD (1999c) Neural mechanisms for processing binocular information II. Complex cells J Neurophysiol 82 909-24 [267]

Apkarian PA (1983) Visual training after long term deprivation: a case report Internat J Neurosci **19** 65-84 [501]

Apkarian PA (1993) Temporal frequency responsivity shows multiple maturation phases: state-dependent visual evoked potential luminance flicker fusion from birth to 9 months Vis Neurosci **10** 1007-18 [471]

Apkarian PA (1996) Chiasmal crossing defects in disorders of binocular vision Eye **10** 222-32 [441, 496]

Apkarian PA, Reits D (1989) Global stereopsis in human albinos Vis Res **29** 1359–70 [496]

Apkarian PA, Shallo-Hoffmann J (1991) VEP projections in congenital nystagmus; VEP asymmetry in albinism: a comparison Invest Ophthal Vis Sci **32** 2653-6 [498]

Apkarian PA, Nakayama K, Tyler CW (1981) Binocularity in the human visual evoked potentials: facilitation summation and suppression EEG Clin Neurophysiol **51** 32–48 [259, 263]

Apkarian PA, Reits D, Spekreijse H (1984) Component specificity in albino VEP asymmetry: maturation of the visual pathway anomaly Exp Brain Res **53** 285-94 [496]

Apkarian PA, Bour L, Barth PG (1994) A unique achiasmatic anomaly detected in non–albinos with misrouted retino–fugal projections Eur J Neurosci **6** 501–7 [441]

Appel MA, Campos JJ (1977) Binocular disparity as a discriminable stimulus parameter for young infants J Exp Psychol **23** 47–56 [482]

Araque A, Paarpura V, Sanzgiri RP, Haydon PG (1999) Trends in Neuroscience **22**, 208-215 [457]

Archer SM, Dubin MW, Stark LA (1982) Abnormal development of kitten retino–geniculate connectivity in the absence of action potentials Science **217** 743–5 [504]

Archie KA, Mel BW (2000) A model for intradendritic computation of binocular disparity Nat Neurosci 3 54-63 [268]

Arditi A (1986) Binocular vision In Handbook of perception and human performance Vol 1 Sensory processes and perception (ed KR Boff, L Kaufman, JP Thomas) Wiley, New York [8]

Arditi A, Anderson PA, Movshon JA (1981b) Monocular and binocular detection of moving sinusoidal gratings Vis Res **21** 329–36 [350]

Arguin M, Cavanagh P, Joanette Y (1994) Visual feature integration with an attention deficit Brain Cogn **27** 44-56 [134]

Ariel M, Daw NW, Rader RK (1983) Rhythmicity in rabbit retinal ganglion cell responses Vis Res **23** 1485–93 [136]

Ariens-Kappers CU, Huber GC, Crosby EC (1936) The comparative anatomy of the nervous system of vertebrates, including man vol 1. MacMillan, New York [459]

Ariotti PE (1973) A little known early seventeenth century treatise on vision: Benedetto Castelli's Discorso sopra la vista. Translation and critical comments Ann Sci **30** 1-30 [21]

Aristotle (1931) Parva naturalia De somni In The works of Aristotle translated into English Vol III Oxford University Press, London [347]

Arterberry ME, Yonas, A (1988) Infants' sensitivity to kinetic information for three-dimensional object shape Percept Psychophys **44** 1-6 [480]

Arterberry ME, Yonas A (2000) Perception of three dimensional shape specified by optic flow by 8-week-old infants Percept Psychophys **62** 550-6 [480]

Ashby FG, Townsend JT (1986) Varieties of perceptual independence Psychol Rev **93** 154-79 [142]

Asher H (1953) Suppression theory of binocular vision Br J Ophthal **37** 37–49 [305]

Aslin RN (1977) Development of binocular fixation in human infants J Exp Child Psychol **23** 133–50]

Aslin RN (1987) Motor aspects of visual development in infancy In Handbook of infant perception Vol 1 From sensation to perception (ed P Salapatek, LB Cohen) pp 43–113 Academic Press, Orlando FL [475]

Aslin RN (1988) Anatomical constraints on oculomotor development: implications for infant perception In Perceptual development in infancy (ed A Yonas) pp 67–104 Erlbaum, Hillsdale N J [484]

Aslin RN, Dobson V (1983) Dark vergence and dark accommodation in human infants Vis Res **32** 1671-8 [476]

Aslin RN, Dumais ST (1980) Binocular vision in infants: a review and a theoretical framework Adv Child Devel Behav **15** 53–94 [485]

Aslin RN, Jackson RW (1979) Accommodative–convergence in young infants: development of a synergistic sensory–motor system Can J Psychol **33** 222–31 [474, 476]

Aslin RN, Johnson SP (1996) Suppression of the optokinetic reflex in human infants: implications for stable fixation and shifts of attention Infant Behav Devel **19** 233-240 [475]

Aslin RN, Salapatek P (1975) Saccadic localization of visual targets by the very young human infant Percept Psychophys **17** 293-302 [475]

Aslin RN, Shea SL (1990) Velocity thresholds in human infants: implications for the perception of motion Devel Psychol **26** 589-98 [472]

Aslin RN, Dobson V, Jackson RW (1982) Dark vergence and dark focus in human infants Invest Ophthal Vis Sci **22** (Abs) 105 [476]

Asztély F, Gustafsson B (1996) Ionotropic glutamate receptors Molec Neurobiol **12** 1-12 [460]

Atkinson J (1972) Visibility of an afterimage in the presence of a second afterimage Percept Psychophys **12** 257–62 [296]

Atkinson J (2000) The developing visual brain Oxford University Press, Oxford [472, 479, 485]

Atkinson J, Braddick O (1976) Stereoscopic discrimination in infants Perception 5 29–38 [482]

Atkinson J, Braddick O (1992) Visual segmentation of oriented textures by infants Behav Brain Res 49 123-31 [471]

Atkinson J, Campbell,FW (1974) The effect of phase on the perception of compound gratings Vis Res 14 159-62 [295]

Atkinson J, Campbell FW, Fiorentini A, Maffei L (1973) The dependence of monocular rivalry on spatial frequency Perception 2 127–33 [295]

Atkinson J, Braddick O, Moar K (1977) Development of contrast sensitivity over the first three months of life in the human infant Vis Res 17 1037–44 [467]

Atkinson J, Hood B, Wattam–Bell J, Anker S, Tricklebank J (1988) Development of orientation discrimination in infancy Perception 17 587–95 [471]

Atkinson J, Smith J, Anker S, Wattam–Bell J, Braddick OJ, Moore AT (1991) Binocularity and amblyopia before and after early strabismus surgery Invest Ophthal Vis Sci 32 820 [518]

Attardi DG, Sperry RW (1963) Preferential selection of central pathways by regenerating optic fibers Exp Neurol 7 46-64 [464]

Au WWL (1993) The sonar of dolphins New York, Springer [548]

Auerbach E, Peachey NS (1984) Interocular transfer and dark adaptation to long-wave test lights Vis Res 27 1043-8 [333]

Averbuch-Heller L, Lewis RF, Zee DS (1999) Disconjugate adaptation of saccades: contribution of binocular and monocular mechanisms Vis Res 39 341-52 [425]

Azzopardi P, Cowey A (1993) Preferential representation of the fovea in the visual cortex Nature 361 719–21 [184]

Azzopardi P, Jones KE, Cowey A (1999) Uneven mapping of magnocellular and parvocellular projections from the lateral geniculate nucleus to the striate cortex in the macaque monkey Vis Res 39 2179-89 [184]

Bach M, Hoffmann MB (2000) Visual motion detection in man is governed by non-retinal mechanisms Vis Res 40 2379-85 [203]

Bach M, Meigen T (1997) Similar electrophysiological correlates of texture segregation induced by luminance orientation motion and stereo Vis Res 37 1409-14 [131]

Bach–y–Rita P (1975) Structural functional correlations in eye muscle fibres Eye muscle proprioception In Basic mechanisms of ocular motility and their clinical implications (ed G Lennerstrand, P Bach–y–Rita) pp 91–108 Pergamon, New York [493]

Backus BT, Fleet DJ, Parker AJ, Heeger, DJ (2001) Human cortical activity correlates with stereoscopic depth perception J Neurophysiol 86 2054-68 [263]

Bacon BA, Villemagne J, Bergeron A, Lepore F, Guillemot JP (1998) Spatial disparity coding in the superior colliculus of the cat Exp Brain Res 119 333-44 [245]

Bacon BA, Lepore F, Guillemot JP (1999) Binocular interactions and spatial disparity sensitivity in the superior colliculus of the Siamese cat Exp Brain Res 124 181-92 [497]

Bacon JH (1976) The interaction of dichoptically presented spatial gratings Vis Res 16 337–44 [332]

Badcock DR, Derrington AM (1987) Detecting the displacements of spatial beats: a monocular capability Vis Res 27 793-7 [282]

Badcock DR, Schor CM (1985) Depth–increment detection function for individual spatial channels J Opt Soc Am A 2 1211–15 [251]

Badcock DR, Westheimer G (1985) Spatial location and hyperacuity: the centre/surround location contribution function has two substrates Vis Res 25 1259-67 [79, 84]

Bagby JW (1957) A cross–cultural study of perceptual predominance in binocular rivalry J Abn Soc Psychol 54 331–4 [309]

Bagley CH (1949) Congenital cataracts Am J Ophthal 32 411-19 [519]

Bagnoli P, Fontanesi G, Casini G, Porciatti V (1990) Binocularity in the little owl *Athene noctua*. I Anatomical investigation of the thalamo-Wulst pathway Brain Behav Evol 35 31-9 [547]

Bahill AT, Ciuffreda KJ, Kenyon R, Stark L (1976) Dynamic and static violations of Hering's law of equal innervation Am J Optom Physiol Opt 53 786–96 [421]

Bahrick LE, Watson JS (1985) Detection of intermodal proprioceptive-visual contingency as a potential basis of self-perception in infants Devel Psychol 21 963-73 [146]

Bailey CH, Kandel ER (1993) Structural changes accompanying memory storage Ann Rev Physiol 55 397-429 [135, 458]

Baitch LW, Levi DM (1988) Evidence for nonlinear binocular interactions in human visual cortex Vis Res 28 1139–43 [260, 282]

Baitch LW, Levi DM (1989) Binocular beats: psychophysical studies of binocular interaction in normal and stereoblind humans Vis Res 29 27–37 [491]

Baitch LW, Ridder WH, Harwerth RS, Smith EL (1991) Binocular beat VEPs: losses of cortical binocularity in monkeys reared with abnormal visual experience Invest Ophthal Vis Sci 32 3096-103 [510]

Baizer JS (1982) Receptive field properties of V3 neurons in monkey Invest Ophthal Vis Sci 23 87-95 [220]

Baizer JS, Ungerleider LG, Desimone R (1991) Organization of visual inputs to the inferior temporal and posterior parietal cortex in macaques J Neurosci 11 168-90 [220, 226]

Baizer JS, Desimone R, Ungerleider LG (1993) Comparison of subcortical connections of inferior temporal and posterior parietal cortex in monkeys Vis Neurosci 10 59–72 [220]

Baker CH (1970) A study of the Sherrington effect Percept Psychophys 8 406–10 [329]

Baker FH, Grigg P, von Noorden GK (1974) Effects of visual deprivation and strabismus on the response of neurons in the visual cortex of the monkey including studies on the striate and prestriate cortex in the normal animal Brain Res 66 185–208 [509]

Baker GE, Reese BE (1993) Chiasmatic course of temporal retinal axons in the developing ferret J Comp Neurol 330 95-104 [440]

Baker R, Highstein SM (1975) Physiological identification of interneurons in the abducens nucleus Brain Res 91 292-8 [429]

Bakin JS, Nakayama K, Gilbert CD (2000) Visual responses in monkey areas V1 and V2 to three-dimensional surface configurations J Neurosci 20 8188-98 [123]

Ball EAW (1952) A study in consensual accommodation Am J Optom Arch Am Acad Optom 29 561-74 [364]

Ball K, Sekuler R (1987) Direction-specific improvement in motion discrimination Vis Res 27 953-65 [354]

Ball W, Tronick E (1971) Infant responses to impending collision: optical and real Science 171 818–20 [478]

Balliet R, Nakayama K (1978) Training of voluntary torsion Invest Ophthal Vis Sci 17 303–14 [411]

Baloh RW, Beykirch K, Honrubia V (1988) Eye movements induced by linear acceleration on a parallel swing J Neurophysiol 60 2000–13 [426]

Baltrusaitis J (1977) Anamorphic art Abrams, New York [40]

Bando T, Yamamoto N, Tsukahara N (1984) Cortical neurons related to lens accommodation in posterior lateral suprasylvian area in cats J Neurophysiol 52 879-91 [365, 431]

Bando T, Hara N, Takagi M, Yamamoto K, Toda H (1996) Roles of the lateral suprasylvian cortex in convergence eye movements in cats Prog Brain Res 112 143-56 [365, 431]

Banks MS (1980) The development of visual accommodation during early infancy Child Devel 51 646–66 [473]

Banks MS (1988) Visual recalibration and the development of contrast and optic flow perception In Perceptual development in infancy (ed A Yonas) pp 145–96 Erlbaum, Hillsdale N J [436]

Banks MS, Bennett PJ (1988) Optical and photoreceptor immaturities limit the spatial and chromatic vision of human neonates J Opt Soc Am A **5** 2059–79 [470]

Banks MS, Salapatek P (1978) Acuity and contrast sensitivity in 1 2 and 3–month–old human infants Invest Ophthal Vis Sci **17** 361–5 [467]

Banks MS, Stevens BR (1985) The development of basic mechanisms of pattern vision: spatial frequency channels J Exp Child Psychol **40** 501-27 [468]

Banks MS, Aslin RN, Letson RD (1975) Sensitive period for the development of human binocular vision Science **190** 675–7 [492, 518]

Banks MS, Geisler WS, Bennett PJ (1987) The physical limits of grating visibility Vis Res **27** 1915–27 [174]

Banton T, Bertenthal, BI (1996) Infants' sensitivity to uniform motion Vis Res **36** 1633-40 [472]

Banton T, Bertenthal BI (1997) Multiple developmental pathways for motion processing Optom Vis Sci **74** 751-60 [472]

Banton T, Levi DM (1991) Binocular summation in vernier acuity J Opt Soc Am A **8** 673-80 [324]

Bapst G (1841) Essai sur l'histoie des panoramas et des dioramas G Masson, Paris [61]

Bárány EH (1946) A theory of binocular visual acuity and an analysis of the variability of visual acuity Acta Ophthal **27** 63–92 [318]

Bárány EH, Halldén U (1948) Phasic inhibition of the light reflex of the pupil during retinal rivalry J Neurophysiol **11** 25–30 [299]

Barbasid M (1994) The Trk family of neurotrophin receptors J Neurobiol **25** 1386-403 [456]

Barbeito R, Tam WJ, Ono H (1986) Two factors affecting saccadic amplitude during vergence: the location of the cyclopean eye and a left-right bias Ophthal Physiol Opt **6** 201-5 [421]

Barlow HB (1958) Temporal and spatial summation in human vision at different background intensities J Physiol **141** 337–50 [331]

Barlow HB (1961) Possible principles underlying the transformations of sensory messages In Sensory communication (ed WA Rosenblith) pp 217-34 MIT Press, Cambridge MA [93, 135, 185]

Barlow HB (1991) Vision tells you more than "what is there" In Representations of vision (ed A Gorea) pp 319–29 Cambridge University Press, New York [135]

Barlow HB, Brindley GS (1963) Inter–ocular transfer of movement aftereffects during pressure blinding of the stimulated eye Nature **200** 1349–50 [347]

Barlow HB, Hill RM (1963) Selective sensitivity to direction of movement in ganglion cells of the rabbit retina Science **139** 412-14 [203]

Barlow HB, Reeves BC (1997) The versatility and absolute efficiency of detecting mirror symmetry in random dot displays Vis Res **19** 783-93 [157]

Barlow HB, Fitzhugh R, Kuffler SW (1957) Change of organization in the receptive fields of the cat's retina during dark adaptation J Physiol **137** 338–54 [298]

Barlow HB, Blakemore C, Pettigrew JD (1967) The neural mechanism of binocular depth discrimination J Physiol **193** 327–42 [59,237, 254, 257]

Barlow HB, Levick WR, Yoon M (1971) Responses to single quanta of light in retinal ganglion cells of the cat Vis Res **11** 87-101 [74]

Barlow HB, Kaushal TP, Hawken M, Parker AJ (1987) Human contrast discrimination and the threshold of cortical neurons J Opt Soc Am A **4** 2366-71 [121]

Barnes RSK (1968) On the evolution of elongated ocular peduncles in the Brachyura System Zool **17** 182-7 [538]

Baro JA, Lehmkuhle S, Kratz KE (1990) Electroretinograms and visual evoked potentials in long–term monocularly deprived cats Invest Ophthal Vis Sci **31** 1405–9 [502, 507]

Barres BA, Raff MC (1993) Proliferation of oligodendrocyte precursor cells depends on electrical activity in axons Nature **361** 258-60 [442]

Barrett BT, Whitaker D, McGraw PV, Herbert AM (1999) Discriminating mirror symmetry in foveal and extra-foveal vision Vis Res **39** 3737-44 [157]

Barris MC, Dawson WW, Trick LR (1981) LASCER Bode plots for normal amblyopic and stereoanomalous observers Doc Ophthal **51** 347-63 [528]

Bartfeld E, Grinvald A (1992) Relationships between orientation-preference pinwheels cytochrome oxidase blobs and ocular-dominance columns in primate striate cortex Proc Nat Acad Sci **89** 11905-9 [209, 211]

Bartlett NR, Eason RG, White CT (1968) Binocular summation in the evoked cortical potential Percept Psychophys **3** 75-6 [328]

Bartmann M, Schaeffel F (1994) A simple mechanism for emmetropization without cues from accommodation or colour Vis Res **34** 873–6 [435]

Barton RA, Harvey PH (2000) Mosaic evolution of brain structure in mammals Nature **405** 1055-8 [445]

Bärtschi WA (1981) Linear perspective (Translated into English by F Bradley) Van Nostrand Reinhold, New York [38]

Batista AP, Andersen RA (2001) The parietal reach region codes the next planned movement in a sequential reach task J Neurophysiol **85** 539-44 [228]

Battersby WS, Wagman IH (1962) Neural limitations of visual excitability IV: spatial determinants of retrochiasmal interaction Am J Physiol **203** 359–65 [334]

Bauer H (1912) Die Psychologie Alhazens Beiträge zur Geschichte der Philosophie des Mittelhalters **5** 1-72 [15, 16, 18]

Bauer HU (1995) Development of oriented ocular dominance bands as a consequence of areal geometry Neural Comput **7** 36-50 [214, 464]

Baylis G, Driver J (2001) Shape-coding in IT cells generalizes over contrast and mirror reversal, but not figure-ground reversal Nature Neurosci **4** 937-42 [222]

Baylor DA, Nunn BJ, Schnapf JL (1984) The photocurrent, noise and spectral sensitivity of rods of the monkey *Macaca fascicularis* J Physiol **357** 575-607 [166]

Baylor DA, Nunn BJ, Schnapf JL (1987) Spectral sensitivities of cones of the monkey *Macaca fascicularis* J Physiol **390** 145–60 [166]

Bear MF, Colman H (1990) Binocular competition in the control of geniculate cell size depends upon cortical N-methyl-d-aspartate receptor activation Proc Nat Acad Sci **87** 9276-9 [504]

Bear MF, Daniels JD (1983) The plastic response to monocular deprivation persists in kitten visual cortex after chronic depletion of norepinephrine J Neurosci **3** 407–16 [513]

Bear MF, Singer W (1986) Modulation of visual cortical plasticity by acetylcholine and noradrenaline Nature **320** 172–6 [513]

Bear MF, Kleinschmidt A, Gu Q, Singer W (1990) Disruption of experience–dependent synaptic modification in striate cortex by infusion of an NMDA receptor antagonist J Neurosci **10** 909–25 [511]

Beard BL, Levi DM, Reich LN (1995) Perceptual learning in parafoveal vision Vis Res **35** 1679-90 [354]

Beare JI (1906) Greek theories of elementary cognition from Alcmaeon to Aristotle Clarendon Press, Oxford [10]

Beare JI (1931) Parva naturalia De Somniis In The works of Aristotle Translated into English (ed WD Ross) Vol 3 pp 461b–462a Oxford University Press, London [10]

Bearse MA, Freeman RD (1994) Binocular summation in orientation discrimination depends on stimulus contrast and duration Vis Res **34** 19–29 [323]

Beasley WC, Peckham RH (1936) An objective study of "cyclotorsion" Psychol Bull **33** 741–2 [278]

Beaver CJ, Ji Q, Fischer QS, Daw NW (2001) Cyclic AMP-dependent protein kinase mediates ocular dominance shifts in cat visual cortex Nature Neurosci **4** 159-63 [511]

Beaver RS, Dunlop SA, Harman AM, Stirling RV, Easter SS., Roberts JD, Beazley S.S (2001) Continued neurogenesis is not a prerequisite for regeneration of a topographic retino-tectal projection Vis Res **41** 1765-70 [447]

Beazley LD (1975) Development of intertectal neuronal connections in *Xenopus*: The effects of contralateral transposition of the eye and eye removal Exp Brain Res **23** 505–18 [541]

Beazley LD, Keating MJ, Gaze RM (1972) The appearance during development of responses in the optic tectum following visual stimulation of the ipsilateral eye in *Xenopus Laevis* Vis Res **12** 407-10 [540]

Bechara A, Tranel D, Damasio H, Adolphs R, Rockland C, Damasio AR (1995) Double dissociation of conditioning and declarative knowledge relative to the amygdala and hippocampus in humans Science **299** 1115-18 [223]

Bechtoldt HP, Hutz CS (1979) Stereopsis in young infants and stereopsis in an infant with congenital esotropia J Ped Ophthal Strab **16** 49–54 [477]

Beck J (1860) Improvements in the stereoscope Photograph J **7** 19-20 [69]

Becker S, Hinton GE (1992) Self-organizing neural network that discovers surfaces in random-dot stereograms Nature **355** 161-3 [268]

Becker W, Jürgens R (1975) Saccadic reactions to double step stimuli: evidence for model feedback and continuous information uptake In Basic mechanisms of ocular motility and their clinical implications (ed G Lennerstrand, P Bach-y-Rita) pp 519-27 Pergamon, Oxford [397]

Bedell HE, Flom MC (1981) Monocular spatial distortion in strabismic amblyopia Invest Ophthal Vis Sci **20** 263–8 [522]

Bedell HE, Flom MC, Barbeito R (1985) Spatial aberrations and acuity in strabismus and amblyopia Invest Ophthal Vis Sci **26** 909–16 [520]

Bedell HE, Yap YL, Flom MC (1990) Fixation drift and nasal-temporal pursuit asymmetries in strabismic amblyopes Invest Ophthal Vis Sci **31** 968-76 [526]

Békésy G von (1967) Sensory inhibition Princeton University Press, Princeton N J [81]

Bell GP, Fenton MB (1986) Visual acuity sensitivity and binocularity in a gleaning insectivorous bat *Macrotus californicus* (Chiroptera: Phyllostomidae) Anim Behav **34** 409-14 [548]

Beller R, Hoyt CS, Marg E, Odom JV (1981) Good visual function after neonatal surgery for congenital monocular cataracts Am J Ophthal **91** 559-65 [519]

Bellocchio EE, Reimer RJ, Fremeau RT, Edwards RE (2000) Uptake of glutamate into synaptic vesicles by an inorganic phosphate transporter Science **289** 957-60 [189]

Bennett AG, Rabbetts RB (1989) Clinical visual optics Butterworth-Heinemann, London [358]

Bennett B (1859) The clairvoyant stereoscope Photograph J **5** 297-8 [69]

Bennett MJ, Smith EL, Harwerth RS, Crawford MLJ (1980) Ocular dominance eye alignment and visual acuity in kittens reared with an optically induced squint Brain Res **193** 33–45 [505]

Bennett PJ, Banks MS (1987) Sensitivity loss in odd-symmetric mechanisms underlies phase anomalies in peripheral vision Nature **326** 873–6 [522]

Bennett RG, Westheimer G (1991) The effect of training on visual alignment discrimination and grating resolution Percept Psychophys **49** 541–6 [87, 354]

Bennett WR (1933) New results in the calculation of modulation products Bell System Technical Journal **12** 228–43 [95]

Benton AL, Hécaen H (1970) Stereoscopic vision in patients with unilateral cerebral disease Neurology **20** 1084–8 [489]

Berardi N, Fiorentini A (1987) Interhemispheric transfer of visual information in humans: spatial characteristics J Physiol **384** 633-47 [232]

Berardi N, Galli L, Maffei L, Siliprandi R (1986) Binocular suppression in cortical neurons Exp Brain Res **63** 581–4 [314]

Berardi N, Bisti S, Maffei L (1987) The transfer of visual information across the corpus callosum: spatial and temporal properties in the cat J Physiol **384** 619-32 [232]

Berardi N, Domenici L, Parisi V, Pizzorusso T, Cellerino A, Maffei L (1993a) Monocular deprivation effects in the rat visual cortex and lateral geniculate nucleus are prevented by nerve growth factor (NGF). I Visual cortex Proc R Soc B **251** 17–23 [512]

Berardi L, Cellerino A, Maffei L (1993b) Monocular deprivation effects in the rat visual cortex and lateral geniculate nucleus are prevented by nerve growth factor (NGF). II Lateral geniculate nucleus Proc R Soc B **251** 25-31 [512]

Bergua A, Skrandies W (2000) An early antecedent to modern random dot stereograms—'the secret stereoscopic writing' of Ramon y Cajal Internat J Psychophysiol **36** 69-72 [27]

Berkeley G (1709) An essay towards a new theory of vision Jeremy Pepyat, Dublin. Reprinted 1922 Dutton, New York. Also in Lindsay AD (Ed) (1910) Theory of vision and other writings by Bishop Berkeley Dent, London [24, 55]

Berlucchi G (1972) Anatomical and physiological aspects of visual functions of corpus callosum Brain Res **37** 371–92 [232]

Berlucchi G, Rizzolatti G (1968) Binocularly driven neurons in visual cortex of split–chiasm cats Science **159** 308–10 [232]

Berlucchi G, Sprague JM, Antonini A, Simoni A (1979) Learning and interhemispheric transfer of visual pattern discriminations following suprasylvian lesions in split-chiasm cats Exp Brain Res **34** 551-74 [497]

Berman N, Cynader M (1972) Comparison of receptive-field organization of the superior colliculus in Siamese and normal cats J Physiol **227** 63-89 [497]

Berman N, Daw NW (1977) Comparison of the critical periods for monocular and directional deprivation in cats J Physiol **265** 279–59 [515]

Berman N, Murphy EH (1982) The critical period for alteration in cortical binocularity resulting from divergent and convergent strabismus Devel Brain Res **2** 181-202 [505, 516]

Berman N, Payne BR (1982) Monocular deprivation in the Siamese cat: development of cortical orientation and direction sensitivity without visual experience Exp Brain Res **46** 147–50 [527]

Berman N, Payne BR (1983) Alterations in connections of the corpus callosum following convergent and divergent strabismus Brain Res **274** 201-12 [453]

Berman N, Blakemore C, Cynader M (1975) Binocular interaction in the cat's superior colliculus J Physiol **276** 595–615 [187, 245]

Berman N, Murphy EH, Salinger WL (1979) Monocular paralysis in the adult cat does not change cortical ocular dominance Brain Res **164** 290–3 [494]

Berman N, Payne BR, Labar DR, Murphy EH (1982) Functional organization of neurons in cat striate cortex: variations in ocular dominance and receptive-field type with cortical laminae and location in the visual field J Neurophysiol **48** 1362-77 [213]

Berman N, Pearson HE, Payne BR (1989) Consequences of visual deprivation in the absence of binocular competitive mechanisms in Siamese cat area 17 Devel Brain Res **50** 69–87 [527]

Berns GS, Dayan P, Sejnowski TJ (1993) A correlational model for the development of disparity selectivity in visual cortex that depends on prenatal and postnatal phases Proc Nat Acad Sci **90** 8277-81 [461]

Berry MJ, Warland DK, Meister M (1997) The structure and precision of retinal spike trains Proc Nat Acad Sci **94** 5411-16 [129]

Berry RN (1948) Quantitative relations among vernier real depth and stereoscopic depth acuities J Exp Psychol **38** 708–21 [324]

Bertenthal BI, Bradbury A (1992) Infants' detection of shearing motion in random-dot displays Devel Psychol **28** 1056-66 [472]

BertrandJ (1889) Calcul des probabilité Chelsea Publishing Co, Bronx, NY [102]

Best PJ, White AM, Minai A (2001) Spatial processing in the brain: the activity of hippocampal place cells Ann Rev Neurosci **24** 459-86 [223]

Bichot NP, Schall JD, Thompson KG (1996) Visual feature selectivity in frontal eye fields induced by experience in mature macaques Nature **381** 697-9 [208]

Bickford ME, Ramcharan E, Godwin DW, Erisir A, Gnadt J, Sherman SM (2000) Neurotransmitters contained in the subcortical extraretinal inputs to the monkey lateral geniculate nucleus J Comp Neurol **427** 701-17 [178]

Bielschowsky A (1898) Über monokuläre Diplopie ohne physikalische Grundlage nebst Bemerkungen über das Sehen Schlielender Graefes Arch klin exp Ophthal **46** 143-83 [296]

Bienenstock EL, Cooper LN, Munro PW (1982) Theory for the development of neuron selectivity: orientation specificity and binocular interaction in visual cortex J Neurosci **2** 32–48 [464]

Biguer B, Prablanc C (1981) Modulation of the vestibulo–ocular reflex in eye–head orientation as a function of target distance in man In Progress in oculomotor research (ed AF Fuchs, W Brecher) pp 525–30 Elsevier, Amsterdam [428]

Bijl P, Koenderink JJ, Kappers AML (1992) Deviations from strict *M* scaling J Opt Soc Am A **9** 1233-9 [184]

Binkofski F, Dohle C, Posse S, Stephan KM, Hefter H, Seitz RJ, Freund HJ (1998) Human anterior intraparietal area subserves prehension. A combined lesion and functional MRI activation study Neurology **50** 1253-9 [227]

Birch DG, Birch EE, Enoch JM (1980) Visual sensitivity resolution and Rayleigh matches following monocular occlusion for one week J Opt Soc Am **70** 954-8 [503]

Birch EE (1985) Infant interocular acuity differences and binocular vision Vis Res **25** 571-6 [481]

Birch EE, Hale LA (1989) Operant assessment of stereoacuity Clin Vis Sci **4** 295–300 [482]

Birch EE, Held R (1983) The development of binocular summation in human infants Invest Ophthal Vis Sci **27** 1103–7 [474]

Birch EE, Petrig B (1996) FPL and VEP measures of fusion stereopsis and stereoacuity in normal infants Vis Res **36** 1321-7 [483]

Birch EE, Stager DR (1985) Monocular acuity and stereopsis in infantile esotropia Invest Ophthal Vis Sci **29** 1627-30 [520]

Birch EE, Swanson WH (2000) Hyperacuity deficits in anisometropic and strabismic amblyopes with known ages of onset Vis Res **40** 1035-40 [521]

Birch EE, Gwiazda J, Held R (1982) Stereoacuity development for crossed and uncrossed disparities in human infants Vis Res **22** 507–13 [481]

Birch EE, Gwiazda J, Held R (1983) The development of vergence does not account for the onset of stereopsis Perception **12** 331–6 [476]

Birch EE, Shimojo S, Held R (1985) Preferential–looking assessment of fusion and stereopsis in infants aged 1–6 months Invest Ophthal Vis Sci **29** 366–70 [482]

Birch EE, Stager DR, Berry P, Everett ME (1990) Prospective assessment of acuity and stereopsis in amblyopic infantile esotropes following early surgery Invest Ophthal Vis Sci **31** 758-65 [518]

Birch EE, Stager DR, Everett ME (1995) Random dot stereoacuity following surgical correction of infantile esotropia J Pediat Ophthal Strabis **32** 231-5 [518]

Birch EE, Fawcett S, Stager DR (2000a) Why does early surgical alignment improve stereoacuity outcomes in infantile esotropia? J Am Assoc Pediat Ophthal Strab **4** 10-14 [374, 518]

Birch EE, Fawcett S, Stager D (2000b) Co-development of VEP motion response and binocular vision in normal infants and infantile esotropes Invest Ophthal Vis Sci **41** 1719-23 [526]

Birkmayer W (1951) Hirnverletzungen Springer–Verlag, Vienna [489]

Birnbaum MH, Koslowe K, Sanet R (1977) Success in amblyopia therapy as a function of age: a literature survey Am J Optom Physiol Opt **54** 299-75 [528]

Birnholz JC (1981) The development of human fetal eye movement patterns Science **213** 679–80 [475]

Bishop PO (1979) Stereopsis and the random element in the organization of the striate cortex Proc R Soc B **204** 415–44 [238, 255]

Bishop PO, Davis R (1953) Bilateral interaction in the lateral geniculate body Science **118** 271-3 [179]

Bishop PO, Henry GH (1971) Spatial vision Ann Rev Psychol **22** 119–60 [233]

Bishop PO, Pettigrew JD (1986) Neural mechanisms of binocular vision Vis Res **29** 1587–600 [59]

Bishop PO, Kozak W, Vakkur GJ (1962) Some quantitative aspects of the cat's eye: axis and plane of reference visual field coordinates and optics J Physiol **163** 466–502 [236]

Bishop PO, Henry GH, Smith CJ (1971) Binocular interaction fields of single units in the cat's striate cortex J Physiol **216** 39–68 [216, 247]

Bishop PO, Coombs JS, Henry GH (1973) Receptive fields of simple cells in the cat striate cortex J Physiol **231** 31–60 [313, 346]

Bisiach E, Vallar G (1988) Hemineglect in humans. In Handbook of Neuropsychology Vol 1. (ed F Boller, J Grafman) pp 195-9. Elsevier, New York [225]

Bisti S, Carmignoto G (1986) Monocular deprivation in kittens differently affects crossed and uncrossed visual pathways Vis Res **29** 875–84 [522]

Bjorklund RA, Magnussen S (1981) A study of interocular transfer of spatial adaptation Perception **10** 511–18 [340]

Black P, Myers RE (1964) Visual functions of the forebrain commissures in the chimpanzee Science **146** 799-800 [355]

Blackie CA, Howland HC (1999) An extension of an accommodation and convergence model of emmetropization to include the effects of illumination intensity Ophthal Physiol Opt **19** 112-125 [435]

Blackwell HR (1952) Studies of psychophysical methods for measuring thresholds J Opt Soc Am **42** 606–16 [77]

Blais BS, Shouval HZ, Cooper LN (1999) The role of presynaptic activity in monocular deprivation: comparison of homosynaptic and heterosynaptic mechanisms Proc Nat Acad Sci **96** 1083-7 [464]

Blake R (1977) Threshold conditions for binocular rivalry J Exp Psychol HPP **3** 251–7 [288]

Blake R (1988) Dichoptic reading: the role of meaning in binocular rivalry Percept Psychophys **44** 133–41 [309]

Blake R (1989) A neural theory of binocular rivalry Psychol Rev **96** 145–67 [311]

Blake R, Boothroyd K (1985) The precedence of binocular fusion over binocular rivalry Percept Psychophys **37** 114–27 [306]

Blake R, Bravo M (1985) Binocular rivalry suppression interferes with phase adaptation Percept Psychophys **38** 277–80 [303]

Blake R, Camisa J (1978) Is binocular vision always monocular? Science **200** 1497–99 [306]

Blake R, Camisa J (1979) On the inhibitory nature of binocular rivalry suppression J Exp Psychol HPP **5** 315–23 [288]

Blake R, Fox R (1972) Interocular transfer of adaptation to spatial frequency during retinal ischaemia Nature New Biol **270** 76–7 [303, 340]

Blake R, Fox R (1973) The psychophysical inquiry into binocular summation Percept Psychophys **14** 161–85 [317, 331]

Blake R, Fox R (1974a) Binocular rivalry suppression: insensitive to spatial frequency and orientation change Vis Res **14** 687–92 [300]

Blake R, Fox R (1974b) Adaptation to invisible gratings and the site of binocular rivalry suppression Nature **279** 488–90 [302]

Blake R, Hirsch HVB (1975) Deficits in binocular depth perception in cats after alternating monocular deprivation Science **190** 1114–16 [513, 520]

Blake R, Lehmkuhle SW (1976) On the site of strabismic suppression Invest Ophthal **15** 660–3 [302]

Blake R, Lema SA (1978) Inhibitory effect of binocular rivalry suppression is independent of orientation Vis Res **18** 541–4 [293]

Blake R, Levinson E (1977) Spatial properties of binocular neurons in the human visual system Exp Brain Res **27** 221–32 [320, 331, 490]

Blake R, O'Shea RP (1988) "Abnormal fusion" of stereopsis and binocular rivalry Psychol Rev **95** 151–4 [307]

Blake R, Overton R (1979) The site of binocular rivalry suppression Perception **8** 143–52 [299, 302]

Blake R, Rush C (1980) Temporal properties of binocular mechanisms in the human visual system Exp Brain Res **3 8** 333–40 [320]

Blake R, Wilson HR (1991) Neural models of stereoscopic vision TINS **14** 445-52 [252]

Blake R, Fox R, McIntyre C (1971) Stochastic properties of stabilized–image binocular rivalry alternations J Exp Psychol **8 8** 327–32 [293, 311]

Blake R, Fox R, Westendorf D (1974) Visual size constancy occurs after binocular rivalry Vis Res **14** 585-6 [297]

Blake R, Breitmeyer B, Green M (1980a) Contrast sensitivity and binocular brightness: dioptic and dichoptic luminance conditions Percept Psychophys **27** 180-1 [301, 307]

Blake R, Westendorf DH, Overton R (1980b) What is suppressed during binocular rivalry? Perception **9** 223–31 [337]

Blake R, Martens W, Di Gianfilippo A (1980c) Reaction time as a measure of binocular interaction in human vision Invest Ophthal Vis Sci **19** 930–41 [490]

Blake R, Sloane M, Fox R (1981a) Further developments in binocular summation Percept Psychophys **30** 296–76 [317]

Blake R, Overton R, Lema–Stern S (1981b) Interocular transfer of visual aftereffects J Exp Psychol HPP **7** 367–81 [343, 345]

Blake R, Zimba L, Williams D (1985) Visual motion binocular correspondence and binocular rivalry Biol Cyber **52** 391–7 [291]

Blake R, Westendorf D, Fox R (1990) Temporal perturbations of binocular rivalry Percept Psychophys **48** 593-602 [311]

Blake R, Yang Y, Westendorf D (1991a) Discriminating binocular fusion from false fusion Invest Ophthal Vis Sci **32** 2821–25 [294]

Blake R, Yang Y, Wilson HR (1991b) On the coexistence of stereopsis and binocular rivalry Vis Res **31** 1191–203 [306]

Blake R, O'Shea RP, Mueller TJ (1992) Spatial zones of binocular rivalry in central and peripheral vision Vis Neurosci **8** 469–78 [297]

Blake R, Yu K, Lokey M, Norman H (1998) Binocular rivalry and motion perception J Cog Neurosci **10** 46-60 [292]

Blakemore C (1969) Binocular depth discrimination and the nasotemporal division J Physiol **205** 471–9 [231]

Blakemore C (1970a) Binocular depth perception and the optic chiasm Vis Res **10** 43–7 [232]

Blakemore C (1970c) The representation of three–dimensional visual space in the cats striate cortex J Physiol **209** 155–78 [238]

Blakemore C (1970d) The range and scope of binocular depth discrimination in man J Physiol **211** 599–622 [280]

Blakemore C (1976) The conditions required for the maintenance of binocularity in the kitten's visual cortex J Physiol **291** 423–44 [507, 513]

Blakemore C, Campbell FW (1969) On the existence of neurones in the human visual system selectively sensitive to the orientation and size of retinal images J Physiol **203** 237–60 [139, 340, 491]

Blakemore C, Cooper GF (1970) Development of the brain depends on the visual environment Nature **228** 477-8 [454]

Blakemore C, Hawken MJ (1982) Rapid restoration of functional input to the visual cortex of the cat after brief monocular deprivation J Physiol **327** 463–87 [508]

Blakemore C, Pettigrew JD (1970) Eye dominance in the visual cortex Nature **225** 429-9 [238]

Blakemore C, Price DJ (1987a) The organization and post–natal development of area 18 of the cat's visual cortex J Physiol **384** 293–92 [454]

Blakemore C, Price DJ (1987b) Effects of dark rearing on the development of area 18 of the cat's visual cortex J Physiol **384** 293–309 [500]

Blakemore C, Sutton P (1969) Size adaptation: a new aftereffect Science **166** 275–7 [350]

Blakemore C, Van Sluyters RC (1974) Reversal of the physiological effects of monocular deprivation in kittens: further evidence for a sensitive period J Physiol **237** 195–216 [517]

Blakemore C, Van Sluyters RC (1975) Innate and environmental factors in the development of the kitten's visual cortex J Physiol **482** 663–716 [453, 508]

Blakemore C, Vital–Durand F (1986a) Effects of visual deprivation on the development of the monkey's lateral geniculate nucleus J Physiol **380** 493–511 [444, 504]

Blakemore C, Vital–Durand F (1986b) Organization and post–natal development of the monkey's lateral geniculate nucleus J Physiol **380** 453–91 [444]

Blakemore C, Fiorentini A, Maffei L (1972) A second neural mechanism of binocular depth discrimination J Physiol **229** 725–49 [255]

Blakemore C, van Sluyters RC, Peck CK, Hein A (1975) Development of cat visual cortex following rotation of one eye Nature **257** 584–7 [513]

Blakemore C, Garey L, Vital–Durand F (1978) The physiological effects of monocular deprivation and their reversal in the monkey's visual cortex J Physiol **283** 223–62 [517]

Blakemore C, Hawken MJ, Mark RF (1982) Brief monocular deprivation leaves subthreshold synaptic input on neurones of the cat's visual cortex J Physiol **327** 489–505 [508, 510]

Blakemore C, Diao Y, Pu M, Wang Y, Xiao Y (1983) Possible functions of the interhemispheric connections between visual cortical areas in the cat J Physiol **337** 331–49 [233]

Blank AA (1953) Luneburg theory of binocular visual space J Opt Soc Am **43** 717–27 [137]

Blank AA (1958) Analysis of experiments in binocular space perception J Opt Soc Am **48** 911-25 [137]

Blasdel GG (1992a) Differential imaging of ocular dominance and orientation selectivity in monkey striate cortex J Neurosci **12** 3115-38 [181, 213]

Blasdel GG (1992b) Orientation selectivity preference and continuity in monkey striate cortex J Neurosci **12** 3139-61 [209, 211]

Blasdel GG, Fitzpatrick D (1984) Physiological organization of layer 4 in macaque striate cortex J Neurosci **4** 880–95 [194, 354]

Blasdel GG, Pettigrew JD (1979) Degree of interocular synchrony required for maintenance of binocularity in kitten's visual cortex J Neurophysiol **42** 1692–710 [514]

Blasdel GG, Salama G (1986) Voltage-sensitive dyes reveal a modular organization in monkey striate cortex Nature **321** 579-85 [209, 211]

Blasdel GG, Lund JS, Fitzpatrick D (1985) Intrinsic connections of macaque striate cortex: axonal projections of cells outside lamina 4C J Neurosci **5** 3350–69 [219]

Blasdel GG, Obermayer K, Kiorpes L (1995) Organization of ocular dominance columns in the striate cortex of neonatal macaque monkeys Vis Neurosci **12** 589-603 [209, 455]

Bloch S, Rivaud S, Martinoya C (1984) Comparing frontal and lateral viewing in the pigeon. III Different patterns of eye movements for binocular and monocular fixation Behav Brain Res **13** 173–82 [545]

Blodi FC, Van Allen MW (1957) Electromyography of the extraocular muscles in fusional movements Am J Ophthal **44** 136–44 [419]

Blough PM (1971) The visual acuity of the pigeon for distant targets J Exp Anal Behav **15** 57–67 [544]

Blum R (1983) The cyclostereoscope Stereo World June 29-31 [72]

Blumenfeld W (1913) Untersuchungen über die scheinbare Grösse im Sehraume Z Psychol **65** 271-404 [137]

Blunt W (1970) The dream king Hamish Hamilton, London [27]

Bobier WR, McRae M (1996) Gain change in the accommodative convergence cross-link Ophthal Physiol Opt **16** 318-25 [390]

Bobier WR, Campbell MC, Hinch M (1992) The influence of chromatic aberration on the static accommodative response Vis Res **32** 823–32 [364]

Bode–Greuel KM, Singer W (1989) The development of N–methyl–D–aspartate receptors in cat visual cortex Devel Brain Res **46** 197–204 [462]

Bodis-Wollner I, Mylin L (1987) Plasticity of monocular and binocular vision following cerebral blindness: evoked potential evidence EEG Clin Neurophysiol **68** 70-4 [488]

Boff KR, Kaufman L, Thomas JP (1986) Handbook of perception and performance. Volume I Sensory processes and perception Wiley New York [8]

Boire D, Morris R, Ptito M, Lepore F, Frost DO (1995) Effects of neonatal splitting of the optic chiasma on the development of feline visual callosal connections Exp Brain Res **104** 275-86 [453]

Bolanowski SJ (1987) Contourless stimuli produce binocular brightness summation Vis Res **27** 1943–51 [328]

Bolanowski SJ, Doty RW (1987) Perceptual "blankout" of monocular homogeneous fields (Ganzfelder) is prevented with binocular viewing Vis Res **27** 967–82 [290]

Boltz RL, Harwerth RS (1979) Fusional vergence ranges of the monkey: a behavioural study Exp Brain Res **37** 87–91 [392]

Boltz RL, Smith EL, Bennett MJ, Harwerth RS (1980) Vertical fusional vergence ranges of the rhesus monkey Vis Res **20** 83-5 [407]

Boman DK, Kertesz AE (1983) Interaction between horizontal and vertical fusional responses Percept Psychophys **33** 565-70 [410]

Boman DK, Kertesz AE (1985) Horizontal fusional responses to stimuli containing artificial scotomas Invest Ophthal Vis Sci **29** 1051–6 [397]

Bomba PC (1984) The development of orientation categories between 2 and 4 months of age J Exp Child Psychol **37** 609–36 [471]

Bonds AB (1989) Role of inhibition in the specification of orientation selectivity of cells in the cat striate cortex Vis Neurosci **2** 41–55 [313, 346]

Bonhoeffer T, Grinvald A (1993) The layout of iso–orientation domains in area 18 of cat visual cortex: optical imaging reveals a pinwheel organization J Neurosci **13** 4157–80 [211]

Bonhoeffer T, Huf J (1985) Position-dependent properties of retinal axons and their growth cones Nature **315** 409-10 [440]

Bonneh Y, Sagi D (1999) Configuration saliency revealed in short duration binocular rivalry Vis Res **39** 271-81 [293]

Bonneh,Y, Sagi D, Karni A (2001) A transition between eye and object rivalry determined by stimulus coherence Vis Res **41** 981-9 [301]

Boothe RG, Williams RA, Kiorpes L, Teller DY (1980) Development of contrast sensitivity in infant *Macaca nemestrina* monkeys Science **208** 1290–2 [467]

Boothe RG, Dobson V, Teller DY (1985) Postnatal development of vision in human and nonhuman primates Ann Rev Neurosci **8** 495–545 [474]

Borg-Graham LJ, Monier C, Frégnac Y (1998) Visual input evokes transient and strong shunting inhibition in visual neurons Nature **393** 369-73 [122, 192]

Boring EG (1942) Sensation and perception in the history of experimental psychology Appleton–Century–Crofts, New York [21]

Boring EG (1950) A history of experimental psychology Appleton–Century–Crofts, New York [117]

Born RT, Tootell BH (1992) Segregation of global and local motion processing in primate middle temporal visual area Nature **357** 497–9 [224]

Bornstein MH, Krinsky SJ, Benasich AA (1986) Fine orientation discrimination and shape constancy in young infants J Exp Child Psychol **41** 49–60 [471]

Bortolotto ZA, Bashir ZI, Davies CH, Collingridge GL (1994) A molecular switch activated by metabotropic glutamate receptors regulates induction of long-term potentiation Nature **368** 740-3 [461]

Bosking WH, Zhang, Y, Schofield B, Fitzpatrick D (1997) Orientation selectivity and the arrangement of horizontal connections in tree shrew striate cortex J Neurosci **17** 2112-7 [196]

Boss VC, Schmidt JT (1984) Activity and the formation of ocular dominance patches in dually innervated tectum of goldfish J Neurosci **4** 2891-905 [464]

Bossink CJH, Stalmeier PF M, de Weert CMM (1993) A test of Levelt's second proposition for binocular rivalry Vis Res **33** 1413–9 [287, 310]

Bossomaier T, Snyder AW (1986) Why spatial frequency processing in the visual cortex? Vis Res **29** 1307-9 [94]

Bough EW (1970) Stereoscopic vision in the macaque monkey Nature **225** 41–2 [548]

Bouman MA (1955) On foveal and peripheral interactions in binocular vision Optica Acta **1** 177–83 [288]

Bouman MA, van den Brink G (1952) On the integrate capacity in time and space of the human peripheral retina J Opt Soc Am **42** 617–20 [78, 331]

Bour LJ (1981) The influence of the spatial distribution of a target on the dynamic response and fluctuations of the accommodation of the human eye Vis Res **21** 1287-96 [361]

Bourassa CM, Rule SJ (1994) Binocular brightness: a suppression-summation trade off Can J Exp Psychol **48** 418-34 [328, 335]

Bourgeois JP, Rakic P (1993) Changes of synaptic density in the primary visual cortex of the macaque monkey from fetal to adult stage J Neurosci **13** 2801–20 [451]

Bourgeois JP, Rakic P (1996) Synaptogenesis in the occipital cortex of macaque monkey devoid of retinal input from early embryonic stages Eur J Neurosci **8** 942-50 [458]

Bourgeois JP, Goldman-Rakic PS, Rakic P (1994) Synaptogenesis in the prefrontal cortex of rhesus monkeys Cerebral Cortex **4** 78-96 [193]

Bovolenta P, Mason C (1987) Growth cone morphology varies with position in the developing mouse visual pathway from retina to first targets J Neurosci **7** 1447-60 [437]

Bowen RW, Wilson HR (1994) A two process analysis of pattern masking Vis Res **34** 645-57 [337]

Bower TGR (1966b) Slant perception and shape constancy in infants Science **151** 832-4 [479]

Bower TGR, Broughton JM, Moore MK (1970a) Infant responses to approaching objects: an indicator of response to distal variables Percept Psychophys **9** 193–6 [478]

Bower TGR, Broughton JM, Moore MK (1970b) Demonstrations of intention in the reaching behavior of neonate humans Nature **228** 679–81 [477]

Bowne SF (1990) Contrast discrimination cannot explain spatial frequency orientation or temporal frequency discrimination Vis Res **30** 449–61 [80, 126]

Boycott B, Wässle H (1999) Parallel processing in the mammalian retina Invest Ophthal Vis Sci **40** 1313-28 [166]

Boyde A (1992) Three–dimensional images of Ramón y Cajal's original preparations as viewed by confocal microscopy TINS **15** 276–8 [186]

Boylan C, Harding GFA (1983) Investigation of visual pathway abnormalities in human albinos Ophthal Physiol Opt **3** 273-85 [496]

Boyle R (1688) A disquisition about the final causes of natural things J Taylor, London See Robert Boyle the works (ed T Birch) Vol 5 Olms, Hildesheim [56]

Boynton RM (1979) Human color vision Holt Rinehart and Winston, New York [283]

Boynton RM, Wisowaty JJ (1984) Selective color effects in dichoptic masking Vis Res **27** 667–75 [334]

Bozzi Y, Pizzorusso T, Cremisi F, Rossi FM, Barsacchi G, Maffei L (1995) Monocular deprivation decreases the expression of messenger RNA for brain-derived neurotrophic factor in the rat visual cortex Neurosci **69** 1133-44 [512]

Braastad BO, Heggelund P (1985) Development of spatial receptive–field organization and orientation selectivity in kitten striate cortex J Neurophysiol **53** 1158–78 [453, 500]

Braccini C, Gambardella G, Suetta G (1980) A noise masking experiment in grating perception at threshold: the implications for binocular summation Vis Res **20** 373–6 [319, 322]

Bracewell RN (1978) The Fourier transform and its applications McGraw–Hill, New York [90, 92]

Braddick OJ (1979) Binocular single vision and perceptual processing Proc R Soc B **204** 503–12 [275]

Braddick OJ, Atkinson J, French J, Howland HC (1979) A photorefractive study of infant accommodation Vis Res **19** 1319–30 [474]

Braddick OJ, Atkinson J, Julesz B, Kropfl W, Bodis–Wollner I, Raab E (1980) Cortical binocularity in infants Nature **288** 363–5 [483]

Braddick OJ, Wattam–Bell J, Day J, Atkinson J (1983) The onset of binocular function in human infants Hum Neurobiol **2** 65–9 [483]

Braddick OJ, Wattam-Bell J, Atkinson J (1986) Orientation-specific cortical responses in early infancy Nature **320** 617-19 [471]

Bradke F, Dotti CG (1999) The role of local actin instability in axon formation Science **283** 1931-4 [448]

Bradley A, Freeman RD (1981) Contrast sensitivity in anisometropic amblyopia Invest Ophthal Vis Sci **21** 467-76 [520]

Bradley A, Freeman RD (1985a) Is reduced vernier acuity in amblyopia due to position contrast or fixation deficits? Vis Res **25** 55–66 [520]

Bradley A, Freeman RD (1985b) Temporal sensitivity in amblyopia: an explanation of conflicting reports Vis Res **25** 39–46 [525]

Bradley A, Freeman RD, Applegate R (1985) Is amblyopia spatial frequency or retinal locus specific? Vis Res **25** 47–54 [521]

Bradley A, Skottun BC, Ohzawa I, Sclar G, Freeman R D (1987) Visual orientation and spatial frequency discrimination: a comparison of single neurons and behavior J Neurophysiol **57** 755-72 [121]

Bradley DC, Andersen RA (1998) Center-surround antagonism based on disparity in primate area MT J Neurosci **18** 7552-65 [243]

Bradley DC, Maxwell M, Andersen RA, Banks MS, Shenoy KV (1996) Mechanisms of heading perception in primate visual cortex Science **273** 1544-7 [224]

Bradley DC, Chang GC, Andersen RA (1998) Encoding of three-dimensional structure-from-motion by primate area MT neurons Nature **392** 714-17 [224]

Bradley DR (1982) Binocular rivalry of real vs subjective contours Percept Psychophys **32** 85-7 [290]

Bradley DV, Fernandes A, Lynn M, Tigges M, Boothe RG (1999) Emmetropization in the rhesus monkey (*Macaca mulatta*): birth to adulthood Invest Ophthal Vis Sci **40** 214-28 [434]

Braendgaard H, Evans SM, Howard CV, Gundersen HJG (1990) The total number of neurons in the human neocortex unbiasedly estimated using optical disectors J Micros **157** 285-304 [446]

Braitenberg V (1985) An isotropic network which implicitly defines orientation columns: discussion of an hypothesis In Models of the visual cortex (ed D Rose, VG Dobson) pp 479–84 Wiley, New York [131]

Braitenberg V, Braitenberg C (1979) Geometry of orientation columns in visual cortex Biol Cyber **33** 179–86 [209]

Bray GM, Villegas-Pérez MP, Vidal–Sanz M, Aguayo AJ (1992) Death and survival of axotomized retinal ganglion cells In Regeneration and plasticity in the mammalian visual system (ed DMK Lam, GM Garth) pp 29–43 MIT Press, Cambridge MA [442]

Brecher GA (1934) Die optokinetische Auslösung von Augenrollung und rotatorischem Nystagmus Pflügers Arch ges Physiol **234** 13–28 [411]

Breese BB (1899) On inhibition Psychol Rev Monogr Supp **3** (whole number 11) [291, 295, 308]

Breese BB (1909) Binocular rivalry Psychol Rev **16** 410–15 [289]

Breinin GM (1955) The nature of vergence revealed by electromyog-raphy Arch Ophthal **54** 407–12 [419]

Breinin GM, Moldaver J (1955) Electromyography of the human extraocular muscles. I Normal kinesiology; divergence mechanism Arch Ophthal **54** 200–10 [429]

Brenner RL, Charles ST, Flynn JT (1969) Pupillary responses in rivalry and amblyopia Arch Ophthal **82** 23-9 [526]

Brewster D (1856) The stereoscope its history theory and construction John Murray, London [64]

Brickley SG, Dawes EA, Keating MJ, Grant S (1998) Synchronizing retinal activity in both eyes disrupts binocular map development in the optic tectum J Neurosci **18** 1491-1504 [540, 541]

Bridge H, Cumming BG (2001) Responses of macaque V1 neurones to binocular orientation diffences J Neurosci **21** 7293-302 [256]

Bridgeman B, Palca J (1980) Role of microsaccades in high acuity observational tasks Vis Res **20** 813–17 [366]

Bridgman CS, Smith KU (1945) Bilateral neural integration in visual perception after section of the corpus callosum J Comp Neurol **83** 57–68 [233]

Briggs F, Callaway EM (2001) Layer-specific input to distinct cell types in layer 6 of monkey primary visual cortex J Neurosci **21** 3600-8 [187]

Briggs W (1676) Ophthalmographia. London [52]

Brigham EO (1974) The fast Fourier transform Prentice–Hall Englewood Cliffs NJ [90, 92]

Brindley GS (1970) Physiology of the retina and visual pathway Williams and Wilkins, Baltimore Md [125]

Brindley GS, Lewin WS (1968) The sensations produced by electrical stimulation of the visual cortex J Physiol **196** 479–93 [184]

Britten KH, van Wezel RJA (1998) Electrical microstimulation of cortical area MST biases heading perception in monkeys Nature Neurosci **1** 59-63 [225]

Britten KH, Newsome WT, Shadlen MN, Celebrini S, Movshon JA (1996) A relationship between behavioral choice and the visual responses of neurons in macaque MT Vis Neurosci **13** 87-100 [208]

Britto J, Tannahill D, Keynes R (2002) A cricical role for sonic hedgehog signaling in the early expansion of the developing brain Nature Neurosci **5** 103-10 [445]

Broadbent H, Westall C (1990) An evaluation of techniques for measuring stereopsis in infants and young children Ophthal Physiol Opt **10** 3-7 [483]

Brodal P (1972) The corticopontine projection from the visual cortex of the cat. I The total projection and the projection from area 17 Brain Res **39** 297–317 [187]

Brodsky MC (1999) Dissociated vertical divergence: a righting reflex gone wrong Arch Ophthal **117** 1216-22 [410]

Brody CD (1998) Slow covariations in neuronal resting potentials can lead to artefactually fast cross-correlations in their spike trains J Neurophysiol **80** 3345-51 [134]

Broerse J, Dodwell PC, Ehrenstein WH (1994) Experiments on the afterimage of stimulus change (Dvorák 1870): a translation with commentary Perception **23** 1135-44 [45]

Bronson GW (1990) Changes in infants visual scanning across the 2- to 14-week period J Exp Child Psychol **49** 101-25 [475]

Brookman KE (1983) Ocular accommodation in human infants Am J Optom Physiol Opt **60** 91-9 [473]

Brosnahan D, Norcia AM, Schor CM, Taylor DG (1998) OKN perceptual and VEP direction biases in strabismus Vis Res **38** 2833-40 [525]

Bross M (1984) Effect of monocular occlusion on lateral phoria Am J Optom Physiol Opt **61** 31-3 [375]

Brotchie PR, Andersen RA, Snyder LH, Goodman SJ (1995) Head position signals used by parietal neurons to encode locations of visual stimuli Nature **375** 232-4 [144, 227]

Brouwer B, Zeeman WPC (1929) The projection of the retina in the primary optic neuron in monkey Brain **49** 1-35 [177]

Brown A, Yates PA, Burrola P, Ortuno D, Vaidya A, Jessell TM, Pfaff SL, O'Leary DDM, Lemke G (2000) Topographic mapping from the retina to the midbrain is controlled by relative but not absolute levels of EphA receptor signalling Cell **102** 77-88 [447]

Brown AM (1990) Development of visual sensitivity to light and color vision in human infants: a critical review Vis Res **30** 1159-88 [471]

Brown AM, Lindsey DT, McSweeney EM, Walters MM (1995) Infant luminance and chromatic contrast sensitivity: optokinetic nystagmus data on 3-month-olds Vis Res **35** 3145-60 [470]

Brown B, Yap MKH, Fan WCS (1993) Decrease in stereoacuity in the seventh decade of life Ophthal Physiol Opt **13** 138-42 [484]

Brown DL, Salinger WL (1975) Loss of X-cells in lateral geniculate nucleus with monocular paralysis: neural plasticity in the adult cat Science **189** 1011-12 [494, 504]

Brown RJ, Norcia AM (1997) A method for investigating binocular rivalry in real-time with the steady-state VEP Vis Res **37** 2701-8 [312]

Brown RJ, Candy TR, Norcia AM (1999) Development of rivalry and dichoptic masking in human infants Invest Ophthal Vis Sci **40** 3327-33 [484]

Bruce AS, Atchinson DA, Bhoola H (1995) Accommodation-convergence relationships and age Invest Ophthal Vis Sci **36** 406-13 [391]

Bruce CJ, Goldberg ME (1985) Primate frontal eye fields: I single neurons discharging before saccades J Neurophysiol **53** 603-35 [227]

Bruce CJ, Isley M, Shinkman PG (1981a) Visual experience and development of interocular orientation disparity in visual cortex J Neurophysiol **46** 215–28 [485]

Bruno N, Cutting JE (1988) Minimodularity and the perception of layout J Exp Psychol Gen **117** 161-70 [148]

Bruno P, Inchingola P, van der Steen J (1995) Unequal saccades produced by aniseikonic patterns: a model approach Vis Res **35** 3473-92 [424]

Brunso-Bechtold JK, Casagrande VA (1985) Presence of retinogeniculate fibres is essential for initiating the formation of each interlaminar space in the lateral geniculate nucleus Devel Brain Res **20** 123-6 [443]

Bryant P (1974) Perception and understanding in young children Methuen London [155]

Bryngdahl O (1976) Characteristics of superposed patterns in optics J Opt Soc Am **66** 87-94 [282]

Bucci MP, Kapoula Z, Eggert T, Garraud L (1997) Deficiency of adaptive control of the binocular coordination of saccades in strabismus Vis Res **37** 2767-77 [421]

Bucci MP, Kapoula Z, Eggert T (1999) Saccade amplitude disconjugacy induced by aniseikonia: role of monocular depth cues Vis Res **39** 3109-22 [425]

Bucci MP, Kapoula Z, Bernotas M, Zamfirescu F (2000) Disconjugate memory-guided saccades to disparate targets: temporal aspects Exp Brain Res **134** 133-8 [424]

Bucci MP, Gomes M, Paris G, Kapoula Z (2001) Disconjugate oculomotor learning caused by feeble image-size inequality: differences between secondary and tertiary positions Vis Res **41** 625-37 [424]

Buchs PA, Muller D (1996) Induction of long-term potentiation is associated with major ultrastructural changes of activated synapses Proc Nat Acad Sci **93** 8040-5 [193]

Buck SL, Pulos E (1987) Rod-cone interaction in monocular but not binocular pathways Vis Res **27** 479-82 [334]

Buckley EG, Seaber JH (1982) The incidence of strabismic amblyopia Invest Ophthal Vis Sci **22** (Abs) 162 [520]

Budden FJ (1972) The fascination of groups Cambridge University Press, London [103, 159]

Buhl EH, Halasy K, Somogi P (1994) Diverse sources of hippocampal unitary inhibitory postsynaptic potentials and the number of synaptic release sties Nature **368** 823-8 [196]

Buisseret P (1995) Influence of extraocular muscle proprioception on vision Physiol Rev **75** 323-38 [495]

Buisseret P, Imbert M (1976) Visual cortical cells: their developmental properties in normal and dark reared kittens J Physiol **255** 511–25 [453, 500]

Buisseret P, Maffei L (1977) Extraocular proprioceptive projections to the visual cortex Exp Brain Res 28 421–5 [494]
Buisseret P, Singer W (1983) Proprioceptive signals from extraocular muscles gate experience–dependent modifications of receptive fields in the kitten visual cortex Exp Brain Res 51 443–50 [495]
Buisseret P, Gary–Bobo E, Imbert M (1978) Ocular motility and recovery of orientational properties of visual cortical neurones in dark–reared kittens Nature 272 816–7 [495]
Buisseret P, Gary-Bobo E, Imbert M (1982) Plasticity in the kitten's visual cortex: effects of the suppression of visual experience upon the orientational properties of visual cortical cells Devel Brain Res 4 417-29 [500]
Buisseret P, Gary-Bobo E, Milleret C (1988) Development of the kitten visual cortex depends on the relationship between the plane of eye movements and visual inputs Exp Brain Res 72 83-94 [495]
Bulfone A, Smiga SM, Shimamua K, Peterson A, Puelles L, Rubenstein JLR (1995) *T-brain-1*: a homolog of *brachyury* whose expression defines molecularly distinct domains within the cerebral cortex Neuron 15 63 [446]
Bullier J Kennedy H (1983) Projection of the lateral geniculate nucleus onto cortical area V2 in the macaque monkey Exp Brain Res 53 168-72 [218, 220]
Bullier J, Girard P, Salin PA (1994) The role of area 17 in the transfer of information to extrastriate visual cortex In Cerebral cortex Volume 10 Primary visual cortex in primates (ed A Peters, KS Rockland) pp 301-30 Plenum, New York [218]
Bullier J, Schall JD, Morel A (1996) Functional streams in occipito-frontal connections in the monkey Behav Brain Res 76 89-97 [228]
Bunim MS (1940) Space in medieval painting and the forerunners of perspective AMS Press, New York [31]
Bunt AH, Minckler DS (1977) Foveal sparing New anatomical evidence for bilateral representation of central retina Arch Ophthal 95 1445–7 [231]
Buonomano DV, Merzenich MM (1998) Cortical plasticity Ann Rev Neurosci 21 149-86 [208]
Buracas GT, Albright TD (1996) Contribution of area MT to perception of three-dimensional shape: A computational study Vis Res 36 869-87 [224]
Burchfiel JL, Duffy FH (1981) Role of intracortical inhibition in deprivation amblyopia: reversal by microinontophoretic bicuculline Brain Res 206 479–84 [510]
Burian HM (1939) Fusional movements: the role of peripheral retinal stimuli Arch Ophthal 21 486–91 [408]
Burke D, Wenderoth P (1989) Cyclopean tilt aftereffects can be induced monocularly: is there a purely binocular process? Perception 18 471–82 [346]
Burke D, Alais D, Wenderoth P (1999) Determinants of fusion of dichoptically presented orthogonal gratings Perception 28 73-88 [289]
Burkhalter A, Van Essen DC (1986) Processing of color form and disparity information in visual areas VP and V2 of ventral extrastriate cortex in the macaque monkey J Neurosci 6 2327–51 [220, 242]
Burkhalter A, Bernardo KL, Charles V (1993) Development of local circuits in human visual cortex J Neurosci 13 1916–31 [452]
Burkhardt DA (1993) Synaptic feedback depolarization and color opponency in cone photoreceptors Vis Neurosci 10 981–9 [167]
Burns BD, Prichard R (1968) Cortical conditions for fused binocular vision J Physiol 197 149–71 [313]
Burr DC, Ross J, Morrone MC (1986) A spatial illusion from motion rivalry Perception 15 59-66 [296]
Burrows RC, Wancio D, Levitt P, Lillien L (1997) Response diversity and the timing of progenitor cell maturation are regulated by developmental changes in EGFR expression in the cortex Neuron 19 251-67 [450]
Burt P, Julesz B (1980) Modifications of the classical notion of Panum's fusional area Perception 9 671–82 [275]
Burton HE (1945) The optics of Euclid J Opt Soc Am 35 357–72 [10]
Busettini C, Miles FA, Schwarz U (1991) Ocular responses to translation and their dependence on viewing distance. II Motion of the scene J Neurophysiol 66 865-78 [427]
Busettini C, Miles FA, Schwarz U, Carl JR (1994) Human ocular responses to translation of the observer and of the scene: dependence on viewing distance Exp Brain Res 100 484-94 [427]
Busettini C, Miles FA, Krauzlis RJ (1996a) Short-latency disparity vergence responses and their dependence on prior saccadic eye movements J Neurophysiol 75 1392-1410 [397, 399]
Busettini C, Masson GS, Miles FA (1997) Radial optic flow induces vergence eye movements with ultra-short latencies Nature 390 512-15 [386]
Busettini C, Fitzgibbon EJ, Miles FA (2001) Short-latency vergence in humans J Neurophysiol 85 1129-52 [397]
Bush GA, van der Steen J, Miles FA (1994) When two eyes see patterns of unequal size they produce saccades of unequal amplitude In Information processing underlying gaze control (ed JM Delgardo-Garcia, E Godaux, PP Vidal) pp 291-7 Pergamon, Oxford [410, 424]
Bushnell MC, Goldberg ME, Robinson DL (1981) Behavioral enhancement of visual responses in monkey cerebral cortex. I Modulation in posterior parietal cortex related to selective visual attention J Neurophysiol 46 755–72 [207, 227]
Butler AK, Uryu K, Rougon G, Chesselet MF (1999) N-Methyl-D-aspartate receptor blockade affects polysialylated neural cell adhesion molecule expression and synaptic density during striatal development Neurosci 89 1169-81 [451]
Büttner–Ennever JA (ed) (1988) Neuroanatomy of the oculomotor system Elsevier, New York [365]
Büttner–Ennever JA, Akert K (1981) Medial rectus subgroups of the oculomotor nucleus and their abducens internuclear input in the monkey J Comp Neurol 197 17–27 [429]
Buzzelli AR (1981) Interocular transfer of a visual aftereffect in different kinds of strabismus Am J Optom Physiol Opt 58 1199-206 [492]
Buzzelli AR (1991) Stereopsis accommodative and vergence facility: do they relate to dyslexia? Optom Vis Sci 68 842-46 [374]
Cabelli RJ, Hohn A, Shatz CJ (1995) Inhibition of ocular dominance column formation by infusion of NT-4/5 or BDNF Science 297 1662-6 [463, 512]
Cabelli RJ, Shelton DL, Segal RA, Shatz CJ (1997) Blockade of endogenous ligands of trkB inhibits formation of ocular dominance columns Neuron 19 63-76 [457, 463]
Cagenello R, Arditi A, Halpern DL (1993) Binocular enhancement of visual acuity J Opt Soc Am A 10 1841-8 [324]
Cai D, DeAngelis GC, Freeman RD (1997) Spatiotemporal receptive field organization in the lateral geniculate nucleus of cats and kittens J Neurophysiol 78 1045-61 [178, 444]
Calford MB, Schmid LM, Rosa MGP (1998) Monocular focal retinal lesions induce short-term topographic plasticity in adult cat visual cortex Proc R Soc B 296 499-507 [198]
Calkins DJ, Sterling P (1996) Absence of spectrally specific lateral inputs to midget ganglion cells in primate retina Nature 381 613-5 [169]
Callaway EM (1998a) Local circuits in primary visual cortex of the macaque monkey Ann Rev Neurosci 21 47-74 [186]

Callaway EM (1998b) Prenatal development of layer-specific local circuits in primary visual cortex of the macaque monkey J Neurosci **18** 1505-27 [451]

Callaway EM, Katz LC (1990) Emergence and refinement of clustered horizontal connections in cat striate cortex J Neurosci **10** 1134–53 [452]

Callaway EM, Katz LC (1991) Effects of binocular deprivation on the development of clustered horizontal connections in cat striate cortex Proc Nat Acad Sci **88** 745-9 [452, 499]

Callaway EM, Katz LC (1992) Development of axonal arbors of layer 4 spiny neurons in cat striate cortex J Neurosci **12** 570-82 [452]

Caminiti R (1995) Spatial vision and movement in the parietal lobe Cerebral Cortex **5** N5 Special Issue [228]

Campbell A (1971) Interocular transfer of mirror-images by goldfish Brain Res **33** 486-90 [355]

Campbell AW (1905) Histological studies on the localization of the cerebral function Cambridge University Press, Cambridge. [28]

Campbell FW (1956) A high–speed infra–red recording optometer J Physiol **133** 31P [358]

Campbell FW (1957) The depth of field of the human eye Optica Acta **4** 157-64 [360, 361]

Campbell FW (1960) Correlation of accommodation between the two eyes J Opt Soc Am **50** 738 [361]

Campbell FW, Green DG (1965) Monocular versus binocular visual acuity Nature **208** 191–2 [318, 320]

Campbell FW, Gubisch RW (1966) Optical quality of the human eye J Physiol **186** 558–78 [171]

Campbell FW, Howell ER (1972) Monocular alternation: a method for the investigation of pattern vision J Physiol **225** 19-21P [295]

Campbell FW, Kulikowski JJ (1966) Orientational selectivity of the human visual system J Physiol **187** 437–45 [491]

Campbell FW, Maffei L (1971) The tilt aftereffect: a fresh look Vis Res **11** 833–40 [345, 492]

Campbell FW, Primrose JAE (1953) The state of accommodation of the human eye in darkness Trans Ophthal Soc U K **73** 353-61 [359]

Campbell FW, Robson JG (1968) Application of Fourier analysis to the visibility of gratings J Physiol **197** 551–66 [93, 138, 201]

Campbell FW, and Westheimer G (1958) Sensitivity of the eye to differences in focus J Physiol **143** 18P [362]

Campbell FW, Westheimer G (1959) Factors influencing accommodation responses of the human eye J Opt Soc Am **49** 568–71 [362, 363]

Campbell FW, Westheimer G (1960) Dynamics of accommodation responses of the human eye J Physiol **151** 285–95 [361, 364]

Campbell FW, Robson JG, Westheimer G (1959) Fluctuations of accommodation under steady viewing conditions J Physiol **145** 579–94 [361]

Campbell FW, Gilinsky AS, Howell ER, Riggs LA, Atkinson J (1973) The dependence of monocular rivalry on orientation Perception **2** 123–5 [295, 296]

Campos EC (1980) Anomalous retinal correspondence Arch Ophthal **98** 299-302 [528]

Campos EC, Chiesi C (1983) Binocularity in comitant strabismus: II Objective evaluation with visual evoked responses Doc Ophthal **55** 277-93 [528]

Campos EC, Bedell HE, Enoch JM, Fitzgerald CR (1978) Retinal receptive field–like properties and Stiles–Crawford effect in a patient with a traumatic choroidal rupture Doc Ophthal 45 381–95 [165]

Campos JJ, Langer A, Crowitz A (1970) Cardiac responses on the visual cliff in prelocomotor human infants Science **170** 196–7 [478]

Cannon MW, Fullenkamp SC (1993) Spatial interactions in apparent contrast: individual differences in enhancement and suppression effects Vis Res **33** 1685-95 [197]

Carandini M, Heeger DJ (1994) Summation and division by neurons in primate visual cortex Science **276** 1333-6 [126, 194, 197, 198]

Carandini M, Barlow HB, O'Keefe LP, Poirson AB, Movshon JA (1997) Adaptation to contingencies in macaque primary visual cortex Philos Trans R Soc **359** 1140-54 [85]

Carder RK, Jones EG, Hendry SHC (1991) Distribution of glutamate neurons and terminals in striate cortex of normal and monocularly deprived monkeys Soc Neurosci Abstr **17** 115 [509]

Cardoso de Oliveira S, Thiele A, Hoffmann PK (1997) Synchronization of neuronal activity during stimulus expectation in a direction discrimination task J Neurosci **17** 9278-60 [132, 134]

Caric D, Price DJ (1999) Evidence that the lateral geniculate nucleus regulates the normal development of visual corticocortical projections in the cat Exp Neurol **156** 353-62 [453]

Carkeet A, Levi, DM, Manny RE (1997a) Development of vernier acuity in childhood Optom Vis Sci **74** 741-50 [470]

Carlson S (1990) Visually guided behavior of monkeys after early binocular visual deprivation Internat J Neurosci **50** 185-94 [501]

Carlson S, Hyvärinen L, Raninen A (1986) Persistent behavioural blindness after early visual deprivation and active visual rehabilitation: a case report Br J Ophthal **70** 607-11 [501]

Carlson S, Pertovaara A, Tanila H (1987) Late effects of early binocular visual deprivation on the function of Brodmann's area 7 of monkeys (*Macaca arctoides*) Devel Brain Res **33** 101-11 [500]

Carlson TA, He S (2000) Visible binocular beats from invisible monocular stimuli during binocular rivalry Curr Biol **10**, 1055-8 [304]

Carlson VR (1962) Size constancy judgments and perceptual compromise J Exp Psychol **63** 68-73 [137]

Carlson WA, Eriksen CW (1966) Dichoptic summation of information in the recognition of briefly presented forms Percept Psychophys **5** 67-8 [324]

Carmignoto G, Vicini S (1992) Activity–dependent decrease in NMDA receptor responses during development of the visual cortex Science **258** 1007–11 [462]

Carmignoto G, Canella R, Candeo P, Comelli MC, Maffei L (1993) Effects of nerve growth factor on neuronal plasticity of the kitten visual cortex J Physiol **464** 343–60 [512]

Carmignoto G, Pizzorusso T, Tia S, Vicini S (1997) Brain-derived neurotrophic factor and nerve growth factor potentiate excitatory synaptic transmission in the rat visual cortex J Physiol **498** 153-64 [456]

Carmon A, Bechtoldt HP (1969) Dominance of the right cerebral hemisphere for stereopsis Neuropsychol **7** 29–39 [489]

Carney T, Shadlen MN, Switkes E (1987) Parallel processing of motion and colour information Nature **328** 647–9 [304]

Caron AJ, Caron RF, Carlson VR (1978) Do infants see objects or retinal images? Shape constancy revisited Infant Behav Devel **1** 229-43 [479]

Caron AJ, Caron RF, Carlson VR (1979) Infant perception of the invariant shape of objects varying in slant Child Devel **50** 716-21 [479]

Carpenter MK, Cui X, Hu Z, Jackson J, Sherman S, Seiger A, and Wahlberg LU (1999) *In vitro* expansion of a multipotent population of human neural progenitor cells Exp Neurol **158** 295-78 [450]

Carpenter RHS (1988) Movements of the eyes Pion, London [365, 368]

Carroll RC, Beattie EC, Zastrow M von, Malenka RC (2001) Role of AMPA receptor endocytosis in synaptic plasticity Nature Rev Neurosci **2** 315-24 [460]

Carroll WM, Jay BS, McDonald WI, Halliday AM (1980) Two distinct patterns of visual evoked response asymmetry in human albinism Nature **286** 604-6 [496]

Carter BAR (1970) Perspective In The Oxford companion to art (ed H Osborne) pp 840-61 Clarendon Press, Oxford [32, 36]

Carter DB (1958) Studies of fixation disparity. II Apparatus, procedure and the problem of constant error Am J Optom Arch Am Acad Optom **35** 590-8 [378]

Carter DB (1963) Effects of prolonged wearing of prism Am J Optom Physiol Opt **40** 295–73 [382]

Carter DB (1964) Fixation disparity with and without foveal fusion contours Am J Optom Arch Am Acad Optom **41** 729-36 [380]

Carter DB (1965) Fixation disparity and heterophoria following prolonged wearing of prisms Am J Optom Arch Am Acad Optom **42** 141–52 [382, 383]

Carterette EC, Friedman MP (1974) Handbook of perception Vol II Psychophysical judgment and measurement Academic Press, New York [73]

Cartmill M (1974) Rethinking primate origins Science **184** 436–43 [532]

Casagrande VA, Brunso–Bechtold JK (1988) Development of lamination in lateral geniculate nucleus: critical factors In Advances in neural and behavioral development (ed PG Shinkman) Vol 3 pp 33–78 Ablex, Norwood NJ [178, 444]

Casagrande VA, Condo GJ (1988) Is binocular competition essential for layer formation in the lateral geniculate nucleus? Brain Behav Evol **31** 198-208 [443]

Casanova C, Freeman RD, Nordmann JP (1989) Monocular and binocular response properties of cells in the striate-recipient zone of the cat's lateral posterior-pulvinar complex J Neurophysiol **62** 544-57 [187, 245, 256]

Casini G, Porciatti V, Fontanesi G, Bagnoli P (1992) Wulst efferents in the little owl *Athene noctua*: an investigation of projections to the optic tectum Brain Behav Evol **39** 101-15 [547]

Castellani V, Boltz J (1997) Membrane-associated molecules regulate the formation of layer-specific cortical circuits Proc Nat Acad Sci **94** 7030-5 [451]

Castelli B (1669) Discorso sopra la vista See Ariotti 1973 [21]

Castrén E, Zafra F, Thoenen H, Lindholm D (1992) Light regulates expression of brain-derived neurotrophic factor mRNA in rat visual cortex Proc Nat Acad Sci **89** 9444-8 [499]

Catalano SM, Shatz CJ (1998) Activity-dependent cortical target selection by thalamic axons Science **281** 559-62 [458]

Catania AC (1965) Interocular transfer of discriminations in the pigeon J Exp Anal Behav **8** 145-55 [354]

Cavanagh P (1987) Reconstructing the third dimension: interactions between color texture motion binocular disparity and shape Comput Vis Gr Im Proc **37** 171-95 [149]

Cavanagh P, Arguin M, Treisman A (1990) Effect of surface medium on visual search for orientation and size features J Exp Psychol HPP **16** 479-91 [124]

Caviness VS, Takahashi T, Nowakowski RS (1995) Numbers time and neocortical neuronogenesis: a general developmental and evolutionary model TINS **18** 379-83 [450]

Cavonius CR (1979) Binocular interactions in flicker Quart J Exp Psychol **31** 273–80 [329, 400]

Celebrini S, Newsome WT (1994) Neuronal and psychophysical sensitivity to motion signals in extrastriate area MST of the macaque monkey J Neurosci **14** 4109-27 [225]

Celebrini S, Thorpe S, Trotter Y, Imbert M (1993) Dynamics of orientation coding in area V1 of the awake primate Vis Neurosci **10** 811–25 [199]

Challacombe JF, Snow DM, Letourneau PC (1996) Role of cytoskeleton in growth cone motility and axonal elongation Sem Neurosci **8** 67-80 [438]

Chalupa LM, Lia B (1991) The nasotemporal division of retinal ganglion cells with crossed and uncrossed projections in the fetal rhesus monkey J Neurosci **11** 191-202 [441]

Chalupa LM, Williams RW, Henderson Z (1984) Binocular interaction in the fetal cat regulates the size of the ganglion cell population Neurosci **12** 1139–46 [442]

Chan SO and Guillery RW (1994) Changes in fiber order in the optic nerve and tract of rat embryos J Comp Neurol **344** 20-32 [439]

Chance FS, Nelson SB, Abbott LF (1999) Complex cells as cortically amplified simple cells Nature Neurosci **2** 277-82 [195]

Chandna A, Pennefather PM, Kovács I, Norcia AM (2001) Contour integration deficits in anisometropic amblyopia Invest Ophthal Vis Sci **42** 875-8 [524]

Chan–Palay V, Palay SL, Billings–Gagliardi SM (1974) Meynert cells in the primate visual cortex J Neurocytol **3** 631–58 [209]

Chao MV (1992) Neurotrophin receptors: a window into neuronal differentiation Neuron **9** 583-93 [456]

Chao-yi L, Creutzfeldt O (1984) The representation of contrast and other stimulus parameters by single neurons in area 17 of the cat Pflügers Arch ges Physiol 401 304-14 [198]

Chapman B, Stryker MP (1993) Development of orientation selectivity in ferret visual cortex and effects of deprivation J Neurosci **13** 5251-62 [500]

Chapman B, Zahs KR, Stryker MP (1991) Relation of cortical cell orientation selectivity to alignment of receptive fields of the geniculocortical afferents that arborize within a single orientation column in ferret visual cortex J Neurosci **11** 1347-58 [199]

Charman WN (1991) Optics of the human eye In Vision and visual dysfunction Vol 1 Visual optics and instrumentation (ed WN Charman) pp 1–29 MacMillan, London [164]

Charman WN, Heron G (1988) Fluctuations in accommodation: a review Ophthal Physiol Opt **8** 153-64 [361]

Charman WN, Tucker J (1977) Dependence of accommodation response on the spatial frequency spectrum of the observed object Vis Res **17** 129-39 [361]

Charman WN, Tucker J (1978) Accommodation as a function of object form Am J Optom Physiol Opt **55** 84-92 [362]

Chatfield C (1997) The analysis of time series Chapman and Hall, London [100]

Chaturvedi V, van Gisbergen JAM (1997) Specificity of saccadic adaptation in three-dimensional space Vis Res **37** 1367-82 [424]

Chaturvedi V, van Gisbergen AM (1998) Shared target selection for combined version-vergence eye movements J Neurophysiol 80 849-62 [419]

Chaturvedi V, van Gisbergen JAM (1999) Perturbation of combined saccade-vergence movements by micro-stimulation in monkey superior colliculus J Neurophysiol **81** 2279-96 [431]

Chaturvedi V, Van Gisbergen JAM (2000) Stimulation in the rostral pole of monkey superior colliculus: effects on vergence eye movements Exp Brain Res **132** 72-8 [431]

Chaudhuri A, Matsubara JA, Cynader MS (1995) Neuronal activity in primate visual cortex assessed by immunostaining for the transcription factor Zif298 Vis Neurosci **12** 35-50 [182, 213, 458]

Chawanya T, Aoyagi T, Nishikawa I, Okuda K, Kuramoto Y (1993) A model for feature linking via collective oscillations in the primary visual cortex Biol Cyber **68** 483–90 [136]

Cheeseman EW, Guyton DL (1999) Vertical fusional vergence: the key to dissociated vertical deviation Arch Ophthal **117** 1188-91 [410, 411]

Chen CC, Kasamatsu Y, Polat U, Norcia AM (2001) Contrast response characteristics of long-range lateral interactions in cat striate cortex Neuro Report 12 655-61 [460]

Chen DF, Schneider GE, Martinou JC, Tonegawa S (1997) Bcl-2 promotes regeneration of severed axons in mammalian CNS Nature 385 434-9 [439, 456]

Chen HX, Otmakhov N, Strack S, Colbran RJ, Lisman J E (2001) Is persistent activity of calcium/calmodulin-dependent kinase required for the maintenance of LTP? J Neurophysiol 851368-76 [460]

Chen Y, Wang Y, Qian N (2001) Disparity tuning to time-varying stimuli J Neurophysiol 86 143-55 [267]

Cheng H, Chino YM, Smith EL, Hamamoto J, Yoshida K (1995) Transfer characteristics of X LGN neurons in cats reared with early discordant binocular vision J Neurophysiol 74 2558-72 [504]

Cheng K, Hasegawa T, Saleem KS, Tanaka K (1997) Comparison of neuronal selectivity for stimulus speed length and contrast in the prestriate visual cortical area V4 and MT of the macaque monkey J Neurophysiol 71 2299-80 [221]

Chen-Huang C, McCrea RA (1998) Viewing distance related sensory processing in the ascending tract of Deiters vestibulo-ocular reflex pathway J Vestib Res 8 175-84 [427]

Cherry EC (1953) Some experiments on the recognition of speech with one and two ears J Acoust Soc Am 25 975–9 [309]

Chérubin d'Orléans P (1671) La dioptique oculaire ou la theorique la positive et la mechanique de l'oculaire dioptique en toutes ses especes Jolly et Benard, Paris [62]

Chérubin d'Orléans P (1677) La vision parfaite ou les concours des deux axes de la vision en un seul point de l'objet Marbre–Cramoisy, Paris [62]

Chesselden W (1728) An account of some observations made by a young gentleman who was born blind or lost his sight so early that he had no remembrance of ever having seen and was couched between 13 and 14 years of age Philos Tr R Soc 35 447-50 [501]

Cheung BSK, Howard IP (1991) Optokinetic torsion: dynamics and relation to circularvection Vis Res 31 1327–36 [411]

Chichilnisky EJ, Baylor DA (1999) Receptive-field microstructure of blue-yellow ganglion cells in primate retina Nature Neurosci 2 889-93 [168]

Chin NB, Breinin GM (1967) Ratio of accommodative convergence to accommodation Arch Ophthal 77 752-6 [390]

Chino YM (1997) Receptive-field plasticity in the adult visual cortex: dynamic signal rerounting or experience-dependent plasticity Sem Neurosci 9 34-46 [197]

Chino YM, Kaplan E (1988) Abnormal orientation bias of LGN neurons in strabismic cats Invest Ophthal Vis Sci 29 644-8 [506]

Chino YM, Shansky MS, Hamasaki DI (1980) Development of receptive field properties of retinal ganglion cells in kittens raised with convergent squint Exp Brain Res 39 313-20 [503]

Chino YM, Shansky MS, Jankowski WL, Banser FA (1983) Effects of rearing kittens with convergent strabismus on the development of receptive field properties in striate cortex neurons J Neurophysiol 50 295–86 [509]

Chino YM, Ridder WH, Czora EP (1988) Effects of convergent strabismus on spatio-temporal response properties of neurons in cat area 18 Exp Brain Res 72 294-78 [505]

Chino YM, Smith EL, Wada H, Ridder WH, Langston AL, Lesher GA (1991) Disruption of binocularly correlated signals alters the postnatal development of spatial properties in cat striate cortical neurons J Neurophysiol 65 841-59 [506]

Chino YM, Kaas JH, Smith EL, Langston AL, Cheng H (1992) Rapid reorganization of cortical maps in adult cats following restricted deafferentation in retina Vis Res 32 789-96 [198]

Chino YM, Cheng H, Smith EL, Garraghty PE, Roe AW, Sur M (1994a) Early discordant binocular vision disrupts signal transfer in the lateral geniculate nucleus Proc Nat Acad Sci 91 6938-42 [504]

Chino YM, Smith EL, Yoshida K, Cheng H, Hamamoto J (1994b) Binocular interactions in striate cortical neurons of cats reared with discordant visual inputs J Neurosci 14 5050-67 [505]

Chino YM, Smith EL, Hatta S, Cheng H (1997) Postnatal development of binocular disparity sensitivity in neurons of the primate visual cortex J Neurosci 17 296-307 [455, 517]

Chino YM, Smith EL, Zhang B, Matsuura K, Mori T, Kaas JH (2001) Recovery of binocular responses by cortical neurons after early monocular lesions Nature Neurosci 4 689-90 [198, 509]

Chklovskii DB (2000) Binocular disparity can explain the orientation of ocular dominance stripes in primate primary visual area (V1) Vis Res 40 1765-73 [215]

Choudhury BP (1980) Binocular depth vision in the rabbit Exp Neurol 68 453-64 [549]

Choudhury BP, Whitteridge D, Wilson ME (1965) The function of the callosal connections of the visual cortex Quart J Exp Physiol 50 215–19 [232]

Chow KL (1973) Neuronal changes in the visual system following visual deprivation In Handbook of sensory physiology (ed R Jung) Vol VII/3A pp 599–630 Springer, New York [499, 504]

Christ RE, Kapadia MK, Westheimer G, Gilbert CD (1998) Perceptual learning of spatial localization: specificity for orientation position and context J Neurophysiol 78 2889-94 [87, 88]

Christakos CN (1994) Analysis of synchrony (correlations) in neural populations by means of unit-to-aggregate coherence computations Neurosci 58 43-57 [136]

Chubb C, Sperling G (1988) Drift-balanced random stimuli: a general basis for studying non-Fourier motion perception J Opt Soc Am A 5 1986-2007 [129]

Churchland PS, Sejnowski TJ (1988) Perspectives on cognitive neuroscience Science 272 741-5 [183]

Cigánek L (1970) Binocular addition of the visually evoked response with different stimulus intensities in man Vis Res 10 479–87 [259]

Ciner EB, Scheiman MM, Schanel–Klitsch E (1989) Stereopsis testing in 18– to 35–month–old children using operant preferential looking Optom Vis Sci 66 782–7 [482]

Ciner EB, Schanel–Klitsch E, Scheiman MM (1991) Stereoacuity development in young children Optom Vis Sci 68 533–6 [482]

Ciner EB, Schanel–Klitsch E, Herzberg C (1996) Stereoacuity development: 6 months to 5 years A new tool for testing and screening Optom Vis Sci 73 43-8 [482]

Ciuffreda KJ (1991) Accommodation and its anomalies In Vision and visual dysfunction Vol 1 Visual optics and instrumentation (ed WN Charman) pp 231–79 MacMillan, London [365]

Ciuffreda KJ, Hokoda SC (1983) Spatial frequency dependence of accommodative responses in amblyopic eyes Vis Res 23 1585-94 [361, 526]

Ciuffreda KJ, Hokoda SC (1985) Subjective vergence error at near during active head rotation Ophthal Physiol Opt 5 411-15 [395]

Ciuffreda KJ, Kenyon RV (1983) Accommodative vergence and accommodation in normals amblyopes and strabismics In Vergence eye movements: Basic and clinical aspects (ed MC Schor, KJ Ciuffreda) pp 99–162 Butterworth, Boston [391]

Ciuffreda KJ, Kenyon RV, Stark L (1980) Increased drift in amblyopic eyes Br J Ophthal 64 7–14 [526]

Ciuffreda KJ, Levi DM, Selenow A (1992) Amblyopia: Basic and clinical aspects Butterworth, Boston [520]

Ciuffreda KJ, Rosenfield M, Chen HW (1997) The AC/A ratio age and presbyopia Ophthal Physiol Opt 17 307-15 [391]

Clare MH, Bishop GH (1954) Responses from an association area secondarily activated from optic cortex J Neurophysiol **17** 271-7 [28]

Clark WEL, Penman GG (1934) The projection of the retina in the lateral geniculate body Proc R Soc B **114** 291-313 [177]

Clarke PGH, Whitteridge D (1976) The cortical visual areas of the sheep J Physiol **256** 497–508 [549]

Clarke PGH, Donaldson IML, Whitteridge D (1976) Binocular visual mechanism in cortical areas I and II of the sheep J Physiol **256** 509–29 [550]

Clarke PGH, Ramachandran VS, Whitteridge D (1979) The development of the binocular depth cells in the secondary visual cortex of the lamb Proc R Soc B **204** 455–65 [550]

Clarke S, Miklossy J (1990) Occipital cortex in man: organization of callosal connections related myelo– and cytoarchitecture and putative boundaries of functional visual areas J Comp Neurol **298** 188–214 [232]

Claudet A (1865) Moving photographic figures J Franklin Inst **50** 346-50 [71]

Cleary M, Houston CA, McFadzean RM, Dutton GN (1998) Recovery in microtropia: implications for aetiology and neurophysiology Br J Ophthal **82** 225-31 [374]

Clelend BG, Mitchell D, Crewther SG, Crewther DP (1980) Visual resolution of retinal ganglion cells in monocularly–deprived cats Brain Res **192** 291–66 [502]

Clelend BG, Crewther DP, Crewther SG, Mitchell DE (1982) Normality of spatial resolution of retinal ganglion cells in cat with strabismus amblyopia J Physiol **329** 235–49 [502]

Cloarec A (1978) Estimation of hit distance by Ranatra Biol Behav **4** 173–91 [532]

Clothiaux EE, Bear MF, Cooper LN (1991) Synaptic plasticity in visual cortex: comparison of theory with experiment J Neurophysiol **66** 1785–804 [135, 514]

Clower DM, Hoffman JM, Votaw JR, Faber TL, Woods RP, Alexander GE (1996) Role of posterior parietal cortex in the recalibration of visually guided reaching Nature **383** 618-21 [144]

Clowes MB (1971a) Picture descriptions In Artificial intelligence and heuristic programming (ed NV Findler, B Meltzer) pp 245-57 Edinburgh University Press, Edinburgh [153]

Cobb WA, Morton HB, Ettlinger G (1967) Cerebral potentials evoked by pattern reversal and their suppression in visual rivalry Nature **216** 1123-5 [312]

Cocker KD, Moseley MJ, Bissenden JG, Fielder AR (1994) Visual acuity and pupillary responses to spatial structure in infants Invest Ophthal Vis Sci **35** 2920-5 [468]

Coe B (1981) The history of movie photography Eastview Editions Westfield NJ [61, 70, 71]

Cogan AI (1979) The relationship between the apparent vertical and the vertical horopter Vis Res **19** 655–65 [412]

Cogan AI (1987) Human binocular interaction: towards a neural model Vis Res **27** 2125–39 [327, 342]

Cogan AI (1989) Do background luminances interact during binocular fusion? Percept Psychophys **46** 560–6 [333]

Cogan AI, Silverman G, Sekuler R (1982) Binocular summation in detection of contrast flashes Percept Psychophys **31** 330–8 [331]

Cogan AI, Clarke M, Chan H, Rossi A (1990) Two–pulse monocular and binocular interactions at the differential luminance threshold Vis Res **30** 1617–30 [331]

Cogan DG (1956) Neurology of the ocular muscles Thomas Springfield Illinois [428]

Cohen J, Burne JF, Winter J, Bartlett P (1986) Retinal ganglion cells lose response to laminin with maturation Nature **322** 465-67 [438]

Cohn TE, Leong H, Lasley DJ (1981) Binocular luminance detection: availability of more than one central interaction Vis Res **21** 1017–23 [327]

Colby CL, Goldberg ME (1999) Space and attention in parietal cortex Ann Rev Neurosci 22 319-49 [228]

Colby CL, Duhamel JR, Goldberg ME (1993) Ventral intraparietal area of the macaque: anatomic location and visual response properties J Neurophysiol **69** 902–14 [226]

Cole RG, Boisvert RP (1974) Effect of fixation disparity on stereo-acuity Am J Optom Physiol Opt **51** 206-13 [377]

Colello RJ, Guillery RW (1992) Observations on the early development of the optic nerve and tract of the mouse J Comp Neurol **317** 357-78 [440]

Colello RJ, Jeffery G (1991) Evaluation of the influence of optic stalk melanin on the chiasmatic pathways in the developing rodent visual system J Comp Neurol **305** 304-12 [496]

Colello RJ, Pott U, Schwab ME (1994) The role of oligodendrocytes and myelin on axon maturation in the developing rat retinofugal pathway J Neurosci **14** 2594-605 [442]

Collett TS (1977) Stereopsis in toads Nature **297** 349–51 [542]

Collett TS (1978) Peering – a locust behaviour pattern for obtaining motion parallax information J Exp Biol **76** 237–41 [534]

Collett TS (1987) Binocular depth vision in arthropods TINS **10** 1–2 [537]

Collett TS, Harkness LIK (1982) Depth vision in animals In Analysis of visual behavior (ed DJ Ingle, MA Goodale, RJW Mansfield) pp 111–76 MIT Press, Cambridge MA [548]

Collett TS, Paterson CJ (1991) Relative motion parallax and target localization in the locust, *Schistocerca gregaria* J Comp Physiol A **169** 615-21 [534]

Collett TS, Udin SB (1988) Frogs use retinal elevation as a cue to distance J Comp Physiol A **163** 677-83 [542]

Collett TS, Udin SB, Finch DJ (1987) A possible mechanism for binocular depth judgments in anurans Exp Brain Res **66** 35–40 [542]

Collewijn H (1970) The normal range of horizontal eye movement in the rabbit Exp Neurol **28** 132-43 [549]

Collewijn H, Erkelens CJ (1990) Binocular eye movements and the perception of depth In Eye movements and their role in visual and cognitive processes: review of oculomotor research (ed E Kowler) pp 213–61 Elsevier, Amsterdam [368]

Collewijn H, van der Mark F, Jansen TC (1975) Precise recording of human eye movements Vis Res **15** 447–50 [365, 413]

Collewijn H, Winterson BJ, Dubois MFH (1978) Optokinetic eye movements in albino rabbits: inversion in anterior visual field Science **199** 1351-3 [496]

Collewijn H, van der Steen J, Ferman L, Jansen TC (1985) Human ocular counterroll: assessment of static and dynamic properties from electromagnetic scleral coil recordings Exp Brain Res **59** 185–96 [410, 411, 413]

Collewijn H, Erkelens CJ, Steinman RM (1988a) Binocular co–ordination of human horizontal saccadic eye movements J Physiol **404** 157–82 [421]

Collewijn H, Erkelens CJ, Steinman RM (1988b) Binocular co–ordination of human vertical saccadic eye movements J Physiol **404** 183–97 [422]

Collewijn H, Erkelens CJ, Steinman RM (1995) Voluntary binocular gaze-shifts in the plane of regard: dynamics of version and vergence Vis Res **35** 3335-3358 [423]

Collewijn H, Erkelens CJ, Steinman RM (1997) Trajectories of the human binocular fixation point during conjugate and non-conjugate gaze-shifts Vis Res **37** 1049-69 [421, 422, 423]

Collins CC, Carlson MR, Scott AB, Jampolsky A (1981) Extraocular muscle forces in normal human subjects Invest Ophthal Vis Sci 20 652–64 [422]

Collins ET (1922) Arboreal life and the evolution of the human eye Lea and Febinger, New York [532]

Collyer SC, Bevan W (1970) Objective measurement of dominance control in binocular rivalry Percept Psychophys 8 437–9 [308]

Coltheart M (1973) Colour–specificity and monocularity in the visual cortex Vis Res 13 2595–8 [344]

Connor CE, Gallant JL, Preddie DC, Van Essen DC (1996) Responses in area V4 depend on the spatial relationship between stimulus and attention J Neurophysiol 75 1306-8 [207]

Connors BW, Gutnick MJ (1990) Intrinsic firing patterns of diverse neocortical neurons TINS 13 99-104 [188]

Constantine–Paton M and Law MI (1978) Eye-specific termination bands in tecta of three-eyed frogs Science 202 639-41 [464]

Constantine–Paton M, Cline HT, Debski E (1990) Patterned activity synaptic convergence and the NMDA receptor in developing visual pathways Ann Rev Neurosci 13 129–54 [462]

Cook EP, Maunsell JHR (2002) Attentional modulation of behavioral performance and neuronal responses in middle temporal and ventral intraparietal areas of macaque monkey J Neurosci 22 1994-2004 [208]

Cook JE, Chalupa LM (2000) Retinal mosaics: new insights into an old concept TINS 23 29-34 [436]

Cook M, Field J, Griffiths K (1978) The perception of solid form in early infancy Child Devel 49 866-9 [479]

Cook PM, Prusky G, Ramoa AS (1999) The role of spontaneous retinal activity before eye opening in the maturation of form and function in the retinogeniculate pathway of the ferret Vis Neurosci 16 491-501 [444]

Cool SJ, Crawford MLJ (1972) Absence of binocular coding in striate cortex units of Siamese cats Vis Res 12 1809-14 [497]

Cooper GR, McGillem CD (1967) Methods of signal and system analysis Holt Rinehart and Winston, New York [92]

Cooper J, Feldman J (1978a) Operant conditioning and assessment of stereopsis in young children Am J Optom Physiol Opt 5 5 532–42 [482]

Cooper J, Feldman J (1978b) Random–dot stereogram performance by strabismic amblyopic and ocular–pathology patients in an operant–discrimination task Am J Optom Physiol Opt 55 599–609 [529]

Cooper J, Feldman J (1979) Assessing the Frisby stereo test under monocular viewing conditions J Am Optom Assoc 50 807–9 [373]

Cooper J, Feldman J (1981) Depth perception in strabismics Br J Ophthal 65 510-11 [373]

Cooper J, Feldman J, Medlin D (1979) Comparing stereoscopic performance of children using the Titmus TNO and Randot stereo tests J Am Optom Assoc 50 821-5 [482]

Cooper J, Feldman J, Horn D, Dibble C (1981) Reliability of fixation disparity curves Am J Optom Physiol Opt 58 960-4 [382]

Cooper ML, Pettigrew JD (1979a) The decussation of the retinothalamic pathway in the cat with a note on the major meridians of the cat's eye J Comp Neurol 187 285-312 [231]

Cooper ML, Pettigrew JD (1979b) The retinothalamic pathways in Siamese cats J Comp Neurol 187 313-48 [497]

Corballis MC, Beale IL (1970a) Bilateral symmetry and behaviour Psychol Rev 77 451-64 [157]

Corballis MC, Beale IL (1970b) Monocular discrimination of mirror-image obliques by pigeons: evidence for lateralized stimulus control Anim Behav 18 563-6 [355]

Corbetta M, Miezin FM, Dobmeyer S, Shulman GL, Petersen SE (1990) Attentional modulation of neural processing of shape color and velocity in humans Science 278 1556-9 [208]

Coren S, Kaplan CP (1973) Patterns of ocular dominance Am J Optom Arch Am Acad Optom 50 283–92 [295]

Cormack LK, Stevenson SB, Schor CM (1993) Disparity-tuned channels of the human visual system Vis Neurosci 10 585-96 [250, 251]

Cornsweet TN (1962) The staircase method in psychophysics Am J Psychol 75 485–91 [76]

Corrette BJ (1989) Prey capture in the praying mantis *Tenodera aridifolia sinensis*: coordination of the capture sequence and strike movements J Exp Biol 148 147–80 [533]

Costa LF (1994) Topographical maps of orientation specificity Biol Cyber 71 537-46 [212]

Costano RM, Gardner EP (1981) Multiple-joint neurons in somatosensory cortex of awake monkeys Brain Res 214 321-33 [143]

Courant R, Robbins H (1941) What is Mathematics? Oxford University Press, London [160]

Cova A, Galiana HL (1995) Providing distinct vergence and version dynamics in a bilateral oculomotor network Vis Res 35 3359-71 [405]

Cova A, Galiana HL (1996) A bilateral model integrating vergence and the vestibulo-ocular reflex Exp Brain Res 107 435-52 [405]

Cowan WM (1973) Neuronal death as a regulative mechanism in the control of cell number in the nervous system Academic Press, New York [442]

Cowan WM, Südhof TC, Stevens CF (2001) Synapses Johns Hopkins University Press, Baltimore [193]

Cowey A (1979) Cortical maps and visual perception: the Grindley Memorial Lecture Quart J Exp Psychol 31 1–17 [218]

Cowey A (1985) Disturbances of stereopsis by brain damage In Brain mechanisms and spatial vision (ed DJ Ingle, M Jeannerod, N Lee) pp 259–78 Nijhoff Dordrecht [233, 488]

Cowey A, Rolls ET (1974) Human cortical magnification factor and its relation to visual acuity Exp Brain Res 21 447–54 [184]

Cowey A, Wilkinson F (1991) The role of the corpus callosum and extrastriate visual areas in stereoacuity in macaque monkeys Neuropsychol 29 465–79 [239]

Cowey A, Parkinson AM, Warnick L (1975) Global stereopsis in monkeys Quart J Exp Psychol 27 93–109 [548]

Coxeter HSM (1961) Non-Euclidean geometry University of Toronto Press Toronto [109]

Coxeter HSM (1964) Projective geometry Blaisdell, New York [105]

Cozzi A, Crespi, B, Valentinotti F, Wörgötter F (1997) Performance of phase-based algorithms for disparity estimation Mach Vis Appl 9 334-40 [266, 268]

Crabus H, Stadler M (1973) Untersuchungen zur Localisierung von Wahrnehmungsprozessen: figurale Nachwirkungen bei binocularen Wettstreit-Bedingungen Perception 2, 67-77 [302]

Crair MC, Ruthazer ES, Gillespie DC, Stryker MP (1997) Ocular dominance peaks at pinwheel center sigularities of the orientation map in cat visual cortex J Neurophysiol 77 3381-5 [209]

Crair MC, Gillespie DC, Stryker MP (1998) The role of visual experience in the development of columns in cat visual cortex Science 279 566-70 [454]

Crair MC, Horton JC, Antonini A, Stryker MP (2001) Emergence of ocular dominance columns in cat visual cortex by 2 weeks of age J Comp Neurol 430 235-49 [454]

Craske B, Crawshaw M (1974) Adaptive changes of opposite sign in the oculomotor systems of the two eyes Quart J Exp Psychol 29 106–13 [151]

Crassini B, Broerse J (1982) Monocular rivalry occurs without eye movements Vis Res 22 203-4 [296]

Craton LG, Yonas A (1988) Infants' sensitivity to boundary flow information for depth at an edge Child Devel 591522-9 [480]

Crawford BH (1940a) Ocular interaction in its relation to measurements of brightness threshold Proc R Soc B **128** 552–9 [320, 333]

Crawford BH (1940b) The effect of field size and pattern on the change of visual sensitivity with time Proc R Soc B **129** 94–106 [333]

Crawford JD, Vilis T (1992) Symmetry of oculomotor burst neuron coordinates about Listing's plane J Neurophysiol **68** 432-48 [432]

Crawford JD, Cadera W, Vilis T (1991) Generation of torsional and vertical eye position signals by the interstitial nucleus of Cajal Science **252** 1551–3 [432]

Crawford MLJ (1998) Column spacing in normal and visually deprived monkeys Exp Brain Res **123** 282-8 [506]

Crawford MLJ, Cool SJ (1970) Binocular stimulation and response variability of striate cortex units in the cat Vis Res **10** 1145–53 [237, 260]

Crawford MLJ, von Noorden GK (1979a) The effects of short-term experimental strabismus on the visual system in *Macaca mulatta* Invest Ophthal Vis Sci **18** 496-504 [503]

Crawford MLJ, von Noorden GK (1979b) Concomitant strabismus and cortical eye dominance in young rhesus monkeys Tr Ophthal Soc UK **99** 369–74 [513]

Crawford MLJ, von Noorden GK (1980) Optically induced concomitant strabismus in monkeys Invest Ophthal Vis Sci **19** 1105–9 [514]

Crawford MLJ, von Noorden GK (1996) Shrinkage and recovery of cells of the lateral geniculate nuclei with prism-rearing in macaques Behav Brain Res **79** 233-8 [503]

Crawford MLJ, von Noorden GK, Meharg LS, Rhodes JW, Harwerth RS, Smith EL, Miller DD (1983) Binocular neurons and binocular function in monkeys and children Invest Ophthal Vis Sci **27** 491–5 [514]

Crawford MLJ, Smith EL, Harwerth RS, von Noorden GK (1984) Stereoblind monkeys have few binocular neurons Invest Ophthal Vis Sci **25** 779–81 [514]

Crawford MLJ, De Faber JT, Harwerth RS, Smith EL, von Noorden GK (1989) The effects of reverse monocular deprivation in monkeys. II Electrophysiological and anatomical studies Exp Brain Res **74** 338–47 [517]

Crawford MLJ, Harwerth RS, Smith EL, von Noorden GK (1996a) Loss of stereopsis in monkeys following prismatic binocular dissociation during infancy Behav Brain Res **79** 207-18 [514]

Crawford MLJ, Pesch TW, von Noorden GK (1996b) Excitatory neurons are lost following prismatic binocular dissociation in infant monkeys Behav Brain Res **79** 227-32 [514]

Crawford MLJ, von Noorden GK, Harwerth RS, Smith EL (1996c) Judgments by monkeys of apparent depth in dynamic random-dot stereograms Behav Brain Res **79** 219-25 [548]

Creed RS (1935) Observations on binocular fusion and rivalry J Physiol **84** 381-92 [300]

Creel D, Witkop C, King RA (1974) Asymmetric visually evoked potentials in human albinos: evidence for visual system abnormalities Invest Ophthal **13** 430-40 [496]

Creel D, O'Donnell FE, Witkop CJ (1978) Visual system anomalies in human ocular albinos Science **201** 931–3 [496]

Creel D, Spekreijse H, Reits D (1981) Evoked potentials in albinos: efficacy of pattern stimuli in determining misrouted optic fibers EEG Clin Neurophysiol **52** 595-603 [496]

Creutzfeldt OD (1977) Generality of the functional structure of the neocortex Naturwissenschaften **64** 507–17 [187]

Crewther DP, Crewther SG, Pettigrew JD (1978) A role for extraocular afferents in post–critical period reversal of monocular deprivation J Physiol **282** 181–95 [495, 508]

Crewther SG, Grewther DP (1993) Amblyopia and suppression in binocular cortical neurones of strabismic cat Neuroreport **4** 1083-6 [505]

Crewther SG, Crewther DP, Peck CK, Pettigrew JD (1980) Visual cortical effects of rearing cats with monocular or binocular cyclotorsion J Neurophysiol **44** 97–118 [485]

Crewther SG, Grewther DP, Mitchell DE (1983) The effects of short–term occlusion therapy on reversal of the anatomical and physiological effects of monocular deprivation in the lateral geniculate nucleus and visual cortex of kittens Exp Brain Res **51** 206–16 [517]

Crewther SG, Grewther DP, Clelland BG (1985) Convergent strabismic amblyopia in cats Exp Brain Res **60** 1-9 [503]

Crick F (1984) Function of the thalamic reticular complex: the searchlight hypothesis Proc Nat Acad Sci **81** 4586–90 [179]

Crick F (1996) Visual perception: rivalry and consciousness Nature **379** 485-6 [314]

Crick F, Koch C (1990) Towards a neurobiological theory of consciousness Seminars in the Neurosciences **2** 293–75 [134]

Cristo GD, Berardi N, Cancedda L, Pizzorusso T, Putignano E, Ratto GM, Maffei L (2001) Requirement of ERK activation for visual cortical plasticity Science **292** 2337-40. [460]

Critchley M (1955) The parietal lobes Arnold, London [135, 229, 488]

Crombie AC (1967) The mechanistic hypothesis and the scientific study of vision: some optical ideas as a background to the invention of the microscope" In Historical aspects of microscopy (ed S Bradbury, GLE Turner) pp 3-113 Heffer Cambridge [15]

Crone RA (1992) The history of stereoscopy Doc Ophthal **81** 1-16 [47, 56]

Crone RA, Everhard–Halm Y (1975) Optically induced eye torsion. I Fusional cyclovergence Graefes Arch klin exp Ophthal **195** 231–9 [413, 416]

Crone RA, Leuridan OMA (1973) Tolerance for aniseikonia. I Diplopia thresholds in the vertical and horizontal meridians of the visual field Graefes Arch klin exp Ophthal **188** 1–16 [273, 278]

Croner LJ, Kaplan E (1995) Receptive fields of P and M ganglion cells across the primate retina Vis Res **35** 7-27 [170]

Cronly–Dillon JR, Glaizner B (1974) Specificity of regenerating optic fibres for left and right optic tecta in goldfish Nature **251** 505-7 [464]

Cronly–Dillon JR, Gregory RL (1991) The evolution of the eye and visual system CRC Press, Boca Raton [485]

Crook JM, Kisvárday ZF, Eysel UT (1996) GABA-induced inactivation of functionally characterized sites in cat striate cortex (area 18): effects on direction selectivity J Neurophysiol **75** 2071-88 [203]

Crook JM, Kisvárday ZF, Eysel UT (1997) GABA-induced inactivation of functionally characterized sites in cat striate cortex: effects on orientation tuning and direction selectivity Vis Neurosci **14** 141-58 [199, 203]

Crovitz HF, Lipscomb DB (1963a) Binasal hemianopia as an early stage in binocular color rivalry Science **139** 596–7 [292]

Crovitz HF, Lipscomb DB (1963b) Dominance of the temporal visual fields at a short duration of stimulation Am J Psychol **76** 631–7 [292]

Crovitz HF, Lockhead GR (1967) Possible monocular predictors of binocular rivalry of contours Percept Psychophys **2** 83–5 [286]

Crowley JC, Katz LC (1999) Development of ocular dominance columns in the absence of retinal input Nature Neurosci **2** 1125-30 [463]

Crowley JC, Katz LC (2000) Early development of ocular dominance columns Science **290** 1321-4 [455]

Crozier WJ, Wolf E (1941) Theory and measurement of visual mechanisms: IV Critical intensities for visual flicker monocular and binocular J Gen Physiol **27** 505–34 [329]

Cudeiro J, Sillito AM (1996) Spatial frequency tuning of orientation-discontinuity-selective corticofugal feedback to the cat lateral geniculate nucleus J Physiol **490** 481-92 [179]

Cudeiro J, González F, Pérez R, Alonso, JM, Acuna C (1989) Does the pulvinar-LP complex contribute to motor programming? Brain Research **484** 367-70 [183]

Culham JC, Dukelow SP, Vilis T, Hassard FA, Gati JS, Menon RS, Goodale,MA (1999) Recovery of fMRI activation in motion area MT following storage of the motion aftereffect J Neurophysiol **81** 388-93 [182]

Cumming BG, DeAngelis GC (2001) The physiology of stereopsis Ann Rev Neurosci **24** 303-38 [237]

Cumming BG, Judge SJ (1986) Disparity–induced and blur–induced convergence eye movement and accommodation in monkey J Neurophysiol **55** 896–914 [389, 397]

Cumming BG, Parker AJ (1997) Responses of primary visual cortical neurons to binocular disparity without depth perception Nature **389** 280-3 [249, 266]

Cumming BG, Parker AJ (1999) Binocular neurons in V1 of awake monkeys are selective for absolute, not relative, disparity J Neurosci **19** 5602-18 [206, 241]

Cumming BG, Parker AJ (2000) Local disparity not perceived depth is signaled by binocular neurons in cortical area V1 of the macaque J Neurosci **20** 4758-67 [241]

Cumming BG, Shapiro SE, Parker AJ (1998) Disparity detection in anticorrelated stereograms Perception **27** 1367-77 [250]

Curcio CA, Sloan KR, Kalina RE, Hendrickson AE (1990) Human photoreceptor topography J Comp Neurol **292** 497–523 [164, 171, 473]

Curtis DW, Rule SJ (1978) Binocular processing of brightness information: a vector–sum model J Exp Psychol: HPP **4** 132–43 [327]

Curtis DW, Rule SJ (1980) Fechner's paradox reflects a nonmonotone relation between binocular brightness and luminance Percept Psychophys **27** 293-6 [335]

Cynader M (1983) Prolonged sensitivity to monocular deprivation in dark-reared cats: effects of age and visual exposure Devel Brain Res **8** 155-64 [516]

Cynader M, Mitchell DE (1980) Prolonged sensitivity to monocular deprivation in dark–reared cats J Neurophysiol **43** 1029–40 [516]

Cynader M, Berman N, Hein A (1976) Recovery of function in cat visual cortex following prolonged deprivation Exp Brain Res **25** 139–56 [500]

Cynader M, Timney BN, Mitchell DE (1980) Period of susceptibility of kitten visual cortex to the effect of monocular deprivation extends beyond 6 months of age Brain Res **191** 545–50 [516]

Cynader M, Gardner JC, Mustari M (1984) Effects of neonatally induced strabismus on binocular responses in cat area 18 Exp Brain Res **53** 384–99 [506]

Cynader M, Gardner JC, Dobbins A, Lepore F, Guillemot JP (1986) Interhemispheric communication and binocular vision: functional and developmental aspects In Two hemispheres – one brain: functions of the corpus callosum (ed F Lepore, M Ptito, HH Jasper) pp 198–209 Liss, New York [232]

Cynader M, Giaschi DE, Douglas RM (1993) Interocular transfer of direction–specific adaptation to motion in cat striate cortex Invest Ophthal Vis Sci **34** (Abs) 1188 [350]

Czepita D, Daw NW (1996) The contribution of NMDA receptors to the visual response in animals that have been partially monocularly deprived Brain Res **728** 7-12 [511]

Czepita D, Reid SNM, Daw NW (1994) Effect of longer periods of dark rearing on NMDA receptors in cat visual cortex J Neurophysiol **72** 1220-6 [462, 500, 511]

D'Azzo JJ, Houpis CH (1995) Linear control system analysis and design conventional and modern McGraw-Hill, New York [98]

da Vinci L (1452) Trattato della pittura (Translated as A treatise on painting by JE Rigaud) London 1802 [52]

Dacey DM, Lee BB, Stafford DK, Pokorny J, Smith VC (1996) Horizontal cells of the primate retina: cone specificity without spectral opponency Science **271** 656-17 [167, 169]

Dailey ME (1964) Dynamic optical imaging of neuronal structure and physiology In Brain Mapping (ed AW Toga, JC Mazziotta) pp 29-46 Academic Press, New York [181]

Dale RT (1982) Fundamentals of ocular motility and strabismus Grune and Stratton, New York [530]

Dalva MB, Katz LC (1994) Rearrangements of synaptic cortical connections in visual cortex revealed by laser photostimulation Science **295** 255-8 [452]

Dalva MB, Ghosh A, Shatz CJ (1994) Independent control of dendritic and axonal form in the developing lateral geniculate nucleus J Neurosci **14** 3588-602 [444]

Damasch H (1994) The origin of perspective (English trans by J Goodman) MIT Press, Cambridge, MA [31]

Damasio AR, Tranel D, Damasio H (1990) Face agnosia and the neural substrates of memory Ann Rev Neurosci **13** 89-109 [222]

Dan Y, Poo M (1992) Hebbian depression of isolated neuromuscular synapses in vitro Science **256** 1570–3 [135]

Dan Y Atick JJ, Reid RC (1996) Efficient coding of neural scenes in the lateral geniculate nucleus: experimental test of a computational theory J Neurosci **16** 3351-62 [135]

Dan Y, Alonso JM, Usrey WM, Reid RC (1998) Coding of visual information by precisely correlated spikes in the lateral geniculate nucleus Nature Neurosci **1** 501-7 [132]

Daniel PM, Whitteridge D (1961) The representation of the visual field on the cerebral cortex in monkeys J Physiol **159** 203–21 [184]

Daniels JD, Norman JL, Pettigrew JD (1977) Biases for oriented moving bars in lateral geniculate nucleus of normal and stripe-reared cats Exp Brain Res **29** 155-72 [179]

Daniels JD, Pettigrew JD, Norman JL (1978) Development of single–neuron responses in kitten's lateral geniculate nucleus J Neurophysiol **41** 1373–93 [444]

Dannemiller JL, Freedland RL (1993) Motion-based detection by 14-week-old infants Vis Res **33** 657-64 [472]

Danta G, Hilton RC, O'Boyle DJ (1978) Hemisphere function and binocular depth perception Brain **101** 569–89 [488, 489]

Dantzker JL, Callaway EM (1998) The development of local, layer-specific visual cortical axons in the absence of extrinsic influences and intrinsic activity J Neurosci **18** 4145-54 [451, 452]

Darian-Smith C, Gilbert CD (1994) Axonal sprouting accompanies functional reorganization in adult cat striate cortex Nature **368** 737-40 [198]

Darian-Smith C, Gilbert CD (1995) Topographic reorganization in the cortex of the adult cat and monkey is cortically mediated J Neurosci **15** 1631-47 [197]

Darrah WC (1964) Stereo views A history of stereographs in America and their collection Times and News Publishing Co, Gettysburg PA [70]

Das A, Gilbert CD (1995a) Long-range horizontal connections and their role in cortical reorganization revealed by optical recording of cat primary visual cortex Nature **375** 780-4 [195, 197]

Das A, Gilbert CD (1995b) Receptive field expansion in adult visual cortex is linked to dynamic changes in strength of cortical connections J Neurophysiol **74** 779-92 [198]

Das A, Gilbert CD (1997) Distortions of visuotopic map match orientation singularities in primary visual cortex Nature **387** 594-8 [185, 214]

Das A, Gilbert CD (1999) Topography of contextual modulations mediated by short-range interactions in primary visual cortex Nature **399** 655-61 [195, 196]

Das I, Sparrow JR, Lin MI, Shih E, Mikawa T, Hempstead BL (2000) Trk C signaling is required for retinal progenitor cell proliferation J Neurosci **20** 2887-95 [456]

Daugman JG (1984) Spatial visual channels in the Fourier plane Vis Res **27** 891-10 [141]

Daugman JG (1985) Uncertainty relation for resolution in space spatial frequency and orientation optimized by two-dimensional visual cortical filters J Opt Soc Am A **2** 1160-9 [141]

Daugman JG (1990) An information–theoretic view of analog representation in striate cortex In Computational neuroscience (ed EL Schwartz) pp 401–23 MIT Press, Cambridge MA [93, 141]

Daugman JG (1991) Self-similar oriented wavelet pyramids: conjectures about neural non–orthogonality In Representations of vision (ed A Gorea) pp 27–46 Cambridge University Press, New York [141]

Daum KM (1982b) The course and effect of visual training on the vergence system Am J Optom Physiol Opt **59** 223-7 [394]

Daum KM (1983) The stability of the fixation disparity curve Ophthal Physiol Opt **3** 13-19 [382]

Daum KM (1988) Characteristics of convergence insufficiency Am J Optom Physiol Opt **65** 429-38 [393]

Daum KM (1989) Evaluation of a new criterion of binocularity Optom Vis Sci **66** 218-28 [389]

Daum KM, Rutstein RP, Cho M, Eskridge JB (1988) Horizontal and vertical vergence training and its effect on vergences and fixation disparity curves: I. Horizontal data Am J Optom Physiol Opt **65** 1-7 [394]

Davenport RW, Thies E, Cohen ML (1999) Neuronal growth cone collapse triggers lateral extensions along trailing axons Nature Neurosci **2** 254-9 [437]

David CT (1982) Compensation for height in the control of groundspeed by *Drosophila* in a new, Barber's pole wind tunnel J Comp Physiol **147** 485-93 [534]

Davis AA, Temple S (1994) A self-renewing multipotential stem cell in embryonic rat cerebral cortex Nature **372** 293-6 [448]

Davison ML (1983) Multidimensional scaling Wiley, New York [144]

Davson H (1962) The eye Academic Press, New York [164, 358]

Daw NW (1995) Visual development Plenum, New York [485, 511]

Daw NW, Videen TO, Rader RK, Robertson TW, Coscia CJ (1985) Substantial reduction of noradrenaline in kitten visual cortex by intraventricular injections of 6–hydroxydopamine does not always prevent ocular dominance shifts after monocular deprivation Exp Brain Res **59** 30–5 [513]

Daw NW, Fox K, Sato H, Czepita D (1992) Critical period for monocular deprivation in the cat visual cortex J Neurophysiol **67** 197–202 [516]

Dawson S (1913) Binocular and uniocular discrimination of brightness Br J Psychol **6** 78–108 [294]

Dawson S (1917) The experimental study of binocular colour mixture I Br J Psychol **8** 510–51 [283, 284, 300]

Day RH (1958) On interocular transfer and the central origin of visual after–effects Am J Psychol **71** 784–9 [344]

Day RH (1961) On the stereoscopic observation of geometrical illusions Percept Mot Skills **13** 277–58 [353]

Day RH, McKenzie BE (1973) Perceptual shape constancy in early infancy Perception **2** 315-20 [479]

Day RH, Wade NJ (1988) Binocular interaction in induced rotary motion Aust J Psychol **40** 159–64 [349]

Day SH, Orel-Bixer DA, Norcia AM (1988) Abnormal acuity development in infantile esotropia Invest Ophthal Vis Sci **29** 327-9 [520]

Dayan P (1998) A hierarchical model of binocular rivalry Neural Comput **10** 1119-35 [311]

De Blas AL (1996) Brain $GABA_A$ receptors studied with subunit-specific antibodies Molec Neurobiol **12** 55-71 [191]

De Carlos JA, O'Leary DDM (1992) Growth and targeting of subplate axons and establishment of major cortical pathways J Neurosci **12** 1194-211 [458]

De Courten C, Garey LJ (1982) Morphology of the neurons in the human lateral geniculate nucleus and their normal development Exp Brain Res **47** 159–171 [443]

De Lange H (1954) Relationship between critical flicker frequency and a set of low–frequency characteristics of the eye J Opt Soc Am **44** 380–9 [329]

De Lange H (1958) Research into the dynamic nature of the fovea–cortex system with intermittent and modulated light J Opt Soc Am **48** 777–89 [93]

De Monasterio FM (1978) Properties of ganglion cells with atypical receptive-field organization in retina of macaques J Neurophysiol **41** 1435-49 [203]

De Ruyter van Steveninck RR, Laughlin SB (1996) The rate of information transfer at graded-potential synapses Nature **379** 642-5 [114]

De Silva HR, Bartley SH (1930) Summation and subtraction of brightness in binocular perception Br J Psychol **20** 271-50 [324]

De Weert CMM, Levelt WJM (1974) Binocular brightness combinations: additive and nonadditive aspects Percept Psychophys **15** 551–62 [325]

De Weert CMM, Levelt WJM (1976a) Comparison of normal and dichoptic color mixing Vis Res **16** 59–70 [284, 300]

De Weert CMM, Levelt WJM (1976b) Dichoptic brightness combination for unequal coloured lights Vis Res **16** 1077-86 [326]

De Weert CMM, Wade NJ (1988) Compound binocular rivalry Vis Res **28** 1031–40 [284, 297]

Dean AF (1981) The relationship between response amplitude and contrast for cat striate cortical neurones J Physiol **318** 413-27 [198]

Dean P, Redgrave P, Westby GWM (1989) Event or emergency? Two response systems in the mammalian superior colliculus TINS **12** 137-47 [257]

DeAngelis GC, Newsome WT (1999) Organization of disparity-selective neurons in macaque area MT J Neurosci **19** 1398-415 [243]

DeAngelis GC, Ohzawa I, Freeman RD (1991) Depth is encoded in the visual cortex by a specialized receptive field structure Nature **352** 156–9 [252, 264]

DeAngelis GC, Robson JG, Ohzawa I, Freeman RD (1992) Organization of suppression in receptive fields of neurons in cat cortex J Neurophysiol **68** 144–163 [200, 313, 346]

DeAngelis GC, Ohzawa I, Freeman RD (1993a) Spatiotemporal organization of simple–cell receptive fields in the cat's striate cortex. I General characteristics and postnatal development J Neurophysiol **69** 1091–117 [202]

DeAngelis GC, Ohzawa I, Freeman RD (1993b) Spatiotemporal organization of simple–cell receptive fields in the cat's striate cortex. II Linearity of temporal and spatial summation J Neurophysiol **69** 1118–35 [202]

DeAngelis GC, Freeman RD, Ohzawa I (1994) Length and width tuning of neurones in the cat's primary visual cortex J Neurophysiol **71** 347–74 [206, 256, 346]

DeAngelis GC, Anzai A, Ohzawa I, Freeman RD (1995) Receptive field structure in the visual cortex: does selective stimulation induce plasticity? Proc Nat Acad Sci **92** 9682-6 [198]

DeAngelis GC, Cumming BG, Newsome WT (1998) Cortical area MT and the perception of stereoscopic depth Nature **394** 677-80 [243]

DeAngelis GC, Ghose GM, OhzawaI, Freeman RD (1999) Functional micro-organization of primary visual cortex: receptive field analysis of nearby neurons J Neurosci **19** 4046-64 [209]

DeAngelis GC, Cumming BG, Newsome WT (2000) A new role for cortical area MT: the perception of stereoscopic depth In The new cognitive neurosciences (ed MS Gazzaniga) pp 305-313 MIT Press, Cambridge MA [243]

DeBruyn B, Rogers BR, Howard IP, Bradshaw MF (1992) Role of positional and orientational disparities in controlling cyclovergent eye movements Invest Ophthal Vis Sci **33** (Abs) 1149 [417]

DeBruyn EJ, Casagrande VA (1981) Demonstration of ocular dominance columns in a New World primate by means of monocular deprivation Brain Res **207** 543-8 [217]

Dehay C, Giroud P, Berland M, Killackey H, Kennedy H (1996) Contribution of thalamic input to the specification of cytoarchitectonic cortical fields in the primate: effects of bilateral enucleation in the fetal monkey on the boundaries, dimensions, and gyrification of striate and extrastriate cortex J Comp Neurol **367** 70-89 [499]

Deiner MS, Kennedy TE, Fazeli A, Serafini T, Tessier-Lavigne M, Sretavan DW (1997) Netrin-1 and DCC mediate axon guidance locally a the optic disc: loss of function leads to optic nerve hypoplasia Neuron **19** 595-89 [436]

Delambre JBJ (1912) The optics of Ptolemy compared with that of Euclid Alhazen and Vitellio Annalen der Physik **40** 371-88 [13]

Delint PJ, Weissenbruch C, Berendschot TTJM, van Norren D (1998) Photoreceptor function in unilateral amblyopia Vis Res **38** 613-17 [519]

Deller M (1988) Why should surgery for early-onset strabismus be postponed? Br J Ophthal **72** 110-115 [518]

Dell'Osso LF, Traccis S, Abel LA (1983) Strabismus—a necessary condition for latent and manifest latent nystagmus Neuro-Ophthal **3** 277-57 [373]

Dell'Osso LF (1996) See-saw nystagmus in dogs and humans Neurology **47** 1372-4 [441]

Dell'Osso LF, Williams RW (1995) Ocular motor abnormalities in achiasmic mutant Belgian sheepdogs: unyoked eye movements in a mammal Vis Res **35** 109-16 [441]

Dell'Osso LF, Schmidt D, Daroff RB (1979) Latent, manifest latent and congenital nystagmus Arch Ophthal **97** 1877-81 [373]

Demanins R, Hess RF (1996a) Effect of exposure duration on spatial uncertainty in normal and amblyopic eyes Vis Res **36** 1189-93 [521]

Demanins R, Hess RF (1996b) Positional loss in strabismic amblyopia: interrelationship of alignment threshold bias spatial scale and eccentricity Vis Res **36** 2771-94 [523]

Demanins R, Wang YZ, Hess RF (1999) The neural deficit in strabismic amblyopia: sampling considerations Vis Res **39** 3573-85 [524]

Demer JL, von Noorden GK (1988) Optokinetic asymmetry in esotropia J Ped Ophthal Strab **25** 286-92 [374, 526]

Demer JL, Zee DS (1984) Vestibulo–ocular and optokinetic deficits in albinos with congenital nystagmus Invest Ophthal Vis Sci **25** 739–45 [498]

Demer JL, Poukens V, Miller JM, Micevych P (1997) Innervation of extraocular pulley smooth muscle in monkeys and humans Invest Ophthal Vis Sci **38** 1774-85 [371]

Denieul P (1982) Effects of stimulus vergence on mean accommodation response microfluctuations of accommodation and optical quality of the human eye Vis Res **22** 561-9 [361]

Denny N, Frumkes TE, Barris MC, Eysteinsson T (1991) Tonic interocular suppression and binocular summation in human vision J Physiol **437** 449-60 [335]

Derrington AM, Hawken MJ (1981) Spatial and temporal properties of cat geniculate neurones after prolonged deprivation J Physiol **314** 107–20 [504]

Derrington AM, Lennie P (1984) Spatial and temporal contrast sensitivities of neurones in lateral geniculate nucleus of macaque J Physiol **357** 219–40 [170]

Des Rosiers MH, Sakurada O, Jehle, Shinohara JJM, Kennedy C, Sokoloff L (1978) Functional plasticity in the immature striate cortex of the monkey shown by the [14C] deoxyglucose method Science **200** 447–9 [455, 509]

Desagulier JT (1716) A plain and easy experiment to confirm Sir Isaac Newton's doctrine of the different refrangibility of the rays of ligh Philos Trans Roy Soc **29** 448-52 [282]

Descargues P (1977) Perspective (Translated from the French by IM Paris) Abrams, New York [38]

Descartes R (1664) Traité de l'homme In Oeuvres de Descartes (ed C Adam, P Tannery) Vol XI 1909 pp 119–215 Cerf, Paris (Translation by TS Hall) Treatise of man Harvard University Press, Cambridge MA 1972 [24]

Desimone R, Schein SJ (1987) Visual properties of neurons in area V4 of the macaque: sensitivity to stimulus form J Neurophysiol **57** 835-68 [221]

Desimone R, Moran J, Schein SJ, Mishkin M (1993) A role for the corpus callosum in visual area V4 of the alert monkey Vis Neurosci **10** 159–71 [232]

Deslandes J (1966) Histoire comparée du cinéma Casterman, Paris [70]

Deubel H (1987) Adaptivity of gain and direction in oblique saccades In Eye movements: From physiology to cognition (ed JK O'Regan, A Levy–Schoen) pp 181–90 Elsevier, Amsterdam [423]

Deubel H, Wolf W, Hauske G (1986) Adaptive gain control of saccadic eye movements Hum Neurobiol **5** 275–53 [423]

DeValois RL (1991) Orientation and spatial frequency selectivity In From pigments to perception (ed A Valberg, BB Lee) pp 291–7 Plenum, New York [205]

DeValois RL, DeValois KK (1988) Spatial vision Oxford University Press, New York [8, 121, 205, 210]

DeValois RL, Walraven J (1967) Monocular and binocular aftereffects of chromatic adaptation Science **155** 463-5 [283, 342]

DeValois RL, Yund EW, Hepler N (1982a) The orientation and direction selectivity of cells in macaque visual cortex Vis Res **22** 531–44 [199, 256]

DeValois RL, Thorell LG, Albrecht DG (1985) Periodicity of striate-cortex-cell receptive fields J Opt Soc Am A **2** 1115-22 [201]

Dews PB, Wiesel TN (1970) Consequences of monocular deprivation on visual behaviour in kittens J Physiol **206** 437–55 [515, 527]

DeYoe EA, Van Essen DC (1985) Segregation of efferent connections and receptive field properties in visual area V2 of the macaque Nature **317** 58–61 [219]

DeYoe EA, Van Essen DC (1988) Concurrent processing streams in monkey visual cortex TINS **11** 219–29 [229]

DeYoe EA, Trusk TC, Wong-Riley MTT (1995) Activity correlates of cytochrome oxidase-defined compartments in granular and supragranular layers of primary visual cortex of the macaque monkey Vis Neurosci **12** 629-39 [213]

Di Stefano M, Bédard S, Marzi CA, Lepore F (1984) Lack of binocular activation of cells in area 19 of the Siamese cat Brain Res **303** 391-5 [497]

Diao YC, Wang YK, Pu ML (1983) Binocular responses of cortical cells and the callosal projection in the albino rat Exp Brain Res **49** 410-18 [496]

Diao YC, Jia WG, Swindale NV, Cynader MS (1990) Functional organization of the cortical 17/18 border region in the cat Exp Brain Res **79** 271-82 [231]

Dias EC, Rocha–Miranda CE, Bernardes RF, Schmidt SL (1991) Disparity selective units in superior colliculus of the opossum Exp Brain Res **87** 546–52 [245]

Diaz–Araya C, Provis JM (1992) Evidence of photoreceptor migration during early foveal development: a quantitative analysis of human fetal retinae Vis Neurosci **8** 505–14 [436]

Diaz-Caneja E (1928) Sur l'alternance binoculaire Annales d'Oculistique **165** 721-31 [299]

DiCarlo JJ, Maunsell JHR (2000) Form representation in monkey inferotemporal cortex is virtually unaltered by free viewing Nature Neurosci **3** 814-21 [222]

Dichgans J, Jung R (1975) Oculomotor abnormalities due to cerebellar lesions In Basic mechanisms of ocular motility and their clinical implications (ed G Lennerstrand, P Bach–y–Rita) pp 281–98 Pergamon, Oxford [498]

Dickinson CM (1986) The elucidation and use of the effect of near fixation in congenital nystagmus Ophthal Physiol Opt **6** 303–11 [498]

Diefenbach TJ, Guthrie PB, Kater SB (2000) Stimulus history alters behavioral responses of neuronal growth cones J Neurosci **20** 1484-94 [438]

Dijkerman HC, Milner AD, Carey DP (1996) The perception and prehension of objects oriented in the depth plane. I. Effects of visual form agnosia Exp Brain Res **112** 442-51 [229]

Dill LM (1975) Predatory behaviour of the zebra spider *Salticus scenicus* (Araneae: Salticidae) Can J Zool **53** 1284–9 [536]

Diner DB, Fender DH (1987) Hysteresis in human binocular fusion: temporalward and nasalward ranges J Opt Soc Am A **4** 1814–19 [279]

Diner DB, Fender DH (1988) Dependence of Panum's fusional area on local retinal stimulation J Opt Soc Am A **5** 1163–9 [279]

Dinse HR, Krüger K, Best J (1990a) A temporal structure of cortical information processing Concepts in Neuroscience **1** 199–238 [131]

Dinse HR, Racanzone GH, Merzenich MM (1990b) Direct observation of neural assemblies during neocortical representational reorganization In Parallel processing in neural systems and computers (ed R Eckmiller, G Hartmann, G Hauske) pp 65–69 Elsevier, Amsterdam [208]

Dobbins AC, Jeo RM, Fiser J, Allman JM (1998) Distance modulation of neural activity in the visual cortex Science **281** 552-5 [243]

Dobkins KR, Teller DY (1996) Infant contrast detectors are selective for direction of motion Vis Res **36** 281-94 [472]

Dobson V, Sebris SL (1989) Longitudinal study of acuity and stereopsis in infants with or at risk for esotropia Invest Ophthal Vis Sci **30** 1146-58 [374, 509]

Dobson V, Teller DY (1978) Visual acuity in human infants: a review and comparison of behavioral and electrophysiological studies Vis Res **18** 1469–83 [468]

Dobson V, Fulton AB, Sebris SL (1984) Cycloplegic refractions of infants and young children: the axis of astigmatism Invest Ophthal Vis Sci **25** 83–7 [474]

Dodd J, Jessell TM (1988) Axon guidance and the patterning of neuronal projections in vertebrates Science **272** 692-99 [438]

Dodd JV, Krug K, Cumming BG, Parker AJ (2001) Perceptually bistable three-dimensional figures evoke high choice probabilities in cortical area MT J Neurosci **21** 4809-21 [224]

Dodwell PC (1983) The Lie transformation group model of visual perception Percept Psychophys **34** 1-16 [136, 156]

Dodwell PC, Muir D, Di Franco D (1976) Responses of infants to visually presented objects Science **194** 209–11 [477]

Domenici L, Berardi N, Carmignoto G, Vantini G, Maffei L (1991) Nerve growth factor prevents the amblyopic effects of monocular deprivation Proc Nat Acad Sci **88** 8811-15 [512]

Domenici L, Cellerino A, Maffei L (1993) Monocular deprivation effects in the rat visual cortex and lateral geniculate nucleus are prevented by nerve growth factor (NGF). II Lateral geniculate nucleus Proc R Soc **251** 25-31 [512]

Domini F, Braunstein M (2001) Influence of a stereo surface on the perceived tilt of a monocular line Percept Psychophys **63** 607-24 [378]

Domini F, Blaser E, Cicerone CM (2000) Color-specific depth mechanisms revealed by a color-contingent depth aftereffect Vis Res **40** 359-64 [353]

Donaldson IML, Long AC (1980) Interactions between extraocular proprioceptive and visual signals in the superior colliculus of the cat J Physiol **298** 85–110 [494]

Donders FC (1864) Accommodation and refraction of the eye The Sydenham Society, London [388]

Donnelly M, Miller RJ (1995) Ingested ethanol and binocular rivalry Invest Ophthal Vis Sci **36** 1548-54 [290]

Donoghue MJ, Rakic P (1999) Molecular evidence for the early specification of presumptive functional domains in the embryonic primate cerebral cortex J Neurosci **19** 5967-79 [446]

Douglas RH, Hawryshyn CW (1990) Behavioural studies of fish vision: an analysis of visual capabilities In The visual system of fish (ed RH Douglas, MBA Djamgoz) pp 373-48 Chapman Hall, London [539]

Douglas RH, Collett TS, Wagner HJ (1986) Accommodation in anuran amphibia and its role in depth vision J Comp Physiol A **158** 133-43 [541]

Douglas RH, Eva J, Guttridge N (1988) Size constancy in goldfish (*Carassius auratus*) Behav Brain Res **30** 37-42 [539]

Douglas RJ, Koch C, Mahowald M, Martin KAC, Suarez HH (1995) Recurrent excitation in neocortical circuits Science **299** 981-5 [188]

Dow BM (1991) Orientation and color columns in monkey striate cortex In From pigments to perception (ed A Valberg, BB Lee) pp 299–74 Plenum, New York [209, 211]

Dowling JE (1987) The retina: An approachable part of the brain Harvard University Press, Cambridge MA [167, 170]

Dowling JE, Boycott BB (1966) Organization of the primate retina Proc R Soc B **166** 80–111 [164]

Downing AC (1972) Optical scanning in the lateral eyes of the copepod *Copilia* Perception **1** 277–61 [536]

Dräger UC (1985) Birth dates of retinal ganglion cells giving rise to the crossed and uncrossed optic projections in the mouse Proc R Soc B **227** 57-77 [439]

Dragoi V, Sur M (2000) Dynamic properties of recurrent inhibition in primary visual cortex: contrast and orientation dependence of contextual effects J Neurophysiol **83** 1019-30 [197]

Drasdo N (1977) The neural representation of visual space Nature **296** 554–6 [184]

Drexler W, Findl O, Schmetterer L, Hitzenberger CK, Fercher AF (1998) Eye elongation during accommodation in humans: differences between emmetropes and myopes Invest Ophthal Vis Sci **39** 2140-7 [359, 435]

Duboscq J (1857) Sur le stéréoscope Bull Soc Fran Photo **3** 77-8 [71]

Dubrovsky BO, Barbas H (1977) Frontal projections to dorsal neck and extraocular muscles Exp Neurol **55** 680–93 [494]

Dudek SM, Bear MF (1989) A biochemical correlate of the critical period for synaptic modification in the visual cortex Science **276** 673-5 [462]

Duffieux PM (1946) L'intégrale Fourier et ses applications a l'optique Imprimeries Oberthur, Rennes [138]

Duffy CJ, Wurtz RH (1991) Sensitivity of MST neurons to optic flow stimuli. II Mechanisms of response selectivity revealed by small–field stimuli J Neurophysiol **65** 1346–59 [225]

Duffy CJ, Wurtz RH (1995) Response of monkey MST neurons to optic flow stimuli with shifted centers of motion J Neurophysiol **15** 5192-208 [225]

Duffy CJ, Wurtz RH (1997) Medial superior temporal area neurons respond to speed patterns in optic flow J Neurosci **17** 2839-51 [225]

Duffy FH, Snodgrass RS, Burchfiel JL, Conway JL (1976) Bicuculline reversal of deprivation amblyopia in the cat Nature **290** 256-7 [510]

Duhamel JR, Colby CL, Goldberg ME (1992) The updating of the representation of visual space in parietal cortex by intended eye movements Science **255** 90–95 [227]

Duhamel JR, Bremmer F, BenHamed S, Graf W (1997) Spatial invariance of visual receptive fields in parietal cortex neurons Nature **389** 845-8 [226]

Duke–Elder S (1961) System of ophthalmology Vol II The anatomy of the visual system Kimpton, London [9, 27]

Duke–Elder S (1968) System of ophthalmology Vol IV The physiology of the eye and of vision Kimpton, London [305]

Dunkeld J, Bower TGR (1980) Infant response to impending collision Perception **9** 549–54 [478]

Dunlap K (1944) Alleged binocular mixing Am J Psychol **57** 559–63 [283]

Dürer A (1525) Underweysung der messung Nuremberg. English translation by WL Strauss, Abaris books, New York [36]

Durgin FH (2001) Texture contrast aftereffects are monocular; texture density aftereffects are binocular Vis Res **41** 2619-30 [351]

Durnford M, Kimura D (1971) Right hemisphere specialization for depth perception reflected in visual field differences Nature **231** 394–5 [489]

Dürsteler MR, von der Heydt R (1983) Plasticity in the binocular correspondence of striate cortical receptive fields in kittens J Physiol **345** 87–105 [485]

Dürsteler MR, Garey LJ, Movshon JA (1976) Reversal of the morphological effects of monocular deprivation in the kitten's lateral geniculate nucleus J Physiol **291** 189–210 [504]

Dürsteler MR, Wurtz RH, Newsome WT (1987) Directional pursuit deficits following lesions of the foveal representation within the superior temporal sulcus of the macaque monkey J Neurophysiol **57** 1292–87 [229]

Dutour EF (1760) Discussion d'un question d'optique L'Académie des Sciences. Mémoires de Mathématique et de physique présentés par Divers Savantes **3** 514-30. An English translation by O'Shea RP (1999) of both Dutour papers is available at http://psy.otago.ac.nz:800/r-oshea.dutour63.html. [62, 282, 350]

Dutour EF (1763) Addition au Mémoire intitulé, Discussion d'un question d'optique L'Académie des Sciences. Mémoires de Mathématique et de physique présentés par Divers Savantes **4** 499-511 [305, 318]

Duwaer AL (1982a) Assessment of retinal image displacement during head movement using an afterimage method Vis Res **22** 1379-88 [395]

Duwaer AL (1982b) Nonmotor component of fusional response to vertical disparity: a second look using an afterimage method J Opt Soc Am **72** 871–7 [407]

Duwaer AL (1983) Patent stereopsis with diplopia in random–dot stereograms Percept Psychophys **33** 443–54 [281]

Duwaer AL, van den Brink G (1981a) Foveal diplopia thresholds and fixation disparities Percept Psychophys **30** 321–9 [380]

Duwaer AL, van den Brink G (1981b) Diplopia thresholds and the initiation of vergence eye–movements Vis Res **21** 1727–37 [394, 407]

Duwaer AL, van den Brink G (1982b) The effect of presentation time on detection and diplopia thresholds for vertical disparities Vis Res **22** 183–9 [277]

Duwaer AL, van den Brink G, van Antwerpen G, Keemink CJ (1982) Comparison of subjective and objective measurements of ocular alignment in the vertical direction Vis Res **22** 983-9 [407]

Duysens J, Maes H, Orban GA (1987) The velocity dependence of direction selectivity of visual cortical neurons in the cat J Physiol **387** 95-113 [203]

Dvorák V (1870) Versuche über Nachbilder von Reizveränderungen Sitzungsbericht der Kaiserlichen Akademie der Wissenschaften: Mathematisch-Naturwissenschaftliche Klasse, II Abteilung (Wein) **61** 257–62. Translation in Broerse *et al.* (1994) [45, 347]

Dwyer P, Wick B (1995) The influence of refractive correction upon disorders of vergence and accommodation Optom Vis Sci **72** 227-32 [394]

Eadie AS, Carlin PJ (1995) Evolution of control system models of ocular accommodation vergence and their interaction Med Biol Engin Comput **33** 517-27 [406]

Eadie AS, Gray LS, Carlin P, Mon-Williams M (2000) Modelling adaptation effects in vergence and accommodation after exposure to a simulated virtual reality stimulus Ophthal Physiol Opt **20** 272-51 [391]

Earle EW (ed) (1979) Points of view The stereograph in America—a cultural history The Book Bus Visual Studies Workshop, Rochester, NY [69]

Easter SS, Schmidt JT (1977) Reversed visuomotor behavior mediated by induced ipsilateral retinal projections in goldfish J Neurosci **40** 1275-54 [464]

Eastwood BS (1982) The elements of vision: the micro-cosmology of Galenic visual theory according to Hunayn ibn Ishäq Tr Am Philos Soc **72** part 5 1-58 [15]

Eastwood BS (1986) Alhazen Leonardo and late-medieval speculation on the inversion of the images in the eye Ann Sci **43** 413-46 [17]

Ebenholtz SM (1970) On the relation between interocular transfer of adaptation and Hering's law of equal innervation Psychol Rev **77** 343-7 [355]

Ebenholtz SM (1983) Accommodative hysteresis: a precursor for induced myopia? Invest Ophthal Vis Sci **27** 513-15 [359]

Ebenholtz SM (1991) Accommodative hysteresis Invest Ophthal Vis Sci **32** 148-53 [359]

Ebenholtz SM, Citek K (1995) Absence of adaptive plasticity after voluntary vergence and accommodation Vis Res **35** 2773-83 [383]

Ebenholtz SM, Fisher SK (1982) Distance adaptation depends upon plasticity in the oculomotor control system Percept Psychophys **31** 551-60 [391]

Eckhorn R, Bauer R, Jordan W, Brosch M, Kruse W, Munk M, Reitboeck HJ (1988) Coherent oscillations: a mechanism for feature linking in the visual cortex? Biol Cyber **60** 121–30 [133]

Eckhorn R, Reitboeck HJ, Arndt M, Dicke P (1990) Feature linking via synchronization among distributed assemblies: simulations of results from cat visual cortex Neural Comput **2** 293–307 [136]

Edgerton SY (1975) The renaissance rediscovery of linear perspective Basic Books, New York [29, 34, 38]

Edwards DP, Purpura, KP, Kaplan E (1995) Contrast sensitivity and spatial frequency responses of primate cortical neurons in and around cytochrome oxidase blobs Vis Res 35 1501-23 [205]

Edwards FA (1995) Anatomy and electrophysiology of fast central synapses lead to a structural model for long-term potentiation Physiol Rev 75 757-87 [189, 193, 458]

Edwards M, Pope DR, Schor CM (1998) Luminance contrast and spatial-frequency tuning of the transient-vergence system Vis Res 38 705-17 [402]

Egelhaaf M, Borst A (1992) Is there a separate control system mediating a "centering response" in honeybees? Naturwissenschaften 79 221–3 [534]

Eggers HM, Blakemore C (1978) Physiological basis of anisometropic amblyopia Science 201 294-7 [505]

Eggert T, Kapoula Z (1995) Position dependency of rapidly induced saccadic disconjugacy Vis Res 35 3493-503 [425]

Ehrenstein W (1925) Versuche über beziehungen zwischen Bewegungs- und Gestaltwahrnehmung. Erste Abhandlung Z Psychol 96 305-52 [347]

Ehrenstein WH (1977) Geometry in visual space—some method-dependent (arti)facts Perception 6 657-60 [137]

Eifuku S, Wurtz RH (1999) Response to motion in extrastriate area MSTl: disparity sensitivity J Neurophysiol 82 2762-75 [225, 243]

Eizenman M, Frecker RC, Hallett PE (1984) Precise non–contacting measurement of eye movements using the corneal reflex Vis Res 27 167–74 [365]

Elberger AJ (1979) The role of the corpus callosum in the development of interocular eye alignment and the organization of the visual field in the cat Exp Brain Res 36 71–85 [233]

Elberger AJ (1980) The effect of neonatal section of the corpus callosum on the development of depth perception in young cats Vis Res 20 177–87 [233]

Elberger AJ (1989) Binocularity and single cell acuity are related in striate cortex of corpus callosum sectioned and normal cats Exp Brain Res 77 213–16 [233]

Elberger AJ (1990) Spatial frequency thresholds of single striate cortical cells in neonatal corpus callosum sectioned cats Exp Brain Res 82 617–27 [233]

Elberger AJ, Smith EL (1983) Binocular properties of lateral suprasylvian cortex are not affected by neonatal corpus callosum section Brain Res 278 259–98 [233]

Elberger AJ, Smith EL (1985) The critical period for corpus callosum section to affect cortical binocularity Exp Brain Res 57 213–23 [233]

Elkins J (1988) Did Leonardo develop a theory of cuvilinear perspective J Warburg and Courtauld Institutes 51 190-6 [37]

Ellard CG, Goodale MA, Timney B (1984) Distance estimation in the Mongolian Gerbil: the role of dynamic depth cues Behav Brain Res 14 29-39 [549]

Ellemberg D, Lewis TL, Liu CH, Maurer D (1999) Development of spatial and temporal vision during childhood Vis Res 39 2325-33 [471]

Ellenberger C, Duane MD, Shuttlesworth E (1978) Electrical correlates of normal binocular vision Arch Neurol 35 834-7 [259]

Ellerbrock VJ (1948) Further study of effects induced by anisometropic corrections Am J Optom Arch Am Acad Optom 25 430–7 [384]

Ellerbrock VJ (1949a) Experimental investigation of vertical fusional movements Part I Am J Optom Arch Am Acad Optom 29 327-37 [407]

Ellerbrock VJ (1949b) Experimental investigation of vertical fusion Part II Am J Optom Arch Am Acad Optom 29 388-399 [407]

Ellerbrock VJ (1952) Effect of aniseikonia on the amplitude of vertical divergence Am J Optom Arch Am Acad Optom 29 403-15 [407]

Ellerbrock VJ (1954) Inducement of cyclofusional movements Am J Optom Arch Am Acad Optom 31 553–66 [412]

Ellerbrock VJ, Fry GA (1941) The after–effect induced by vertical divergence Am J Optom Arch Am Acad Optom 18 450–4 [384]

Ellerbrock VJ, Fry GA (1942) Effects induced by anisometropic corrections Am J Optom Arch Am Acad Optom 19 444–59 [384]

Elliot J (1852) Letter to the Editor Lon Edin Dub Philos Mag J Sci 3 397 [64]

Elliott T, Shadbolt NR (1996) A mathematical model of activity-dependent anatomical segregation induced by competition for neurotrophic support Biol Cyber 75 463-70 [464]

Elliott T, Shadbolt NR (1999) A neurotrophic model of the development of the retinogeniculocortical pathway induced by spontaneous retina waves J Neurosci 19 7951-70 [464]

Elliott T, Shadbolt NR (2002) Dissociating ocular dominance column development and ocular dominance plasticity: a neurotrophic model Biol Cybern 86 281-92 [464]

Elliott T, Howarth, CI, Shadbolt NR (1997) Axonal processes and neural plasticity. III Competition for dendrites Philos Tr R Soc B 352 1975-83 [464]

Elston GN, Tweedale R, Rosa MGP (1999) Cortical integration in the visual system of the macaque monkey: large-scale morphological differences in the pyramidal neurons in the occipital, parietal and temporal lobes Proc R Soc B 296 1367-74 [197, 222]

Emerson RC (1997) Quadrature subunits in directionally selective simple cells: spatiotemporal interactions Vis Neurosci 14 357-71 [203]

Emerson RC, Bergen JR, Adelson EH (1992) Directionally selective complex cells and the computation of motion energy in cat visual cortex Vis Res 32 203-18 [202]

Emmerton J (1983) Functional morphology of the visual system In Physiology and behaviour of the pigeon (ed M Abs) pp 221–66 Academic Press, London [545]

Emsley HH (1952) Visual optics Hatton Press, London [171]

Endo M, Kaas JH., Jain N, Smith EL, Chino YM (2000) Binocular cross-orientation suppression in the primary visual cortex (V1) of infant rhesus monkeys Invest Ophthal Vis Sci 41 4022-31 [484]

Engel AK, König P, Singer W (1991) Direct physiological evidence for scene segmentation by temporal coding Proc Nat Acad Sci 88 9136–40 [133]

Engel AK, König P, Kreiter AK, Schillen TB, Singer W (1992) Temporal coding in the visual cortex: new vistas on integration in the nervous system TINS 15 218–29 [133, 134]

Engel E (1956) The role of content in binocular resolution Am J Psychol 69 87–91 [309]

Engel GR (1967) The visual processes underlying binocular brightness summation Vis Res 7 753–67 [326]

Engel GR (1969) The autocorrelation function and binocular brightness mixing Vis Res 9 1111–30 [326]

Engel GR (1970b) Tests of a model of binocular brightness Can J Psychol 27 335–52 [326]

Engel SA (1996) Looking into the black box: new directions in neuroimaging Neuron 17 375-8 [182]

Engert F, Bonheoffer T (1997) Synaptic specificity of long-term potentiation breaks down at short distances Nature 388 279-84 [460]

Enoch JM,Lakshminarayanan V (2000) Duplication of unique optical effects of ancient Egyptian lenses from the IV/V dynasties: lenses fabricated ca 2920-2700 BC or roughly 4600 years ago Ophthal Physiol Opt 20 129-30 [19]

Enoch JM, Rabinowicz IM (1976) Early surgery and visual correction of an infant born with unilateral eye lens opacity Doc Ophthal **41** 371-82 [518]

Enoch JM, Tobey FL (ed) (1981) Retinal photoreceptor optics Springer Verlag, Berlin [165]

Enoch JM, Birch DG, Birch EE (1979) Monocular light exclusion for a period of days reduces directional sensitivity of the human retina Science **206** 705–7 [165]

Enoksson P (1963) Binocular rivalry and monocular dominance studied with optokinetic nystagmus Acta Ophthal **41** 544–63 [287]

Enright JT (1980) Ocular translation and cyclotorsion due to changes in fixation distance Vis Res **20** 595–601 [420]

Enright JT (1984) Changes in vergence mediated by saccades J Physiol **350** 9–31 [422]

Enright JT (1986a) The aftermath of horizontal saccades: saccadic retraction and cyclotorsion Vis Res **29** 1807-14 [370]

Enright JT (1986b) Facilitation of vergence changes by saccades: influences of misfocused images and of disparity stimuli in man J Physiol **371** 69–87 [422]

Enright JT (1987a) Perspective vergence: oculomotor responses to line drawings Vis Res **27** 1513–29 [386]

Enright JT (1987b) Art and the oculomotor system: perspective illustrations evoke vergence changes Perception **16** 731–46 [386]

Enright JT (1989a) Convergence during human vertical saccades: probable causes and perceptual consequences J Physiol **410** 45-65 [422]

Enright JT (1990) Stereopsis cyclotorsional "noise" and the apparent vertical Vis Res **30** 1487–97 [415]

Enright JT (1992a) The remarkable saccades of asymmetrical vergence Vis Res **32** 2291–76 [421, 423]

Enright JT (1992b) Unexpected role of the oblique muscles in the human vertical fusional reflex J Physiol **451** 279–93 [410, 411]

Enright JT (1996b) Slow velocity asymmetrical convergence: a decisive failure of "Hering's law" Vis Res **36** 3667-84 [421]

Enright JT (1998a) On the "cyclopean eye": saccadic asymmetry and the reliability of perceived straight-ahead Vis Res **38** 459-69 [421]

Enright JT (1998b) Monocularly programmed human saccades during vergence changes? J Physiol **512** 235-50 [423]

Enroth–Cugell C, Robson JG (1966) The contrast sensitivity of ganglion cells of the cat J Physiol **187** 517–52 [168]

Eriksen CW (1966) Independence of successive inputs and uncorrelated error in visual form perception J Exp Psychol **72** 29–35 [318]

Eriksen CW, Greenspon TS (1968) Binocular summation over time in the perception of form at brief durations J Exp Psychol **76** 331-6 [324]

Eriksen CW, Greenspon TS, Lappin J, Carlson WA (1966) Binocular summation in the perception of form at brief durations Percept Psychophys **1** 415–9 [318, 324]

Eriksson ES (1980) Movement parallax and distance perception in the grasshopper (*Phaulacridium vittatum*) J Exp Biol **86** 337-40 [534]

Eriksson L, Dahlbom M, Widén L (1990) Positron emission tomography—a new technique for studies of the central nervous system J Micros **157** 305-33 [182]

Erkelens CJ (1987) Adaptation of ocular vergence to stimulation with large disparities Exp Brain Res **66** 507–16 [399, 401, 404]

Erkelens CJ (1988) Fusional limits for a large random–dot stereogram Vis Res **28** 345–53 [281]

Erkelens CJ, Collewijn H (1985a) Motion perception during dichoptic viewing of moving random–dot stereograms Vis Res **25** 583–8 [395]

Erkelens CJ, Collewijn H (1985b) Eye movements in relation to loss and regaining of fusion of disjunctively moving random–dot stereograms Hum Neurobiol **4** 181–8 [393, 395]

Erkelens CJ, Collewijn H (1985c) Eye movements and stereopsis during dichoptic viewing of moving random–dot stereograms Vis Res **25** 1689–700 [400]

Erkelens CJ, Collewijn H (1991) Control of vergence: gating among disparity inputs by voluntary target selection Exp Brain Res **87** 671–78 [401]

Erkelens CJ, Regan D (1986) Human ocular vergence movements induced by changing size and disparity J Physiol **379** 145–69 [386]

Erkelens CJ, van der Steen J, Steinman RM, Collewijn H (1989a) Ocular vergence under natural conditions. I Continuous changes of target distance along the median plane Proc R Soc B **236** 417–40 [398, 399]

Erkelens CJ, Steinman RM, Collewijn H (1989b) Ocular vergence under natural conditions. II Gaze shifts between real targets differing in distance and direction Proc R Soc B **236** 441–65 [387, 399, 421, 423]

Erkelens CJ, Collewijn H, Steinman RM (1989c) Asymmetrical adaptation of human saccades to anisometropic spectacles Invest Ophthal Vis Sci **30** 1132–45 [424]

Erskine L, Williams SE, Brose K, Kidd T, Rachel RA, Goodman CS, Tessier Lavigne M, Mason CA (2000) Retinal ganglion cell axon guidance in the mouse optic chiasm: expression and function of Robos and Slits J Neurosci **20** 4975-82 [447]

Erwin E, Miller KD (1998) Correlation-based development of ocularly matched orientation and ocular dominance maps: determination of required input activities J Neurosci **18** 9870-95 [464]

Erwin E, Miller, KD (1999) The subregion correspondence model of binocular simple cells J Neurosci **19** 7212-29 [253, 266]

Erwin E, Obermeyer K, Schulten K (1995) Models of orientation and ocular dominance columns in the visual cortex: a critical comparison Neural Comput **7** 425-68 [212]

Eskandar EN, Assad JA (1999) Dissociation of visual , motor and predictive signals in parietal cortex during visual guidance Nature Neurosci **2** 88-93 [227]

Eskandar EN, Richmond BJ, Optican LM (1992) Role of inferior temporal neurons in visual memory. I Temporal encoding of information about visual images recalled images and behavioral context J Neurophysiol **68** 1277-95 [223]

Euclid (300 BC/1945) Optics (Translated by HE Burton) J Opt Soc Am **35** 357–72 [10]

Evans BJW (1997) Pickwell's binocular vision anomalies Investigation and treatment Butterworth-Heinemann, London [358, 376]

Evans BJW, Drasco N, Richards IL (1994) Investigation of accommodative and binocular function in dyslexia Ophthal Physiol Opt **14** 5-19 [374]

Everson RM, Prashanth AK, Gabbay M, Knight BW, Sirovich L, Kaplan E (1998) Representation of spatial frequency and orientation in the visual cortex Proc Nat Acad Sci **95** 8334-8 [210]

Evinger C (1988) Extraocular motor nuclei: location morphology and afferents In Neuroanatomy of the oculomotor system (ed JA Büttner–Ennever) pp 81–118 Elsevier, New York [419]

Evinger LC, Fuchs AF, Baker R (1977) Bilateral lesions of the medial longitudinal fasciculus in monkeys: effects on the horizontal and vertical components of voluntary and vestibular induced eye movements Exp Brain Res **28** 1–20 [429]

Exner S (1868) Über die zu einer Gesichtswahrnemung Nöthige Zeit Sitzungsbericht der Akademie Wissenschaft Wien **58** 601–32 [340]

Exner S (1891) The physiology of the compound eyes of insects and crustaceans (Translated in 1989 by RC Hardie) Springer, New York [536, 537]

Eyre MB, Schmeeckle MM (1933) A study of handedness eyedness and footedness Child Devel 4 73–8 [295]

Eysel UT, Shevelev IA, Lazareva NA, Sharaev GA (1998) Orientation tuning and receptive field structure in cat striate neurons during local blockade of intracortical inhibition Neurosci 84 25-36 [199]

Eysteinsson T, Barris MC, Denny N, Frumkes TE (1993) Tonic interocular suppression binocular summation and the evoked potential Invest Ophthal Vis Sci 34 2743–8 [333]

Fagiolini M, Pizzorusso T, Berardi N, Domenici L, Maffei L (1994) Functional postnatal development of the rat primary visual cortex and the role of visual experience: dark rearing and monocular deprivation Vis Res 34 709-20 [515]

Fahle M (1982a) Cooperation between different spatial frequencies in binocular rivalry Biol Cyber 44 27–9 [288]

Fahle M (1982b) Binocular rivalry: suppression depends on orientation and spatial frequency Vis Res 22 787–800 [288]

Fahle M (1983) Non–fusible stimuli and the role of binocular inhibition in normal and pathologic vision especially strabismus Doc Ophthal 55 323–40 [528]

Fahle M (1987) Naso-temporal asymmetry of binocular inhibition Invest Ophthal Vis Sci 28 1016-17 [292, 520]

Fahle M (1993a) Figure–ground discrimination from temporal information Proc R Soc B 254 199–203 [131]

Fahle M (1994) Human pattern recognition: parallel processing and perceptual learning Perception 23 411-27 [354]

Fahle M (1997) Specificity of learning curvature orientation and vernier discriminations Vis Res 37 1885-95 [88, 528]

Fahle M, Edelman S (1993) Long–term learning in vernier acuity: effects of stimulus orientation range and of feedback Vis Res 33 397–412 [88, 354]

Fahle M, Palm G (1991) Perceptual rivalry between illusory and real contours Biol Cyber 66 1-8 [290]

Fahle M, Schmid M (1988) Naso–temporal asymmetry of visual perception and of the visual cortex Vis Res 28 293–300 [473]

Falmagne JC (1985) Elements of psychophysical theory Oxford University Press, New York [73, 77]

Fantz R (1965) Visual perception from birth as shown by pattern selectivity Ann N Y Acad Sci 118 793–814 [476]

Faraday M (1831) On a peculiar class of optical deception Journal of the Royal Institution February [70]

Farge M, Hunt J, Vassilicos S (Eds) (1993) Wavelets, fractals and fourier transforms: new developments and new applications Oxford University Press, Oxford [141]

Farid H, Adelson EH (2001) Synchrony does not promote grouping in temporally structured displays Nature Neurosci 4 875-6 [134]

Farina WM, Varjú D, Zhou Y (1994) The regulation of distance to dummy flowers during hovering flight in the hawk moth *Macroglossum stellatarum* J Comp Physiol A 174 239-47 [535]

Faugeras O (1993) Three-dimensional computer vision MIT Press, Cambridge MA [109]

Favreau OE (1978) Interocular transfer of color–contingent motion aftereffects; positive aftereffects Vis Res 18 841–4 [353]

Favreau OE (1979) Persistence of simple and contingent motion aftereffects Percept Psychophys 29 187–94 [85]

Favreau OE, Cavanagh P (1983) Interocular transfer of a chromatic frequency shift Vis Res 23 951–7 [351]

Favreau OE, Cavanagh P (1984) Interocular transfer of a chromatic frequency shift: temporal constraints Vis Res 2 7 1799—804 [351]

Fawcett JW (1993) Growth–cone collapse: too much of a good thing? TINS 16 165–7 [438]

Fawcett JW, Willshaw DJ (1982) Compound eyes project stripes on the optic tectum in Xenopus Nature 296 350-2 [465]

Fawcett SL, Birch EE (2000) Motion VEPs, stereopsis, and bifoveal fusion in children with strabismus Invest Ophthal Vis Sci 41 411-17 [526]

Fawcett SL, Raymond JE, Astle WF, Skov CMB (1998) Anomalies of motion perception in infantile esotropia Invest Ophthal Vis Sci 39 727-35 [525]

Fazeli MS (1992) Synaptic plasticity: on the trail of the retrograde messenger TINS 15 115–17 [461]

Fechner GT (1860) Uber einige Verhältnisse des binokularen Sehens Berichte Sächs gesamte Wissenschaft 7 337-564 [287]

Feldman DE (2000) Inhibition and plasticity Nature Neurosc 3 303-4 [515]

Feldman DE, Knudsen EI (1998) Experience-dependent plasticity and the maturation of glutamatergic synapses Neuron 20 1067-71 [462]

Feldman DE, Brainard MS, Knudsen EI (1996) Newly learned auditory responses mediated by NMDA receptors in the owl inferior colliculus Science 271 525-8 [546]

Feldman JM, Cooper J (1980) Rapid assessment of stereopsis in pre-verbal children using operant techniques: a preliminary study J Am Optom Assoc 51 767-71 [482]

Feldman JM, Cooper J, Carniglia P, Schiff FM, Skeetes JN (1989) Comparison of fusional ranges measured by Risley prisms vectograms and computer orthopter Optom Vis Sci 66 375-82 [392, 393]

Feldman M, Cohen B (1968) Electrical activity in the geniculate body of the alert monkey associated with eye movements J Neurophysiol 31 455–66 [179]

Felleman DJ, Van Essen DC (1987) Receptive field properties of neurons in area V3 of macaque monkey extrastriate cortex J Neurophysiol 57 889–920 [220, 229, 242, 243]

Felleman DJ, Van Essen DC (1991) Distributed hierarchical processing in the primate cerebral cortex Cerebral Cortex 1 1–47 [218, 220, 230]

Felleman DJ, Burkhalter A,Van Essen DC (1997a) Cortical connections of areas V3 and VP of macaque monkey extrastriate visual cortex J Comp Neurol 379 21-47 [220, 221]

Felleman DJ, Xiao Y, McClendon E (1997b) Modular organization of occipito-temporal pathways: cortical connections between visual area 4 and visual area 2 and posterior inferotemporal ventral area in macaque monkeys J Neurosci 17 3185-200 [220]

Feller MB, Wellis DP, Stellwagen D, Werblin FS, Shatz CJ (1996) Requirement for cholinergic synaptic transmission in the propagation of spontaneous retinal waves Science 272 1182-7 [459]

Felton TB, Richards W, Smith RA (1972) Disparity processing of spatial frequencies in man J Physiol 225 349–62 [251]

Fender D, Julesz B (1967) Extension of Panum's fusional area in binocularly stabilized vision J Opt Soc Am 57 819–30 [279, 280, 395]

Fenelon B, Neill RA, White CT (1986) Evoked potentials to dynamic random dot stereograms in upper centre and lower fields Doc Ophthal 63 151-6 [262]

Ferguson ES (1977) The mind's eye: nonverbal thought in technology Science 197 827-36 [37, 160]

Ferman L, Collewijn H, Van den Berg AV (1987a) A direct test of Listing's law. I Human ocular torsion measured in static tertiary positions Vis Res 27 929–38 [370]

Ferman L, Collewijn H, Jansen TC, van den Berg AV (1987b) Human gaze stability in horizontal vertical and torsional direction during voluntary head movements evaluated with a three–dimensional scleral induction coil technique Vis Res 27 811–28 [411, 413, 415]

Fernald RD (1985) Eye movements in the African cichlid fish *Haplochromis burtoni* J Comp Physiol A**156** 199–208 [539]

Ferraina S, Paré M, Wurtz RH (2000) Disparity sensitivity of frontal eye field neurons J Neurophysiol **83** 625-9 [244, 432]

Ferrera VP, Nealey TA, Maunsell JHR (1992) Mixed parvocellular and magnocellular geniculate signals in visual area V4 Nature **358** 756–8 [229]

Ferrera VP, Nealey TA, Maunsell JHR (1994) Responses in macaque visual area V4 following inactivation of the parvocellular and magnocellular LGN pathways J Neurosci **14** 2080-88 [229]

Ferster D (1981) A comparison of binocular depth mechanisms in areas 17 and 18 of the cat visual cortex J Physiol **311** 623–55 [238, 247, 248, 252, 313]

Ferster D (1987) Origin of orientation selective EPSP's in simple cells of cat visual cortex J Neurosci **7** 1780–91 [313]

Ferster D (1988) Spatially opponent excitation and inhibition in simple cells of the cat visual cortex J Neurosci **8** 1172-80 [194]

Ferster D (1990) Binocular convergence of excitatory and inhibitory synaptic pathways onto neurons of cat visual cortex Vis Neurosci **4** 625–9 [214]

Ferster D, Miller KD (2000) Neural mechanisms of orientation selectivity in the visual cortex Ann Rev Neurosci **23** 441-71 [199]

Ferster D, Chung S, Wheat H (1996) Orientation selectivity of thalamic input to simple cells of cat visual cortex Nature **380** 279-52 [199]

Fiala JC, Feinberg M, Popov V, Harris KM (1998) Synaptogenesis via dendritic filopodia in developing hippocampal area CA1 J Neurosci **18** 8900-11 [192]

Field DJ (1987) Relations between the statistics of natural images and the response properties of cortical cells J Opt Soc Am A **4** 2379-94 [94]

Field DJ, Hayes A, Hess RF (1993) Contour integration by the human visual system: evidence for a local "association field" Vis Res **33** 173–93 [124]

Field JV (1986) Piero della Francesca's treatment of edge distortion J Warb Court Inst **49** 66-90 [35]

Filipek PA, Richelme C, Kennedy DN, Caviness VS (1994) The young adult human brain: an MRI-based morphometric analysis Cerebral Cortex **4** 344-60 [446]

Finch DJ, Collett TS (1983) Small–field binocular neurons in the superficial layers of the frog optic tectum Proc R Soc B **217** 491–7 [540]

Fincham EF (1951) The accommodation reflex and its stimulus Br J Ophthal **35** 381–93 [361, 363, 364, 387]

Fincham EF (1955) The proportion of ciliary muscular force required for accommodation and convergence J Physiol **128** 99-112 [390]

Fincham EF (1962) Accommodation and convergence in the absence of retinal images Vis Res **1** 425–40 [387]

Fincham EF, Walton J (1957) The reciprocal actions of accommodation and convergence J Physiol **137** 488–508 [390]

Findlay JM, Harris LR (1993) Horizontal saccades to dichoptically presented targets of differing disparities Vis Res **33** 1001-10 [421]

Findler N (1979) Associative networks New York, Academic press [152]

Finger S (2000) Minds behind the brain Oxford University Press, New York [27]

Fink GR, Dolan RJ, Halligan PW, Marshall JC, Frith CD (1997) Space-based and object-based visual attention: shared and specific domains Brain **120** 2013-28 [207]

Finney DJ (1971) Probit analysis Cambridge University Press, London [75]

Fioravanti F, Inchingolo P, Pensiero S, Spanio M (1995) Saccadic eye movement conjugation in children Vis Res **35** 3217-28 [421]

Fiorentini A, Berardi N (1981) Learning in grating waveform discrimination: specificity for orientation and spatial frequency Vis Res **21** 1149-58 [354]

Fiorentini A, Maffei L (1970) Electrophysiological evidence for disparity detectors in human visual system Science **169** 208–9 [261]

Fiorentini A, Maffei L (1974) Change of binocular properties of the simple cells of the cortex in adult cats following immobilization of one eye Vis Res **14** 217–8 [495]

Fiorentini A, Bayly EJ, Madei L (1972) Peripheral and central contributions to psychophysical spatial interactions Vis Res **12** 253-9 [334]

Fiorentin A, Sireteanu R, Spinelli D (1976) Lines and gratings: different interocular after-effects Vis Res **16** 1303-9 [340]

Fiorentini A, Maffei L, Bisti S (1979) Change of binocular properties of cortical cells in the central and paracentral visual field projections of monocularly paralyzed cats Brain Res **171** 541-44 [495]

Fiorentini A, Maffei L, Cenni MC, Tacchi A (1985) Deafferentation of oculomotor proprioception affects depth discrimination in adult cats Exp Brain Res **59** 296-301 [494]

Fiorentini A, Berardi N, Maffei L (1995) Nerve growth factor preserves behavioral visual acuity in monocularly deprived kittens Vis Neurosci **12** 51-5 [512]

Firth A Y (1999) Adie syndrome: evidence for refractive error and accommodative asymmetry as the cause of amblyopia Am J Ophthal **128**118-9 [520]

Fischer AJ, Reh TA (2001) Müller glia are a potential source of neural regeneration in the postnatal chicken retina Nature Neurosci **4** 247-52 [187, 450]

Fischer B, Boch R (1981) Enhanced activation of neurones in prelunate cortex before visually guided saccades of trained rhesus monkeys Exp Brain Res **44** 129–37 [207]

Fischer B, Krüger J (1979) Disparity tuning and binocularity of single neurons in cat visual cortex Exp Brain Res **35** 1–8 [238]

Fischer KF, Lukasiewicz PD, Wong RO (1998) Age-dependent and cell class-specific modulation of retinal cell bursting activity by GABA J Neurosci **18** 3767-8 [444]

Fischer M, Kaech S, Wagner U, Brinkhaus H, Matus A (2000) Glutamate receptors regulate actin-based plasticity in dendritic spines Nature Neurosci **3** 887-94 [193]

Fischer M, Kaech S, Knutti D, Matus A (1998) Rapid actin-based plasticity in dendritic spines Neuron **20** 847-54 [438]

Fisher CB, Bornstein MH (1982) Identification of symmetry: effects of stimulus orientation and head position Percept Psychophys **32** 443-8 [102]

Fisher RA (1966) The design of experiments Oliver and Boyd, London [148]

Fisher SK, Ciuffreda KJ, Hammer, S (1987) Interocular equality of tonic accommodation and consenuality of accommodative hysteresis Ophthal Physiol Opt **7** 17-20 [359]

Fisher SK, Ciuffreda KJ, Tannen B, Super P (1988a) Stability of tonic vergence Invest Ophthal Vis Sci **29** 1577–81 [372]

Fisher SK, Ciuffreda KJ, Bird JE (1988b) The effect of monocular versus binocular fixation on accommodation hysteresis Ophthal Physiol Opt **8** 438-42 [359]

Fisher SK, Ciuffreda KJ, Bird JE (1990) The effect of stimulus duration on tonic accommodation and tonic vergence Optom Vis Sci **67** 441-9 [359]

Fite KV (1969) Single unit analysis of binocular neurons in the frog optic tectum Exp Neurol **27** 475–86 [540]

Fite KV (1973) The visual fields of the frog and toad: a comparative study Behav Biol **9** 707–18 [539, 541]

Fite KV, Rosenfield–Wessels S (1975) A comparative study of deep avian foveas Brain Behav Evol **12** 97–115 [546]

Fitzpatrick D, Itoh K, Diamond IT (1983) The laminar organization of the lateral geniculate body and the striate cortex of the squirrel monkey (*Saimiri sciureus*) J Neurosci **3** 673-702 [186]

Fitzpatrick D, Lund JS, Blasdel GG (1985) Intrinsic connections of macaque striate cortex: afferent and efferent connections of lamina 4C J Neurosci **5** 3329–49 [186, 188]

Fitzpatrick D, Usrey WM, Schofield BR, Einstein G (1994) The sublaminar organization of corticogeniculate neurons in layer 6 of macaque striate cortex Vis Neurosci **11** 307-15 [178, 187]

Fitzsimonds RM, Poo MM (1998) Retrograde signaling in the development and modification of synapses Physiol Rev **78** 143-170 [457]

Flanagan JR, Rao AK (1995) Trajectory adaptation to a nonlinear visuomotor transformation: Evidence of motion planning in visually perceived space J Neurophysiol **74** 2174-8 [144]

Fleagle JG (1988) Primate adaptation and evolution Academic Press, San Diego [216]

Fleet DJ, Jepson AD, Jenkin M (1991) Phase-based disparity measurement Comput Vis Gr Im Proc **53** 198-210 [266, 268, 269]

Fleet DJ, Wagner H, Heeger DJ (1996a) Neural encoding of binocular disparity: energy models position shifts and phase shifts Vis Res **36** 1839-57 [253, 268]

Fleet DJ, Wagner H, Heeger DJ (1996b) Modelling binocular neurons in the primary visual cortex In Computational and biological mechanisms of visual coding (ed M. Jenkin, L Harris) Cambridge University Press, London [249, 268]

Flitcroft DI (1990) A neural and computational model for the chromatic control of accommodation Vis Neurosci **5** 547-555 [363]

Flitcroft DI(1998) A model of the contribution of oculomotor and optical factors to emmetropization and myopia Vis Res **3 8** 2869-79 [435]

Flitcroft DI, Morley JW (1997) Accommodation in binocular contour rivalry Vis Res **37** 121-5 [299, 365]

Flitcroft DI, Judge SJ, Morley JW (1992) Binocular interactions in accommodation control: effects of anisometropic stimuli J Neurosci **12** 188-203 [299, 364]

Flom MC (1960a) On the relationship between accommodation and accommodative vergence Part I. Linearity Am J Optom Arch Am Acad Optom **37** 474-82 [388]

Flom MC (1960b) On the relationship between accommodation an accommodative convergence. Part II: Stability Am J Optom Arch Am Acad Optom **37** 517-23 [388]

Flom MC (1960c) On the relationship between accommodation an accommodative convergence. Part III: Effects of orthoptics Am J Optom Arch Am Acad Optom **37** 619-32 [390]

Flom MC (1963) Treatment of binocular anomalies in children In Vision of children an optometric symposium (ed MJ Hirsch, RE Wick) pp 197–228 Chilton, Philadelphia [374]

Flom MC, Heath GG, Takahashi E (1963) Contour interaction and visual resolution: contralateral effects Science **142** 979–89 [339]

Florence SL, Kaas JH (1992) Ocular dominance columns in area 17 of Old World macaque and talapoin monkeys: complete reconstructions and quantitative analysis Vis Neurosci **8** 449-62 [213]

Florence SL, Conley M, Casagrande VA (1986) Ocular dominance columns and retinal projections in New World spider monkeys (*Ateles ater*) J Comp Neurol **273** 234-48 [216]

Fogt N, Jones R (1997) Comparison of the monocular occlusion and a direct method for objective measurement of fixation disparity Optom Vis Sci **74** 43-50 [379]

Fogt N, Jones R (1998a) Comparison of fixation disparities obtained by objective and subjective methods Vis Res **38** 411-21 [379]

Fogt N, Jones R (1998b) The effect of forced vergence on retinal correspondence Vis Res **38** 2711-19 [379]

Foley JD, van Dam A, Feiner SK, Hughes JF (1990) Computer graphics Addison-Wesley New York Foley JD, van Dam A, Feiner SK, Hughes JF (1990) Computer graphics Addison-Wesley New York [109]

Foley JE (1974) Factors governing interocular transfer of prism adaptation Psychol Rev **81** 183-6 [355]

Foley JE, Miyanshi K (1969) Interocular effects in prism adaptation Science **165** 311-12 [355]

Foley JM (1964) Desarguesian property in visual space J Opt Soc Am **54** 684–92 [137]

Foley JM, Richards W (1974) Improvement in stereoanomaly with practice Am J Optom Physiol Opt **51** 935–8 [488]

Forster LM (1979) Visual mechanisms of hunting behaviour in *Trite planiceps* a jumping spider (Araneae: Salticidae) NZ J Zool **6** 79–93 [536]

Foster DH, Mason, RJ (1977) Interaction between rod and cone systems in dichoptic masking Neurosci Lett **4** 39-42 [341]

Fotheringhame D, Baddeley R (1997) Nonlinear principal components analysis of neuronal spike train data Biol Cyber **77** 283-8 [130]

Fourcade F (1962) La peinture murale de Touen Houang Editions Cercle d'Art, Paris [31]

Fox K, Daw NW (1993) Do NMDA receptors have a critical function in visual cortical plasticity? TINS **16** 116–22 [461]

Fox K, Sato H, Daw N (1989) The location and function of NMDA receptors in cat and kitten visual cortex J Neurosci **9** 2743–54 [462]

Fox K, Daw N, Sato H, Czepita D (1992) The effect of visual experience on development of NMDA receptor synaptic transmission in kitten visual cortex J Neurosci **12** 2972-84 [511]

Fox R (1991) Binocular rivalry In Vision and visual dysfunction Vol 9 Binocular vision (ed D Regan) pp 93–110 MacMillan, London [287]

Fox R, Blake RR (1971) Stereoscopic vision in the cat Nature **233** 55–6 [548]

Fox R, Check R (1966a) Binocular fusion: a test of the suppression theory Percept Psychophys **1** 331–4 [306]

Fox R, Check R (1966b) Forced–choice form recognition during binocular rivalry Psychonom Sci **6** 471–2 [306, 308]

Fox R, Check R (1968) Detection of motion during binocular rivalry suppression J Exp Psychol **78** 388–95 [303]

Fox R, Check R (1972) Independence between binocular rivalry suppression duration and magnitude of suppression J Exp Psychol **93** 283–9 [310]

Fox R, Herrmann J (1967) Stochastic properties of binocular rivalry alternations Percept Psychophys **2** 432–6 [311]

Fox R, Patterson R (1981) Depth separation and lateral interference Percept Psychophys **30** 513–20 [338]

Fox R, Rasche F (1969) Binocular rivalry and reciprocal inhibition Percept Psychophys **5** 215–17 [287, 310]

Fox R, Todd S, Bettinger LA (1975) Optokinetic nystagmus as an objective indicator of binocular rivalry Vis Res **15** 849–53 [287]

Fox R, Lehmkuhle SW, Westendorf DH (1976) Falcon visual acuity Science **192** 293–5 [546]

Fox R, Lehmkuhle SW, Bush RC (1977) Stereopsis in the falcon Science **197** 79–81 [546]

Fox R, Aslin RN, Shea SL, Dumais ST (1980) Stereopsis in human infants Science **207** 323–4 [482]

Fox R, Patterson R, Francis EL (1986) Stereoacuity in young children Invest Ophthal Vis Sci **27** 598–609 [483]

Fraenkel GS, Gunn DL (1940) The orientation of animals Oxford University Press, Oxford [112]

France TD Ver Hoeve JN (1994) VECP evidence for binocular function in infantile esotropia J Ped Ophthal Strab **31** 225-31 [260]

Franceschetti AT, Burian HM (1971) Visually evoked responses in alternating strabismus Am J Ophthal **71** 1292-7 [527]

Francis EL, Owens DA (1983) The accuracy of binocular vergence for peripheral stimuli Vis Res **23** 13–19 [380, 397]

Frangenberg T (1986) The image and the moving eye. Jean Pélerin (Viator) to Guidoboldo del Monte J Warburg and Courtland Institutes **49** 150-71 [36]

Frangenberg T (1991) Perspectivist Aristotelianism: three case-studies of cinquecento visual theory J Warb Court Inst **54** 137-58 [10]

Fraser J (1908) A new visual illusion of direction Br J Psychol **2** 307–20 [124]

Fraser SE (1991) Relative roles of positional cues and activity–based cues in the patterning of the retinotectal projection In Development of the visual system (ed DMK Lam, CJ Shatz) pp 123–32 MIT Press, Cambridge MA [442]

Fredenburg P, Harwerth RS (2001) The relative sensitivities of sensory and motor fusion to small binocular disparities Vis Res **41** 1969-79 [104]

Freedman DJ, Riesenhuber M, Poggio T, Miller EK (2001) Categorical representation of visual stimuli in the primate prefrontal cortex Science **291** 312-16 [153, 228]

Freeman AW, Jolly N (1994) Visual loss during interocular suppression in normal and strabismic subjects Vis Res **34** 2043-50 [333]

Freeman AW, Nguyen VA (2001) Controlling binocular rivalry Vis Res **41** 2943-50 [301]

Freeman AW, Nguyen VA, Jolly N (1996) Components of visual acuity loss in strabismics Vis Res **36** 765--74 [520]

Freeman RD, Bonds A (1979) Cortical plasticity in monocularly deprived immobilized kittens depends on eye movement Science **206** 1093–5 [495]

Freeman RD, Bradley A (1980) Monocularly deprived humans: nondeprived eye has supernormal vernier acuity J Neurophysiol **43** 1645-53 [509]

Freeman RD, Ohzawa I (1988) Monocularly deprived cats: binocular tests of cortical cells reveal functional connections from the deprived eye J Neurosci **8** 2791–506 [509]

Freeman RD, Ohzawa I (1990) On the neurophysiological organization of binocular vision Vis Res **30** 1661–76 [248, 252]

Freeman RD, Ohzawa I (1992) Development of binocular vision in the kitten's striate cortex J Neurosci **12** 4721–36 [453]

Freeman RD, Tsumoto T (1983) An electrophysiological comparison of convergent and divergent strabismus in the cat: electrical and visual activation of single cortical cells J Neurophysiol **49** 238-53 [505]

Freeman W (1975) Mass action in the nervous system Academic Press, New York [136]

Freeman, WT (1994) The generic viewpoint assumption in a framework for visual perception Nature **368**, 542-5 [161]

Frégnac Y, Imbert M (1978) Early development of visual cortical cells in normal and dark-reared kittens: relationship between orientation selectivity and ocular dominance J Physiol **278** 27-44 [453, 500]

Frégnac Y, Imbert M (1984) Development of neuronal selectivity in the primary visual cortex of the cat Physiol Rev **64** 325–434 [458, 513]

Frégnac Y, Shulz D, Thorpe S, Bienenstock E (1988) A cellular analogue of visual cortical plasticity Nature **333** 368-70 [463]

Frégnac Y, Burke JP Smith D, Friedlander MJ (1994) Temporal covariance of pre- and postsynaptic activity regulates functional connectivity in the visual cortex J Neurophysiol **71** 1403-21 [135]

Freier BE, Pickwell LD (1983) Physiological exophoria Ophthal Physiol Opt **3** 297-72 [375]

Freiwald WA, Kreiter AK, Singer W (1995) Stimulus dependent intercolumnar synchronization of single unit responses in cat area 17 Neuroreport **6** 2348-52 [133]

Frens MA, Beuzekom AD von Sándor PS, Henn V (1998) Binocular coupling in chameleon saccade generation Biol Cyber **78** 57-61 [543]

Freud SL (1964) The physiological locus of the spiral aftereffect Am J Psychol **77** 422–8 [347]

Freund TF, Martin KA C, Soltesz I, Somogyi P, Whitteridge D (1989) Arborization pattern and postsynaptic targets of physiologically identified thalamocortical afferents in striate cortex of the macaque monkey J Comp Neurol **289** 315-36 [188]

Friauf E, Shatz CJ (1991) Changing patterns of synaptic input to subplate and cortical plate during development of visual cortex J Neurophysiol **66** 2059–71 [458]

Fricke C, Lee JS, Geiger-Rudolph S, Bonhoeffer F, Chien CB (2001) *astray*, a Zebrafish *roundabout* homolog required for retinal axon guidance Science **292** 507-10 [440]

Friedenwald JS (1936) Diagnosis and treatment of anisophoria Arch Ophthal **15** 283–307 [375]

Friedlander MJ, Stanford LR (1984) Effects of monocular deprivation on the distribution of cell types in the LGN: a sampling study with fine-tipped micropipettes Exp Brain Res **53** 451-61 [504]

Friedlander MJ, Martin KAC, Wassenhove-McCarthy D (1991) Effects of monocular visual deprivation on geniculocortical innervation of area 18 in cat J Neurosci **11** 3298-88 [506]

Friedman Z Neumann E, Hyams SW, Peleg B (1980) Ophthalmic screening of 38,000 children age 1 to 21/2 years in child welfare clinics J Ped Ophthal Strab**17** 291-7 [372]

Friedman-Hill SR, Robertson LC, Treisman A (1995) Parietal contributions to visual feature binding: evidence from a patient with bilateral lesions Science **299** 853-5 [134]

Fries P, Roelfsema PR, Engel AK, König P, Singer W (1997) Synchronization of oscillatory responses in visual cortex correlates with perception in interocular rivalry Proc Nat Acad Sci **94** 12999-704 [314]

Fries P, Reynolds JH, Rorie AE, Desimone R (2001) Modulation of oscillatory neuronal synchronization by selective visual attention Science **291** 1560-63 [134]

Frisby JP, Mayhew JEW (1980b) The role of spatial frequency tuned channels in vergence control Vis Res **20** 727–32 [393]

Frisby JP, Buckley D, Bergin L, Hill L (1993) Cyclotorsion to slanted surfaces with consistent and conflicting stereo and texture slant cues Perception **22** (ECVP Abs) 115 [416]

Frisén L, Lindblom B (1988) Binocular summation in humans: evidence for a hierarchical model J Physiol **402** 773–82 [324]

Frith C, and Dolan RJ (1997) Bran mechanisms associated with top-down processes in perception Philos Trans R Soc **352** 1221-30 [208]

Fronius M, Sireteanu R (1989) Monocular geometry is selectively distorted in the central visual field of strabismic amblyopes Invest Ophthal Vis Sci **30** 2034–44 [522, 523]

Fronius M, Sireteanu R (1994) Pointing errors in strabismics: complex patterns of distorted visuomotor coordination Vis Res **34** 689–707 [527]

Frost BJ, Scilley PL, Wong SCP (1981) Moving background patterns reveal double–opponency of directionally specific pigeon tectal neurons Exp Brain Res **43** 173–85 [224]

Frost BJ, Goodale MA, Pettigrew JD (1983) A search for functional binocularity in the pigeon Proc Soc Neurosci **9** 823 [545]

Frost BJ, Wise LZ, Morgan B, Bird D (1990) Retinotopic representation of the bifoveate eye of the kestrel (Falco sparverius) on the optic tectum Vis Neurosci **5** 231-9 [546]

Frost DO, Metin C (1985) Induction of functional retinal projections to the somatosensory system Nature **317** 162-4 [447]

Frostig RD (1994) What does *in vivo* optical imaging tell us about the primary visual cortex in primates In Cerebral cortex Volume 10 Primary visual cortex in primates (ed A Peters, KS Rockland) pp 331-58 Plenum, New York [181]

Fry GA (1939) Further experiments on the accommodation–convergence relationship Am J Optom Arch Am Acad Optom **16** 325–34 [392]

Fry GA (1950) Visual perception of space Am J Optom Arch Am Acad Optom **27** 531–53 [367]

Fry GA (1968) Nomograms for torsion and direction of regard Am J Optom Arch Am Acad Optom **45** 631–41 [370]

Fry GA (1983) Basic concepts underlying graphical analysis In Vergence eye movements: Basic and clinical aspects (ed MC Schor, KJ Ciuffreda) pp 403–38 Butterworth, Boston [391]

Fry GA, Bartley SH (1933) The brilliance of an object seen binocularly Am J Ophthal **16** 687–93 [325]

Fry GA, Treleaven CL, Baxter RC (1945) Specification of the direction of regard Am J Optom Arch Am Acad Optom **2 2** 351–60 [370]

Fujita I, Tanaka K, Ito M, Cheng K (1992) Columns for visual features of objects in monkey inferotemporal cortex Nature **360** 343–6 [445]

Fujita S (1990) Morphogenesis of the human brain as studied by 3-D computer graphics simulation J Micros **157** 259-69 [222]

Fukuda H, Blake R (1992) Spatial interactions in binocular rivalry J Exp Psychol HPP **18** 362–70 [299]

Fulton AB, Albert DM, Craft JL (1978) Human albinism Light and electron microscope study Arch Ophthal **96** 305-10 [496]

Funahashi S, Bruce CJ, Goldman–Rakic PS (1989) Mnemonic coding of visual space in the monkey's dorsolateral prefrontal cortex J Neurophysiol **61** 331–49 [228]

Furchner CS, Ginsburg AP (1978) "Monocular rivalry" of a complex waveform Vis Res **18** 1641-8 [296]

Fuster JM, Jervey JP (1981) Inferotemporal neurons distinguish and retain behaviorally relevant features of visual stimuli Science **212** 952–5 [208]

Gabbiani F, Krapp HG, Laurent G (1999) Computation of object approach by a wide-field, motion-sensitive neuron J Neurosci **19** 1122-41 [534]

Gabor D (1946) Theory of communication J Inst Elec Engin (London) **93** 429–57 [139]

Gabrieli DE, Brewer JB, Desmond JE, Glover GH (1997) Separate neural bases of two fundamental processes in the human medial temporal lobe Science **276** 294-6 [222]

Gaffan D (1994) Scene–specific memory for objects: a model of episodic memory impairment in monkeys with fornix transection J Cog Neurosci **6** 305-20 [223]

Gaillard F (1985) Binocularly driven neurons in the rostral part of the frog optic tectum J Comp Physiol A **157** 47–55 [540]

Galand G, Liege B (1974) Réponses visuelle unitaires chez la troute In Vision in fishes new approaches in research (ed MA Ali) pp 127-36 Plenum, New York [539]

Galarreta M, Hestrin S (2001) Electrical synapses between GABA-releasing interneurones Nature Rev **2** 425-33 [190]

Galen C (175) De usa partium corporis humani (Translated in 1968 by MT May) Cornell University Press, Ithaca NY [13]

Gallant JL, Braun J, Van Essen DC (1993) Selectivity for polar hyperbolic and Cartesian gratings in macaque visual cortex Science **259** 100-3 [136, 212, 221]

Gallant JL, Connor CE, Rakshit S, Lewis JW, van Essen DC (1996) Neural responses to polar, hyperbolic, and cartesian gratings in area V4 of the macaque monkey J Neurophysiol **76** 2718-39 [212]

Galletti C, Battaglini PP (1989) Gaze-dependent visual neurons in area V3A of monkey prestriate cortex J Neurosci **9** 1112-1125 [220]

Galletti C, Battaglini PP and Fattori P (1993) Parietal neurons encoding spatial locations in craniotopic coordinates Exp Brain Res **96** 221–9 [228]

Galletti C, Battaglini PP and Fattori P (1997) The posterior parietal cortex in humans and monkeys News Physiol Sci **12** 166-71 [228]

Galli L, Maffei L (1988) Spontaneous impulse activity of rat retinal ganglion cells in prenatal life Science **272** 90–1 [436, 459]

Gallo G, Lefcort FB, Letourneau PC (1997) The trkA receptor mediates growth cone turning toward a localized source of nerve growth factor J Neurosci **17** 5445-54 [457]

Gamlin PDR, Clarke RJ (1995) Single-unit activity in the primate nucleus reticularis tegmenti pontis related to vergence and ocular accommodation J Neurophysiol **73** 2115-9 [431]

Gamlin PDR, Mays LE (1992) Dynamic properties of medial rectus motoneurons during vergence eye movements J Neurophysiol **67** 64–74 [430]

Gamlin PDR, Yoon K (2000) An area for vergence eye movement in primate frontal cortex Nature **407** 10037 [432]

Gamlin PDR, Gnadt JW, Mays LE (1989) Abducens internuclear neurones carry an inappropriate signal for ocular convergence J Neurophysiol **62** 70-81 [419, 431]

Gamlin PDR, Zhang Y, Clendaniel RA, Mays LE (1994) Behavior of identified Edinger-Westphal neurons during ocular accommodation J Neurophysiol **72** 2368-82 [365]

Gamlin PDR, Yoon K, Zhang H (1996) The role of cerebro-ponto-cerebellar pathways in the control of vergence eye movements Eye **10** 167-71 [431, 432]

Ganguly K, Kiss L, Poo MM (2000) Enhancement of presynaptic neuronal excitability by correlated presynaptic and postsynaptic spiking Nature Neurosci **3** 1018-29 [461]

Ganz L, Hirsch HVB, Tieman SB (1972) The nature of perceptual deficits in visually deprived cats Brain Res **44** 547-68 [524]

Garaschuk O, Linn J, Eilers J, Konnerth A (2000) Large-scale oscillatory calcium waves in the immature cortex Nature Neurosci **3** 452-9 [459]

Gardner JC, Cynader MS (1987) Mechanisms for binocular depth sensitivity along the vertical meridian of the visual field Brain Res **413** 60–74 [233]

Gardner JC, Raiten EJ (1986) Ocular dominance and disparity–sensitivity: why there are cells in the visual cortex driven unequally by the two eyes Exp Brain Res **64** 505–14 [215, 238]

Garey LJ (1984) Structural development of the visual system of man Hum Neurobiol **3** 75-80 [443]

Garey LJ, Blakemore C (1977) The effects of monocular deprivation on different neuronal classes in the lateral geniculate nucleus of the cat Exp Brain Res **28** 259-78 [503]

Garey LJ, de Courten C (1983) Structural development of the lateral geniculate nucleus and cortex in monkey and man Behav Brain Res **10** 3–13 [451]

Garey LJ, Dreher B, Robinson SR (1991) The organization of the visual thalamus In Neuroanatomy of the visual pathways and their development (ed B Dreher, SR Robinson) pp 176–234 CRC Press, Boston [178]

Garner WR (1962) Uncertainty and structure as psychological concepts Wiley, New York [77]

Garner WR (1974) The processing of information and structure Erlbaum, New York [142]

Garraghty PE, Salinger WL, MacAvoy MG, Schroeder CE, Guido W (1982) The shift in X/Y ratio after chronic monocular paralysis: a binocularly mediated barbiturate-sensitive effect in the adult lateral geniculate nucleus Exp Brain Res **47** 301-8 [494]

Garraghty PE, Roe AW, Chino YM, Sur M (1989) Effect of convergent strabismus on the development of physiologically identified retinogeniculate axons in cats J Comp Neurol **289** 202-12 [504]

Gary-Bobo E, Milleret C, Buisseret P (1986) Role of eye movements in developmental processes of orientation selectivity in the kitten visual cortex Vis Res **29** 557-67 [495]

Garzia RP (1987) Efficacy of vision therapy in amblyopia: a literature review Am J Optom Physiol Opt **64** 393–404 [528]

Garzia RP, Nicholson SB (1988) The effect of volition on the horizontal forced-vergence fixation disparity curve Am J Optom Physiol Opt **65** 61-3 [378]

Gaska JP, Jacobson LD, Pollen DA (1988) Spatial and temporal frequency selectivity of neurons in visual cortical area V3A of the macaque monkey Vis Res **28** 1179-91 [220]

Gaska JP, Jacobson LD, Chen HW, Pollen DA (1994) Space-time spectra of complex cell filters in the macaque monkey: a comparison of results obtained with pseudowhite noise and grating stimuli Vis Neurosci **11** 805-21 [203]

Gassendi P (1658) Gassendi; opera omnia Vol 2 p 395 Lyon [305]

Gattass R, Oswaldo–Cruz E, Sousa APB (1979) Visual receptive fields of units in the pulvinar of cebus monkey Brain Res **160** 413–30 [187]

Gattass R, Sousa APB, Gross CG (1988) Visuotopic organization and extent of V3 and V4 of the macaque J Neurosci **8** 1831-45 [221]

Gawne TJ (2000) The simultaneous coding of orientation and contrast in the responses of V1 complex cells Exp Brain Res **133** 293-302 [129]

Gawne TJ, Richmond BJ, Optican LM (1991) Interactive effects among several stimulus parameters on the responses of striate cortical complex cells J Neurophysiol **66** 379–89 [129]

Gawne TJ, Kjaer TW, Richmond BJ (1996) Latency: another potential code for feature binding in striate cortex J Neurophysiol **76** 1356-60 [136]

Gaze RM, Jacobson M (1962) The projection of the binocular visual field on the optic tecta of the frog Quart J Exp Physiol **47** 273–80 [540]

Gaze RM, Keating MJ (1970) The restoration of the ipsilateral visual projection following regeneration of the optic nerve in the frog Brain Res **21** 207-16 [464]

Gaze RM, Sharma SC (1970) Axial differences in the reinnervation of the goldfish optic tectum by regenerating optic nerve fibres Exp Brain Res **10** 171-81 [464]

Gazzaniga MS, LeDoux JE (1978) The integrated mind Plenum, New York [489]

Geesaman BJ, Andersen RA (1966) The analysis of complex motion patterns by form/cue invariant MSTd neurons J Neurosci **16** 4716-32 [225]

Gegenfurtner KR, Kiper DC, Fenstemaker SB (1996) Processing of color form and motion in macaque area V2 Vis Neurosci **13** 161-72 [219]

Gegenfurtner KR, Kiper DC, Levitt JB (1997) Functional properties of neurons in macaque area V3 J Neurophysiol **77** 1906-23 [220]

Geisler WS (1984) Physical limits of acuity and hyperacuity J Opt Soc Am A **1** 775-82 [79, 81, 470]

Georgeson MA (1984) Eye movements, afterimages and monocular rivalry Vis Res **27** 1311-19 [296]

Georgeson MA (1988) Spatial phase dependence and the role of motion detection in monocular and dichoptic forward masking Vis Res **28** 1193–1205 [332]

Georgeson MA, Phillips R (1980) Angular selectivity of monocular rivalry: experiment and computer simulation Vis Res **20** 1007–13 [296]

Georgeson MA, Sullivan GD (1975) Contrast constancy: deblurring in human vision by spatial frequency J Physiol **252** 627-56 [93]

Georgopoulos AP (1991) Higher order motor control Ann Rev Neurosci **14** 361-77 [144]

Gernsheim H (1969) History of photography McGraw-Hill, New York [61, 62, 69, 71]

Gerstein GL, Aertsen AMH (1985) Representation of cooperative firing activity among simultaneously recorded neurons J Neurophysiol **54** 1513-28 [131]

Gescheider GA (1976) Psychophysics: Method and theory Erlbaum, Hillsdale NJ [73]

Gestrin PJ, Teller DY (1969) Interocular hue shifts and pressure blindness Vis Res **9** 1297-71 [283]

Ghose GM, Freeman RD (1992) Oscillatory discharge in the visual system: does it have a functional role? J Neurophysiol **6 8** 1558–74 [133, 136]

Ghose GM, Freeman RD (1997) Intracortical connections are not required for oscillatory activity in the visual cortex Vis Neurosci **14** 963-79 [136]

Ghose GM, Ts'o DY (1997) Form processing modules in primate area V4 J Neurophysiol **77** 2191-96 [221]

Ghose GM, Ohzawa I, Freeman RD (1994a) Receptive–field maps of correlated discharge between pairs of neurons in the cat's visual cortex J Neurophysiol **71** 330–46 [132]

Ghose GM, Freeman RD, Ohzawa I (1994b) Local intracortical connections in the cat's visual cortex: postnatal development and plasticity J Neurophysiol **72** 1290-303 [195]

Ghose GM., Yang T, Maunsell JHR (2002) Physiological correlates of perceptual learning in monkey V1 and V2 J Neurophysiol **87** 1867-88 [208]

Ghosh A, Greenberg ME (1995) Calcium signaling in neurons: molecular mechanisms and cellular consequences Science **298** 239-47 [460]

Ghosh A, Shatz CJ (1992a) Involvement of subplate neurons in the formation of ocular dominance columns Science **255** 1441–3 [451]

Ghosh A, Shatz CJ (1992b) Pathfinding and target selection by geniculocortical axons J Neurosci **12** 39-55 [458]

Ghosh A, Shatz CJ (1994) Segregation of geniculocortical afferents during the critical period: a role for subplate neurons J Neurophysiol **14** 3862-80 [458]

Gibson EJ (1967) Principles of perceptual learning and development Appleton-Century-Crofts, London [153]

Gibson EJ, Walk RD (1960) The visual cliff Sci Am **202** 64–71 [478]

Gibson EJ, Owsley CJ, Walker A, Megaw-Nyce J (1979) Development of the perception of invariants: substance and shape Perception **8** 609-19 [480]

Gibson JJ (1933) Adaptation after–effect and contrast in the perception of curved lines J Exp Psychol **16** 1–31 [345, 351]

Gibson JJ (1937) Adaptation aftereffect and contrast in the perception of tilted lines. II Simultaneous contrast and the areal restriction of the aftereffect J Exp Psychol **20** 553-69 [345, 492]

Gibson JJ, Kaplan GA, Reynolds HN, Wheeler K (1969) The change from visible to invisible: a study of optical transitions Percept Psychophys **5** 113-16 [480]

Gilbert CD, Wiesel TN (1979) Morphology and intracortical projections of functionally characterized neurones in the cat visual cortex Nature **280** 120–5 [186, 195, 452]

Gilbert CD, Wiesel TN (1985) Intrinsic connectivity and receptive field properties in visual cortex Vis Res **25** 365–74 [195]

Gilbert CD, Wiesel TN (1989) Columnar specificity of intrinsic horizontal and connections in cat visual cortex J Neurophysiol **9** 2732–42 [196]

Gilbert CD, Wiesel TN (1990) The influence of contextual stimuli on the orientation selectivity of cells in primary visual cortex of the cat Vis Res **30** 1689–1701 [197]

Gilbert CD, Wiesel TN (1992) Receptive field dynamics in adult primary visual cortex Nature **356** 150-2 [197]

Gilbert CD, Ts'o DY, Wiesel TN (1991) Lateral interactions in visual cortex In From pigments to perception (ed A Valberg, BB Lee) pp 239–47 Plenum, New York [197, 314]

Gilbert DS, Fender DH (1969) Contrast thresholds measured with stabilized and non–stabilized sine–wave gratings Optica Acta **16** 191–204 [366]

Gilchrist J, Pardhan S (1987) Binocular contrast detection with unequal monocular illuminance Ophthal Physiol Opt **7** 373-7 [322]

Gilinsky AS, Doherty RS (1969) Interocular transfer of orientational effects Science **164** 454–5 [336]

Gillard-Crewther S, Crewther DP (1988) Neural site of strabismic amblyopia in cats: X-cell acuities in the LGN Exp Brain Res **72** 503-9 [502]

Gillespie DC, Crair MC, Stryker MP (2000) Neurotrophin-4/5 alters responses and blocks the effect of monocular deprivation in cat visual cortex during the critical period J Neurosc **20** 9174-86 [463, 512]

Gioanni H, Bennis M, Sansonetti A (1993) Visual and vestibular reflexes that stabilize gaze in the chameleon Vis Neurosci **10** 947–56 [543]

Giolli RA, Guthrie MD (1969) The primary optic projections in the rabbit An experimental degeneration study J Comp Neurol **136** 99–116 [549]

Glantz RM (1974) Defense reflex and motion detector responsiveness to approaching targets: the motion detector trigger to the defense reflex pathway J Comp Physiol **95** 297-314 [538]

Glasser S, Ingle D (1978) The nucleus isthmus as a relay station in the ipsilateral visual projection to the frog's optic tectum Brain Res **159** 214-18 [540]

Glazewski S, Giese KP, Silva A, Fox K (2000) The role of a-CaMKII autophosphorylation in neocortical experience-dependent plasticity Nature Neurosci **3** 911-18 [460]

Gleason G, Schor C, Lunn R, Maxwell J (1992) Directionally selective short–term nonconjugate adaptation of vertical pursuits Vis Res **33** 33–46 [385]

Glendenning KK, Kofron EA, Diamond IT (1976) Laminar organization of projections of the lateral geniculate nucleus to the striate cortex in *Galago* Brain Res **105** 538-46 [216]

Glickstein M, Miller J, Smith OA (1964) Lateral geniculate nucleus and cerebral cortex: evidence for a crossed pathway Science **145** 159–61 [233]

Glickstein M, Stein J, King RA (1972) Visual input to the pontine nucleus Science **178** 1110-11 [365]

Gnadt JW, Beyer J (1998) Eye movements in depth: what does the monkey's parietal cortex tell the superior colliculus? Neuro Report **29** 233-8 [432]

Gnadt JW, Mays LE (1995) Neurons in monkey parietal area LIP are tuned for eye-movement parameters in three-dimensional space J Neurophysiol **73** 280-97 [243, 432]

Gochin PM (1996) The representation of shape in the temporal lobe Behav Brain Res **76** 99-116 [222]

Gödecke I, Bonhoeffer T (1996) Development of identical orientation maps for two eyes without common visual experience Nature **379** 251-54 [471]

Godement P, Salaun J, Mason CA (1990) Retinal axon pathfinding in the optic chiasma Neuron **5** 173–86 [440]

Godement P, Wang LC, Mason CA (1994) Retinal axon divergence in the optic chiasm: dynamics of growth cone behavior at the midline J Neurosci **14** 7027-39 [440]

Goldsmith TH, Cronin TW (1993) Retinoids of seven species of mantis shrimp Vis Neurosci **10** 915–20 [538]

Goldstein B (ed) (2001) Blackwell handbook of perception Blackwell Publisher, Oxford [8]

Golomb D, Kleinfeld D, Reid RC, Shapley RM, Shraiman BI (1994) On temporal codes and the spatiotemporal response of neurons in the lateral geniculate nucleus J Neurophysiol **72** 2990-3003 [130]

Gonzalez F, Perez R (1998a) Modulation of cell responses to horizontal disparities by ocular vergence in the visual cortex of the awake *macaca mulatta* monkey Neurosci Lett **275** 101-4 [254]

Gonzalez F, Perez R (1998b) Neural mechanisms underlying stereoscopic vision Prog Neurobiol **55** 191-227 [237]

Gonzalez F, Krause F, Perez R, Alonso JM, Acuna C (1993a) Binocular matching in monkey visual cortex: single cell responses to correlated and uncorrelated dynamic random dot stereograms Neurosci **52** 933–9 [240]

Gonzalez F, Revola JL, Perez R, Acuna C, Alonso JM (1993b) Cell responses to vertical and horizontal retinal disparities in the monkey visual cortex Neurosci Lett **160** 167-70 [257]

Good WV, da Sa LCF, Lyons CJ, Hoyt CS (1993) Monocular visual outcome in untreated early onset esotropia Br J Ophthal **77** 492-4 [528]

Goodale MA (1983) Visually guided pecking in the pigeon (*Columba livia*) Brain Behav Evol **22** 22–40 [545]

Goodale MA, Ellard CG, Booth L (1990) The role of image size and retinal motion in the computation of absolute distance by the Mongolian gerbil (*Meriones unguiculatus*) Vis Res **30** 399-413 [549]

Goodenough DR, Sigman E, Oltman PK Rosso J, Mertz H (1979) Eye torsion in response to a tilted visual stimulus Vis Res **19** 1177–9 [411]

Goodhill GJ (1993) Topography and ocular dominance: a model exploring positive correlations Biol Cyber **69** 109-18 [464]

Goodhill GJ, Löwel S (1996) Theory meets experiment: correlated neural activity helps determine ocular dominance column periodicity TINS **18** 437-9 [506]

Goodhill GJ, Willshaw DJ (1990) Application of the elastic net algorithm to the formation of ocular dominance stripes Network **1** 41-59 [214]

Gordon B, Gummow L (1975) Effects of extraocular muscle section on receptive fields in cat superior colliculus Vis Res **15** 1011-19 [505]

Gordon B, Presson J (1977) Effects of alternating occlusion on receptive fields in cat superior colliculus J Neurophysiol **40** 1406–14 [505]

Gordon FR, Yonas A (1976) Sensitivity to binocular depth information in infants J Exp Child Psychol **22** 413–22 [477]

Gordon JA, Stryker MP (1996) Experience-dependent plasticity of binocular responses in the primary visual cortex of the mouse J Neurosci **16** 3274-86 [507]

Goss DA (1995) Accommodation convergence and fixation disparity: A manual of clinical analysis Professional Press/Fairchild Publications, New York [8, 376, 378, 392]

Goss DA, Zhai H (1994) Clinical and laboratory investigations of the relationship of accommodation and convergence function with refractive error: a literature review Doc Ophthal **86** 349-80 [359]

Gosser HM (1977) Selected attempts at stereoscopic moving pictures and their relationship to the development of motion picture technology 1852-1903 Arno Press, New York [70, 71]

Gottmann K, Mehrie A, Gisselmann G, Hatt H (1997) Presynaptic control of subunit composition of NMDA receptors mediating synaptic plasticity J Neurosci 17 2766-74 [460]

Gould E, Reeves AJ, Graziano MSA, Gross CG (1999) Neurogenesis in the neocortex of adult primates Science 286 548-52 [193, 446, 450]

Gouras P (1991) Cortical mechanisms of colour vision In Vision and visual dysfunction Vol 6 The perception of colour (ed P Gouras) pp 179-97 MacMillan, London [126, 204]

Gouras P, Link K (1966) Rod and cone interaction in dark adapted monkey ganglion cells J Physiol 184 499-510 [334]

Govi G (1885) L'Ottica di Claudio Tolomeo Turin [12]

Grafstein B, Laureno R (1973) Transport of radioactivity from eye to visual cortex in the mouse Exp Neurol 39 44–57 [212]

Graham N, Nachmias J (1971) Detection of grating patterns containing two spatial frequencies: a comparison of single-channel and multiple-channel models Vis Res 11 251-9 [139]

Graham NVS (1989) Visual pattern analyzers Oxford University Press, New York [84, 141]

Graham PA (1974) The epidemiology of strabismus Br J Ophthal 58 227–31 [373, 374]

Granrud CE (1986) Binocular vision and spatial perception in 4– and 5–month–old infants J Exp Psychol HPP 12 36–49 [477]

Granrud CE, Yonas A, Smith IM, Arterberry ME, Glicksman ML, Sorknes AC (1984) Infants' sensitivity to accretion and deletion of texture as information for depth at an edge Child Devel 55 1630-6 [480]

Granrud CE, Yonas A, Opland EA (1985) Infants' sensitivity to the depth cue of shading Percept Psychophys 37 415-19 [481]

Grant S, Keating MJ (1989a) Changing patterns of binocular visual connections in the intertectal system during development of the frog *Xenopus laevis.* I Normal maturational changes in response to changing binocular geometry Exp Brain Res 75 99–116 [540]

Grant S, Keating MJ (1989b) Changing patterns of binocular visual connections in the intertectal system during development of the frog *Xenopus laevis.* II Abnormalities following early visual deprivation Exp Brain Res 75 117–32 [541]

Grant S, Keating MJ (1992) Changing patterns of binocular visual connections in the intertectal system during development of the frog *Xenopus laevis.* III Modifications following early eye rotation Exp Brain Res 89 383–96 [541]

Grasse KL (1994) Positional disparity sensitivity of neurons in the cat accessory optic system Vis Res 13 1673-89 [245, 256]

Gratiolet LP (1854) Note sur les expansions des racines cérébrales du nerf optique et sur leur terminaison dans une région déterminée de l'écorce des hémisphères Comp Rendu Acad Sci 39 274-7 [23]

Graves AL, Trotter Y, Frégnac Y (1987) Role of extraocular muscle proprioception in the development of depth perception in cats J Neurophysiol 58 816–31 [494]

Gray CM, Di Prisco GV (1997) Stimulus-dependent neuronal oscillations and local synchronization in striate cortex of the alert cat J Neurosci 17 3239-53 [132]

Gray CM, McCormick DA (1996) Chatering cells: superficial pyramidal neurons contributing to the generation of synchronous oscillations in the visual cortex Science 274 109-13 [131, 136, 188]

Gray CM, Singer W (1989) Stimulus–specific neuronal oscillations in orientation columns of cat visual cortex Proc Nat Acad Sci 86 1698–1702 [132]

Gray CM, Engel AK, König P, Singer W (1991) Synchronous neuronal oscillations in cat visual cortex: functional implications In Representations of vision (ed A Gorea) pp 83–96 Cambridge University Press, New York [133]

Gray GE, Sanes JR (1991) Migratory paths and phenotypic choices of clonally related cells in the avian central nervous system Neuron 6 211-25 [450]

Gray JR, Lee JK, Robertson RM (2001) Activity of descending contralateral movement detector neurons and collision avoidance behaviour in response to head-on visual stimuli in locusts J Comp Physiol A 187 115-29 [534]

Gray LS, Winn B, Gilmartin B (1993a) Accommodative microfluctuations and pupil diameter Vis Res 33 2083-90 [362]

Gray LS, Winn B, Gilmartin B (1993b) Effect of target luminance on microfluctuations of accommodation Ophthal Physiol Opt 13 258-65 [362]

Gray LS, Winn B, Gilmartin B, Eadie AS (1993c) Objective concurrent measures of open–loop accommodation under photopic conditions Invest Ophthal Vis Sci 34 2996–3003 [390]

Gray, M.S, Pouget A, Zemel RS, Nowlan SJ. Sejnowski TJ (1998) Reliable disparity estimation through selective integration Vis Neurosci 15 511-28 [269]

Gray R, Regan D (1996) Cyclopean motion perception produced by oscillations of size disparity and location Vis Res 36 655-65 [149]

Graziano MSA, Gross CG (1993) A bimodal map of space: somatosensory receptive fields in the macaque putamen with corresponding visual receptive fields Exp Brain Res 97 96-109 [229]

Graziano MSA, Andersen RA, Snowden RJ (1994) Tuning of MST neurons to spiral motions J Neurosci 14 54-67 [225]

Graziano MSA, Hu XT, Gross CG (1997) Visuospatial properties of ventral premotor cortex J Neurophysiol 77 2298-92 [229]

Green DG, Powers MK, Banks MS (1980) Depth of focus, eye size and visual acuity Vis Res 20 827-35 [474]

Green DM, Swets JA (1966) Signal detection theory and psychophysics Wiley, New York [73, 76, 310]

Green M (1991) Visual search visual streams and visual architectures Percept Psychophys 50 388-403 [134]

Green M (1992a) Visual search: detection identification and localization Perception 21 765-777 [134]

Green M, Blake R (1981) Phase effects in monoptic and dichoptic temporal integration: flicker and motion detection Vis Res 21 365–72 [331]

Green M, Odom JV (1984) Comparison of monoptic and dichoptic masking by light Percept Psychophys 35 265-8 [338]

Greene HA, Madden DJ (1987) Adult age differences in visual acuity stereopsis and contrast sensitivity Am J Optom Physiol Opt 64 749-53 [484]

Greenhut A, Young FA (1953) Visual depth perception in the rat J GenetPsychol 82 155-82 [549]

Greenspon TS, Eriksen CW (1968) Interocular nonindependence Percept Psychophys 3 93–6 [324]

Gregg FM, Parks MM (1992) Stereopsis after congenital monocular cataract extraction Am J Ophthal 114 314-17 [502]

Gregory RL, Wallace JG (1963) Recovery from early blindness A case study Exp Psychol Monogr No 2 Heffer, London [501]

Gregory RL, Ross HE, Moray N (1964) The curious eye of *Copilia* Nature 201 1166–8 [536]

Gresty MA, Bronstein AM, Barratt H (1987) Eye movement responses to combined linear and angular head movement Exp Brain Res 65 377–84 [427]

Greuel JM, Luhmann HJ, Singer W (1988) Pharmacological induction of use-dependent receptive field modifications in the visual cortex Science 272 74-7 [462]

Grieve KL, Acuna C, Cudeiro J (2000) The primate pulvinar nuclei: vision and action TINS 23 35-9 [183]

Griffin DR (1958) Listening in the dark Yale University Press, Yale [548]

Griffin JR, Grisham JD (1995) Binocular anomalies Diagnosis and vision therapy Butterworth-Heinemann, Boston [376]

Grimsley G (1943) A study of individual differences in binocular color fusion J Exp Psychol **32** 82–7 [284]

Grinvald A, Lieke EE, Frostig RD, Hildesheim R (1994) Cortical point-spread function and range lateral interactions revealed by real-time optical imaging of macaque monkey primary visual cortex J Neurosci **14** 2545-68 [195, 197]

Grinvald A, Slovin H, Vanzetta I (2000) Non-invasive visualization of cortical columns by fMRI Nature Neurosci **3** 105-7 [183, 213]

Grobstein P, Comer C, Kostyk S (1980) The potential binocular field and its tectal representation in *Rana pipiens* J Comp Neurol **190** 175–85 [539, 541]

Grobstein P, Reyes A, Zwanziger L, Kostyk SK (1985) Prey orienting in frogs: accounting for variations in output with stimulus distance J Comp Physiol A **156** 775-85 [541]

Gronwall DMA, Sampson H (1971) Ocular dominance: a test of two hypotheses Br J Psychol **62** 175–85 [295]

Grosof DH, Shapley RM, Hawken MJ (1993) Macaque V1 neurons can signal 'illusory' contours Nature **365** 550-2 [123]

Gross CG (1973) Visual functions of inferotemporal cortex In Handbook of sensory physiology (ed R Jung) Vol VII/3B pp 451–82 Springer, New York [218]

Gross CG (1992) Representation of visual stimuli in inferior temporal cortex Philos Tr R Soc B **335** 3-10 [222]

Gross CG, Rocha-Miranda CE, Bender DB (1972). Visual properties of neurons in inferotemporal cortex of the macaque J Neurophysiol **35** 96-111 [222]

Gross KJ, Hickey TL (1980) Abnormal laminar patterns in the lateral geniculate nucleus of an albino monkey Brain Res **190** 231-7 [496]

Grossberg S (1990) A model cortical architecture for the preattentive perception of 3–D form In Computational neuroscience (ed ER Schwartz) pp 117–38 MIT Press, Cambridge MA [230]

Grossberg S, Kelly F (1999) Neural dynamics of binocular brightness perception Vis Res **39** 3796-816 [328]

Grossberg S, Marshall JA (1989) Stereo boundary fusion by cortical complex cells: a system of maps, filters, and feedback networks for multiplexing distributed data Neural Networks **2** 29-51 [268]

Grossberg S, Mcloughlin NP (1997) Cortical dysnamocs of three-dimensional surface perception: binocular and half-occluded scenic images Neural Networks, **10**, 1583-605 [268]

Grossberg S, Somers D (1991) Synchronized oscillations during cooperative feature linking in a cortical model of visual perception Neural Networks **4** 453–66 [136]

Grosslight JH, Fletcher HJ, Masterton RB, Hagen R (1978) Monocular vision and landing performance in general aviation pilots: cyclops revisited Hum Factors **20** 27-33 [5]

Grossman I, Magnus SW (1964) Groups and their graphs Random House, New York [103]

Grosvenor T (1987) Reduction in axial length with age: an emmetropizing mechanism for the adult eye Am J Optom Physiol Opt **64** 657-63 [435]

Grove AC (1991) An introduction to the Laplace transform and the z transform Prentice Hall, New York [97]

Grüber HE, Fink CD, Damm V (1957) Effects of experience on perception of causality J Exp Psychol **53** 89-93 [160]

Gruberg ER, Lettvin JY (1980) Anatomy and physiology of a binocular system in the frog *Rana pipiens* Brain Res **192** 313–25 [540]

Grunewald A, Mingolla E (1998) Motion after-effect due to binocular sum of adaptation to linear motion Vis Res **38** 2963-71 [349]

Grüsser OJ, Grüsser–Cornehls U (1965) Neurophysiological Grundlagen des Binocularsehens Arch Psychiat Z ges Neurol **207** 296–317 [215, 216, 319]

Grüsser OJ, Grüsser–Cornehls U (1976) Neurophysiology of the anuran visual system In Progress in neurobiology (ed R Llinás, W Precht) pp 297–385 Springer, New York [539, 541]

Gu Q (1995) Involvement of nerve growth factor in visual cortex plasticity Rev Neurosci **6** 329-51 [512]

Gu Q, Bear MF, Singer W (1989) Blockade of NMDA–receptors prevents ocularity changes in kitten cortex after reversed monocular deprivation Devel Brain Res **47** 281–88 [511]

Gu Q, Patel B, Singer W (1990) The laminar distribution and postnatal development of serotonin-immunoreactive axons in the cat primary visual cortex Exp Brain Res **81** 257-66 [192]

Gu Q, Liu Y, Cynader MS (1994) Nerve growth factor-induced ocular dominance plasticity in adult cat visual cortex Proc Nat Acad Sci **91** 8408-12 [512]

Gu X, Spitzer NC (1995) Distinct aspects of neuronal differentiation encoded by frequency of spontaneous Ca^{2+} transients Nature **375** 784-7 [436]

Guido W, Salinger WL (1989) 6-hydroxydopamine treatment blocks the effects of chronic monocular paralysis in the cat's lateral geniculate nucleus Brain Res **501** 397-400 [504]

Guido W, Salinger WL, Schroeder CE (1988) Binocular interactions in the dorsal lateral geniculate nucleus of monocularly paralyzed cats: extraretinal and retinal influences Exp Brain Res **70** 417-28 [494]

Guido W, Tumosa N, Spear PD (1989) Binocular interactions in the cat's dorsal lateral geniculate nucleus: I Spatial-frequency analysis of responses of X Y and W cells to nondominant-eye stimulation J Neurophysiol **62** 529-43 [179]

Guilford JP (1954) Psychometric methods McGraw–Hill, New York [73]

Guillemot JP, Paradis MC, Samson A, Ptito M, Richer L, Lepore F (1993) Binocular interaction and disparity coding in area 19 of visual cortex in normal and split–chiasm cats Exp Brain Res **94** 405–17 [232, 233]

Guillery RW (1969) An abnormal retinogeniculate projection in Siamese cats Brain Res **14** 739-41 [496]

Guillery RW (1972) Binocular competition in the control of geniculate cell growth J Comp Neurol **144** 117–30 [504]

Guillery RW (1986) Neural abnormalities in albinos TINS **9** 364–7 [496]

Guillery RW, Kaas JH (1971) A study of normal and congenitally abnormal retinogeniculate projections in cats J Comp Neurol **143** 73-100 [496]

Guillery RW, Stelzner DJ (1970) The differential effects of unilateral lid closure upon the monocular and binocular segments of the dorsal lateral geniculate nucleus of the cat J Comp Neurol **139** 413-22 [504]

Guillery RW, Casagrande VA, Oberdorfer MD (1974) Congenitally abnormal vision in Siamese cats Nature **252** 195-99 [497]

Guillery RW, Okoro AN, Witkop CJ (1975) Abnormal visual pathways in the brain of a human albino Brain Res **96** 373–7 [496]

Guillery RW, Ombrellaro M, LaMantia AL (1985) The organization of the lateral geniculate nucleus and of the geniculocortical pathway that develops without retinal afferents Devel Brain Res **20** 221-3 [443]

Guillery RW, Mason CA, Taylor JSH (1995) Developmental determinants at the mammalian optic chiasm J Neurosci **15** 4727-37]

Guillin O, Diaz J, Carroll P, Griffon N, Schwartz JC, Sokoloff P (2001) BDNF controls dopamine D3 receptor expression and triggers behavioural sensitization Nature **411** 86-9 [441]

Guirao A, González C, Redondo M, Geraghty E, Norrby S, Artal P (1999) Average optical performance of the human eye as a function of age in a normal population Invest Ophthal Vis Sci **40** 203-13 [484]

Guire ES, Lickey ME, Gordon B (1999) Critical period for the monocular deprivation effect in rats: assessment with sweep visually evoked potentials J Neurophysiol **81** 121-8 [515]

Guitton D (1999) Gaze shifts in three-dimensional space: a closer look at the superior colliculus J Comp Neurol **413** 77-82 [431]

Güldenagel M, Ammermüller J, Feigenspan A, Teubner B, Degen J, Söhl G, Willecke K, Weiler R (2001) Visual transmission deficits in mice with targeted disruption of the gap junction gene connexin36 J Neurosci **21** 6036-44 [167]

Gulick WL, Lawson RB (1976) Human stereopsis Oxford University Press, New York [6, 8, 59]

Gulyás B, Roland PE (1994) Binocular disparity discrimination in human cerebral cortex: functional anatomy by positron emission tomography Proc Nat Acad Sci **91** 1239-43 [263]

Gunter R (1951) Binocular fusion of colours Br J Psychol **42** 363–72 [283, 284]

Guo S, Reinecke RD, Fendick M, Calhoun JH (1989) Visual pathway abnormalities in albinism and infantile nystagmus: VECPs and stereoacuity measurements J Ped Ophthal Strab 29 97 104 [496]

Guo Y, Udin B (2000) The development of abnormal axon trajectories after rotation of one eye in *Xenopus* J Neurosci **20** 4189-97 [541]

Gur M (1991) Perceptual fade–out occurs in the binocularly viewed Ganzfeld Perception **20** 645–54 [290]

Gur M, Snodderly DM (1997) Visual receptive fields of neurons in primary visual cortex (V1) move in space with the eye movements of fixation Vis Res **37** 257-65 [395]

Gur M, Beylin A, Snodderly DM (1997) Response variability of neurons in primary visual cortex (V1) of alert monkeys J Neurosci **17** 2914-20 [251]

Gutfreund Y, Yarom Y, Segev I (1995) Subthreshold oscillations and resonant frequency in guinea-pig cortical neurons: physiology and modelling J Physiol **483** 621-40 [136]

Guth SL (1971) On probability summation Vis Res **11** 747–50 [319]

Guthrie DM (1990) The physiology of the optic tectum In The visual system of fish (ed RH Douglas, MBA Djamgoz) pp 300-42 Chapman Hall, London [539]

Gutierrez C, Cola MG, Seltzer B, Cusick C (2000) Neurochemical and connectional organization of the dorsal pulvinar complex in monkeys J Comp Neurol **419** 61-86 [183]

Gutiérrez-Igarza K, Fogarty DJ, Pérez-Cerdá F, Doñate-Oliver F, Albus K, Matute C (1996) Localization of AMPA-selective glutamate receptor subunits in the adult cat visual cortex Vis Neurosci **13** 61-72 [191]

Guyton DL (2000) Dissociated vertical deviation: etiology, mechanism, and associated phenomena J Am Ass Ped Ophthal Strab **4** 131-41 {9} [410]

Gwiazda J, Scheiman M, Mohindra I, Held R (1984) Astigmatism in children: changes in axis and amount from birth to six years Invest Ophthal Vis Sci **25** 88–92 [474]

Gwiazda J, Bauer J, Held R (1989) Binocular function in human infants: correlation of stereoptic and fusion-rivalry discriminations J Ped Ophthal Strab **26** 128-32 [482]

Gwiazda J, Bauer J, Thorn F, Held R (1997) Development of spatial contrast sensitivity from infancy to adulthood: psychophysical data Optom Vis Sci **74** 785-9 [468, 484]

Gwiazda J, Grice K, Thorn F (1999) Response AC/A ratios are elevated in myopic children Ophthal Physiol Opt **19** 173-9 [390, 435]

Haas HJ (1980) Binoculare Korrektion. Die Methodeik und Theorie von HJ Haas Willy Schrinkel, Düsseldorf [382]

Haenny PE, Schiller PH (1988) State dependent activity in monkey visual cortex. I Single cell activity in V1 and V4 on visual tasks Exp Brain Res **69** 225–44 [208]

Hagemans KH, Wildt GJ von der (1979) The influence of stimulus width on the contrast sensitivity function in amblyopia Invest Ophthal **18** 842-7 [521]

Haggard P, Hutchinson K, Stein J (1995) Patterns of coordinated multi-joint movement Exp Brain Res **107** 254-66 [143]

Hahm DE (1978) Early Hellenistic theories of vision and the perception of color In Studies in perception (ed PK Machamer, RG Turnbull) pp 61-85 Ohio State University Press, Columbus Ohio [10, 12]

Hahm JO, Langdon RB, Sur M (1991) Disruption of retinogeniculate afferent segregation by antagonists to NMDA receptors Nature **351** 568–70 [444]

Hahmann U, Güntürkün O (1993) The visual acuity for the lateral visual field of the pigeon Vis Res **33** 1659–64 [544]

Haig ND (1993) Why is the retina capable of resolving finer detail than the eye's optical or neural systems? Spatial Vis **7** 257–73 [174]

Hain TC, Luebke AE (1990) Phoria adaptation in patients with cerebellar dysfunction Invest Ophthal Vis Sci **31** 1394-7 [431]

Haines RF (1977) Visual response time to colored stimuli in peripheral retina: evidence for binocular summation Am J Optom Physiol Opt **54** 387–98 [328]

Hainline L (1993) Conjugate eye movements of infants In Early visual development normal and abnormal (ed K Simons) pp 47-79 Oxford University Press, New York [475]

Hainline L, Riddell PM (1995) Binocular alignment and vergence in early infancy Vis Res **35** 3229-36 [476]

Hainline L, Turkel J, Abramov I, Lemerise E, Harris CM (1984) Characteristics of saccades in human infants Vis Res **27** 1771-80 [475]

Hainline L, Riddell PM, Grose-Fifer J, Abramov I (1992) Development of accommodation and vergence in infancy Behav Brain Res **49** 33-50 [473]

Hajos A (1968) Sensumotorische Koordinationsprozesse bei Richtungslokalisation Z Exp Angew Psychol **15** 435-61 [355]

Hajos A, Ritter M (1965) Experiments to the problem of interocular transfer Acta Psychol**27** 81–90 [345, 351, 355]

Haldat C (1806) Expériences sur la double vision J de Physique **63** 387–401 [282]

Hall TS (1972) Treatise of man René Descartes Harvard University Press, Cambridge MA [25]

Halloran MC, Kalil K (1994) Dynamic behaviors of growth cones extending in the corpus callosum of living cortical brain slices observed with video microscopy J Neurosci **14** 2161-77 [437, 453]

Halmagyi GM, Gresty MA, Leech J (1980) Reversed optokinetic nystagmus (OKN): mechanism and clinical significance Ann Neurol **7** 429–35 [498]

Halpern DL, Patterson R, Blake R (1987a) Are stereoacuity and binocular rivalry related? Am J Optom Physiol Opt **64** 41-4 [290]

Hamasaki DI, Flynn JT (1981) Amblyopic eyes have longer reaction times Invest Ophthal Vis Sci **21** 846-53 [525]

Hamer RD, Norcia AM (1994) The development of motion sensitivity during the first year of life Vis Res **34** 2387-402 [472]

Hamer RD, Schneck ME (1984) Spatial summation in dark–adapted human infants Vis Res **27** 77–85 [434]

Hamer RD, Norcia AM, Tyler CW, Hsu–Winges C (1989) The development of monocular and binocular VEP acuity Vis Res **29** 397–408 [469]

Hamilton CR, Tieman SB, Winter HL (1973) Optic chiasm section affects discriminability of asymmetric patterns by monkeys Brain Res **49** 427-31 [355]

Hamilton DB, Albrecht DG, Geisler WS (1989) Visual cortical receptive fields in monkey and cat: spatial and temporal phase transfer function Vis Res **29** 1285-308 [201]

Hammond JH (1981) The camera obscura Hilkger, Bristol [39, 60]

Hammond JH (1987) The camera lucida Hilger, Bristol [39]

Hammond P (1979) Stimulus dependence of ocular dominance of complex cells in area 17 of the feline visual cortex Exp Brain Res **35** 583–9 [215]

Hammond P (1981) Non-stationarity of ocular dominance in cat striate cortex Exp Brain Res **42** 189–95 [215, 248]

Hammond P (1991) Binocular phase specificity of striate corticotectal neurones Exp Brain Res **87** 615–23 [291]

Hammond P, Fothergill LK (1991) Interocular comparison of length summation and end-inhibition in striate cortical neurones of the anaesthetized cat Proc Physiol Soc **446** 232P [214]

Hammond P, Kim JN (1996) Role of suppression in shaping orientation and direction selectivity of complex neurons in cat striate cortex J Neurophysiol **75** 1163-76 [216]

Hammond P, MacKay DM (1981) Modulatory influences of moving textured backgrounds on responsiveness of simple cells in feline striate cortex J Physiol **319** 431–42 [193]

Hammond P, Mouat GSV (1988) Neural correlates of motion after-effects in cat striate cortical neurones: interocular transfer Exp Brain Res **72** 21–8 [350]

Hammond P, Pomfrett CJD (1991) Interocular mismatch in spatial frequency and directionality characteristics of striate cortical neurones Exp Brain Res **85** 631–40 [255]

Hammond P, Mouat GSV, Smith AT (1988) Neural correlates of motion after-effects in cat striate cortical neurones: monocular adaptation Exp Brain Res **72** 1–20 [350]

Hampton DR, Kertesz AE (1982) Human response to cyclofusional stimuli containing depth cues Am J Optom Physiol Opt **59** 21–7 [412]

Hampton DR, Kertesz AE (1983a) Fusional vergence response to local peripheral stimulation J Opt Soc Am **73** 7–10 [397]

Hampton DR, Kertesz AE (1983b) The extent of Panum's area and the human cortical magnification factor Perception **12** 161–65 [273]

Hamsher K de S (1978) Stereopsis and unilateral brain disease Invest Ophthal Vis Sci **17** 336–43 [489]

Hänny P, von der Heydt R (1982) The effect of horizontal-plane environment on the development of binocular receptive fields of cells in cat visual cortex J Physiol **329** 75–92 [485]

Hänny P, von der Heydt R, Poggio GF (1980) Binocular neuron responses to tilt in the monkey visual cortex Evidence for orientation disparity processing Exp Brain Res **41** A29 [255]

Hara N, Steffen H, Roberts DC, Zee DS (1998) Effects of horizontal vergence on the motor and sensory components of vertical fusion Invest Ophthal Vis Sci **39** 2298-76 [407]

Harker GS (1960) Two stereoscopic measures of cyclorotation of the eyes Am J Optom Arch Am Acad Optom **37** 461–73 [412]

Harkness L (1977) Chameleons use accommodation cues to judge distance Nature **297** 346–9 [543]

Harkness L, Bennet-Clark HC (1978) The deep fovea as a focus indicator Nature **272** 814-16 [544]

Harrad RA, Hess RF (1992) Binocular integration of contrast information in amblyopia Vis Res **32** 2135–50 [491]

Harrad RA, McKee SP, Blake R, Yang Y (1994) Binocular rivalry disrupts stereopsis Perception **23** 15-28 [307]

Harris AE, Ermentrout GB, Small SL (1997) A model of ocular dominance column development by competition for trophic factor Proc Nat Acad Sci **94** 9944-9 [464]

Harris GG (1960) Binaural interactions of impulsive stimuli and pure tones J Acoust Soc Am **32** 685-92 [87, 149]

Harris J (1775) A treatise of optics B White, London p 171 [56]

Harris L, Jenkin M (2002) Levels of perception New York, Springer [74]

Harris JM, Willis A (2001) A binocular site for contrast-modulated masking Vis Res **41** 873-81 [336]

Harris P, MacFarlane A (1974) The growth of the effective visual field from birth to seven weeks J Exp Child Psychol **18** 340-8 [475]

Harris VA, Hayes W, Gleason JM (1974) The horizontal–vertical illusion: binocular and dichoptic investigations of bisection and verticality components Vis Res **14** 1323–6 [354]

Harris WA (1984) Axonal pathfinding in the absence of normal pathways and impulse activity J Neurosci **4** 1153-62 [438]

Harter MR, Seiple WH, Salmon L (1973) Binocular summation of visually evoked responses to pattern stimuli in humans Vis Res **13** 1433–46 [258, 259]

Harter MR, Seiple WH, Musso M (1974) Binocular summation and suppression: visually evoked cortical responses to dichoptically presented patterns of different spatial frequency Vis Res **14** 1169–80 [328]

Harter MR, Towle VL, Zakrzewski M, Moyer SM (1977) An objective indicant of binocular vision in humans: size-specific interocular suppression of visual evoked potentials EEG Clin Neurophysiol **43** 825-36 [337]

Hartline HK (1938) The response of single optic nerve fibres of the vertebrate eye to illumination of the retina Am J Physiol **121** 400–15 [168]

Hartline HK, Graham CH (1932) Nerve impulses from single receptors in the eye J Cell Comp Physiol **1** 277-95 [168]

Hartmann EE, Banks MS (1992) Temporal contrast sensitivity in human infants Vis Res **32** 1163-8 [472]

Hartmann EU, Succop A, Buck SL, Weiss AH, Teller DY (1993) Quantification of monocular optokinetic nystagmus asymmetries and motion perception with motion-nulling techniques J Opt Soc Am A **10** 1835-40 [526]

Hartveit E, Ramberg SI, Heggelund P (1993) Brain stem modulation of spatial receptive field properties of single cells in the dorsal lateral geniculate nucleus of the cat J Neurophysiol **70** 1644–55 [179]

Harvey AR (1980) A physiological analysis of subcortical and commissural projections of areas 17 and 18 of the cat J Physiol **302** 507–34 [232]

Harvey LO (1978) Single representation of the visual midline in humans Neuropsychol **16** 601-10 [232]

Harwerth RS, Boltz RL (1979b) Behavioral measures of stereopsis in monkeys using random dot stereograms Physiol Behav **22** 229-234 [548]

Harwerth RS, Smith EL (1985) Binocular summation in man and monkey Am J Optom Physiol Opt **62** 439–46 [320]

Harwerth RS, Smith EL, Levi DM (1980) Suprathreshold binocular interactions for grating patterns Percept Psychophys **27** 43-50 [320]

Harwerth RS, Smith EL, Boltz RL, Crawford MLJ, von Noorden GK (1983a) Behavioral studies on the effect of abnormal early visual experience in monkeys: spatial modulation sensitivity Vis Res **23** 1501–10 [520]

Harwerth RS, Smith EL, Okundaye OJ (1983b) Oblique effects vertical effects and meridional amblyopia in monkeys Exp Brain Res **53** 142-50 [522]

Harwerth RS, Smith EL, Crawford MLJ, von Noorden GK (1984) Effects of enucleation of the nondeprived eye on stimulus deprivation amblyopia in monkeys Invest Ophthal Vis Sci **25** 10-25 [517]

Harwerth RS, Smith EL, Duncan GC, Crawford MLJ, von Noorden GK (1986a) Effects of enucleation of the fixating eye on strabismic amblyopia in monkey Invest Ophthal Vis Sci **27** 276–54 [527]

Harwerth RS, Smith EL, Duncan GC, Crawford MLJ, von Noorden GK (1986b) Multiple sensitive periods in the development of the primate visual system Science **232** 235–8 [515]

Harwerth RS, Smith EL, Crawford MLJ, von Noorden GK (1990) Behavioral studies of the sensitive period of development of visual functions in monkeys Behav Brain Res **41** 179–98 [518]

Harwerth RS, Smith EL, Paul AD, Crawford MLJ, von Noorden GK (1991) Functional effects of bilateral form deprivation in monkeys Invest Ophthal Vis Sci **32** 2311–27 [501]

Harwerth RS, Smith EL, Siderov J (1995) Behavioral studies of local stereopsis and disparity vergence in monkeys Vis Res **35** 1755-70 [392, 548]

Harwerth RS, Smith EL, Crawford MLJ, von Noorden GK (1997) Stereopsis and disparity vergence in monkeys with subnormal binocular vision Vis Res **37** 483-93 [518]

Hasebe H, Ohtsuki H, Kono R, Nakahira Y (1998) Biometric confirmation of the Hirschberg ratio in strabismic children Invest Ophthal Vis Sci **39** 2782-5 [375]

Hasebe H Oyamada H, Kinomura S, Kawashima R, Ouchi Y, Nobezawa S, Tsukada H, Yoshikawa E, Ukai K, Takada R, Takagi M, Abe H, Fukuda H, Bando T (1999) Human cortical areas activated in relation to vergence eye movements—a PET study Neuroimage **10** 200-8 [432]

Haslwanter T (1995) Mathematics of three-dimensional eye rotations Vis Res **35** 1727-39 [370]

Hastorf AH, Myro G (1959) The effect of meaning on binocular rivalry Am J Psychol **72** 393–400 [309]

Hata Y, Stryker MP (1994) Control of thalamocortical afferent rearrangement by postsynaptic activity in developing visual cortex Science **295** 1732-5 [511]

Hata Y, Tsumoto T, Sato H, Hagihara K, Tamura H (1988) Inhibition contributes to orientation selectivity in visual cortex of cat Nature **335** 815-7 [199]

Hata Y, Tsumoto T, Sato H, Hagihara K, Tamura H (1993) Development of local interactions in cat visual cortex studied by cross–correlation analysis J Neurophysiol **69** 40–56 [452]

Hata Y, Ohshima M, Ichisaka S, Wakita M, Fukuda M, Tsumoto T (2000) Brain-derived neurotrophic factor expands ocular dominance columns in visual cortex in monocularly deprived and nondeprived kittens but does not in adult cats J Neurosci **20** RC57 [463, 512]

Hattori M, Osterfield M, Flanagan JG (2000) Regulated cleavage of a contact-mediated axon repellent Science **289** 1360-5 [447]

Hawken MJ, Parker AJ (1984) Contrast sensitivity and orientation selectivity in lamina IV of the striate cortex of old world monkeys Exp Brain Res **54** 367-72 [199]

Hawken MJ, Shapley RM, Grosof DH (1996) Temporal-frequency selectivity in monkey visual cortex Vis Neurosci **13** 477-92 [201]

Hawken MJ, Blakemore C, Morley JW (1997) Development of contrast sensitivity and temporal-frequency selectivity in primate lateral geniculate nucleus Exp Brain Res **114** 86-98 [444]

Hay JC, Pick HL, Rosser E (1963) Adaptation to chromatic aberration by the human visual system Science **141** 167–9 [351]

Haydon PG (2001) Glia: listening and talking Nature Rev Neurosci **2** 185-93 [188]

Hayes RM (1989) 3-D movies A history and filmography of stereoscopic cinema McFarland, London [72å]

Hearn D, Baker MP (1986) Computer graphics Prentice-Hall, Englewood Cliffs NJ [109]

Heath G, Hofstetter HW (1952) The effect of orthoptics on the zone of binocular vision in intermittent exotropia – a case study Am J Optom Arch Am Acad Optom **29** 12–31 [392]

Hebb DO (1949) The organization of behavior Wiley, New York [135, 459, 514]

Hebbard FW (1962) Comparison of subjective and objective measurements of fixation disparity J Opt Soc Am **52** 706–12 [379]

Hebbard FW (1964) Effect of blur on fixation disparity Am J Optom Arch Am Acad Optom **41** 540–48 [380]

Hecht H, van Doorn A, Koenderink JJ (1999) Compression of visual space in natural scenes and their photographic counterparts Percept Psychophys **61** 1299-86 [137]

Hecht S (1928) On the binocular fusion of colors and its relation to theories of color vision Proc Nat Acad Sci **14** 237–41 [283]

Hecht S, Mintz EU (1939) The visibility of single lines at various illuminations and the retinal basis of visual resolution J Gen Physiol **22** 593–612 [78]

Heckmann T, Schor CM (1989b) Panum's fusional area estimated with a criterion–free technique Percept Psychophys **45** 297–306 [272, 274]

Heeger DJ (1992a) Normalization of cell responses in cat striate cortex Vis Neurosci **9** 181-97 [126, 203]

Heeger DJ (1992b) Half-squaring in responses of cat striate cells Vis Neurosci **9** 427-43 [194]

Heeger DJ (1999) Linking visual perception with human brain activity Curr Opin Neurobiol **9** 474-9 [182]

Heeley DW, Buchanan-Smith HM (1996) Mechanisms specialized for the perception of image geometry Vis Res **36** 3607-27 [200]

Hegde J, Van Essen DC (2000) Selectivity for complex shapes in primate visual area V2 J Neurosci **20** RC61 [220]

Heggelund P, Albus K (1978) Orientation selectivity of single cells in striate cortex of cat: the shape of orientation tuning curves Vis Res **18** 1067–71 [199]

Hein A, Diamond R (1983) Contributions of eye movement to the representation of space In Spatially oriented behavior (ed M Jeannerod) pp 119–33 Springer, Berlin [494]

Held R (1981) Acuity in infants with normal and anomalous visual experience In The development of perception: Psychobiological perspectives (ed RN Aslin, JR Roberts, MR Petersen) Vol 2 Academic Press, New York [471, 486]

Held R (1991) Development of binocular vision and stereopsis In Vision and visual dysfunction Vol 9 Binocular vision (ed D Regan) pp 170–8 MacMillan, London [485]

Held R, Bauer JA (1967) Visually guided reaching in infant monkeys after restricted rearing Science **155** 718-20 [146]

Held R, Leibowitz, HW, Teuber HL (1978) Handbook of sensory physiology, Volume I Perception Springer-Verlag, Heidelberg [8]

Held R, Birch EE, Gwiazda J (1980) Stereoacuity of human infants Proc Nat Acad Sci **77** 5572–4 [469, 481]

Heller D (1988) History of eye movements In Eye movement research (ed G Lüer U Lass, J Shallo-Hoffmann) pp 37-54 Hogrefe, Toronto [51]

Heller J, Hertz JA, Kjaer TW, Richmond BJ (1995) Information flow and temporal coding in primate pattern vision J Comput Neurosci **2** 175-93 [130]

Helmchen C, Rambold H, Fuhry L, Büttner U (1998) Deficits in vertical and torsional eye movements after uni- and bilateral muscimol inactivation of the interstitial nucleus of Cajal of the alert monkey Exp Brain Res **119** 436-52 [432]

Helmholtz H von (1864) Uber den horopter Graefes Arch klin exp Ophthal **10** 1-60 [58]

Helmholtz H von (1893) Popular lectures on scientific subjects (Translated by E Atkinson) Longmans Green, London [58, 305]

Helmholtz H von (1910) Physiological optics Dover, New York 1962 (Translation by JPC Southall from the 3rd German edition of Handbuch der Physiologischen Optik) Vos Hamburg [45, 71, 150, 275, 278, 283, 287, 372, 411, 412]

Helveston EM, von Noorden GK (1967) Microtropia Arch Ophthal **78** 272-81 [381]

Hendricks JM, Holliday IE, Ruddock KH (1981) A new class of visual defect: spreading inhibition elicited by chromatic light stimuli Brain **104** 813-40 [221]

Hendrickson AE, Boothe R (1976) Morphology of the retina and dorsal lateral geniculate nucleus in dark–reared monkeys (Macaca nemestrina) Vis Res **16** 517–21 [499, 504]

Hendrickson AE, Drucker D (1992) The development of parafoveal and mid-peripheral human retina Behav Brain Res **49** 21-31 [436]

Hendrickson AE, Kupfer C (1976) The histogenesis of the fovea in the macaque monkey Invest Ophthal Vis Sci **15** 746–56 [436]

Hendrickson AE, Tigges M (1985) Enucleation demonstrates ocular dominance columns in Old World macaque but not in New World squirrel monkey visual cortex Brain Res **333** 340-4 [216]

Hendrickson AE, Yuodelis C (1984) The morphological development of the human fovea Ophthalmology **91** 603–12 [436]

Hendrickson AE, Wilson JR, Ogren MP (1978) The neuroanatomical organization of pathways between the dorsal lateral geniculate nucleus and visual cortex in old world and new world primates J Comp Neurol **182** 123-36 [217]

Hendrickson AE, Movshon JA, Eggers HM, Gizzi MS, Boothe RG, Kiorpes L (1987) Effects of early unilateral blur on the macaque's visual system. II Anatomical observations J Neurosci **7** 1327-39 [509]

Hendry SHC, Jones EG (1986) Reduction in number of immunostained GABAergic neurones in deprived-eye dominance columns of monkey area 17 Nature **320** 750-3 [502]

Hendry SHC, Kennedy MB (1986) Immunoreactivity for a cadmodulin-dependent protein kinase is selectively increased in macaque striate cortex after monocular deprivation Proc Nat Acad Sci **83** 1536-40 [181, 213]

Hendry SHC, Reid RC (2000) The koniocellular pathway in primate vision Ann Rev Neurosci **23** 127-53 [178]

Hendry SHC, Yoshioka T (1994) A neurochemically distinct third channel in the macaque dorsal lateral geniculate nucleus Science **294** 575-7 [178]

Hendry SHC, Huntsman MM, Vinuela A, Mohler H, de Blas AL, Jones EG (1994) $GABA_A$ receptor subunit immunoreactivity in primate visual cortex: distribution in macaques and humans and regulation by visual input in adulthood J Neurosci **14** 2383-401 [181, 213]

Hennessy RT, Iida T, Shina K, Leibowitz HW (1976) The effect of pupil size on accommodation Vis Res **16** 587-9 [361, 364]

Henning GB, Hertz BG (1973) Binocular masking level differences in sinusoidal grating detection Vis Res **13** 2755–63 [338]

Henning GB, Hertz BG (1977) The influence of bandwidth and temporal properties of spatial noise on binocular masking–level differences Vis Res **17** 399–402 [338]

Henry GH (1991) Afferent inputs receptive field properties and morphological cell types in different laminae of the striate cortex In Vision and visual dysfunction Volume 4 The neural basis of visual function (ed AG Leventhal) pp 223-270 MacMillan, London [186]

Henry GH, Vidyasagar TR (1991) The evolution of visual pathways In The evolution of the eye and visual system (ed JR Cronly–Dillon, RL Gregory) pp 442–65 CRC Press, Boca Raton [444]

Henry GH, Bishop PO, Coombs JS (1969) Inhibitory and sub–liminal excitatory receptive fields of simple units in cat striate cortex Vis Res **9** 1289–96 [216]

Henson DB (1978) Corrective saccades: effects of altering visual feedback Vis Res **18** 63–7 [423]

Henson DB, Dharamshi BG (1982) Oculomotor adaptation to induced heterophoria and anisometropia Invest Ophthal Vis Sci **22** 234–40 [385]

Henson DB, North RE (1980) Adaptation to prism–induced heterophoria Am J Optom Physiol Opt **57** 129–37 [382]

Henson DB, Williams DE (1980) Depth perception in strabismus Br J Ophthal **64** 349-53 [373]

Hepler N (1968) Color: a motion–contingent aftereffect Science **162** 376–7 [352]

Hergueta S, Ward R, Lemire M, Rio JP, Reperant J, Weidner C (1992) Overlapping visual fields and ipsilateral retinal projections in turtles Brain Res **29** 427-33 [175, 532, 544]

Hering E (1861) Beitrage zur Physiologie Vol 5. Engelmann, Leipzig [283, 284, 287]

Hering E (1864) Allgemeine geometrische Auslösung des Horopterproblems In Beitrage zur Physiologie Part 4, pp 225-86 Engelmann, Leipzig [45]

Hering E (1865) Die Gesetze der binocularen Tiefenwahrnehmung Arch für Anat Physiol Wissen Med 152–165 [45, 49]

Hering E (1868) Die Lehre vom Binocularen Sehen Engelmann, Leipzig English Translation by B. Brideman and L. Stark, Plenum [418]

Hering E (1874) Outlines of a theory of the light sense (Translated by L Hurvich, D Jameson) Harvard University Press, Cambridge MA 1964 [294]

Hering E (1879) Spatial sense and movements of the eye (Translated by CA Radde) Am Acad Optom, Baltimore 1942 [52, 283]

Hermann JS, Samson CR (1967) Critical detection of the accommodative convergence to accommodation ratio by photosensor-oculograph Arch Ophthal **78** 427-30 [390]

Hermans TG (1943) Torsion in persons with no known eye defect J Exp Psychol **32** 307–27 [413]

Hernandez-Gonzalez A, Cavada C, Reinoso-Suárez F (1994) The lateral geniculate nucleus projects to the inferior temporal cortex in the macaque monkey Neuroreport **5** 2993-6 [222]

Heron G, Winn B (1989) Binocular accommodation reaction and response times for normal observers Ophthal Physiol Opt **9** 176-83 [361]

Heron G, Dholakia S, Collins DE, McLaughlan H (1985) Stereoscopic threshold in children and adults Am J Optom Physiol Opt **62** 505-15 [483]

Heron G, Charman WN, Gray LS (1999) Accommodation responses and ageing Invest Ophthal Vis Sci **40** 2872-83 [358]

Herpers MJ, Caberg HB, Mol JMF (1981) Human cerebral potentials evoked by moving dynamic random dot stereograms EEG Clin Neurophysiol **52** 50-6 [262]

Hershberger W (1970) Attached-shadow orientation perceived as depth by chickens reared in an environment illuminated from below J Comp Physiol Psychol **73** 407-11 [481]

Herz A, Sulzer B, Küne R, van Hemmen JL (1989) Hebbian learning reconsidered: representation of static and dynamic objects in associative nets Biol Cyber **60** 457-67 [135]

Hesler J, Pickwell D, Gilchrist J (1989) The accommodative contribution to binocular vergence eye movements Ophthal Physiol Opt **9** 379-84 [384]

Hess DT, Edwards MA (1987) Anatomical demonstration of ocular segregation in the retinogeniculocortical pathway of the New World capuchin monkey (*Cebus apella*) J Comp Neurol **294** 409-20 [217]

Hess RF (1978) Interocular transfer in individuals with strabismic amblyopia: a cautionary note Perception **7** 201–5 [340, 491]

Hess RF (1980) A preliminary investigation of neural function and dysfunction in amblyopia I. Size-selective channels Vis Res **20** 749-54 [523]

Hess RF (1991) The site and nature of suppression in squint amblyopia Vis Res 31 111-17 [492]

Hess RF, Anderson SJ (1993) Motion sensitivity and spatial under-sampling in amblyopia Vis Res 33 881-96 [525]

Hess RF, Baker CL (1984) Assessment of retinal function in severely amblyopic individuals Vis Res 27 1367-76 [519]

Hess RF, Bradley A (1980) Contrast perception above threshold is only minimally impaired in human amblyopia Nature 287 463-4 [521]

Hess RF, Demanins R (1998) Contour integration in anisometropic amblyopia Vis Res 38 889-94 [523]

Hess RF, Field DJ (1994) Is the spatial deficit in strabismic amblyopia due to loss of cells or an uncalibrated disarray of cells Vis Res 34 3397-406 [523]

Hess RF, Field DJ (1995) Contour integration across depth Vis Res 35 1699-711 [124]

Hess RF, Holliday I (1992b) The spatial localization deficit in amblyopia Vis Res 32 1319-39 [523]

Hess RF, Howell ER (1977) The threshold contrast sensitivity function in strabismic amblyopia: evidence for a two type classification Vis Res 17 1049–55 [520]

Hess RF, Pointer JS (1985) Differences in the neural basis of human amblyopia: the distribution of the anomaly across the visual field Vis Res 25 1577–94 [527, 528]

Hess RF, Campbell FW, Greenhalgh T (1978) On the nature of the neural abnormality in human amblyopia; neural aberrations and neural sensitivity loss Pflügers Arch ges Physiol 377 201–7 [522, 523]

Hess RF, Campbell FW, Zimmern R (1980) Differences in the neural basis of human amblyopia: the effect of mean luminance Vis Res 20 295-305 [520, 521]

Hess RF, France TD, Tulunay-Keesey U (1981) Residual vision in humans who have been monocularly deprived of pattern stimulation in early life Exp Brain Res 44 295-311 [529]

Hess RF, Hayes A, Kingdom FAA (1997a) Integrating contours within and through depth Vis Res 37 691-6 [124]

Hess RF, McIlhagga W, Field DJ (1997b) Contour integration in strabismic amblyopia: the sufficiency of an explanation based on positional uncertainty Vis Res 37 3145-61 [524]

Hess RF, Demanins R, Bex PJ (1997c) A reduced motion aftereffect in strabismic amblyopia Vis Res 37 1303-11 [348, 493, 525]

Hess RF, Wang YZ, Demanins R, Wilkinson F, Wilson HR (1999b) A deficit in strabismic amblyopia for global shape detection Vis Res 39 901-14 [521]

Hetherington PA, Swindale NV (1999) Receptive field and orientation scatter studied by tetrode recordings in cat area 17 Vis Neurosci 16 637-52 [209, 237, 255]

Heuer H, Owens DA (1989) Vertical gaze direction and the resting posture of the eyes Perception 18 363-77 [372]

Heuer H, Dunkel-Abels G, Brüwer M, Kröger H, Römer T, Wischmeier E (1988) The effects of sustained vertical gaze deviation on the resting state of the vergence system Vis Res 28 1337–44 [372]

Hevner RF, Wong-Riley MTT (1993) Mitochrondrial and nuclear gene expression for cytochrome oxidase subunits are disproportionately regulated by functional activity in neurones J Neurosci 13 1805-19 [510]

Heywood CA, Gadotti A, Cowey A (1992) Cortical area V4 and its role in the perception of color J Neurosci 12 4056-65 [222]

Hickey TL (1977) Postnatal development of the human lateral geniculate nucleus: relationship to a critical period for the visual system Science 198 836–8 [443]

Hickey TL, Spear PD, Kratz AE (1977) Quantitative studies of cell size in the cat's dorsal lateral geniculate nucleus following visual deprivation J Comp Neurol 172 295–82 [503]

Hietanen JK, Perrett DI (1996) Motion sensitive cells in the macaque superior temporal polysensory area: response discrimination between self-generated and externally generated pattern motion Behav Brain Res 76 155-67 [223]

Higgins KE, Daugman JG, Mansfield RJW (1982) Amblyopic contrast sensitivity: insensitivity to unsteady fixation Invest Ophthal Vis Sci 23 113-20 [520]

Higo N, Oishi T, Yamashita A, Matsuda K, Hayashi M (2000) Expression of GAP-43 and SCG10 mRNAs in lateral geniculate nucleus of normal and monocularly deprived macaque monkeys J Neurosci 20 6030-8 [512]

Hine T, Thorn F (1987) Compensatory eye movements during active head rotation for near targets: effects of imagination rapid head oscillation and vergence Vis Res 27 1639–57 [428]

Hinkle DA, Connor, CE (2001) Disparity tuning in macaque area V4 NeuroReport 12 365-9 [243]

Hinton GE (1987) The horizontal-vertical delusion Perception 16 667-80 [160]

Hinton GE (1989) Connectionist learning procedures Artificial Intelligence 40 185–234 [95, 269]

Hinton GE, McClelland JL, Rumelhart DE (1986) Distributed representations In Parallel distributed processing (ed DE Rumelhart, JL McClelland) Vol 1 pp 77–109 MIT Press, Boston [132]

Hirsch HVB, Spinelli DN (1971) Modification of the distribution of receptive field orientation in cats by selective visual exposure during development Exp Brain Res 13 509-27 [454]

Hirsch J, Curcio CA (1989) The spatial resolution capacity of the human foveal retina Vis Res 29 1095–101 [174]

Hirsch J, Hylton R (1982) Limits of spatial–frequency discrimination as evidence of neural interpolation J Opt Soc Am 72 1367 71 [81]

Hirsch JA, Gilbert CD (1991) Synaptic physiology of horizontal connections in the cat's visual cortex J Neurosci 11 1800–9 [196]

Hirsch JA, Gilbert CD (1993) Long–term changes in synaptic strength along specific intrinsic pathways in the cat visual cortex J Physiol 461 277–62 [198, 208, 452]

Hirsch JA, Alonso JM, Reid RC, Martinez LM (1998) Synaptic integration in striate cortical simple cells J Neurosci 18 9517-28 [199]

Hirschberg J (1982-92) The history of ophthalmology 11 Volumes (Translated by FC Blodi, JP Bonn) Wayenborgh [59]

Hitchcock PF, Hickey TL (1980) Ocular dominance columns: evidence for their presence in humans Brain Res 82 176–9 [213]

Hockney D (2001) Secret knowledge. Rediscovering the lost techniques of the old masters Thames and Hudson, London [39]

Hoekstra J, Van der Groot DPJ, Van den Brink G, Bilsen FA (1974) The influence of the number of cycles upon the visual contrast threshold for spatial sine wave patterns Vis Res 14, 365-8 [521]

Hofer MM, Barde YA (1988) Brain-derived neurotrophic factor prevents neuronal death *in vivo* Nature 331 291-2 [456]

Hoffman CS (1962) Comparison of monocular and binocular color matching J Opt Soc Am 52 75–80 [284]

Hoffman CS, Price AC, Garrett ES, Rothstein W (1959) Effect of age and brain damage on depth perception Percept Mot Skills 9 283-6 [484]

Hoffman WC (1966) The Lie algebra of visual perception J Math Psychol 3 65-98 [103, 109, 136]

Hoffmann KP, Lippert P (1982) Recovery of vision with the deprived eye after the loss of the non-deprived eye in cats Hum Neurobiol 1 45-8 [508, 527]

Hoffmann KP, Schoppmann A (1984) Shortage of binocular cells in area 17 of visual cortex in cats with congenital strabismus Exp Brain Res 55 470-82 [505]

Hoffmann KP, Sherman SM (1974) Effects of early monocular deprivation on visual input to cat superior colliculus J Neurophysiol **37** 1276–86 [505]

Hofman MA (1985) Size and shape of the cerebral cortex in mammals. I. The cortical surface Brain Behav Evol **27** 28-40 [445]

Hofman MA (1989) On the evolution and geometry of the brain in mammals Prog Neurobiol **32** 137-58 [445]

Hofmann FB, Bielschowsky A (1900) Über die der Willkür entzogenen Fusionsbewegungen der Augen Pflügers Arch ges Physiol **80** 1–40 [377, 413]

Hofstetter HW (1945) The zone of clear single binocular vision Am J Optom Arch Am Acad Optom **22** 301–33 and 361–84 [390, 392]

Hofstetter HW (1983) Graphical analysis In Vergence eye movements: Basic and clinical aspects (ed CM Schor, KJ Ciuffreda) pp 439-64 Butterworth-Heinemann, Boston [392]

Hofstetter HW, Bertsch JD (1976) Does stereopsis change with age? Am J Optom Physiol Opt **53** 664-7 [484]

Hohmann A, Creutzfeldt OD (1975) Squint and the development of binocularity in humans Nature **254** 613–14 [492]

Hohmann CF, Berger-Sweeney J (1998) Cholinergic regulation of cortical development and plasticity Perspectives in Developmental Neurobiology **5** 401-25 [457]

Holden AL, Low JC (1989) Binocular fields with lateral-eyed vision Vis Res **29** 361-67 [544]

Holland HC (1965) The spiral after–effect Pergamon, Oxford [347]

Holländer H, Vanegas H (1977) The projection from the lateral geniculate nucleus onto the visual cortex in the cat A quantitative study with horseradish-peroxidase J Comp Neurol **173** 519-36 [186]

Hollins M (1980) The effect of contrast on the completeness of binocular rivalry suppression Percept Psychophys **27** 550–6 [297]

Hollins M, Bailey GW (1981) Rivalry target luminance does not affect suppression depth Percept Psychophys **30** 201–3 [288]

Hollins M, Leung EHL (1978) The influence of color on binocular rivalry In Visual psychophysics and physiology (ed JC Armington, J Krausfopf, BR Wooten) pp 181–90 Academic Press, New York [290]

Hollmann M, Heinemann S (1994) Cloned glutamate receptors Ann Rev Neurosci **17** 31-108 [191]

Holmes EJ, Gross CG (1984) Effects of inferior temporal lesions on discrimination of stimuli differing in orientation J Neurosci **4** 3063-8 [222]

Holmes G, Horrax G (1919) Disturbances of spatial orientation and visual attention with loss of stereoscopic vision Arch Neurol Psychiat **1** 385–407 [228, 488]

Holmes OW (1859) The stereoscope and the stereograph Atlantic Monthly **3** 738-48. Also (1861) **5** 13-29; (1863) **7** 1-15 [69]

Holopigian K (1989) Clinical suppression and binocular rivalry suppression: the effects of stimulus strength on the depth of suppression Vis Res **29** 1325-33 [288]

Holopigian K, Blake R (1983) Spatial vision in strabismic cats J Neurophysiol **50** 287–96 [509]

Holopigian K, Blake R, Greenwald M (1986) Selective losses in binocular vision in anisometropic amblyopes Vis Res **29** 621–30 [529]

Holopigian K, Blake R, Greenwald MJ (1988) Clinical suppression and amblyopia Invest Ophthal Vis Sci **29** 444–51 [530]

Hölscher C (1997) Nitric oxide the enigmatic neuronal messenger: its role in synaptic plasticity TINS **20** 298-303 [461]

Holt CE, Harris WA (1993) Position guidance and mapping in the developing visual system J Neurobiol **27** 1400-22 [437]

Home R (1984) Binocular summation: a study of contrast sensitivity visual acuity and recognition Vis Res **18** 579-85 [324]

Honda H, Findlay JM (1992) Saccades to targets in three–dimensional space: dependence of saccadic latency on target location Percept Psychophys **52** 167–74 [423]

Hooge ITC van den Berg AV (2000) Visually evoked cyclovergence and extended Listing's law J Neurophysiol **8 3** 2757-75 [370]

Hooten K, Myers E, Worrall R, Stark L (1979) Cyclovergence: the motor response to cyclodisparity Graefes Arch klin exp Ophthal **210** 65–8 [413]

Hopkins AA (1898) Magic. Stage illusions, special effects and trick photography Reprinted by Dover, New York, 1976 [61]

Horng JL, Semmlow JL, Hung GK, Ciuffreda KJ (1998a) Initial component control in disparity vergence: A model-based study IEEE Tr Biomed Engin **45** 279-57 [405]

Horng JL, Semmlow JL, Hung GK, Ciuffreda KJ (1998b) Dynamic asymmetries in disparity convergence eye movements Vis Res **38** 2761-8 [406, 423]

Horowitz MW (1949) An analysis of the superiority of binocular over monocular visual acuity J Exp Psychol **39** 581–96 [317]

Horridge GA (1978) The separation of visual axes in apposition compound eyes Philos Tr R Soc B **285** 1–59 [532]

Horsten GPM, Winkelman JE (1962) Electrical activity of the retina in relation to histological differentiation in infants born prematurely and at full term Vis Res **2** 299-76 [471]

Horton JC (1984) Cytochrome oxidase patches: a new cytoarchitectonic feature of monkey visual cortex Philos Tr R Soc B **304** 199–253 [205, 455]

Horton JC, Hedley-White ET (1984) Mapping of cytochrome oxidase patches and ocular dominance columns in human visual cortex Philos Tr R Soc B **304** 255-72 [204, 510]

Horton JC, Hocking DR (1996a) Intrinsic variability of ocular dominance column periodicity in normal macaque monkeys J Neurosci **16** 7228-39 [214]

Horton JC, Hocking DR (1996b) Anatomical demonstration of ocular dominance in striate cortex of the squirrel monkey J Neurosci **16** 5510-22 [217]

Horton JC, Hocking DR (1996c) An adult-like pattern of ocular dominance columns in striate cortex of newborn monkeys prior to visual experience J Neurosci **16** 1791-1807 [455]

Horton JC, Hocking DR (1997) Timing of the critical period for plasticity of ocular dominance columns in macaque striate cortex J Neurosci **17** 3684-709 [517]

Horton JC, Hocking DR (1998a) Monocular core zones and binocular strips in primate striate cortex revealed by the contrasting effects of enucleation, eyelid suture, and retinal laser lesions on cytochrome oxidase activity J Neurosci **18** 5433-55 [205]

Horton JC, Hocking DR (1998b) Effect of early monocular enucleation upon ocular dominance columns and cytochrome oxidase activity in monkey and human visual cortex Vis Neurosci **15** 289-303 [215, 517]

Horton JC, Hubel DH (1981) Regular patchy distribution of cytochrome oxidase staining in primary visual cortex of the macaque monkey Nature **292** 762–4 [203]

Horton JC, Stryker MP (1993) Amblyopia induced by anisometropia without shrinkage of ocular dominance columns in human striate cortex Proc Nat Acad Sci **90** 5494-8 [509]

Horton JC, Dagi LR, McCrane EP, de Monasterio FM (1990) Arrangement of ocular dominance columns in human visual cortex Arch Ophthal **108** 1025-31 [213, 214]

Horton JC, Hocking DR, Kiorpes L (1997) Pattern of ocular dominance columns and cytochrome oxidase in a macaque monkey with naturally occurring anisometropic amblyopia Vis Neurosci **14** 681-9 [509]

Horton JC, Hocking DR, Adams DL (1999) Metabolic mapping of suppression scotomas in striate cortex of macaques with experimental strabismus J Neurosci **15** 7111-29 [509]

Hosoba M, Bando T, Tsukahara N (1978) The cerebellar control of accommodation of the eye in the cat Brain Res **153** 495-505 [365]

Hotta T, Kamena K (1963) Interactions between somatic and visual or auditory responses in the thalamus of the cat Exp Neurol **8** 1–13 [179]

House D (1982) Depth perception in frogs and toads Lecture notes in biomathematics Number 80 Springer, New York [539]

Houston CA, Cleary M, Dutton GN, McFadzean RM (1998) Clinical characteristics of microtropia--is microtropia a fixed phenomenon? Brit J Ophthal **82** 219-27 [374]

Houtman WA, van der Pol BAE (1982a) Fusional movements by peripheral retinal stimulation Graefes Arch klin exp Ophthal **218** 218-20 [408]

Houtman WA, van der Pol BAE (1982b) Fixation disparity in vertical vergence Ophthalmologica **185** 220-5 [409]

Houtman WA, Roze JH, Scheper W (1981) Vertical vergence movements Doc Ophthal **51** 199-207 [409]

Hovis JK (1989) Review of dichoptic color mixing Optom Vis Sci **66** 181–90 [284]

Hovis JK, Guth SL (1989a) Dichoptic opponent hue cancellations Optom Vis Sci **66** 304–19 [284]

Hovis JK, Guth SL (1989b) Changes in luminance affect dichoptic unique yellow J Opt Soc Am A **6** 1297-301 [284]

Howard IP (1959) Some new subjective phenomena apparently due to interocular transfer Nature **184** 1516–17 [290]

Howard IP (1960) Attneave's interocular color–effect Am J Psychol **73** 151–2 [352]

Howard IP (1961) An investigation of a satiation process in the reversible perspective of a revolving skeletal cube Quart J Exp Psychol **13** 19–33 [149, 150]

Howard IP (1974) Proposals for the study of adaptation to anomalous schemata Perception **3** 497-513 [161]

Howard IP (1978) Recognition and knowledge of the water-level principle Perception **7** 151-60 [160]

Howard IP (1982) Human visual orientation Wiley, Chichester [24, 126, 144, 197, 344, 355, 454, 471, 522]

Howard IP (1986) The vestibular system In Handbook of perception and performance (ed KR Boff, L Kaufman, JP Thomas) Chap 11 Wiley, New York [365, 426]

Howard IP (1993a) The optokinetic system In The vestibulo–ocular reflex nystagmus and vertigo (ed JA Sharpe, HO Barber) pp 163–84 Raven Press, New York [365, 490]

Howard IP (1996) Alhazen's neglected discoveries of visual phenomena Perception **25** 1203-18 [18]

Howard IP (1997a) Interactions within and between the spatial senses J Vestib Res **7** 311-45 [151]

Howard IP (1997b) Seeing in reverse Nature **389** 235-6 [403]

Howard IP, Evans J (1963) The measurement of eye torsion Vis Res **3** 447–55 [413]

Howard IP, Kaneko H (1994) Relative shear disparities and the perception of surface inclination Vis Res **34** 2505-17 [418]

Howard IP, Ohmi M (1984) The efficiency of the central and peripheral retina in driving human optokinetic nystagmus Vis Res **27** 969–76 [417]

Howard IP, Rogers BJ (1995) Binocular vision and stereopsis Oxford University Press, New York. [8]

Howard IP, Templeton WB (1966) Human spatial orientation Chichester, Wiley [369]

Howard IP, Wade N (1996) Ptolemy on binocular vision Perception **25** 1189-203 [47]

Howard IP, Zacher JE (1991) Human cyclovergence as a function of stimulus frequency and amplitude Exp Brain Res **85** 445–50 [413]

Howard IP, Bergström SS, Ohmi M (1990) Shape from shading in different frames of reference Perception **19** 523–30 [78, 481]

Howard IP, Ohmi M, Sun L (1993) Cyclovergence: a comparison of objective and psychophysical measurements Exp Brain Res **97** 349–55 [413]

Howard IP, Sun L, Shen X (1994) Cycloversion and cyclovergence: the effects of the area and position of the visual display Exp Brain Res **100** 509-14 [417]

Howard IP, Allison RS, Zacher JE (1997) The dynamics of vertical vergence Exp Brain Res **116** 153-9 [410]

Howard, IP, Fang X, Allison RS, Zacher JE (2000) Effects of stimulus size and eccentricity on horizontal and vertical vergence Exp Brain Res **130** 124-32 [396, 408]

Howell ER, Hess RF (1978) The functional area for summation to threshold for sinusoidal gratings Vis Res **18** 369-74 [521]

Howell ER, Mitchell DE, Keith CG (1983) Contrast thresholds for sign gratings of children with amblyopia Invest Ophthal Vis Sci **27** 782–7 [522]

Howland HC (1991) Determination of ocular refraction In Vision and visual dysfunction Vol 1 Visual optics and instrumentation (ed WN Charman) pp 399– 414 MacMillan, London [358]

Howland HC (1993) Early refractive development In Early visual development, normal and abnormal (ed K Simons) pp 5-13 Oxford University Press, New York [434]

Howland HC, Howland B (1974) Photorefraction: a technique for the study of refractive state at a distance J Opt Soc Am **64** 270–9 [473]

Howland HC, Sayles N (1984) Photorefractive measurements of astigmatism in infants and young children Invest Ophthal Vis Sci **25** 93–102 [474]

Howland HC, Dobson V, Sayles N (1987) Accommodation in infants as measured by photorefraction Vis Res **27** 2141-52 [173]

Hoyt CS, Stone RD, Fromer C (1981) Monocular axial myopia associated with neonatal eyelid closure in human infants Am J Ophthal **91** 197-200 [435]

Huang EJ, Reichardt LF (2001) Neurotrophins: roles in neuronal development and function Ann Rev Neurosci **24** 677-36 [457]

Hubel DH, Livingstone MS (1987) Segregation of form color and stereopsis in primate area 18 J Neurosci **7** 3378–415 [219, 239]

Hubel DH, Wiesel TN (1959) Receptive fields of single neurones in the cat's visual cortex J Physiol **148** 574–91 [58, 213, 235]

Hubel DH, Wiesel TN (1962) Receptive fields binocular interaction and functional architecture in the cat's visual cortex J Physiol **160** 106–54 [58, 213-215, 235, 255]

Hubel DH, Wiesel TN (1963) Receptive fields of cells in striate cortex of very young visually inexperienced kittens J Neurophysiol **29** 994–1002 [454]

Hubel DH, Wiesel TN (1965) Binocular interaction in striate cortex of kittens reared with artificial squint J Neurophysiol **28** 1041–59 [505, 508, 513]

Hubel DH, Wiesel TN (1967) Cortical and callosal connections concerned with the vertical meridian in the cat J Neurophysiol **30** 1561–73 [232]

Hubel DH, Wiesel TN (1968) Receptive fields and functional architecture of monkey striate cortex J Physiol **195** 215–43 [214]

Hubel DH, Wiesel TN (1969) Anatomical demonstration of columns in the monkey striate cortex Nature **221** 747–50 [212]

Hubel DH, Wiesel TN (1970a) Stereoscopic vision in macaque monkey Nature **225** 41–2 [236, 239]

Hubel DH, Wiesel TN (1970b) The period of susceptibility to the physiological effects of unilateral eye closure in kittens J Physiol **206** 419–36 [515]

Hubel DH, Wiesel TN (1971) Aberrant visual projections in the Siamese cat J Physiol **218** 33–62 [497]

Hubel DH, Wiesel TN (1973) A re-examination of stereoscopic mechanisms in area 17 of the cat J Physiol **232** 29-30P [255]

Hubel DH, Wiesel TN (1974a) Uniformity of monkey striate cortex: a parallel relationship between field size scatter and magnification factor J Comp Neurol **158** 295–306 [184, 214]

Hubel DH, Wiesel TN (1974b) Sequence regularity and geometry of orientation columns in the monkey striate cortex J Comp Neurol **158** 297-93 [209]

Hubel DH, Wiesel TN (1977) Functional architecture of macaque monkey visual cortex Proc R Soc B **198** 1–59 [193, 214]

Hübener M, Shoham D, Grinvald A, Bonhoeffer T (1997) Spatial relationships among three columnar systems in cat area 17 J Neurosci **17** 9270-84 [210]

Huffman DA (1971) Impossible objects as nonsense sentences In Machine Intelligence **6** (ed B Meltzer, D Michie) pp 295-327 Edinburgh University Press, Edinburgh [30]

Hughes A (1971) Topographical relationships between the anatomy and physiology of the rabbit visual system Doc Ophthal **30** 33–159 [549]

Hughes A (1972) Vergence in the cat Vis Res **12** 1961–94 [366]

Hughes A (1977) The topography of vision in mammals of contrasting life style: comparative optics and retinal organization In Handbook of sensory physiology Vol VII/5 (ed F Crescitelli) pp 615–756 Springer, New York [532, 548]

Hughes A, Vaney DI (1982) The organization of binocular cortex in the primary visual area of the rabbit J Comp Neurol **294** 151–64 [549]

Hultén KG (1952) A peep show by Carel Fabritius Art Quart **15** 279-90 [40]

Humphrey AL, Hendrickson AE (1983) Background and stimulus-induced patterns of high metabolic activity in the visual cortex (area 17) of the squirrel and macaque monkey J Neurosci **3** 345-58 [217]

Humphriss D (1982) The psychological septum An investigation into its function Am J Optom Physiol Opt **59** 639-41 [287]

Hung GK (1992a) Quantitative analysis of associated and disassociated phorias: linear and nonlinear static models IEEE Tr Biomed Engin **39** 135-45 [380]

Hung GK (1992b) Adaptation model of accommodation and vergence Ophthal Physiol Opt **12** 319-29 [384]

Hung GK (1997) Quantitative analysis of the accommodative convergence to accommodation ratio: linear and nonlinear static models IEEE Tr Biomed Engin 44 306-16 [389, 391]

Hung GK (1998) Dynamic model of saccade-vergence interactions Med Sci Res **26** 9-14 [423]

Hung GK, Ciuffreda KJ (1988) Dual-mode behaviour in the human accommodation system Ophthal Physiol Opt **8** 327-32 [358]

Hung GK, Semmlow JL, Ciuffreda KJ (1983) Identification of accommodative vergence contribution to the near response using response variance Invest Ophthal Vis Sci **27** 772–7 [389]

Hung GK, Semmlow JL, Ciuffreda KJ (1986) A dual-mode dynamic model of the vergence eye movement system IEEE Tr Biomed Engin **BME-33** 1021–36 [405]

Hung GK, Wang T, Ciuffreda KJ, Semmlow JL (1989) Suppression of sensitivity to surround displacement during vergence eye movements Exp Neurol **105** 300–5 [368]

Hung GK, Sun L, Semmlow JL, Ciuffreda KJ (1990) Suppression of sensitivity to change in target disparity during vergence eye movements Exp Neurol **110** 291–7 [368]

Hung GK, Semmlow JL, Sun L, Ciuffreda KJ (1991) Vergence control of central and peripheral disparities Exp Neurol **113** 202–11 [396]

Hung GK, Ciuffreda KJ, Semmlow JL, Horng JL (1994) Vergence eye movements under natural viewing conditions Invest Ophthal Vis Sci **35** 3486-92 [398]

Hung GK, Ciuffreda KJ, Rosenfield M (1996) Proximal contribution to a linear static model of accommodation and vergence Ophthal Physiol Opt **16** 31-41 [387]

Hung GK, Zhu H, Ciuffreda KJ (1997) Convergence and divergence exhibit different response characteristics to symmetric stimuli Vis Res **37** 1197-1205 [397, 398]

Hung LF, Crawford MLJ, Smith EL (1995) Spectacle lenses alter eye growth and the refractive status of young monkeys Nature Med **1** 761-5 [435]

Hupé JM, James AC, Payne BR, Lomber SG, Girard P, Bullier J (1998) Cortical feedback improves discrimination between figure and background by V1, V2 and V3 neurones Nature **394** 784-7 [188]

Hupé JM, James AC, Girard P, Lomber SG, Payne BR, Bullier J (2001) Feedback connections act on the early part of the responses in monkey visual cortex J Neurophysiol **85** 134-145 [206]

Hurvich LM, Jameson D (1951) The binocular fusion of yellow in relation to color theories Science **114** 199–202 [283]

Husi H, Ward MA, Choudhary JS, Blackstock WB, Grant SGN (2000) Proteomic analysis of NMDA receptor-adhesion protein signaling complexes Nature Neurosci **3** 661-9 [193]

Huttenlocher PR, de Courten C (1987) The development of synapses in striate cortex of man Hum Neurobiol **6** 1–9 [451]

Huxley JS (1932) Problems of relative growth Methuen, London [107]

Hyman I (Ed) (1974) Brunelleschi in perspective Prentice-Hall, Englewood Cliffs, NJ [32]

Hyson MT, Julesz B, Fender DH (1983) Eye movements and neural remapping during fusion of misaligned random–dot stereograms J Opt Soc Am **73** 1665–73 [280]

Hyvärinen J, Poranen A (1974) Function of the parietal associative area 7 as revealed from cellular discharges in alert monkeys Brain **97** 673-92 [227]

Hyvärinen J, Hyvärinen L, Linnankoski I (1981) Modification of parietal association cortex and functional blindness after binocular deprivation in young monkeys Exp Brain Res **42** 1-8 [500]

Ide CF, Fraser SE, Meyer RL (1983) Eye dominance columns from an isogenic double-nasal frog eye Science **221** 292-5 [465]

Ikeda H, Jacobson SG (1977) Nasal field loss in cats reared with convergent squint: behavioural studies J Physiol **270** 367–81 [522]

Ikeda H, Tremain KE (1979) Amblyopia occurs in retinal ganglion cells in cats reared with convergent squint without alternating fixation Exp Brain Res **35** 559-82 [503]

Ikeda H, Wright MJ (1976) Properties of LGN cells in kittens reared with convergent squint: a neurophysiological demonstration of amblyopia Exp Brain Res **25** 63-77 [503]

Ikeda H, Plant GT, Tremain KE (1977) Nasal field loss in kittens reared with convergent squint: neurophysiological and morphological studies of the lateral geniculate nucleus J Physiol **270** 345-66 [503]

Ikeda M (1965) Temporal summation of positive and negative flashes in the visual system J Opt Soc Am **55** 1527–34 [331]

Ikeda M, Nakashima Y (1980) Wavelength difference limit for binocular color fusion Vis Res **20** 693–7 [283]

Ikeda M, Sagawa K (1979) Binocular color fusion limit J Opt Soc Am **69** 316–20 [284]

Ikonomidou C, Bosch F, Miksa M, Bitttigau P, Vöckler J, Dikranian K, Tenkova TI, Stefovska V, Turski L, Olney JW (1999) Blockade of NMDA receptors and apoptotic neurodegeneration in the developing brain Science **283** 70-4 [461]

Illing RB, Wässle H (1981) The retinal projection to the thalamus in the cat: a quantitative investigation and a comparison with the retinotectal pathway J Comp Neurol **202** 295-85 [177]

Imamizu H, Uno Y, Kawato M (1995) Internal representations of the motor apparatus: Implications from generalization in visuomotor learning J Exp Psychol HPP**21** 1174-98 [143]

Imamura K, Kasamatsu T (1989) Interaction of noradrenergic and cholinergic systems in regulation of ocular dominance plasticity Neurosci Res **6** 519-36 [513]

Imamura K, Kasamatsu T (1991) Ocular dominance plasticity restored by NA infusion to aplastic visual cortex of anesthetized and paralyzed kittens Exp Brain Res **87** 309-18 [512]

Imamura K, Kasamatsu T, Shirokawa T, Ohashi T (1999) Restoration of ocular dominance plasticity mediated by adenosine 3',5' monophosphate in adult visual cortex Proc Roy Soc B **296** 1507-16 [513]

Imbert M, Buisseret P (1975) Receptive field characteristics and phasic properties of visual cortical cells in kittens reared with or without visual experience Exp Brain Res **22** 25–36 [500]

Inchingolo P, Accardo A, Pozzo SD, Pensiero S, Perissutti P (1996) Cyclopean and disconjugate adaptive recovery from post-saccadic drift in strabismic children before and after surgery Vis Res **36** 2897-913 [374]

Indow T, Watanabe T (1984) Parallel–alleys and distance–alleys on horopter plane in the dark Perception **13** 165–82 [137]

Ingle D (1972) Depth vision in monocular frogs Psychonom Sci **29** 37-8 [541]

Ingle D (1976) Spatial vision in anurans In The amphibian visual system – a multidisciplinary approach (ed K Fite) Academic Press, New York [541]

Ingle D, Cook J (1977) The effect of viewing distance upon size preference of frogs for prey Vis Res **17** 1009–13 [539, 541]

Ingling CR (1991) Psychophysical correlates of parvo channel function In From pigments to perception (ed A Valberg, BB Lee) pp 413–27 Plenum, New York [244]

Ingling CR, Grigsby SS (1990) Perceptual correlates of magnocellular and parvocellular channels: seeing form and depth in afterimages Vis Res **30** 823–8 [245]

Ingling CR, Martinez–Uriegas E (1985) The spatiotemporal properties of the r–g X–cell channel Vis Res **25** 33–8 [244]

Ingram RM, Walker C (1979) Refraction as a means of predicting squint or amblyopia in preschool siblings of children known to have these defects Br J Ophthal **63** 238-42 [473, 522]

Innocenti GM (1981) Growth and reshaping of axons in the establishment of visual callosal connections Science **212** 827-7 [453]

Innocenti GM, Frost DO (1979) Effects of visual experience on the maturation of the efferent system to the corpus callosum Nature **280** 231-3 [453]

Innocenti GM, Frost DO (1980) The postnatal development of visual callosal connections in the absence of visual experience or of the eyes Exp Brain Res **39** 365-75 [232, 500]

Innocenti GM, Frost DO, Illes J (1985) Maturation of visual callosal connections in visually deprived kittens: a challenging critical period J Neurosci **5** 255-67 [453]

Inoue Y, Takemura A, Kawano K, Kitama T, Miles FA (1998) Dependence of short-latency ocular following and associated activity in the medial superior temporal area (MST) on ocular vergence Exp Brain Res **121** 135-44 [427]

Ireland FH (1950) A comparison of critical flicker frequencies under conditions of monocular and binocular stimulation J Exp Psychol **40** 282–6 [329]

Irving EL, Robertson KM (1991) Monocular components of the fixation disparity curve Optom Vis Sci **68** 117-29 [379]

Irving EL, Robertson KM (1996) Influences of monocular image degradation on the monocular components of fixation disparity Ophthal Physiol Opt **16** 329-35 [379]

Isley MR, Rogers-Ramachandran DC, Shinkman PG (1990) Interocular torsional disparity and visual cortical development in the cat J Neurophysiol **64** 1352-60 [485]

Issa NP, Trachtenberg JT, Chapman B, Zahs KR, Stryker MP (1999) The critical period for ocular dominance plasticity in the Ferret's visual cortex J Neurosci **19** 6965-78 [515]

Issa NP, Trepel C, Stryker MP (2000) Spatial frequency maps in cat visual cortex J Neurosci **20** 8504-14 [210]

Ito M, Sanides D, Creutzfeldt OD (1977) A study of binocular convergence in cat visual cortex neurons Exp Brain Res **2 8** 21–35 [214]

Ito M, Tamura H, Fujita I, Tanaka K (1995) Size and position invariance of neuronal responses in monkey inferotemporal cortex J Neurophysiol **73** 218-29 [222]

Ittelson WH, Ames A (1950) Accommodation convergence and their relation to apparent distance J Psychol **30** 43–62 [386]

Ivins JP, Porrill J, Frisby JP (1999) Instability of torsion during smooth asymmetric vergence Vis Res **39** 993-1009 [416]

Ivins WM (1973) On the rationalization of sight Da Capo Press, New York [34, 36]

Jacobs DG, Blakemore C (1988) Factors limiting the postnatal development of visual acuity in the monkey Vis Res **28** 947–58 [470]

Jacobson LD, Gaska JP, Chen HW, Pollen DA (1993) Structural testing of multi-input linear-nonlinear cascade models for cells in macaque striate cortex Vis Res **33** 609-29 [203]

Jacobson M, Gaze RM (1965) Selection of appropriate tectal connections by regenerating optic nerve fibers in adult goldfish Exp Neurol **13** 418-30 [464]

Jacobson M, Hirsch HVB (1973) Development and maintenance of connectivity in the visual system of the frog. I The effects of eye rotation and visual deprivation Brain Res **49** 47–65 [541]

Jacobson SG, Mohindra I, Held R (1981) Age of onset of amblyopia in infants with esotropia Doc Ophthal **30** 210-23 [518]

Jacobson SG, Mohindra I, Held R (1982) Visual acuity in infants with ocular diseases Am J Ophthal **93** 198–209 [468]

Jagadeesh B, Gray CM, Ferster D (1992) Visually evoked oscillations of membrane potential in cells of visual cortex Science **257** 552–4 [133]

Jagadeesh B, Wheat HS, Kontsevich LL, Tyler CW, Ferster D (1997) Direction selectivity of synaptic potentials in simple cells of the cat visual cortex J Neurophysiol **78** 2772-89 [203]

Jampel RS (1960) Convergence divergence pupillary reactions and accommodation of the eyes from faradic stimulation of the macaque brain J Comp Neurol **115** 371-400 [365, 432]

Jampel RS (1967) Multiple motor systems in the extraocular muscles of man Invest Ophthal **6** 288–93 [429]

Jampolsky A (1956) Esotropia and convergent fixation disparity of small degree: differential diagnosis and management Am J Ophthal **41** 825-33 [381]

Jampolsky A (1962) Management of exodeviations In Strabismus. Symposium of the New Orleans Academy of Ophthalmology Mosby, St. Louis [381]

Jampolsky A, Flom BC, Freid AN (1957) Fixation disparity in relation to heterophoria Am J Ophthal **43** 97–106 [379]

Janssen P, Vogels R, Orban GA (2000) Selectivity for 3D shape that reveals distinct areas within macaque inferior temporal cortex Science **288** 2054-6 [222, 243]

Janssen P, Vogels R, Orban GA (2000b) Three-dimensional shape coding in inferior temporal cortex Neuron **27** 385-97 [244]

Janssen P, Vogels R, Liu Y, Orban GA (2001) Macaque inferior temporal neurons are selective for three-dimensional boundaries and surfaces J Neuroci **21** 9419-29 [243]

Jarvis CR, Xiong ZG, Plant JR, Churchill D, Lu WY, MacVicar BA, MacDonald JF (1997) Neurotrophin modulation of NMDA receptors in cultured murine and isolated rat neurons J Neurophysiol **78**, 2363-71 [456]

Jaschinski W (1997) Fixation disparity and accommodation as a function of viewing distance and prism load Ophthal Physiol Opt **17** 327-39 [380, 383]

Jaschinski W (1998) Fixation disparity at different viewing distances and the preferred viewing distance in a laboratory near-vision task Ophthal Physiol Opt **18** 30-39 [380]

Jaschinski W (2001) Fixation disparity and accommodation for stimuli closer and more distant than oculomotor tonic positions Vis Res **41** 923-33 [380]

Jaschinski W, Koitcheva V, Heuer H (1998) Fixation disparity accommodation dark vergence and dark focus during inclined gaze Ophthal Physiol Opt **18** 351-9 [377]

Jaschinski W, Bröde P, Griefahn B (1999) Fixation disparity and nonius bias Vis Res **39** 669-77 [378]

Jaschinski-Kruza W (1990) Effects of stimulus distance on measurements of dark convergence Ophthal Physiol Opt **10** 273-51 [372]

Jaschinski-Kruza W (1993) Fixation disparity at different viewing distances of a visual display unit Ophthal Physiol Opt **13** 27-34 [377]

Jaschinski–Kruza W (1994) Dark vergence in relation to fixation disparity at different luminance and blur levels Vis Res **3 4** 1197–204 [380]

Jay B, Witkop CJ, King RA (1982) Albinism in England Birth Defects **18** 319–25 [495]

Jeeves MA (1991) Stereo perception in callosal agenesis and partial callosotomy Neuropsychol **29** 19-34 [233]

Jeffery G (1984) Retinal ganglion cell death and terminal field retraction in the developing rodent visual system Devel Brain Res **13** 81-96 [440]

Jeffery G (1989) Shifting retinal maps in the development of the lateral geniculate nucleus Devel Brain Res **46** 187–96 [444]

Jeffery G (1990) The topographic relationship between shifting binocular maps in the developing dorsal lateral geniculate nucleus Exp Brain Res **82** 408–16 [440, 444]

Jeffery G (1997) The albino retina: an abnormality that provides insight into normal retinal development TINS **20** 165-9 [496]

Jeffery G, Perry VH (1982) Evidence for ganglion cell death during development of the ipsilateral retinal projection in the rat Devel Brain Res **2** 176–80 [442]

Jeffery G, Schütz G, Montoliu L (1994) Correction of abnormal retinal pathways found with albinism by introduction of a functional tyrosinase gene in transgenic mice Devel Biol **166**, 460-4 [495]

Jenkin MR, Jepson AD, Tsotsos JK (1991) Techniques for disparity measurement Comp Vis Im Proc: Unstand **53** 14-30 [265]

Jenkin MR, Jepson AD (1988) The measurement of binocular disparity In Computational processes in human vision (ed ZW Pylyshyn) pp 69-98 Ablex Publishing, Norwood NJ [252]

Jenkins TCA, Pickwell LD, Abd-Manan F (1992) Effect of induced fixation on binocular visual acuity Ophthal Physiol Opt **12** 299-301 [317]

Jenkins TCA, Abd-Manan F, Pardhan S, Murgatroyd RN (1994) Effect of fixation disparity on distance binocular visual acuity Ophthal Physiol Opt **14** 129-31 [317]

Jennings JAM, Charman WN (1981) Off-axis image quality in the human eye Vis Res **21** 445–55 [173]

Jennings WW, Vanet P (1952) New direct-vision stereo-projection screen J Motion Pict Televis Engin **59** 22-7 [72]

Jensen RJ, Devoe RD (1983) Comparisons of directionally sensitive with other ganglion cells of the turtle retina: intracellular recording and staining J Comp Neurol **217** 271-87 [203]

Jepson A, Richards W, Knill DC (1996) Model structure and reliable inference. In Perception as Bayesian inference (ed DC Knill, W Richards) pp 63-92 Cambridge, Cambridge University Press [102]

Jiang BC (1995) Parameters of accommodative and vergence systems and the development of late-onset myopia Invest Ophthal Vis Sci **36** 1737-42 [390, 435]

Jiang BC (1996) Accommodative vergence is driven by the phasic component of the accommodative controller Vis Res **36** 97-102 [391]

Jiang BC, Woessner WM (1996) Dark focus and dark vergence: an experimental verification of the configuration of the dual-interactive feedback model Ophthal Physiol Opt **16** 342-7 [390]

Jiang BC, Gish KW, Leibowitz HW (1991) Effect of luminance on the relation between accommodation and convergence Optom Vis Sci **68** 220-5 [359]

Jiao SL, Han C, Jing QC, Over R (1984) Monocular–contingent and binocular–contingent aftereffects Percept Psychophys **3 5** 105–10 [349]

Jiménez JR, Olivares JL, Pérez-Ocón F, del Barco LJ (2000a) Associated phoria in relation to stereopsis with random-dot stereograms Optom Visi Sci **77** 47-50 [381]

Johannsen DE (1930) A quantitative study of binocular color vision J Gen Psychol **4** 282–308 [283]

Johansson CB, Momma S, Clarke DL, Risling M, Lendahl U, Frisén J (1999) Identification of a neural stem cell in the adult mammalian central nervous system Cell **96** 25-34 [450]

Johansson G (1973) Visual perception of biological motion and a model for its analysis Percept Psychophys **14** 201-11 [145]

Johnson CA (1976) Effects of luminance and stimulus distance on accommodation and visual resolution J Opt Soc Am **66** 138-42 [359, 361]

Johnson CA, Post RB, Chalupa LM, Lee TJ (1982) Monocular deprivation in humans: a study of identical twins Invest Ophthal Vis Sci **23** 135-8 [509]

Johnson RR, Burkhalter A (1997) A polysynaptic feedback circuit in rat visual cortex J Neurosci **17** 7129-40 [187]

Jones DG, van Sluyters RC, Murphy KM (1991) A computational model for the overall pattern of ocular dominance J Neurosci **11** 3794-808 [214]

Jones JP, Palmer LA (1987) An evaluation of the two-dimensional Gabor filter model of simple receptive fields in cat striate cortex J Neurophysiol **58** 1233-58 [141]

Jones KR, Kalil RE, Spear PD (1984a) Effects of strabismus on responsivity spatial resolution and contrast sensitivity of cat lateral geniculate neurons J Neurophysiol **52** 538–52 [504]

Jones KR, Spear PD, Tong L (1984b) Critical periods for effects of monocular deprivation: differences between striate and extrastriate cortex J Neurosci **4** 2543–52 [516]

Jones LS (1996) Integrins: possible functions in the adult CNS TINS **19** 68-72 [438]

Jones R (1977) Anomalies of disparity detection in the human visual system J Physiol **294** 621–40 [404]

Jones R (1980) Fusional vergence: sustained and transient components Am J Optom Physiol Opt **57** 640–4 [400, 402]

Jones R, Kerr KE (1972) Vergence eye movements to pairs of disparity stimuli with shape selection cues Vis Res **12** 1425–30 [402]

Jones R, Stephens GL (1989) Horizontal fusional amplitudes Invest Ophthal Vis Sci **30** 1638–42 [392, 396]

Jones RK, Lee DN (1981) Why two eyes are better than one: the two views of binocular vision J Exp Psychol HPP **7** 30–40 [4]

Jordan M Luthardt G, Meyer-Naujoks C, Roth G (1980) The role of eye accommodation in the depth perception of common toads Z Naturforsch **35c** 851-2 [542]

Joshua DE, Bishop PO (1970) Binocular single vision and depth discrimination Receptive field disparities for central and peripheral vision and binocular interaction on peripheral single units in cat striate cortex Exp Brain Res **10** 389–416 [238, 253]

Joynson RB (1971) Michotte's experimental methods Brit J Psychol **62** 293-302 [159]

Judd CH (1907) Photographic records of convergence and divergence Psychol Rev Psychol Monogr **8** 370-423 [377]

Judge AW (1950) Stereoscopic photography Chapman Hall, London [70]

Judge SJ (1985) Can current models of accommodation and vergence control account for the discrepancies between AC/A measurements made by the fixation disparity and phoria methods Vis Res **25** 1999–2001 [389]

Judge SJ (1987) Optically–induced changes in tonic vergence and AC/A ratio in normal monkeys and monkeys with lesions of the flocculus and ventral paraflocculus Exp Brain Res **66** 1–9 [389]

Judge SJ (1991) Vergence In Vision and visual dysfunction Vol 8 Eye movements (ed RHS Carpenter) pp 157–72 MacMillan, London [368]

Judge SJ, Cumming BG (1986) Neurons in monkey midbrain with activity related to vergence eye movement and accommodation J Neurophysiol **55** 915–30 [365, 430, 431]

Judge SJ, Miles FA (1985) Changes in the coupling between accommodation and vergence eye movements induced in human subjects by altering the effective interocular distance Perception **14** 617–29 [389]

Julesz B (1971) Foundations of cyclopean perception University of Chicago Press, Chicago [8, 82, 86, 532]

Julesz B, Miller JE (1975) Independent spatial frequency tuned channels in binocular fusion and rivalry Perception **4** 125–43 [306, 320]

Julesz B, Tyler CW (1976) Neurontropy an entropy–like measure of neural correlation in binocular fusion and rivalry Biol Cyber **22** 107–19 [262]

Julesz B, Breitmeyer B, Kropfl W (1976a) Binocular-disparity–dependent upper–lower hemifield anisotropy and left–right hemifield isotropy as revealed by dynamic random–dot stereograms Perception **5** 129–41 [489]

Julesz B, Petrig B, Buttner U (1976b) Fast determination of stereopsis in rhesus monkey using dynamic random dot stereograms J Opt Soc Am **66** 1090 [548]

Julesz B, Kropfl W, Petrig B (1980) Large evoked potentials to dynamic random–dot correlograms permit quick determination of stereopsis Proc Nat Acad Sci **77** 2348–51 [262]

Kaas JH, Guillery RW (1973) The transfer of abnormal visual field representations from the dorsal lateral geniculate nucleus to the visual cortex in Siamese cats Brain Res **59** 61–95 [496, 497]

Kaas JH, Guillery RW, Allman JM (1972) Some principles of organization in the dorsal lateral geniculate nucleus Brain Behav Evol **6** 253–99 [177, 178]

Kaas JH, Harting JK, Guillery RW (1974) Representation of the compete retina in the contralateral superior colliculus of some animals Brain Res **65** 343-6 [174]

Kaas JH, Lin CS, Casagrande VA (1976) The relay of ipsilateral and contralateral retinal inputs from the lateral geniculate nucleus to striate cortex in the owl monkey: a transneuronal transport study Brain Res **106** 371-8 [217]

Kaczmarek L, Kossut M, Skangiel-Kramska J (1997) Glutamate receptors in cortical plasticity: molecular and cellular biology Physiol Rev **77** 217-55 [191, 461]

Kaernbach C, Schröger E, Jacobsen T, Roeber U (1999) Effects of consciousness on human brain waves following binocular rivalry Neuroreport **10** 713-6 [312]

Kahn JI, Foster DH (1981) Visual comparison of rotated and reflected random-dot patterns as a function of their positional symmetry and separation in the field Quart J Exp Psychol **33A** 155-66 [102]

Kahneman D (1968) Methods findings and theory in studies of visual masking Psychol Bull **70** 693–7 [341]

Kaiser P, Boynton RM (1985) Role of the blue mechanism in wavelength discrimination Vis Res **25** 523-9 [290]

Kalarickal GJ, Marshall JA (1999) Models of receptive-field dynamics in visual cortex Vis Neurosci **16** 1055-81 [198]

Kalil RE (1980) A quantitative study of the effects of monocular enucleation and deprivation on cell growth in the dorsal lateral geniculate nucleus of the cat J Comp Neurol **189** 483-527 [503]

Kalil RE (1990) The influence of action potentials on the development of the central visual pathway in mammals J Exp Biol **153** 291–76 [444]

Kalil RE, Jhaveri SR, Richards W (1971) Anomalous retinal pathways in the Siamese cat: an inadequate substrate for normal binocular vision Science **174** 302-5 [496]

Kalil RE, Spear PD, Langsetmo A (1984) Response properties of striate cortex neurons in cats raised with divergent or convergent strabismus J Neurophysiol **52** 514-37 [522]

Kaminska B, Kaczmarek L, Chaudhuri A (1996) Visual stimulation regulates the expression of transcription factors and modulates the composition of AP-1 in visual cortex J Neurosci **16** 3968-78 [458]

Kaminska B, Kaczmareki L, Chaudhuri A (1997) Activity-dependent regulation of cytochrome b gene expression in monkey visual cortex J Comp Neurol **379** 271-82 [510]

Kandler K, Katz LC (1998) Coordination of neuronal activity in developing visual cortex by gap junction-mediated biochemical communication J Neurosci **18** 1419-27 [459]

Kanizsa G (1979) Organization in vision: Essays on Gestalt perception Praeger, New York [123]

Kapadia MK, Gilbert CD, Westheimer G (1994) A quantitative measure for short–term cortical plasticity in human vision J Neurosci **14** 451-7 [198]

Kapadia MK, Ito M, Gilbert CD, Westheimer G (1995) Improvement in visual sensitivity by changes in local context: parallel studies in human observers and in V1 of alert monkeys Neuron **15** 843-56 [124, 196]

Kapadia MK, Westheimer G, Gilbert CD (2000) Spatial distribution of contextual interactions in primary visual cortex and in visual perception J Neurophysiol **84** 2048-62 [196, 197]

Kaplan D, Glass L (1995) Understanding nonlinear dynamics Springer, New York [100]

Kaplan E, Shapley RM (1986) The primate retina contains two types of ganglion cells with high and low contrast sensitivity Proc Nat Acad Sci **83** 2755-7 [177]

Kaplan IT, Metlay W (1964) Light intensity and binocular rivalry J Exp Psychol **67** 22-6 [289]

Kapoula Z, Hain TC, Zee DS, Robinson DA (1987) Adaptive changes in post–saccadic drift induced by patching one eye Vis Res **27** 1299–307 [422]

Kapoula Z, Optican LM, Robinson DA (1990) Retinal image motion alone does not control disconjugate postsaccadic eye drift J Neurophysiol **63** 1000-9 [424]

Kapoula Z, Eggert T, Bucci MP (1995) Immediate saccade amplitude disconjugacy induced by unequal images Vis Res **35** 3505-18 [422]

Kapoula Z, Eggert T, Bucci MP (1996a) Disconjugate adaptation of the vertical oculomotor system Vis Res **36** 2735-45 [424]

Kapoula Z, Bucci MP, Eggert T, Zamfirescu F (1996b) Fast disconjugate adaptations of saccades in microstrabismic subjects Vis Res **36** 103-8 [425]

Kapoula Z, Bucci MP, Eggert T, Garraud L (1997) Impairment of the binocular coordination of saccades in strabismus Vis Res **37** 2757-66 [422]

Kapoula Z, Bucci MP, Lavigne-Tomps F, Zamfirescu F (1998) Disconjugate memory-guided saccades to disparate targets: evidence for 3D sensitivity Exp Brain Res **122** 413-23 [425]

Kapoula Z, Bernotas M, Haslwanter T (1999) Listing's plane rotation with convergence: role of disparity, accommodation, and depth perception Exp Brain Res **129** 175-86 [370]

Kapoula Z, Bucci MP, Bernotas M, Zamfirescu F (2000) Motor execution is necessary to memorize disparity Exp Brain Res **131** 500-10 [425]

Kapral R, Showalter K (1995) Chemical waves and patterns Kluwer Norwell MA [212]

Karnath HO (2001) New insights into the functions of the superior temporal cortex Nature Rev Neurosci **2** 569-76 [225]

Karnath HO, Ferber S, Himmelbach M, (2001) Spatial awareness is a function of the temporal not the posterior parietal lobe Nature **411** 950-3 [225]

Karten HJ, Hodos W, Nauta WJH, Revzin AM (1973) Neural connections of the "visual Wulst" of the avian telencephalon Experimental studies in the pigeon (*Columbia livia*) and owl (*Speotyto cunicularia*) J Comp Neurol **150** 253–77 [545]

Kasai T, Morotomi T (2001) Event-related potentials during selective attention to depth and form in global stereopsis Vis Res **41** 1379-88 [263]

Kasamatsu T (1991) Adrenergic regulation of visuocortical plasticity: a role of the locus coeruleus system. Prog Brain Res **88** 599-616 [513]

Kasamatsu T, Pettigrew JD (1979) Preservation of binocularity after monocular deprivation in the striate cortex of kittens treated with 6–hydroxydopamine J Comp Neurol **185** 139–61 [512]

Kasamatsu T, Pettigrew JD, Ary M (1979) Restoration of visual cortical plasticity by local microperfusion of norepinephrine J Comp Neurol **185** 163–82 [513]

Kasamatsu T, Pettigrew JD, Ary M (1981) Cortical recovery from effects of monocular deprivation: acceleration with norepinephrine and suppression with 6-hydroxydopamine J Neurophysiol **45** 254-66 [512]

Kasamatsu T, Watabe K, Heggelund P, Schöller E (1985) Plasticity in cat visual cortex restored by electrical stimulation of the locus coeruleus Neurosci Res **2** 365–86 [513]

Kasamatsu T, Kitano M, Sutter EE, Norcia AM (1998a) Lack of lateral inhibitory interactions in visual cortex of monocularly deprived cats Vis Res **38** 1-12 [507]

Kasamatsu T, Imamura K, Mataga N, Hartveit E, Heggelund U, Heggelund P (1998b) Roles of N-methyl-D-aspartate receptors in ocular dominance plasticity in developing visual cortex: re-evaluation Neuroscience **82** 687-700 [511]

Kasamatsu T, Polat U, Pettet MW, Norcia AM (2001) Colinear facilitation promotes reliability of single-cell responses in cat striate cortex Exp Brain Res **138** 2, 163-72 [117, 196]

Kastner S, Ungerleider LG (2000) Mechanisms of visual attention in the human cortex Ann Rev Neurosci **23** 315-41 [208]

Kastner S, de Weerd P, Desimone R, Ungerleider LG (1998) Mechanisms of directed attention in the human extrastriate cortex as revealed by functional MRI Science **282** 108-11 [182, 207]

Kato H, Bishop PO, Orban GA (1981) Binocular interaction on monocularly discharged lateral geniculate and striate neurons in the cat J Neurophysiol **46** 932–51 [178, 216]

Katsumi O, Tsuyoshi T, Hirose T (1986) Effect of aniseikonia on binocular function Invest Ophthal Vis Sci **27** 601–4 [260]

Katz LC, Callaway EM (1992) Development of local circuits in mammalian visual cortex Ann Rev Neurosci **15** 31–56 [456]

Katz LC, Gilbert CD, Wiesel TN (1989) Local circuits and ocular dominance columns in monkey striate cortex J Neurosci **9** 1389–99 [186]

Katz LM, Levi DM, Bedell, HE (1984) Central and peripheral contrast sensitivity in amblyopia with varying field size Doc Ophthal **58** 351-73 [522]

Katz PS, Frost WN (1996) Intrinsic neuromodulation: altering neuronal circuits from within TINS **19** 54-61 [114]

Kaufman L (1963) On the spread of suppression and binocular rivalry Vis Res **3** 401–15 [293, 294, 297]

Kaufman L (1964) Suppression and fusion in viewing complex stereograms Am J Psychol **77** 193–205 [305, 306]

Kaufman L (1974) Sight and mind An introduction to visual perception Oxford University Press, London [282]

Kaufmann R, Maland J, Yonas A (1981) Sensitivity of 5- and 7-month-old infants to pictorial depth information J Exp Child Psychol **32** 162-8 [481]

Kaufmann-Hayoz R, Kaufmann F, Stucki M (1986) Kinetic contours in infants' visual perception Child Devel **57** 292-9 [479]

Kavadellas A, Held R (1977) Monocularity of color–contingent tilt aftereffects Percept Psychophys **21** 12–14 [352]

Kaye M, Mitchell DE, Cynader M (1982) Depth perception eye alignment and cortical ocular dominance of dark–reared cats Devel Brain Res **2** 37–53 [508]

Keating MJ, Feldman JD (1975) Visual deprivation and intertectal neuronal connexions in *Xenopus laevis* Proc R Soc B **191** 467–74 [541]

Keating MJ, Gaze RM (1970a) The ipsilateral retinotectal pathway in the frog Quart J Exp Physiol **55** 284–92 [540]

Keating MJ, Gaze RM (1970b) The depth distribution of visual units in the contralateral optic tectum following regeneration of the optic nerve in the frog Brain Res **21** 197-206 [541]

Keating MJ, Kennard C (1976) Binocular visual neurones in the frog thalamus Proc Physiol Soc **258** 69P [540]

Keating MJ, Dawes EA, Grant S (1992) Plasticity of binocular visual connections in the frog *Xenopus Laevis*: reversibility of effects of early visual deprivation Exp Brain Res **90** 121–8 [541]

Keck MJ, Price RL (1982) Interocular transfer of the motion aftereffect in strabismus Vis Res **22** 55–60 [493]

Keele KD (1955) Leonardo da Vinci on vision Proc R Soc Med **48** 384–90 [19, 20, 38, 52]

Keenan JM, Willshaw HE (1992) Outcome of strabismus surgery in congenital esotropia Br J Ophthal **76** 342-5 [374]

Keller EL (1973) Accommodative vergence in the alert monkey Motor unit analysis Vis Res **13** 1565–75 [430]

Keller EL (1989) The cerebellum In The neurobiology of saccadic eye movements (ed RH Wurtz, ME Goldberg) pp 391-411 Elsevier, Amsterdam [431]

Keller EL, Robinson DA (1972) Abducens unit behaviour in the monkey during vergence eye movements Vis Res **12** 369–82 [429, 430]

Kelley WVD (1924) Stereoscopic pictures Tr Soc Motion Pict Engin **17** 149-52 [72]

Kellman PJ (1984) Perception of three-dimensional form by human infants Percept Psychophys **36** 353-8 [480]

Kellman PJ, Short KR (1987) Development of three-dimensional form perception J Exp Psychol: HPP **13** 545-57 [480]

Kelly SA, Chino YM, Cotter SA, Knuth J (1997) Orientation anisotropy and strabismus Vis Res **37** 151-63 [522]

Kemp M (1978) Science, non-science and nonsense: the interpretation of Brunelleschi's perspective Art History 1 134-61 [31]

Kemp M (1990) The science of art Yale University Press, New Haven [40]

Kennedy H, Bullier J, Dehay C (1985) Cytochrome oxidase activity in the striate cortex and lateral geniculate nucleus of the newborn and adult macaque monkey Exp Brain Res 6 1 204–9 [455]

Kennedy H, Dehay C, Bullier J (1986) Organization of the callosal connections of visual areas V1 and V2 in the macaque monkey J Comp Neurol 277 398–415 [232]

Kennedy H, Meissirel C, Dehay C (1991) Callosal pathways and their compliancy to general rules governing the organization of corticocortical connectivity In Neuroanatomy of the visual pathways and their development (ed B Dreher, SR Robinson) pp 327–59 CRC Press, Boston [232]

Kenyon RV, Ciuffreda KJ, Stark L (1978) Binocular eye movements during accommodative vergence Vis Res 18 545-55 [421]

Kenyon RV, Ciuffreda KJ, Stark L (1980a) Dynamic vergence eye movements in strabismus and amblyopia: symmetric vergence Invest Ophthal Vis Sci 19 60–74 [390]

Kenyon RV, Ciuffreda KJ, Stark L (1980b) Unequal saccades during vergence Am J Optom Physiol Opt 57 586–94 [421, 423]

Kenyon RV, Ciuffreda KJ, Stark L (1981) Dynamic vergence eye movements in strabismus and amblyopia: asymmetric vergence Br J Ophthal 65 167–76 [527]

Kepler J (1604) Ad vitellionem paralipomena Marnium and Haer, Frankfurt [23]

Kepler J (1611) Dioptrice Vindelicorum, Augsburg [23]

Kerr KE (1998) Anomalous correspondence—the cause or consequence of strabismus Optom Vis Sci 75 17-22 [374]

Kersten D, Legge GE (1983) Convergence accommodation J Opt Soc Am 73 332–8 [390]

Kertesz AE (1972) The effect of stimulus complexity on human cyclofusional response Vis Res 12 699–704 [411, 413]

Kertesz AE (1973) Disparity detection within Panum's fusional areas Vis Res 13 1537–43 [278]

Kertesz AE (1981) Effect of stimulus size on fusion and vergence J Opt Soc Am 71 289-93 [278, 392, 407]

Kertesz AE, Lee HJ (1987) Comparison of simultaneously obtained objective and subjective measurements of fixation disparity Am J Optom Physiol Opt 64 734–8 [379]

Kertesz AE, Sullivan MJ (1978) The effect of stimulus size on human cyclofusional response Vis Res 18 567–71 [411, 413, 417]

Kertesz AE, Hampton DR, Sabrin HW (1983) The unreliability of nonius line estimates of vertical fusional vergence performance Vis Res 23 295-97 [378, 407]

Khan AZ, Crawford JD (2001) Ocular dominance reverses as a function of horizontal gaze angle Vis Res 41 1743-8 [295]

Kicliter E, Northcutt RG (1975) Ascending afferents to telencephalon of ranid frogs: an anterograde degeneration study J Comp Neurol 161 239–54 [540]

Kidd T, Bland KS, Goodman CS (1999) Slit is the midline repellent for the Robo receptor in *Drosophila* Cell 96 785-94 [440]

Kim DS, Bonhoeffer T (1994) Reverse occlusion leads to a precise restoration of orientation preference maps in visual cortex Nature 370 370-2 [517]

Kim DS, Duong RQ, Kim SG (2000) High resolution mapping of iso-orientation columns by fMRI Nature Neurosci 3 164-8 [183]

Kim JN, Shadlen MN (1998) Neural correlates of a decision in the dorsolateral prefrontal cortex of the macaque Nature Neurosci 2 176-85 [228]

Kimura D, Durnford M (1974) Normal studies on the function of the right hemisphere in vision In Hemisphere function in the human brain (ed SJ Dimond, JG Beaumont) pp 25-47 Elek, London [489]

Kimura F, Nishigori A, Shirokawa T, Tsumoto T (1989) Long–term potentiation and N–methyl–D–aspartate receptors in the visual cortex of young rats J Physiol 414 125–44 [461]

King SM, Cowey A (1992) Defensive responses to looming visual stimuli in monkeys with unilateral striate cortex ablation Neuropsychol 30 1017-27 [183, 257]

King WM, Zhou W (1995) Initiation of disjunctive smooth pursuit in monkeys: evidence that Hering's law of equal innervation in not obeyed by the smooth pursuit system Vis Res 35 3389-400 [419]

King WM, Zhou W, Tomlinson RD, McConville KMV, Page WK, Paige GD, Maxwell JS (1994) Eye position signals in the abducens and oculomotor nuclei of monkeys during ocular convergence J Vestib Res 4 401-8 [429]

Kingdom FAA (1997) Simultaneous contrast: the legacies of Hering and Helmholtz Perception 29 673-7 [44]

Kinnear PE, Jay B, Witcop CJ (1985) Albinism Survey Ophthal 30 75-101 [495]

Kiorpes L (1992) Effect of strabismus on the development of vernier acuity and grating acuity in monkeys Vis Neurosci 9 253–9 [522]

Kiorpes L, Boothe RG (1980) The time course for the development of strabismus amblyopia in infant monkeys (*Macaca nemestrina*) Invest Ophthal Vis Sci 19 841-5 [527]

Kiorpes L, Wallman J (1995) Does experimentally-induced myopia cause hyperopia in monkeys? Vis Res 35 1289-97 [435]

Kiorpes L, Kipper DC, Movshon JA (1993) Contrast sensitivity and vernier acuity in amblyopic monkeys Vis Res 33 2301–11 [522]

Kiorpes L, Walton PJ, O'Toole LP, Movshon JA, Lisberger SG (1996) Effects of early-onset strabismus on pursuit eye movements and on neuronal responses in area MT of macaque monkeys J Neurosci 16 6537-53 [526]

Kiorpes L, Kiper DC, O'Keefe LP, Cavanaugh JR, Movshon JA (1998) Neuronal correlates of amblyopia in the visual cortex of macaque monkeys with experimental strabismus and anisometropia J Neurosci 18 6411-27 [510]

Kiorpes L, Tang C, Movshon JA (1999) Factors limiting contrast sensitivity in experimentally amblyopic macaque monkeys Vis Res 39 4152-60 [519]

Kiper DC (1994) Spatial phase discrimination in monkeys with experimental strabismus Vis Res 34 437–47 [522]

Kiper DC, Gegenfurtner KR, Movshon JA (1996) Cortical oscillatory responses do not affect visual segmentation Vis Res 36 539-44 [133]

Kirby MA, Steineke TC (1992) Morphogenesis of retinal ganglion cells during formation of the fovea in the Rhesus macaque Vis Neurosci 9 603-16 [436]

Kirk DL, Levick WR, Cleland BG, Wässle H (1976a) Crossed and uncrossed representation of the visual field by brisk–sustained and brisk–transient cat retinal ganglion cells Vis Res 16 225–31 [231]

Kirk DL, Levick WR, Cleland BG (1976b) The crossed or uncrossed destination of axons of sluggish–concentric and non–concentric cat retinal ganglion cells with an overall synthesis of the visual field representation Vis Res 16 233–6 [231]

Kirkwood A, Bear MF (1994a) Homosynaptic long-term depression in the visual cortex J Neurosci 14 3404-12 [460]

Kirkwood A, Bear MF (1994b) Hebbian synapses in visual cortex J Neurosci 14 1634-45 [461]

Kirkwood A, Dudek SM, Gold JT, Aizenman CD, Bear MF (1993) Common forms of synaptic plasticity in the hippocampus and neocortex in vitro Science 290 1518–21 [461]

Kirkwood A Lee HK, Bear MF (1995) Co-regulation of long-term potentiation and experience dependent synaptic plasticity in visual cortex by age and experience Nature **375** 328-31 [462]

Kirkwood A, Rioult MG, Bear MF (1996) Experience-dependent modification of synaptic plasticity in visual cortex Nature **381** 529-8 [460]

Kirmse W, Kirmse R, Milev E (1994) Visuomotor operation in transition from object fixation to prey shooting in chameleons Biol Cyber **71** 209-14 [544]

Kirsten C (1986) Dokumente einer Freundschaft. Briefwechsel zwischen Hermann von Helmholtz und Emil du Bois-Reymond 1846-1894. Akademie Verlag, Berlin [46]

Kitaoji H, Toyama K (1987) Preservation of position and motion stereopsis in strabismic subjects Invest Ophthal Vis Sci **28** 1290–67 [529]

Kitterle FL, Thomas J (1980) The effects of spatial frequency orientation and color upon binocular rivalry and monocular pattern alternation Bull Psychonom Soc **16** 405–7 [295]

Kitterle FL, Kaye RS, Nixon H (1974) Pattern alternation: effects of spatial frequency and orientation Percept Psychophys **16** 543-6 [295]

Klein R (1961) Pomponius Gauricus on perspective Art Bulletin **43** 211-30 [31]

Klein R (1977) Stereopsis and the representation of space Perception **6** 327–32 [5]

Klein SA, Levi DM (1985) Hyperacuity thresholds of 1.0 second: theoretical predictions and empirical validation J Opt Soc Am A **2** 1170–90 [81, 87, 142]

Klein SA, Levi DM (1986) Local multipoles for measuring contrast and phase sensitivity Invest Ophthal Vis Sci **27** (Abs) 225 [142]

Klein SA, Casson E, Carney T (1990) Vernier acuity as line and dipole detection Vis Res **30** 1703–19 [142]

Kleinschmidt A, Bear MF, Singer W (1987) Blockade of "NMDA" receptors disrupts experience dependent plasticity of kitten striate cortex Science **238** 355-7 [504]

Kleinschmidt A, Büchel C, Zeki S, Frackowiak RSJ (1998) Human brain activity during spontaneously reversing perception of ambiguous figures Proc R Soc B **295** 2727-33 [162, 182]

Knierim JJ, Van Essen DC (1992) Neuronal responses to static texture patterns in area V1 of the alert monkey J Neurophysiol **67** 961–80 [197, 200, 206]

Knill DC, Richards W (1996) Perception as Bayesian inference Cambridge, Cambridge University Press [102]

Knorr WR (1992) When circles don't look like circles: An optical theorem in Euclid and Pappus Arch Hist Exact Sci **44** 287-328 [28]

Knudsen EI (1982) Auditory and visual maps of space in the optic tectum of the owl J Neurosci **2** 1177–94 [546]

Knudsen EI (1989) Fused binocular vision is required for development of proper eye alignment in barn owls Vis Neurosci **2** 35–40 [547]

Koch C, Poggio T (1992) Multiplying with synapses and neurones In Single neuron computation (ed T McKenna, J Davis, SF Zornetzer) pp 315-45 Academic Press, New York [128]

Koch C, Ullman S (1985) Shifts in selective visual attention: towards the underlying neural circuitry Hum Neurobiol **4** 219–27 [162]

Koch C, Marroquin J, Yuille A (1986) Analog "neuronal" networks in early vision Proc Nat Acad Sci **83** 4293–7 [114]

Koenderink JJ (1982) Different concepts of "ray" in optics: link between resolving power and radiometry Am J Physics **50** 1012-15 [11]

Koenderink JJ (1990) Solid shape MIT Press, Cambridge Mass [109, 141]

Koenderink JJ, van Doorn AJ, Lappin JS (2000) Direct measurement of the curvature of visual space Perception **29** 69-79 [138]

Koenigsberger L (1902) Hermann von Helmholtz. F Vieweg und Sohn, Braunschweig [42]

Koestler A (1960) The Watershead. A biography of Johannes Kepler Doubleday, New York [23]

Koffka K (1935) Principles of Gestalt psychology Harcourt Brace, New York [123]

Köhler W, Wallach H (1944) Figural aftereffects: an investigation of visual processes Proc Am Philos Soc **88** 299–357 [84]

Kohly RP, Regan D (2000) Coincidence detectors: visual processing of a pair of lines and implications for shape discrimination Vis Res **40** 2291-306 [145]

Kohn H (1960) Some personality variables associated with binocular rivalry Psychol Rec **10** 9–13 [309]

Kohonen T (1995) Self organization and associative memory New York, Springer [454]

Kojic L, Dyck RH, Douglas RM, Matsubara J, Cynader MS (2000) Columnar distribution of serotonin-dependent plasticity within kitten striate cortex Proc Nat Acad Sci **97** 1841-4 [462]

Koken PW, Erkelens CJ (1993) Simultaneous hand tracking does not affect human vergence pursuit Exp Brain Res **96** 494–500 [398]

Kolb FC, Braun J (1995) Blindsight in normal observers Nature **377** 336-8 [293]

Kolb H (1970) Organization of the outer plexiform layer of the primate retina: electron microscopy of Golgi–impregnated cells Philos Tr R Soc B **258** 291–83 [166, 169]

Kolers PA, Rosner BS (1960) On visual masking (metacontrast): dichoptic observation Am J Psychol **73** 2–21 [341]

Köllner H (1914) Das funktionelle Überwiegen der nasalen Netzhauthälften im gemeinschaftlichen Sehfeld Arch Augenheilk **76** 153–64 [292]

Komatsu H, Roy JP, Wurtz RH (1988) Binocular disparity sensitivity of cells in area MST of the monkey Soc Neurosci Abstr **14** 202 [256]

Kommerell G (1996) The relationship between infantile strabismus and latent nystagmus Eye **10** 274-81 [526]

Kommerell G, Mehdorn E (1982) Is an optokinetic defect the cause of congenital and latent nystagmus? In Functional basis of ocular motility disorders (ed G Lennerstrand DS Zee, EL Keller) pp 159-67 Pergamon, New York [373]

Kommerell G, Olivier D, Theopold H (1976) Adaptive programming of phasic and tonic components in saccadic eye movements Investigations in patients with abducens palsy Invest Ophthal **15** 657–60 [425]

Kommerell G, Ullrich D, Gilles U, Bach M (1995) Asymmetry of motion VEP in infantile strabismus and in central vestibular nystagmus Doc Ophthal **89** 373-81 [472]

Kondo T, Raff M (2000) Oligodendrocyte precursor cells reprogrammed to become multipotential CNC stem cells Science **289** 1754-6 [448]

König P, Schillen TB (1991) Stimulus–dependent assembly formation of oscillatory responses: I Synchronization Neural Comput **3** 155–66 [136]

König P, Engel AK, Löwel S, Singer W (1993) Squint affects synchronization of oscillatory responses in cat visual cortex Eur J Neurosci **5** 501-8 [506]

König P, Engel AK, Roelfsema PR, Singer W (1995) How precise is neuronal synchronization? Neural Comput **7** 469-85 [131]

Konishi M (1993) Listening with two ears Sci Am **298** 66–73 [546]

Kontsevich LL, Tyler CW (2000) Relative contributions of sustained and transient pathways to human stereoprocessing Vis Res **40** 3275-55 [245]

Kooi FL, Toet A, Tripathy SP, Levi DM (1994) The effect of similarity and attention on contour interaction in peripheral vision Spatial Vis 8 255-79 [197, 339]

Korenberg MJ, Hunter IW (1986) The identification of nonlinear biological systems: LNL cascade models Biol Cybern 55 125-34 [95]

Kori AA, Schmid-Priscoveanu A, Straumann D (2001) Vertical divergence and counterroll eye movements evoked by whole-body position steps about the roll axis of the head in humans J Neurophysiol 85 671-8 [410]

Kornack DR, Radic P (1995) Radial and horizontal deployment of clonally related cells in the primate neocortex: relationship to distinct mitotic lineages Neuron 15 311-21 [450]

Korr H, Schmitz C (1999) Facts and fictions regarding post-natal neurogenesis in the developing human cerebral cortex J Theor Biol 200 291-7 [446, 450]

Korsching S (1993) The neurotrophic factor: a reexamination J Neurosci 13 2739–48 [442, 456]

Kossel A, Löwel S, Bolz J (1995) Relationships between dendritic fields and functional architecture in striate cortex of normal and visually deprived cats J Neurosci 15 3913-29 [506]

Kotulak JC, Morse SE (1995) The effect of perceived distance on accommodation under binocular steady-state conditions Vis Res 35 791-5 [360]

Kotulak JC, Schor CM (1986a) The dissociability of accommodation from vergence in the dark Invest Ophthal Vis Sci 27 544–51 [387]

Kotulak JC, Schor CM (1986b) Temporal variations in accommodation during steady-state conditions J Opt Soc Am A 3 223-7 [362]

Kotulak JC, Schor CM (1986c) The accommodative response to subthreshold blur and to perceptual fading during the Troxler phenomenon Perception 15 7-15 [360]

Kotulak JC, Schor CM (1986d) A computational model of the error detector of human visual accommodation Biol Cybern 54 189-94 [363]

Kotulak JC, Schor CM (1987) The effects of optical vergence, contrast, and luminance on the accommodative response to spatially bandpass filtered targets Vis Res 27 1797-806 [361]

Kourtzi Z, Kanwisher N (2001) Representation of perceived object shape by the human lateral occipital complex Science 293 1506-9 [206]

Kouyama N, Marshak DW (1992) Bipolar cells specific for blue cones in the macaque retina J Neurosci 12 1233-52 [166]

Kovács G, Vogels R, Orban GA (1995) Selectivity of macaque inferior temporal neurons for partially occluded shapes J Neurosci 15 1984-97 [124]

Kovács I, Papathomas TV, Yang M, Fehér A (1996) When the brain changes its mind: interocular grouping during binocular rivalry Proc Nat Acad Sci 93 15508-11 [299]

Kovács I, Polat U, Pennefather PM, Chandna A, Norcia AM (2000) A new test of contour integration deficits in patients with a history of disrupted binocular experience during visual development Vis Res 40 1775-83 [524]

Kran BS, Ciuffreda KJ (1988) Noncongruent stimuli and tonic adaptation Am J Optom Physiol Opt 65 703-9 [383]

Krapp HG, Hengstenberg R, Egelhaaf M (2001) Binocular contributions to optic flow processing in the fly visual system J Neurophysiol 85 724-34 [535]

Kratz KE, Lehmkuhle S (1983) Spatial contrast sensitivity of monocularly deprived cats after removal of the non–deprived eye Behav Brain Res 7 291–6 [520]

Kratz KE, Spear PD, Smith DC (1976) Postcritical–period reversal of effects of monocular deprivation on striate cortex cells of the cat J Neurophysiol 39 501–11 [508]

Krautheimer R, Krautheimer-Hess T (1982) Lorenzo Ghiberti Princeton University Press, Princeton, NJ [32, 33]

Kreiman G, Koch C, Fried I (2000) Category-specific visual responses of single neurons in the human medial temporal lobe Nature Neurosci 3 946-53 [223]

Kreiter AK, Singer W (1996) Stimulus-dependent synchronization of neuronal responses in the visual cortex of the awake macaque monkey J Neurosci 16 2381-96 [133]

Krekling S (1973a) Comments on cyclofusional eye movements Graefes Arch klin exp Ophthal 188 231–8 [411]

Krekling S, Blika S (1983b) Meridional anisotropia in cyclofusion Percept Psychophys 34 299-300 [279]

Kremenitzer JP, Vaughan HG, Kurzberg D, Dowling K (1979) Smooth-pursuit eye movements in the newborn infant Child Devel 50 442-8 [475]

Krimsky E (1972) The corneal light reflex Charles Thomas Springfield Illinois [375]

Krishnan VV, Stark L (1983) Model of the disparity vergence system In Vergence eye movements: Basic and clinical aspects (ed MC Schor, KJ Ciuffreda) pp 349–72 Butterworth, Boston [404]

Krishnan VV, Phillips S, Stark L (1973a) Frequency analysis of accommodation accommodative vergence and disparity vergence Vis Res 13 1545–54 [397, 400]

Krishnan VV, Farazian F, Stark L (1973b) An analysis of latencies and prediction in the fusional vergence system Internat J Optom 50 933–9 [397]

Krishnan VV, Shirachi D, Stark L (1977) Dynamic measures of vergence accommodation Am J Optom Physiol Opt 54 470-3 [358]

Kritzer MF, Cowey A, Somogyi P (1992) Patterns of inter- and intralaminar GABAergic connections distinguish striate (V1) and extrastriate (V2 V4) visual cortices and their functionally specialized subdivisions in the rhesus monkey J Neurosci 12 4545-64 [188, 196]

Krommenhoek KP, Van Gisbergen JAM (1994) Evidence for nonretinal feedback in combined version-vergence movements Exp Brain Res 102 95-109 [432]

Krubitzer LA, Kaas JH (1990) Cortical connections of MT in four species of primates: areal modular and retinotopic patterns Vis Neurosci 5 165–204 [223]

Krug K, Akerman CJ, Thompson ID (2001) Responses of neurons in neonatal cortex and thalamus to patterned visual stimulation through the naturally closed lids J Neurophysiol 851436-43 [454]

Krüger K, Kiefer W, Groh A, Dinse HR, von Seelen W (1993) The role of the lateral suprasylvian visual cortex of the cat in object–background interactions: permanent deficits following lesions Exp Brain Res 97 40–60 [224]

Kruger K, Tam AS, Lu C, Sretavan DW (1998) Retinal ganglion cell axon progression from the optic chiasm to initiate optic tract development requires cell autonomous function of GAP-43 J Neurosci 18 5692-705 [441]

Kruger PB, Pola J (1986) Stimuli for accommodation: blur chromatic aberration and size Vis Res 29 957–71 [364]

Kruger PB, Pola J (1987) Dioptric and non-dioptric stimuli for accommodation: target size alone and with blur and chromatic aberration Vis Res 27 555-67 [360]

Kruger PB, Nowbotsing S, Aggarwala KR, Mathews S (1995a) Small amounts of chromatic aberration influence dynamic accommodation Optom Vis Sci 72 656-66 [364]

Kruger PB, Mathews S, Aggarwala KR, Yager D, Kruger ES (1995b) Accommodation responds to changing contrast of long middle and short spectral-waveband components of the retinal image Vis Res 35 2715-29 [364]

Kruger PB, Mathews S, Katz M, Aggarwala KR, Nowbotsing S (1997) Accommodation without feedback suggests directional signals specify ocular focus Vis Res **37** 2511-29 [363, 364]

Krumholz DM, Fox RS, Ciuffreda KJ (1986) Short-term changes in tonic accommodation Invest Ophthal Vis Sci **27** 552-7 [359]

Kubová Z, Kuba M, Juran J, Blakemore C (1996) Is the motion system relatively spared in amblyopia? Evidence from cortical evoked responses Vis Res **36** 181-90 [525]

Kubovy M (1986) The psychology of perspective and Renaissance art Cambridge University Press, Cambridge [38]

Kuffler SW (1953) Discharge patterns and functional organization of mammalian retina J Neurophysiol **16** 37–68 [168]

Kulikowski JJ (1978) Limit of single vision in stereopsis depends on contour sharpness Nature **275** 129-7 [273]

Kulikowski JJ (1980) Processing of patterns by simple cells in the cat visual cortex Neurosci Lett Supplement S95 [140]

Kulikowski JJ (1992) Binocular chromatic rivalry and single vision Ophthal Physiol Opt **12** 168-70 [299]

Kuljis RO, Rakic P (1990) Hypercolumns in primate visual cortex can develop in the absence of cues from photoreceptors Proc Nat Acad Sci **87** 5303-6 [455]

Kumagami T, Zhang B, Smith EL, Chino YM (2000) Effect of onset age of strabismus on the binocular response properties of neurons in the monkey visual cortex Invest Ophthal Vis Sci **41** 948-54 [517]

Kunishima N, Shimada Y, Tsuji Y, Sato T, Yamamoto M, Kumasaka T, Nakanishi S, Jingami H, Morikawa K (2000) Structural basis of glutamate recognition by a dimeric metabotropic glutamate receptor Nature **407**, 971-7 [191]

Kuppermann BD, Kasamatsu T (1983) Changes in geniculate cell size following brief monocular blockade of retinal activity in kittens Nature **306** 465-8 [503]

Kuppermann BD, Kasamatsu T (1984) Enhanced binocular interaction in the visual cortex of normal kittens subjected to intracortical norepinephrine perfusion Brain Res **302** 91-9 [512]

Kwee IL, Fujii Y, Matsuzawa H, Nakada T (1999) Perceptual processing of stereopsis in humans: high-field (3.0-tesla) functional MRI study Neurology **53** 1599-601 [263]

Lachica EA, Beck, PD, Casagrande VA (1992) Parallel pathways in macaque monkey striate cortex: anatomically defined columns in layer III Proc Nat Acad Sci **89** 3566-70 [205]

Lack LC (1969) The effect of practice on binocular rivalry control Percept Psychophys **6** 397–400 [308]

Lack LC (1971) The role of accommodation in the control of rivalry Percept Psychophys **10** 38-42 [308]

Lack LC (1978) Selective attention and the control of binocular rivalry Mouton, New York [303]

Lagae L, Raiguel S, Orban GA (1993) Speed and direction selectivity of macaque middle temporal neurons J Neurophysiol **69** 19–39 [223]

Lagae L, Maes H, Raiguel S, Xiao DK, Orban GA (1994) Responses of macaque STS neurons to optic flow components: a comparison of areas MT and MST J Neurophysiol **71** 1597-29 [225]

Lagréze WD, Sireteanu R (1991) Two–dimensional spatial distortions in human strabismic amblyopia Vis Res **31** 1271–88 [522]

Lal R, Friedlander MJ (1990) Effect of passive eye position changes on retinogeniculate transmission in the cat J Neurophysiol **63** 502-22 [494]

Lam AKC, Chau ASY, Lam WY, Leung GYO, Man BSH (1996) Effects of naturally occurring visual acuity differences between two eyes in stereoacuity Ophthal Physiol Opt **16** 189-95 [484]

Lam DMK, Shatz CJ (1991) Development of the visual system MIT Press, Cambridge MA [485]

Lam K, Sefton AJ, Bennett MR (1982) Loss of axons from the optic nerve of the rat during early postnatal development Devel Brain Res **3** 487–91 [442]

Lamb TD (1987) Sources of noise in photoreceptor transduction J Opt Soc Am A **4** 2295-300 [166]

Lamme VAF (1995) The neurophysiology of figure-ground segregation in primary visual cortex J Neurosci **15** 1605-15 [206]

Lamme VAF, Spekreijse H (1998) Neuronal synchrony does not represent texture segregation Nature **396** 362-6 [133]

Lamme VAF, van Dijk BW, Spekreijse H (1993) Organization of texture segregation processing in primate visual cortex Vis Neurosci **10** 781–90 [131]

Land MF (1969) Movements of the retinae of jumping spiders (Salticidae: Dendryphantinae) in response to visual stimuli J Exp Biol **51** 471–93 [536]

Land MF (1988) The functions of eye and body movements in *Labidocera* and other copepods J Exp Biol **140** 381–91 [536]

Land MF (1995) Fast-focus telephoto eye Nature **373** 658-9 [544]

Land MF, Marshall JN, Brownless D, Cronin TW (1990) The eye–movements of the mantis shrimp *Odontodactylus scyllarus* (Crustacea: Stomatopoda) J Comp Physiol **167** 155–66 [537, 538]

Land PW, Lund RD (1979) Development of the rat's uncrossed retinotectal pathway and its relation to plasticity studies Science **205** 698-700 [444]

Lange–Malecki B, Creutzfeldt OD, Hinse P (1985) Haploscopic colour mixture with and without contours in subjects with normal and disturbed binocular vision Perception **14** 587–600 [284]

Lansford TG, Baker HD (1969) Dark adaptation: an interocular light–adaptation effect Science **164** 1307–9 [333]

Lansing RW (1964) Electroencephalographic correlates of binocular rivalry in man Science **146** 1325–7 [312]

Larkum ME, Zhu JJ, Sakmann B (1999) A new cellular mechanism for coupling inputs arriving at different cortical layers Nature **398** 338-41 , [114, 187]

Larsen JS (1971) The sagittal growth of the eye IV Ultrasonic measurement of the axial length of the eye from birth to puberty Acta Ophthal **49** 873–86 [434]

Larson WL (1982) A technique to measure accommodative convergence heterophoria and the AC/A during single binocular vision Am J Optom Physiol Opt **59** 111-15 [389]

Larson-Prior LJ, Ulinski PS, Slater NT (1991) Excitatory amino acid receptor-mediated transmission in geniculocortical and intracortical pathways within visual cortex J Neurophysiol **66** 293-306 [191]

Lashley KS, Russell JT (1934) The mechanism of vison. XI A preliminary test of innate organization J Genet Psychol **45** 136-44 [549]

Lau E (1921) Neue Untersuchungen über das Tiefen- und Ebenensehen Z Sinnesphysiol **53** 1-35 [377]

Law MI, Constantine-Paton M (1980) Right and left eye bands in frogs with unilateral tectal ablations Proc Nat Acad Sci **77** 2314-18 [464]

Law MI, Constantine-Paton M (1981) Anatomy and physiology of experimentally produced striped tecta J Neurosci **1** 741-59 [464]

Lawden MC, Hess RF, Campbell FW (1982) The discrimination of spatial phase relationships in amblyopia Vis Res **22** 1005–16 [522]

Lawler KA, Cowey A (1986) The effects of pretectal and superior collicular lesions on binocular vision Exp Brain Res **63** 402-8 [431]

Lawwill T, Biersdorf WR (1968) Binocular rivalry and visual evoked responses Invest Ophthal **7** 378-85 [312]

Le Clerc S (1679) Discours touchant de point de veue dans lequel il est prouvé que les chose qu'on voit distinctement ne sont veues que d'un oeil Jolly, Paris [56]

Le Conte J (1881) Sight Kegan Paul, London [412]
Le Gros Clark WE (1970) History of the primates British Museum (Natural History), London [532]
LeDoux JE, Deutsch G, Wilson DH, Gazzaniga MS (1977) Binocular stereopsis and the anterior commissure in man The Physiologist 20 55 [233]
Lee BB (1996) Receptive field structure in the primate retina Vis Res 36 631-44 [169, 170]
Lee D, Malpeli JG (1994) Global form and singularity: modeling the blind spot's role in lateral geniculate morphogenesis Science 293 1292–4 [177]
Lee SH, Blake R (1999a) Visual form created solely from temporal structure Science 284 1165-8 [133]
Lee SH, Blake R (1999b) Rival ideas about rivalry Vis Res 3 9 1447—54 [301]
Lee TS, Mumford D, Romero R, Lamme VAF (1998) The role of the primary visual cortex in higher level vision Vis Res 38 2429-54 [206]
Lee WCA, Nedivi E (2002) Extended plasticity of visuaol cortex in dark-reared animals may result from prolonged expression of cpg15-like genes J Neurosci 22 1807-15 [516]
Leeman F (1976) Hidden images Abrams, New York [40]
Legge GE (1979) Spatial frequency masking in human vision: binocular interactions J Opt Soc Am 69 838–47 [336, 491]
Legge GE (1984a) Binocular contrast summation. I Detection and discrimination Vis Res 27 373–83 [313, 321, 323]
Legge GE (1984b) Binocular contrast summation. II Quadratic summation Vis Res 27 385–94 [321]
Legge GE, Rubin GS (1981) Binocular interactions in suprathreshold contrast perception Percept Psychophys 3 0 49–61 [326]
Leguire LE, Blake R, Sloane M (1982) The square wave illusion and phase anisotropy of the human visual system Perception 11 547–56 [303]
Leguire LE, Rogers GL, Bremer DL (1991) Visual-evoked response binocular summation in normal and strabismic infants Invest Ophthal Vis Sci 32 129-33 [483]
Lehky SR (1983) A model of binocular brightness and binaural loudness perception in humans with general applications to nonlinear summation of sensory inputs Biol Cyber 49 89–97 [327]
Lehky SR (1988) An astable multivibrator model of binocular rivalry Perception 17 215–28 [311]
Lehky SR (1995) Binocular rivalry is not chaotic Proc R Soc B 259 71-6 [311]
Lehky SR, Blake R (1991) Organization of binocular pathways: modeling and data related to rivalry Neural Comput 3 44-53 [312]
Lehky SR, Maunsell JHR (1996) No binocular rivalry in the LGN of alert macaque monkeys Vis Res 36 1225-34 [312]
Lehky SR, Sejnowski TJ (1990) Neural model of stereoacuity and depth interpolation based on a distributed representation of stereo disparity J Neurosci 10 2281–99 [241]
Lehky SR, Sejnowski TJ, Desimone R (1992) Predicting responses of nonlinear neurons in monkey striate cortex to complex patterns J Neurosci 12 3568–81 [95]
Lehman RA W, Spencer DD (1973) Mirror-image shape discrimination: interocular reversal of responses in the optic chiasm sectioned monkey Brain Res 52 23-41 [354, 355]
Lehmann D, Fender DH (1967) Monocularly evoked electroencephalogram potentials : influence of target structure presented to the other eye Nature 215 204-5 [312]
Lehmann D, Fender DH (1968) Component analysis of human averaged evoked potentials: dichoptic stimuli using different target structure EEG Clin Neurophysiol 27 542-53 [312]
Lehmann D, Fender DH (1969) Averaged visual evoked potenials in humans: mechanism of dichoptic interaction studied in a subject with a split chiasma EEG Clin Neurophysiol 27 142-45 [312]
Lehmann D, Julesz B (1977) Human average evoked potentials elicited by dynamic random-dot stereograms EEG Clin Neurophysiol 43 469 [262]
Lehmann D, Julesz B (1978) Lateralized cortical potentials evoked in humans by dynamic random–dot stereograms Vis Res 18 1295–71 [262, 489]
Lehmann D, Wälchli P (1975) Depth perception and localization of brain lesions J Neurol 209 157–64 [489]
Lehmkuhle SW, Fox R (1975) Effect of binocular rivalry suppression on the motion aftereffect Vis Res 15 855–9 [303]
Lehmkuhle SW, Fox R (1976) On measuring interocular transfer Vis Res 16 428–30 [347]
Lehmkuhle SW, Fox R (1980) Effect of depth separation on metacontrast masking J Exp Psychol HPP 6 605–21 [340]
Lehmkuhle SW, Kratz KE, Mangel SC, Sherman SM (1980) Effects of early monocular lid suture on spatial and temporal sensitivity of neurons in dorsal lateral geniculate nucleus of the cat J Neurophysiol 43 542–56 [504]
Lehmkuhle SW, Kratz KE, Sherman SM (1982) Spatial and temporal sensitivity of normal and amblyopic cats J Neurophysiol 48 372–87 [527]
Lehmkuhle SW, Sherman SM, Kratz KE, (1984) Spatial contrast sensitivity of dark–reared cats with striate cortex lesions J Neurosci 4 2719–27 [527]
Lehrer M, Srinivasan MV, Zhang SW (1988) Motion cues provide the bee's visual world with a third dimension Nature 332 356–7 [534]
Leibowitz H, Owens DA (1975) Night myopia and the intermediate dark focus of accommodation J Opt Soc Am 65 1121-8 [359]
Leibowitz H, Walker L (1956) Effect of field size and luminance on the binocular summation of suprathreshold stimuli J Opt Soc Am 46 171–2 [328]
Leicester J (1968) Projection of the vertical meridian to cerebral cortex of the cat J Neurophysiol 31 371–82 [231]
Leigh RJ, Rushton DN, Thurston SE, Hertle RW, Yaniglos SS (1988) Effects of retinal image stabilization in acquired nystagmus due to neurologic disease Neurology 38 122–7 [498]
Leigh RJ, Maas EF, Grossman GE, Robinson DA (1989) Visual cancellation of the torsional vestibulo–ocular reflex in humans Exp Brain Res 75 221–6 [411]
Lein ES, Shatz CJ (2000) Rapid regulation of brain-derived neurotrophic factor MRNA within eye–specific circuits during ocular dominance column formation J Neurosci 20 1470-83 [463, 512]
Lein ES, Finney EM, McQuillen PS, Shatz CJ (1999) Subplate neuron ablation alters neurotrophin expression and ocular dominance column formation Proc Nat Acad Sci 96 13491-5 [458]
Lein ES, Hohn A, Shatz CJ (2000) Dynamic regulation of BDNF and NT-3 expression during visual system development J Comp Neurol 420 1-18 [456, 457]
Leinonen L, Hyvärinen J, Nyman G, Linnankoski I (1979) Functional properties of neurons in lateral part of association area 7 in awake monkeys Exp Brain Res 34 299-320 [227]
Lejeune A (1956) L'Optique de Claude Ptolémée (Latin version translated from Arabic by the Emir Eugène of Sicile) University of Louvain, Louvain [12, 13]
Lejeune A (1989) L'Optique de Claude Ptolémée dans la version latine d'apres l'arabe de l'emir Eugene de Sicile With a translation in French. Brill, Leiden [12]

Lema SA, Blake R (1977) Binocular summation in normal and stereoblind humans Vis Res **17** 691–5 [490, 491]

Lemij HG, Collewijn H (1991a) Long–term nonconjugate adaptation of human saccades to anisometropic spectacles Vis Res **31** 1939–54 [424]

Lemij HG, Collewijn H (1991b) Short–term nonconjugate adaptation of human saccades to anisometropic spectacles Vis Res **31** 1955–66 [424]

Lemij HG, Collewijn H (1992) Nonconjugate adaptation of human saccades to anisometropic spectacles: meridian–specificity Vis Res **32** 453–64 [425]

Lemke G (2001) Glial control of neuronal development Ann Rev Neurosci **24** 87-105 [457]

Lennerstrand G, von Noorden GK, Campos EC (1988) Strabismus and amblyopia Plenum, New York [374]

Lennie P (1998) Single units and visual cortical organization Perception **27** 889-935 [219]

Lennie P, Haake PW, Williams DR (1991) The design of chromatically opponent receptive fields In Computational models of visual processing (ed MS Landy, JA Movshon) pp 71–82 MIT Press, Cambridge MA [169]

Leonard CM, Rolls ET, Wilson FAW, Baylis GC (1985) Neurons in the amygdala of the monkey with responses selective for faces Behav Brain Res **15** 159-76 [223]

Leonards U, Sireteanu R (1993) Interocular suppression in normal and amblyopic subjects: the effect of unilateral attenuation with neutral density filters Percept Psychophys **54** 65-74 [294]

Leonards U, Singer W, Fahle M (1996) The influence of temporal phase differences on texture segmentation Vis Res **36** 2989-97 [133]

Leopold DA, Logothetis NK (1996) Activity changes in early visual cortex reflect monkeys' percepts during binocular rivalry Nature **379** 549-53 [314]

Lepore F, Guillemot JP (1982) Visual receptive field properties of cells innervated through the corpus callosum in the cat Exp Brain Res **46** 413–27 [233]

Lepore F, Samson A, Molotchnikoff S (1983) Effects on binocular activation of cells in visual cortex of the cat following the transection of the optic tract Exp Brain Res **50** 392–6 [232]

Lepore F, Ptito M, Lassonde M (1986) Stereoperception in cats following section of the corpus callosum and/or the optic chiasma Exp Brain Res **61** 258–64 [232, 233]

Lepore F, Samson A, Paradis MC, Ptito M (1992) Binocular interaction and disparity coding at the 17–18 border: contribution of the corpus callosum Exp Brain Res **90** 129–40 [232]

Letourneau PC, Condic ML, Snow DM (1994) Interactions of developing neurons with the extracellular matrix J Neurosci **14** 915-28 [438]

Leuba G, Garey LJ (1987) Evolution of neuronal numerical density in the developing and aging human visual cortex Hum Neurobiol **6** 11–18 [446]

LeVay S (1986) Synaptic organization of claustral and geniculate afferents to the visual cortex of the cat J Neurosci **6** 3564-75 [183]

LeVay S, Ferster D (1977) Relay cell classes in the lateral geniculate nucleus of the cat and the effects of visual deprivation J Comp Neurol **172** 563-84 [504]

LeVay S, Gilbert CD (1976) Laminar patterns of geniculocortical projection in the cat Brain Res **113** 1-19 [527]

LeVay S, Voigt T (1988) Ocular dominance and disparity coding in cat visual cortex Vis Neurosci **1** 395–414 [214, 236, 238, 254]

LeVay S, Hubel DH, Wiesel TN (1975) The pattern of ocular dominance columns in macaque visual cortex revealed by a reduced silver stain J Comp Neurol **159** 559-75 [212]

LeVay S, Stryker MP, Shatz CJ (1978) Ocular dominance columns and their development in layer IV of the cat's visual cortex: a quantitative study J Comp Neurol **179** 223–44 [454, 455]

LeVay S, Wiesel TN, Hubel DH (1980) The development of ocular dominance columns in normal and visually deprived monkeys J Comp Neurol **191** 1–51 [455, 463, 500, 509, 517]

LeVay S, Connolly M, Houde J, Van Essen DC (1985) The complete pattern of ocular dominance stripes in the striate cortex and visual field of the macaque monkey J Neurosci **5** 486–501 [214, 339, 473]

Levelt WJM (1965a) Binocular brightness averaging and contour information Br J Psychol **56** 1–13 [324, 325]

Levelt WJM (1965b) On binocular rivalry Institute for Perception, Soesterberg, The Netherlands [287]

Levelt WJM (1966) The alternation process in binocular rivalry Br J Psychol **57** 225–38 [287, 310]

Levelt WJM (1967) Note on the distribution of dominance times in binocular rivalry Br J Psychol **58** 143–5 [293]

Leventhal AG, Ault SJ, Vitek DJ (1988) The nasotemporal division in primate retina: the neural basis of macular sparing and splitting Science **270** 66-7 [176, 231]

Leventhal AG, Wang Y, Schmolesky MT, Zhou Y (1998) Neural correlates of boundary perception Vis Neurosci **15** 1107-18 [124]

Levi DM (1991) Spatial vision in amblyopia In Vision and visual dysfunction Vol 10 Spatial vision (ed D Regan) pp 212–38 CRC Press, Boca Raton [520, 521]

Levi DM, Klein S (1982a) Hyperacuity and amblyopia Nature **298** 298–9 [520]

Levi DM, Klein S (1982b) Differences in vernier discrimination for gratings between strabismic and anisometropic amblyopes Invest Ophthal Vis Sci **23** 389–407 [522]

Levi DM, Klein S (1983) Spatial localization in normal and amblyopic vision Vis Res **23** 1005–17 [522]

Levi DM, Klein S (1985) Vernier acuity, crowding and amblyopia Vis Res **25** 979–91 [509, 521, 523]

Levi DM, Klein S (1990a) The role of separation and eccentricity in encoding position Vis Res **30** 557-85 [273]

Levi DM, Klein S (1990b) Equivalent intrinsic blur in amblyopia Vis Res **30** 1995-2022 [522]

Levi DM, Sharma V (1998) Integration of local orientation in strabismic amblyopia Vis Res **38** 775-81 [524]

Levi DM, Harwerth R, Manny RE (1979a) Suprathreshold spatial frequency detection and binocular interaction in strabismic and anisometropic amblyopia Invest Ophthal Vis Sci **18** 714–25 [529]

Levi DM, Harwerth RS, Smith EL (1979b) Humans deprived of normal binocular vision have binocular interactions tuned to size and orientation Science **206** 852–3 [492]

Levi DM, Harwerth RS, Smith EL (1980) Binocular interactions in normal and anomalous binocular vision Doc Ophthal **49** 303–27 [490]

Levi DM, Pass AF, Manny RE (1982) Binocular interactions in normal and anomalous binocular vision: effects of flicker Br J Ophthal **66** 57–63 [329, 490]

Levi DM, Klein S, Aitsebaomo AP (1984) Detection and discrimination of the direction of motion in central and peripheral vision of normal and amblyopic observers Vis Res **27** 789-800 [525]

Levi DM, Klein S, Aitsebaomo AP (1985) Vernier acuity crowding and cortical magnification Vis Res **25** 963–77 [184, 339]

Levi DM, Klein SA, Yap YL (1987) Positional uncertainty in peripheral and amblyopic vision Vis Res **27** 581-97 [523]

Levi DM, Klein SA, Wang H (1994a) Discrimination of position and contrast in amblyopic and peripheral vision Vis Res **34** 3293-313 [522]

Levi DM, Waugh SJ, Beard BL (1994b) Spatial scale shifts in amblyopia Vis Res **34** 3315-33 [522]

Levi DM, Polat U, Hu YS (1997) Improvement in vernier acuity in adults with amblyopia Invest Ophthal Vis Sci **38** 1493-1509 [528]

Levi DM, Klein SA, Sharma V (1999) Position jitter and undersampling in pattern perception Vis Res **39** 445-65 [524]

Levi L, Zee DS, Hain TC (1987) Disjunctive and disconjugate saccades during symmetrical vergence Invest Ophthal Vis Sci **28** (Abs) 332 [422]

Levick WR, Kirk DL, Wagner HG (1981) Neurophysiological tracing of a projection from temporal retina to contralateral visual cortex of cat Vis Res **21** 1677–9 [231]

Levine RL, Jacobson M (1975) Discontinuous mapping of retina onto tectum innervated by both eyes Brain Res **98** 172-6 [464]

Levinson E, Sekuler R (1975) The independence of channels in human vision selective for direction of movement J Physiol **250** 347–66 [347]

Levitt FB, Van Sluyters RC (1982) The sensitivity period for strabismus in the kitten Devel Brain Res **3** 323-7 [516]

Levitt H (1971) Transformed up-down methods in psychoacoustics J Acoust Soc Am **49** 467-77 [76]

Levitt JB, Lund JS (1997) Contrast dependence of contextual effects in primate visual cortex Nature **387** 73-6 [197]

Levitt JB, Kiper DC, Movshon A (1994) Receptive fields and functional architecture of macaque V2 J Neurophysiol **71** 2517-42 [219]

Levitt JB, Yoshioka T, Lund JS (1995) Connections between the pulvinar complex and cytochrome oxidase-defined compartments in visual area V2 of macaque monkey Exp Brain Res **104** 419-30 [183, 219, 245]

Levitt JB, Schumer RA, Sherman SM, Spear PD, Movshon JA (2001) Visual response properties of neurons in the LGN of normally reared and visually deprived macaque monkeys J Neurophysiol **85** 2111-29 [179, 504]

Levitt P, Moore RY (1979) Development of the noradrenergic innervation of neocortex Brain Res **162** 273–59 [192, 512]

Levitt P, Barbe MF, Eagleson KL (1997) Patterning and specification of the cerebral cortex Ann Rev Neurosci 20 1-27 [447]

Levy WB, Baxter RA (1996) Energy efficient neural codes Neural Comput **8** 531-43 [122]

Lewin GR, Barde YA (1996) Physiology of the neurotrophins Ann Rev Neurosci **19** 289-317 [456]

Lewis CE, Blakeley WR, Swaroop R, Masters RL, McMurty TC (1973) Landing performance by low-time private pilots after the sudden loss of binocular vision: cyclops II Aviat Space Environ Med **44** 1271-45 [5]

Lewis J (1996) Neurogenic genes and vertebrate neurogenesis Curr Opin Neurobiol **6** 3-10 [446]

Lewis JL (1970) Semantic processing of unattended messages using dichotic listening J Exp Psychol **85** 225–8 [309]

Lewis RF, Zee DS, Gaymard BM, Guthrie BL (1994) Extraocular muscle proprioception functions in the control of ocular alignment and eye conjugacy J Neurophysiol **72** 1028-31 [494]

Lewis TL, Maurer D (1992) The development of the temporal and nasal visual fields during infancy Vis Res **32** 903–11 [473]

Lewis TL, Maurer D, Chung JYY, Holmes-Shannon R, Schaik CS (2000) The development of symmetrical OKN in infants: quantification based on OKN acuity for nasalward versus temporalward motion Vis Res **40** 445-53 [475]

Li W, Thier P, Wehrhahn C (2000) Contextual influence on orientation discrimination of humans and responses of neurons in V1 of alert monkeys J Neurophysiol **83** 941-54 [200]

Li XC, Jarvis ED, Alvarez-Borda B, Lim DA, Nottebohm F (2000) A relationship between behavior, neurotrophin expression, and new neuron survival Proc Nat Acad Sci **97** 8584-9 [458]

Lie I, Opheim A (1990) Long-term stability of prism correction of heterophorics and heterotropics; a 5 year follow-up J Am Optom Assoc **61** 491-8 [382]

Liege B, Galand G (1972) Single–unit visual responses in the frog's brain Vis Res **12** 609–22 [540]

Lilien J, Arregui C, Li H, Balsamo J (1999) The juxtamembrane domain of cadherin regulates integrin-mediated adhesion and neurite outgrowth J Neurosci Res **58** 727-34 [438]

Lin CH, Forscher P (1995) Growth cone advance is inversely proportional to retrograde F-actin flow Neuron **14**, 763-71 [438]

Lindberg DC (1976) Theories of vision from Al–Kindi to Kepler University of Chicago Press, Chicago [15, 17, 19, 59]

Lindberg DC (1978) The intromission-extramission controversy in Islamic visual theory: Alkindi versus Avicenna In Studies in perception (ed PK Machamer, RG Turnbull) pp 136-59 Ohio University Press, Columbus Ohio [10, 15]

Lindberg DC (1983) Studies in the history of medieval optics Valorium, London [18, 20, 22]

Lindblom B, Westheimer G (1989) Binocular summation of hyperacuity tasks J Opt Soc Am A **6** 585-9 [324]

Lindsay RM, Wiegand SJ, Alter CA, DiStefano PS (1994) Neurotrophic factors: from molecules to man TINS **17** 182-90 [456]

Lines CR, Milner AD (1983) Nasotemporal overlap in the human retina investigated by means of simple reaction time to lateral light flash Exp Brain Res **50** 166–72 [232]

Linksz A (1952) Physiology of the eye Vol II Vision Grune and Stratton, New York [231]

Linsker R (1986) From basic network principles to neural architecture: emergence of orientation columns Proc Nat Acad Sci **83** 8779-83 [212]

Lippert J, Fleet DJ, Wagner H (2000) Disparity tuning as simulated by a neural net Biol Cybern **83** 61-72 [269]

Lisman JE (1989) A mechanism for the Hebb and the anti-Hebb processes underlying learning and memory Proc Nat Acad Sci **86** 9574-8 [460]

Lisman JE (1997) Bursts as a unit of neural information: making unreliable synapses reliable TINS **20** 38-43 [131]

Little AMG (1971) Roman perspective painting and the ancient stage Star Press, Kennebunk Maine [28]

Liu L, Schor CM (1994) The spatial properties of binocular suppression zone Vis Res **34** 937–47 [298]

Liu L, Schor CM (1995) Binocular combination of contrast signals from orthogonal orientation channels Vis Res **35** 2559-67 [328]

Liu L, Tyler CW, Schor CM (1992a) Failure of rivalry at low contrast: evidence of a suprathreshold binocular summation process Vis Res **32** 1471–9 [253]

Liu L Tyler CW, Schor CM, Ramachandran VS (1992b) Position disparity is more efficient in coding depth than phase disparity Invest Ophthal Vis Sci **33** (Abs) 1373 [289]

Liu Y, Meiri KF, Cynader MS, Gu Q (1996) Nerve growth factor induced modification of presynaptic elements in adult visual cortex in vivo Brain Res **732** 36-42 [457, 511]

Livingstone MS (1996a) Ocular dominance columns in New World monkeys J Neurosci **16** 2086-96 [216]

Livingstone MS (1996b) Differences between stereopsis interocular correlation and binocularity Vis Res **36** 1127-40 [260]

Livingstone MS, Hubel DH (1981) Effects of sleep and arousal on the processing of visual information Nature **291** 554–61 [179, 207]

Livingstone MS, Hubel DH (1984) Anatomy and physiology of a color system in the primate visual cortex J Neurosci **4** 309–56 [196, 204]

Livingstone MS, Hubel DH (1987a) Connections between layer 4B of area 17 and the thick cytochrome oxidase stripes of area 18 in the squirrel monkey J Neurosci **7** 3371–7 [219]

Livingstone MS, Hubel DH (1988) Segregation of form, color movement and depth: anatomy physiology and perception Science **270** 740–9 [229, 244]

Livingstone MS, Tsao DY (1999) Receptive fields of disparity selective neurons in macaque striate cortex Nature Neurosci **2** 825-32 [248, 268]

Livingstone MS, Nori S, Freeman DC, Hubel DH (1995) Stereopsis and binocularity in the squirrel monkey Vis Res **35** 345-54 [217]

Ljubinkovic R (1964) Medieval art in Yugoslavia. The church of the Apostles in the partriarchate of Pec Publishing House Jugoslavija, Beograd, Yugoslavia [31]

Llinás R, Grace AA, Yarom Y (1991) In vitro neurons in mammalian cortical layer 4 exhibit intrinsic oscillatory activity in the 40 to 50 Hz frequency range Proc Nat Acad Sci **88** 897-901 [136]

Lock A, Collett T (1979) A toad's devious approach to its prey: a study of some complex uses of depth vision J Comp Physiol **131** 179–89 [541]

Lockett A (1913) The evolution of the modern stereoscope Sci Am Supplement Number 76 276-9 [64]

Lodovichi C, Berardi N, Pizzorusso T, Maffei L (2000) Effects of neurotrophins on cortical plasticity: same or different? J Neurosci **20** 2155-65 [463]

Logothetis NK (1998) Single units and conscious vision Philos Tr R Soc **353** 1801-18 [313]

Logothetis NK, Schall JD (1989) Neuronal correlates of subjective visual perception Science **275** 761–3 [314]

Logothetis NK, Schall JD (1990) Binocular motion rivalry in macaque monkeys: eye dominance and tracking eye movements Vis Res **30** 1409–19 [287]

Logothetis NK, Leopold DA, Sheinberg DL (1996) What is rivalling during binocular rivalry Nature **380** 621-4 [301]

Lom B, Cohen-Cory S (1999) Brain-derived neurotrophic factor differentially regulates retinal ganglion cell dendrites and axonal arborization *in vivo* J Neurosci **19** 9928-38 [457]

London R, Wick BC (1982) Changes in angle lambda during growth: theory and clinical applications Am J Optom Physiol Opt **59** 568-72 [475]

Long GM (1979) The dichoptic viewing paradigm: do the eyes have it Psychol Bull **86** 391–403 [344]

Longuet–Higgins HC (1981) A computer algorithm for reconstructing a scene from two projections Nature **293** 133–5 [102]

Loomis JM, Collins CC (1978) Sensitivity to shifts of a point stimulus: an instance of tactile hyperacuity Percept Psychophys **27** 487–92 [81]

Löpping B, Weale RA (1965) Changes in corneal curvature following ocular convergence Vis Res **5** 207–15 [390]

Lore R, Sawatski D (1969) Performance of binocular and monocular infant rats on the visual cliff J Comp Physiol Psychol **67** 177-81 [548]

Lorente de Nó R (1949) Cerebral cortex: architecture intracortical connections motor projections In Physiology of the nervous system 3rd edn (ed JF Fulton) Chap 15 Oxford University Press, London [208]

Lotmar W (1976) A theoretical model for the eye of new-born infants Graefes Arch klin exp Ophthal **198** 179–85 [434]

Lotze H (1852) Medicinische Psychologie Weidmann, Leipzig. Part translation in Herrnstein and Boring (1965) [45]

Lowe KN, Ogle KN (1966) Dynamics of the pupil during binocular rivalry Arch Ophthal **75** 395–403 [299]

Löwel S (1994) Ocular dominance column development: strabismus changes the spacing of adjacent columns in cat visual cortex J Neurosci **14** 7451-68 [505]

Löwel S, Singer W (1992) Selection of intrinsic horizontal connections in the visual cortex by correlated neuronal activity Science **255** 209-12 [506]

Lu C, Fender DH (1972) The interaction of color and luminance in stereoscopic vision Invest Ophthal **11** 482–90 [244]

Luck SJ, Chelazzi L, Hillyard SA, Desimone R (1997) Neural mechanisms of spatial selective attention in areas V1 V2 and V4 of macaque visual cortex J Neurophysiol **77** 27-42 [208]

Ludvigh E, McKinnon P (1966) Relative effectivity of foveal and parafoveal stimuli in eliciting fusion movements of small amplitude Arch Ophthal **76** 443–9 [397]

Ludvigh E, McKinnon P (1968) Dependence of the amplitude of fusional convergence movements on the velocity of the eliciting stimulus Invest Ophthal **7** 347–52 [399]

Ludvigh E, McKinnon P, Zaitzeff L (1964) Temporal course of the relaxation of binocular duction (fusion) movements Arch Ophthal **71** 389-399 [383]

Ludvigh E, McKinnon P, Zaitzeff L (1965) Relative effectivity of foveal and parafoveal stimuli in eliciting fusion movements Arch Ophthal **73** 115–21 [396]

Lueck CJ, Hamlyn P, Crawford TJ, Levy IS, Brindley GS, Watkins ES, Kennard C (1991) A case of ocular tilt reaction and torsional nystagmus due to direct stimulation of the midbrain in man Brain **114** 2069-79 [432]

Luhmann HJ, Millán LM, Singer W (1986) Development of horizontal intrinsic connections in the cat striate cortex Exp Brain Res **63** 443–8 [452]

Lumer ED, Friston KK, Rees G (1998) Neural correlates of perceptual rivalry in the human brain Science **280** 1930-4 [312]

Lund JS, Boothe RG (1975) Interlaminar connections and pyramidal neuron organization in the visual cortex area 17 of the macaque monkey J Comp Neurol **159** 305-34 [178]

Lund JS, Boothe RG, Lund RD (1977) Development of neurons in the visual cortex of the monkey (Macaca nemestrina): a Golgi study from fetal day 127 to postnatal maturity J Comp Neurol **176** 149–88 [451]

Lund RD (1965) Uncrossed visual pathways of hooded and albino rats Science **149** 1506-7 [496]

Lund RD, Mitchell DE (1979a) Plasticity of visual callosal projections Soc Neurosci Symp **4** 142–52 [453]

Lund RD, Mitchell DE (1979b) Asymmetry in the callosal connections of strabismic cats Brain Res **167** 176-9 [453]

Lund RD, Mustari MJ (1977) Development of the geniculocortical pathway in rats J Comp Neurol **173** 289-305 [451]

Lund RD, Mitchell DE, Henry GH (1978) Squint-induced modification of callosal connections in cats Brain Res **144** 169-72 [453]

Luneburg RK (1947) Mathematical analysis of binocular vision Edwards, Ann Arbor Michigan [137]

Luneburg PK (1950) The metric of binocular visual space J Op Soc Am **40** 627-42 [137]

Luo L, Jan LY, Jan YN (1997) Rho family small GTP-binding proteins in growth cone signalling Curr Opin Neurobiol **7** 81-6 [438]

Luskin MB, Shatz CJ (1985) Neurogenesis of the cat's primary visual cortex J Comp Neurol **272** 611–31 [448]

Luthardt-Laimer G (1983) Distance estimation in binocular and monocular salamanders Z Tierpsychol **63** 233-40 [543]

Lynch JC, Mountcastle VB, Talbot WH, Yin TCT (1977) Parietal lobe mechanisms for directed visual attention J Neurophysiol **40** 362–89 [207]

Lynes JA (1980) Brunelleschi's perspectives reconsidered Perception 9 87–99 [32]

Lyon DC, Kaas JH (2001) Connectional and architectonic evidence for dorsal and ventral V3, and dorsomedial area in marmoset monkeys J Neurosci **21** 249-61 [220]

Lythgoe RJ, Phillips LR (1938) Binocular summation during dark adaptation J Physiol **91** 427-36 [319]

MacDermott AB, Role LW, Siegelbaum SA (1999) Presynaptic ionotropic receptors and the control of transmitter release Annn Rev Neurosci **22** 443-85 [192]

Mach E (1929) The principles of physical optics An historical and philosophical treatment (Translated by JS, Anderson, AFA Young) Methuen, London [50]

MacKay DM (1961) Interactive processes in visual perception In Sensory communication (ed WA Rosenblith) pp 339-355 Wiley, New York [136]

MacKay DM (1968) Evoked potentials reflecting interocular and monocular suppression Nature **217** 81-3 [312]

MacKay DM, MacKay V (1975) Dichoptic induction of McColloughtype effects Quart J Exp Psychol **27** 225–33 [352]

Mackensen G (1958) Reaktionszeitmessungen bei Amblyopie Graefes Arch klin exp Ophthal **159** 636-42 [525]

Macknik SL, Haglund MM (1999) Optical images of visible and invisible percepts in the primary visual cortex of primates Proc Nat Acad Sci **96** 15208-10 [340]

MacLeod DIA (1972) The Schrödinger equation in binocular brightness combination Perception **1** 321–4 [326]

Macmillan NA, Creelman CD (1991) Detection theory: A users guide Cambridge University Press, New York [77]

MacNeil MA, Masland RH (1998) Extreme diversity among amacrine cells: implications for function Neuron **20** 971-82 [168]

Macy A, Ohzawa I, Freeman RD (1982) A quantitative study of the classification and stability of ocular dominance in the cat's visual cortex Exp Brain Res **48** 401-8 [215]

Maddox EE (1893) The clinical use of prisms and the decentering of lenses John Wright, Bristol England [368, 372, 376, 388]

Maeda M, Sato M, Ohmura T, Miyazaki Y, Wang AH, Awaya S (1999) Binocular depth-from-motion in infantile and late-onset esotropia patients with poor stereopsis Invest Ophthal Vis Sci **40** 3031-6 [525]

Maffei L, Bisti S (1976) Binocular interaction in strabismic kittens deprived of vision Science **191** 279–80 [494]

Maffei L, Fiorentini A (1976a) Monocular deprivation in kittens impairs spatial resolution of geniculate neurones Nature **294** 754–5 [504]

Maffei L, Fiorentini A (1976b) Asymmetry of motility of the eyes and change of binocular properties of cortical cells in adult cats Brain Res **105** 73-8 [495]

Maffei L, Berardi N, Bisti S (1986) Interocular transfer of adaptation after effect in neurons of area 17 and 18 of split chiasm cats J Neurophyiol **55** 966-76 [340]

Maffei L, Berardi N, Domenici L, Parisi V, Pizzorusso T (1992) Nerve growth factor (NGF) prevents the shift in ocular dominance distribution of visual cortical neurons in monocularly deprived rats J Neurosci **12** 4651-62 [512]

Magee JC, Johnston D (1997) A synaptically controlled associative signal for Hebbian plasticity in hippocampal neurons Science **275** 209-13 [113]

Magnin M, Cooper HM, Mick G (1989) Retinohypothalamic pathway: a breach in the law of Newton-Müller-Budden Brain Res **488** 390-7 [176]

Magoon EH, Robb RM (1981) Development of myelin in human optic nerve and tract Arch Ophthal **99** 655–9 [442]

Makous N, Boothe R (1974) Cones block signals from rods Vis Res **14** 285-94 [334]

Makous W, Sanders RK (1978) Suppressive interactions between fused patterns In Visual psychophysics and physiology (ed JC Armington, J Krausfopf, BR Wooten) pp 167–79 Academic Press, New York [288, 306]

Makous W, Teller D, Boothe R (1976) Binocular interaction in the dark Vis Res **16** 473–6 [335]

Malach R, Van Sluyters RC (1989) Strabismus does not prevent recovery from monocular deprivation: a challenge for simple Hebbian models of synaptic modification Vision Neurosci **3** 297–73 [514]

Malach R, Ebert R, Van Sluyters RC (1984) Recovery from effects of brief monocular deprivation in the kitten J Neurophysiol **51** 538-51 [515]

Malach R, Amir Y, Harel M, Grinvald A (1993) Relationship between intrinsic connections and functional architecture revealed by optical imaging and *in vivo* targeted biocytin injections in primate striate cortex Proc Nat Acad Sci **90** 10469-73 [196]

Maldonado H, Rodriguez E (1972) Depth perception in the praying mantis Physiol Behav **8** 751-9 [532]

Maldonado PE, Gödecke I, Gray CM, Bonhoeffer T (1997) Orientation selectivity in pinwheel centres in cat striate cortex Science **276** 1551-5 [212]

Malebranche N (1674) De la recherche de la verité Preland, Paris [56]

Maletic-Savatic M, Malinow R, Svoboda K (1999) Rapid dendritic morphogenesis in CA1 hippocampal dendrites induced by synaptic activity Science **283** 1923 [193]

Malgaroli A, Ting AE, Wendland B, Bergamaschi A, Villa A, Tsien RW, Scheller RH (1995) Presynaptic components of long-term potentiation visualized at individual hippocampal synapses Science **298** 1627-8 [136]

Mallett RFJ (1964) The investigation of heterophoria at near and a new fixation disparity technique The Optician **148** 547-551 574-81 [377]

Mallot HA, Roll A, Arndt PA (1996c) Disparity-evoked vergence is driven by interocular correlation Vis Res **36** 2925-37 [396]

Malmstrom FW, Randle RJ (1976) Effects of visual imagery on the accommodation response Percept Psychophys **19** 450-3 [359]

Malpeli JG, Baker FH (1975) The representation of the visual field in the lateral geniculate nucleus of *Macaca mulatta* J Comp Neurol **161** 569-94 [177]

Malpeli JG, Schiller PH, Colby CL (1981) Response properties of single cells in monkey striate cortex during reversible inactivation of individual lateral geniculate laminae J Neurophysiol **46** 1102-9 [194, 195]

Malpeli JG, Lee D, Baker FH (1996) Laminar and retinotopic organization of the macaque lateral geniculate nucleus: magnocellular and parvocellular functions J Comp Neurol **375** 363-77 [177]

Malsburg C von der (1979) Development of ocularity domains and growth behaviour of axon terminals Biol Cyber **32** 49-62 [464]

Malsburg C von der and Schneider W (1986) A neural cocktail–party processor Biol Cyber **54** 29–40 [132]

Manas L (1958) The effect of visual training upon the ACA ratio Am J Optom Arch Am Acad Optom **35** 428-37 [390]

Mandelbaum J (1960) An accommodation phenomenon Arch Ophthal **63** 923-6 [361]

Mangel SC, Wilson JR, Sherman SM (1983) Development of neuronal response properties in the cat dorsal lateral geniculate nucleus during monocular deprivation J Neurophysiol **5 0** 270–64 [504]

Mann I (1964) The development of the human eye Grune and Stratton, New York [485]

Mann VA (1978) Different loci suggested to mediate tilt and spiral motion aftereffects Invest Ophthal Vis Sci **17** 903–9 [492]

Mann VA, Hein A, Diamond R (1979a) Patterns of interocular transfer of visuomotor coordination reveal differences in the representation of visual space Percept Psychophys **25** 35-41 [356]

Manning KA, Riggs LA (1984) Vergence eye movements and visual suppression Vis Res **27** 521–29 [368]

Manning ML, Finlay DC, Dewis SAM, Dunlop DB (1992) Detection duration thresholds and evoked potential measures of stereosensitivity Doc Ophthal **79** 161-75 [262]

Manny RE (1980) Monocular vergence movements produced by external visual feedback Am J Optom Physiol Opt **57** 236-44 [419]

Manny RE, Klein SA (1985) A three alternative tracking paradigm to measure vernier acuity of older infants Vis Res **25** 1275-52 [469]

Manny RE, Levi DM (1982) Psychophysical investigations of the temporal modulation sensitivity function in amblyopia: spatiotemporal interactions Invest Ophthal Vis Sci **22** 525–34 [525]

Mapperson B, Lovegrove W (1991) Orientation and spatial-frequency-specific surround effects on binocular rivalry Bull Psychonom Soc **29** 95-7 [293]

Mapperson B, Bowling A, Lovegrove W (1982) Problems for an after-image explanation of monocular rivalry Vis Res **22** 1233-4 [296]

Maraini G, Porta R (1978) Interocular transfer of a visual aftereffect in early–onset esotropia Arch Ophthal **96** 1853–6 [492]

Marcelja S (1980) Mathematical description of the responses of simple cortical cells J Opt Soc Am **70** 1297-1300 [140]

Marchiafava PL (1966) Binocular reciprocal interaction upon optic fiber endings in the lateral geniculate nucleus of the rat Brain Res **2** 188-92 [179]

Marcos S, Navarro R (1997) Determination of the foveal cone spacing by ocular speckle interferometry: limiting factors and acuity predictions J Opt Soc Am A **14** 731-40 [174]

Marcos S, Moreno E, Navarro R (1999) The depth-of-field of the human eye from objective and subjective measurements Vis Res **39** 2039-49 [361]

Marcus RC, Mason CA (1995) The first retinal axon growth in the mouse optic chiasm: axon patterning and the cellular environment J Neurosci **15** 6389-402 [439]

Marcus RC, Blazeski R, Godement P, Mason CA (1995) Retinal axon divergence in the optic chiasm: uncrossed axons diverge from crossed axons within a midline glial specialization J Neurosci **15** 3716-29 [440]

Marcus RC, Wang LC, Mason CA (1996) Retinal axon divergence in the optic chiasma: midline cells are unaffected by the albino mutation Development **122** 859-68 [496]

Marcus RC, Shimamura K, Sretavan D, Lai E, Rubenstein JLR, Mason CA (1999) Domains of regulatory gene expression and the developing optic chiasm: correspondence with retinal axon paths and candidate signaling cells J Comp Neurol **403** 346-58 [441]

Mareschal I, Baker CL (1998) A cortical locus for the processing of contrast-defined contours Nature Neurosci **1** 150-4 [220]

Marey EJM (1895) Le mouvement (Translated by E Pritchard) Appleton, New York 1972 [71]

Mark RF, Davidson TM (1966) Unit responses from commissural fibres of optic lobes of fish Science **152** 797-9 [539]

Markoff JI, Sturr JF (1971) Spatial and luminance determinants of the increment threshold under monoptic and dichoptic viewing J Opt Soc Am **61** 1530–7 [334]

Markram H, Lübke J, Frotscher M, Sakmann B (1997) Regulation of synaptic efficiency by coincidence of postsynaptic APs and EPSPs Science **275** 213-15 [113, 460]

Marmarelis PZ, Marmarelis VZ (1978) Analysis of physiological systems: the white noise approach Plenum, New York [94]

Marr D (1982) Vision Freeman San Francisco [89, 121, 142]

Marran L, Schor CM (1998) Lens induced aniso-accommodation Vis Res **38** 3601-19 [364]

Marrocco RT, McClurkin JW (1979) Binocular interaction in the lateral geniculate nucleus of the monkey Brain Res **168** 633–7 [178, 180]

Marrocco RT, Carpenter MA, Wright SE (1985) Spatial contrast sensitivity: effects of peripheral field stimulation during monocular and dichoptic viewing Vis Res **25** 917–27 [339]

Marrocco RT, McClurkin JW, Alkirc MT (1996) The influence of the visual cortex on the spatiotemporal response properties of lateral geniculate nucleus cells Brain Res **737** 110-18 [179, 180]

Marrs GS, Green SH, Dailey ME (2001) Rapid formation and remodelling of postsynaptic densities in developing dendrites Nature Neurosci **4** 1006-13 [189]

Marshall NJ, Land MF, King CA, Cronin TW (1991) The compound eyes of mantis shrimps (Crustacea Hoplocarida Stomatopoda). II Colour pigments in the eyes of stomatopod crustaceans: polychromatic vision by serial and lateral filtering Philos Tr R Soc B **334** 57–84 [538]

Marshall WH, Talbot SA (1940) Recovery cycle of the lateral geniculate of the nembutalized cat Am J Physiol **129** P417-18 [179]

Marshall WH, Talbot SA (1942) Recent evidence for neurological mechanisms in vision leading to a general theory of sensory acuity Biological Symposium **7** 117–64 [366]

Martens TG, Ogle KN (1959) Observations on accommodative convergence Am J Ophthal **47** 455-63 [388]

Martenson C, Stone K, Reedy M, Sheetz M (1993) Fast axonal transport is required for growth cone advance Nature **366** 66–9 [437]

Martin GR (1984) The visual fields of the tawny owl *Strix aluco* L Vis Res **27** 1739–51 [546]

Martin GR, Young SR (1983) The retinal binocular field of the pigeon (*Columbia livia*: English racing homer) Vis Res **23** 911–15 [544]

Martin JI (1970) Effects of binocular fusion and binocular rivalry on cortically evoked potentials EEG Clin Neurophysiol **28** 190-201 [312]

Martin KAC, Ramachandran VS, Rao VM, Whitteridge D (1979) Changes in ocular dominance induced in monocularly deprived lambs by stimulation with rotation gratings Nature **277** 391–3 [549]

Martin PR (1986) The projection of different retinal ganglion cell classes to the dorsal lateral geniculate nucleus in the hooded rat Exp Brain Res **62** 77-88 [177]

Martinoya C, Rey J, Bloch S (1981) Limits of the pigeon's binocular field and direction for best binocular viewing Vis Res **21** 1197–200 [544]

Martinoya C, Le Houezec J, Bloch S (1988) Depth resolution in the pigeon J Comp Physiol A **163** 33–42 [545]

Marzi CA, Simoni A, di Stefano M (1976) Lack of binocularly driven neurones in the Siamese cat's visual cortex does not prevent successful interocular transfer of visual form discriminations Brain Res **105** 353-7 [497]

Marzi CA, Antonini A, Legg CR (1980) Callosum-dependent binocular interactions in the lateral suprasylvian area of Siamese cats which lack binocular neurons in areas 17 and 18 Brain Res **197** 230-5 [497]

Marzi CA, Antonini A, Di Stefano M, Legg CR (1982) The contribution of the corpus callosum to receptive fields in the lateral suprasylvian visual areas of the cat Behav Brain Res **4** 155–76 [233]

Marzi CA, Antonucci G, Pizzamiglio L, Santillo C (1986) Simultaneous binocular integration of the visual tilt effect in normal and stereoblind observers Vis Res **29** 477–83 [344, 493]

Maske R, Yamane S, Bishop PO (1984) Binocular simple cells for local stereopsis: a comparison of receptive field organizations for the two eyes Vis Res **27** 1921–9 [214, 255]

Maske R, Yamane S, Bishop PO (1986a) End–stopped and binocular depth discrimination in the striate cortex of cats Proc R Soc B **229** 257–76 [236, 238, 253, 254]

Maske R, Yamane S, Bishop PO (1986b) Stereoscopic mechanisms: binocular responses of the striate cells of cats to moving light and dark bars Proc R Soc B **229** 227–56 [238]

Masland RH, Raviola E (2000) Confronting complexity: strategies for understanding the microcircuity of the retina Ann Rev Neurosci **23** 279-84 [181]

Mason CA, Wang LC (1997) Growth cone form is behavior-specific and consequently position-specific along the retinal axon pathway J Neurosci **17** 1086-100 [437]

Massaro DW (1988) Ambiguity in perception and experimentation J Exp Psychol Gen **117** 417-21 [148]

Masson GS, Busettini CM, Miles FA (1997) Vergence eye movements in response to binocular disparity without depth perception Nature **389** 283-6 [402]

Mastai ML d'Otrange (1976) Illusion in art Secker and Warburg, London [40]

Masters WM, Raver KAS (2000) Range discrimination by big brown bats (*Eptesicus fuscus*) using altered model echoes: implications for signal processing J Acoust Soc Am **107** 625-37 [548]

Mastronarde DN (1983) Correlated firing of cat retinal ganglion cells. I Spontaneously active inputs to X– and Y–cells J Neurophysiol **49** 303–27 [168]

Mateeff S, Yakimoff N, Hohnsbein J, Ehrenstein WH, Bohdanecky Z, Radil T (1991) Selective directional sensitivity in visual motion perception Vis Res **31** 131-8 [525]

Mates JWB (1978) Eye movements of African chameleons: spontaneous saccadic timing Science **199** 1087–9 [543]

Mather G, Verstraten F, Anstis S (1998) The motion aftereffect MIT Press Cambridge, MA Mather et al. 1998 [347]

Mathis U, Eschbach S, Rossel S (1992) Functional binocular vision is not dependent on visual experience in the praying mantis Vis Neurosci **9** 199-203 [533]

Matin E (1974) Saccadic suppression: a review and an analysis Psychol Bull **81** 899-917 [368]

Matin L (1962) Binocular summation at the absolute threshold for peripheral vision J Opt Soc Am **52** 1276–86 [318, 331]

Matsubara J, Cynader M, Swindale NV, Stryker MP (1985) Intrinsic projections within visual cortex: evidence for orientation–specific local connections Proc Nat Acad Sci **8 2** 935–9 [196]

Matsuoka K (1984) The dynamic model of binocular rivalry Biol Cyber **49** 201-8 [310]

Matthews N, Geesaman BJ, Qian N (2000) The dependence of motion repulsion and rivalry on the distance between moving elements Vis Res **40** 2025-36 [291]

Matthews PB (1988) Proprioceptors and their contribution to somatosensory mapping: complex messages require complex processing Can J Physiol Pharmacol **66** 430-8 [143]

Mattingley JB, Davis G, Driver J (1997) Preattentive filling-in of visual surfaces in parietal extinction Science **275** 671-4 [124]

Maturana HR, Frenk S (1963) Directional movement and horizontal edge detectors in the pigeon retina Science **142** 977-9 [203]

Maturana HR, Lettvin JY, McCullach WS, Pitts WH (1960) Anatomy and physiology of vision in the frog J Gen Physiol **43** 129-75 [203]

Mauch DH, Nägler K, Schumacher S, Göritz C, Müller EC, Otto A, Pfrieger FW (2001) CNS synaptogenesis promoted by glial-derived cholesterol Science **294** 1354-7 [457]

Maude N (1978) Stereo photography its inception rise and fall Br J Photog Annual [66]

Maunsell JHR, McAdams CJ (2000) Effects of attention on neuronal response properties in visual cerebral cortex In The new cognitive neurosciences (ed MS Gazzaniga) pp 315-27 MIT Press, Cambridge MA. [207]

Maunsell JHR, Van Essen DC (1983a) The connections of the middle temporal visual area (MT) and their relationship to a cortical hierarchy in the macaque monkey J Neurosci **3** 2563–86 [217]

Maunsell JHR, Van Essen DC (1983b) Functional properties of neurons in middle temporal visual area of the macaque monkey. II Binocular interactions and sensitivity to binocular disparity J Neurophysiol **49** 1148–67 [224, 229, 243, 256, 257]

Maunsell JHR, Van Essen DC (1987) Topographic organization of the middle temporal visual area in the macaque monkey: representational biases and the relationship to callosal connections and myeloarchitectonic boundaries J Comp Neurol **296** 535–55 [223, 232]

Maunsell JHR, Nealey TA, DePriest DD (1990) Magnocellular and parvocellular contributions to responses in the middle temporal visual area (MT) of the macaque monkey J Neurosci **10** 3323-34 [223, 229]

Maunsell JHR, Ghose GM, Assad JA, McAdams CJ, Boudreau CE, Noerager BD (1999) Visual response latencies of magnocellular and parvocellular LGN neurons in macaque monkeys Vis Neurosci **16** 1-14 [178]

Maurer D, Martello M (1980) The discrimination of orientation by young infants Vis Res **20** 201–4 [471]

Maurer D, Lewis TL, Brent HP, Levin AV (1999) Rapid improvement in the acuity of infants after visual input Science, **286** 108-10 [468]

Maxwell GF, Lemij HG, Collewijn H (1995) Conjugacy of saccades in deep amblyopia Invest Ophthal Vis Sci **36** 2514-22 [526]

Maxwell JS, King WM (1992) Dynamics and efficacy of saccade-facilitated vergence eye movements in monkeys J Neurophysiol **68** 1278–59 [423]

Maxwell JS, Schor CM (1994) Mechanisms of vertical phoria adaptation revealed by time–course and two–dimensional spatiotopic maps Vis Res **34** 2741–51 [385]

Maxwell JS, Schor CM (1996) Adaptation of vertical eye alignment in relation to head tilt Vis Res **36** 1195-6 [385]

Maxwell JS, Schor CM (1997) Head-position-dependent adaptation of noncomitant vertical skew Vis Res **37** 441-6 [385]

Maxwell JS, Schor CM (1999) Adaptation of torsional eye alignment in relation to head roll Vis Res **39** 4192-9 [385]

Mayer DL, Dobson V (1982) Visual acuity development in infants and young children as assessed by operant preferential looking Vis Res **22** 1141-51 [468]

Mayer ML, Westbrook GL, Guthrie PB (1984) Voltage–dependent block by Mg2+ of NMDA responses in spinal cord neurones Nature **309** 291–3 [460]

Mayhew JEW, Anstis SM (1972) Movement aftereffects contingent on color intensity and pattern Percept Psychophys **12** 77–85 [353]

Mays LE (1984) Neural control of vergence eye movements: convergence and divergence neurons in midbrain J Neurophysiol **51** 1091–108 [430]

Mays LE, Porter JD (1984) Neural control of vergence eye movements: activity of abducens and oculomotor neurons J Neurophysiol **52** 743–61 [430]

Mays LE, Sparks DL (1980) Dissociation of visual and saccade-related responses in superior colliculus neurones J Neurophysiol **43** 207-32 [246]

Mays LE, Porter JD, Gamlin PDR, Tello CA (1986) Neural control of vergence eye movements: neurons encoding vergence velocity J Neurophysiol **56** 1007–21 [398, 430]

Mays LE, Zhang Y, Thorstad MH, Gamlin PDR (1991) Trochlear unit activity during ocular convergence J Neurophysiol **6 5** 1484–91 [430]

McAdams CJ, Maunsell JHR (1999) Effects of attention on orientation-tuning functions of single neurons in macaque cortical area V4 J Neurosci **19** 431-41 [207]

McAllister AK, Lo DC, Katz LC (1995) Neurotrophins regulate dendritic growth in developing visual cortex Neuron **15** 791-803 [457]

McAllister AK, Katz LC, Lo DC (1997) Opposing roles for endogenous BDNF and NT-3 in regulating cortical dendritic growth Neuron **18** 767-78 [457]

McAllister AK, Katz LC, Lo DC (1999) Neurotrophins and synaptic plasticity Ann Rev Neurosci **22** 295-318 [457]

McBrien NA, Millodot M (1987) The relationship between tonic accommodation and refractive error Invest Ophthal Vis Sci **28** 997-1004 [359]

McCandless JW, Schor CM, Maxwell JS (1996) A cross-coupling model of vertical vergence adaptation IEEE Tr Biomed Engin **43** 27-34 [385]

McCarthy JE (1993) Directional adaptation effects with contrast modulated stimuli Vis Res **33** 2653-62 [348]

McCarty JA, Demer JL, Hovis LA, Nuwer MR (1992) Ocular motility anomalies in developmental misdirection of the optic chiasma Am J Ophthal **113** 86-95 [441]

McClurkin JW, Optican LM (1996) Primate striate and prestriate cortical neurons during discrimination. I Simultaneous temporal encoding of information about color and pattern J Neurophysiol **75** 481-95 [129]

McClurkin JW, Optican LM, Richmond BJ (1994) Cortical feedback increases visual information transmitted by monkey parvocellular lateral geniculate nucleus neurons Vis Neurosci **11** 601-17 [180]

McColl SL, Mitchell DE (1998) Stereodeficient subjects show substantial differences in interocular transfer of two motion adaptation aftereffects Vis Res **38** 1889-900 [493]

McColl SL, Ziegler L, Hess RF (2000) Stereodeficient subjects demonstrate non-linear stereopsis Vis Res **40**1167-77 [529]

McCollough C (1965) Colour adaptation of edge–detectors in the human visual system Science **149** 1115–16 [85, 351]

McConnell SK (1995a) Constructing the cerebral cortex: neurogenesis and fate determination Neuron **15** 761-8 [450]

McConnell SK (1995b) Strategies for the generation of neuronal diversity in the developing central nervous system J Neurosci **15** 6987-8 [451]

McConnell SK, Kaznowski CE (1991) Cell cycle dependence of laminar determination in developing neocortex Science **254** 282–5 [451]

McConnell SK, Ghosh A, Shatz CJ (1989) Subplate neurons pioneer the first axon pathway from the cerebral cortex Science **275** 978–2 [458]

McConnell SK, Ghosh A, Shatz CJ (1994) Subplate pioneers and the formation of descending connections from cerebral cortex J Neurosci **14** 1892-1907 [458]

McCormack G, Fisher SK (1996) The source of disparity vergence innervation determines prism adaptation Ophthal Physiol Opt **16** 73-82 [383]

McCormack G, Fisher SK, Wolf K (1991) Retinal eccentricity of fusion detail affects vergence adaptation Optom Vis Sci **68** 711–17 [383]

McCullough RW (1978) The fixation disparity-heterophoria relationship J Am Optom Assoc **49** 369-72 [379]

McFadden SA (1987) The binocular depth stereoacuity of the pigeon and its relation to the anatomical resolving power of the eye Vis Res **27** 1967–80 [545]

McFadden SA, Reymond L (1985) A further look at the binocular visual field of the pigeon (*Columba livia*) Vis Res **25** 1741–6 [544]

McFarlane S, Pollock NS (2000) A role for voltage-gated potassium channels in the growth of retinal axons in the developing visual system J Neurosci **20** 1-20-9 [439]

McGehee DS, Role LW (1996) Presynaptic ionotropic receptors Curr Opin Neurobiol **6** 342-9 [192]

McGuire BA, Gilbert CD, Rivlin PK, Wiesel TN (1991) Targets of horizontal connections in macaque primary visual cortex J Comp Neurol 305 370-92 [195]

McKay R (1997) Stem cells in the central nervous system Science **276** 66-71 [450]

McKee MC, Young DA, Kohl P, Reinke R, Yolton RL (1987) Effect of head and eye positions on fixation disparities phoria and ductions at near Am J Optom Physiol Opt **64** 909–15 [378]

McKee SP, Harrad RA (1993) Fusional suppression in normal and stereoanomalous observers Vis Res **33** 1645–58 [530]

McKee SP, Mitchison GJ (1988) The role of retinal correspondence in stereoscopic matching Vis Res **28** 1001–12 [268]

McKee SP, Westheimer G (1970) Specificity of cone mechanisms in lateral interaction J Physiol **206** 117-28 [341]

McKee SP, Klein SA, Teller DY (1985) Statistical properties of forcedchoice psychometric functions: implications of probit analysis Percept Psychophys **37** 286–98 [77]

McKee SP, Silverman GH, Nakayama K (1986) Precise velocity disparitydiscrimination despite random variations in temporal frequency and contrast Vis Res **29** 609–19 [126]

McKee SP, Bravo MJ, Taylor DG, Legge GE (1994) Stereo matching precedes dichoptic masking Vis Res **34** 1047–60 [338]

McKenzie BE, Day RH (1972) Object distance as a determinant of visual fixation in early infancy Science **178** 1108-10 [478]

McKinney RA, Capagna M, Dürr R, Gähwiler BH, Thompson SM (1999) Miniature synaptic events maintain dendritic spines via AMPA receptor activation Nature Neurosci 2 44-9 [193]

McLaughlin D, Shapley R, Shelley M, Wielaard DJ (2000) A neuronal network model of macaque primary visual cortex (V1): orientation selectivity and dynamics in the input layer 4Ca Proc Nat Acad Sci **97** 8087-92 [200]

McLaughlin SC (1967) Parametric adjustment in saccadic eye movements Percept Psychophys **2** 359–62 [423]

McLean A (1972) Humanism and the rise of science in Tudor England Heinemann, London [13]

McLean J, Palmer LA (1989) Contribution of linear spatiotemporal receptive field structure to velocity selectivity of simple cells in area 17 of cat Vis Res **29** 675-9 [202]

McLean J, Palmer LA (1994) Organization of simple cell responses in the three-dimensional (3-D) frequency domain Vis Neurosci **11** 295-306 [202]

McLean J, Palmer LA (1998) Plasticity of neuronal response properties in adult cat striate cortex Vis Neurosci **15** 177-96 [208]

McLin LN, Schor CM (1988) Voluntary effort as a stimulus to accommodation and vergence Invest Ophthal Vis Sci **29** 1739–46 [387]

McLin LN, Schor CM, Kruger PB (1988) Changing size (looming) as a stimulus to accommodation and vergence Vis Res 28 883–98 [360, 386]

Meakin SO, Shooter EM (1992) The nerve growth factor family of receptors TINS 15 323–31 [512]

Meenes M (1930) A phenomenological description of retinal rivalry Am J Psychol 42 290-9 [287]

Meissirel C, Chalupa LM (1994) Organization of pioneer retinal axons within the optic tract of the rhesus monkey Proc Nat Acad Sci 91 3906-10 [439]

Meissirel C, Wikler KC, Chalupa LM, Rakic P (1997) Early divergence of magnocellular and parvocellular functional subsystems in the embryonic primate visual system Proc Nat Acad Sci 94 5900-5 [443]

Meister M (1996) Multineuronal codes in retinal signalling Proc Nat Acad Sci 93 609-14 [132, 199]

Meister M, Wong RO L, Baylor DA, Shatz CJ (1991) Synchronous bursts of action potentials in ganglion cells of the developing mammalian retina Science 252 939–43 [443]

Mel BW (1994) Information processing in dendritic trees Neural Comput 6 1031-85 [128]

Mello NK (1966) Concerning inter-hemispheric transfer of mirror-image patterns in pigeon Physiol Behav 1 293-300 [355]

Menon RS, Ogawa S, Strupp JP, Ugurbil K (1997) Ocular dominance in human V1 demonstrated by functional magnetic resonance imaging J Neurophysiol 77 2780-7 [213]

Menon RS, Gati JS, Goodyear BG, Luknowsky DC, Thomas CG (1998a) Spatial and temporal resolution of functional magnetic resonance imaging Biochem Cell Biol 76 560-71 [182]

Menon RS, Luknowsky DC, Gati JS (1998b) Mental chronometry using latency-resolved functional MRI Proc Nat Acad Sci 95 10902-7 [183]

Meredith GM, Meredith CGW (1962) Effect of instructional conditions on rate of binocular rivalry Percept Mot Skills 15 655–64 [308]

Merigan WH (1989) Chromatic and achromatic vision of macaques: role of the P pathway J Neurosci 9 776-83 [229]

Merigan WH (1996) Basic visual capacities and shape discrimination after lesions of extrastriate area V4 in macaques Vis Neurosci 13 51-60 [221]

Merigan WH, Maunsell JHR (1990) Macaque vision after magnocellular lateral geniculate lesions Vis Neurosci 5 347–52 [229]

Mershon DH, Amerson TL (1980) Stability of the dark focus of accommodation Invest Ophthal Vis Sci 19 217-221 [359]

Merzenich MM, Kaas JH, Wall J, Nelson RJ, Sur M, Fellemen D (1983) Topographic reorganization of somatosensory cortical areas 3B and 1 in adult monkeys following restricted deafferentation Neurosci 8 33-55 [197]

Mestre DR, Brouchon M, Ceccaldi M, Poncet M (1992) Perception of optical flow in cortical blindness: a case report Neuropsychol 30 783-95 [183]

Meyer GE (1974) Pressure blindness and the interocular transfer of size aftereffects Percept Psychophys 16 222-4 [351]

Meyer RL (1982) Tetrodotoxin blocks the formation of ocular dominance columns in goldfish Science 218 589–91 [466]

Meyer-Franke A, Kaplan MR, Pfrieger FW, Barres BA (1995) Characterization of the signaling interactions that promote the survival and growth of developing retinal ganglion cells in culture Neuron 15 805-19 [442]

Michael CR (1968) Receptive fields of single optic nerve fibers in a mammal with an all cone retina. II. Directionally selective units J Neurophysiol 31 257-67 [203]

Michotte A (1963) La perception de la causalité Louvain: Publications Universitaires de Louvain, English translation, Methuen, London [160]

Miezin FM, Myerson J, Julesz B, Allman JM (1981) Evoked potentials to dynamic random–dot correlograms in monkey and man: a test for cyclopean perception Vis Res 21 177–9 [262]

Mijovoc P (1966) Medieval art in Yugoslavia. Decani. Publishing House Jugoslavija, Beograd, Yugoslavia [31]

Mikaelian HH (1975) Interocular generalization of orientation specific color aftereffects Vis Res 15 661–3 [351]

Mikhael S, Nicolle D, Vilis T (1995) Rotation of Listing's plane by horizontal vertical and oblique prism-induced vergence Vis Res 35 3273-54 [370, 410, 411]

Milder DG, Reinecke RD (1983) Phoria adaptation to prisms Arch Ophthal 40 339-42 [431]

Miles FA, Judge SJ, Optican LM (1987) Optically induced changes in the couplings between vergence and accommodation J Neurosci 7 2576–89 [391]

Miles WR (1930) Ocular dominance in human adults J Gen Psychol 3 412–30 [295]

Miller EK, Li L, Desimone R (1993) Activity of neurons in anterior inferior temporal cortex during a short-term memory task J Neurosci 13 1460-78 [222]

Miller EK, Erickson CA, Desimone R (1996) Neural mechanisms of visual working memory in prefrontal cortex of the macaque J Neurosci 16 5154-67 [222]

Miller JM, Robins D (1987) Extraocular muscle sideslip and geometry in monkeys Vis Res 27 381-92 [371]

Miller JM, Ono H, Steinbach MJ (1980) Additivity of fusional vergence and pursuit eye movements Vis Res 20 43–8 [420]

Miller JM, Anstis T, Templeton WB (1981) Saccadic plasticity: parametric adaptive control by retinal feedback J Exp Psychol HPP 7 356–66 [423]

Miller KD (1994) A model for the development of simple cell receptive fields and the ordered arrangement of orientation columns through activity–dependent competition between on– and offcenter inputs J Neurosci 14 409–41 [454]

Miller KD, Chapman, B, Stryker MP (1989a) Visual responses in adult cat visual cortex depend on N-methyl-D-aspartate receptors Proc Nat Acad Sci 86 5183-7 [462, 511]

Miller KD, Keller JB, Stryker MP (1989b) Ocular dominance column development: analysis and simulation Science 275 605–15 [464]

Miller RJ (1978) Temporal stability of the dark focus of accommodation Am J Optom Physiol Opt 55 447-50 [359]

Miller RJ (1980) Ocular vergence-induced accommodation and its relation to dark focus Percept Psychophys 28 125-32 [360]

Miller SM, Liu BB, Ngo TT, Hooper G, Riek S, Carson RG, Pettigrew JD (2000) Interhemispheric switching mediates perceptual rivalry Curr Biol 10 383-92 [314]

Miller WT, Sutton RS, Werbos PJ (1991) Neural networks for control MIT Press, Cambridge MA [95, 269]

Milleret C, Buser P (1984) Receptive field sizes and responsiveness to light in area 18 of the adult cat after chiasmotomy Post–operative evolution; role of visual experience Exp Brain Res 57 73–81 [232]

Milleret C, Gary-Bobo E, Buisseret P (1988) Comparative development of cell properties in cortical area 18 of normal and dark-reared kittens Exp Brain Res 71 8-20 [453]

Milner AD (1997) Vision without knowledge Philos Trans R Soc 352 1279-56 [229]

Milner AD, Goodale MA (1995) The visual brain in action Oxford University Press, Oxford [229]

Milner B (1974) Hemispheric specialization: scope and limits In The neurosciences: Third study programme (ed FO Schmitt, FG Worden) pp 75–89 MIT Press, Cambridge MA [489]

Milner PM (1974) A model for visual shape recognition Psychol Rev 81 521–35 [132]

Minciacchi D, Antonini A (1984) Binocularity in the visual cortex of the adult cat does not depend on the integrity of the corpus callosum Behav Brain Res **13** 183–92 [233]

Minken AWH, van Gisbergen JAM (1994) A three-dimensional analysis of vergence movements at various levels of elevation Exp Brain Res **101** 331-45 [415]

Minken AWH, van Gisbergen JAM (1996) Dynamical version-vergence interactions for a binocular implementation of Donder's law Vis Res **36** 853-67 [370]

Minken AWH, Gielen CCAM, van Gisbergen JAM (1995) An alternative three-dimensional interpretation of Hering's equal-innervation law for version and vergence eye movements Vis Res **35** 93-102 [415]

Minsky M (1961) Steps toward artificial intelligence Proceedings of the Institute of Radio Engineering **49** 8-30 [153]

Minsky M (1975) A framework for representing knowledge In The psychology of computer vision (ed PH Winston) pp McGraw-Hill, New York [100]

Minsky M , Papert E (1969) Perceptrons MIT Press, Cambridge, MA [105]

Minucci PK, Connors MM (1964) Reaction time under three viewing conditions: binocular dominant eye and nondominant eye J Exp Psychol **67** 298–75 [328]

Mioche L, Perenin MT (1986) Central and peripheral residual vision in humans with bilateral deprivation amblyopia Exp Brain Res **62** 259–72 [501]

Mioche L, Singer W (1989) Chronic recordings from single sites of kitten striate cortex during experience-dependent modifications of receptive-field properties J Neurophysiol **62** 185-97 [507]

Misantone LJ, Stelzner DJ (1974) Behavioural manifestations of competition of retinal endings for sites in doubly innervated frog optic tectum Exp Neurol **45** 364-76 [464]

Mishkin M (1982) A memory system in the monkey Philos Trans Roy Soc B **289** 85-95 [222]

Mitchell AM, Ellerbrock VJ (1955) Fixation disparity and the maintenance of fusion in the horizontal meridian Am J Optom Arch Am Acad Optom **32** 520–34 [382, 383]

Mitchell DE (1966a) Retinal disparity and diplopia Vis Res **6** 441–51 [274, 277]

Mitchell DE (1966b) A review of the concept of "Panum's fusional areas" Am J Optom Arch Am Acad Optom **43** 387–401 [272, 273]

Mitchell DE (1970) Properties of stimuli eliciting vergence eye movements and stereopsis Vis Res **10** 145–62 [397, 400, 403]

Mitchell DE (1988a) The extent of visual recovery from early monocular or binocular visual deprivation in kittens J Physiol **395** 639–60 [516, 517]

Mitchell DE (1988b) The recovery from early monocular visual deprivation in kittens In Perceptual development in infancy (ed A Yonas) pp 1–34 Erlbaum, Hillsdale N J [517, 528]

Mitchell DE (1988c) Animal models of human strabismic amblyopia In Advances in neural and behavioral development (ed PG Shinkman) Vol 3 pp 209–69 Ablex, Norwood NJ [528]

Mitchell DE (1991) The long-term effectiveness of different regimens of occlusion on recovery from early monocular deprivation in kittens Philos Tr R Soc B **333** 51–79 [528]

Mitchell DE, Blakemore C (1970) Binocular depth perception and the corpus callosum Vis Res **10** 49–54 [233]

Mitchell DE, Timney B (1982) Behavioural measurement of normal and abnormal development of vision in the cat In Analysis of visual behavior (ed DJ Ingle, MA Goodale, RJW Mansfield) pp 483–523 MIT Press, Cambridge MA [482, 501, 516]

Mitchell DE, Ware C (1974) Interocular transfer of visual after-effect in normal and stereoblind humans J Physiol **236** 707–21 [492]

Mitchell DE, Reardon J, Muir DW (1975) Interocular transfer of the motion aftereffect in normal and stereoblind observers Exp Brain Res **22** 163–75 [493]

Mitchell DE, Cynader M, Movshon JA (1977) Recovery from the effects of monocular deprivation J Comp Neurol **176** 53–63 [517]

Mitchell DE, Kaye M, Timney B (1979) Assessment of depth perception in cats Perception **8** 389–96 [548]

Mitchell DE, Ruck M, Kaye MG, Kirby S (1984) Immediate and long-term effects on visual acuity of surgically induced strabismus in kittens Exp Brain Res **55** 420-30 [503, 505]

Mitchell DE, Murphy KM, Dzioba HA, Horne JA (1986) Optimization of visual recovery from early monocular deprivation in kittens: implications for occlusion therapy in the treatment of amblyopia Clin Vis Sci **1** 173–7 [528]

Mitchell DE, Gingras G, Kind PC (2001) Initial recovery of vision after early monocular deprivation in kittens is faster when both eyes are open Proc Nat Acad Sci **98** 11662-67 [517]

Mitchell RT, Liaudansky LH (1955) Effect of differential adaptation of the eyes upon threshold sensitivity J Opt Soc Am **45** 831–4 [333]

Mitchison GJ, Crick F (1982) Long axons within the striate cortex: their distribution orientation and patterns of connection Proc Nat Acad Sci **79** 3661–5 [196]

Mitrofanis J, Guillery RW (1993) New views of the thalamic reticular nucleus in the adult and the developing brain TINS **16** 270-5 [458]

Miyashita EM, Hevner R, Wasserman KM, Martinez S, Rubenstein LR (1999) Early neocortical regionalization in the absence of thalamic innervation Science **285** 906-9 [447]

Miyashita Y (1988) Neuronal correlate of visual associative long-term memory in the primate temporal cortex Nature **335** 817-20 [222]

Mize RR, Marc RE (ed) (1992) GABA in the retina and central visual system Prog Brain Res Vol 90 Elsevier, London [178, 191]

Mohindra I (1975) A technique for infant vision examination Am J Optom Physiol Opt **52** 867-70 [473]

Mohindra I, Zwaan J, Held R, Brill S, Zwaan F (1985) Development of acuity and stereopsis in infants with esotropia Ophthal **92** 691-7 [519]

Mohn G, Van Hof-van Duin J (1983) On the relation of stereoacuity to interocular transfer of the motion and the tilt aftereffects Vis Res **23** 1087–96 [492, 493]

Mohn G, Van Hof-van Duin J (1986) Development of the binocular and monocular visual fields of human infants during the first year of life Clin Vis Sci **1** 51–64 [473]

Mohn G, Van Hof-van Duin J (1991) Development of spatial vision In Spatial vision (ed D Regan) pp 179–211 CRC Press, Boca Raton [473]

Mok D, Ro A, Cadera W, Crawford JD, Vilis T (1992) Rotation of Listing's plane during vergence Vis Res **32** 2055–64 [370, 415]

Molnár Z, Blakemore C (1999) Development of signals influencing the growth and termination of thalamocortical axons in organotypic culture Exp Neurol 156 363-93 [451]

Molnár Z, Adams R, Blakemore C (1998) Mechanisms underlying the early establishment of thalamocortical connections in the rat J Neurosci **18** 5723-45 [458]

Molyneux W (1692) A treatise of Dioptricks B Tooke, London [23]

Montague PR, Gancayco CD, Winn MJ, Marchase RB, Friedlander MJ (1994) Role of NO production in NMDA receptor–mediated neurotransmitter release in cerebral cortex Science **293** 973–7 [461]

Montero VM (1992) A quantitative study of synaptic contacts on interneurons and relay cells of the cat lateral geniculate nucleus Exp Brain Res **86** 257–70 [178]

Monyer H, Sprengel R, Schoepfer R, Herb A, Higuchi M, Lomeli H, Burnashev N, Sakmann B, Seeburg PH (1992) Heteromeric NMDA receptors: molecular and functional distinction of subtypes Science **256** 1217-22 [190]

Moody SA (2000) Cell fate and cell migration in the developing cerebral cortex Academic Press, New York [451]

Mooney R, Madison DV, Shatz CJ (1993) Enhancement of transmission at the developing retinogeniculate synapse Neuron **10** 815-25 [444]

Moore RJ, Spear PD, Kim CBY, Xue JT (1992) Binocular processing in the cat's dorsal lateral geniculate nucleus. III Spatial frequency orientation and direction sensitivity of nondominant–eye influences Exp Brain Res **89** 588–98 [180, 312]

Moores E, Frisby JP, Buckley DE, Fawcett A (1998) Vergence control across saccades in dyslexic adults Ophthal Physiol Opt **18** 452-62 [374]

Moraglia G, Schneider B (1990) Effects of direction and magnitude of horizontal disparities on binocular unmasking Perception **19** 581–93 [338]

Moraglia G, Schneider B (1991) Binocular unmasking with vertical disparity Can J Psychol **45** 353–66 [338]

Moran G, Timney B, Sorenson L, Desrochers B (1983) Binocular depth perception in the Meerkat (*Suricata suricatta*) Vis Res **23** 965–9 [548]

Moran J, Desimone R (1985) Selective attention gates visual processing in the extrastriate cortex Science **229** 782–4 [207]

Morgan H, Symmes D (1982) Amazing 3-D Little Brown Co, Boston [70]

Morgan MJ, Hotopf WHN (1989) Perceived diagonals in grids and lattices Vis Res **29** 1005–15 [124]

Morgan MJ, Regan D (1987) Opponent models for line interval discrimination: interval and vernier performance compared Vis Res **27** 107-18 [145]

Morgan MJ, Mason AJS, Solomon JA (1997) Blindsight in normal subjects Nature **385** 401-2 [293]

Morgan MW (1944) Accommodation and its relationship to convergence Am J Optom Arch Am Acad Optom 21 183-95 [388]

Morgan MW (1968) Accommodation and vergence Am J Optom Arch Am Acad Optom **45** 417–53 [360, 391]

Morley JW, Lindsey JW, Judge SJ (1988) Prism–adaptation in a strabismic monkey Clin Vis Sci **3** 1–8 [383]

Morley JW, Judge SJ, Lindsey JW (1992) Role of monkey midbrain near–response neurons in phoria adaptation J Neurophysiol **67** 1475–92 [431]

Morris JS, Friston KJ, Dolan RJ (1997) Neural responses to salient visual stimuli Proc R Soc **294** 769-75 [245]

Morrone MC, Burr DC, Maffei L (1982) Functional implications of cross–orientation inhibition of cortical visual cells. I Neurophysiological evidence Proc R Soc B **216** 335–54 [313]

Morrone MC, Burr DC, Speed HD (1987) Cross-orientation inhibition in cats is GABA mediated Exp Brain Res **67** 635-44 [313]

Morrone MC, Burr DC, Fiorentini A (1993) Development of infant contrast sensitivity to chromatic stimuli Vis Res **33** 2535-52 [470]

Morrone MC, Tosetti M, Montanaro D, Fiorentini A, Cioni G, Burr DC (2000) A cortical area that responds specifically to optic flow, revealed by fMRI Nature Neurosci **3**,1322-8 [224]

Moschovakis AK (1995) Are laws that govern behavior embedded in the structure of the CNS? The case of Hering's law Vis Res **35** 3207-16 [419, 429]

Moschovakis AK, Scudder CA, Highstein SM (1990) A structural basis for Hering's law: projections to extraocular motoneurons Science **278** 1118–19 [419]

Moss SJ, Smart TG (2001) Constructing inhibitory synapses Nature Rev Neurosci **2** 240-50 [191]

Motter BC (1991) Beyond extrastriate cortex: the parietal visual system In Vision and visual disfunction (ed AL Leventhal) Vol IV pp 371–87 MacMillan, London [223]

Motter BC (1993) Focal attention produces spatially selective processing in visual cortical areas V1 V2 and V4 in the presence of competing stimuli J Neurophysiol **70** 909–19 [207]

Motter BC (1994) Neural correlates of attentive selection for color or luminance in extrastriate area V4 J Neurosci **14** 2178-89 [207]

Motter BC, Poggio GF (1984) Binocular fixation in the rhesus monkey: spatial and temporal characteristics Exp Brain Res **54** 304–14 [395]

Moulden BP (1980) After–effects and the integration of patterns of neural activity within a channel Philos Tr R Soc B **290** 39–55 [342, 345]

Mountcastle VB (1957) Modality and topographic properties of single neurons of cat's somatic sensory cortex J Neurophysiol **20** 408–34 [209]

Mountcastle VB (1997) The columnar organization of the neocortex Brain **120** 701-22 [209]

Mountcastle VB (1998) Perceptual neuroscience. The cerebral cortex Harvard University Press, Cambridge, MA [446]

Mountcastle VB, Lynch JC, Georgopoulos A, Sakata H, Acuna C (1975) Posterior parietal association cortex of the monkey Command functions for operations within extrapersonal space J Neurophysiol **38** 871-908 [143, 227]

Mousavi MS, Schalkoff RJ (1994) ANN impementation fo stereo vision using a muliti-layer feedback architecture IEEE Tr Man Mach Cyber **24** 1220-38 [268]

Movshon JA (1976) Reversal of the behavioural effects of monocular deprivation in the kitten J Physiol **291** 175–87 [517]

Movshon JA, Blakemore C (1974) Functional reinnervation in kitten visual cortex Nature **251** 504-5 [516]

Movshon JA, Dürsteler MR (1977) Effects of brief periods of unilateral eye closure on the kitten's visual system J Neurophysiol **40** 1255-65 [507]

Movshon JA, Kiorpes L (1988) Analysis of the development of spatial contrast sensitivity in monkey and human infants J Opt Soc Am A **5** 2166-72 [467]

Movshon JA, Lennie P (1979) Pattern–selective adaptation in visual cortical neurones Nature New Biol **278** 850–2 [303]

Movshon JA, Newsome WT (1996) Visual response properties of striate cortical neurons projecting to area MT in macaque monkey J Neurosci **16** 7733-41 [224]

Movshon JA, Chambers BEI, Blakemore C (1972) Interocular transfer in normal humans and those who lack stereopsis Perception **1** 483–90 [492]

Movshon JA, Thompson ID, Tolhurst DJ (1978) Spatial and temporal contrast sensitivity of neurones in areas 17 and 18 of the cat's visual cortex J Physiol **283** 101–20 [201, 202, 248]

Movshon JA, Eggers HM, Gizzi MS, Hendrickson AE, Kiorpes L, Boothe RG (1987) Effects of early unilateral blur of the macaque's visual system. III Physiological observations J Neurosci **7** 1340-51 [509]

Mower GD, Christen WG (1985) Role of visual experience in activating critical period in cat visual cortex J Neurophysiol **53** 572-89 [516]

Mower GD, Christen WG (1989) Evidence for an enhanced role of GABA inhibition in visual cortical dominance of cats reared with abnormal monocular experience Devel Brain Res **45** 211-18 [510]

Mower GD, Burchfiel JL, Duffy FH (1981a) The effects of dark rearing on the development and plasticity of the lateral geniculate nucleus Devel Brain Res **1** 418-27 [499]

Mower GD, Berry D, Burchfiel JL, Duffy FH (1981b) Comparison of the effects of dark-rearing and binocular suture on development and plasticity of cat visual cortex Brain Res **220** 255-67 [516]

Mower GD, Caplan CJ, Letsou G (1982) Behavioral recovery from binocular deprivation in the cat Behav Brain Res **4** 209-15 [501]

Mower GD Christen WG, Caplan CJ (1983) Very brief exposure eliminates plasticity in the cat visual cortex Science **221** 178-80 [516]

Mower GD, Caplan CJ, Christen WG, Duffy FH (1985) Dark rearing prolongs physiological but not anatomical plasticity of the cat visual cortex J Comp Neurol **235** 448-66 [500]

Mowforth P, Mayhew JEW, Frisby JP (1981) Vergence eye movements made in response to spatial–frequency–filtered random–dot stereograms Perception **10** 299–304 [393]

Mueller BK (1999) Growth cone guidance Ann Rev Neurosci **22** 351-88 [438]

Mueller TJ (1990) A physiological model of binocular rivalry Vis Neurosci **4** 63–73 [311]

Mueller TJ, Blake R (1989) A fresh look at the temporal dynamics of binocular rivalry Biol Cyber **61** 223-32 [287, 310]

Müller CM, Best J (1989) Ocular dominance plasticity in adult cat visual cortex after transplantation of cultured astrocytes Nature **342** 427–30 [457]

Müller H (1854) Über die entoptische Wahrnehmung der Netzhautgefässe insbesondere als Beweismittel für die Lichtperception durch die nach hinten gelegenen Netztzhautelemente Verhandlungen der Physiologischen Medizin Gesellschaft Würzburg **5** 411 [26, 164]

Müller J (1829) Zur Vergleichenden Physiologie des Gesichtssinnes des Menschen und der Thiere Cnobloch Leipzig [58, 421]

Müller J (1843) Elements of physiology Vol 2 pp 1147–8 (Translated by W Baly) Tayler and Walton, London [117, 388]

Müller JR, Metha AB, Krauskopf J, Lennie P (1999) Rapid adaptation in visual cortex to the structure of images Science **285** 1405-8 [121]

Munk H (1879) Physiologie der Sehsphäre der Grosshirnrinde Centralblatt für praktische Augenheilkunde **3** 255-66 [27]

Munk MHJ, Roelfsema PR, König P, Engel AK, Singer W (1996) Role of reticular activation in the modulation of intracortical synchronization Science **272** 271-4 [132, 134]

Munoz P, Semmlow JL, Yuan, W, Alvarez TL (1999) Short term modification of disparity vergence eye movements Vis Res **39** 1695-705 [425]

Muntz WRA (1961) Interocular transfer in *Octopus vulgaris* J Comp Physiol Psychol **54** 49-55 [354]

Murasugi CM, Salzman CD, Newsome WT (1993) Microstimulation in visual area MT: effects of varying pulse amplitude and frequency J Neurosci **13** 1719-29 [348]

Murata A, Gallese V, Luppino G, Kaseda M, Sakata H (2000) Selectivity for the shape, size, and orientation of objects for grasping in neurons of monkey parietal area AIP J Neurophysiol **83** 2580-601 [227]

Murata T, Shimizu H (1993) Oscillatory binocular system and temporal segmentation of stereoscopic depth surfaces Biol Cyber **68** 381–91 [131]

Murch GM (1972) Binocular relationships in a size and color orientation specific aftereffect J Exp Psychol **93** 30–4 [350]

Murch GM (1974) Color contingent motion aftereffects: single or multiple levels of processing Vis Res **14** 1181–4 [351]

Murdoch JR, McGhee CNJ, Glover V (1991) The relationship between stereopsis and fine manual dexterity: pilot study of a new instrument Eye **5** 642-43 [5]

Murphy CJ, Howland HC (1983) Owl eyes: accommodation corneal curvature and refractive state J Comp Physiol **151** 277-84 [547]

Murphy KM, Mitchell DE (1987) Reduced visual acuity in both eyes of monocularly deprived kittens following a short or long period of reverse occlusion J Neurosci **7** 1529–36 [528]

Murphy KM, Mitchell DE (1991) Vernier acuity of normal and visually deprived cats Vis Res **31** 253–66 [522]

Murphy KM, Jones DG, Van Sluyters RC (1995) Cytochrome-oxidase blobs in cat primary visual cortex J Neurosci **15** 4196-208 [204]

Murphy PC, Sillito AM (1989) The binocular input to cells in the feline dorsal lateral geniculate nucleus (dLGN) J Physiol **415** 393-408 [180]

Murphy PC, Duckett SG, Sillito AM (1999) Feedback connections to the lateral geniculate nucleus and cortical response properties Science **286** 1552-4 [179]

Murray ADN, Calcutt C (1990) The incidence of amblyopia in long-standing untreated infantile esotropia In Strabismus and early ocular motility disorders (ed EC Campos) pp 45-9 MacMillan, London [374, 528]

Murray E (1939) Binocular fusion and the locus of 'yellow' Am J Psychol **52** 117–21 [283]

Mussap AJ, Levi DM (1995) Binocular processes in vernier acuity J Opt Soc Am A **12** 225-33 [337]

Mussap AJ, Levi DM (1999) Orientation-based segmentation in strabismic amblyopia Vis Res **39** 411-18 [524]

Mussap AJ, Levi DM (2000) Amblyopic deficits in detecting a dotted line in noise Vis Res **40** 3297-309 [524]

Mustari MJ, Cynader M (1981) Prior strabismus protects kitten cortical neurons from the effects of monocular deprivation Brain Res **211** 165-70 [516]

Muybridge E (1899) Animals in motion Chapman Hall, London [71]

Myers RE (1955) Interocular transfer of pattern discrimination in cats following section of crossed optic fibres J Comp Physiol Psychol **48** 470-3 [354]

Nachmias J, Sansbury RV (1974) Grating contrast: discrimination may be better than detection Vis Res **14** 1039–42 [323]

Nachmias J, Weber A (1975) Discrimination of simple and complex gratings Vis Res **15** 217-23 [126]

Nadarajah B, Brunstrom JE, Grutzendler J, Wong ROL (20001) Two modes of radial migration in early development of the cerebral cortex Nature Neurosci **4** 143-9 [450]

Nagel A (1868) Über das Vorkommen von wahren Rollungen des Auge um die Gesichtslinie Arch für Ophthal **14** 228–46 [416]

Nakagawa Y, Johnson JE, O'Leary DDM (1999) Graded and areal expression patterns of regulatory genes and cadherins in embryonic neocortex independent of thalamocortical input J Neurosci 15 10877-85 . [447]

Nakamura H, Gattass R, Desimone R, Ungerleider LG (1993) The modular organization of projections from areas V1 and V2 to areas V4 and TEO in macaques J Neurosci **13** 3681-91 [220]

Nakamura H, Kuroda T, Wakita M, Kusunoki M, Kato A, Mikami A, Sakata H, Itoh K (2001) From three-dimensional space vision to prehensile hand movements: the lateral intraparietal area links the area V3A and the anterior intraparietal area in macaques J Neurosci **21** 8174-87 [227]

Nakayama K (1975) Coordination of extraocular muscles In Basic mechanisms of ocular motility and their clinical implications (ed B Lennerstrand, P Bach–y–Rita) pp 193–208 Pergamon, New York [371]

Nakayama K (1978) A new method of determining the primary position of the eye using Listings law Am J Optom Arch Am Acad Optom **55** 331–6 [370]

Nánez JE (1988) Perception of impending collision in 3-to 6-week-old human infants Infant Behav Devel **11** 447-63 [478]

Nánez JE, Yonas A (1994) Effects of luminance and texture motion on infant defensive reactions to optical collision Infant Behav Devel **17** 165-74 [479]

Narasimhan PT, Jacobs RE (1964) Neuroanatomical micro-magnetic resonance imaging In Brain Mapping (ed AW Toga, JC Mazziotta) pp 147-59 Academic Press, New York [182]

Naya Y, Yoshida M, Miyashita Y (2001) Backward spreading of memory-retrieval signal in the primate temporal cortex Science **291** 661-4 [222]

Nealey TA, Maunsell JHR (1994) Magnocellular and parvocellular contributions to the responses of neurons in macaque striate cortex J Neurosci **14** 2069-79 [219]

Nedergaard M (1994) Direct signaling from astrocytes to neurons in cultures of mammalian brain cells Science **293** 1768-71 [114]

Needham J (1962) Science and civilization in China Vol 4 Part 1 Cambridge University Press, London [18, 19]

Neill RA, Fenelon B (1988) Scalp response topography to dynamic random dot stereograms EEG Clin Neurophysiol **69** 209-217 [262]

Neisser U (1967) Cognitive psychology Appleton–Century–Crofts, New York [87, 154, 155]

Nelson JI, Frost BJ (1978) Orientation selective inhibition from beyond the classic visual receptive field Brain Res **139** 359–65 [197, 200, 206]

Nelson JI, Frost BJ (1985) Intracortical facilitation among co–oriented co–axially aligned simple cells in cat striate cortex Exp Brain Res **61** 54 61 [124, 193, 196, 200]

Nelson JI, Kato H, Bishop PO (1977) Discrimination of orientation and position disparities by binocularly activated neurons in cat striate cortex J Neurophysiol **40** 290–83 [255]

Neuenschwander S, Singer W (1996) Long-range synchronization of oscillatory light responses in geniculate nucleus Nature **379** 728-33 [133, 136]

Newell A, Simon HA (1972) Human problem solving Prentice-Hall, Englewood Cliffs, NJ [155]

Newhouse M, Uttal WR (1982) Distribution of stereoanomalies in the general population Bull Psychonom Soc **20** 48–50 [487]

Newsome WT, Paré EB (1988) A selective impairment of motion perception following lesions of the middle temporal visual area (MT) J Neurosci **8** 2201-11 [224]

Newsome WT, Wurtz RH, Komatsu H (1988) Relation of cortical areas MT and MST to pursuit eye movements. II Differentiation of retinal from extraretinal inputs J Neurophysiol **60** 604–20 [224]

Newton I (1704) Opticks Smith and Walford, London. 1979 printing, Dover, New York [55]

Ng AYK, Stone J (1982) The optic nerve of the cat: appearance and loss of axons during normal development Devel Brain Res **5** 293–71 [442]

Ngo TT, Miller, SM, Liu GB, Pettigrew JD (2000) Binocular rivalry and perceptual coherence Curr Biol **10** R134-6 [299]

Nguyen VA, Freeman A, Wenderoth P (2001) The depth and selectivity of suppression in binocular rivalry. Percept Psychophys **63** 348-60 [301]

Nicholson W (1802) Narrative and explanation of the appearance of phantoms and other figures in the exhibition of the phantasmagoria: with remarks on the philosophical use of common occurrences J Nat Philosoph Arts 1 147-51 [61]

Niclou SP, Jia L, Raper JA (2000) Slit2 is a repellent for retinal ganglion cell axons J Neurosci **20** 4962-74 [447]

Nicol JAC (1989) The eyes of fishes Clarendon Press, Oxford [539]

Niebur E, Koch C, Rosen C (1993) An oscillation–based model for the neuronal basis of attention Vis Res **33** 2798–802 [136]

Nieder A, Wagner H (2000) Horizontal-disparity tuning of neurons in the visual forebrain of the behaving barn owl J Neurophysiol **83** 2967-79 [547]

Nieder A, Wagner H (2001) Hierarchical processing of horizontal disparity information in the visual forebrain of behaving owls J Neurosci **21** 4514-22 [547]

Nikara T, Bishop PO, Pettigrew JD (1968) Analysis of retinal correspondence by studying receptive fields of binocular single units in cat striate cortex Exp Brain Res **6** 353–72 [231, 236, 238]

Nishida S, Ashida H (2000) A hierarchical structure of motion system revealed by interocular transfer of flicker motion aftereffects Vis Res **40** 295-78 [348]

Nishida S, Ashida H (2001) A motion aftereffect seen more strongly by the non-adapted eye: evidence of multistage adaptation in visual motion processing Vis Res **41** 561-70 [348]

Nishida S, Sato T (1995) Motion aftereffect with flickering test patterns reveals higher stages of motion processing Vis Res **35** 477-90 [348]

Nishida S, Ashida H, Sato T (1994) Complete transfer of motion aftereffect with flickering test Vis Res **34** 2707-16 [348]

Nixon RB, Helveston EM, Miller K, Archer SM, Ellis FD (1985) Incidence of strabismus in neonates Am J Ophthal **100** 798-801 [374]

Noble J (1966) Mirror-images and the forebrain commissures of the monkey Nature **211** 1293-6 [355]

Noble J (1968) Paradoxical interocular transfer of mirror-image discrimination in the optic chiasm sectioned monkey Brain Res **10** 127-51 [355]

Noctor SC, Flint AC, Weissman TA, Dammerman RS, Kriegstein AR (2001) Neurons derived from radial glial cells establish units in neocortex Nature **409** 714-20 [448]

Noda H, Tamaki Y, Iwama K (1972) Binocular units in the lateral geniculate nucleus of chronic cats Brain Res **41** 81-99 [179]

Nomura M (1993) A model for neural representation of binocular disparity in striate cortex: distributed representation and veto mechanism Biol Cyber **69** 165–71 [248]

Nomura M, Matsumoto G, Fugiwara S (1990) A binocular model for the simple cell Biol Cyber **63** 237–42 [248]

Norcia AM (1996) Abnormal motion processing and binocularity: infantile esotropia as a model system for effects of early interruptions of binocularity Eye **10** 259-65 [526]

Norcia AM, Tyler CW (1985) Spatial frequency sweep VEP: visual acuity during the first year of life Vis Res **25** 1399–408 [259, 468]

Norcia AM, Sutter EE, Tyler CW (1985) Electrophysiological evidence for the existence of coarse and fine disparity mechanisms in human Vis Res **25** 1603 11 [263]

Norcia AM, Tyler CW, Allen D (1986) Electrophysiological assessment of contrast sensitivity in human infants Am J Optom Physiol Opt **63** 12–15 [467]

Norcia AM, Tyler CW, Hamer RD (1988) High visual contrast sensitivity in the young human infant Invest Ophthal Vis Sci **29** 44–9 [467]

Norcia AM, Tyler CW, Hamer RD (1990) Development of contrast sensitivity in the human infant Vis Res **30** 1475-86 [467, 468]

Norcia AM, Garcia H, Humphry R, Holmes A, Hamer RD, Orel-Bixler D (1991) Anomalous motion VEPs in infants and in infantile esotropia Invest Ophthal Vis Sci **32** 436-9 [472, 526]

Norcia AM, Hamer RD, Jampolsky A, Orel-Bixler D (1995) Plasticity of human motion processing mechanisms following surgery for infantile esotropia Vis Res **35** 3279-96 [526]

Norling JA (1939) Three-dimensional motion pictures J Soc Motion Pict Engin **33** 612-34 [71]

Norman HF, Norman, JF, Bilotta J (2000) The temporal course of suppression during binocular rivalry Perception **29** 831-41 [291, 310]

Norman JF, Dawson TE, Buler AK (2000) The effect of age upon the perception of depth and 3-D shape from differential motion and binocular disparity Perception **29** 1335-9 [484]

Norren DV, Vos JJ (1974) Spectral transmission of the human ocular media Vis Res **14** 1237-44 [163]

Norris CR, Kalil K (1991) Guidance of callosal axons by radial glia in the developing cerebral cortex J Neurosci **11** 3481-92 [453]

Norris KS (1968) The echolocation of marine mammals. In The biology of marine mammals (ed TH Anderson) pp 391-423 New York, Academic Press [548]

North RV, Sethi B, Henson DB (1986) Effects of prolonged forced vergence upon the adaptation system Ophthal Physiol Opt **6** 391-6 [383]

North RV, Sethi B, Owen K (1990) Prism adaptation and viewing distance Ophthal Physiol Opt **10** 81-5 [383]

North RV, Henson DB, Smith TJ (1993) Influence of proximal accommodative and disparity stimuli upon the vergence system Ophthal Physiol Opt **13** 239-43 [387]

Nothdurft HC, Gallant JL, Van Essen DC (2000) Response profiles to texture border patterns in area V1 Vis Neurosci **17** 421-36 [197]

Nuzzi G, Franchi A (1983) Binocular interaction in visual–evoked responses: summation facilitation and inhibition in a clinical study of binocular vision Ophthal Res **15** 291–82 [483]

Nye PW (1968) The binocular acuity of the pigeon measured in terms of the modulation transfer function Vis Res **8** 1041–53 [544]

O'Brien B (1951) Vision and resolution in the central retina J Opt Soc Am **41** 882-94 [171, 173]

O'Dell C, Boothe RG (1997) The development of stereoacuity in infant rhesus monkeys Vis Res **37** 2975--84 [481]

O'Donovan M (1999) The origin of spontaneous activity in developing networks of the vertebrate nervous system Curr Opin Neurobiol **9** 94-104 [459]

O'Keefe J, Nadel L (1978) The hippocampus as a cognitive map Clarendon Press, Oxford [223]

O'Keefe LP, Berkley MA (1991) Binocular immobilization induced by paralysis of the extraocular muscles of one eye: Evidence for an interocular proprioceptive mechanism J Neurophysiol **66** 2022–33 [420]

O'Keefe LP, Movshon JA (1998) Processing of first- and second-order motion signals by neurons in area MT of the macaque monkey Vis Neurosci **15** 305-17 [224]

O'Kusky J, Colonnier M (1982) Postnatal changes in the number of neurons and analysis in normal and monocularly deprived animals J Comp Neurol **210** 291–306 [186, 451, 509]

O'Leary A, Wallach H (1980) Familiar size and linear perspective as distance cues in stereoscopic depth constancy Percept Psychophys **27** 131–5 [151]

O'Malley CD (1964) Andreas Vesalius of Brussels Cambridge University Press, London [22]

O'Rourke NA, Dailey ME, Smith SJ, McConnell SK (1992) Diverse migratory pathways in the developing cerebral cortex Science **258** 299–302 [450]

O'Shea RP (1987) Chronometric analysis supports fusion rather than suppression theory of binocular vision Vis Res **27** 781–91 [306]

O'Shea RP, Corballis PM (2000) Binocular rivalry in a split-brain observer Invest Ophthal Vis Sci **41** S732 [314]

O'Shea RP, Wilson RG, and Duckett A (1993) The effects of contrast reversal on the direct, indirect, and interocularly-transferred tilt aftereffect NZ J Psychol **22** 94-100 [345]

O'Shea RP, Blake R (1986) Dichoptic temporal frequency differences do not lead to binocular rivalry Percept Psychophys **39** 59–63 [294]

O'Shea RP, Crassini B (1981a) The sensitivity of binocular rivalry suppression to changes in orientation assessed by reaction–time and forced–choice techniques Perception **1 0** 283–93 [301]

O'Shea RP, Crassini B (1981b) Interocular transfer of the motion after–effect is not reduced by binocular rivalry Vis Res **21** 801–4 [303]

O'Shea RP, Crassini B (1982) The dependence of cyclofusion on orientation Percept Psychophys **32** 195–6 [278]

O'Shea RP, Crassini B (1984) Binocular rivalry occurs without simultaneous presentation of rival stimuli Percept Psychophys **36** 296–76 [294]

O'Shea RP, Williams DR (1996) Binocular rivalry with isoluminant stimuli visible only via short-wavelength-sensitive cones Vis Res **36** 1561-71 [291]

O'Shea RP, Blake R, Wolfe JM (1994a) Binocular rivalry and fusion under scotopic luminances Perception **23** 771-84 [289]

O'Shea RP, McDonald AA, Cumming A, Peart D, Sanderson G, Moltenov CB (1994b) Interocular transfer of the movement aftereffect in central and peripheral vision of people with strabismus Invest Ophthal Vis Sci **35** 313–17 [493]

O'Shea RP, Simms AJH, Govan DG (1997) The effect of spatial frequency and field size on the spread of exclusive visibility in binocular rivalry Vis Res **37** 175-83 [297]

O'Shea WF, Ciuffreda KJ, Fisher SK, Tannen B, Super P (1988) Relation between distance heterophoria and tonic vergence Am J Optom Physiol Opt **65** 787–93 [298, 376]

Obermayer K, Blasdel GG (1993) Geometry of orientation and ocular dominance columns in monkey striate cortex J Neurosci **13** 4114-29 [209, 211]

Odom JV, Chao GM (1995) Models of binocular luminance interaction evaluated using visually evoked potential and psychophysical measures: a tribute to M Russell Harter Internat J Neurosci **80** 255-80 [332]

Odom JV, Hoyt CS, Marg E (1981) Effects of natural deprivation and unilateral eye patching on visual acuity of infants and children Arch Ophthal **99** 1412–16 [528]

Oettermann S (1997) The panorama. History of a mass medium New York, Zone Books [61]

Ogawa S, Tank DW, Menon R, Ellermann JM, Kim SG, Merkle H, Ugurbil K (1992) Intrinsic signal changes accompanying sensory stimulation: functional brain mapping with magnetic resonance imaging Proc Nat Acad Sci **89** 5951-5 [182]

Ogle KN (1939c) Relative sizes of ocular images of the two eyes in asymmetrical convergence Arch Ophthal **22** 1046-67 [406]

Ogle KN (1964) Researches in binocular vision Hafner, New York [8, 273]

Ogle KN, Ellerbrock VJ (1946) Cyclofusional movements Arch Ophthal **36** 700–35 [411, 416]

Ogle KN, Prangen A de H (1951) Further considerations of fixation disparity and the binocular fusional processes Am J Ophthal **34** 57–72 [379]

Ogle KN, Prangen A de H (1953) Observations on vertical divergences and hyperphorias Arch Ophthal **49** 313–34 [272, 273, 379, 381, 383, 384, 406]

Ogle KN, Schwartz JT (1959) Depth of focus of the human eye J Opt Soc Ame **49** 273-80 [360]

Ogle KN, Wakefield JM (1967) Stereoscopic depth and binocular rivalry Vis Res **7** 89–98 [306]

Ogle KN, Mussey F, Prangen A de H (1949) Fixation disparity and the fusional processes in binocular single vision Am J Ophthal **32** 1069–87 [379]

Ogle KN, Martens TG, Dyer JA (1967) Oculomotor imbalance in binocular vision and fixation disparity Lea and Febiger, Philadelphia [8, 376, 379, 380, 381, 383, 389]

Ohmi M, Howard IP, Everleigh B (1986) Directional preponderance in human optokinetic nystagmus Exp Brain Res **63** 387-94 [525]

Ohtsuka K, Nagasaka Y (1999) Divergent axon collaterals from the rostral superior colliculus to the pretectal accommodation-related areas and the omnipause neuron area in the cat J Comp Neurol 413 68-76 [431]

Ohtsuki H (1994) Critical period for restoration of normal stereoacuity in acute-onset comitant esotropia Am J Ophthal **118** 502-8 [374]

Ohwaki S (1960) On the destruction of geometrical illusions in stereoscopic observation Tohoku Psychol Folia **29** 27–36 [353]

Ohzawa I, Freeman RD (1986a) The binocular organization of simple cells in the cat's visual cortex J Neurophysiol **56** 221–42 [247, 254, 260]

Ohzawa I, Freeman RD (1986b) The binocular organization of complex cells in the cat's visual cortex J Neurophysiol **5 6** 273–60 [248, 260]

Ohzawa I, Freeman RD (1988) Cyclopean visual evoked potentials: a new test of binocular vision Vis Res **28** 1167–70 [260]

Ohzawa I, Sclar G, Freeman RD (1985) Contrast gain control in the cat's visual system J Neurophysiol **54** 651-67 [248, 340]

Ohzawa I, DeAngelis GC, Freeman RD (1990) Stereoscopic depth discrimination in the visual cortex: neurons ideally suited as disparity detectors Science **279** 1037–41 [237, 248, 249]

Ohzawa I, DeAngelis GC, Freeman RD (1996) Encoding of binocular disparity by simple cells in the cat's visual cortex J Neurophysiol **75** 1779-805 [252, 254]

Ohzawa I, DeAngelis GC, Freeman RD (1997) Encoding of binocular disparity by simple cells in the cat's visual cortex J Neurophysiol **77** 2879-910 [249, 250]

Okada A, Lansford R, Weimann JM, Fraser SE, McConnell SK (1999) Imaging cells in the developing nervous system with retrovirus expressing modified green fluorescent protein Exp Neurol **156** 394-406 [433]

Okada M, Erickson A, Hendrickson, AE (1994) Light and electron microscopic analysis of synaptic development in Macaca monkey retina as described by immunocytochemical labeling for the synaptic vesicle protein J Comp Neurol **339** 535-58 [436]

Olberg RM, Worthington AH, Venator KR (2000) Prey pursuit and interception in dragonflies J Comp Physiol A, **186** 155-62 [534]

Oliet SHR, Piet R, Poulain DA (2001) Control of glutamate clearance and synaptic efficiency by glial coverage of neurons Science **292**, 923-6 [188]

Oliver G, Gruss P (1997) Current views on eye development TINS **20** 415-21 [434]

Olshausen BA, Field DJ (1996) Emergence of simple-cell receptive field properties by learning a sparse code for natural images Nature **381** 607-9 [94]

Olson CR (1980) Spatial localization in cats reared with strabismus J Neurophysiol **43** 792-806 [527]

Olson CR, Freeman RD (1975) Progressive changes in kitten striate cortex during monocular vision J Neurophysiol **38** 29-32 [507]

Olson CR, Freeman RD (1978a) Eye alignment in kittens J Neurophysiol **41** 848-59 [474]

Olson CR, Freeman RD (1978b) Monocular deprivation and recovery during sensitive period in kittens J Neurophysiol **4 1** 65-74 [507, 516]

Olson CR, Freeman RD (1980a) Profile of the sensitivity period for monocular deprivation in kittens Exp Brain Res **39** 17-21 [515]

Olson CR, Freeman RD (1980b) Cumulative effect of brief daily periods of monocular vision on kittens striate cortex Exp Brain Res **38** 53–6 [507]

Olum V (1956) Developmental differences in the perception of causality Am J Psychol **69** 417-23 [160]

Ono H (1980) Hering's law of equal innervation and vergence eye movement Am J Optom Physiol Opt **57** 578–85 [418]

Ono H (1983) The combination of version and vergence In Vergence eye movements: Basic and clinical aspects (ed MC Schor, KJ Ciuffreda) pp 373–400 Butterworth, Boston [421]

Ono H, Mapp AP (1995) A restatement and modification of Wells-Hering's laws of visual direction Perception **27** 237-52 [378]

Ono H, Nakamizo S (1978) Changing fixation in the transverse plane at eye level and Hering's law of equal innervation Vis Res **18** 511–19 [421]

Ono H, Tam WJ (1981) Asymmetrical vergence and multiple saccades Vis Res **21** 739–43 [421]

Ono H, Wade NJ (1985) Resolving discrepant results of the Wheatstone experiment Psychol Res **47** 135–42 [65]

Ono H, Hasdorf A, Osgood CE (1966) Binocular rivalry as a function of incongruity of meaning Scand J Psychol 7 225-33 [309]

Ono H, Komoda M, Mueller ER (1971a) Intermittent stimulation of binocular disparate colors and central fusion Percept Psychophys 9 343–7 [283]

Ono H, Nakamizo S, Steinbach MJ (1978) Nonadditivity of vergence and saccadic eye movement Vis Res **18** 735–39 [421]

Oohira A, Zee DS (1992) Disconjugate ocular motor adaptation in rhesus monkey Vis Res **32** 489–97 [424]

Oohira A, Zee DS, Guyton DL (1991) Disconjugate adaptation to long–standing large–amplitude spectacle–corrected anisometropia Invest Ophthal Vis Sci **32** 1693–703 [424]

Ooi TL, He ZJ (1999) Binocular rivalry and visual awareness: the role of attention Perception **28** 551-74 [293, 308]

Ooi TL, Loop MS (1994) Visual suppression and its effect upon color and luminance sensitivity Vis Res **34** 2997-3003 [300]

Optican LM (1982) Saccadic dysmetria In Functional basis of ocular motility disorders (ed G Lennerstrand, DS Zee, EL Keller) pp 441–51 Pergamon, New York [426]

Optican LM, Robinson DA (1980) Cerebellar–dependent adaptive control of primate saccadic system J Neurophysiol **44** 1058–76 [426]

Optican LM, Zee DS, Chu FC (1985) Adaptive response to ocular muscle weakness in human pursuit and saccadic eye movements J Neurophysiol **54** 110–22 [425]

Oram MW, Perrett DI (1992) Time course of neural responses discriminating different views of the face and head J Neurophysiol **68** 70–84 [123]

Oram MW, Perrett DI (1996) Integration of form and motion in the anterior superior temporal polysensory area (STPa) of the macaque monkey J Neurophysiol **76** 109-29 [229]

Oram MW, Wiener MC, Lestienne R, Richmond BJ (1999) Stochastic nature of precisely timed spike patterns in visual system neuronal responses J Neurophysiol **81** 3021-33 [130]

Orbach HS, Van Essen DC (1993) *In vivo* tracing of pathways and spatio–temporal activity patterns in rat visual cortex using voltage sensitive dyes Exp Brain Res **94** 371–92 [181]

Orban GA, Sunaert S, Todd JT, Van Hecke P, Marchal G (1999) Human cortical regions involved in extracting depth from motion Neuron **27** 929-40 [224]

Oster G (1965) Optical art App Optics **4** 1359-69 [282]

Osterberg G (1935) Topography of the layer of rods and cones in the human retina Acta Ophthal Supp 6 1–103 [164]

Oswald I (1957) After–images from retina and brain Quart J Exp Psychol **9** 88–100 [86, 302, 344]

Otsu Y, Kimura F, Tsumoto T (1995) Hebbian induction of LTP in visual cortex: perforated patch-clamp study in cultured neurons J Neurophysiol **74** 2737-44 [461]

Ott M (2001) Chameleons have independent eye movements but synchronize both eyes during saccadic prey tracking Exp Brain Res **139** 173-9 [543]

Ott D, Seidman SH, Leigh RJ (1992) The stability of human eye orientation during visual fixation Neurosci Lett **142** 183–6 [365, 415]

Ott M, Schaeffel F (1995) A negatively powered lens in the chameleon Nature **373** 692-4 [543]

Ott M, Schaeffel F, Kirmse W (1998) Binocular vision and accommodation in prey-catching chameleons J Comp Physiol **182** 319-30 [543]

Over R, Long N, Lovegrove W (1973) Absence of binocular interaction between spatial and color attributes of visual stimuli Percept Psychophys **13** 534–40 [352]

Owens DA (1980) A comparison of accommodative responsiveness and contrast sensitivity for sinusoidal gratings Vis Res **20** 159-67 [361]

Owens DA, Leibowitz HW (1980) Accommodation convergence and distance perception in low illumination Am J Optom Physiol Opt **57** 540–50 [372, 390]

Owens DA, Leibowitz HW (1983) Perceptual and motor consequences of tonic vergence In Vergence eye movements: Basic and clinical aspects (ed MC Schor, KJ Ciuffreda) pp 25–98 Butterworth, Boston [372, 380]

Owens DA, Tyrrell RA (1992) Lateral phoria at distance: contributions of accommodation Invest Ophthal Vis Sci **33** 2733–43 [376]

Owens RL, Higgins KE (1983) Long-term stability of the dark focus of accommodation Am J Optom Physiol Opt **60** 32-8 [359]

Owens WC, Hughes WF (1948) Results of surgical treatment of congenital cataract Arch Ophthal **39** 339-50 [519]

Owsley C (1983) The role of motion in infants' perception of solid shape Perception **12** 707-17 [480]

Owsley C, Sloane ME (1990) Vision and aging In Handbook of neuropsychology (ed F Boller, J Grafman) Vol 4 pp 229-49 Elsevier, Amsterdam [484]

Oyster CW (1999) The human eye Sinauer, Sunderland MA [164]

Paap KR, Ebenholtz SM (1977) Concomitant direction and distance aftereffects of sustained convergence: a muscle potentiation explanation for eye–specific adaptation Percept Psychophys **21** 307–14 [355]

Packwood J, Gordon B (1975) Stereopsis in normal domestic cat Siamese cat and cat raised with alternating monocular occlusion J Neurophysiol **38** 1485-99 [497]

Pagano CC, Turvey MT (1995) The inertial tensor as a basis for the perception of limb orientation J Exp Psychol HPP **21** 1070-87 [143]

Page WK, Duffy CJ (1999) MST neuronal responses to heading direction during pursuit eye movements J Neurophysiol **81** 596-610 [225]

Paige GD (1989) The influence of target distance on eye movement responses during vertical linear motion Exp Brain Res **77** 585–93 [427]

Paige GD (1991) Linear vestibulo–ocular (LVOR) and modulation by vergence Acta Otolaryngol **48** 282–6 [427]

Paige GD, Telford L, Seidman SH, Barnes GR (1998) Human vestibuloocular reflex and its interactions with vision and fixation distance during linear and angular head movement J Neurophysiol **80** 2391-404 [427, 428]

Palecek SP, Loftus JC, Ginsberg MH, Lauffenburger DA, Horwitz AF (1997) Integrin-ligand binding properties govern cell migration speed through cell-substratum adhesiveness Nature **385** 537-40 [437]

Palmer DA (1961) Measurement of the horizontal extent of Panum's area by a method of constant stimuli Optical Acta **8** 151–9 [273, 277]

Palmer EA, von Noorden GK (1978) The relationship between fixation disparity and heterophoria Am J Ophthal **86** 172-5 [379]

Panofsky E (1927) Die Perspektive als symbolische Form Vorträge der Bibliothek Warburg 1925-25 Leipzig and Berlin. [31]

Panofsky E (1940) The Codex Huygens and Leonardo da Vinci's art theory The Warburg Institute, London [35]

Panum PL (1858) Physiologische Untersuchungen über das Sehen mit zwei Augen Schwers, Keil [57, 272]

Paolini M, Distler C, Bremmer F, Lappe M, Hoffmann CP (2000) Response to continuously changing optic flow in area MST J Neurophysiol **84** 730-43 [225]

Pape HC, Eysel UT (1986) Binocular interactions in the lateral geniculate nucleus of the cat: GABAergic inhibition reduced by dominant afferent activity Exp Brain Res **61** 295-71 [180]

Papert S (1964) Stereoscopic synthesis as a technique for locating visual mechanisms MIT Quart Prog Rep **73** 239–43 [349]

Paradiso MA, Shimojo S, Nakayama K (1989) Subjective contours tilt aftereffects and visual cortical organization Vis Res **29** 1205–13 [346]

Pardhan S, Rose D (1999) Binocular and monocular detection of Gabor patches in binocular two-dimensional noise Perception **28** 203-15 [322]

Pardhan S, Gilchrist J, Douthwaite W (1989) The effect of spatial frequency on binocular contrast inhibition Ophthal Physiol Opt **9** 46–9 [322]

Paris J, Prestrude AM (1975) On the mechanism of the interocular light adaptation effect Vis Res **15** 595–603 [333]

Park JK, Williams BP, Alberta JA, Stiles CD (1999) Bipotent progenitor cells process conflicting cues for neurons and glia in a hierarchical manner J Neurosci **19** 10383-9 [448]

Park RS, Park GE (1933) The center of ocular rotation in the horizontal plane J Physiol **104** 545-52 [368]

Parker A, Hawken M (1985) Capabilities of monkey cortical cells in spatial-resolution tasks J Opt Soc Am A **2** 1101-14 [201]

Parker AJ, Newsome WT (1998) Sense and the single neuron Ann Rev Neurosci **21** 227-77 [77, 114, 121, 250]

Parker J (1858) To make stereoscopic spectacles Photograph J **5** 69 [69]

Parkes L, Lund J, Angelucci A, Solomon JA, Morgan M (2001) Compulsory averaging of crowded orientation signals in human vision Nature Neurosci **4** 739-44 [79]

Parks MM (1969) The monofixation syndrome Tr Am Ophthal Soc **67** 608-56 [381]

Parks TE (1984) Illusory figures: a (mostly) atheoretical review Psychol Bull **95** 282–300 [123]

Parnavelas JG (1999) Glial cell lineages in the rat cerebral cortex Exp Neurol **156** 418-29 [450]

Parnavelas JG (2000) The origin and migration of cortical neurones: new vistas TINS **23** 129-31 [450]

Parodi O, Combe P, Ducom JC (1996) Temporal coding in vision: coding by the spike arrival times leads to oscillations in the case of moving targets Biol Cyber **74** 497-509 [136]

Parra P, Gulyas AI, Miles R (1998) How many subtypes of inhibitory cells in the hippocampus? Neuron **20** 983-93 [192]

Parronchi A (1964) Studi su la prospettiva Aldo Martello, Milan [33]

Parsons S, Jones G (2000) Acoustic identification of twelve species of echolocating bat by discriminant function analysis and artificial neural networks J Exp Biol **203** 2941-56 [548]

Pass AF, Levi DM (1982) Spatial processing of complex stimuli in the amblyopic visual system Invest Ophthal Vis Sci **23** 780-6 [522]

Pasternak T, Tompkins J, Olson CR (1995) The role of striate cortex in visual function of the cat J Neurosci **15** 1940-50 [168, 186]

Pasupathy A, Connor CE (1999) Responses to contour features in macaque area V4 J Neurophysiol **82** 2790-502 [221]

Patel SS, Ogmen H, White JM, Jiang BC (1997) Neural network model of short-term horizontal disparity vergence dynamics Vis Res **37** 1383-99 [405]

Patel SS, Jiang BC, White JM, Ogmen H (1999) Nonlinear alteration of transient vergence dynamics after sustained convergence Optom Vis Sci **76** 656-63 [383]

Patterson R, Fox R (1984a) Stereopsis during continuous head motion Vis Res **27** 2001–3 [395]

Patterson R, Fox R (1984b) The effect of testing method on stereoanomaly Vis Res **27** 403–8 [487]

Patterson R, Fox R (1990) Metacontrast masking between cyclopean and luminance stimuli Vis Res **30** 439–48 [340]

Patterson R, Martin WL (1992) Human stereopsis Hum Factors **34** 669–92 [8]

Payne BR (1990) Representation of the ipsilateral visual field in the transition zone between areas 17 and 18 of the cat's cerebral cortex Vis Neurosci **4** 445–74 [231]

Payne BR, Berman N (1983) Functional organization of neurons in cat striate cortex: variations in preferred orientation and orientation selectivity with receptive-field type ocular dominance and location in the visual-field map J Neurophysiol **49** 1051-68 [214]

Payne BR, Berman N, Murphy EH (1981) A quantitative assessment of eye alignment in cats after corpus callosum transection Exp Brain Res **43** 371–6 [429]

Payne BR, Pearson HE, Berman N (1984a) Role of corpus callosum in functional organization of cat striate cortex J Neurophysiol **52** 570–94 [233]

Payne BR, Pearson HE, Berman N (1984b) Deafferentation and axotomy of neurons in cat striate cortex: time course of changes in binocularity following corpus callosum transection Brain Res **307** 201-15 [233]

Payne WH (1967) Visual reaction times on a circle about the fovea Science **155** 481-82 [473]

Peck CK, Blakemore C (1975) Modification of single neurons in the kitten's visual cortex after brief periods of monocular visual experience Exp Brain Res **22** 57-68 [515]

Peckham RH, Hart WM (1960) Binocular summation of subliminal repetitive visual stimulation Am J Ophthal **49** 1121–5 [329]

Peiper A (1963) Cerebral function in infancy and childhood Pitman, London [478]

Peli E (1983) Normal stereo acuity despite anisometropic-amblyopia J Am Optom Assoc **54** 919-21 [529]

Peng G, Qiu F, Ginzburg VV, Jasnow D, Balazs AC (2000) Forming supramolecular networks from nonoscale rods in binary, phase-separating mixtures Science **288** 1802-4 [212]

Penn AA, Riquelme PA, Feller MB, Shatz CJ (1998) Competition in retinogeniculate pattering driven by spontaneous activity Science **279** 2108-12 [444]

Penne A, Baraldi P, Fonda S, Ferrar F (1987) Incremental binocular amplitude of the pattern visual evoked potential during the first five months of life: electrophysiological evidence of the development of binocularity Doc Ophthal **65** 15-23 [483]

Penrose R (1979) The topology of ridge systems Ann Hum Genet **42** 435-44 [211]

Perenin MT, Vighetto A (1988) Optic ataxia: a specific disruption in visuomotor mechanisms Brain **111** 643-74 [228]

Perkel DH, Gerstein GL, Moore GP (1967) Neuronal spike trains and stochastic point processes. II Simultaneous spike trains Biophys J **7** 419–40 [452]

Perlmutter AL, Kertesz AE (1982) Human vertical fusional response under open and closed loop stimulation to predictable and unpredictable disparity presentations IEEE Tr Biomed Engin **29** 57-61 [409]

Perrett DI, Rolls ET, Caan W (1982) Visual neurones responsive to faces in the monkey temporal cortex Exp Brain Res **47** 329-42 [222]

Perry E, Walker M, Grace J, Perry R (1999) Acetylcholine in mind" a neurotransmitter correlate of consciousness TINS **22** 273-8 [192]

Perry VH, Cowey A (1985) The ganglion cell and cone distributions in the monkey's retina: implications for central magnification factors Vis Res **25** 1795–1810 [184, 442]

Perry VH, Henderson Z, Linden R (1983) Postnatal changes in retinal ganglion cell and optic axon populations in the pigmented rat J Comp Neurol **219** 356–68 [442]

Perry VH, Oehler R, Cowey A (1984) Retinal ganglion cells that project to the dorsal lateral geniculate nucleus in the macaque monkey Neurosci **12** 1101–23 [177]

Peters A, Rockland KS (1994) Cerebral cortex Volume 10 Primary visual cortex in primates Plenum, New York [186, 187]

Petrig B, Julesz B, Kropfl W, Baumgartner G, Anliker M (1981) Development of stereopsis and cortical binocularity in human infants: electrophysiological evidence Science **213** 1402-5 [483]

Pettet MW, Gilbert CD (1992) Dynamic changes in receptive–field size in cat primary visual cortex Proc Nat Acad Sci **89** 8366–70 [198]

Pettigrew JD (1974) The effect of visual experience on the development of stimulus specificity by kitten cortical neurones J Physiol **237** 49–74 [453, 500]

Pettigrew JD (1978) A comparison of the visual-field representation of the visual Wulst of *Falco* and *Tyto* with a note on the evolution of frontal vision In Frontiers of visual science (ed S Cool, E Smith) pp 328-35 Springer, New York [546]

Pettigrew JD (1979) Binocular visual processing in the owl's telencephalon Proc R Soc B **204** 435–54 [547]

Pettigrew JD (1986) The evolution of binocular vision In Vis Neurosci (ed JD Pettigrew, KJ Sanderson, WR Levick) pp 208–222 Cambridge University Press, London [531, 547]

Pettigrew JD (1991) Evolution of binocular vision In The evolution of the eye and visual system (ed JR Cronly–Dillon, RL Gregory) pp 271–83 CRC Press, Boca Raton [548]

Pettigrew JD, Dreher B (1987) Parallel processing of binocular disparity in the cat's retinogeniculate pathways Proc R Soc B **232** 297–321 [231, 238]

Pettigrew JD, Konishi M (1976a) Neurons selective for orientation and binocular disparity in the visual Wulst of the barn owl Science **193** 675–8 [547]

Pettigrew JD, Konishi M (1976b) Effect of monocular deprivation on binocular neurones in the owl's visual Wulst Nature **294** 753-4 [547]

Pettigrew JD, Nikara T, Bishop PO (1968) Binocular interaction on single units in cat striate cortex: simultaneous stimulation by single moving slit with receptive fields in correspondence Exp Brain Res **6** 391–410 [59, 236, 237, 251, 500]

Pettigrew JD, Ramachandran VS, Bravo H (1984) Some neural connections subserving binocular vision in ungulates Brain Behav Evol **27** 65–93 [549]

Pham TA, Impey S, Storm DR, Stryker, MP (1999) CRE-mediated gene transcription in neocortical neuronal plasticity during the developmental critical period Neuron **22** 63-72 [512]

Phillips GC, Wilson HR (1984) Orientation bandwidths of spatial mechanisms measured by masking J Opt Soc Am A **1** 229-32 [199]

Phillips S, Stark L (1977) Blur: a sufficient accommodative stimulus Doc Ophthal **43** 65–89 [360]

Philpot BD, Sekhar AK, Shouval HZ, Bear MF (2001) Visual experience and deprivation bidirectionally modify the composition and function of NMDA receptors in visual cortex Neuron **29** 157-69 [462]

Piantanida TP (1986) Stereo hysteresis revisited Vis Res **29** 431–7 [280, 281]

Pick HL, Hay JC, Willoughby RH (1966) Interocular transfer of adaptation to prismatic distortion Percept Mot Skills **23** 131-5 [355]

Pickwell LD (1972) Hering's law of equal innervation and the position of the binoculus Vis Res **12** 1499–1507 [421]

Pickwell LD (1985) The increase in convergence inadequacy with age Ophthal Physiol Opt **5** 347-8 [393, 484]

Pickwell LD (1989) Binocular vision anomalies Butterworth-Heinemann, Oxford [8]

Pickwell LD, Hampshire R (1981) The significance of inadequate convergence Ophthal Physiol Opt **1** 13-18 [484]

Pickwell LD, Gilchrist JM, Hesler J (1988) Comparison of associated heterophoria measurements using the Mallett test for near vision and the Sheedy disparometer Ophthal Physiol Opt **8** 19-25 [377]

Pimentel B, Sanz C, Varela-Nieto I, Rapp UR, De Pablo F, de la Rosa EJ (2000) c-Raf regulates cell survival and retinal ganglion cell morphogenesis during neurogenesis J Neurosci **20** 3254-62 [456]

Pinckney GA (1964) Reliability of duration as a measure of the spiral aftereffect Percept Mot Skills **18** 375–6 [347]

Pini A (1993) Chemorepulsion of axons in the developing mammalian central nervous system Science **291** 95–9 [438]

Pinter RB, Nabet B (1992) Nonlinear vision: Determination of neural receptive fields function and networks CRC Press, London [95]

Pirchio M, Spinelli D, Fiorentini A, Maffei L (1978) Infant contrast sensitivity evaluated by evoked potentials Brain Res **141** 179–84 [467]

Pirenne MH (1943) Binocular and uniocular threshold of vision Nature **152** 698–9 [83, 318]

Pirenne MH (1970) Optics painting and photography Cambridge University Press, Cambridge [38, 40]

Pitblado CB (1979) Cerebral asymmetries in random-dot stereopsis: reversal of direction with changes in dot size Perception **8** 683–90 [489]

Pittman RN (1985) Release of plasminogen activator and a calcium-dependent metalloprotease from cultured sympathetic and sensory neurons Devel Biology **110** 91-101 [438]

Plateau J AF (1849) Troisiemè note sur de nouvelles applications curieuse de la persistance des impressions de la rétine Bull Acad R Sci Bel **16** 37-9 [71]

Plateau JAF (1850) Vierte Notiz über neue sonderbare Anwenduggen des Verweilens der Eindrücke auf die Netzhaut Poggendorff's Ann Physik Chem **80** 287–92 [347]

Platter F (1583) De corporis humani structura et usu König, Basel [22]

Plug C, Ross HE (1994) The natural moon illusion: a multifactor angular account Perception **23** 321-33 [18]

Pobuda M, Erkelens CJ (1993) The relationship between absolute disparity and ocular vergence Biol Cyber **68** 221–8 [398, 404, 405]

Poggio GF (1991) Physiological basis of stereoscopic vision In Vision and vision dysfunction Vol 9 Binocular vision (ed D Regan) pp 227–38 MacMillan, London [240]

Poggio GF, Fischer B (1977) Binocular interaction and depth sensitivity in striate and prestriate cortex of behaving rhesus monkey J Neurophysiol **40** 1392–405 [216, 239, 256]

Poggio GF, Poggio T (1984) The analysis of stereopsis Ann Rev Neurosci **7** 379–412 [239]

Poggio GF, Talbot WH (1981) Mechanisms of static and dynamic stereopsis in foveal cortex of the rhesus monkey J Physiol **315** 469–92 [240, 256]

Poggio GF, Motter BC, Squatrito S, Trotter Y (1985) Responses of neurons in visual cortex (VI and V2) of the alert Macaque to dynamic random–dot stereograms Vis Res **25** 397–406 [240]

Poggio GF, Gonzalez F, Krause F (1988) Stereoscopic mechanisms in monkey visual cortex: binocular correlation and disparity selectivity J Neurosci **8** 4531–50 [240]

Poggio T, Fahle M, Edelman S (1992) Fast perceptual learning in visual hyperacuity Science **256** 1018–21 [87, 354]

Pohl W (1973) Dissociation of spatial discrimination deficits following frontal and parietal lesions in monkeys J Comp Physiol Psychol **82** 227–39 [229]

Polak NA, Jones R (1990) Dynamic interactions between accommodation and convergence IEEE Tr Biomed Engin **3 7** 1011–14 [391]

Polat U (1999) Functional architecture of long-range perceptual interactions Spatial Vis **12** 143-62 [524]

Polat U, Norcia AM (1996) Neurophysiological evidence for contrast dependent long-range facilitation and suppression in the human cortex Vis Res **36** 2099-109 [196]

Polat U, Sagi D (1993) Lateral interactions between spatial channels: suppression and facilitation revealed by lateral masking experiments Vis Res **33** 993-9 [196]

Polat U, Sagi D (1994) The architecture of perceptual spatial interactions Vis Res **34** 73–8 [124, 196]

Polat U, Sagi D, Norcia AM (1997) Abnormal long-range spatial interactions in amblyopia Vis Res **37** 737-44 [524]

Pollack P (1977) The picture history of photography Adams, New York [65]

Pollen DA (1981) Phase relationships between adjacent simple cells in the visual cortex of the cat Science 212 1409-11 [141, 209]

Pollen DA, Feldon SE (1979) Spatial periodicities of periodic complex cells in the visual cortex cluster at one-half octave intervals Invest Ophthal Vis Sci **18** 429-34 [201]

Polonsky A, Blake R, Braun J, Heeger DJ (2000) Neuronal activity in human primary visual cortex correlates with perception during binocular rivalry Nature Neurosci **3** 1153-9 [312]

Polyak S (1941) The retina University of Chicago Press, Chicago [16, 164, 170]

Polyak S (1957) The vertebrate visual system pp 109–110 University of Chicago Press, Chicago [19, 22, 28, 59164, 445]

Pope DR, Edwards M, Schor CM (1999a) Orientation and luminance polarity tuning of the transient-vergence system Vis Res **39** 575-84 [402]

Popovic Z, Sjöstrand J (2001) Resolution, separation of retinal ganglion cells, and cortical magnification in humans Vis Res **41** 1313-19 [184]

Popple AV, Levi DM (2000) Amblyopes see true alignment where normal observers see illusory tilt Proc Nat Acad Sci **97** 11667-72 [524]

Popple AV, Smallman HS, Findlay JM (1998) The area of spatial integration for initial horizontal disparity vergence Vis Res **3 8** 319-29 [396]

Porciatti V, Fontanesi G, Raffaelli A, Bagnoli P (1990) Binocularity in the little owl *Athene nocta.* II Properties of visually evoked potentials from the Wulst in response to monocular and binocular stimulation with sine wave gratings Brain Behav Evol **35** 40-8 [547]

Porrill J, Ivins JP, Frisby JP (1999b) The variation of torsion with vergence and elevation Vis Res **39** 3934-50 [415]

Porta GB della (1558) Magiae naturalis English edition of 1658 reprinted by Basic Books, New York 1957 [20]

Porta GB della (1593) De refractione Optices Parte Carlinum and Pacem Naples [20, 305]

Porter JD, Baker RS (1992) Prenatal morphogenesis of primate extraocular muscle: neuromuscular junction formation and fiber type differentiation Invest Ophthal Vis Sci **33** 657–70 [429]

Porter JD, Guthrie BL, Sparks DL (1983) Innervation of monkey extraocular muscles: localization of sensory and motor neurons by retrograde transport of horseradish peroxidase J Comp Neurol **218** 208–19 [428, 429]

Porterfield W (1738) An essay concerning the motion of our eyes Edinburgh Medical Essays and Observations **4** 127-294 [24]

Porterfield W (1759) A treatise on the eye: The manner and phenomena of vision A Miller, London [56]

Poteser M, Kral K (1995) Visual distance discrimination between stationary targets in praying mantis: an index of the use of motion parallax J Exp Biol **198** 2127-37 [533]

Potts MJ, Harris JP (1979) Dichoptic induction of movement aftereffects contingent on color and on orientation Percept Psychophys **29** 25–31 [352]

Pouget A, Sejnowski TJ (1994) A neural model of the cortical representation of egocentric distance Cerebral Cortex **4** 314-29 [254]

Pozzo A (1593) Rules and examples of perspective First published in Rome in Latin. An English translation appeared in London in 1707, which was reproduced in 1971 by Benjamin Blom, Inc., London [36]

Prablanc C, Tzavaras A, Jeannerod M (1975) Adaptation of the two arms to opposite prism displacements Quart J Exp Psychol **27** 667-71 [355]

Pratt-Johnson JA, Tillson G (1981) Visual results after removal of congenital cataracts before the age of 1 year Can J Ophthal **16** 19-21 [519]

Pratt-Johnson JA, Tillson G (1983) Sensory results following treatment of infantile esotropia Can J Ophthal **18** 175-7 [374, 528]

Pratt-Johnson JA, Tillson G (1989) Unilateral congenital cataract: binocular status after treatment J Ped Ophthal Strab **29** 72-75 [502]

Prazdny K (1985c) On the disparity gradient limit for binocular fusion Percept Psychophys **37** 81–3 [276]

Predebon J (1994) Convergence responses to monocularly viewed objects: implications for distance perception Perception **23** 303-19 [386]

Prentice WCH (1948) New observations of binocular yellow J Exp Psychol **38** 284–8 [283]

Prévost A (1843) Essai sur la theorie de la vision binoculaire Ramboz, Geneva [58]

Prévost P (1804) Essais de philosophie ou étude de l'esprit humain Paschoud, Geneva [57]

Price DJ, Blakemore C (1985) The postnatal development of the association projection from visual cortical area 17 to area 18 in the cat J Neurosci **5** 2743–52 [453]

Price DJ, Ferrer JMR, Blakemore C, Kato N (1994) Postnatal development and plasticity of corticocortical projections from area 17 to area 18 in the cat's visual cortex J Neurosci **14** 2747-62 [453]

Priestley J (1772) The history and present state of discoveries relating to vision light and colours Johnson, London [18, 21, 24]

Prince SJD, Eagle RA (2000a) Weighted directional energy model of human stereo correspondence Vis Res **40** 1143-55 [268]

Prince SJD, Pointon AD, Cumming BG, Parker AJ (2002a) Quantitative analysis of the responses of V1 neurons to horizontal disparity in dynamic random-dot stereograms J Neurophysiol **87** 191-208 [242, 246, 249]

Prince SJD, Cumming BG, Parker AJ (2002b) Range and mechanism of encoding of horizontal disparity in macaque V1 J Neurophysiol **87** 209-21 [252]

Prince SJD, Pointon AD, Cumming BG, Parker AJ (2000c) The precision of single neuron responses in cortical area V1 during stereoscopic depth judgments J Neurosci **20** 3387-3400 [250]

Proffitt DR, Gilden, DL (1989) Understanding natural dynamics J Exp Psychol: HPP **15** 284-93 [160]

Provine RR, Enoch JM (1975) On voluntary ocular accommodation Percept Psychophys **17** 209-12 [359]

Provine RR, Westerman JA (1979) Crossing the midline: limits of early eye-hand behavior Child Devel **50** 437-41 [477]

Ptito A, Zatorre RJ, Larson WL, Tosoni C (1991) Stereopsis after unilateral anterior temporal lobectomy Dissociation between local and global measures Brain **114** 1323-33 [489]

Ptito M, Lepore F, Guillemot JP (1992) Loss of stereopsis following lesions of cortical areas 17–18 in the cat Exp Brain Res **89** 521–30 [238, 488]

Ptolemy C Optics See Lejeune (1956 1989) and Smith (1996) [12]

Pugh M (1958) Visual distortion in amblyopia Br J Ophthal **42** 449–60 [522, 523]

Purkinje J (1825) Beobachtungen und Versuche zur Physiologie der Sinne Vol 2 p 60 JG Calve, Prague [347]

Purves D, Lichtman JW (1985) Principles of neural development Sinauer, Sunderland MA [485]

Pylyshyn ZW (1973) What the mind's eye tells the mind's brain Psychol Bull **80** 1-24 [154]

Qian N (1994) Computing stereo disparity and motion with known binocular cell properties Neural Comput **6** 390-404 [266, 268]

Qian N, Andersen RA (1997) A physiological model for motion-stereo integration and a unified explanation of Pulfrich-like phenomena Vis Res **37** 1683-98 [268]

Qian N, Zhu Y (1997) Physiological computation of binocular disparity Vis Res **37** 1811-27 [249, 266, 268]

Quereau J (1954) Some aspect of torsion Arch Ophthal **5** 783–8 [370]

Quick MW, Boothe RG (1989) Measurement of binocular alignment in normal monkeys and in monkeys with strabismus Invest Ophthal Vis Sci **30** 1159-68 [375]

Quick MW, Tigges M, Gammon JA, Boothe RG (1989) Early abnormal visual experience induces strabismus in infant monkeys Invest Ophthal Vis Sci **30** 1012–17 [374]

Quick RF (1974) A vector–magnitude model of contrast detection Kybernetik **16** 65–7 [83]

Quinlan EM, Philpot BD, Huganir RL, Bear MF (1999) Rapid, experience-dependent expression of synaptic NMDA receptors in visual cortex *in vivo* Nature Neurosci **2** 352-7 [462]

Raff MC, Barres BA, Burne JF, Coles HS, Ishizaki Y, Jacobson MD (1993) Programmed cell death and the control of cell survival: lessons from the nervous system Science **292** 695–700 [442]

Rainey BB (2000) The effect of prism adaptation on the response AC/A ratio Ophthal Physiol Opt **20** 199-206 [389]

Rakic P (1974) Neurons in rhesus monkey visual cortex: systematic relation between time of origin and eventual disposition Science **183** 425–7 [448]

Rakic P (1976) Prenatal genesis of connections subserving ocular dominance in the rhesus monkey Nature **291** 467–71 [443, 455]

Rakic P (1981) Development of visual centers in the primate brain depends on binocular competition before birth Science **214** 928–31 [444, 503]

Rakic P (1988) Specification of cerebral cortical areas Science **271** 170–6 [209, 448, 451]

Rakic P, Riley KP (1983) Overproduction and elimination of retinal axons in the fetal rhesus monkey Science **219** 1441–4 [442]

Rakic P, Bourgeois JP, Eckenhoff MF, Zecevic N, Goldman–Rakic PS (1986) Concurrent overproduction of synapses in diverse regions of the primate cerebral cortex Science **232** 232–5 [451]

Ramachandran VS (1975) Suppression of apparent movement during binocular rivalry Nature **256** 122–3 [304]

Ramachandran VS (1991) Form motion and binocular rivalry Science **251** 950–1 [349]

Ramachandran VS, Sriram S (1972) Stereopsis generated with Julesz patterns in spite of rivalry imposed by colour filters Nature **237** 347–8 [304]

Ramachandran VS, Cobb S, Levi L (1994a) The neural locus of binocular rivalry and monocular diplopia in intermittent exotropes Neuroreport **5** 1141-44 [296]

Ramachandran VS, Cobb S, Levi L (1994b) Monocular double vision in strabismus Neuroreport **5** 1418 [296]

Ramoa AS, McCormick DA (1994) Enhanced activation of NMDA receptor responses at the immature retinogeniculate synapse J Neurosci **14** 2098-105 [444]

Ramoa AS, Freeman RD, Macy A (1985) Comparison of response properties of cells in the cat's visual cortex at high and low luminance levels J Neurophysiol **54** 61-72 [168, 200]

Ramoa AS, Shadlen M, Freeman RD (1987) Dark–reared cats: unresponsive cells become visually responsive with microiontophoresis of an excitatory amino acid Exp Brain Res **65** 658–65 [499]

Ramoa AS, Paradiso MA, Freeman RD (1988) Blockade of intracortical inhibition in kitten striate cortex: effects on receptive field properties and associated loss of ocular dominance plasticity Exp Brain Res **73** 285-96 [510]

Ramoa AS, Mower AF, Liao D, Jafri SIA (2001) Suppression of cortical NMDA receptor function prevents development of orientation selectivity in the primary visual corte J Neurosci **21** 4299-309 [454, 511]

Ramón y Cajal S (1901) Recreaciones estereoscópicas y binoculares La Fotgrapfía **27** 41-8. English translation available from Ian Howard. [27]

Ramón y Cajal S (1911) Histologie du system nerveux de l'homme et des vertébrés A Maloine, Paris [58, 175, 235]

Ramón y Cajal S (1937) Recollections of my life (Translated by EH Graige) Mem Am Philos Soc 3 [27]

Ransom–Hogg A, Spillmann L (1980) Perceptive field size in fovea and periphery of the light–and dark–adapted retina Vis Res **20** 221–8 [184]

Rao RPN, Ballard DH (1999) Predictive coding in the visual cortex: functional interpretation of some extra-classical receptive-field effects Nature Neurosci **2** 79-87 [123]

Rao SC, Rainer G, Miller EK (1997) Integration of what and where in the primate prefrontal cortex Science **276** 821-4 [228]

Rao VM (1977) Tilt illusion during binocular rivalry Vis Res **17** 327–8 [303]

Raper JA, Grunewald EB (1990) Temporal retinal growth cones collapse on contact with nasal retinal axons Exp Neurol **109** 70-4 [440]

Raphan T (1998) Modeling control of eye orientation in three dimensions. I Role of muscle pulleys in determining saccadic trajectory J Neurophysiol **79** 2953-67 [371]

Rasengane TA, Allen D, Manny RE (1997) Development of temporal contrast sensitivity in human infants Vis Res 37 1747-54 [471]

Rashbass C (1970) The visibility of transient changes of luminance J Physiol **210** 165–86 [331]

Rashbass C (1981) Reflections on the control of vergence In Models of oculomotor behavior and control (ed BL Zuber) pp 139–48 CRC Press, Boca Raton [405]

Rashbass C, Westheimer G (1961a) Disjunctive eye movements J Physiol **159** 339–60 [397, 398, 404, 409]

Rashbass C, Westheimer G (1961b) Independence of conjugate and disjunctive eye movements J Physiol **159** 361-4 [420]

Ratcliff F (1965) Mach Bands: Quantitative studies on neural networks in the retina Holden–Day, San Francisco [122]

Ratcliff G, Davies-Jones GAB (1972) Defective visual localization in focal brain wounds Brain **95** 49-60 [228]

Rauschecker JP (1991) Mechanisms of visual plasticity: Hebb synapses NMDA receptors and beyond Physiol Rev **71** 587–615 [458]

Rauschecker JP, Singer W (1981) The effects of early visual experience on the cat's visual cortex and their possible explanation by Hebb synapses J Physiol **310** 215–39 [508]

Rauschecker JP, Campbell FW, Atkinson J (1973) Colour opponent neurones in the human visual system Nature **275** 42–3 [295]

Rauschecker JP, von Grünau MW, Poulin C (1987) Centrifugal organization of direcfion preferences in the cat's lateral suprasylvian cortex and its relation to flow field processing J Neurosci 7 943-58 [224]

Raybourn MS (1975) Spatial and temporal organization of the input to frog optic tectum Brain Behav Evol **11** 161–78 [540]

Raymond JE (1993) Complete interocular transfer of motion adaptation effects on motion coherence thresholds Vis Res **33** 1865–70 [347, 348, 493]

Read JCA (2002) A Bayesian model of stereopsis depth and motion direction discrimination Biol Cybern **86** 117-36 [102]

Read JCA, Cumming B, and Parker AJ (2000). Local models can explain the reduced response of disparity-tuned V1 neurons to anti-correlated images. Soc Neurosci Abstr **26** 1845 [249]

Reading RW (1983a) Binocular vision—foundations and applications Butterworth, London [8]

Reading RW (1992) Vergence errors: some hitherto unreported aspects of fixation disparity Optom Vis Sci **69** 538-43 [379]

Reading RW (1994) Variations in the monocular components of fixation disparity Optom Vis Sci **71** 371-6 [379]

Recanzone GH, Schreiner CE, Merzenich MM (1993) Plasticity in the frequency representation of primary auditory cortex following discrimination training in adult owl monkeys J Neurosci **13** 87–103 [208]

Rees G, Friston K, Koch C (2000) A direct quantitative relationship between the functional properties of human and macaque V5 Nature Neurosci **3** 716-23 [224]

Reese BE, Baker GE (1992) Changes in fiber organization within the chiasmatic region of mammals Vis Neurosci **9** 527–33 [440]

Reese BE, Maynard TM, Hocking DR (1994) Glial domains and axonal reordering in the chiasmatic region of the developing ferret J Comp Neurol **349** 303-27 [441]

Reeves A, Peachey NS, Auerbach E (1986) Interocular sensitization to a rod-detected test Vis Res **29** 1119-27 [333]

Regal DM (1981) Development of critical flicker frequency in human infants Vis Res **21** 549-55 [471]

Regal DM, Boothe R, Teller DY, Sackett GP (1976) Visual acuity and visual responsiveness in dark–reared monkeys (Macaca Nemestrina) Vis Res **16** 523–30 [501]

Regan D (1973) Rapid objective refraction using evoked potentials Invest Ophthal **12** 669–79 [259]

Regan D (1977) Speedy assessment of visual acuity in amblyopia by the evoked potential method Ophthalmologica **175** 159–64 [259]

Regan D (1982) Visual information channeling in normal and disordered vision Psychol Rev **89** 407–44 [126]

Regan D (1989a) Human brain electrophysiology Evoked potentials and evoked magnetic fields in science and medicine Elsevier, New York [182, 258, 263]

Regan D (1991a) Vision and visual dysfunction Vol 9 Binocular vision (ed D Regan) MacMillan, London [8]

Regan D (1991b) Objects described by colour disparity and motion In Vision and visual dysfunction Vol 10 Spatial vision (ed D Regan) pp 135-78 MacMillan, London [148]

Regan D (1991c) Vision and visual dysfunction Vol 10 Spatial vision MacMillan, London [8]

Regan D, Beverley KI (1985) Postadaptation orientation discrimination J Opt Soc Am A **2** 147–55 [126]

Regan D, Price P (1986) Periodicity in orientation discrimination and the unconfounding of visual information Vis Res **2 9** 1299–302 [81]

Regan D, Spekreijse H (1970) Electrophysiological correlate of binocular depth perception in man Nature **225** 92–4 [258, 261]

Regan D, Erkelens CJ, Collewijn H (1986a) Necessary conditions for the perception of motion in depth Invest Ophthal Vis Sci **27** 584–97 [395]

Regan D, Gray R, Hamstra SJ (1996) Evidence for a neural mechanism that encodes angles Vis Res **36** 323-30 [200]

Regan MP, Regan D (1988) A frequency domain technique for characterizing nonlinearities in biological systems J Theor Biol **133** 293–317 [95, 128, 260, 282]

Regan MP, Regan D (1989) Objective investigation of visual function using a nondestructive zoom–FFT technique for evoked potential analysis Can J Neurol Sci **16** 168–79 [95, 260, 282]

Reh TA, Constantine-Paton M (1985) Eye-specific segregation requires neural activity in three-eyed *Rana pipiens*. J Neurosci **5** 1132-43 [465]

Reichardt W (1987) Evaluation of optical motion information by movement detectors J Comp Physiol A **161** 533-47 [127]

Reichardt LF (1992) Neuronal interactions with the extracellular matrix that regulate axon growth In Regeneration and plasticity in the mammalian visual system (ed DMK Lam, GM Garth) pp 59–70 MIT Press, Cambridge MA [437]

Reid CB, Liang I, Walsh C (1995) Systematic widespread clonal organization in cerebral cortex Neuron **15** 299-310 [450]

Reid RC, Alonso JM (1995) Specificity of monosynaptic connections from thalamus to visual cortex Nature **378** 281-4 [199]

Reid RC, Shapley RM (1992) Spatial structure of cone inputs to receptive fields in primate lateral geniculate nucleus Nature **356** 716-7 [169]

Reid RC, Soodak RE, Shapley RM (1991) Directional selectivity and spatiotemporal structure of receptive fields of simple cells in cat striate cortex J Neurophysiol **66** 505-29 [203]

Reid RC, Victor JD, Shapley RM (1997) The use of m-sequences in the analysis of visual neurons: linear receptive field properties Vis Neurosci **14** 1015-27 [95]

Reid SNM (1995) Immunohistochemical study of two phosphoinositidelinked metabotropic glutamate receptors (mGluR1a and mGluR5) in the cat visual cortex before during and after the peak of the critical period for eye-specific connections J Comp Neurol **355** 470-7 [461]

Reid SNM, Daw NW, Czepita D, Flavin HJ, Sessa WC (1996) Inhibition of nitric oxide synthesis does not alter ocular dominance shifts in kitten visual cortex J Physiol **494** 511-17 [461]

Reid T (1764) An inquiry into the human mind On the principles of common sense Miller Kinnaird and Bell, Edinburgh [62, 519]

Reinagel P, Reid RC (2000) Temporal coding of visual information in the thalamus J Neurosci **20** 5392-400 [130, 131]

Reiter HO, Stryker MP (1988) Neural plasticity without postsynaptic action potentials: less–active inputs become dominant when kitten visual cortical cells are pharmacologically inhibited Proc Nat Acad Sci **85** 3627–37 [511]

Reiter HO, Waitzman DM, Stryker MP (1986) Cortical activity blockade prevents ocular dominance plasticity in the kitten visual cortex Exp Brain Res **65** 182–8 [511]

Remole A (1984) Binocular fixation misalignment measured by border enhancement: a simplified technique Am J Optom Physiol Opt **61** 118–27 [379]

Remole A (1985) Fixation disparity vs binocular fixation misalignment Am J Optom Physiol Opt **62** 25–34 [379]

Remole A, Code SM, Matyas CE, McLeod MA, White DJ (1986) Objective measurement of binocular fixation misalignment Am J Optom Physiol Opt **63** 63–8 [379]

Rentschler I, Hilz R (1985) Amblyopic processing of positional information Part I: Vernier acuity Exp Brain Res **60** 270-8 [521, 523]

Ress D, Backus BT, Heeger DJ (2000) Activity in primary visual cortex predicts performance in a visual detection task Nature Neurosci **3** 940-5 [207]

Rettig G, Roth G (1986) Retinofugal projections in salamanders of the family Plethodontidae Cell Tissue Res **273** 385-96 [542]

Reynolds BA, Weiss S (1992) Generation of neurons and astrocytes from isolated cells of the adult mammalian nervous system Science **255** 1707-10 [450]

Reynolds JH, Chelazzi L, Desimone R (1999) Competitive mechanisms subserve attention in macaque areas V2 and V4 J Neurosci **19** 1736-53 [208]

Richards W (1966) Attenuation of the pupil response during binocular rivalry Vis Res **6** 239–40 [299]

Richards W (1970) Stereopsis and stereoblindness Exp Brain Res **10** 380–8 [487]

Richards W (1971a) Independence of Panum's near and far limits Am J Optom Arch Am Acad Optom **48** 103–9 [272]

Richards W (1971b) Anomalous stereoscopic depth perception J Opt Soc Am **61** 410–14 [487]

Richards W (1972) Response functions for sine– and square–wave modulations of disparity J Opt Soc Am **62** 907–11 [251]

Richards W (1973) Reversal in stereo discrimination by contrast reversal Am J Optom Arch Am Acad Optom **50** 853–62 [487]

Richards W, Regan D (1973) A stereo field map with implications for disparity processing Invest Ophthal Vis Sci **12** 904–9 [488]

Richmond BJ, Optican LM (1987) Temporal encoding of two-dimensional patterns by single units in primate inferior temporal cortex. II Quantification of response waveforms J Neurophysiol **57** 147 61 [129]

Richmond BJ, Optican LM (1990) Temporal encoding of two-dimensional patterns by single units in primate primary visual cortex. II Information transmission J Neurophysiol **64** 370–80 [129]

Richmond BJ, Optican LM, Spitzer H (1990) Temporal encoding of two–dimensional patterns by single units in primate primary visual cortex. I Stimulus-response relations J Neurophysiol **64** 351-369 [129]

Richmond FJR, Johnston WSW, Baker RS, Steinbach MJ (1984) Palisade endings in human extraocular muscles Invest Ophthal Vis Sci **25** 471–6 [493]

Richter GAM (1970) Perspective in Greek and Roman art Phaidon, London [28]

Richter JP (1970) The notebooks of Leonardo da Vinci Dover, New York [19]

Riddell JL (1853) Notice of a binocular microscope Am J Sci Arts **15** 68 [62]

Ridder WH, Smith EL, Manny RE, Harwerth RS, Kato K (1992) Effects of interocular suppression on spectral sensitivity Optom Vis Sci **69** 227-36 [300]

Riddle DR, Lo DC, Katz LC (1995) NT-4-mediated rescue of lateral geniculate neurons from effects of monocular deprivation Nature **378** 189-91 [504, 512]

Riddoch G (1917) Dissociation of visual perceptions due to occipital injuries with especial reference to appreciation of movement Brain **40** 15–57 [488]

Riedel G (1996) Function of metabotropic glutamate receptors in learning and memory TINS **19** 214-23 [191, 461]

Riesenhuber M, Poggio T (1999) Hierarchical models of object recognition in cortex Nature Neurosci **2** 1019-25 [146]

Riggs LA, Niehl EW (1960) Eye movements recorded during convergence and divergence J Opt Soc Am **50** 913–20 [401]

Riggs LA, Whittle P (1967) Human occipital and retinal potentials evoked by subjectively faded visual stimuli Vis Res **7** 441-51 [312]

Rind FC (1996) Intracellular characterization of neurons in the locust brain signaling impending collision J Neurophysiol **7 5** 986-95 [534]

Ringach DL, Hawken MJ, Shapley R (1996) Binocular movements caused by the perception of three-dimensional structure from motion Vis Res **36** 1479-92 [387]

Ringach DL, Hawken MJ, Shapley R (1997) Dynamics of orientation tuning in macaque primary visual cortex Nature **387** 281-4 [200]

Ringstedt T, Braisted JE, Brose K, Kidd T, Goodman C, Tessier-Lavigne M, O'Leary DDM (2000) Slit inhibition of retinal axon growth and its role in retinal axon pathfinding and innervation patterns in the diencephalon J Neurosci **20** 4983-91 [447]

Ripps H, Chin NB, Siegel IM, Breinen GM (1962) The effect of pupil size on accommodation convergence and the AC/A ratio Invest Ophthal **1** 127–35 [389]

Rittenhouse CD, Shouval HZ, Paradiso MA, Bear MF (1999) Monocular deprivation induces homosynaptic long-term depression in visual cortex Nature **397** 347-50 [511]

Ritz R, Gerstner W, Fuentes U, van Hemmen JL (1994) A biological motivated and analytically soluble model of collective oscillations in the cortex Biol Cyber **71** 349-58 [136]

Rivest J, Cavanagh P (1996) Localizing contours defined by more than one attribute Vis Res **36** 53-66 [148]

Rivest J, Cavanagh P, Lassonde M (1994) Interhemispheric depth judgments Neuropsychol **32** 69-76 [233]

Rivest J, Boutet I, Intriligator J (1997) Perceptual learning of orientation discrimination by more than one attribute Vis Res **37** 273-81 [148]

Rizzo M, Damasio H (1985) Impairment of stereopsis with focal brain lesions Ann Neurol **18** 147 [489]

Rizzo M, Nawrot M, Blake R, Damasio A (1992) A human visual disorder resembling area V4 dysfunction in the monkey Neurology **42** 1175-80 [221]

Robb RM, Mayer DL, Moore BD (1987) Results of early treatment of unilateral congenital cataracts J Ped Ophthal Strab **27** 178–81 [519]

Roberts EB, Ramoa AS (1999) Enhanced NR2A subunit expression and decreased NMDA decay time at the onset of ocular dominance plasticity in the ferret J Neurophysiol **81** 2587-91 [462]

Roberts EB, Meredith MA, Ramoa AS (1998) Suppression of NMDA receptor function using antisense DNA blocks ocular dominance plasticity while preserving visual responses J Neurophysiol **80** 1021-32 [511]

Roberts N, Westall C (1990) OKN asymmetries in amblyopia and their effect on velocity perception Clinical Visual Science **5** 383-9 [526]

Robertson VM, Fry GA (1937) After-images observed in complete darkness Am J Psychol **49** 295–76 [352]

Robinson DA (1963) A method of measuring eye movement using a scleral search coil in a magnetic field IEEE Tr Biomed Engin **BME-10** 137–45 [413]

Robinson DL, Kertzman C (1995) Covert orienting of attention in macaques. III Contributions of the superior colliculus J Neurophysiol **74** 713-721 [207]

Robinson DL, Petersen SE (1992) The pulvinar and visual salience TINS **15** 127-32 [183, 207, 245]

Robinson DL, Goldberg ME, Stanton GB (1978) Parietal association cortex in the primate: sensory mechanisms and behavioral modifications J Neurophysiol **41** 910-32 [207]

Robinson DL, Baizer JS, Dow BM (1980) Behavioral enhancement of visual responses of prestriate neurons of the rhesus monkey Invest Ophthal Vis Sci **9** 1120–3 [207]

Robinson DL, McClurkin JW, Kertzman C (1990) Orbital position and eye movement influences on visual responses in the pulvinar nuclei of the behaving macaque Exp Brain Res **82** 235-46 [227]

Robinson DL, Bowman EM, Kertzman C (1995) Covert orienting of attention in macaques. II Contributions of parietal cortex J Neurophysiol **74** 698-712 [207]

Robinson DN (1968) Visual disinhibition with binocular and interocular presentation J Opt Soc Am **58** 254–7 [341]

Robinson JD (2001) Mechanisms of synaptic transmission Oxford University Press, Oxford [193, 456]

Robinson SR (1991) Development of the mammalian retina In Neuroanatomy of the visual pathways and their development (ed B Dreher, SR Robinson) pp 69–128 CRC Press, Boston, and in Vision and visual dysfunction (ed JR Cronly-Dillon) Vol 3 pp 69-128 MacMillan, London [436, 485]

Robinson TR (1895) Experiments with Fechner's paradoxon Am J Psychol **7** 9-23 [324]

Robson JG, Troy JB (1987) Nature of the maintained discharge of Q, X and Y retinal ganglion cells of the cat J Opt Soc Am A **4** 23017 [170]

Rockland KS (1985) A reticular pattern of intrinsic connections in primate area V2 (area 18) J Comp Neurol **235** 467-78 [195]

Rockland KS, Lund JS (1982) Widespread periodic intrinsic connections in the tree shrew visual cortex Science **215** 1532–4 [195, 452]

Rockland KS, Lund JS (1983) Intrinsic laminar lattice connections in primate visual cortex J Comp Neurol **216** 303-18 [195]

Rockland KS, Van Hoesen GW (1994) Direct temporal-occipital feedback connections to striate cortex (V1) in the macaque monkey Cerebral Cortex **4** 300-13 [230]

Rodieck RW (1965) Quantitative analysis of cat retinal ganglion cell response to visual stimuli Vis Res **5** 583-601 [141]

Rodieck RW, Dreher B (1979) Visual suppression from nondominant eye in the lateral geniculate nucleus: a comparison of cat and monkey Exp Brain Res **35** 465–77 [180]

Rodriguez E, George N, Lachaux JP, Martinerie J, Renault B, Varela FJ (1999) Perception's shadow: long distance synchronization of human brain activity Nature, **397** 430-3 [133]

Rodriguez-Tébar A, Dechant G, Barde YA (1990) Binding of brain-derived neurotrophic factor to the nerve growth factor receptor Neuron **4** 487-92 [456]

Roe AW, Ts'o DY (1995) Visual topology in primate V2: multiple representation across functional stripes J Neurosci **15** 3689-715 [219]

Roe AW, Pallas SL, Kwon YH, Sur M (1992) Visual projections routed to the auditory pathway in ferrets: receptive fields of visual neurons in primary auditory cortex J Neurosci **12** 3651-64 [447]

Roelfsema PR, Engel AK, König P, Singer W (1997) Visuomotor integration is associated with zero time-lag synchronization among cortical areas Nature **385** 157-60 [134]

Roelfsema PR, Lamme VAF, Spekreijse H (1998) Object-based attention in the primary visual cortex of the macaque monkey Nature **395** 376-81 [207]

Rogers BJ (1992) The perception and representation of depth and slant in stereoscopic surfaces In Artificial and biological vision systems (ed GA Orban, HH Nagel) pp 271-296 Springer-Verlag, Berlin [416]

Rogers BJ, Bradshaw MF (1999) Disparity minimisation, cyclovergence, and the validity of nonius lines as a technique for measuring torsional alignment Perception **28** 127-41 [417]

Rogers BJ, Cagenello R (1989) Disparity curvature and the perception of three–dimensional surfaces Nature **339** 135–7 [256]

Rogers BJ, Howard IP (1991) Differences in the mechanisms used to extract 3–D slant from disparity and motion parallax cues Invest Ophthal Vis Sci **32** (Abs) 695 [416]

Rogers DC, Hollins M (1982) Is the binocular rivalry mechanism tritanopic? Vis Res **22** 515-20 [291]

Rohault J (1671) Traité de Physique Savreux, Paris [24, 26, 56]

Rolls ET (1992) Neurophysiological mechanisms underlying face processing within and beyond the temporal cortical visual areas Philos Tr R Soc B **335** 11–21 [130]

Rolls ET (1994) Brain mechanisms for invariant visual recognition and learning Behav Proc **33** 113-38 [160]

Rolls ET, Cowey A (1970) Topography of the retina and striate cortex and its relationship to visual acuity in rhesus and squirrel monkeys Exp Brain Res **10** 298–310 [184]

Rolls ET, Tovee MJ (1995) Sparseness of the neural representation of stimuli in the primate temporal visual cortex J Electrophysiol 73 713-29 [160]

Rolls ET, Tovee MJ, Purcell DG, Stewart AL, Azzopardi P (1994) The responses of neurons in the temporal cortex of primates and face identification and detection Exp Brain Res **101** 473-84 [222]

Rolls ET, Treves A, Tovee MJ (1997) The representational capacity of the distributed encoding of information provided by populations of neurons in primate temporal visual cortex Exp Brain Res **114** 149-62 [161]

Rolls ET, Treves A, Robertson RG, Georges-Francois P, Panzeri S (1998) Information about spatial view in an ensemble of primate hippocampal cells J Neurophysiol **79** 1797-813 [223]

Romano PE, von Noorden GK (1971) Limitations of cover test in detecting strabismus Am J Ophthal **72** 10-12 [375]

Romano PE, Romano JA, Puklin JE (1975) Stereoacuity development in children with normal binocular single vision Am J Ophthal **79** 966–71 [482]

Rommetveit R, Toch H, Svendsen D (1968) Semantic syntactic and associative context effects in a stereoscopic rivalry situation Scand J Psychol **9** 145–9 [308]

Ronan CA (1978) The shorter science and civilization in China An abridgement of Joseph Needham's book Science and civilization in China Vol 2 p 351 Cambridge University Press, Cambridge [16]

Roorda A Williams DR (1999) The arrangement of the three cone classes in the living eye Nature **397** 520-22 [164]

Rosa MGP, Gattass R, Fiorani M, Soares JGM (1992) Laminar columnar and topographic aspects of ocular dominance in the primary visual cortex of *Cebus* monkeys Exp Brain Res **8 8** 279–64 [217]

Rosa MGP, Schmid LM, Calford MB (1995) Responsiveness of cat area 17 after monocular inactivation: limitation of topographic plasticity in adult cortex J Physiol **482** 589-608 [198, 509]

Rosar WH (1985) Visual space as physical geometry Perception **14** 403–25 [137]

Rose A (1948) The sensitivity performance of the human eye on an absolute scale J Opt Soc Am **38** 196–208 [166]

Rose D (1978) Monocular versus binocular contrast thresholds for movement and pattern Perception **7** 195–200 [331]

Rose D (1980) The binocular: monocular sensitivity ratio for movement detection varies with temporal frequency Perception **9** 577-80 [331]

Rose D (1999) The historical roots of the theories of local sign Perception **28** 675-85 [45, 117]

Rose D, Blake R, Halpern DL (1988) Disparity range for binocular summation Invest Ophthal Vis Sci **29** 283–90 [320]

Rosen E (1956) The invention of eyeglasses J Hist Med Allied Sci **11** 13–46; 183-218 [19]

Rosenberg R, Flax N, Brodsky B, Abelman L (1953) Accommodative levels under conditions of asymmetric convergence Am J Optom Arch Am Acad Optom **30** 274-54 [364]

Rosenfield M (1997) Tonic vergence and vergence adaptation Optom Vis Sci **74** 303-28 [372]

Rosenfield M, Ciuffreda KJ (1990) Distance heterophoria and tonic vergence Optom Vis Sci **67** 667–9 [376]

Rosenfield M, Ciuffreda KJ (1991a) Accommodation responses to conflicting stimuli J Opt Soc Am A **8** 422-7 [361]

Rosenfield M, Ciuffreda KJ (1991b) Effect of surround propinquity on the open-loop accommodative response Invest Ophthal Vis Sci **32** 142-7 [359]

Rosenfield M, Gilmartin B (1988a) Accommodative adaptation induced by sustained disparity vergence Am J Optom Physiol Opt **65** 118-29 [391]

Rosenfield M, Gilmartin B (1988b) The effect of vergence adaptation on convergent accommodation Ophthal Physiol Opt **8** 172-7 [391]

Rosenfield M, Ciuffreda KJ, Hung GK (1991) The linearity of proximally–induced accommodation and vergence Invest Ophthal Vis Sci **32** 2985–91 [386]

Rosenfield M, Ciuffreda KJ, Hung GK, Gilmartin B (1993) Tonic accommodation: a review. I Basic aspects Ophthal Physiol Opt **13** 296-84 [359]

Rosenfield M, Ciuffreda KJ, Hung GK, Gilmartin B (1994) Tonic accommodation: a review. II Accommodative adaptation and clinical aspects Ophthal Physiol Opt 14 295-77 [360]

Rosenfield M, Ciuffreda KJ, Chen HW (1995a) Effect of age on the interaction between the AC/A and CA/C ratios Ophthal Physiol Opt **15** 451-5 [391]

Rosenfield M, Ciuffreda KJ, Ong E, Super S (1995b) Vergence adaptation and the order of clinical vergence range testing Optom Vis Sci **72** 219-223 [392]

Rosenfield M, Rappon JM, Carrel MF (2000) Vergence adaptation and the clinical AC/A ratio Ophthal Physiol Opt **20** 207-11 [389]

Ross HE (2000) Cleomedes (c. 1st century AD) on the celestial illusion, atmospheric enlargement, and size-distance invariance Perception **29**, 863-71 [18]

Ross HE, Ross GM (1976) Did Ptolemy understand the moon illusion? Perception **5** 377-85 [18]

Ross JE (1983) Disturbance of stereoscopic vision in patients with unilateral stroke Behav Brain Res **7** 99–112 [489]

Ross-Dommasch E, Morris E (1990) What are we doing when we occlude infantile esotropes? Am Orthopt J **40** 80-7 [528]

Rossel S (1980) Foveal fixation and tracking in the praying mantis J Comp Physiol **139** 307–31 [533]

Rossel S (1983) Binocular stereopsis in an insect Nature **302** 821–2 [533]

Rossel S (1986) Binocular spatial localization in the praying mantis J Exp Biol **120** 295–81 [532]

Rossel S (1991) Spatial vision in the praying mantis: is distance implicated in size detection? J Comp Physiol **169** 101–8 [533]

Rossel S (1996) Binocular vision in insects: how mantids solve the correspondence problem Proc Nat Acad Sci **93** 13229-32 [533]

Rossel S, Mathis U, Collett T (1992) Vertical disparity and binocular vision in the praying mantis Vis Neurosci **8** 165–70 [532, 533]

Rossetti Y, Tadary B, Prablanc C (1994) Optimal contributions of head and eye positions to spatial accuracy in man tested by visually directed pointing Exp Brain Res **97** 487-96 [144]

Rossi AF, Paradiso MA (1999) Neural correlates of perceived brightness in the retina, lateral geniculate nucleus, and striate cortex J Neurosci **19** 6145-56 [197, 206]

Rossi AF, Desimone R, Ungerleider LG (2001) Contextual modulation in primary visual cortex of macaques J Neurosci **21** 1698-709 [206]

Roth G (1987) Visual behavior in salamanders Springer, Berlin [542]

Rothstein TB, Sacks JG (1972) Defective stereopsis in lesions of the parietal lobe Am J Ophthal **73** 281–4 [489]

Roumes C, Planter J, Menu JP, Thorpe S (1997) The effects of spatial frequency on binocular fusion: from elementary to complex images Hum Factors **39** 359-73 [274]

Rouse MW, Tittle JS, Braunstein ML (1989) Stereoscopic depth perception by static stereo–deficient observers in dynamic displays with constant and changing disparity Optom Vis Sci **66** 355–62 [488]

Rovamo J, Virsu V (1979) An estimation and application of the human cortical magnification factor Exp Brain Res **37** 495–510 [184, 273, 473]

Rowe JB, Toni I, Josephs O, Frackowiak RSJ, Passingham RE (2000) The prefrontal cortex: response selection or maintenance without working memory Science **288** 1656-60 [228]

Rowe MH (1991) Functional organization of the retina In Neuroanatomy of the visual pathways and their development (ed B Dreher, RS Robinson) pp 1–58 CRC Press, Boston [168]

Roy JP, Komatsu H, Wurtz RH (1992) Disparity sensitivity of neurons in monkey extrastriate area MST J Neurosci **12** 2778-92 [254, 256]

Rozhkova GI, Nickolayev PP, Shchadrin VE (1982) Perception of stabilized retinal stimuli in dichoptic viewing conditions Vis Res **22** 293–302 [290]

Rucker FJ, Kruger PB (2001) Isolated short-wavelength sensitive cones can mediate a reflex accommodation response Vis Res **41** 911-22 [363]

Ruddock KH, Wigley E (1976) Inhibitory binocular interaction in human vision and a possible mechanism subserving stereoscopic fusion Nature **290** 604–6 [340]

Ruddock KH, Waterfield VA, Wigley E (1979) The response characteristics of an inhibitory binocular interaction in human vision J Physiol **290** 37–49 [340]

Rumelhart DE, McClelland JL (1986) Parallel distributed processing MIT Press, Cambridge MA [95, 269]

Rushton WAH (1961) Peripheral coding in the nervous system In Sensory communication (ed WA Rosenblith) pp 169-82 Wiley, London [154]

Rushworth MFS, Paus T, Sipila PK (2001) Attentional systems and the organization of the human parietal cortex J Neurosci **21** 5262-71 [227]

Russell JT (1932) Depth discrimination in the rat J Genet Psychol **40** 136-59 [548]

Ruthazer ES, Stryker MP (1996) The role of activity in the development of long-range horizontal connections in area 17 of the ferret J Neurosci **16** 7253-69 [452]

Ruthazer ES, Baker GE, Stryker MP (1999) Development and organization of ocular dominance bands in primary visual cortex of the sable ferret J Comp Neurol **407** 151-65 [454]

Rutstein RP (1977) Fixation disparity and stereopsis Am J Optom Physiol Opt **54** 550-5 [381]

Rutstein RP, Corliss D (1999) Relationship between anisometropia, amblyopia, and binocularity Optom Vis Sci **76** 229-33 [519]

Rutstein RP, Eskridge JB (1984) Stereopsis in small-angle strabismus Am J Optom Physiol Opt **61** 491-8 [373]

Rutstein RP, Fuhr PS (1992) Efficacy and stability of amblyopia therapy Optom Vis Sci **69** 747–54 [528]

Rutstein RP, Daum KM, Cho M, Eskridge JB (1988) Horizontal and vertical vergence training and its effect on vergences fixation disparity curves and prism adaptation: II vertical data Am J Optom Physiol Opt **65** 8-13 [407]

Ruttum M, von Noorden GK (1983) Adaptation to tilting of the visual environment in cyclotropia Am J Ophthal **96** 229-37 [411]

Sabatini BL, Regehr WG (1999) Timing of synaptic transmission Ann Rev Physiol **61** 521-42 [134]

Sabra AI (1966) Ibn al-Haytham's criticisms of Ptolemy's Optics J Hist Philos **4** 145-49 [52]

Sabra AI (1978) Sensation and inference in Alhazen's theory of visual perception In Studies in perception (ed PK Machamer, RG Turnbull) pp 160-84 Ohio State University Press, Columbus Ohio [17]

Sabra AI (1987a) Islamic optics In Dictionary of the middle ages (ed JR Strayer) Vol 9 pp 270-7 Scribner, New York [15]

Sabra AI (1987b) Psychology versus mathematics: Ptolemy and Alhazen on the moon illusion In Mathematics and its application to science and natural philosophy in the Middle Ages (ed E Grant, JE Murdoch) pp 217-47 Cambridge University Press, London [18]

Sabra AI (1989) Form in ibn al-Haytham's theory of vision Z Gesch Arabi-Islam Wissen **5** 115-40 [17]

Sabra AI (1996) On seeing the stars. II Ibn al-Haytham's "answers" to the "doubts" raised by Ibn Ma'dän Z Gesch Arab-Islam Wissen **10** 1-60 [18]

Sabrin HW, Kertesz AE (1983) The effect of imposed fixational eye movements on binocular rivalry Percept Psychophys **34** 155–7 [293]

Sachs MB, Nachmias J, Robson JG (1971) Spatial-frequency channels in human vision J Opt Soc Am **61** 1176-86 [139]

Sagawa K (1981) Minimum light intensity required for color rivalry Vis Res **21** 1467-74 [291]

Sagawa K (1982) Dichoptic color fusion studied with wavelength discrimination Vis Res **22** 945–52 [284]

Sagi D, Hochstein S (1985) Lateral inhibition between spatially adjacent spatial–frequency channels? Percept Psychophys **37** 315–22 [122]

Saida S, Ono H (1984) Interaction between saccade and tracking vergence Vis Res **27** 1289–94 [421]

Saida S, Ono H, Mapp AP (2001) Closed-loop and open-loop accommodative vergence eye movements Vis Res **41** 77-86 [421]

Saito H, Yukie M, Tanaka K, Hikosaka K, Fukada Y, Iwai E (1986) Integration of direction signals of image motion in the superior temporal sulcus of the macaque monkey J Neurosci **6** 145-57 [225, 256]

Sakata H, Shibutani H, Kawano K (1980) Spatial properties of visual fixation neurons in posterior parietal association cortex of the monkey J Neurophysiol **43** 1654-72 [227]

Sakata H, Shibutani H, Tsurugai K (1986) Parietal cortical neurons responding to rotary movement of visual stimulus in space Exp Brain Res **61** 658–63 [225]

Sakata H, Shibutani H, Ito Y, Tsurugai K, Mine S, Kusunoki M (1994) Functional properties of rotation-sensitive neurons in the posterior parietal association cortex of the monkey Exp Brain Res **101** 183-202 [228]

Sakata H, Kusunoki TM, Murata A, Tanaka Y (1997) The parietal association cortex in depth perception and visual control of hand action TINS **20** 350-6 [228]

Sakata H, Taira M, Kusunoki TM, Murata A, Tsutsui K, Tanaka Y Shein W, Miyashita Y (1999) Neural representation of three-dimensional features of manipulation objects with stereopsis Exp Brain Res **128** 160-9 [226, 243]

Sakitt B (1982) Why the cortical magnification factor in rhesus can not be isotropic Vis Res **22** 417-21 [184]

Sakitt B, Barlow HB (1982) A model for the economic encoding of the visual image in cerebral cortex Biol Cyber **43** 97-108 [94, 141]

Saladin JJ, Carr LW (1983) Fusion lock diameter and the forced vergence fixation disparity curve Am J Optom Physiol Opt **60** 933–43 [380]

Saladin JJ, Sheedy JE (1978) Population study of fixation disparity heterophoria and vergence Am J Optom Physiol Opt **55** 744-50 [379, 383]

Salapatek P, Cohen L (1987) Handbook of infant perception Academic Press, New York [185]

Salin PA, Bullier J (1995) Corticocortical connections in the visual system: structure and function Physiol Rev **75** 107-54 [187]

Samorajski T, Keefe JR, Ordy JM (1965) Morphogenesis of photoreceptor and retinal ultrastructure in a sub–human primate Vis Res **5** 639–48 [436]

Sanderson KJ, Sherman M (1971) Nasotemporal overlap in visual field projected to lateral geniculate nucleus in the cat J Neurophysiol **34** 453–66 [231]

Sanderson KJ, Darian–Smith I, Bishop PO (1969) Binocular corresponding receptive fields of single units in the cat dorsal lateral geniculate nucleus Vis Res **9** 1297–303 [179]

Sanderson KJ, Bishop PO, Darian–Smith I (1971) The properties of the binocular receptive fields of lateral geniculate neurons Exp Brain Res **13** 178--207 [179, 180]

Sanderson KJ, Guillery RW, Shackelford RM (1974) Congenitally abnormal visual pathways in mink (*Mustela vison*)with reduced retinal pigment J Comp Neurol **154** 225–48 [496]

Sandor PS, Frens MA, Henn V (2001) Chameleon eye position obeys Listing's law Vis Res **41** 2245-51 [371]

Sanes JR (1989) Extracellular matrix molecules that influence neural development Ann Rev Neurosci **12** 491–516 [437]

Sanger TD (1988) Stereo disparity computation using Gabor filters Biol Cyber **59** 405-18 [252, 268]

Sarmiento RF (1975) The stereoacuity of macaque monkey Vis Res **15** 493–8 [548]

Sáry G, Vogels R, Orban GA (1993) Cue-invariant shape selectivity of macaque inferior temporal neurons Science **290** 995-7 [222, 244]

Sasaki Y, Cheng H, Smith EL, Chino Y (1998) Effects of early discordant binocular vision on the postnatal development of parvocellular neurons in the monkey lateral geniculate nucleus Exp Brain Res **118** 341-51 [504]

Sato M, Edwards M, Schor CM (2001) Envelope size-tuning for transient disparity vergence Vis Res **41** 1695-707 [402]

Sato T, Kawamura T, Iwai E (1980) Responsiveness of inferotemporal single units to visual pattern stimuli in monkeys performing discrimination Exp Brain Res **38** 313-19 [222]

Sauer B, Kammradt G, Krauthausen I, Kretschmann HJ, Lange HW, Wingert F (1983) Qualitative and quantitative development of the visual cortex in man J Comp Neurol **214** 441–50 [446]

Saul AB, Cynader MS (1989) Adaptation in single units in visual cortex: the tuning of aftereffects in the spatial domain Vis Neurosci **2** 593–607 [84]

Saunders JB De CM, O'Malley CD (1950) The illustrations from the works of Andreas Vesalius of Brussels World Publishing, New York [22]

Saunders KJ (1995) Early refractive development in humans Survey Ophthal **40** 207-16 [474]

Savoy RL (1984) "Extinction" of the McCollough effect does not transfer interocularly Percept Psychophys **36** 571–6 [352]

Sawa M, Ohtsuka K (1994) Lens accommodation evoked by microstimulation of the superior colliculus in the cat Vis Res **34** 975-81 [365]

Sawatari A, Calloway EM (1996) Convergence of magno- and parvocellular pathways in layer 4B of macaque primary visual cortex Nature **380** 442-6 [186]

Sawtell NB, Huber KM, Roder JC, Bear MF (1999) Induction of NMDA receptor-dependent long-term depression in visual cortex does not require metabotropic glutamate receptors J Neurophysiol **82** 3594-3597 [511]

Scammon RE, Wilmer HA (1950) Growth of the components of the human eyeball. II Comparison of the calculated volumes of the eyes of the newborn and of adults and their components Arch Ophthal **43** 620–37 [434]

Scanziani M, Malenka RC, Nicoll RA (1996) Role of intercellular interactions in heterosynaptic long-term depression Nature **380** 446-50 [135, 460]

Schaafsma SJ, Duysens J (1996) Neurons in the ventral intraparietal area of awake macaque monkey closely resemble neurons in the dorsal part of the medial superior temporal area in their responses to optic flow patterns J Neurophysiol **76** 4056-68 [226]

Schaafsma SJ, Duysens J, Gielen CCAM (1997) Responses in ventral intraparietal area of awake macaque monkey to optic flow patterns corresponding to rotation of planes in depth can be explained by translation and expansion effects Vis Neurosci **14** 633-46 [226]

Schacter DL, Relman E, Uecker A, Polster MR, Yun LS, Cooper LA (1995) Brain regions associated with retrieval of structurally coherent visual information Nature **376** 587-90 [223]

Schade OH (1956) Optical and photoelectric analog of the eye J Opt Soc Am **46** 721-39 [41, 138]

Schaeffel F, Diether S (1999) The growing eye: an autofocus system that works on very poor images Vis Res **39** 1585-9 [435]

Schaeffel F, Howland HC (1988) Mathematical model of emmetropization in the chicken J Opt Soc Am **5** 2080-6 [435]

Schaeffel F, Howland HC (1991) Properties of the feedback loops controlling eye growth and refractive state in the chicken Vis Res **31** 717-34 [435]

Schaeffel F, Howland HC (1995) Myopia Vis Res **35** Number 9 Special Issue [435]

Schaeffel F, Wagner H (1992) Barn owls have symmetrical accommodation in both eyes but independent pupillary responses to light Vis Res **32** 1149-55 [548]

Schaeffel F, Glasser A, Howland HC (1988) Accommodation refractive error and eye growth in chickens Vis Res **28** 639–57 [434]

Schaeffel F, Troilo D, Wallman J, Howland HC (1990) Developing eyes that lack accommodation grow to compensate for imposed defocus Vis Neurosci **4** 177-83 [434]

Schapero M, Levy M (1953) The variation of proximal convergence with change in distance Am J Optom Arch Am Acad Optom **30** 403-16 [386]

Scharff LFV (1997) Decreases in the critical disparity gradient with eccentricity may reflect the size-disparity correlation J Opt Soc Am A **14** 1205-12 [277]

Scharre JE, Cotter SA, Block SS, Kelly SA (1990) Normative contrast sensitivity data for young children Optom Vis Sci **67** 829-32 [468]

Schechter PB, Murphy EH (1976) Brief monocular visual experience and kitten cortical binocularity Brain Res **109** 165-8 [507]

Scheetz AJ, Williams RW, Dubin MW (1995) Severity of ganglion cell death during early postnatal development is modulated by both neuronal activity and binocular competition Vis Neurosci **12** 605-10 [442]

Scheetz AJ, Nairn AC, Constanine-Paton M (2000) NMDA receptor-mediated control of protein synthesis at developing synapses Nature Neurosci **3** 211-16 [458]

Scheidt RA, Kertesz AE (1993) Temporal and spatial aspects of sensory interactions during human fusional response Vis Res **33** 1259–70 [277]

Scheiffele P, Fan J, Choih J, Fetter R, Serafini T (2000) Neuroligin expressed in nonneuronal cells triggers presynaptic development in contacting axons Cell **101** 657-69 [457]

Scheiman M, Wick B (1994) Clinical management of binocular vision Philadelphia, Lippincott [8]

Schein SJ, De Monasterio FM (1987) Mapping of retinal and geniculate neurons onto striate cortex of macaque J Neurosci **7** 996–1009 [177, 178, 185, 204, 297]

Scheiner C (1619) Oculus hoc est fundamentum opticum Agricola Innsbruck [24]

Scheiner C (1630) Rosa Ursina Phaeum Bracciani [24]

Schellart NAM (1990) The visual pathways and central non-tectal processing In The visual system of fish (ed RH Douglas, MBA Djamgoz) pp 345-71 Chapman Hall, London [539]

Scherer WJ, Udin SB (1989) N-methyl-D-aspartate antagonists prevent interaction of binocular maps in *Xenopus* tectum J Neurosci **9** 3837-43 [462]

Schiff ND, Purpura KP, Victor JD (1999) Gating of local network signals appears as stimulus-dependent activity envelopes in striate cortex J Neurophysiol **82** 2182-96 [99, 258]

Schillen TB, König P (1994) Binding by temporal structure in multiple feature domains of an oscillatory neuronal network Biol Cyber **70** 397-405 [136]

Schiller PH (1965) Monoptic and dichoptic visual masking by patterns and flashes J Exp Psychol **69** 193-9 [341]

Schiller PH (1970) The discharge characteristics of single units in the oculomotor and abducens nuclei of the unanesthetized monkey Exp Brain Res **10** 347–62 [430]

Schiller PH (1992) The on and off channels of the visual system TINS **15** 86–92 [168]

Schiller PH (1993) The effects of V4 and middle temporal (MT) area lesions on visual performance in the rhesus monkey Vis Neurosci **10** 717-46 [221, 243]

Schiller PH, Dolan RP (1994) Visual aftereffects and the consequences of visual system lesions on their perception in the rhesus monkey Vis Neurosci **11** 643-65 [245]

Schiller PH, Greenfield A (1969) Visual masking and the recovery phenomenon Percept Psychophys **6** 182–4 [341]

Schiller PH, Lee K (1991) The role of the primate extrastriate area V4 in vision Science **251** 1251-3 [208]

Schiller PH, Smith M (1968) Monoptic and dichoptic metacontrast Percept Psychophys **3** 237-9 [341]

Schiller PH, Wiener M (1962) Binocular and stereoscopic viewing of geometrical illusions Percept Mot Skills **15** 739-47 [354]

Schiller PH, Wiener M (1963) Monoptic and dichoptic masking J Exp Psychol **66** 386-93 [341]

Schiller PH, Finlay BL, Volman SF (1976) Quantitative studies of single cells in monkey striate cortex. II Orientation specificity and ocular dominance J Neurophysiol **39** 1320–33 [215]

Schiller PH, Sandell JH, Maunsell HR (1986) Functions of the on and off channels of the visual system Nature **322** 827–5 [168]

Schiller PH, Logothetis NK, Charles ER (1990) Role of the color-opponent and broad-band channels in vision Vis Neurosci **5** 321-46 [223, 229, 245]

Schlaggar BL, O'Leary DDM (1991) Potential of visual cortex to develop an array of functional units unique to somatosensory cortex Science **522** 1556–60 [447]

Schlossman A, Priestley BS (1952) Role of heredity in etiology and treatment of strabismus Arch Ophthal **47** 1-20 [374]

Schmid LM, Rosa MGP, Calford MB, Ambler JS (1996) Visuotopic reorganization in the primary visual cortex of adult cats following monocular and binocular retinal lesions Cerebral Cortex **6** 388-405 [198]

Schmidt KE, Kim DS, Singer W, Bonhoeffer T, Löwel S (1997) Functional specificity of long-range intrinsic and interhemispheric connections in the visual cortex of strabismic cats J Neurosci **17** 5480-92 [506]

Schmielau F, Singer W (1977) The role of visual cortex for binocular interactions in the cat lateral geniculate nucleus Brain Res **120** 354–61 [180]

Schmolesky MT, Wang Y, Hanes DP, Thompson KG, Leutgeb S, Schall JD, Leventhal AG (1998) Signal timing across the macaque visual system J Neurophysiol **79** 3272-8 [178]

Schnapf JL, Nunn BJ, Meister M, Baylor DA (1990) Visual transduction in cones of the monkey *Macaca fascicularis* J Physiol **427** 681-713 [166]

Schneck ME, Haegerstrom-Portnoy G, Lott LA, Brabyn JA (2000) Ocular contributions to age-related loss in coarse stereopsis Optome Vis Sci **77** 531-6 [484]

Schneider B, Moraglia G (1992) Binocular unmasking with unequal interocular contrast: the case for multiple cyclopean eyes Percept Psychophys **52** 639-60 [338]

Schneider B, Moraglia G (1994) Binocular vision enhances target detection by filtering the background Perception **23** 1297-86 [323, 532]

Schneider B, Moraglia G, Jepson A (1989) Binocular unmasking: an analogue to binaural unmasking Science **273** 1479–81 [338]

Schneider B, Moraglia G, Speranza F (1999) Binocular vision enhances phase discrimination by filtering the background Percept Psychophys **61** 468-89 [338]

Schor CM (1975) A directional impairment of eye movement control in strabismus amblyopia Invest Ophthal Vis Sci **14** 692–7 [526]

Schor CM (1977) Visual stimuli for strabismic suppression Perception **6** 583–93 [530]

Schor CM (1979a) The influence of rapid prism adaptation upon fixation disparity Vis Res **19** 757–65 [375, 382, 405]

Schor CM (1979b) The relationship between fusional vergence eye movements and fixation disparity Vis Res **19** 1359–67 [381, 382]

Schor CM (1983a) Fixation disparity and vergence adaptation In Vergence eye movements: Basic and clinical aspects (ed MC Schor, KJ Ciuffreda) pp 465–516 Butterworth, Boston [381, 383]

Schor CM (1983b) Subcortical binocular suppression affects the development of latent and optokinetic nystagmus Am J Optom Physiol Opt **60** 481-502 [526]

Schor CM (1986) The Glen A Adaptive regulation of accommodative vergence and vergence accommodation Am J Optom Physiol Opt **63** 587-609 [391]

Schor CM (1988) Imbalanced adaptation of accommodation and vergence produces opposite extremes of the AC/A and CA/C ratios Am J Optom Physiol Opt **65** 341-48 [391]

Schor CM (1992) A dynamic model of cross-coupling between accommodation and convergence: simulations of step and frequency responses Optom Vis Sci **69** 258–69 [391]

Schor CM (1999) The influence of interactions between accommodation and convergence on the lag of accommodation Ophthal Physiol Opt **19** 134-50 [390, 435]
Schor CM, Ciuffreda KJ (1983) Vergence eye movements: Basic and clinical aspects Butterworth, Boston [8, 368]
Schor CM, Hallmark W (1978) Slow control of eye position in strabismic amblyopia Invest Ophthal Vis Sci **17** 577-81 [526]
Schor CM, Horner D (1989) Adaptive disorders of accommodation and vergence in binocular dysfunction Ophthal Physiol Opt **9** 294–8 [391]
Schor CM, Kotulak JC (1986) Dynamic interactions between accommodation and convergence are velocity sensitive Vis Res **29** 927–42 [391, 405]
Schor CM, Levi DM (1980) Disturbances of small–field horizontal and vertical optokinetic nystagmus in amblyopia Invest Ophthal Vis Sci **19** 668–83 [373, 526]
Schor CM, McCandless JW (1995a) Distance cues for vertical vergence adaptation Optom Vis Sci **72** 478-86 [385]
Schor CM, McCandless JW (1995b) An adaptable association between vertical and horizontal vergence Vis Res **35** 3519-27 [385]
Schor CM, McCandless JW (1997) Context-specific adaptation of vertical vergence to correlates of eye position Vis Res **37** 1929-37 [385]
Schor CM, Narayan V (1982) Graphical analysis of prism adaptation convergence accommodation and accommodative convergence Am J Optom Physiol Opt **59** 774-84 [380]
Schor CM, Tsuetaki TK (1987) Fatigue of accommodation and vergence modifies their mutual interactions Invest Ophthal Vis Sci **28** 1250–9 [391]
Schor CM, Tyler CW (1981) Spatio–temporal properties of Panum's fusional area Vis Res **21** 683–92 [278]
Schor CM, Bridgeman B, Tyler CW (1983) Spatial characteristics of static and dynamic stereoacuity in strabismus Invest Ophthal Vis Sci **27** 1572-9 [529]
Schor CM, Wood IC, Ogawa J (1984b) Binocular sensory fusion is limited by spatial resolution Vis Res **27** 661–5 [273, 298]
Schor CM, Robertson KM, Wesson M (1986a) Disparity vergence dynamics and fixation disparity Am J Optom Physiol Opt **63** 611–18 [380]
Schor CM, Kotulak JC, Tsuetaki T (1986b) Adaptation of tonic accommodation reduces accommodative lag and is masked in darkness Invest Ophthal Vis Sci **27** 820-7 [359]
Schor CM, Wesson M, Robertson KM (1986c) Combined effects of spatial frequency and retinal eccentricity upon fixation disparity Am J Optom Physiol Opt **63** 619–29 [399]
Schor CM, Landsman L, Erickson P (1987) Ocular dominance and the interocular suppression of blur in monovision Am J Optom Physiol Opt **64** 723–30 [289]
Schor CM, Heckmann T, Tyler CW (1989) Binocular fusion limits are independent of contrast luminance gradient and component phases Vis Res **29** 821–35 [274]
Schor CM, Gleason G, Horner D (1990) Selective nonconjugate binocular adaptation of vertical saccades and pursuits Vis Res **30** 1827–44 [425]
Schor CM, Alexander J, Cormack L, Stevenson S (1992) Negative feedback control model of proximal convergence and accommodation Ophthal Physiol Opt **12** 307–18 [387, 391]
Schor CM, Gleason G, Maxwell J, Lunn R (1993a) Spatial aspects of vertical phoria adaptation Vis Res **33** 73–84 [385]
Schor CM, Gleason G, Lunn R (1993b) Interactions between short-term vertical phoria adaptation and nonconjugate adaptation of vertical pursuits Vis Res **33** 55–64 [385]
Schor CM, Maxwell JS, Stevenson SB (1994) Isovergence surfaces: the conjugacy of vertical eye movements in tertiary positions of gaze Ophthal Physiol Opt **14** 279-85 [367, 384, 409]
Schor CM, Fusaro RE, Wilson N, McKee SP (1997) Prediction of early-onset esotropia from components of the infantile squint syndrome Invest Ophthal Vis Sci **38** 719-40 [374]
Schor CM, Maxwell JS, Graf EW (2001) Plasticity of convergence-dependent variations of cyclovergence with vertical gaze Vis Res **41** 3353-69 [415]
Schoups AA, Orban GA (1996) Interocular transfer in perceptual learning of a pop-out discrimination task Proc Nat Acad Sci **93** 7358-62 [354]
Schoups AA, Vogels R, Orban GA (1995a) Human perceptual learning in identifying the oblique orientation: retinotopy orientation specificity and monocularity J Physiol **483** 797-810 [354]
Schoups AA, Elliott RC, Friedman WJ, Black IB (1995b) NGF and BDNF are differentially modulated by visual experience in the developing geniculocortical pathway Devel Brain Res **86** 329-34 [456, 463, 499, 512]
Schrödinger E (1926) Die Gesichtempfindungen In Mueller–Pouillet's Lehrbuch der Physik (11th edn) Book 2 Part 1 pp 456–560 Vieweg, Braunschweig [326]
Schroeder CE, Tenke CE, Arezzo JC, Vaughan HG (1990) Binocularity in the lateral geniculate nucleus of the alert monkey Brain Res **521** 303–10 [180]
Schroeder TL, Rainey BB, Goss DA, Grosvenor TP (1996) Reliability of and comparisons among methods of measuring dissociated phoria Optom Vis Sci **73** 389-97 [376]
Schultze M (1866) Zur Anatomie und Physiologie der Retina Arch Mikros Anat Ent **2** 165-286 [26]
Schuman EM, Madison DV (1991) A requirement for the intercellular messenger nitric oxide in long-term potentiation Science **254** 1503-6 [461]
Schuster HG, Wagner P (1990) A model for neuronal oscillations in the visual cortex Biol Cyber **64** 77–82 [136]
Schwartz EL (1980) Computational anatomy and functional architecture of striate cortex: a spatial mapping approach to perceptual coding Vis Res **20** 645–70 [185]
Schwartz TL, Dobson V, Sandstrom DJ, Van Hof–van Duin J (1987) Kinetic perimetry assessment of binocular visual field shape and size in young infants Vis Res **27** 2163–75 [436, 473]
Schwarz U, Miles FA (1991) Ocular responses to translation and their dependence on viewing distance. I Motion of the observer J Neurophysiol **66** 851-64 [427]
Schwarz W (1993) Coincidence detectors and two–pulse visual temporal integration: new theoretical results and comparison data Biol Cyber **69** 173–82 [78, 319]
Schwassmann HO (1968) Visual projection upon the optic tectum in foveate marine teleosts Vis Res **8** 1337–48 [539]
Sclar G, Freeman RD (1982) Orientation selectivity in the cat's striate cortex is invariant with stimulus contrast Exp Brain Res **46** 457-61 [200]
Sclar G, Ohzawa I, Freeman RD (1986) Binocular summation in normal, monocularly deprived, and strabismic cats: visual evoked potentials Exp Brain Res **62** 1-10 [490, 527]
Sclar G, Maunsell JHR, Lennie P (1990) Coding of image contrast in central visual pathways of the macaque monkey Vis Res **30** 1-10 [199, 219]
Scobey RP, Gabor AJ (1989) Orientation discrimination sensitivity of single units in cat primary visual cortex Exp Brain Res **77** 398-406 [81, 199]
Scott AB (1981) Botulinum toxin injection of eye muscles to correct strabismus Tr Am Ophthal Soc **79** 734–70 [374]
Scott TR, Wood DZ (1966) Retinal anoxia and the locus of the aftereffect of motion Am J Psychol **79** 435–42 [347]
Sefton AJ (1986) The regulation of cell numbers in the developing visual system In Vis Neurosci (ed JD Pettigrew, WR Levick) pp 2145–56 Cambridge University Press, London [442]

Seidemann E, Zohary E, Newsome WT (1998) Temporal gating of neural signals during performance of a visual discrimination task Nature **394** 72-5 [224]

Sekuler R, Pantle A, Levinson E (1978) Physiological basis of motion perception In Handbook of sensory physiology (ed R Held, HW Leibowitz, HL Teuber) Vol VIII pp 67–98 Springer–Verlag, Berlin [350]

Selby SA, Woodhouse JM (1981) The spatial frequency dependence of interocular transfer in amblyopes Vis Res **21** 1401–8 [492]

Selfridge O (1959) Pandemonium: a paradigm for learning. In Symposium on the mechanisation of though processes H.M. Stationary Office, London [153]

Semmlow JL, Heerema D (1979a) The synkinetic interaction of convergence accommodation and accommodative convergence Vis Res **19** 1237-42 [380]

Semmlow JL, Heerema D (1979b) The role of accommodative convergence at the limits of fusional vergence Invest Ophthal Vis Sci **18** 970–6 [392]

Semmlow JL, Hung G (1979) Accommodative and fusional components of fixation disparity Invest Ophthal Vis Sci **18** 1082–6 [380, 384]

Semmlow JL, Hung GK (1980) Binocular interactions of vergence components Am J Optom Physiol Opt **57** 559-65 [380]

Semmlow JL, Hung G (1983) The near response Theories of control In Vergence eye movements: Basic and clinical aspects (ed C Schor, K Ciuffreda) pp 175–95 Butterworth, Boston [387]

Semmlow JL, Venkiteswaran N (1976) Dynamic accommodative vergence components in binocular vision Vis Res **16** 403–10 [391]

Semmlow JL, Wetzel P (1979) Dynamic contributions of the components of binocular vergence J Opt Soc Am **69** 639–45 [389, 397]

Semmlow JL, Hung G, Ciuffreda KJ (1986) Quantitative assessment of disparity vergence components Invest Ophthal Vis Sci **27** 558–64 [402, 403, 405]

Semmlow JL, Hung G, Horng JL, Ciuffreda KJ (1993) Initial control component in disparity vergence eye movements Ophthal Physiol Opt **13** 48-55 [401]

Semmlow JL, Hung G, Horng JL, Ciuffreda KJ (1994) Disparity vergence eye movements exhibit preprogrammed motor control Vis Res **34** 335-43 [406]

Semmlow JL, Yuan W, Alvarez TL (1998) Evidence for separate control of slow version and vergence eye movements: Support for Hering's Law Vis Res **38** 1145-52 [420]

Sen DK, Singh B, Shroff NM (1977) Diagnosis and measurement of cyclodeviation Br J Ophthal **61** 690-2 [413]

Sen DK, Singh B, Mathur GP (1980) Torsional fusional vergences and assessment of cyclodeviation by synoptophore method Br J Ophthal **64** 354-7 [278]

Sengelaub DR, Finlay BL (1981) Early removal of one eye reduces normally occurring cell death in the remaining eye Science **213** 573-4 [442]

Sengpiel F, Blakemore C (1994) Interocular control of neuronal responsiveness in cat visual cortex Nature **368** 847-50 [313]

Sengpiel F, Blakemore C (1996) The neural basis of suppression and amblyopia in strabismus Eye **10** 250-8 [530]

Sengpiel F, Blakemore C, Kind PC, Harrad R (1994) Interocular suppression in the visual cortex of strabismic cats J Neurosci **14** 6855-71 [313]

Sengpiel F, Blakemore C, Harrad R (1995a) Interocular suppression in the primary visual cortex: a possible neural basis of binocular rivalry Vis Res **35** 179-95 [313]

Sengpiel F, Freeman TCB, Blakemore C (1995b) Interocular suppression in cat striate cortex is not orientation selective Neuroreport **6** 2235-9 [312, 313]

Sengpiel F, Troilo D, Kind PC, Graham B, Blakemore C (1996) Functional architecture of area 17 in normal and monocularly deprived marmosets (*Callithrix jacchus*) Vis Neurosci **13** 135-60 [216, 217]

Sengpiel F, Sen A, Blakemore C (1997) Characteristics of surround inhibition in cat area 17 Exp Brain Res **116** 216-28 [196]

Sengpiel F, Baddeley RJ Freeman TCB, Harrad R, Blakemore C (1998) Different mechanisms underlie three inhibitory phenomena in cat area 17 Vis Res **38** 2067-80 [313]

Serafini T, Colamarino SA, Leonardo ED, Wang H, Beddington R, Skarnes WC, Tessier-Lavigne M (1996) Netrin-1 is required for commissural axon guidance in the developing vertebrate nervous system Cell **87**, 1001-14 [453]

Sereno AB, Maunsell JHR (1998) Shape selectivity in primate lateral intraparietal cortex Nature **395** 500-3 [229]

Sereno MI, Dale AM, Reppas JB, Kwong KK, Belliveau JW, Brady TJ, Rosen BR, Tootell RBH (1995) Borders of multiple visual areas in humans revealed by functional magnetic resonance imaging Science **298** 889-93 [218]

Sermasi E, TropeaD, Domenici L (1999) A new form of synaptic plasticity is transiently expressed in the developing rat visual cortex: a modulatory role for visual experience and brain-derived neurotrophic factor Neurosci **91** 163-73 [462]

Servos P, Goodale MA (1994) Binocular vision and the on-line control of human prehension Exp Brain Res **98** 119-27 [4]

Servos P, Goodale MA, Jakobson LS (1992) The role of binocular vision in prehension: a kinematic analysis Vis Res **32** 1513–21 [4]

Sethi B (1986a) Vergence adaptation: a review Doc Ophthal **63** 277–63 [381]

Sethi B (1986b) Heterophoria: a vergence adaptive position Ophthal Physiol Opt **6** 151-6 [383]

Sethi B, Henson DB (1984) Adaptive changes with prolonged effect of comitant and noncomitant vergence disparities Am J Optom Physiol Opt **61** 506–12 [385]

Sethi B, North RV (1987) Vergence adaptive changes with varying magnitudes of prism–induced disparities and fusional amplitudes Am J Optom Physiol Opt **64** 293–8 [383]

Shadlen M, Carney T (1986) Mechanisms of human motion perception revealed by a new cyclopean illusion Science **232** 95–7 [304]

Shallo-Hoffmann J, Faldon M, Hague S, Riordan-Eva P, Fells P, Gresty M (1997) Motion detection deficits in infantile esotropia without nystagmus Invest Ophthal Vis Sci **38** 219-29 [525, 526]

Shapley R (1991) Receptive field structure of P and M cells in the monkey retina In From pigments to perception (ed A Valberg, BB Lee) pp 95–104 Plenum, New York [170]

Shapley R, So YT (1980) Is there an effect of monocular deprivation on the proportions of X and Y cells in the cat lateral geniculate nucleus? Exp Brain Res **39** 41–8 [504]

Shapley RM, Victor JD (1978) The effect of contrast on the transfer properties of cat retinal ganglion cells J Physiol **285** 275-98. [248]

Sharif MM (1966) A history of Muslim philosophy Otto Harrassowitz, Wiesbaden [14]

Sharma K, Abdul-Rahim AS (1992) Vertical fusion amplitude in normal adults Am J Ophthal **114** 636-7 [407]

Sharma V, Levi DM, Coletta NJ (1999) Sparse-sampling in the visual cortex of strabismic amblyopes Vis Res **39** 3529-36 [522, 523]

Sharma V, Levi DM, Klein SA (2000) Undercounting features and missing features: evidence for a high-level deficit in strabismic amblyopia Nature Neurosci **3** 496-501 [524]

Shattuck S, Held R (1975) Color and edge sensitive channels converge on stereo-depth analyzers Vis Res **15** 309-11 [352]

Shatz CJ (1977a) Abnormal interhemispheric connections in the visual system of Boston Siamese cats; A physiological study J Comp Neurol **171** 229-46 [497]

Shatz CJ (1977b) Anatomy of interhemispheric connections in the visual system of Boston Siamese and ordinary cats J Comp Neurol **173** 497-518 [497]

Shatz CJ (1983) The prenatal development of the cat's retinogeniculate pathway J Neurosci **3** 482–99 [443]

Shatz CJ (1990a) Impulse activity and the patterning of connections during CNS development Neuron **5** 745-56 [439]

Shatz CJ (1990b) Competitive interactions between retinal ganglion cells during prenatal development J Neurobiol **2 1** 197–211 [443]

Shatz CJ, LeVay S (1979) Siamese cat: altered connections of visual cortex Science **204** 328-30 [497]

Shatz CJ, Sretavan DW (1986) Interactions between retinal ganglion cells during development of the mammalian visual system Ann Rev Neurosci **9** 171-207 [436]

Shatz CJ, Stryker MP (1978) Ocular dominance columns in layer IV of the cat's visual cortex and the effects of monocular deprivation J Physiol **281** 297-83 [507]

Shatz CJ, Stryker MP (1988) Prenatal tetrodotoxin infusion blocks segregation of retinogeniculate afferents Science **272** 87–9 [444]

Shatz CJ, Lindstöm S, Wiesel TN (1977) The distribution of afferents representing the right and left eyes in the cat's visual cortex Brain Res **131** 103-16 [214]

Shatz CJ, Ghosh A, McKonnell SK, Allendoerfer KL, Friauf E, Antonini A (1991) Subplate neurons and the development of neocortical connections In Development of the visual system (ed DMK Lam, CJ Shatz) pp 175–96 MIT Press, Cambridge MA [451]

Shaw C, Cynader M (1984) Disruption of cortical activity prevents ocular dominance changes in monocularly deprived kittens Nature **308** 731–4 [511]

Shaw L, Roder B, Bushnell EW (1986) Infants' identification of three-dimensional form from transformations of linear perspective Percept Psychophys **40** 301-10 [480]

Shaw WT (1861) Description of a new optical instrument called the "Stereotrope" J Franklin Inst **42** 128-30 Also Proc R Soc **11** 70-72 [71]

Shea SL, Fox R, Aslin RN, Dumas ST (1980) Assessment of stereopsis in human infants Invest Ophthal Vis Sci **19** 1400-4 [477]

Shea SL, Doussard–Roosevelt JA, Aslin RN (1985) Pupillary measures of binocular luminance summation in infants and stereoblind adults Invest Ophthal Vis Sci **29** 1064–70 [474]

Shea SL, Aslin RN, McCulloch D (1987) Binocular VEP summation in infants and adults with abnormal binocular histories Invest Ophthal Vis Sci **28** 356–65 [483]

Sheedy JE (1980) Actual measurement of fixation disparity and its use in diagnosis and treatment J Am Optom Assoc **51** 1079–84 [377]

Sheedy JE, Bailey IL Buri M, Bass E (1986) Binocular vs monocular performance Am J Optom Physiol Opt **63** 839–46 [4]

Sheinberg DL, Logothetis NK (1997) The role of temporal cortical areas in perceptual organization Proc Nat Acad Sci **94** 3408-13 [314]

Shelhamer M, Merfeld DM, Mendoza JC (1995) Effect of vergence on the gain of the linear vestibulo-ocular reflex Acta Otolaryngologica Supplement 520 72-76 [427]

Sheng M, Cummings J, Roldan LA, Jan YN, Jan LY (1994) Changing subunit composition of heteromeric NMDA receptors during development of rat cortex Nature **368** 144-7 [460]

Sheppard AM, Pearlman AL (1997) Abnormal reorganization of preplate neurons and their associated development in the reeler mutant mouse J Comp Neurol **378** 173-9 [449]

Sherk H, LeVay S (1983) Contributions of the cortico-claustral loop to receptive field properties in area 17 of the cat J Neurosci **3** 2121-27 [183]

Sherman SM (1972) Development of interocular alignment in cats Brain Res **37** 187-203 [474]

Sherman SM (1973) Visual field defects in monocularly and binocularly deprived cats Brain Res **49** 25–45 [522]

Sherman SM (1974) Monocularly deprived cats: improvement of the deprived eye's vision by visual decortication Science **186** 297-9 [527]

Sherman SM (1996) Dual response modes in lateral geniculate neurons: Mechanisms and functions Vis Neurosci **13** 205-13 [179]

Sherman SM, Koch C (1986) The control of retinogeniculate transmission in the mammalian lateral geniculate nucleus Exp Brain Res **63** 1–20 [178, 179]

Sherman SM, Spear PD (1982) Organization of visual pathways in normal and visually deprived cats Physiol Rev **62** 738-854 [508, 525]

Sherman SM, Stone J (1973) Physiological normality of the retina in visually deprived cats Brain Res **60** 227–30 [502]

Sherman SM, Wilson JR (1975) Behavioral and morphological evidence for binocular competition in the postnatal development of the dog's visual system J Comp Neurol **161** 183-96 [504]

Sherman SM, Hoffmann KP, Stone J (1972) Loss of a specific cell type from the dorsal lateral geniculate nucleus in visually deprived cats J Neurophysiol **35** 532-41 [504]

Sherman SM, Wilson JR, Guillery RW (1975) Evidence that binocular competition affects the postnatal development of Y-cells in the cat's lateral geniculate nucleus Brain Res **100** 441-4 [504]

Sherrington CS (1904) On binocular flicker and the correlation of activity of corresponding retinal points Br J Psychol **1** 29–60 [305, 324, 329]

Sherry DF, Jacobs LF, Gaulin SJC (1992) Spatial memory and adaptive specialization of the hippocampus TINS **15** 298–303 [461]

Sheth BR, Sharma J, Rao SC, Sur M (1996) Orientation maps of subjective contours in visual cortex Science **274** 2110-16 [124]

Shevelev IA, Sharaev GA, Lazareva NA, Novikova RV, Tikhomirov AS (1993) Dynamics of orientation tuning in the cat striate cortex neurons Neurosci **56** 865–76 [199]

Shevelev IA, Eysel UT, Lazareva NA, Sharaev GA (1998) The contribution of intracortical inhibition to dynamics of orientation tuning in cat striate cortex neurons Neurosci **84** 11-23 [199]

Shikata E, Hamzei F, Glauche V, Knab R, Dettmers C, Weiller C, Büchel C (2001) Surface orientation discrimination activates caudal and anterior intraparietal sulcus in humans: an event-related fMRI study J Neurophysiol **85** 1309-14 [226]

Shimojo S, Held R (1987) Vernier acuity is less than grating acuity in 2- and 3-month-olds Vis Res **27** 77-86 [469]

Shimojo S, Birch EE, Gwiazda J, Held R (1984) Development of vernier acuity in infants Vis Res **27** 721-8 [469]

Shimojo S, Bauer J, O'Connell KM, Held R (1986) Pre–stereoptic binocular vision in infants Vis Res **29** 501–10 [482]

Shimono K, Ono H, Saida S, Mapp AP (1998) Methodological caveats for monitoring binocular eye position with nonius stimuli Vis Res **38** 591-600 [378]

Shinkman PG, Bruce CJ (1977) Binocular differences in cortical receptive fields of kittens after rotationally disparate binocular experience Science **197** 285–7 [485]

Shinkman PG, Isley MR, Rogers DC (1983) Prolonged dark rearing and development of interocular orientation disparity in visual cortex J Neurophysiol **49** 717–29 [485]

Shinkman PG, Timney B, Isley MR (1992) Binocular depth perception following early experience with interocular torsional disparity Vis Neurosci **9** 303–12 [485]

Shipley T (1957a) Convergence function in binocular visual space. I. A note on theory J Opt Soc Am **47** 795-803 [137]

Shipley T (1957b) Convergence function in binocular visual space. II. Experimental report J Opt Soc Am **47**, 804-21 [137]

Shipley T, Rawlings SC (1970a) The nonius horopter. I History and theory Vis Res **10** 1225–62 [57, 58]

Shipp S, Zeki S (1985) Segregation of pathways leading from area V2 to areas V4 and V5 of macaque monkey visual cortex Nature **315** 322–5 [223]

Shirke AM, Malinow R (1997) Mechanisms of potentiation by calcium-cadmodulin kinase II of postsynaptic sensitivity in rat hippocampal CA1 neurons J Neurophysiol **78** 2682-92 [460]

Shirokawa T, Kasamatsu T, Kuppermann BD, Ramachandran VS (1989) Noradrenergic control of ocular dominance plasticity in the visual cortex of dark-reared cats Devel Brain Res **47** 303-8 [512]

Shlaer R (1971) Shift in binocular disparity causes compensatory change in the cortical structure of kittens Science **173** 638–41 [484]

Shlaer R (1972) An eagle's eye: quality of the retinal image Science **176** 920–22 [545]

Shoham D, Hübener M, Schulze S, Grinvald A, Bonhoeffer T (1997) Spatio-temporal frequency domains and their relation to cytochrome oxidase staining in cat visual cortex Nature **385** 529-33 [205]

Shors TJ, Miesegaes G, Beylin A, Zhao M, Rydel T, Gould E (2001) Neurogenesis in the adult is involved in the formation of trace memories Nature **410** 372-5 [450]

Shou T, Leventhal AG (1989) Organized arrangement of orientation-sensitive relay cells in the cat's dorsal lateral geniculate nucleus J Neurosci **9** 4287-302 [179]

Shouval H, Intrator N, Law CC, Cooper LN (1996) Effect of binocular cortical misalignment on ocular dominance and orientation selectivity Neural Comput **8** 1021-40 [464]

Shu T, Richards LJ (2001) Cortical axon guidance by the glial wedge during the development of the corpus callosum J Neurosci **21** 2749-58 [453]

Shubnikov AV, Koptsik VA (1974) Symmetry in science and art Plenum, New York [103, 159]

Siegel H, Duncan CP (1960) Retinal disparity and diplopia vs luminance and size of target Am J Psychol **73** 280–4 [274]

Siegel RE (1970) Galen on sense perception S Kager, Basel [10, 14]

Sillito AM, Kemp JA, Milson JA, Berardi N (1980a) A re–evaluation of the mechanisms underlying simple cell orientation selectivity Brain Res **194** 517–20 [199]

Sillito AM, Kemp JA, Patel H (1980b) Inhibitory interactions contributing to the ocular dominance of monocularly dominated cells in the normal cat striate cortex Exp Brain Res **41** 1-10 [216]

Sillito AM, Kemp JA, Blakemore C (1981) The role of GABAergic inhibition in the cortical effects of monocular deprivation Nature **291** 318–20 [510]

Sillito AM, Jones HE, Gerstein GL, West DC (1994) Feature-linked synchronization of thalamic relay cell firing induced by feedback from the visual cortex Nature **369** 479-82 [134]

Sillito AM, Grieve KL, Jones HE, Cudeiro J, Davis J (1995) Visual cortical mechanisms detecting focal orientation discontinuities Nature **378** 492-96 [200]

Silva LR, Amitai Y, Connors BW (1991) Intrinsic oscillations of neocortex generated by layer 5 pyramidal neurons Science **251** 432–5 [136]

Silver MA, Stryker MP (1999) Synaptic density in geniculocortical afferents remains constant after monocular deprivation in the cat J Neurosci **19** 10829-42 [507]

Silver MA, Stryker MP (2000) Distributions of synaptic vesicle proteins and GAD65 in deprived and nondeprived ocular dominance columns in layer IV of kitten primary visual cortex are unaffected by monocular deprivation J Comp Neurol **422** 652-64 [510]

Simmers AJ, Gray LS, McGraw PV, Winn B (1999) Functional visual loss in amblyopia and the effect of occlusion therapy Invest Ophthal Vis Sci **40** 2859-71 [528]

Simmons DR, Kingdom FAA (1998) On the binocular summation of chromatic contrast Vis Res **38** 1063-71 [323]

Simmons JA, Stein RA (1980) Acoustic imaging in bat sonar: echlocation signals and the evolution of echlocation J Comp Physiol A **135** 61-84 [548]

Simons K (1981b) Stereoacuity norms in young children Arch Ophthal **99** 439–45 [482]

Simons K (1993) Early visual development normal and abnormal Oxford University Press, New York [485]

Simonsz HJ, Harting F, de Waal BJ, Verbeeten BWJM (1985) Sideways displacement and curved path of recti eye muscles Arch Ophthal **103** 127-8 [371]

Simpson T (1991) The suppression effect of simulated anisometropia Ophthal Physiol Opt **11** 350-8 [287]

Sincich LC, Blasdel GG (2001) Oriented projections in primary visual cortex of the monkey J Neurosci **21** 4416-26 [196]

Sindermann F, Lüddeke H (1972) Monocular analogues to binocular contour rivalry Vis Res **12** 763–72 [295]

Singer W (1970) Inhibitory binocular interaction in the lateral geniculate body of the cat Brain Res **18** 165–70 [179]

Singer W (1977) Effects of monocular deprivation on excitatory and inhibitory pathways in cat striate cortex Exp Brain Res **30** 25-41 [508]

Singer W (1978) The effect of monocular deprivation on cat parastriate cortex: asymmetry between crossed and uncrossed pathways Brain Res **157** 351–5 [522]

Singer W (1990) The formation of cooperative cell assemblies in the visual cortex J Exp Biol **153** 177–97 [460]

Singer W, Gray CM (1995) Visual feature integration and the temporal correlation hypothesis Ann Rev Neurosci **18** 555-86 [132]

Singer W, Tretter F (1976) Receptive–field properties and neural connectivity in striate and parastriate cortex of contour–deprived cats J Neurophysiol **39** 613–30 [500]

Singer W, Rauschecker J, Werth R (1977) The effect of monocular exposure to temporal contrasts on ocular dominance in kittens Brain Res **134** 568–72 [513]

Singer W, Yinon, U, Tretter F (1979) Inverted monocular vision prevents ocular dominance shift in kittens and impairs the functional state of visual cortex in adult cats Brain Res **164** 294-9 [507]

Singer W, Freeman B, Rauschecker J (1981) Restriction of visual experience to a single orientation affects the organization of orientation columns in cat visual cortex Exp Brain Res **41** 199-215 [454]

Sireteanu R (1982) Binocular vision in strabismic humans with alternating fixation Vis Res **22** 889-896 [529]

Sireteanu R (1987) Binocular luminance summation in humans with defective binocular vision Invest Ophthal Vis Sci **28** 349-55 [474]

Sireteanu R (1996) Development of the visual field: results from human and animal studies In Infant vision (ed F Vital-Durand, J Atkinson, OJ Braddick) pp 17-31 Oxford University Press, Oxford [473]

Sireteanu R, Fronius M (1981) Naso-temporal asymmetries in human amblyopia: consequences of long-term interocular suppression Vis Res **21** 1055-63 [522]

Sireteanu R, Fronius M (1990) Human amblyopia: structure of the visual field Exp Brain Res **79** 603–14 [522]

Sireteanu R, Hoffmann KP (1979) Relative frequency and visual resolution of X– and Y–cells in the LGN of normal and monocularly deprived cats: interlaminar differences Exp Brain Res **34** 591–603 [504]

Sireteanu R, Maurer D (1982) The development of the kitten's visual field Vis Res **22** 1105–11 [473]

Sireteanu R, Singer W (1980) The "vertical effect" in human squint amblyopia Exp Brain Res **40** 354-57 [522]

Sireteanu R, Fronius M, Singer W (1981) Binocular interaction in the peripheral visual field of humans with strabismic and anisometropic amblyopia Vis Res **21** 1065–74 [492]

Sireteanu R, Singer W, Fronius M, Greuel JM, Best J, Fiorentini A, Bisti S, Schiavi C, Campos EC (1993a) Eye alignment and cortical binocularity in strabismic kittens: a comparison between tenotomy and recession Vis Neurosci **10** 541-9 [505]

Sireteanu R, Lagreze WD, Constantinescu DH (1993b) Distortions in two-dimensional visual space perception in strabismic observers Vis Res **33** 677-90 [524]

Sireteanu R, Fronius M, Constantinescu DH (1994) The development of visual acuity in the peripheral visual field of human infants: binocular and monocular measurements Vis Res **34** 1659-71 [473]

Skaliora I, Scobey RP, Chalupa LM (1993) Prenatal development of excitability in cat retinal ganglion cells: action potentials and sodium currents J Neurosci **13** 313–23 [436, 444, 459]

Skarf B (1973) Development of binocular single units in the optic tectum of frogs raised with disparate stimulation to the eyes Brain Res **51** 352–7 [541]

Skarf B, Eisenman M, Katz LM, Bachynsnki B, Klein R (1993) A new VEP system for studying binocular single vision in human infants J Ped Ophthal Strab **30** 237-42 [483]

Skoczenski AM, Aslin RN (1992) Spatiotemporal factors in infant position sensitivity: single bar stimuli Vis Res **32** 1761-9 [469]

Skoczenski AM, Aslin RN (1995) Assessment of vernier acuity development using the "equivalent intrinsic blur" paradigm Vis Res **35** 1879-88 [470]

Skoczenski AM, Norcia AM (1998) Neural noise limitations on infant visual sensitivity Nature **391** 697-700 [470]

Skottun BC, Freeman RD (1984) Stimulus specificity of binocular cells in the cat's visual cortex: ocular dominance and the matching of left and right eyes Exp Brain Res **56** 206–16 [214]

Skottun BC, Bradley A, Romoa AS (1986a) Effect of contrast on spatial frequency tuning of neurones in area 17 of cat's visual cortex Exp Brain Res **63** 431-5 [199, 201]

Skottun BC, Bradley A, Freeman RD (1986b) Orientation discrimination in amblyopia Invest Ophthal Vis Sci **27** 532-7 [522]

Skottun BC, Bradley A, Sclar G, Ohzawa I, Freeman RD (1987) The effects of contrast on visual orientation and spatial frequency discrimination: a comparison of single cells and behavior J Neurophysiol **57** 773-86 [198, 201]

Skrandies W (1991) Contrast and stereoscopic visual stimuli yield lateralized scalp potential fields associated with different neural generators EEG Clin Neurophysiol **78** 274-83 [262]

Skrandies W (1997) Depth perception and evoked brain activity: the influence of horizontal disparity and visual field location Vis Neurosci **14** 527-32 [262]

Skrandies W, Vomberg HE (1985) Stereoscopic stimuli activate different cortical neurones in man: electrophysiological evidence Internat J Psychophysiol **2** 293-6 [263]

Slater AM (ed) (1998) Perceptual development Psychology Press, Hove UK [485]

Slater AM, Findlay JM (1975a) The corneal reflection technique and the visual preference method: sources of error J Exp Child Psychol **20** 270–7 [475]

Slater AM, Findlay JM (1975b) Binocular fixation in the newborn baby J Exp Child Psychol **20** 278–73 [476]

Slater AM, Kirby R (1998) Innate and learned perceptual abilities in the newborn infant Exp Brain Res **123** 90-94 [479]

Slater AM, Morison V (1985) Shape constancy and slant perception at birth Perception **14** 337-44 [479]

Slater AM, Morison V, Somers M (1988) Orientation discrimination and cortical function in the human newborn Perception **17** 597–602 [471]

Sloane ME, Blake R (1984) Selective adaptation of monocular and binocular neurons in human vision J Exp Psychol HPP **1 0** 406–42 [340]

Sloper JJ (1972) Gap junctions between dendrites in the primate neocortex Brain Res **44** 641-6 [190]

Smallman HS, MacLeod DIA, He S, Kentridge RW (1996) Fine grain of the neural representation of human spatial vision J Neurosci **16** 1852-9 [174]

Smart TG (1997) Regulation of excitatory and inhibitory neurotransmitter ion channels by protein phosphorylation Curr Opin Neurobiol **7** 358-67 [191]

Smith AM (1996) Ptolemy's theory of visual perception (English translation of the *Optics*) Tr Am Philos Soc **86** Part 2 [12]

Smith AT (1983) Interocular transfer of colour–contingent threshold elevation Vis Res **23** 729–34 [353]

Smith AT, Edgar GK (1994) Antagonistic comparison of temporal frequency filter outputs as a basis for speed perception Vis Res **34** 253-65 [203]

Smith AT, Jeffreys DA (1979) Evoked potential evidence for differences in binocularity between striate and prestriate regions of human visual cortex Exp Brain Res **36** 375-80 [260]

Smith DC, Holdefer RN (1985) Binocular competitive interactions and recovery of visual acuity in long–term monocularly deprived cats Vis Res **25** 1783–94 [527]

Smith DC, Spear PD, Kratz KE (1978) Role of visual experience in postcritical–period reversal of effects of monocular deprivation in cat striate cortex J Comp Neurol **178** 313–28 [508]

Smith DC, Lorber S, Stanford LR, Loop MS (1980) Visual acuity following binocular deprivation in the cat Brain Res **183** 1-11 [501]

Smith EL, Bennett MJ, Harwerth RS, Crawford MLJ (1979) Binocularity in kittens reared with optically induced squint Science **204** 875–7 [513]

Smith EL, Harwerth RS, Browning R, Jacobson D, Kircher ML (1980) Stereopsis and fixation disparity following prolonged monocular occlusion Am J Optom Physiol Opt **57** 413-19 [518]

Smith EL, Levi DM, Harwerth RS, White JM (1982) Color vision is altered during the suppression phase of binocular rivalry Science **218** 802–4 [300]

Smith EL, Levi DM, Manny RE, Harwerth RS, White JM (1985a) The relationship between binocular rivalry and strabismic suppression Invest Ophthal Vis Sci **29** 80–7 [530]

Smith EL, Harwerth RS, Crawford MLJ (1985b) Spatial contrast sensitivity deficits in monkeys produced by optically induced anisometropia Invest Ophthal Vis Sci **29** 330–42 [518]

Smith EL, Chino YM, Ridder WH, Kitagawa K, Langston A (1990) Orientation bias of neurons in the lateral geniculate nucleus of macaque monkeys Vis Neurosci **5** 525-45 [179]

Smith EL, Harwerth RS, Siderow J, Wingard M, Crawford MLJ, von Noorden GK (1992) Prior binocular dissociation reduces monocular form deprivation amblyopia in monkeys Invest Ophthal Vis Sci **33** 1804–10 [527]

Smith EL, Hung LF, Harwerth RS (1994) Effects of optically induced blur on the refractive status of young monkeys Vis Res **34** 293–301 [435]

Smith EL, Chino YM, Ni J, Ridder WH, Crawford MLJ (1997a) Binocular spatial phase tuning characteristics of neurons in the macaque striate cortex J Neurophysiol **78** 351-65 [248]

Smith EL, Chino YM, Ni J, Cheng H (1997b) Binocular combination of contrast signals by striate cortical neurons in the monkey J Neurophysiol **78** 366-82 [248]

Smith EL, Chino YM, Ni J, Cheng H, Crawford MLJ, Harwerth RS (1997c) Residual binocular interactions in the striate cortex of monkeys reared with abnormal binocular vision J Neurophysiol **78** 1353-62 [509]

Smith JC, Atkinson J, Anker S, Moore AT (1991) A prospective study of binocularity and amblyopia in strabismic infants before and after corrective surgery: implications for the human critical period Clin Vis Sci **6** 335–53 [519]

Smith KU, Schmidt J, Putz V (1970) Binocular coordination: feedback of synchronization of eye movements for space perception Am J Optom Arch Am Acad Optom **47** 679-89 [421]

Smith R (1738) A compleat system of opticks in four books Cambridge [56]

Smith R (1985) Vergence eye-movement responses to whole-body linear acceleration stimuli in man Ophthal Physiol Opt **5** 303-11 [426]

Snider CA, Dehay C, Berland M, Kennedy H, Chalupa LM (1999) Prenatal development of retinogeniculate axons in the macaque monkey during segregation of binocular inputs J Neurosci **19** 220-8 [443]

Snippe HP, Koenderink JJ (1992) Discrimination thresholds for channel–coded systems Biol Cyber **66** 543–51 [79]

Snodderly DM, Gur M (1995) Organization of striate cortex of alert trained monkeys (Macaca fascicularis): ongoing activity stimulus selectivity and widths of receptive field activating regions J Neurophysiol **74** 2100-25 [178, 204, 213]

Snow R, Hore J, Vilis T (1985) Adaptation of saccadic and vestibulo-ocular systems after extraocular muscle tenectomy Invest Ophthal Vis Sci **29** 927–31 [426]

Snowden RJ, (1992) Sensitivity to relative and absolute motion Perception **21** 563-8 [350]

Snowden RJ, Hammett ST (1992) Subtractive and divisive adaptation in the human visual system Nature **355** 278-50 [313]

Snowden RJ, Treue S, Erickson RG, Andersen RA (1991) The response of area MT and V1 neurons to transparent motion J Neurosci **11** 2768-85 [224]

Snowden RJ, Treue S, Andersen RA (1992) The response of neurons in areas V1 and MT of the alert rhesus monkey to moving random dot patterns Exp Brain Res **88** 389–400 [61, 203]

Snyder AW, Miller WH (1977) Photoreceptor diameter and spacing for highest resolution J Opt Soc Am **67** 696-8 [165, 173]

Snyder AW, Miller WH (1978) Telephoto lens system of falconiform eyes Nature **273** 127–9 [545]

Snyder AW, Bossomaier TRJ, Hughes A (1986) Optical image quality and the cone mosaic Science **231** 499-501 [173]

Snyder LH, Lawrence DM, King WM (1992) Changes in vestibuloocular reflex (VOR) anticipate changes in vergence angle in monkey Vis Res **32** 569–75 [428]

Snyder LH, Grieve KL, Brotchie P, Andersen RA (1998) Separate body- and world-referenced representations of visual space in parietal cortex Nature **394** 887-91 [228]

Sobel EC (1990) Depth perception by motion parallax and paradoxical parallax in the locust Naturwissenschaften **77** 271–3 [534]

Soechting JF, Terzuolo CA (1988) Sensorimotor transformations underlying the organization of arm movements in three-dimensional space Can J Physiol Pharmacol **66** 502-7 [143]

Softky W (1994) Sub-millisecond coincidence detection in active dendritic trees Neurosci **58** 13-41 [135]

Sokol S, Nadler D (1979) Simultaneous electroretinograms and visually evoked potentials from adult amblyopes in response to a pattern stimulus Invest Ophthal Vis Sci **18** 848-55 [525, 527]

Sokoloff L, Reivich M, Kennedy C, Des Rosiers MH, Patlak CS, Pettigrew KD, Sakurada O, Shinohara M (1977) The [14C] deoxyglucose method for measurement of local cerebral glucose utilization procedure and normal values in the conscious and anesthetized rat J Neurochem **28** 897–916 [212]

Solomon JA, Morgan MJ (1999) Dichoptically cancelled motion Vis Res **39** 2293-7 [293]

Solomon JS, Doyle JF, Burkhalter A, Nerbonne JM (1993) Differential expression of hyperpolarization-activated currents reveals distinct classes of visual cortical projection neurons J Neurosci **13** 5082-91 [188]

Solomon Sg, White AJR, Martin PR (2002) Extraclassical receptive field properties of parvocellular, magnocellular, and koniocellular cells in the primate lateral geniculate nucleus J Neurosci **22** 338-49 [178]

Solomons H (1978) Binocular vision, a programmed text Heinemann, London [8]

Somani RAB, Desouza JFX, Tweed D, Vilis T (1998) Visual test of Listing's law during vergence Vis Res **38** 911-23 [370]

Song H, Ming G, Poo M (1997) cAMP-induced switching in turning direction of nerve growth cones Nature **388** 275-9 [438]

Soodak RE, Shapley RM, Kaplan E (1987) Linear mechanism of orientation tuning in the retina and lateral geniculate nucleus of the cat J Neurophysiol **58** 297-75 [179]

Sparks DL, Porter JD (1983) Spatial localization of saccade targets. II Activity of superior colliculus neurons preceding compensatory saccades J Neurophysiol **49** 64-74 [227]

Sparks DL, Mays LE, Gurski MR, Hickey TL (1986) Long– and short-term monocular deprivation in the rhesus monkey: effects on visual fields and optokinetic nystagmus J Neurosci **6** 1771–80 [510, 526]

Spatz WB (1989) Loss of ocular dominance columns with maturity in the monkey *Callithrix jacchus* Brain Res **488** 376-80 [217]

Spear PD, Tong L (1980) Effects of monocular deprivation on neurons in cat's lateral suprasylvian visual area. I Comparison of binocular and monocular segments J Neurophysiol **4 4** 568–84 [507]

Spear PD, Langsetmo A, Smith DC (1980) Age-related changes in effects of monocular deprivation on cat striate cortex neurons J Neurophysiol **43** 559–80 [508]

Spekreijse H, Reits D (1982) Sequential analysis of the visual evoked potential system in man: nonlinear analysis of a sandwich system Ann N Y Acad Sci **388** 72-97 [95]

Spekreijse H, van der Tweel LH, Regan D (1972) Interocular sustained suppression: correlations with evoked potential amplitude and distribution Vis Res **12** 521–6 [259]

Spencer JB (1858) Stereoscopes for books Photograph J **5** 79 [69]

Spencer RF, Porter JD (1988) Structural organization of the extraocular muscles In Neuroanatomy of the oculomotor system (ed JA Büttner–Ennever) pp 33–80 Elsevier, New York [429]

Spencer RW, Wilson WK (1954) Accommodative response in asymmetric convergence Am J Optom Arch Am Acad Optom **31** 498-505 [364]

Spencer S, Willard MB (1992) GAP–43 and regrowth of retinal ganglion cell axons In Regeneration and plasticity in the mammalian visual system (ed DMK Lam, GM Garth) pp 97–105 MIT Press, Cambridge MA [442]

Sperling G (1965) Temporal and spatial masking. I. Masking by impulse flashes J Op Soc Am **55** 541-59 [338]

Sperling G (1970) Binocular vision: a physical and a neural theory Am J Psychol **83** 461–534 [268]

Sperling G, Dosher BA (1995) Depth from motion In Early vision and beyond (ed TV Papathomas) pp 133-43 MIT Press, Cambridge Mass [149, 150]

Sperry RW (1945) Restoration of vision after crossing of optic nerves and after contralateral transplantation of the eye J Neurophysiol **8** 15-28 [448]

Sperry RW (1951) Mechanisms of neural maturation In Handbook of experimental psychology (ed SS Stevens) pp 236-80 Wiley, New York [433, 447, 464]

Sperry RW, Clark E (1949) Interocular transfer of visual discrimination habits in a teleost fish Physiol Zool **22** 372-8 [354]

Sperry RW, Minor N, Myers RE (1955) Visual pattern perception following subpial slicing and tantalum wire implantations in the visual cortex J Comp Physiol Psychol **48** 50–8 [209]

Spillmann L (1993) The perception of movement and depth in moiré patterns Perception **22** 287-308 [282]

Spillmann L, Dresp B (1995) Phenomena of illusory form: can we bridge the gap between levels of explanation? Perception **27** 1333-64 [123]

Sporns O, Tononi G, Edelman GM (1991) Modeling perceptual grouping and figure–ground segregation by means of active reentrant connections Proc Nat Acad Sci **88** 129–33 [136]

Springbett BM (1961) Some stereoscopic phenomena and their implications Br J Psychol **52** 105–9 [353]

Springer AD, Cohen SM (1981) Optic fiber segregation in goldfish with two eyes innervating one tectual lobe Brain Res **225** 23-36 [464]

Springer AD, Gaffney JS (1981) Retinal projections in the goldfish: a study using cobaltous lysine J Comp Neurol **203** 401-27 [539]

Spruston N, Schiller Y, Stuart G, Sakmann B (1995) Activity-dependent action potential invasion and calcium influx into hippocampal CA1 dendrites Science **298** 297-300 [115]

Squatrito S, Maioli MG (1996) Gaze field properties of eye position neurones in areas MST and 7a of the macaque monkey Vis Neurosci **13** 385-98 [225]

Srebro R (1978) The visually evoked response: binocular facilitation and failure when binocular vision is disturbed Arch Ophthal **96** 839–44 [259, 527]

Sretavan DW (1990) Specific routing of retinal ganglion cell axons at the mammalian optic chiasm during embryonic development J Neurosci **10** 1995–2007 [439, 440]

Sretavan DW, Reichardt LF (1993) Time-lapse video analysis of retinal ganglion cell axon pathfinding at the mammalian optic chiasm: growth cone guidance using intrinsic chiasm cues Neuron **10** 761-77 [440]

Sretavan DW, Shatz CJ (1986a) Prenatal development of retinal ganglion cell axons: segregation into eye-specific layers within the cat's lateral geniculate nucleus J Neurosci **6** 234-51 [443]

Sretavan DW, Shatz CJ (1986b) Prenatal development of cat retinogeniculate axon arbors in the absence of binocular interactions J Neurosci **6** 990-1003 [443]

Sretavan DW, Shatz CJ (1987) Axon trajectories and pattern of terminal arborization during the prenatal development of the cat's retinogeniculate pathway J Comp Neurol **255** 386-400 [443]

Sretavan DW, Feng L, Puré E, Reichardt LF (1994) Embryonic neurons of the developing optic chiasm express L1 and CD44 cell surface molecules with opposing effects on retinal axon growth Neuron **12** 957-75 [440]

Sretavan DW, Puré E, Siegel MW, Reichardt LF (1995) Disruption of retinal axon ingrowth by ablation of embryonic mouse optic chiasm neurons Science **299** 98-101 [440]

Srinivasan MV, Venkatesh S (1997) From living eyes to seeing machines Oxford University Press, Oxford [535]

Srinivasan MV, Zhang SW (1993) Evidence for two distinct movement–detecting mechanisms in insect vision Naturwissenschaften **80** 38–41 [534]

Srinivasan MV, Lehrer M, Zhang SW, Horridge GA (1989) How honeybees measure their distance from objects of unknown size J Comp Physiol A **165** 605-13 [535]

Srinivasan MV, Zhang SW, Lehrer M, Collett TS (1996) Honeybee navigation *en route* to the goal: visual flight control and odometry J Exp Biol **199** 237-44 [534]

Srinivasan MV, Poteser M, Kral K (1999) Motion detection in insect orientation and navigation Vis Res **39** 2749-66 [535]

Srinivasan MV, Zhang SW, Altwein M, Tautz J (2000) Honeybee navigation: nature and calibration of the "odometer" Science **287** 851-3 [535]

Srinivasan R, Russell DP, Edelman GM, Tonini G (1999) Increased synchronization of neuromagnetic responses during conscious perception J Neurosci **19** 5435-48 [312]

St Cyr GF, Fender DH (1969) The interplay of drifts and flicks in binocular fixation Vis Res **9** 275–65 [395]

Stalmeier PFM, de Weert CMM (1988) Binocular rivalry with chromatic contours Percept Psychophys **44** 456-62 [291]

Stanford LR (1987) Conduction velocity variations minimize conduction time differences among retinal ganglion cell axons Science **238** 358-60 [168]

Star EN, Kwiatkowski DJ, Murphy N (2002) Rapid turnover of actin in dendritic spines and its regulation by activity Nature Neurosci **5** 239-46 [438]

Starr BS (1971) Veridical and paradoxical interocular transfer of left/right mirror image discriminations Brain Res **31** 377 [355]

Steadman P (2001) Vermeer's camera Oxford University Press, Oxford [39]

Steeves JKE, Reed MJ, Steinbach MJ, Kraft SP (1999) Monocular horizontal OKN in observers with early- and late-onset strabismus Behav Brain Res **103** 135-43 [526]

Steffen H, Walker MF, Zee DS (2000) Rotation of Listing's plane with convergence: independence from eye position Invest Ophthal Vis Sci **41** 715-21 [370]

Stein JF (1991) Space and the parietal association areas In Brain and space (ed J Paillard) pp 185-222 Oxford University Press, New York [228]

Steinbach MJ, Money KE (1973) Eye movements of the owl Vis Res **13** 889–91 [547]

Steinbuch JG (1811) Beitrag zur Physiologie der Sinne Schragg, Nurnberg [45]

Steiner V, Blake R, Rose D (1994) Interocular transfer of expansion rotation and translation motion aftereffects Perception **23** 1197-202 [348]

Steinman RM, Collewijn H (1980) Binocular retinal image motion during active head rotation Vis Res **20** 415–29 [395]

Steinman RM, Cunitz RJ, Timberlake GT, Herman M (1967) Voluntary control of microsaccades during maintained monocular fixation Science **155** 1577–9 [366]

Steinman RM, Haddad GM, Skavenski AA, Wyman D (1973) Miniature eye movements Science **181** 810–19 [365]

Steinman RM, Levinson JZ, Collewijn H, van der Steen J (1985) Vision in the presence of known natural retinal image motion J Opt Soc Am A **2** 229–33 [395]

Stemmler M, Usher M, Niebur E (1995) Lateral interactions in primary visual cortex: a model bridging physiology and psychophysics Science **299** 1877-80 [197]

Stephens GL, Jones R (1990) Horizontal fusional amplitudes after adaptation to prism Ophthal Physiol Opt **10** 25–8 [383]

Sterling P (1990) Retina In Synaptic organization of the brain (ed G Shepard) Chap 6 Oxford University Press, Oxford [168]

Stetter M, Bartsch H, Obermayer K (2000) A mean-field model for orientation tuning, contrast saturation, and contextual effects in the primary visual cortex Biol Cybern **82** 291-304 [200]

Stevens CF (1995) Cortical synaptic transmission In The cortical neuron (ed MJ Gutnick, I Mody) pp 27-32 Oxford University Press, Oxford [189]

Stevens CF (2001) An evolutionary scaling law for the primate visual system and its basis in cortical function Nature **411** 193-5 [446]

Stevenson SB, Cormack LK, Schor CM, Tyler CW (1992) Disparity tuning in mechanisms of human stereopsis Vis Res **32** 1685–94 [251]

Stevenson SB, Cormack LK, Schor CM (1994) The effect of stimulus contrast and interocular correlation on disparity vergence Vis Res **34** 383–96 [394]

Stevenson SB, Lott LA, Yang J (1997) The influence of subject instruction on horizontal and vertical vergence tracking Vis Res **37** 2891-8 [409, 410]

Stevenson SB, Reed PE, Yang J (1999) The effect of target size and eccentricity on reflex disparity vergence Vis Res **39** 823-32 [396, 409]

Stidwill D (1997) Clinical survey: epidemiology of strabismus Ophthal Physiol Opt **17** 536-9 [373]

Stiles WS (1939) The directional sensitivity of the retina and the spectral sensitivities of the rods and cones Proc R Soc B **127** 64-105 [341]

Stiles WS, Crawford BH (1933) The luminance efficiency of rays entering the eye pupil at different points Proc R Soc B **112** 428-50 [165]

Stoerig P, Cowey A (1989) Wavelength sensitivity in blindsight Nature **342** 916-18 [220]

Stoerig P, Cowey A (1997) Blindsight in man and monkey Brain **120** 535-9 [218]

Stone D, Mathews S, Kruger PB (1993) Accommodation and chromatic aberration: effect of spatial frequency Ophthal Physiol Opt **13** 274-52 [364]

Stone J (1966) The naso-temporal division of the cat's retina J Comp Neurol **129** 585–600 [231]

Stone J, Fabian M (1966) Specialized receptive fields of the cat's retina Science,**152**, 1277-9 [203]

Stone J, Fukuda Y (1974) The naso-temporal division of the cat's retina re-examined in terms of Y- X- and W-cells J Comp Neurol **155** 377-94 ` [231]

Stone J, Rapaport DH (1986) The role of cell death in shaping the ganglion cell population of the adult cat retina In Vis Neurosci (ed JD Pettigrew, WR Levick) pp 157–65 Cambridge University Press, London [442]

Stone J, Leicester J, Sherman SM (1973) The naso-temporal division of the monkey's retina J Comp Neurol **150** 333-48 [231]

Stoner GR, Albright TD (1993) Image segmentation cues in motion processing: implications for modularity in vision J Cog Neurosci **5** 129-49 [160]

Stork DG, Wilson HR (1990) Do Gabor functions provide appropriate descriptions of visual cortical receptive fields J Opt Soc Am A **7** 1362-73 [142]

Stratton GM (1917) Theophrastus and the Greek physiological psychology before Aristotle Macmillan, New York [10]

Straumann D, Zee DS, Solomon D, Lasker AG, Roberts DC (1995) Transient torsion during and after saccades Vis Res **35** 3321-34 [370]

Strauss WL (1977) Commentary to the translation of Dürer's The painter's manual Abaris Books New York [36]

Straznicky C, Tay D, Hiscock J (1980) Segregation of optic fibre projections into eye-specific bands in dually innervated tecta in xenopus Neurosci Lett **19** 131-6 [464]

Stromeyer CF (1978) Form–color aftereffects in human vision In Handbook of sensory physiology (ed H Teuber, R Held) Vol VII pp 97–142 Springer, New York [351]

Stromeyer CF, Julesz B (1972) Spatial-frequency masking in vision: critical bands and spread of masking J Opt Soc Am **62** 1221-32 [139]

Stromeyer CF, Mansfield RJW (1970) Colored aftereffects produced with moving edges Percept Psychophys **7** 108–14 [353]

Strong DS (1979) Leonardo on the eye Garland, New York [20, 35, 52]

Struck MC, Ver Hoeve JN, France TD (1996) Binocular cortical interactions in the monofixation syndrome J Ped Ophthal Strab **33** 291-7 [381]

Stryker MP (1991) Activity–dependent reorganization of afferents in the developing mammalian visual system In Development of the visual system (ed DMK Lam, CJ, Shatz) pp 297–87 MIT Press, Cambridge MA [199]

Stryker MP, Blakemore C (1972) Saccadic and disjunctive eye movements in cats Vis Res **12** 2005-13 [392]

Stryker MP, Harris WA (1986) Binocular impulse blockage prevents the formation of ocular dominance columns in cat visual cortex J Neurosci **6** 2117–33 [463]

Stryker MP, Sherk H, Leventhal AG, Hirsch HVB (1978) Physiological consequences for the cats visual cortex of effectively restricting early visual experience with oriented contours J Neurophysiol **41** 896-909 [454]

Stuart GJ, Sakmann B (1994) Active propagation of somatic action potentials into neocortical pyramidal cell dendrites Nature **367** 69-72 [113]

Stuckmann I, Weigmann A, Shevchenko A, Mann M, Hutter WB (2001) Ephrin B1 is expressed on neuroepithelial cells in correlation with neocortical neurogenesis J Neurosci **21** 2726-37 [447]

Sturm AK, König P (2001) Mechanisms to synchronize neuronal activity Biol Cyber **84** 153-72 [136]

Sturr JF, Teller DY (1973) Sensitization by annular surrounds: dichoptic properties Vis Res **13** 909-18 [334]

Sugie N (1982) Neural models of brightness perception and retinal rivalry in binocular vision Biol Cyber **43** 13–21 [310, 327]

Sugihara H, Murakami I, Shenoy KV, Andersen RA, Komatsu H (2002) Response of MSTd neurons to simulated 3D orientation of rotating planes J Neurophysiol **87** 273-85 [225]

Sumner FC, Watts FP (1936) Rivalry between uniocular negative after–images and the vision of the other eye Am J Psychol **48** 109–16 [352]

Supér H, Spekreijse H, Lamme VAF (2001) Two distinct modes of sensory processing observed in monkey primary visual cortex Nature Neurosc **4** 304-9 [206]

Sur M, Humphrey AL, Sherman SM (1982) Monocular deprivation affects X- and Y-cell retinogeniculate terminations in cats Nature **300** 183-5 [504]

Sur M, Pallas SL, Roe AW (1990) Cross-modal plasticity in cortical development: differentiation and specification of sensory neocortex TINS **13** 227-33 [447]

Suter DM, Errante LD, Belotserkovsky V, Forscher P (1998) The Ig superfamily cell adhesion molecule, apCAM, mediates growth cone steering by substrate-cytoskeletal coupling J Cell Biol **141** 227-40 [438]

Sutter EE (1992) A deterministic approach to nonlinear systems analysis. In Nonlinear vision: Determination of neural receptive fields function and networks (ed RB Pinter, B Nabet) pp 171-220 CRC Press, London [95]

Sutterlin AM, Prosser CL (1970) Electrical properties of goldfish optic tectum J Electrophysiol **33** 36-45 [539]

Suzuki H, Kato E (1966) Binocular interaction at cat's lateral geniculate body J Neurophysiol **29** 909–20 [179]

Suzuki H, Takahashi M (1970) Organization of lateral geniculate neurons in binocular inhibition Brain Res **23** 291-4 [180]

Suzuki WA, Miller EK, Desimone R (1997) Object and place memory in the macaque entorhinal cortex J Neurophysiol **7 8** 1062-81 [223]

Suzuki Y, Büttner-Ennever JA, Straumann D, Hepp K, Hess BJM, Henn V (1995) Deficits in torsional and vertical rapid eye movements and shift of Listing's plane after uni- and bilateral lesions of the rostral interstitial nucleus of the medial longitudinal fasciculus Exp Brain Res **106** 215-32 [432]

Swadlow HA (1983) Efferent systems of primary visual cortex: a review of structure and function Brain Res Rev **6** 1–27 [187]

Swan H (1863) On a new kind of miniature Photograph J **8** 351 [69]

Swanson WH, Birch E (1990) Infant spatiotemporal vision: dependence of spatial contrast sensitivity on temporal frequency Vis Res **30** 1033-48 [471]

Swanston MT, Wade NJ (1985) Binocular interaction in induced line rotation Percept Psychophys **37** 363–8 [349]

Swets JA (1964) Signal detection and recognition by human observers Wiley, New York [73]

Swindale NV (1980) A model for the formation of ocular dominance stripes Proc R Soc B **208** 273-64 [212]

Swindale NV (1981) Absence of ocular dominance patches in dark-reared cats Nature **290** 332-333 [463, 500]

Swindale NV (1982) A model for the formation of orientation columns Proc R Soc B **215** 211-30 [212, 464]

Swindale NV (1988) Role of visual experience in promoting segregation of eye dominance patches in the visual cortex of the cat J Comp Neurol **297** 472-88 [500]

Swindale NV (1992) A model for the coordinated development of columnar systems in primate striate cortex Biol Cyber **66** 217-30 [212]

Swindale NV (1996) The development of topography in the visual cortex: a review of models Network: Computation in Neural Systems 7 161-247 [212]

Swindale NV, Cynader MS (1986) Physiological segregation of geniculo-cortical afferents in the visual cortex of dark-reared cats Brain Res **362** 281-6 [500]

Swindale NV, Mitchell DE (1994) Comparison of receptive field properties of neurons in area 17 of normal and bilaterally amblyopic cats Exp Brain Res **99** 399-10 [522, 525]

Swindale NV, Vital-Durand F, Blakemore C (1981) Recovery from monocular deprivation in the monkey. III Reversal of anatomical effects in the visual cortex Proc R Soc B **213** 435-50 [517]

Swindale NV, Matsubara JA, Cynader MS (1987) Surface organization of orientation and direction selectivity in cat area 18 J Neurosci **7** 1414–27 [211]

Swindale NV, Shoham D, Grinvald A, Bonhoeffer T, Hübener M (2000) Visual cortex maps are optimized for uniform coverage Nature Neurosci **3** 822-6 [185]

Switkes E, Bradley A, Schor C (1990) Readily visible changes in color contrast are insufficient to stimulate accommodation Vis Res **30** 1367–76 [364]

Sylvestre PA, Galiana HL, Cullen KE (2002) Conjugate and vergence oscillations during saccades and gaze shifts: implications for integrated control of binocular movement J Neurophysiol **87** 257-72 [422]

Symons LA, Pearson PM, Timney B (1996) The aftereffect to relative motion does not show interocular transfer Perception 25 651-60 [348]

Szabadics J, Lorincz A, Tamas G (2001) _ and _ frequency synchronization by dendritic GABAergic synapses an gap junctions in a network of cortical interneurones Journal of Neurosci 21 5824-31 [196]

Szebenyi G, Dent EW, Callaway JL, Seys C, Lueth H, Kalil K (2001) Fibroblast growth factor–2 promotes axon branching of cortical neurons by influencing morphology and behaviour of the primary growth cone J Neurosci **21** 3932-41 [457]

Székely G (1971) The mesencephalic and diencephalic optic centres in the frog Vis Res Supplement 3 299–79 [540]

Szentágothai J (1973) Neuronal and synaptic architecture of the lateral geniculate nucleus In Handbook of sensory physiology (ed R Jung) Vol VII/3B pp 141–76 Springer–Verlag, New York [186]

Taira M, Tsutsui KI, Jiang M, Yara K, Sakata H (2000) Parietal neurons represent surface orientation from the gradient of binocular disparity J Neurophysiol **83** 3140-46 [226, 243]

Takada R, Hara N, Yamamoto K, Toda H, Ando T, Hasebe H, Abe H, Bando T (2000) Effects of localized lesions in the lateral suprasylvian cortex on convergence eye movement in cat Neurosci Res **36** 275-83 [431]

Takagi M, Toda H, Bando T (1993) Extrastriate cortical neurons correlated with ocular convergence in the cat Neurosci Res **17** 141-58 [431]

Takagi M, Oyamada H, Abe H, Zee DS, Haswbe H, Miki A, Usui T, Hasegawa S, Bando T (2001) Adaptive changes in dynamic properties of human disparity-induced vergence Invest Ophthal Visual Sci **42** 1479-86 [425]

Takai K, Shin RM, Inoue T, Kato K, Mikoshiba K (1998) Regulation of nerve growth mediated by inositol 1,4,5-triphosphate receptors in growth cones Science **282** 1705-8 [439]

Takayama Y, Sugishita M, Kido T, Ogawa M, Fukuyama H, Akigushi I (1994) Impaired stereoacuity due to a lesion in the left pulvinar J Neurol Neurosurg Psychiat **57** 652-4 [245]

Takeda T, Hashimoto K, Hiruma N, Fukui Y (1999) Characteristics of accommodation toward apparent depth Vis Res 39 2087-97 [360, 386]

Takemura A, Inoue Y, Kawano K, Quaia C, Miles FA (2001) Single-unit activity in cortical area MST associated with disparity-vergence eye movements: evidence for population coding J Neurophysiol **85** 2245-66 [431]

Tallon-Baudry C, Bertrand O, Delpuech C, Pernier J (1996) Stimulus specificity of phase-locked and non-phase-locked 40 Hz visual responses in human J Neurosci **16** 4270-9 [133]

Tallon-Baudry C, Bertrand O, Fischer C (2001) Oscillatory synchrony between human extrastriate areas during visual short term memory maintenance J Neurosci **21** 1-5 [135]

Tam WJ, Stelmach LB (1998) Display duration and stereoscopic depth discrimination Can J Exp Psychol **52** 56-61 [487]

Tamler E, Jampolsky A, Marg E (1958) An electromyographic study of asymmetric convergence Am J Ophthal **46** 174–82 [419]

Tan RKT, O'Leary DJ (1986) Stability of the accommodative dark focus after periods of maintained accommodation Invest Ophthal Vis Sci **27** 1414-7 [359]

Tanaka H, Uka T, Yoshiyama K, Kato M, Fujita I (2001) Processing of shape defined by disparity in monkey inferior temporal cortex J Neurophysiol **85** 735-44 [244]

Tanaka K, Fukada Y, Saito HA (1989) Underlying mechanisms of the response specificity of expansion/contraction and rotation cells in the dorsal part of the medial superior temporal area of the monkey J Neurophysiol **62** 642-56 [225]

Tanaka K, Saito HA, Fukada Y, Moriya M (1991) Coding visual images of objects in the inferotemporal cortex of the macaque monkey J Neurophysiol **66** 170–89 [222]

Tanaka S (1991) Theory of ocular dominance column formation Biol Cyber **64** 293-72 [212]

Tanner WP (1956) Theory of recognition J Acoust Soc Am **28** 882-888 [81, 327]

Tansley BW, Boynton RM (1978) Chromatic border perception: the role of red- and green-sensitive cones Vis Res **18** 683-97 [290]

Tao W, Lai E (1992) Telencephalon-restricted expression of BF-1 a new member of the HNF-3/*fork head* gene family in the developing rat brain Neuron **8** 957-66 [450]

Taylor J (1738) Le mechanisme ou le nouveau Traité de l'anatomie du globe de l'oeil avec l'usage de ses différentes paries et de celles qui lui sont contigues David, Paris [56, 282]

Taylor JSH, Guillery RW (1995) The effect of a very early monocular enucleation upon the development of the uncrossed retinofugal pathway in ferrets J Comp Neurol **357** 331-40 [440]

Taylor MJ, Roberts DC, Zee DS (2000) Effect of sustained cyclovergence on eye alignment: rapid torsional phoria adaptation Invest Ophthal Vis Sci **41** 1076-83 [414, 416, 418]

Taylor MM (1963) Tracking the neutralization of seen rotary movement Percept Mot Skills **16** 513–19 [347]

Taylor MM, Creelman CD (1967) PEST: efficient estimates on probability functions J Acoust Soc Am **41** 782–7 [76]

Teller DY (1979) A forced–choice preferential looking procedure: a psychophysical technique for use with human infants Infant Behav Devel **2** 135–53 [477]

Teller DY, Gallanter E (1967) Brightness luminances and Fechner's paradox Percept Psychophys **2** 297–300 [326]

ten Doesschate G (1962) Oxford and the revival of optics in the thirteenth century Vis Res **1** 313–42 [15, 18]

ten Doesschate G (1964) Perspective. Fundamentals, contoversials, history B de Graaf, Nieuwkoop [32]

ten Doesschate G, Alpern M (1967) Influence of asymmetrical photo excitation of the two retinas on pupil size J Neurophysiol **30** 577–85 [474]

Teping C, Silny J (1987) evidence of pericentral stereopsis in random dot VECP Doc Ophthal **66** 291-66 [262]

Tessier-Lavigne M, Goodman CS (1996) The molecular biology of axon guidance TINS **274** 1123-33 [438]

Thibos LN, Still DL, Bradley A (1996) Characterization of spatial aliasing and contrast sensitivity in peripheral vision Vis Res **36** 279-58 [173]

Thiele A, Bremmer F, Ilg UJ, Hoffmann KP (1997) Visual responses of neurons from areas V1 and MT in a monkey with late onset strabismus: a case study Vis Res **37** 853-63 [530]

Thoenen H (1995) Neurotrophins and neuronal plasticity Science **270** 593-8 [456, 463, 511]

Thomas C, Moss C, Vater M (2000) Advances in the study of echolocation in bats and dolphins University of Chicago Press, Chicago [548]

Thomas FH, Dimmick FL, Luria SM (1961) A study of binocular color mixture Vis Res **1** 108–20 [283, 284]

Thomas GJ (1956) Effect of contours on binocular CFF obtained with synchronous and alternate flashes Am J Psychol **69** 369–77 [329]

Thomas J (1977) A reciprocal model for monocular pattern alternation Percept Psychophys **22** 310-12 [295]

Thomas J (1978a) Binocular rivalry: the effects of orientation and pattern color arrangement Percept Psychophys **23** 360-2 [293]

Thomas J (1978b) Normal and amblyopic contrast sensitivity functions in central and peripheral retinas Invest Ophthal **17** 746-53 [521]

Thomas JP, Gille J (1979) Bandwidths of orientation channels in human vision J Opt Soc Am **69** 652–60 [84]

Thompson AM, Nawrot M (1999) Abnormal depth perception form motion parallax in amblyopic observers Vis Res **39** 1407-13 [529]

Thompson DW (1952) On growth and form Cambridge University Press, Cambridge [107, 157, 159]

Thompson I, Graham M (1995) The representation of the visual field in the visual cortex of anaesthetized albino ferrets J Physiol **489** 10P [496]

Thompson KG, Zhou Y, Leventhal AG (1994a) Direction-sensitive X and Y cells within the A laminae of the cat's LGNd Vis Neurosci **11** 927-38 [179]

Thompson KG, Leventhal AG, Zhou Y, Liu D (1994b) Stimulus dependence of orientation and direction sensitivity of cat LGNd relay cells without cortical inputs: a comparison with area 17 cells Vis Neurosci **11** 939-51 [179]

Thomson LC (1947) Binocular summation within the nervous pathway of the pupillary light reflex J Physiol **106** 59-65 [317]

Thorn F, Boynton RM (1974) Human binocular summation at absolute threshold Vis Res **14** 445–58 [318, 331]

Thorn F, Gwiazda J, Cruz AAV, Bauer JA, Held R (1994) The development of eye alignment convergence and sensory binocularity in young infants Invest Ophthal Vis Sci **35** 544–53 [476]

Thorpe PA, Blakemore C (1975) Evidence for a loss of afferent axons in the visual cortex of monocularly deprived cats Neurosci Lett **1** 271–6 [513]

Thorpe S, Fize D, Marlot C (1996) Speed of processing in the human visual system Nature **381** 520-2 [130]

Thorpe SJ, Celebrini S, Trotter Y, Imbert M (1991) Dynamics of stereo processing in area V1 of the awake primate J Neurosci **4** (Supp) 83 [251]

Thurman J, David J (1978) The magic lantern Atheneum New York [59]

Tieman DG, Tumosa N, Tieman SB (1983a) Behavioral and physiological effects of monocular deprivation: a comparison of rearing with diffusion and occlusion Brain Res **280** 41–50 [510, 514]

Tieman DG, McCall MA, Hirsch HVB (1983b) Physiological effects of unequal alternating monocular exposure J Neurophysiol **49** 804-18 [514]

Tieman SB (1984) Effects of monocular deprivation on geniculocortical synapses in the cat J Comp Neurol **222** 166-76 [507]

Tieman SB, Tumosa N (1997) Alternating monocular exposure increases the spacing of ocularity domains in area 17 of cats Vis Neurosci **14** 929-39 [505]

Tigges J, Tigges M (1979) Ocular dominance columns in the striate cortex of chimpanzee (*Pan troglodytes*) Brain Res **166** 386-90 [216]

Tigges M, Hendrickson AE, Tigges J (1984) Anatomical consequences of long-term monocular eyelid closure on lateral geniculate nucleus and striate cortex in squirrel monkey J Comp Neurol **227** 1-13 [216, 503]

Tigges M, Tigges J, Fernandes A, Eggers HM, Gammon JA (1990) Postnatal axial eye elongation in normal and visually deprived rhesus monkeys Invest Ophthal Visual Sci **31** 1035-46 [435]

Tigges M, Boothe RG, Tigges J, Wilson JR (1992) Competition between an aphakic and an occluded eye for territory in striate cortex of developing rhesus monkeys: cytochrome oxidase histochemistry in layer 4C J Comp Neurol **316** 173-86 [181, 213]

Timney B (1981) Development of binocular depth perception in kittens Invest Ophthal Vis Sci **21** 493-6 [478]

Timney B (1983) The effects of early and late monocular deprivation on binocular depth perception in cats Devel Brain Res 7 235-43 [516]

Timney B (1988) The development of depth perception In Advances in neural and behavioral development Vol 3 (ed PG Shinkman) pp 153-208 Ablex, Norwood NJ [485]

Timney B (1990) Effects of brief monocular deprivation on binocular depth perception in the cat: a sensitive period for the loss of stereopsis Vis Neurosci **5** 273-80 [515]

Timney B, Keil K (1994) Local and global stereopsis in the horse Invest Ophthal Vis Sci **35** (Abs) 2110 [550]

Timney B, Keil K (1996) Horses are sensitive to pictorial depth cues Perception **25** 1121-8 [550]

Timney B, Mitchell DE, Giffin F (1978) The development of vision in cats after extended periods of dark–rearing Exp Brain Res **31** 547–60 [501]

Timney B, Mitchell DE, Cynader M (1980) Behavioral evidence for prolonged sensitivity to effects of monocular deprivation in dark–reared cats J Neurophysiol **43** 1041–54 [516]

Timney B, Elberger AJ, Vandewanter ML (1985) Binocular depth perception in the cat following early corpus callosum section Exp Brain Res **60** 19-29 [233]

Timney B, Wilcox LM, St John R (1989) On the evidence for a 'pure' binocular process in human vision Spatial Vis **4** 1–15 [307]

Timney B, Symons LA, Wilcox LM, O'Shea RP (1996) The effect of dark and equiluminant occlusion on the interocular transfer of visual aftereffects Vis Res **36** 707-15 [347]

Toates FM (1970) A model of accommodation Vis Res **10** 1069-76 [358]

Toates FM (1974) Vergence eye movements Doc Ophthal **37** 153-214 [368]

Toates FM (1975) Control theory in biology and experimental psychology Hutchinson, London [92]

Toda H, Takagi M, Yoshizawa T, Bando T (1993) Disjunctive eye movement evoked by microstimulation in an extrastriate cortical area of the cat Neurosci Res **12** 300-6 [431]

Todd EM (1991) The neuroanatomy of Leonardo da Vinci American Association of Neurological Surgeons, Park Ridge, Illinois [20]

Toet A, Levi DM (1992) The two–dimensional shape of spatial interaction zones in the parafovea Vis Res **32** 1349–57 [339]

Tolhurst DJ, Movshon JA, Thompson ID (1981) The dependence of response amplitude and variance of cat visual cortical neurons on stimulus contrast Exp Brain Res **41** 414-19 [198]

Tolhurst DJ, Movshon JA, Dean AF (1983) The statistical reliability of signals in single neurons in cat and monkey visual cortex Vis Res **23** 775-85 [251]

Tomlinson RD, Schwarz DWF, Fredrickson JM (1978) Cerebellar and brainstorm responses to eye muscle stretch in the cat In Vestibular mechanisms in health and disease (ed JD Hood) pp 45–51 Academic Press, New York [494]

Tong F, Engel SA (2001) Interocular rivalry revealed in the human cortical blind-spot representation Nature **411** 195-9 [313]

Tong F, Nakayama K, Vaughan JT, Kanwisher N (1998) Binocular rivalry and visual awareness in human extrastriate cortex Neuron **21** 753-9 [312]

Tong L, Guido W, Tumosa N, Spear PD, Heidenreich S (1992) Binocular interactions in the cat's dorsal lateral geniculate nucleus. II Effects on dominant–eye spatial–frequency and contrast processing Vis Neurosci **8** 557–66 [180, 311]

Toni N, Buchs PA, Nikonenko I, Bron CR, Muller D (1999) LTP promotes formation of multiple spine synapses between a single axon terminal and a dendrite Nature **402** 421-5 [193]

Tootell RBH, Silverman MS, De Valois RL, Jacobs GH (1983) Functional organization of the second cortical visual area in primates Science **220** 737–9 [219]

Tootell RBH, Hamilton SL, Switkes E (1988a) Functional anatomy of macaque striate cortex. IV. Contrast and magno-parvo streams J Neurophysiol **8** 1594–1609 [186]

Tootell RBH, Silverman MS, Hamilton SL, Switkes E, De Valois RL (1988b) Functional anatomy of macaque striate cortex. V. Spatial–frequency J Neurosci **8** 1610–27 [204]

Tootell RBH, Hamilton SL, Silverman MS, Switkes E (1988c) Functional anatomy of macaque striate cortex. I. Ocular dominance binocular interactions and baseline conditions J Neurosci **8** 1500–30 [131, 214]

Tootell RBH, Switkes E, Silverman MS, Hamilton SL (1988d) Functional anatomy of macaque striate cortex. II. Retinotopic organization J Neurosci **8** 1531-68 [184]

Tootell RBH, Silverman MS, Hamilton SL, DeValois RL, Switkes E (1988e) Functional anatomy of macaque striate cortex. III. Color J Neurosci **8** 1569-93 [205]

Tootell RBH, Reppas JB, Kwong KK, Malach R, Born RT, Brady TJ, Rosen BR, Belliveau JW (1995) Functional analysis of human MT and related visual cortical areas using magnetic resonance imaging J Neurosci **15** 3215-30 [223]

Tootell RBH, Dale AM, Sereno MI, Malach R (1996) New images from the human cortex TINS **19** 481-9 [219]

Tootell RBH, Mendola JD, Hadjikhani NK, Ledden PJ, Liu AK, Reppas JB, Sereno MI, Dale AM (1997) Functional analysis of V3A and related areas in human visual cortex J Neurosci **17** 7060-78 [220]

Tootle JS (1993) Early postnatal development of visual function in ganglion cells of the cat retina J Neurophysiol **69** 1645-60 [437]

Torgerson WS (1958) Theory and methods of scaling Wiley, New York [73, 77]

Tourtual CT (1827) Die Sinne des Menschen Regensberg, Münster [45]

Tovée MJ, Rolls ET, Treves A, Bellis RP (1993) Information encoding and the responses of single neurons in the primate temporal visual cortex J Neurophysiol **70** 640–54 [130]

Towle VL, Harter MR, Previc FH (1980) Binocular interaction of orientation and spatial frequency channels: evoked potentials and observer sensitivity Percept Psychophys **27** 351-60 [337]

Townsend JT (1968) Binocular information summation and the serial processing model Percept Psychophys **4** 125–8 [324]

Toyama K, Fugii K, Kasai S, Maeda K (1986a) The responsiveness of Clare-Bishop neurons to size cues for motion stereopsis Neurosci Res **4** 110-28 [431]

Toyama K, Kitaoji H, Umetani K (1991) Binocular neuronal responsiveness in Clare-Bishop cortex Exp Brain Res **86** 471-82 [497]

Trachtenberg JT, Trepel C, Stryker MP (2000) Rapid extragranular plasticity in the absence of thalamocortical plasticity in the developing primary visual cortex Science **287** 2029-32 [506]

Trachtenberg JT, Stryker MP (2001) Rapid anatomical plasticity of horizontal connections in the developing visual cortex J Neurosci **21** 3476-82 [506]

Traub RD, Whittington MA, Stanford IM, Jeffreys JGR (1996) A mechanism for generation of long-range synchronous fast oscillations in the cortex Nature **383** 621-4 [136]

Treisman A (1988) Features and objects Quart J Exp Psychol **40A** 201–38 [134]

Treisman A, Schmidt H (1982) Illusory conjunctions in the perception of objects Cog Psychol **14** 107-141 [134]

Tremain KE, Ikeda H (1982) Relationship between amblyopia LGN cell 'shrinkage' and cortical ocular dominance in cats Exp Brain Res **45** 273–52 [503]

Treue S, Anderson RA (1996) Neural responses to velocity gradients in macaque cortical area MT Vis Neurosci **13** 797-804 [224]

Treue S, Maunsell JHR (1996) Attentional modulation of visual motion processing in cortical areas MT and MST Nature **382** 539-41 [207, 208]

Trick GL, Compton JR (1982) Analysis of the effect of temporal frequency on the dichoptic visual-evoked response Am J Optom Physiol Opt **59** 155-61 [259]

Trick GL, Guth SL (1980) The effect of wavelength on binocular summation Vis Res **20** 975–80 [320]

Tripathy SP, Levi DM (1994) Long–range dichoptic interactions in the human visual cortex in the region of the blind spot Vis Res **34** 1127–38 [339]

Troelstra A, Zuber BL, Miller D, Stark L (1964) Accommodation tracking: a trial–and–error function Vis Res **4** 585–94 [362, 363]

Troilo D, Wallman J (1991) The regulation of eye growth and refractive state: an experimental study of emmetropization Vis Res **31** 1237-50 [434]

Trotter Y, Celebrini S (1999) Gaze direction controls response gain in primary visual-cortex neurons Nature **398** 239-42 [254]

Trotter Y, Frégnac Y, Buisseret P (1987) The period of susceptibility of visual cortical binocularity to unilateral proprioceptive deafferentation of extraocular muscles J Neurophysiol **58** 795–815 [494]

Trotter Y, Celebrini S, Stricanne B, Thorpe S, Imbert M (1992) Modulation of neural stereoscopic processing in primate area V1 by the viewing distance Science **257** 1279–81 [134, 254]

Trotter Y, Celebrini S, Beaux JC, Grandjean B, Imbert M (1993) Long–term dysfunctions of neural stereoscopic mechanisms after extraocular muscle proprioceptive deafferentation J Neurophysiol **69** 1513–29 [494]

Trotter Y, Celebrini S, Stricanne B, Thorpe S, Imbert M (1996) Neural processing of stereopsis as a function of viewing distance in primate visual cortical area V1 J Neurophysiol **76** 2872-85 [254]

Troyer TW, Krukowski AE, Priebe NJ, Miller KD (1998) Contrast-invariant orientation tuning in cat visual cortex: thalamocortical input tuning and correlation-based intracortical connectivity J Neurosci **18** 5908-27 [200]

Truchard AM, Ohzawa I, Freeman RD (2000) Contrast gain control in the visual cortex: monocular versus binocular mechanisms J Neurosci **20** 3017-32 [248]

Trueswell JC, Hayhoe MM (1993) Surface segmentation mechanisms and motion perception Vis Res **33** 313–28 [149]

Trusk TC, Kaboord WS, Wong–Riley MTT (1990) Effects of monocular enucleation tetrodotoxin and lid suture on cytochrome–oxidase reactivity in supragranular puffs of adult macaque striate cortex Vis Neurosci **4** 185–204 [509]

Ts'o DY, Gilbert CD (1988) The organization of chromatic and spatial interactions in the primate striate cortex J Neurosci **8** 1712–27 [205]

Ts'o DY, Gilbert CD, Wiesel TN (1986) Relationships between horizontal interactions and functional architecture in cat striate cortex as revealed by cross–correlation analysis J Neurosci **6** 1160–70 [196]

Ts'o DY, Gilbert CD, Frostig RD, Grinvald A, Wiesel TN (1989) Functional architecture of visual area 18 of Macaque monkey Soc Neurosci Abstr **15** 161 [219]

Ts'o DY, Frostig RD, Lieke E, Grinvald A (1990) Functional organization of primate visual cortex revealed by high resolution optical imaging Science **279** 417–19 [181, 205, 211, 213]

Ts'o DY, Roe AW, Gilbert CD (2001) A hierarchy of the functional organization for color, form, and disparity in primate visual area V2 Vis Res **41** 1333-49 [220]

Tsai RYL, McKay RDG (2000) Cell contact regulates fate choice by cortical stem cells J Neurosci **20** 3725-35 [450]

Tsal Y (1989) Do illusory conjunctions support the feature integration theory? A critical review of theory and findings J Exp Psychol HPP **15** 394-400 [134]

Tsuetaki TK, Schor LM (1987) Clinical method for measuring adaptation of tonic accommodation and vergence accommodation Am J Optom Physiol Opt **64** 437-49 [390]

Tsumoto T, Suda K (1978) Evidence for excitatory connections from deprived eye to the visual cortex in monocularly deprived kittens Brain Res **153** 150-6 [509]

Tsumoto T, Suda K (1980) Three groups of cortico-geniculate neurons and their distribution in binocular and monocular segments of cat striate cortex J Comp Neurol **193** 223-6 [180]

Tsutsui KI, Jiang M, Yara K, Sakata H, Taira M (2001) Integration of perspective and disparity cues in surface-orientation selective neurons of area CIP J Neurophysiol **86** 2856-67 [243]

Tucker J, Charman WN (1979) Reaction and response times for accommodation Am J Optom Physiol Opt **56** 490–503 [361]

Tucker J, Charman WN, Ward PA (1986) Modulation dependence of the accommodation response to sinusoidal gratings Vis Res **29** 1693-1707 [361]

Tulunay Keesey U (1960) Effects of involuntary eye movements on visual acuity J Opt Soc Am **50** 769–74 [366]

Tumosa N, Tieman SB, Hirsch HVB (1982) Visual field deficits in cats reared with unequal alternating monocular exposure Exp Brain Res **47** 119-29 [514]

Tumosa N, Nunberg S, Hirsch HVB, Tieman SB (1983) Binocular exposure causes suppression of the less experienced eye in cats previously reared with unequal alternating monocular exposure Invest Ophthal Vis Sci **27** 496-5006 [514]

Tumosa N, McCall MA, Guido W, Spear PD (1989a) Responses of lateral geniculate neurons that survive long–term visual cortex damage in kittens and adult cats J Neurosci **9** 280–98 [180, 311]

Tumosa N, Tieman SB, Tieman DG (1989b) Binocular competition affects the pattern and intensity of ocular activation columns in the visual cortex of cats Vis Neurosci **2** 391-407 [508, 509]

Turing AM (1952) The chemical basis of morphogenesis Philos Tr R Soc B **237** 37-72 [212]

Turlejski K, Kossut M (1985) Decrease in the number of synapses formed by subcortical inputs to the striate cortex of binocularly deprived cats Brain Res **331** 115-25 [499]

Turner KM, Burgoyne RD, Morgan A (1999) Protein phosphorylation and the regulation of synaptic membrane traffic TINS **22** 459-64 [189]

Turner RS (1993) Consensus and controversy. Helmholtz and the visual perception of space In Hermann von Helmholtz (ed D Cahan) pp 154-204 University of California Press, London [42, 65]

Turner RS (1994) In the eye's mind Princeton University Press, Princeton NJ [42]

Tweed D (1997) Visual-motor optimization in binocular control Vis Res **37** 1939-51 [371]

Twitchell TE (1970) Reflex mechanisms and the development of prehension In Mechanisms of motor skill development (ed K Connolly) pp 25–37 Academic Press, London [477]

Tychsen L, Boothe RG (1996) Latent fixation nystagmus and nasotemporal asymmetries of motion visually evoked potentials in naturally strabismic primate J Ped Ophthal Strab **33** 148-52 [347]

Tychsen L, Burkhalter A (1997) Nasotemporal asymmetries in V1: ocular dominance columns of infant, adult, and strabismic monkeys J Comp Neurol **388** 32-46 [214, 455]

Tychsen L, Lisberger SG (1986) Maldevelopment of visual motion processing in humans who had strabismus with onset in infancy J Neurosci **6** 2795–508 [525]

Tychsen L, Hurtig RR, Scott WE (1985) Pursuit is impaired but the vestibulo-ocular reflex is normal in infantile strabismus Arch Ophthal **103** 536-9 [526]

Tyler CW (1971) Stereoscopic depth movement: two eyes less sensitive than one Science **174** 958–61 [349]

Tyler CW (1973) Stereoscopic vision: cortical limitations and a disparity scaling effect Science **181** 276–8 [275]

Tyler CW (1983) Sensory processing of binocular disparity In Vergence eye movements: Basic and clinical aspects (ed MC Schor, KJ Ciuffreda) pp 199–296 Butterworth, Boston [8]

Tyler CW (1990) A stereoscopic view of visual processing streams Vis Res **30** 1877-95 [245]

Tyler CW (1991a) Cyclopean vision In Vision and visual dysfunction Vol 9 Binocular Vision (ed D Regan) pp 38–74 MacMillan, London [8]

Tyler CW (1997) On Ptolemy's geometry of binocular vision Perception **29** 1579-81 [49]

Tyler CW (1999) Human symmetry detection exhibits reverse eccentricity scaling Vis Neurosci **16** 919-22 [157]

Tyler CW, Apkarian PA (1985) Effects of contrast orientation and binocularity in the pattern evoked potential Vis Res **25** 755-66 [260]

Tyler CW, Kaitz MF (1977) Binocular interactions in the human visual evoked potential after short-term occlusion and anisometropia Invest Ophthal Vis Sci **16** 1070-3 [510]

Tyler CW, Sutter EE (1979) Depth from spatial frequency difference: an old kind of stereopsis? Vis Res **19** 859–65 [306, 307]

Tyler WJ, Pozzo-Miller LD (2001) BDNF enhances quantal neurotransmitter release and increases the number of docked vesicles at the active zones of hippocampal excitatory synapses J Neurosci **21** 4249-58 [457]

Tytla ME, Lewis TL, Maurer D, Brent HP (1993) Stereopsis after congenital cataract Invest Ophthal Vis Sci **34** 1767–73 [502]

Udin SB (1989) Development of the nucleus isthmi in Xenopus. II Branching patterns of contralaterally projecting isthmotectal axons during maturation of binocular maps Vis Neurosci **2** 153-63 [540]

Uhlarik JJ, Canon LK (1971) Influence of concurrent and terminal exposure conditions on the nature of perceptual adaptation J Exp Psychol **91** 233-9 [355]

Uka T, Tanaka H, Yoshiyama K, Kato M, Fujita I (2000) Disparity selectivity of neurons in monkey inferior temporal cortex J Neurophysiol **84** 120-32 [244]

Ukwade MT (2000) Effects of nonius line and fusion lock parameters on fixation disparity Optom Vis Sci **77** 309-20 [377]

Ullian EM, Sapperstein SK, Christopherson KS, Barres BA (2001) Control of synapse number by glia Science **291** 657-61 [457]

Ullman S (1979) The interpretation of visual motion MIT Press, Cambridge MA [102]

Ungerleider LG, Desimone R (1986) Cortical connections of visual area MT in the macaque J Comp Neurol **278** 190-222 [224]

Ungerleider LG, Mishkin M (1982) Two cortical visual systems In Analysis of visual behavior (ed DJ Ingle, MA Goodale, RJW Mansfield) pp 549–86 MIT Press, Cambridge MA [223]

Ungerleider LG, Gaffan D, Pelak VS (1989) Projections from inferior temporal cortex to prefrontal cortex via the uncinate fascicle in rhesus monkey Exp Brain Res **76** 473-84 [228]

Usher M, Donnelly N (1998) Visual synchrony affects binding and segmentation in perception Nature **394** 179-82 [133]

Usrey WM, Alonso JM, Reid RC (2000) Synaptic interactions between thalamic inputs to simple cells in cat visual cortex J Neurosci **20** 5461-7 [135]

Uttal WR (1983) Visual form detection in 3-dimensional space Erlbaum, Hillsdale NJ124]

Uttal WR, Fitzgerald J, Eskin TE (1975a) Parameters of tachistoscopic stereopsis Vis Res **15** 705–12 [341]

Uttal WR, Baruch T, Allen L (1995) Dichoptic and physical information combination: a comparison Perception **27** 351-62 [324]

Uylings HBM (1990) The prenatal and postnatal development of the rat cerebral cortex. In The cerebral cortex of the rat (ed B Kolb, RC Tees) pp 35-76 MIT Press, Cambridge MA [448]

Vaadia E, Benson DA, Hienz RD, Goldstein MH (1986) Unit study of monkey frontal cortex: active localization of auditory and of visual stimuli J Neurophysiol **56** 934-52 [207]

Vaadia E, Haalman I, Abeles M, Bergman H, Prut Y, Slovin H, Aertsen A (1995) Dynamics of neuronal interactions in monkey cortex in relation to behavioural events Nature **373** 515-8 [131, 133]

Vaina LM, Solomon J, Chowdhury S, Sinha P, Belliveau JW (2001) Functional neuroanatomy of biological motion perception in humans Proc Nat Acadf Sci **98** 11656-61 [226]

Valtschanoff JG, Weinberg RJ (2001) Laminar organization of the NMDA receptor complex within the postsynaptic density J Neurosci **15** 1211-17 [191]

Valverde F (1991) The organization of the striate cortex In Neuroanatomy of the visual pathways and their development (ed B Dreher, SR Robinson) pp 235–77 CRC Press, Boston [186]

Van Bergeijk WA (1962) Variations on a theme of Békésy: A model of binaural interaction J Acoust Soc Am **34** 1431-7 [149]

Van de Castle RL (1960) Perceptual defense in a binocular-rivalry situation J Person **28** 448–62 [309]

Van de Geer JP, De Natris PJA (1962) Dutch distorted rooms from the seventeenth century Acta Psychol **20** 101-104 [40]

Van de Grind WA, van Hof P, van der Smagt MJ, Verstraten FA (2001) Slow and fast visual motion channels have independent binocular rivalry stages Proc Roy Soc B **268** 437-43 [348]

Van der Smagt MJ, Verstraten FA, van der Grind WA (1999) A new transparent motion aftereffect Nature Neurosc **2** 595-6 [348]

Van der Steen J, Bruno P (1995) Unequal amplitude saccades produced by aniseikonic patterns: effects of viewing distance Vis Res **35** 3459-71 [410, 424]

Van der Tweel LH, Estévez O (1974) Subjective and objective evaluation of flicker Ophthalmologica **169** 70–81 [329, 490]

Van der Willigen RF, Frost B, Wagner H (1998) Stereoscopic depth perception in the owl Neuroreport **9** 1233-7 [547]

Van der Willigen RF, Frost B, Wagner H (2002) Depth generalization from stereo to motion parallax in the owl. J Comp Physiol A **187** 997-1007 [547]

Van der Zwan R, Wenderoth P, Alais D (1993) Reduction of a pattern-induced motion aftereffect by binocular rivalry suggests the involvement of extrastriate mechanisms Vis Neurosci **10** 703-9 [303]

Van Essen DC, Zeki SM (1978) The topographic organization of rhesus monkey prestriate cortex J Physiol **227** 193–229 [221, 232]

Van Essen DC, Maunsell JHR, Bixby JL (1981) The middle temporal visual area in the macaque: myeloarchitecture connections functional properties and topographic organization J Comp Neurol **199** 293-329 [223]

Van Essen DC, Newsome WT, Bixby JL (1982) The pattern of interhemispheric connections and its relationship to extrastriate visual areas in the macaque monkey J Neurosci **2** 295-83 [232]

Van Essen DC, Newsome WT, Maunsell HR (1984) The visual field representation in striate cortex of the macaque monkey: asymmetries, anisotropies, and individual variability Vis Res 27 429-48 [184]

Van Essen DC, Anderson CH, Felleman DJ (1992) Information processing in the primate visual system: an integrated systems perspective TINS 255 419–23 [218]

Van Essen DC, Lewis JW, Drury HA, Hadjikhani N, TootellRBH, Bakircioglu M, Miller MI (2001) Mapping visual cortex in monkeys and humans using surface-based atlases Vis Res 41 1359-78 [183]

Van Hearingen R, McClurg P, Cameron KD (1986) Comparison of Wesson and modified Sheedy fixation disparity tests Do fixation disparity measures relate to normal binocular status? Ophthal Physiol Opt 6 397-400 [377]

Van Hof MW, Steele Russell I (1977) Binocular vision in the rabbit Physiol Behav 19 121–8 [549]

Van Hof–van Duin J (1976a) Development of visuomotor behavior in normal and dark–reared cats Brain Res 104 233–41 [501]

Van Hof–van Duin J (1976b) Early and permanent effects of monocular deprivation on pattern discrimination and visuomotor behavior in cats Brain Res 111 291–76 [517]

Van Kruysbergen NAWH, de Weert CMM (1993) Apparent motion perception: the contribution of the binocular and monocular systems An improved test based on motion aftereffects Perception 22 771–84 [349]

Van Kruysbergen NAWH, de Weert CMM (1994) Aftereffects of apparent motion: the existence of an AND-type binocular system in human vision Perception 23 1069-83 [349]

Van Leeuwen AF, Collewijn H, Erkelens CJ (1998) Dynamics of horizontal vergence movements: interaction with horizontal and vertical saccades and relation with monocular preferences Vis Res 38 3943-54 [423]

Van Meeteren A (1978) On the effective quantum efficiency of the human eye Vis Res 18 257-67 [166]

Van Opstal AJ, Hepp K, Hess BJM, Straumann D, Henn V (1991) Two- rather than three-dimensional representation of saccades in monkey superior colliculus Science 252 1313-15 [371]

Van Rijn LJ, Collewijn H (1994) Eye torsion associated with disparity-induced vertical vergence in humans Vis Res 34 2307-16 [410]

Van Rijn LJ, van den Berg AV (1993) Binocular eye orientation during fixation: Listing's law extended to include eye vergence Vis Res 33 691-708 [371, 415]

Van Rijn LJ, van der Steen J, Collewijn H (1992) Visually induced cycloversion and cyclovergence Vis Res 32 1875–83 [414]

Van Rijn LJ, van der Steen J, Collewijn H (1994a) Eye torsion elicited by oscillating gratings: effects of orientation wavelength and stationary contours Vis Res 34 533–40 [414, 415]

Van Rijn LJ, van der Steen J, Collewijn H (1994b) Instability of ocular torsion during fixation: cyclovergence is more stable than cycloversion Vis Res 34 1077–87 [416]

Van Rijn LJ, Simonsz HJ, ten Tussher MPM (1997) Dissociated vertical deviation: relation to disparity-induced vertical vergence Strabismus 5 13-20 [410]

Van Rijn LJ, ten Tusscher MPM, de Jong I, Hendrikse F (1998) Asymmetrical vertical phorias indicating dissociated vertical deviation in subjects with normal binocular vision Vis Res 38 2973-78 [410]

Van Sluyters RC (1978) Reversal of the physiological effects of brief periods of monocular deprivation in the kitten J Physiol 284 1–17 [516]

Van Sluyters RC, Levitt FB (1980) Experimental strabismus in the kitten J Neurophysiol 43 686–99 [494]

Van Sluyters RC, Stewart DL (1974a) Binocular neurons of the rabbit's visual cortex: receptive field characteristics Exp Brain Res 19 166–95 [549]

Van Sluyters RC, Stewart DL (1974b) Binocular neurons of the rabbit's visual cortex: effects of monocular sensory deprivation Exp Brain Res 19 196–204 [549]

Van Vreeswijk C, Sompolinsky H (1996) Chaos in neuronal networks with balanced excitatory and inhibitory activity Science 274 1727-6 [100, 130]

Vancura KL, Jay DG (1998) G proteins and axon growth Sem Neurosci 9 209-19 [439]

Vandenbussche E, Vogels R, Orban GA (1986) Human orientation discrimination: changes with eccentricity in normal and amblyopic vision Invest Ophthal Vis Sci 27 237-45 [522]

Vanzetta I, Grinvald A (1999) Increased cortical oxidative metabolism due to sensory stimulation: implications for functional brain imaging Science 286 1555-8 [182]

Varela FJ, Singer W (1987) Neuronal dynamics in the visual corticothalamic pathway revealed through binocular rivalry Exp Brain Res 66 10–20 [180, 311]

Varley FH (1890) A camera for taking consecutive pictures of objects in motion Br J Photog 37 777-80 [71]

Varner D, Cook JE, Schneck ME, McDonald MA, Teller DY (1985) Tritan discrimination by 1– and 2–month–old human infants Vis Res 25 821–31 [470]

Varoqui H, Erickson JD (1997) Vesicular neurotransmitter transporters Molec Neurobiol 15 165-92 [190]

Vautin RG, Berkley MA (1977) Responses of single cells in cat visual cortex to prolonged stimulus movement: neural correlates of visual aftereffects J Neurophysiol 40 1051–65 [350]

Vereecken EP, Brabant P (1984) Prognosis for vision in amblyopia after the loss of the good eye Arch Ophthal 102 220–4 [527]

Verhage M, Maia AS, Plomp JJ, *et al.* (2000) Synaptic assembly of the brain in the absence of neurotransmitter secretion Science 287 864-9 [439]

Verhoeff FH (1934) Cycloduction Tr Am Ophthal Soc 32 208–28 [411, 413, 416]

Verhoeff FH (1935) A new theory of binocular vision Arch Ophthal 13 151–75 [305]

Verhoeff FH (1959) Fixation disparity Am J Ophthal 48 339–41 [378]

Verkhrastsky A, Orkand RK, Kettenmann H (1998) Glial calcium: homeostasis and signalling function Physiol Rev 78 1-42 [188]

Versino M, Hurko O, Zee DS (1996) Disorders of binocular control of eye movements in patients with cerebellar dysfunction Brain 119 1933-50 [431]

Verstraten FAJ, Fredericksen RE, van de Grind WA (1994a) Movement aftereffect of bi-vectorial transparent motion Vis Res 34 349-58 [348]

Verstraten FA, van der Smagt MJ, van de Grind WA (1998) Aftereffect of high speed motion Perception 27 1055-66 [348]

Via SE (1977) Visually mediated snapping in the bulldog ant: a perceptual ambiguity between size and distance J Comp Physiol 121 33-51 [533]

Vickery RM, Morley JW (1999) Binocular phase interactions in area 21a of the cat J Physiol 514 541-549 [239]

Victor JD (1979) Nonlinear systems analysis: comparison of white noise and sum of sinusoids in a biological system Proc Nat Acad Sci 76 996–8 [95]

Victor JD, Purpura KP (1996) Nature and precision of temporal coding in visual cortex: a metric-space analysis J Neurophysiol 76 1310-29 [130]

Videen TO, Daw NW, Rader RK (1984) The effect of norepinephrine on visual cortical neurons in kittens and adult cats J Neurosci 4 1607–17 [513]

Vidyasagar TR (1976) Orientation specific colour adaptation at a binocular site Nature **291** 39-40 [352]

Vidyasagar TR, Mueller A (1994) Function of GABA inhibition in specifying spatial frequency and orientation selectivities in cat striate cortex Exp Brain Res **98** 31-8 [199]

Vidyasagar TR, Stuart GW (1993) Perceptual learning in seeing form from motion Proc R Soc B **254** 271–4 [88, 354]

Vidyasagar TR, Urbas JV (1982) Orientation sensitivity of cat LGN neurones with and without inputs from visual cortical areas 17 and 18 Exp Brain Res **46** 157-69 [179]

Vidyasagar TR Pei X, Volgushev M (1996) Multiple mechanisms underlying the orientation selectivity of visual cortical neurones TINS **19** 272-7 [199]

Vieth GAU (1818) Über die Richtung der Augen Ann Physik **28** 233–53 [57]

Viirre E, Tweed D, Milner K, Vilis TA (1986) Reexamination of the gain of the vestibuloocular reflex J Neurophysiol **56** 439–50 [428]

Viirre E, Cadera W, Vilis T (1987) The pattern of changes produced in the saccadic system and vestibuloocular reflex by visually patching one eye J Neurophysiol **57** 92–103 [426]

Viirre E, Cadera W, Vilis T (1988) Monocular adaptation of the saccadic system and vestibulo–ocular reflex Invest Ophthal Vis Sci **29** 1339–47 [426]

Vilis T, Yates S, Hore J (1985) Visual patching of one eye produces changes in saccadic properties in the unseeing eye Devel Brain Res **17** 290–2 [426]

Vilis T, Hepp K, Schwarz U, Henn V (1989) On the generation of vertical and torsional rapid eye movements in the monkey Exp Brain Res **77** 1–11 [432]

Vincent A, Regan D (1995) Parallel independent encoding of orientation spatial frequency and contrast Perception **27** 491-9 [126]

Vinje WE, Gallant JL (2000) Sparse coding and decorrelation in primary visual cortex during natural vision Science **287** 1273-75 [122]

Virsu V, Rovamo J (1979) Visual resolution, contrast sensitivity, and the cortical magnification factor Exp Brain Res **37** 475-94 [184]

Virsu V, Taskinen H (1975) Central inhibitory interactions in human vision Exp Brain Res **23** 65–74 [344]

Vital–Durand F, Putkonen PTS, Jeannerod M (1974) Motion detection and optokinetic responses in dark-reared kittens Vis Res **14** 141-2 [501]

Vital–Durand F, Garey LJ, Blakemore C (1978) Monocular and binocular deprivation in the monkey: morphological effects and reversibility Brain Res **158** 45–64 [504]

Vital Durand F, Atkinson J, Braddick O (ed) (1996) Infant vision Oxford University Press, Oxford [485]

Vogels R, Orban GA (1990) How well do response changes of striate neurons signal differences in orientation: a study in the discriminating monkey J Neurosci **10** 3543-58 [121]

Volchan E, Gilbert CD (1995) Interocular transfer of receptive field expansion in cat visual cortex Vis Res **35** 1-6 [198]

Volgushev M, Pei X, Vidyasagar TR, Creutzfeldt OD (1993) Excitation and inhibition in orientation selectivity of cat visual cortex neurons revealed by whole-cell recordings *in vivo* Vis Neurosci **10** 1151-5 [199]

Volgushev M, Vidyasagar TR, Pei X (1995) Dynamics of the orientation tuning of postsynaptic potentials in the cat visual cortex Vis Neurosci **12** 621-8 [199]

Volkmann AW (1836) Neue Beiträge zur Physiologie des Gesichtssinnes Breitkopft, Leipzig [58, 287]

Volkmann FC, Riggs LA, Moore RK, White KD (1978) Central and peripheral determinants of saccadic suppression In Eye movements and the higher psychological functions (ed JW Senders, DA Fisher, RA Monty) pp 35–54 Erlbaum, Hillsdale NJ [368]

Von der Heydt R, Peterhans E (1989) Mechanisms of contour perception in monkey visual cortex. I Lines of pattern discontinuity J Neurosci **9** 1731-48 [124, 197, 346]

Von der Heydt R, Adorjani CS, Hänny P, Baumgartner G (1978) Disparity sensitivity and receptive field incongruity of units in the cat striate cortex Exp Brain Res **31** 523–45 [238, 251, 254]

Von der Heydt R, Hänny P, Dürsteler MR, Poggio GF (1982) Neuronal responses to stereoscopic tilt in the visual cortex of the behaving monkey Invest Ophthal Vis Sci **22** (Abs) 12 [256]

Von der Heydt R, Peterhans E, Baumgartner G (1984) Illusory contours and cortical neuron responses Science **227** 1290-2 [220]

Von der Heydt R, Zhou H, Friedman HS (2000) Representation of stereoscopic edges in monkey visual cortex Vis Res **40** 1995-67 [242]

Von Frisch K (1993) The dance and orientation of bees Harvard University Press, Cambridge MA [535]

Von Grünau M (1979a) The role of maturation and visual experience in the development of eye alignment in cats Exp Brain Res **37** 41-7 [474]

Von Grünau M (1979b) Binocular summation and the binocularity of cat visual cortex Vis Res **19** 813-16 [490]

Von Grünau M (1982) Comparison of the effects of induced strabismus on binocularity in areas 17 and the LS area in the cat Brain Res **276** 325-9 [507]

Von Grünau M, Rauschecker JP (1983) Natural strabismus in non-Siamese cats: lack of binocularity in the striate cortex Exp Brain Res **52** 307-10 [505]

Von Hofsten C (1977) Binocular convergence as a determinant of reaching behavior in infants Perception **6** 139-44 [478]

Von Hofsten C, Fazel-Zandy S (1984) Development of visually guided hand orientation in reaching J Exp Child Psychol **38** 208-19 [478]

Von Hofsten C, Ronqvist L (1988) Preparation for grasping an object: a developmental study J Exp Psychol: HPP **14** 610-21 [478]

Von Hofsten C, Rosander K (1996) The development of gaze control and predictive tracking in young infants Vis Res **36** 81-96 [475]

Von Holst E (1954) Relations between the central nervous system and the peripheral organs Br J Anim Behav **2** 89-94 [146]

Von Holst E, Mittelstaedt H (1950) Das Reafferenzprinzip Naturwissenschaften **37** 464-76 [145]

Von Noorden GK (1961) Reaction time in normal and amblyopic eyes Arch Ophthal **66** 695-701 [525]

Von Noorden GK (1990) Binocular vision and ocular motility Mosby, St Louis MO [8, 374, 393]

Von Noorden GK, Crawford MLJ (1977) Form deprivation without light deprivation produces the visual deprivation syndrome in Macaca mulatta Brain Res **129** 37-44 [504]

Von Noorden GK, Crawford MLJ (1978) Lid closure and refractive error in macaque monkey Nature **272** 53-5 [435]

Von Noorden GK, Crawford MLJ (1981) Failure to preserve cortical binocularity in strabismic monkeys raised in a unidirectional visual environment Invest Ophthal Vis Sci **20** 665–70 [514]

Von Noorden GK, Crawford MLJ (1992) The lateral geniculate nucleus in human strabismic amblyopia Invest Ophthal Vis Sci **33** 2729–32 [503]

Von Noorden GK, Leffler MB (1966) Visual acuity in strabismic amblyopia under monocular and binocular conditions Arch Ophthal **76** 172–7 [521]

Von Noorden GK, Middleditch PR (1975) Histology of the monkey lateral geniculate nucleus after unilateral lid closure and experimental strabismus: further observations Invest Ophthal 14 674–83 [503]

Von Noorden GK, Crawford MLJ, Middleditch PR (1977) Effect of lid suture on retinal ganglion cells in *Macaca mulatta* Brain Res 122 437-44 [503]

Von Seelen W (1970) Zur Informationsverarbeitung im visuellen System der Wirbeltiere Kybernetik 7 89–106 [209]

Von Senden M (1960) Space and sight Methuen, London [501]

Von Stein A, Chiang C, König P (2000) Top-down processing mediated by interareal synchronization Proc Nat Acad Sci 97 14748-53 [134]

Vu ET, Krasne FB (1992) Evidence for a computational distinction between proximal and distal neuronal inhibition Science 255 1710-12 , [123, 192]

Wade NJ (1973) Binocular rivalry and binocular fusion of after–images Vis Res 13 999–1000 [294]

Wade NJ (1975a) Binocular rivalry between single lines viewed as real images and afterimages Percept Psychophys 17 571–7 [311]

Wade NJ (1975b) Monocular and binocular rivalry between contours Perception 4 85–95 [290, 295]

Wade NJ (1976a) Monocular and dichoptic interaction between afterimages Percept Psychophys 19 149-54 [296]

Wade NJ (1976b) On interocular transfer of the movement aftereffect in individuals with and without normal binocular vision Perception 5 113–18 [493]

Wade NJ (1977) Binocular rivalry between after-images illuminated intermittently Vis Res 17 310-12 [311]

Wade NJ (1978) Why do patterned afterimages fluctuate in visibility? Psychol Bull 85 338–52 [290]

Wade NJ (1980) The influence of colour and contour rivalry on the magnitude of the tilt illusion Vis Res 20 229–33 [303]

Wade NJ (1981) A note on the history of binocular microscopes Perception 10 591–2 [62]

Wade NJ (1983) Brewster and Wheatstone on vision Academic Press, New York [65]

Wade NJ (1987) On the late invention of the stereoscope Perception 16 785–818 [59]

Wade NJ (1998a) Light and sight since antiquity Perception 27 633-58 [26]

Wade NJ (1998b) Early studies of eye dominance Laterality 3 97-108 [20]

Wade NJ (2000) Jean Théopile Desaguliers (1683-1744) and eighteenth century vision research Brit J Psychol 91 275-85 [56]

Wade NJ, de Weert CMM (1986) Aftereffects in binocular rivalry Perception 15 419–34 [287]

Wade NJ, Heller D (1997) Scopes of perception: the experimental manipulation of space and time Psychol Res 60 227-37 [41]

Wade NJ, Hughes P (1999) Fooling the eyes: *trompe l'oeil* and reverse perspective Perception 28 115-19 [40]

Wade NJ, Ono H (1985) The stereoscopic views of Wheatstone and Brewster Psychol Res 47 125–33 [65]

Wade NJ, Wenderoth P (1978) The influence of colour and contour rivalry on the magnitude of the tilt after–effect Vis Res 18 827–35 [303]

Wade NJ, De Weert CMM, Swanston MT (1984) Binocular rivalry with moving patterns Percept Psychophys 35 111–22 [291]

Wade NJ, Swanston MT, de Weert CMM (1993) On interocular transfer of motion aftereffects Perception 22 1365-80 [347]

Wagner H, Frost B (1993) Disparity–sensitive cells in the owl have a characteristic disparity Nature 364 796–7 [547]

Wagner H, Frost B (1994) Binocular responses of neurons in the barn owl's visual Wulst J Comp Physiol A 174 661-70 [547]

Wagner H, Schaeffel F (1991) Barn owls (*Tyto alba*) use accommodation as a distance cue J Comp Physiol 169 515–21 [547]

Wagner M (1985) The metric of visual space Percept Psychophys 38 483–95 [137]

Wales R, Fox R (1970) Increment detection thresholds during binocular rivalry suppression Percept Psychophys 8 90–4 [305]

Walk RD, Gibson EJ (1961) A comparative and analytical study of visual depth perception Psychol Monogr 75 (Whole No 15) 1–44 [478]

Walker GA, Ohzawa I, Freeman RD (1998) Binocular cross-orientation suppression in the cat's striate cortex J Neurophysiol 79 227-39 [313]

Walker GA, Ohzawa I, Freeman RD (1999) Asymmetric suppression outside the classical receptive field of the visual cortex J Neurosci 19 10536-53 [197, 200]

Walker GA, Ohzawa I, Freeman RD (2000) Suppression outside the classical cortical receptive field Vis Neurosci 17 369-79 [195]

Walker P (1975) Stochastic properties of binocular rivalry alternations Percept Psychophys 18 467–73 [311]

Walker P (1978a) Orientation–selective inhibition and binocular rivalry Perception 7 207–14 [344]

Walker P (1978b) Binocular rivalry: central or peripheral selective processes? Psychol Bull 85 376–89 [311]

Walker P, Powell DJ (1979) The sensitivity of binocular rivalry to changes in the nondominant stimulus Vis Res 19 277–9 [301]

Wallace GK (1959) Visual scanning in the desert locust *Schistocera gregaria Forskål* J Exp Biol 36 512–25 [534]

Wallach H, Frey KJ (1972) Adaptation in distance perception based on oculomotor cues Percept Psychophys 11 77-83 [355]

Wallach H, O'Connell DN (1953) The kinetic depth effect J Exp Psychol 45 205-17 [480]

Walline JJ, Mutti DO, Zadnik K, Jones LA (1998) Development of phoria in children Optom Vis Sci 75 605-10 [476]

Wallis G, Rolls ET (1997) Invariant face and object recognition in the visual system Prog Neurobiol 51 167-94 [132, 222]

Wallman J, Adams JI (1987) Developmental aspects of experimental myopia in chicks: susceptibility recovery and relation to emmetropization Vis Res 27 1139–63 [434]

Wallman J, Turkel J (1978) Extreme myopia produced by modest change in early visual experience Science 201 1279-51 [435]

Walls GL (1953) Interocular transfer of after–images Am J Optom Arch Am Acad Optom 30 57–64 [343, 347]

Walls GL (1963) The vertebrate eye and its adaptive radiations Hafner, New York [175, 531, 539, 543, 548]

Walsh C (1988) Age-related fiber order in the ferret's optic nerve and optic tract J Neurosci 6 1635-42 [437, 439]

Walsh C, Guillery RW (1985) Age-related order in the optic tract of the ferret J Neurosci 5 3061-9 [437]

Walsh G, Charman WN (1988) Visual sensitivity to temporal change in focus and its relevance to the accommodation response Vis Res 28 1207-21 [362]

Walsh V, Butler SR (1996) The effects of visual cortex lesions on the perception of rotated shapes Behav Brain Res 76 127-42 [229]

Wandell BA (1999) Computational neuroimaging of human visual cortex Ann Rev Neurosci 22 145-73 [182]

Wang C, Dreher B (1996) Binocular interactions and disparity coding in area 21a of cat extrastriate visual cortex Exp Brain Res 108 257-72 [239]

Wang C, Dreher B, Burke W (1994) Non-dominant suppression in the dorsal lateral geniculate nucleus of the cat: laminar differences and class specificity Exp Brain Res 97 451-65 [180]

Wang GY, Liets LC, Chalupa LM (2001) Unique functional properties of On and Off pathways in the developing mammalian retina J Neurophysiol 21 4310-17 [437]

Wang H, Levi DM, Klein SA (1998) Spatial uncertainty and sampling efficiency in amblyopic position acuity Vis Res **38** 1239-51 [523]

Wang LC, Dani J, Godement P, Marcus RC, Mason CA (1995) Crossed and uncrossed retinal axons respond differently to cells of the optic chiasm midline in vitro Neuron **15** 1349-64 [440]

Wang Q, Cavanagh P, Green M (1994) Familiarity and pop-out in visual search Percept Psychophys **56** 495-500 [17]

Wang XF, Daw NW (1996) Metabotropic glutamate receptors potentiate responses to NMDA and AMPA from layer V cells in rat visual cortex J Neurophysiol **76** 808-15 [461]

Wang YC, Jiang S, Frost BJ (1993) Visual processing in pigeon nucleus rotundus: luminance color motion and looming subdivisions Vis Neurosci **10** 21–30 [545]

Ward PA (1987) The effect of stimulus contrast on the accommodation response Ophthal Physiol Opt **7** 9-15 [359]

Ward PA, Charman WN (1985) Effect of pupil size on steady state accommodation Vis Res **25** 1317-29 [361]

Ward R, Repérant J, Hergueta S, Miceli D, Lemire M (1995) Ipsilateral visual projections in non-eutherian species: random variation in the central nervous system? Brain Res Rev **20** 155-70 [531, 532]

Ware C, Mitchell DE (1974a) On interocular transfer of various visual aftereffects in normal and stereoblind observers Vis Res **14** 731–4 [491, 492]

Warland DK, Reinagel P, Meister M (1997) Decoding information from a population of retinal ganglion cells J Neurophysiol **78** 2336-50 [127]

Washburn MF (1933) Retinal rivalry as a neglected factor in stereoscopic vision Proc Nat Acad Sci **19** 773–7 [305]

Washburn MF, Faison C, Scott R (1934) A comparison between the Miles A–B–C method and retinal rivalry as tests of ocular dominance Am J Psychol **46** 633–6 [295]

Wässle H, Boycott BB (1991) Functional architecture of the mammalian retina Physiol Rev **71** 447–80 [170]

Wässle H, Peichl L, Boycott BB (1981) Dendritic territories of cat retinal ganglion cells Nature **292** 344–5 [170]

Wässle H, Grünert U, Röhrenbeck J, Boycott BB (1990) Retinal ganglion cell density and cortical magnification factor in the primate Vis Res **30** 1897–911 [170, 184, 442]

Watamaniuk SNJ, Duchon A (1992) The human visual system averages speed information Vis Res **32** 931–41 [80]

Watamaniuk SNJ, Sekuler R, Williams DW (1989) Direction perception in complex dynamic displays: the integration of direction information Vis Res **29** 47–59 [80]

Watanabe M (1996) Reward expectancy in primate prefrontal neurons Nature **382** 629-32 [208]

Watson AB (1986) Temporal sensitivity In Handbook of perception and human performance Vol 1: Sensory processes and perception (ed K Boff, L Kaufman, JP Thomas) chapter 6 Wiley, New York [471]

Watson AB, Nachmias J (1977) Patterns of temporal interaction in the detection of gratings Vis Res **17** 893–902 [331]

Watson AB, Pelli DG (1983) QUEST: a Bayesian adaptive psychometric method Percept Psychophys **33** 113–20 [76]

Watson AB, Robson JG (1981) Discrimination at threshold: labelled line detectors in human vision Vis Res **21** 1115-22 [139]

Watt RJ, Hess RF (1987) Spatial information and uncertainty in anisometropic amblyopia Vis Res **27** 661-74 [522]

Watt RJ, Morgan MJ (1983) The recognition and representation of edge blur: evidence for spatial primitives in human vision Vis Res **23** 1465-77 [362]

Watt RJ, Morgan MJ, Ward RM (1983) Stimulus features that determine the visual location of a bright bar Invest Ophthal Vis Sci **27** 66–71 [79]

Wattam–Bell J (1991) Development of motion–specific cortical responses in infancy Vis Res **31** 287–97 [472]

Wattam–Bell J (1992) The development of maximum displacement limits for discrimination of motion direction in infancy Vis Res **32** 621–30 [472]

Wattam-Bell J (1994) Coherence thresholds for discrimination of motion direction in infants Vis Res **34** 877-83 [472]

Wattam–Bell J (1996a) Visual motion processing in one-month-old infants: preferential looking experiments Vis Res **36** 1671-7 [472]

Wattam-Bell J (1996b) Development of visual motion processing In Infant vision (ed F Vital-Durand, J Atkinson, OJ Braddick) pp 80-94 Oxford University Press, Oxford [473]

Webster MJ, Shatz CJ, Kliot M, Silver J (1988) Abnormal pigmentation and unusual morphogenesis of the optic stalk may be correlated with retinal axon misguidance in embryonic Siamese cats J Comp Neurol **299** 592-611 [497]

Webster MJ, Bachevalier J, Ungerleider LG (1994) Connections of inferior temporal area TEO and TE with parietal and frontal cortex in macaque monkeys Cerebral Cortex **4** 470-83 [222, 228, 229]

Wedgwood CV (1995) The thirty years war. Random House, London. [23]

Wehner R (1981) Spatial vision in arthropods In Handbook of sensory physiology Vol VIIC (ed H Autrum) pp 287-616 Springer, Berlin [537]

Wehrhahn C, Westheimer G, Abulencia A (1990) Binocular summation in temporal-order detection J Opt Soc Am A **7** 731-2 [332]

Weiskrantz L (1987) Blindsight: a case study and implications Oxford University Press, London [218, 257]

Weiss S, Hochman D, MacVicar BA (1993) Repeated NMDA receptor activation induces distinct intracellular calcium changes in subpopulations of striatal neurons in vitro Brain Res **627** 63–71 [461]

Weisstein N (1972) Metacontrast In Handbook of sensory physiology (ed D Jameson, LM Hurvich) Vol VII/4 pp 233–72 Springer, New York [340]

Wekiky M, Katz LC (1997) Disruption of orientation tuning in visual cortex by artificially correlated neuronal activity Nature **386** 680-5 [454]

Weliky M, Bosking WH, Fitzpatrick D (1996) A systematic map of direction preference in primary visual cortex Nature **379** 725-8 [211]

Wells WC (1792) An essay upon single vision with two eyes; together with experiments and observations on several other subjects in optics T Cadell, London [49, 52]

Welpe E, von Seelen W, Fahle M (1980) A dichoptic edge effect resulting from binocular contour dominance Perception **9** 683–93 [286]

Wensveen J M, Harwerth RS, Smith EL (2001) Clinical suppression in monkeys reared with abnormal binocular visual experience Vis Res **41** 1593-608 [373]

Werkhoven P, Koenderink JJ (1990) Extraction of motion parallax structure in the visual system Biol Cyber **63** 185-91 [109]

Werner C, Himstedt W (1984) Eye accommodation during prey capture behaviour in salamanders (*Salamandra salamandra* L) Behav Brain Res **12** 69-73 [542]

Werner H (1935) Studies on contour: I Qualitative analysis Am J Psychol **47** 40–64 [340]

Werner H (1940) Studies on contour: strobostereoscopic phenomena Am J Psychol **53** 418–22 [340, 341]

Werner O (1910) Zur Physik Leonardo da Vinci Magdeburg [16]

Westall CA (1986) Binocular vision: its influence on the development of visual and postural reflex eye movements Ophthal Physiol Opt **6** 139-43 [475]

Westall CA, Woodhouse JM, Brown VA (1989) OKN asymmetries and binocular function in amblyopia Ophthal Physiol Opt **9** 299-76 [526]

Westall CA, Eizenman M, Kraft SP, Panton CM, Chatterjee S, Sigesmund D (1998) Cortical binocularity and monocular optokinetic asymmetry in early-onset esotropia Invest Ophthal Vis Sci **39** 1352-9 [526]

Westendorf DH (1989) Binocular rivalry and dichoptic masking: suppressed stimuli do not mask stimuli in a dominating eye J Exp Psychol HPP **15** 485–92 [339]

Westendorf DH, Fox R (1974) Binocular detection of positive and negative flashes Percept Psychophys **15** 61–5 [330]

Westendorf DH, Fox R (1975) Binocular detection of vertical and horizontal line segments Vis Res **15** 471–76 [320, 330]

Westendorf DH, Fox R (1977) Binocular detection of disparate light flashes Vis Res **17** 697–702 [320]

Westendorf DH, Blake R, Fox R (1972) Binocular summation of equal energy flashes of unequal duration Percept Psychophys **12** 445–8 [331]

Westendorf DH, Langston A, Chambers D, Allegretti C (1978) Binocular detection by normal and stereoblind observers Percept Psychophys **27** 209–14 [490]

Westendorf DH, Blake R, Sloane M, Chambers D (1982) Binocular summation occurs during interocular suppression J Exp Psychol HPP **8** 81–90 [300]

Westheimer G (1955) The relationship between accommodation and accommodative convergence Am J Optom Arch Am Acad Optom **32** 206-12 [333, 388]

Westheimer G (1965) Spatial interaction in the human retina during scotopic vision J Physiol **181** 881–94 [333]

Westheimer G (1967) Spatial interaction in human cone vision J Physiol **190** 139-54 [334]

Westheimer G (1972) Optical properties of vertebrate eyes. In Handbook of sensory physiology Vol. VII2 (ed MGF Fuortes) pp 449-82. Springer, New York [173]

Westheimer G (1984a) Line–separation discrimination curve in the human fovea: smooth or segmented? J Opt Soc Am A **1** 683–4 [81]

Westheimer G (2001) Is peripheral visual acuity susceptible to perceptual learning in the adult? Vis Res **41** 47-52 [87]

Westheimer G, Blair SM (1975) The ocular tilt reaction—a brainstem oculomotor routine Invest Ophthal Vis Sci **14** 833- [432]

Westheimer G, Hauske G (1975) Temporal and spatial interference with vernier acuity Vis Res **15** 1137–41 [197, 339]

Westheimer G, McKee SP (1975) Visual acuity in the presence of retinal–image motion J Opt Soc Am **65** 847–50 [366]

Westheimer G, McKee SP (1977) Integration regions for visual hyperacuity Vis Res **17** 89–93 [366]

Westheimer G, Mitchell AM (1956) Eye movement responses to convergence stimuli Arch Ophthal **55** 848–56 [398, 399, 421]

Westheimer G, Mitchell DE (1969) The sensory stimulus for disjunctive eye movements Vis Res **9** 749–55 [402, 429]

Westheimer G, Shimamura K, McKee SP (1976) Interference with line-orientation sensitivity J Opt Soc Am **66** 332–8 [339]

Weyland TG, Malpeli JG (1993) Responses of neurons in primary visual cortex are modulated by eye position J Neurophysiol **69** 2258-60 [185]

Wheatstone C (1838) Contributions to the physiology of vision – Part the first On some remarkable and hitherto unobserved phenomena of binocular vision Philos Tr R Soc **128** 371–94 [62, 272, 287]

Wheatstone C (1853) On the binocular microscope and on stereoscopic pictures of microscopic objects Tr Microscop Soc Lon **1** 99–102 [64]

White BL, Castle P, Held R (1964) Observations on the development of visually–directed reaching Child Devel **35** 349–63 [477]

White CT, Bonelli L (1970) Binocular summation in the evoked potential as a function of image quality Am J Optom Arch Am Acad Optom **47** 304-9 [328]

White J (1967) The birth and rebirth of pictorial space Farber and Farber, London [38]

White KD, Petry HM, Riggs LA, Miller J (1978) Binocular interactions during establishment of McCollough effects Vis Res **18** 1201–15 [303, 351, 352]

White LE, Bosking WH, Williams SM, Fitzpatrick D (1999) Maps of central visual space in ferret V1 and V2 lack matching inputs from the two eyes J Neurosci **19** 7089-99 [531]

White LE, Coppola DM, Fitzpatrick D (2001) The contribution of sensory experience to the maturation of orientation selectivity in ferret visual cortex Nature **411** 1049-52 [508]

Whitefoot H, Charman WN (1992) Dynamic retinoscopy and accommodation Ophthal Physiol Opt **12** 8-17 [359]

Whitten DN, Brown KT (1973) Photopic suppression of monkey's rod receptor potential, apparently by a cone-initiated lateral inhibition Vis Res **13** 1629-58 [334]

Whittle EJ, Bryans C, Timney B (1987) Visual acuity in isotropic cats following occlusion of the non-deviating eye Behav Brain Res **27** 101-9 [508]

Whittle P (1965) Binocular rivalry and the contrast at contours J Exp Psychol **17** 217–29 [287]

Whittle P, Challands PDC (1969) The effect of background luminance on the brightness of flashes Vis Res **9** 1095-1110 [333]

Whittle P, Bloor DC, Pocock S (1968) Some experiments on figural effects in binocular rivalry Percept Psychophys **4** 183–8 [299]

Wick B (1985) Forced vergence fixation disparity at distance and near in an asymptomatic young adult population Am J Optom Physiol Opt **62** 591–9 [377, 383]

Wick B, Bedell HE (1989) Magnitude and velocity of proximal vergence Invest Ophthal Vis Sci **30** 755–60 [386]

Wick B, Cook D (1987) Management of anomalous correspondence: efficacy of therapy Am J Optom Physiol Opt **64** 405–10 [374]

Wick B, Currie D (1991) Dynamic demonstration of proximal vergence and proximal accommodation Optom Vis Sci **68** 163-7 [386]

Wick B, Ryan JB (1982) Clinical aspects of cyclophoria: definition diagnosis therapy J Am Optom Assoc **53** 987-95 [411]

Wickelgren LW (1967) Convergence in the human newborn J Exp Child Psychol **5** 74–85 [475]

Wicklein M, Strausfeld NJ (2000) Organization and significance of neurons that detect change of visual depth in the hawk moth *manduca sexta* J Comp Neurol **427** 356-76 [535]

Wiemer J, Burwick T, von Seelen W (2000) Self-organizing maps for visual feature representation based on natural binocular stimuli Biol Cybern **82** 97-110 [454]

Wieniawa-Narkiewicz BM, Wimborne BM, Michalski A, Henry GH (1992) Area 21a in the cat and the detection of binocular orientation disparity Ophthal Physiol Opt **12** 299-72 [255]

Wiesel TN, Hubel DH (1963a) Single cell responses in striate cortex of kittens deprived of vision in one eye J Neurophysiol **29** 1003–17 [507]

Wiesel TN, Hubel DH (1963b) Effects of visual deprivation on morphology of cells in the cat's lateral geniculate body J Neurophysiol **29** 978–93 [503]

Wiesel TN, Hubel DH (1965a) Extent of recovery from the effects of visual deprivation in kittens J Neurophysiol **28** 1060–72 [508]

Wiesel TN, Hubel DH (1965b) Comparison of the effects of unilateral and bilateral eye closure on cortical unit responses in kittens J Neurophysiol **28** 1029–40 [508]

Wiesel TN, Hubel DH (1974) Ordered arrangement of orientation columns in monkeys lacking visual experience J Comp Neurol **158** 307–18 [455]

Wiesel TN, Raviola E (1979) Increase in axial length of the macaque monkey eye after corneal opacification Invest Ophthal Vis Sci **18** 1232–6 [435]

Wiesel TN, Hubel DH, Lam DMK (1974) Autoradiographic demonstration of ocular–dominance columns in the monkey striate cortex by means of transneuronal transport Brain Res **79** 273–9 [212]

Wiesenfelder H, Blake R (1990) The neural site of binocular rivalry relative to the analysis of motion in the human visual system J Neurosci **10** 3880–8 [303]

Wiesenfelder H, Blake R (1991) Apparent motion can survive binocular rivalry suppression Vis Res **31** 1589–99 [304]

Wiesenfelder H, Blake R (1992) Binocular rivalry suppression disrupts recovery from motion adaptation Vis Neurosci **9** 143-8 [304]

Wiggers W, Roth G, Eurich C, Straub A (1995) Binocular depth perception mechanisms in tongue-projecting salamanders J Comp Physiol **176** 365-77 [542]

Wilcox LM, Timney B, St John R (1990) Measurement of visual aftereffects and inferences about binocular mechanisms in human vision Perception **19** 43–55 [343, 345]

Wilcox LM, Timney B, Girash M (1994) On the contribution of a binocular 'AND' channel at contrast threshold Perception **23** 659-69 [345]

Wildsoet CF, Cameron KD (1985) The effect of illumination and foveal fusion lock on clinical fixation disparity measurements with the Sheedy disparometer Ophthal Physiol Opt **5** 171-8 [377]

Wildsoet CF, Howland HC, Falconer S, Dick K (1993) Chromatic aberration and accommodation: their role in emmetropization in the chick Vis Res **33** 1593-603 [434]

Williams D, Gassel MM (1962) Visual function in patients with homonymous hemianopia: Part I The visual fields Brain **85** 175-250 [231]

Williams DR (1988) Topography of the foveal cone mosaic in the living human eye Vis Res **28** 433–54 [164, 174]

Williams R (1974) The effect of strabismus on dichoptic summation of form information Vis Res **14** 307-9 [324]

Williams RA, Boothe RG (1981) Development of optical quality in the infant monkey (*Macaca nemestrina*) eye Invest Ophthal Vis Sci **21** 728-36 [470]

Williams RA, Fender DH (1977) The synchrony of binocular saccadic eye movements Vis Res **17** 303-6 [421]

Williams RW, Bastiani MJ, Lia B, Chalupa LM (1986) Growth cones, dying axons and developmental fluctuations in the fiber population of the cat's optic nerve J Comp Neurol **276** 32-69 [442]

Williams RW, Garraghty PE, Goldowitz D (1991) A new visual system mutation Achiasmatic dogs with congenital nystagmus Soc Neurosci Abstr **17** 187 [441, 497]

Williams RW, Hogan D, Garraghty PE (1994) Target recognition and visual maps in the thalamus of achiasmatic dogs Nature **367** 637–9 [441, 497]

Wilson FAW, Scalaidhe PO, Goldman–Rakic PS (1993) Dissociation of object and spatial processing domains in primate prefrontal cortex Science **290** 1955–7 [228]

Wilson HR (1986) Responses of spatial mechanisms can explain hyperacuity Vis Res **29** 453–69 [81]

Wilson HR (1988) Development of spatiotemporal mechanisms in infant vision Vis Res **28** 611–28 [470]

Wilson HR (1991a) Psychophysical models of spatial vision and hyperacuity In Spatial vision (ed D Regan) pp 64–86 CRC Press, Boca Raton [139]

Wilson HR (1991b) Model of peripheral and amblyopic hyperacuity Vis Res **31** 967-82 [522]

Wilson HR (1999) Simplified dynamics of human and mammalian neocortical neurons J Theor Biol **200** 375-88 [188]

Wilson HR, Gelb DJ (1984) Modified line element theory for spatial–frequency and width discrimination J Opt Soc Am A **1** 127–31 [84]

Wilson HR, Kim J (1994) A model for motion coherence and transparency Vis Neurosci **11** 1205-20 [129]

Wilson HR, McFarlane DK, Phillips GC (1983) Spatial frequency tuning of orientation selective units by oblique masking Vis Res **23** 873-82 [139]

Wilson HR, Blake R, Pokorny J (1988) Limits of binocular fusion in the short wave sensitive ("blue") cones Vis Res **28** 555-62 [277]

Wilson HR, Blake R, Halpern DL (1991) Coarse spatial scales constrain the range of binocular fusion on fine scales J Opt Soc Am A **8** 229–36 [276]

Wilson HR, Blake R, Lee SH (2001) Dynamics of travelling waves in visual perception Nature **412** 907-10 [298]

Wilson JR, Nevins CL (1991) Effects of monocular deprivation on the visual fields of squirrel monkeys Behav Brain Res **44** 129-31 [510]

Wilson JR, Sherman SM (1977) Differential effects of early monocular derivation on binocular and monocular segments of cat striate cortex J Neurophysiol **40** 891–903 [527]

Wilson JR, Webb SW, Sherman SM (1977) Conditions for dominance of one eye during competitive development of central connections in visually deprived cats Brain Res **136** 277–87 [514]

Wilson JR, Lavallee KA, Joosse MV, Hendrickson AE, Boothe RG, Harwerth RS (1989) Visual fields of monocularly deprived macaque monkeys Behav Brain Res **33** 13-22 [510]

Wilson MA, Bower JM (1991) A computer simulation of oscillatory behavior in primate visual cortex Neural Comput **3** 498–509 [136]

Wilson MA, McNaughton BL (1993) Dynamics of the hippocampal ensemble code for space Science **291** 1055-8 [223]

Wilson ME, Cragg BG (1967) Projections from the lateral geniculate nucleus in the cat and monkey J Anat **101** 677–92 [233]

Wilson ME, Bluestein EC, Parks MM (1993) Binocularity in accommodative esotropia J Ped Ophthal Strab **30** 233-6 [374]

Wimbauer S, Wenisch OG, Miller KD, van Hemmen JL (1997) Development of spatiotemporal receptive fields of simple cells: I Model formulation Biol Cyber **77** 453-61 [454]

Winkelman JE (1951) Peripheral fusion Arch Ophthal **45** 425–30 [396]

Winkelman JE (1953) Central and peripheral fusion Arch Ophthal **50** 179–83 [401]

Winn B (2000) Accommodative microfluctuations: a mechanism for steady-state control of accommodation In Accommodation and vergence mechanisms in the visual system (ed O Franzén, H Richter, L Stark) pp 129-40 Birkhäuser Verlag, Basel [361]

Winn B, Gilmartin B, Sculfor DL, Bamford JC (1994) Vergence adaptation and senescence Optom Vis Sci **71** 797-800 [383]

Winter HJJ (1954) The optical researches of Ibn Al-Haitham Centaurus **3** 190-210 [15]

Winterkorn JMS, Shapley R, Kaplan E (1981) The effect of monocular paralysis on the lateral geniculate nucleus of the cat Exp Brain Res **42** 117–21 [494]

Winterson BJ, Collewijn H (1976) Microsaccades during finely guided visuomotor tasks Vis Res **16** 1387–90 [366]

Wiser AK, Callaway EM (1997) Ocular dominance columns and local projections of layer 6 pyramidal neurons in macaque primary visual cortex Vis Neurosci 14 271-51 [213]

Witasek St (1899) Über die Natur der geometrisch-optischen Täuschungen Z Psychol Physiol Sinnesorg 19 81–174 [353]

Witkin A (1981) Recovering surface shape and orientation from texture Artif Intell 17 17-45 [102]

Witkop CJ, Nance WE, Rawls RF, White JG (1970) Autosomal recessive oculocutaneous albinism in man: evidence for genetic heterogeneity Am J Hum Genet 22 55-74 [495]

Wizenmann A, Thanos S, Boxberg Y von and Bonhoeffer F (1993) Differential reaction of crossing and non-crossing rat retinal axons on cell membrane preparations from the chiasma midline: an in vitro study Development 117 725-35 [440]

Wohlgemuth A (1911) On the after–effect of seen movement Br J Psychol Monogr Supp No 1 1–117 [347, 349, 350]

Wojciulik E, Kanwisher N, Driver J (1998) Modulation of activity in the fusiform face area by covert attention: an MRI study J Neurophysiol 79 1574-8 [313]

Wolf E, Zigler MJ (1955) Course of dark adaptation under various conditions of pre–exposure and testing J Opt Soc Am 45 696–702 [333]

Wolf E, Zigler MJ (1963) Effects of uniocular and binocular excitation of the peripheral retina with test fields of various shapes on binocular summation J Opt Soc Am 53 1199–205 [320]

Wolf E, Zigler MJ (1965) Excitation of the peripheral retina with coincident and disparate test fields J Opt Soc Am 55 1517–19 [320]

Wolf KS, Bedell HE, Pedersen SB (1990) Relations between accommodation and vergence in darkness Optom Vis Sci 67 89–93 [390]

Wolfe JM (1983a) Afterimages binocular rivalry and the temporal properties of dominance and suppression Perception 12 439–45 [294]

Wolfe JM (1983b) Influence of spatial frequency luminance and duration on binocular rivalry and abnormal fusion of briefly presented dichoptic stimuli Perception 12 447–56 [294]

Wolfe JM (1984a) Short test flashes produce large tilt aftereffects Vis Res 27 1959–64 [84]

Wolfe JM (1984b) Reversing ocular dominance and suppression in a single flash Vis Res 27 471-8 [301]

Wolfe JM (1986a) Briefly presented stimuli can disrupt constant suppression and binocular rivalry suppression Perception 1 5 413–17 [294]

Wolfe JM (1986b) Stereopsis and binocular rivalry Psychol Rev 93 299–82 [305, 306, 307]

Wolfe JM, Held R (1981) A purely binocular mechanism in human vision Vis Res 21 1755–9 [342, 343, 345]

Wolfe JM, Held R (1982) Binocular adaptation that cannot be measured monocularly Perception 11 287–95 [345]

Wolfe JM, Held R (1983) Shared characteristics of stereopsis and the purely binocular process Vis Res 23 217–27 [342, 346]

Wolfe JM, O'Connell M (1986) Fatigue and structural change: two consequences of visual pattern adaptation Invest Ophthal Vis Sci 27 538–43 [85]

Wolfe JM, O'Connell M (1987) Adaptation of the resting state of accommodation Invest Ophthal Vis Sci 28 992-6 [359]

Wolfe JM, Owens DA (1981) Is accommodation colorblind? Focusing chromatic contours Perception 10 53–62 [361, 364]

Wolosker H, Blackshaw S, Snyder SH (1999) Serine racemase: a glial enzyme synthesizing D-serine to regulate glutamate-N-Methyl-D-asparate neurotransmission Proc Nat Acad Sci 9 6 13409-14 [461]

Wolpert DM, Gharhramani Z, Jordon MI (1994) Perceptual distortion contributes to the curvature of human reaching movements Exp Brain Res 98 153-6 [144]

Wolpert DM, Gharhramani Z, Jordon MI (1995) Are arm trajectories planned in kinematic of dynamic coordinates? An adaptation study Exp Brain Res 103 460-70 [143, 144]

Wolpert DM, Goodbody SJ, Husain M (1998) Maintaining internal representations: the role of the human superior parietal lobe Nature Neurosci 1 529-33 [228]

Wong ROL (1999) Retinal waves and visual system development Ann Rev Neurosci 22 29-47 [459]

Wong ROL, Oakley DM (1996) Changing patterns of spontaneous bursting activity of On and Off retinal ganglion cells during development Neuron 16 1087-95 [459]

Wong ROL, Chernjavsky A, Smith SJ, Shatz CJ (1995) Early functional neural networks in the developing retina Nature 374 716-18 [436]

Wong–Riley MTT (1979a) Changes in the visual system of monocularly sutured or enucleated cats demonstrable with cytochrome oxidase histochemistry Brain Res 171 11–28 [203, 504]

Wong–Riley MTT (1979b) Columnar cortico-cortical interconnections within the visual system of the squirrel and macaque monkeys Brain Res 162 201–17 [217, 504]

Wong–Riley MTT (1989) Cytochrome oxidase: an endogenous metabolic marker for neuronal activity TINS 12 94–101 [203]

Wong–Riley MTT (1994) Primate visual cortex Dynamic metabolic organization and plasticity revealed by cytochrome oxidase In Cerebral cortex Volume 10 Primary visual cortex in primates (ed A Peters, KS Rockland) pp 141-200 Plenum, New York [204, 509]

Wong–Riley MTT, Hevner RF, Cutlan R, Earnest M, Egan R, Frost J, Nguyen T (1993) Cytochrome oxidase in the human visual cortex: distribution in the developing and the adult brain Vis Neurosci 10 41–58 [446, 455]

Woo GCS (1974a) The effect of exposure time on the foveal size of Panum's area Vis Res 14 473–80 [277]

Woo GCS (1974b) Temporal tolerance of the foveal size of Panum's area Vis Res 14 633–5 [278]

Woo GCS, Reading RW (1978) Panum's area explained in terms of known acuity mechanism Br J Physiol Opt 32 30–7 [274]

Wood ER, Dudchenko PA, Eichenbaum H (1999) The global record of memory in hippocampal neuronal activity Nature 397 613-16 [223]

Wood ICJ, Fox JA, Stevenson MG (1978) Contrast threshold of random dot stereograms in anisometropic amblyopia: a clinical investigation Br J Ophthal 62 34–8 [529]

Wood JM, Collins MJ, Carkeet A (1992) Regional variations in binocular summation across the visual field Ophthal Physiol Opt 12 46-51 [321]

Wright J, Bourke PD, Chapman CL (2000) Synchronous oscillation in the cerebral cortex and object coherence: simulation of basic electrophysiological findings Biol Cybern 83 341-53 [136]

Wright KW (1996) Clinical optokinetic nystagmus asymmetry in treated esotropes J Ped Ophthal Strab 33 153-5 [526]

Wright KW, Matsumoto E, Edelman PM (1992) Binocular fusion and stereopsis associated with early surgery for monocular congenital cataracts Arch Ophthal 110 1607-9 [519]

Wright KW, Edelman PM, McVey JH, Terry AP, Lin M (1994) High-grade stereo acuity after early surgery for congenital esotropia Arch Ophthal 112 913-9 [518]

Wright L (1983) Perspective in perspective Routledge and Kegan Paul, London [38, 39]

Wright LA, Wormald RPL (1992) Stereopsis and ageing Eye 6 473-6 [484]

Wright MJ (1986) Apparent velocity of motion aftereffects in central and peripheral vision Perception **15** 603–12 [347]

Wu HP, Williams CV, McLoon SC (1994) Involvement of nitric oxide in the elimination of a transient retinotectal projection in development Science **295** 1593-6 [461]

Wu Q, Maniatis T (1999) A striking organization of a large family of human neural cadherin-like adhesion genes Cell, **97** 779-90 [437]

Wundt W (1894) Lectures on human and animal psychology MacMillan, New York [58]

Wurtz RH, Goldberg ME, Robinson DL (1980) Behavioural modulation of visual responses in the monkey: stimulus selection for attention and movement Prog Psychobiol Physiol Psychol **9** 43–83 [207]

Xiang Z, Huguenard JR, Prince DA (1998) Cholinergic switching within neocortical inhibitory networks Science **281** 985-8 [192]

Xiao DK, Raiguel S, Koenderink J, Orban GA (1995) Spatial heterogeneity of inhibitory surrounds in the middle temporal visual area Proc Nat Acad Sci **92** 11303-6 [224]

Xue JT, Ramoa AS, Carney T, Freeman RD (1987) Binocular interaction in the dorsal lateral geniculate nucleus of the cat Exp Brain Res **68** 305-10 [180, 245]

Xue JT, Carney T, Ramoa AS, Freeman RD (1988) Binocular interaction in the perigeniculate nucleus of the cat Exp Brain Res **69** 497-508 [245]

Yabuta NH, Callaway EM (1998) Functional streams and local connections of layer 4C neurons in primary visual cortex of the macaque monkey J Neurosci **18** 9489-99 [186]

Yabuta NH, Sawatari A, Callaway EM (2001) Two functional channels from primary visual cortex to dorsal visual cortical areas Science **292** 297-10 [186]

Yakovlev PI, Lecours A (1967) The myelogenetic cycles of regional maturation of the brain In Regional development of the brain in early life (ed A Minkowski) Blackwell, Oxford [442]

Yan HQ, Mazow ML, Dafny N (1996) NFG prevents the changes induced by monocular deprivation during the critical period in rats Brain Res **706** 318-22 [512]

Yan XX, Zheng DS, Garey LJ (1992) Prenatal development of GABA–immunoreactive neurons in the human striate cortex Devel Brain Res **65** 191–204 [448, 449]

Yang DS, Fitzgibbon EJ, Miles FA (1999) Short-latency vergence eye movements induced by radial optic flow in humans: dependence on ambient vergence level J Neurophysiol **81** 945-9 [386]

Yang J, Stevenson SB (1999) Post retinal processing of background luminance Vis Res **39** 4045-51 [337]

Yang Y, Rose D, Blake R (1992) On the variety of percepts associated with dichoptic viewing of dissimilar monocular stimuli Perception **21** 47–62 [286]

Yap M, Brown B, Clarke J (1994) Reduction in stereoacuity with age and reduced illuminance Ophthal Physiol Opt **14** 298-301 [484]

Yarbus AL (1967) Eye movements and vision (Translated by LA Riggs) Plenum, New York [421]

Yates FA (1964) Giordano Bruno and the hermetic tradition Routledge and Kegan Paul, London [18, 23]

Yau KW, Baylor DA (1989) Cyclic GMP–activated conductance of retinal photoreceptor cells Ann Rev Neurosci **12** 289–327 [165]

Yee RD, Baloh RW, Honrubia V (1980) A study of congenital nystagmus: optokinetic nystagmus Br J Ophthal **64** 929–32 [498]

Yekta AA, Pickwell LD, Jenkins TCA (1989) Binocular vision age and symptoms Ophthal Physiol Opt **9** 115-20 [484]

Yellott JI, Wandell BA (1976) Color properties of the contrast flash effect: monoptic vs dichoptic comparisons Vis Res **16** 1275-80 [341]

Yeshurun Y, Schwartz EL (1990) Neural maps as data structures Fast segmentation of binocular images In Computational neuroscience (ed EL Schwartz) pp 256–66 MIT Press, Cambridge MA [142, 251]

Yeshurun Y, Schwartz EL (1999) Cortical hypercolumn size determines stereo fusion limits Biol Cyber **80** 117-29 [273]

Ygge J, Zee DS (1995) Control of vertical eye alignment in three-dimensional space Vis Res **35** 3169-81 [424]

Yildirim C, Mutlu FM, Chen Y, Altinsoy HI (1999) Assessment of central and peripheral fusion and near and distance stereoacuity in intermittent exotropic patients before and after strabismus surgery Am J Ophthal **128** 222-30 [374]

Yinon U (1975) Eye rotation in developing kittens: the effect on ocular dominance and receptive field organization of cortical cells Exp Brain Res **27** 215-18 [541]

Yinon U (1978) Chronic asymmetry in the extraocular muscles of adult cats: stability of cortical neurons Exp Brain Res **32** 275–85 [495]

Yinon U, Auerbach E, Blank M, Friesenhausen J (1975) The ocular dominance of cortical neurones in cats developed with divergent and convergent squint Vis Res **15** 1251–6 [505]

Yonas A, Owsley C (1987) Development of visual space perception In Handbook of infant perception (ed P Salapetek, L Cohen) Vol 3 pp 79–122 Academic Press, London [485]

Yonas A, Bechtold AG, Frankel D, Gordon FR, McRoberts G, Norcia A, Sternfels S (1977) Development of sensitivity to information for impending collision Percept Psychophys **21** 97–104 [478]

Yonas A, Oberg C, Norcia A (1978) Development of sensitivity to binocular information for the approach of an object Devel Psychol **14** 147–52 [478]

Yonas, A, Kuskowski M, Sternfels S (1979) The role of frames of reference in the development of responsiveness to shading information Child Devel **50** 495-500 [481]

Yonas A, Pettersen L, Granrud CE (1982) Infants' sensitivity to familiar size as information for distance Child Devel **53** 1285-90 [479]

Yonas A, Craton LG, Thompson WB (1987a). Relative motion—kinetic information for the order of depth at an edge Percept Psychopys **41** 53-9 [480]

Yonas A, Arterberry ME, Granrud CE (1987b) Four-month-old infants' sensitivity to binocular and kinetic information for three-dimensional-object shape Child Devel **58** 910-17 [480]

Yoon M (1971) Reorganization of retinotectal projection following surgical operations on the optic tectum in goldfish Exp Neurol **33** 395-411 [464]

Yoon M (1972) Reversibility of the reorganization of retinotectal projection in goldfish Exp Neurol **35** 565-77 [464]

Yoshioka T, Levitt JB, Lund JS (1994) Independence and merger of thalamocortical channels within macaque monkey primary visual cortex: anatomy of interlaminar projections Vis Neurosci **11** 467-89 [186, 205]

Yoshioka T, Blasdel GG, Levitt JB and Lund JS (1996) Relation between patterns of intrinsic lateral connectivity ocular dominance and cytochrome oxidase-reactive regions in macaque monkey striate cortex Cerebral Cortex **6** 297-310 [196]

Young FA, Leary GA (1991) Refractive error in relation to the development of the eye In Vision and visual dysfunction Vol 1 Visual optics and instrumentation (ed WN Charman) pp 29–44 MacMillan, London [435]

Young LR, Sheena D (1975) Survey of eye movement recording methods Behav Res Meth Instrum **7** 397–429 [365]

Young MP, Tanaka K, Yamane S (1992) On oscillating neuronal responses in the visual cortex of the monkey J Neurophysiol **67** 1464–74 [133]

Young RA (1987) The Gaussian derivative model for spatial vision Spatial Vis **2** 273-93 [141]

Young RW (1971) Shedding of discs from rod outer segments in the rhesus monkey J Ultrastruct Res **34** 190–203 [165]

Young T (1801) On the mechanism of the eye Philos Tr R Soc **91** 23-88 [26]

Young T (1802) On the theory of light and colours Philos Tr R Soc **92** 12–48 [118]

Yu K, Blake R (1992) Do recognizable figures enjoy an advantage in binocular rivalry? J Exp Psychol **18** 1158–73 [309]

Yu TW, Bargmann CI (2001) Dynamic regulation of xon guidance Nature Neurosci **4** 1169-76 [447]

Yuan W, Semmlow JL (2000) The influence of repetitive eye movements on vergence performance Vis Res **40** 3089-98 [400]

Yuan W, Semmlow JL, Munoz P (2000) Effects of prediction on timing and dynamics of vergence eye movements Ophthal Physiol Opt **20** 298-305 [398]

Yuasa J, Hirano S, Yamagata M, Noda M (1996) Visual projection map specified by topographic expression of transcription factors in the retina Nature **382** 632-5 [439, 457]

Yukie M, Iwai E (1981) Direct projection from the dorsal lateral geniculate nucleus to the prestriate cortex in macaque monkeys J Comp Neurol **201** 81-97 [220]

Yuodelis C, Hendrickson A (1986) A qualitative and quantitative analysis of the human fovea during development Vis Res **29** 847–55 [436]

Yuste R, Bonhoeffer T (2001) Morphological changes in dendritic spines associated with long-term synaptic plasticity Ann Rev Neurosci **24** 1071-89 [193]

Yuste R, Denk W (1995) Dendritic spines as basic functional units of neuronal integration Nature **357** 682-4 [115]

Yuste R, Nelson DA, Rubin WW, Katz LC (1995) Neuronal domains in developing neocortex: mechanisms of coactivation Neuron **14** 7-17 [459]

Zacharias GL, Young LR (1981) Influence of combined visual and vestibular cues on human perception and control of horizontal rotation Exp Brain Res **41** 159-71 [149]

Zajaczkowska A (1956) Experimental test of Luneburg's theory. Horopter and Alley experiments J Opt Soc Am **46** 514-27 [137]

Zakharenko SS, Zablow L, Siegelbaum SA (2001) Visualization of changes in presynaptic function during long-term synaptic plasticity Nature Neurosc **4** 711-17 [461]

Zanker J, Mohn G, Weber U, Zeitler-Driess K, Fahle M (1992) The development of vernier acuity in human infants Vis Res **32** 1557-64 [469]

Zee DS, Levi L (1989) Neurological aspects of vergence eye movements Rev Neurol **145** 613–20 [405]

Zee DS, Fitzgibbon EJ, Optican LM (1992) Saccade–vergence interactions in humans J Neurophysiol **68** 1627–42 [422, 423]

Zeil J, Nalbach G, Nalbach HO (1986) Eyes, eye stalks and the visual world of semi-terrestrial crabs J Comp Physiol A **159** 801-11 [538]

Zeki SM (1974b) Functional organization at a visual area in the posterior bank of the superior sulcus of the rhesus monkey J Physiol **236** 549–73 [217]

Zeki SM (1978) Uniformity and diversity of structure and function in rhesus monkey prestriate visual cortex J Physiol **277** 273-90 [220, 260]

Zeki SM (1979) Functional specialization and binocular interactions in the visual areas of rhesus monkey prestriate cortex Proc R Soc B **204** 379-97 [217]

Zeki SM (1990) A century of achromatopsia Brain **113** 1721-77 [221]

Zeki S (2001) Localization and globalization in conscious vision Ann Rev Neurosci **24** 57-86 [230]

Zeki SM, Fries W (1980) A function of the corpus callosum in the Siamese cat Proc R Soc B **207** 279–58 [233]

Zeki S, Shipp S (1988) The functional logic of cortical connections Nature **335** 311–17 [132]

Zeki SM, Watson JDG, Lueck CJ, Friston KJ, Kennard C, Frackowiak RSJ (1991) A direct demonstration of functional specialization in human visual cortex J Neurosci **11** 641–9 [221, 224]

Zemon V, Pinkhasov E, Gordon J (1993) Electrophysiological tests of neural models: evidence for nonlinear binocular interactions in humans Proc Nat Acad Sci **90** 2975-8 [260]

Zenisek D, Steyer JA, Almers W (2000) Transport, capture and exocytosis of single synaptic vesicles at active zones Nature **406** 849-54 [189]

Zetlan SR, Spear PD, Geisert EE (1981) The role of corticogeniculate projections in the loss of Y-cells in monocularly deprived cats Vis Res **21** 1035-9 [504]

Zhang H, Gamlin PDR (1998) Neurons in the posterior interposed nucleus of the cerebellum related to vergence and accommodation. I Steady-state characteristics J Neurophysiol **79** 1255-69 [365, 431]

Zhang Y, Gamlin PDR, Mays LE (1991) Antidromic identification of midbrain near response cells projecting to the oculomotor nucleus Exp Brain Res **84** 525–8 [430]

Zhang Y, Mays LE, Gamlin PDR (1992) Characteristics of near response cells projecting to the oculomotor nucleus J Neurophysiol **67** 944–60 [430]

Zhang YZ, Moheban DB, Conway BR, Bhattacharyya A, Segal RA (2000) Cell surface Trk receptors mediate NGF-induced survival while internalized receptors regulate NGF-induced differentiation J Neurosci **20** 5671-8 [456]

Zheng JQ, Felder M, Connor JA, Poo M (1994) Turning of nerve growth cones induced by neurotransmitters Nature **368** 140–4 [439]

Zhou H, Friedman HS, von der Heydt R (2000) Coding of border ownership in monkey visual cortex J Neurosci **20** 6594-611 [206]

Zhou W, King WM (1998) Premotor commands encode monocular eye movements Nature **393** 692-5 [429]

Zhou Y, Leventhal AG, Thompson KG (1995) Visual deprivation does not affect the orientation and direction sensitivity of relay cells in the lateral geniculate nucleus of the cat J Neurosci **15** 689-98 [499]

Zhou YX, Baker CL (1994) Envelope-responsive neurons in areas 17 and 18 of cat J Neurophysiol **72** 2134-50 [201]

Zhou YX, Baker CL (1996) Spatial properties of envelope-responsive cells in area 17 and 18 neurons of the cat J Neurophysiol **75** 1038-50 [128, 201]

Zhu Q, Julien JP (1999) A key role for GAP-43 in the retinotectal topographic organization Exp Neurol **155** 228-42 [439, 440]

Zhu Y, Qian N (1996) Binocular receptive field models, disparity tuning and characteristic disparity Neural Comput **8** 1647-77 [253, 547]

Zielinski BS, Hendrickson AE (1992) Development of synapses in macaque monkey striate cortex Vis Neurosci **8** 491–504 [451]

Zihl J, Cramon D von and Mai N (1983) Selective disturbance of movement vision after bilateral brain damage Brain **106** 313-40 [224]

Zimba LD, Blake R (1983) Binocular rivalry and semantic processing: out of sight out of mind J Exp Psychol HPP **9** 807–15 [308]

Zipser K, Lamme VAF, Schiller PH (1996) Contextual modulation in primary visual cortex J Neurosci **16** 7376-89 [206]

Zohary E (1992) Population coding of visual stimuli by cortical neurons tuned to more than one dimension Biol Cyber **66** 295-72 [122]

Zohary E, Celebrini S, Britten KH, Newsome WT (1994a) Neuronal plasticity that underlies improvement in perceptual performance Science **293** 1289–92 [208]

Zohary E, Shadlen MN, Newsome WT (1994b) Correlated neuronal discharge rate and its implications for psychophysical performance Nature **370** 140-3 [318]

Zuber BL, Stark L (1968) Dynamical characteristics of the fusional vergence eye–movement system IEEE Tr Syst Sci Cyber **4** 72–9 [397]

Zuidam I, Collewijn H (1979) Vergence eye movements of the rabbit in visuomotor behavior Vis Res **19** 185–94 [549]

Index of Portraits in Volume 1

Subject Index Volume 1